Contemporary Systems Thinking

Series Editor

Robert L. Flood
Maastricht School of Management, Maastricht, The Netherlands

More information about this series at http://www.springer.com/series/5807

Piero Mella

The Magic Ring

Systems Thinking Approach to Control Systems

Second Edition

Piero Mella
Department of Economics and Management
University of Pavia
Pavia, Italy

ISSN 1568-2846
Contemporary Systems Thinking
ISBN 978-3-030-64196-2 ISBN 978-3-030-64194-8 (eBook)
https://doi.org/10.1007/978-3-030-64194-8

This Springer imprint is published by the registered company Springer Nature Switzerland AG
The registered company address is: Gewerbestrasse 11, 6330 Cham, Switzerland

Preface to the First Edition

Thinking is easy, acting is difficult, and to put one's thoughts into action is the most difficult thing in the world. – Johann Wolfgang Von Goethe

The origins – This book represents the completion and evolution of my book *Systems Thinking. Intelligence in Action* (Springer 2012), which presented the general view of the "world" considered as a "system of systems" of interconnected variables. That book presented the theory, language and basic rules of *Systems Thinking* while this book develops the theory, language and rules behind the functioning of *Control Systems*, which were presented in a single short chapter. Though this book is the logical development of the previous book, it is a self-contained work compared to *Systems Thinking* since it does not suppose any prior knowledge of Systems Thinking – whose basic concepts are presented at the beginning – but gradually introduces a true Discipline of Control Systems that is accessible to a large group of non-specialists. The choice of topics and examples – in this endless field of knowledge – derives from the need to make the Discipline of Control accessible to "everyone".

The assumptions – The idea of developing a Discipline of Control that stands on its own with regard to Systems Thinking is based on the hypothesis – whose validity I have tried to demonstrate throughout the book – that among all types of systems Control Systems occupies an absolutely preeminent position. Even if we are not accustomed to "seeing them", "recognizing them" or "designing them", they are everywhere inside and around us. Only their ubiquitous presence makes possible our world, life, society and existence, constructing an ordered and liveable world, erecting barriers against disorder, and directing the irregular dynamics of the world toward states of homeostatic equilibrium in all its vital and social environments. It is for this reason I have tried to make clear the need to gradually accustom ourselves to thinking not only in terms of systems but above all in terms of Control Systems: "*from Systems Thinking to Control Thinking*"; this is my epistemological proposal.

The objectives – This book presents a gradual path toward "educating" the reader in understanding how Control Systems truly operate and in recognizing, simulating and improving them in all fields of activity. Starting from the hypothesis that knowledge of Control Systems is not only a technical fact but also represents a discipline – that is, "A discipline is a developmental path for acquiring certain skills or competencies. (…) To practice a discipline is to be a lifelong learner. You "never arrive"; you spend your life mastering disciplines." (Senge, 1992, p. 12) – I have set the objective of making Control Systems a topic that is, in a certain sense, simple and attractive by turning to the effective symbolism typical of Systems Thinking models and avoiding too technical and formal a treatment of the subject. Thus, the reader should know that this is not an engineering, physics, biology or economics text, nor a mathematics one either. Technical or mathematical tools are not used to construct Control Systems; instead, I use the highly simple and universal logic behind the notion itself of *control process* and the simple and universal action of the Control Systems that produce this process.

The title – Why "Ring"? – Control Systems (in their most simple form) are particular apparatuses (logical or technical) that cause a variable *Y*t to attain an objective *Y** by gradually reducing the error (gap, distance) *E*t = *Y** – *Y*t through a series of "adjustments" of *Y*t produced by acting on a *control variable*, or *lever*, *X*t, so that the values of *Y*t approach *Y**; the iterations of this process continue and, if the system is well designed, *E*t becomes zero, even in the presence of external disturbances, *D*t, which cannot be controlled by the system. Using the logical language of Systems Thinking we can easily see that a control system is a typical balancing loop that connects the three variables [*X*t, *Y*t and *E*t]; this loop is represented by a typical circular graphic model that in fact resembles a *Ring*. We can also imagine that each iteration to reduce *E*t is like a *rotation* of the *Ring*.

The title – Why "Magic"? – It is probably difficult to realize that you can read these sentences *only* because at least five fundamental automatic Control Systems, which exist in our bodies to produce vision, are acting simultaneously; that is, five *Rings* that rotate at an incredible speed. I will briefly mention these, and without any particular order (for simplicity's sake I shall omit the time reference in the variables).

The first *Ring* is the focus Control System, which works through the crystalline lens which, by varying its shape using the tensor muscles (*X*), varies the focal distance in order to eliminate the gap between the actual focus (*Y*) and that needed to read the letter (*Y**), even when the distance between the text and the reader's eye (*D*) varies. The second *Ring* is the control system which, by changing the diameter of the pupil (*X*), varies the quantity of light (*Y*) that strikes the retina (cones and rods) to maintain (*Y**) constant even with variations in the ambient light (*D*). The third *Ring* is the control system that, by using the motor eye muscles (*X*), produces the macro and micro movements in the eyeball (*Y*) in order to read the words in the line (*Y**), if necessary aided by the rotation of the head (controlled by an additional *Ring*). The fourth *Ring* is the control system that produces tears (*X*) to keep the cornea wet (*Y*) in an optimal manner (*Y**), even when there is a change in evaporation due to heat at various times from various sources (*D*). The fifth *Ring* (the final one I shall mention) is perhaps the most important: the Control System that allows us to *recognize* signs written on a page as *letters*, and the latter as *words* given rhythm by punctuation marks, which leads to the *recognition* of sentences.

Assuming that these five *Rings alone* "rotate" *only* 20 times per second, then in order to read this paragraph in, let's say 60 seconds, we need 6000 total iterations

(5 × 20 × 60), that is, 6000 control actions of which we are normally unaware: the number, in fact, required to read second by second (obviously ignoring whether or not the reader understands what he has read). Does this number of iterations seem too small? Consider that obviously these five *Rings* function continuously all day long our entire lives: let us assume on average 14 hours for 80 years. During this period the five Control Systems produce more than (145 × 109) iterations. Our five *Rings* rotate at an incredible speed only to guarantee vision, one of the five senses. Just think about the others!

Does it seem like a small number now?

The most complex "machine" on earth is probably the human brain, formed on average by (100 × 109 = 1011) neurons (in addition to an even larger number of auxiliary cells). The complexity of this machine derives from the fact that the neurons are connected by means of various types of synapses, whose overall number is estimated to be 1014. By using the simple model of the neural networks we can imagine that the synapses have a particular "weight" which determines the strength of the signal received by each transmitter neuron which is then sent to the receptor neurons, and that each neuron in turn has a specific threshold value below which the neuron does not "fire" the signals received, causing the connection with the receptor neurons down the line not to take place. It is immediately clear that for each neuron there must correspond "at least one" Control System for identifying the threshold value and that for each synapsis "at least one" Control System for sending the signals with their specific weights. At this moment our brain is working thanks to the action of (1011 + 1014) *Rings* rotating thousands of times per second, and for the entire human population [(1011 + 1014) × 6 × 109] neural *Rings* are operating (in the brain). We may not realize it, but this is "magic".

Does this seem too small a number?

The average number of cells in the body of a man weighing 70 kilograms is estimated to be 1014. Each cell, as an autopoietic and thus homeostatic system, can exist thanks to the action of a large number of Control Systems that govern the cell cycle, instant by instant, and to the other Control Systems that regulate cellular interactions. Estimating that there are *only* 100 Control Systems for each cell, at this moment your body can remain alive thanks to the action of 1016 invisible *Rings* that rotate uninterruptedly for your entire life. At this moment the human population exists thanks to the action of [6 × 109 × 1016] cellular Control Systems in addition to [(1011 + 1014) × 6 × 109] neural Control Systems, trillions of trillions of invisible *Rings* that rotate incessantly, untiringly, second after second, year after year. I am sure few people have considered this.

Do you still think this is not significant?

Without exaggerating and becoming tiresome, I would point out that the cellular control characterizes all cells of all living beings. Ants alone are estimated to number from 1015 to 1016. How many insects are there? How many birds? How many mammals? How many fish and molluscs? It is estimated that there are more than (400 ×106) units of phytoplankton in a cubic meter of water. The oceans and seas are made up of about (1.34 × 109) km3 of water, the equivalent of (1.34 × 1018) cubic meters. Overall there are (400 × 1.34 × 1024) individual plankton, single- or multi-celled, which at this moment are kept alive by ($k \times 400 \times 1.34 \times 1024$) invisible *Rings* that rotate in whirling fashion second after second. It is too difficult, useless even, to estimate the number of vegetal units – plants, shrubs, blades of grass, mushrooms, lichen, mould, etc. – and the number of cells of each individual. However, there is no need for such an estimate to grasp the impressive number of *Rings* necessary to keep the planet's flora alive.

And what about the billions of man-made Control Systems? And the trillions of Control Systems that regulate relationships between individuals, populations and species? And the endless number of Control Systems that operate in our atmosphere, seas, and in all physical phenomena?

There is a lot still to discover about how Control Systems function as motors for learning and knowledge; I will deal with this in Chapters 9 on.

If there is something magical in life, society and organization, in orders and evolution, this something is represented by the "magic" action of the "rings" which, through their ubiquitous presence and variety of forms, account for individuality, maintain the order of things, restore equilibrium and regulate interactions.

Why "magic *Ring*"? Independently of its nature, function and physical structure, each Control System always has the same structure of a *Ring*. A "magic *Ring*", in fact.

The content – The book is divided into 10 Chapters.

Chapter 1 seeks to review the basic language of Systems Thinking and the models it allows us to create. This language and these models will be the instruments for presenting the logic of Control Systems in a manner that is understandable to everyone. It is suggested that the study of Control Systems be considered as a true discipline, according to Peter Senge's conception, a true Sixth Discipline.

Chapter 2 introduces the control process, presenting the *theoretical structure* of three simple Control Systems we all can observe: those that permit us to regulate the sound volume (radio, TV, iPod, stereo, computer, etc.), the water temperature (shower, radiator, boiler, etc.) and the room temperature (air conditioner, refrigerator, etc.). Despite the simplicity of the three kinds of systems, we can gain some fundamental knowledge from them about the basic structure of a Control System by learning the basic concepts: *manager, governor, action rate, reaction rate, reaction time, delays* and *disturbances*. After presenting the general theoretical model of this system, we examine several new concepts: (1) the *chain of control*, which represents the technical structure of the process; (2) the various types of *delays* in control, extending the considerations already begun in Chapter 2; and (3) *interferences* among the Control Systems which can give rise to disturbance dynamics.

Chapter 3 proposes a *general typology* of Control Systems with examples taken from observations of reality, distinguishing among: (a) systems of *attainment* and *recognition*; (b) *continuous-* or *discrete-lever systems*; (c) *fixed* or *variable objective* systems; (d) *steering* and *halt* Control Systems; (e) systems of *collision, anti-collision* and *queuing* systems; and (f) *tendential* and *combinatory* systems; to name but a few.

Chapter 4 broadens the view of Control Systems by introducing two important generalizations: (1) *multilever* Control Systems, with levers that are independent or dependent of each other; (2) *multi-objective* systems, with independent or interdependent objectives. In multilever systems, it is fundamental to understand the importance of the control *strategy*, that is, of the specification of the priorities for the activation of the different levers used to eliminate the deviation from the objective. In multi-objective systems, on the other hand, it is fundamental to understand the importance of the control *policy*, which decides on the order of priorities for intervening on the various objectives. To further generalize, we shall examine the notion of *impulse systems*, which play a fundamental role in life.

Chapter 5 outlines the guidelines for recognizing, observing or designing Control Systems and presents the problems that arise regarding their logical realization. It introduces the fundamental distinction between symptomatic and structural control.

Chapters 6, 7, 8, and 9 undertake a "mental journey" through various "environments", increasingly broader in scope, suggesting to the reader how to recognize therein Control Systems that, by their ubiquitous presence, make the world possible in all its manifestations.

Chapter 6 considers daily individual actions – the domestic and civic environment (the most familiar ones) – in order to arrive at Control Systems that maintain the psychic-physical system.

Chapter 7 guides the reader through the physical, biological and social environments. Particular attention is given to Control Systems that operate in Societies and in *combinatory systems*, a special class of systems which has not yet been adequately examined.

Chapter 8 considers the environment of organizations, trying to demonstrate how organizations are themselves Control Systems by presenting two basic models: Stafford Beer's Viable System Model and my own Model of the Organization as an Efficient System of Transformation (MOEST). The chapter also illustrates the logic behind the many Control Systems that operate in the network of processes in organizations and which are necessary to maintain autopoiesis and increase the efficiency of the entire organization.

Chapter 9 deals with three specific topics: (1) the *cognition process*, understood as the result of the action of Control Systems that recognize differences and order them based on Gregory Bateson's notion of "mind"; (2) the *signification process*, which allows man to translate cognitive content into signs and to form languages; and (3) the *scientific process* and the explanatory power of models.

Chapter 10, the final chapter, deals with two topics: (1) some ideas about a Discipline of Control Systems are proposed; (2) the human aspects of control. A reflection on the content and limits of this book are presented as FAQs in the final pages of the chapter.

The style and method – In order to facilitate the understanding, acquisition and practice of the discipline, Control Systems have been presented in a logical way by using the symbols of Systems Thinking to produce easy-to-construct models. Thus, I have intentionally avoided expressing Control Systems in a typically mathematical form, which would have used differential equations or differences, preferring instead to use simulation examples with "everyday resources", in particular Excel files accompanied (in a few cases) by Powersim, to demonstrate to the reader how relatively easy it is to do the calculations to achieve the control. Moreover, I have chosen particularly simple and clear examples by presenting very common Control Systems we all can observe and experiment with in order to draw lessons in general theory from them.

The readership – The book is "for everyone". There is no prerequisite required to read and understand it, in particular math and computer knowledge. The use of the simplest tools of Excel helps in creating models for basic simulations of simple systems in order to improve understanding. However, I have felt it useful to also present several simple models in Powersim. A number of examples are aimed at systems found in organizations and companies; thus, the text can aid in the professional growth of managers, consultants and corporate employees in general. Teachers, professionals, and educated people in general can also find sources for reflection and occupational tools from the text. A word of caution for the reader: *in order to master the Control Discipline it is necessary to read this book* "slowly" *and refer the standard models presented to one's own world and experiences.*

Piero Mella Pavia, Italy

Preface to the Second Edition

Six years have passed since the publication of *The Magic Ring*. During this period, I have received many comments: some of appreciation, others of constructive criticism, and others, the most numerous and useful, with observations, suggestions, ideas, and starting points for reflection. In light of these comments, I started editing, integrating, and expanding my private files from the first edition for my own update. When I realized that the changes and additions had become significant, I sent the publisher a proposal, immediately accepted, to prepare a second edition.

To guide the reader, it is useful to summarize the main differences between the two editions, following the structure of the new edition.

Part I: Discovering the "Ring"

The five chapters of Part I have undergone limited changes. The comments I had received confirmed the appreciation for the chosen approach: to present the logical structure of Control Systems, adopting the language of Systems Thinking without employing the mathematical formalism of differential equations, and translating the logic of Control Systems into balancing loops that could be easily understood even by readers without a specific mathematical or engineering background.

Chapter 1. Two sections were added that developed two very simple simulation models of the explosive dynamics of a virus both before and after measures taken to counteract its spread. I have also added a section containing a list of systemic archetypes proposed by Peter Senge (1990), supplemented by a number of other "new" archetypes that expand on Senge's original list.

Chapters 2, 3, 4, and 5. Apart from a few minor changes, these chapters, which represent the core of Control Theory, have remained unchanged in both structure and content. Only Chapter 5 has been enriched with two additions: the first, new Sect. 5.6, highlights the "Risks of Failure of the Control Process Due to the Variables to be Controlled"; the second presents "The Principles of Systems Thinking Applied to Problem Solving."

Part II: The Magic of the Ring

The chapters of Part II present the ubiquitous, irreplaceable action of Control Systems in all the observable "environments" and "contexts": personal, family, social, biological, organizational, productive, and cognitive. The most significant changes have been made to Part II, the most relevant of which are the following:

Chapter 6. There are two main additions. In Sect. 6.6, "Control System for Global Warming," four reinforcement loops caused by global warming that lead to its increase are mentioned; Sect. 6.7, "Rings Acting on Earthquakes and Tsunamis," has been added. Moreover, Maslow's model of the hierarchy of needs has been updated.

Chapter 7 (New). The "old" Chapter 7 "The Magic Ring in Action: Life Environments" has been split into two distinct chapters. The "new" Chapter 7 highlights the Magic Rings that act for the balance between the actors in the social environment and for sustainability. Many additions have been made to the chapter: a more careful examination of the control of "population longevity" and of the "dynamics of the world population"; integration of the theory of Combinatory Systems and the Tragedy of the Common Resources; Sect. 7.9, "Change Management in a Complex World", has been greatly expanded; and Sect. 7.10.2 has been added, highlighting the "Modus Operandi of Some Relevant Social Phenomena Following the Combinatory Systems Model," along with Sect. 7.10.5, "The PSC Model Applied to Stereotypes and Gender Discrimination."

Chapter 8 (New). The part of "old" Chapter 7 regarding the biological environment has been entirely rewritten and considerably expanded and supplemented, becoming the "new" Chapter 8, which deals with the vast topic of population dynamics and control from both a quantitative view, the change in the number of individuals in a population, and a qualitative one, the variation or change in the phenotypes in evolutionary processes. Several simple simulation models of population dynamics are examined in this chapter, which also considers the evolution of populations of non-biological entities, particularly robots and organizations that form the nodes of production networks.

Chapter 9 (New). This chapter considers the action of the Magic Rings, without which organizations could not exist. Modest variations have been made to several sections; nevertheless, the most important variation concerns the six new sections added to "Complementary Material,", which deal with control through Budgeting and Reporting, Cost Measurement and Control through Standard Costs, and the "External Control of the Company's Economic and Financial Efficiency" by means of Position Analysis.

Chapter 10 (New). This is a completely new chapter whose subject is the Rings that control the two fundamental variables – quality and productivity – which are vital both for organizations and for the global economic system. The chapter presents the numerous techniques for evaluating the "quality levels" together with the levers to be used to control these variables. Moreover, after presenting a coherent frame of reference, the chapter examines the drivers of productivity before discussing the consequences of continual growth in productivity and quality. The Control of Plant Efficiency through Total Productive Maintenance and Terborgh's MAPI formula conclude the additions.

Chapter 11 (New). This, too, is a new chapter dealing with warehouse control and production processes, which are typical forms of management control. A number of calculation models have been developed for warehouse and production control. The logic of the "global production control methods" is then examined: MRP, JIT, OPT, concluding with FMS and HMS. The final section of the chapter addresses the problem of project control by examining PERT/CPM and ALTAI not only as network programming techniques but also as optimization and control tools.

Chapters 12 and 13 (New). These correspond to the "old" Chapters 9 and 10 and have undergone only marginal changes.

Final considerations The new edition has been enriched by a substantial number of models and a significant bibliography. I have also augmented the text by providing numerous quotes that probably may present a possible source of distraction for the hasty reader, yet are essential to "give the floor" to the authors of the concepts that I have taken up and used. I therefore found it useful to report their thoughts directly without filtering them through my interpretations, paraphrasing, or summaries.

Special thanks go to the publisher, for having given me the opportunity of this second edition, and the editorial staff for their invaluable effort in preparing the book.

Pavia, Italy Piero Mella

September 2020

Contents

List of Figures

List of Tables

Part I
Discovering the "Ring"

Chapter 1
The Language of Systems Thinking for Control Systems

Abstract We are about to begin a wonderful trip in the world of control systems, which are all around us. We are not used to seeing them, but their presence is indispensable for the existence of the world and of life. We need to gradually understand how they really operate in order to recognize, simulate, and improve them. The reader should know that this is not a mathematics, engineering, physics, biology, economics, or sociology book. I have chosen to present the logic of control systems using the simple but powerful and clear language of systems thinking for interpreting the world as a system of dynamic systems of interconnected and interacting variables.

Keywords Fifth discipline · Systems thinking · Learning organization · Sixth discipline · Causal chain · Causal loops · CLD, Causal loop diagram · Dynamic instability · Virus diffusion · Law of Accelerating Returns

> Systems thinking plays a dominant role in a wide range of fields from industrial enterprise and armaments to esoteric topics of pure science. Innumerable publications, conferences, symposia and courses are devoted to it. Professions and jobs have appeared in recent years which, unknown a short while ago, go under names such as systems design, systems analysis, systems engineering and others. They are the very nucleus of a new technology and technocracy; their practitioners are the "new utopians" of our time [...] who—in contrast to the classic breed whose ideas remained between the covers of books-are at work creating a New World, brave or otherwise (Ludwig von Bertalanffy 1968, p. 3).

> The great shock of twentieth-century science has been that systems cannot be understood by analysis. The properties of the parts are not intrinsic properties but can be understood only within the context of the larger whole. Thus the relationship between the parts and the whole has been reversed. In the systems approach the properties of the parts can be understood only from the organization of the whole. Accordingly, systems thinking concentrates not on basic building blocks, but on basic principles of organization. Systems thinking is "contextual", which is the opposite of analytical thinking. Analysis means taking something apart in order to understand it; systems thinking means putting it into the context of a larger whole (Fritjof Capra 1996a, pp. 29–30).

P. Mella, *The Magic Ring*, Contemporary Systems Thinking,
https://doi.org/10.1007/978-3-030-64194-8_1

We are about to begin a wonderful trip in the world of control systems, which are all around us. We are not used to seeing them, but their presence is indispensable for the existence of the world and of life. We need to gradually understand how they really operate in order to recognize, simulate, and improve them. The reader should know that this is not a mathematics, engineering, physics, biology, economics, or sociology book. It does not treat mathematical or technical tools used to produce or even simulate factual control systems.

I have chosen to present the logic of control systems using the simple but powerful and clear language of systems thinking (as proposed by Peter Senge 1990) for interpreting the world as a system of dynamic systems of interconnected and interacting variables. The first chapter thus serves to recall the principles of systems thinking and presents the rules of the formal language used to represent dynamic systems—the language we will also use to construct models of the control systems that will be developed in subsequent chapters. The pervasiveness of the logic, notions, and techniques that can be applied to the multiform variety of control systems allows us to develop a true "art" of control—a "discipline"—which I shall call the *Sixth Discipline, since it represents the evolution of the way* systems thinking (the "fifth discipline") *observes the world.*

1.1 The Sixth Discipline: The Discipline of Control Systems

In his important work, *The Fifth Discipline: The Art and Practice of the Learning Organization* (Senge 1990, 2006; Senge et al. 1994), which is aimed mainly at managers, professionals, and educated people in general, Peter Senge seeks to spread systems thinking by applying a simple and clear methodology and symbology that everyone can easily understand and that permits us to build qualitative models of the world understood as the interpenetration—to different extents—of systems of interacting variables (Mella 2012) of all types of structure and complexity that often form holarchies or holonic networks (Mella 2009).

Senge indicates to managers the "superiority" of systems thinking not in order to propose a new form of thinking but to emphasize the need for organizations and managers to think systematically. Precisely for this reason, he considers systems thinking as a discipline—the fifth discipline—everyone should follow in order to get accustomed to succinct thinking, the only kind that can understand the properties of a whole without, however, forgetting about those of the parts.

In fact, the history of systems thinking, as a form of holistic thinking, is very old, as Fritjof Capra shows in his important work *The Web of Life*:

> The ideas set forth by organismic biologists during the first half of the century helped to give birth to a new way of thinking—"systems thinking"—in terms of connectedness, relationships, context. According to the systems view, the essential properties of an organism, or living system, are properties of the whole, which none of the parts have. They arise from the interactions and relationships among the parts (Capra 1996b, p. 30).

In this sense, Senge views systems thinking as an efficient method for abandoning reductionist thinking, which blocks us from seeing the emergent properties of systems—above all, dynamic systems—and from learning to think in a succinct manner as well.

Rather than concentrate on mathematical and technical formalism—which can be found, if necessary, in specialist texts on systems theory (von Bertalanffy 1968; Boulding 1956, Klir 1969; Casti 1985; Skyttner 2005), systems science (László 1983, Sandquist 1985; Klir 1991), systems analysis and design (Wasson 2006; Gopal 2002), control theory and cybernetics (Ashby 1957; Wiener 1961; Johnson et al. 1963; Leigh 2004), and other fields—Senge wished to codify a way of thinking about dynamics and complexity—that is, a systems thinking—that presents an effective approach for observing and modeling reality, forcing the observer to search for the circular connections, the loops among the interrelated variables of which reality is composed. More than a technique, systems thinking is thus a mental disposition (Anderson and Johnson 1997, p. 20), an attitude, a language, and a paradigm.

> Systems Thinking [is] a way of thinking about, and a language for describing and understanding, the forces and interrelationships that shape the behavior of Systems. This discipline helps us see how to change systems more effectively, and to act more in tune with the larger processes of the natural and economic world (Senge et al. 1994, p. 6).

> Systems Thinking is a Paradigm and a Learning Method. The first conditions the second. The second supports the first. The two parts form a synergistic whole (Richmond 1994, online).

> "Systems Thinking" has long been a cornerstone in our work on organizational learning, but the term often seems more daunting (it can easily sound like an intellectual task reserved for Ph.D.'s) than helpful. In fact, systems thinking is not about fighting complexity with more complexity. It simply means stepping back and seeing patterns that are, when seen clearly, intuitive and easy to grasp (Senge et al. 2008, p. 23).

It is for this reason that it must be gradually learned and systematically applied; it must become a true discipline that every company man, as well as each of us, must master through experience.

> Systems thinking is a discipline for seeing wholes, recognizing patterns and interrelationships, and learning how to structure those interrelationships in more effective, efficient ways (Senge and Lannon-Kim 1991, p. 24).

Other terms can also be used to indicate this new way of thinking. Barry Richmond, who was one of the most renowned experts in this discipline (he was the founder, in 1984, of the High-Performance System), stated:

> Systems Thinking, a Systems Approach, Systems Dynamics, Systems Theory and just plain ol' "Systems" are but a few of the many names commonly attached to a field of endeavor that most people have heard something about, many seem to feel a need for, and few really understand. [...] As I prefer the term "Systems Thinking", I'll use it throughout as the single descriptor for this field of endeavor (Richmond 1991, p. 1).

Systems thinking has been proposed as the "fifth discipline" precisely because, together with four other disciplines, it is presented as a means for building learning organizations, which are organizations that develop a continual collective learning by putting all their members in a position to learn together while supplying them with the instruments for such collective learning.

> Systems thinking is the fifth discipline. [...] It is the discipline that integrates the [other four] disciplines, fusing them into a coherent body of theory and practice. [...] Without a systemic orientation, there is no motivation to look at how the disciplines interrelate. By enhancing each of the other disciplines, it continually reminds us that the whole can exceed the sum of its parts (Senge 2006, p. 12).

Among the most notable results of systems thinking, the most important in my view is its ability to allow us to understand, explain, and simulate the *modus operandi* of control systems, which are the basic elements of biological and social life; of populations, societies, and organizations; and even of most physical phenomena. It is not difficult to see that without the action of control systems, which, though by nature extremely varied, can all be represented by the same models, the "world" itself—understood as a system of increasingly more extensive systems—would probably not exist, if the variables it is made up of were not in some way limited by constraints and objectives that keep the system "what it is" within a time horizon that preserves the relations among the variables.

Why, then, do we need a new book exclusively dedicated to control systems? The reason I propose this book is to demonstrate, using the logical and symbolic apparatus of systems thinking, that all control systems, no matter the context they operate in, can be represented by the same "logical models" and obey a relatively simple theory whose understanding does not require a particular knowledge of math, physics, or engineering. Thus, this is not an engineering, physics, biology, or economics book, nor for that matter a math book; neither technical tools (Sontag 1998; de Carvalho 1993; Goodwin et al. 2001; Leigh 2004) nor mathematical equations (Vanecek and Celikovsky 1996; Imboden and Pfenninger 2013) necessary to construct sophisticated control systems are considered. My sole aim is to present the highly simple logic underlying the concept of control, which is understood as the gradual approximation of the movement of a variable toward desired values (objectives), made possible by the action of one or more additional variables—the control levers—whose values are recalculated, based on some specific time scale, in order to gradually counter any movement away from the objective.

Thus, I also propose a more ambitious objective: to arouse in the reader curiosity about control systems and help him or her recognize these in the various environments they operate in, thereby providing a guide to the construction of models useful for understanding and simulating their behavior. I will not follow particular techniques but merely help the reader get accustomed to the logic of control and to work toward mastering them, slowly guiding him or her toward increasingly vaster systems. In other words, I will introduce him or her to what I feel could become a true discipline of control systems—or, more simply, "control discipline"—in line with Peter Senge's understanding of the term "discipline."

> By "discipline", I do not mean an "enforced order" or "means of punishment", but a body of theory and technique that must be studied and mastered to be put into practice. A discipline is a developmental path for acquiring certain skills or competencies.
>
> A discipline is a developmental path for acquiring certain skills or competencies. As with any discipline, from playing the piano to electrical engineering, some people have an innate "gift", but anyone can develop proficiency through practice.
>
> To practice a discipline is to be a lifelong learner. You "never arrive"; you spend your life mastering disciplines (Senge 2006, p. 10).

Following Senge's fifth discipline, I hope not to be too pretentious in calling the discipline of control systems the sixth discipline, that is, the discipline of the control of the individual, the collectivity, the organizations, and the ecosystem—in other words, the discipline of our world's present and future. In fact, in Chap. 19 (entitled appropriately enough "A Sixth Discipline?") of the first edition of *The Fifth Discipline* Senge, in the specific context of organizations and companies, writes:

> The five disciplines now converging appear to comprise a critical mass. They make building learning organizations a systematic undertaking, rather than a matter of happenstance. But there will be other innovations in the future. [...] perhaps one or two developments emerging in seemingly unlikely places, will lead to a wholly new discipline that we cannot even grasp today. [...] Likewise, the immediate task is to master the possibilities presented by the present learning discipline, to establish a foundation for the future (Senge 1990, p. 363).

It truly is difficult to think we can lay the foundations for the future without understanding the logic of the control systems that allow us to maintain intact over time our organism, society, and the organizations we are a part of, which, through their processes, generate stability and order as well as development and innovation. I believe that the need for a new culture of control is also recognized by Peter Senge in his recent book *The Necessary Revolution*, which applies systems thinking to the socioeconomic system in order to explore the paths of change in industrial civilizations—a change held to be indispensable for guaranteeing the *sustainability* of our social and economic system in a context where common exhaustible resources are becoming increasingly scarcer.

> A sustainable world, too, will only be possible by thinking differently. With nature and not machines as their inspiration, today's innovators are showing how to create a different future by learning how to see the larger systems of which they are a part and to foster collaboration across every imaginable boundary. These core capabilities—seeing systems, collaborating across boundaries, and creating versus problem solving—form the underpinnings, and ultimately the tools and methods, for this shift in thinking (Senge et al. 2008, p. 11).

Since change is only possible by applying the logic of control systems, *change management*, which decides how to regulate the system in order to achieve the new objectives, must necessarily identify the structure and state of the actual socioeconomic system along with the levers to modify the objectives through the levers of cultural change, which lead to a *working together to create a sustainable world* (for more on the processes of change management, see Chap. 7, Sect. 7.9).

As Fritjof Capra states, evolution, innovation, and learning are also typical of control processes as regards the vision provided by cybernetics as well as the more comprehensive view of self-organizing systems:

> For Ashby [in his Cybernetic view] all possible structural changes take place within a given "variety pool" of structures, and the survival chances of the system depend on the richness, or "requisite variety", of that pool. There is no creativity, no development, no evolution. The later models, [Self-organization view] by contrast, include the creation of novel structures and modes of behavior in the processes of development, learning, and evolution (Capra 1996b, p. 85).

How can we consciously guide the present toward the future, proactively face the unavoidable consequences of the processes under way—global warming, extinction, population growth, spread of epidemics, arms escalation, space exploration, biological and nanotechnological research, inflation, poverty, etc.—and consciously react to new events that we cannot foresee today (cataclysms, harmful innovations, etc.) if we are not able to understand the logic, power, or networks of control systems in action and, above all, to design and build effective control systems? How can we innovate, create, and manage new forms of organization, new machines, and new processes without also conceiving new control systems? If we think about it carefully, even the purely evolutionary process "produces" nothing other than a continuous innovation in control systems by "producing" new sensors (antennae, eyes, electrical discharge sensors, etc.), new regulatory organs (the nervous system and brain in particular), and new effectors of movement and of the recognition of forms (see Chap. 2, Sect. 2.13). Where are the control systems? They are within us and everywhere around us. We must be able to identify them by zooming out to catch the extreme variety, richness, and importance of the macro control systems as well as zooming in to catch the infallible effectiveness of the micro control systems, which are so essential to life.

Though different, all control systems can be referred to a few model types that are particularly useful in all fields of research. We need to get used to observing them and learn to recognize and master them by trying to understand their *modus operandi*. We will realize that we are formed by control systems, and surrounded by control systems, that we can exist and live thanks only to the control systems that regulate our environment and entire ecosystem. When we apply the sixth discipline, our life will no longer seem the same, and perhaps—with the same certainty as John von Neumann (the first citation at the beginning of Chap. 2)—we shall succeed in controlling—after having understood and identified the most suitable control systems—even the most unstable phenomena that are leading us toward the threshold of chaos.

Before examining the logic of control and control systems it is useful to recall the basic principles of the fifth discipline, systems thinking (Sects. 1.2, 1.3, and 1.4), and touch on the other four disciplines (Sect. 1.7.9). At this point the importance and pervasiveness of control systems and the relevance of the sixth discipline will be even clearer.

1.2 The Fifth Discipline: The Five Basic Rules of Systems Thinking

I am convinced that intelligence depends on the ability to construct coherent and sensible models (Mella 2012) in order to acknowledge and understand the world. I believe that:

> *Intelligent* persons are those who understand (and comprehend) quickly and effectively; who are not content to "look at the world with their eyes" (objects, facts, phenomena and processes) but who are able "to see the world with their minds" by constructing models to "understand" how the world is (description), how it functions (simulation), and how we can act as part of it (decision and planning), even without having the need, or possibility, of "looking at everything" (Mella 2012, p. 3).

> If the capacity to *see* and not simply look at [the world] depends on the ability to construct models to understand, explain and simulate the world, then the most useful and effective models to strengthen our intelligence are the systems ones based on the logic of Systems Thinking (*ibidem*, p. 6),

since these models allow us to represent complexity, dynamics, and change (the relevance of models for our knowledge and intelligence is shown in Chap. 9, Sect. 9.14). The cognitive efficiency of models derived from systems thinking owes to their ease of construction; they require only perspicacity and acumen; rely on elementary techniques; are understandable even for non-experts; can easily be communicated, debated, and improved; and without too much difficulty turned into quantitative simulation models based on system dynamics logic. Of course we should not be content only with systems thinking models, but for those with little time or resources to construct more sophisticated (though less immediate) models the following proverb always applies: *Beati monoculi in terra caecorum* [blessed are those who see with one eye in the land of the blind]; that is, in a dynamic and complex world blessed are those who, knowing how to construct systems thinking models, have at least one eye in a land of blind people (Mella 2012: Sect. 4.13). However, before examining the technique for constructing models it is useful to recall what systems thinking entails and what its logical and theoretical bases are. Even if Peter Senge has not explicitly stated them, five fundamental rules are at the basis of systems thinking in my opinion, as I have tried to demonstrate in my book *Systems Thinking* (2012, Chap. 1).

FIRST RULE: *If we want to understand the world, we must be able to "see the trees and the forest"; we must develop the capacity to "zoom" from the whole to the parts, from systems to components*, and vice versa.

This rule, which is at the basis of systems thinking, can be translated as follows: Reality is *permeated with systems*, increasingly vaster in scope, which form a global structure that creates a global process that cannot be *understood only* from the *outside* or the *inside*. This rule, whose mastery requires constant practice, can be translated: If we want to broaden our intelligence, we *must develop the capacity to "zoom" from* parts *to* the whole *and from* entities *to* components. In this sense we can say (Mella 2009) that this FIRST RULE of systems thinking "operationalizes" the *holonic view* introduced in 1967 by Arthur Koestler in his book *The Ghost in the Machine*.

> Parts and wholes in an absolute sense do not exist in the domain of life [...] The organism is to be regarded as a multi-levelled hierarchy of semi-autonomous sub-wholes, branching into sub-wholes of a lower order, and so on. Sub-wholes on any level of the hierarchy are referred to as holons [...] The concept of holon is intended to reconcile the atomistic and holistic approaches (Koestler 1967, p. 341).

To understand the dynamics of the world we must always remember that each "object" to which our observation is directed is at the same time a whole—composed of smaller parts—and part of a larger whole: a holon! Ken Wilber, another contemporary proponent of the holonic view (Wilber 2000), is even more explicit, stating: "*The world is not composed of atoms or symbols or cells or concepts. It is composed of holons*" (Wilber 2001, p. 21). I shall return to the holonic view of the world in more detail in Chap. 3, Sect. 3.8.

SECOND RULE: *We must not limit our observation to that which appears constant but "search for what varies"; the variables and their variations are what interest the systems thinker.*

This rule, equally as important as the preceding one, imposes greater attention and a more intense discipline, since systems thinking tells us to shift from a "world of objects" (people, processes, concepts, etc.)—whether trees or forests—to a "world of variables" that connote those objects. By definition a variable X may assume different values (measurements, states, quantity, etc.), x_0, x_1, x_2, x_3, etc., which we can assign through a succession of measurements. Let $x(t_0)$, $x(t_1)$, $x(t_2)$, $x(t_3)$, etc. represent the values of X measured at the end of regular intervals, $[t_1, t_2, t_3, t_4$, etc.], all within a defined period of reference T. In this sense, X is a (discrete) time variable and its measured values, arranged along a time axis, form the dynamics (motion, trajectory, evolution, trend, etc.) of X with respect to T. Systems thinking is interested above all in the variations they reflect: $\Delta x(t_1) = x(t_1) - x(t_0)$; $\Delta x(t_2) = x(t_2) - x(t_1)$; $\Delta x(t_3) = x(t_3) - x(t_2)$; etc. This seems easier than it really is. One needs sensitivity and experience to select the truly significant variables. Systems thinking explores and models systems of interconnected and interacting variables that produce the complexity we observe in our world.

> Senge sorts two types of complexity from this—detail complexity and dynamic complexity. Detail complexity arises where there are many variables, which are difficult, if not impossible, to hold in the mind at once and appreciate as a whole. Dynamic complexity arises where effects over time of interrelatedness are subtle and the results of actions are not obvious; or where short term and long term effects are significantly different; [...] (Flood 1999, p. 13).

However, we must not limit ourselves to explicitly stating the variables we consider useful but must be able to measure the "variations" they undergo over time. For systems thinking, each object must be observed as a "vector of variables" and the dynamics of the objects are perceived only if we are able to observe and measure the dynamics of the variables they represent. Systems thinking also requires us to *select* the relevant variables and *restrict* their number so as to consider only those most relevant for the construction of models. In general, a few attempts are sufficient to narrow down the most interesting variables.

THIRD RULE: *If we truly wish to understand reality and change, we must make an effort "to understand the cause of the variations in the variables we observe" by forming chains of causal relationships among the connected variables while identifying and specifying:*

1. The *processes* that "produce" the dynamics in the variables and the *machines* (or *systemic structures*) that "produce" those *processes*

2. The *variables* that "carry out" those processes (causes or inputs) and those that "derive" from the processes (effects or outputs)

For simplicity's sake we could even call the input and output variables "causes" (causal variables) and "effects" (caused variables), respectively. However, we must always remember that the processes—to the extent they are considered *black boxes*—always play the role of producer of the *effects*, given the *causes* and the more or less broad set of "initial" and "boundary or surrounding" conditions. This third rule is perhaps the most important. It assumes the constant exercise of the two preceding ones while also completing them, forming together a very potent logical system: *if you want to understand the world you mustn't be content with only observing variables and their variations, but you must search for the "cause" of the variations in the variables you observe.* This seems quite obvious, but the fact that normally we do not observe, or else fail to consider, that the variables change their values due to some process carried out by some "machine" makes it equally important to practice applying this rule, which has three corollaries:

(a) Every observable variable is a *cause* or an *effect* of other variables.
(b) Cause and effect variables are linked to form causal chains (A causes B, B causes C, and so on) where each variable (A, B, C, etc.) is both the effect of variables up the line and a cause for those down the line.
(c) It is not possible to understand the world's dynamics without identifying and studying the causal chains that make up the world.

FOURTH RULE: *It is not enough to search for the causes of the variations we observe by only searching for the causal chains among variables; we must also "link together the variables in order to specify the* loops *among all the variations."*

A loop is formed when two variables, *X* and *Y*, are linked in a *dual direction*, *X* causes *Y* and *Y* causes *X* (naturally we must take into account the time sequence of the observations), so that the concept of *cause* and *effect*—which holds between two (or more) linked variables forming a causal chain—loses its significance when we consider those variables linked together by one or more *loops* (Richardson 1991). For this reason systems thinking states: if we really want to "understand" the world and its changes, it is not enough to reason in terms of chains of causes and effects between variables; we must instead get used to *circular* thinking. In other words, we must be aware not only that every variable is the cause of variations in variables connected down the line in the causal chain but also that the causal chain can become a *loop*, since the last variable in the chain can be the cause of the first variable, which thus becomes "the effect of its effects," thereby creating a circular connection. In conclusion, this rule obliges us to make an effort to link together the variables until we obtain a loop among their variations. In other words, we must move from the causal chains to the systemic *interconnections* and from the *linear variations* to the *systemic interactions* among the variables of interest. In brief, we must see the world in terms of circular processes, or feedback loops, abandoning the "linear thinking" ("laundry list thinking") that only considers chains of causes and effects and getting accustomed to "circular thinking" (loops and causal loop diagrams, CLDs), identifying the loops that connect the variables.

> [...] If you took the time to record your thoughts, I'll bet they took the form of a [...] "laundry list". I like to refer to the mental modelling process that produces such lists as laundry list thinking. I believe it to be the dominant thinking paradigm in most of the Western world today. [...] Notice that the implicit assumptions in the laundry list thinking process are that (1) each factor contributes as a cause to the effect, i.e., causality runs one way; (2) each factor acts independently; (3) the weighting factor of each is fixed; and (4) the way in which each factor works to cause the effect is left implicit (represented only by the sign of the coefficients, i.e., this factor has a positive or a negative influence). The systems thinking paradigm offers alternatives to each of these assumptions. First, according to this paradigm, each of the causes is linked in a circular process to both the effect and to each of the other causes. Systems thinkers refer to such circular processes as feedback loops (Richmond 1993, p. 117).

A set of variables linked in various ways by loops so that the behavior of any variable depends on that of all the others forms a *dynamic system*, according to systems thinking. Systems thinking defines "system" as a unitary set of interconnected variables—capable of producing emerging macro-dynamics that do not coincide with any of the micro-dynamics of the individual variables or their partial subsystems—whose logical structure it investigates and represents (Mella 2012, p. 21). The chains and loops that bind together the variables form the *logical structure of the system*. Systems thinking states that to "describe, model, and control" the world it is enough to understand the logical structure of the systems of variables it is composed of, leaving to engineers, physicists, biologists, doctors, economists, sociologists, psychologists, and other specialized scientists the task of investigating the operating structure of those systems.

At this point we can now finally derive a general law of systems thinking: *in order to explain the dynamics of a variable, do not search for its immediate cause but rather define and investigate the logical structure of the dynamic system it is a part of.* But how large must a dynamic system be? How many variables do we have to connect? To answer these questions, systems thinking proposes this final rule.

FIFTH RULE: *When we observe the world, we must "always specify the boundaries of the system we wish to investigate."*

In fact, the FIRST RULE of systems thinking obliges us to zoom inside a system—thereby identifying increasingly smaller subsystems—as well as outside a system to identify ever larger supersystems. Are we thus destined (or "condemned") to having a holistic view without limits? Do we have to zoom ad infinitum? Certainly not! Systems thinking is the art of "seeing" the world, and in order for what we see to have a true meaning it must depend on our cognitive interests. Heinz Von Foerster, one of the founders of "radical constructivism" in the transition to second-order cybernetics, recalled that the observer cannot separate himself from the observed universe; therefore, he must delimit (build) his own field of observation and decide how far the observed system should extend, tracing its boundary with reference to the environment: "The world, as we perceive it, is our own invention" (von Foerster Page, online). We cannot have a forest without limits. For this reason the fifth rule of systems thinking obliges us to always specify the boundaries of the system being investigated, since our capacity to understand a system is diminished when it becomes too large. In fact there are two boundaries: an *external* boundary which delimits the system when we zoom from the parts to the whole and an *internal* one when we zoom from the whole to the parts. It is not easy to identify or set the system's boundaries; fortunately, the more we apply ourselves in the discipline of systems thinking, the more the solution to this problem becomes easy, almost spontaneous.

1.3 The Construction of Models Based on Systems Thinking: The *Rings*

As I have observed, systems thinking, or the fifth discipline, has the great merit of making immediately and easily understandable the *causal* interrelationships among the variables that make up a *dynamic* system, understood as a *pattern* of *interconnected* and *interacting* variables of whatever type, size, and structure. In order to make the causal relations among the variables in a dynamic system immediately and easily understandable, systems thinking constructs models by following five operational rules, which can be summed up as follows (Mella 2012).

1. Once the cause and effect links between two variables (for example, X and Y) have been determined, the link is represented by simply using an arrow (of whatever shape, thickness, or color) that unequivocally correlates their variations. The cause variable (input) is written at the tail of the arrow and the effect variable (output) at the head. To indicate that X causes Y, systems thinking writes:

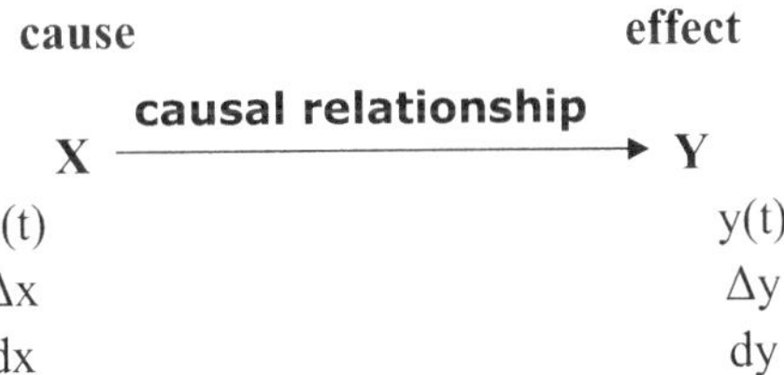

   The direction of the arrow indicates the direction of the causal link, so that

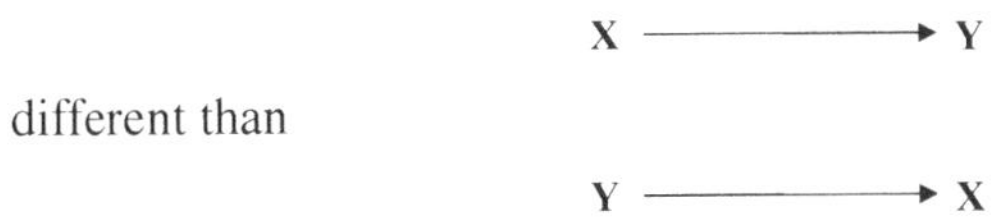

   is different than

   Y ⟶ X

2. After having specified the direction of the cause and effect link we need to understand the direction of variation of this link; two variables have the *same direction* of variation (**s**) if increases or decreases (+/−) in the former result in corresponding increases or decreases (+/−) in the latter. They have the *opposite*

direction (**o**) if increases or decreases in the former (+/−) result in corresponding decreases or increases (−/+) in the latter. The direction of variation is indicated by placing (**s**) or (**o**) next to the arrow, as shown in the following models:

3. Identify the causal chains, the significant variables these are composed of, and the direction of variation of the entire chain. Each causal chain has a direction of variation **s** or **o** that depends on the directions of variation of the causal links they are made up of. If there is an odd number of variations **o** among the variables in the chain, then the entire chain has a direction of variation **o**; an increase (decrease) in the *first* variable (+/−) produces a decrease (increase) in the *last* one (−/+); if there are no **o** variations, or an even number of them, then the causal chain has direction **s**.
4. In many cases two variables—let us suppose *X* and *Y*—can be linked by cause and effect relations that act contemporaneously in two opposite directions. We must recognize and represent the *two-way* circular cause and effect links, that is, those between *X* and *Y* and, at the same time, between *Y* and *X*. This possible dual relation is represented by two opposite arrows—from *X* to *Y* and *Y* to *X*— which form the simplest model of a dynamic system, called the CLD or causal diagram. Loops can be *basic*, when there are only two variables, or *compound*, when more than two variables are joined in a circular link. The loops are distinguished by the effect produced on the variables they are composed of, as a result of the direction of variation of the variables themselves. There are only two basic types of loop:

   (a) *Reinforcing* loops [**R**], which produce a *reciprocal increase* or *reduction*— in successive repetitions of the system's cycle—in the values of the two variables having an identical *direction of variation*: "**s** and **s**" or "**o** and **o**"; a basic reinforcing loop can be represented by the following model:

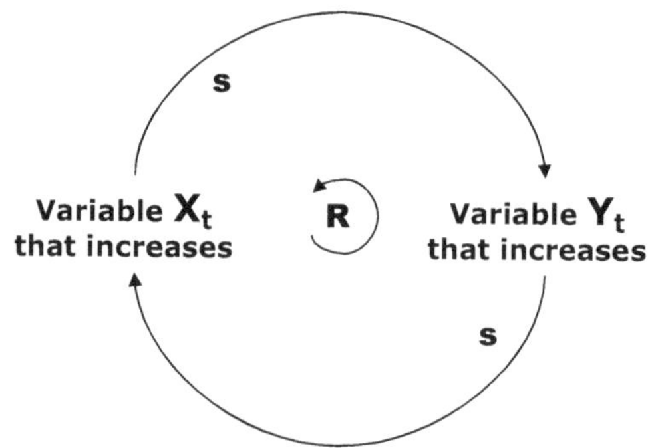

(b) *Balancing* loops [**B**], which *maintain relatively stable* the values of the connected variables, which are connected by a different *direction of variation*: "**s** and **o**" or "**o** and **s**"; a basic balancing loop can be represented as follows:

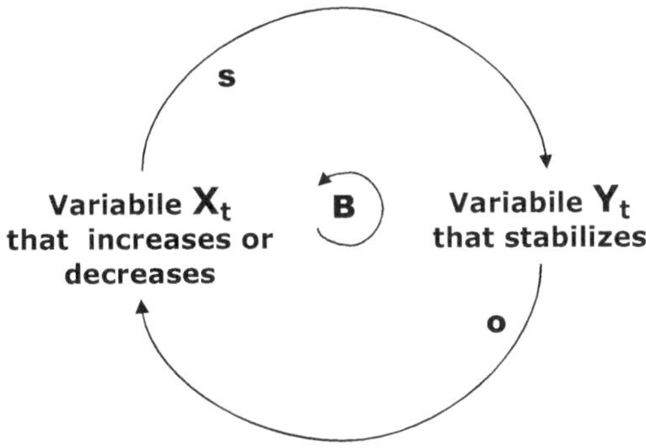

I propose referring to a *basic* or a *compound loop* with the more evocative term "*Ring*."

5. Join together the various reinforcing and balancing loops in order to represent any type of reality formed by the interconnected variables included within the system's boundary. *A system of loops in which all variables are linked by arrows, without there being an initial and final variable, is defined as a general CLD.*

By applying these simple operational rules, systems thinking states that a dynamic world, no matter how large and complex, can be effectively described and modeled through various combinations of loops [**R**] and [**B**], even by inserting in each of the loops a large number of variables linked by causal relations (zooming in) or by adding any number of loops through some variables (zooming out). Each general CLD represents a model of a *dynamic system* in which the dynamics of each variable depends on the dynamics of the others, in addition to variables outside the system, which, in the CLD, are represented with arrows whose origin is not linked to other arrows. To clarify operating rules 1–5 above, it is useful to present some simple models of dynamic systems.

The price of a good has a direction of variation contrary to that of the volumes of supply (downward arrow, **o**), but the price variations produce a variation of same direction in the production volumes (upward arrow, **s**), as shown in Fig. 1.1. The causal model that links price and supply shows how, in successive *cycles of the system*, an equilibrium [**B**] is produced among the variables; if the price falls too much then supply falls; if supply declines then the price increases again.

The model in Fig. 1.2 shows that it is possible to join two balancing loops to form a system that describes the well-known law of the market under perfect competition; price represents the key variable that produces an equilibrium between supply and demand.

Observing the two joint *Rings* in Fig. 1.2, we can easily see the dynamics of the variables: if the price falls too much, then supply will also fall and demand rise, and

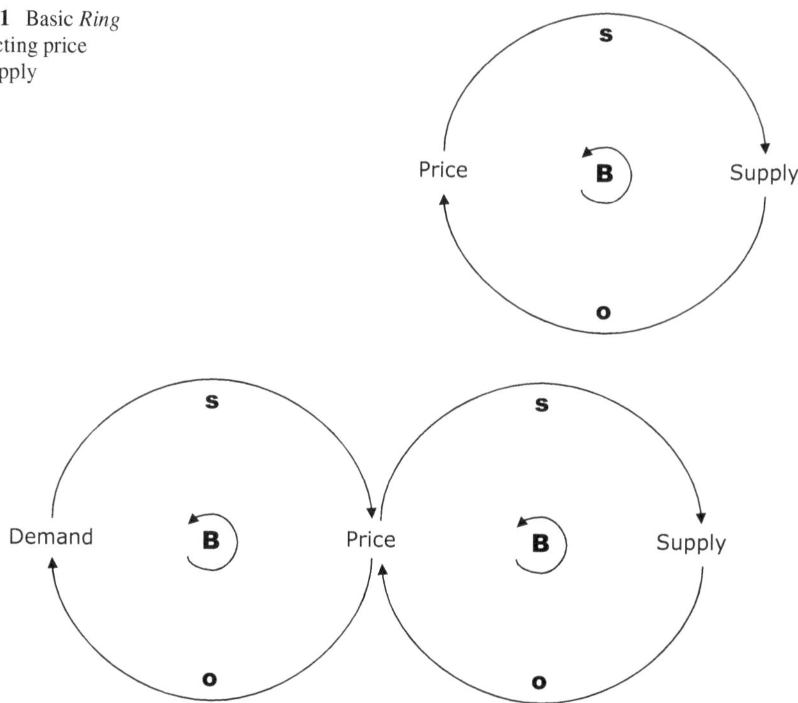

Fig. 1.1 Basic *Ring* connecting price and supply

Fig. 1.2 The law of the market under perfect competition

so the price increases again, inverting the direction of variation in demand and supply. The CLD in Fig. 1.3 instead represents a reinforcing loop [**R**], since each of the variables causes an increase in the second one (this example was presented in Senge); in successive repetitions of the process, the variation in both variables is accentuated.

The *Ring* in Fig. 1.4 represents a composite, though simplified, model of the system that regulates the struggle for life among populations, groups, and individuals.

The CLD in Fig. 1.5 connects the main variables that drive the marketing strategy.

Figure 1.6 shows the structure that regulates our job and wealth aspirations.

Figure 1.7 illustrates the system that describes burnout from stress; the search for short-term symptomatic solutions for stress can aggravate the problem in the long term rather than solve it.

Now that we have summarized the five rules of systems thinking and represented the structure of some simple systems, we can understand the approach of the fifth discipline in observing the interconnected systems of variables. As we know, Ludwig von Bertalanffy provided the following general definition of system:

> A system can be defined as a complex of interacting elements. Interaction means that the elements, p, stand in relations, R, so that the behavior of an element p in R is different from its behavior in another relation R′. If the behaviors of R and R′ are not different, there is no interaction and the elements behave independently with respect to the relations R and R′ (von Bertalanffy 1968, p. 55).

Fig. 1.3 Arms escalation

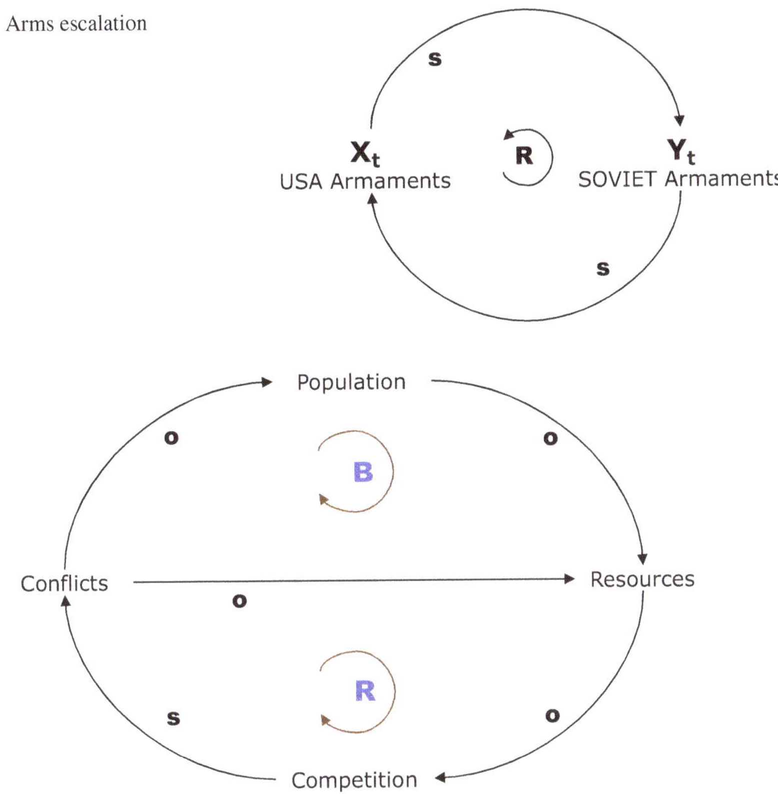

Fig. 1.4 Struggle for life

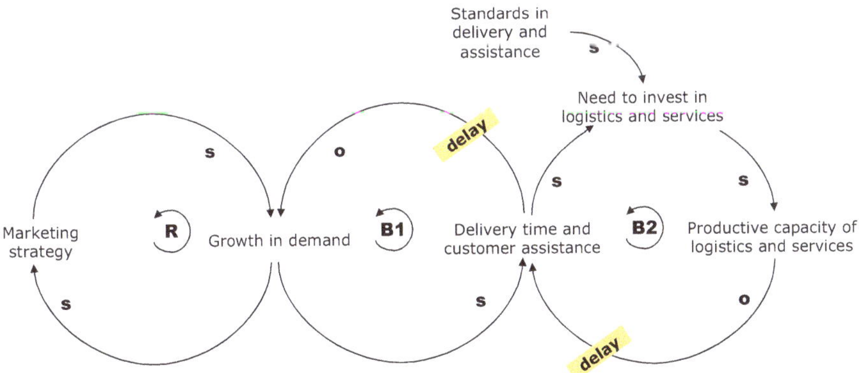

Fig. 1.5 Marketing strategy

This definition implies that each system has its own emergent characteristics compared to those of its constituent parts. Thus, the definition of dynamic system as a dynamic unity composed of interconnecting and interacting variables is justified.

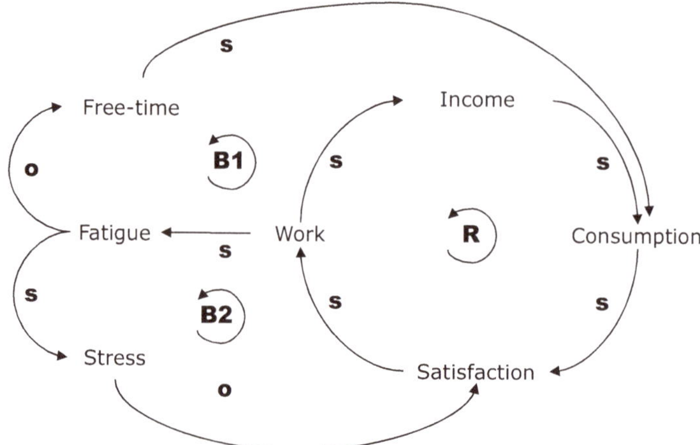

Fig. 1.6 Job satisfaction

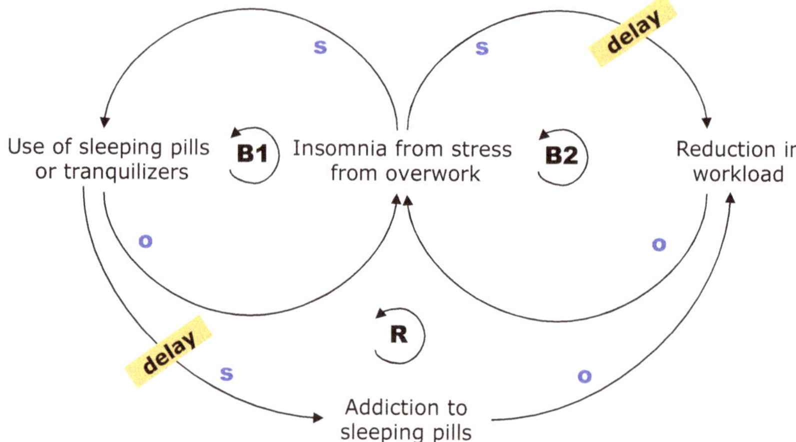

Fig. 1.7 Burnout from stress

Naturally, a dynamic system derives from a system of processors that produce the dynamics in the input (causes) and output (effects) variables.

Along with dynamic systems, there are *structural* systems and *ordinal* systems. *Structural systems* are operational systems made up of elements (usually processors) connected by a map of defined and stable relations forming the invariant organization of the system, according to Maturana and Varela's interpretation (1980) so clearly described by Stafford Beer in the Preface to their work.

> The relations between components that define a composite unity (system) as a composite unity of a particular kind, constitute its organization. In this definition of organization the components are viewed only in relation to their participation in the constitution of the unity (whole) that they integrate. This is why nothing is said in it about the properties that the

components of a particular unity may have other than those required by the realization of the organization of the unity (Beer 1992, p. XIX).

For example, systems of production facilities in a factory, organs that are part of the organization of corporations, battalions, companies, and platoons in a regiment are all structural. Clearly every dynamic system has its own structural system, concisely indicated as the "technical structure" of the system. *Ordinal systems* are units made up of non-dynamic elements (objects, concepts, words, etc.) arranged in an ordered map: the Eiffel Tower, made up of the ordered arrangement of structural components; this book, composed of the ordered arrangement of chapters, sections, sentences, and words; and the "afterlife" system described in Dante's *Divine Comedy*, with its ordered arrangement of the nine circles of hell, the seven terraces of purgatory, and the nine celestial spheres of Paradise. Every dynamic system is characterized by its own *structural system* that defines the system's "logical structure" and is represented by the CLDs.

Systems can be observed from an *external point of view* (forest) following a typical synthetic, teleological, or instrumental approach, according to which these systems, viewed as a whole, in many cases denote an object-oriented macro behavior, in that they serve a particular aim, or from an *internal point of view* (trees), following a structuralist, analytical approach, according to which it detects the macro behavior of the unitary system as the emergent effect of the micro behaviors.

The rules of systems thinking and the simple examples presented clearly show that the *fifth discipline has mainly an analytical, internal approach*; the system derives from the interconnected actions of the variables that form the structure. However, it is equally apparent that, after having identified the "trees," we must be able to observe the "forest" by identifying the boundary of the system observed as a whole, according to the synthetic approach, trying to analyze the emerging effects of the overall system. Many systems in systems thinking have no objective and no clear-cut instrumentality; but they are nevertheless wholes that possess their own logical structure and macro behavior and that are investigated following a holistic and holonic approach.

1.4 From Systems Thinking to System Dynamics: A Simulation of a Dynamic System

From the brief considerations in the preceding sections, it should be clear that systems thinking, or the fifth discipline, considers solely *dynamic systems* that have loops among the variables, so that the latter, being mutually linked, form "a whole" that presents its own dynamics, *new* and *emergent*, with respect to the dynamics of the component variables.

Systems thinking models are typically suited to qualitative research into the systems represented; they offer an understanding of their structure and allow us to examine the direction of the dynamics of the variables. These same models can, however, be translated into simulation models that allow us to numerically calculate and represent in tables and graphs the succession of values of the variables produced by the action and structure of the system under study. Expressed quantitatively, systems thinking adopts the simulation procedures of system dynamics, a discipline and technique that unquestionably goes back to Jay Forrester and his fundamental book *Industrial Dynamics* (Forrester 1961). In recent works, Forrester defines systems dynamics as follows:

> By "systems thinking" I mean the very popular process of talking about systems, agreeing that systems are important, and believing that intuition will lead to effective decisions. Without a foundation of systems principles, simulation, and an experimental approach, systems thinking runs the risk of being superficial, ineffective, and prone to arriving at counterproductive conclusions…
>
> "System dynamics" is a professional field that deals with the complexity of systems. System dynamics is the necessary foundation underlying effective thinking about systems. System dynamics deals with how things change through time, which covers most of what most people find important. System dynamics involves interpreting real life systems into computer simulation models that allow one to see how the structure and decision-making policies in a system create its behavior (Forrester 1999, p. 1).

> System dynamics combines the theory, methods, and philosophy needed to analyze the behavior of systems not only in management, but also in environmental change, politics, economic behavior, medicine, engineering, and other fields. System dynamics provides a common foundation that can be applied wherever we want to understand and influence how things change through time. The system dynamics process starts from a problem to be solved—a situation that needs to be better understood, or an undesirable behavior that is to be corrected or avoided. The first step is to tap the wealth of information that people possess in their heads. […] System dynamics uses concepts drawn from the field of feedback control to organize available information into computer simulation models (Forrester 1991, p. 5).

Let us consider the *basic loops*, or *Rings*, connecting two variables—for example, X, cause, and Y, effect—which change their values over a time interval $T = [t = 0, t = 1, t = 2$, etc.] divided into intervals of equal length, thereby producing reciprocal differences—ΔX_t and ΔY_t—through two opposing and linked *processes.* There are different ways to specify the *functions* needed to simulate the joint dynamics of X and Y (in both directions), since it is possible to variously define the equations that link the variations of X to those of Y.

First, the time sequence of the variations must be accurately determined; for example, the values ΔX_t could produce a variation ΔY_t at the same instant t (which is the normal assumption) or a variation ΔY_{t+1} quantified at instant $t + 1$, a variation ΔY_{t+2}, or even other variants; in the opposite direction, a variation ΔY_t could quantify a variation ΔX_{t+1} (as is normally the case) or present other variants.

Second, systems thinking reminds us that the processes that produce the variations in X and Y are carried out by a set of interconnected operative "apparatuses," or "machines," of a particular type and nature, each of which can be viewed as a black box whose *modus operandi* can even be ignored, unless we wish to zoom inside them. To simulate the dynamics of X_t and Y_t, we must know how each "operative machine" transforms the variations in the cause variable, X, and the variations in the other effect variable, Y. Therefore, we must assume that each "operative machine" generates the dynamics of the variables by means of a *variation function* that makes the values of the effect variable, Y_t, depend on those of the cause variable, X_t. This implies that for each simulation we must define the two *variation functions*:

1. The function for variations in Y due to variations in X, which we will call $g(Y/X)$, so that an input ΔX_t *always* produces, for example, the output:

$$\Delta Y_{t+1} = \left[g(Y/X)\Delta X_t\right] + DY_t$$

2. The function for variations in X due to variations in Y, which we will call $h(X/Y)$, so that an input ΔY_t *always* produces, for example, the output:

$$\Delta X_{t+1} = \left[h(X/Y)\Delta Y_t\right] + DX_t$$

The variables DY_t and DX_t indicate "disturbances" to the two variables from the environment outside the loop. We can represent the loop connecting X and Y in a dual direction, for example, using the two joint equations, which show a process length of one period:

$$\begin{cases} \Delta Y_t = \left[g(Y/X)\Delta X_t\right] + DY_t \\ \Delta X_{t+1} = \left[h(X/Y)\Delta Y_t\right] + DX_{t+1} \end{cases} \tag{1.1}$$

If we assume that the structure of the "machine" does not vary over time and does not depend on the values of X or Y, then we can also assume, by simplifying, that the activity of the "machines" that connect X_t and Y_t is represented by an "action parameter," which we indicate by "g = const" for X and "h = const" for Y, so that we can represent the loop (at least in the simplest cases) using two joint equations:

$$\begin{cases} \Delta Y_t = \left[g\,\Delta X_t\right] + DY_t \\ \Delta X_{t+1} = \left[h\,\Delta Y_t\right] + DX_{t+1} \end{cases} \tag{1.2}$$

I propose to call g the *rate of action* and h the *rate of reaction*. However, to carry out the simulations we must introduce into the model, in addition to the action parameters g and h (or their more complex formulations), the *initial values* X_0 and Y_0 of the two variables, along with any limits of variation admissible for them, or *variation constraints*. Figures 1.8 and 1.9 illustrate how the reinforcing "rings" and balancing "rings" can be structured to carry out simple simulations.

To make clearer the dynamics of the reinforcing loops we can assume that, in a simplified way, X and Y of the loop model [**R**] in Fig. 1.8 are linked by the following equations:

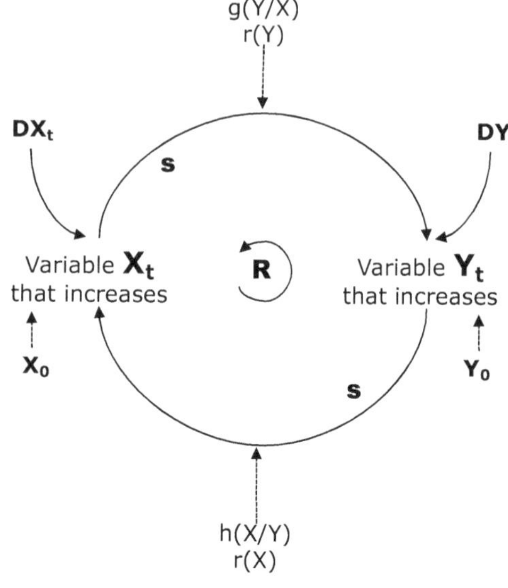

Fig. 1.8 Structure of a reinforcing *Ring* for simulations

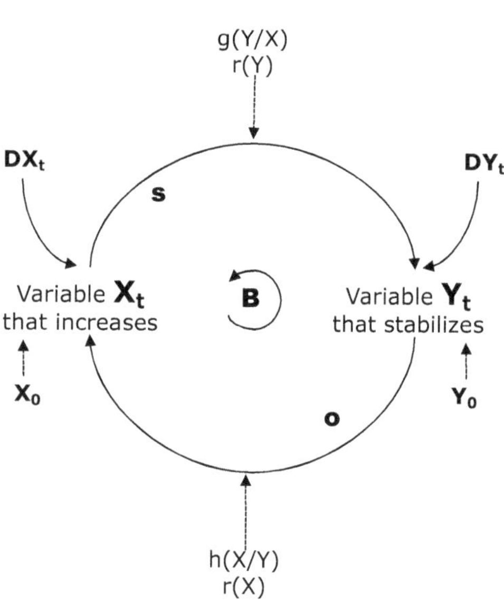

Fig. 1.9 Structure of a balancing *Ring* for simulations

$$\begin{cases} X_{t+1} = X_t + \dfrac{Y_t h(Y/X)}{r(Y/X)} + DX_t \\ Y_{t+1} = Y_t + \dfrac{X_{t+1} g(X/Y)}{r(X/Y)} + DY_t \end{cases} \tag{1.3}$$

$t = 1, 2, \ldots$

where

$$\begin{cases} \Delta Y_{t+1} = Y_{t+1} - Y_t \\ \Delta X_{t+1} = X_{t+1} - X_t \end{cases} \tag{1.4}$$

In addition to the functions describing the variation of the first variable as a function of the second, expressions in Eq. (1.3) introduce the variable r to indicate the *reaction time*, which represents the number of time scales over which the adjustment of variations in Y due to variations in X (of variations in X due to variations in Y) takes place. Referring to Y, a value $r = 2$ means that the variation in Y is halved in each interval since it is distributed over two intervals (I will take this up in more detail below). If g and h represent the *action* and *reaction rates* (which do not depend on the value of the causal variables X and Y), $r = 1$, and there are no disturbances, then the system normally has a linear dynamics; if, instead, g and h are themselves functions of X and Y and/or of time and/or $r = 2, 3, \ldots$, then the dynamics of the system will very likely be nonlinear.

A loop [**B**] can be represented in many ways, either by acting on the signs of the action rates g and h or on the signs of the causal links; in any event, we can construct the following recursive functions:

$$\begin{cases} X_{t+1} = X_t - \dfrac{Y_t h(Y/X)}{r(Y/X)} + DX_t \\ \\ Y_{t+1} = Y_t + \dfrac{X_{t+1} g(X/Y)}{r(X/Y)} + DY_t \end{cases} \tag{1.5}$$
$$t = 1, 2, \ldots$$

As Fig. 1.9 clearly shows, a process in the balancing *loops* leads to a variation in the direction **s** of the Y_t, but this causes a variation in the opposite direction, **o**, in X_t; in this case, an impulse (an increase or a decrease in X_t) is "balanced" by the opposite variation in Y_t. The two variables *balance* each other; in other words, a positive impulse ΔX_t causes an incremental variation ΔY_t. The new value of Y_{t+1} produces a decreasing variation in X_{t+1}, so that if the processes occur over several cycles, the two variables will tend to reciprocally compensate their variations and the balancing loop will produce cyclical dynamics in the variables.

There are no other elementary *Rings* since a loop where both variables vary in direction **o** would still be a reinforcing loop entirely similar to that in the model in Fig. 1.8.

1.5 Two Fundamental Laws of Systems Thinking

Given that we can represent a dynamic world of interconnected variables, building CLDs makes clear that *systems thinking* represents a very powerful and efficient way of thinking that enhances our intelligence. Systems thinking must be considered, however, not only as a technique but also as a *discipline* for efficient and effective thinking, in that it proposes:

- To train us to observe reality as composed of dynamic systems
- To provide us powerful models of description and simulation
- To improve our ability to acquire knowledge, that is, to learn
- To develop our intelligence

Systems thinking, precisely because it is a tool for developing our intelligence, must be learned gradually through practice and constant improvement (Checkland 1999). It is a discipline that requires the systems thinker to have a deep knowledge and to constantly apply its rules as well as to have the willingness to continually improve. This objective can be achieved by developing the necessary competencies in order to:

- Perceive and recognize the circularity of phenomena
- See how systems really function
- Intuit the effects of actions over time
- Feel responsible for the system's performance
- Predict the future
- Control dynamic processes

It is important to clarify which systems thinking examines and what types of models can thereby be obtained. As we have clearly specified, due to its intrinsic logic systems thinking considers *dynamic systems* of any kind in any field, building models of a world of incessant movement in continual transformation and evolution. This discipline considers not only dynamic but also *repetitive* systems, which are able to repeat their processes over time, as well as *recursive* systems, capable of interacting with themselves, in the sense that their output, entirely or in part, becomes their own inputs. Even if we are not used to observing them, repetitive and recursive systems are all around us; they are the typical essence of nature. Life itself is recursive in its typical process of birth, reproduction, and death, which is destined to repeat itself again and again.

Systems thinking is particularly sensitive to *systems with memory*; it forces us to consider the connections between variables, always zooming between high-level variables, which accumulate variation over time, and more detailed (state) variables, which cause variations over time; it forces us to observe the dynamics of recursive processes and not only individual pairs of values to consider the loops and not only the pure causal connections (von Foerster 2003, p. 143).

With its core of general principles and concepts, the systems thinking discipline makes it possible (though not easy, since it is a discipline) "to see" an interconnected, dynamic, and recursive world, that is, to identify the structure that determines the trends in phenomena and produces complexity, nonlinearity, and memory in systems that operate in all fields of human action—biological, economic, and social. A good model of the world under examination must always foresee balancing factors for the reinforcing loops and reinforcing factors for the balancing loops.

The systems thinking models are certainly not the only ones capable of increasing our knowledge of the world, but in my view their cognitive effectiveness owes to their ease of construction and communication (Nonaka 1994). The only skills they require are perspicacity and insight; they use elementary techniques, are understandable even to non-experts, and can easily be communicated, examined, and improved. They allow us to learn to collectively improve our understanding of the world and can be easily translated into quantitative simulation models.

The concepts introduced so far allow us to present a *first fundamental law* of systems thinking:

1. Law of Structure and Component Interaction. *In order to understand and control the dynamics in the world it is necessary to identify the systemic structures that make up this world.*

   (a) On the one hand, the behavior of a variable depends on the system in which it is included.
   (b) On the other hand, the behavior of the entire system depends on both its *logical* and *technical* structures; that is, on the interconnections among its component variables (causal chains and causal loops) and on the "machines" producing their variations.
   This law has several *corollaries*:

      (i) There are no "isolated" and "free" variables, only interconnected ones.
      (ii) Each variable has a *cause* for its variations and produces an *effect* on the other variables.
      (iii) Analyzing the dynamics of a variable (e.g., analysis as a historical series of values) does not mean *explaining* or *understanding* the dynamics but simply *describing them*.
      (iv) It is useless to try to modify the values of a variable if first we do not understand the systemic structure of which it is a part, since the balancing loops will restore its value and the reinforcing loops will increase it.
      (v) In observing a dynamic world, the ceteris paribus assumption is never valid.

   Connected to the preceding fundamental law is a *second fundamental law* of systems thinking, which I shall name as:

2. Law of Dynamic Instability. *Expansion and equilibrium are processes that do not last forever; they are not propagated* ad infinitum. *Sooner or later stability is disturbed. Sooner or later the dynamics are stabilized* (see Sect. 1.7.4). Even if we are not aware of it, in every systemic context the reinforcing loops are always linked to some balancing loop and vice versa.

1.6 Systems Archetypes: Three Relevant Structures of Human Behavior—The Archetypes of the Three Myopias

A particularly useful class of systems that systems thinking deals with are *systems archetypes*, or standard structures, that is, *simple general* and *stable* models of relations that frequently recur in various situations in any type of organization, public as well as private companies, and in different environments.

> One of the most important, and potentially more empowering, insights to come from the young field of systems thinking is that certain patterns of structure recur again and again. This "systems archetypes" or "generic structures" embody the key to learning to see structures in our personal and organizational lives. [...] Because they are subtle, when the archetypes arise in a family, an ecosystem, a news story, or a corporation, you often don't see them as so much as feel them. Sometimes they produce a sense of deja vu, a hunch that you've seen this pattern of forces before. "There it is again" you say to yourself (Senge 2006, p. 93).

The aim of archetypes is to rapidly increase the capacity of the observer, the manager, the governor, or the decision-maker to see the systemic problems, recognize the structures that determine these problems (see Chap. 5, Sect. 5.9), and identify leverage effect of these structures for the purpose of formulating definitive solutions. Senge (1990) describes ten archetypes that have been adopted by all of his followers and that by now have become part of the shared heritage of knowledge of this field. In his book (1990, in particular, Appendix 2), Peter Senge presented ten archetypes, which, in his "original list," are as follows:

1. Balancing process with delay
2. Limits to growth (also known as Limits to Success)
3. Shifting the burden
4. Special case: shifting the burden to the intervenor
5. Eroding goals (also known as Drifting Goals)
6. Escalation
7. Success to the successful (also known as Path Dependence)
8. Tragedy of the commons (Hardin, 1968; Ostrom, 1990)
9. Fixes that fail
10. Growth and underinvestment

The literature on systems thinking, along with many consultants, have normally considered Senge's archetypes as a finite set (e.g., Wolstenholme and Corben 1993; Dowling et al. 1995; Kim 2000; Kim and Anderson 2011; Braun 2002; Heaven & Earth 2000; Continuous-Improvement-Ass. 2003; Bellinger 2004; Dempsey 2015; Insight-Maker 2017), even if other authors had presented their own list of system archetypes that could be applied in specific contexts, for example, Jay Forrester (1970) and Dana Meadows (1982). Nevertheless, Senge's list represents the most frequent and visible archetypes in organizations, even though it can be extended. Above all, most of Senge's archetypes present a mirror-image archetype, to which new important and significant archetypes can be added. I hold that, along with Senge's *ten standard archetypes*, we can identify other archetypes, equally important and widespread (Mella 2012), several of which we shall examine in subsequent chapters (Chap. 5, in particular).

In my opinion there are three basic *archetypes*, which usually occur together and act in a unitary manner, which account for a vast variety of behaviors by human agents and which also justify the validity of many of Senge's standard archetypes (Sect. 1.7.8). These basic structures describe man's natural instinct, reinforced by the prevailing educational and cultural models, not to give enough importance to the harm his behavior can produce; as a result, in the face of such harmful consequences the human agents refrain from modifying their cultural attitudes and applying systems thinking to assess the disastrous effects of their behavior.

The first archetype, which I propose to call the "short-term preference archetype," or "Temporal Myopia" archetype, is innate in all men who adopt a "short-sighted" rationality principle. This archetype manifests itself when a particular behavior is repeated over time and produces, *at the same time*, immediate, *short-term advantages* (benefits, or pleasures) and *disadvantages* (sacrifices, or harm) that occur over the *long run* with a significant delay. Here it is not a question of choosing between two or more advantages and two or more disadvantages, distributed differently over time, but of understanding how an immediate, or short-term, certain *advantage* is evaluated when accompanied by a long-term, probable, even notable *disadvantage* under the assumption that these are produced by the *same repetitive behavior.*

Since the short-term *advantages* are immediately perceived, it is "natural" to attribute high value to them, and this leads the agents to prefer and repeat the behavior that produces them. At the same time, the perception of the short-term advantages is accompanied by a diminished perception (probability, weight, judgment) of the long-term *disadvantages*, which makes it likely the agent will repeat the behavior that produces the short term advantages, even if long-term disadvantages are inexorably associated with these. Figure 1.10 shows the simple structure of the "short-term preference archetype": the repetitive behavior produces short-term advantages (vertical arrow, **s**) that, in turn, encourage the continuation of the behavior, according to loop [**R1**].

The current advantages reduce the perception of the long-term disadvantages—which are thus produced with a considerable *delay*—and this encourages even more of the same behavior in order to gain present advantages, as shown in loop [**R2**], so that the long-term disadvantages do not condition behavior but represent only its effect (no loop).

The two loops act together to increase *short-term preferences*, which guide behavior and inexorably create long-term problems when the *disadvantages* caused by the repeated behavior appear. It is useful to point out that the terms *advantages* and *disadvantages* must be taken broadly. The chance to avoid a disadvantage (harm, an unpleasant event) could be perceived as an advantage, and a missed advantage could be perceived as a disadvantage. In any case, the structure in Fig. 1.10 would not change much. The *short-term* advantage prevails over the

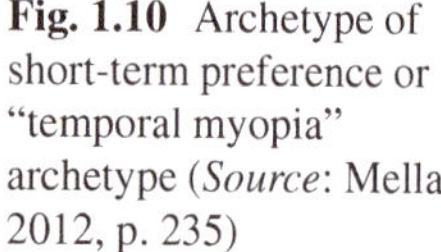

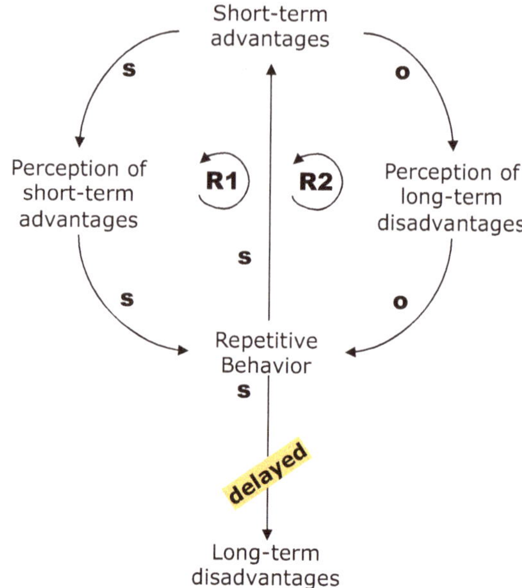

Fig. 1.10 Archetype of short-term preference or "temporal myopia" archetype (*Source*: Mella 2012, p. 235)

long-term disadvantage in conditioning behavior. There is no way out: this archetype is always in action, and when it is not it is always lurking.

Let us reflect on the behavior of parents who prefer to overfeed their children and allow them to watch TV or play with the PC for hours and hours in order to avoid any disturbance or complaint today, not thinking about the long-term physical and psychological consequences. Consider pollution "today" and the lack of interest in the damage "tomorrow." The behavior of the smoker who prefers the short-term pleasure of thousands of cigarettes to the danger from future harm, perceived as remote, is exactly the same as that of the person who prefers a convenient disposal of waste, accumulating this in landfills or burning it in incinerators, thereby demonstrating a lack of interest in the damage "tomorrow." Let me take another example: spending cuts for present academic research or health care to favor a balanced budget "today" will cause relevant damage to the state of "medium- to long-term" competitiveness and health status of the population. There is a popular saying: "The ant is always wiser than the cicada," but the cicada prefers "today" the pleasure that "tomorrow" will lead to its death. Not to mention the behavior of the careless driver ….

The leverage effect that derives from this ARCHETYPE is fundamental: when we perceive its action, it is useless to try and oppose loop [**R1**] (anti-smoking laws, emission standards); instead, it is necessary to weaken loop [**R2**] through a convincing information and training campaign, day after day, to make the decision-maker aware of the disadvantages the behavior can cause (awareness campaign, company training). The lack of respect for the Kyoto protocol is mainly to be attributed to those countries that do not want to limit their emissions in order not to undermine growth "today," not worrying about the damage global warming will cause in upcoming decades.

The *archetype* of short-term preference has two parallel variants that are equally widespread and ruinous:

1. The agent prefers the objectives that bring *local* advantages over those that would provide *global* ones; this is the "local preference archetype", or "Spatial Myopia" archetype (Fig. 1.11 left).
2. The agent prefers the objectives that produce *individual* advantages over those for larger *groups*; this is the "individual preference archetype" or "Social Myopia" archetype (Fig. 1.11 right).

Because the local and "partisan" advantages are perceived in the short term while the global and collective ones are usually noted in the long term, these structures can be considered as particular cases of the preceding one. Nevertheless, we can distinguish them by referring to them as the *archetypes* of "individual preference" (Mella 2012, p. 236).

We are all inclined to give priority to objectives that bring advantages to our family, club, or association rather than to those that bring advantages to the collectivity. Anyone working in a department tries to make it function in the best way possible, even if this conflicts with the functioning of other departments. Judgments on the local disadvantages from public works (rubbish dumps, incinerators, dams, highways, etc.) prevail over those regarding the global advantages they will provide to the larger territory. Even if we are aware that public landfills or incinerators are necessary for the common good, we steadfastly refuse to allow these to be placed in our towns, since local interests prevail over the global advantages they will provide to the larger territory. The NIMBY effect is always present. "Not in my back yard" is the most common individual response to the community's request to put up with the construction of public works for the advantage of the collectivity.

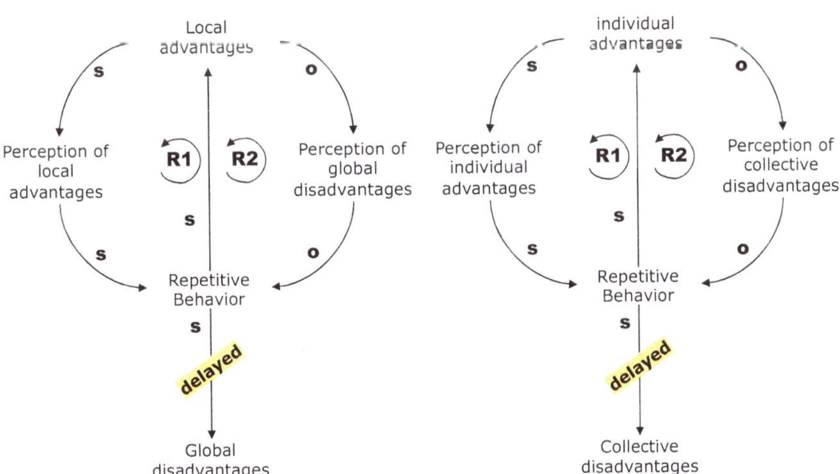

Fig. 1.11 Archetype of local (*left*) and individual (*right*) preference or "spatial and social myopia" archetypes. (*Source*: Mella 2012, p. 236)

It is easy to see that the three archetypes often overlap; thus, it is useful to present them using a single model, as shown in Fig. 1.12.

The three archetypes that dominate the prevailing form of individual–agent behavior are very powerful, since the individual that carries out the archetypes sees them as entirely rational. In fact, every agent naturally behaves based on his or her own individual, local, and short-term rationalities, not "willingly" accepting being guided by a global rationality. The short-term, local, and individual advantages prevail over the long-term, global, and collective ones in conditioning behavior. Therefore, if we wish to avoid the harmful consequences of the actions of the three archetypes, it is necessary to act to counter the reinforcing effect of loops [**R1**] and [**R2**]. There are three important weakening actions for such loops; to better understand these we can refer to Fig. 1.13, where the three archetypes have been grouped into a single model of equal significance.

The first solution is to try to weaken or eliminate loop [**R1**]. Since the local advantages would immediately reveal the benefit of the individual action, an appropriate system of *disincentives*, even in the form of sanctions or the imposition of costs and taxes, proportional in some way to the amount of short-term local advantages, would lessen the perception of the convenience from repeating the current behavior, thus forming the balancing loop [**B1**], which could compensate the effects of loop [**R1**].

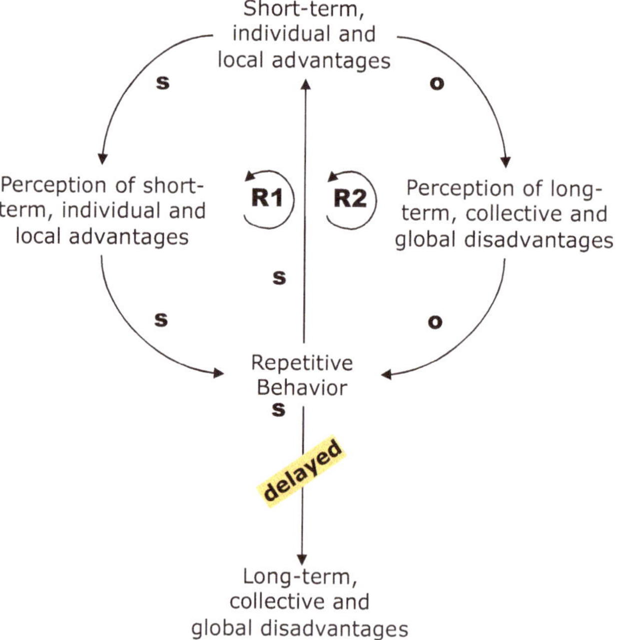

Fig. 1.12 Archetype of short-term, local, and individual preference or "the three myopias" archetype

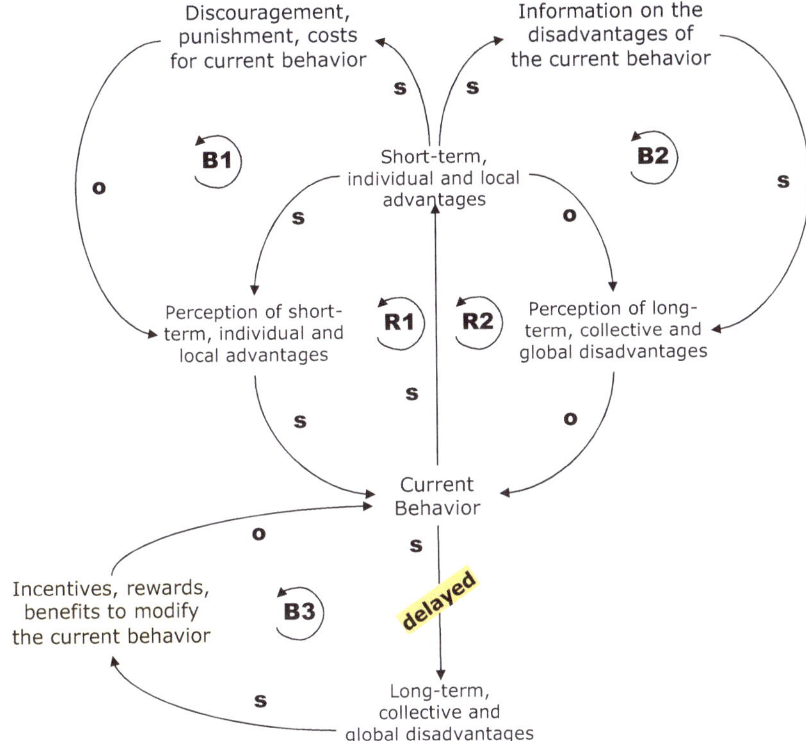

Fig. 1.13 Actions to weaken the behavioral myopia archetypes

A second way to counter the archetype is to weaken loop [**R2**], which produces an underestimate by the agent of the long-run disadvantages. An incisive, widespread, and continual *stream of information* about the negative long-run global effects due to short-term, local, and individualistic behaviors would lead to a conscious awareness by the agent of the global harm from his or her behavior. The persuasive capacity of this *information* must be proportional to the amount of short-term local advantages. The more relevant these are, the more the information must be insistent, repeated, uniform, credible, convincing, and diffused over vast territories.

The third form of intervention to reduce the negative effects from the action of the archetype in Fig. 1.12 is to act directly on the agent's current behavior through incentives, [**B3**], even in the form of social and economic *benefits*, rewards, tax breaks, etc., to induce the agent to modify his or her current behavior by reducing its intensity and frequency, adopting processes and technologies that reduce the long-term global disadvantages.

The *modus operandi* of the three archetypes will be taken up in Chap. 6 and in Sect. 7.7, Chap. 7.

1.7 Complementary Material

1.7.1 Reinforcing Loop: Arms Escalation

Reinforcing loops represent many typical processes in which the loop variables reciprocally increase their values. This section presents basic reinforcing loops to represent escalation processes where X and Y continually reinforce each other. The subsequent section will consider a "ring" that shows a typical explosive process, in which a variable reinforces itself through increases generated by the variable itself. To simulate a "ring" that represents an escalation process we go back to the model in Fig. 1.3, which describes the arms race, simulating this with expressions in Eq. (1.2) using Excel, whose use is particularly effective and simple to use for "small" loops formed by a limited number of variables (no more than eight or ten, in my experience). The simulation model that reveals the arms dynamics of two nations, X and Y, is illustrated in Fig. 1.14.

In the stylized loop that serves as a *control panel*, we can immediately read the values assigned to the action and reaction rates as well as the initial values of the two variables. I have introduced the assumption that X, after observing the low initial level of Y's armaments (equal to $Y_0 = 50$), compared to its own level of $X_0 = 200$, overvalues its initial superiority. I also assume that X has a low action rate, $h(X/Y) = 10\%$, compared to an action rate for Y of $g(X/Y) = 15\%$, as well as a higher reaction time of $rX = 2$. This means that while Y reacts more energetically to variations in X's armaments, since $g = 15\%$, X reacts less intensely ($h = 10\%$) and more slowly ($r = 2$). The arms dynamics (Fig. 1.10) show that X's initial overestimation of its own strength allows Y to achieve a superior level of armaments starting already in the seventh year.

It is clear that the model does not explicitly represent the generating processes of the arms dynamics of X and Y. These dynamics are in fact produced when, due to processes fed by psychological and social motivations considered as a black box, increases in the armaments of one side, say X, increase the fear of Y's leaders and vice versa. This fear moves Y's leaders to increase their arms, but, following a typical escalation process, the increase in Y's arms causes an increase in X's arsenal; loop [**R**] acts over many periods. To take these variables into account we need a new model, which we shall examine in Sect. 1.7.5.

1.7.2 Reinforcing Loops: Virus Explosion

The power of reinforcing loops as models to represent expansive phenomena can be illustrated by a simple example presenting several simulations of the *dynamics of individuals in a population infected by a pathogen* for a period $T = 100$ weeks (or some other period). The model is "extremely simplified" in that it considers only two variables:

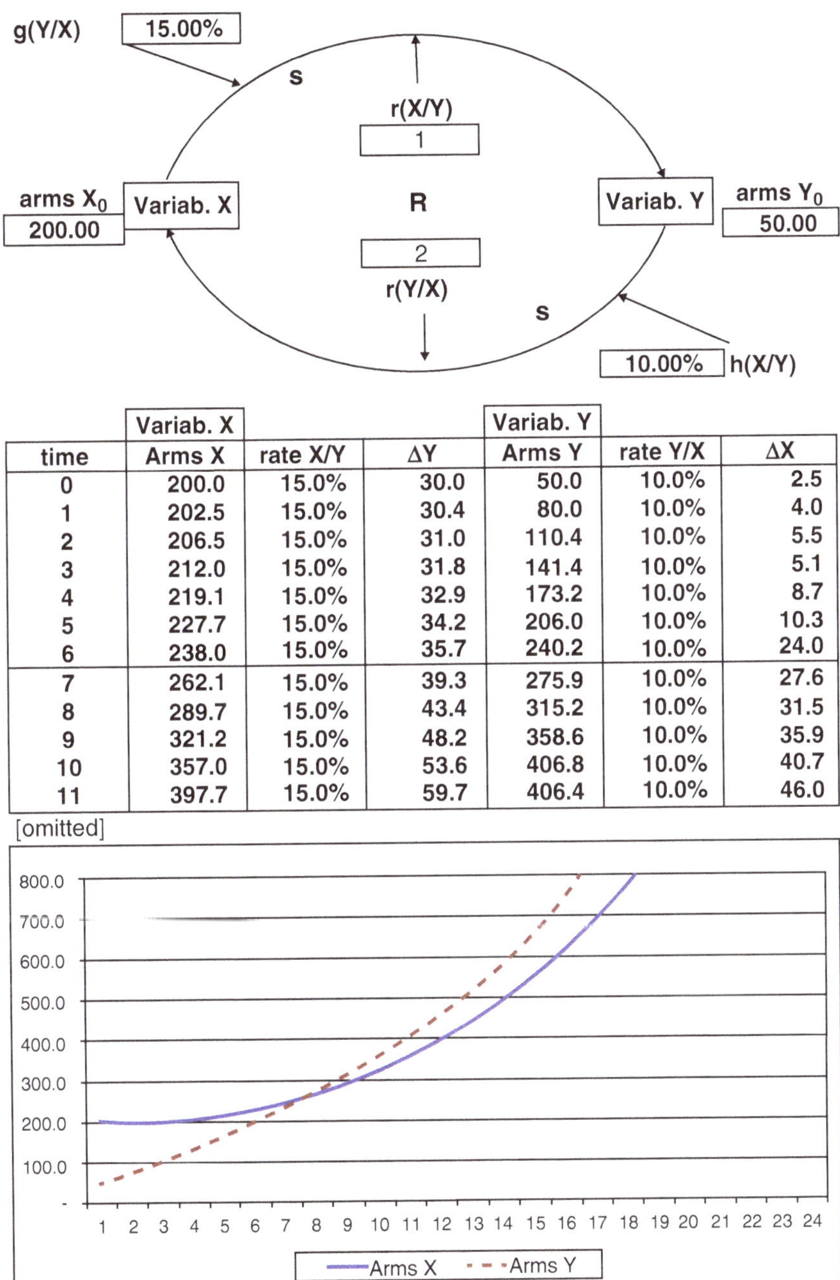

	Variab. X			Variab. Y		
time	Arms X	rate X/Y	ΔY	Arms Y	rate Y/X	ΔX
0	200.0	15.0%	30.0	50.0	10.0%	2.5
1	202.5	15.0%	30.4	80.0	10.0%	4.0
2	206.5	15.0%	31.0	110.4	10.0%	5.5
3	212.0	15.0%	31.8	141.4	10.0%	5.1
4	219.1	15.0%	32.9	173.2	10.0%	8.7
5	227.7	15.0%	34.2	206.0	10.0%	10.3
6	238.0	15.0%	35.7	240.2	10.0%	24.0
7	262.1	15.0%	39.3	275.9	10.0%	27.6
8	289.7	15.0%	43.4	315.2	10.0%	31.5
9	321.2	15.0%	48.2	358.6	10.0%	35.9
10	357.0	15.0%	53.6	406.8	10.0%	40.7
11	397.7	15.0%	59.7	406.4	10.0%	46.0

Fig. 1.14 Arms escalation (see Fig. 1.3)

- Variable X = number of infected individuals
- Variable Y = number of infected individuals
- Initial value of variable $X_0 = 1$; the epidemic starts with a single "infected individual zero"
- Maximum value of variable X limited to $X_{max} = 1$ billion units

In the *first simulation*, the following rates are assumed:

- $g(Y/X)$, *contagion rate* (*reproduction rate* or *transmission rate*)—called $R0$ in the early stages of the pandemic and Rt after containment actions—is indicated as a percentage: $g(Y/X) = 200\%$; this means that each element of the population infects two elements at each step (delays are not introduced in the simulation).
- $h(X/Y)$, rate of increase of infected individuals in the population, $h(X/Y) = 100\%$: this rate indicates how many of the newly infected individuals increase the infected population.
- The reaction times of the two variables (see Chap. 2, Sect. 2.9) are $rX = rY = 1$, as no delays have been assumed.

As shown in the simulation in Fig. 1.15, with such very high rates (which are unrealistic in human populations), the entire population would become infected toward the end of Week 19. The assumptions behind the model can be made more

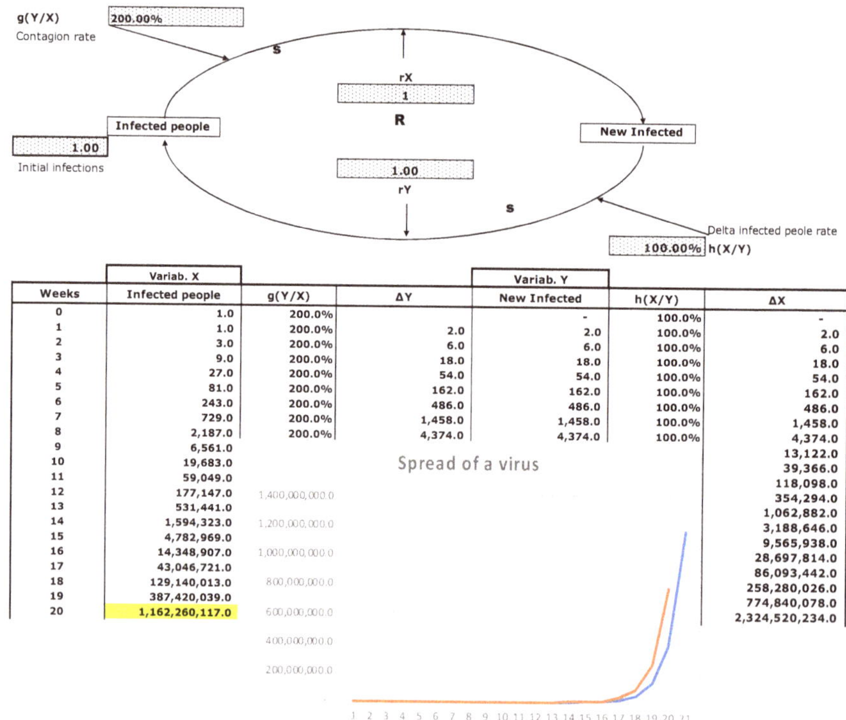

Weeks	Variab. X Infected people	g(Y/X)	ΔY	Variab. Y New Infected	h(X/Y)	ΔX
0	1.0	200.0%		-	100.0%	-
1	1.0	200.0%	2.0	2.0	100.0%	2.0
2	3.0	200.0%	6.0	6.0	100.0%	6.0
3	9.0	200.0%	18.0	18.0	100.0%	18.0
4	27.0	200.0%	54.0	54.0	100.0%	54.0
5	81.0	200.0%	162.0	162.0	100.0%	162.0
6	243.0	200.0%	486.0	486.0	100.0%	486.0
7	729.0	200.0%	1,458.0	1,458.0	100.0%	1,458.0
8	2,187.0	200.0%	4,374.0	4,374.0	100.0%	4,374.0
9	6,561.0					13,122.0
10	19,683.0					39,366.0
11	59,049.0					118,098.0
12	177,147.0					354,294.0
13	531,441.0					1,062,882.0
14	1,594,323.0					3,188,646.0
15	4,782,969.0					9,565,938.0
16	14,348,907.0					28,697,814.0
17	43,046,721.0					86,093,442.0
18	129,140,013.0					258,280,026.0
19	387,420,039.0					774,840,078.0
20	1,162,260,117.0					2,324,520,234.0

Fig. 1.15 Spread of a virus: number of infected cases

realistic by taking into account the reactions of the authorities to the contagion data. First of all, we can assume that there are traffic bans, mandatory transmission containment instruments (masks, for example), and social distancing measures, and that this reduces the rate of contagion in 6 weeks. In parallel, it can be assumed that hospitals have equipped themselves to care for the infected and that, therefore, the rate of increase in the number of infected individuals is also significantly reduced. In the *second simulation*, the following lower rates are assumed:

- Starting from Week 7 the following rates apply:
- $g(Y/X)$, contagion rate (in percentage) $g(Y/X) = 100\%$. Each individual in the population infects only one other individual.
- $h(X/Y)$, rate of increase of infected individuals in the population, $h(X/Y) = 20\%$. Four out of five infected individuals are cured after treatment.

With these more realistic rates, the contagion of the entire population takes place in Week 49. What is more interesting is that the number of infected individuals exceeds 300,000 units only after 40 weeks.

The simulation model allows us to hypothesize other, even more realistic, rates that slow down the dynamics of contagion even further; for example, we could set $g(Y/X) = 80\%$ and $h(X/Y) = 10\%$. In Sect. 1.7.7, the model will be modified to also introduce the hypothesis that the number of infected individuals is reduced because they are cured either through treatment or spontaneously.

1.7.3 Multiple Loops: The Law of Accelerating Returns

A significant example of a multiple reinforcing loop is the one that produces the model in Ray Kurzweil's example (2001, 2005) of his Law of Accelerating Returns. After presenting the dynamics of qualitative and quantitative growth of the main environmental and technological variables, particularly over the last few thousand years, and observing that the data show exponential growth, Kurzweil proposes the theory of the acceleration of growth or "theory of accelerating returns": changes in the environmental variables that, until yesterday, were studied in terms of *velocity* must now be interpreted in terms of *acceleration*, a characteristic that affects all aspects of our world: technology, public life, political movements and wars, economic dynamics (Evans 1991), as well as social aspects. As recognized by Ray Kurzweil, during the age of the Pharaohs or, 2000 years later, the Romans, technological progress, military conquests, power dynamics, changes in the borders between states, and other macro phenomena were interconnected but followed an understandable dynamics in terms of a slow *linear evolution* based on *constant rates of change*. For some years now, such linear dynamics have given way to the gradual "reinforcement" of the speed of change of all phenomena; that is, an *acceleration* of change has caused an *increase in the rates of change* of the different variables in the system.

An analysis of the history of technology shows that technological change is exponential, contrary to the common-sense "intuitive linear" view. So we won't experience 100 years of progress in the 21st century — it will be more like 20,000 years of progress (at today's rate) (Kurzweil 2001).

Adopting the logic and language of systems thinking, we can easily recognize that the *acceleration* in the variations pointed out by Kurzweil occurs when two (or more) variables, X_t and Y_t, which form a *reinforcing loop*, not only produce the reciprocal variations but also lead to a gradual *increase in the variation rates*—$g(Y/X)$ and $h(X/Y)$—of the variables (or at least in one of them), as shown in the model in Fig. 1.16, where the variable Y also impacts the rate $g(Y/X)$, based on the function $k = G(Y)$, and the variable X impacts the function $h(X/Y)$, based on the function $h = F(X)$, thereby *accelerating* the variations in the interacting variables.

In the economic and technical world, the *modus operandi* of the model in Fig. 1.16 is not an exception but the rule (Castera, 2016); Fig. 1.17 shows an example of how the Law of Accelerating Returns impacts the growth in electronics and informatics.

What causes a slow dynamic process to be transformed into a fast, and at times explosive, one? Systems thinking offers a very simple formal explanation. The slow dynamics is almost always due to a simple law of *linear accumulation* of the type $Y = (1 + r\ n)$, $r > 0$; if r is negative, there is a decrease. When Y increases (decreases) beyond a certain limit, the dynamics become *nonlinear* since one or more reinforcement loops between Y and other variables that interconnect and interact with it come into play. The increase in Y leads, for example, to an increase in a variable Z, which interacts with Y in a reinforcing loop, which leads to a further increase in Y, which once again increases Z, and so on, for several recursions typical of the systems in Systems Thinking. As further detailed in Chap. 6, Sect. 6.6, after a period of slow growth, the increase in the "greenhouse gases" as a result of

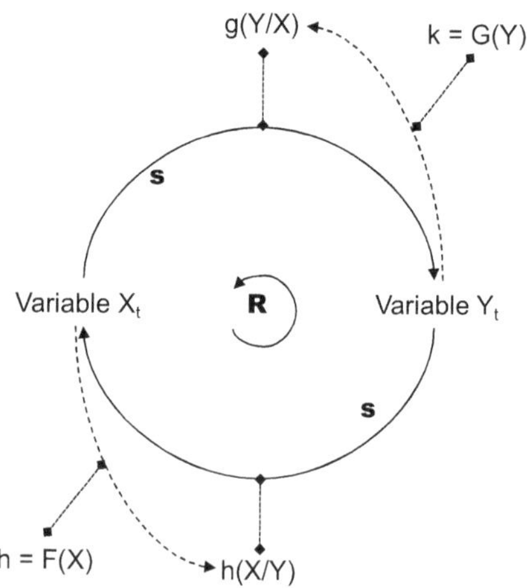

Fig. 1.16 Loop showing the action of accelerating returns

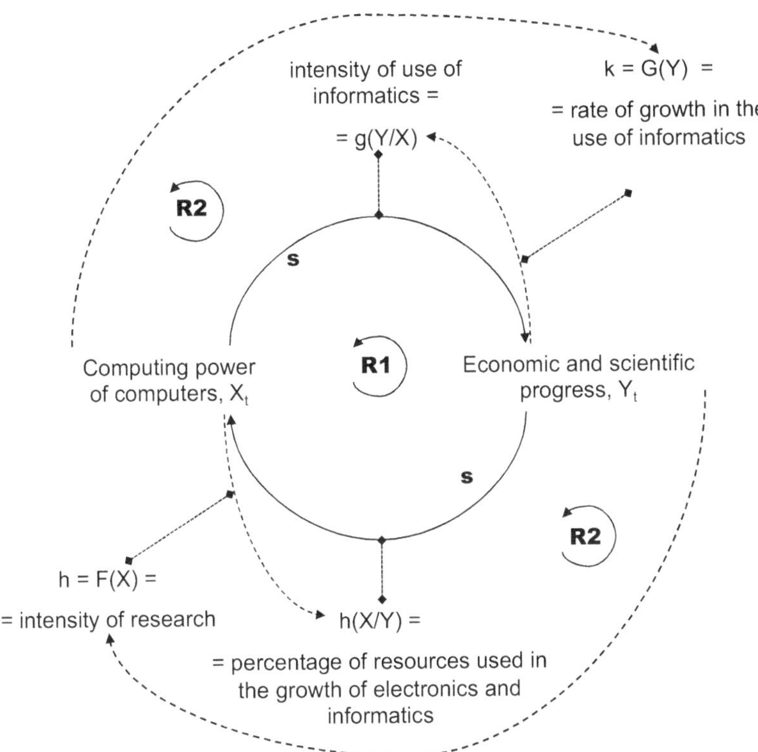

Fig. 1.17 The action of the Law of Accelerating Returns on the growth in electronics and informatics. (*Source*: my elaboration of the model in Kurzweil, 2001, online)

man's production and consumption activity becomes significant and preoccupying. By trapping the heat from the sun's rays, the greenhouse gases become a determining factor in "global warming," which leads to the gradual warming of the soil, causing the growth of trees that retain heat, and of the seas, causing an increase in dark algae that produces the same effect. Moreover, global warming causes the melting of glaciers at the North and South Poles, thereby reducing the "albedo effect" and leading to a new increase in the Earth's temperature, and in the tundra, causing the escape of methane bubbles (which before were trapped in the ice) that further increase the greenhouse gases. By the time we become aware of global warming it is too late to take remedial action due to the action of these and other reinforcing loops that are difficult to control. As observed in Sect. 1.5, the dynamics of a variable are never isolated, since every variable is contained in dynamic systems of interacting variables.

To demonstrate how the Law of Accelerating Returns works, Kurzweil provides a formal example that I have translated into the model in Fig. 1.18, which, after what was said above, has no need of further comment.

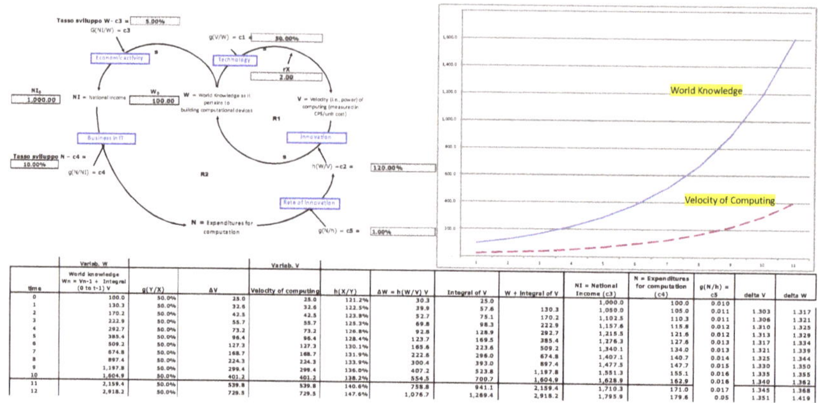

	Variab. W			Variab. V									
time	World knowledge Wn = Vn-1 + Integral (0 to t-1) V	g(Y/X)	ΔV	Velocity of computing	h(X/Y)	ΔW = h(W/V) V	Integral of V	W + Integral of V	NI = National Income (c3)	N = Expenditures for computation (c4)	g(N/h) = c5	delta V	delta W
0	100.0	50.0%	25.0	25.0	121.2%	30.3	25.0		1,000.0	100.0	0.010		
1	130.3	50.0%	32.6	32.6	122.5%	39.9	57.6	130.3	1,050.0	105.0	0.011	1.303	1.317
2	170.2	50.0%	42.5	42.5	123.8%	52.7	75.1	170.2	1,102.5	110.3	0.011	1.306	1.321
3	222.9	50.0%	55.7	55.7	125.3%	69.8	98.3	222.9	1,157.6	115.8	0.012	1.310	1.325
4	292.7	50.0%	73.2	73.2	126.8%	92.8	128.9	292.7	1,215.5	121.6	0.012	1.313	1.329
5	385.4	50.0%	96.4	96.4	128.4%	123.7	169.5	385.4	1,276.3	127.6	0.013	1.317	1.334
6	509.2	50.0%	127.3	127.3	130.1%	165.6	223.6	509.2	1,340.1	134.0	0.013	1.321	1.339
7	674.8	50.0%	168.7	168.7	131.9%	222.6	296.0	674.8	1,407.1	140.7	0.014	1.325	1.344
8	897.4	50.0%	224.3	224.3	133.9%	300.4	393.0	897.4	1,477.5	147.7	0.015	1.330	1.350
9	1,197.8	50.0%	299.4	299.4	136.0%	407.2	523.8	1,197.8	1,551.3	155.1	0.016	1.335	1.355
10	1,604.9	50.0%	401.2	401.2	138.2%	554.5	700.7	1,604.9	1,628.9	162.9	0.016	1.340	1.362
11	2,159.4	50.0%	539.8	539.8	140.6%	758.8	941.1	2,159.4	1,710.3	171.0	0.017	1.345	1.368
12	2,918.2	50.0%	729.5	729.5	147.6%	1,076.7	1,269.4	2,918.2	1,795.9	179.6	0.05	1.351	1.419

Fig. 1.18 Model (derived from Kurzweil, 2001) highlighting the variables that produce an accelerating return on the growth in electronics and informatics (see Fig. 1.17)

1.7.4 The General Law of Dynamic Instability

At the end of Sect. 1.5, I mentioned a general and intuitive law of systems thinking, which I have called as follows:

LAW OF DYNAMIC INSTABILITY: *Expansion and equilibrium are processes that do not last forever; they are not propagated* ad infinitum. *Sooner or later stability is disturbed. Sooner or later the dynamics are stabilized.*

Operationally speaking, this law affirms in a simple way that, though we are unaware or unable to observe this, every expansion loop is always associated with a balancing loop that dampens the expansion dynamics; and vice versa, every balancing loop is associated with some type of expansion loop that counters the balancing effect. Moreover, external disturbances can come from the environment in the form of braking variables that counter the expansion or acceleration variables that eliminate the balancing effect, as shown in the model in Fig. 1.19.

Paraphrasing Newton's first law of mechanics: "Every object remains in its state of rest or uniform motion in a straight line unless a force intervenes to modify this state", Systems Thinking could instead state: "Every repetitive system does not endlessly produce its own reinforcing or balancing processes because other processes intervene to reverse the dynamics". It seems impossible to respect the wise motto: "Quieta non movere, mota quietare!" (Mella 2012, p. 73).

A simple and interesting numerical simulation—in one of its many possible variants—is presented in Fig. 1.20. Several simulation tests are presented in Fig. 1.21. The diagram shows that the curve in bold, which depicts the dynamics of the stabilized variable Y_t, has a cyclical dynamics precisely due to the reinforcement from the curve for variable X_t and the balancing effects from the curve for variable Z_t.

By modifying the values in the control panel (directly shown in the CLD at the top of the table), we obtain the explosive dynamics of growth in Y if variable X prevails (see test 4 in Fig. 1.21) or the explosive decreasing dynamics if Z prevails. By

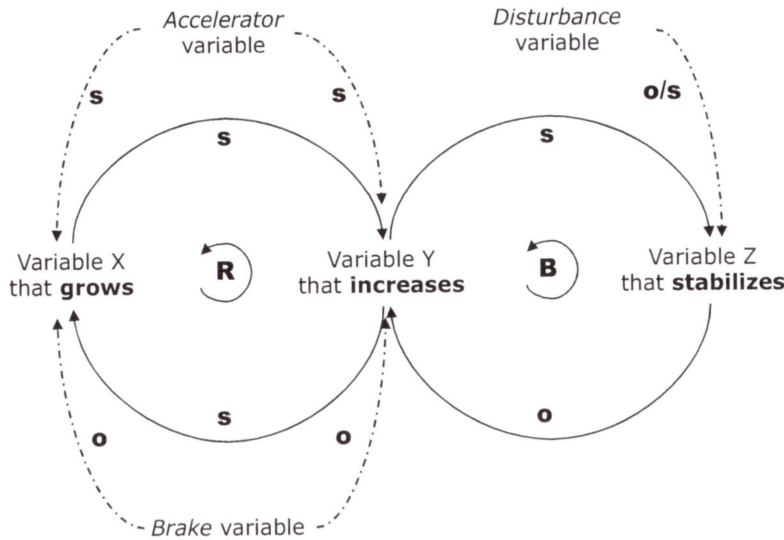

Fig. 1.19 The model representing the general law of dynamic instability

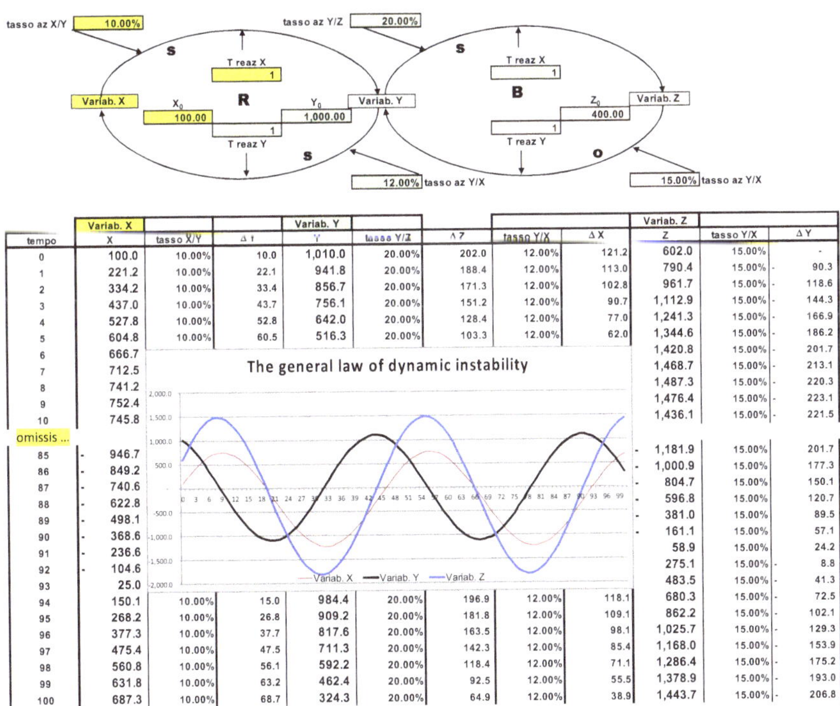

tempo	Variab. X: X	tasso X/Y	Δ Y	Variab. Y: Y	tasso Y/Z	Δ Z	tasso Y/X	Δ X	Variab. Z: Z	tasso Y/X	Δ Y
0	100.0	10.00%	10.0	1,010.0	20.00%	202.0	12.00%	121.2	602.0	15.00%	-
1	221.2	10.00%	22.1	941.8	20.00%	188.4	12.00%	113.0	790.4	15.00%	- 90.3
2	334.2	10.00%	33.4	856.7	20.00%	171.3	12.00%	102.8	961.7	15.00%	- 118.6
3	437.0	10.00%	43.7	756.1	20.00%	151.2	12.00%	90.7	1,112.9	15.00%	- 144.3
4	527.8	10.00%	52.8	642.0	20.00%	128.4	12.00%	77.0	1,241.3	15.00%	- 166.9
5	604.8	10.00%	60.5	516.3	20.00%	103.3	12.00%	62.0	1,344.6	15.00%	- 186.2
6	666.7								1,420.8	15.00%	- 201.7
7	712.5								1,468.7	15.00%	- 213.1
8	741.2								1,487.3	15.00%	- 220.3
9	752.4								1,476.4	15.00%	- 223.1
10	745.8								1,436.1	15.00%	- 221.5
omissis ...											
85	- 946.7								- 1,181.9	15.00%	201.7
86	- 849.2								- 1,000.9	15.00%	177.3
87	- 740.6								- 804.7	15.00%	150.1
88	- 622.8								- 596.8	15.00%	120.7
89	- 498.1								- 381.0	15.00%	89.5
90	- 368.6								- 161.1	15.00%	57.1
91	- 236.6								58.9	15.00%	24.2
92	- 104.6								275.1	15.00%	- 8.8
93	25.0								483.5	15.00%	- 41.3
94	150.1	10.00%	15.0	984.4	20.00%	196.9	12.00%	118.1	680.3	15.00%	- 72.5
95	268.2	10.00%	26.8	909.2	20.00%	181.8	12.00%	109.1	862.2	15.00%	- 102.1
96	377.3	10.00%	37.7	817.6	20.00%	163.5	12.00%	98.1	1,025.7	15.00%	- 129.3
97	475.4	10.00%	47.5	711.3	20.00%	142.3	12.00%	85.4	1,168.0	15.00%	- 153.9
98	560.8	10.00%	56.1	592.2	20.00%	118.4	12.00%	71.1	1,286.4	15.00%	- 175.2
99	631.8	10.00%	63.2	462.4	20.00%	92.5	12.00%	55.5	1,378.9	15.00%	- 193.0
100	687.3	10.00%	68.7	324.3	20.00%	64.9	12.00%	38.9	1,443.7	15.00%	- 206.8

Fig. 1.20 The general law of dynamic instability: a simulation

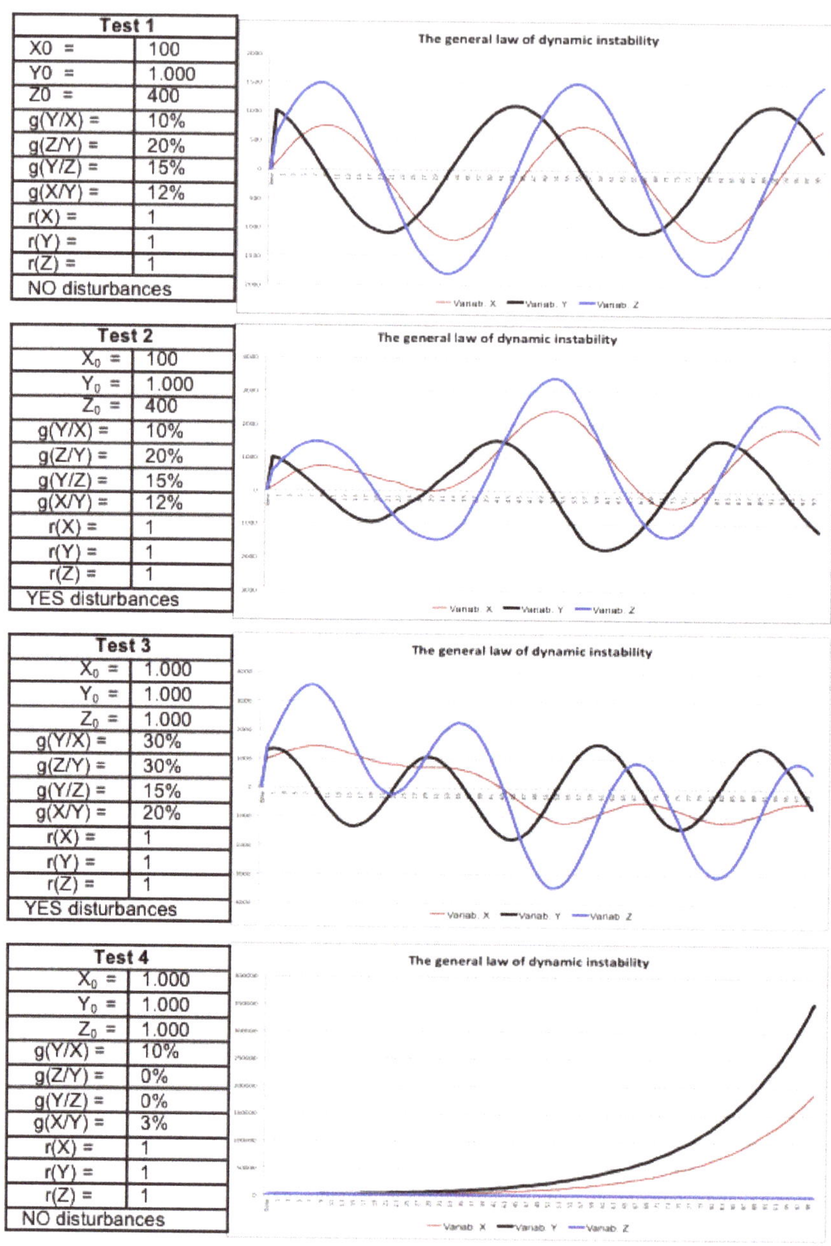

Fig. 1.21 The general law of dynamic instability: simulation test (the *bold line* represents the stabilized variable *Y*)

introducing outside disturbances, the dynamics can become quite irregular. I will not add any comments since the loops produced by the systems thinking technique are clearly representative and understandable. By adopting the simulation model behind Fig. 1.20 we can simulate other dynamics, introducing new parameters or even disturbance elements for one or even all three of the variables.

The first three tests reveal the relative stability in the dynamics of variable *Y* (the thicker curve), which is stabilized even in the presence of disturbances; in fact, the latter are "neutralized" by the joint action of the variables *X* and *Z*. Test 4 shows that by deactivating loop [**B**] we get only the reinforcing effect of loop [**R**].

1.7.5 The Law of Dynamic Instability: Richardson's Model

An initial example of how we can apply the law of dynamic instability to give a more correct representation of dynamic systems comes from reformulating the arms escalation model in Fig. 1.10, which only contained a single loop [**R**] that produced an apparently infinite growth in *X*'s and *Y*'s arms.

To avoid this infinite growth, which is entirely unrealistic, we must introduce, in addition to the *action* and *reaction rates* *g* and *h*, respectively—which we can call "coefficients of defense," since they indicate to what extent fear moves *X* and *Y* to increase their arms—two new coefficients as well, called the "coefficients of saturation," since they express the degree of tolerability in the population for the economic sacrifices it must make to strengthen their country's armaments. We then get the model in Fig. 1.22, which adds two loops [**B**] to loop [**R**]; these loops, as stated in the law of dynamic instability, impede an uncontrolled growth in arms. Figure 1.15 is based on the model by Lewis Richardson (1949), considered to be the founder of the scientific analysis of conflicts.

> Lewis Fry Richardson (1881-1953) was one of the most original thinkers ever to apply their minds to the study of war and its causes. In a strict sense he was an amateur in that he held no position of profit dedicated to the pursuit of these researches. He was always employed to do something else and did the research at week-ends and in his retirement. … It was after his death that his work had its impact (Nicholson 1999, p. 541).

Not taking into account, for simplicity's sake, other variables introduced by Richardson, I have reformulated the model by assuming that *Y*'s arms increase due to fear of the size of *X*'s arsenal, according to a (positive) *coefficient of defense*, and undergo a slowing down proportionate to the size of the armaments themselves, according to a (negative) *coefficient of saturation*. Since *X* starts with an arsenal five times that of *Y*'s ($X_0 = 500$ against $Y_0 = 100$), I have set a low *coefficient of defense* ($h = 5\%$ against $g = 15\%$ for *Y*) as well as a high *rate of saturation* (15% against 5% for *Y*), since we can assume that the population of *X* is intolerant toward additional sacrifices to increase its already high arms level. On the other hand, *Y*'s citizens, conscious of their initial inferiority, are very frightened by *X*'s arsenal and will tolerate even substantial sacrifices to shore up their security. We can observe these dynamics in many contexts, even in the recent developments in the conflicts in the Middle and Far East.

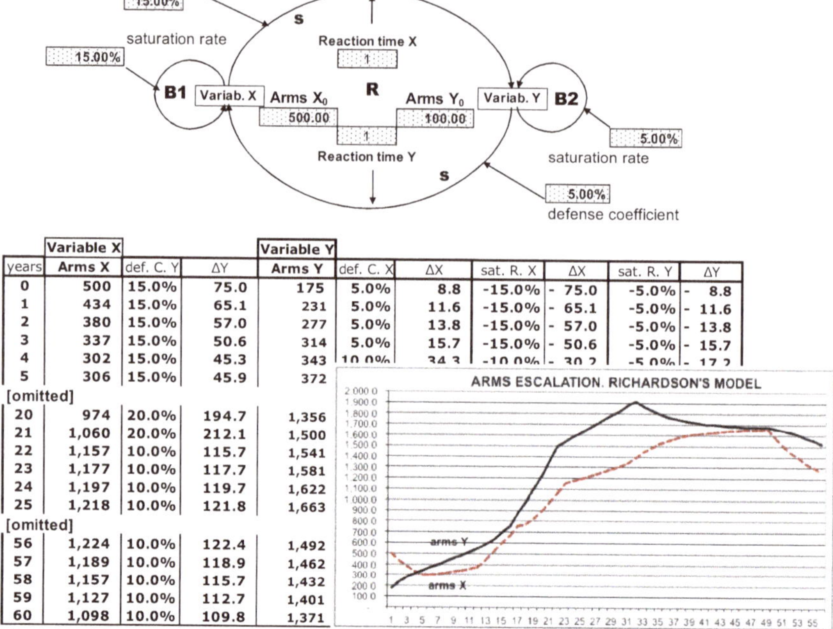

	Variable X			Variable Y						
years	Arms X	def. C. Y	ΔY	Arms Y	def. C. X	ΔX	sat. R. X	ΔX	sat. R. Y	ΔY
0	500	15.0%	75.0	175	5.0%	8.8	-15.0%	- 75.0	-5.0%	- 8.8
1	434	15.0%	65.1	231	5.0%	11.6	-15.0%	- 65.1	-5.0%	- 11.6
2	380	15.0%	57.0	277	5.0%	13.8	-15.0%	- 57.0	-5.0%	- 13.8
3	337	15.0%	50.6	314	5.0%	15.7	-15.0%	- 50.6	-5.0%	- 15.7
4	302	15.0%	45.3	343	10.0%	34.3	-10.0%	- 30.2	-5.0%	- 17.2
5	306	15.0%	45.9	372						
[omitted]										
20	974	20.0%	194.7	1,356						
21	1,060	20.0%	212.1	1,500						
22	1,157	10.0%	115.7	1,541						
23	1,177	10.0%	117.7	1,581						
24	1,197	10.0%	119.7	1,622						
25	1,218	10.0%	121.8	1,663						
[omitted]										
56	1,224	10.0%	122.4	1,492						
57	1,189	10.0%	118.9	1,462						
58	1,157	10.0%	115.7	1,432						
59	1,127	10.0%	112.7	1,401						
60	1,098	10.0%	109.8	1,371						

Fig. 1.22 Richardson's model

Fig. 1.22 Richardson's model

The model in Fig. 1.22 is useful for understanding the opposite phenomenon of a gradual reduction in recent years in the nuclear arms of the two greatest world powers, which has occurred due to both the increase in the saturation rate and the decrease in the defense coefficient. This tendency has continued even recently, and according to what was published in America24 on March 24, 2010:

> Washington and Moscow are close to signing a new nuclear treaty. U.S. president Barack Obama briefed members of Congress on the state of negotiations with Russia on the reduction of nuclear arms. All aspects of the agreements are said to have been defined and the formal announcement could come in the next few days. Negotiations between American and Russian representatives have been going on for months with the aim of signing a new treaty to replace the 1991 treaty, which expired in December 2009 (www.america24.com).

1.7.6 The Law of Dynamic Instability: Prey–Predator Populations

No population grows infinitely, since growth is always balanced by at least three factors operating individually or jointly: a maximum limit of available resources and the presence of predators and human controls. I propose the very simple model in Fig. 1.23, which represents the dynamics of two populations, sardines and sharks, which, being preys and predators, limit in turn their growth.

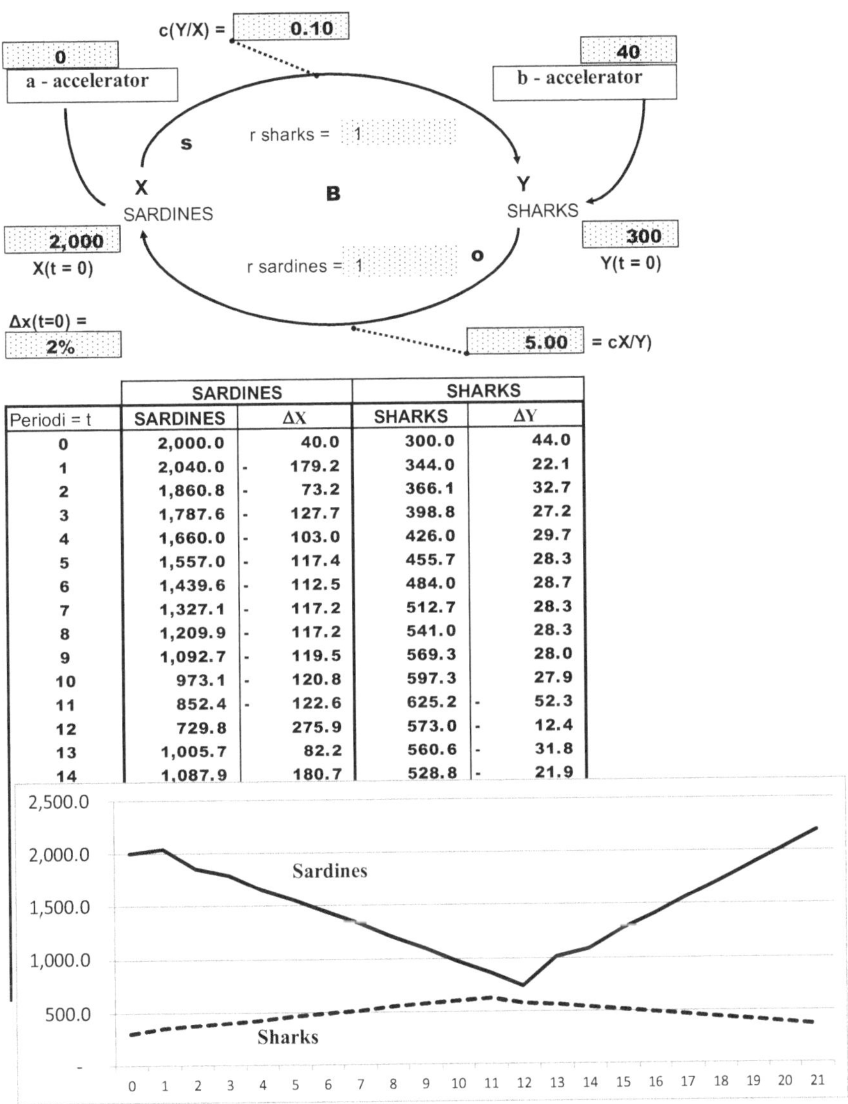

	SARDINES		SHARKS	
Periodi = t	SARDINES	ΔX	SHARKS	ΔY
0	2,000.0	40.0	300.0	44.0
1	2,040.0	- 179.2	344.0	22.1
2	1,860.8	- 73.2	366.1	32.7
3	1,787.6	- 127.7	398.8	27.2
4	1,660.0	- 103.0	426.0	29.7
5	1,557.0	- 117.4	455.7	28.3
6	1,439.6	- 112.5	484.0	28.7
7	1,327.1	- 117.2	512.7	28.3
8	1,209.9	- 117.2	541.0	28.3
9	1,092.7	- 119.5	569.3	28.0
10	973.1	- 120.8	597.3	27.9
11	852.4	- 122.6	625.2	- 52.3
12	729.8	275.9	573.0	- 12.4
13	1,005.7	82.2	560.6	- 31.8
14	1,087.9	180.7	528.8	- 21.9

Fig. 1.23 Prey–predator dynamics (the model is a very simplified application of the Volterra–Lotka equations)

The more sardines increase in number, the greater the increase in the shark population due to the abundance of food; this increase in shark numbers reduces the number of sardines, which causes a reduction in the number of predators. These dynamics are described by a balancing loop that produces oscillating dynamics for the two populations normally limited in size. I would note that the study of population dynamics has a solid tradition going back to Vito Volterra and Alfred Lotka and to their famous equations that describe the coevolution of prey–predator populations. Note that the model in Fig. 1.23 does not derive from these equations but is presented with the sole aim of illustrating in a simplified way how the law of dynamic instability works (the original equations are examined in Chap. 8, Sect. 8.5).

1.7.7 The Law of Dynamic Instability: Counteracting the Spread of a Virus

Section 1.7.2 presented a model of explosive dynamics regarding individuals infected by the spread of a virus. The model appeared unrealistic as it did not cover the possibility of containment of the virus and the reduction in the number of infected individuals implemented by public and private structures (Burgiel 2020). In fact, the General Law of Dynamic Instability states that any model developed with Systems Thinking logic cannot include only reinforcement (or balancing) loops without also including balancing (or reinforcement) loops that counteract the dynamics of the main loop. Figure 1.15, however, indicates that to counteract expansive dynamics some external intervention may be sufficient to produce a "braking action" that:

- Reduces the *contagion* rate, $g(Y/X)$, or Rt, (red zones, social distancing, immunizing equipment, masks, tampons, etc.).
- Reduces the *infected* rate, $h(X/Y)$ (enhanced immune defenses, a spontaneous cure, etc.).
- Introduces the *cure rate*, j, which expresses the effort to enhance and improve *health care* (s), which leads to infected individuals being cured (s), so that the increase in *cured individuals* reduces the number of the *infected*. This important "braking factor" has been taken into account by adding loop B to the model.

I have modified the simulation in Sect. 1.7.2 by introducing the following hypotheses (Fig. 1.24):

- The *contagion*, of infection, rate, $g(Y/X)$, which initially had a value of 2 (each infected person in turn infected two healthy individuals), has been gradually reduced to the values that appear in the third column.
- The *infected* rate, which had a value of 100% (all the newly infected increased the number of infected persons), has gradually been reduced to the values seen in the third column.
- The *cure* rate has the values indicated in the penultimate column.

In the simulation, these hypotheses produce dynamics of the infected population that fall to zero in Week 40.

Note 1: The model was presented only to theoretically demonstrate how the General Law of Dynamic Instability behaves. A simulation of real cases could also take into account other factors that affect the dynamics of infection; but, above all, it should include growth rates for contagion, infected individuals, and cured individuals that have been determined through a careful epidemiological and health study.

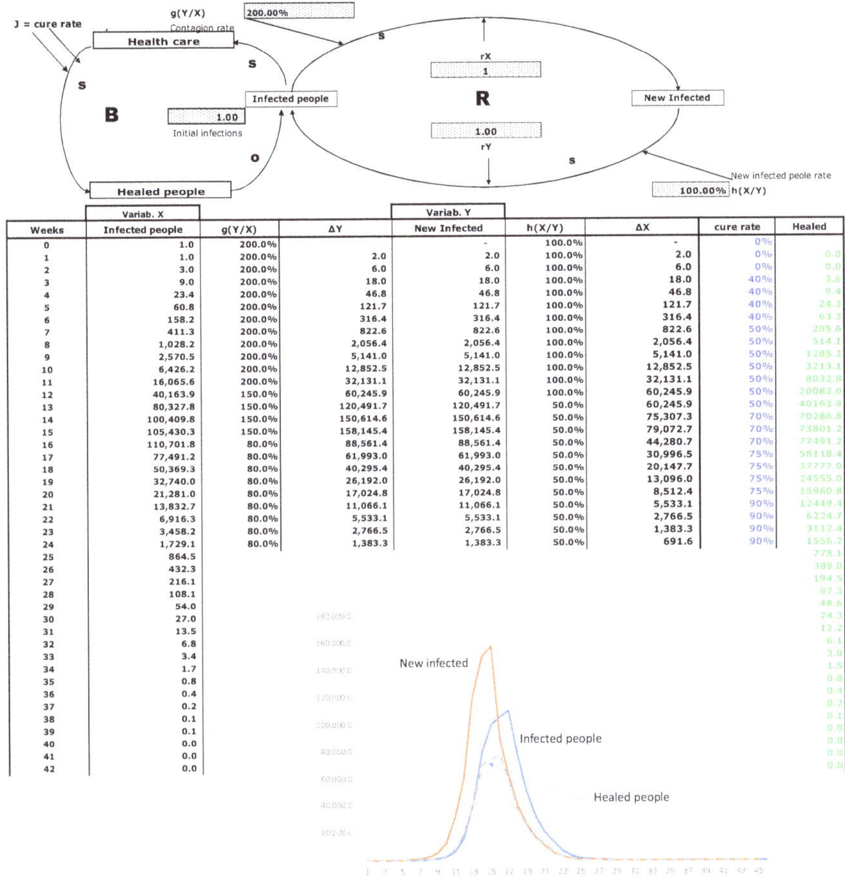

Weeks	Variab. X Infected people	g(Y/X)	ΔY	Variab. Y New Infected	h(X/Y)	ΔX	cure rate	Healed
0	1.0	200.0%		-	100.0%	-	0%	
1	1.0	200.0%	2.0	2.0	100.0%	2.0	0%	0.0
2	3.0	200.0%	6.0	6.0	100.0%	6.0	0%	0.0
3	9.0	200.0%	18.0	18.0	100.0%	18.0	40%	3.6
4	23.4	200.0%	46.8	46.8	100.0%	46.8	40%	9.4
5	60.8	200.0%	121.7	121.7	100.0%	121.7	40%	24.3
6	158.2	200.0%	316.4	316.4	100.0%	316.4	40%	63.3
7	411.3	200.0%	822.6	822.6	100.0%	822.6	50%	205.6
8	1,028.2	200.0%	2,056.4	2,056.4	100.0%	2,056.4	50%	514.1
9	2,570.5	200.0%	5,141.0	5,141.0	100.0%	5,141.0	50%	1285.2
10	6,426.2	200.0%	12,852.5	12,852.5	100.0%	12,852.5	50%	3213.1
11	16,065.6	200.0%	32,131.1	32,131.1	100.0%	32,131.1	50%	8032.8
12	40,163.9	150.0%	60,245.9	60,245.9	100.0%	60,245.9	50%	20082.0
13	80,327.8	150.0%	120,491.7	120,491.7	50.0%	60,245.9	50%	40163.9
14	100,409.8	150.0%	150,614.6	150,614.6	50.0%	75,307.3	70%	70286.8
15	105,430.3	150.0%	158,145.4	158,145.4	50.0%	79,072.7	70%	73801.2
16	110,701.8	80.0%	88,561.4	88,561.4	50.0%	44,280.7	70%	77491.2
17	77,491.2	80.0%	61,993.0	61,993.0	50.0%	30,996.5	75%	58118.4
18	50,369.3	80.0%	40,295.4	40,295.4	50.0%	20,147.7	75%	37777.0
19	32,740.0	80.0%	26,192.0	26,192.0	50.0%	13,096.0	75%	24555.0
20	21,281.0	80.0%	17,024.8	17,024.8	50.0%	8,512.4	75%	15960.8
21	13,832.7	80.0%	11,066.1	11,066.1	50.0%	5,533.1	90%	12449.4
22	6,916.3	80.0%	5,533.1	5,533.1	50.0%	2,766.5	90%	6224.7
23	3,458.2	80.0%	2,766.5	2,766.5	50.0%	1,383.3	90%	3112.4
24	1,729.1	80.0%	1,383.3	1,383.3	50.0%	691.6	90%	1556.2
25	864.5							778.1
26	432.3							389.0
27	216.1							194.5
28	108.1							97.3
29	54.0							48.6
30	27.0							24.3
31	13.5							12.2
32	6.8							6.1
33	3.4							3.0
34	1.7							1.5
35	0.8							0.8
36	0.4							0.4
37	0.2							0.2
38	0.1							0.1
39	0.1							0.0
40	0.0							0.0
41	0.0							0.0
42	0.0							0.0

Fig. 1.24 Counteracting the spread of a virus

Note 2: In its simplicity, the model can be applied to other types of pandemics, even non-viral ones; for example, the plague, cholera, and measles epidemics, lice epidemics at school, the *Xylella fastidiosa*, which has devastated ancient olive trees in Italy's southern Apulia region and which has no known cure, as well as many others.

1.7.8 Senge's Archetypes List Expanded

Space constraints make it impossible to present Senge's Archetypes in detail. Senge's list represents the most frequent and visible archetypes in organizations, even though it can be extended. Above all, most of Senge's archetypes present a mirror-image archetype, to which new important and significant archetypes can be added. Instead, the most interesting ones will be briefly described, in addition to the most relevant ones in Senge's list, leaving the reader to refer to Chap. 5 (and Mella 2012, Chap. 4), for a more in-depth treatment with examples.

ARCHETYPE of the FIXES THAT FAIL: Urgent and short-term problem-solving strategies can be adopted in dealing with certain problems that nevertheless cause unforeseen consequences—which will appear later—that make continual readjustments of the solution necessary (*Senge*).

SHIFTING THE BURDEN: When a symptom arises, symptomatic solutions can delay the search for the problem and the adoption of definitive solutions; such a delay can lead to collateral effects, which further impede the search for long-term solutions. Systems thinking favors and accelerates the search for a definitive solution (*Senge*).

ERODING GOALS: The manager/decision-maker, perceiving the difficulties in achieving the objective (symptoms), often responds by "adjusting" the objective—that is, by reducing the system's performance expectations—rather than intervening on the operational conditions that give rise to the problem itself (*Senge*).

STRENGTHENING GOALS or INSATIABILITY: When the manager/decision-maker obtains a result above the standard, instead of reducing the action to achieve the objective he continually *raises* the level of the standards (*mirror-image archetype*).

DEGRADATION OF THE ERROR ASSESSMENT: An altered perception of the "distance" from the objective induces the decision-maker to undersize the *decisional lever*, thereby producing a new deviation (*mirror-image archetype*).

PERSISTENCE: The wrong perception of the deviation induces the decision-maker to *overestimate the error* and to activate every possible lever to eliminate it, showing by such intolerance toward any slight deviation a true *persistence*, or even implacability, in the control action (*mirror-image archetype*).

SUCCESS TO THE SUCCESSFUL: There is a tendency to assign more resources to those who successfully employ them. The subject that receives more resources has a greater probability of a better performance compared to subjects whose performance remains stable over time. Thus, an initial distribution of resources produces a permanent performance differential (*Senge*).

PUNISHMENT FOR SUCCESS: The person who works better is always given more work; the shirkers are instead very often underutilized. It almost seems as if the successful person, the one with the better performance, is punished by being given more to do (*mirror-image archetype*).

TRAGEDY OF THE COMMONS: When two or more subjects share a common resource, sooner or later it will be depleted (see Chap. 7, Sect. 7.10). *"Freedom in a commons brings ruin to all"* (Hardin 1968, p. 1244) (*Senge*).

LIMITS TO GROWTH: Nothing grows ad infinitum. Every growth process that entails a growth condition available in limited quantities is destined sooner or later to end (*Senge*).

GROWTH AND UNDERINVESTMENT: When growth is slowed by the gradual saturation of the productive capacity, and this saturation could, in turn, be eliminated through additional investment in capacity, normally the capacity saturation impedes new investment as well (*Senge*).

ACCIDENTAL ADVERSARIES: When two or more subjects cooperate for their mutual success, parallel to or subsequent to actions to collaborate and support one another to achieve such success they also act to pursue their own personal and individual success, continually impeding one another and lowering the collective and individual performances (*Senge*).

SHORT-TERM, INDIVIDUAL, AND LOCAL PREFERENCES: Preference for short-term, individual, or local advantages hides the collective and global disadvantages that will occur over the long term; for this reason, short-term, individual, and local advantages are preferred to long-term and global disadvantages (*new archetype*). This archetype is also known as the archetype of the three myopias: temporal, social, and spatial (Fig. 1.12).

1.7.9 The "Six" Disciplines of Learning Organizations

Peter Senge introduced systems thinking not as a new theory of systems but as the fifth of the disciplines needed to construct a *learning organization*.

Referring the interested reader to Senge's work and to the vast literature on the topic (Mella 2012), here I will merely summarize Senge's theory for those who do not need to study it in greater detail.

The first discipline is "personal mastery." In order to improve the organization's performance it is necessary to develop the discipline of personal mastery, which, bringing out in each individual of the organization the desire to learn and create, teaches him or her to face life in a creative and nonreactive way, which favors individual growth through personal learning and improvement, thereby contributing to clarifying the individual's view about his life and work.

> Personal mastery is the discipline of continually clarifying and deepening our personal vision, of focusing our energies, of developing patience, and of seeing reality objectively. As such, it is an essential cornerstone of the learning organization—the learning organization's spiritual foundation (Senge 2006, p. 7).

> The discipline of personal mastery [...] starts with clarifying the things that really matter to us, of living our lives in the service of our highest aspirations (*ibidem*, p. 8).

The second discipline, referred to as "mental models," is particularly relevant because it connects with the ability to clarify and improve the frames and repertoires that affect attention and decision-making and thus to ameliorate the action individuals undertake in their organization. In fact, a mental model is a "pattern" or a "theory" that guides a person in the decisions and choices he or she makes when he or she has to act. If the results of the model are positive, the model is strengthened; otherwise, it is "set aside" and another model of behavior is sought (I shall deal more amply with the mental models in Chap. 12, Sect. 12.13).

> 'Mental models' are deeply ingrained assumptions, generalizations, or even pictures or images that influence how we understand the world and how we take action (Senge 2006, p. 8).

Mental models influence in a crucial way the quality of thinking of individuals, but they are also pervasive in their effect on organizations, in the form of operational procedures (I do this because it is written so in the manual), widely accepted organizational practices (I act like this because everyone else does), standardized decisional rules (it might be wrong, but that is what I have been told to do), and so on. The mental models discipline is fundamental for organizational learning since it not only increases the group's or the individual's capacity to form a stock of shared knowledge but also facilitates the process for recognizing and modifying the group mental models in order to collectively decide in an effective way, as if the decision came from a single individual.

> The discipline of working with mental models starts with turning the mirror inward; learning to unearth our internal pictures of the world, to bring them to the surface and hold them rigorously to scrutiny (Senge 2006, p. 8).

"Building shared vision" is the third discipline. In order to build a shared vision, the heads and leaders must continually analyze and share their personal visions with the other members of the organization and have these accepted by the latter, who then, in turn, spread them through personal commitment. As Senge observes, in most modern-day organizations relatively few people can be said to be enrolled and even a lesser amount to be committed. Most are in a state of conformism, whether formal or genuine. The enrolled people assume the role of gaining the attention of other members in order to improve their attention and knowledge. The discipline of creating a shared vision has always been fundamental for the long-term survival of organizations, but it is even more essential today for building learning organizations.

> If any one idea about leadership has inspired organizations for thousands of years, it's the capacity to hold a shared picture of the future we seek to create. […] The practice of shared vision involves the skills of unearthing shared 'pictures of the future' that foster genuine commitment and enrollment rather than compliance (Senge 2006, p. 9).

The fourth discipline is referred to as "team learning," that is, to say, the process that seeks to create and develop the team's capacity to focus the attention of all participants toward stimuli, categories, and objectives in order to work together in a coordinated manner to obtain results its members truly desire, perhaps in order to achieve a shared vision through dialogue, in order to listen to the different points of view, and discussion, in order to search for the best point of view to support the decisions that must be made.

> The discipline of team learning starts with "dialogue", the capacity of members of a team to suspend assumptions and enter into a genuine "thinking together". […] "The discipline of dialogue also involves learning how to recognize the patterns of interaction in teams that undermine learning" (Senge 2006, p. 10).

Alignment is a necessary condition in order that the power given to an individual can increase the power of the entire team.

> […] there are striking examples where the intelligence of the team exceeds the intelligence of the individuals in the team, and where teams develop extraordinary capacities for coordinated action (Senge 2006, p. 9).

The fifth discipline is "systems thinking." In order to develop a learning organization the four disciplines must, according to Senge, be jointly and coherently applied. This requires a fifth discipline, which serves as a unifier and coordinator for the other four, and it is this that justifies the role of systems thinking.

The role of the sixth discipline, "control thinking," is also clear. If learning organizations increase their survival capacity to the extent they are able to successfully react to environmental disturbances and produce a change in their position to achieve states more favorable to their survival—if the capacity to react and change depends on the extent to which they learn quickly as unitary systems, thus making the network of their processes even more efficient—then it is clear that they learn to the extent to which they can set their own objectives, translate these into coherent and shared individual objectives, verify their achievement, and take the necessary collective and individual actions to identify and eliminate deviations from the achievement of the objectives. As the following chapter will show, this is precisely the logic behind control.

The role of the sixth discipline is to contribute to the construction of learning organizations by promoting an education and discipline of control at all levels in order to transform the organizations themselves into networks of control systems, as I shall try to demonstrate in subsequent chapters.

1.8 Summary

Systems thinking—which Peter Senge considers the fifth discipline for forming learning organizations—represents an effective language and a powerful logical tool for understanding and describing dynamic systems as well as control systems, as we shall see in the next chapter (Sect. 1.1).

There are five basic rules of systems thinking (Sect. 1.2).

1. If we want to understand the world, we must be able to "see the trees and the forest"; we must develop the capacity to "zoom" from the whole to the parts, from systems to components, and vice versa.
2. We must not limit our observation to that which appears constant but "search for what varies"; the variables are what interest the systems thinker.
3. If we truly wish to understand reality and change, we must make an effort "to understand the cause of the variations in the variables we observe."
4. It is not enough to search for the causes of the variations we observe only by searching for the causal chains among the variables; we must also "link together the variables in order to specify the loops among all the variations."
5. When we observe the world, we must "always specify the boundaries of the system we wish to investigate."

The simplest models that describe dynamic systems are the reinforcing loops [**R**] and balancing loops [**B**]. Dynamic systems of whatever size are constructed by joining together an appropriate number of loops [**R**] and [**B**]. The basic balancing or reinforcing loops that connect two variables can also be called *Rings* since they usually assume a circular representation (Sect. 1.3).

Dynamic systems are repetitive systems equipped with memory and closed with respect to the environment. They can be simulated using system dynamics techniques (Sect. 1.4).

Two important laws of systems thinking have been presented (Sect. 1.5) along with three fundamental "archetypes" describing the human behavioral trait of preferring short-term benefits, even if these produce long-term damage (Sect. 1.6).

Chapter 2
The Ring: The General Structure of Control Systems

Abstract Chapter 1 presented the principles of systems thinking along with the rules and formal language, and this discipline uses to represent dynamic systems. This chapter deals with the feedback control problem as a general form of efficient control. By adopting the systems thinking language we shall construct the general *logical model* of a *closed-loop control system*, the "Ring," which allows the process control to take place (Sects. 2.1, 2.2, 2.3, 2.4, 2.5, 2.6, 2.7, and 2.8). This approach allows us to develop a true "art" of control—a *discipline* based on systems thinking logic that enables us to observe, understand, and modify the world and our future. I shall define the *logical* and the *technical structure* of the *Ring* along with the *recursive equations* that determine its dynamics.

Keywords Control system · The Ring · Logical structure of control systems · Control levers · External disturbance · Delays · Action rate · Reaction rate · Reaction time · Management and governance of control systems · Precision of control systems · Effector · Sensor · Regulator · Continuous and discrete control systems · Connections and interferences · Non-symmetrical systems

> All processes that are stable we shall predict. All processes that are unstable we shall control… This was John von Neumann's dream (John von Neumann's dream, from Dyson 1988, p. 182).

> In case a system approaches a stationary state, changes occurring may be expressed not only in terms of actual conditions, but also in terms of the distance from the equilibrium state; the system seems to "aim" at an equilibrium to be reached only in the future. Or else, the happenings may be expressed as depending on a future final state (von Bertalanffy 1968, p. 75).

> Stability is commonly thought of as desirable, for its presence enables the system to combine flexibility and activity in performance with something of permanence. Behaviour that is goal-seeking is an example of behaviour that is stable around a state of equilibrium. Nevertheless, stability is not always good, for a system may persist in returning to some state that, for other reasons, is considered undesirable (Ashby 1957, p. 61).

P. Mella, *The Magic Ring*, Contemporary Systems Thinking,
https://doi.org/10.1007/978-3-030-64194-8_2

Chapter 1 presented the principles of systems thinking along with the rules and formal language, and this discipline uses to represent dynamic systems. This chapter deals with the feedback control problem as a general form of efficient control. By adopting the systems thinking language we shall construct the general *logical model* of a *closed-loop control system*, the "Ring," which allows the process control to take place (Sects. 2.1, 2.2, 2.3, 2.4, 2.5, 2.6, 2.7, and 2.8). The feedback control system model represents the general foundation for every ordered and stable dynamic in the world, and thus the fundamental instrument by which the world itself maintains its existence. This approach allows us to develop a true "art" of control—a *discipline* based on systems thinking logic that enables us to observe, understand, and modify the world and our future. I shall define the *logical structure* of the *Ring* along with the *recursive equations* that determine its dynamics. The general *logical structure* of control systems is integrated with its *technical structure*, which is made up of four apparatuses or machines—effector, sensor, regulator, and information network—which, in fact, make up the *chain of control* (Sect. 2.9). The model also includes three types of *delays* and *disturbances*—action, detection, and regulation—which characterize the processes carried out by the apparatuses. Other new and important concepts are introduced: (1) the management and governance processes; (2) the interferences between control systems; and (3) different areas of application of the general model.

2.1 The Control Process

As I have pointed out in Chap. 1, systems thinking views the world as a system of systems of interconnected variables, thereby constructing a "world of interacting variables" linked by causal chains, by reinforcing and balancing loops. Since the world is made up of variables, we might expect incredible disorder—variables that grow and diminish and that, through their dynamics, impact other variables, which in turn are connected to others, as part of a vortex of dynamics that combines in a disorderly fashion, as if we were observing the turbulent dynamics of the sun's surface, the atmosphere, or the oceans. Yet the world normally appears ordered, except for some exceptions in the context of complex systems (CSS 2013); the dynamics of variables seems regulated over time and in space, and any excessive variation is brought back to normality through special systems known as control systems.

In fact, control systems are systems that control the world's fundamental variables. Cells, as well as all living beings, can remain alive for a long time thanks to the order that stabilizes the multiple interactions among their constituent variables. If the disturbances are not so intense as to be destructive, then autopoiesis can continue over time, allowing the living being to exist. It is clear that when the temperature falls below zero, men normally do not die of exposure, and that when the climate turns hot, they do not die scorched and dehydrated. They are able to maintain their temperature within the standard vital parameters, except for very extreme variations in the outside temperature. If it is cold, one dresses more warmly, lights a fire, lives in the barn with other animals, finds refuge in a cave, builds earthen houses that insulate better, even adopts modern temperature regulators and solar paneling, modern insulating fabric, and air conditioners, and constructs thermoregulated skyscrapers.

What is true for intelligent man is also true for other species: winter animals go into hiding, hibernate, or gather together in small herds. Penguins resist polar temperatures by forming large compact flocks and allowing the outside members to rotate from the outside to the inside of the flock to allow every member to be protected from the cold. In summer, animals adopt strategies against the heat, for example, by looking for watercourses or going under the sand. Evolution has provided many ways to regulate temperature, equipping animals with fur and layers of fat, even developing cold-blooded animals. The vegetable kingdom also has a wide variety of strategies against excessive heat or cold.

As subsequent chapters provide a wide range of examples, these few examples above will suffice to show the importance of control and control systems. Though there are several definitions of control and control systems, it is useful to start with the simple and significant one by Michael Arbib:

> In general terms, therefore, a control problem is to choose the input to some system in such a way as to cause its output to behave in some desired way, whether to stay near a set reference value (the regulator problem), or to follow close upon some desired trajectory (the tracking problem) (Arbib 1987, online).

Starting from the above quote, I shall define as "controllable" with respect to time a variable, Y_t, if it can take on in a succession of periods, or instants, $t = 1, 2, \ldots$ (based on a temporal scale to be specified) a desired or established value, Y^*. The instant t^* in which the objective is achieved defines the *control instant.*

The value Y^* that the variable Y_t must attain can represent a *constraint* (e.g., do not go below 10,000 ft; do not exceed 50 km per hour; emissions not to exceed 150 parts per million; costs not to exceed $120 per unit of product), a *limit* (break the high jump record; the temperature of dew at which fog forms; a body temperature of 36 °C), and an *objective* or *goal* (raise the temperature to 24 °C; reach 12,000 m of altitude; turn left; etc.). The distinction between *constraints*, *limits*, and *objectives* is not always clear; for this reason, unless clearly indicated I will use Y^* (set point) to indicate an *objective* to be reached (as well as a *constraint* or a *limit*).

The graph in Fig. 2.1 clearly represents the control problem for a variable Y_t (curve a) whose dynamics must converge toward the objective Y^*, indicated by a half line and assumed to remain constant over time.

When the values for Y_t move away from those for Y^*, *gaps* develop, which must be eliminated. The *gap* (Δ) is calculated as the difference $\Delta(Y)_t = Y^* - Y_t$, so that it is immediately clear that if $\Delta(Y)_t < 0$, then the dynamics of Y_t must be reduced, and vice versa if $\Delta(Y)_t > 0$. The *gap* can be given various names (deviation, difference, etc.), but henceforth I shall use the general term "error," indicating this by $E(Y)_t$, which is the equivalent of $\Delta(Y)_t$. An attempt at control Y_t noticeably reduces the error (curve b), though not entirely. The control is successful when the dynamics of Y_t at instant t^* coincide with the value Y^* (dashed line c) and $\Delta(Y)_t = 0$.

We must specify how it is possible to act on the variable Y_t so that, at successive instants t_1, t_2, t_3, etc., we can modify its values and make these move toward Y^*. The answer is clear: to control Y_t we always require the action of a control system.

2.2 The Logical Structure of Control Systems in the Language of Systems Thinking

After these introductory comments and with the aid of Fig. 2.1, we can easily introduce the concept of a control system by adopting the logic and language of systems thinking as shown in Fig. 2.2.

We know from systems thinking that the dynamics of Y_t can be considered the effect of one or more variables, X_t, that causes those dynamics. I shall define a *control variable* (*action* or *active* variable) as any variable X_t on whose values those of Y_t depend based on the causal relation, so that the variations in Y_t can be considered produced by variations in X_t. For this reason, the variable X_t can also be referred to with the more evocative term *control lever*. The values assigned to X_t in successive instants are not "arbitrary" but are *calculated* and *readjusted* to *gradually* eliminate the error between X^* and the values of Y_t. I shall define as *manager* of the system the person who can calculate and modify the values of X_t, that is, in figurative language, "activate the control lever." This simple logic is translated into the *standard model* of Fig. 2.2, which shows the *logical structure* of the *single-lever* and *single-objective* control systems, *without delays*, able to control only a single variable, as shown in Fig. 2.1 (*multilever* and *multiobjective* control systems will be examined in Chap. 4).

Without discussing for now the internal operating mechanisms (these will be taken up in Sect. 2.9), I shall define as a control system any *logical* or *technical* structure (algorithm or machine, rule or apparatus, etc.) that, for a set of instants,

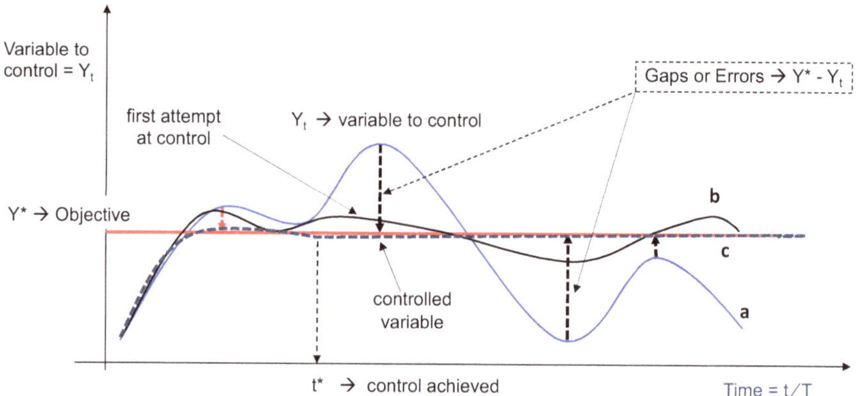

Fig. 2.1 Control problem

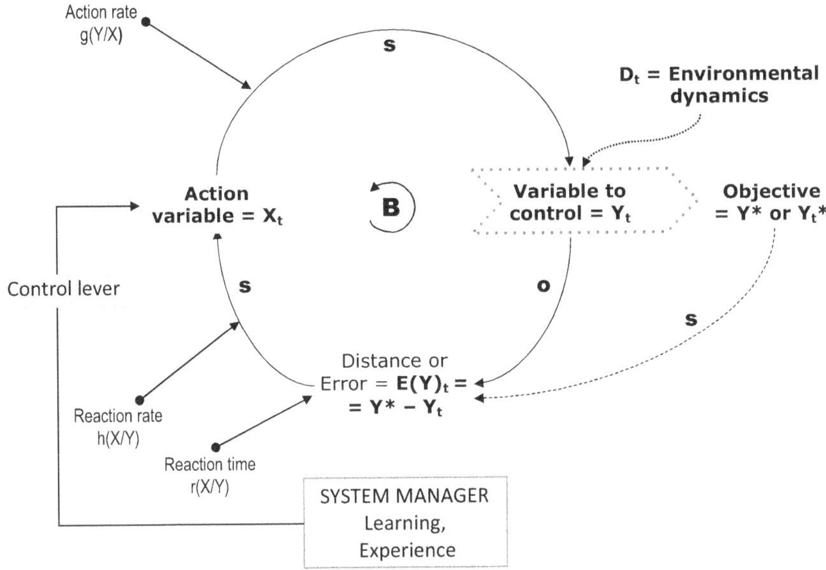

Fig. 2.2 Standard model of a single-lever control system

perceives $E(Y)_t = Y^* - Y_t$, calculates and assigns the values of the lever X_t, and produces the appropriate Y_t to gradually eliminate, when possible, the error $E(Y)$, thereby achieving the equality $Y_{t^*} = Y^*$ at instant t^*.

Since the system in fact "perceives and evaluates" the errors with regard to the objective, it operates according to the logic referred to as *feedback control*.

The control system is *repetitive* and functions by means of *action* (X acts on Y) and *reaction* ($E(Y)$ acts on X); through a certain number of interventions (iterations) on the *control lever*, the control system tries to achieve an objective (*goal-seeking systems*) or respect a constraint or limit (*constraint-keeping systems*).

By observing the model in Fig. 2.2 according to systems thinking logic, it is clear that the control system is a *Ring*; that is, it produces a balancing loop through a logical structure of the model containing five fundamental variables and three functions.

The five variables of the model in Fig. 2.2 are as follows:

1. The *variable to control* Y_t, or *measured output*, which represents the *governed* variable (the *passive* variable), which characterizes the behavior of the entire system as it drives this variable to achieve the objective.
2. The *objective*, or *set reference*, which may be a time variable Y_t^* (tracking objective) or, as a particular case, a constant value Y^* (goal or constraint) that must be reached and possibly maintained by the variable to control; without specifications, I shall consider the objective as a constant value over time: Y^*.
3. The error (deviation, gap, variance), $E(Y)_t = Y^*_t - Y_t$, which represents the *distance* between the values of the objective Y^* and those of Y_t; the causal links between Y and $E(Y)$ have direction **o** since, if we increase the value of Y, the error

with respect to Y^* decreases; the causal links between Y^* and $E(Y)$ have direction **s** since, if we increase the level of the objective, we also increase the distance from Y.

4. The *action variable* or the *control lever* X_t (*governing* or *active* variable), which must be operated to modify Y_t; the causal link between X and Y has direction **s** since we assume that, in order to increase Y, we must increase X (the opposite relation will be considered below).
5. The variable D_t, which indicates the possible external Disturbances connected to environmental Dynamics (D could also be a disturbance constant that acts one or more times); Fig. 2.2 does not indicate any direction of variation of the causal link between D and Y, since this direction can be either **s** or **o** based on the nature of D.

Note. In the figures in this book, I shall use the symbols $\Delta(Y)$ or $E(Y)$ indifferently. Moreover, in order not to complicate the use of symbols, when it is understood that we are talking about dynamic variables (over time), I shall write X, Y, E, and D, without indicating the variable t as a subscript.

Taking into account the direction of variation between the constituent variables, I shall define the model in Fig. 2.2 as a control system with structure [**s-o-s**].

The three functions needed to determine the dynamics of the system are the following:

1. The *action function* $g(Y/X)$ (gain) indicates the measurement of the variation ΔY_t for each variation ΔX_t; if $g(Y/X) = g$, a constant, then the action function becomes simply an *action rate*, which indicates the variation in Y for each unit of X; the accuracy of the *action function* or *rate* is fundamental for the efficiency of the control system and depends on the instruments available in relation to the type of system and on the control objectives.
2. The *reaction function* $h(X/Y)$, or *reaction rate* h, indicates the variation in X for each variation in E, and thus the variation in Y; usually, $h = 1/g$; in this case the system is perfectly *symmetrical* to the control; in general, however, the system can be controlled even if $h \neq 1/g$; in general the lower the h is with respect to $1/g$, the easier is the control, since the variations of X as a function of $\Delta(Y)$ are attenuated, thereby producing more dampened impulses for X; if $h > 1/g$, the system can produce increasingly greater oscillations. Section 2.16.11 considers the case of asymmetrical control systems.
3. The *reaction time* $r(X/Y)$, or $r(X)$, represent the *time to eliminate the error* (or the percentage of error, which is eliminated in each step); this function indicates the speed with which the control system moves toward the objective; if $r(X/Y) = r$, a constant, a reaction time of $r = 1$ indicates an *immediate* but *sudden* control; a reaction time $r = 2, 3, \ldots$ makes the control *less fast* but *smoother*; $r = 2$ means that in each step a percentage of $(100/r)$ is eliminated. In order not to produce sudden adjustments, the variations of the error should be gradual and $r > 1$. The *reaction time* plays a fundamental role in control systems that produce their

effects on people; the manager must therefore balance the need for rapid control (if a pedestrian is crossing the street, a sudden braking is required) with that of graduality (if you are giving a ride to your old granny, it is better to avoid sudden braking).

Notice that the assumptions that $g(Y/X)$ and $h(X/Y)$ are *constant coefficients* (i.e., *action rates*), for example, $g = 4$ and $h = 0.25$, do not represent a serious limitation for control systems; quite the opposite, the presence of constant coefficients are found in a large variety of control systems with linear dynamics. In many other cases, however, $g(Y/X)$ and $h(X/Y)$ are functions of X_t and/or Y_t and/or E_t, so that the system can also have nonlinear dynamics, as occurs when, for example, $g(Y/X) = 1.5–0.1X_{t-1}$; in this case the dynamics of Y_t would depend not only on the constant *action rate* ($g = 1.5$) but also on the new variable coefficient $0.1X_{t-1}$.

What matters most is to understand that the dynamics of the control lever—and as a consequence of the entire system—depends not so much on the value of Y_t as on the "distance" between Y^* and Y_t. The concept of "distance," or "Error," is fundamental in all feedback control systems; in fact, Norbert Wiener, the founder of cybernetics, proposes this concept as the basic mechanism of such a form of control:

> Now, suppose that I pick up a lead pencil. To do this, I have to move certain muscles. However, for all of us but a few expert anatomists, we do not know what these muscles are; and even among the anatomists, there are few, if any, who can perform the act by a conscious willing in succession of the contradiction of each muscle concerned. On the contrary, what we will is to pick the pencil up. Once we have determined on this, our motion proceeds in such a way that we may say roughly that the amount by which the pencil is not yet picked up is decreased at each stage. This part of the action is not in full consciousness. To perform an action in such a manner, there must be a report to the nervous system, conscious or unconscious, of the amount by which we have failed to pick up the pencil at each instant (Wiener 1961, p. 7).

Systems thinking has made the concept of distance operative simply by creating the new variable $E(Y)_t = Y^* - Y_t$, which tells us "by how much we have exceeded Y^*" or "how close we are to Y^*" (Fig. 2.1).

Let us agree to denote systems that function according to the model in Fig. 2.2 also as *Decision and control systems*, since the control of Y_t implies that explicit decisions regarding changes in X_t are made by the system manager. A control system is defined as *automatic*, or *cybernetic*, if the entire system can be structured and programmed so that its functioning is independent of human decisions and interventions; the *cybernetic* system appears to be closed within itself, and the only link with the environment is the specification of the objective Y^* by the *manager*. We shall return to these aspects of control systems in Chap. 3. We should note at this point that control systems operate automatically for the most part, at microscopic and macroscopic levels; for this reason they are not easily observed or understood.

2.3 The *Ring* in Action: The Heuristic Model of a Control System

On the basis of this minimal set of variables and functions, represented by the logical model in Fig. 2.2, we can derive the simple *heuristic model* that describes the recursive equations producing the behavior of the system (assumed to be without delays) through the dynamics of its variables.

For simplicity's sake, we assume that (a) the time scale is represented by $t = 0, 1, 2, \ldots n, \ldots$; (b) $g(Y/X) = g$ is constant; (c) the system is symmetrical and $h = 1/g$; and (d) $r(X/Y) = r$, that is, $r = 1$ or $r = 2$, etc.

- At time ($t = 0$), detect the *initial state*, Y_0, of the variable to control, and the objective, Y*.
- At time ($t = 0$), quantify the "distance," or error, between the initial state and the objective to achieve, calculating the initial deviation:

$$E(Y)_0 = Y^* - Y_0 = \Delta(Y)_0 \tag{2.0}$$

- At time ($t = 0$), determine the value X_1 to be assigned to the action variable in the subsequent moment $t = 1$; this value can be assigned by the manager of the system at his discretion, based on his experience, or it can be determined more efficiently by multiplying the deviation by the *reaction time* h (assumed to be constant):

$$X_1 = X_0 + E(Y)_0 \; h \tag{2.1}$$

- At time ($t = 1$), determine, using the *action rate* g (assumed constant), the first value Y_1 for the variable to control:

$$Y_1 = gX_1 + D_1 \tag{2.2}$$

- Again at time ($t = 1$), determine the new error:

$$E(Y)_1 = Y^* - Y_1 \tag{2.3}$$

  where $E(Y)_1 < E(Y)_0$
- At time ($t = 2$), the action variable is adjusted by adding the correction factor, calculated using the *reaction rate* h, according to the equation:

$$X_2 = X_1 + E(Y)_1 \; h \tag{2.4}$$

- As there are *no delays*, again at ($t = 2$) calculate the new value for the variable to control, using the equation:

$$Y_2 = gX_2 + D_2 \tag{2.5}$$

- Proceeding recursively, at discrete moments ($t = n$) and ($t = n + 1$), we calculate the new values for Y and X from the equations:

$$Y_n = gX_n + D_n \tag{2.6}$$

$$E(Y)_n = Y^* - Y_n \tag{2.7}$$

$$X_{n+1} = X_n + E(Y)_n \ h \tag{2.8}$$

- Having assumed that the values of the reaction time are $r = 1$ or $r = 2$, etc., we can modify (2.8) (also modifying the preceding relations) to get the following relation:

$$X_{n+1} = X_n + \frac{(Y^* - Y_n)h}{r} \tag{2.9}$$

- In this sense, r signifies *time required to eliminate error*. Note that since (2.9) allows us to adjust the variables with respect to the errors of the preceding instant by multiplying the error by ($1/r$), the system in theory will never achieve equilibrium. It is thus necessary to define a *tolerance* that, when reached, will lead the system to automatically ignore the residual error and consider the objective value to have been reached.

The system achieves a *stable equilibrium* when the deviation $E(Y)_{n^*} = (Y^* - Y_{n^*})$ is zero for all moments subsequent to n^* and the variable Y_{n^*} reaches and maintains the value Y^*; or when there is a deviation opposite in sign that the system eliminates by acting appropriately with new values for X. This stability can be *long-lasting* but, due to the action of the *law of dynamic instability* that characterizes all systems from the system thinking perspective (see Chap. 1, Sect. 1.6.3), this cannot be indefinitely maintained, since normally some disturbance, D_n, occurs that requires new adjustments. The above elements are usually sufficient by themselves to describe the general model of system control without delays, using the symbols of systems thinking as shown in Fig. 2.2.

Regarding the way the values of the variables X_n and Y_n are calculated, I would note that, through (2.6) and (2.8), the cumulative values of variables X_n and Y_n are always determined. One alternative is possible: calculating the cumulative values for X_n or for Y_n alone. An equivalent alternative is possible: calculating the difference values: $\Delta X_n = X_{n+1} - X_n$ and $\Delta Y_n = g\Delta X_n$.

Naturally, in both cases the dynamics always leads to equilibrium, though the way in which it is represented changes; for example, in the numerical tables for the

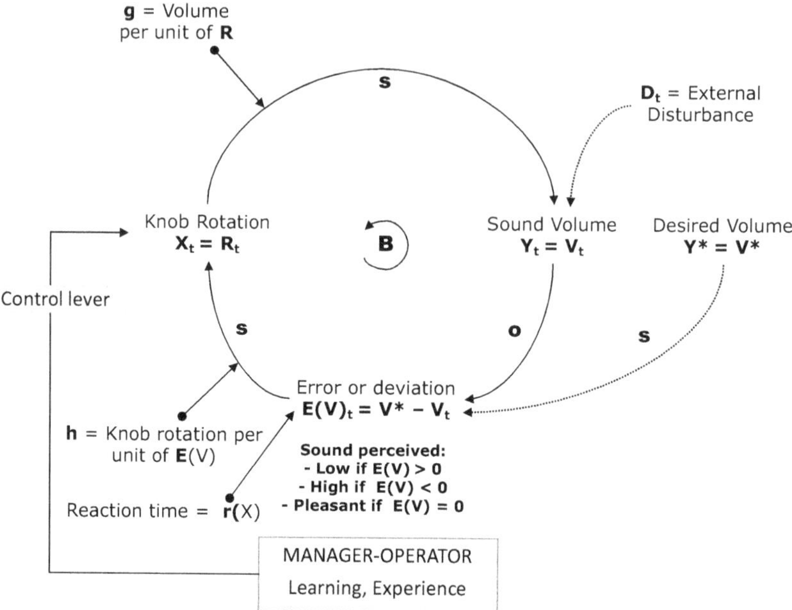

Fig. 2.3 Audio control system with one continuous lever without delay

audio and shower temperature models shown in Figs. 2.3 and 2.9 below, in order to facilitate the understanding of the dynamics I have chosen to represent the dynamics of Y_n and only the variations in X_n for each n.

2.4 Some Technical Notes

Even if the control system model has been completely outlined, some technical notes are useful to clarify or supplement several elements of the model in Fig. 2.2.

First technical note A control system is structured to *achieve* the value Y^* through a succession of "adjustments" to X; therefore, there can be no control system if we do not have some type of objective. Whatever the nature of the objective, it can be made explicit or implicit. Y^* represents an *implicit* objective if it entails a constraint or limit that, though not defined, is the result of the "nature of things." Our normal body temperature is an implicit objective (constraint) of our physiological system, which maintains the temperature over time by acting on "sweating" and "chilling." The temperature of 22 °C we want to maintain in our office with a heater is an *explicit* objective. Thus, we can define a control system in simple terms as a *Ring* that operates repetitively to allow the Y_t to achieve the *explicit* or *implicit* objective independently of its significance.

Second technical note I observe that in very large systems composed of many subsystems—whose causal loop diagram (CLD) is made up of a large number of variables—it is not always easy to recognize the existence of control systems. There is, however, a simple *rule*: a control system exists whenever we can recognize in the general system we are observing and modeling the presence of a variable that takes on the significance of an error with respect to a value Y^*, appropriately defined, that serves as an implicit or explicit objective.

Third technical note Control systems are dynamic systems that achieve or maintain the objective in a *succession of iterations*. Nevertheless, there are also control systems for which the objective must be achieved in a *single* cycle, as for example, in hunting, an activity where normally the hunter has only one shot to catch the prey (objective), or in industrial or artisan activities, where a given worker must carry out an activity that leads to irreversible effects. Also falling within this category are all those systems where a decision must be made that produces a single, non-modifiable effect. I shall call these control systems *one-shot systems* or simply *decisional systems*. Decisional processes, in the traditional sense of the term—choosing the best alternative for achieving an objective—can thus be considered as special cases of control processes (see Chap. 5, Sect. 5.9). Henceforth, unless otherwise specified, the term control system will mean a *pluricycle system of decision-making and control*: a *Ring* that rotates several times.

Fourth technical note The control system in Fig. 2.2 has a simple feature: the action on X_t is proportionate to the error (in the various ways that will be examined below). These systems can be defined as *proportional control systems* in that they correct X_t, and thus Y_t, in a measure that is proportionate to $E(Y_t)$. In some cases it can be useful to quantify the action on X_t not so much in terms of the error but of the variation in $E(Y_t)$; the corrective measure is proportionate to the rate of variation more so than to the amount of the error. Systems that adopt this form of control can be called *derivative control systems*; they permit a more "precise" regulation of the variable Y_t: the smaller is the reduction in $E(Y)_t$, the more dampened is the variation in X_t (e.g., approaching the dock the speed of the boat's motor is reduced). There is also a third possibility: quantifying the reaction on X_t not based on the error dynamics, at every instant, but on the dynamics of the accumulated error, from the beginning of the control and continuing on through each successive instant. For this reason these systems are known as *integral control systems*, and they can be applied to phenomena for which the error indicates an overall magnitude to be "bridged" (volume, distance, etc.); in this case the intervention on X must be proportionate to the magnitude that is "narrowed." In many cases it is also necessary to maneuver the control lever to take into account possible future errors, in order to bring forward the control and make it more efficient. A control system that seeks to control Y by setting values for the X lever so as to simultaneously take into account three measures—(1) the *size* of the error at every instant (P, *proportional* to the present error), (2) the total (integral) of the past errors (I, integral of past errors), and (3) the prediction of future errors (D, *derivative* of the error; that is, the slope of the error over time)—is called a proportional-integral-derivative control system or PID control system (Sontag 1998; Zhong 2006).

Fifth technical note The terms control and regulation are often used with various meanings. For example, in *control theory* (Ogata 2002; Sontag 1998, p. 372) the term *regulation* indicates the form of control typical of systems that must maintain the variable Y stable *around* the values of an objective (*set point*) that is set once and only varied occasionally by the system's *manager*. Along with *systems of regulation* do we have *position control systems* (or servo-mechanisms), whose function is to modify the variable Y through X in order to reach an objective that can also vary over time (I shall deal with these systems in Chap. 3, Sect. 3.4). A second meaning, in addition to the two just mentioned, is based on the distinction between the control process and the technical system ("machine") that carries this out, in the sense that: "I cannot control the temperature because the mixer faucet is not well regulated." Here the *control* regards the directional process of Y_t in moving toward the objective; the *regulation*, on the other hand, concerns the parameters for the functioning of the "machine" that carries out the *control*. It is clear that these two concepts are linked, in that the process depends on the "machine." *Ergo*: the *control* depends on the *regulation*. In order to avoid ambiguity, I shall use the term *calibration* rather than *regulation* to refer to the specific techniques of the "machine" that carry out the control process. According to a third meaning, control implies the attainment of an objective set by man; regulation instead represents a "natural" control, without human intervention, as in the case of a mutual regulation of populations (see Chap. 8, Sect. 8.2). In this book, to avoid ambiguity I shall use the term "control" with the more general meaning of indicating the entire *logical* process that leads Y to achieve the objective Y^*, even assuming deliberate or systematic variations in the latter. I shall use the term "regulation" to indicate the choice of values for X that are necessary to modify Y in order to eliminate the error, as in the expressions "to control the temperature (variable Y), regulate the mixer faucet lever (control lever for X)"; "regulate the speed (control lever for X), otherwise we'll go off the road (variable Y)"; "get going on the rudder (control lever for X) and change direction (variable Y), otherwise we won't enter the harbor (objective)." I shall use the term "calibration" to indicate the control on the parameters of the "machine" that implements the control process.

Sixth technical note The logic of a *closed-loop* or *feedback control* described so far is fundamental for the arguments in this book. Nevertheless, we can also design control systems defined as *open-loop*, *feedforward*, or even *commands* where the value of the control lever is determined only at $t = 0$ based on a *model known to the manager* by which he regulates the causal relations between X and Y, without further interventions on the trajectory of Y, in order to eliminate any error (Sect. 2.16.9 will deal with this topic further).

The next sections present several models typical of control systems and describe simple systems we have all encountered in our daily lives, in order that the reader can verify the *modus operandi* of those systems and understand how they perfectly achieve the control process.

2.5 Continuous Single-Lever Control System Without Delays

It is useful to begin by presenting the simplest control system, with a single control lever, X_t, with continuous variation within a preestablished range of values. This system regulates a variable, Y_t, which is also characterized by continuous variation and which reacts instantly—that is, without delay—to the variations of the control lever in order to achieve an objective, X^*, fixed over time and with a constant and known action rate, g, that regulates the variation of Y_t, assuming the reaction time is $r = 1$.

The continuity of the variables, the absence of delays, a fixed objective, and a fixed action rate are all features that make the control very simple. Knowing g and observing the amount of $E(Y)$, the manager assigns values to X_t in successive instants ($t = 1, 2, \ldots$), which allow Y_t to reduce the distance from Y^* and to eliminate $E(Y)$ at a certain instant t^*. This control system can be easily simulated, and if the manager had access to a precise error *detector*, while knowing $E(Y)_0$ and the action rate g, he would be able to calculate the new value X_1 to the extent necessary to obtain the perfect control already by instant one, since the reaction time is $r = 1$. The error elimination would be immediate, and thus very abrupt. To make the control less abrupt we could introduce a reaction time of $r > 1$.

Though it has a quite elementary logical structure, this type of system is very common, as our daily experience will confirm. We find such a control system when we have to adjust the volume on our radio, TV, stereo, cell phone, iPod, etc., using a knob with a continuous rotation. We observe the same logic when we have to change the direction of our car by turning the steering wheel or when we wish to change the water flow by turning the faucet, and so on.

When we listen to the radio we can judge the volume level as being pleasant, too high, or too low; thus, there is an optimal, subjective level V^* that represents the objective of the control. Given the physical structure of the "machine" producing it—which we assume to be a black box—we regulate the sound volume by rotating the regulating knob by a certain number of "degrees" or "notches" displayed on a graduated scale, as shown in Fig. 2.3.

The knob rotations, R_t, represent the action variable; that is, the *control lever*. The sound volume, V_t, is the variable to control in order to achieve the objective V^*. A clockwise rotation (cause) produces the effect of "increasing the sound volume"; a counterclockwise rotation "reduces the volume." If we define a clockwise rotation as positive (+) and a counterclockwise one as negative (−), then a causal relation in direction **s** exists between R and V.

The sound control is achieved through a succession of decisions that transmits to R the impulses $\Delta^+ R_t$= "turn right" or $\Delta^- R_t$= "turn left," which depend on the sign and perceived entity of the error, $E(V)_t$, of the actual volume compared to the desired one. Once the objective (V^*) of a pleasant sound is reached, the system remains stable, in the sense that no further rotation ΔR_t is applied to the knob, unless the user modifies the objective V^* or some *external disturbance* to the sound volume makes a correction necessary. If the control occurs *manually*, "by ear," a normal manager–

operator would be able to achieve the objective within a few time units, though with a certain tolerance. If we assume that the control system is *automatic* and that the manager–operator only sets the objective V^*, then the dynamics of the variables that make up the logical structure of the control system for the volume can be derived from (2.0, 2.1, 2.2, 2.3, 2.4, 2.5, 2.6, 2.7, and 2.8).

Let us assume the following initial inputs:

- $R_0 = V_0 = 0$; the initial values of the control lever and the volume are zero.
- $V^* = 40$ dB.
- $g = ½$; if we assume the knob rotates only a maximum of 120° and that the maximum attainable volume is 60 dB, then the action rate $g = ½$ means we have to rotate the knob 2° to obtain a volume increase of 1 dB.
- $h = 2 = 1/g$; this means that each unit of error $E(V)$ requires a correction of 2° to the knob.
- Reaction time $r = 1$.

These values allow us to produce the following iterations that carry out a control process in a single step:

- At time ($t = 0$), quantify the "distance" between the initial state and the objective to achieve, calculating the initial deviation:

$$E(V)_0 = Y^* - Y_1 = 40 - 0 = 40 \quad \text{decibels}$$

- At time ($t = 0$), determine the value R_1 to be assigned to the control lever at $t = 1$; this value can be assigned by the manager–operator of the system at his discretion, based on his experience, or it can be determined more efficiently, and automatically, by multiplying the error $E(V)_0$ by the *reaction rate*, $h = 2$:

$$R_1 = R_0 + E(V)_0 \ h = 0 + (40 \times 2) = 80 \quad \text{degrees of rotation of the knob}$$

- At time ($t = 1$), determine, using the *action rate* $g = 1/2$, the first value V_1 for the sound volume:

$$V_1 = R_1 g + D_1 = \left(80 \times \frac{1}{2}\right) + 0 = 40 \quad \text{decibels}$$

- At time ($t = 1$) determine the new deviation:

$$E(V)_1 = V^* - V_1 = 40 - 40 = 0 (\text{control achieved}).$$

- Observe that the automatic control of the sound volume has led to the immediate attainment of the objective at instant $t^* = 1$. If we desire a more gradual increase in volume, we could construct the control system using a reaction time of $r(R) = 2$. Figure 2.4 shows the numerical progression of the dynamics of the variables under this hypothesis. The "control panel" at the top of the table indicates the parameters the manager–operator must specify, including the volume objective and the action rate g. For convenience sake, the error is shown in the last column. It is important to realize at this point that the discomfort we experience when the volume is too high or too low is the *signal* that there is a *gap* with respect to the level we judge to be pleasant; this sense of discomfort is a *symptom* that creates the need to control V; thus it is subjective, often depending on environmental circumstances as well.

As shown in the system's control panel in Fig. 2.4 (first two lines at the top), we have assumed that the *reaction time* is $r(R) = 2$ time units; this means that rotating the dial 80° cannot be done in a single action, since the error reduction must occur over 2 time units. As a result, and as we can read from the line corresponding to $t = 1$, the rotation is not $R_1 = 80°$ but $R_1 = 40$, which produces an increase of only $V_1 = 20$ dB of sound (third column, $t = 1$), thereby reducing $E(V)_1$ by half. At $t = 2$, a 40° rotation of the dial is to occur in order to eliminate the error. Since $r(R) = 2$, the dial rotation is only $R_2 = 20°$, a volume increase of only 10 dB, which makes the perceived volume $V_2 = 30$ dB (third column, $t = 2$); the system continues on in this way until instant $t^* = 10$, when the error is eliminated. There is thus a nonlinear control, as we can easily see from the graph in Fig. 2.4.

By observing the values from column 2 (X_t= "cumulative rotation"), we note that the sum of the partial rotations from $t = 1$ to $t = 10$ is precisely $X_{10} = 80°$ and that corresponding to this value of the action lever is a volume level of $V_{10} = 40$ dB. At $t = 10$, the objective has been achieved and the error eliminated. If the reaction time had been $r(R) = 4$, the adjustment would have occurred for one-fourth of $E(V)$ for each period.

The concept of *reaction time* may seem unrealistic since we are not used to thinking that a control system does not immediately adjust X to eliminate the error and achieve the objective Y^*. However, the opposite is true. Experience shows that normally the corrections made by the control system are not violent but occur gradually over several time instants. If you are not convinced of this, consider how irritating it would be if a friend of yours turned on the radio and the radio suddenly rose from zero to 60 dB, making those present jump. And consider how annoying, or even harmful, it would be if, by turning the shower mixer, the water immediately became hot or cold; you would jump out of the shower. Or if, when you get into your friend's car, it immediately reaches cruising speed, slamming your body against the seat. However, with single-lever control systems, especially those that gradually produce their effects on people, the adjustments are normally fairly gradual, and this gradualism depends on the non-instantaneous *reaction time*, which indicates by how much the error must be reduced in each interval.

Returning to the model in Fig. 2.4, the penultimate column shows that from $t = 12$ to $t = 15$, there is an external disturbance (e.g., the nearby church bells are ringing) that reduces the perceived volume by 10 dB. I have therefore assumed that at $t = 12$ the manager–operator decides to increase the objective from $Y^* = 40$ to $Y^* = 50$ dB in order to counter the disturbance effect, and the volume V rapidly increases as the graph shows in Fig. 2.4. Once the disturbance ceases, the audio system produces a decibel level that is so annoying that the manager–operator decides to once again reduce the objective to $V^* = 40$ dB. The control system gradually reduces the dial rotation, returning the decibel value to that of the objective, which is reached at $t = 24$, with a minimal tolerance.

The continuous-lever control system described in the model in Fig. 2.3 is, despite its simplicity, entirely general and capable of representing numerous regulation processes. I shall here mention a few of these processes we all have experienced.

System name	Y = variable to control	X = continuous control lever
Stereo system	Volume level	Dial or regulator
Lamp	Level of light	Electricity regulator
Roller blind	Amount of light that enters	Lift cord
Eye	Amount of light that enters	Pupil size
Automobile	Direction of movement	Steering wheel rotation
Automobile	Speed (increase)	Pressure on gas pedal
Sailboat	Direction of movement	Rudder (angle of rudder)
Sailing ship	Direction of movement	Steering wheel (rotation)
Military drone	Direction of movement	Rotation of joystick
Nuclear engine	Power produced	Number of uranium bars
Automobile	Arrival time	Average speed

2.6 Discrete Single-Lever Control System Without Delay

Other control dynamics are possible, which depend on the type of *physical apparatus* used in the regulation process. In fact, along with *continuous-lever* systems there are *discrete-lever* control systems, which are equally simple and widespread.

Figure 2.5 presents a variant of the audio control in which the control lever is not a continuous rotating dial but the button on a remote control device (that can be pressed in two directions for several successive instants for a certain period) to increase or decrease the audio by a *constant amount* so that, at the same time, the error $E(V)_t$ falls at each instant by the same discrete amount at each subsequent pressing of the button. The control lever with discrete variations that, at each variation, varies the control variable by a discrete quantity (the movement of the lever) represents a "fixed-turning" lever. When there is a discrete and fixed-turning lever, the reaction time is not set, since the speed of reaction is directly established by the manager–operator. As shown in Fig. 2.5, the dynamics of V_t, while linear, gives similar results to those in Fig. 2.4.

We can consider as *discrete* or *fixed-turning lever* systems all those where the lever causes a variation in Y_t proportional to the action time, so that each action time unit of the lever is entirely equivalent to the pressure of a regulator button. If a tap produces a constant water flow so that the water volume is, for example, $g = 1$ L per

g = decibels/dial	0,5	V_0 =	0	r(R) =	2
h = 1/g = dial/decibels	2	V^*_0 =	40	tolerance	0

t	delta R_t = rotation	X_t = cumulative rotation	V_t = volume/ decibels	V^*_t = objective	variation in objective	D = disturbanc e	DEVIATION $E(V)_t=V^*-V_t$
0	0,00	0,0	0,0	40,0	0	0	40,0
1	40,00	40,0	20,0	40,0	0	0	20,0
2	20,00	60,0	30,0	40,0	0	0	10,0
3	10,00	70,0	35,0	40,0	0	0	5,0
4	5,00	75,0	37,5	40,0	0	0	2,5
5	2,50	77,5	38,8	40,0	0	0	1,3
6	1,25	78,8	39,4	40,0	0	0	0,6
7	0,63	79,4	39,7	40,0	0	0	0,3
8	0,31	79,7	39,8	40,0	0	0	0,2
9	0,16	79,8	39,9	40,0	0	0	0,1
10	0,08	79,9	40,0	40,0	0	0	0,0
11	0,04	80,0	40,0	40,0	0	0	0,0
12	0,02	80,0	30,0	50,0	10	-10	20,0
13	20,01	100,0	40,0	50,0	0	-10	10,0
14	10,00	110,0	45,0	50,0	0	-10	5,0
15					0	-10	2,5
16					-10	0	-18,7
17					0	0	-9,4
18					0	0	-4,7
19					0	0	-2,3
20					0	0	-1,2
21					0	0	-0,6
22					0	0	-0,3
23					0	0	-0,1
24					0	0	-0,1
25					0	0	0,0

AUDIO VOLUME CONTROL

Objective

Volume/decibel

Fig. 2.4 Dynamics of the variables in the volume control with a continuous lever

second, then I can fill a container with a volume of $V^* = 10$ L by keeping the lever open (i.e., the constant opening of the tap) for a period $T = 10$ s, which would be equivalent to producing the water volume for 1 s ten times. The similarity to the discrete button-lever is totally evident. We can include in this class all control systems whose lever is a constant flow per unit of time, $f_t = f =$ "constant flow/time unit," where $t = 0, 1, 2, \ldots$ (1 L per second, 1 m^3 per day, 100 m per minute, etc.) and the variable to control is a flow integral in a defined interval T, "$Y_t = \int f dt = (f \times T)$": liters accumulated in 20 s; gas consumed in 40 days; distance covered in 30 min, etc. Given these characteristics, the variations of the control lever can, in fact, correspond to the instants $X_t = t$, which refer to the use of the lever; the higher X_t is (the number of instants of flow emission), the greater Y_t becomes (overall flow volume).

Even the system in Fig. 2.5 follows this rule. In fact, we can observe that the discrete lever was pressed four times, or four instants, and that V_t takes on the value $V_t = 4 \times 10 = 40$. This interpretation allows us to broaden considerably the number of *discrete* systems with a *fixed-turning lever* that we can encounter in the real world, which can be represented by the model in Fig. 2.5. Some of these systems are indicated in the following table.

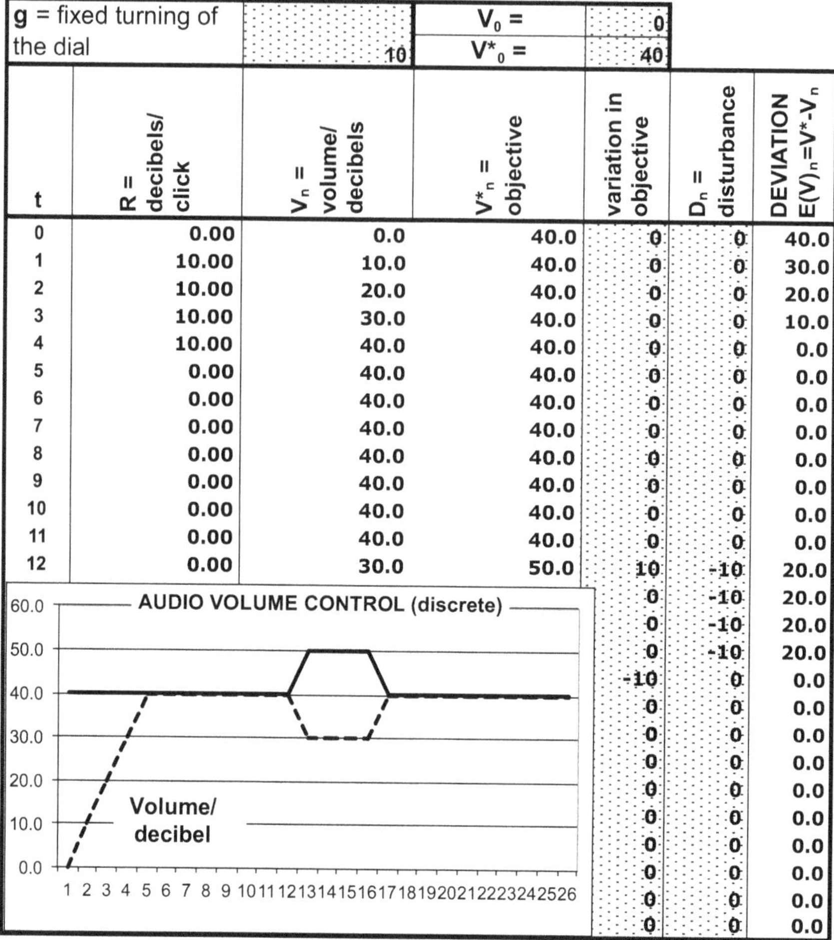

g = fixed turning of the dial	10	V_0 =	0			
		V^*_0 =	40			

t	R = decibels/click	V_n = volume/decibels	V^*_n = objective	variation in objective	D_n = disturbance	DEVIATION $E(V)_n=V^*-V_n$
0	0.00	0.0	40.0	0	0	40.0
1	10.00	10.0	40.0	0	0	30.0
2	10.00	20.0	40.0	0	0	20.0
3	10.00	30.0	40.0	0	0	10.0
4	10.00	40.0	40.0	0	0	0.0
5	0.00	40.0	40.0	0	0	0.0
6	0.00	40.0	40.0	0	0	0.0
7	0.00	40.0	40.0	0	0	0.0
8	0.00	40.0	40.0	0	0	0.0
9	0.00	40.0	40.0	0	0	0.0
10	0.00	40.0	40.0	0	0	0.0
11	0.00	40.0	40.0	0	0	0.0
12	0.00	30.0	50.0	10	-10	20.0
				0	-10	20.0
				0	-10	20.0
				0	-10	20.0
				-10	0	0.0
				0	0	0.0
				0	0	0.0
				0	0	0.0
				0	0	0.0
				0	0	0.0
				0	0	0.0
				0	0	0.0
				0	0	0.0
				0	0	0.0

Fig. 2.5 Sound volume control with discrete control lever

System name	Y = variable to control	X = discrete control lever
Stereo system	Sound level	Pressing the remote control
Lamp	Level of light	Variator with relay
Sailboat	Direction of movement	Rudder angle
Cylindrical cistern	Amount/depth of liquid	Flow emission time
Automobile	Distance traveled	Hours at constant average speed

2.7 Control System with On–Off Lever

We can further widen the class of systems that clearly follows the *modus operandi* described in Sect. 2.6 by including as well the type of control system defined as *on–off*. In fact, for such systems we can assume that the action of the lever corresponds to a duration T formed by a succession of consecutive instants, $X_t = t$, in each of which the variable Y_t undergoes a variation (continuous or fixed-turning) until the objective is achieved. The control lever is activated (*on*) when the error occurs and is deactivated (*off*) when the error is eliminated. If an external disturbance should produce a new error, the lever will once again activate (*on*) until the new error reaches zero (*off*). Thus, an external observer would perceive the activity of these systems as a succession of activations and deactivations of the control process; that is, a succession of periods T where the lever is operating, in each of which there is a variation in Y_t that is necessary to eliminate the error, following a typical *on–off* behavior, which accounts for the term *on–off* systems.

Due to the simplicity of their logic, *on–off* systems are quite common. Some of those that are immediately recognizable in our home environments are presented in the following table.

System name	Y = variable to control	X = continuous control lever
Air conditioner	Environmental temperature	On–off compressor
Refrigerator	Temperature below zero	On–off compressor
Radiator fan	Heat of radiator	On–off fan motor
Forehead sweating	Body temperature	On–off sweat gland
Chills	Body coldness	On–off muscular vibration
Windshield wipers	Transparency of windshield	On–off wiper oscillator
Brightness regulator	Brightness monitor	On–off led power
Fireplace	Room temperature	I/O feeding fire
Fan	Air temperature	I/O fan power supply

On–off systems usually function *automatically*, since the variable to regulate undergoes variations from external disturbances whose intensity, duration, and starting time cannot be easily predicted. For this reason, with *on–off* systems—for example, air conditioners—the activation and deactivation of X_t does not depend on decisions by the manager but on the magnitude of $E(Y)_t$ and the control systems operating program.

Figure 2.6 presents an initial standard model to examine an automatic *on–off* control system for regulating room temperature by means of a basic single-lever air conditioner. We immediately observe that this model is entirely similar to that in Fig. 2.5, except for the fact that the *reaction rate*, h, corresponds to the command to turn on the conditioner, which functions automatically for a number of instants $X_t = t$, in each of which cold air is emitted at the temperature indicated by the *action rate*, g; the actual set-point temperature is gradually lowered to the objective A^*.

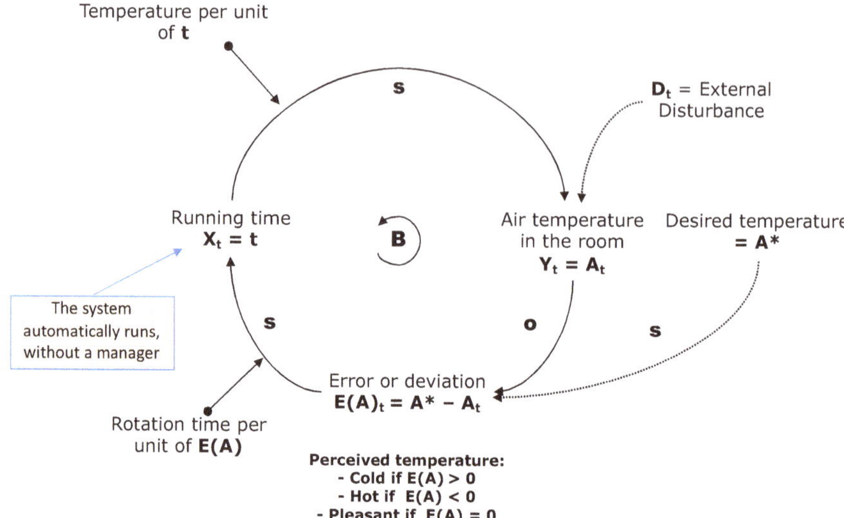

Fig. 2.6 Model of an automatic on–off air conditioner

The desired temperature is memorized through an appropriate graduated dial (or a digital display) indicating the various temperatures that, when set at a fixed point, memorize the objective Y^*.

If outside disturbances should once again raise the temperature, the regulator would again turn on the compressor for a sufficient interval to reach the objective. It is clear that over a sufficiently long period, the control system produces a succession of compressor activations (on) and deactivations (off). Obviously when producing an air conditioner, as well as other *on–off* systems, account is taken of a tolerance interval for Y_t with respect to the objective, so that the compressor does not have to turn on and off too frequently.

On–off systems can also operate with a *delay*, though the delayed start is usually compensated for by an equally delayed turning off. The *reaction time* also appears not to be significant since its effect is compensated for, once again by the length of the activation period.

A non-automatic *on–off* system—that is, one where the activations and deactivations of X_t (which determine Y) depend directly on the manager's decisions—is often indicated by the symbol I/O: "I" for *on*, "O" for *off*. We can find such non-automatic systems in, for example, country homes where a room is heated by adding logs to the fireplace for a certain period or cooled down by turning a fan on or off; or in vintage cars without radiator fans, which must therefore stop from time to time to cool down the liquid in the radiator.

2.8 Continuous Single-Lever Control System with Delay

This section takes up control systems with a *delay* between the variation of X_t and that of Y_t. We shall construct a new standard model that allows us to introduce the concept of delay and to clarify how delays condition the behavior of control systems. Unless stated otherwise, it is assumed at all times that the variations in Y_t resulting from the variations in X_t are instantaneous; this means that the system takes no time to carry out its process or that the amount of time is negligible, exactly as in the example of turning off the radio dial, which seems to "instantaneously" change the volume, at most taking account of the gradualness resulting from the reaction time. When instead there is a *relatively long* and *unexpected* interval between the assignment of values to X_t, and the "answer" comes from $Y_{t+\Delta t}$, and not Y_t—with a non-negligible Δt—then the control system operates with a delay of Δt.

In practical terms familiar to all, we can state that Δt represents a delay when the interval is so long that the *manager* can decide to vary X *at least twice* before the *first* variation is produced in Y. Precisely for this reason, *delays* make it quite complicated for control systems to achieve their objective since, if we attribute two impulses, ΔX_1 and ΔX_2, to X in rapid succession, we can have an answer, $\Delta Y_{\Delta t}$, that is twice that of the desired one, which would then require a new impulse of opposite sign to X_t.

The presence of *delays* in models is indicated by the word "delay" appearing on the arrow as many times as there are impulses from X before the first response from Y, or by placing two short crosswise segments on the arrow.

It is important to realize that *delays* do not always depend on functional defects of the control system but, in most cases, are innate to the functioning of the "machines" that activate control processes (Sect. 2.9). Thus *delays cannot be eliminated*, arbitrarily reduced, or even ignored. Once cognizant of the existence of delays and the impossibility of eliminating them, *there is only one strategy open to us*: learning *to identify them* and, through experience, *reducing them in number* and *length*. Delays can also arise in the recursive equations in Sect. 2.3. A single *delay* means that the value of a variable to control, Y_n, at a given instant, $t = n$, depends on the value X_{n-1} of a control lever rather than on X_n. The recursive Eqs. (2.6, 2.7, and 2.8) become:

$$Y_n = X_{n-1}\ g + D \tag{2.6/1}$$

$$E(Y)_n = Y^* - Y_n \tag{2.7/1}$$

$$X_{n+1} = X_n + E(Y)_n\ h. \tag{2.8/1}$$

Two *delays* mean that at $t = n$ the value of a variable to control, Y_n, depends on the value X_{n-2} of a control lever rather than on X_n.

The recursive eqs. (2.6/1, 2.7/1, and 2.8/1) become:

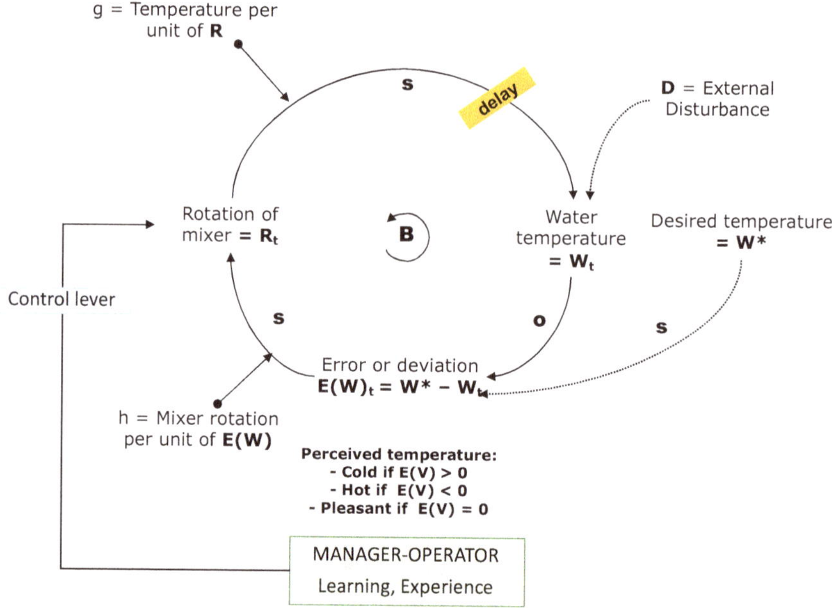

Fig. 2.7 Control system for regulating the shower temperature using a mixer

$$Y_n = X_{n-2}\ g + D_n \tag{2.6/2}$$

$$E(Y)_n = Y^* - Y_n \tag{2.7/2}$$

$$X_{n+1} = X_n + E(Y)_n\ h \tag{2.8/2}$$

To understand the effect of *delays* we can consider, for example, the system that regulates the water temperature of a shower with a mixer (single-lever control), as shown in the model in Fig. 2.7, which represents a theoretical shower with a delay between the regulation of the mixer and the variation in the water temperature, exactly similar to the sound regulation model in Fig. 2.4.

Figure 2.8 presents a *numerical simulation*, using Excel, of the system in Fig. 2.7, assuming the "delay = 0." In Fig. 2.8, I have assumed that the shower temperature is regulated by a mixer that, for each degree of rotation ($R = X$), varies the water temperature ($W = Y$) according to the constant rate "$g = 0.5$ degrees/notches," and that the system is perfectly symmetrical, so that "$h = 2$ notches/degrees." Let us assume the objective is a temperature "$W^* = 28°$" and that, as the shower is symmetric, for each degree by which the temperature *differs* from the objective the regulator must rotate the mixer 2 notches ($h = 2$ notch/degrees) to vary the water temperature by one degree (condition of symmetry); or, what is the same thing, that for each notch of rotation of the regulator the temperature rises "$g = 0.5$ degrees/notches."

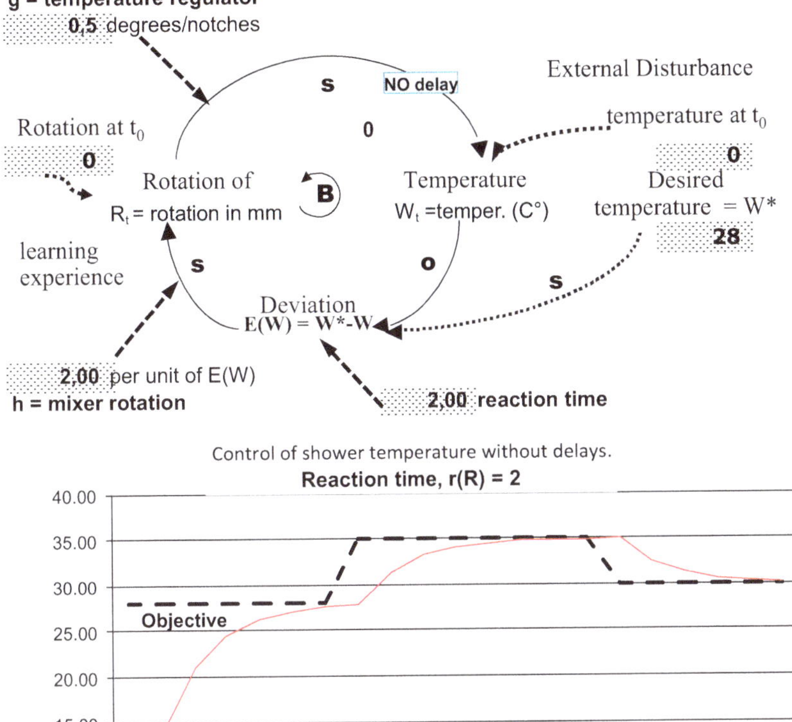

Fig. 2.8 Shower temperature control, without delay

Figure 2.8 shows that if the system is *without delays* then only a few regulations are needed to reach a stable temperature of $W^* = 28$, even if the manager thinks it opportune to vary the temperature objective twice (at $t = 7$ and $t = 15$). These assumptions can obviously vary from shower to shower, and the model can easily take account of different values. However, the model in Fig. 2.8 is not realistic. How many times, in fact, have we turned on the mixer and not felt the expected variation in temperature, so that we have to turn the mixer some more, with the "devastating effect" of ice-cold or boiling water. The water temperature control system normally has at least one *delay*.

Here are some difficulties we may encounter in controlling the shower temperature when there is a delay: we are under the shower and, feeling cold (this is the signal that accompanies the "error temperature"), we turn the mixer (initial impulse) to increase the temperature. Since the water has a rather long route to follow, from the boiler to the shower, some seconds, Δt, are required for the temperature to adjust to the rotation of the mixer. The cold is unpleasant and we are not willing to put up with it, so we turn the regulator further (second impulse); finally, after an interval Δt (which depends on the physical structure of the shower, the length of the tubes that bring the water, the wall temperature, etc.) the temperature turns pleasant, as a response to the initial rotation of the mixer (initial impulse).

However, after another interval the effect of the second impulse kicks in and the water begins to scald us. We rush to reverse the rotation of the regulator but nothing happens, because the delay also operates in reverse; so there is a new impulse, with the result that we might produce too strong a variation with a consequent discharge of ice-cold water, as we have all experienced. If we do not "jump out" of the shower or turn off the water, but instead patiently continue regulating the mixer, we usually end up with a pleasant temperature. This phenomenon occurs most often in hotel rooms when we shower for the first time; thereafter the regulation is easier, because we have learned how the water temperature "responds"—in gradations and with a reaction *delay*—to the rotation of the regulator.

Figure 2.9 shows the dynamics of the values of the same shower in Fig. 2.8, in which I have assumed *a single delay*. The first graph indicates the dynamics of W_t and R_t under the assumption of a *reaction time* of one period: that is, an instantaneous adjustment of the temperature following the rotation of the mixer. The second graph presents the dynamics for the same variables with a *reaction time* of two periods.

We immediately observe that, while the instantly adjustable shower (first diagram above) does not allow us to achieve the desired objective, but leaves the poor manager–user continuously going from hot to cold, the presence of a *reaction time* of $r(R) = 2$ (second diagram down) makes the water temperature adjustment slower and more gradual; the oscillations are less accentuated and quickly "attenuate" to allow the objective to be reached. This strange behavior shows that delays make it difficult to set up an automatic control that tends to reach and maintain an objective through a mechanical regulator. However, the different dynamics show us why, despite the fact all showers always have some built-in delay, we are always able to take a shower: the presence of a reaction time $r(R) > 1$ produces a gradual adjustment of the temperature that makes slight oscillations "acceptable" while we wait for the desired objective.

Our simulation model applies naturally to non-automatic systems as well, even though it is obvious that a manual regulation to adjust the shower temperature would have required a different number of attempts based on the manager–operator's experience.

This *conclusion* is general in nature: as the *existence of a delay* causes serious regulation problems, since it makes it difficult to reach and maintain the objective, the manager–operator of the system develops a *learning process* that, by learning from errors and delays, makes the control more effective. A second *conclusion* is also clear: in order to describe and simulate a single-lever control system with delay,

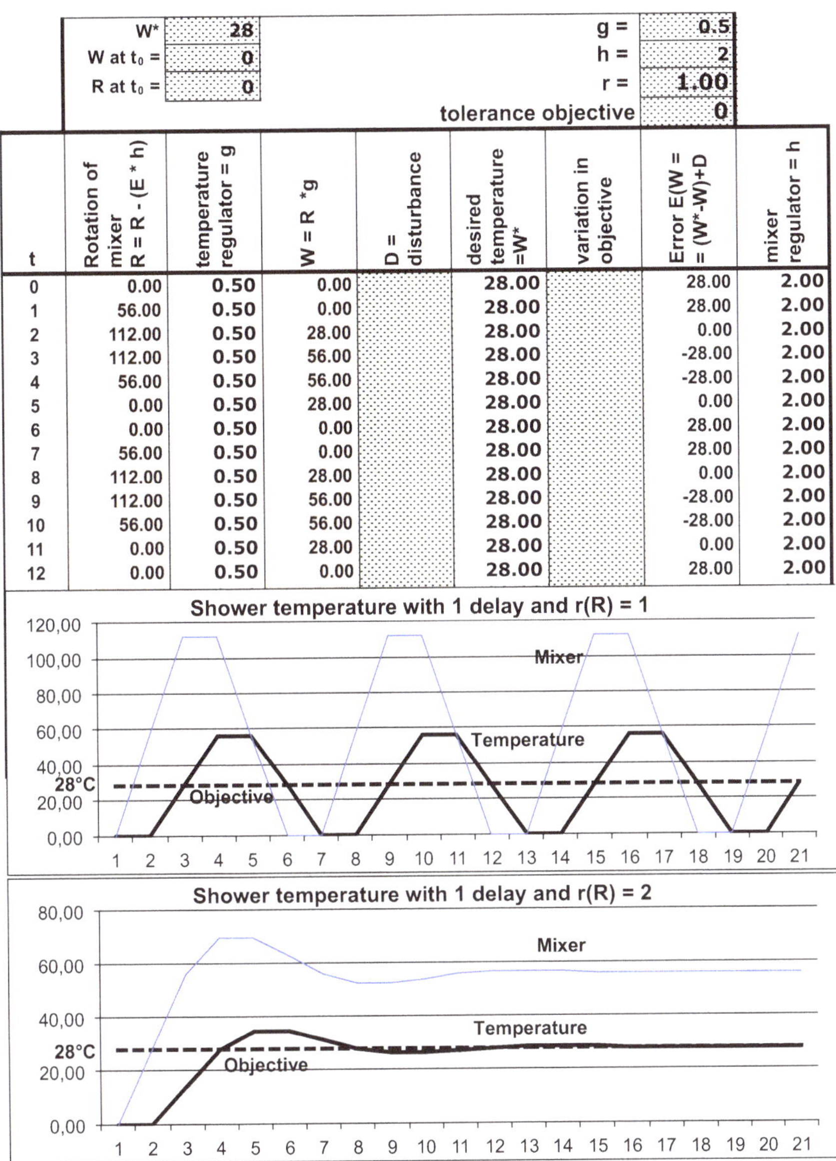

W*	28	g =	0.5
W at t_0 =	0	h =	2
R at t_0 =	0	r =	1.00
		tolerance objective	0

t	Rotation of mixer R = R - (E * h)	temperature regulator = g	W = R *g	D = disturbance	desired temperature =W*	variation in objective	Error E(W = = (W*-W)+D	mixer regulator = h
0	0.00	0.50	0.00		28.00		28.00	2.00
1	56.00	0.50	0.00		28.00		28.00	2.00
2	112.00	0.50	28.00		28.00		0.00	2.00
3	112.00	0.50	56.00		28.00		-28.00	2.00
4	56.00	0.50	56.00		28.00		-28.00	2.00
5	0.00	0.50	28.00		28.00		0.00	2.00
6	0.00	0.50	0.00		28.00		28.00	2.00
7	56.00	0.50	0.00		28.00		28.00	2.00
8	112.00	0.50	28.00		28.00		0.00	2.00
9	112.00	0.50	56.00		28.00		-28.00	2.00
10	56.00	0.50	56.00		28.00		-28.00	2.00
11	0.00	0.50	28.00		28.00		0.00	2.00
12	0.00	0.50	0.00		28.00		28.00	2.00

Fig. 2.9 Shower temperature control: one delay and two alternative reaction times

not only is it necessary to specify the action g and reaction h rates, the reaction time r, and the *initial* values of both X and Y, but also it is indispensable to determine the number of delays that intervene between a variation in X and the corresponding variation in Y.

2.9 The Technical Structure of a Single-Lever Control System: The Chain of Control

Despite their apparent simplicity, the control system models illustrated in previous sections have allowed us to make some general observations. Those models need to be further developed, but even the basic ideas presented so far allow us to understand the full power of control systems and the fact that without the concept of *control process* and without control systems as instruments for realizing such a process, we cannot fully understand the dynamics in our world. One thing is certain: without the action of billions upon billions of *Rings*, of varying hierarchical levels, life itself could not exist, neither single-celled life nor, even more so, the life of organisms, man as an individual, organizations, society, and ecosystems, as I shall demonstrate in Part 2 of this book.

In order to fully grasp the action of control systems we must always remember that the dynamics of the variables shown in the model in Fig. 2.2 are *produced* by interconnected *machines*—that is, by physical, biological, social "apparatuses," or those of other types and configurations—that develop the processes on which the dynamics of the variables depend. Through their interconnections these *machines*, whatever their nature, form the *technical structure* (or *real system*) on whose processes the *logical structure* (or *formal* system) of the control system depends.

By integrating the *logical structure* presented in the standard model of Fig. 2.2, this chapter proposes to define the *general technical structure* of control systems. Figure 2.10 shows a *complete model* that applies to any single-lever control system since it includes the "apparatuses" that constitute the *technical structure* of control systems, which is usually investigated by engineers, biologists, physiologists, economists, businessmen, etc. Knowledge of the *machines* and *processes* that produce the variations in X_t and Y_t is fundamental; unfortunately, in many cases their nature and behavior cannot be observed or understood, in which case they are considered *black boxes* and are not represented in the model; in such cases only the logical model can be represented.

We can generalize: all control systems depend on a *technical structure* made up of three basic process-producing "machines" linked together by an *information network* (not shown in Fig. 2.10) that processes the information (mechanical, electrical, electronic, human, etc.) necessary for the functioning of the apparatuses, transmitting this information to them.

(a) The *effector* (or *operator*, or *engine*), which represents the *apparatus* (natural or artificial), in fact enables a variation, ΔX_t, to be transformed into a corresponding variation ΔY_t, taking account of the *action function*, "$g(Y/X)$," and the values of the disturbance variables "D_t"; according to the engineering logic of the control (Sect. 2.16.10), this apparatus is considered as a physical system to be controlled.

(b) The *sensor* (or *error detector*), which represents the *apparatus* (natural or artificial) whose function is to quantify the deviation or error by measuring the value of Y_t in a form useful for comparing it with the objective Y^* and determin-

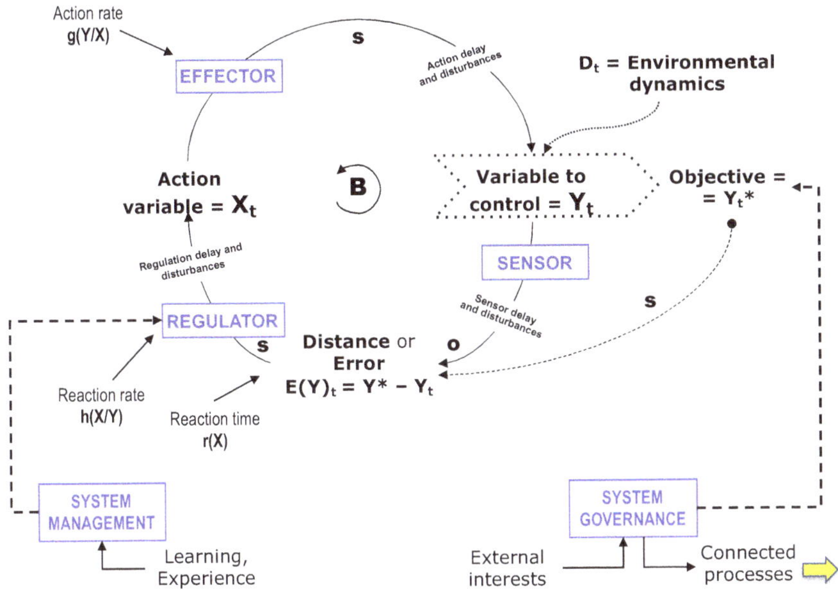

Fig. 2.10 Complete model of a single-lever control system

ing the error $E(Y)$; without a precise calculation of the error we cannot regulate X_t; from a technical point of view, the sensor can be thought of as composed of a *detector*, the apparatus that measures the values of Y_t, as well as a *comparator* that, through appropriate mechanisms (transducers and actuators), compares at each moment the values of Y_t with those of Y^*, calculating the deviations E_t.

(c) The *regulator* (*controller* or *compensator*), which represents the *apparatus* capable of "operating" the action variable—that is, calculating the new value of X_t, taking account of $E(Y)$, based on a *reaction function* $h(X/Y)$; technically the regulator is the controller apparatus in a strict sense; by using the data provided by the detector it acts on the effector.

(d) The *information network*, which is made up of various types of information transmission channels linked in various ways and which transforms the three apparatuses above—together with the *processes* carried out—into a "real" control system. This transmission network is not shown in Fig. 2.10, since it can be considered an integral part of the three "machines" that make up the real *control chain*.

Note that the concepts of network and *information transmission channels* concern not only the mechanical, electrical, electronic, or biological links among the apparatuses in the control chain but also they must be taken in a wider sense as a component of the technical (physical) structure of control systems—taking into account the various types of such systems. If the connecting chain (information transmission apparatus) between the rudder wheel (control lever) and the rudder (effector) should break, then it would be impossible to control the ship's direction;

similarly, a trauma that damaged the nerves in a part of our body (information transmission apparatus) would impede the action of the muscles (control lever) and inhibit any postural control by our body.

If the shower temperature (Y) is kept at the desired value (Y^*), this is due to the control system formed, for example, by the skin (sensor), the manual maneuvering (regulator) of the mixer tap (control lever), and the time needed for the hot water heater to turn on (effector), on the condition that the nervous system correctly transmits the sensorial impulses. If the stereo volume (Y) does not exceed the levels that allow us to pleasantly listen to music (Y^*), this is because of a control system composed of the ear (detector), the regulation of the sound level (regulator) by rotating the knob (control lever), and the internal apparatus that produces the sound (effector), on the condition that the network of electrical connections allows the apparatuses to function properly. If the quantity of light reaching the retina (Y) is relatively constant throughout the day (Y^*), this owes to the action (control lever) of a muscular apparatus (effector) that modifies the opening of the pupil (effector), with the eventual aid of eyeglasses and artificial light (voluntary control levers adopted by the manager). If our body temperature (Y) is maintained within its physiological limits (Y^*), this is due to a control system made up of the skin (detector), which transmits the sensation of hot/cold (error) to the brain (regulator) so that it can regulate the temperature through the biological apparatuses (effectors) of sweating and chilling (control levers), aided by clothes, heating, cooling, and transferal to environments with favorable climates (effectors).

No part of our world can exist without a control system, in the absence of which the variables take on unacceptable values and a catastrophe occurs; the system that variable is a part of is disturbed and even destroyed. It is all too clear that when arms escalation and increased violence occur, the tragedy of war cannot be avoided without the help of a control system. Nevertheless, no war drags on forever, since sooner or later a political and social control system is created that stops the escalation. Control systems are often difficult to perceive since they are so microscopic (control systems in cells, the eye, neurons), vast (control systems for climate, population dynamics, society), or complex (control systems for neural activity, the immune system, the economic system) as to make their representation and understanding difficult.

There is no doubt that if man has survived for millennia and attained his present level of prosperity, this owes to his success in controlling the waters, fertility, productivity, epidemics, and wars. As yet no effective control systems for births, global temperature, desertification, typhoons, and new epidemics have been conceived, but everything leads us to believe that such systems will be devised and constructed as soon as effective apparatuses that include these phenomena are devised.

First technical note Figure 2.10 presents a general model of a control system where the control lever, X_t, acts directly on the variable to control, Y_t. This represents a so-called perfect *Ring*, since the three fundamental variables—X_t, Y_t, and $E(Y)_t$—are directly linked through a single complex "apparatus." The FIRST GENERAL RULE of systems thinking (see Chap. 1, Sect. 1.5) teaches us that if we wish to construct useful systems models we must develop the capacity to "zoom" from the whole to the parts, from systems to components, and vice versa. We can apply this rule to control systems as well; if we want to have an in-depth understanding of how one variable acts on another, we can zoom in and break up the "apparatuses" into smaller components, thereby specifying the links between them.

Second technical note Normally we assume a control system is formed by an *unvarying* structure when the apparatuses that make up the *chain of control* are characterized by action parameters whose values are unequivocally determined and do not vary as a function of X, Y, or $E(Y)$. I shall define a *varying structure* control system as one whose structure (and thus capacity for control) is varied by the apparatuses as a function of one or more of the systems' fundamental variables. Operationally speaking, in control systems with an *unvarying structure*, $g(Y/X) = g=$"constant" is an action *rate*; if, instead, we have an action *function* $g(Y/X) = g(X, Y, E, D)$, then the system would have a *varying* structure and the effector might even produce a nonlinear dynamics in Y_t as X_t varies. The possibility of using the apparatuses in the chain of control to vary the action function usually derives from the system manager's ability to strengthen, when necessary, the power of control, as we shall see in Chap. 5, Sect. 5.4 (Fig. 5.2).

2.10 Management and Governance of the Control System

Figure 2.10 indicates two fundamental actors that in various ways normally intervene in any control system: the *manager* and the *governor*.

Above all, I would like to reflect on the *regulation mechanism* and *process* whose function is to decide which new value (or variation) to give the action variable, X_t, taking into account the amount of error, $E(Y)$, in order to enable the system to drive Y_t to reach Y^*. The *regulation process* is carried out by an actor I shall define as the *system manager*. More precisely, the *manager* of the control system (in the broadest sense of the term) is defined as the subject (individual or group, organ or organization) that, by means of a *series of decisions*, operates on the *regulator* to vary X_t in order to modify Y_t, assuming that the functioning of the *apparatus* does not vary over time. The *manager* must almost always directly intervene on the *effector* and on the *sensor apparatus*, taking into account the values of the functions $g(Y/X)$, $h(X/Y)$, and $r(X)$. Precisely for this reason the *manager's* activity—the *management process*—is based on appropriate knowledge (together with sufficient experience) of the entire *control chain*; such knowledge must allow the manager to judge the *appropriateness of the objectives* to be achieved, to carry out the *information-gathering* procedures to determine the error and quantify the *delays*, and to grasp the functioning of the *effector*, as well as to gain mastery over the operational mechanisms of the *regulator* and the *information network*.

A *cybernetic system* (or a system with *internal* regulation) is a control system whose *manager* is an integral part of the chain of control. No management intervention in a cybernetic system comes from outside. The system is self-sufficient in producing its own dynamics in order to achieve the objective. If the manager was instead outside the chain of control, separated from the regulator, then the system would be *externally regulated.*

Under a normal shower we are the *managers* of the control system, detecting the temperature with our skin and regulating the mixer lever with our hands, even if we cannot control the boiler that heats the water, the length of the tubes it runs through, or the external temperature. If we include within the chain of control interventions by the user–manager as well, then a normal shower becomes a typical closed-loop control system; to an outside observer it would be considered a cybernetic system in all regards. On the other hand, if we use a modern shower, which allows us to set the desired temperature on a thermostat, then we have a cybernetic system, and we are only users but not managers of the system. If the control levers are external to the shower and, when we want to vary the water temperature, we have to ask an outside manager to regulate the mixer, then the control system cannot be defined as cybernetic, even though it maintains the features of a control system with feedback. Similarly, when we drive we are the internal managers that control our cars, since we influence a certain number of action variables (brakes, gas pedal, gear shift) in order to regulate speed (variable under control) so as not to exceed an external limit (constraint). For an outside observer, on the other hand, a car with its driver (e.g., the taxi that must take us to our destination) is in all respects a cybernetic system.

In our offices the temperature is automatically regulated by a thermostat whose desired temperature (objective) we determine; this is an automatic cybernetic system. The cook who regulates the gas to maintain the oven temperature stable is an external manager; in order to avoid burnt or undercooked cakes, kitchens have for some time now been equipped with an automatic control consisting of internal thermostats that signal an error in temperature to a regulator that then varies the flow of gas. We can thus reach the following conclusion: a system is "cybernetic" *to the outside observer* when the manager of the system is an integral part of the chain of control, so that the system tries to achieve the objective set by the *governance* without interventions from outside control.

Referring precisely to control systems with feedback, an *initial question* arises: if the manager is part of the control system (internal or external), then who sets the *objective*? I shall define *governor* as the actor (individual or group, organ or organization) that sets the objective, Y^*, which management must achieve. Thus the governor is not part of the control system but an *external* agent that considers the control system as *instrumental* for achieving the objective Y^*, which is necessary for him/it to achieve his/its own personal interests or produce other *processes down the line*, as shown by the arrow in Fig. 2.10 leading from the box labeled "governance" (lower right of figure). The control system for the order of finish of the competitors in a race is instrumental for the regularity of the race itself; the control system for the production specifications of a machine is instrumental for obtaining the desired quality components to be used in the production process; the control of the stereo volume or shower temperature is necessary in order to allow oneself several minutes of relaxation. The *governance* process can also be complex, particularly in the many cases where the system is viewed as *instrumental* by various actors at the same time, each of whom proposes a different objective in order to attain its own individual interests. Thus, the *system objective* is defined through a "political negotiation" among the actors on the basis of the latter's relative power. We know this all too well from having to reach an agreement with our office colleagues about the temperature to set on the thermostat, or when we are driving other passengers and have different opinions on the proper speed, or when we have to discuss with other family members the best place to go for our picnic.

2.11 Design and Realization of the Control System

A *new question* now arises: who in fact constructs the control system to achieve the objective set by the *governance*? To answer this, we need to consider two new actors external to the system: the *designer* and the *constructor*, understood in the broadest sense. These are the actors who, after assessing the objectives, identify the most suitable chain of control structure and determine the functions $g(Y/X)$, $h(X/Y)$, and $r(Y)$.

The *manager* limits himself to making a certain number of decisions involving the regulation of an already-formed chain of control (whose functioning he probably is not familiar with) that is developed by the *designer* and built by the *constructor*, both of whom are external subjects, concrete or ideal, who make the system to control available to the manager.

There is no doubt that we are the managers of the car we are driving through city traffic or that our car is the product of the engineers that designed it and the firm that built it. Today the car industry is designing cars with automatic controls that, by means of systems for determining position through General Position Systems (GPS) *devices*, will enable a destination (spatial objective) to be reached in the most efficient way possible through an increasingly integrated control of the vehicle (see Chap. 6, Sect. 6.2).

It should be noted that it is not easy to design and construct a control system. The designer and constructor must take into account all the factors that can alter the values of Y_t. Some of these are *external*, being linked to "environmental variety," while others are internal and innate to the "endogenous variety" that characterizes the technical functioning of the apparatuses making up the system's physical structure. For example, it is clear that a control system for an airplane's speed must function whether the plane is flying through calm air or in the midst of turbulence, near sea level at 20 °C or at 36,000 ft at −30 °C, etc. (environmental variety); or flying on a full or semi-full tank (the weight decreases along with the distance traveled); with efficient engines or less efficient ones, etc. (endogenous variety).

I would like to point out that most systems that operate in our body to keep us alive are automatic control systems. Among the simplest and most evident are those that regulate our heart beat, or arterial pressure, in order to adjust this to the movement of our body so as to maintain a constant supply of blood; those that control body temperature through sweating or chills; those that ward off fatigue by requiring the body to rest; and those that regulate the intensity of movement and generate tiredness. It is equally clear that such systems must function efficiently, both in young and not-so-young people (endogenous variety), in warm or cold environments, at sea level or in the high mountains (environmental variety).

At times physiological control systems are not only completely automatic. We become the external manager of a chain of control; if sweating is not sufficient to lower our body temperature, we take off our clothes; if we still feel hot, we take a cool shower, and if the heat persists for a long time, we take a holiday in the mountains.

2.12 Delays and Disturbances in the Control: Variants of the Control Model

The model in Fig. 2.10 also shows us that in the *logical structure* of control systems delays are inevitable, since they occur in the *technical structure.*

> Delays between actions and consequences are everywhere in human systems … Unrecognized delays can also lead to instability and breakdown, especially when they are long (Senge 2006, p. 88).

> If a government manages to cut back on births today, it will affect school sizes in about 10 years, the labor force in 20 years, the size of the next generation in about 30 years, and the number of retirees in about 60 years. Mathematically, this is very much like trying to control a space probe far out in the solar system, where your commands take hours to reach it, or like trying to control the temperature of your shower when there's a half minute delay between adjusting the tap and the hot water reaching you. If you don't take the delay into account properly you can get scalded (William Brian Arthur 1973, in Waldrop 1992, p. 27).

There are three types of delay (Coyle 1977; Roberts et al. 1983):

(a) The most frequent is the *action delay*—or output delay—which depends on the effector and slows down the response of Y to an impulse from X; this delay depends on the *effector.*
(b) The second is the *delay in detection*, or information delay; this is a deceptive delay, since it acts on the perception and measurement of the error; if the error is not quickly detected, the regulator can produce a new and dangerous impulse from X.
(c) The third is the *regulation delay*, which occurs when the *regulator* does not promptly respond to the error.

These delays are always lying in ambush. They are inevitable, and only when they are negligible, with a normal standard of tolerance, can the control system be defined as *without delays*. Often we are not aware of their presence because the system admits a certain *tolerance* in its functioning, so that slight asynchronies in the variation of the variables do not noticeably influence its behavior.

Figure 2.10 shows the *disturbances* along with the *delays*, since the former can also create or accentuate the delays. Usually the disturbances act on the variable to control, *Y*.

> Disturbances: Disturbance is a signal which tends to adversely affect the value of the output of a system. If such a disturbance is generated within the system itself, it is called an internal disturbance. The disturbance generated outside the system acting as an extra input to the system in addition to its normal input, affecting the output adversely is called an external disturbance (Bakshi and Bakshi 2009, pp. 1–3).

More generally, by *disturbance* I mean any phenomenon, internal or external to the system, which in some way makes the activity of a given apparatus inefficient; in fact, along with delays, disturbances also act on each of the apparatuses that make up the chain of control.

In the examples of the control systems for water temperature and audio volume, we can easily recognize the close connection between delays and disturbances. A body that has been in the cold perceives a higher temperature with a longer delay than one that has been in normal conditions (*detector delay*), just as a person who for hours has listened to music at full volume in a discotheque perceives a radio volume that is too high with a delay, thereby angering his or her neighbors. Those who have soap in their eyes and feel the scalding water have a difficult time finding the temperature regulator (*regulation delay*), just as someone listening to music in the dark has difficulty reaching the volume control on the radio.

You have invited friends over and want to amaze them with your homemade pizza. After preparing the pizza with the dough, sauce, and mozzarella cheese, you put it in the electric oven (*effector*) because the instructions say to cook it at 180 °C (objective: *Y**). Even if you are not an expert cook, you can regulate the appropriate dial (*control lever X*)—based on your acquired experience—so that it provides a temperature of 180 °C (controlled variable *Y*). If the oven has an internal thermometer that gives the temperature (*sensor*), the temperature will not cause problems, and you need only check the cooking time (a second variable to control). If the oven does not have an automatic *sensor* for the temperature, or you do not trust the one it has, then you become the *manager* of the system and must periodically visually check (*detecting*) the cooking progress of the pizza. If you see that the pizza is cooking at too low or too high a heat (error between *Y** and *Y*), you intervene to regulate the dial (*control lever*) several notches to vary the oven temperature in the desired direction. The three delays are waiting in ambush: the oven can adjust the temperature more slowly than planned (*action delay*); distracted by the arrival of your guests, you do not check how the pizza is cooking at frequent enough intervals (*detection delay*); and even though you detect an error you might wait several minutes before regulating the gas knob (*regulation delay*). Making a pizza is not only a question of the right recipe (productive combination) but also of the friends waiting to have it for dinner (users or clients); above all, it represents a *managerial problem regarding the control of a process*.

Delays in detecting the error or in the action of the control levers are often due to physical limits, or limits related to obtaining and transmitting the information. Our car thermometer always shows with a delay the outside temperature that indicates the presence of ice, and we could decide to slow down after having gone some kilometers unaware of the icy roads. Moreover, if we are not careful we might not notice the temperature signal and employ the levers to reduce speed with a delay. For this reason, many cars even adopt an acoustic signal to alert the driver when the temperature is reached at which ice forms. This reduces the risk, as long as we are not driving with the stereo going full blast, which would interfere with our hearing the signal. To control the movement of the Mars probe, it is necessary to detect its position and send signals with information about the new objectives and the tasks to carry out. This requires several minutes between the reception and transmission of the signals, and it is possible that a probe approaching a gulley could receive a message to halt after it has already fallen in.

First technical note Although delays are linked to the time it takes for the processes to be carried out, these should nevertheless be identified or confused with the *normal* functioning times (even when lengthy) of the apparatuses carrying out those processes. The delay must be considered as such by management when it perceives an apparatus response time *greater than what would be expected under normal operating conditions*. We all know that by turning the mixer faucet, we do not produce an instantaneous variation in temperature; if our experience has memorized an action time of, for example, 4 s, then we will start to perceive a real delay after 6 or 8 s, which leads us to turn the dial further. A delay *beyond the normal functioning times* can result from our error in maneuvering the levers, from a functioning disturbance in the apparatuses, or even from external events. Only delays should be considered as anomalous, and there is no remedy for them except to trust in our experience and attentively and regularly observe the environment.

Second technical note Since delays are always held to be a destabilizing factor, they are the main focus of the analysis and design of control systems. However, we must not underestimate the effect on control systems from the opposite situation: a *premature occurrence* that is manifested when apparatuses respond in a shorter time than management expects under normal operating conditions. Whether or not Y_t varies earlier than our experience indicates is normal or with a delay, our reaction is the same: we try as soon as possible to reset the values on the control lever, X_t. We can deduce that a premature occurrence produces control problems, which in a certain sense are the same as those resulting from a delay. In fact, a premature occurrence forces the manager to reduce the action and reaction times with respect to the normal (longer) ones that regulate the action of the technical apparatuses. For the manager it is as if the system responded "with a delay" to the premature action of the levers.

2.13 Strengthening and Precision of Control Systems

Let us return to the *general model* in Fig. 2.10 in order to deal with a technical topic that will allow us to reflect on the historical evolution and the evolutionary tendencies of control systems, while also making some brief considerations that will be further explored in subsequent chapters.

It seems that the most relevant progress in control systems, of whatever type, has come about through the improvement in the power of the *effectors* that control the variable Y through the control lever X (these considerations also hold for the multi-lever systems we shall examine in Chap. 4). There is clearly a greater capacity of control of the route of a boat that has powerful turbine motors equipped with stabilizers controlled by electric motors than one powered by the wind and steered by a manually operated rudder.

But while modern devices that produce the effect of X on Y have considerable power and incomparable precision, the greatest progress in control systems must be attributed to improvements in *detection* and *regulation* apparatuses.

Who tunes his radio or television today by means of manually operated dials and knobs, as was the case at the dawn of radio transmission technology? Nobody thinks twice about this, since now there is automatically and electronically controlled tuning, which is much more precise and faster. From regulating routes using an oar we have evolved to the bar rudder, the circular rudder, and then to the modern rudder with servo-control. Today at home we no longer measure our child's temperature by placing the palm of our hand over his forehead but by using modern laser ray devices that do not even have to be in contact with the skin.

The lack of some devices in the chain of control prohibits us from controlling a number of variables that we could instead influence. While city authorities try to keep under control the amount of fine dust in the air through measurements at a detection facility and by banning traffic, we have yet to find appropriate effectors to control global warming, even though this phenomenon can be measured with precise instruments. It is not science fiction to imagine a future with powerful suction devices that can suck in icy air from the higher strata of the atmosphere to bring to the warmer zones to lower the temperature, or powerful purifiers that could eliminate air pollution in cities. And how will it be possible to control viral epidemics if we lack detection systems to measure the epidemic situation and effector systems to act against the spread of the virus? How can companies activate the market control levers or those for the demand for its products if the marketing departments do not have the proper instruments for detecting the main market variables on which to intervene? These few examples should be enough to give an idea of the progress of control systems, regarding both the precision of error measurements and the development of true control devices.

In the area of mechanical and biological controls, we seem to have reached a state of absolute precision: zero tolerance with respect to the functioning, pollution, and maintenance objectives. Evolution in these areas has involved a multitude of attempts at making more efficient apparatuses through which organisms try to control their main processes. Even a quick survey reveals the evolution in the limbs of terrestrial animals; that is, in the effectors that produce speed and direction (see Chap. 8, Sect. 8.8). We can also observe the many forms of propulsion in marine animals—from tail fins to forceful water expulsion—and of direction in the water—from rear to lateral to front fins. Birds and many insects are equipped with three-dimensional wings to move in space, and all reptiles have developed a totally or partially crawling movement.

Even the *effector organs* to grasp and capture objects have evolved in quite different ways. The hands, along with the bones and muscular apparatuses of the neck and arm, are the effectors *par excellence* in man and in many primates, while the trunk enables the elephant to grasp things with extreme precision. In order to capture insects by eliminating the distance from their prey, frogs, toads, and chameleons can quickly flick out their tongues. The woodpecker also sticks out its tongue to capture insects inside tree trunks, and we find similar effector organs in mammals that feed on ants or termites, such as the anteater. Many classes of mollusks and polyps have developed very elaborate tentacles they can use to grab prey; many insects and crustaceans (e.g., ants, beetles, lobsters, and shrimp) are equipped with more or less efficient claws to capture their prey.

What variety of *sensors* has evolution produced in nature? In addition to the five human senses of which we are all familiar, evolution has produced many other types of sensors. A very efficient one is the biosonar of bats and dolphins and porpoises, which enables them to perceive the dynamics of many environmental variables and determine the gap between their position and an object. Even Lorenzini's ampullae, which function as electroreceptors of the prey's magnetic field, and the lateral line system, employing neuromasts, are precise receptors that allow sharks and torpedoes to locate the position of external objects and determine the distance to these objects. I will also cite the fantastic world of insects: ants have senses for pheromones, through which they can recognize others of their species and determine the presence of a pheromone trail, which they can follow by calculating the distance between this trail and their position. Bees have compound eyes that enable them to locate the hive at a distance of many kilometers; they are also equipped with tactile "sensiles," olfactory systems, thermoreceptors, and hygroreceptors capable of identifying many chemical and physical variables. Many insect species are also equipped with mechanical receptors that permit them to get information about the position and state of various parts of their body and many winged insects have receptors on their heads to measure wind speed.

Even decision-making and control apparatuses have been refined through evolution. Evolution has progressed from strictly reactive automatic regulators in many living beings to the nervous system (which also acts as an information transmission system) and then to the brain, the most evolved regulatory apparatus, and finally the human brain, which has enabled man to gain knowledge and develop language, science, and technology. Through the cognitive power of his brain man has substantially expanded his apparatuses, inventing machinery to produce things more efficiently, to detect phenomena in an extremely precise way, and to support his control decisions. These aspects will be developed further in subsequent chapters in Part 2.

2.14 Connections and Interferences Among Single-Lever Control Systems

In this section, I would like to introduce the important new concept of the *connection* and *interference* between single-lever control systems.

Two (or more) control systems can be *connected* in the sense that the dynamics of a system's controlled variable depends not only on the values of its control lever but also on the values of the controlled variable of another system. Two (or more) control systems *interfere* if the connection is *reciprocal*. Thus, interference occurs when two or more control systems are interrelated in some way so that the variation in the controlled variable of one system also produces changes in the controlled variable of the other and vice versa.

When the systems interfere with each other, the simple dynamics of each system considered on its own can become complex. Let us assume we have two single-lever control systems that obey the following equations (see (2.6) and (2.7) in Sect. 2.3):

$$B_n = A_n \ g_1 + C_n \tag{2.10}$$

$$A_{n+1} = A_n + \left(B^* - B_n\right) \ h_1; \tag{2.11}$$

and

$$Q_n = P_n \, g_2 + R_n \tag{2.12}$$

$$P_{n+1} = P_n + \left(Q^* - Q_n\right) \ h_2; \tag{2.13}$$

In order to connect two showers and create *interference*, we can consider the first shower as a disturbance to the other and vice versa. We can thus transform (2.10) and (2.12) by assuming that the values of the disturbance variable, C_n, of the first system are a function of the value of the controlled variable, Q_n, of the second system (or, in many cases, of the $E(Y)$), and that this interference is based on a parameter i_Q; at the same time we assume that the values of R_n of the second system are a function of the value B_n (or of the error of the first system) according to a parameter i_B.

Taking appropriate account of the time displacement, Eqs. (2.10) and (2.12) can be rewritten as shown below, by directly including as disturbances the temperatures of the connected shower, while (2.11) and (2.13) would not change.

$$B_n = A_n \ g_1 + i_Q \ Q_{n-1} \tag{2.10/1}$$

$$Q_n = P_n \ g_2 + i_B \ B_{n-1} \tag{2.12/1}$$

Figure 2.11 shows the dynamics of the temperatures and regulations of two showers in the same apartment that interfere with each other (since they are connected to the same hydraulic system), assuming the same *reaction time* for both showers, $r(X) = 2$. The first shower is opened at the initial instant; the second begins to function after six periods. To emphasize the interferences, I have chosen high values for the coefficients "$i_Q = 30\%$" and "$i_B = 20\%$."

Equations (2.10/1) and (2.12/1) clearly illustrate the interference process; in fact, we can see that the opening of the second shower produces a lowering of the temperature of the first, which forces the manager of the first shower, in the succeeding instant, to readjust his temperature; however, this causes a variation in the temperature of the second, which is readjusted, only to again interfere with the temperature of the first shower. The dynamics are complicated due to the presence of a *delay* that, in the initial phases, produces an increase in the temperatures of the two showers above their respective objectives. Nevertheless, a reaction time of $r(X) = 2$ protects the control system from breaking down by closing the mixers. A reaction time of $r(X) > 2$ would make the delay even more tolerable.

When the parameters i_Q and i_B are close to 0, the interferences are modest and the reciprocal variations of the temperatures are not too high, so that the managers of the two showers can quickly succeed, with only a few adjustments, to restore the desired temperature. It is even easier to achieve the objectives without any large oscillations by assuming a reaction time of 2 (or even higher).

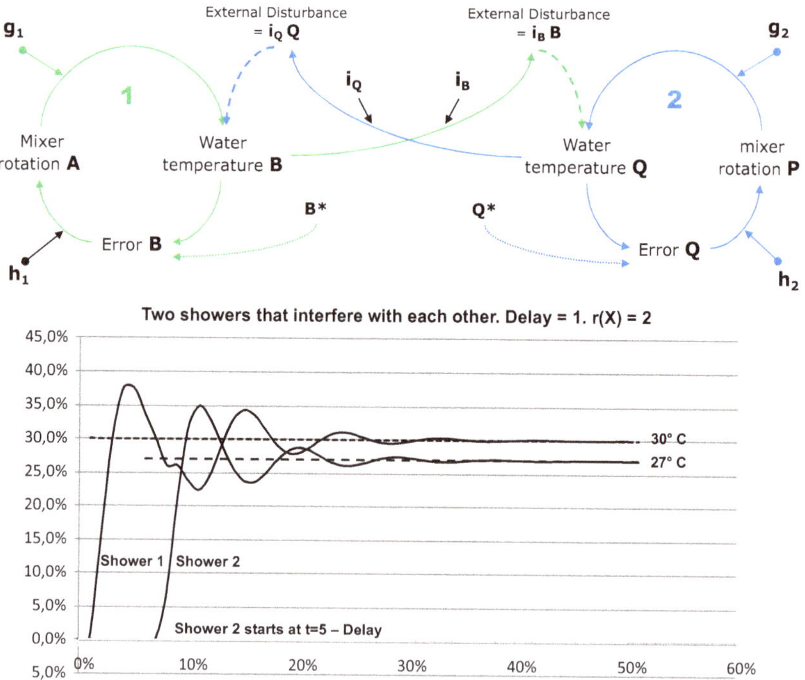

Fig. 2.11 Dynamics of the temperatures of two showers that interfere with each other through Y_t, with $i_Q = -30\%$ and $i_B = -20\%$ and with delay = 1 and reaction time $r(X) = 2$

Figure 2.12 extends the example in Fig. 2.11 by presenting the dynamics of three interfering showers.

In order to work through the example and make the graph understandable, the interference percentages have been set as follows: Shower 1 ($i_2 = -25\%$, $i_3 = -25\%$); Shower 2 ($i_1 = -25\%$, $i_3 = -25\%$); Shower 3 ($i_1 = -25\%$, $i_2 = -80\%$); the symbol i_k, with $k = 1, 2$, or 3, indicates the percentage at which each shower would interfere with showers to which it is connected. The temperature dynamics of the three showers becomes even more complex by the existence of interference together with the presence of a delay. In any event, in this example all the control systems enable the temperature to reach the objective set by the governance.

Equally complex dynamics would occur if we assume interference quantified not on the basis of the temperature but on the error. In this case, (2.10/1) and (2.12/1) would become:

$$B_n = A_n \; g_1 + i_Q \left(Q^* - Q_{n-1}\right) \qquad (2.10/2)$$

$$Q_n = P_n \; g_2 + i_B \left(B^* - B_{n-1}\right) \qquad (2.12/2)$$

Figure 2.13 indicates the temperature dynamics of the two showers determined from the last two equations.

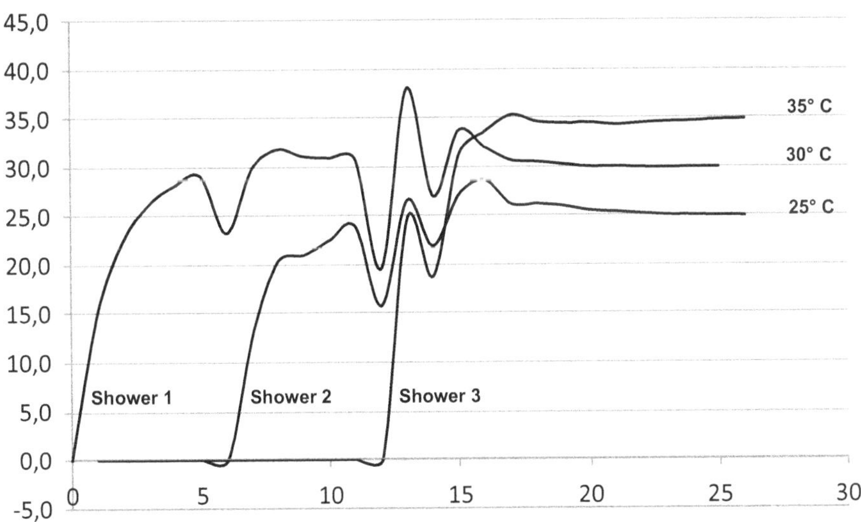

Fig. 2.12 Dynamics of the temperatures of three showers that interfere with each other through Y_t: delay = 2 and reaction time $r(Y) = 2$

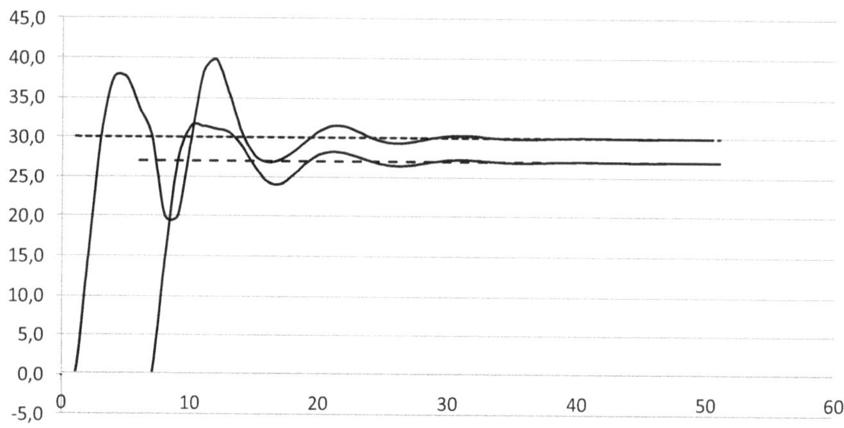

Fig. 2.13 Dynamics of the temperatures of two showers that interfere with each other through $E(Y)_t$: delay = 2 and reaction time $r(Y) = 2$ (compare with Fig. 2.11)

2.15 Areas of Application of the General Model

The general model of a control system in Fig. 2.2 can be found, or applied, in all situations where we wish to modify a variable in order to achieve an objective. It is useful to briefly mention some of these situations, which will be dealt with in more detail in subsequent chapters.

The system we apply in *problem solving* is certainly a control system, as is shown in Fig. 2.14. Solving a problem means achieving a state Y^* considered optimal, starting from a non-optimal state, Y. The problem is perceived as an error between the two states, $E(Y) = Y^* - Y =$ PROBLEM. The problem solver (i.e., the manager of the control system) must eliminate the error through decisions involving the activation of the control levers. The model can be applied to both artificial and natural systems.

For example, thirst represents the error $E(Y)$ between an optimal salt/water density (D^*) and the actual density (D); this is the problem we all experience when subject to scorching heat. To get rid of the thirst, we must restore the correct saline density by ingesting water (drinking). In all respects, we have a control system here that detects the problem (error) and solves it (activation of the lever). This is a natural system for most higher-order animals, who are the managers that decide which levers to activate and to what degree (see Chap. 4, Sect. 4.2, Fig. 4.6).

Equally clear, in my view, is the representation of the decision-making process as a control system, as shown in the model in Fig. 2.15.

Decision-making represents the process that leads to the choice of some alternative in order to achieve some objective, understood in the broadest sense. The need to arrive at a decision comes from the perception of a state of dissatisfaction regarding a given situation that contrasts with a situation we hold to be better and which thus represents the objective to achieve. This dissatisfaction is a symptom of the error between "where we would like to be" and "where we are," and it represents the

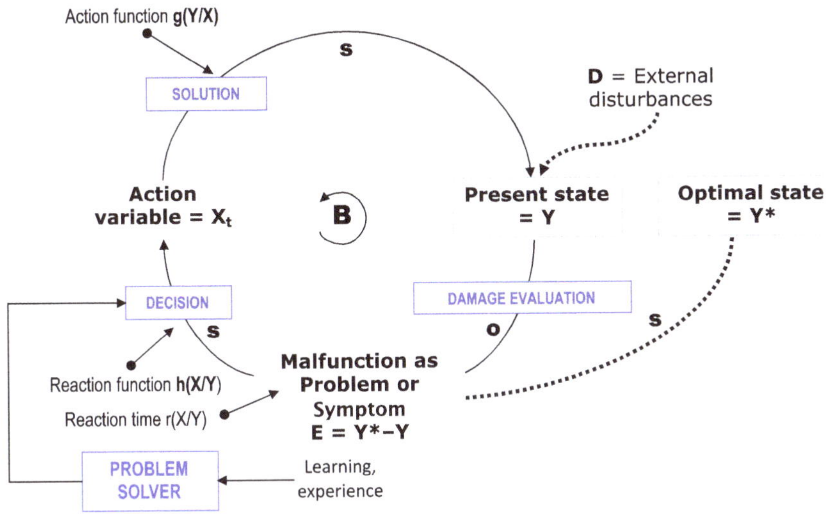

Fig. 2.14 Problem solving as a control system

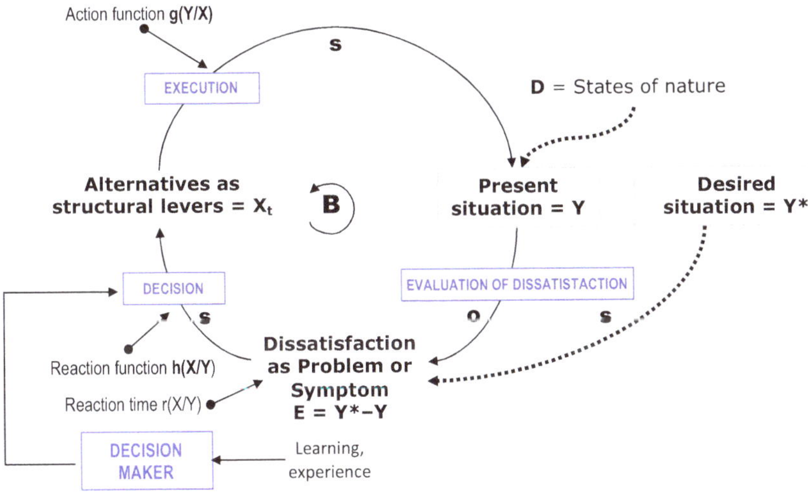

Fig. 2.15 Decision-making as a control system

problem; that is, the error $E(Y)$ to eliminate. From experience, the decision-maker identifies, evaluates, and activates the decisional levers, that is, the various action alternatives, to modify the present situation in an attempt to attain the desired one, taking into consideration external disturbances, or states of nature (see Chap. 5, Sect. 5.9). An efficient information system for evaluating the attainable results and the probabilities of success regarding the various alternatives is activated by the decision-maker while regulating the control levers, X, resulting in the decision-maker choosing the best alternative.

Change is a natural part of the life of every individual, society, or organization. Change requires a modification in one's *present* state, which must be transformed into the *desired* one to be attained; once more the logic of control systems is appropriate, as shown in the model in Fig. 2.16. The management of the control system that produces change can also be referred to as *change management*. Through some *change management* apparatus, the need for change is perceived. Change management searches for the *change drivers*; that is, the levers to activate in order to trigger the process of change that should transform the present state into the desired one (see Chap. 7, Sect. 7.9).

In many cases this change is obligatory, being the consequence of the action of biological programs (physical and mental development) or social ones (learning and education) that define how the change is to come about and at what speed. In many other cases the change is voluntary and motivated by the need to avoid disadvantages or to gain better living conditions. In addition to obligatory change, individuals, organizations, and societies apply decision-making processes whenever they need to produce or face processes of change, whether these processes derive from the need to face environmental situations or are directly activated to produce an improvement in one's situation. These processes of change must be managed so as to maximize benefits and minimize costs, and they can be viewed as a generalization of problem solving, if the change comes from *external pressures*, and therefore must be endured, or of decision-making, if the change is actively sought after.

The general model in Fig. 2.10 can also be interpreted as a system for controlling that the performance of any activity is satisfactory, as shown in Fig. 2.17.

When the *performance measurement* reveals a performance gap, $E(Y)$, between the actual performance, (Y), and the desired one, (Y^*), the performance manager intervenes and assigns the most appropriate values to the *performance driver*, X, or

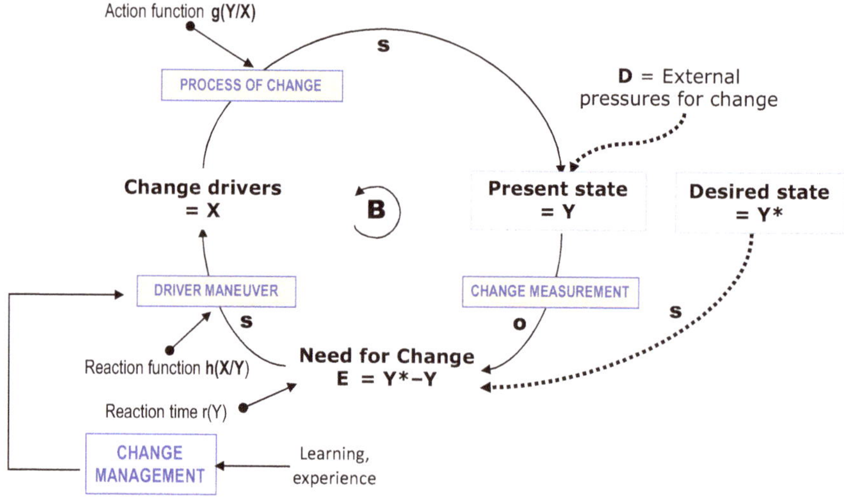

Fig. 2.16 Change management process as a control system

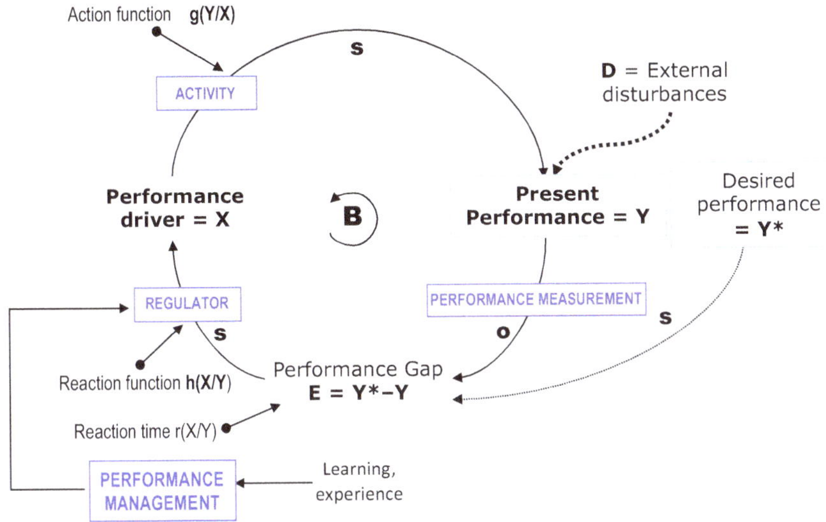

Fig. 2.17 Performance control systems

the performance control lever, so as to enable the activity to produce a new current performance that will reduce or eliminate the performance gap. The model in Fig. 2.17 has two very general variants depending on the meaning we attribute to the performance concept adopted.

In the first variant, the performance driver X represents the *levels* or *inputs* of an activity or process, while Y indicates the obtainable *results* that must move toward the result objective Y^* by means of the regulation of X. In the second variant the variable X represents the specific techniques of an activity, or process on which performance depends; however, this variant differs from the first in that the variable to control, Y, is not a result but a *performance*, or quality *standard* defined from time to time for the activity or process. For both variants, the *performance management* must achieve a desired performance *level*, Y^*, which, in the first case, represents a *performance objective*, while in the second a *performance standard* (see Chap. 8, Sects. 8.1 and 8.4).

2.16 Complementary Material

2.16.1 Simulation Tools

In Sect. 2.8, the simulation model for the shower temperature using a mixer (Fig. 2.8) was developed using Excel, since a spreadsheet can rapidly and efficiently calculate the necessary iterations to produce the system dynamics. As the reader can observe, even those without a sophisticated knowledge of the topic can develop an automatic

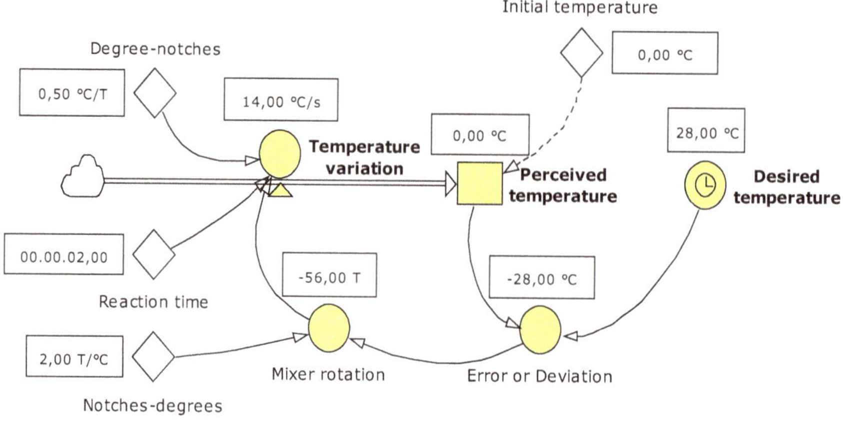

Fig. 2.18 Shower temperature control, without delay, simulated by Powersim (see Fig. 2.8)

control system by simply respecting the rules indicated in the iterative equation models (2.0, 2.1, 2.2, 2.3, 2.4, 2.5, 2.6, 2.7, 2.8, and 2.9) of Sect. 2.3, with appropriate adjustments to take into account any operating constraints for the system.

To simulate more complex control systems other software can be used, such as *Mathematics*—the computational program developed by *Wolfram Research*, which has a section specifically dedicated to control systems—or the simulation software based on relations that link *stock*, or *level* variables with *flow*, or *rate* variables, which change the makeup of the stock. Figure 2.18 illustrates the Stock and Flow Diagram (SFD) that corresponds to the example in Fig. 2.8 for the shower temperature control, without delay, with a mixer functioning as a continuous control lever.

From a mathematical point of view, *stock* is a variable that derives from the accumulation or integration of *flows* (with the appropriate sign), while *flows* are variables that add (inflows = input) or subtract (outflows = output) to the amount of the *stock*. The intensity of the *flows* depends on the variation rates (converters) indicated (in various ways) in the corresponding labels. The simulation model that results is called a Stock and Flow Diagram (SFD) or a Level and Flow Structure (LFS). It is obviously necessary to define the temporal scale that determines the dynamics of *flows* and *stocks* (Coyle 1977; Roberts 1978; Sterman 2000, 2001; Mella 2012). Among the many types of simulation software, all of which are relatively simple, we can mention the following:

- Powersim (http://www.powersim.com).
- MyStrategy (http://www.strategydynamics.com/mystrategy/default.asp).
- ithink and stella (http://www.iseesystems.com/index.aspx),
- Vensim (http://www.vensim.com).
- Excel Software (http://www.excelsoftware.com/).

I would not be surprised if the reader finds the model in Fig. 2.18 quite different and less significant than the one in Fig. 2.7, which is then simulated in Fig. 2.8, if only because the SFD in Fig. 2.18 does not show the direction of variation and the *loops*, which make the model easy to understand.

In fact, this is the main "limit" of SFD simulation models; but this is balanced by the immeasurable advantage of being able to quickly develop models with hundreds of variables, which is almost impossible with Excel. Nevertheless, if we do not allow ourselves to be confused by the direction of the arrow, we can easily recognize that the SFD in Fig. 2.18 corresponds completely to the CLD in Fig. 2.7; in Fig. 2.18, the variable Y_t, which indicates the "perceived temperature," represents a *stock* that is accumulated based on the values of the "temperature variation" variable, which depends on the "mixer rotation" and is equivalent to X_t. All the other elements in Fig. 2.18, in particular the graph showing the dynamics of the temperatures, can be easily recognized in Fig. 2.8.

2.16.2 Control of an Elevator

The simple operation of taking an elevator is made possible by the simple control system in Fig. 2.19.

Figure 2.20 numerically illustrates the dynamics of the elevator, represented by discrete time intervals. When we get on an elevator we want to go to a certain floor (objective), going up or down according to our initial position. To achieve our objective, we simply press the button that indicates the floor we want to go to and think no more of it. The pressure from the button indicating the floor is interpreted by the machine (effector) as the objective to achieve, determining the displacement relative to the initial floor and turning on the elevator motor for the time necessary to eliminate the gap. The elevator machine begins to move past the floors that separate us from the one we wish to go to; when the objective is reached, it stops.

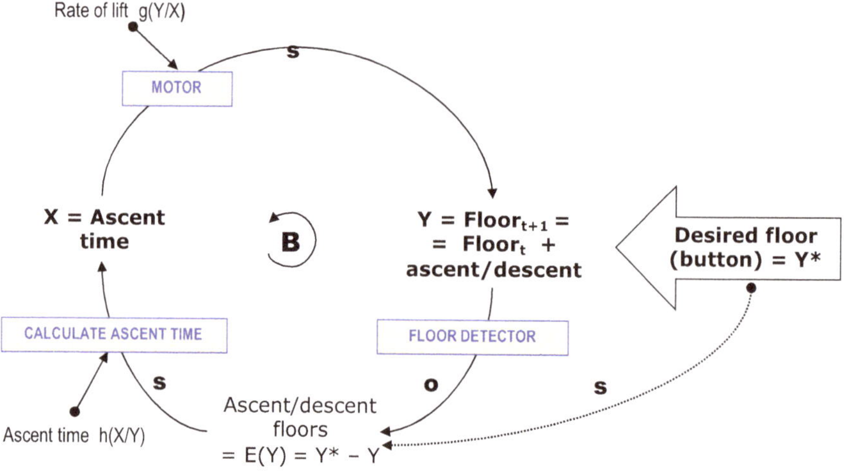

Fig. 2.19 The control system for an elevator

2.16.3 *Searching for a Street, a Page, and a Word*

The functioning logic behind the above-mentioned simple control system for an elevator can be applied to other systems that regulate quite different processes. Above all, the system that allows us to identify a certain street address starting from either end of the street. If, for example, we are at number 5 and want to go to number 11, we only need to behave like the elevator; obviously our trajectory will be horizontal rather than vertical, but the process of eliminating the gap is entirely analogous. We all use the same control system when we look for the page number in a book or newspaper. If we are on page 5 and want to get to page 125, the elevator model is still perfectly applicable.

The model can also be extended to looking up a word in a dictionary. Here we are not controlling the gap between two numbers but that between two words in the dictionary, as shown in Fig. 2.21. As humans we are too used to thinking in terms of letters, that is, symbols; but we should not forget that for computers, letters are merely codified representations of numbers. We can generalize these observations by remembering that as humans we can recognize letters and search for a word in a dictionary because we have learned a catalogue of letters that does not change over time; that is, an alphabet. The recognition of a letter is carried out through a control system that identifies the letters from a catalogue and scrolls from letter to letter, just like an elevator scrolls through its "catalogue of floors." The recognizing systems will be taken up in Chap. 3, Sects. 3.1 and 3.9.5.

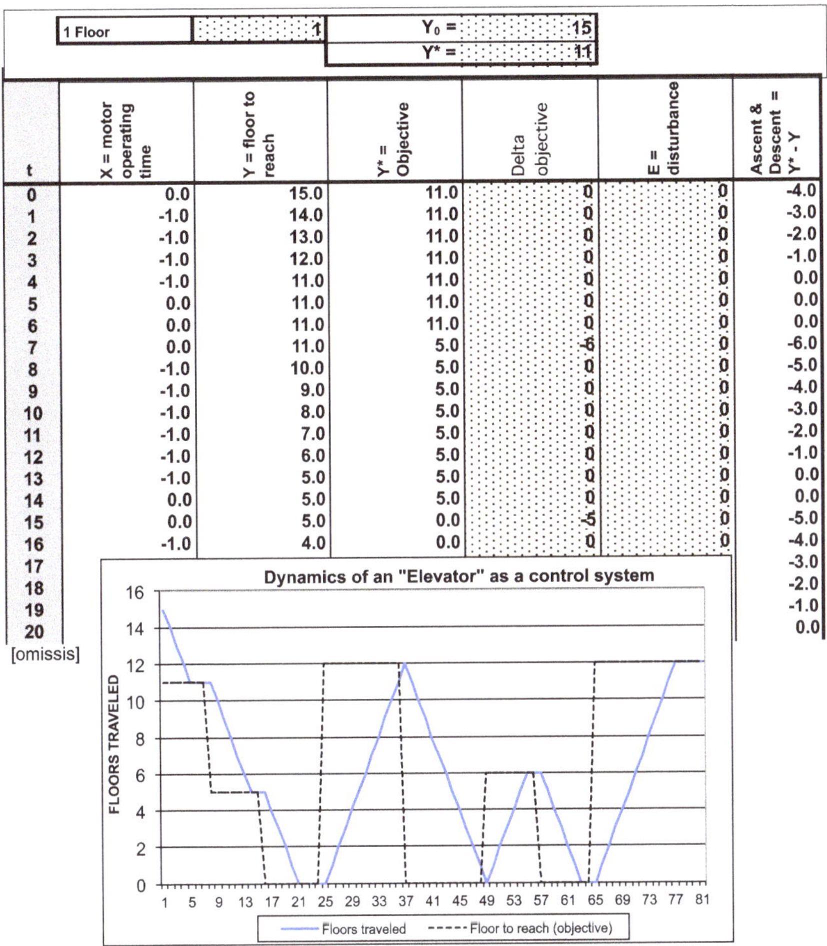

1 Floor	1	Y_0 = 15
		Y* = 11

t	X = motor operating time	Y = floor to reach	Y* = Objective	Delta objective	E = disturbance	Ascent & Descent = Y* - Y
0	0.0	15.0	11.0	0	0	-4.0
1	-1.0	14.0	11.0	0	0	-3.0
2	-1.0	13.0	11.0	0	0	-2.0
3	-1.0	12.0	11.0	0	0	-1.0
4	-1.0	11.0	11.0	0	0	0.0
5	0.0	11.0	11.0	0	0	0.0
6	0.0	11.0	11.0	0	0	0.0
7	0.0	11.0	5.0	-6	0	-6.0
8	-1.0	10.0	5.0	0	0	-5.0
9	-1.0	9.0	5.0	0	0	-4.0
10	-1.0	8.0	5.0	0	0	-3.0
11	-1.0	7.0	5.0	0	0	-2.0
12	-1.0	6.0	5.0	0	0	-1.0
13	-1.0	5.0	5.0	0	0	0.0
14	0.0	5.0	5.0	0	0	0.0
15	0.0	5.0	0.0	-5	0	-5.0
16	-1.0	4.0	0.0	0	0	-4.0
17						-3.0
18						-2.0
19						-1.0
20						0.0
[omissis]						

Fig. 2.20 Dynamics of a control system for an elevator

2.16.4 *The Trajectories of a Car and a Boat*

The system for regulating the audio level (Sect. 2.5 and Fig. 2.3) can similarly be applied to a vast array of seemingly diverse phenomena. In particular, it is entirely analogous to the control system for the route of any type of car and vehicle that uses a steering wheel to change direction, as shown in the model in Fig. 2.22. The system for regulating a *shower* can similarly be applied to the control system for the route of any type of boat in a sea or lake (river navigation must also take into account the single-directional current) by means of a rudder wheel. The trajectory of a boat differs from that of a car for two reasons. Above all, the dynamics of a boat is subject to *inertia*; secondly, the direction of a boat is to a great extent autonomous while

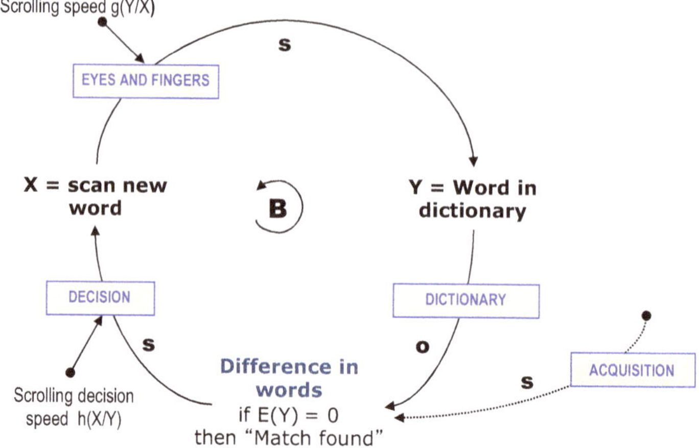

Fig. 2.21 Control system for recognizing a word in a dictionary

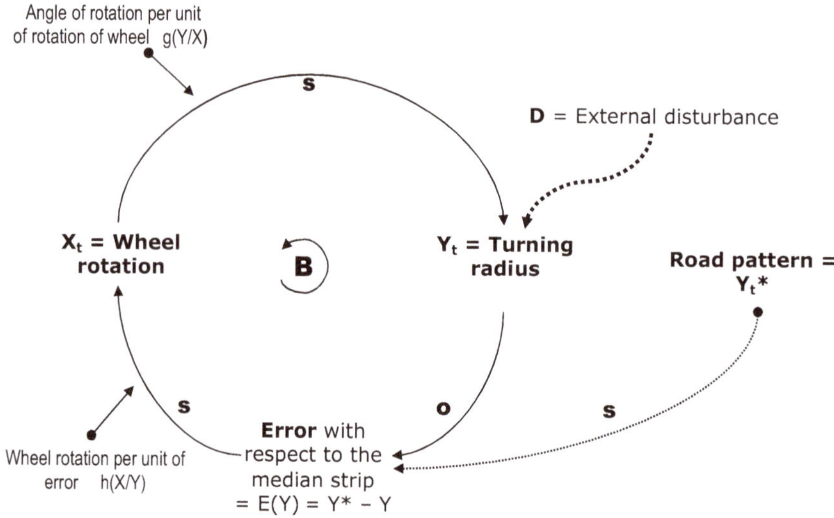

Fig. 2.22 Control system for the trajectory of a car

that of a car is conditioned by the types of curves on the road. In fact, in order to steer a boat toward point Y^*, defined as a geographical location, we turn the rudder a certain number of degrees; the rudder transmits the signal—with appropriate amplifications—to the real rudder, at the stern of the boat, which imparts the appropriate rotation. Since movement in a fluid must overcome inertia, unless we correctly calculate the rotation of the rudder wheel we run the risk of not achieving the objective.

Modern automatic control systems already take into account *inertia* when determining the correct trajectory; however, to simplify things we can imagine that the control of a boat under the assumption of a manual control is similar to that of the shower temperature when there are *delays*. The helmsman sets the direction but the boat responds with a *delay* due to *inertia*, so that most probably there is a gradually attenuating oscillating movement—entirely like that in the graph at the bottom of Fig. 2.9, with a reaction time of two periods—which enables the desired direction to be maintained. The helmsman of a sailboat or a boat without automatic controls must be quite experienced; however, the manual control often leads to failures, as the tragedy of the Titanic reminds us.

2.16.5 Shower with Two Delays

Figure 2.23 presents the same shower as in Fig. 2.9, but assuming *two delays*. Note that in the graph at the top (reaction time of one period) the temperature dynamics inevitably leads the user to "jump out" of the shower almost immediately; the reaction time of three periods means the 28 °C temperature can be reached with some scalding water.

It is intuitively clear that in any type of control system a reaction time greater than one period does not represent an inconvenience; on the contrary, delays are fundamental for the regulation of Y_t even when, without a reaction time of two or more periods, they could lead the system to destruction. With a reaction time of four periods, and with the data in Fig. 2.9, the shower would not produce oscillations and thus enable the objective to be achieved and stably maintained.

2.16.6 Direct and Inverse Control

I shall define *direct* control as the control carried out by a control system according to the [**s-o-s**] logic shown in Fig. 2.2. Many control systems operate in an *inverse* manner according to the sequence [**o-o-o**]. A simple inverse control system is the cooling system in our refrigerator. The time the compressor is turned on, *X*, influences in direction (**o**) the temperature of the refrigerating cell, *Y*, in order to adjust this in line with the temperature objective, *Y**, we have set with the memory apparatus (dial with numbered notches).

In control systems for a supermarket trolley, a hang glider, a bicycle, or a sailboat with tiller, the relations between the shifting of the angle of the bar and the angle of direction of the trolley, hang glider, bicycle, or boat are in direction (**o**). Figure 2.24 shows the inverse control system of a small motor boat with a tiller.

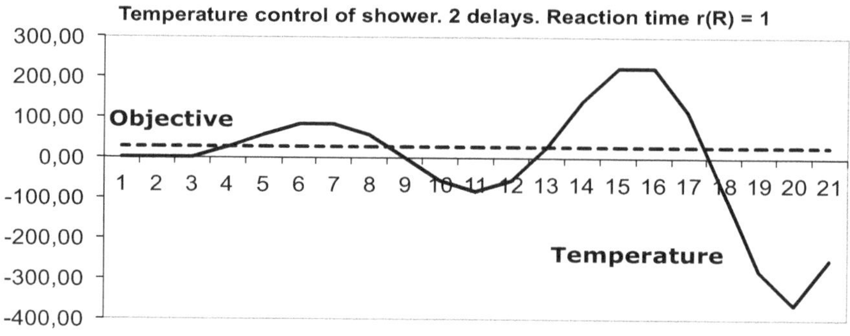

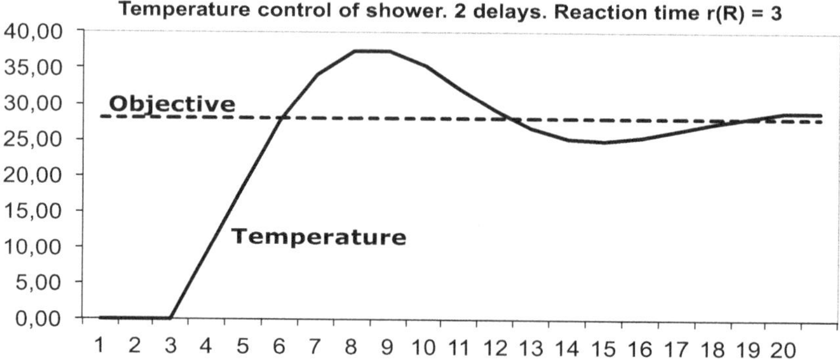

Fig. 2.23 Regulation of a shower with mixer and two delays

2.16.7 Simulation of an On–Off System: The Hot and Cold Air Conditioner

Figure 2.25 presents a basic numerical simulation of a hot and cold air conditioner viewed as an on–off system.

2.16.8 Simulation of Two Interfering Showers Using Powersim

The model of a control system for two interfering showers presented in Fig. 2.13 can be translated using Powersim into the SFD in Fig. 2.26.

2.16.9 Feedforward Control

The control systems described so far—and which are the subject of this book—act according to a *closed-loop* or *feedback* logic, in that X_{t+1} is recalculated based on the error $E(Y)_t$ and influences Y_{t+1} so as to gradually reduce the error. Nevertheless,

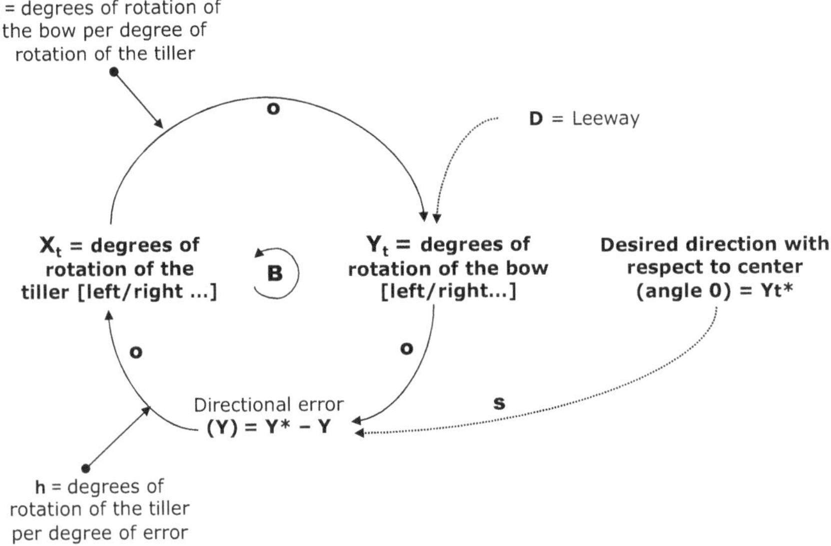

Fig. 2.24 Inverse control system [**o-o-o**]: control of a boat's direction

it is possible to also plan *open-loop*, *feedforward*, or *command control systems* where, based on a *well-known model* regulating the causal relations between X and Y, the manager calculates a one-time initial value, X_0, for the control lever to trigger the dynamics of Y_t in order to achieve the objective Y^*. Once the system starts up, it is not possible to adjust Y_t, since the manager cannot modify the control lever's initial value. For this reason, *feedforward* systems have no need to calculate the error in order to achieve the objective but only to "command" X appropriately in order to eliminate beforehand the error, which is already known or predictable.

Formally speaking, the *feedforward* control system has a causal *chain*, non-repetitive structure that tries to achieve the objective Y^* by preemptively setting X_0. In fact, such a system is missing an internal *controlling system* capable of intervening on the trajectory of Y_t, while this is being produced; it can only "command" the control lever X_0 in order to program the trajectory of Y_t toward Y^*, which explains why it is also called a *system of command* or *commands* (Fig. 2.27).

The feedback control systems in Fig. 2.10 can also be called *decision and control systems*, since the control of Y_t is achieved through decisions to vary X_t over a succession of repetitions of the control cycle. Feedforward systems can also be called *decision* or *one-shot control systems* since they try to reach the objective through a single cycle and thus through a single *initial decision*. They are typically systems that "fire a single shot" to achieve the objective. The decisions are made based on precise calculations; however, when an error is detected in the attempt to achieve Y^* the manager cannot correct this through other regulating decisions to eliminate the error.

The archer takes account of the features of the bow and arrow, also estimating the distance, wind, and other disturbances, before precisely aiming his arrow, transmitting a direction, X_t, deemed useful for hitting the target at point Y^*. He then releases

Increase in temperature C°/t	4	$Y_{t=0}$ =	40
Tolerance precision in degrees C°	1.5	Y* =	25
Δ Outside temperature	0.5		

t	action rate degrees/time period	Y = inside temperature	Y* = OBJECTIVE	Variation in objective	D = disturbance	Error = Y* - Y
0	0.0	40.5	25.0	0	0.5	-15.5
1	0.0	41.0	25.0	0	0.50	-16.0
2	0.0	41.5	25.0	0	0.50	-16.5
3	-4.0	38.0	25.0	0	0.50	-13.0
4	-4.0	34.5	25.0	0	0.50	-9.5
5	-4.0	31.0	25.0	0	0.50	-6.0
6	-4.0	27.5	25.0	0	0.50	-2.5
7	-4.0	24.0	25.0	0	0.50	1.0
8	-4.0	20.5	25.0	0	0.50	4.5
9	-2.5	18.5	25.0	0	0.50	6.5
10	0.0	19.0	25.0	0	0.50	6.0
11	4.0	23.5	20.0	-5	0.50	-3.5
12	4.0	28.0	20.0	0	0.50	-8.0
13	4.0	32.5	20.0	0	0.50	-12.5
14	-3.5	29.5	20.0	0	0.50	-9.5
15	-4.0	26.0	20.0	0	0.50	-6.0

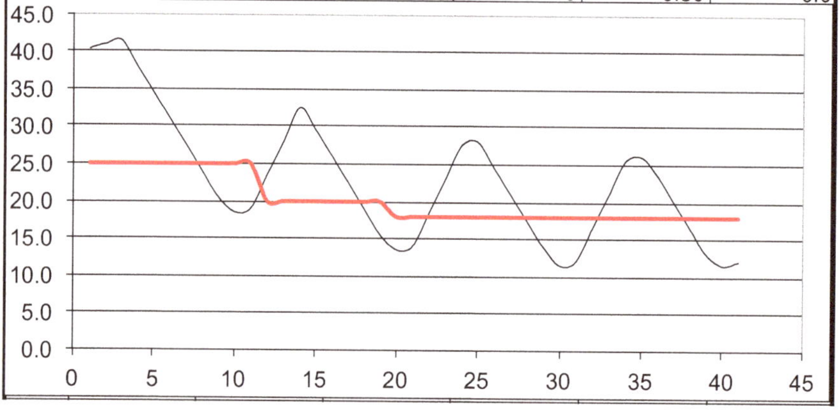

Fig. 2.25 Simulation of an automatic on–off air conditioner

the arrow and observes its flight. He cannot do anything further since, once the arrow has left (X_0), the trajectory Y_t is predetermined and no intervention is possible to eliminate any possible error $E(Y)$. The same is true for the hunter who fires his rifle, the space center that launches an engine-less probe, the worker who ignites a blast furnace in which he places what he deems to be the proper mixture to produce an alloy, the firm launching an ad campaign in newspapers, the cook placing what is believed to be the right amount of rice into boiling water to feed his dining companions, or the farmer spreading what he considers the right amount of fertilizer for a bountiful harvest.

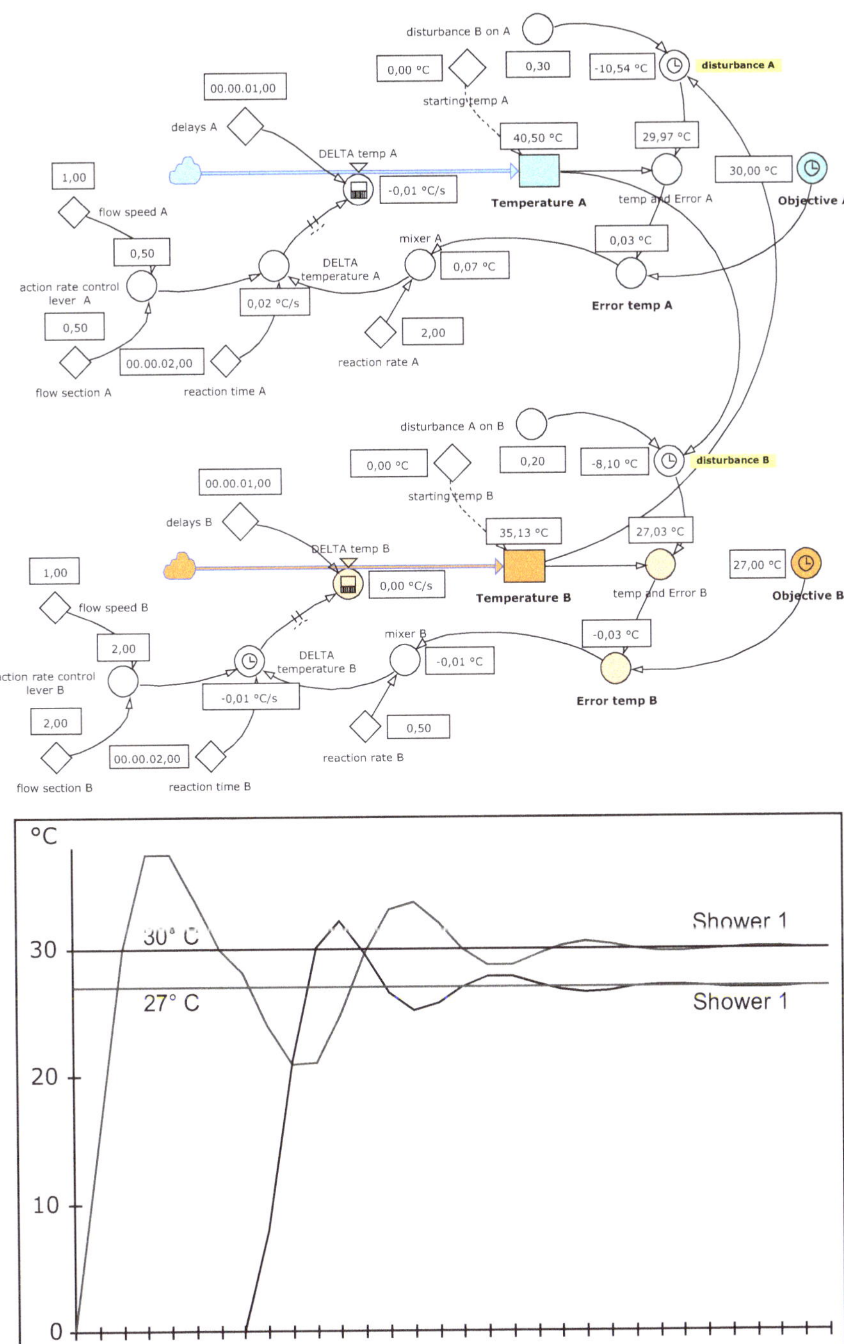

Fig. 2.26 Simulation using Powersim of two interfering showers (reaction time equal to 2 and delay equal to 1)

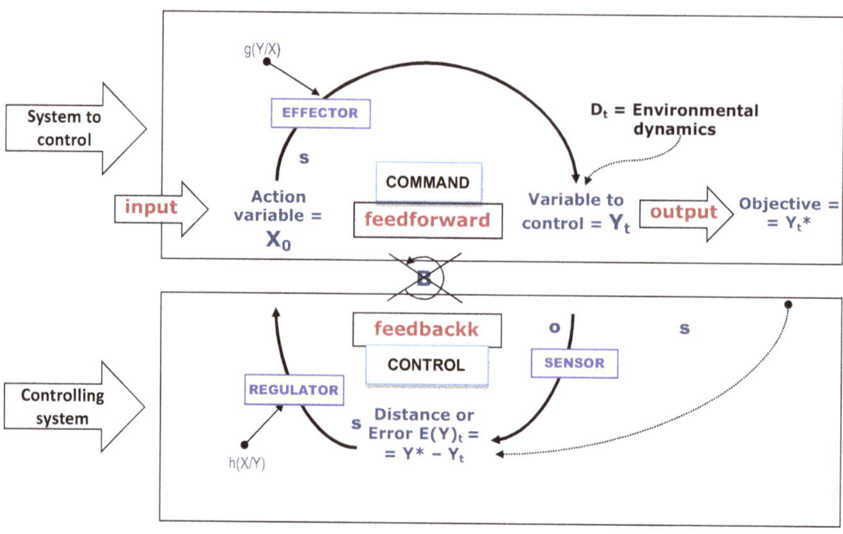

Fig. 2.27 Feedforward control system

We can recognize *feedforward* control systems all around us, since they derive from man's innate need to achieve objectives even without being able to activate a system that can detect and eliminate the errors. Equally evident is the fact that man tends to replace feedforward control with feedback control whenever possible, especially when the costs of failure in achieving the objective are very high.

The advantages of feedback control systems can be summarized in the table of Fig. 2.28.

2.16.10 The Engineering Definition of Control Systems

Figure 2.27 allows us also to point out that, according to engineering theory about control systems, the effector is considered to be a *system to control* and the detector (sensor), together with the regulator, a control system that, connected by feedback to the system to control, transforms the latter into a *system under control*, as the following precise definition explains:

> The Control System is that means by which any quantity of interest in a machine, mechanism or other equipment is maintained or altered in accordance with a desired manner. Consider, for example, the driving system of an automobile. The speed of the automobile is a function of the position of its accelerator. The desired speed can be maintained (or desired change in speed can be achieved) by controlling pressure on the accelerator pedal. The automobile driving system (accelerator, carburetor and engine-vehicle) constitutes a Control System. For the automobile driving system the input (command) signal is the force on the accelerator pedal which, through linkages, causes the carburetor valve to open (close) so as to increase or decrease fuel (liquid form) flow to the engine, bringing the engine-vehicle speed (controlled variable) to the desired value (Nagrath and Gopal 2008, p. 2).

No.	Open Loop	Closed Loop
1	Any change in output has no effect on the input; i.e., feedback does not exists.	Changes in output, affect the input which is possible by use of feedback.
2	Output measurement is not requirad for operation of system.	Output measurament is necessary.
3	Feedback element is absent.	Feedback element is present.
4	Error detector is absent.	Error detector is necessary.
5	It is inaccurate and unreliable.	Highly accurate and reliable
6.	Highly sensitive to the disturbances.	Less sensitive to the disturbances.
7	Highly sensitive to the environmental changes.	Less sensitive to the environmental changes.
8	Bandwidth is small.	Bandwidth is large.
9	Simple to construct and cheap.	Complicated to design and hence costly.
10	Generaly are stable in nature.	Stability is the major consideration while designing.
11.	Highty affected by nonlinearities.	Reduced effecy of nonlinearities.

Fig. 2.28 Comparison of open-loop and closed-loop control systems. (*Source*: Bakshi and Bakshi 2009, p. 16)

> [...] The general block diagram of an automatic Control System, which is characterized by a feedback loop, is shown in Figure 1.4. An error detector compares a signal obtained through feedback elements, which is a function of the output response, with the reference input. Any difference between these two signals constitutes an error or actuating signal, which actuates the control elements. The control elements in turn alter the conditions in the plant (controlled member) in such a manner as to reduce the original error (*ibidem*, p. 5).

The Fig. 1.4 mentioned by Nagrath and Gopal is reproduced in Fig. 2.29; as we can see, the feedback control system perfectly corresponds to the one in Fig. 2.10, which can be arrived at through simple position changes of the various elements.

Figure 2.30 illustrates a more analytical model that perfectly corresponds to that in Fig. 2.29 (taken from Varmah 2010), which also highlights the input transducer, that is, the apparatus that transforms the command input (e.g., the indication of the desired temperature on a graduated scale) into the reference input, that is, the objective in terms that can be read from the control system. These typically engineering apparatuses enrich the technical model of a control system in Fig. 2.10 while not reducing its generality.

In this book I have adopted the *language* of systems thinking, feeling it appropriate to focus on the controlled variable, Y_t, and to assume that, if its dynamics lead it to converge to a value Y^*, defined as a constraint or objective, then its dynamics are

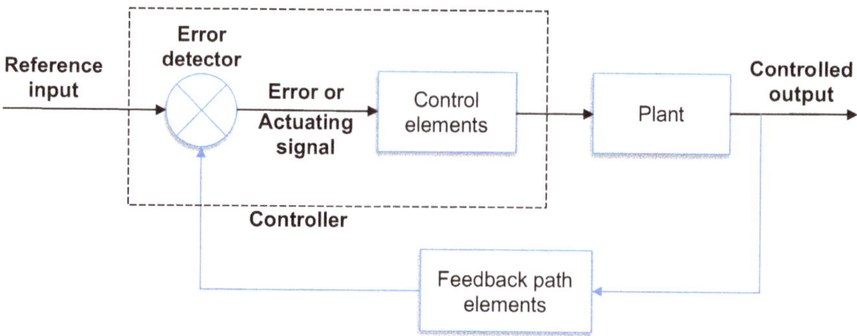

Fig. 2.29 Model of feedback control systems from an engineering perspective (*Source*: Nagrath and Gopal 2008, "Fig. 1.4. General block diagram of an automatic Control System," p. 5)

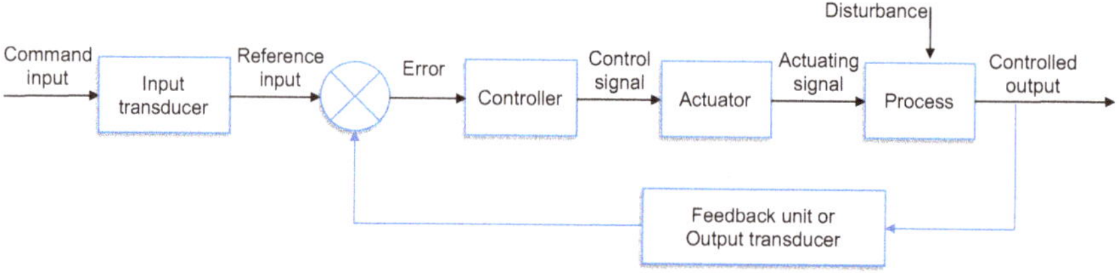

Fig. 2.30 Engineering model of feedback control systems with the essential terminology (*Source*: Varmah 2010, "Fig. 1.2 Closed-loop Control System," p. 4)

governed by a control system. Thus, the term "control system" indicates the entire logical *loop* between the controlled variable Y_t and the control levers X_t, through the error $E(Y)$, independently of the physical, biological, or social structure of the mechanisms that make Y_t tend toward $Y*$.

2.16.11 An Analytical Appendix: The Optimal Value of the Lever and the Optimal Control Period in Non-symmetrical Systems

Sect. 2.3 presented the general model of a control system in the form of recursive equations, as these can be immediately grasped and easily applied using both spreadsheets and general programming language. Nevertheless, it is appropriate, at least for the sake of thoroughness, to provide a brief analytical discussion that can also help in understanding the operational logic of simulation programs for stocks and flows. *In any event, the concepts in this section are not indispensable, and the reader can even go back to them later.*

I shall start with the simplest model (the mono-lever one) of a symmetrical control system ($h = 1/g$) in which the action equation, which enables us to calculate the optimal value $Y*$, has the following simple form:

$$Y^* = g\ X_{n*} + D \tag{2.14}$$

where X_{n*} represents the optimal value of the control lever and D the sum of the external disturbances. Let us assume the initial value of the system is:

$$Y_0 = g\ X_0 \tag{2.15}$$

Subtracting (2.15) from (2.14) we obtain:

$$Y^* - Y_0 = g\ X_{n*} - g\ X_0 + D$$

which immediately allows us to determine the value of X_{n*}, since $E = Y^* - Y_0$:

$$\frac{Y^* - Y_0 - D}{g} = X_{n*} - X_0 \tag{2.16}$$

Example 1: The objective is $Y^* = 40$, with $Y_0 = 0$, $g = 2$, $h = 1/2$, and $D = 0$; from (2.16) we immediately obtain: $X_{n*} = 20$, since $X_0 = 0$.

Example 2: The objective is $Y^* = 80$, with $g = 6$, $h = 1/6$, and $D = 20$; referring to (2.16) we get $X^* = 10$.

Example 3: The same values as in *Example* 2, but with $D = -20$; from (2.16) we obtain $X^* = 16.67$ as the optimal value of the lever, as the reader can verify.

Equation (2.16) clearly shows that the reaction rate, h, is not an essential datum in the model in Fig. 2.2, since it corresponds to the reciprocal of g; this relationship creates a system of symmetrical control. Nevertheless, its introduction in models is useful for giving a logical completeness to the loop in Fig. 2.2, since it enables the optimal value of the control lever to be calculated immediately in one sole step.

In order to identify a general solution for calculating the optimal value, X_{n*}, for the control lever in the case of an *asymmetrical* system, I propose this simple line of reasoning.

Let us go back to (2.16) and, to simplify things, assume that $X_0 = 0$, $Y_0 = 0$, and $D = 0$, and that $h° \neq (1/g)$; let us also assume that $g° = (1/h°)$ is the action rate corresponding to $h°$, so that we can write the control system, in its minimal asymmetrical form, as follows (omitting time references):

$$\begin{cases} g\ X = Y \\ E + Y^* - Y \\ X_n = (1/g°)E \end{cases} \tag{2.17}$$

Adopting the value for the lever X_n calculated in the third equation, if this value is optimal, we should obtain the objective value, Y^*, calculated in the first equation in (2.17):

$$g\frac{1}{g^{\circ}}E = Y^{*} \tag{2.18}$$

However, Eq. (2.18) shows that the lack of symmetry affects the dynamics of the system when the initial value of X is different from the optimal one:

- If $g < g^{\circ}$, then the system underestimates X_n and does not achieve Y^*, thereby producing a new error to be eliminated through a new calculation of X_n; the system achieves its objective through a succession of iterations.
- If $g > g^{\circ}$, the system overestimates the value of the lever and exceeds the objective; depending on the value of g°, the system may converge or diverge.

Figure 2.31, which is derived from Fig. 2.8, represents the results from four tests carried out on the temperature system, without delays and with a variation in the objective at $t = 9$. In general:

- Test 1: For $g = g^{\circ}$, the system is perfectly symmetrical.
- Test 2: For $g = 2\ g^{\circ}$, the system oscillates over constant periods.
- Test 3: For $g < 2\ g^{\circ}$, the system converges with diminishing oscillations.
- Test 4: For $g > 2\ g^{\circ}$, the system diverges.

A final observation: Considering—for simplicity's sake—only symmetrical systems, (2.16) not only allows us to quantify the optimal value of X in *continuous phase* control systems—as the previous examples have shown—but is also equally useful for *discrete step* control systems, since it permits us to determine the *control period*, that is, the number of periods at the end of which Y attains Y^*. Let us assume x^* to be the rate of variation of X; from (2.16) above, $X_{n^*} = n^*x^*$, with n^* being the number of times step x^* must be repeated in order to obtain the optimal value of the lever X, which gives us the value Y^*. The number n^*, which represents the number of periods the step must be repeated, expresses the control period needed to bring Y to the objective value Y^*.

From (2.16), we automatically obtain n^*:

$$\frac{Y^{*} - Y_0 - D}{g} + X_0 = n^{*}x^{*} \tag{2.19}$$

Example 4: The objective is $Y^* = 80$, with $g = 5$, $h = 1/5$, and $D = -20$, assuming a fixed step, $x^* = 1$; we immediately get $n^* = 20$, which reduces to 10 when $x^* = 2$.

Example 5: The objective is $Y^* = 120$, with $g = 10$, $h = 1/10$, and $D = 20$, assuming a fixed step, $x^* = 2$; we immediately get $n^* = 5$.

Equation (2.19) allows us to make another simple, though important, calculation. Let us assume we have set the condition that the control must be carried out during a predetermined period, n^*, which represents a second time objective. It is clear that in order to simultaneously achieve the objectives Y^* and n^*, the manager must be able to set the *action rate*, g (the capacity), for the lever X.

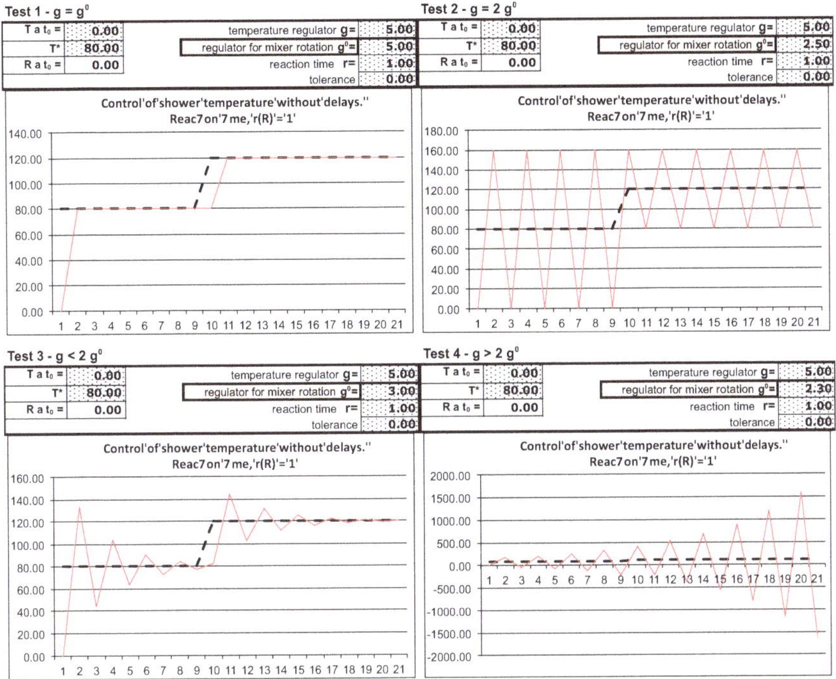

Fig. 2.31 Dynamics of non-symmetrical temperature control systems

Thus, Eq. (2.19) allows us to calculate the action rate, g^*, that enables Y to attain the value Y^* during a period of n^* instants. Setting $X_0 = 0$, we immediately get:

$$\frac{Y^* - Y_0 - D}{n^* x^*} = g^* \tag{2.20}$$

Example 6: The objective is $Y^* = 120$, $D = 20$, $x^* = 2$, and $n^* = 5$; we immediately get $g^* = 10$, which decreases to $g^* = 4$ if the step increases to $x^* = 5$.

2.17 Summary

We have learned that:

1. The *logical structure* of single-lever control systems (Sect. 2.2) always includes the following elements (Fig. 2.2): X_t; Y_t; Y^*; $E_t = Y^* - Y$; D_t; $g(Y/X)$; $h(X/Y)$; $r(X/Y)$; and delays.
2. The dynamics of the variables X_t and Y_t can be determined with the recursive equations, without delay, indicated in Sect. 2.3, or with delay, as indicated in Sect. 2.8.

3. Every control system has its own *logical* and *technical structure* (Sect. 2.9 and Fig. 2.10).
4. The *technical structure*—or *chain of control*—is a "machine" composed of mechanisms (or apparatuses) that develop the processes that produce the variations in the variables X, Y, and E; it includes: the *effector*, the *sensor*, the *regulator*, the *information network*.
5. Single-lever control systems can interfere or be linked (Sect. 2.14) in order to form a wider system; if there are delays, the global system can produce oscillating dynamics (Figs. 2.11, 2.12, and 2.13); nevertheless, the presence of a reaction time of 2, 3, or even more time units makes it possible to achieve Y^*.
6. In any control system we can find a *manager* and a *governor* (Sect. 2.10).
7. A control system is developed by the *designer* and built by the *constructor* (Sect. 2.11).
8. The general model of control system can represent all situations of change (Sect. 2.15), in particular:

   - The problem-solving process (Fig. 2.14)
   - The decision-making process (Fig. 2.15)
   - The change management process (Fig. 2.16)
   - The performance control systems (Fig. 2.17)

Chapter 3
The Ring Variety: A Basic Typology

Abstract Chapter 2 presented the logic of control systems and described the technical structure that produces the general logical structure. I have thought it useful to begin with the simplest control systems, single-objective (a single Y and single Y^*) and single-lever (a single X) ones, in which the action function determining the dynamics of the variables (g and h) are represented as simple constants. This chapter presents several fundamental classes of control systems, among which are natural and artificial systems, manual or automatic cybernetic systems, quantitative and qualitative systems, and attainment and recognition systems. Finally, I shall consider various forms of interconnections among control systems and the holarchies of control systems.

Keywords Control system · Recognition system · Automatic control system · Manual control system · Steering control systems · Halt control systems · Systems of "Pursuit;" · Collision and anticollision systems · Parallel and serial connections · Explorative systems · Zeno's paradox · Watt's centrifugal regulator · Floating ball regulator

> It seems that cybernetics is many different things to many different people, but this is because of the richness of its conceptual base. And this is, I believe, very good, otherwise cybernetics would become a somewhat boring exercise. However, all of those perspectives arise from one central theme, and that is that of circularity [...] While this was going on, something strange evolved among the philosophers, the epistemologists and the theoreticians: they began to see themselves included in a larger circularity, maybe within the circularity of their family, or that of their society and culture, or being included in a circularity of even cosmic proportions. What appears to us today most natural to see and to think, was then not only hard to see, it was even not allowed to think! (Heinz Von Foerster 1990, online).

Chapter 2 presented the logic of control systems and described the technical structure that produces the general logical structure. I have thought it useful to begin with the simplest control systems, single-objective (a single Y and single Y^*), and single-lever (a single X) ones, in order to present and simulate several elementary systems

P. Mella, *The Magic Ring*, Contemporary Systems Thinking,
https://doi.org/10.1007/978-3-030-64194-8_3

drawn from our daily experiences, in which the action function determining the dynamics of the variables (g and h) are represented as simple constants. This chapter presents several fundamental classes of control systems, distinguished by the type of human intervention in the control or by the nature of the objective: natural and artificial systems, manual or automatic cybernetic systems, quantitative and qualitative systems, attainment and recognition systems, steering and halt control systems, fixed- and variable-objective systems (or systems of pursuit), tendential and combinatory systems, as well as systems distinguished by other features. Finally, we shall consider various forms of interconnections among control systems and the holarchies of control systems.

3.1 Manual and Automatic Control Systems: Cybernetic Systems

The first useful distinction to make—of an engineering nature—is that between *manually controlled systems* and *automatic control systems.* This distinction is based on the presence or lack thereof of human intervention in the chain of control

> Systems [which] involve continuous manual control by a human operator … are classified as manually controlled systems. In many complex and fast-acting systems, the presence of the human element in the control loop is undesirable because the system response may be too rapid for an operator or the demand on the operator's skill may be unreasonably high. … Even in situations where manual control could be possible, an economic case can often be made for reducing human supervision. Thus in most situations the use of some equipment which performs the same intended function as a continuously employed human operator is preferred. A system incorporating such equipment is known as an automatic Control System (Nagrath and Gopal 2008, pp. 3–4).

> In short, a Control System is, in the broadest sense, an interconnection of the physical components to provide a desired function, involving some kind of controlling action in the system (Bakshi and Bakshi 2009, pp. 1–3).

The above-mentioned distinction is appropriate but, precisely because it derives from an engineering view of control systems, which focuses on the control of systems understood as mechanical apparatuses, we need to make several observations and additions. Above all, recall that the Preface of this book presented several examples that show that, in terms of quantity and quality, most of the *automatic control systems* are *natural*, biological, or even social systems where human intervention is not possible. In fact, all biological systems that control the existence of all living beings, from the tiniest cell to organizations and society, are natural systems. Such systems act on organs of all types and varieties that usually cannot be directly controlled (breathing, heartbeat, immune system, etc.), on the body's structural components, even on the biological clocks that regulate most of our biological cycles, etc. Control systems that operate in the physical and meteorological world to produce clouds or dew or in ecological systems to restore balance in prey, and predator populations also represent natural systems.

In his daily activities, man represents a formidable control system that produces hundreds of control activities each minute. None of us bump into other passers-by when we encounter them or cross the street without first observing the traffic light or how far away the moving cars are. We check our position and that of our limbs and set objectives for our behavior. We drive our car to the desired destination and in doing so we grab and maneuver the driving wheel, reach and operate the various levers, adjust from time to time the direction of the car, the speed, and the distance from other cars, etc. I shall consider these systems in more detail in the second part of the book.

All control systems designed and constructed by man, whether or not *manually controlled* or *automatic* control systems, are *manmade control systems*. The distinction between *manual* or *automatic* systems is not simple to make, since human intervention in *artificial* control systems can take quite different forms and intensities and act differently on the "apparatuses" that make up the chain of control.

It can be useful to refer to Fig. 3.1, which is directly derived from Fig. 2.10. We can define those control systems as *automatic* in which human intervention is limited to externally setting the objectives, Y^*, which the variable to control, Y, must achieve (point ①). The objective represents the only "information" for the system regarding the dynamics to set for Y; *sensor*, *regulator*, and *effector* are natural or artificial apparatuses on which man does not intervene, unless possibly to supply energy and for maintenance. Thus, man assumes the role of *governor* of the control system; by setting the latter's objectives, he guides the system in the action space of Y but does not take on the role of *manager*, since management is an integral part of the regulatory apparatus. We can define those control systems as *manually guided* where man becomes the manager of the system and intervenes through his own decisions and actions on the regulator in order to determine the values of the control lever X (point ②). The objectives can be assigned by the manager himself or by an outside governor.

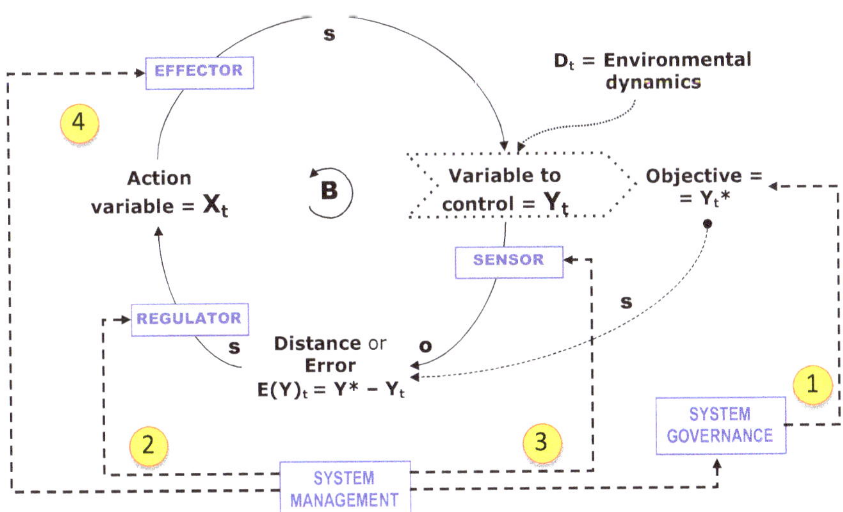

Fig. 3.1 Types of human intervention in artificial control systems

An *automatic* control system guided toward a fixed objective determined by an outside *governor*, but guided by a *manager within the chain of* control, is also defined as a *cybernetic system*; the *cybernetic system* appears to be closed within itself, and the only link with the environment is represented by the specification of the objective *Y** by the governor.

The definition of *cybernetic system* can be linked to the idea of "guiding" a system in time/space (e.g., a ship); that is, of controlling its movement toward an objective (port of arrival) by means of an automatic control system in which the human manager (helmsman) is replaced by a mechanical manager (GPS-guided rudder), thereby creating an automatic *Ring*. The science that studies how to produce automatic systems was founded by Norbert Wiener, who defined this as the science of the control (and guidance) of systems. In fact, Wiener's book is entitled: *Cybernetics, or control and communication in the animal and the machine* (1961).

> We have decided to call the entire field of control and communication theory, whether in the machine or the animal, by the name *Cybernetics*, which we form from the Greek χυβερνητητ [sic] or *steersman*. In choosing this term we wish to recognize that the first significant paper on feedback mechanisms is an article on governors, which was published by Clerk Maxwell in 1868, and that *governor* is derived from a Latin corruption of χυβερνητητ [sic]. We also wish to refer to the fact that the steering engines of a ship are indeed one of the earliest and best-developed forms of feedback mechanisms (Wiener 1961, p. 11).

Norbert Wiener—precisely because he viewed cybernetics as the *theory of the guidance* or *governance* of systems; that is, the study of the *logic* according to which the helmsman (manager) guides the ship to the port (objective *Y**) indicated by the ship owner—assigned a fundamental role to both the *communication* of the objective to the manager and the latter's orders (control) to modify the route (*Y*) by means of the rudder (*X*). For cybernetics, the true input of the control system is the information on the amount of error (*E*(*Y*)) with respect to the objective, so that the manager can rectify the value of the control lever to guide the "ship" to port. In another important book, *Cybernetics and Society*, Norbert Wiener wrote:

> In giving the definition of Cybernetics in the original book, I classed communication and control together. Why did I do this? [...] When I communicate with another person, I impart a message to him [...] Furthermore, if my control is to be effective I must take cognizance of any message from him which may indicate that the order is understood and has been obeyed. [...] When I give an order to a machine, the situation is not essentially different from that which arises when I give an order to a person. [...] Naturally there are detailed differences in messages and in problems of control, not only between a living organism and a machine, but within each narrower class of beings (Wiener 1954).

Precisely to the extent it is a science of guidance and communication, cybernetics today is mainly associated with the control of machines and mechanical systems. None other than Ludwig von Bertalanffy, considered to be the founder of General Systems Theory, recognized that:

> [A] great variety of systems in technology and in living nature follow the feedback scheme, and it is well-known that a new discipline, called Cybernetics, was introduced by Norbert Wiener to deal with this phenomena. The theory tries to show that mechanisms of a feedback nature are the bases of teleological or purposeful behaviour in man-made machines as well as living organisms and in social systems (von Bertalanffy 1968, p. 44).

Nevertheless, as Ross Ashby, another founder of cybernetics, clearly states that what cybernetics examines are not machines but the control systems that allow these to function in order to produce the desired dynamics.

> Many a book has borne the title "Theory of Machines", but it usually contains information about *mechanical* things, about levers and cogs. Cybernetics, too, is a "theory of machines", but it treats, not things but *ways of behaving*. It does not ask "what *is* this thing?" but "*what does it do*?" [...] It is thus essentially functional and behaviouristic. [...] Cybernetics deals with all forms of behaviour in so far as they are regular, or determinate, or reproducible. The materiality is irrelevant, and so is the holding or not of the ordinary laws of physics. [...] The truths of cybernetics are not conditional on their being derived from some other branch of science. Cybernetics has its own foundations. [..] Cybernetics stands to the real machine—electronic, mechanical, neural, or economic—much as geometry stands to a real object in our terrestrial space (Ashby 1957, pp. 1–2).

Even more recent authors have recognized that cybernetics does not have a single, so-to-speak practical, function but represents instead a general *discipline* for observing and interpreting the world:

> Cybernetics is not exactly a discipline. Rather, it should be thought of as an approach which helps thinkers and practitioners across a whole range of different disciplines secure a vantage from which they are likely to gain a more complete understanding of their own area of practice. Its focus is upon meaningful and effective action and behaviour in whatever domain of human activity (Velentzas and Broni 2011, p. 737).

I have provided the above quotes because they show that even cybernetics as a *discipline* studies the *logical models* of control systems not necessarily their *technical realization*. I would like to specify that this book is not a cybernetics text, since it does not investigate how control is effected in systems containing an objective—machines, organisms, societies, organizations, etc.—but how such systems can be interpreted as Control Systems. I shall not attempt to clarify how the eye focuses on objects but how the eye is "in and of itself" the Control System that focuses on objects. Nor shall we investigate how firms can be controlled but how they are Control Systems "in and of themselves."

I wish to make also clear that this book considers the general theory of control systems in a different, probably broader, way than cybernetics; I shall not consider only automatic–cybernetic systems but *natural systems* as well, where human intervention is absent, and *manual systems*, where human intervention not only is connected to the communication of orders and the reception of answers but also can directly intervene on the chain of control itself. I would note that *natural systems* are not only cybernetic. A control system can be defined as *natural* if the following twofold condition applies: (1) it is *automatic* and (2) it has no *outside governor*, since the objectives are in the form of constraints, limits, or endogenous conditions. The eye is a natural control system, since it tries to control the quantity of light that must reach the retina (constraint Y^*) through the automatic control system of the pupil, which, through small, speedily contracting muscles, rapidly varies its diameter. All (human) eyes function in this manner.

The general theory of control systems does not consider only the logic behind the functioning of control systems—also proposing a significant typology of such systems—but becomes a true control discipline whose objective is to demonstrate how fundamental control systems are for the existence of our planet, aiding the reader to identify which control systems are operating in each sector of our knowledge. As a discipline, it also seeks to study how every *automatic* control system can grow and become *manually* guided, and how every *manually* guided system can transform itself into an automatic system in which management can become part of the control apparatus itself.

I would also note that the *natural* control systems for our physiological control are not only automatic; we can become the outside managers of the chain of control. If the light is too strong, we lower our eyelids, wear sunglasses, or seek out the shade. If it is hot, the natural body temperature control system activates the sweating process; however, if the heat is still too high and sweating does not sufficiently lower our body temperature, we take off some of our clothes. If we still feel hot, we take a cool shower; and if the heat persists for a long time we activate the air conditioner or take a holiday in the mountains.

Going back to Fig. 3.1, we can finally observe that man can direct the control process by intervening in various ways on the technical structure of control systems: (a) he can function as a *sensor* (point ③) by detecting the errors directly with his own organs: by feeling when the water or air temperature is too hot; by perceiving a too high or too low sound volume; by noticing that the automobile is veering dangerously toward the guard rail; by tasting the risotto to see if it is done, etc.; (b) he can function as the *effector* (point ④) by stepping on the brake pedal, turning the handle bars on the bicycle, opening/closing a valve, etc.; (c) he can be the regulator (point ②) that, based on his knowledge, decides when or to what extent to maneuver the control lever; d) finally, he can intervene globally (points ① + ② + ③ + ④) by forming the entire chain of control, where man becomes "the" control system. To convince yourselves of this, just "look around."

Furthermore, the control discipline reveals the importance of the *observer's point of view* in recognizing and constructing models and simulating the behavior of control systems. As taxi passengers, we view ourselves as the *governor* of the transportation system (we give the destination) and observe the car as a control system (machine) whose manager-controller is the taxi driver; if we are driving our car, we believe we ourselves are the control system for speed and direction, which we perceive with our own senses and modify with our own limbs, considering the car as the effector toward the desired destination; if an alien intelligence observes the Earth from a UFO several miles above, and sees a car winding through a network of roads, it would probably think it was seeing natural control systems operating, changing direction and speed without any objective or human intervention, without ever colliding or veering dangerously away from the flow of traffic, exactly as we imagine the ants do that are marching along the plants in our garden.

3.2 Quantitative and Qualitative Control Systems: Attainment and Recognition Control Systems

The general model of a control system shown in Fig. 3.1 can be applied to both *quantitative* and *qualitative* control. The control is *quantitative* if the variable Y_t represents a *quantitative measure* (in any of its multiform expressions) and the objective is a value that Y_t must reach (*set reference*) or a set of values to be achieved (*track reference*).

I shall define *attainment control systems* as those control systems that act to "attain" a *quantitative* objective, independently of the control context. All the control systems models presented so far—audio, temperature, directional, speed control, etc.—are cases of *attainment control systems*. A simple observation will immediately reveal that *attainment control systems* are all around us, since we encounter them whenever both the variable "Y_t" and the objective "Y^*" can be quantified or are separated by a distance that is in some way measurable. Figure 3.2 illustrates the typical example of the control system we activate when we want to grab something with our hand; for example, the gesture of grabbing a cigar or a pencil as described by Norbert Wiener:

> If I pick up my cigar, I do not will to move any specific muscles. Indeed in many cases, I do not know what those muscles are. What I do is to turn into action a certain feedback mechanism; namely, a reflex in which the amount by which I have failed to pick up the cigar is turned into a new and increased order to the logging muscles, whichever they may be (Wiener 1954, p. 26).

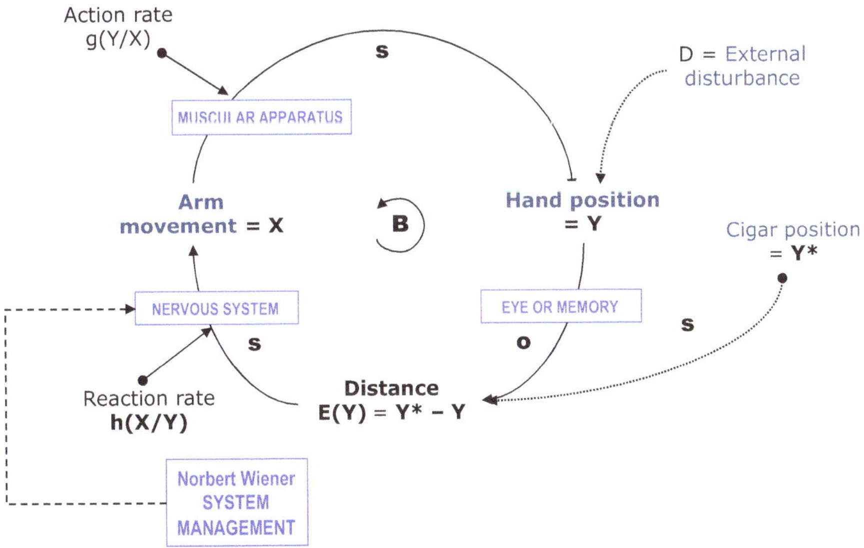

Fig. 3.2 Attainment control system. Norbert Wiener picks up his cigar

Other less obvious examples of such systems are those structured to control that given levels of average hourly productivity in a factory are achieved, that a desired monthly level of *operating cash flow* in a certain department is achieved, that the public debt does not exceed $2 trillion, that a given threshold of dust particles in the air is not exceeded, that our body temperature does not go beyond its normal limits, that microwave ovens operate for the desired length of time, or that our eye or our reflex camera can bring images into focus, etc.

We can distinguish among the *attainment* control systems based on the nature of the objective:

1. *Goal-seeking systems*: those in which Y^* represents a quantitative *objective*: achieve a result, or attain or maintain operating standards.
2. *Constraint-keeping systems*: in which Y^* represents a *constraint* to respect or a *limit* not to be exceeded; these are usually *halt systems* (see Sect. 3.3).

Recalling the concept of a control problem (Sect. 2.1), we can further distinguish among control systems as follows:

1. *Regulator systems*, where Y^* (objective or constraint) is a specific value (goal, constraint, or limit) that Y_t must reach and maintain over time.
2. *Tracking systems*, or *path systems*, where Y^* (goal, constraint, or limit) is a *trajectory*, that is, a sequence of values, Y_t^*, however formed, which Y_t must follow.

Despite the abundance and ease of observation of quantitative control systems, we must nevertheless be aware that there exists an equally broad and important class of *qualitative control systems* in which both the variable Y_t and the objective Y^* are qualitative in the broadest sense of the term (objects, colors, flavors, forms, etc.). If we want a white wash load (Y^*, qualitative objective) we must modify the state of cleanliness (Y_t, qualitative variable) until we are satisfied ($E(Y)$ reduced to zero). If we want a perfectly cooked roast (Y^*, qualitative objective) we must modify the cooking state (Y_t, qualitative variable) until we are satisfied ($E(Y)$ reduced to zero).

It is immediately clear that in qualitative control systems, the manager–governor faces the problem of coherently expressing the qualitative variables based on a scale of differentiated states so that the detector apparatus can compare the present state, Y_t, and that representing the objective, Y^*, in order to determine the variance. How do we determine the state of "cleanliness" of laundry? How can we detect the cooking state of food? The problem of how to graduate the "states of variety" in the qualitative dimensions is often left to the experience of the manager. The cook knows if the food is well flavored, not salty enough or almost done; every housewife knows when the laundry or dishes are cleaned to the right extent. Nevertheless, procedures to determine the "states of variety" of the qualitative variables have been worked out, the most important of which are presented in Sect. 3.9.8. However, these examples are not enough to make us aware of the generality of *qualitative control systems*; in order to fully understand their *modus operandi* and to introduce the very relevant class of *control systems of recognition*, I now propose four other examples.

Let us first imagine we have to buy a button to replace the one we lost on our gray suit (Y^*). We go to the store that sells buttons to buy the one we need and are immediately embarrassed because the shop owner asks us the button size. We remember it is "around" 2 cm in diameter. Then, when we are asked to specify the color, we panic because the owner presents us with dozens of buttons (Y_n) of that diameter in various shades of gray, and we do not know how to *recognize* (identify) the one closest to the desired shade, assuming the size is correct. There are four possibilities for arriving at the optimal choice; the first is to have at hand the gray suit with the lost button, so that we can compare the remaining buttons (Y^*) with those the shop owner has in stock. If we are lucky, we will find a button that we *recognize* as identical to the desired one. If we do not have the suit with us, we can, as an alternative, at least have a photograph of the button to replace (Y^*). We have had the foresight and wisdom to have taken a photo of the suit and its buttons on our SmartPhone before leaving home, which we now show to the shop owner. In this case as well we can easily *recognize* the button to purchase, thereby solving our problem. As a third alternative, before leaving home we could have compared our suit buttons with a color chart (catalog) containing a wide variety of shades of gray, numbered progressively (or by name), and identified the number that corresponds to the shade of the suit buttons. Here, too, if the shop owner has the same chart, we can precisely *recognize* the button. It is perfectly clear that if we cannot compare the buttons on the shop counter with an identical copy of the original one, using a graphical model or the shades of gray on a chart, the only chance left to recognize the desired button is to hope in our prodigious visual memory, together with a good deal of luck, in order to choose without having the actual sample at hand.

Let us now consider a second example. Our university mate, George, who I have not seen for years, telephones to invite me to a party he will be attending. I am happy for the chance to get together with my old friend. I arrive at the party only to realize there are around 300 people in a large reception room; I start looking for my friend George based on my recollection of his facial features. Walking around the room, I observe one person after another, obviously only the males. "This isn't George; it resembles him, but it's not him …." After numerous attempts, I spot someone at the end of the room who *could* be my friend; however, when I approach him I realize he is not George, though he quite resembles him. I continue searching for several minutes until I see a person who, despite his hair color and some wrinkles, appears to be my friend. I have finally *recognized* George. The same recognition would probably have been made by someone to whom I had assigned the task and to whom I had given a photo of George. Of course, in both cases it would not have been necessary or appropriate to compare the fingerprints of those present at the party with those the army had of George. But in other cases, which I shall leave to the reader to imagine, fingerprints would be useful for recognition (Sect. 3.9.3).

As a third example, consider an eagle that, in its majestic flight, crosses a valley in search of food (Y^*). Observing the ground from a certain height with his sharp eyes, he notices many animals (Y_t) that would be suitable as food. He identifies a huge animal and immediately compares it with the *catalog* of possible prey his life experience has compiled year after year in his memory; in this catalog (memory) that animal is *identified* as a bison, and his memory reminds him that it is too large and cannot be lifted into the air; thus it does not represent a prey. He then *identifies* a snake, but it is too close to the rocks and thus difficult to capture. In the middle of the alley, he observes an animal of suitable body mass, and his catalog (memory) *identifies* it as a small deer that seems not to have noticed the approaching eagle. The eagle quickly swoops down and captures the deer and takes it to its nest.

We have taken a course for *sommelier*, tasting dozens of different wines, memorizing for each its taste, color, aroma, and other organoleptic features, resulting in our compiling a catalog. At the final exam to earn qualification as a *Sommelier*, we are given a wine (Y^*) and asked to indicate the name, year of production, and other characteristics determined by the jury. We taste the wine (Y^*) and immediately our mind scans the "catalogue" of stored "wine models" (Y_n), calling them up one by one until we can identify the wine objective as a 2009 *Brunello di Montalcino*.

On careful reflection, the *identification* process of animals carried out by the eagle, or of wines by the sommelier, is different from the *recognition* of my friend George at the party or of the right button for my gray suit. In *recognizing* George (or the button), I compare "real objects" that *scroll by* before me (Y_n), $n = 1, 2, \ldots$, with a "model-objective" (Y^*) and, when there is a correspondence, I am able to *recognize* a friend (a button). The eagle (the aspiring sommelier), on the other hand, from time to time compares the "stored models, or catalogue," Y_n, that *scrolls by* in his mind one after the other with a "real object" (Y^*); that is, with the animal it observes (the wine he tastes). Only when it/he succeeds in *identifying* a prey (or a type of wine) is the qualitative objective (prey, wine) achieved.

I can now generalize. In all situations, where we have to *recognize* some "model-objective" (George, button) by comparing it with "real objects" from a set that is sequenced (invitations to the party, buttons on the counter) we activate a *system of recognition* where the qualitative variable to control, Y_n, represents the "real objects" that we *scan* in order to recognize the "model-objective" Y^*. In situations where we instead have to *identify* a "real-object" set as an *objective* Y^* by *scanning through* a "catalogue of models", Y_n, we activate an *identification system.*

It is not always easy to distinguish between these two types of system since, despite their operational differences, systems of *recognition* and of *identification* follow the same logic and can be described in a similar way, if we consider that, in a strict sense, all observed *objects* can be considered, from a technical point of view, as models of *objects* or as *technical descriptions* of objects (Sect. 9.4).

In order to *recognize* a model-objective or *identify* a real-object-objective, Y^*, we require:

- An *archive* of "*N*" *objects* (buttons, party invitations) or *models* (preys, wines tasted, and memorized) represented by a variable Y_n, $n = 1, 2, \ldots, n, \ldots, N$.
- An *effector* apparatus that scans the *archive* by changing the value of Y_n over a time sequence based on some predetermined criterion (alphabetical order, size, sex, location, route, etc.).
- A *sensor* apparatus to compare the value of Y_n with the object/model-objective to recognize, Y^*, in order to determine, through preestablished rules, if $Y_n = Y^*$ or if there is an error $E(Y)$.

- A *regulatory* apparatus that evaluates $E(Y)$ and decides whether or not to continue with the *scrolling* of the catalog ($n + 1$, $n + 2$, etc.).
- An *action lever*, X_n, that orders the effector to *scroll through* the catalog for successive "n" if $E(Y) \neq 0$ (according to the predetermined rules of comparison).

I thus propose to consider *recognition systems* as a general class of qualitative control systems and *identification systems* as a subclass. In any event, it is clear that the process of *recognition* or of *identification* just outlined corresponds perfectly with the general logic of control examined in Chap. 2 and that this process is carried out by a qualitative control system.

The model in Fig. 3.3 illustrates the logic of *systems of recognition*, which can be applied, for example, to a person, animal, or machine that explores the environment—that is, the archive of possible objects or models—in order to recognize an object or model-objective and continue the exploration until the error is eliminated.

As Fig. 3.3 shows, all *recognition control systems* must possess special *apparatuses* in their chain of control. The *sensor* organs must be guided by special *algorithms* that permit us to make a description of the observed model. The *sensors* can make a comparison with the model-objective only if they are guided by some comparison *algorithm* capable of carrying out the comparison between the observed description and the objective one. Finally, the *regulation* apparatus must be guided by some error evaluation and decision-making algorithm in order to decide how to act on the control lever (Zhao and Chellappa 2006).

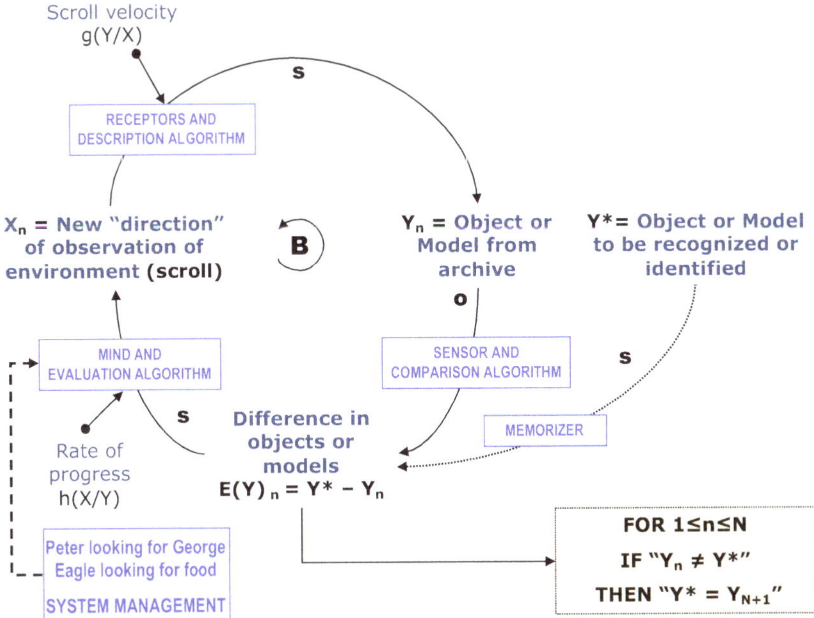

Fig. 3.3 Control system of recognition-identification

Figure 3.3 allows us to explore the infinite world of *qualitative control systems of recognition* that permeate our life as well as that of all animals equipped with sensor organs that make it possible to compare models and objects. In living systems the model-objective is memorized mentally by both the predator, which must identify a prey, and the prey, which must recognize the predator; by both the pollinating bee that must identify a flower to pollinate and the warrior who must distinguish friends from enemies; by both the infant that must recognize his parents among a multitude of people and those who must recognize the person calling on the phone. How would our life be if we could not recognize people, objects, words, etc.? Consider how many people we can recognize on a daily basis from facial features as well as from tones of voice, the way of striking keys on a keyboard or of stopping a car in front of a house, etc.; not to mention the ability to read a daily newspaper and recognize keys in a larger set, a badge, a credit card, the telephone number of friends, etc.

Many *control systems of identification* are artificial and can scan different types of "catalogue" at high speed in order to recognize the model-objective, as in the case of the facial and expression recognition systems used by the police or by intelligence agencies in airports (Sect. 3.9.3). If there is a substantial inherent difference between asking my son: "let me know when you see (recognize) mom arriving" and ordering the policeman: "check if this digital fingerprint is in (recognized by) the AFIS," from a formal point of view these are two typical recognition systems: the former one human and the latter computerized.

Most *control systems of recognition–identification* particularly in the animal kingdom are usually linked to some *attainment control system* with a spatial objective; when the chameleon recognizes its prey and locates it at a given spatial point, it tries to *reach it* with the control system that uses an extendable tongue; when the hawk recognizes a mouse it dives to reach it; when a doctor recognizes a pathology he starts to initiate a control system to eliminate it.

The *control systems of recognition–identification* are fundamental for the survival of individuals and species. They operate not only at a macrolevel but a micro one as well, inside our bodies. Our immune system represents the most powerful *recognition–identification* system in our bodies, without which our existence would not be possible. The immune system protects the organism from aggression from pathogenic agents, recognizing in a highly specific manner millions of substances extraneous to the organism: the antigens, viruses, or bacteria. This defense is based on the action of lymphocytes that produce the antibodies that recognize the antigens; the lymphocytes then attach to the antigens in order to facilitate the work of the macrophages that then proceed to destroy them. The system acts at the microlevel through the continual production (X) of different antibodies (Y_n) until one of these couples to (recognizes) the intruder (the antigen Y^*), thus activating the control system that leads, when possible, to the latter's neutralization. When the recognition fails the organism undergoes aggression from the antigens and often dies. When there is no recognition capacity the so-called autoimmune diseases can occur.

The box at the bottom right in Fig. 3.3 contains a conditional clause that allows the archive of objects models to be updated in the event the object model set as the objective is not recognized. Regarding this specific aspect, see Sect. 3.9.4.

3.3 Steering and Halt Control Systems

All the control systems we have examined so far have two important characteristics: they were constructed to achieve a fixed objective, Y^*, over the entire duration of the control—except in the case of occasional changes introduced by the governance—and the values of Y_t could be greater than or lesser than the objective. We can now relax these two conditions. Let us begin by considering the second condition.

Control systems—represented by the model in Fig. 3.1—in which Y can take on values that are above (e.g., water too hot) or below (water too cold) the objective Y^* (pleasant temperature) can be called "steering" control systems, or even "two-way" systems; the system acts to achieve the objective through positive or negative adjustments that converge toward Y^*, independently of the initial value of Y. Control systems for sound volume, temperature, direction, and speed are typically "two-way" control systems. The models developed for "two-way" systems can also be applied to "one-way" systems, or "halt" systems, where the values of Y must tend toward but not exceed Y^*, in which case the control process would fail. The error can never change sign.

Let us consider, for example, having to pour a liquid, X_t, into a container (a glass, artificial basin) until level Y_t reaches the value objective Y^*. It is clear that for values $Y_t < Y^*$ the control system continues to increase the level until $Y_t = Y^*$; if, however, we do not stop the flow when the objective is achieved but allow $Y_t > Y^*$, then the system has not attained the desired control (the glass overflows and the basin floods the surrounding area).

"One-way" control is very common in all control contexts and is found in all situations where Y^* is by nature a nonexceedable limit or constraint. If by supplying light and heat to a greenhouse we can control the ripening state of strawberries before picking them, then if we exceed the objective we lose the harvest. If we want to reduce the content of a tank by removing the drain plug, we cannot exceed the lower limit that is set as an objective. If we want to catch the bus that leaves at a certain hour and the control fails, the failure is signaled by the fact we have to wait for the next bus.

"Halt" control systems can also be represented with the same models as the others; the uniqueness of such systems merely requires a "precise regulation" of the chain of control of Y so that it does not exceed Y^* before we stop the control process. This implies the need to *measure* the error at the end of short intervals, and without delays, and to carry out the *regulation* without delays as well, thereby reducing the action rate "*g*" when the error falls below the critical level, thus moving from a normal to a "fine tuning" control. In fact, we all reduce the speed of our car to a minimum when we must park it near the wall in our garage in order to avoid dents, just as we reduce the flow of champagne (by adjusting the inclination of the bottle) when we reach the capacity level of the glass, or lower the flame when the roast appears to be done. Even my Mac possesses a "fine tuning" utility for brightness, called "Shades," which, contrary to the classical system of control, allows me to proceed in smaller increments to have a better control on the level of brightness.

In order not to risk exceeding the objective or limit, control systems, especially "Halt" control systems, may call for particular apparatuses that send the manager a particular "warning" of "nearness to the objective," above all when it is difficult to detect the error. We are all familiar with the beep–beep sound that warns us when we are near an obstacle while we are backing up, signaling a need for "fine tuning" our speed.

These apparatuses can be considered part of the broad class of *algedonic alerts* (from the Greek αλγος, pain and ηδος, pleasure), which associates a signal of *urgency* to the error $E(Y)$, even when this is not directly detected; the signal is usually accompanied by a sensation of pleasure or pain, so that the manager is moved to quickly eliminate it by controlling Y_t. If we touch an object which is too hot, a signal of pain makes us immediately withdraw our hand from the object, even if a thermometer to detect the temperature would have stopped us from touching the object in the first place.

The term *algedonic signal* has been used by Stafford Beer to indicate a form of control exercised by the environment on our behavior and objectives, through a system of reward or punishment. Each reward or punishment spurs us to modify our former objectives, and thus our behavior, which subsequently will again be judged by the environment to determine whether it merits new rewards or punishments (Widrow et al. 1973). Thus, algedonic alerts are particularly useful for understanding that the outside environment has detected errors, $E(Y)$, in our behavior and that we must urgently change our objectives and perhaps even the entire control system.

> Algedonic Regulation: literally, regulation by pain and pleasure, more generally, by rewards and punishments for the products rather than the behavior leading to it. E.g., people may be trained to perform a task by explaining to them the role their task plays within the larger system of which they are a part, but they may also be trained algedonically by a series of rewards and punishments that offer no such explanations. The algedonic regulator must have an image of the expected system of behavior but it restricts fluctuations not in the behavior of its parts but in their outputs (Krippendorff 1986).

> There is a last feature of the VSM to incorporate: it is extremely important, but there is not much to be said—so don't flick the page too soon. We were speaking of somnolence. It is an occupational hazard of System Five. After all, all those filters on the main axis ... maybe Five will hear the whole organism droning on, and simply 'fall asleep'. For this reason, a special signal (I call it algedonic, for pain and pleasure) is always identifiable in viable systems. It divides the ascending signal—which we know to be entering the metasystemic filtration arrangements—coming from System One, and uses its own algedonic filter to decide whether or not *to alert system five*. The cry is 'wake up—danger!' (Beer 1995, p. 126).

3.4 Fixed- and Variable-Objective Systems (or Systems of "Pursuit")

Turning to the other condition regarding the type of variability of the objective (Sect. 3.3), we have already seen that we can also achieve objectives that vary over time in an intermittent and occasional manner, so as to form a path (Y_t*) which the variable Y_t must follow, as shown, for example, in Figs. 2.4 and 2.8. I propose the following generalization: control systems can also be structured to achieve objectives that *systematically vary* over time: that is, *dynamic objectives*.

If the dynamics of the system occur at discrete instants, then the variation in the objective always precedes the variation in X_t and Y_t, so that, in principle, the system might not be able to achieve its objective. However, we can distinguish two cases:

1. The objective varies independently of the variables X_t and Y_t and has the following form: $Y_{t+1}^* = H(Y_t^*)$, where H has to be specified.

   Under these assumptions, and agreeing that, for simplicity's sake, $H(Y_t^*) = Y_t^* + H_t$, the control system can be defined as *evolving*, since it tends to achieve an objective that varies, or evolves, for *exogenous reasons*; for example, in the control of the level of liquid in a container, which changes with each fill up, or in many games where the players modify the position of an object-objective (a ball that changes direction when hit by other players).

   The system will achieve its objective if the dynamics of the latter stabilizes around a constant value for some period of time. Evolving systems are typical in living and social systems, in individuals and organizations, since they characterize the development phase. Figure 3.4 shows the dynamics of a single-lever system with a variable objective having the form:

$$Y_{t+1}^* = Y_t^* + H_t,$$

   with "H" taking the values of $H = 0$ to $H = 10$, for $t = 0$ until $t = 10$, and the values from $H = 10$ to $H = 0$ for $t = 11$ until $t = 21$.

   Since at $t = 0$, $Y_0^* = 60$, and since for $t > 21$ the objective has a constant value of 170, the system succeeds in carrying out the control with an assumed reaction time of $r(Y) = 1$; the longer the reaction time is, the more the variable representing the objective moves away from Y.

2. The objective varies as a function of $E(Y)_t$ (or X_t) according to the following form: $Y_{t+1}^* = G(Y_t^*, X_t, E(Y)_t)$, with G representing the "speed of flight" of Y_t^*(G can also be negative), which needs to be specified.

   Under these assumptions, and also assuming, for simplicity's sake, that

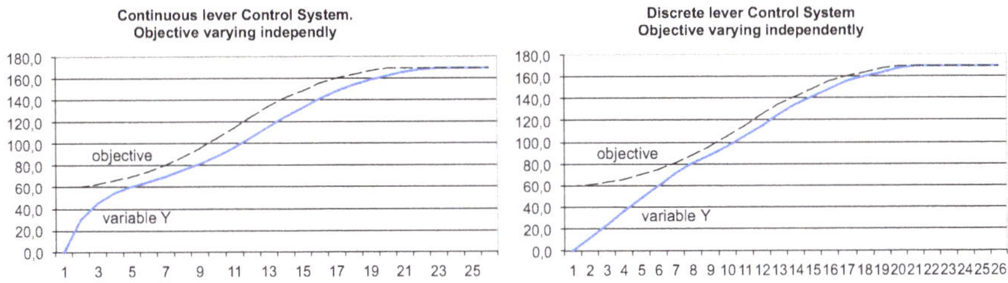

Fig. 3.4 Dynamics of a single-lever system of pursuit with an independently varying objective

$$Y_{t+1}^{*} = Y_{t}^{*} + GE(Y)_{t}$$

such systems can be defined as *systems of pursuit*, since it seems that Y_t tries to pursue the objective by reducing the error $E(Y)_t$, and that Y_t* "observes" the controlled variable and adjusts its values to those of this variable, trying to "flee" from the pursuer by further increasing the distance by $GE(Y)_t$. On the one hand, the size of the variable Y_t depends on the overall amount of $E(Y)_t$, and on the other the size of the objective is, in turn, modified based on the coefficient G applied to $E(Y)_t$. The conditions that enable the pursuer, Y_t, to reach the pursued, Y_t*, can easily be calculated, though they differ according to the type of *control lever*:

1. If the lever X_t varies *continuously* according to the fixed *action rate* "g," and the system operates with a fixed *reaction time* "r," then Y_t reaches Y_t* if $G < 1/r$; in fact, recalling (2.6) and (2.9) from Sect. 2.3 (omitting the time reference), Y increases by $\Delta Y = Xg/r = Ehg/r = Eg$, while Y^* by $\Delta Y^* = EG$; if we want $\Delta Y > \Delta Y^*$, then we must have $E/r > EG$; it follows that if $G < 1/r$, then the pursuer will reach the pursued; if $G > 1/r$, the pursued flees and is never reached; if $G = 1/r$ there is a position of deadlock.
2. If the lever X_t varies *discretely*, with a *fixed rate* "g," and the system has a reaction time $r = 1$, then Y_t reaches Y_t* if $G < g/E_0$; the demonstration of this is analogous to the previous point.

Figure 3.5 shows various tests. In both cases we can modify both "g" and "G" so as to produce dynamics of pursuit with a variable "distance," as shown in Test 3, where we assume that, upon seeing Y_t approach, Y^* increases his speed of flight from $G = 33\%$ to $G = 60\%$.

The cases I have presented, as abstract as they may seem, are significant regarding the situations of escalation we frequently encounter in real life: from the pursuits and flights of cars or Olympic races to the pursuits and flights involving predators and their prey in the struggle for survival (Sect. 8.4).

In fact, observing such processes, it is natural to think they are regulated by *systems of pursuit*. The lion pursues the gazelle, but if this has a constant competitive advantage the lion will never catch up to it, as the following African proverb clearly attests to:

> Every morning in Africa, a gazelle wakes up. It knows it must outrun the fastest lion or it will be killed. Every morning in Africa, a lion wakes up. It knows it must run faster than the slowest gazelle, or it will starve. It doesn't matter whether you're the lion or a gazelle: when the sun comes up, you'd better be running.

Here is a significant variant of the above:

> A lion wakes up each morning thinking, "All I've got to do today is run faster than the slowest antelope". An antelope wakes up thinking, "All I've got to do today is run faster than the fastest lion". A human wakes up thinking, "To hell with who's fastest, I'll outlast the bastards".

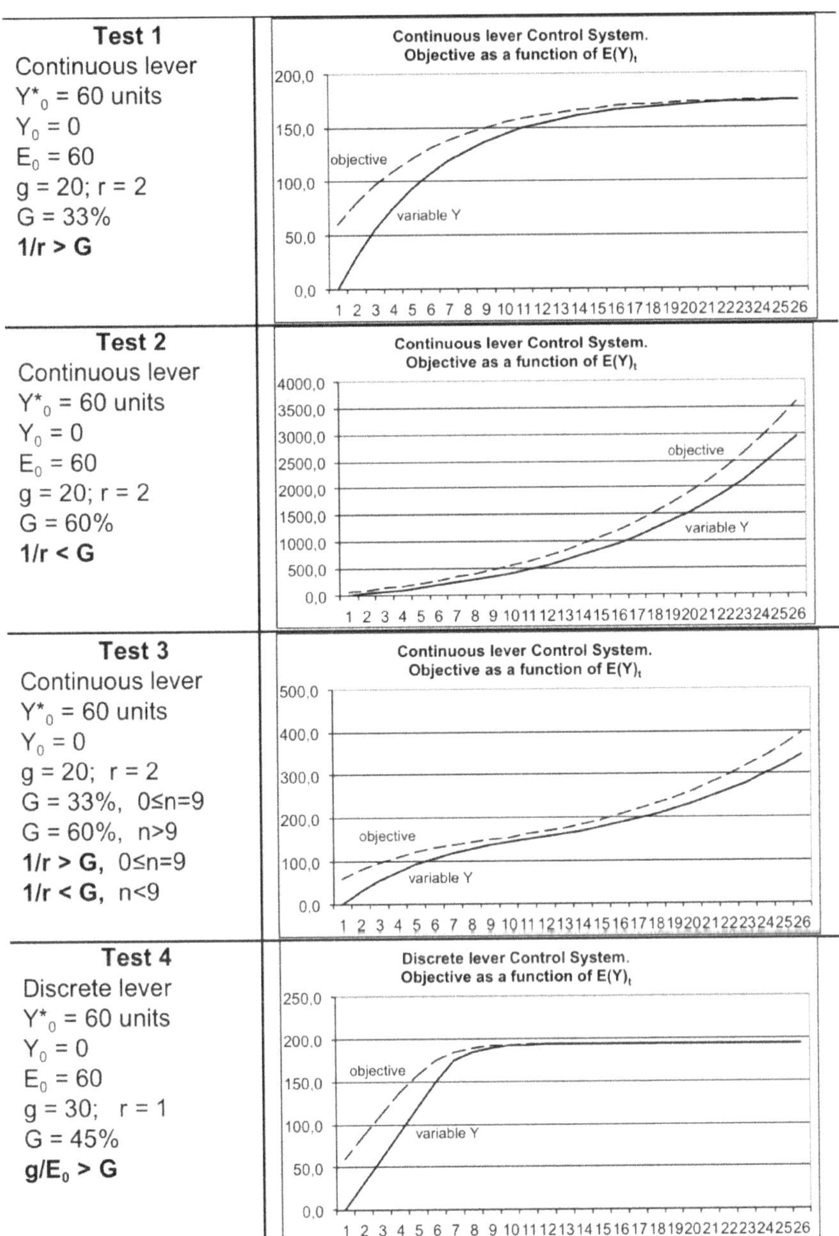

Fig. 3.5 Dynamics of a single-lever system of pursuit with an objective varying as a function of $E(Y)_t$

Systems of pursuit can give rise to various forms of *escalation*, both in a biological context—as occurs, for example, in the evolutionary escalation of the systems of attack and defense of predators and prey, respectively, as Darwin so well described—and in a sociopolitical context. We can observe these tendencies, for example, in arms escalations, feuds, wars, and violence, as well as in the economic field, when firms develop their strategies to improve quality and productivity to win the "war" of competition.

A special *system of pursuit* is the one that leads to continual progress in all the scientific, technical, economic, and business sectors. A sports record (Y^*) is set and immediately thousands of athletes train and compete to break it (objective); the average level of preparation (variable X) and performance (variable Y) increases, so that, sooner or later, someone sets a new record, thereby motivating thousands of athletes to train and compete in the continual pursuit of an objective that always seems to be just out of reach.

A similar process occurs when a new scientific discovery (Y^*) induces thousands of researchers to broaden their research (variable X), which soon gives rise to new discoveries (variable Y). The level of scientific knowledge (objective) continually progresses. How else can we explain the dynamics of progress and well-being if not as the result of the action of control systems with variable objectives? I shall return to a discussion of the above-mentioned aspects in Sect. 5.7.

3.5 Collision, Anticollision, and Alignment Systems

Systems of pursuit can be seen in systems of *collision, anticollision*, and *alignment*, whose logic can be summed up as follows. Assume there are two Control Systems, A and B, which interfere with each other, in the sense that one of the two, say B, has the objective of achieving and maintaining a certain velocity, while A's objective is to achieve and maintain B's velocity, increased or reduced by a constant P (in the simplest-case scenario). If $P > 0$, then A will catch up to and exceed B's speed. If A and B are in single file (same route objective) a collision will occur. If A is designed to collide with B then we have a *collision system*. If A is following B, then A is a system of pursuit with respect to B. If A precedes B, then A is a system of flight from B. Clear examples of such systems are predators that must pursue their prey and catch up to it, that is, collide with the prey, or prey that must flee from their predators by increasing their speed. If $P = 0$ then A will maintain the same speed as B; if they are vertically aligned we get the *alignment* of A and B, where A follows B. If $P < 0$ then A will maintain its speed below that of B by the constant P, which has a negative value. System A is an anticollision system, since A is lined up behind B but, in order to avoid colliding with B, A reduces its speed with respect to B.

We are used to considering collision or anticollision processes as spatial processes carried out by systems whose objectives are defined as dynamics in space. There is no contradiction here with the model just described, since, if we assume that the systems A and B are vertically aligned—that is, they are on a trajectory along a curve; however, this is defined—then it is easy to translate their differences in speed (as well as the constant P) in terms of distances along the spatial trajectory the systems themselves are moving along. It would also be possible to carry out a control of the spatial trajectory as well as the velocities in order to guarantee a collision or to avoid one. Nevertheless, we would have to consider two objectives contemporaneously (spatial position and speed) and several levers. We shall discuss these aspects in Sect. 4.4.

3.6 Tendential and Combinatory Control Systems

I shall conclude this *short typology* by mentioning two special classes of control system that will be useful when we examine combinatory systems in some detail (Sect. 7.5).

The first class are systems that can be called *tendential* control systems. In many feedforward control systems, the control variable X acts *randomly*, in the sense that, although it is calculated in order not to produce a gap between Y and the objective, Y^* (which, for simplicity's sake, we shall consider constant over time), it produces values of Y_t that do not permit the objective to be pursued. This is due to the work of unknown and uncontrollable *disturbance* factors that intervene, which we can define as *causal* in all respects. If the feedforward control system operates over a certain number of cycles—for example, an archery champion trying to hit the bull's-eye—we can assume that the system manager (the athlete) gains more and more experience and determines the values of X in an increasingly more accurate fashion, thereby increasing as well the precision of Y with respect to the Y^*.

To evaluate the amount of control of such a system—that is, to understand if over time the system will move closer to the objective—we must evaluate not the gaps between each single Y_t and Y^*, but the *trend* calculated using a set of values, Y_t, based on a static procedure we have chosen to describe the dynamics, since Y_t can be treated as a *time series* in all respects. I will only briefly mention the following possibilities, suggesting the reader consult a statistics text for the applied techniques:

(a) Evaluate the *trend* by simply calculating the moving arithmetic average for all values of Y_t from $t = 0$ to the interval of the last observation; the closer the average is to the objective, the more we can assume that the system is "tendentially" reaching the objective.
(b) If we need to make a progressive evaluation, then rather than using the average for all values of Y, it could be useful to estimate the trend by calculating the *moving average* using an appropriate interval of values centered on Y_t; since there are many types of moving averages—*simple moving average*, *weighted moving average*, *exponential moving average*, to name but a few—we must choose the type of moving average that is most significant for expressing the *speed of adjustment* of Y to the objective, taking into account the nature of the variables and the length of the various control cycles of the system;

(c) Evaluate the *trend* using least-square techniques; in the simplest case, the trend may be expressed by estimating the line which minimizes the square of the variance of each Y_t with respect to Y^* for the interval of observation and repositions the slope of the trend line through the given data points of Y_t.

These statistical techniques to detect the dynamics of the error and its movement toward zero are widely used in performance quality control for sports, organizations, and companies.

Figure 3.6 represents the control chart of a system of regulation that, acting on a *control lever X*, seeks to maintain the dynamics of the values of the controlled variable Y_t, which expresses the quality of a generic production process, within the objective values $Y^* = 130 \pm 5$. The control chart indicates the moving averages of 5 and 9 values to test whether or not there is movement toward the objective. We see from an initial visual analysis that from the period $t = 16$ onward the values of the variable Y_t tend to stabilize around values close to the objective; this shows that the control actions on the technical variables X were successful.

The second type of system I shall refer to as *combinatory* control systems. Their unique characteristic comes from the fact that the controlled variable Y is the composite result of the combined action (the meaning of which will be defined below) of a group of individuals (animals, persons, or other) whose collective action, X, produces the values of Y that permit the objective, Y^*, to be achieved.

The gap between Y and Y^* results in a change in the behavior of the individuals—no matter whether uniform or varied, contemporaneous, synchronized, or at different time periods—so that the "combination of the behaviors", X, produces a new value for Y.

Combinatory systems usually also act as *tendential systems*, in that in the combined action of various individuals the values of the variable Y must be

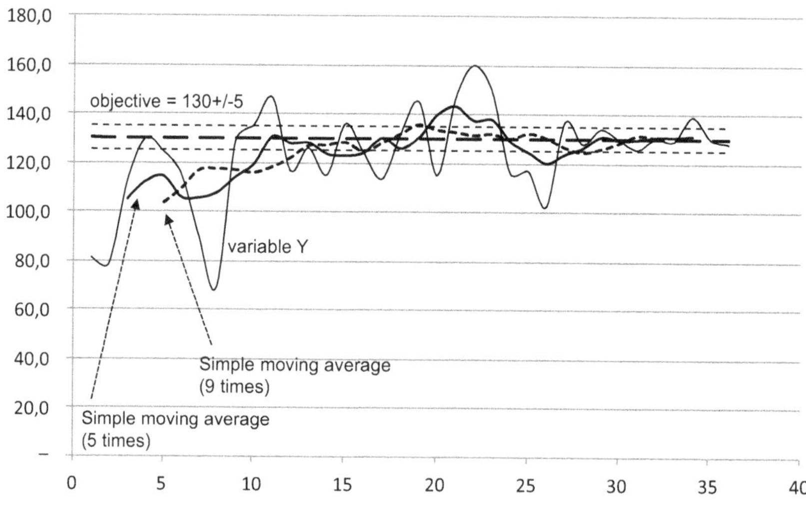

Fig. 3.6 Dynamics of an abstract tendential system

cycles	Agents A(1)	A(2)	A(3)	A(4)	A(5)	A(6)	A(7)	A(8)	A(9)	A(10)	mean = Y	objective = Y*	Error = Y* - Y	
t=1	153	155	151	150	155	150	155	149	148	151	152	160	- 8	Action on X
t=2	154	155	149	155	154	152	156	149	149	153	153	160	- 7	
t=3	154	159	149	152	158	152	156	147	149	153	153	160	- 7	
t=4	156	159	152	156	158	152	156	150	151	154	154	160	- 6	
t=5	156	160	153	156	158	152	156	153	151	154	155	160	- 5	
t=6	156	160	153	156	158	155	156	155	151	158	156	160	- 4	
t=7	157	160	155	156	162	157	156	155	151	158	157	160	- 3	
t=8	159	163	155	157	162	157	157	157	151	158	158	160	- 2	
t=9	159	165	157	159	162	157	158	157	155	158	159	160	- 1	
t=10	163	165	160	159	162	160	158	157	155	158	160	160	- 0	maintaining control
t=11	165	168	160	159	162	160	161	158	156	159	161	160	1	
t=12	162	160	164	161	162	160	163	158	156	163	161	160	1	
t=13	160	160	162	161	162	160	160	160	162	163	161	160	1	
t=14	158	158	162	161	162	162	158	158	162	163	160	160	0	
t=15	157	165	157	158	167	159	157	158	162	166	161	160	1	
t=16	157	161	157	160	159	159	159	160	161	166	160	160	0	
t=17	155	163	160	158	159	158	157	155	157	166	159	160	- 1	
t=18	159	162	161	160	161	160	159	155	154	166	160	160	- 0	
t=19	161	162	162	160	159	162	159	155	154	166	160	160	-	
t=20	162	164	157	157	157	163	158	159	156	160	159	160	- 1	

Fig. 3.7 Logic of combinatory control systems

tendentially evaluated in their movement toward the objective (a broad discussion will be developed in Sect. 7.5).

To understand the logic of combinatory systems it is useful to refer to Fig. 3.7, which shows the values $[y_t^1, y_t^2, \ldots, y_t^N]$ lined up in rows, obtained from $N = 10$ agents over $T = 20$ cycles as a function of a control lever X (not shown in the table). The system's response in terms of Y_t is determined as the average of the individual values of the $N = 10$ agents and is presented in the "average" column. Even without a detailed examination, a simple reading of the data indicates that, starting from period $t = 10$, a specific control has been carried out on X (not shown in the table) that gradually eliminated the variance. Naturally, since we are dealing with a *tendential system* the variance in this specific example is never completely eliminated; although it returns in subsequent periods, it nevertheless remains at moderate values. Combinatory control systems are quite common, and they will be dealt with more at length in Sect. 7.6.

3.7 Parallel or Serial Connections

Until now, we have considered the control of the variable Y_t in its movement toward the objective Y^* as a process carried out by a single control system. We can immediately extend this logic to cases in which the control of Y_t is carried out by a plurality of control systems connected *in parallel* or *serially*.

The *parallel control* occurs when changes in Y_t require the action of a lever, X_t, which cannot be activated by a single control system. Thus, *partial* control systems are set in motion which are small in scale and capable of producing an overall effect equal to that of a larger-sized control system.

For simplicity' sake, let us assume that only *two* partial systems, A1 and A2, are activated; then, the *parallel* connection can be achieved in two ways.

(a) *Decomposition*: the objective Y^{A*} is decomposed (i.e., divided) into two subobjectives, Y^{A1*} and Y^{A2*} (which are not necessarily equal), so that $Y^{A1*} + Y^{A2*} = i^{A*}$; these partial objectives are assigned to two *parallel partial systems*, A1 and A2.
(b) *Composition* of the lever X^{A} that controls Y^{A} by summing the values of the levers, X^{A1} and X^{A2}, of the two subsystems.

Form (a) of parallel control is purely additive, since we sum the effects produced by the partial objectives. Figure 3.8 shows a possible simple parallel control system based on the decomposition approach; in this case the system A must control the variable Y^{A} through the *partial* systems A1 and A2, which are connected systems (Sect. 2.14).

The objective of *A*, $Y^{A*} = 350$, is divided into two subobjectives, $Y^{A1*} = 210$ and $Y^{A2*} = 140$ (other rules for subdivision would be equally admissible). The system A2 is "slower" than A1 since it is characterized by a reaction time of $R = 3$ and an action (power) rate equal to 1.

Parallel control through the *decomposition* of the objective is rather common in both the socioeconomic and corporate environments. A firm could divide up the

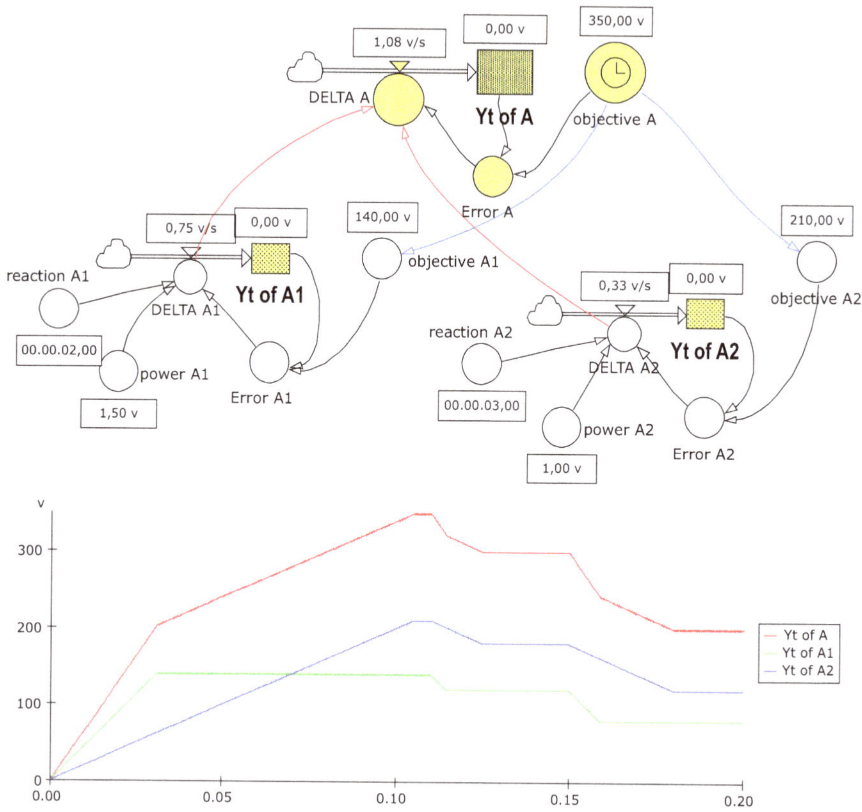

Fig. 3.8 Parallel systems based on the decomposition of Y^* (simulated with Powersim)

general revenue objective into partial objectives to assign to the various geographic divisions. The government, whose objective is to guarantee its citizens a certain quantity of health and educational services, should necessarily assign parts of these objectives to the various health and educational units in the country.

Control by linking partial parallel systems can also occur through the *composition* of the control lever X, as shown in the second case above. Unlike the first case, here the objective Y^{A*} is not broken up, since we assume it must be achieved only through system A, on the condition the latter is able to assign a sufficient value to its own lever, X_t^A. When it is not possible or convenient to directly give a value to X_t^A of system A, then two partial systems, A1 and A2, could be activated in order to independently produce the partial values of the lever X_t^{A1} and X_t^{A2}, so that their sum (composition) result in: $X_t^A = X_t^{A1} + X_t^{A2}$.

This form of segmentation is also quite common. If a force of 100 men is needed to move an obelisk, each of these men represents a partial control system that does not move the obelisk by even a hundredth part, but simply produces one-hundredth of the force necessary to place the obelisk in the correct position.

However, with parallel control systems through *composition* of the X_t^A we need to clarify what objectives Y^{A1*} and Y^{A2*} to assign to A1 and A1.

A simple solution is to assign to A1 and A2 an objective corresponding to a share, E^{A1} and E^{A2}, of the error of the main system, $E^A(Y) = Y^{A*} - Y_t^A$. As a result, in trying to produce values Y^{A1*} and Y^{A2*} to achieve E^{A1} and E^{A2}, A1 and A2 in fact end up producing a value of the A lever, $X_t^A = X^{A1*} + X^{A2*}$, which is necessary to achieve the general objective Y^{A*}, thereby eliminating the overall system error $E^A(Y)$, as can be seen in Fig. 3.9. In this figure, even with the same parameters of the model in Fig. 3.8, we note a different movement toward the achievement of Y^{A*}. In fact, the two systems, A1 and A2, in Fig. 3.9 achieve their objectives contemporaneously, since they uniformly eliminate over time the shares E^{A1} and E^{A2} assigned to them; precisely for this reason the control ends when E^{A1} and E^{A2} become zero.

Let us now consider the structures that are formed when several control systems operate *serially*. Two systems, A (upstream) and B (downstream) are serially connected when the variable controlled by A, Y^A, becomes the lever X^B of the second system in order to achieve the objective Y^{B*}. Thus, the following causal chain is formed: $[X^A \rightarrow Y^A = X^B \rightarrow Y^B \rightarrow Y^{B*}]$, which clearly shows that Y^B is in fact controlled by the lever X^A of the system upstream. The objective to assign to system A can be represented by the values that must be produced in B's lever: that is, $Y^{A*} = X^{B*}$.

If it is possible to calculate in advance the value X^{B*} needed to achieve the objective of B, then Y^{A*} is a *fixed* objective; otherwise it is *variable*, and A continues to operate until the system *downstream* has not eliminated its error. In both cases the objective of the system *upstream* is assigned by the control apparatus of the system *downstream*.

Serial control is the rule, not the exception. If we apply the FIRST RULE of systems thinking (Sect. 1.2) and zoom in, we can easily see that most control systems can be *serially* coupled to another instrumental system whose function is to produce the variations in the first system's lever.

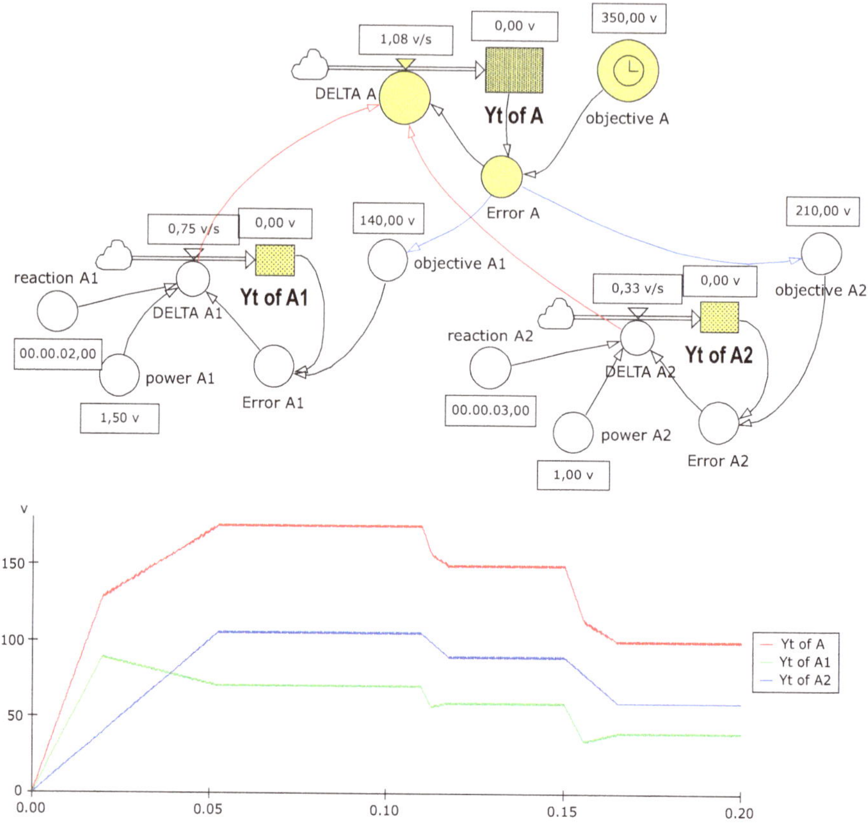

Fig. 3.9 Parallel systems from the composition of *X*

If we instead zoom out we can easily realize that each control system can be an element, in *parallel* or *serially*, for a larger system. In order to drive a car we have to turn the front wheel (downstream system) appropriately; however, this depends on the rotation of the steering wheel, which in turn depends on the "torque" produced by our arm (upstream system). When we, along with our colleagues, achieve the objectives assigned to us at work, the department also achieves its objectives; if all the departments achieve their objectives then the organization as a whole does so as well.

When we observe control systems from a systems thinking perspective and try to identify the relations between *X* and *Y*, having in mind only a single objective *Y**, we are at a summary level; that is, we do not zoom in and explicitly consider whether this system derives from the composition of parallel partial control systems or whether other variables operate between *X* and *Y*, each of which is controlled by serially arranged partial systems. In order to treat the topic in a manageable way, in this *section* we have only considered two partial subsystems, A1 and A2 or B. More generally, we have to keep in mind that both *parallel* and *serial* arrangements can be formed by any number of partial systems.

Finally, we must consider that *parallel* and *serial* structures can be joined together to form complex control structures that characterize biological, social, and organizational contexts. These aspects will be taken up in Part II of the book.

3.8 Holarchies of Control Systems

The points developed in the previous *section* can be generalized from a holonic perspective (Mella 2009). When we observe control systems from a systems thinking point of view and try to identify the relations between X and Y, having a single objective, Y^*, in mind, we are treating the topic in a concise manner; the control system is viewed as an autonomous entity. Systems thinking teaches us, however, that we must get accustomed to thinking of every control system as interacting with other systems at the same level, or at a superordinate or subordinate one. In addition to analyzing the *serial* and *parallel* connections, which produce a monodirectional interaction among control systems, we can and must extend our observation and consider control systems as classes of observed entities arranged in *hierarchical levels*. This necessity is made clear by reflecting on the fact that many control systems, though apparently acting independently, are nevertheless necessary for the functioning of others at both a superior and inferior level; this is true for any level of observation. In other words, if we zoom in and out at the same time it appears that everywhere there is a holonic hierarchy, a *holarchy*, among control systems.

We become aware of the various levels of control systems as soon as we realize we can control the actions of our body (move ourselves, feed ourselves, etc.) because our organs function correctly thanks to their control systems; these organs, in turn, function without errors thanks to the control systems in the cells that make up the tissue in our bodies; the cells function as a result of the control systems which regulate their autopoiesis.

It is clear that I can control my fingers on a keyboard to write this sentence, since I control my arm, shoulder, and entire muscular–skeletal apparatus; moreover, the flow of words can pour out from the keyboard because the recognition system, thanks to the neural system, identifies the appropriate words from the catalog of my knowledge (grammar and vocabulary), arranging them in the correct sequence. However, this implies that the areas of the brain that concern memory and various other functions are being controlled; and, at a lower level, the individual neurons are controlled, which in turn control the impulses that arrive from "upstream" to discharge other impulses "downstream." It is equally clear that the control of the direction and speed of our car is possible since its various organs and components—amount of gas in the tank, turbochargers, electronic system, tires, etc.,—are linked to specific control systems.

Even the control system that seeks to keep pollution under certain thresholds around the globe implies emissions control in large urban areas, and, at increasingly lower levels, in cities and energy-providing companies, which limit individual emissions by controlling the energy consumption of plants and the maintenance state of facilities. How can we control infant mortality rates if we do not activate controls on research and health structures (research labs, obstetric and neonatology departments) and on families (hygiene and diet)? How can we control health and family-related structures if we do not control the level of preparation and education of their personnel? In short, while taking nothing away from the circular interconnections among control systems, it is clear that there are various levels of control systems that constitute a *holarchy*.

The notion of *holarchy* comes from the concept of *holon*, coined by Arthur Koestler (1967, 1978), who viewed the *holon* as a *Janus-faced entity*: if it observes its own *interior* it considers itself a *whole* formed by (containing) subordinate *parts*; if it observes its *exterior* it considers itself a *part* or *element* of (contained in) a vaster *whole*. In any event, it sees itself as a *self-reliant* and *unique* entity that tries both to survive (it is a *viable system*) and to integrate with other holons:

> These sub-wholes—or 'holons', as I have proposed to call them—are Janus-faced entities which display both the independent properties of wholes and the dependent properties of parts. Each holon must preserve and assert its autonomy, otherwise the organism would lose its articulation and dissolve into an amorphous mass—but at the same time the holon must remain subordinate to the demands of the (existing or evolving) whole (Koestler 1972, pp. 111–112).

Each holon includes those from lower levels, but it cannot be reduced to these; it transcends them at the same time that it includes them, and it has emerging properties (Edwards 2003).

Due to their Janus-faced nature holons must necessarily be connected to other holons in a typical *vertical arborizing structure* known as a *holarchy*, which can be viewed as a *multilayer* system (multistrata) (in the sense of Mesarovic et al. 1970) or *multilevel* system, with a *tree structure* (Pichler 2000). Each holon is a *head* holon for the subtended part of the branch and a *member* holon for the upper part. It is relevant to observe that holarchies are not holons but arrangements of holons that represent conceptual entities whose function is to bring out the essentiality of the vertical interactions among holons.

The completeness principle (Mella 2009) must, in any event, apply in the multilayer holarchy: each subordinate level represents holons which are less extensive and recompressed into the holons at the superordinate level (arborization effect), with the understanding that all the base holons must be included in the final holon.

Koestler (1967, p. 344) defines *output holarchies*—or *descendent* holarchies—as those that operate according to the *trigger-release principle*. In such holarchies the *top holon* carries out its own processes thanks solely to the activities of the subordinate holons, which the top holon itself coordinates by sending information about the activities the subordinate ones must undertake. He defines *input holarchies*—or *ascendant* holarchies—as those holarchies that operate according to the *filtering principle*; they produce progressive syntheses from the subordinate to the superordinate levels, as if, at each level, the holons filtered and synthesized the inputs from the subordinate holons.

We can easily note that the holarchies of control systems are at the same time both input and output holarchies; *they are descendent relative to the objectives and ascendant relative to the variables under control.* The top-holon control system "launches" its objective, *Y**, as information to lower-level control systems which, in turn, "launch" subobjectives to even lower levels. On the other hand, control system holarchies are also *ascendant*, in the sense that the values for *Y* (or *X*) at a certain level are syntheses of lower levels that are filtered (synthesized) to higher levels. The close connection between control systems for the *Y* variables at different levels (ascendant connection) and the *Y** objectives (descendent connection) is a general feature of control system holarchies; this connection fully reflects the general principle, well outlined by Koestler, of the behavioral complexity of holons:

> Hierarchies can be regarded as 'vertically arborising structures whose branches interlock with those of other hierarchies at a multiplicity of levels and form 'horizontal' networks: arborization and reticulation are complementary principles in the architecture of organisms and societies (Koestler 1967, p. 345).

This feature is shown also in the model in Fig. 3.10; while the basic control systems have irregular dynamics, their aggregation produces less regular dynamics as we proceed upward toward higher-level control systems. Figure 3.10 presents a simple control system holarchy formed by partial systems, arranged over three levels according to a typically holonic arrangement set out in a descending *series*; the main objective of the *top holon* (A), *Y** = 360 units, is broken up into several percentages (which I have introduced for the simulation); the percentage shares of the main objective are assigned to the subsystems, which are arranged in *parallel* at each level. The control system at level A represents an accumulator of the variable *Y* for the sublevels A1 and A2, which in turn accumulate the values of the variable *Y* from the five base holons (A11, A12, A21, A22, A23). Each variable *Y* of each system and each level acts to achieve its particular objective and, moving up through the holarchy, to allow the objective of the top holon (A) to be achieved. The *upper* part of Fig. 3.10 indicates the initial situation and the subdivision of the main objective; the *lower* part shows the results of the simulation after 60 iterations. Figure 3.11 presents *Y*'s movement toward the objectives; we can see that, despite the fact the initial objective was subject to two exogenous variations, the system correctly controls the *Y* variables.

From a *bottom-up* perspective, the values of the *Y* variable at level A derive from the values produced at the lower levels; moreover, these values, from a top-down perspective, are conditioned by the holons at the superordinate levels, since level A indicates the objectives to achieve at the lower levels, which for simplicity's sake are set equal to a proportional share of the objective of the higher-level holon. In this twofold relation, descendent (*Y**) and ascendant (*Y*), we obtain a coevolving dynamics among the control systems at the various levels, following the typical relation that Ken Wilber, in his *Twenty Tenets* (Meyerhoff 2005), clearly describes:

> 10. Holarchies co-evolve. The micro is always within the macro (all agency is agency in communion). 11. The micro is in relational exchange with macro at all levels of its depth (Wilber 2000, pp. 34–35).

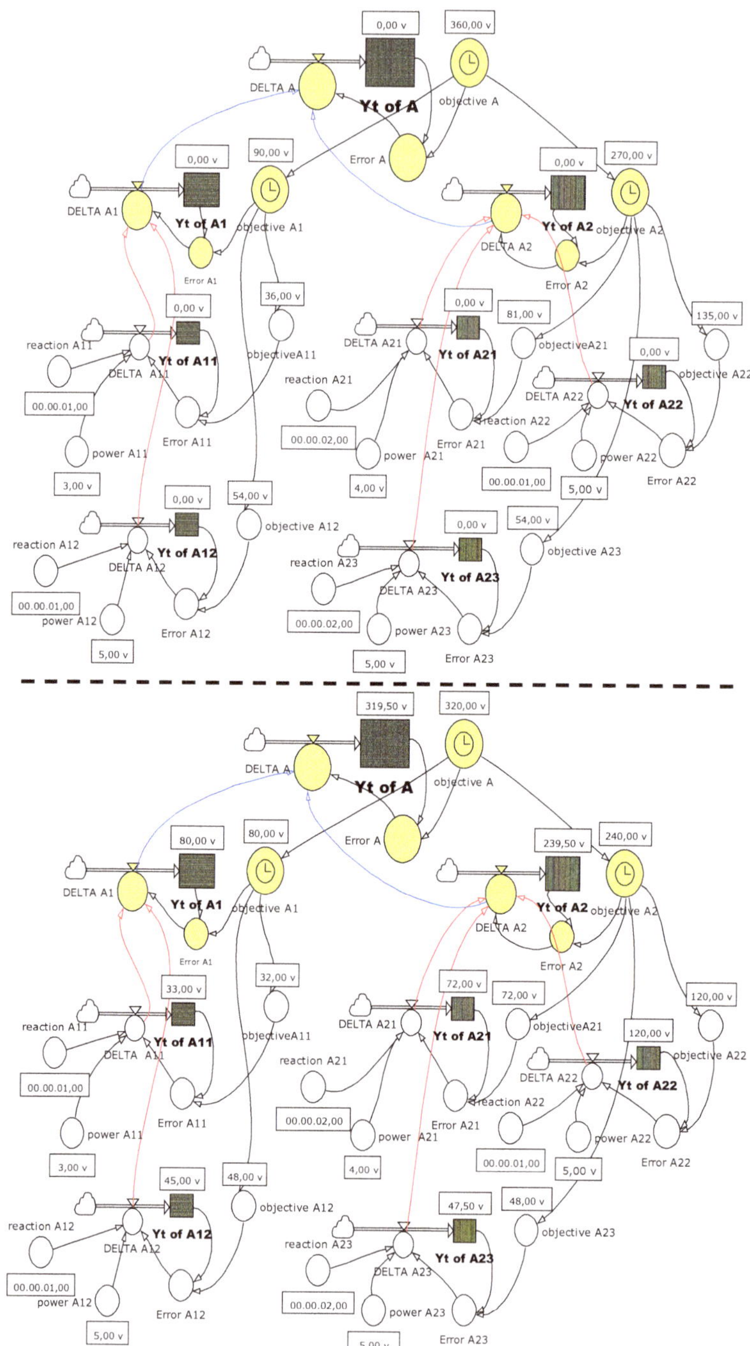

Fig. 3.10 Control system with three levels, based on a holonic arrangement

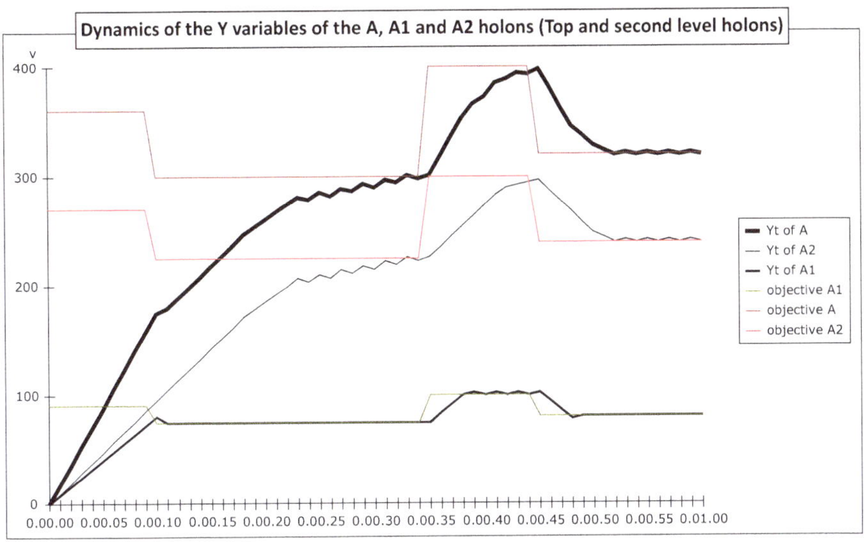

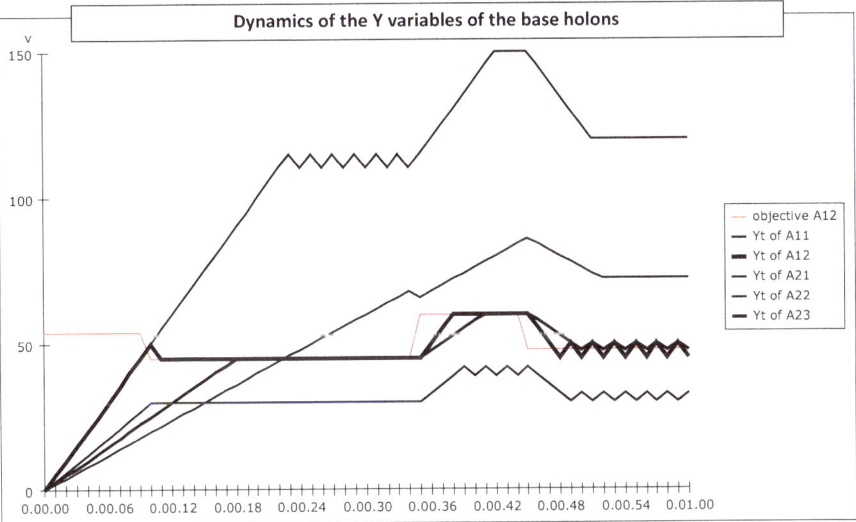

Fig. 3.11 Dynamics of the variables to control (*Y*) in a holonic system

Such conditionings not only act in an ascending direction but reveal their effects in a descending one as well. This means that in a control system holarchy each system at a given level acts together to condition the higher-level systems (following one of the arrangements in Figs. 3.8 and 3.9); however, at the same time it receives from the latter appropriate inputs that guide its own behavior. These inputs are then also transmitted in descending fashion along the lower branch, thanks to the coordinating capacity the system has in relation to those systems composing the subtended branch.

Moreover, as Wilber explicitly notes, since every holon at a certain level (and this is particularly true for holonically ordered control systems) includes subordinate holons in its own structure, even while it transcends them, in order to survive it must preserve and regenerate these, since it depends on them for its existence. This makes it more likely that lower-level systems will be maintained and consolidated over time.

From an *ascendant* observational perspective, by gradually zooming in we can verify that a miniscule pixel is functioning because a microcontrol system provides it with a certain state, since a superordinate control system makes this state necessary in order to adjust the state of the screen. However, the state of the screen derives in turn from the control objectives of the software operating at that moment, which in turn responds to the control needs of the person using it, who in turn is influenced by the control needs of the group in which he provides his activities, which in turn are influenced by the control system represented by the operating center in which the group functions. This control system is part of a vaster control system of the department, the division, and the entire organization, which in turn is under the control of other superordinate systems.

We can make similar considerations for all living organisms and for social organizations. For example, in living organisms the macrocontrol of organs depends on the micro controls of the constituent cells; in a plant, each cell in the root produces a process controlled from a higher level, but the Y^{ij} variables of the control systems of cells are synthesized into the Y^{j} variables in the (increasingly larger) branching in the roots, eventually becoming the Y variable of the entire organ, represented by the roots as a whole. It is clear that the activities in the roots condition those in the other parts of the plant; however, it is equally clear that the operating objectives of the roots and branches, down to the individual cells, do not determine the higher-level objectives but, on the contrary, derive from the dietary and survival constraints of the plant as a whole.

Another example: the control of global warming implies emissions control by individual countries, which reverberates down to the control of the individual regions, companies or families, and the individual machines and their components that produce the substances feeding the *greenhouse effect*.

It surely is not the faster beating of the heart that spurs us to run in order to increase our blood pressure, blood circulation, and oxygenation; rather the opposite is true. Our objective to run imposes a control on the organs; the muscles need energy, there must be ample blood flow, the lungs must produce more oxygen, and thus the heart must beat faster, at levels determined by our physiological programming. Thus, the heart is regulated by higher-level controls and, as a single unit, controls the contraction of its muscle fibers, which determine the movement of the cells of which it is composed. This may be a simplified view of things, but it is obvious that it is never the heart beating fast that forces us to run, but just the opposite: it is our running that forces our organism to also control our heartbeat. Only if the latter becomes anomalous and is not due to any special physiological need does the heart "spur" us to set in motion pharmacological or electronic control systems to bring our heartbeat "under control" according to our physiological objectives or constraints.

3.9 Complementary Material

3.9.1 *Some Well-Known Cybernetic Systems*

As observed in Sect. 3.1, cybernetics deals with automatic mechanical, biological, and social systems. In the collective mind, cybernetic systems are usually thought of as mechanical systems that can control the variable *Y* by automatically regulating *X*, as we can clearly understand from the following quote:

> Watt's governor was a superb example of feedback control. Feedback controllers, mechanisms that sense a discrepancy and correct it, are absolutely shot through our world today. We go through hardly an hour of any day without using feedback devices—float valves in our toilets, thermostats in our rooms, pressure-control valves and carburetion electronics in our cars.
>
> [...] That sort of thing was common in the Hellenistic world. One of the first feedback devices was the water-clock flow regulator. The 3rd-century BC engineer Ktsebios made the ancient water-clock into an accurate timekeeper by inventing a float stopper to regulate a constant flow of water into the indicator tank (Lienhard 1994).

There has been rapid progress in mechanical cybernetic systems thanks to the ideas from engineers and technicians, which have succeeded in replacing the human manager with a mechanical one. The history of control systems and cybernetic systems has mainly been "written" by engineers who have designed and produced automatically controlled mechanical systems. Among the many examples presented below are those of models of the main automatic control apparatuses, indicated in the quote, which are now part of the history of cybernetic systems.

Figure 3.12 shows Watt's *Centrifugal Regulator* (otherwise known as the Centrifugal Speed Control Device, Flyball Governor, or Watt's valve) invented by Matthew Boulton and James Watt and produced in 1788; it served to automatically control the speed of the steam locomotive and was applied to the control of the rotation of the driveshaft in all steam engines as well as windmills.

Watt's regulator is a valve that regulates the on/off state of the flow of steam that generates pressure in the piston. This is coupled to a mechanical speed detector made up of two oscillating spheres mounted on the spindle that rotates when activated by the driveshaft. When the speed of the motor increases, the oscillating and rotating spheres rise due to the action of the centrifugal force, thereby raising the axes linked to the on/off valve that closes off the steam when the speed raises the levers (due to centrifugal force) to the limit the governor has chosen as the speed objective. The lack of steam diminishes the pressure in the piston, which reduces the speed, thus lowering the spheres and once again opening the steam valve to increase the pressure and speed.

Figure 3.13 shows a system for the automatic regulation of the level of liquid in a container based on a principle entirely similar to that in Fig. 3.12. The level detector is represented by a bob that is pushed up by Archimedes' force, while in Fig. 3.12 the two rotating balls were pushed up by a centrifugal force. A flow of liquid, constant or irregular, raises the level in the container, which in turn raises the bob. When

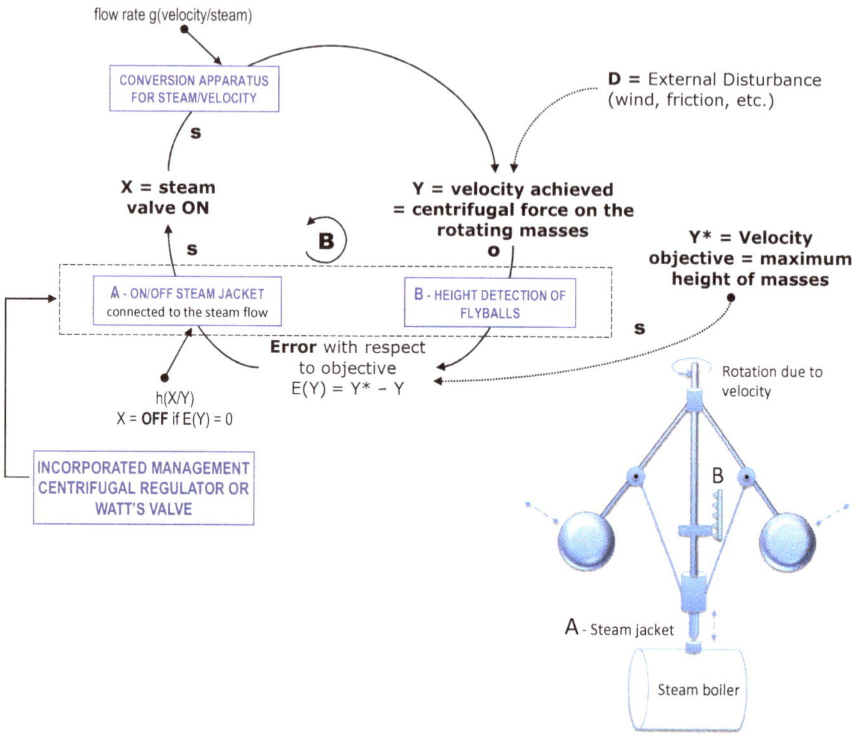

Fig. 3.12 Automatic control system. Watt's centrifugal regulator

the bob reaches the maximum desired level, a butterfly valve is closed by the force of buoyancy, which stops the flow of liquid. The automatic system in Fig. 3.13 can be improved by also controlling the outflow of the liquid in the tank through a valve opener to eliminate the excess liquid in the event liquid flows in from an outside disturbance. This system will be examined in the next chapter, as the addition of this second control lever leads to a two-lever system (Fig. 4.1).

Since these systems were conceived and created in previous centuries, it would seem they would be outdated. Yet their main principle is still valid; the automatic control is achieved through more modern detection and regulation systems, which are supported by electric engines commanded by relays. The *Flyball Governor* has been greatly simplified, since the operation of the *flyballs* has been replaced by pressure valves which open when the steam pressure, or in general the gas pressure, exceeds a certain level, so as to reduce the pressure, precisely as the *flyballs* functioned.

A simple modification transforms the automatic control system in Fig. 3.13 into a water clock, one of the simplest automatic systems for measuring the flow of time. A straw introduces regular flows of water, or even single drops, into a container or tank (even a large one) that can operate for many hours. The floating ball flows vertically, while attached to it is an indicator; on the wall of the container are notches

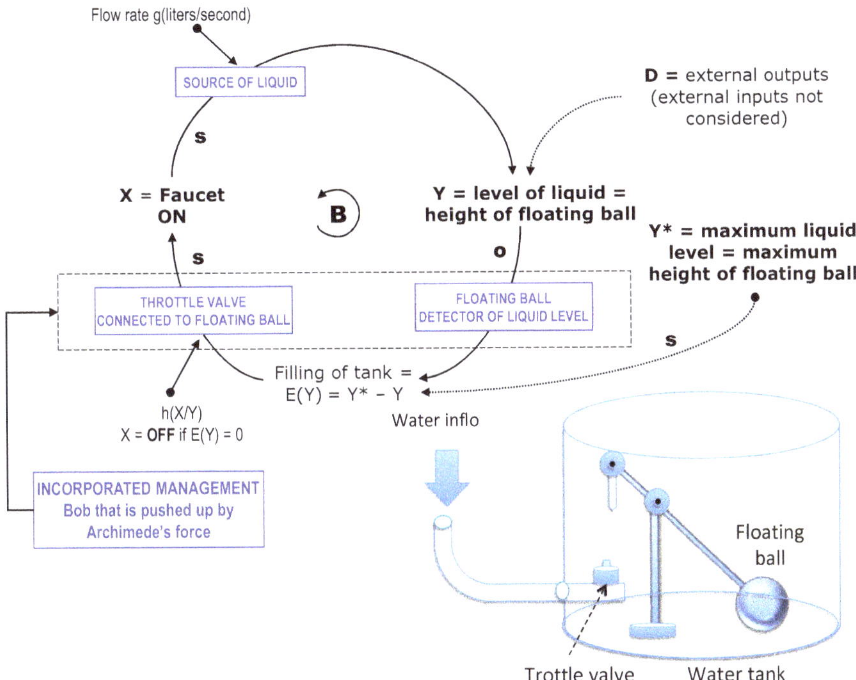

Fig. 3.13 Automatic control system. Floating ball regulator

indicating units of time. The gradual filling of the container, which is assumed to be regular, raises the floating ball to increasingly higher notches, and this measures the flow of time in the unit chosen for the notches: 15, 30, 60 min, etc., according to the experience of the clockmaker. When the container is completely filled a control mechanism sometimes operates in a manner opposite to that in Fig. 3.13. When the floating ball reaches its maximum level a butterfly valve is pushed that quickly empties the container, thus permitting the clock to begin operating again (this is a typical CI-PO impulse system, which will be illustrated in Sect. 4.3). Hourglasses function similarly: the upper part releases the sand to the lower part; when the latter is full, the manager turns over the hourglass and the cycle starts again.

A variant which simplifies the preceding mechanism (by eliminating the floating ball) consists simply in constructing a container—preferably of glass and cylinder shaped, with a narrow base and sufficient height—whose walls have regularly distanced notches with a predetermined time scale. The container is filled with water which is then let out through a small valve at its base. This regular outflow lowers the level of the liquid to increasingly lower notches, which allows the time flow to be measured and the time to be known by simply observing the notch corresponding to the water level. Though simpler than the preceding example, this clock has the disadvantage that the water outflow is never regular, since the pressure of the remaining liquid varies as the water flows out.

Even Galileo Galilei invented a similar mechanism, equally simple though not entirely automatic. Instead of filling up a graduated container, his design involved filling a container, of whatever form, and having the water regularly flow out through a straw at the bottom. The water would be collected in a second container at the base of the first and then weighed when it was time to measure how much time had passed. The weight is, in effect, converted into time using a simple scale derived from experience: for example, a kilogram of water could correspond to 10 min or an hour in relation to the amount of outflow in the upper container. If the weight-to-time conversion scales are accurate, this measure not only will provide more precise results but also have the advantage of permitting the calculation of the submultiples: if a kilogram of water equaled 1 h, then 100 g of water would be equivalent to 6 min, etc.

I do not wish to always backtrack, but it is important to note that the first modern clocks to precisely measure time flow were the *pendulum clocks*, which function in a similar fashion to the flyballs in the system in Fig. 3.12, not to the floating ball in the model in Fig. 3.13, followed by the water clocks. We have to go back to Galileo, in the seventeenth century, to find the isochronism of the mechanical pendulum. The simplest mechanism to measure time is by a weight, a sphere, or a cylinder placed at the lower end of a rigid shaft with a pivot at the top end. A spring, or some similar device, moves the oscillating shaft while an instrument measures the number of oscillations, each of which is accomplished in the same period of time. The clock-maker measures the time for an oscillation based on his models, thus producing a unit of time; the duration of a period of time can thus be determined by multiplying the unit of time by the number of oscillations detected. If the time of an oscillation were 4 s, then 30 oscillations would measure 2 min. The addition of a quadrant provided the finishing touch for the pendulum clock we are all familiar with. The functioning similarity to the other mechanisms we have examined is clear. In Watt's *Centrifugal Regulator*, the flyballs were moved by the centrifugal force generated by the rotating speed of the driveshaft; in the pendulum clock the weight was triggered by the force of gravity.

We can immediately observe that all clocks, water, sand, or pendulums, as well as modern wristwatches, are not control systems but "machines"; the variable that measures time flows—the height of the floating ball, the level and number of oscillations, or another internal "engine"—is not a variable that is controlled but simply the variable Y_t produced by the effector-clock. This is why all clocks, old as well as modern ones, must be periodically checked to eliminate the error, $E(Y)$, between the actual time (Y_t) and the time of reference (Y^*). The lever X on our clock (depending on the model) allows us to become the manager of the *manual* control system of time. For many digital clocks, this *manual* control system can become entirely *automatic* if the clock has an internal detector apparatus to capture the international standard time or the UTC/GMT (coordinated universal time), calculated using a weighted average of signals from a certain number of atomic clocks, located in several countries.

3.9.2 Halt Control Systems

"One-way" control systems are very common, whether or not the objective is in the form of a constraint that cannot be exceeded, a limit that is insurmountable, or a variable objective, as we typically see in systems referred to as "constraint-keeping" control systems. Figure 3.14 shows the model of a control system that allows us to park our car without crashing into the front wall of our garage.

"One-way" control systems are also common in our simplest daily activities and more numerous than steering control systems. It is true that, when we measure the quantity of rice for a meal for four people, we can calculate the right amount by incremental adjustments (up or down) in the rice before adding it to the water. However, when the rice is cooking and we wish to add the right amount of salt, it is better to do so in small, gradual quantities, since if we add too much the rice will become inedible. The rice also has to be cooked just right, and it must be tasted (detect the error) several times in order not to be overcooked.

In all mixtures and blends with several components (risotto, for example) the right amount of each component can never exceed the limit of the other components otherwise the product must be definitively eliminated.

It is not even necessary to give overly complicated examples. Figure 3.15 represents the control system that enables a *sommelier* to pour champagne always to the same (approximate) level in flute glasses.

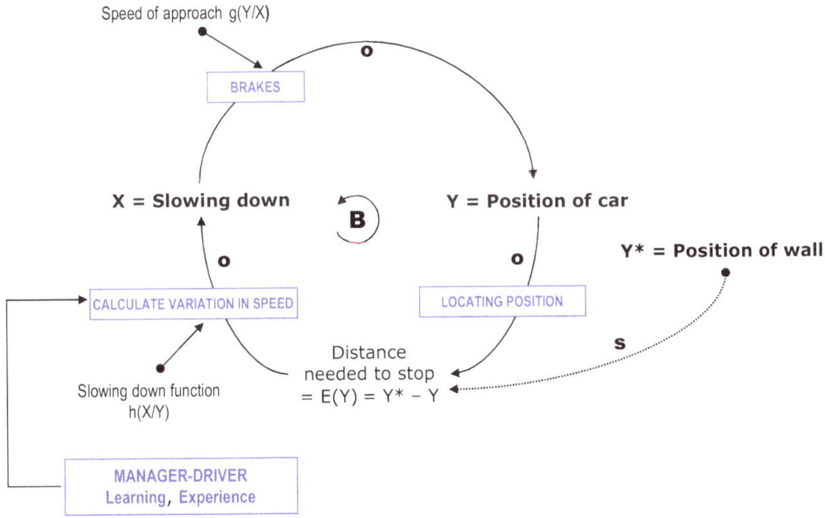

Fig. 3.14 Halt control system. Parking against a wall

3.9.3 Biometric Systems of Recognition and Identification

Going back to the distinction between systems of *attainment* and recognition (Sect. 3.2), let us consider *biometric systems*; that is, artificial *systems of recognition–identification* of a person based on several biological characteristics, which can be divided into physiological and behavioral aspects.

The physiological features refer to data that differentiates individuals in a singular way (static data), such as fingerprints, the shape of the hand, the form of the iris, the network of veins, or even the facial image. The behavioral characteristics seek to identify an individual based on the way the individual carries out at action, such as his handwriting, his style of walking, vocal features, and in general the way he performs an activity. Among the highly precise automatic systems of recognition and identification are the systems of recognition of *fingerprints* from a continually updated archive, as shown in Fig. 3.16, which is derived from Fig. 3.3.

The fingerprint to recognize is obtained from an optical scanner and set as the objective (Y^*). The effector apparatus scans the archived fingerprints and the detector compares these and identifies the error (no match found) or the absence of error (match found).

The system of fingerprint recognition used today by the police is automatic (AFIS, automated fingerprint identification system) and the high speed of recognition allows authorities to produce very large archives that can be quickly updated. In order to speed up the recognition of a fingerprint, AFIS uses classification algorithms of the fingerprint pattern and of the sampling of parts of the fingerprint itself or of systems of points. In addition to AFIS, fingerprint recognition is also wide-

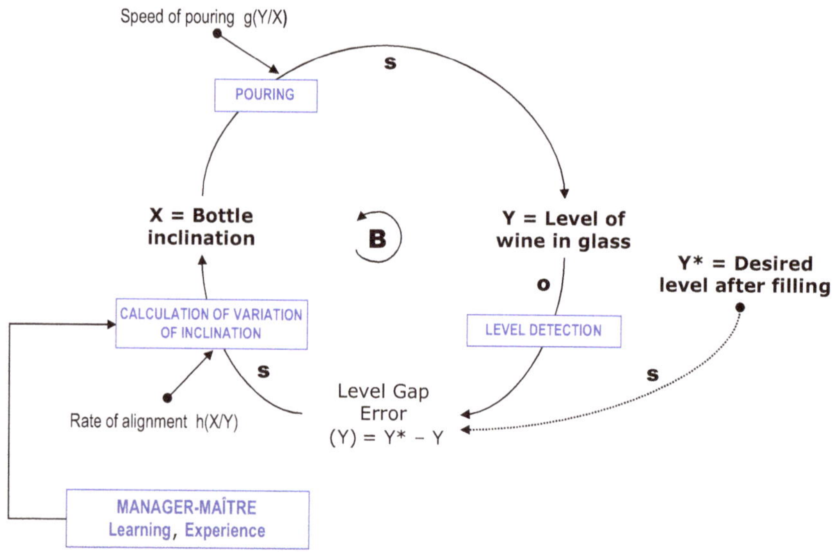

Fig. 3.15 Halt control system. Pouring champagne to the same level

spread in many personal electronic products, ranging from laptop computers to cell phones, and is replacing recognition by means of badges. Another recognition system, similar to AFIS, is *hand* recognition; an appropriately arranged scanner obtains an image of the hand using an infrared camera (which captures the image in black and white, ignoring surface details) and sends the scanned profile of the hand to the sensor apparatus, which compares it to the image from a previously assembled archive.

The high speed and reliability of recognition systems has made possible more ambitious *pattern recognition* techniques applied to the face or physique of individuals. Such systems are based on the fact that the face (or body structure) of a person can be recognized not by observing it in all its details but by referring to certain fundamental characteristics; for example, the color of the eyes or the distance between them; the shape and size of the nose and ears; the broadness of the forehead; the shape of the chin; hair color and styling.

An automatic *pattern recognition* system for the face is a control system that acquires images from some scanning system and identifies the proportions, distances, and ratios between the various somatic features. It forms a model that represents the objective (Y^*) to recognize and compares this with the models in the archive. Systems that recognize people by their retina or iris are very accurate. We have all seen machines that scan the retina to identify a person in order to authorize access to some device.

A DNA recognition system was created at the start of the 1990s involving data banks containing DNA profiles which, under the law, can be accessed and preserved, for example, the DNA of people who have committed certain types of crimes

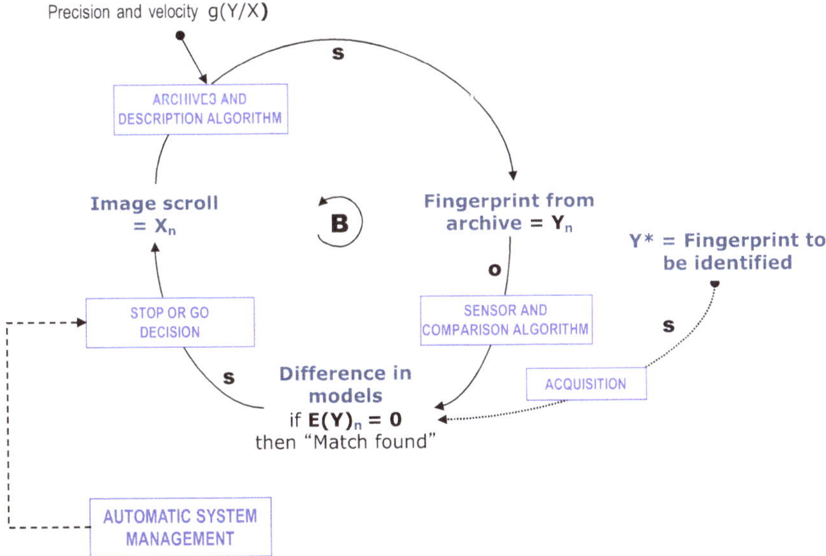

Fig. 3.16 Model of automated fingerprint identification system

and those who have had restrictions placed on their personal freedoms; DNA samples from biological evidence acquired during criminal proceedings; the DNA of missing people or their relatives and of unidentified corpses or body remains. A DNA sample taken at some site is analyzed and the profile is compared with those in the data base; if recognized, then the person it belongs to is identified. Not all countries have DNA data banks, which is understandable given the delicacy of the problem of the privacy protection of citizens and of the safekeeping of samples and access to them.

This type of control system functions just like that in the model in Fig. 3.16.

3.9.4 Explorative Systems

The systems of recognition presented in Sect. 3.2 also allow us to describe the functioning of *explorative systems*. Exploring an unknown environment of whatever kind, from a territory to the web, a bookstore to a just-purchased book, from craters on Mars to the starry firmament, signifies, in fact, scanning the objects of this environment and comparing these with an archive of models. The more the objects in the environment are identified, the better known the explored environment becomes. Since they are not in the archive, the unidentified objects are transformed into models (description, designs, photos, films, etc.) and then *included* in the archive, which gradually grows to make future explorations easier.

In order to understand the exploration process as an activity of a control system of *recognition*, let us imagine we have a dark room of unknown content which we are forced to explore; we move around without being able to see anything, only perceiving tactile sensations, for example, when we run into something, or sounds connected to these impacts. Our mind tries to identify the "objects" that could have caused these perceptions, and we can thus identify a sharp-cornered piece of furniture or the sharp edge itself, a glass that has fallen and shattered, a soft cushion, a carpet, etc.

The Center for Space Observation is exploring a swath of the night sky and "strange signals" arrive from space, which are captured by instruments. The astronomers immediately analyze these signals and look in their "atlas of celestial bodies" to identify objects that could be the source of the signals. Let us assume we are able to identify a pulsar located in the *XYZ* coordinates, which seems to be the source. The exploration has then finished. In the event no cataloged "object" can be linked to those signals, the astronomers would presumably add a report to their catalog: "Today at MM hours signals were identified with the following characteristics (position, mass, waves emitted, etc.) which we have named "H3b12." From that moment on, any astronomer identifying similar signals can use the catalog to recognize the sender of the signal.

If we substitute "astronomer" with "doctor" and "strange signals from space" with "strange symptoms of patient," you will find an explanation for the many papers found in medical journals. Likewise, if we substitute "astronomer" with "nature explorer in new lands" and "strange signals from space" with "unknown fauna and flora" we will find ourselves going back to Charles Darwin's *Voyage of the Beagle*. We have represented the *explorative* activity as the system of recognition and identification shown in Fig. 3.3, including, in the box at the lower right, the "instructions for updating the catalogue"; for example, in the following way:

$$\text{for } 1 \le n \le N, \quad \text{if } Y_n \ne Y^*, \quad \text{then } Y^* = Y_{N+1}.. \tag{3.1}$$

A variant of explorative systems are systems of *improvement*, characterized by an increase in the objective when the exploration identifies values for Y_n which are higher than those of the objective. Belonging to this type of system are *selecting systems*: an exploration of the environment leads to selecting the "best" object in terms of one of its qualities: the highest, sweetest, most flavorful, coldest, most stocked with merchandise, wealthiest, etc. This selection must occur through a comparison between the objects to be selected, which are identified by exploring the environment, and the temporary object considered to be the "best" and set as the objective. In order to transform a system of recognition and identification into a *selecting system* we must replace (3.1) in Fig. 3.3 by the following instruction:

$$\text{for } 1 \le n \le N, \quad \text{if } Y_n > Y^*, \quad \text{then } Y^* = Y_n. \tag{3.2}$$

This instruction differs from (3.1) in that it produces a continual increase in the objective in selecting other "best" objects.

3.9.5 Looking Up a Word in the Dictionary

The control system for looking up the word *Y** (Sect. 2.16.3) can operate by directly comparing the entire word with every word in the dictionary using the model in Fig. 2.21; however, as shown in Fig. 3.17, in order to do so it must read the first letter of *Y** and start scanning through the dictionary, page after page, until it comes to the words whose first letter is the same as that in *Y**.

It then reads the second letter and continues scanning until it arrives at the words whose second letters are the same as in *Y**, proceeding in this manner until it arrives at the last letter in *Y**. If the dictionary contains a word whose last letter is the same as the last letter of *Y**, then the search has ended and we can read the definition or translation.

3.9.6 Achilles and the Tortoise: Zeno's Paradox

Aristotle writes that, in order to support the idea that reality is made up of a single and immutable Being, Zeno of Elea proposed several paradoxes to show, absurdly, the illusion of both plurality and motion, despite what our daily lives clearly reveal to us.

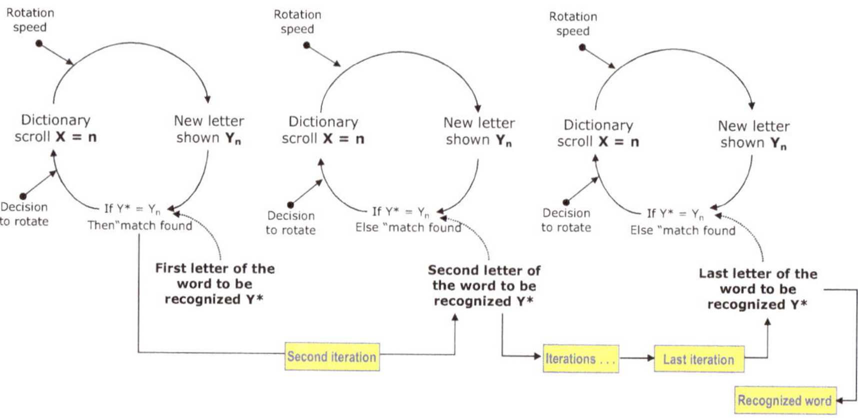

Fig. 3.17 Control system for recognizing a word in the dictionary by sequencing its component letters

The most famous of Zeno's paradoxes is probably that which argues against motion, whereby the philosopher wants to show that, despite the fact Achilles was faster than a tortoise, he could never have caught up to and passed it in a race if the latter had had a head start (the paradox is summarized by Smith (2013), in Prime). As we can see in the model in Fig. 3.18, the action variable, X_t, represents the amount of time Achilles decides to run after having observed the position of the tortoise at time "t"; the variable Y_t, on the other hand, is his position after having run the amount of time decided at the previous instant.

In fact, starting after an amount of time, let us say $t = n$, Achilles, in order to overtake the tortoise, would have had to arrive at where the latter was positioned at "n"; however, in the meantime the tortoise advances an additional, shorter distance, which Achilles must bridge. When Achilles reaches this new position the tortoise advances again some distance; this continuous pursuit translates into an infinite series of small mutual advancements, increasingly shorter in distance, so that from this perspective the gap between Achilles and the tortoise could never have been eliminated.

This is a classic example control systems of *pursuit*; in this case Achilles, the system's manager, has the objective of reaching the observed position of the tortoise (Y_t*) at every "t" by changing his position by means of decisions to run for a sufficient length of time to eliminate the gap.

Figure 3.19—even though it involves a somewhat reduced number of control cycles and of decimal places—shows that Zeno's example calls for its effects to occur punctually. However, this occurs because the system in Fig. 3.18 (as conceived of by Zeno) *is poorly designed*.

In order to modify the regulation decisions Achilles must rationally make, we need only transform the control system from a *variable objective* (the tortoise who is fleeing) into a *fixed objective* (the *finish line*) system.

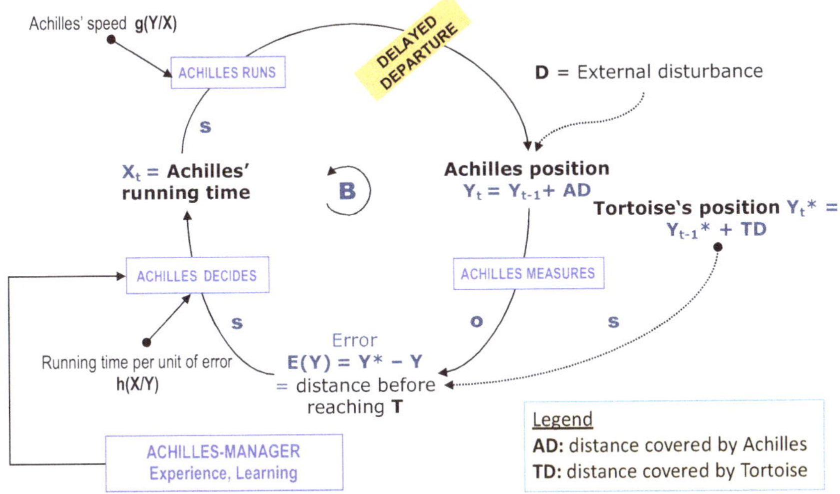

Fig. 3.18 Control system which Zeno had in mind. Achilles fails to pass the turtle

In this new system Achilles does not have to decide the *running time* needed to catch up to the tortoise but the *time needed to reach the finish line*, as illustrated in the model in Fig. 3.20. As we can easily see from the calculations, it is guaranteed that Achilles will surpass the tortoise, on the condition that his speed is greater than his adversary's and there is a sufficient distance to the finish line.

In fact, what is truly different in this second system with respect to the previous one is the procedure for quantifying the error: in the first case this is calculated with respect to the *variable objective* of the tortoise's position, while in the second in terms of the *fixed finish line* (objective).

3.9.7 Serial Systems: The Oven and the Boiler

Figure 3.21 represents a control system for how long food is cooked in an oven that can be regulated in terms of temperature and cooking time.

The oven temperature is, in turn, a variable controlled by a second control system represented by the boiler, which provides the oven temperature through two levers: the boiler temperature and the amount of time the boiler is turned on. It is not possible to cook food by lengthening the operating time of the oven, *tF*, if the oven temperature, *TF*, is too low; thus, in order to cook food we need a minimum oven temperature, *TF**, which is set as the objective of the second system (the boiler) which, through its two levers, allows the oven temperature, *TF*, to reach the minimum necessary to cook the food.

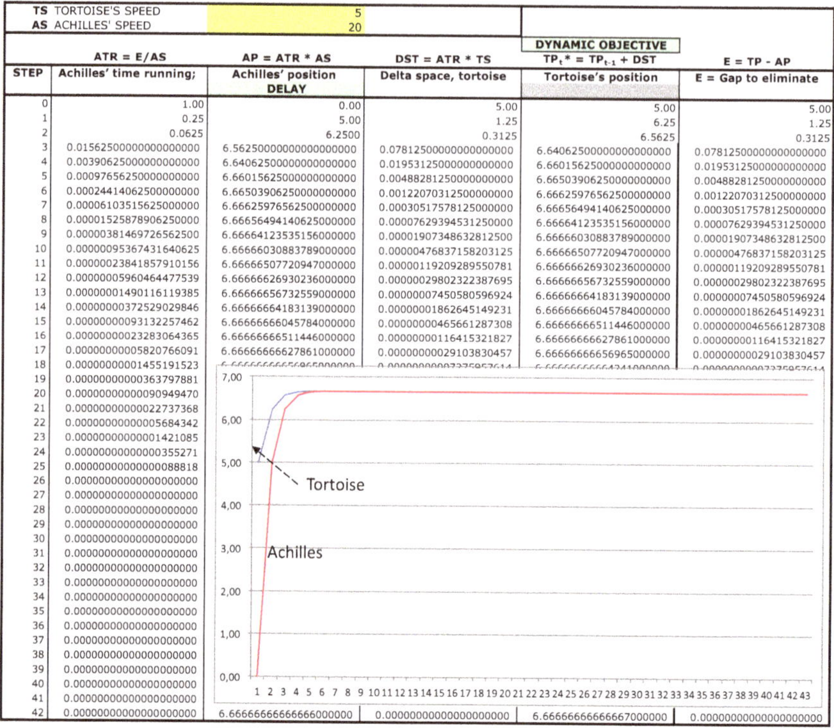

	TS TORTOISE'S SPEED	5			
	AS ACHILLES' SPEED	20			
	ATR = E/AS	AP = ATR * AS	DST = ATR * TS	DYNAMIC OBJECTIVE TP$_t$* = TP$_{t-1}$ + DST	E = TP - AP
STEP	**Achilles' time running;**	**Achilles' position DELAY**	**Delta space, tortoise**	**Tortoise's position**	**E = Gap to eliminate**
0	1.00	0.00	5.00	5.00	5.00
1	0.25	5.00	1.25	6.25	1.25
2	0.0625	6.2500	0.3125	6.5625	0.3125
3	0.01562500000000000000	6.56250000000000000000	0.07812500000000000000	6.64062500000000000000	0.07812500000000000000
4	0.00390625000000000000	6.64062500000000000000	0.01953125000000000000	6.66015625000000000000	0.01953125000000000000
5	0.00097656250000000000	6.66015625000000000000	0.00488281250000000000	6.66503906250000000000	0.00488281250000000000
6	0.00024414062500000000	6.66503906250000000000	0.00122070312500000000	6.66625976562500000000	0.00122070312500000000
7	0.00006103515625000000	6.66625976562500000000	0.00030517578125000000	6.66656494140625000000	0.00030517578125000000
8	0.00001525878906250000	6.66656494140625000000	0.00007629394531250000	6.66664123535156000000	0.00007629394531250000
9	0.00000381469726562500	6.66664123535156000000	0.00001907348632812500	6.66666030883789000000	0.00001907348632812500
10	0.00000095367431640625	6.66666030883789000000	0.00000476837158203125	6.66666507720947000000	0.00000476837158203125
11	0.00000023841857910156	6.66666507720947000000	0.00000119209289550781	6.66666626930236000000	0.00000119209289550781
12	0.00000005960464477539	6.66666626930236000000	0.00000029802322387695	6.66666656732559000000	0.00000029802322387695
13	0.00000001490116119385	6.66666656732559000000	0.00000007450580596924	6.66666664183139000000	0.00000007450580596924
14	0.00000000372529029846	6.66666664183139000000	0.00000001862645149231	6.66666666045784000000	0.00000001862645149231
15	0.00000000093132257462	6.66666666045784000000	0.00000000465661287308	6.66666666511446000000	0.00000000465661287308
16	0.00000000023283064365	6.66666666511446000000	0.00000000116415321827	6.66666666627861000000	0.00000000116415321827
17	0.00000000005820766091	6.66666666627861000000	0.00000000029103830457	6.66666666656965000000	0.00000000029103830457
18	0.00000000001455191523	[illegible]	[illegible]	[illegible]	[illegible]
19	0.00000000000363797881				
20	0.00000000000090949470				
21	0.00000000000022737368				
22	0.00000000000005684342				
23	0.00000000000001421085				
24	0.00000000000000355271				
25	0.00000000000000088818				
26	0.00000000000000000000				
27	0.00000000000000000000				
28	0.00000000000000000000				
29	0.00000000000000000000				
30	0.00000000000000000000				
31	0.00000000000000000000				
32	0.00000000000000000000				
33	0.00000000000000000000				
34	0.00000000000000000000				
35	0.00000000000000000000				
36	0.00000000000000000000				
37	0.00000000000000000000				
38	0.00000000000000000000				
39	0.00000000000000000000				
40	0.00000000000000000000				
41	0.00000000000000000000				
42	0.00000000000000000000	6.66666666666666000000	0.00000000000000000000	6.66666666666667000000	0.00000000000000000000

Fig. 3.19 A simulation of Zeno's paradox

3.9.8 *Qualitative Control Systems: Procedure to Determine the States of Variety of the Qualitative Variables*

In order for the *qualitative* control systems in Sect. 3.2 to function, it is necessary to identify or define appropriate "*scales of variety*," or *qualitative scales*, for the qualitative dimensions of color, flavor, smell, tactile sensation, etc., which enables a *qualitative state* of the variable to control, Y_t, to be compared with a *qualitative state* set as the objective, Y^*, as we clearly saw in the example of recognizing the replacement button for our clothing or the wine at a wine-tasting competition. The determination of the "states of variety" is not easy since, in undertaking the qualitative determination, it is not possible to apply procedures of enumeration, measurement, and metricization. We start by distinguishing between *direct* and *indirect* qualitative determination; in the first case the process to determine each qualitative dimension specified several elementary states, while indirect determination derives the states using operations carried out on elementary states.

Specifying the name, order of finish, license plate, or identity card number falls under direct determination. Specifying the state of purity of a diamond instead comes under indirect determination, since the state of purity is obtained by, for

TS	TORTOISE'S SPEED	10	Y* = Distance to the finish line	500	
AS	ACHILLES' SPEED	20			
	DELAY			FIXED OBJECTIVE	
	AD = t AS	TD = t TS	GAP = DT - DA	Y*	E = Y* - AD
STEP	DISTANCE TRAVELED BY ACHILLES DELAY	DISTANCE TRAVELED BY TORTOISE	Gap	Finish line	E = ACHILLES' GAP TO FINISH LINE
0	0.0	0.0	0.0	500.0	500.0
1	0.0	10.0	-10.0	500.0	500.0
2	0.0	20.0	-20.0	500.0	500.0
3	0.0	30.0	-30.0	500.0	500.0
4	0.0	40.0	-40.0	500.0	500.0
5	0.0	50.0	-50.0	500.0	500.0
6	20.0	60.0	-40.0	500.0	480.0
7	40.0	70.0	-30.0	500.0	460.0
8	60.0	80.0	-20.0	500.0	440.0
9	80.0	90.0	-10.0	500.0	420.0
10	**100.0**	**100.0**	0.0	500.0	400.0
11	120.0	110.0	10.0	500.0	380.0
12	140.0				360.0
13	160.0				340.0
14	180.0				320.0
15	200.0				300.0
16	220.0				280.0
17	240.0				260.0
18	260.0				240.0
19	280.0				220.0
20	300.0				200.0
21	320.0				180.0
22	340.0				160.0
23	360.0				140.0
24	380.0				120.0
25	400.0				100.0
26	420.0				80.0
27	440.0				60.0
28	460.0				40.0
29	480.0	290.0	190.0	500.0	20.0
30	500.0	300.0	200.0	500.0	0.0
31	520.0	310.0	210.0	500.0	-20.0

Fig. 3.20 Control system Achilles employs. Achilles surpasses the tortoise

example, combining the quality of the color, the degree of impurity, and the transparency of the diamond. While the color of a button is determined in a direct manner based on "color scales," the flavor of a wine represents a mix of a number of qualitative states which the wine taster is able to recognize.

Among the possible determination procedures, I will only mention here the following, all of which are general and widely applicable:

- Specific procedure
- Analogic procedure
- Diagnostic procedure
- Defining or conventional procedure

The *specific procedure* searches for the qualitative state of the dimension to determine through observations or specific research. It is applied, for example, in attributing and searching for names or the specific signs of the objects of observation, in identifying their composition, specifying their structure, and so on. This procedure is not simple at all. Consider the difficulties in identifying the name of a crime suspect who refuses to give information about himself, or in identifying the structure of a biological system (cellular, the human brain) or the composition of a chemical mix or compound.

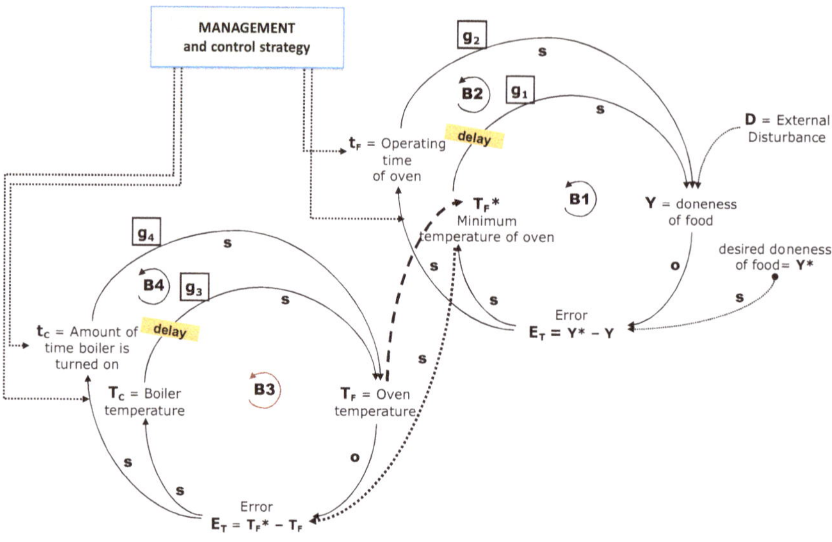

Fig. 3.21 Serial systems. The oven and boiler

The *analogic procedure* involves specifying the quality of a given object or phenomena by comparing this to a predetermined and known scale of varieties. This is the preferred procedure for qualifying colors, flavors, odors, noises, and so on. We have all learned the simplest scales of variety through education, which we refer every day. We say our car is red, that we have tasted food with a prevalently sour flavor, or that we have bought lavender soap. When we wish to be more precise we may specify that our car is "Ferrari red" or "blood red" and that the food tastes of bitter lemon or of lemon mixed with orange. The analogic procedure makes the determinations more precise by using a *standard scale* (chromatic, balance of flavors, olfactory, acoustic, etc., recognized in the context of a group of operators) which is referred to in order to specify, with the maximum precision that is possible, admissible and practical, to which element in the scale the dimension to determine can be analogically compared. The difficulties in doing so are well known to manufacturers of colors, paints, colored fabric, restorers, cooks, wine producers, perfume manufacturers, musicians, piano tuners, and so on.

A particular type of analogic procedure is undertaking a *comparison with a sample or prototype*," which is applied to specify the quality of complex objects (facial shape of a person, esthetic aspect of an object, structure, etc.) through a "line by line" comparison with a sample object, which represents the only state of comparison of a hypothetical scale of reference and which is set as the objective of the qualitative control system.

The *diagnostic procedure* occurs in all cases in which the "way of being" of a qualitative dimension is spelled out by a definition that indicates the element or elements characterizing the various states of the scale of variety. In order to specify the "way of being" of a qualitative dimension, specific surveys can be undertaken to identify "symptoms" or "signs" that can distinguish between the various states of the dimension. This procedure is adopted to identify the state of illness of a given individual or the state of insolvency of an enterprise. The diagnostic procedure is also widely used in the judicial sphere. In fact, many norms contain definitions of "status" (parent, owner, partner, firm, etc.) or define particular types of action (declaring, registering, informing, adopting, selling, paying off debt, etc.).

In many cases the qualitative dimensions appear so detailed and complex that their quality is conventionally decreed through appropriate *circumscribing*, or *conventional definitions.* In this case the determination of the "states of variety" of a dimension consist in verifying whether or not the determination conforms to the conventional definition. Consider how difficult it still is to ascertain whether or not a biological entity possesses the quality of "belonging to the living sphere." When is a biological structure deemed to be living (animated)? How can an exploratory system—a space probe on a new planet—recognize when it encounters a living biological being? Biologists have provided circumscribed definitions for the notion of "living," definitions that change along with scientific and technical progress which increase our capacity to understand phenomena and refine our observations.

The "macroscopic living being" (perceivable to the naked eye or by using modest magnifying lenses) gave way to Galileo's invention of the microscope that enabled man to recognize the "microscopic living being": single-cell organisms, bacteria, viruses, and so on. This made it necessary to deal with the problem of their inclusion in the sphere of the living (animated) being, which entailed modifying or refining the definition of "living." Equally as complex is ascertaining the quality "belonging to the category '*alive*'." When a biological structure belongs to the category of the living, is it *alive*? When does it cease being alive? This determination is particularly difficult in pathological or "cutting-edge" cases we can all imagine.

The above considerations and examples reveal how the procedures for qualitative determinations are not always individually applicable: instead the methods of qualification must often be used in combination. Moreover, the methods of qualitative determination, especially the definition procedure, often refer to specific ordinal or cardinal scales that allow the states of the scales of variety to be more accurately determined.

3.10 Summary

Single-lever and single-objective control systems can be divided into several interesting categories, of which the following are particularly relevant:

(a) Artificial and Natural Control Systems (Sect. 3.1)
(b) Manually Controlled systems, Automatic Control Systems, or Cybernetic Systems (Sect. 3.1)
(c) Quantitative and Qualitative Control Systems (Sect. 3.2)
(d) Attainment and Recognizing Systems (Sect. 3.2)

(e) Steering and Halt Control Systems (Sect. 3.3)
(f) Fixed- and Variable-Objective Systems (or systems of "pursuit") (Sect. 3.4)
(g) Collision, Anticollision, and Alignment Systems (Sect. 3.5)
(h) Tendential and Combinatory Control Systems (Sect. 3.6)
(i) Parallel or Serially Connected Systems (Sect. 3.7)
(j) Control Systems forming an Holarchy (Sect. 3.8)

Chapter 4
The Ring Completed: Multilever and Multiobjective Control Systems

Abstract Chapter 2 presented the general notion of a single-lever control system, illustrating the logical and technical structures of such a system. In addition, Chap. 3 proposed several important types of control systems that, though dissimilar in appearance, satisfy the definition and general logic of single-lever control systems. This chapter presents a broader view of control systems by introducing two important generalizations: (1) *multilever* control systems and (2) *multiobjective* control systems. Multilever control systems can have dependent or independent levers. Multiobjective control systems can be apparent or effective. Specifying the control strategies requires introducing the concept of *cost–benefit analysis* applied to the various levers. Specifying the control policies brings up the notion of *scale of priorities* regarding the various objectives.

Keywords Multilever control system · Multilayer control system · Multiobjective control systems · Control strategy · Control policy · Cost–benefit analysis · The direct comparison procedure · The standard gamble method

> There once were two watchmakers, named Hora and Tempus, who manufactured very fine watches. Both of them were highly regarded, and the phones in their workshops rang frequently—new customers were constantly calling them. However, Hora prospered, while Tempus became poorer and poorer and finally lost his shop. The watches the men made consisted of about 1,000 parts each. Tempus had so constructed his that if he had one partly assembled and had to put it down—to answer the phone, say—it immediately fell to pieces and had to be reassembled from the elements. [...] The watches that Hora made were no less complex than those of Tempus. But he had designed them so that he could put together subassemblies of about ten elements each. Ten of these subassemblies, again, could be put together into a larger subassembly; and a system of ten of the latter subassemblies constituted a whole watch. Hence, when Hora had to put down a partly assembled watch in order to answer the phone, he lost only a small part of his work, and he assembled his watches in only a fraction of the man-hours it took Tempus (Herbert Simon 1962, p. 90).

Chapter 2 presented the general notion of a single-lever control system, illustrating the logical and technical structures of such a system. In addition, Chap. 3 proposed several important types of control systems that, though dissimilar in

P. Mella, *The Magic Ring*, Contemporary Systems Thinking,
https://doi.org/10.1007/978-3-030-64194-8_4

appearance, satisfy the definition and general logic of single-lever control systems. This chapter presents a broader view of control systems by introducing two important generalizations: (1) *multilever* control systems and (2) *multiobjective* control systems. Multilever control systems can have dependent or independent levers. Multiobjective control systems can be apparent or effective. Generalizing further, we will examine the notion of *impulse systems*, which play a fundamental role in life.

In *multilever* control systems, especially those where the control levers can be activated independently of one another, the *manager* must first define the *control strategy* to be adopted—that is, programming the activation of the various levers to achieve the objective—by choosing the most appropriate levers in terms of the objectives to be achieved and their urgency, together with their activation order. In *multiobjective* systems, the choice of strategy is coupled with the definition of the *control policy*, which chooses the order of priorities regarding actions on the various objectives. Specifying the control strategies requires introducing the concept of *cost–benefit analysis* applied to the various levers. Specifying the control policies brings up the notion of *scale of priorities* regarding the various objectives.

4.1 Dual-Lever Control System with Mutually Dependent Levers

In order to make it easy to understand the general logic of control, the previous chapters presented *single-lever* control systems—where a single lever, X_t, is sufficient to direct the trajectory of the variable to control, Y_t, toward the objective Y^*—since these are relatively simple to illustrate, even though they often reveal complex dynamics.

It is time now to abandon the limitation of a single control lever and assume that the *passive* variable, Y_t, can be controlled through "*N*" *active* variables: $X1_t$, $X2_t$, …, Xn_t, … XN_t. We thus can form control systems with two, three, or multiple control "levers," which I shall define as *pluri-lever* or *multilever* systems.

As in single-lever systems, in multilever ones the values of the control levers depend on the amount of error $E(Y)$. Based on the manner in which the levers depend on $E(Y)$, we can divide multilever systems into two general categories: *monolayer* and *multilayer* systems.

To simplify the analysis of multilever control systems, it is appropriate to first examine *monolayer* systems, postponing until Sect. 4.5 the examination of *multilayer* ones. I shall begin with the simplest case of the *dual-lever* control system, where Y_t (effect) is controlled by $X1_t$ and $X2_t$ (cause).

I define *monolayer* control systems as those where all the levers are regulated by the manager based on the "general," or *first-level*, error: $E^1(Y)_t = Y^* - Y_t$ (we assume that there is a fixed objective). This means that the system can be described "simply" by generalizing the recursive eqs. (2.6), (2.7), and (2.8), which were introduced in Sect. 2.3. In order to simplify the notation (until the discussion of *multilayer* systems in Sect. 4.5), the first-level error will always be indicated by $E(Y)_t$ to avoid confusion.

Assuming that the *reaction times* are equal to "1" for both levers, the action rate "g" and reaction rate "h" are constant, $t = 0, 1, 2, \ldots$, and that there are no delays, then the new recursive equations for the *dual-lever* systems become

$$Y_n = [X1_n\ g_1 + X2_n\ g_2] + D_n \tag{4.1}$$

$$E(Y)_n = Y^* - Y_n \tag{4.2}$$

$$X1_{n+1} = X1_n + E(Y)_n\ h_1; \tag{4.3}$$

$$X2_{n+1} = X2_n + E(Y)_n\ h_2. \tag{4.4}$$

It is immediately clear from the above equations that the manager must carefully decide how to regulate the value of the levers; if he or she applies (4.3) and (4.4) in a cumulative manner (taking into account for both the amount of $E(Y)_n$) and also applies (4.1) with the new values for the levers, he or she could cause Y_{n+1} to become too large, thus making the control difficult, if not impossible.

Thus, the manager must decide in what *proportion* to regulate the two control levers by determining their *relative weights*, s_1 and s_2—$0 \le s_1 \le 1$, $0 \le s_2 \le 1$, $s_1 + s_2 = 1$—, in order to complete (4.1) as follows:

$$Y_n = [s_1(X1_n\ g_1) + s_2(X2_n\ g_2)] + D_n$$

Clearly, in order to control Y_t the manager must carry out a convex linear combination of the two levers. The vector $[s_1$ and $s_2]$ represents the *control strategy*.

Two cases are possible: the *levers* can be *mutually dependent* or *independent*:

(a) $X1_t$ and $X2_t$ are *mutually dependent* if the manager cannot regulate the former independently of the latter because the two levers vary in the "opposite" direction with respect to the variations in Y_t. Thus, a variation in $X1_t$ in sense "**s**" normally implies as well a variation in $X2_t$ in the opposite *direction* "**o**," even if the variations are different in size; this type of control is defined as a *constrained-variable* (or opposed) control.

(b) $X1_t$ and $X2_t$ are *independent* when management can decide whether or not to modify only the first, second, or both levers; there is thus a *free variable* (or concurrent) control.

The simplest example of a *mutually dependent dual-lever* control system is that which regulates the water level (Y_t) in a tub or a tank by the use of two opposite control "levers", as illustrated in Fig. 4.1. Lever $X1_t$ indicates the width of the drain opening and the amount of time it remains open; lever $X2_t$ is the rotation of the water faucet that regulates the water flow input. If the manager is rational, then it is

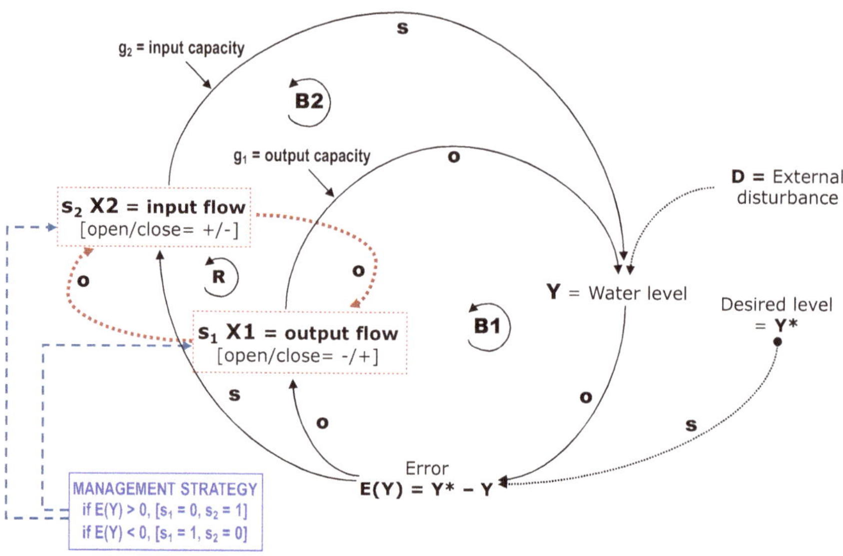

Fig. 4.1 Control of the water level with two *mutually dependent* levers

intuitively clear that when there is an inflow of water from an opening of the faucet there cannot at the same time be an outflow through the drain (unless there is some distraction). Similarly, when the water drains out, it is useless to open the water inflow faucet.

Thus, in this control system the manager must manipulate the faucet and drain in the *opposite direction of variation*, until one of the two is closed. To indicate this constraint, I have inserted in the CLD in Fig. 4.1 the *pair of dashed arrows*, in direction "**o**," representing the reinforcing *loop* indicated by [**R**], whose effect is to indicate to the manager that he or she must rotate the two control levers in the *opposite* direction.

The water level dynamics in the bathtub, with correlated inflows and outflows and discrete-variation levers, following the strategy shown in Fig. 4.1 of manipulating the input and output levers in the opposite direction, is illustrated in Fig. 4.2. The control panel at the top of Fig. 4.2 shows the parameters that indicate the construction specifications chosen by the designer (the speed and section of X1 and X2).

I define the output lever X1 that causes a decrease in Y_t as a *negative* lever; each notch of the output lever means that X1 reduces the water level with a negative outflow of $g_1 = 5$ units of water (maximum); this value derives from the product of two parameters: the *speed*, $vg_1 = 1$, that is, the intensity of the liquid flowing out for each notch in the open direction, and the *section*, $sg_1 = 5$, which instead indicates how many notches on the output lever can be kept open at any instant, thus expressing the maximum *section* of the drain hole. The product of the *section* and the *velocity*, $g_1 = 5$, quantifies the maximum *capacity* flow of the lever output, that is, the maximum amount of the outflow that reduces the water level at each instant.

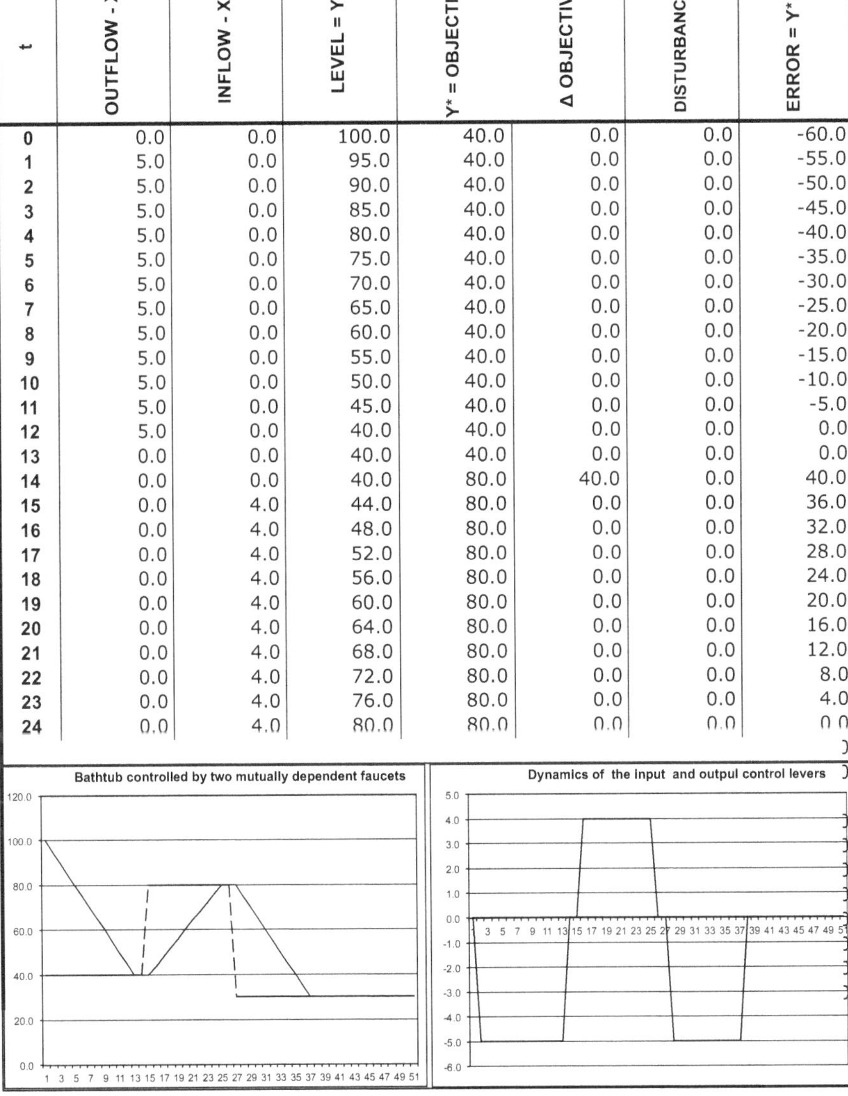

Levers			
output lever X1		**input lever X2**	
section	5	section	2
speed	1	speed	2
capacity	5	capacity	4

Y_0 =	100
Y* =	40

t	OUTFLOW - X1	INFLOW - X2	LEVEL = Y	Y* = OBJECTIVE	Δ OBJECTIVE	DISTURBANCES	ERROR = Y* - Y
0	0.0	0.0	100.0	40.0	0.0	0.0	-60.0
1	5.0	0.0	95.0	40.0	0.0	0.0	-55.0
2	5.0	0.0	90.0	40.0	0.0	0.0	-50.0
3	5.0	0.0	85.0	40.0	0.0	0.0	-45.0
4	5.0	0.0	80.0	40.0	0.0	0.0	-40.0
5	5.0	0.0	75.0	40.0	0.0	0.0	-35.0
6	5.0	0.0	70.0	40.0	0.0	0.0	-30.0
7	5.0	0.0	65.0	40.0	0.0	0.0	-25.0
8	5.0	0.0	60.0	40.0	0.0	0.0	-20.0
9	5.0	0.0	55.0	40.0	0.0	0.0	-15.0
10	5.0	0.0	50.0	40.0	0.0	0.0	-10.0
11	5.0	0.0	45.0	40.0	0.0	0.0	-5.0
12	5.0	0.0	40.0	40.0	0.0	0.0	0.0
13	0.0	0.0	40.0	40.0	0.0	0.0	0.0
14	0.0	0.0	40.0	80.0	40.0	0.0	40.0
15	0.0	4.0	44.0	80.0	0.0	0.0	36.0
16	0.0	4.0	48.0	80.0	0.0	0.0	32.0
17	0.0	4.0	52.0	80.0	0.0	0.0	28.0
18	0.0	4.0	56.0	80.0	0.0	0.0	24.0
19	0.0	4.0	60.0	80.0	0.0	0.0	20.0
20	0.0	4.0	64.0	80.0	0.0	0.0	16.0
21	0.0	4.0	68.0	80.0	0.0	0.0	12.0
22	0.0	4.0	72.0	80.0	0.0	0.0	8.0
23	0.0	4.0	76.0	80.0	0.0	0.0	4.0
24	0.0	4.0	80.0	80.0	0.0	0.0	0.0

Fig. 4.2 Dynamics of a bathtub controlled by two *mutually dependent* faucets

The parameters for the input lever X2—which produces the inflow—have a similar meaning. Since it represents a *positive* value, each notch on the input lever increases the water level by $g_2 = 4$ units, since X2 has a water *speed* of $vg_2 = 2$; the lever can be opened as far as the lever *section* $sg_2 = 2$ at each instant. The maximum *capacity* flow of the lever input is thus $g_2 = 4$ units per instant.

Since in Fig. 4.2 I have assumed the water level must be reduced from $Y_0 = 100$ to $Y^* = 40$ units (initial objective), the system manager acts rationally if, at $t = 1$, he or she opens the lever $X1_1$ to its maximum capacity, $g_1 = 5$ units (maximum admissible section and maximum admissible velocity); as a consequence, at $t = 1$ the level falls from $Y_0 = 100$ to $Y_1 = 95$ units of water.

The manager could have opened lever $X1_1$ by only one notch or chosen a lower velocity; but this would not have been rational, since it would have slowed down the achievement of the objective.

The level $Y^* = 40$ units is reached at $t = 12$. At $t = 14$ the governance changes the objective, which makes it necessary to increase the level from $Y^* = 40$ to $Y^* = 80$ units. Thus, the inflow lever must be used, which has a *water flow capacity* of $g_2 = 4$ units, thereby filling the tub at instant $t = 24$. To complete this simple simulation, we assume that at $t = 26$ the governance decides to decrease the water level from $Y^* = 80$ to $Y^* = 30$ units; this new objective is achieved at $t = 36$.

The graph of the water level dynamics can easily be produced, as shown in Fig. 4.2; the graph reveals the best *strategy* (see also Fig. 4.1) for varying the water level in the tub: opening only the output lever for water *outflow* or the input lever for water *inflow*. Nevertheless, the manager could have made other, illogical choices; for example, he or she could have tried to empty or fill the tub by simultaneously opening, for several periods, both the outflow and inflow levers. To avoid the risk of overflowing, designers usually give the outflow lever a greater overall capacity than the inflow one, as in the example in Fig. 4.2. To an outside observer such choices would appear irrational, because they are inefficient. Thus, the dynamics illustrated in Fig. 4.2 can be considered to be produced by an *automatic* control system that adopts the best *strategy* for achieving its objectives.

The simple control system for a tub with interdependent levers can be generalized in a number of ways, since it can be compared to a vast number of systems that entail a level that varies on the basis of inputs and outputs over different periods of time: warehouses, ships, stocks, etc. In Sect. 4.3, we will use this control system to simulate the behavior of other systems.

Technical note The control system in Fig. 4.2 is a *dual-lever system*, but if it had been designed to allow the manager to control both the *speed* and *section* of the two levers, then the system would have become a *four-lever system* composed of two *mutually dependent* pairs of levers: one pair that regulates the input flows and the other the output flows. However, within each pair, the levers of speed and section would be *independent* of each other, and the definition of the strategy to adopt would be less simple.

4.2 Dual-Lever Control System with Independent Levers: Control Strategy

The simplest example of *independent* "dual-lever" control system—as defined in point (b) of previous section—is the regulation of the shower temperature using two separate faucets: one, ($X1_t = H_t$), to regulate the flow of hot water, and the other, ($X2_t = C_t$), to regulate the cold water flow. The water temperature ($Y_t = W_t$) derives from a *mix* of H_t and C_t. If the water is too cold, the manager opens H_t or closes C_t; if it becomes too hot, he or she can close H_t or open C_t or adjust both faucets simultaneously until he or she obtains the desired temperature H^*.

Figure 4.3 represents the *logical structure* of a shower with two separate faucets to control water temperature, which is a general version of the model in Fig. 2.7; the hot and cold sensations felt by the person taking the shower generate the error temperature, which I have indicated as $E(W) = W^* - W_t$. The *governor-manager* of the system (the person taking the shower) can decide, after calculating and evaluating the temperature error (using his or her own skin), how to vary the rotation of the faucets, taking into account the temperature deviation rates, g_1 and g_2 (assumed to be constant), produced by each degree of rotation of the faucets.

Let us assume (though each of us may choose differently) that the *governor-manager*, upon feeling the water too hot, prefers to play it safe and *always* close the hot water faucet first, only subsequently opening the cold water one, if the first decision does not solve the problem; when the water is too cold, the manager chooses to open the hot water faucet and then regulate the cold water one. Thus, the manager has chosen a precise *control strategy* that can be expressed as follows: "if you feel a

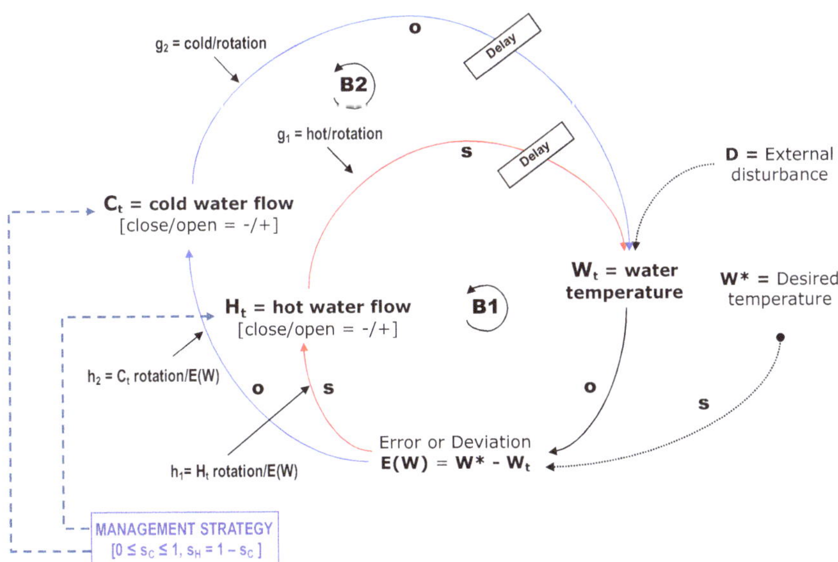

Fig. 4.3 Dual-lever temperature control

significant difference between the water temperature detected by your skin (W) and the optimal temperature (W^*), first regulate faucet H_t and only subsequently faucet C_t." How many of us would share this strategy? How many would prefer to act simultaneously on the two faucets so as to speed up the regulation process? Conclusion: The control with two *free variables always* implies a *strategy* that defines an *order of priorities* of action on the *control levers* or the choice of a *mix* of the two levers based on a specific vector $[s_1, s_2]$ that operationalizes the *control strategy*; these options do not depend on the structure of the control system but on the information available to the manager and on his or her experience.

The term *strategy* used in the context of control refers to "a pattern of regulation of the control levers established by the manager based on his experience," after having determined that what was successful in the past can lead to success in the future. This definition is perfectly in line with the classical term of strategy as "*the employment of battles to gain the end of war*" coined by the Prussian general, Carl von Clausewitz. According to this view, *strategy* is the set of actions (control levers) by which the army commander (manager), based on his knowledge and experience, uses the available arms and his soldier's skills (effector) to defeat the enemy in battle (control lever) in order to defend or conquer a territory for the well-being of his country (objectives).

Today, the term is used in a more general sense in all fields where it is necessary to decide how to activate a group of action variables (Xn) in order to achieve a set of objectives (Y^*) given the presence of adversaries or competitors or even other outside disturbances. It is used in this sense in *game theory* to indicate the *rule* or the *frequency* with which the various courses of action must be chosen during the successive repetitions of a game in order to maximize the winnings. The management of multilever systems must therefore develop strategic thinking.

> Strategic thinking is an intentional process easily lost amid the pressures of operational decision-making and tactical leadership. Strategic thinking focuses on finding and developing unique opportunities to create value by enabling a provocative and creative dialogue among people who can affect a company's direction. Strategic thinking is a way of understanding the fundamental drivers of a business and rigorously challenging conventional thinking about them, in conversation with others (Reynolds 2013, online).

In management theory, Johnson and Scholes define strategy, as follows:

> Strategy is the direction and scope of an organisation over the long-term: which achieves advantage for the organisation through its configuration of resources within a challenging environment, to meet the needs of markets and to fulfill stakeholder expectations (Johnson and Scholes 2002, p. 3).

Chandler also includes the determination of the objectives in the definition of strategy:

> Strategy can be defined as the determination of the basic long-term goals and objectives of an enterprise, and the adoption of courses of action and the allocation of resources necessary for carrying out these goals (Chandler Jr. 1962, p. 13).

The term is well suited to multilever control systems, since there is a group of control levers that must be manipulated in a sequence of "moves and countermoves" by assigning positive or negative values, with a certain frequency and with successive repetitions of the control process, so as to achieve the objective with the maximum efficiency, in the minimum amount of time, and with the minimum use of resources.

While in *manual* control systems the strategy must be decided each time and changed during the period of control, the creation of an *automatic* control system with *two independent levers* requires by nature that the *control strategy* adopted be defined *in advance.* An example of an *automatic* control system of water temperature is shown in the Excel simulation in Fig. 4.4, which presents a segment of the numerical dynamics—for 35 instants—of the water temperature variable, W_t, for a perfectly *symmetrical* shower, assuming two delays in adjusting the temperature and a reaction time of 1.

The dynamics are determined by the choices indicated in the initial control panel. As the control panel shows, I have assumed that each notch of the hot water faucet increases (if opened) or reduces (if closed) the water temperature by $g_1 = 1°$; likewise, the cold water faucet reduces (if opened) or increases (if closed) the temperature by $g_2 = 1°$. The control *strategy* was determined in this example as follows: since the two faucets can be simultaneously regulated, if the *governor-manager* wants to rapidly reduce the temperature from $W_0 = 35$ °C (initial temperature) to $W^* = 10$ °C (objective), he or she uses both the hot water "lever" (second column), which is closed 2.5 notches (with each notch having a range of 10°), and the cold water one, which instead is open 2.5 notches. A similar strategy is used if he or she needs to increase the temperature.

Nevertheless, we can observe that the use of the two levers is not perfectly symmetrical since, in this specific example, the user-manager prefers to regulate both levers to rapidly lower the temperature when it is too high and must be lowered; but when the temperature, at $t = 7$, reaches the "cold threshold" of $E(W)_7 = 3°$, he or she regulates the temperature by only acting on the hot water lever (closing it slowly) to avoid the risk of lowering the temperature below the objective.

Vice versa, when the temperature must be raised (which occurs in period $t = 18$), the manager manipulates both levers so that the water quickly heats up; but when the "hot threshold" of $E(W)_{24} \approx 5°$ is reached ($t = 24$) he or she prefers to reach the objective by using the cold water lever (closing it slowly) so as not to risk exceeding the objective. Of course, this is not the only possible strategy, but it reflects our natural cautionary tendency, in this case not to risk "exceeding" the tolerance limits of hot and cold.

The left-hand table in Fig. 4.5 indicates the dynamics of the water temperature. The right-hand table indicates the trends in the opening and closing of the faucets, which represent the two control levers.

I think it is important to emphasize that in many cases dual-lever control systems are "*badly designed,*" since they are programmed to always give preference to a single control lever, which causes errors in regulation; only intervention by an outside manager with considerable experience and knowledge can avoid the error that leads to irreversible damage to the behavior of the entire system.

H hot lever	1	C cold lever	-1	W_0 =	35
hot treshold	5	cold treshold	3	W* =	10
r(X)	1	capacity	10		

t	H = HOT LEVER	C = COLD LEVER	WATER TEMPERATURE	W* = OBJECTIVE	Δ OBJECTIVE	DISTURBANCES	ERROR = W* - W
0	0.0	0.0	35.0	10.0	0.0	0.0	-25.0
1	-2.5	2.5	35.0	10.0	0.0	0.0	-25.0
2	-2.5	2.5	35.0	10.0	0.0	0.0	-25.0
3	-2.5	2.5	30.0	10.0	0.0	0.0	-20.0
4	-2.0	2.0	25.0	10.0	0.0	0.0	-15.0
5	-1.5	1.5	20.0	10.0	0.0	0.0	-10.0
6	-1.0	1.0	16.0	10.0	0.0	0.0	-6.0
7	-0.6	0.6	13.0	10.0	0.0	0.0	-3.0
8	-0.3	0.0	11.0	10.0	0.0	0.0	-1.0
9	-0.1	0.0	9.8	10.0	0.0	0.0	0.2
10	0.0	0.0	9.5	10.0	0.0	0.0	0.5
11	0.0	-0.1	9.4	10.0	0.0	0.0	0.6
12	0.0	-0.1	9.4	10.0	0.0	0.0	0.6
13	0.0	-0.1	9.5	10.0	0.0	0.0	0.5
14	0.0	-0.1	9.5	10.0	0.0	0.0	0.5
15	0.0	0.0	9.6	10.0	0.0	0.0	0.4
16	0.0	0.0	9.6	10.0	0.0	0.0	0.4
17	0.0	0.0	9.7	10.0	0.0	0.0	0.3
18	0.0	0.0	9.7	30.0	20.0	0.0	20.3
19	2.0	-2.0	9.8	30.0	0.0	0.0	20.2
20	2.0	-2.0	9.8	30.0	0.0	0.0	20.2
21	2.0	-2.0	13.9	30.0	0.0	0.0	16.1
22	1.6	-1.6	17.9	30.0	0.0	0.0	12.1
23	1.2	-1.2	21.9	30.0	0.0	0.0	8.1
24	0.8	-0.8	25.2	30.0	0.0	0.0	4.8
25	0.0	-0.5	27.6	30.0	0.0	0.0	2.4
26	0.0	-0.2	29.2	30.0	0.0	0.0	0.8
27	0.0	-0.1	29.7	30.0	0.0	0.0	0.3
28	0.0	0.0	29.9	30.0	0.0	0.0	0.1
29	0.0	0.0	30.0	30.0	0.0	0.0	0.0
30	0.0	0.0	30.0	30.0	0.0	0.0	0.0
31	0.0	0.0	30.0	20.0	-10.0	0.0	-10.0
32	-1.0	1.0	30.0	20.0	0.0	0.0	-10.0
33	-1.0	1.0	30.0	20.0	0.0	0.0	-10.0
34	-1.0	1.0	28.0	20.0	0.0	0.0	-8.0
35	-0.8	0.8	26.0	20.0	0.0	0.0	-6.0

Fig. 4.4 Shower controlled by two independent and continuous faucets with two delays

We can become aware of this problem by considering, for example, a biological control system we constantly have to deal with, often in a rather problematic way: the control system of *thirst*, which can be represented by the model in Fig. 4.6. Among the many biological constraints in our body is that of the optimal or the normal *density standard* of salt, S, with respect to the volume of water, W, which is indicated in Fig. 4.6 by $D^* = (S/W)^*$; this represents the *objective* (or constraint) the system must achieve and maintain.

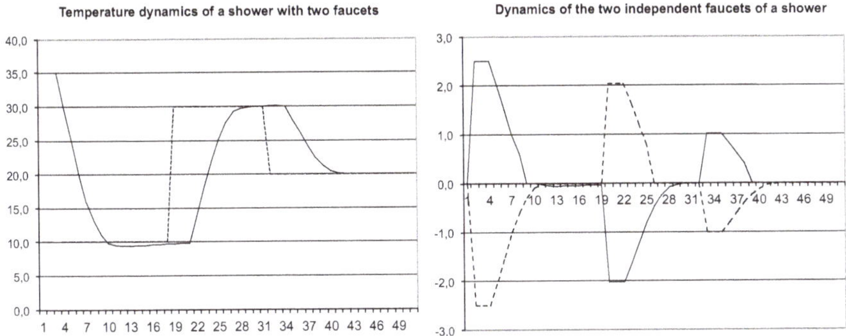

Fig. 4.5 Dynamics of the variables of the control system for the shower in Fig. 4.4

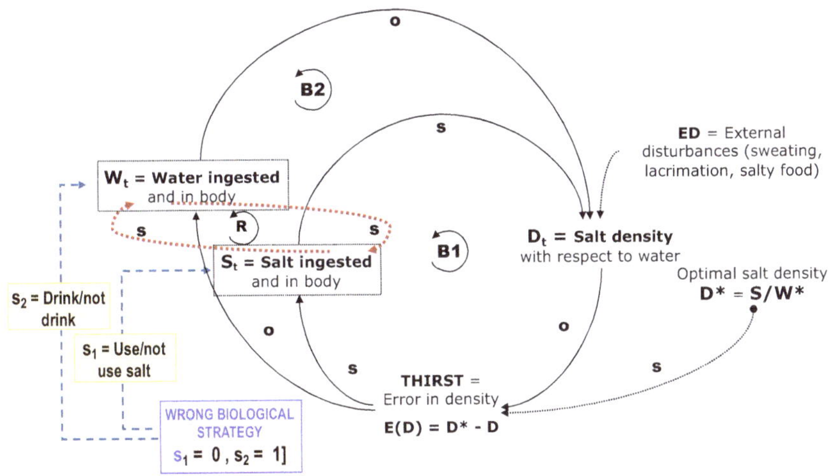

Fig. 4.6 Dual-lever salt water density control

When the effective salt density (D_t) does not coincide with the standard one (D^*), the error $E(D) = D^* - D_t$ manifests itself in our organism as a *sensation of thirst*. The higher the difference ($D^* - D_t$), the more intense is the *thirst stimulus*. Thus, Fig. 4.6 represents a *dual-lever* control system. The ratio D_t between salt and water can be controlled by regulating the lever S_t for the amount of salt ingested, thereby forming the *loop* [**B1**], or the lever for the quantity of water ingested, W_t, or drinking lever, thereby forming the *loop* [**B2**].

In many cases, the control system fails because it is "badly designed." In fact, the system cannot determine whether the *thirst* is caused by a scarcity/excess of water or by a scarcity/excess of salt. Thirst derives from $E(D)$, whether or not it is caused by the values of W or S.

Thus, thinking of our own experience, when we are thirsty we always use the single control lever W: "drink water." If the error, $E(D)$, is due to an excess of salt in the numerator (too much salt and too little water, because we have eaten salty foods), then by acting on the W_t lever we can reduce and eliminate thirst (we must take into account delays). However, if the sensation of burning thirst is from a lack of salt (we have sweated or cried too much, causing the salt to be expelled from our body along with the liquids), then the signal from the error $E(D)$ (too little salt and too much water) leads us, unfortunately, to drink again in order to try and eliminate our thirst. The intake of liquids further worsens D_t and increases the error, causing greater thirst and desire to consume liquids. This goes on until the system breaks down with … water indigestion. This is well known by people who live in warm-weather countries or who work in the sun; these managers know this simple strategy: "if, even when drinking, our burning thirst becomes unbearable, it is useless to insist on drinking; it is better to take a packet of minerals salts."

In the animal kingdom, this system appears to be well designed in some cases. Nocturnal porcupines, that well-known rodent with long, pointy spines, search out salty foods, licking and eating anything covered with salt, from household goods to tires that have gone over salt spread on the road! Elephants consume grasses, small plants, bushes, fruit, twigs, tree bark, and roots. To supplement the diet, elephants will dig up earth to obtain salt and minerals. Over time, African elephants have hollowed out deep caverns in a volcano mountainside on the Ugandan border to obtain salt licks and minerals (http://www.seaworld.org/animal-info/info-books/elephants/diet.htm). Finally, we can note the periodic migration of zebras in Africa: each year they cover hundreds of kilometers to reach the Botswana's salt pans in order to eat salt and stock up for the rest of the year.

Another control system that, though badly designed, operates in a perfectly logical manner is the one that regulates our body temperature when we have a fever. An increase in temperature should produce significant sweating; however, as we have all observed and experienced, a fever makes us feel cold and gives us chills, which, in serious cases in young children, even leads to feverish convulsions.

It appears that the control system for body temperature is not well designed. In fact, however, from a logical point of view it functions perfectly. We know that chills are one of the physiological levers to combat outside cold with respect to the normal body temperature. Chills serve to produce heat in the muscles when our sensors perceive too low an environmental temperature. If we reflect on it, even when we are feverish our temperature sensors perceive an environmental temperature below our body temperature, which has risen due to the fever; as a result the body chill lever is triggered along with the one that causes us to look for blankets to reduce the sensation of cold. Thus, when we have a fever the natural control system for body temperature appears to be badly designed, since it cannot distinguish whether or not the difference $E(Y)$ between the body temperature, Y^*, and the environmental temperature, Y, is due to the fact that the latter is too low with respect to our equilibrium temperature or that our body temperature is too high with respect to the environmental one. In order to overcome the poor functioning of this natural control system, the outside manager (doctor, family member, etc.) must intervene to activate the auxiliary levers to rectify the natural ones, for example, by placing cold water packs on the fevered body, taking off clothing, or taking medicine to lower the fever. Obviously, when the fever falls rapidly the natural control system acts in the opposite way to produce sweating and the sensation of heat, which causes the fevered individual to throw off his blankets.

4.3 Impulse Control Systems

The water tanks and shower operate with *two levers*, X1 and X2, that control Y through gradual variations over time; even if measured at discrete instants, the levers usually take on definite values for each $t = 0, 1, 2, \ldots$, which are included in the observation interval T. Such systems can be defined as *gradual* or *continuous functioning* control systems, depending on whether the control levers admit a discrete or a continuous functioning.

Let me now introduce another important class of dual-lever control systems: those where one control lever produces a sequence of inputs or outputs for Y_t over a certain period, while the other produces a single value, opposite in direction of variation and resembling an *impulse*, at the beginning or the end of the period. More precisely, a system operates by *impulses* when a control lever—for example, X1—has a nonzero value as soon as Y_t, or the variance $E(Y)_t$, reaches a given level (previously defined as the *objective*) while maintaining the value $X1_t = 0$ for every other level of Y_t or E_t. The second control lever, X2, instead takes on the value zero if the other variable has a nonzero value and has a value opposite to that of the objective when X1 becomes zero. This joint behavior produces a periodic variation of Y from zero to the value of the objective. To show this, let us take as an example an ideal warehouse where stocks of whatever kind are accumulated; these stocks increase by means of new inputs and decrease as a result of the output.

We let Y_t indicate the variable that expresses the stock's trends, X1 the *output* variable, and X2 the *input* variable. We assume that the two levers are interdependent, as in the case of the bathtub. Impulses can function in two ways: the initial stock is at a level equal to $Y_{\min}$ and is increased gradually by the X2 lever until it reaches the objective Y^*; the stock, through the action of the X1 lever, then takes on the level $Y_t = 0$, from which it begins to gradually increase until it once again reaches the objective Y^*. The cycle repeats itself until period T is over. We can call this ideal *impulse control system*, CI-PO, an acronym for *continuous input–point output*, which immediately calls to mind its operational method.

Figure 4.7 shows numerically the first two cycles of a control system that functions as I have just described. I have not included variations in objectives or disturbances, but these would only produce a temporary alteration of the cycles, which would regularly repeat themselves once the disturbance has ended. We observe from the control panel that the objective is $Y^* = 100$ and that the input variables X2, with a *velocity* of $v_2 = 2$ and a *section* of $s_2 = 10$, resupply the warehouses with shipments of *capacity* $g_2 = 20$ units, which produces the same increase in the stock for each t_h. When Y_t reaches $Y^* = 100$ units it is "discharged" with a single output produced by the X1 lever, whose capacity is $g_1 = 100$ units; the stock is eliminated and starts to increase anew.

Figure 4.8 illustrates numerically the consequences of raising the objective, which occur at $t = 9$, as well as of decreasing disturbances in the instants from $t = 18$ to $t = 22$. Despite these disturbances, the control system always succeeds in achieving the value $Y^* = 100$.

OUT	X1 velocity	10	X1 section	10	initial stock	Y_0 =	0
IN	X2 velocity	2	X2 section	10	point output	Y^* =	100

t	OUTPUT LEVER X1	INPUT LEVER X2	STOCK Y_T	OBJECTIVE Y^*	Δ OBJECTIVE	D = DISTURBANCES	ERROR $E(Y)_t = Y^* - Y^t$
0	0.0	0.0	0.0	100.0	0.0	0.0	100.0
1	0.0	20.0	20.0	100.0	0.0	0.0	80.0
2	0.0	20.0	40.0	100.0	0.0	0.0	60.0
3	0.0	20.0	60.0	100.0	0.0	0.0	40.0
4	0.0	20.0	80.0	100.0	0.0	0.0	20.0
5	0.0	20.0	100.0	100.0	0.0	0.0	0.0
6	-100.0	0.0	0.0	100.0	0.0	0.0	100.0
7	0.0	20.0	20.0	100.0	0.0	0.0	80.0
8	0.0	20.0	40.0	100.0	0.0	0.0	60.0
9	0.0	20.0	60.0	100.0	0.0	0.0	40.0
10	0.0	20.0	80.0	100.0	0.0	0.0	20.0
11	0.0	20.0	100.0	100.0	0.0	0.0	0.0
12	-100.0	0.0	0.0	100.0	0.0	0.0	100.0
13	0.0	20.0	20.0	100.0	0.0	0.0	80.0
14	0.0	20.0	40.0	100.0	0.0	0.0	60.0
15	0.0	20.0	60.0	100.0	0.0	0.0	40.0
16	0.0	20.0	80.0	100.0	0.0	0.0	20.0
17	0.0	20.0	100.0	100.0	0.0	0.0	0.0
18	-100.0	0.0	0.0	100.0	0.0	0.0	100.0

IMPULSE CONTROL SYSTEM. CI-PO

stock

point output

continuous input

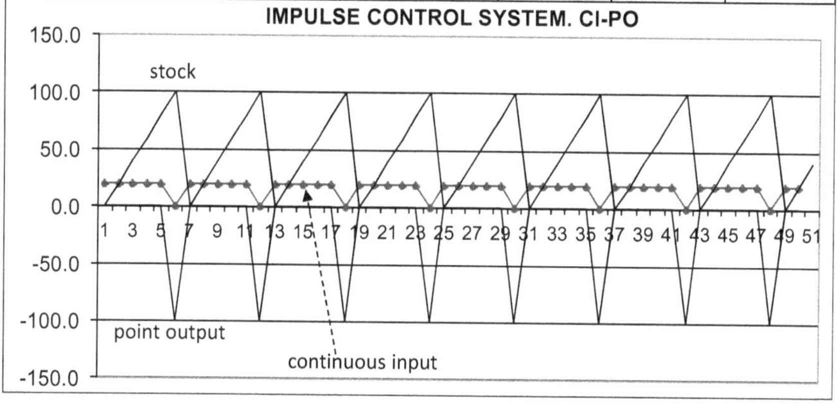

Fig. 4.7 Impulse control system with CI–PO functioning, discrete levers, and no disturbances

We can imagine an alternative form of functioning of the ideal warehouse viewed as a control system: the warehouse starts at a certain level and is gradually decreased through the X1 lever until it reaches the value $Y_t = 0$, after which it is increased by the X2 lever until it reaches the restocking value $Y_0 = 100$. After the increase, the stock then gradually diminishes through the action of the X1 lever. I propose naming this type of ideal control system a PI–CO system, which stands for *point input–continuous output*, in order to succinctly highlight its operational logic.

We can improve the model of the PI–CO system by adding the constraint that the resupply occurs not at stock level $Y_t = 0$ but at a positive level, $Y_{min} > 0$, which I call the *security level*; this takes on the significance of the objective Y^* that the system

OUT	X1 velocity	10	X1 section	10	initial stock	Y_0 =	0
IN	X2 velocity	2	X2 section	10	point output	Y* =	100

t	OUTPUT LEVER X1	INPUT LEVER X2	STOCK Y_T	OBJECTIVE Y*	Δ OBJECTIVE	D = DISTURBANCES	ERROR E(Y)$_t$ = Y* - Y^t
0	0.0	0.0	0.0	100.0	0.0	0.0	100.0
1	0.0	20.0	20.0	100.0	0.0	0.0	80.0
2	0.0	20.0	40.0	100.0	0.0	0.0	60.0
3	0.0	20.0	60.0	100.0	0.0	0.0	40.0
4	0.0	20.0	80.0	100.0	0.0	0.0	20.0
5	0.0	20.0	100.0	100.0	0.0	0.0	0.0
6	-100.0	0.0	0.0	100.0	0.0	0.0	100.0
7	0.0	20.0	20.0	100.0	0.0	0.0	80.0
8	0.0	20.0	40.0	100.0	0.0	0.0	60.0
9	0.0	20.0	60.0	150.0	50.0	0.0	90.0
10	0.0	20.0	80.0	150.0	0.0	0.0	70.0
11	0.0	20.0	100.0	150.0	0.0	0.0	50.0
12	0.0	20.0	120.0	150.0	0.0	0.0	30.0
13	0.0	20.0	140.0	150.0	0.0	0.0	10.0
14	0.0	10.0	150.0	150.0	0.0	0.0	0.0
15	-150.0	0.0	0.0	150.0	0.0	0.0	150.0
16	0.0	20.0	20.0	150.0	0.0	0.0	130.0
17	0.0	20.0	40.0	150.0	0.0	0.0	110.0
18	0.0	20.0	50.0	150.0	0.0	-10.0	100.0
19	0.0	20.0	50.0	150.0	0.0	-20.0	100.0
20	0.0	20.0	40.0	150.0	0.0	-30.0	110.0
21	0.0	20.0	40.0	150.0	0.0	-20.0	110.0
22	0.0	20.0	50.0	150.0	0.0	-10.0	100.0
23	0.0	20.0	70.0	150.0	0.0	0.0	80.0
24	0.0	20.0	90.0	150.0	0.0	0.0	60.0

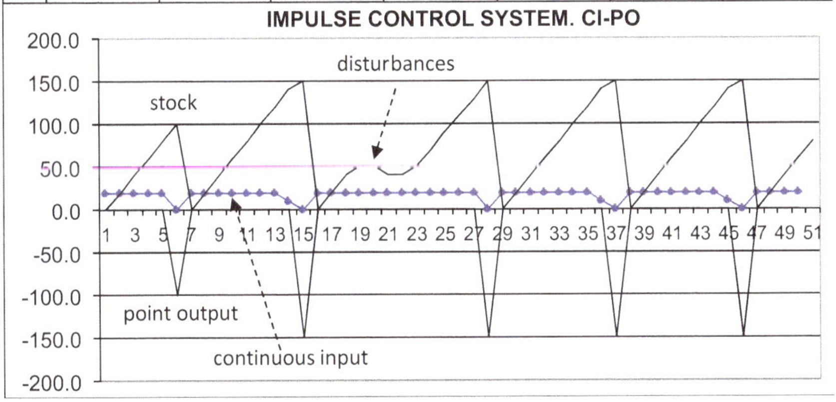

Fig. 4.8 Control system with CI–PO functioning, with disturbances

must achieve and maintain as a minimum level. Figure 4.9 shows numerically the first two cycles of a warehouse that functions in this way; it also assumes a variation in the objective in period $t = 18$.

At first sight, control systems that operate *by impulses* may seem somewhat abstract; but even after quick reflection we can easily realize how common they are. A typical CI–PO control system is our toilet. When we push the button to flush the

OUT	X1 velocity	3	X1 section	5	initial stock Y_0 =		100
IN	X2 velocity	10	X2 section	10	security level Y^* =		20
t	OUTPUT LEVER X1	INPUT LEVER X2	STOCK Y_T	OBJECTIVE Y^*	Δ OBJECTIVE	D = DISTURBANCES	ERROR $E(Y)_t = Y^* - Y^t$
0	0.0	0.0	100.0	20.0	0.0	0.0	-80.0
1	-15.0	0.0	85.0	20.0	0.0	0.0	-65.0
2	-15.0	0.0	70.0	20.0	0.0	0.0	-50.0
3	-15.0	0.0	55.0	20.0	0.0	0.0	-35.0
4	-15.0	0.0	40.0	20.0	0.0	0.0	-20.0
5	-15.0	0.0	25.0	20.0	0.0	0.0	-5.0
6	-5.0	0.0	20.0	20.0	0.0	0.0	0.0
7	0.0	100.0	120.0	20.0	0.0	0.0	-100.0
8	-15.0	0.0	105.0	20.0	0.0	0.0	-85.0
9	-15.0	0.0	90.0	20.0	0.0	0.0	-70.0
10	-15.0	0.0	75.0	20.0	0.0	0.0	-55.0
11	-15.0	0.0	60.0	20.0	0.0	0.0	-40.0
12	-15.0	0.0	45.0	20.0	0.0	0.0	-25.0
13	-15.0	0.0	30.0	20.0	0.0	0.0	-10.0
14	-10.0	0.0	20.0	20.0	0.0	0.0	0.0
15	0.0	100.0	120.0	20.0	0.0	0.0	-100.0
16	-15.0	0.0	105.0	20.0	0.0	0.0	-85.0
17	-15.0	0.0	90.0	20.0	0.0	0.0	-70.0
18	-15.0	0.0	75.0	40.0	20.0	0.0	-35.0
19	-15.0	0.0	60.0	40.0	0.0	0.0	-20.0
20	-15.0	0.0	45.0	40.0	0.0	0.0	-5.0
21	-5.0	0.0	40.0	40.0	0.0	0.0	0.0
22	0.0	100.0	140.0	40.0	0.0	0.0	-100.0
23	-15.0	0.0	125.0	40.0	0.0	0.0	-85.0
24	-15.0	0.0	110.0	40.0	0.0	0.0	-70.0

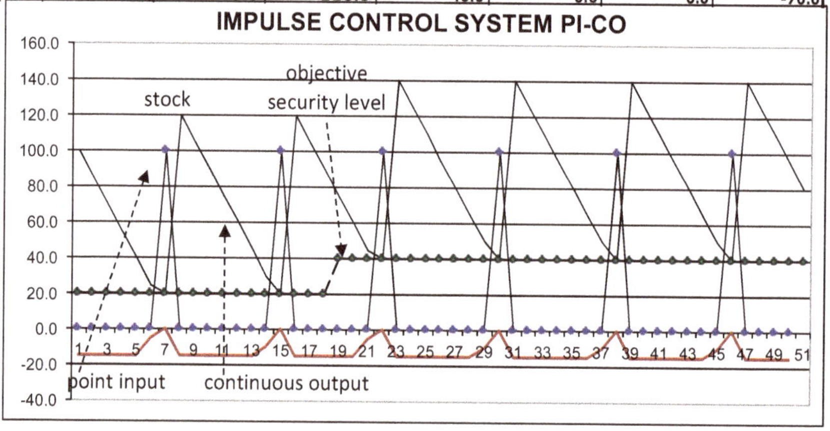

IMPULSE CONTROL SYSTEM PI-CO

Fig. 4.9 Control system with PI–CO functioning, without disturbances

toilet the water drains out almost instantaneously (PO); after the tank is emptied it starts to fill again (CI), until the sensor and regulation apparatus, already shown in Fig. 3.13, stop the water level from rising after reaching the maximum level for the tank.

The velocity with which the tank fills up depends, as always, on the *capacity* of lever X1, which in turn depends on the tube *section* and the water *velocity*. We can also view plans for capital accumulation as CI–PO control systems. Most CI–PO control systems are observed in the living world; fruit slowly grows (CI) with the gradual help of heat and nutrients (variable X2); it then matures (Y^*), falls from the tree, or ends up on some Table (PO). But the control system in the next seasonal cycle will once again transform the input of heat and nutrients into a new fruit.

The fact that, as a rule, pregnancy in humans lasts about 40 weeks is the clearest demonstration of the action of such systems, even if in this case the cycles are not automatic or regular.

PI–CO control systems are equally as common. Our pantry, fridge, and the gas tank in our car are under our daily observation. Unless we constantly visit hypermarkets or gas stations, we all supply ourselves with a stock of food, home products, and gas (PI) to last for several days; we use these goods regularly (CO), resupplying ourselves when their levels fall "dangerously" close to the security level, Y^*.

A good meal supplies us with the energy (PI) for several hours of work, during which it is used up (CO), thus making us desire a new meal (PI). This cycle repeats itself throughout our lifetime.

Systems that function not by impulses but gradually can, by extension and when appropriately construed, be viewed as *continuous input–continuous output* (CI–CO) systems.

4.4 Multilever Control Systems

The considerations in the preceding *sections* for *dual-lever* control systems are also valid for *multilever* systems with any number of levers, all or in part *mutually dependent* or *independent*. Obviously, the greater the number of levers, the more the rational manager must determine an effective *strategy* that takes into account the urgency of achieving the objective and the *capacity* of each lever, in addition to the cost of regulation (for these aspects, see Sect. 4.7). Control without a strategy means "blindly navigating"; control is by "trial and error," and it is not easy to optimize the time required to achieve the objective. Often control systems are multilever because the levers have a limited range of action and, when the error is large, must be employed contemporaneously (Fig. 4.10) or in succession (Fig. 4.11) to achieve the objective. In these cases, the strategy depends on the possibility of regulating the levers.

An example of this type of control is the *spontaneous control system* for focusing our eyes on objects in a horizontal plane with respect to the normal vertical position of the bust, based on the model in Fig. 4.10. This system is also suited to objects in any area that can be taken in by our gaze; however, if we wish to limit ourselves to the horizontal plane, the control system in Fig. 4.10 requires the action of only four control levers.

As we can all verify, when we have to focus on an object in a horizontal plane with about a 20° radius to the left or the right with respect to the center, we are

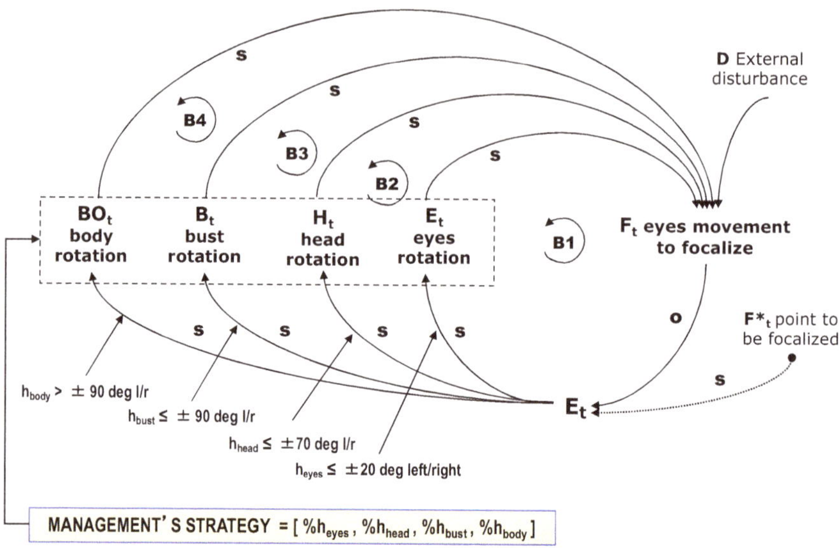

Fig. 4.10 Focusing on an object in a horizontal plane with a multilever control system

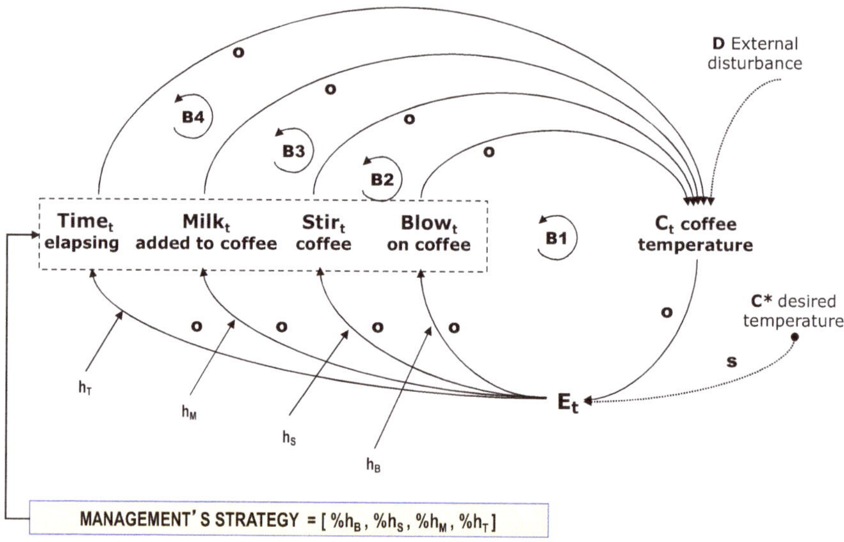

Fig. 4.11 Reducing the temperature of a liquid with a multilever control system

"naturally" inclined to horizontally rotate our eyeballs (E_t); for objects further to the outside, within about a 70° radius, focusing also requires the rotation of the head (H_t). For objects even further to the outside, up to about a 90° radius left or right, we must rotate our bust (B_t). Beyond this limit, if we want to focus our gaze on an object "behind" us we are forced to turn our entire body (BO_t). Under normal conditions we all adopt this *natural strategy* of four-lever regulation: first rotating the eyes, then the head, then the bust, and finally, if necessary, our body, according to the percentages to apply to the reaction rates "h" to regulate the various levers, as shown in the lower part of Fig. 4.10. In the model in Fig. 4.10, the levers are normally regulated in *cumulative fashion* by acting jointly in the order indicated by the strategy in relation to the amount of the error.

However, if the manager tries to regulate the system under "non-normal" conditions, he or she may be forced to find a different strategy. When I have a stiff neck, I try to rotate my eyes as much as possible and then rotate my upper body, thereby avoiding rotating my head so as not to feel cervical pain. When I have a bad back, I try not to activate the lever for rotating my bust, preferring instead to rotate my body by moving my feet. When I am driving my car, I am quite limited in regulating the lever that rotates my bust, and the lever for rotating my body is completely deactivated. Fortunately, the latter lever is usually substituted by the rotation of the eyes, which is possible, thanks to the *supplementary lever* of the rear-view mirror, which produces a change in perspective and scale.

Another multilever system that we use daily is the control of the temperature of a liquid or of food that is too hot to be ingested, since its temperature exceeds that which our organism recognizes as the tolerance limit. Figure 4.11 represents a possible configuration of the control system for lowering the temperature of coffee served to us boiling hot.

When we bring the cup to our mouth and our sensors perceive the temperature of the coffee (C_t), which is too high compared to the desired temperature (C^*), we spontaneously activate an initial lever: blowing on the coffee to cool it (BLOW lever); a second, alternative lever is to stir the coffee several times with a cold spoon, thus exploiting the lower ambient temperature (STIR lever). If we are in a rush, we can also add cold milk (MILK lever). On the contrary, if time allows we can simply wait for the necessary amount of time for the coffee to dissipate its incorporated heat to the atmosphere, thus lowering the temperature (TIME lever). These four levers can be used in succession, or one can be used as an alternative to another, or in a *mix* decided on by the manager-user. During a quick coffee break, we would normally activate the BLOW and MILK levers. During a gala, no one would dream of blowing on his or her coffee or stirring it repeatedly. The preferred lever could be to allow the time to go by.

We should note the difference between the control system in Fig. 4.10 and that in Fig. 4.11. The system in Fig. 4.10 is a typical *attainment* control system, whose levers can have positive and negative values. The system in Fig. 4.11 is a typical *halt* system; if the coffee becomes too cold, it is undrinkable and the levers can be regulated, individually or jointly, in a single direction only (once the cold milk is added it cannot be removed).

A typical multilever control system that is particularly interesting because it is common in nature and has an infinite number of applications is that which controls the movement of a point toward an objective in an "N-dimensional space." For a space of $N = 4$ dimensions, defined by the coordinates $S = (x, y, z, w)$ placed over four theoretical axes, the system can be described as a control system to shift a point

$P_t = (x_t, y_t, z_t, w_t)$, defined in S, toward a different point, assumed to be the objective, $P^* = (x^*, y^*, z^*, w^*)$, assumed to be fixed. In fact, the shifting of P toward P^* in the four-dimensional space entails the gradual elimination of the distance between P^* and P, according to the typical logic of control (Sect. 2.1). In order to achieve the objective, the control system must modify the position of the "intercepts" in each of the four theoretical axes.

A relatively simple, holonic-like process (Sect. 3.8) to produce the dynamics of the coordinates of P is to transform the single control system into four second-level control systems, considering each of the coordinates of the primary system objective, P^*, as a basic objective of the second-level systems. The control system for shifting P_t toward P^* is thus divided into four single-lever control systems, each of which is able to control the dynamics of a single coordinate of P_t, thereby determining and eliminating the error of the coordinate with respect to the value of the objective embodied in P^*. The result is the gradual shifting of P toward P^*, as shown in Fig. 4.12, which takes place over 101 periods with the parameters indicated in the control panel.

We must decide if and when P will reach P^* in order to allow the lower level systems to stop acting. For this we need to only go back to the first-level control system and calculate, for each instant, the Euclidean distance between P^* and P, representing the error $E(P)$, which is defined as

$$P^* - P = \sqrt{\left(x^* - x_t\right)^2 + \left(y^* - y_t\right)^2 + \left(w^* - w_t\right)^2 + \left(z^* - z_t\right)^2}$$

The *four-lever* first-level control system stops when $E(P) = P^* - P = 0$, which occurs at $t = 100$, when the four *single-lever*, second-level control systems, each acting independently, eliminate the distances in each coordinate (not shown in Fig. 4.12), thereby allowing the first-level system to shift the point from P to P^*.

In fact, Fig. 4.12 shows how the control system shifts the point of coordinates $P_0 = (0, -20, -50, -20)$ toward the *objective* point of coordinates $P^* = (180, 160, 150, 100)$ by means of the four second-level systems, each of which produces a shift along the axes, with the action rates "g" specified in the system's control panel; different action rates have been chosen for each pair of axes. The graph at the bottom of the table shows the dynamics of point P_t as it moves toward P^*, which is reached when the control system eliminates the distance between the two points. This movement is also shown by the curve representing the distance of P_t, calculated with *respect to the origin*. With these settings, the four variables—x_t, y_t, z_t, w_t—are, in all respects, the four control levers the manager of the first-level control system regulates to produce the movement of P_t toward P^* by activating the second-level systems.

What strategy can the manager use to regulate the four levers (x_t, y_t, z_t, w_t) that control P? There are a number of possible strategies. Management could decide to modify the position of P by acting first on the "x" coordinate, then on "y," and subsequently on the other coordinates. It could also begin the movement of "w" and then act on "z," and so on. In the system in Fig. 4.12, it was decided to activate all

		x	y	z	w	
	P	0	-20	-50	-20	initial position of point P
	P*	180	160	150	100	position of the objective point P*
action rates -	g	4	4	2	2	

	Translators				Point coordinates				Objective coordinates				Distance of coordinates				Dynamics		
t	g_X = traslat X	g_Y = traslat Y	g_Z = traslat Z	g_W = traslat W	x	y	z	w	x*	y*	z*	w*	dx=x*-x	dy=y*-y	dy=y*-y	dw=w*-w	distance of P from P*	Distance of P* from origin	Distance of P from origin
0					0	-20	-50	-20	180	160	150	100	180	180	200	120	345.25	300.83	57.45
1	4	4	2	2	4	-16	-48	-18	180	160	150	100	176	176	198	118	339.23	300.83	53.85
2	4	4	2	2	8	-12	-46	-16	180	160	150	100	172	172	196	116	333.23	300.83	50.79
3	4	4	2	2	12	-8	-44	-14	180	160	150	100	168	168	194	114	327.23	300.83	48.37
4	4	4	2	2	16	-4	-42	-12	180	160	150	100	164	164	192	112	321.25	300.83	46.69
5	4	4	2	2	20	0	-40	-10	180	160	150	100	160	160	190	110	315.28	300.83	45.83
6	4	4	2	2	24	4	-38	-8	180	160	150	100	156	156	188	108	309.32	300.83	45.83
	[omitted]																		
44	4	4	2	2	176	156	38	68	180	160	150	100	4	4	112	32	116.62	300.83	247.75
45	4	4	2	2	180	160	40	70	180	160	150	100	0	0	110	30	114.02	300.83	253.97
46	0	0	2	2	180	160	42	72	180	160	150	100	0	0	108	28	111.57	300.83	254.85
47	0	0	2	2	180	160	44	74	180	160	150	100	0	0	106	26	109.14	300.83	255.76
48	0	0	2	2	180	160	46	76	180	160	150	100	0	0	104	24	106.73	300.83	256.69
49	0	0	2	2	180	160	48	78	180	160	150	100	0	0	102	22	104.35	300.83	257.66
50	0	0	2	2	180	160	50	80	180	160	150	100	0	0	100	20	101.98	300.83	258.65
	[omitted]																		
60	0	0	2	2	180	160	70	100	180	160	150	100	0	0	80	0	80.00	300.83	270.00
61	0	0	2	0	180	160	72	100	180	160	150	100	0	0	78	0	78.00	300.83	270.53
62	0	0	2	0	180	160	74	100	180	160	150	100	0	0	76	0	76.00	300.83	271.06
	[omitted]																		
96	0	0	2	0	180	160	142	100	180	160	150	100	0	0	8	0	8.00	300.83	296.92
97	0	0	2	0	180	160	144	100	180	160	150	100	0	0	6	0	6.00	300.83	297.89
98	0	0	2	0	180	160	146	100	180	160	150	100	0	0	4	0	4.00	300.83	298.86
99	0	0	2	0	180	160	148	100	180	160	150	100	0	0	2	0	2.00	300.83	299.84
100	0	0	2	0	180	160	150	100	180	160	150	100	0	0	0	0	0.00	300.83	300.83

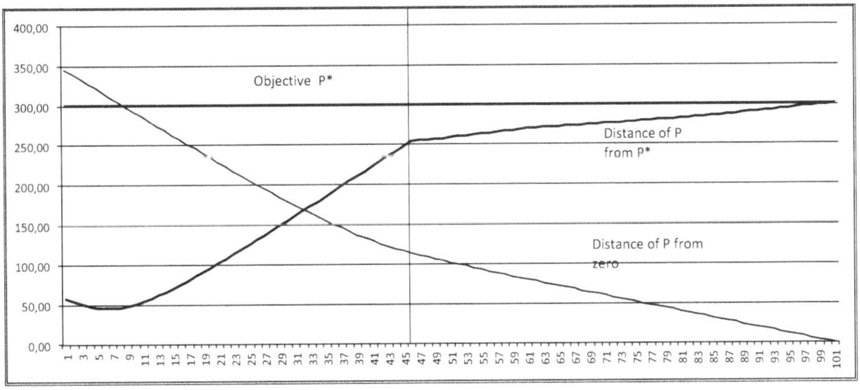

Fig. 4.12 Dynamics of a point P that shifts toward an objective P^*

four of the levers of P using the maximum value of "g" permitted for each (last line in the control panel). At time $t = 45$ the variables "x" and "y," which have higher *action rates* than the others ($g_x = 4$ and $g_y = 4$, as shown in the control panel), reach the objective values $x^* = 180$ and $y^* = 160$ (see the horizontal line at line $t = 45$). Thus, in periods $t > 45$ their value is held at "0," since no further shift is necessary. The other two levers, with lower *action rates* ($g_z = 2$ and $g_w = 2$, as shown in the control panel), continue their trajectory until $t = 100$, when they reach their objective value, $w^* = 150$ and $z^* = 100$ (objective coordinates at P^*). Since the four second-level systems have attained their objectives, even the *four levers* stop acting. Precisely because of the differing dynamics of the four levers of the second-level systems, at $t = 45$ the curve representing the distance of P from zero, together with that for the distance of P^* from P, shows a sudden variation.

If all the coordinates had infinitesimal and equal action rates, the optimal strategy would equate to the shifting of P_t toward P^* with a continuous straight movement along a straight line passing from P to P^*. The presence of "obstacles" between P and P^* and the use of different action rates might require the adoption of other strategies.

A concrete example of a control system that operates like the one in Fig. 4.12, in only a *three-dimensional* space, is the control of the position of the "movable point" of a crane that must move a load of bricks from point $P_0 = (x_0, y_0, z_0)$ to point $P^* = (x^*, y^*, z^*)$, for example, to the top of a roof. The moving point, $P_t = (x_t, y_t, z_t)$, shifts from P_0 to P^* by acting on the three coordinates through the angular rotation of the crane shaft (x_t), the horizontal shifting of the carriage (y_t), and the vertical shifting of the pulley (z_t). The control of these coordinates is carried out through the three single-lever, second-level control systems, which produce the variation in the P_t coordinates. Each of these is activated by a different effector (motor) and has a different *action rate*: the rotation of the crane is slow while that of the shifter of the horizontal and vertical axes is faster. The control *strategy* adopted by the crane operator to shift the load depends on the starting and ending position, but above all on his or her personal preferences and the constraints imposed by the "operations manual," which indicates the safety procedures to follow.

A new variant is the four-dimensional control system, which adds a fourth time dimension to the three spatial ones (w = time). Any vehicle that travels in three-dimensional space in order to move from one point in space to another in a fixed time interval acts in this manner (Sect. 4.8.6).

4.5 Multilayer Control Systems

The preceding *sections* presented *multilever* control systems whose multiple levers are regulated by the manager to eliminate the distance $E(Y)$ between Y_t and the objective Y^* through an appropriate *strategy*. The systems we have studied so far are *single-layer* ones (Sect. 4.1) in that all the levers have the capability to eliminate the "general" or *first-level* error, $E^1(Y)_t = Y^* - Y_t$. I shall define these levers as *first-order levers*, as they can all be regulated to eliminate $E^1(Y)_t$ (henceforth I shall also indicate the first-level error simply as $E(Y)_t$, since the "general" error is, by definition, a *first-level* error).

There are many cases where the *first-level* error is so large that the action of the *first-order* levers at our disposal is not able to eliminate it, thus leaving behind a residual error I shall define as a *second-level* error: $E^2(Y)_t = 0 - E^1(Y)_t$. This expression clearly indicates that if $E^1(Y)_t$ is permanently different from zero, the "distance" needed to completely eliminate the error is $E^2(Y)_t = 0 - E^1(Y)_t$, a value that represents, in fact, the *second-level* error. In order to eliminate $E^2(Y)_t$, we must activate an *additional* lever, which I define as the *second-order* levers, which are added to the *first-order* levers and are regulated only when the latter are insufficient. If there is still a residual *third-level error*, $E^3(Y)_t = 0 - E^2(Y)_t$, after activating the *second-order* levers, then it is necessary to activate *third-order* levers, continuing on in this manner until the residual error is eliminated, if possible.

I shall define a *multilayer* control system as one setup to eliminate the overall error, $E^1(Y)$, by dividing it into different layers, $E^2(Y)_t$, $E^3(Y)_t$, etc., each controlled by levers at successive orders; in this way the control system is divided into various *layers*, each of which use successive order levers to eliminate a portion of the error, as shown in the general model in Fig. 4.13.

Figure 4.13 illustrates a multilayer control model that, though abstract, can be characterized as follows:

(a) The model includes a first-level control obtained through the two action variables, X1a and X1b, which I have assumed to have the opposite dynamic effect on the variable to control, Y_t; two balancing loops are thus produced: [**B1a**] and [**B1b**].
(b) The model includes successive second-, third-, and fourth-level controls that are activated when management observes that the control at the preceding level is not able to eliminate the error; each error of a certain level derives from the error of the preceding level, when the latter does not reach the desired value of "zero error"; appropriate order levers are activated at each level control.
(c) The controls at the subsequent levels produce the balancing loops [**B2**], [**B3**], and [**B4**]; I have indicated two possible directions of variation, "**s/o,**" on the arrows that start from the successive deviations; these arrows must connect with the adjacent ones with direction "**s/o.**"
(d) We can define as an *ordinary* or an *operational* control that which is carried out at the *first level* through the *first-order* levers, X1a and X1b, which are *ordinary* levers.
(e) We can define as a *reinforcing* control that which is carried out at the *second level* through the *second-order* lever, X2.
(f) The *third-order* levers, X3, at the *third level*, are usually levers that produce an *extraordinary* control made necessary when the levers from the preceding levels are insufficient.
(g) The levers from the other, successive levels are usually those that modify the structure itself of the apparatuses that produce the control and for this reason are defined as *structural* levers.
(h) There are many possible delays for each control variable (which I have not indicated in Fig. 4.13) whose presence can make the control system complex, in the sense that the behavior of the variable to control can have unpredictable, even chaotic dynamics.

Therefore, in order to define the control strategy, the manager must decide to which level the control must be extended, also taking into account the fact that the

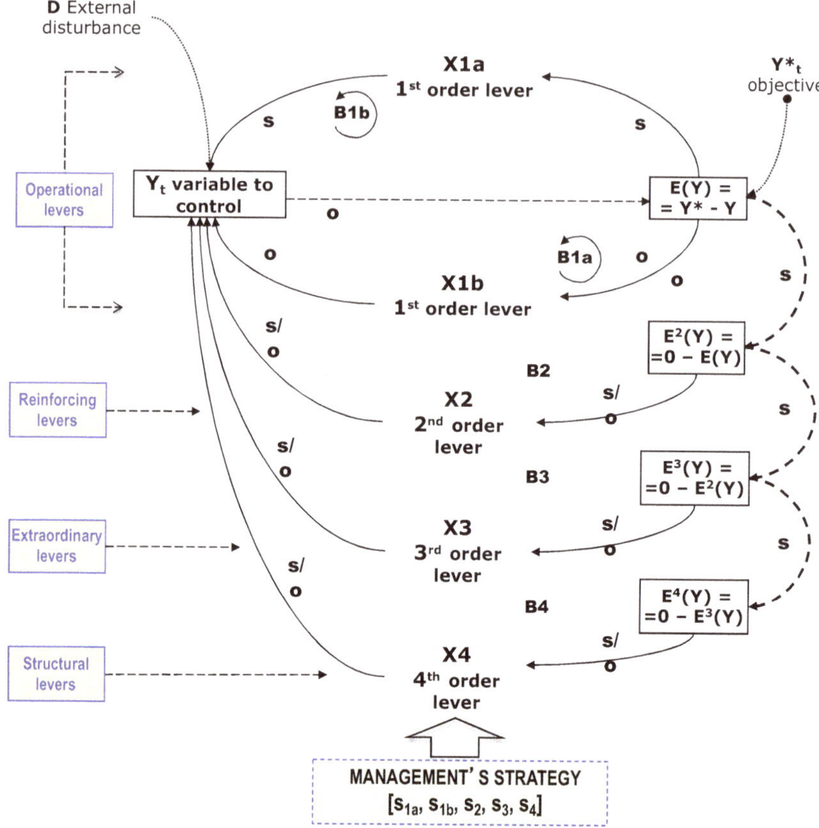

Fig. 4.13 A multilever and multilayer control system. General model

more the *control level* increases, the more the levers become difficult to activate, thus requiring a longer time to produce their effects on Y_t.

Two immediate examples will clarify the logic of *multilayer* control systems. Several times I have observed that the control of our body temperature is fundamental to our survival. This control can be achieved through a multilayer system similar to that in Fig. 4.14. When we are in a state of normal health there are two physiological levers nature has provided us with: chills and movement help us to raise our body temperature (lever X1a); sweating and resting in the shade allow us to lower it (lever X1b). However, these levers have a *capacity* of action limited to only a few degrees of body temperature, spurring man, beginning in the prehistoric era, to search for strengthening levers in the form of clothing of any kind (lever X2), from primitive furs to modern artificial fibers that nearly completely insulate us from heat and cold. If clothing is not enough, man tries to regulate his body temperature through *extraordinary levers*, so called because they depend on the environment in which he lives and the materials he may find there; these levers include all forms of

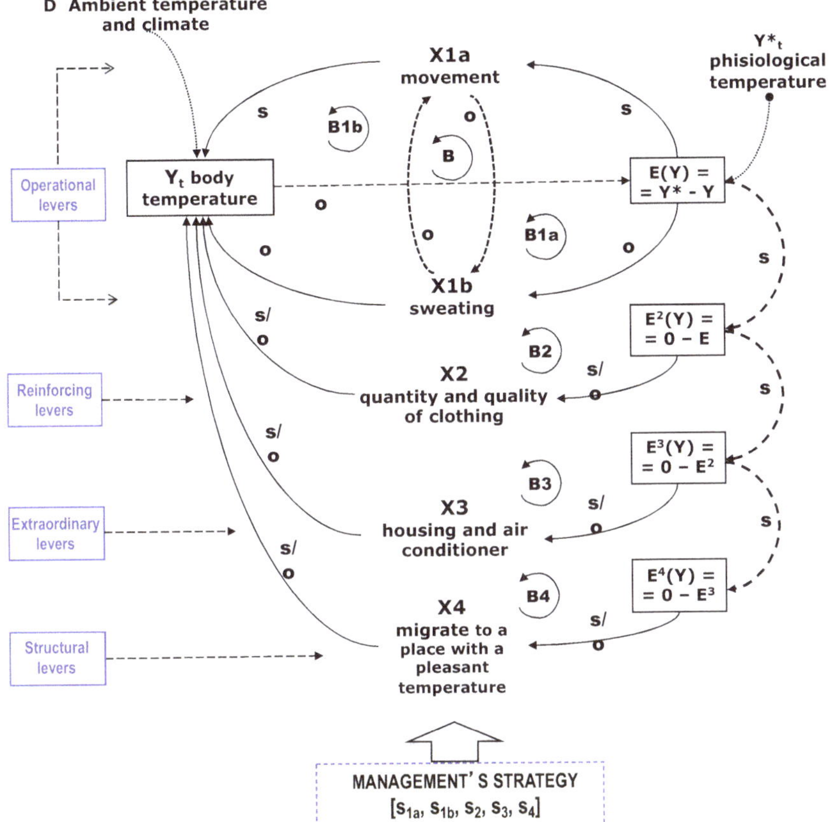

Fig. 4.14 Multilayer control system for body temperature

habitation that insulate him from the outside temperature and from the inside one by preserving the artificial heat produced by fireplaces, stoves, and even modern inverter air conditioners for hot or cold air (lever X3). We could stop here at this three-level system; however, recalling the first rule of systems thinking, we could also zoom out further and consider other levers, such as "going to the mountains," "going to live near the poles," and so on.

As a second example, let us consider the familiar system presented in Fig. 4.15: a car (effector) whose driver (manager) wants to control its cruising speed (variable to control) in order to maintain an average speed (objective) or to remain within the speed limit (constraint), even if the steepness of the road as well as other external disturbances, such as traffic and the environment, can alter the speed of the car (Mella 2012, pp. 146–148).

Under normal driving conditions the driver-manager has two main "levers" to control velocity (Y_t): the accelerator (X1a) and the brakes (X1b), which are *linked* and operate in the opposite direction. In fact, the pedals are usually controlled by the

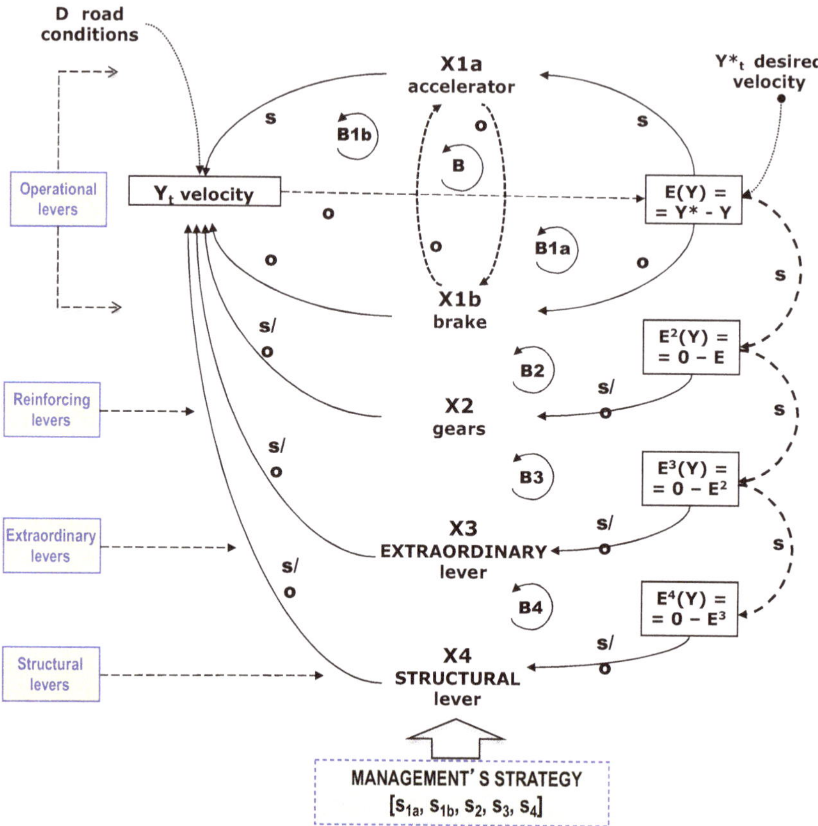

Fig. 4.15 Multilayer control system for car velocity. (Adapted from Mella 2012, p. 147)

same foot to make it impossible to use them simultaneously, which in practical terms fixes the relation in direction "**o**." The accelerator and brake "levers" are *first-order operational* levers, since they are regulated by the manager-driver based on the amount of the *first-level* error, $R(V) = Y^* - Y_t$, which is calculated when the speed differs significantly from the objective or the speed limit.

As a *reinforcement* of these two *first-order* levers, the gear shift (X2) can be used along with the action on X1a and X1b as a third operational (reinforcing) lever. Today almost all car manufacturers equip their cars with automatic transmission, which automatically reinforces, without the manager's intervention, the action of the X1a and X1b levers. However, many car models still have manual transmission, which must be activated by the manager-driver, as every driver has his or her own strategy for changing gears. Some use it before the other two levers, while others, to avoid the risk of breaking the engine, prefer to use the gear shift only when they feel that the action of the other two levers is insufficient. However, in general most people choose the strategy of first reducing the pressure on the accelerator lever, X1a, to zero, then acting on the gear lever, X2 (hitting the brakes), and finally the brake lever, X1b (downshifting). If, when going uphill, the driver sees that the speed has been reduced too much, he or she eliminates the pressure on the brake, steps on the accelerator, and shifts into the appropriate gear. In fact, this is the strategy adopted by car manufacturers for cars with automatic transmissions.

The brake and accelerator are *first-order* levers for a *first-level control*, since they are almost without delay and must be acted on first; the action on the gear shift can be considered as having a delayed effect (even if a slight one), and thus this lever is a *second-order* one for a *second-level* control; this means that the system has an additional *objective* of "zero deviation": $E^2(Y) = 0$. If the action on the accelerator, brake, and gear is not sufficient for the control (considering the time required to reach Y^*), then the manager-driver will detect a further "nonzero error," which conflicts with the second objective of "zero error." There is a *third-level error*, $E^3(Y) \neq 0$, which requires the use of a *third-order* control lever, such as driving zigzag or going onto the shoulder and even rubbing the side of the car against a wall; in extreme cases the driver might even decide to stop the car suddenly by crashing into a tree rather than going headlong into a gulley. These levers can be defined as *extraordinary control levers* in order to distinguish them from the *operational* levers used under normal conditions.

If the control system, through the use of the *extraordinary* levers, permits the car to stop without causing a deadly accident, then the management could modify some parts of the car in order to better control it in future (remember that control systems are recursive), such as by adjusting the brakes, increasing the power of the engine, or changing the gears. It is clear that these control measures can be viewed as *structural* control levers, to distinguish them from *operational* and *extraordinary* control levers, which act on the available levers and not on the structure.

The example reveals a *general conclusion*: in the case of *multilayer* control by means of multiple levers the definition of an order of *priority of actions*, that is, of the *strategy*, for the control lever is crucial and depends on the knowledge (through experience or by referring to the driver's manual) of both the *rate of variation* of each lever—accelerator, brake, and gear shift—with regard to the velocity and the *inverse rate of maneuver* of the levers as a function of the amount of error, taking account of the different *delays* in their "response to the maneuver."

The role of strategy is fundamental in *multilayer* control systems. The strategy affects the efficiency and speed of the control; the experienced manager, following a strategy that has proven successful in past control cycles, is advantaged with respect to one who must choose a control strategy for the first time. This statement is always true, as we can easily verify through observation. The great strategists are the warlords, skilled managers who have gained the necessary knowledge to maneuver their levers, represented by their armed divisions. The great entrepreneurs are skilled managers who know how to maneuver their company's financial, economic, and production levers to achieve success. We ourselves know which levers to use to cook a successful meal, drive in city traffic, find a parking place during rush hour, paginate a book on the computer, or convince someone to see a film and not go to a football match.

I shall define *plan of action*, or *routine* or *consolidated strategy*, as a strategy the manager feels can be effectively applied in all *similar* situations requiring the activation of a control system and which he or she thus *normally* applies in the manual control of all problems requiring analogous control processes. Through *routines* even manual control tends to become transformed into an automatic control, since the manager no longer has to decide on an ad hoc strategy each time but rather to carry out a *routine* deemed effective.

Surely none of us has to decide each time which muscles to activate to grab a pencil, to write, to remain in equilibrium, or to walk or which levers to activate to control our car during normal driving conditions. And my wife certainly does not have to establish an ad hoc strategy each day to activate the levers needed to prepare dinner. She is quite capable of cooking even while looking at the TV and monitoring her children's homework, because she follows well-tested *action plans*, well-established *routines*.

The concept of *routine*, that is, a *consolidated strategy*, is mentioned often in studies on organizations, in which a large number of individuals activate control systems to regulate their behavior and the organizational processes without having to select a new strategy each time, instead following well-established routines, procedures, and rules that are adapted to the changing environmental or internal conditions. Routines are referred to by different names. Argyris and Schön, for example, refer to them as "theories of action":

> When someone is asked how he would behave under certain circumstances, the answer he usually gives is his espoused theory of action for that situation. This is the theory of action to which he gives allegiance, and which, upon request, he communicates to others. However the theory that actually governs his actions is his theory-in-use, which may or may not be compatible with his espoused theory; furthermore, the individual may or may not be aware of incompatibility of the two theories (Argyris and Schön 1974, pp. 6–7).

Technical Note The systems in Figs. 4.14 and 4.15 were created using the same number of levers only in order to better clarify the concept of *multilayer* control system, since it is obvious that systems can exist with a different number of levels as well as levers of the same order.

4.6 Multiobjective Control Systems and Control Policies

All the systems presented in the preceding *sections* have in common the fact they concern a single objective: water temperature or level, velocity, and so on. These limitations are not at all necessary; we can consider even more general models where the *manager* must control several objectives simultaneously by employing an appropriate number of levers with various characteristics.

I shall define as a *multiobjective* (or *pluri-objective*) control system, *in the true sense*, one characterized by a set of objective variables: Y1, Y2, …, YM, to be controlled simultaneously in order to achieve an equal number of objectives, under the *condition* that the *M* variables representing the objective are *interdependent*. The contemporaneous presence of many objectives does not always signify that the control system is *multiobjective* in the *true sense*; in fact, we can find multiobjective systems that are totally *apparent* or even *improper*. In order to better clarify the behavioral logic of multiobjective systems in the *true sense*, it is appropriate to begin with an examination of *apparent* and *improper* multiobjective systems.

Assuming that the *M* variables, Y1, Y2, ..., YM, though controlled by the same system, are *independent*, each regulated by independent levers that do not interact with the control of other objectives, then the multiobjective system must be considered only *apparent*, since it derives from a simple *operational union* of *M* single-objective systems. A typical *apparent* pluri-objective system appears during a parade: each soldier that marches must control, at the same time, a multitude of variables: the vertical alignment (with respect to the parading soldiers in front), the horizontal alignment (with respect to the soldiers on each side), the cadenced step, the speed, and, if it is the first row, the marching direction. Even if all these variables must be *jointly* controlled by each soldier in order to have a perfect parade alignment, each of these does not interfere with the others (unless the soldier is distracted) and can be controlled through the appropriate levers by employing different sensors (sight and sound) to eliminate errors with respect to the objectives of a cadenced and aligned march. Each marching soldier can thus be viewed as an autonomous, *apparent*, multiobjective control system. Nevertheless, this system interferes with other control systems, represented by parading companions, and these interferences can involve various control variables. The interferences are evident: for example, when a soldier is "out of step" and causes disorder his fellow soldiers, rather than continuing to march in the same step, try to adjust, causing a series of cyclical oscillations that are clear to everyone that has witnessed a parade. The result is that the soldiers become unaligned and change their marching speed, even only for brief moments, until the control systems (the soldiers) regain control of this variable.

Another *apparent*, multiobjective control system is one we have all experienced when driving in city traffic. Similar to parading soldiers, we have to remain aligned vertically and horizontally by following the flow of traffic surrounding us, which can change speed at any moment. However, in addition we must check the traffic lights, speed limit, route to take to reach our destination, gas level, car temperature, and directional changes, along with many other variables. We realize that managing our car is a difficult task indeed; fortunately, we can control one variable at a time by acting alternately on one and then the other, trusting the speed of our sensors and effectors, but especially the promptness of regulation which experience gives us. Driving slow in city traffic is the sign of a new driver who lacks experience.

A very general case of *apparent*, multiobjective systems is *alternating objectives* control systems, which occur when the manager can alternately achieve different objectives, one at a time, without the momentary objective interfering with the others. As an example, let us assume once again that we are driving our car along a curvy road. The final objective is to keep the car within the driving lane; but to do so we have to, without probably even realizing it, translate this general objective into two *specific alternating objectives*: maintain a constant distance from the lane dividing line to the left or a constant distance from the right border line. Normally we move from one to the other of the two objectives in continuation depending on the direction of the road. When there is a curve to the left, many prefer to follow the "distance from the lane dividing line" objective, ignoring the distance from the right side; when there is curve to the right the objective focuses on the right border line, not considering the dividing line. Those who try to achieve the *final objective* of staying in the center of the lane by ignoring the specific objectives regarding the distances from the left and right will drive in a swerving manner, as we have all had occasion to observe.

The *alternating objectives* control system is even more evident when we are driving along a narrow road. We need to only recall how often our gaze (detector) shifts to the left or the right of the garage door when we are parking our car. If you need more proof, then consider how our eyes shift continuously when we align letters and lines while writing with a pen, and on the way we pay attention to the two sides of the street when, as pedestrians, we have to cross.

The second type of multiobjective system, which I have defined as *improper*, occurs when the *M* variables-objectives are *interfering* but we can nevertheless control the achievement of each one; the control system becomes in this case an *improper multiobjective system*, since it is formed by the union of *M interfering* single-objective systems, each of which, however, is able to eliminate the error with regard to its individual objective, unless the interactions among the systems are too strong.

This is the case with a familiar control system: the user-manager who is showering and wants, *at the same time*, to control the water *temperature* and *flow* variables through the use of two levers, the *temperature mixer* (hot–cold) and *faucet*, to regulate the water *flow* (not too strong–very strong). If the mixer can regulate the temperature without affecting the flow, or if the faucet can regulate the flow without interfering with the temperature, then the multiobjective control is only *apparent*, since there is a union of two single-objective systems. The manager must only decide which control variable to use first; in other words, he must establish the priorities for achieving the two objectives (policy). However, if as often occurs the variation in flow also causes a variation in temperature, which will then be corrected using the mixer, then the system can be considered as the *union* of two single-objective *interfering* systems, since the control of flow interferes with the water temperature, even if the reverse is not true (Sect. 2.14).

Another multiobjective system with *interfering objectives* can be observed in an airplane that must fly to a certain destination. The control system represented by the airplane is much more complicated than that of the shower; nevertheless, it has quite similar operating features. The *manager* of this system—the pilot together with the ground staff—must simultaneously control at least three main *operational variables*—direction, speed, and altitude—by using a certain number of levers, as shown in the model in Fig. 4.16. The pilot can *achieve* the final objective, *represented by the flight destination, only by maintaining control of these variables for the entire length of the flight.* I shall leave it to the reader's intuition to interpret the control system in Fig. 4.16, though it is quite clear that what makes the control complicated is the fact that the three variables indicated are interconnected, since many control levers are common to at least two of them, so that it is not possible to regulate one variable without also regulating the others (see Mella 2012, Sect. 3.7 for a more detailed description).

By observing the rules for the control of speed and altitude indicated in Fig. 4.16, we see that there are no direct constraints among these variables, even though there are indirect ones; this is due to the fact that the *governed* variables are controlled by common commands (levers): an increase in speed cannot be obtained by simply giving the plane the throttle (increase in the number of revolutions of the propeller), since this lever also controls altitude; giving the throttle increases not only velocity but altitude as well. To gain speed the action on the throttle must be accompanied by the maintenance of altitude. Vice versa, by using the vertical control stick to reduce altitude the pilot also increases speed; thus, to avoid this increase he or she must let up on the throttle. In conclusion, the objectives of velocity and altitude are linked by

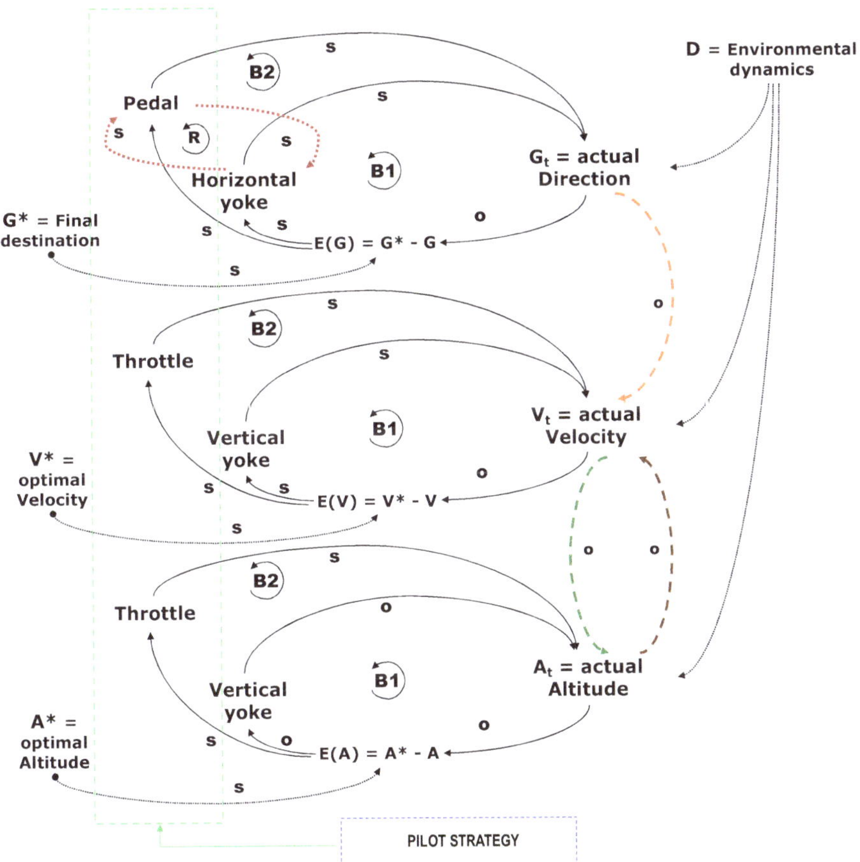

Fig. 4.16 Model of the multiobjective control of a plane

relations in direction "**o**." This is a novelty with respect to the preceding models, since now *the links are placed on the objectives* and not on the control levers of the variables.

Examining these dynamics, we see that the objectives indicated in Fig. 4.16—taking into account the variations adopted by the pilot (manager) during the trip—are always achieved, even if the reciprocal interactions among these variables necessarily lead to adjustment oscillations. Precisely in order to avoid these oscillations, which would be unpleasant for the passengers, the *designers* of modern aircraft add other control levers to the fundamental ones, which leads to a "precise control" of the *governed* variables: additional ailerons and rudders that automatically intervene to avoid oscillations, control of the engine's revolutions without having to use the throttle, and so on.

In conclusion, even if the objectives interfere with each other the pilot-manager is still able to control the system by bringing all the variables to the value required by the objectives. Thus, a plane represents an *improper multiobjective* control system in which the pilot-manager always achieves his or her objectives, even if the *interferences* among the variables to control necessarily cause adjustment oscillations.

What makes a system *effectively multiobjective* in a *true sense* is the *condition of interdependence* among the M variables to control, so that a control lever regulating a variable Y_i *inevitably* regulates at least one other variable Y_j, whether or not the direction is [**s**] or [**o**]. If both objectives vary in direction [**s**] with respect to the same control *lever*, then the objectives are *cumulative* or *complementary*; if they vary in direction [**o**] they are *concurrent* or *antagonistic*.

It is not difficult to find examples of such systems, even if it is not often easy to describe them effectively. As always I shall begin by proposing a *multiobjective, multilever* control system we have all experienced when, as passengers-managers, we have had to regulate the position of the car seat when the position can be controlled through separate *levers* activated by buttons that function by means of electric effectors.

For simplicity's sake, I assume that we wish to achieve four objectives: (1) travel safety, (2) passenger-manager comfort, (3) guest passenger comfort, and (4) outside visibility of manager. To achieve these objectives, we use three control levers that act on the car seat position by modifying (a) *height*, (b) *distance* from windshield, and (c) *inclination*, according to the following rules:

1. Safety:
    - Low seat
    - Seat set all the way back
    - Slight inclination
2. Passenger-manager comfort:
    - Seat in middle position
    - Seat all the way back
    - Strong inclination

3. Comfort of rear-seat passenger for reasons of courtesy:
   - Low seat
   - Seat all the way forward
   - Strong inclination
4. Outside visibility:
   - High seat
   - Seat all the way forward
   - Medium inclination

A quick look at this situation immediately makes clear that it is not possible to achieve all the objectives contemporaneously. If we use the lever to move the seat forward and backward, we are affecting at the same time safety, manager comfort, passenger comfort, and visibility. Whatever lever we use, all the objectives come into conflict with one another. The system is *truly* multiobjective and the objectives *concurrent*.

Here is a second example. We are preparing dinner for friends and want to control the variety and tastiness of the dishes but at the same time the amount of work needed, time, cost, healthiness of the meal, and so on. It is impossible to totally control all *governed* variables contemporaneously and to reach our objectives at the same time. The control system we are the *manager* of is *multiobjective in the true sense*, since the variables are interdependent and we must prioritize the objectives to optimize. The *governance* of the system must choose, for example, whether to favor quality, by preparing caviar and champagne, or to minimize cost, by serving shrimp and mineral water; whether to cook a pot roast for 12 h or a pasta with sauce in 10 min; whether to use truffles or butter; and whether to worry about cholesterol or flavor the dishes with egg and cheese.

All multiobjective systems, from *apparent* ones with independent levers (shower, car) to *improper* ones with only interfering ones (airplanes), even *multiobjective systems in the true sense*, with interdependent objectives (car seat, dinner), require the manager to ask the *governor* (if the latter is not also the manager) to decide the *order in which the objectives must be achieved.*

When driving over a narrow road with curves the governor-manager must choose if and when to observe the distance from the left and right sides of the road. Under the shower the governor-manager must decide whether to regulate first temperature and then water flow or vice versa. In short, the *governor-manager* would do well to determine an order of importance of the objectives, that is, establish a "control policy" in order to decide which objectives to give priority to. If he does not choose the policy before initiating the control, the policy will be decided "on the field" in a *contingent* manner.

Since controlling a plane toward its destination implies the contemporaneous presence of several partial objectives (direction, velocity, altitude) that must be simultaneously controlled by acting on the specific control variables, it is clear that the *manager*-pilot must decide on a *control policy* that defines the *order* and *priority* of the variables to control and thus the objectives to achieve. In case of a deviation from the route due to lateral wind, he or she could first of all correct the direction and subsequently regulate the altitude or he or she could maintain the actual route and try to reach a better altitude. In the case of headwinds, he or she could use the velocity control while maintaining the actual route or he or she could try to maintain the plane's velocity while deviating from his or her route in order to avoid the headwinds. In the case of tailwinds, he or she could maintain the throttle constant to arrive ahead of time or he or she could let up on the throttle to take advantage of the "push" from the wind and arrive on time. Another alternative would be to change his or her route in order to arrive on time without reducing the throttle. The need to decide on a *priority of objectives* in a multiobjective system with *interdependent* and *concurrent* (seat regulation) objectives is much more apparent.

The outline for the general model of control is thus the following: a multiobjective system can be controlled only if the *manager-governor* defines the *set of objectives* to pursue (and the *policy* that determines their relative priority) and if, at the same time, there are sufficient control "levers," which are compatible and coordinated to use based on a defined *strategy*. Several methods for assigning a relative weight of importance, or urgency, to the objectives are presented in Sects. 4.8.8 and 4.8.9.

4.7 Optimal Strategies and Policies: Two General Preference Models

"Control," viewed as a concrete *managerial* action to guide the system toward its objectives, is always an "online" activity, that is, one *concomitant* to the system's dynamics. As such it must be regulated by a *policy* and *strategy* of control the *governor* and *manager* must specify beforehand or while the system is in action. If the *policy* and *strategy* are defined before the system's trajectory begins, the control is defined as *formalized*. If these are decided while the trajectory is unfolding, the control is *contingent*.

The formulation of a control *strategy* implies that the *manager*-user be able to activate the control levers based on an *order of preference* by searching whenever possible for the most efficient and effective order, that is, the *optimal strategy*. The choice of order of preference is not always easy. This problem would require a good deal of treatment, and here I can only present some general, brief indications. I shall start by considering that several *levers* can be activated at no cost (with whatever meaning we wish to attribute to that term, even in the sense of damage, disadvantage, sacrifice, etc.), while others can be activated only with an *explicit* cost (direct sacrifices from activating the lever) or an *implicit* one (lost advantage from activating the other levers).

There is no cost in adjusting the stereo volume; when showering there is no cost in increasing the water temperature by reducing the cold water flow, while there is a greater cost if we increase the flow of hot water (fuel, electricity, etc.); there is no explicit cost to reducing the speed of a car by letting up on the gas pedal (in fact, this reduces fuel consumption), while there is a cost to braking (brake disks are consumed), especially if we take into account the cost due to the damage caused by not braking and having an accident.

In choosing the control levers *cost* plays an important role, though it is not the only parameter to consider. Clearly, we must also take account of the *amount of error* and the *importance of the objective*. If there is a speed limit and I note the presence of the highway patrol, I will not undertake an economic calculation of the costs and benefits of braking but try instead to reduce "at all costs" my speed using every available lever, so as to avoid a ticket. Even less would I calculate the costs if my car was going down a steep incline with narrow curves; if my brakes were not sufficient and the speed increased to as to endanger my physical well-being, I would instinctively activate the extraordinary levers (steer into a mound of dirt, scrape the side of the car against a wall, etc.) independently of the cost of activation, even if I know a few people who would rather risk their lives than ruin the body work on their car.

Therefore, the idea of an *optimal strategy* is a relative one and depends on the type of objective and advantages in play and on the amount of error and the costs associated with each lever. When there is time to act rationally by programming the activation of the control levers without being pressured by the urgency of the situation or the gravity of the control, then we can determine the *optimum strategy* through a cost–benefit analysis (CBA), evaluating all the cost factors associated with the levers (direct costs and indirect damage) as well as all the benefits in terms of the advantages from reducing the error and achieving the objective.

In many cases applying CBA techniques encounters considerable difficulties since, in evaluating the different control levers, the manager takes into account monetarily nonquantifiable factors of psychological and emotional origin or factors linked to culture and preferences, which at times are even prejudicial. As proof of this consider the simple control system of the blood pressure of someone who has suffered from "high blood pressure" for some time. To reduce the pressure to its physiological limits, different levers can be activated, among which skipping meals and losing weight, avoiding the use of alcohol, giving up smoking, living a healthier lifestyle, exercising, and taking blood pressure pills. How should a rational person behave in determining an *optimal strategy*? Each of us can provide our own answer, but it is very likely that most people would, irrationally, prefer the blood pressure pill lever to giving up smoking, food, and alcohol (Sect. 5.7).

Similar considerations can be made to define the *optimal policy* of a multiobjective system. The choice of the order to achieve the different objectives depends in part on the costs/benefits associated with the available levers, but even more on the *importance* and *urgency* of the objectives the system must achieve. Usually, the first objective to achieve should be the one deemed most urgent or important, since a high benefit (if achieved) or a high fallout (if not achieved) is associated with it. If the control is not uniquely tied to the urgency of the objectives, then we can add techniques for ordering the objectives on a "pseudo-cardinal scale" to the definition of *optimal policy*, so that the policy choice can entail metrics that assign objective *priority indicators* that are coherent with the idea of the distance between urgency, importance, etc. Several methods have been devised for assigning pseudo-cardinal scales in order to prioritize the objectives of the observed control system.

In *automatic systems* these priority scales are determined by the system's designer and incorporated in the strategy that activates the levers in the most appropriate way to achieve the most important objectives first, followed by the others. For example, a digital reflex camera in the "auto" setting usually prioritizes the objective of optimizing the quantity of light reaching the sensor in order to allow photos to be taken with little light, even if this is at the expense of the objectives of depth of field and still image. In order to allow the manager to better adjust the levers, other operational modes are provided that favor different objectives, which the manager-photographer can choose using simple commands according to the objective deemed most important at a certain moment. A well-known reflex camera provides the following operational modes for the manager to choose based on his or her objectives: "Automatic mode with detection of 32 scenes, steady photos in any situation, gorgeous photos in low light conditions, high-speed, high-resolution photos." Clearly each mode favors a particular objective while ensuring that the other objectives are gradated along a scale of decreasing importance.

In a social context the scale of priorities of objectives is established by norms, customs, or beliefs of various types. Should a control system for saving passengers on a ship privilege saving goods or people? Or saving an adult male, adult woman, a sick person, or a baby? The phrase "priority for women and children" helps to establish the scale of objectives. In setting up control systems for human behavior, the laws of nations and religions often define several scales of objectives that guide the control processes: For example, should we first control tax evasion or privacy? Murder or theft? Theft or corruption? Lying or a code of silence? Prison overcrowding or an increase in crime? Educational levels or resources for schools, etc.?

Scales to order objectives based on urgency, importance, utility, etc. are rarely *quantitative*. Though it is theoretically possible to imagine an objective being doubly urgent or useful compared to another, it is not normally possible to determine or define a *measuring procedure*, since any measurement requires that an objective that can serve as a base unit of a quantitative scale be identified or defined. In order to assign an *index of priority* (urgency, importance, utility, etc.) to each objective, we need a procedure that, borrowing a term from topology, I shall define as *metrics procedure*—which differs from measurement—capable of creating *pseudo-cardinal scales*. I shall define *pseudo-cardinal* as a scale whose numbers are not derived from measurement but from procedures which assign *coherent indices* to the various states of a qualitative variable, so that we can compare the dimensions of these states unambiguously by employing the assigned indices.

It is useful to recall two metrics procedures that will be described in Sects. 4.8.8 and 4.8.9:

- The *direct comparison* procedure
- The *linear combination* procedure in the *standard gamble* method variant

However, even the definition of the optimal policy is affected (perhaps to a large extent) by the *personal equation* of the governor-manager, whose scales of preference reflect cultural factors ("women and children first"), ethical factors ("before food we require safety"), political factors ("first aid to workers then to firms"), and so on. I would also note that the definition of the optimal policy is also affected by several general psychological models that, often unconsciously, influence the choices and limit the rationality of the decision-maker.

It is natural that the rational person has a greater preference for short-term advantages than long-term ones and, conversely, long-term disadvantages are preferred to short-term ones, even if the amount of short-term advantages or disadvantages is less than the respective long-term ones. As we know, these principles are the basis for the financial calculation that provides the techniques for comparing *alternative* financial advantages and disadvantages at different "times."

Nevertheless, Sect. 1.7.8 highlights three systems archetypes (Fig. 1.12) that describe three general structures of human behavior which reveal the widespread tendency for man to behave in an apparent rational manner in an individual and local context, if viewed in the short term, but in an irrational manner if we consider the long-term effects this behavior produces, in other words, at the collective and global level. If we do not apply systems thinking to choose and order the objectives, we run the risk of being "swallowed up" by the short-term archetypes (individual and local preference) and of not properly evaluating the disadvantages this produces in policy formation by the system's governance, with the result that, even though the system permits the objectives to be achieved, this outcome will generate even more serious future problems, which will require even more powerful control systems.

4.8 Complementary Material

4.8.1 Flying in a Hot Air Balloon

We have all imagined how fascinating a flight in a hot air balloon could be, even if most of us have not had this experience. Rising among the clouds and being taken away by the wind must be an extremely pleasant sensation.

Let us try to understand how to take off and land a hot air balloon, that is, how to control the *altitude* of this aircraft. We all know that the upward thrust is supplied by Archimedes' law: *a body receives a push from the bottom upwards equal to the weight of the volume of fluid displaced*. Since hot air is lighter than cold air, in order to overcome the weight of the nylon envelope, the gondola, and the harness, it is necessary for the ball's volume to be sufficiently large to displace an external volume of air capable of supplying the thrust. Figure 4.17 makes it easier to understand the logic of the system to control altitude.

The hot air is not injected into the envelope. A burner—which represents the *control lever* for climbing—in the lower mouth of the envelope heats the air in the

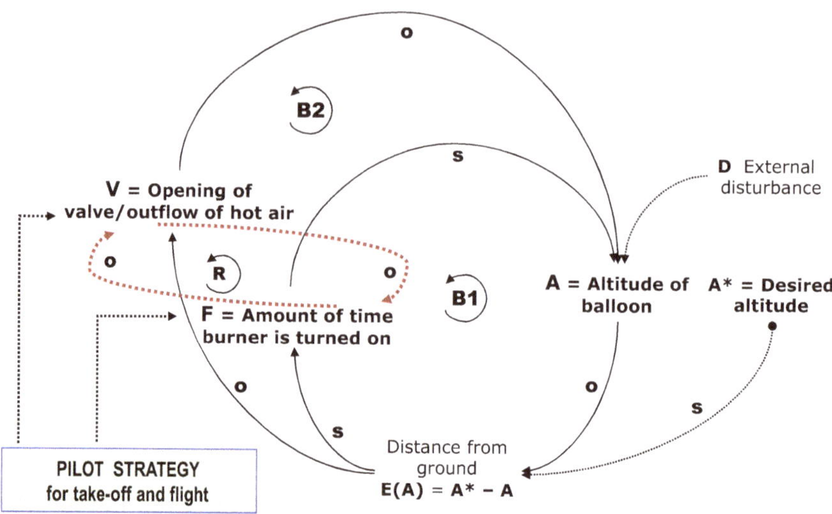

Fig. 4.17 The control system for a hot air balloon

balloon; thus, in order to take off the burner needs to only operate for the time needed to heat the internal air mass, based on a heating parameter that depends on the device used. How do we bring it down? It would be enough just to wait for the internal air to cool down, but this usually takes too long to be efficient. There is a *second lever* for landing (the one that controls the descent) represented by a simple valve at the top of the balloon. When opened the hot internal air is released, thereby leaving room for the cooler air that is injected through the lower opening of the balloon. Figure 4.18 shows the general Powersim simulation model for a hot air balloon.

The simulation assumes that initially the balloon is at 100 m of altitude, at which point the pilot ignites the burner to make the aircraft rise to 300 m; after 40 s of flight the pilot orders the air valve to be opened to descend back down to 100 m. After 25 s the balloon lands. The bold line traces its trajectory. The line at the bottom represents the ground level. We can observe that, due to the time needed to heat the air, the ascent does not stop precisely at 300 m but continues on for some meters due to inertia, after which it settles at the desired altitude. For the descent, the adjustment of the valve permits a more precise control, and if the pilot is experienced the airship gently touches down.

4.8.2 *Submerging in a Submarine*

Even if submerging in a submarine is not as thrilling as a flight in a hot air balloon, it may nevertheless be an equally interesting experience. We are not interested here in the visual sensations but in the control system of this vessel for underwater navi-

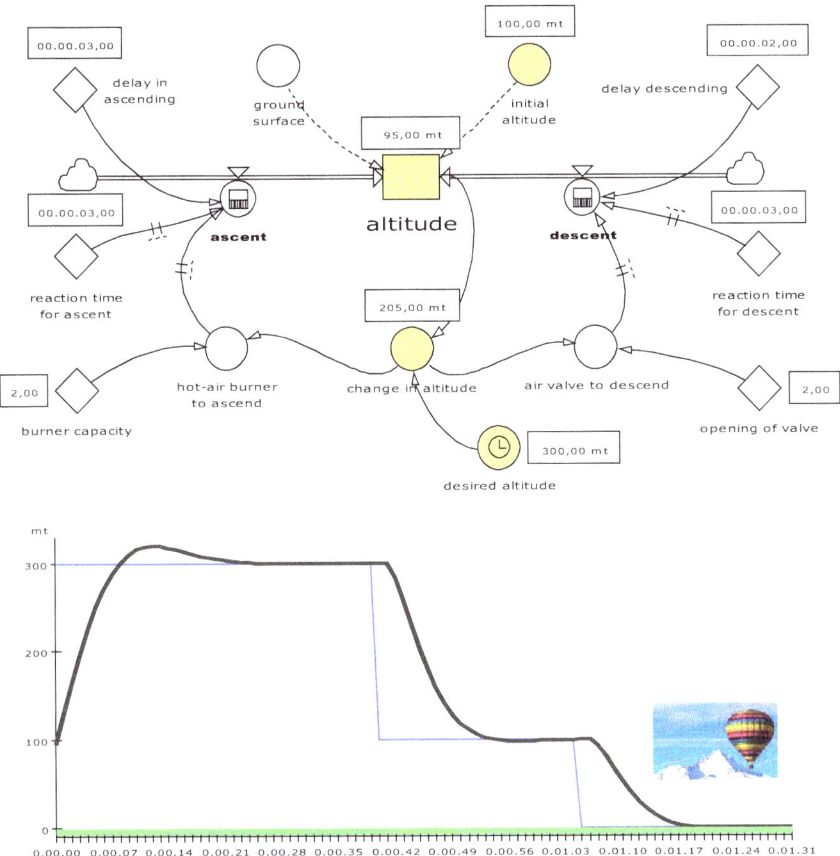

Fig. 4.18 Two-lever control of a hot air balloon, simulated using Powersim

gation. It should be absolutely clear that the control system for the depth of the submarine—illustrated in Fig. 4.19—is the same as that which regulates the altitude of a hot air balloon in Fig. 4.17. The submarine tends to float as a result of Archimedes' law of buoyancy. It can descend to the deepest depths of the sea only if its mass is increased through the emission of seawater through an open valve that allows the water to gradually fill a certain number of ballast tanks.

When the mass of injected water, added to the weight of the submarine, exceeds the weight of the water it displaces, then the submarine can submerge. In order to surface, a pump injects pressurized air into the ballast tanks; the pressure of air forces the water to flow from the valve. Gradually the total mass of the submarine becomes less than that of the water displaced by its overall volume, and the submarine inexorably rises until it surfaces. In order for the submarine to remain stably at a certain depth, it is necessary to adjust its mass—through the emission or the expulsion of water from the internal chambers (tanks)—so that it is equal to the mass of displaced water. As in the case of the hot air balloon, the commander-manager of the

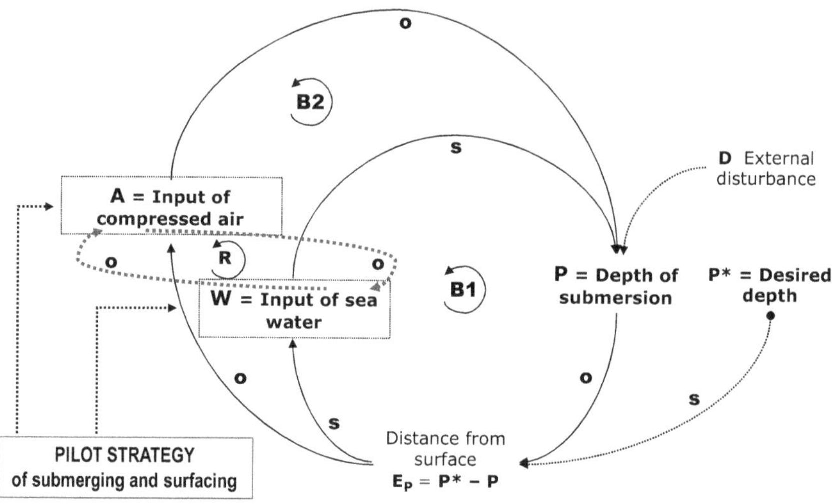

Fig. 4.19 The control system of a submarine

submarine uses the input control levers (valve for injection of water) and the output ones (compressor for the compressed air used to expel water) in an interdependent way, activating one at a time. Figure 4.20 presents the Powersim simulation of the control system. Initially the submarine is 100 m below sea level and starts to descend to −300 m. After 40 s of descent the commander gives the order to rise back up to 100 m and 25 s after that to resurface. The bold line traces the trajectory.

Note that, due to the time needed to empty tubs and the delays caused by the water's viscosity, the descent does not stop at −300 m but continues due to inertia for some additional meters, after which the submarine rises again and stabilizes at the desired depth. The same dynamics hold for resurfacing. Inertia causes the submarine to exceed the limit ordered by the commander, and when the vessel comes to the surface it jumps out of the water, like a dolphin. The only obvious difference with the hot air balloon is that for the latter the ground represents the lower limit of descent, while for the submarine the upper level of ascent is the sea level. On the other hand, the hot air balloon has an altitude limit that cannot be exceeded; this limit is defined on each occasion by the air density. The submarine has a lower depth limit, which is a function of the vessel's resistance to outside water pressure. The similarities between the two control systems, independently of the differences in the "machines" they are applied to, are immediately clear when we compare Figs. 4.18 and 4.20, which coincide except for the inversion in the direction of motion.

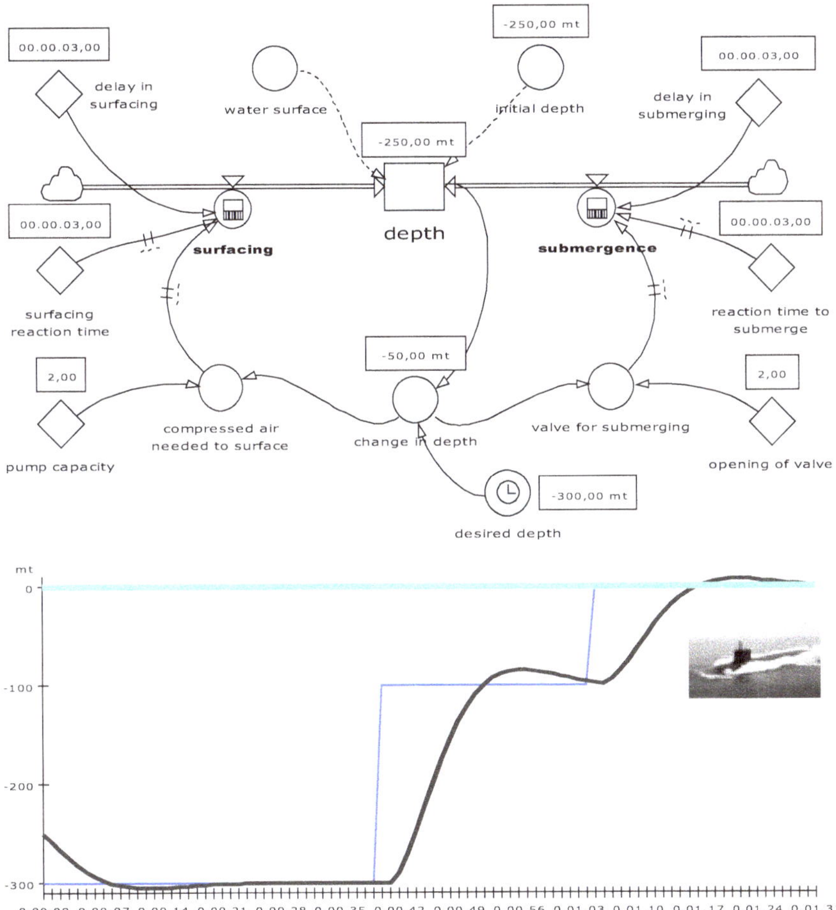

Fig. 4.20 Two-lever control of a submarine, simulated with Powersim

4.8.3 A Multilever System: Mix of N *Components*

Section 4.4 presented an example of the spatial movement of a point in an N-dimensional space, based on the logic presented in Fig. 4.12. A system directly derived from the preceding one enables us to create and maintain a given mix, MIX_t, among N components $[x^1, x^2, x^n, \ldots, x^N]$, with any N (whole, finite). Each component, x_n, must enter the mix in an "ideal" quantity, x^{*n}, which represents the objective; the mix is perfectly formed when all N components have the values $x^n = x^{*n}$. The values of the N quantities making up the mix at instant "t," $MIX_t = [x^1{}_t, x^2{}_t, \ldots, x^n{}_t, \ldots, x^N{}_t]$, form a point in the N-component space; the control system must shift this point to the point formed by the N quantities-objectives: $M^* = [x^{*1}, x^{*2}, \ldots, x^{*n}, \ldots, x^{*N}]$.

The system can act based on constant *action rates* applied over a given time interval or in a manner proportional to the error. These methods of control must be well designed and take into account the nature of the mix, the relative quantities of the components, the reaction times, and the maximum allowable time for completing the control.

4.8.4 Control of a Mobile Platform

The control system to shift a point toward another point, which is described in Sect. 4.4, can be applied in many other ways. One of these is to control the horizontal equilibrium of a platform placed on irregular ground, for example, the platform of a crane, a scaffolding, and a marine platform, where the platform is set on four adjustable legs, moved by hydraulic pistons, which can be set at different heights so as to counter the roughness and unevenness of the ground.

Observing the model in Fig. 4.21, we can describe how the control functions as follows: first of all, the height of the rough ground is measured to form point P_0, and then the maximum height, P^*, is calculated, that is, the height-objective to which the position of the entire horizontal platform must refer.

Given the initial position of the legs, P_0, the system calculates the necessary extension of each leg with respect to the maximum height, P^*, required to bring the horizontal platform to that maximum height. Referring again to Fig. 4.21, we see from the control panel that, for example, leg X has an initial extension of $x_0 = 3$ units of measure (centimeters, inches, etc.); it must rest at a ground level of 15, which corresponds to the maximum ground height. Thus, the value $x_0 = 3$ must be adjusted and brought to $x^* = 0$ by using a small motor that, for each "t," shifts the leg $g_x = 1$ unit. Leg X is level at $t = 0$. Leg W has an initial position of $w_0 = 1$ and must be level at ground height 15. Taking into consideration that the ground on which W rests already has a height of 2, then leg W must be extended to $w^* = 13$, which, since $g_w = 1$, is reached at $t = 11$. The same procedure can be observed for the other two legs. At time $t = 11$ all the legs are extended so as to level the platform at height 15, thus terminating the control, since the distance from the objective is eliminated (line 11 in the right-hand column). Note that this system also acts to level a mobile platform anchored in another platform, as occurs, for example, with platforms anchored to the ceiling of a cave or a building.

4.8.5 Industrial Robots and Movement Systems

The multilever systems in Sect. 4.4 can operate contemporaneously on two *movement systems*, whose dynamics interact according to given rules. Let us first consider the case of two systems, A and B, and apply the rule that A is independent and follows a path toward its spatial objectives, indicated by the coordinates X, Y, and Z,

names of legs		X	Y	Z	W	
initial level of legs	P_0	3	2	2	1	INITIAL POSITION
height of the roughness of ground	P_{terra}	15	12	8	2	15 MAXIMUM HEIGHT
delta extension of legs	ΔP_0	-3	1	5	12	
extension of legs	P*	0	3	7	13	OBJECTIVE
		1	1	1	1	r = reaction times
		1	1	1	1	g = action rates = translators

t	g_x	g_y	g_z	g_w	x	y	z	w	x*	y*	z*	w*	dx=x*-x	dy=y*-y	dz=z*-z	dw=w*-w	distance
0					3	2	2	1	0	3	7	13	-3	1	5	12	13.38
1	-1	1	1	1	2	3	3	2	0	3	7	13	-2	0	4	11	11.87
2	-1	0	1	1	1	3	4	3	0	3	7	12	-1	0	3	9	9.54
3	-1	0	1	1	0	3	5	4	0	3	7	12	0	0	2	8	8.25
4	0	0	1	1	0	3	6	5	0	3	7	12	0	0	1	7	7.07
5	0	0	1	1	0	3	7	6	0	3	7	12	0	0	0	6	6.00
6	0	0	0	1	0	3	7	7	0	3	7	12	0	0	0	5	5.00
7	0	0	0	1	0	3	7	8	0	3	7	12	0	0	0	4	4.00
8	0	0	0	1	0	3	7	9	0	3	7	12	0	0	0	3	3.00
9	0	0	0	1	0	3	7	10	0	3	7	12	0	0	0	2	2.00
10	0	0	0	1	0	3	7	11	0	3	7	12	0	0	0	1	1.00
11	0	0	0	1	0	3	7	12	0	3	7	12	0	0	0	0	0.00
12	0	0	0	0	0	3	7	12	0	3	7	12	0	0	0	0	0.00
13	0	0	0												0	0	0.00
14	0	0	0												0	0	0.00
15	0	0	0												0	0	0.00
16	0	0	0												0	0	0.00
17	0	0	0												0	0	0.00
18	0	0	0												0	0	0.00
19	0	0	0												0	0	0.00
20	0	0	0												0	0	0.00

Fig. 4.21 Horizontal leveling of a platform supported by four legs

and its temporal ones (coordinate *T*), while B must follow A and have the objective of reaching, for every "*t*," A's position, as shown in Fig. 4.22.

A's initial point is $PA_0 = (0, 0, 0)$, which must shift to the objective PA* = (40, 40, 40). B's initial point is $PB_0 = (40, 40, 40)$, and B must follow A to the distance-objective PB* = (30, 30, 30). These values indicate the distance each coordinate of B must reach with reference to the respective coordinate of A. The distance of pursuit, calculated from the origin, is 51.96 (placed in the cell positioned between the two systems). This value must be reached by B's trajectory, which follows that of A, and it can be read from the penultimate column of system B (second table). Obviously, this value must also be determined by the difference between the distance from the origin of A (last column in the first table) and B (third-to-last column in the second table). As Fig. 4.23 also shows, B's objective of pursuit, PB*, is achieved at $t = 9$, when the values of the last two columns of system B coincide. Clearly, B is a typical *control system of pursuit*, since B tries to pursue A, whose trajectory represents B's variable objective.

SYSTEM "A"	X	Y	Z	T	
P = (x, y, z, t) =	0	0	0	0	position
P* = (x*, y*, z*, t*) =	40	40	40	50	objective
action rates-translators	5.0	5.0	5.0	1	

t	g_x	g_y	g_w	gt	x	y	z	t	x*	y*	z*	t*	dx=x*-x	dy=y*-y	dy=y*-y	dt=t*-t	Distance of A from objective	Distance of A from ORIGIN
0					0	0	0	0	40	40	40	50	40	40	40	50	69.28	0.00
1	5	5	5	1	5	5	5	1	40	40	40	50	35	35	35	49	60.62	8.66
2	5	5	5	1	10	10	10	2	40	40	40	50	30	30	30	48	51.96	17.32
3	5	5	5	1	15	15	15	3	40	40	40	50	25	25	25	47	43.30	25.98
4	5	5	5	1	20	20	20	4	40	40	40	50	20	20	20	46	34.64	34.64
5	5	5	5	1	25	25	25	5	40	40	40	50	15	15	15	45	25.98	43.30
6	5	5	5	1	30	30	30	6	40	40	40	50	10	10	10	44	17.32	51.96
7	5	5	5	1	35	35	35	7	40	40	40	50	5	5	5	43	8.66	60.62
8	5	5	5	1	40	40	40	8	40	40	40	50	0	0	0	42	0.00	69.28
9	0	0	0	1	40	40	40	9	40	40	40	50	0	0	0	41	0.00	69.28
10	0	0	0	1	40	40	40	10	40	40	40	50	0	0	0	40	0.00	69.28
11	0	0	0	1	40	40	40	11	40	40	40	50	0	0	0	39	0.00	69.28
12	0	0	0	1	40	40	40	12	40	40	40	50	0	0	0	38	0.00	69.28
13	0	0	0	1	40	40	40	13	40	40	40	50	0	0	0	37	0.00	69.28
14	0	0	0	1	40	40	40	14	40	40	40	50	0	0	0	36	0.00	69.28
15	0	0	0	1	40	40	40	15	40	40	40	50	0	0	0	35	0.00	69.28

SYSTEM "B"	X	Y	Z	T			
P = (x, y, z, t) =	40	40	40	0	position		
Distance of B from A	30	30	30	50	objective	Distance/pursuit	51.96
action rates-translators	5.0	5.0	5.0	1			

t	g_x	g_y	g_w	gt	x	y	z	t	x*	y*	z*	t*	dx=x*-x	dy=y*-y	dy=y*-y	dt=t*-t	Distance of B from objective	Distance of B from ORIGIN	Distance of B from A	Objective of distance
0					40	40	40	0	0	0	0	50	40	40	40	50	69.28	69.28	69.28	51.96
1	-5	-5	-5	1	35	35	35	1	5	5	5	50	30	30	30	49	51.96	60.62	51.96	51.96
2	0	0	0	1	35	35	35	2	10	10	10	50	25	25	25	48	43.30	60.62	43.30	51.96
3	5	5	5	1	40	40	40	3	15	15	15	50	25	25	25	47	43.30	69.28	43.30	51.96
4	5	5	5	1	45	45	45	4	20	20	20	50	25	25	25	46	43.30	77.94	43.30	51.96
5	5	5	5	1	50	50	50	5	25	25	25	50	25	25	25	45	43.30	86.60	43.30	51.96
6	5	5	5	1	55	55	55	6	30	30	30	50	25	25	25	44	43.30	95.26	43.30	51.96
7	5	5	5	1	60	60	60	7	35	35	35	50	25	25	25	43	43.30	103.92	43.30	51.96
8	5	5	5	1	65	65	65	8	40	40	40	50	25	25	25	42	43.30	112.58	43.30	51.96
9	5	5	5	1	70	70	70	9	40	40	40	50	30	30	30	41	51.96	121.24	51.96	51.96
10	0	0	0	1	70	70	70	10	40	40	40	50	30	30	30	40	51.96	121.24	51.96	51.96
11	0	0	0	1	70	70	70	11	40	40	40	50	30	30	30	39	51.96	121.24	51.96	51.96
12	0	0	0	1	70	70	70	12	40	40	40	50	30	30	30	38	51.96	121.24	51.96	51.96
13	0	0	0	1	70	70	70	13	40	40	40	50	30	30	30	37	51.96	121.24	51.96	51.96
14	0	0	0	1	70	70	70	14	40	40	40	50	30	30	30	36	51.96	121.24	51.96	51.96
15	0	0	0	1	70	70	70	15	40	40	40	50	30	30	30	35	51.96	121.24	51.96	51.96

Fig. 4.22 Pursuit control systems

According to the models above, the movement systems with multilevel control are the typical systems studied and produced by *robotics*. In particular, Cartesian robots (which, for example, move a utensil placed at the end of a mobile arm from one point to another in three-dimensional Cartesian space) are referred to as typical movement systems. Two robots programmed to transfer an object between them (a utensil or a component to work on) can be fully conceived of as *collision* systems. However, if they must place the object at a certain distance from the other, they are conceived of as alignment systems.

Cartesian robots can operate in more than three dimensions. For example, let us take an industrial robot that must work on a component in a fixed period of time by carrying out the following program: grab the object; turn it along a vertical axis; choose, one after the other, three tools from its "catalogue"; carry out three successive processes in a fixed amount of time; and, finally, rotate the object and move it forward to permit other objects to be worked on. In this case the robot is in all respects a seven-dimensional (seven-lever) control system: the three spatial dimensions, the speed of the moving objects (shifting time), the degrees of rotation of the transporting head, the manufacturing time, and the catalog of tools.

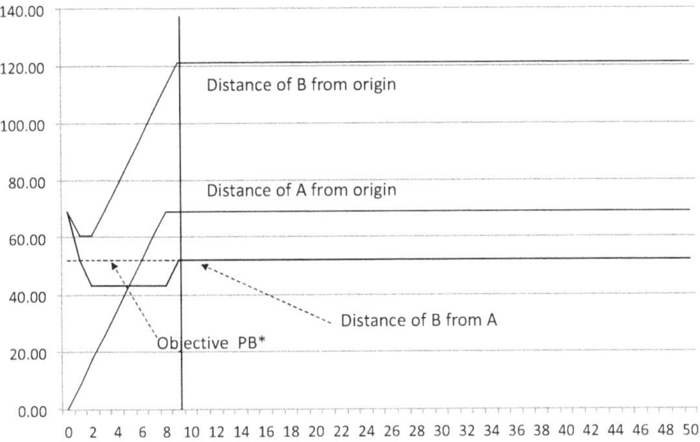

Fig. 4.23 Dynamics of two systems of pursuit in a three-dimensional space

4.8.6 *Focusing*

Figure 4.10 presents a simple multilever control system that regulates focusing on an object placed on a horizontal plane by rotating the eyes, neck, upper body, and entire body. Figure 4.24 constructs a simple simulation model of this system.

The control panel indicates the action rates for each of the four levers and the maximum range of variation of each lever. These ranges are quantified in geometric degrees with respect to the central focus point, which is equal to "0"; in other words, they are quantified for each semi-plane to the right or the left of the line that is perpendicular to the vertical horizon line. The objective as well is set based on the same logic, with the object to be focused on positioned in the semi-plane to the right or the left of center. I have assumed that the management tries to minimize the time needed to focus on the object and thus adopts the strategy for activating the four levers in parallel, each up to its maximum radius of action.

4.8.7 *Demand, Supply, and Price: Dual-Objective Control System*

Figure 1.2 in Chap. 1 presented a typical model of the "law of supply and demand," expressed in systems thinking language. According to Walras' classical theory of equilibrium (Walras 1874), the price of a good can be considered a variable controlled by the demand and supply of that good, and not as a control lever. However, Fig. 1.2 shows that, if demand and supply influence price, then price in turn influences demand and supply. If demand rises, price rises; if the price rises then so does supply. Vice versa, if supply falls, price increases, which reduces demand and once again increases supply.

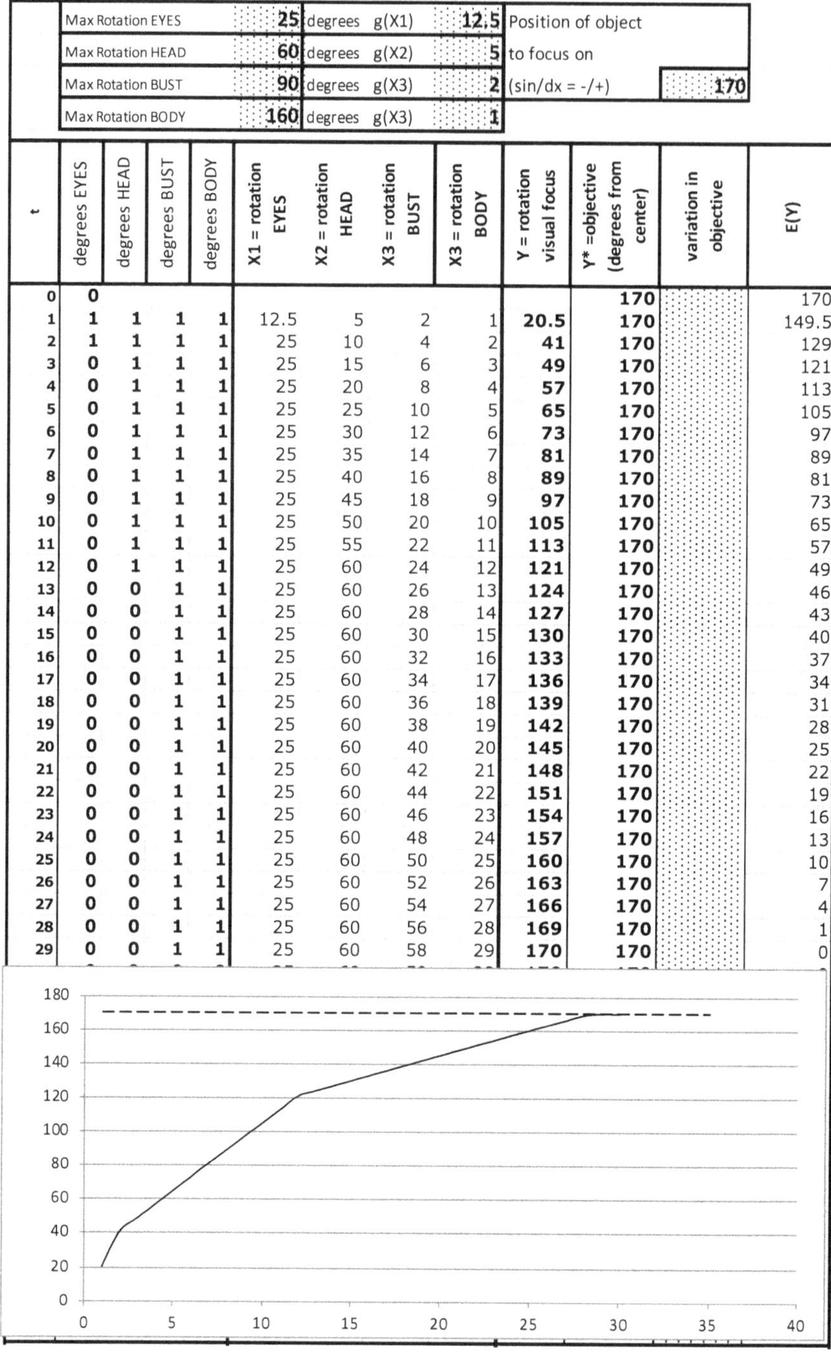

Max Rotation EYES	25	degrees	g(X1)	12.5	Position of object	
Max Rotation HEAD	60	degrees	g(X2)	5	to focus on	
Max Rotation BUST	90	degrees	g(X3)	2	(sin/dx = -/+)	170
Max Rotation BODY	160	degrees	g(X3)	1		

t	degrees EYES	degrees HEAD	degrees BUST	degrees BODY	X1 = rotation EYES	X2 = rotation HEAD	X3 = rotation BUST	X3 = rotation BODY	Y = rotation visual focus	Y* =objective (degrees from center)	variation in objective	E(Y)
0	0									170		170
1	1	1	1	1	12.5	5	2	1	20.5	170		149.5
2	1	1	1	1	25	10	4	2	41	170		129
3	0	1	1	1	25	15	6	3	49	170		121
4	0	1	1	1	25	20	8	4	57	170		113
5	0	1	1	1	25	25	10	5	65	170		105
6	0	1	1	1	25	30	12	6	73	170		97
7	0	1	1	1	25	35	14	7	81	170		89
8	0	1	1	1	25	40	16	8	89	170		81
9	0	1	1	1	25	45	18	9	97	170		73
10	0	1	1	1	25	50	20	10	105	170		65
11	0	1	1	1	25	55	22	11	113	170		57
12	0	1	1	1	25	60	24	12	121	170		49
13	0	0	1	1	25	60	26	13	124	170		46
14	0	0	1	1	25	60	28	14	127	170		43
15	0	0	1	1	25	60	30	15	130	170		40
16	0	0	1	1	25	60	32	16	133	170		37
17	0	0	1	1	25	60	34	17	136	170		34
18	0	0	1	1	25	60	36	18	139	170		31
19	0	0	1	1	25	60	38	19	142	170		28
20	0	0	1	1	25	60	40	20	145	170		25
21	0	0	1	1	25	60	42	21	148	170		22
22	0	0	1	1	25	60	44	22	151	170		19
23	0	0	1	1	25	60	46	23	154	170		16
24	0	0	1	1	25	60	48	24	157	170		13
25	0	0	1	1	25	60	50	25	160	170		10
26	0	0	1	1	25	60	52	26	163	170		7
27	0	0	1	1	25	60	54	27	166	170		4
28	0	0	1	1	25	60	56	28	169	170		1
29	0	0	1	1	25	60	58	29	170	170		0

Fig. 4.24 Control of horizontal focus

As Mordecai Ezekiel demonstrated in his famous diagram (Ezekiel 1938, pp. 262, 264), which derives from the "cobweb theory," the search for equilibrium between demand and supply and the determination of an equilibrium price can be viewed as the result of a series of repositionings "on" the demand and supply curves, based on the typical logic of dynamic systems, which, in relation to the slope of those curves, causes prices to converge toward an equilibrium price (in a typical cobweb representation) or to periodically oscillate.

In a simplified attempt at explanation, referring to the model in Fig. 1.2, if $[D - S > 0]$, then the consumers whose demand is left unsatisfied will exercise pressure on the producers, causing them to raise prices in the attempt to increase profits, as they are sure that they will sell their production. If $[D - S < 0]$, then the producers will note an increase in stocks due to lower sales, which will move them to reduce prices in the attempt to eliminate the stocks, thereby providing incentives to consumers to buy their goods.

The model in Fig. 1.2, though interesting from the point of view of a qualitative understanding of the dynamics of the variables, can be improved as soon as we translate it into a control system that transforms the model in Fig. 1.2 into the more efficient one in Fig. 4.25, where the "true" variable to control is the difference between demand and supply $[D - S]$. The model also indicates the action rates "g" and reaction rates "h." The former express the price elasticity of demand and supply. If these rates are constant, then demand and supply have a straight-line trend; if, on the other hand, "g" varies with the size of D and S, then these variables have nonlinear dynamics. The reaction rate could be differentiated for D and S, in which case we shall indicate them as "g_D" and "g_S."

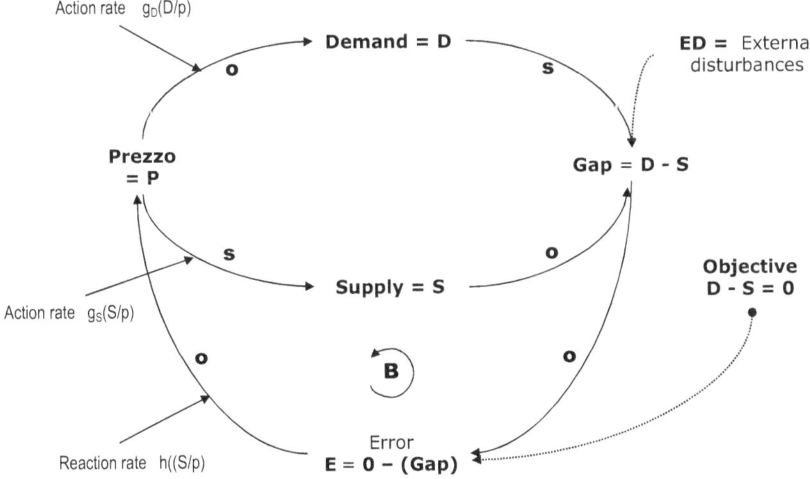

Fig. 4.25 The control system for demand and supply through price

Figure 4.26 presents a very simple teaching simulation model that assumes that:

1. For each unit of price, demand and supply vary by ten units.
2. At $p = 0$ demand is $D = 100$ and supply $O = 0$.
3. For every ten units of gap, price varies by one unit.
4. At $t = 0$ price is $p_0 = 0$.

We can observe that, under the simplified conditions assumed in the model, the system attains equilibrium at $t = 5$, when demand and supply are equal, at price $p = 5$. The result is easy to calculate, even intuitively, since the initial difference between D and S is $E = 100$; this causes the price to increase, and since price increases by one unit at each successive instant, after five instants demand falls to $D = 50$ and supply increases to $S = 50$.

To make the example more significant, we assume that for the successive 10 periods after $t = 5$, demand increases by 10 each period ("Disturbance D" column), which is equivalent to an overall increase in demand in the decade of 100 units, so that at $t = 15$ total demand is $D_{15} = 200$ units. Since supply at $t = 0$ is $S_0 = 0$, under the hypothesized conditions the equilibrium price becomes $p_{\text{equilibrium}} = 10$.

At $t = 18$ supply begins to fall over 6 periods ("Disturbance S" column); this is equivalent to a situation where maximum demand is 200, after the increase occurring over 10 periods, while total supply falls to 40. The control system inevitably brings the price to $p = 13$, at which level demand and supply are $D_{23} = S_{23} = 70$.

While we can easily intuit that the price rises to $p = 10$ to equilibrate demand and supply at 100, the increase in price to $p = 13$, with demand and supply in equilibrium at 70, represents a counterintuitive result.

There is no error here; at $t = 18$ the equilibrium $D_{17} = S_{17} = 100$ is altered by the fall in supply by 10 units. The system must readjust to increase price to $p_{19} = 11$ in order to reduce demand and, at the same time, provide the incentive to supply more.

Nevertheless, a new reduction in supply occurs that leads to a further increase in price. This pattern goes on until $t = 23$, when a stable equilibrium is restored at price $p_{23} = 13$. As proof of the correctness in the behavior of the control system of price in Fig. 4.26, we assume that the reduction in supply is eliminated by an annual increase of ten units over six periods ("Disturbance S" column), once again bringing total supply to $S_{29} = 100$. At the end of the decreasing dynamics and adjustment of the system, price returns to $p_{30} = 10$ and equilibrium is again achieved at $D_{30} = S_{30} = 100$.

4.8.8 Ordering of the Objectives to Define Control Policy: The Direct Comparison Procedure

As we observed in Sect. 4.7, in multiobjective control systems the definition of a *policy* assumes the possibility of the *governor* identifying the most "important," "urgent," or "useful" objectives. In order to assign to each objective, a *priority index*

D_0	Starting Demand	100	g_D = Action rate(price/demand)	10
O_0	Starting Supply	0	g_S = Action rate (price/supply)	10
p_0	Starting price	0	Δ p	1

t	Price	Demand D	Supply S	disturbance D	disturbance S	Error = = (D-S)	Δ p
0	0	100	0			100	1
1	1	90	10			80	1
2	2	80	20			60	1
3	3	70	30			40	1
4	4	60	40			20	1
5	5	50	50			0	0
6	5	60	50	10		10	1
7	6	60	60	10		0	0
8	6	70	60	10		10	1
9	7	70	70	10		0	0
10	7	80	70	10		10	1
11	8	80	80	10		0	0
12	8	90	80	10		10	1
13	9	90	90	10		0	0
14	9	100	90	10		10	1
15	10	100	100	10		0	0
16	10	100	100			0	0
17	10	100	100			0	0
18	10	100	90		-10	10	1
19	11	90	90		-10	0	0
20	11	90	80		-10	10	1
21	12	80	80		-10	0	0
22	12	80	70		-10	10	1
23	13	70	70		-10	0	0
24	13	70	70			0	0
25	13	70	80		10	-10	-1
26	12	80	80		10	0	0
27	12	80	90		10	-10	-1
28	11	90	90		10	0	0
29	11	90	100		10	-10	-1
30	10	100	100		10	0	0
31	10	100	100			0	0
32	10	100	100			0	0

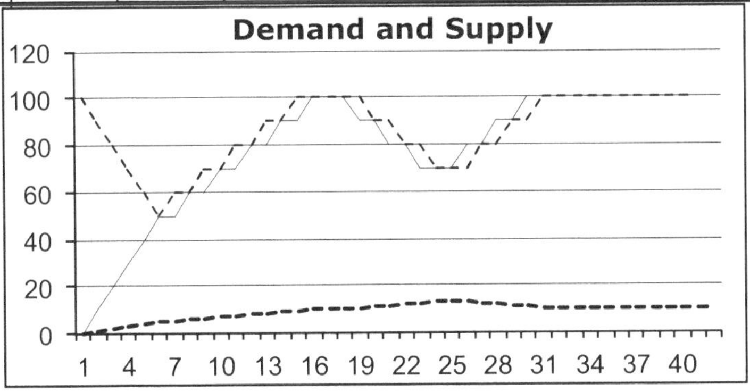

Fig. 4.26 Simulation of a control system of demand and supply (teaching model)

(urgency, importance, utility, etc.), it is necessary to rely on some *metrics procedure* (which is different from measurement) that can produce *pseudo-cardinal scales* capable of indicating the "distance" between the values assigned to the objectives. There are two procedures that can be applied:

- The *direct comparison* procedure
- The *linear combination* procedure using the *standard gamble* method

I shall define direct comparison as the procedure that seeks to define the pseudo-cardinal scales by deriving the priority indices through repeated comparisons among the various objectives. This method was originally applied in decision-making theory and in operational research to determine the weights of the objectives or a pay-off measure of a given set of predicted results from various courses of action. In this context the method must be attributed to Churchman and Ackoff, in their work entitled *An Approximate Measure of Value* (1954, pp. 172–180); see also Churchman, Ackoff, and Arnoff (1957, pp. 132–139); Ackoff and Sasieni (1968, pp. 51–54), describe the assumptions for the application of the method; a critical analysis can be found in Fishburn (1964) and in Fishburn (1967b, pp. 69–82). The method is applied not only to the objectives but also to all the objects to which no measurement procedure can be applied but for which the notion of distance or relation between their states makes sense.

In order to understand the logic of this method, which is applied to general objects, an analogy is useful. Let us assume that we have collected five objects, for example, five rocks, and that we want to determine their relative weights in the absence of a measurement procedure. We could first of all order the five rocks according to an estimate of their weight using our sense organs. Let us say that the order goes from O1, the heaviest object, to O5, the lightest. The following question arises: How heavier is O1 than O2, O3, O4, and O5? A scale would have allowed us to give a precise answer, since it *measures* the weight of the different rocks. If O1 weighed 1000 g, O2 500, O3 400, O4 300, and O5 100, then we could state that O1 is ten times heavier than O5, but also that O1 is twice as heavy as O2 or that O4 is triple the weight of O1, etc.

When it is not possible to measure objects, we can turn to the *direct comparison* procedure, which allows us to construct a *relative and approximate* cardinal scale, ad hoc, even without being able to measure the objects precisely. The procedure involves three phases:

1. The objects are ordered in a decreasing scale, from the heaviest to the lightest, using some instrument which is easy to construct or simply by using our senses.
2. Each object is assigned a quantitative index, or approximate measure, which, based on our senses, could come close to the weight (which we do not know) of the objects.
3. The first approximation measures are subjected to a series of adjustments so that, at the end of the process, they are consistent with the rational estimate of the differences of weights and relations among the objects; thus, we can state that if O1 is assigned an indicator of 100 and O2 of 50, then O1 is supposed to have double

the weight of O2, with respect to the indicators assigned in a coherent manner. The observer must be able to carry out an operational procedure for the ordinal comparison of any subsystems of the objects of observation.

To better understand the above procedure, let us apply it to the five objects mentioned above, assuming obviously that we do not have a graduated scale to weigh them. I suggest that the reader imagine carrying out the metricizing process using a common two pan balance, without, of course, having the weights against which to weigh the objects (e.g., 10, 100, 200, 500-g weights), which would have permitted an accurate measurement. We can proceed with the *systematic* and *repeated* comparison of the five objects by placing them in various subsystems in the two dishes.

Let us assume that we have obtained the results shown in the table in Fig. 4.27. The way the comparison works is simple: the heaviest object, O1 (left dish, for example), is compared to the group of remaining objects (right dish). Then O1 is compared to the remaining objects, excluding O5, which is the lightest; then O1 with the remaining objects, excluding O4 as well, and so on until O1 exceeds the weight of the remaining objects in the right-hand dish, which in our example occurs in the fourth comparison (see column 1).

As shown in phase (2), we assume that the five objects have been assigned the *first approximation indices* shown in column [2] of Fig. 4.28. As is immediately clear, apart from the change in scale, the first approximation indices are far from the measured weights of the objects obtained with a graduated scale. A comparison with the table in Fig. 4.27 allows us to verify the adequacy of the indices assigned as a first approximation and to proceed with their gradual adjustment. Let us begin with comparison (8) in column 3 of Fig. 4.27. Since we have to verify that

Column 1	Column 2	Column 3
START		
(1) O1 < O2 + O3 + O4 + O5		
(2) O1 < O2 + O3 + O4		
(3) O1 < O2 + O3 + O5		
(4) O1 > O2 + O3 OK		
	(5) O2 < O3 + O4 + O5	
	(6) O2 < O3 + O4	
	(7) O2 > O3 + O5 OK	
		(8) O3 > O4 + O5
		END

Fig. 4.27 Direct comparison procedure. The systematic comparison

O3 > O4 + O5, then it also must be the case that n3 > n4 + n5. However, we instead observe from the indices in column [2] of Fig. 4.28 that n3 = n4 + n5 = 5. Thus, it is necessary to adjust the index assigned to O3 to obtain the desired inequality. If we change n3 = 5 to n3 = 5.5, then the indices of O3, O4, and O5 are congruent; they are written in column [3] of Fig. 4.28.

We then proceed with comparison (7) in Fig.4.27, from which we get O2 > O3 + O5. As a result, we should get n2 > n3 + n5; since we can see from column [3] of Fig. 4.28 that n3 + n5 = 6.5, we see the inconsistency of n2 = 5,5. We can remove this inconsistency, for example, by setting n2 = 7. This new value is written in column (4) of Fig. 4.28, along with that for column (3).

We proceed in the same manner for the analysis of the other comparisons. From comparison (6) we should have n2 < n3 + n4; since, with the indices already present in column [4], the inequality is verified, the adequacy of n2 = 7 is further confirmed.

Comparison (5) should result in n2 < n3 + n4 + n5; since the inequality is confirmed, the adequacy of n2 = 7 is definitively verified (this comparison is superfluous, with the validity of (6) already having been confirmed).

From comparison (4) we should have n1 > n2 + n3; the data from column [4] in Fig. 4.28, where n1 = 8 and n2 + n3 = 12.5, reveals the nonadequacy of n1 = 8. Thus, we could, for example, change n1 = 8 to n1 = 13; this value, along with the preceding ones, appears in column [5] of Fig. 4.28.

Comparison (3) indicates that the index n1 = 13 is consistent; the index n1 > 13.5 would not have been so, however.

The index n1 = 13 in comparison (2) is appropriate; however, n1 > 16.5 would not have been adequate.

The index n1 = 13 in comparison (1) is verified as being consistent.

Objects	First Approximation Indices	Initial Adjustment	Second Adjustment	Final Adjustment	Definitive Indices	Normalized Indices
[1]	[2]	[3]	[4]	[5]	[6]	[7]
O1	n1 = 8	n1 = 8	n1 = 8	*n1 = 13*	n1 = 13	N1 = 0,43
O2	n2 = 5,5	n2 = 5,5	*n2 = 7*	*n2 = 7*	n2 = 7	N2 = 0,23
O3	n3 = 5	*n3 = 5,5*	n3 = 5,5	n3 = 5,5	n3 = 5,5	N3 = 0,18
O4	n4 = 4	n4 = 4	n4 = 4	n4 = 4	n4 = 4	N4 = 0,13
O5	n5 = 1	n5 = 1	n5 = 1	n5 = 1	n5 = 1	N5 = 0,03
TOT					30,5	1

Fig. 4.28 Direct comparison procedure. Adjustment of the indicators

After the verification of congruence has been completed, the first approximation measures in column [5] of Fig. 4.28 can be considered as definitive, forming a pseudo-cardinal scale. These measures can then be rewritten in column [6] and normalized, as indicated in column [7], by calculating the ratio of each measure in column [6] to the total for the measures.

We observe that the procedure would have provided equally consistent, though different, results even if the verification process for consistency had started with comparison (1) rather than (8) in Fig. 4.27. However, in this case, as the reader can readily see, based on the numerical data from the example the procedure would have been less efficient, requiring a greater number of "revisions" of the first approximation measures.

The direct comparison procedure is particularly simple and well suited to evaluating the importance or the urgency of the objectives of the multiobjective control system. However, there are a number of limits to the practical application of this method; above all, the governor-manager must allocate the objectives according to an ordinal scale, which is not always possible, especially when there is a circularity of preferences.

Secondly, this technique does not provide measures of importance or urgency but coherent indices assuming a correct observation of the distances. In effect, each measure can be considered as an element of a given interval of uncertainty. It is easy to check that the scale of indicators in Fig. 4.28 would be coherent even if the index n4 = 4 rather than n4 = 5 is assigned to object O4. Thus, the indicators obtained are subjective and are only valid within more or less uncertain boundaries and coherent only when using the judgment scales of the observer.

4.8.9 The Standard Gamble Method

The standard gamble method, or linear combination procedure (Sect. 4.8.8), can also be appropriately applied whenever it is necessary to determine the indices of the preferences of objectives or, more generally, of objects with respect to nonquantitative measures. The method is based on the hypothesis that an individual is capable of expressing coherent preferences among "linear combinations" of objects evaluated with respect to a qualitative dimension. The method was proposed in this context by John von Neumann and Oskar Morgenstern, in Theory of Games and Economic Behavior (1953, p. 26 and following). The authors assembled a system of axioms that guarantee the coherence of the determinations. In particular, the measure for utility must exclude any complementarity or incompatibility among the objects, whether the latter be monetary in nature, objectives, or courses of action. For a general axiomatization, see also Luce and Raiffa (1967), p. 23 and following. A complete axiomatization in the context of a complex formalization of the method can be found in Champernowne (1969, Vol. 1, p. 9), as well as in Blackwell and Girshich (1954, p. 104).

Assume that there are three objects, O1, O2, and O3, observed in a nonquantitative dimension D (importance, urgency, etc.), and that the observer can order them, for example, in the following way: O1 > O2 > O3.

We also assume that the evaluator-individual can determine a *coefficient of equivalence*, "c," from "0" to "1," so that O2 (intermediate object) is held to be equivalent (~) to the "linear combination" of the other two:

$$\mathrm{O2} \sim c\mathrm{O1} + (1 - c)\mathrm{O3}.$$

As the objects have already been ordered as O1 > O2 > O3, it follows that the evaluator-individual:

(a) Holds that if $c = 1$, O2 is then *less* than the linear combination of O1 and O3, with respect to the qualitative dimension D.
(b) Holds that if $c = 0$, O2 is instead *greater* than the linear combination of the other two.

Once "c" has been determined, from *personal choice* or through some *experiment*, the evaluator-individual can assign the following indices to the three objects:

$$\mathrm{n1} = 1, \mathrm{n2} = c, \mathrm{n3} = 0.$$

These indices form a *pseudo-cardinal* scale of order. The measures assigned are coherent and reflect the conditions of observation of the distances among the objects, though this relies on the subjectivity of the observer's choices. This method can also be applied to a number of objects of whatever kind, by combining them in groups of three, if the following conditions, or axioms, apply:

1. Condition of coherence: The observer must be able to order the objects coherently based on an ordinal scale.
2. Condition of combinability: The observer must be able to form linear combinations between any pair of coherently ordered objects.
3. Condition of existence: The observer must be able to determine the coefficient of equivalence for every imaginable linear combination; this is the most difficult condition to find.
4. Condition of comparability: The observer must be able to compare objects and linear combinations and to coherently order them.
5. Composition/decomposition: It is possible to combine several combinations into one or to break up one combination into others based on the elementary rules of algebra.

If the above conditions hold, the linear combination method is equally as useful as the *direct comparison* one in metricizing aspects such as utility or the importance of objectives or objects in any other nonmeasurable domain.

In fact, the method was originally created to assign measures to the utility of given objects in order to determine the payoff measures in the field of decision-making and game theory. In this context the fundamental problem with the linear combination method (to determine the coefficients of equivalency, "*c*") is immediately solved by calculating the probabilities the observer (the observer is a decision-making subject) must be capable of quantifying. In this way the linear combinations become "lotteries," or "standard games," that involve the various objects; in this case the objects were payoffs deriving from the carrying out of courses of action or strategies. For this reason, in the context of decision-making and games theory the procedure is also referred to as the *standard gamble method.*

The method can be applied with several variations to illustrate that I shall turn to a simple numerical example applied in decision-making theory. If a decision-maker, in hypothetically carrying out four alternatives of action, expects to obtain the following results, R1 > R2 > R3 > R4, which follow a decreasing order in terms of preferences, we can also assume that the results are objectives of a control system or represent other nonmeasurable objects of observation. We need to assume that the decision-maker can create "lotteries" where those results must be obtained with the given probabilities "*p*," which play the role of the coefficient of equivalence "*c*" of the general method.

Let us consider a *first variant* of the *standard gamble* in which the decision-maker creates the following pairs of lotteries:

LOTTERY A: [R1 with probability p1 + R4 with probability (1 − p1)] or [R2 certain]
LOTTERY B: [R1 with probability p′1 + R4 with probability (1 − p′1)] or [R3 certain]

Assume that the decision-maker is "forced" to choose between the lotteries (which correspond to the linear combinations among the objects of observation) and is indifferent in his or her evaluation of the two terms of LOTTERY A, if $p = 30/45$, and the two terms of LOTTERY B, if $p' = 15/45$. These probability valuations give us the following preference indices:

$$\mathrm{n1} = 10, \mathrm{n2} = 6.7, \mathrm{n3} = 3.3, \mathrm{n4} = 0.$$

These indices are coherent with the observation of the distances among the preferences toward the various results, even if result n4 is difficult to manage with a ratio scale, thereby making the method cumbersome in many situations.

Now let us consider a *second variant* of the *standard gamble* that overcomes this limitation; in this case the decision-maker has to choose among three lotteries (there are four results):

LOTTERY A: [R1 with probability p1] or [R2 certain]
LOTTERY B: [R1 with probability p′1] or [R3 certain]
LOTTERY C: [R1 with probability p″1] or [R4 certain]

It is assumed now that the decision-maker, forced to choose among the lotteries, evaluates indifferently the two terms in LOTTERY A, if p1 = 80%, the two terms in LOTTERY B, if p′ = 50%, and the two terms in LOTTERY C, if p″1 = 15%. Using R1 as the term of comparison, we can then assign the following preference indices:

$$n1 = 10, n2 = 8, n3 = 5, n4 = 1.5.$$

These indices do not coincide with those assigned in the first variant of the *standard gamble*, since the lotteries whose indifference probabilities have to be evaluated have changed; however, in this case as well the method guarantees the coherence of the resulting scale.

There is a *third variant* of the *standard gamble* we can consider, in which, in place of the preceding lotteries, the decision-maker is faced with the following ones, whose main term of comparison is R4 instead of R1:

LOTTERY A: [R1 with probability p1] or [R4 certain]
LOTTERY B: [R2 with probability p2] or [R4 certain]
LOTTERY C: [R3 with probability p3] or [R4 certain]

Assuming that the decision-maker, forced to evaluate the lotteries, has chosen the following probabilities, p1 = 19/200, p2 = 1/8, and p3 = 2/7, we can assign the following preference indices:

$$n1 = 200 / 19 \approx 10.5, n2 = 8.0, n3 = 3.5, n4 = 1.$$

This third variant appears, at least in theory, to be preferable since, like the *direct comparison* method, it takes as its term of comparison, or *scale unit*, the "lesser" object (R4, in our example).

The limitations of the *linear combination* procedure are evident: first and foremost, the assumption of the comparability of the linear combinations and, secondly, the choice of the *coefficient* of indifference. For this reason, the *direct comparison* method is perhaps more "operational," as this method involves only comparisons among the objects and aggregations of objects (sums, noncombinations of the dimensional states).

4.9 Summary

We have learned that:

1. Systems can be multilever (Sect. 4.4), multilayer (Sect. 4.5), and multiobjective (Sect. 4.6).
2. The simplest multilever systems are the two-levered ones: one lever, X1, to produce increases (decreases) in *Y*, and another, X2, to produce variations of equal or opposite sign in *Y*. The levers can be:

   (a) *Mutually dependent* (Sect. 4.1), or linked, if X1 and X2 can only produce variations of the opposite sign as *Y* (Fig. 4.1).
   (b) *Independent* (Sect. 4.2), or free, if both X1 and X2 can produce variations of the same sign as *Y* or variations of the opposite sign (Fig. 4.3).

3. A particular type of two-lever system is an impulse system (Sect. 4.3), typically used to control warehouses and biological life; such systems can be:
   (a) CI–PO functioning (Figs. 4.7 and 4.8).
   (b) PI–CO functioning (Fig. 4.9).
4. Multilayer control systems (Sect. 4.5) can have levers of various orders that act at different levels of control (Fig. 4.13), specifically:
   (a) *Operational*, or first-order, levers; these are used immediately (they can be free or linked).
   (b) *Reinforcing*, or second-order, levers if they are complementary to the operational levers.
   (c) *Extraordinary*, or third-order, levers; these are used when the first-level levers are insufficient to eliminate the variance.
   (d) *Structural*, or fourth-order, levers; these are used even if the extraordinary levers are not sufficient to achieve the objective.
5. In multilever systems, the manager must determine the order of activation of the levers, that is, the *control strategy* (Sect. 4.5), by carrying out a CBA to identify the most effective and efficient levers (Sect. 4.7).
6. The multiobjective systems (Sect. 4.6) are those where the manager simultaneously controls different objectives by employing levers of varying levels (Fig. 4.16); these can be of three types:
   (a) *Apparent*, if they can be conceived of as the conjunction of several independent single-objective systems.
   (b) *Improper*, if they derive from various interfering single-objective systems.
   (c) *Proper*, if they must achieve multiple interdependent objectives.
7. In multiobjective systems, the manager must define the order of priority of the objectives, that is, the control *policy* (Sect. 4.7), which must be linked to the control *strategy*, which as we have seen concerns the choice of the control levers.
8. The definition of the control policy must occur on the basis of accurate techniques to order the objectives, which involve defining their scale of priorities, according to metricization procedures that determine pseudo-cardinal scales (Sects. 4.8.8 and 4.8.9).

Chapter 5
The Ring: Observation and Design

Abstract In this chapter, I shall outline the guidelines for recognizing, observing, or designing control systems and the problems that arise regarding their logical realization. I shall introduce the fundamental distinction between symptomatic and structural control. I shall then examine the concepts of the *efficiency* and *effectiveness* of control systems and the various techniques for increasing the effectiveness of control through the use of internal and external interventions to strengthen the system. A review of the various possible risks of failure of the control process due to structural causes will then be presented. The last part of the chapter considers the *human aspects of control*, which involves several ontological aspects of control: discouragement, insatiability, persistence, and underestimation. The chapter concludes by presenting the logic behind the decision-making process.

Keywords Design control system · Pathologies of control · Symptomatic control · Structural control · Effectiveness of control systems · Efficiency of control systems · Risk of failure of the control process · Archetypes · Problem solving · Leverage effect · Definitive solution · Multicriteria decision-making

> Many workers in the biological sciences—physiologists, psychologists, sociologists—are interested in cybernetics and would like to apply its methods and techniques to their own speciality. Many have, however, been prevented from taking up the subject by an impression that its use must be preceded by a long study of electronics and advanced pure mathematics; for they have formed the impression that cybernetics and these subjects are inseparable.
>
> The author is convinced, however, that this impression is false. The basic ideas of cybernetics can be treated without reference to electronics, and they are fundamentally simple; so although advanced techniques may be necessary for advanced applications, a great deal can be done, especially in the biological sciences, by the use of quite simple techniques, provided they are used with a clear and deep understanding of the principles involved (Ross Ashby 1957, Preface, p. v).

The previous chapter completed the presentation of the logical structure of control systems by introducing two important generalizations: (1) *multilever* control systems and the concept of *control strategy* and (2) *multiobjective* control systems

P. Mella, *The Magic Ring*, Contemporary Systems Thinking,
https://doi.org/10.1007/978-3-030-64194-8_5

and the concept of *control policy*. The chapter also mentioned the concepts of *cost–benefit analysis* and *scale of priorities* as basic elements in determining control strategies and policies. In this chapter, which generalizes the considerations presented in the preceding ones, I shall outline the guidelines for recognizing, observing, or designing control systems and the problems that arise regarding their logical realization. I shall introduce the fundamental distinction between symptomatic and structural control. A symptomatic control is short-term and tends to eliminate the error by acting only on the error itself through symptomatic levers. A structural control acts on the effective causal structure between *X* and *Y*; it is long term and tries to permanently eliminate the error by acting on *Y*. I shall then examine the concepts of the efficiency and effectiveness of control systems and the various techniques for increasing the effectiveness of control through the use of internal and external interventions to strengthen the system. A review of the various possible risks of failure of the control process due to structural causes will then be presented, with particular reference to failure caused by the use of the wrong levers. Two fundamental archetypes, "Fixes That fail" and "Shifting the Burden," will clarify how these risks manifest themselves in a widespread and recurring manner. The last part of the chapter considers the *human aspects of control*, which involves several ontological aspects of control: discouragement, insatiability, persistence, and underestimation. The chapter concludes by presenting the logic behind the decision-making process or behind the replacement of the present states for desired future ones; the decision-making process can, in all respects, be viewed as a unique control process.

5.1 How to Recognize or Design the Logical Structure of a Control System

After presenting the theory of *multilever* and *multiobjective* control systems, and in order to be prepared to recognize the vastness of the array of control systems in all observational and operational contexts that will be examined in Part II, the reader will have the opportunity to apply himself in the control discipline. In order to fully achieve the objective set out in Sect. 1.1—to systematize the control discipline and favor its application and spread—it is necessary to complete the task by proposing several guidelines for the correct *observation*, analysis or *design* of the *logical* structure of control systems. It is useful to recall in this *section* some of the many aspects and problems we have encountered in the analysis of control systems in previous chapters.

We must always keep in mind that, even before we realize the *physical* structure of the system, *observing* or *designing* a control system means understanding or designing the *logical* structure of the control process. The first step in configuring the *logical* structure of a control system is to identify the *governed* variable *Y* (or variables, in the case of multiobjective systems) to control. This is not a free choice, since obviously the control is necessary when the manager of the system identifies or sets an objective or constraint, Y^*. The *objectives* or *constraints* determine the choice of the variable whose dynamics allows the objectives or constraints to be achieved.

In the examples illustrated in the preceding *sections*, many control systems were prearranged for the achievement of an objective represented by a specific value of the variable Y_t. We define these objectives as *goal* or *punctual* objectives. Many control systems have a more complex functioning, since they must produce a dynamics in Y_t that conforms to a succession of differently timed values. A succession of values of Y_t that the control system must achieve is defined as a *path* or *tendential objective*, and the system is a *tracking* control system. It is one thing to reach a destination—for example, a city street—and another to do so by taking a certain route with constraints, for example, passing through the center or by your aunt's house. This second type of objective is more difficult to identify, and often the observer views it as a simple succession of punctual objectives or as variations of an original *goal* objective.

Objectives can be further broken down into *static* (or fixed) objectives, which have a sole value (goal)—or a fixed sequence of values (tendential objectives)—to achieve, or as *dynamic* objectives, which vary according to a function that depends on the values of the variable Y and/or X, as analytically illustrated in Sect. 3.4, which can be applied to the case of two systems.

It is obvious that a hungry lion has the objective of reaching the gazelle that is calmly nibbling the grass in the savanna; or, better yet, to reach the point where the gazelle is at that moment. If the lion is downwind and the gazelle is relatively still while nibbling, then its position represents a *static goal* for the approaching lion, who can easily activate a system of collision (Sect. 4.8.5). Since it is well known that lions do not prefer that tactic but instead approach their prey trusting in their speed and strength, the *position* of the gazelle, as we can easily imagine, almost never represents a static objective. The gazelle, perceiving the danger from the predator, starts to flee, so that its position becomes a *dynamic* objective that the pursuing lion's trajectory must reach.

This *dynamic* objective—the gazelle's trajectory Y^*—that the lion's trajectory must reach can be considered a function of the lion's trajectory, Y_t, and thus its exertions, X_t, as well as of the gazelle's position in the chase environment, Y^*_t; thus we can more generally write: $Y^*_{t+1} = F(Y_t, X_t, Y^*_t)$, using one of the forms presented in Sect. 3.4. On the other hand, the position of the flower the bee wants to reach represents a clear example of a *static* objective. We realize that in the struggle for existence all prey are the objectives of predators; for vegetarian predators, the prey (their position) generally represents a *static* objective; for carnivorous predators the position of a live prey is vital to reach and normally represents a *dynamic* objective. We must recognize that these two types of objective are present in all aspects of life, even the sporting one. In fact, many interesting control systems in the world of sports have variable objectives. Though I enjoy watching the high jump, long jump, the running and walking competitions, the field events with their many types of equipment, and cycling events (all sports that have a fixed objective), I find it much more pleasing to watch baseball, rugby or soccer, all sports where each player is obviously a control system trying to achieve a variable objective, with variations that depend in part on external disturbances and in part on actions by the player himself.

Many objectives are *quantitative* and represent desired numerical values for specific quantitative variables, which can be measured using a given procedure that is defined from time to time. Along with these we should not underestimate the broad category of *qualitative* objectives (flavors, colors, forms, etc.) set by the manager-governor based on his needs and experience.

The producers of food and drinks (wines, cheeses, cookies, soft drinks, etc.) or those that prepare them (chefs, restaurateurs, etc.) well know how difficult it is to maintain a long-standing flavor standard, even with changes in the quality of the raw materials, and for this reason they set up sophisticated control systems to guarantee constant quality over time.

Artists also "know" how their work will look in the end before producing it, and this knowledge guides their work toward achieving their vision (achievement); the great Michelangelo clearly expressed this idea when he said that *the statue already exists in the block of marble, and that the sculptor only has to free it* by eliminating the excess marble. A quote normally attributed to Leonardo clarifies this concept.

> In every block of marble I see a statue as plain as though it stood before me, shaped and perfect in attitude and action. I have only to hew away the rough walls that imprison the lovely apparition to reveal it to the other eyes as mine see it (Shaikh and Leonard-Amodeo 2005).

Perhaps most of the control systems in operation every day, and which allow us to have a normal existence, are those characterized by qualitative objectives. We ourselves change restaurants or cafés when we feel that the qualitative objectives of flavor, courtesy, and environment are not met. Similarly, my mother-in-law knows when the wash is not perfectly clean when it comes out of the machine, just as my mechanic knows that the noise from my motorcycle's engine is different from what one would expect.

The presence of qualitative objectives may make the formalization of control systems more complicated, but this in no way changes their logic. The most difficult *qualitative* objectives are forms (faces, environments, models, fingerprints, stripes on a blanket, etc.), which all animals equipped with sight, including humans, learn to recognize better from repetitions of the recognition cycle, time after time, which allows them to construct a catalog of forms. This process helps them to live and survive.

As *recognizing systems* (Sect. 3.2), we recognize the face, voice, posture, walk, and handwriting of an incredibly high number of people and animals. We recognize the attitude, gestures, clothing of friends and enemies, just as animals recognize the odors, footprints, plumage, etc., of friends and enemies. In many cases, it is not always easy to identify the control objectives and variables, since a variable to control can sometimes be only apparent. This occurs whenever Y_t is part of a causal chain (e.g., *A*–*B*–*C*–*Y*). Since *Y* varies as a result of the variations of the variables preceding it in the causal chain, the "true" variable to control may not be *Y* but *A*, making it then necessary to identify an objective to assign to A, say A^*, which is entirely equivalent to the objective Y^* we desire for Y, so that when A reaches A^*, Y reaches Y^*. It may be possible to control Y with ad hoc levers, but this control would only be temporary since the subsequent variations in A would take Y to nonoptimal values. I shall define this situation as *apparent control*, since there is an attempt to constrain variable Y without, however, controlling the variable A, from which Y's dynamics structurally derive.

I would like now to consider a new question: how many levers must we trigger in order to achieve a complete recognition or to construct a meaningful control system? If we go back to the example of the two-lever control system of water temperature in Fig. 4.3, we immediately realize that in the case of a control with two independent levers, the variable to control (the water temperature) derives, in fact, from a *mix* of hot and cold water. What our skin perceives (sensor apparatus) as a symptom is not so much the *volume* of hot or cold water but the sum of the temperatures of the two water flows, obviously weighted based on the size of the flows. Thus, it is possible to maintain this ratio by varying both the temperature of the flows and their size.

In our showers, we do not have the problem of varying the temperature of the water that flows from the two taps (or the problem arises only in the event of outside disturbances) since usually the temperature of the hot and cold water is maintained constant and the perceived temperature—from the *mix* of the two flows—depends only on the size of the two flows, given the constant water temperatures. If it were possible to vary both the size and temperature of the flows, then the control system of the shower would become a *four-lever system*. Since it is much more efficient to regulate water temperature using a mixer, as in the model in Fig. 2.7, showers with two separate taps have almost disappeared.

We can generalize: if in many cases control systems can be *observed* or *designed* with a *few* or with *many* control levers, then the *observers* or *designers* are torn between two tendencies: *on the one hand*, whether or not to increase the number of control *levers* [Xn], in order to allow a "precise control" of the *variable Y*, and on the other whether or not to reduce the number of *levers* [Xn], through appropriate *mixes* of their values using special apparatuses, in order to make the "control quick" and easy for the management of the system.

How many of us would be content to take a shower in a multilever control system where you had to control both the water flow and temperature? That is why modern mixers today regulate the shower very quickly and more comfortably. A chemical plant where the regulation of the mixture temperatures must be quite precise is another thing altogether. How many of the readers remember all the *manual* regulations that were necessary to take a good photograph with the "old" reflex cameras? How many of us hurried to replace our cameras with one that automatically regulated light and focus? How many of us who now have digital cameras that are completely "automatic" still remember the problems we had with manual settings for the shutter opening and shutter speed?

How should the observers and designers of control systems *behave*? Should they increase or decrease the control levers? Should they search for "precision" or instead aim for the "speed and simplicity" of regulation? We can only answer in general: observers and designers must above all determine the function the control system must carry out and whether or not the system is part of a larger system; subsequently they must evaluate whether or not to *zoom out*, toward a more general perspective (by reducing the number of control levers with appropriate *mixes* of action variables) or *in*, taking in more detail (by increasing the number of [Xn]).

The above observations pose a more general question: are there some criteria that suggest how many objectives [Ym] to control simultaneously and how to identify the optimal number of control levers [Xn]? Recalling the general rules of systems thinking proposed in Sect. 1.2, in particular the *first* (zoom in and out) and the *fifth* (specify the boundaries), the answer is a number of jointly controlled objectives and a number of control levers jointly considered depend on how the manager of the control observes reality.

We must always keep in mind (Mella 2012, p. 101) that the models that represent control systems must also (or above all) determine the essential aspects of the logical reality they represent, and thus must identify and connect all, and only those, levers [Xn] which the observer holds to be truly significant for his control objectives [Ym]. The more we zoom in "toward the smallest details," the more the number of levers are reduced; the more we zoom out "toward the larger picture," broadening the observed environment, the more the control systems will involve multiple objectives and imply multiple levers.

The eye system alone (zooming in on a narrower perspective) can automatically control the act of focusing on various objects (Fig. 4.10), even in movement, through the process of "accommodation"; that is, the rapid contraction of the crystalline lens (effector) using the ciliary muscle (regulator) which, by adding diopters for close-in focusing, guarantees a clear image (focus) even for objects very close to the eye. There are various disturbances to the focusing capacity, the most common of which are, as we know, myopia (the difficulty in focusing on distant objects), presbyopia (which inhibits our ability to focus on nearby objects), and hypermetropia (which makes it difficult to focus on both distant and nearby objects). An auxiliary lever that facilitates the ability to focus on objects is an increase in sunlight, natural, or artificial. If we zoom out and also consider the individual during his lifetime, we can add a third control lever: eyeglasses or contact lenses, which adjust our focus artificially. There are also structural levers that correct visual disturbances by intervening on the structure of the eye or by using excimer laser treatment or other surgical interventions deemed appropriate for more serious cases.

We have reached a *general conclusion*: the number of objectives and control levers of any specific control system does not depend so much on the "nature" of the system as on the extension of the *boundary* we place to circumscribe the reality we are observing or designing. In other words, every control system—and the model built to represent it—is not absolute and objective but relative and subjective. We can further *generalize*: once we have set up a control system, we can usually set new objectives and add other levers by zooming out to get "a broader picture," considering that an increase in the number of objectives to achieve also means increasing the number of control levers and the number of layers needed to achieve full control.

5.2 Symptomatic and Structural Control

After having recognized (observation) or defined (designed) the variable to be controlled, Y_t, the objective to attain, Y^*, and the control lever, X_t, it is important to reflect on the significance of the error $E(Y)$ (I am referring to single-lever and single-objective systems, but all of the following considerations hold for any more general control system). In formal terms, for the control system as such this variable has no other significance than that of representing the *distance* (gap) from Y^* for each instant of detection of the values of Y_t. For the system *manager* this variable has, however, a completely different meaning, which is connected to the manager's physical shape and experiences. In many cases, $E(Y)$ is identified by the manager–user as a *symptom* that is physically associated with distance.

In the shower, $E(Y)$ indicates if the water temperature is below or above the desired temperature, while for the manager–user it signifies the sensation of cold, hot, or well-being. For the manager–user of the radio system, $E(Y)$ signifies deafening noise or imperceptible whispering. In the system that controls the human body's state of nutrition or hydration, $E(Y)$ represents the sensation of hunger, thirst, or satiety. Tiredness is the physiological sensation that derives from the difference between the tolerated level of lactic acid in the muscles and the actual level.

We need to get accustomed to the fact that almost all our sensations of *need* or *satisfaction*, of *dissatisfaction* or *satiety*, can be interpreted as the meaning we attribute to the perception of a deviation between a normal physiological or mental state Y^* (objective or limit) and an actual state Y. They are the *symptoms* of disequilibrium between the actual situation and normality (objective or limit). These considerations are important when we face the next step in identifying the control levers that enable Y to reach Y^*, thereby eliminating $E(Y)$.

The search for the control levers must meet the general requirement of each *Causal Loop Diagram*; the X must have a causal relationship with the Y. Before making any further considerations, and going back to what we have just observed about the meaning of the error, we can state that, rather than finding a causal link with Y, the observer or designer often identifies an immediate link between X and $E(Y)$. For example, thirst, as we see in Fig. 4.6, is not a symptom of the lack of liquid but of the alteration of the salt/water ratio with respect to a normal value; it is a value of $E(Y)$ and not of Y. Thus, it should be clear that the causal link is between the water intake and the salt/water ratio. It would be a mistake to see a direct causal relation between the amount of water drunk and the intensity of thirst, which could lead to catastrophic consequences for our organism.

The same is true for pain, which in general does not represent the variable to control but the *symptom* of a difference (error) between a normal and an altered state of some part or organ in our body. Yet many of us think that pain is cured with analgesics. It is not pharmacologically wrong to take a painkiller for a headache, but the cause of the headache is to be found in some physiological alteration (stress, indigestion, cold, etc.); the control levers for a headache must try to eliminate this alteration, not only the symptom; X must act on Y and not on $E(Y)$.

However, in many circumstances when we cannot identify the lever that acts directly on Y—since the causal link between Y and its causes is unknown—we have no choice but to settle for acting on $E(Y)$, as if this were the true variable to control. For that matter, the curative power of some herbs for a headache has been known for centuries, even if it is only today that we know about many of the *levers* to use to eliminate the true physiological causes of headaches, which act on the chemistry of the brain or on the stress produced from the action of other organs (digestion, liver, etc.). Unfortunately not all *levers* that could control headaches are known, and the spread of headache centers is proof of this.

These considerations allow us to understand the distinction between *symptomatic* and *structural* control systems. Taking doses of analgesics to keep a headache under control, just like spreading on cream to fight muscle pain after a workout, eating to satisfy our hunger, or a digestive against overeating are merely attempts to find the *levers* that produce a *symptomatic* control.

A *symptomatic* control is short term; it eliminates the error by acting only on the error itself by using *symptomatic levers* not on its causes. In subsequent repetitions of the cycle the error is inevitably destined to return, since its causes have not been eliminated. On the other hand, a *structural control*, precisely because it acts on the effective *causal structure* between X and Y, is long term and tries to permanently eliminate the deviation by acting on Y.

Therefore, a *rule* to follow in the discipline of control is the following: *in observing or designing control systems we must identify the levers that can intervene on the structural variables that produce the symptoms, and not only on those acting on the symptoms*; in other words, we must identify a link between the *structural levers* X and Y and not between the *symptomatic lever W*, which acts on the symptom, $E(Y)$. Nevertheless, many persons, through ignorance or obstinacy, favor symptomatic control, thereby often aggravating the initial control problem. This is such a common error that it represents an obstacle in many organizations and in many control contexts, so much so that Peter Senge considers it a true *systems* ARCHETYPE, which he calls "Shifting the burden," which will be illustrated in Sect. 5.7.

5.3 Effectiveness and Efficiency of Control Systems

The general *conclusion* presented at the end of Sect. 5.1 leads to another question: is there some criterion for assessing the *effectiveness* of a control system; that is, its capacity to achieve its objectives or to respect its constraints, with the maximum precision and the minimum risk of failure? And is there also some criterion for testing the *efficiency* of a control system; that is, the capacity of the system to be effective by utilizing the *minimum number of control levers* activated by processes requiring the *minimum use of resources*?

To answer these questions, it is especially important to observe that control systems have their own *utility* and that only the governor–manager can determine the overall *utility* of the system by evaluating its effectiveness and efficiency. To assess the *utility* of the control system it is natural to consider the *effectiveness* requirement as having priority over *efficiency*, since an ineffective system does not allow the variable Y to achieve its objective; thus, *effectiveness* must be viewed as a *vital feature*, often to be achieved "at all costs." Only in the case of the design of alternative control systems—all capable of achieving the same objectives, with the same risk of failure—must the criterion of efficiency have equal importance to effectiveness in the evaluation. Nevertheless, we must consider that a control system is effective not only if properly sized—including all the relevant variables that influence the objectives—but also if the *levers* have a *range of variation* suitable for the admissible variations of the variables to control.

We must remember a *key principle of cybernetics*, which all evaluations of the effectiveness of control systems must adhere to: the *law of the necessary variety*, formulated by Ross Ashby (1957), according to which the "variety" of admissible states of the Y variables of a control system must be equal to or greater than the "variety" of the reality (disturbances, $E(Y)$ variables), which can alter the values of the variables to control, X, to such an extent as to not permit the objectives to be achieved. From this it follows that an effective control system must possess control levers, X, in *number*, *variety*, and *potency* so as to allow the error to be eliminated, despite the vastness of the environmental "variety" "[Only] *variety can destroy variety*" (Ashby 1957, p. 207). Ashby provides a formal demonstration, but I think it is sufficient to remind the reader how Ashby applies the law of requisite variety for the control of living systems:

> … the law of Lienhard Variety enables us to apply a measure to regulation. Let us go back and reconsider what is meant, essentially, by "regulation". There is first a set of disturbances D, that start in the world outside the organism, often far from it, and that threaten, if the regulator R does nothing, to drive the essential variables E outside their proper range of values. The values of E correspond to the "outcomes" [of the regulator]. Of all these E-values only a few (η) are compatible with the organism's life, or are unobjectionable, so that the regulator R, to be successful, must take its value in a way so related to that of D that the outcome is, if possible, always within the acceptable set … i.e., within physiological limits (Ashby 1957, p. 209).

In any case, the larger the variety of actions available (X) to a control system, the larger the variety of disturbances it is able to compensate for.

> Ashby's law is perhaps the most famous (some would say the only successful) principle of cybernetics recognized by the whole Cybernetics and Systems Science community. The Law has many forms, but it is very simple and commonsensical: a model system or controller can only model or control something to the extent that it has sufficient internal variety to represent it. For example, in order to make a choice between two alternatives, the controller must be able to represent at least two possibilities, and thus one distinction. From an alternative perspective, the quantity of variety that the model system or controller possesses provides an upper bound for the quantity of variety that can be controlled or modelled (Heylighen and Joslyn 2001, online).

Following Fig. 2.10, if $[X]$ stands for the vector of "active" variables and $[Y(X, D)]$ the "passive" ones, which suffer the "variety" of external disturbance variables $[D]$, the *control domain* is defined as the admissible range of the variables to control:

$$\left[Y(X,D)_{\min} \le Y \le Y(X,D)_{\max}\right]$$

This domain is obtained as functions of the values of the levers $[X]$ and the disturbances $[D]$ (to simplify the symbols, the variables in square brackets signify vectors of variables). The law of Requisite Variety enables us to specify the *General Criterion of Effectiveness*: a control system is *effective* in controlling $[Y(X, D)]$ if it includes a number and quality of control levers, $[X]$, capable of producing, overall, a sufficiently large *control domain* to achieve the objectives $[Y^*]$; that is, to eliminate the errors $[Y^* - Y(X, D)]$, producing values such that:

$$\left[Y(X,D)_{\min} \le Y^* \le Y(X,D)_{\max}\right]. \tag{5.1}$$

If the *control domain* is always $[Y^* > Y(X, D)_{max}]$, then the control system is *undersized*: if it is always $[Y^* < Y(X, D)_{min}]$, the system is *oversized*; in either case it is ineffective.

The *General Criterion of Effectiveness* produces the following actions to restore effectiveness:

1. Every *undersized* control system can always be adequately sized by means of two actions capable of raising $Y(X, D)_{max}$:
   (a) Inserting new control levers in $[X]$ and/or strengthening the range of admissible variation of $[X]$ by acting on $g(Y/X)$ and $h(X/Y)$ and/or.
   (b) Controlling the disturbance variables $[D]$, which will result in a rise in the maximum values of $[Y]$.
2. Every *oversized* control system can always be adequately sized by means of two mirror-image actions that can reduce $Y(X, D)_{min}$:
   (a) Reducing the number of control levers in $[X]$ (deactivation) or weakening some of them by reducing the range of admissible variation.
   (b) Controlling the disturbance variables $[D]$ to reduce the minimum values of $[Y]$.

   Control actions 1(a) and 2(a) can be defined as *structural* or *internal* because they impact on effectiveness by influencing the internal structure of the control system. Actions 1(b) and 2(b) act instead on the external environment, which represents the source of the disturbances $[D]$; therefore, I shall call these *environmental* or *external* control actions.

Structural actions depend on the nature of the control system and the available technology, thus on design potential and managerial quality. *Environmental* actions depend, on the other hand, on the type of disturbance that alters the dynamics of $[Y]$; they can also be carried out by additional autonomous control systems that combine with the main control system in a complementary way.

We must now reflect on which form of control action—structural or environmental—to carry out. I must first note that the general criterion of *Effectiveness* allows us to state the following *General Criterion of Efficiency*: in general, control systems are efficient to the extent they are designed using the *minimum number* of control levers, [*X*], necessary to produce an adequate control domain for the objectives [*Y**], both in terms of *precision* and the *cost* of control, thereby reducing as much as possible the *external control* of the disturbance variables *D*. Second, it is appropriate to point out that a number of jointly controlled objectives and a number of control levers simultaneously considered depend on how the *designer* and the *manager* of the control observe reality; the more they zoom in "on a narrower perspective," the fewer control levers there are. The more they zoom out "toward a broader perspective" of the observed environment, the more the control will involve multiple objectives and imply multiple levers.

The eye system taken by itself (zooming in on a narrower perspective) can control the quantity of light that reaches the retina through the contraction of the pupil (first lever) and the closing of the eyelids (second lever); if we also consider the individual, we can add a third control lever: dark eyeglass lenses. If we zoom in to observe an individual in a room we can implement the control with *environmental actions* as well, such as the opening or closing of the shutters and the switching on or off of the lamp. If we consider that the individual must also have the sharpness of vision necessary to read tiny characters, then we cannot limit the control to only the quantity of light but must also include contrast as a further control objective. This control system is entirely similar to that for the control of the exposure in manual reflex cameras, which is shown in Fig. 5.1.

Generalizing, the more we zoom out (a broader perspective) the more necessary or useful it becomes to increase the *number* of [*Y*] variables that the control system must consider simultaneously. However, we must at the same time consider an increasingly higher *number* of disturbance variables [*D*], with the inevitable result of an increase in the *number* of control levers [*X*] needed to achieve the objectives.

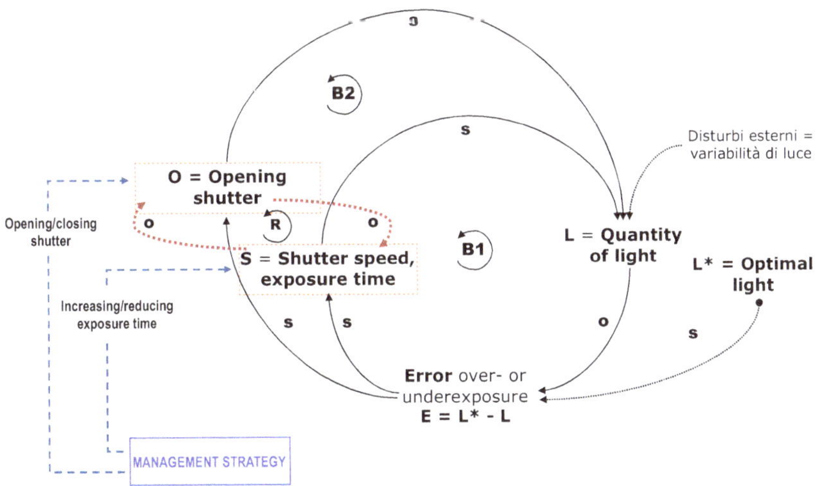

Fig. 5.1 Model of a control system for the exposure in a manual reflex camera

It follows that the larger the environment in which the control takes place, the more numerous and complex are the *structural* and *environmental* control actions necessary to make the system *effective*.

We thus have an initial answer to the questions we started with: the size of the control does not depend so much on the "nature" of the situation we are examining as it does on the extension of the *boundary* that, in delimiting this reality, defines the control objectives the observer can pursue. In other words, every control system is not absolute and objective but *relative* and *subjective*.

5.4 Strengtheners, Turbos, and Multilevers

It is important here to introduce another important point: when *observing* or *designing* control systems it is necessary to carefully analyze the *nature* of the control levers [*X*] in order to see whether or not they can vary continuously and for discrete values, and especially if they have a *capacity*, which derives from a *section* and a *speed of transition*. This necessitates the determination of values for the system's rates of action, "$g(Y/[X])$," reaction, "$h([X]/Y)$," and the characteristics of symmetry, ($h = 1/g$). Only knowledge of these structural elements allows us to study the system's behavior and thus make calculations using the recursive model that produces it or to simulate its dynamics by using some specific tool.

In order to further consider control mechanisms, let us return to the control system for a *bathtub controlled by two independent faucets*, similar to the one in Fig. 4.1. We assume that the system has a positive lever, X2, that increases the water level at a *section* sX2 = 2 and a *velocity* of vX2 = 2; assuming also that the system is perfectly linear and symmetrical, this means that the *capacity* is gX2 = 4 units ($s \times v$) and the water level rises by that amount for each period of measurement.

Let us consider an operator–manager who, finding the tub empty, sets the objective of reaching a level of $Y^* = 500$ units in a given period, $T = (0–50)$. We can easily verify that the objective cannot be achieved *in the time available* since, with a capacity of gX2 = 4 units for each "t," the level would only rise by $Y_{50} = 200$ units after 50 instants. The system is *undersized* in terms of the objective; or, from a different perspective, the objective is *oversized* in relation to the available effectiveness of the control system. In this specific example, in order to strengthen the control system we can carry out several *actions* on the system's *effector* mechanism, which we define as *structural actions*; for instance (simplifying the symbols) we can:

1. Increase the *velocity* of the outflow; if the velocity is increased from $v = 2$ to $v' = 5$ per unit of section, with a section lever equal to $s = 2$, there would be an inflow of $g = 10$ units per instant; the objective could be reached in 50 periods. This action can occur through a *structural* intervention that installs a pump to increase the water *pressure* and thus the *velocity* of inflow as well.
2. Increase the section of the inflow lever while maintaining the same flow velocity; this adjustment assumes that it is possible to restructure the physical system by increasing the section of the water pipes.

3. Add a *new inflow lever* capable of supplying additional water; this implies a more detailed restructuring of the system, which then would have two independent levers for the water inflow in addition to a dependent lever for the outflow.

Environmental actions can also be taken that do not concern the *structure* but the external *environment*. These can again involve the *velocity* of the water inflow by means of a lever or a "device" that is not part of the system; for example, a pump placed in the tank where the water comes from through the inflow lever or a process that brings the objectives to a more manageable level for the system.

Mutatis mutandis, the situation is similar for the outflow process which, if undersized, can be regulated through *structural* and *environmental* interventions.

In general, we can define *boosting* as any *structural* intervention that increases the *capacity* of the lever $g(Y/X)$. *Boosting* can occur through two *internal* interventions, which are individual or combined:

- An *amplifying intervention*, if it acts on $g(Y/X)$ to produce an increase in the section, "*s*," of the lever (the meaning of which is to be specified for any concrete system).
- An accelerator or *turbo intervention*, if it acts to increase the flow velocity, "*v*."

Amplifying and turbo processes can depend on both the controlled variable X and the error $E(Y)$; in this case the control system frequently becomes multilayered.

The *apparatuses* that produce these effects are internal *servomechanisms* that can produce linear dynamics, if their effects do not depend on the amount of error, or nonlinear dynamics, if in some way they are a function of the $E(Y)$ or its components, the level of Y and/or the objective. Nonlinear boosting processes are more familiar to us from our experiences since, as we know, a turbo has stronger effects when there is a large variance with respect to the objective; the effects are gradually reduced when the dynamics of X approach the objective.

A *new control lever*, V_t (itself dependent on the error, $E(Y)_t$), which produces a further flow with respect to the lever X_t, can be defined as an *auxiliary lever* that is *internal* to the structure; if, instead, V_t is not regulated by $E(Y)$, then the *servomechanism* is *external* or environmental.

Applying these internal and external actions to the general model of a control system in Fig. 2.10—assuming only one lever, X, to make the graph simpler—we get the *undersized* control system model, with the possibility of *booster* adjustments, as shown in Fig. 5.2, which, although it seems complicated, includes all the booster elements mentioned above. The initial system is represented by the *loop* [**B1**], to which is added the second lever V_t, which starts the loop [**B2**]; this lever represents an internal *auxiliary lever* which can booster the action of X on the variable Y.

The *amplifier* and the *turbo* are represented instead as links between the error and the action parameter, $g(Y/X)$, which is divided into its two elements: section and *velocity*. The external disturbance, D, which disrupts the value of the variable, Y, is kept under control by an *environmental* control system that, by means of the external

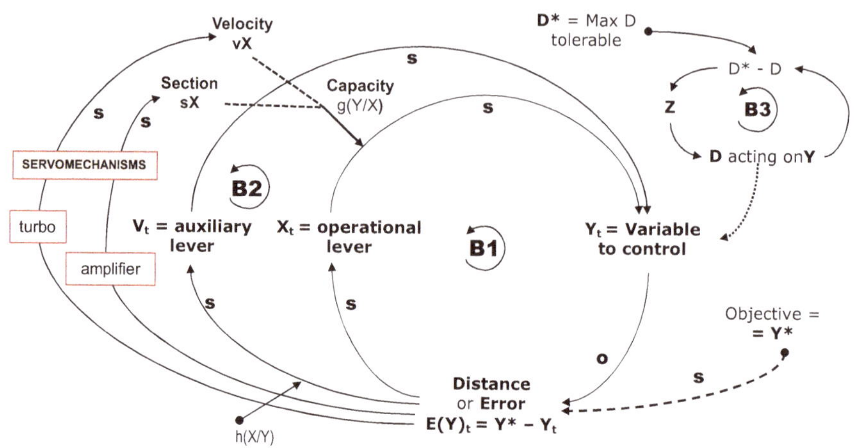

Fig. 5.2 Single-lever control system with booster adjustments

lever "Z," carries out an external control process to reduce the undersizing of the initial system. By definition, the *environmental* control is *external* to the main control system. Lever "Z" could also be viewed as a second- or third-order *structural* lever (Fig. 4.13), and thus placed inside the main control system, which then becomes a *multilayer* control system (Sect. 4.5).

Let us not, however, get too far ahead of ourselves. Today all cars have the *turbo*, the well-known servomechanism that increases the pressure of the mixture that enters the cylinders to increase their power in order to increase the car's velocity through pressure on the gas pedal. An external *amplification* increases the car's piston displacement; but given equal piston displacements, cars with a turbo with more thrust are more desirable. A bicycle's amplification mechanism is its derailleur, which allows the gears to be changed so that, with the same pedaling frequency, the speed increases. The turbo effect can only come from the outside, by increasing the muscular strength of the cyclist's legs. An auxiliary lever could be represented by an electric motor that would make the bicycle go faster. The amplification effect is achieved in a sailboat by varying the number and surface area of the sails; the turbo effect cannot be achieved structurally, but depends instead on the wind velocity. An auxiliary lever could be represented by a motor that added its effect to that of the sails.

5.5 Risk of Failure of the Control Process Due to Structural Causes

After having presented some guidelines for observing and designing control systems (Sect. 5.1) we need to deal with the other side of the coin: providing indications for recognizing the many *risks of failure* of the control process. We must begin by considering several errors that very frequently lead to an inadequate control system that risks failing as soon as it starts up. The most frequent of these concern the definition of the variable to control [*Y*], the control levers [*X*], and the objectives [*Y**]; in particular, the inability to correctly *identify* or *design* the link between the variables [*Y*] and the levers [*X*] that control the former in order to achieve the chosen objective. This problem is evident when we consider, for example, the biological control system for thirst that we examined in Sect. 4.2.

Remaining on the topic of objectives, it is useful to consider a risk factor that is often underestimated but that can cause catastrophic damage to the system due to the inability to connect levers to objectives when the levers produce cumulative or contrasting effects. Many problems in piloting an aircraft result from the beginner pilot's unawareness that it is not possible to increase altitude by only using the lever represented by the horizontal tail rudder without at the same time using the throttle lever. When the plane "raises its nose," its speed is reduced; if the engine power is not increased the plane risks stalling. Similarly it is useless to try to treat liver disease by using pharmacological levers that damage the kidneys; try to lose weight with intensive sports activities without considering the effect of this on your heart; or try to increase national output through levers that tend to increase savings, since postponing consumer purchases results in lower and postponed demand, with the opposite effect on the desired objective. A lack of awareness of and respect for the interrelations between the objectives and the related control levers is one of the most serious types of failure of any control process.

A second factor that can lead to the ineffectiveness of the system in carrying out the objective and achieving control is linked to the insufficient *potency* of the control levers; that is, the incapacity of these to influence *Y*, taking into account the nature and potency of the *effector* apparatus. As observed in Sect. 5.4, if actions are not taken to increase the potency of the levers that are undersized, the control system will inevitably fail to achieve the objective.

If a river regularly overflows, we can control the flood waters by building fixed embankments; this lever only has a section, in this case represented by the height of the embankment. It is not possible to also determine a *velocity* for this lever. Even if we build an outlet canal for the water, which activates when the water level rises, we still have a lever with only a section. However, if we use a water-scooping system that is activated when the water level reaches the flood level, then we have a control lever whose *capacity* or *potency* derives from both the section (number of water-scoopers, diameter of the tubes) and the *velocity*, that is, the pumping capacity. Examples of failures in the controls due to insufficient *capacity* of the control levers are so evident that we need not mention them; we all know it is impossible for a ship to leave the port if the tugboat is not powerful enough, just as it is impossible to quench your thirst on a torrid day with a drop of water per minute. It is nearly impossible for a goal-keeper to stop all the penalty shots; his speed and arm length are too limited with respect to the speed of the ball, and he can only stop the shot when it is too central or slow.

Linked to insufficient *potency* of the levers is another risk factor that can lead to the failure of a control: the incapacity or impossibility of *observing* or *designing* an adequate number of control levers and of ensuring a quality suitable for the variable to control. How can we control the continual rising temperature on our planet (Sect. 6.6 below)? How can we stem the gradual desertification around the world? How can we eliminate the periodic invasions of locusts? How do we control the reproductive rhythm of cancerous cells before there is an explosive and lethal increase in them? Based on our current knowledge, these all seem to be uncontrollable systems, even if the efforts to create adequate *levers* for these situations are well known. Some of these levers have been discovered (pollution, forest fires, the spread of algae), some have been developed conceptually (radiotherapy, receptor inhibitors in cells), and others are being experimented (new food plants derived from genetic engineering, education of the population, and food and medical assistance). One thing, however, is clear: the variables to control will always be out of control until the basic *levers* that affect these dynamics are not completely included in an effective control system.

Last but not least, we shall consider an important structural factor almost certain to produce failure in the control, even when this is well designed. Sect. 2.9 showed that as the *chain of control* is made up of three distinct apparatuses—*effector*, *detector*, and *regulator*—it is necessary to have a *transmission network* for information, however constructed, which connects the three apparatuses and must be considered complementary to the three "machines" making up the *true* control system. Nevertheless, clearly this network must be considered, in all regards, as a specific *autonomous* apparatus, on whose correct functioning depends that of the entire control system. The concepts of *network* and *transmission channels* for information not only concerns mechanical, electrical, or electronic connections among the apparatuses in the chain of control but must also be viewed in a broader sense in relation to the nature of control systems. Here are some examples.

Some hold that the battle of Waterloo was probably lost by Napoleon because he could not activate an important control lever: the positioning of General Emmanuel de Grouchy's 30,000 troops, which represented the right wing of the French alignment. The reason for this was because it was not possible to communicate orders to them and thus carry out the strategy; the channel of communication (the dispatch riders) between the regulator (Napoleon) and the effector of the lever (de Grouchy's troops on the right wing) did not function perfectly. Moreover, it seems that Napoleon advised de Grouchy to listen for the sound of the cannons to understand in which direction to march to bring support; however, de Grouchy stuck to the previous orders to pursue the Prussians rather than maneuver in aid of Napoleon.

A cataclysm that physically separates two populations, prey and predator, would represent a malfunctioning of communications between the effector and detector, impeding the biological interactions among individuals and upsetting the biological control system. No matter how relevant the gap is between actual sales data and the budgeted data, no measure (regulation) could be taken with regard to the sales force (effector) if the head of periodic sales reporting (detector) did not transmit the noted variances to the marketing director (regulator) in a timely fashion. No control of the punctuality of urban buses could be carried out if the reports by the route inspectors (detectors) did not arrive at the municipal transport command center (regulator).

A final striking example reveals the insidious nature of human error regarding information *transmission channels*. As we know, the most accredited theory about the sinking of the Titanic in April 1912, which caused more than 1500 victims, is that the transatlantic liner collided with an iceberg due to a defect in the rudder (effector), which was not adequately sized and equipped for the rapid evasive action required by the late sighting of the obstacle (error detection). Recently, however, another fact has emerged. The tragedy was not caused by the delay in detecting the iceberg or by the inadequately sized rudder, but probably was due to a simple error in communicating the control decisions for the evasive action. We read from the diary (only now made public) of the second officer, Charles Lightoller, the ship's fourth most senior officer (who survived the sinking):

> The error on the ship's maiden voyage between Southampton and New York in 1912 happened because at the time seagoing was undergoing enormous upheaval because of the conversion from sail to steam ships. The change meant there were two different steering systems and different commands attached to them. Some of the crew on the Titanic were used to the archaic Tiller Orders associated with sailing ships and some to the more modern Rudder Orders. Crucially, the two steering systems were the complete opposite of one another. So a command to turn "hard a starboard" meant turn the wheel right under the Tiller system and left under the Rudder. When First Officer William Murdoch spotted the iceberg two miles away, his "hard a-starboard" order was misinterpreted by the Quartermaster Robert Hitchins. He turned the ship right instead of left and, even though he was almost immediately told to correct it, it was too late and the side of the starboard bow was ripped out by the iceberg. The steersman panicked and the real reason why Titanic hit the iceberg, which has never come to light before, is because he turned the wheel the wrong way", said Lady Patten who is the wife of former Tory Education minister, Lord (John) Patten (Alleyne 2010, online).

The quote speaks for itself.

There are many other risk factors of control; however, I wish to conclude this *section* by pointing out another element that often makes control problematic. Though taken up at the end, it is certainly not last in terms of its importance and the disastrous effects it can have on the control process.

We saw in Sect. 2.9 that an essential component of the chain of control is the process of *detecting error*—that is, the deviation $E(Y)$—which is a process set off by a *sensor* apparatus that is often very complex. We need to realize that without an efficient detection process for error, or when the process is unreliable and imprecise and operates with a nonadmissible tolerance, it is not possible to calibrate the regulation of X so that it acts appropriately on Y; as a result control is not possible or is extremely imprecise. This risk emerges not only in cases of possible malfunctioning of the mechanical or electrical instruments for error detection but also, and above all, when the detection is carried out by the operator–manager using his own senses and knowledge. There is no need here to use overly sophisticated examples such as the lack of control of solar batteries on the space probe due to the malfunctioning of the charge detection; we need only point out all the times our car tires have gone soft because, being distracted, we had failed to check the pressure; what a relief today to have an automatic pressure gauge that signals any problem, thus allowing us to promptly carry out the necessary control.

5.6 Risks of Failure of the Control Process Due to some Characteristics of the Variables to be Controlled

As we have seen in Sect. 1.5, the systems observed and built based on the logic of Systems Thinking are systems of *variables*, not systems of objects. For this reason, Systems Thinking considers all types of dynamic systems in all fields, building models to understand, simulate, and above all "control" a complex world of incessant movement in continual transformation and evolution. In many situations, no matter how much time and energy we dedicate to it, the construction of effective control systems models is not possible or is impeded by particular characteristics of the variables to be controlled. I will mention seven of the conditions that make it difficult, if not impossible, to perceive "dynamics," thus impeding the control manager from "understanding" and "controlling" those variables.

Note. These difficulties in building effective control systems are in part suggested by Senge's book (1990) and by my book on Systems Thinking (2012). A more in-depth analysis can be found in *Obstacles to Managing Dynamic Systems. The Systems Thinking Approach* (Rangone and Mella 2019).

1. *Temporal slowness* or the metaphor of the *boiled frog*. This obstacle derives from the extreme difficulty of observing the dynamics of variables that change very slowly so that, when we realize that a control is necessary, the variable has assumed values so high that no control system can change them. This difficulty is associated with a boiled frog since, when a frog is immersed in a pot of cold water that is slowly heated by a flame, it will attempt to jump out of the water as the temperature slowly, but inexorably rises. Because his limbs are numb from the heat, the frog remains in the water and is boiled, thus becoming a boiled frog.

   > Though there is nothing restraining him, the frog will sit there and boil. Why? Because the frog's internal apparatus for sensing threats to survival is geared to sudden changes in his environment, not to slow, gradual changes (Senge, 1990, p. 17).

   The *boiled frog metaphor* can be observed in many real-life situations: individual, organizational, and social (Teal & Co. 2016). The situation described in the metaphor of the *boiled frog* applies, though with inverted dynamics, to many phenomena marked by a gradual reduction in some variable; a frog that is immersed in water that becomes increasingly cold will not be able to jump out because it is *frozen*. The decline in the birth rate in Japan and in other countries can be described as a frozen frog phenomenon (McCurri, 2018, online).
2. *Speed of processes* or metaphor of the *networking effect*. This obstacle is linked to the speed of processes and phenomena (usually involving accumulation and propagation based on exponential laws) that are so rapid we are not able to "recognize" their evolution until they have already produced their effects on the system, as occurs with the explosion of a virus (Sect. 1.7.2) or in cases of accelerating returns (Kurzweil 2001) (Sect. 1.7.3).

   > Consider this parable: a lake owner wants to stay at home totend to the lake's fish and make certain that the lake itself will not become covered with lily pads, which are said to double

their number every few days. Month after month, he patiently waits, yet only tiny patches of lily pads can be discerned, and they don't seem to be expanding in any noticeable way. With the lily pads covering less than 1 percent of the lake, the owner figures that it's safe to take a vacation and leaves with his family. When he returns a few weeks later, he's shocked to discover that the entire lake has become covered with the pads, and his fish have perished. By doubling their number every few days, the last seven doublings were sufficient to extend the pads' coverage to the entire lake. (Seven doublings extended their reach 128-fold.) This is the nature of exponential growth (Kurzweil 2005, p. 4).

3. Another typical case is the so-called *networking effect*, which operates especially in *networks* of elements that spread some information, or effect, at too high a speed to be observed, as in the case of phenomena spreading by word of mouth or pandemics. Systems Thinking observes that the only defense against the speed of processes is a careful monitoring of the causes of the variation, along with an increase in the resistance and resilience and a reduction in the vulnerability (Ellison et al. 1997; Bianchi 2019) of the structures that would suffer the effects of this obstacle.
4. *Spatial distance*, or the metaphor of the *butterfly effect*, also known as the *Turing effect* (Turing 1950), the obstacle from *distance in space*. There are systems—composed of a very *high number* of variables that interact, being linked by *nested loops* contained in other *loops*—so complex that even a slight variation in the value of one of the variables is enough to produce cumulative effects that spread throughout the entire system, thereby altering other variables distant in time and space (Andrews, 2011). The term *butterfly effect* derives from the physicist Edward Lorenz, who, in 1979, stated that if the theories of complex systems and chaos were correct, then the fluttering wings of a butterfly in Brazil would be enough to alter climate patterns, even permanently, and to generate a tornado in the Caribbean. Conversely, "*If the flap of a butterfly's wings can be instrumental in generating a tornado, it can equally well be instrumental in preventing a tornado*" (Lorenz, 1972, p. 1).
5. *Temporal distance* or the metaphor of *Temporal Myopia*. As illustrated in Sect. 1.6, preference for short-term, individual, or local advantages hide the collective and global disadvantages that occur over the long term; this, however, inexorably creates long-term problems when the disadvantages caused by the repeated behavior appear so that the long-term disadvantages do not condition behavior but represent only its effect.
6. *Observational arrow* or the metaphor of the *monodirectional view*. People often "*look*" in one direction only and do not "*perceive*" the variables taking place at their backs or in other directions, at times even choosing to ignore these so that it is impossible to control them. In particular, the observational arrow obstacle to control occurs when the observer merely considers causal chains, ignoring the existence of loops and avoiding searching for them. This is the typical behavior of those who do not see the harmful implications of their actions and continue undeterred in behaving according to cause–effect logic, in a typical linear and noncircular view. A typical example of a "monodirectional view" is described by Peter Senge as "I am my position," which he considered as a learning disability:

> Most see themselves within a "system" over which they have little or no influence. They "do their job," put in their time, and try to cope with the forces outside of their control. Consequently, they tend to see their responsibilities as limited to the boundaries of their position (Senge 1990, p. 17).

More generally, the observer adopting a monodirectional view is inside a limited system and ignores, or wants to ignore, the fact that this system is part of a larger system.

7. *Complexity of structure.* This difficulty is linked to the structural, computational, and temporal complexity of *systems with memory*. This form of complexity has been well described by Heinz von Foerster (2003, p. 143), the father of "second-order cybernetics," who views a system (machine) with *memory*—defined as a nontrivial machine (von Foerster 1984, p. 10)—as a *complex system* deriving from the interconnection of systems (machines) without memory, or trivial machines, which in Systems Thinking represent the elementary processes between two variables based on a cause–effect relation. In many cases, it is impossible to understand and control the complex behavior of a system with memory because of the incredible number of behaviors [input-states-output] that a system with *memory* can produce.

The seven metaphors touched on above represent difficulties that are not easy to overcome regarding the perception, and therefore the control, of many environmental variables and of social, economic, and political systems. The only remedy is to become aware of their existence and hone even more our attention and sensitivity to them. To this end, we have also indicated for each of these difficulties some strategies proposed by Systems Thinking to deal with these problems. In particular, we want to tackle a stimulating challenge: setting up "sensors" that signal in advance the starting up of the action of the seven obstacles that make it difficult to observe and understand the dynamics of systems so that the systems thinker will not be unprepared when he has to suffer their effects.

5.7 Risks of Failure of the Control Process Due to Improper Levers: "Shifting the Burden" ARCHETYPE

Much more subtle are the risks from choosing the *wrong lever*, which is not able to control the variable to be kept under control. There are several variations of this situation. The most evident is choosing levers that only act on the *symptoms*. The action of symptomatic levers, *X*, starts the control positively; but since *Y* is not directly controlled, the error is not eliminated but instead returns even stronger later on. This unsuccessful control takes on two forms that are so frequent they are recognized as widespread recurring *archetypes* in every context (Sect. 1.7.8). The first variant of a failure due to the erroneous identification of the control levers can be represented by two archetypes similar in their effects: the ARCHETYPE of the "Fixes That Fail," which operates in all contexts where control favors short-term levers, and the more general ARCHETYPE of "Shifting the Burden," which acts in situations where control is based on symptomatic levers (this can be viewed as a generalization of the previous archetype).

The "Fixes That Fail" ARCHETYPE is applied to control systems that can have several levers. Some of these are symptomatic levers, which act immediately on Y or directly on $E(Y)$, though their effect can be short term so that, once eliminated, the deviation returns after a short period. Other *levers* are instead *structural* and long term and often act with a delay; however, their effect is permanent. Considering the period needed for them to act, the former can be called *contingent* levers, while the latter can also be called *strategic* levers. Figure 5.3 describes the archetype of the "Fixes That Fail."

Pollution has to be reduced? The simplest *symptomatic* lever is to block cars from circulating. This lever has an immediate effect on pollution, as stopping cars from circulating reduces dangerous emissions. However, clearly this affect is *short term* because the economics of a region is based on the rapid displacement of people, and thus cars cannot be blocked from circulating for a long period of time. Quite different is the *structural* lever that rationalizes traffic by redesigning the road system to avoid traffic jams; or the lever, even longer term, whereby the subway or rail system is improved. An even stronger effect would be produced by the *lever*

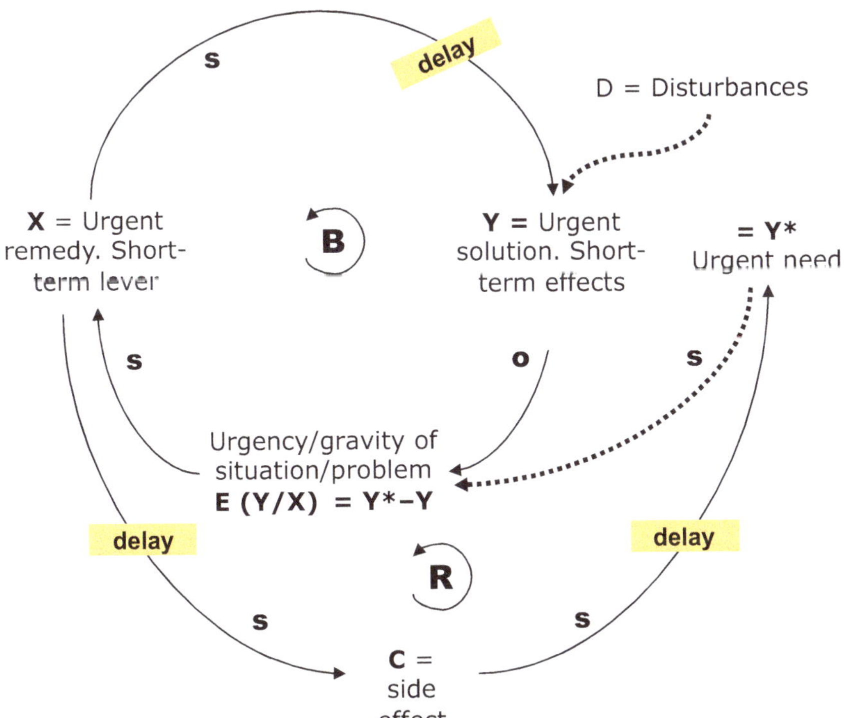

Fig. 5.3 "Fixes That Fail" ARCHETYPE applied to the control process

represented by the spread of electric automobiles. These are long-term levers not only on account of the time needed to produce the necessary infrastructures but also because of the need to modify citizens' attitudes toward using their cars.

Drug consumption must be controlled? The shortest-term *symptomatic* lever is the usual one: try to arrest drug traffickers and drug dealers; a similar measure would produce a short-term improvement since, as we observe in real life, the spread of drugs almost always immediately recurs in other ways. This depends not so much on the possibility of business activity in this area but on demand. A *structural* lever with permanent effects, but which is applicable only in the long term, would be to improve education at all levels about the harmful effects of drugs, accompanied by social conditions that provide more positive incentives.

When a control system is structured only with contingent, short-term levers the control can fail, because in the long-term unforeseen consequences always arise that manifest themselves as disturbances that do not permit the objective to be achieved and stably maintained, thereby generating new, more serious deviations that make a continual readjustment of the solution necessary. The "Shifting the Burden" ARCHETYPE shows the recurring risk of control failure when the control system is designed solely to eliminate the symptom, so that the user–manager, perceiving a temporary improvement in the unpleasant sensations linked to the symptom, neglects to activate the levers that would lead to a long-lasting control of Y, thereby causing a net worsening in the system's performance, with often dramatic consequences (Kim 1999). If we try to control a fever with ice on our forehead without treating the infection with antibiotics, the system will improve for a few hours but the infection will spread. If we try to resupply a warehouse with extraordinary production, without replanning production and sales, the rundown of stocks will inevitably resurface.

Figure 5.4 shows the structure of the "Shifting the Burden" ARCHETYPE applied to the control process. When the control system is structured to reduce the intensity of $E(Y)$, but the structural levers have a long operating time or involve delays, then $E(Y)$ appears as a serious symptom to be eliminated through symptomatic levers, by forming loop [**B2**], which can cause a delay in the activation of the strategic levers for the purpose of the structural control [**B1**]; this delay can lead to collateral effects that further hinder the search for strategic levers, thereby producing a new and more serious variance through the strengthening action [**R**]; systems thinking—precisely because it brings out the perverse *modus operandi* of this archetype—favors and accelerates the search for strategic levers for an effective control. Fortunately this archetype does not always act in all control contexts, its action is more easily encountered the more intense the unpleasant effect associated with the symptom is so that in order to eliminate it we end up by reducing the efforts to activate structural levers, with the consequent delay in achieving the objectives.

My father has a bad back because he weights too much, and the weight compresses the vertebrae in the spinal column. The pain he feels, E^1, comes from the excessive weight (Y) compared to the weight the spinal column can support (Y^*). The structural solution would be for my father to reduce the amount of food (X) he eats through an appropriate diet in order to reduce his weight (Y), since obviously the excess weight impedes him from moving freely. However, the effects of the diet

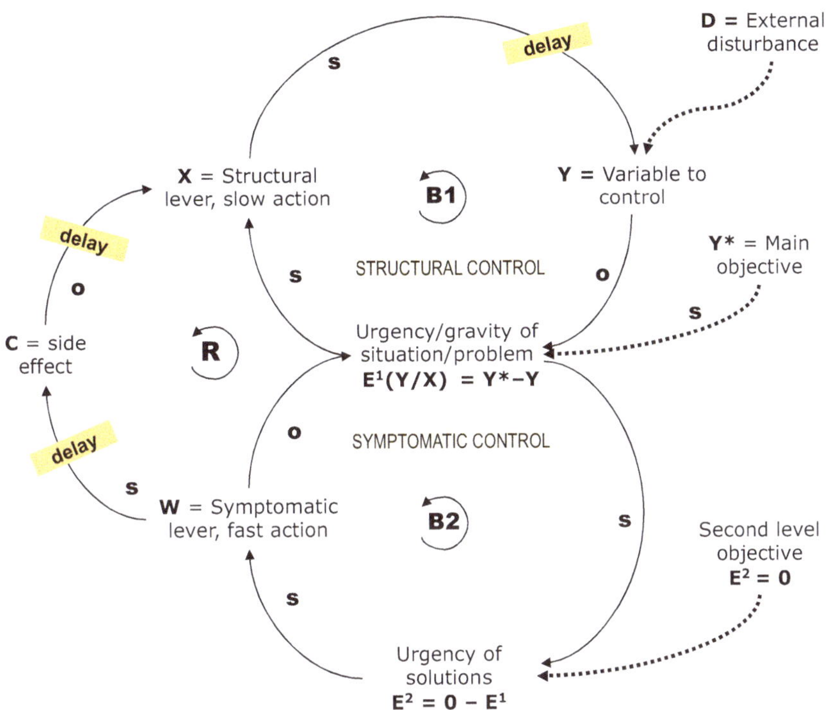

Fig. 5.4 "Shifting the Burden" ARCHETYPE applied to the control process

are seen only after a long delay, and thus the excess weight ($Y > Y^*$) does not diminish fast enough. The elimination of back pain thus becomes the overriding variable, E^2, which requires that E^1 be urgently eliminated; for this reason my father turns to a symptomatic lever: painkillers and physiotherapy (the structural lever in the form of a spinal column operation is discarded a priori). The symptomatic levers act quickly so that the desire to eat less (X) diminishes, and therefore the pain reappears. The continued use of pain killers produces dangerous side effects on the liver as well as reduces the ability to concentrate, thereby making work more difficult.

The "Shifting the Burden" ARCHETYPE, which generalizes, so to speak, the "Fixes That Fail" ARCHETYPE, has particular importance, since it favors the tendency to prefer short-term symptomatic solutions that immediately produce visible effects, in this way putting off strategic, definitive solutions, thus laying the basis for the production of permanent negative effects. Therefore, this archetype reveals our natural tendency to prefer present, individual or local advantages that can be immediately enjoyed to future ones we hold as too far off in time, as shown by the "Short-term Preference" ARCHETYPE (Sect. 1.6.6). The "Shifting the Burden" ARCHETYPE is quite common, and it produces its negative effects in quite different situations, individual, social, or organizational. Fig. 1.7 (Sect. 1.3) presented the burnout-from-stress problem, which can be translated into the model in Fig. 5.5.

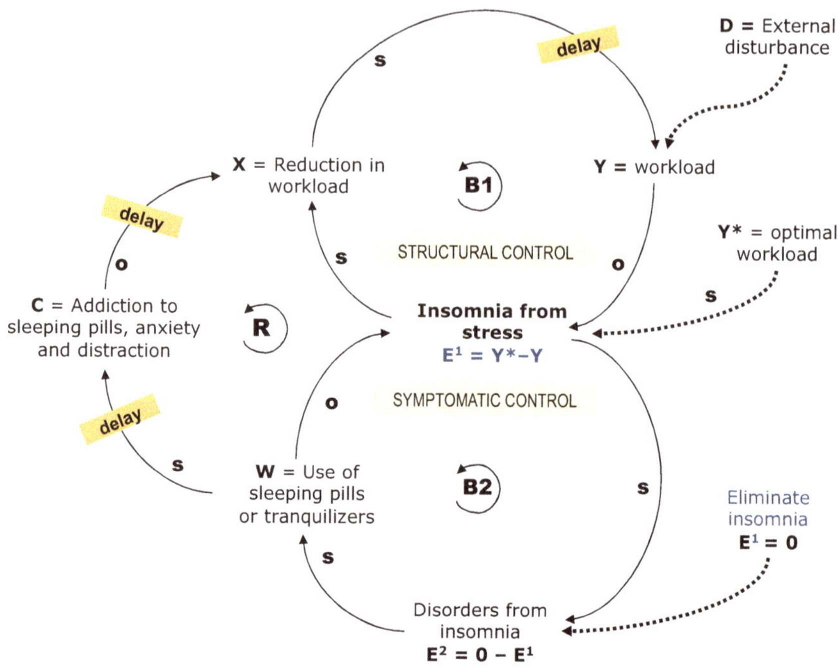

Fig. 5.5 Burnout from stress revisited (see Sect. 1.3, Fig. 1.7)

Many growing companies are faced by a strong rise in demand but are not able to provide this through present production. The optimal solution would be to increase the productive capacity through appropriate investments. Since the demand pressure is always higher and there is an increased risk of losing customers, and since the growth in the productive capacity can occur only with a long delay and a substantial investment of financial capital, companies find it much easier to supply themselves externally, by providing other firms with the know-how to produce their products. This problem is so widespread that it can be described as "Shifting the Burden," as shown in the model in Fig. 5.6.

Outsourcing "wins" over increased productive capacity, but it has two negative consequences: the first is an increased dependence on the outside supplier; the second, a reduced urgency to raise capital to finance the increase in productive capacity. These two effects, acting jointly, postpone for a long time the company's productive growth, thereby increasing the risk linked to outside supply.

This archetype inexorably appears in many other situations involving organizations and firms. Figure 5.7 represents the model that explains the delays in organizational growth. In fact, managerial problems are often resolved by recourse to outside consulting which, on the one hand, can solve the problem in the short term, but on the other inevitably provides a symptomatic solution, thereby delaying the true long-term strategic solution, which consists in training managers with ad hoc in-house courses to favor their professional development.

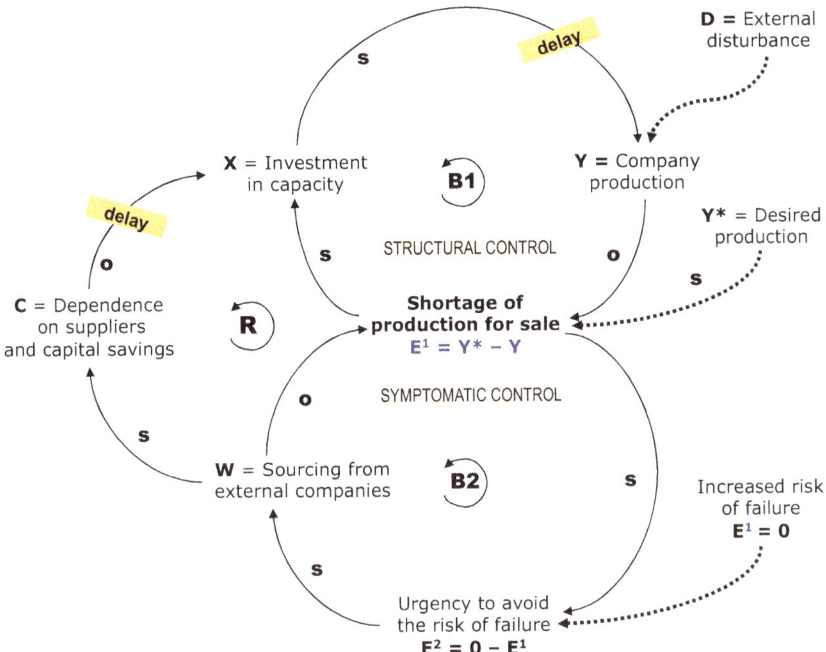

Fig. 5.6 Victory of external sourcing

The "shifting the burden" ARCHETYPE has an interesting extension. We know that control systems normally try to lead Y to the values Y^* by means of the levers X; however, in this attempt the elimination of the error becomes difficult when the system undergoes strong disturbances from the external variables, D, which impede the maintenance of equilibrium. Every attempt to readjust the equilibrium, disturbed by D, through the use of the levers X ends up being only temporary, and readjustments are then continually needed.

It is thus natural to ask if the variable Y that the main control system tries to control cannot be considered, in all respects, a *symptomatic variable* controlled by the *symptomatic lever* X; and if, in order to achieve Y^* and maintain a stable equilibrium, it is not necessary to instead intervene on the external disturbances D, which represent the true *strategic variable*, whose control is essential for achieving the control of Y. Based on what we have seen in Sect. 5.4, this would necessitate the activation of a second external control system on D through the structural levers Z, which are able to act on the disturbance variables, as illustrated in Fig. 5.2. Figure 5.4 would still be appropriate, with the variant that the loop [**B2**] would constitute the long-term structural control system of D, while loop [**B1**] would represent the system that initiates the short-term symptomatic control of Y, as better shown in the model in Fig. 5.8.

The "Shifting the Burden" ARCHETYPE can thus operate in a descending manner, zooming in, when the structural control of Y is replaced by a symptomatic control,

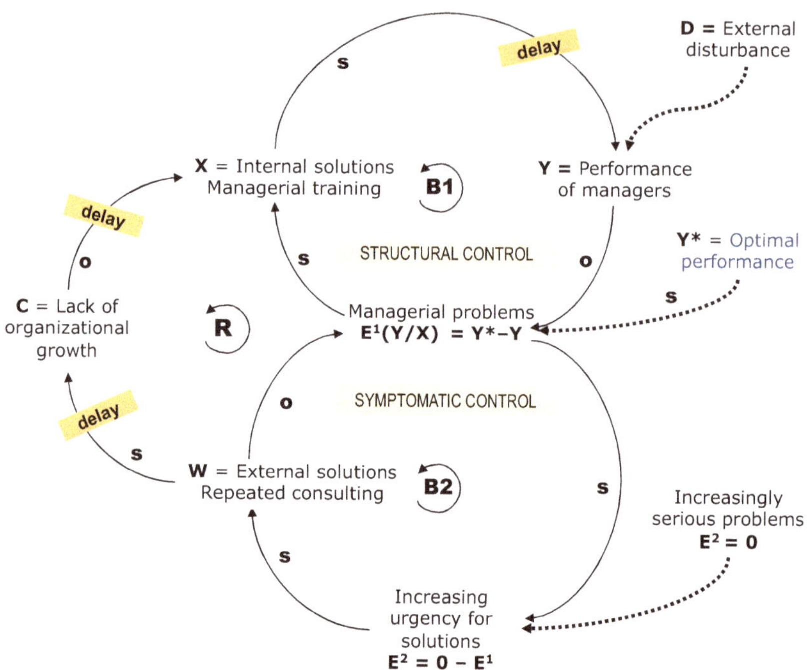

Fig. 5.7 Victory of consultants

as in the model in Fig. 5.4; or it can operate in an ascending manner, zooming out, when the original control on Y is viewed as symptomatic and is supported by the structural control on D, as illustrated in Fig. 5.8. We can also create a *control chain*, where the base-level disturbances are controlled by a higher level system. Any disturbances at the new level could be controlled at an even higher level, all the way up to the maximum level of admissible structural control, after which no new structural level can be employed.

There are an endless number of examples of this expansion of the original control system's confines. If the control of the level of rice fields is compromised by the variability in the river level, it is necessary to construct embankments; if these are not sufficient then the course of the river must be changed or a dam constructed; and if even this is not enough, then the course of another river can be changed to keep the artificial basin adequately supplied with water. If I cannot control the quiet in my home because of noisy neighbors (external disturbance), I can ask them to behave in an educated manner or I can even call in the police. In conclusion, the more the system is expanded, the more the control levers of a given level become symptomatic levers for a higher level of control; the higher level levers are always more effective, but also slower and more costly, so that it is fundamental to carry out a cost–benefit analysis (Sect. 4.7).

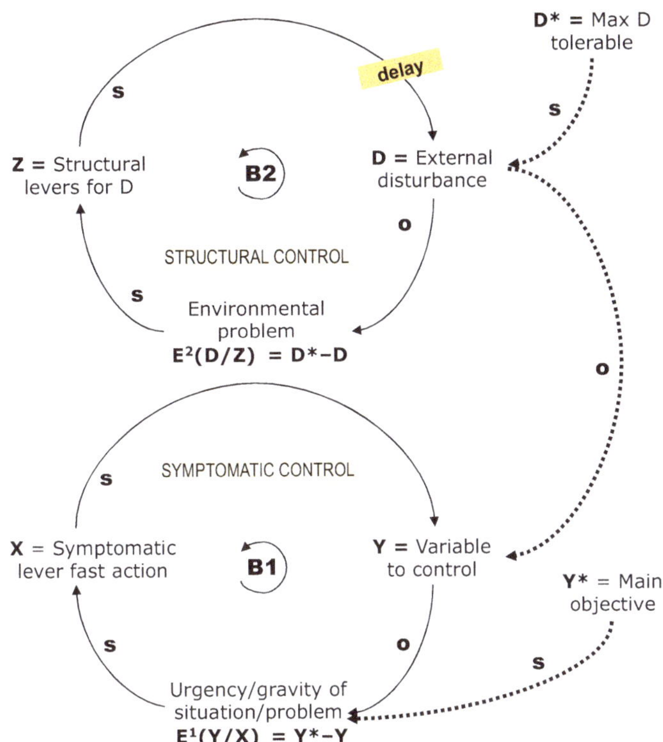

Fig. 5.8 Structural control of external disturbances

Linked to the risk of failure due to the wrong choice of levers is the wrong specification of the action function $g(Y/X)$ and of the complementary reaction function $h(X/Y)$, taking into account that not all control systems are perfectly symmetrical (Sect. 2.3). In all the examples in the preceding chapters, I have chosen to represent "g" and "h" as constants, in order to simplify the simulations and make them readily understandable. It is intuitively clear that, even with this simplified assumption, the quantification of the rates requires particular attention, as regards both the design and realization of the system and the construction of its model. If we abandon this simplifying assumption and assume instead that $g(Y/X)$ and $h(X/Y)$ are functions of both X and Y, then the control system can also take on a rather complicated mathematical structure, since the relations between the variations in X_t, Y_t, and E_t must be described by differential or difference equations. In any event, we must always remember that without the correct formulation of the functional links among the fundamental variables of the control system, and without a precise quantification of the parameters included in these functions, there is a high risk of failure for the control system.

5.8 Pathologies of Control: Discouragement, Insatiability, Persistence, and Underestimation

I shall conclude this examination of the characteristics of human intervention in control processes by presenting three recurring situations, three ARCHETYPES, which reflect so-called *pathological situations* regarding the use of control systems. The first situation is the "discouragement" that often overcomes the governor–manager when he is not able to achieve the objectives he has set for the system. We know from Sect. 5.4 that control systems are designed and created to achieve the objectives, Y^*, and that the governor–manager first turns to the operational levers; if the objectives cannot be achieved by those alone, he then turns to the auxiliary, extraordinary, and structural levers; and in some cases he even uses symptomatic and short-term levers when the action times of the others are too long. Nevertheless, in many cases where the manager–governor cannot completely control Y and the error is not eliminated, he becomes discouraged and turns to a fallback solution: lowering the level of the objective Y^* so as to eliminate $E(Y)$ without achieving the optimum value for Y.

This failure of the control is very common and can be observed in many different situations in which man manages the control systems. For this reason this situation represents a true systems ARCHETYPE, called "Eroding goals." This archetype is illustrated in the model in Fig. 5.9, where the loop [**B1**] represents the control system that must achieve the objective Y^*; loop [**B2**] represents the process by which the manager–governor decides to achieve less lofty objectives rather than the chosen one, thereby creating an "alibi" for not making the correct regulations and allowing the system to degrade. If "Eroding goals" is allowed to act undisturbed, a "spiral of degradation" inevitably is triggered regarding the global performance of the control system. The erosion of the *goals* or of the *performance standards* usually is accompanied by harmful side effects, which are not always easy to identify because they appear with a delay in the long term. When it is suspected that this archetype is operating, it is thus always necessary to make an effort to identify the side effects, at least the most evident short-term ones. The lower box in Fig. 5.9 represents the possibility of side effects, such as the occurrence or worsening of problems. The side effects could also be viewed as external inflows or disturbances; in this way specific control systems could be identified to mitigate these harmful side effects, as illustrated in the diagram in Fig. 5.2.

Nevertheless, despite its clear-cut danger this archetype is found very frequently in decision-making contexts, especially when the manager–governor of the control system is unable to achieve the objectives, judging them to have been too lofty in the first place and being content to accept more modest objectives in the original control system. It is incredible to witness in the papers how this archetype is at play even on a daily basis in families when the behavior of relatives and children is allowed to deteriorate to unsupportable levels of boorishness.

Air pollution has exceeded the tolerance threshold? Then raise the threshold through legislation. The high level of coli bacteria in our waters is endangering

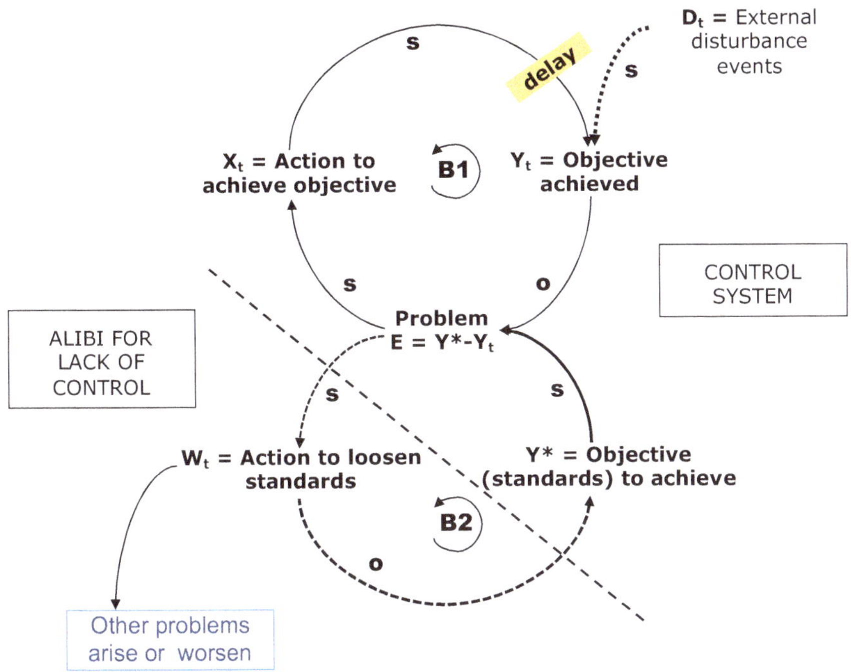

Fig. 5.9 Archetype of "Eroding Goals"

bathing? Then raise the tolerance limits with a local decree. Air emissions are raising the temperatures on the planet to dangerous levels because the Kyoto protocol is not enforced? Then let us put off applying it for a few years. Students are too rowdy (they curse, rebel against their teachers, are badly mannered at home)? "They are only kids" ... but then we should not be surprised if schools are out of control and bullying has become an ordinary occurrence. Our prisons are overflowing and the conditions of inmates are inhumane? Then we can raise the social tolerance for small thefts, scams, violent episodes, and grant pardons and amnesties. The prisons will empty (the system initially improving) but they will soon become overcrowded again. Some youth have caused damage during the procession (or even game, film, and class outing)? "These are only small fringes" ... but let us not be surprised if the areas around stadiums or movie houses become ungovernable (for more, see Mella 2012, pp. 240–241). In our daily lives, we all fall victim to the temptation to become discouraged when faced with control difficulties. Instead of remaining firm in our aspirations and objectives, we often prefer to sacrifice our ambitions rather than make the effort needed to achieve our objectives regarding sports, study, or work. Figure 5.10 illustrates the tragic case of a decrease in motivation for personal improvement, with the risk of a loss in self-confidence.

The model in Fig. 5.11 shows how the Eroding Goals archetype functions in a common situation that leads to a decline in the quality of services. When the gap between the desired and actual quality is too large, quality control activities are

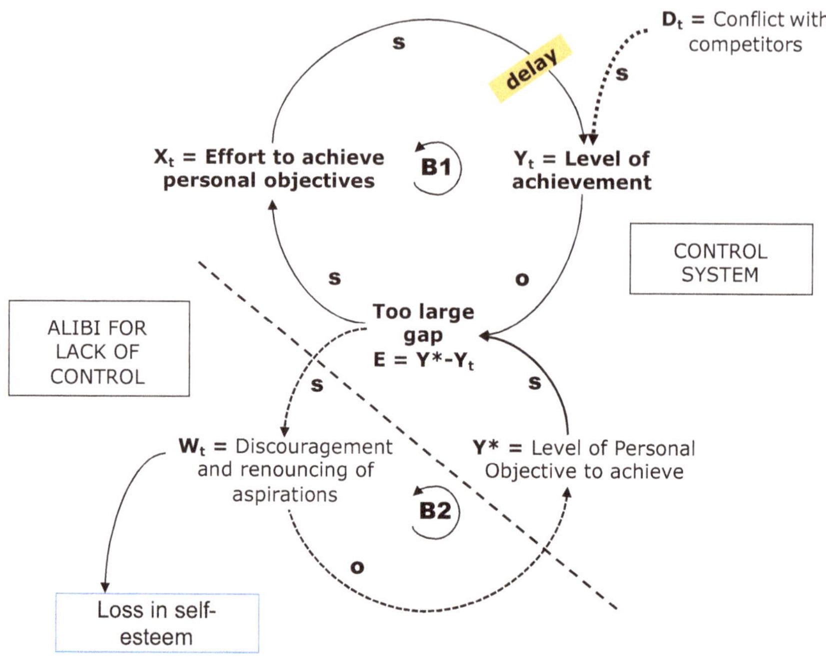

Fig. 5.10 Eroding individual goals and discouragement

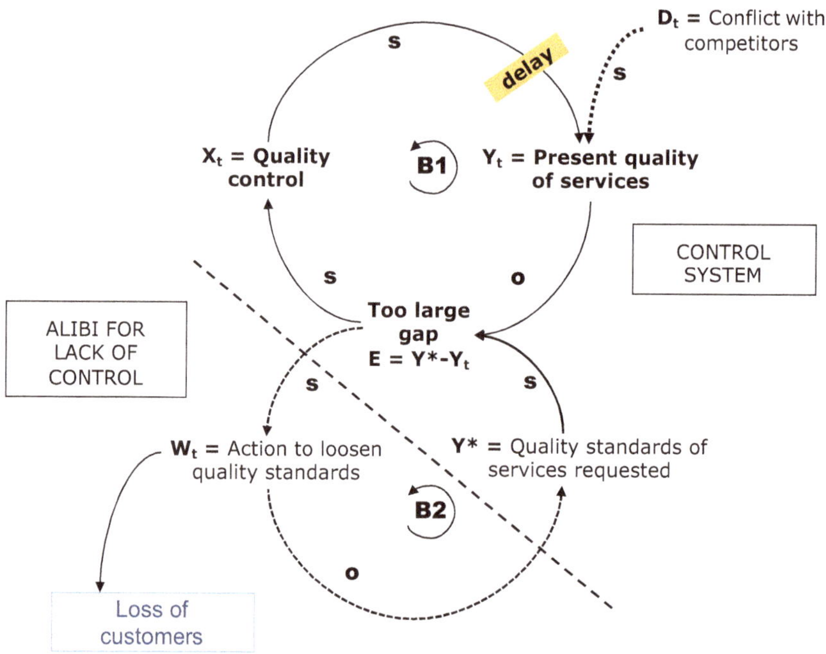

Fig. 5.11 Eroding goals and degradation in quality

insufficient. Rather than insist in using all the levers available to maintain the original standards, the firm prefers to reduce the acceptable standards. This situation occurs very often in public companies that provide services to customers; however, when this occurs in private companies, which are subject to competition, the decline in the objectives leads to the side effect of a loss of business from unsatisfied clients.

How should the governor–manager behave when he becomes aware of eroding objectives? Here are some possible actions he can take:

(a) First of all, it must be clear that the problem of achieving an objective or a given standard always arises in control systems.
(b) It is then necessary to evaluate whether achievable objectives and realistic performance standards have been set.
(c) A constant focus on the objective and standard must be imposed, and steps must be taken to react to achievement errors, however modest.
(d) It is not enough to manipulate the short-term decision-making levers but to work to identify and utilize the long-term levers which, by modifying the generating system, can definitively solve the problem.

It is useful to note that the ARCHETYPE of "Eroding Goals" has a mirror-image variant that we can define as the "Strengthening Goals or Insatiability" ARCHETYPE. This variant is very important, even if in the context of control systems it has not yet received the proper attention. This archetype (illustrated in Fig. 5.12) acts whenever a better result than the objective is achieved and the governor–manager decides not to eliminate the gap by reducing the action of the control lever (thereby producing the loop [**B1**]), but instead to raise the level of standards (loop [**B2**]), thus once again increasing the distance, $E(Y)$, and forcing loop [**B1**] to act more intensively.

Those who are subject to this archetype have no way out. No matter how hard they try they will *never be satisfied*; and even if they do not become frustrated themselves, they end up subjecting their colleagues to an unbearable performance stress. The harmful effects of this archetype are instead felt in organizations whose managers continually raise the performance objectives, as well as in families whose children continually pester their parents for more free time, or whose parents badger their children to get better grades or do better in sports.

Anyone obsessed with exercising an increasingly precise control and who is never satisfied should remember Voltaire's motto: "*the best is the enemy of the good.*" Nevertheless, in many cases "Insatiability" produces positive effects; for example, it characterizes most of the work of scientists, who are never satisfied and fulfilled by their knowledge, of researchers, always thirsting for new discoveries, and of athletes, who are always struggling to set new records.

Figure 5.13 shows how this archetype works for athletes and their trainers, both of whom are searching for continually improving results, which at times harms the athletes themselves, who are subjected to more and more intense training sessions.

The ARCHETYPE of "Insatiability" is often accompanied by another type of equally serious pathological behavior: the *persistence of the controller* with regard to the *controlled subject*. I shall call this the ARCHETYPE of "Persistence," which is

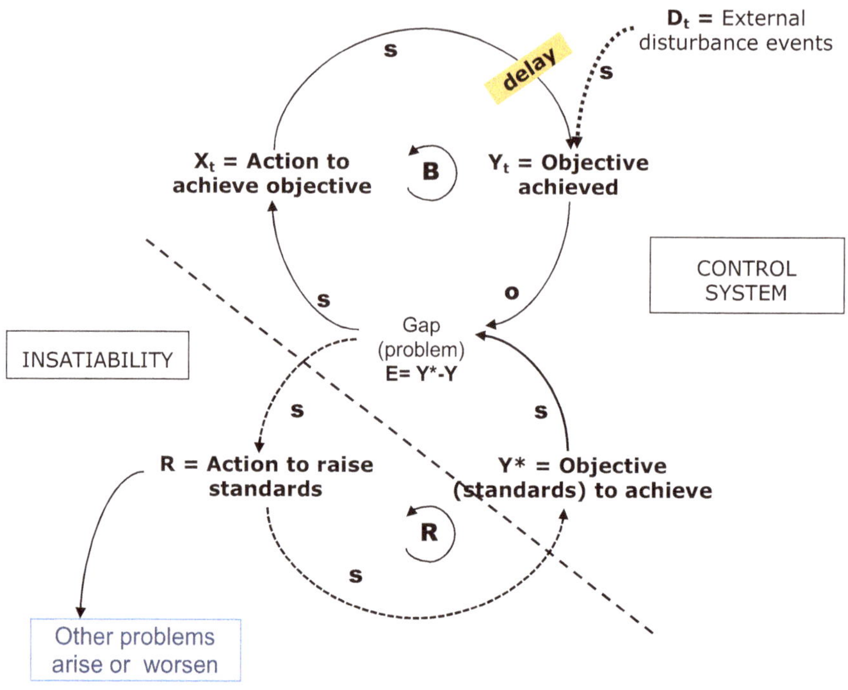

Fig. 5.12 ARCHETYPE of "Strengthening Goals or Insatiability"

shown in Fig. 5.14. It is as equally frequent as the INSATIABILITY archetype and perhaps even more dangerous.

The *pathological* aspect consists in the erroneous perception of the size of the error by the manager, who overestimates the error and brings into play any lever that could possibly eliminate it, demonstrating by this behavior his intolerance toward even the slightest deviation. This behavior represents a true, often *pathological persistence* in the control action. If the objective of the control, Y^*, is the behavior or the performance, Y, of some individual, and the *controlled* subject produces even a minimum error, $E(Y) = Y^* - Y$, the manager immediately punishes him, thereby demotivating, frustrating, and, in extreme cases, even harming the controlled subject. Children are the first to be affected by this archetype when parents, teachers, or instructors (governors) say: "Do this!" (objective), and then remain there observing the child (manager) trying to achieve the objective, persisting in punishing him if he cannot "maneuver the levers" to achieve the goal.

Examples of how this archetype operates are clear to all of us: the smallest grammatical error is punished with a blue line underscore or a bad mark; the slightest formal mistake in a tax return leads to a stiff penalty; even a minimal excess of speed is sanctioned, even harshly, by the highway police who, perhaps from a hideout, try to catch even the most insignificant infraction. These exaggerated examples are meant to show clearly that this archetype reveals a control pathology whose use should be eliminated.

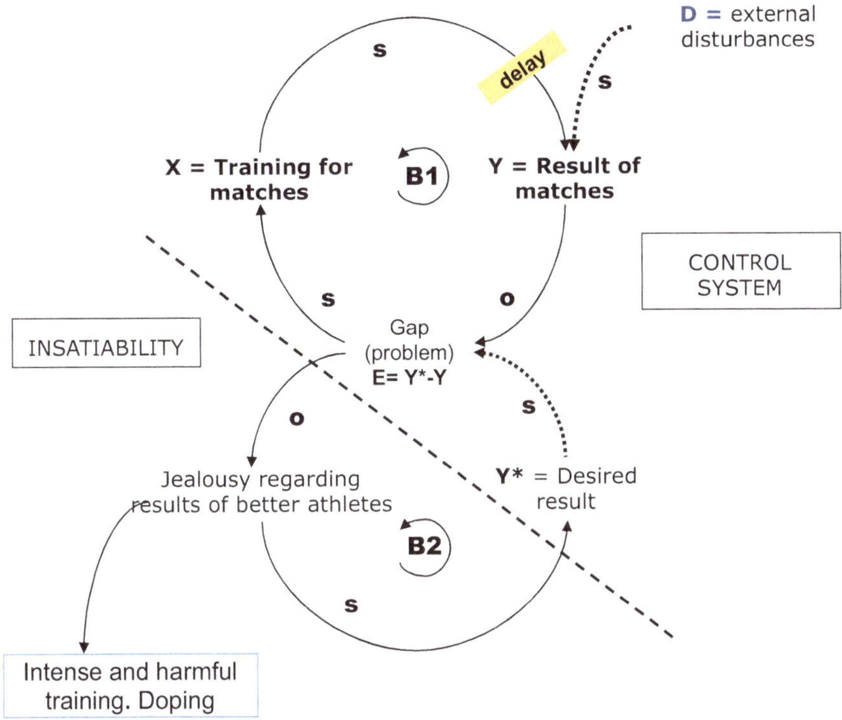

Fig. 5.13 Athletes' Insatiability for victory

With reference again to the educational system, Fig. 5.15 illustrates how the archetype of persistence acts in obsessively attempting to eliminate academic errors, punishing students for even the slightest mistake. The students' insatisfaction leads in many cases to students dropping out of school, and even to more serious situations, which often end up in the crime section of newspapers. The two archetypes described above also produce harmful effects in universities. Students know all too well that their inability to pass certain exams is not due to the difficulty of the subject-matter but to the intransigence of the teacher. Who does not remember at least one personal instance of this?

To conclude I shall present a final anomalous, even *pathological* aspect to human intervention in control, which represents the mirror image of the ARCHETYPE of "Persistence." This aspect is equally frequent and perhaps even more dangerous. While "Persistence" leads to an overestimation of the error, leading to the use of punitive levers that are often exaggerated, in many other cases we often observe the manager *underestimating* the amount of error and *acting with a delay* in *regulating* the levers, resulting in the failure to achieve the objective or to achieving it with a delay, thereby causing (even serious) damage.

There is always the same traffic on the road, and if we want to arrive in time at the airport, we have to leave the office at 6 p.m.; but there is always someone who

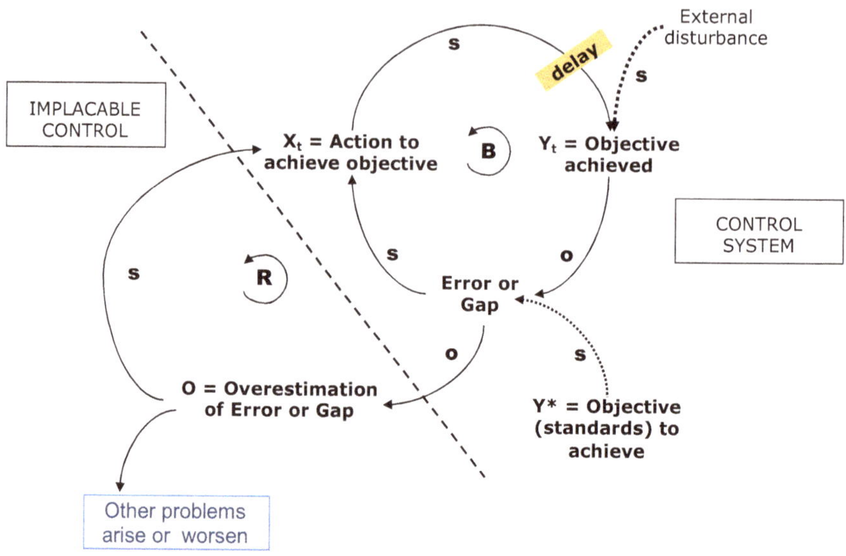

Fig. 5.14 ARCHETYPE of "Persistence"

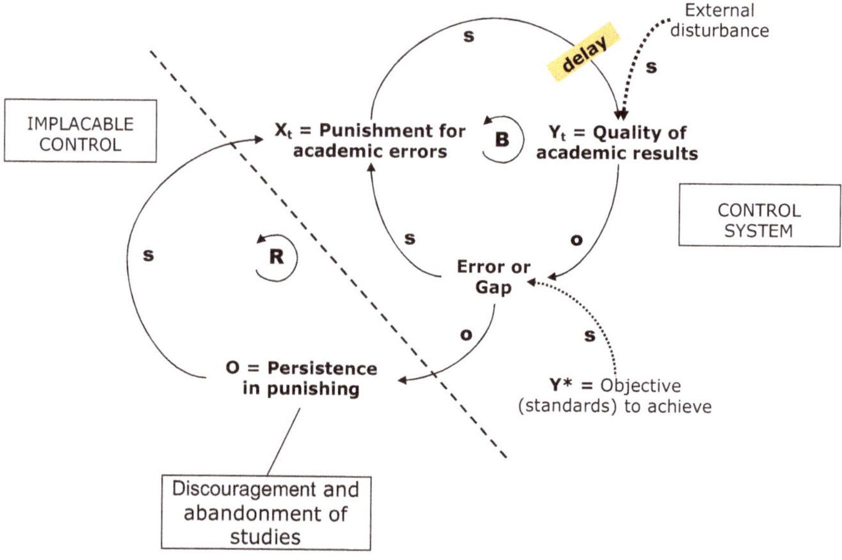

Fig. 5.15 "Persistence" in trying to eliminate academic errors

is not ready: "We'll get there in time just the same …" Instead, we arrive when the check-in has already closed. Are you always late for appointments even though you feel you have left in time and there is nothing stopping you from leaving a few minutes earlier? Even though you could be more careful, do you always wait too long to send your purchase order so that your supplier is forced to deliver it late, each time causing you to stop production?

Such behaviors are not due to bad faith, ignorance, incompetence, unwillingness, or irrationality but to the malfunctioning of the sensor and regulation apparatuses, which are not able to correctly judge the distance, $E(Y)$, between the present position, Y, and the objective, Y^*, and to properly regulate the system's levers. I propose calling this general phenomenon the ARCHETYPE of the "Degradation of the error Assessment," which is illustrated in Fig. 5.16.

This archetype is clearly operating when there is a temporal objective; the erroneous assessment of the "time remaining" leads to delays and to even more serious damage. However, this archetype has a more general action that concerns all types of variables to control. Do these words sound familiar: "Check how much gas we have"; "There's enough still to go a long way." The result: waiting for the tow truck to arrive. "Take the dog out to pee"; "In a moment, there's still time." The result: you have to clean the floor. "It's time to leave"; "In a little while, we'll make it in time." The result: you arrive after the film has started. "Study because your exam is coming up"; "Don't worry! I'll manage without any problem." The result: you fail the exam.

I shall conclude this brief presentation of the pathologies of control with the model in Fig. 5.17, which highlights why many people are latecomers "by nature." This is not a question of a lack of volition or of scant respect for timetables; rather it represents the systematic underestimation of the distance between the actual time and the time of the appointment so that there is delayed or insufficient action on the levers that could accelerate the actions needed to arrive promptly. I would finally point out that persistence as well as the underestimation of the error gives rise to side effects that can aggravate the situation, as is explicitly indicated in the lower part of the various figures showing these archetypes.

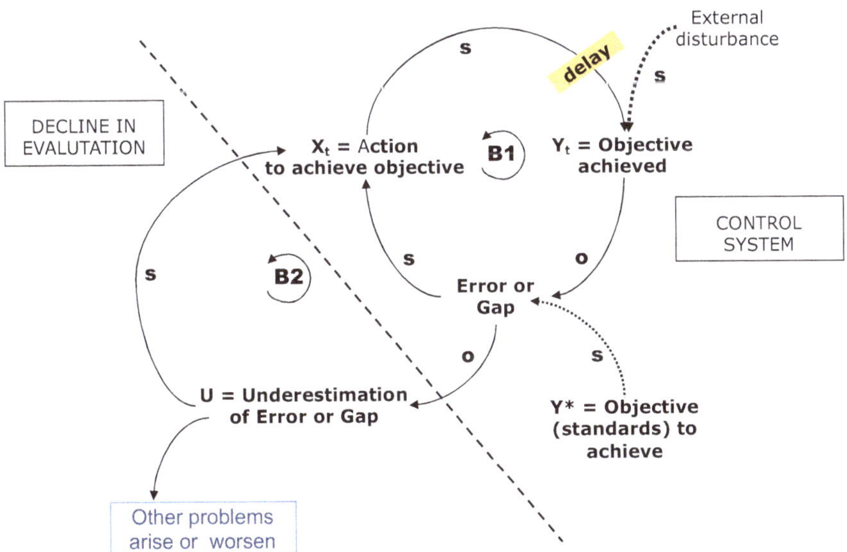

Fig. 5.16 "Degradation of the error assessment" ARCHETYPE

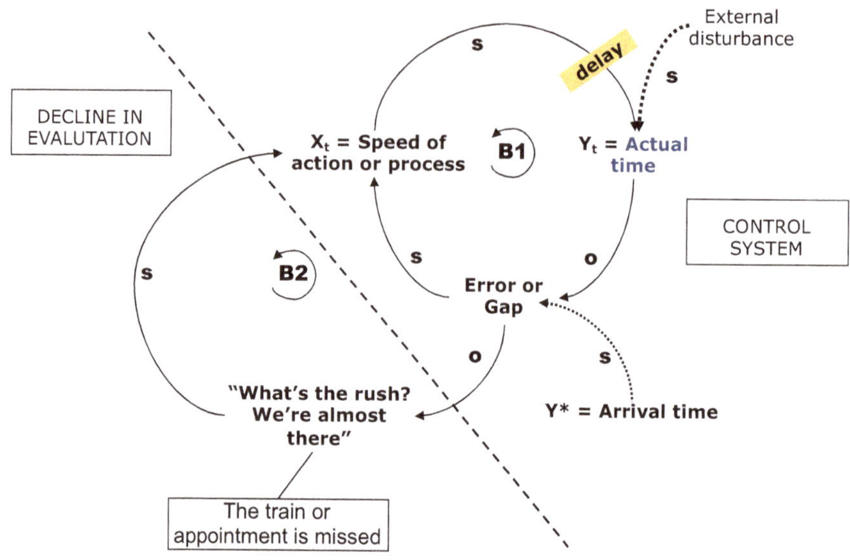

Fig. 5.17 "Degradation" of the error Assessment in the habitual latecomer

5.9 Problem Solving and Control Systems

To broaden our understanding of control systems that every day influence our behavior, it is useful to emphasize the relation between the discipline of control and problem solving, the process by which individuals and organizations recognize and solve their problems (Mayer 1992). We know that "every day and all through the day" we are bombarded by all types of problems to solve in order to live a dignified life (Robertson 2001). Whatever the problem—where to eat and what to wear, not enough money for a project, what name to give our second child, the flat tire to fix on our car, the choice of a new job, a stopped-up sink, etc.—they all have one thing in common: they describe a gap between our present state and the one we have lost or are trying to reach.

> In other words, a distinguishing feature of a problem is that there is a goal to be reached and how you get there is not immediately obvious. To put it at its simplest, you have a problem when you are required to act but you don't know what to do (Robertson 2001, p. 2).

Technically speaking, a *problem* is the perception of a gap between the present situation, which we are not satisfied with and would like to improve, and a desired situation, an objective to achieve; or it reveals a worrisome situation we would like to bring back to optimal conditions: "*A problem exists when someone identifies a desired change in a situation*" (Harris 2002, p. 25).

In this sense, the solution to a problem is equivalent to producing a change, and managing solutions to problems can be conceived of as a specific and circumscribed process of change management (Sect. 7.9). In fact, the *solution* of the problem implies the choice of some action to take to try and eliminate the gap between the present and desired situations. The actions we can undertake—called the *courses of action*, action alternatives or, simply, *alternatives*—must be carefully evaluated in order to estimate, with the maximum precision possible, the results, effects, and consequences of each action in terms of the objective, taking into account the possible *states of nature* (i.e., the external events)—which do not depend on the decision maker—that can create disturbances in reaching the desired state. The act of deciding is not a simple procedure for choosing an alternative but rather a complex process made up of several phases:

(a) Identifying the existence of a problem and specifying the objectives to pursue to modify the actual state (problem finding):

> Many organizations and their managers drive toward the future while looking through the rear-view-mirror. They manage in relation to events that have already occurred, rather than anticipate and confront the challenge of the future (Morgan 1988, p. 4).

(b) Recognizing the type of problem to solve (problem setting).
(c) Redefining the problem; that is, restructuring the assumptions of the problem and gaining a new perspective on it:

> Problem finding is concerned with identifying the problem. Many problems seemed to be apparent: lots of complaints about pay rates, absenteeism, high turnover of personnel, low standards of efficiency, marked aggression against management, etc.
>
> Redefining the problem: the boundary-examination technique was used to restructure the assumptions of the problem and produce a new perspective on it (Proctor 1999, p. 85).

(d) Searching for information (even if it is forecasted information) compatible with the cost of the search and, based on the information, carrying out the subsequent steps.
(e) Searching for technical, legal, and economic constraints to the admissible alternatives, considering the external environment and the internal resources, and delimiting the area of admissible solutions; there is need for *awareness* of the system of *values*, *rules*, and *policies* used in evaluating the identified alternatives.
(f) Choosing evaluation criteria for each alternative based on rules (subjective-intuitive or objective-formal). If a single criterion is chosen, the decision is *single-criterion* or *single-objective*; if there are multiple criteria, the decision is *multicriteria* or *multiobjective* (Sect. 5.12.1).
(g) Identifying and evaluating the various alternatives based on the chosen criteria, taking into account the results each alternative permits, based on a given process of forecasting and calculation.
(h) Choosing the best alternative, anticipating the circumstances that most likely could condition the results (noncontrollable variables or states of nature); ensuring that the choice is *compatible* with the system of values, policies, and rules shared within the organization.
(i) Implementing the chosen alternative; that is, carrying out a concrete process of action that modifies the present situation in order to achieve the desired one.

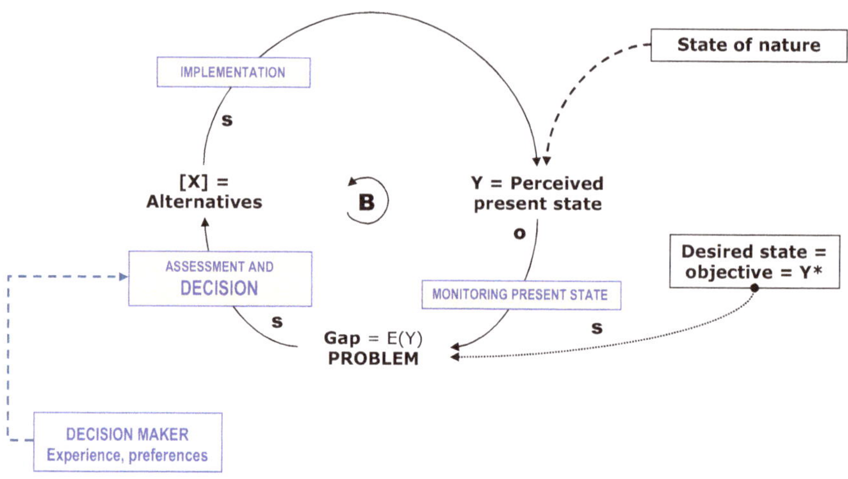

Fig. 5.18 Model of problem solving as a control system

Along with implementing the alternatives, the gaps between states and objectives must be controlled; this represents the phase where we evaluate whether the process of change is capable of modifying the present situation in the desired manner.

When the implementation is not capable of attaining the desired results, new problems arise and the process becomes cyclical. The process by which solutions are sought for problems is called problem solving (Shibata 1998). Using the terminology, we have just presented, problem solving can be represented as a control system by means of the model in Fig. 5.18.

The reason for this identification is immediately clear if we consider that a problem is nothing other than the gap (deviation, error, and variance) between two states—actual and desired—that we judge to be useless or harmful and wish to eliminate or reduce as much as possible. Resolving the problem means eliminating the gap, which in fact implies designing the control system so that it identifies the lever [*X*] that, by producing the results [*Y*], taking into account the states of nature [*D*], eliminates the gap with respect to the desired value [*Y**].

Naturally we must first verify if the problem (gap) is real—that it derives from an actual, undesired state—or imaginary—that it derives from an unrealistic objective. If we specify the proper elements in the model in Fig. 5.18 we can obtain some interesting types of problems.

(a) *Problems of attainment* and *restoration* (or *maintenance*). This classification concerns the type of objective; if the objective is *Y** = "a future state we want to reach" through alternatives that guide the present state toward that objective, then the problem is to produce the *change* that eliminates the gap, thereby *attaining* the objective. What stock should I invest in? Should I introduce a new product? Should I replace the machinery? Which TV channel should I watch tonight? If instead the objective is an historical state, we wish to *restore* by bringing the present state back to the previous situation, or if it is the present state that we wish to maintain despite external disturbances, then the problem is one of *restoration or maintenance*. I have to change my tire, but how do I do this in the dark? The sink is stopped up and I desperately have to wash the dishes. Product quality is deteriorating and I have to block deliveries. *Attainment* problems are typically problems of result; *restoration and maintenance* problems are typically problems of control.

(b) *Isolated problems* and *recurring problems*. The distinction here concerns how often the problem comes up. The former occur in a restricted context and/or come up only once and/or can be quickly and definitively resolved. Recurring problems arise several times, are linked to previous problems, and their resolution each time influences subsequent problems. How often should I go to the hypermarket? I'm stressed out; I need something to calm me down. These are typical recurring problems because they can arise every day, week after week. Many species are becoming extinct. What can we do? This can also be considered a recurring problem since, for example, if we try to avoid the extinction of a species (panda) the problem arises for other species (whales, monkeys, parrots, etc.). It is not always easy to distinguish between the two types of problems; of course no problem is completely one-off, just as no problem is definitely recurring. We have to assess whether or not to maintain or break the links with other problems.

(c) *Problems which are deterministic, probabilistic, and uncertain in nature*. Here the distinction refers to the type of knowledge of the states of nature that can emerge. The problem is in a deterministic context if only one state of nature is foreseen for each alternative, so that for each X we can calculate the result obtained in terms of Y and choose the X necessary to produce $Y = Y^*$. Technical, engineering or simple financing problems are deterministic in nature. In many circumstances, for each alternative we can foresee alternative events whose probability of occurrence can be estimated; for each X we can calculate Y as the expected value of the random variable of possible results, thereby choosing X based on criteria that take into account the result and the *risk* associated with the attainment of the various possible values of Y. These problems are probabilistic in nature and are typical of the biological and social fields. If alternative events are foreseen but the probabilities of their occurrence are not known, then the problem is *indeterminate* and the choice of X must be based on the results estimated with diverse criteria that take into account the *preference for risk* (e.g., the *min–max* criterion or the *regret* criterion), caution (the max–min criterion), or absolute uncertainty. These problems frequently occur in all aspects of our daily life (what road to take to avoid traffic?) as well as in the physical, biological, and social worlds.

(d) *Problems with the states of nature* or *with rational adversaries*. When the values of the action variables Y do not depend solely on the X levers or on "natural" outside disturbances but instead are conditioned by countermoves by rational adversaries, then the problem can be compared to a game and be dealt with using decisional criteria from *game theory*.

From this distinction, it is clear that problem solving is identified with a control system when it typically deals with *maintenance* problems, *recurring* and *recursive*, that are deterministic or probabilistic in nature. Even when problem solving deals with attainment and "one-shot" problems, which do not have repetitive decision-making cycles, it is still a control system in which the decision must lead to the elimination of the gap—and thus to a solution to the problem—in one cycle only. In these circumstances the problem-solving process is also known as the decision-making process (Simon et al., 1986; Harris 1998, 2002; Shaw 2001).

Problem solving is not only an application of the logic of control systems; it is a very complex process that utilizes a corpus of methods and techniques considered effective for solving problems in the various contexts we operate in as individuals, members of an organization, or as organizations.

> The work of managers, of scientists, of engineers, of lawyers—the work that steers the course of society and its economic and governmental organizations—is largely the work of making decisions and solving problems. It is the work of choosing issues that require attention, setting goals, finding or designing suitable courses of action, and evaluating and choosing among alternative actions. The first three of these activities—fixing agendas, setting goals, and designing actions—are usually called problem solving; the last, evaluating and choosing, is usually called decision making (Simon and Associates 1986, online).

5.10 Problem Solving and the Leverage Effect

After this brief survey, I would like to reflect on the important contributions systems thinking and control thinking can make to understanding the logic of problem solving, beginning with a particularly important aspect.

If we represent problem solving as a control system, it is clear that, formally speaking, the problem appears as a "gap" between the objective and the present state we wish to modify through the choice of appropriate decision-making levers; in fact, *the problem is a symptom*, but of what?

We must search for the answer in the basic logic of systems thinking according to which every state, event, or situation must be viewed as the result of the action of some system operating on one or more variables. By following this rule of systems thinking we are warned that a problem must not be identified with the *evident symptoms*—that is, the perceived gaps in the values of some variable—that require urgent measures: the *symptom* is not "the" problem because "the" problem is in the *structure* of the entire system and its whole dynamics. In order to solve "the" problem it is not enough to remove the *symptom*—settling for short-term symptomatic solutions (Sect. 5.2) with equally short-term effects—but to intervene on the *structure* producing that symptom.

> The definition of a problem and the action taken to solve it largely depend on the view which the individuals or groups that discovered the problem have of the system to which it refers. A problem may thus find itself defined as a badly interpreted output, or as a faulty output of a faulty output device, or as a faulty output due to a malfunction in an otherwise faultless system, or as a correct but undesired output from a faultless and thus undesirable

system. All definitions but the last suggest corrective action; only the last definition suggests change, and so presents an unsolvable problem to anyone opposed to change (Brun 1970, online)

Employing the logic of control systems, we can redefine a problem as an anomaly, inconvenience, or difficulty in *carrying out systemic processes* or in *achieving given programmed objectives* which the manager of some management control system perceives when the control system displays an error, $E(Y) = Y^* - Y$. Systems thinking goes deeper and teaches us that a problem concerning the values of the variable Y arises precisely because of the bad—or undesired—functioning of the *generating system* of Y (normally considered a black box), which disturbs the optimal state, Y^*; the overall action of these disturbances can be considered the result of the action of external events. In order to solve the problem—that is, to eliminate the gap—we must reconstruct the *generating system* of Y (transform this into a white box) and determine the systemic levers, that is, the courses of action X, that operate to produce the desired Y^* in spite of D.

In other words, the consequences of each course of action do not derive from some *immediate* effect $X \rightarrow Y$ but from the overall *lever effect* ($X \rightarrow$ Generating System $\rightarrow Y$) that intervenes on the *causal map* of the system to control and on the *loops* on which the systemic lever operates. Thus, the *definitive solution* of a problem almost never comes from variables directly connected to the symptom but from variables that are often distant, located along causal chains and loops in the generating system of Y; these variables must be recognized and their connections accurately described.

The model in Fig. 5.18 must be completed as shown in the model in Fig. 5.19.

The model demonstrates the important general principle that derives from the foundations of systems thinking we have just examined: we must not limit ourselves to considering problems only as undesired *symptoms* of *immediate causes* that are to be uncovered and eliminated (symptomatic solution), but rather as the undesired effects of the functioning of some *generating system* that must be recognized, specified, and controlled by means of control systems provided with appropriate systemic levers (definitive solution).

Peter Senge, for that matter, has been clear on this point. He defines a *definitive solution* that exploits the potential of the system structure and its loops—not limiting itself to symptomatic interventions on individual variables—as the *leverage effect*.

> The advantage of systems thinking derives from the leverage effect—seeing in what way the actions and changes in the structures can lead to long-lasting, meaningful improvements. Often the leverage effect follows the principle of the economy of means, according to which the best results do not come from large-scale efforts but from well-concentrated small actions. Our non-system way of thinking causes significant specific damage because it continually leads us to concentrate on low leverage effect changes: we concentrate on symptoms of higher stress. We correct and improve the symptoms: but such efforts are limited, when things go well, to improving short-term factors, while worsening the situation in the long run (Senge 1990, p. 131).

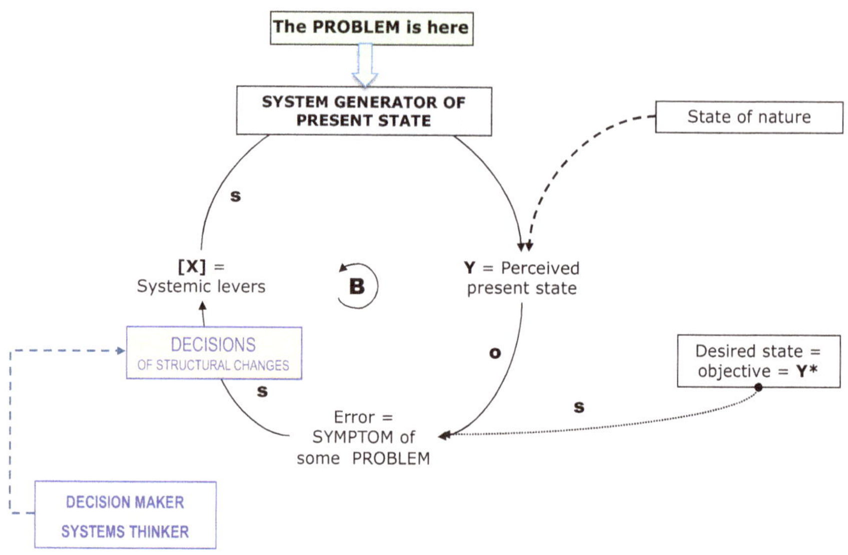

Fig. 5.19 Problem-solving model and the system generating the problem-causing state

Obviously, as the quote clearly demonstrates, and Fig. 5.20 shows, exploiting the *leverage effect* means identifying the *loops* in the structure of the system—that is, the subsystems—which allow for *greater beneficial effects* on the symptoms, with a *minimum use of resources*, taking into account the time needed for the *leverage effect* to come into effect.

In fact, the causal diagrams do not automatically *highlight* any solution; instead, they must be carefully studied in order to identify one or more linked *loops*—called *system* or *structural levers*—which, by being acted upon, can cause the *leverage effect*. This represents the most difficult moment; if we cannot recognize the *system lever* that can bring about the *leverage effect*, then no decision can be said to be consciously taken and no problem definitively resolved.

Once again, the fundamental instrument for reaching a definitive solution is the *structural map* of the situation; in problem solving the construction of the structural map must *start from the detected symptom*, from the difficulty encountered, in order to arrive, through successive connections, to the formation of the map of the processes. The problem emerges when we find anomalous relations or noncoherent trends in elementary processes or in chains of processes, which reveal unwanted reinforcements or undesired balancings, or when appropriate variable-generating programs are not applied.

The construction of the structural map makes it easier to *identify the alternatives* and assess the consequences, as shown in Fig. 5.20. There are three types of systemic levers [*X*] (i.e., the action alternatives):

- [X1]. Measures to increase the *performance of the personnel* that act through the physical structure of the processes that determine the dynamics to be modified.

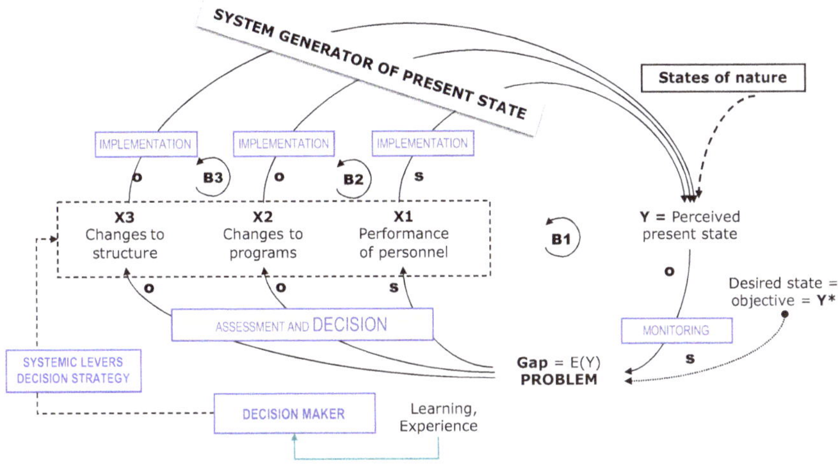

Fig. 5.20 Problem-solving model as a control system

- [X2]. Measures that lead to *improvements in the action programs* (rules, routines, etc.) that activate the processes and determine their dynamics.
- [X3]. Measures to change the *system structure*.

In order to identify the alternatives, the decision-maker must behave like the manager of a control system, evaluating the error and formulating *strategy* by adopting the well-known technique of "*what would happen if* …; *otherwise* …" Forecasting the consequences of each alternative means being able to simulate the changes the processes have undergone through their hypothetical implementation.

The structural map allows us to highlight the results obtainable in both the short and medium-long term. This *simulation* reveals all the usefulness of systems thinking; in fact, we must remember that every action has *immediate* consequences for the processes it modifies as well as long-term consequences, since every modification of given processes spreads to all the others linked to them in the structural map.

In order to identify the effective control levers the loops must be examined. In the medium term a *strengthening loop* can amplify, even unexpectedly, modest effects that appear in the short term. On the other hand, *balancing loops* can render useless positive short-term results, even eliminating them in the long term. Finally, the chain of processes in the opposite direction can even create loops capable of inverting the direction of the expected results so that the alternative we wish to introduce to eliminate a problem could actually worsen it.

Here are four important lessons from system thinking applied to problem solving (Sect. 5.11):

1. If you want to solve a problem through alternatives that change the system's organization or structure, then redesign the map of the processes and analyze the loops; the new network of processes will permit an assessment of the risk that

the consequences of the alternatives turn out to be the opposite of what was desired.
2. For a correct identification of the problem we must distinguish between process problems (often linked to operational programs) and structure problems; it is then necessary to evaluate the alternatives for structural changes through a loops analysis carried out over the medium-long term.
3. If you wish to solve a problem using alternatives that only modify the processes (thus employing new operational programs), pay attention to the alternatives that must produce an expanding or contracting dynamics, keeping in mind that, according to the Law of Dynamic Instability (Sect. 1.5), no expansion or contraction can be maintained for long; every reinforcing loop always produces a balancing loop which, in the long run, can change the direction of the dynamics.
4. In addition, be careful about the alternatives that eliminate an expanding or contracting dynamics; every balancing generates contrasting processes that can prevail and trigger the dynamics once again, often in the opposite direction.

This theoretical perspective which considers the solution of a problem as the search for the most appropriate control system to activate a process of change to modify the present state and bring about a future state (set as the objective) appears more useful the more frequently problems arise in *large systems* in which the *fundamental laws* of systems thinking are particularly evident and always act. Such systems are usually studied within the context of information systems and networks; however, their significance is more general, and they can be applied to various fields: social systems, economic systems, environmental systems, computer networks, logistics and transportation systems, large manufacturing factories, and so on (Dewan 1969), even if the concept has not yet been precisely defined.

> Regulation and control in the very large system is of peculiar interest to the worker in any of the biological sciences, for most of the systems he deals with are complex and composed of almost uncountably many parts.
>
> The ecologist may want to regulate the incidence of an infection in a biological system of great size and complexity, with climate, soil, host's reactions, predators, competitors, and many other factors playing a part.
>
> The economist may want to regulate against a tendency to slump in a system in which prices, availability of labour, consumer demands, costs of raw materials, are only a few of the factors that play some part.
>
> The sociologist faces a similar situation. And the psychotherapist attempts to regulate the working of a sick brain that is of the same order of size as his own, and of fearful complexity.
>
> These regulations are obviously very different from those considered in the simple mechanisms of the previous chapter. At first sight they look so different that one may well wonder whether what has been said so far is not essentially inapplicable. This, however, is not so. To repeat what was said … many of the propositions established earlier are stated in a form that leaves the size of the system irrelevant. […]
>
> Before we proceed we should notice that when the system is very large the distinction between D, the source of the disturbances, and T, the system that yields the outcome, may be somewhat vague, in the sense that the boundary can often be drawn in a variety of ways that are equally satisfactory (Ashby 1957, pp. 244–245).

5.11 The Principles of Systems Thinking Applied to Problem Solving

Referring to Fig. 5.19, solving a problem means achieving a state Y* considered "optimal," starting from a present, nonoptimal state Y. The Error E(Y) = Y* − Y must be conceived of as the evident "SYMPTOM" of the incorrect dynamic of a state variable, which has departed from the optimal value. The problem-solver (i.e., the manager of the Control System) must eliminate the error through decisions involving the activation of the systemic levers to modify the variable Y, which is out of control. Systems Thinking warns us that a problem must not be identified with the evident symptoms, proposing the following general rule:

GENERAL RULE: We must not limit ourselves to viewing problems as undesired symptoms of immediate causes (Ishikawa and Kondo 1969; Ishikawa 1976) to discover and eliminate (symptomatic solution) but as the undesired effects of the functioning of some system—or subsystem—that must be recognized, specified and controlled (definitive solution). From this general rule, some specific principles can be derived.

FIRST PRINCIPLE: Apply Systems Thinking above all to avoid problems from arising or appearing too soon, when the symptom is not yet high.

Systems Thinking helps us with a second very relevant principle:

SECOND PRINCIPLE: The symptoms are not the problem but a sign that some problem has caused them. Do not stop at the symptom but apply Systems Thinking to recognize the true nature of the problem and avoid symptomatic solutions which, by not solving the problem but only eliminating the symptoms, often aggravate it.

COROLLARY: Try to determine if, by your behavior, you "are part of" the problem or "the problem itself."

> Albert Einstein reputedly said, "If I had an hour to save the world, I would spend 59 minutes defining the problem and one minute finding solutions". Unfortunately, most organizations are running around spending 60 minutes finding solutions to problems that do not matter. …. One of the greatest skills an organization can possess is the ability to ask the right question in the right way (Shapiro 2011, online).

The pressing outbreak of a symptom—perhaps as the consequence of inefficient problem solving—requires us to understand the system from which it has originated (Fig. 5.19). We know that systems structures are found in various contexts and produce similar effects. Knowledge of the system structure is indispensable to avoid activating processes that could produce undesirable effects. In particular, we must pay attention to the creation of reinforcing loops which, if triggered, would lead to devastating effects. We can therefore formulate these additional principles:

THIRD PRINCIPLE: In problem solving, it is necessary to always consider the decision in the context of some systems process—which characterizes a given situation of an individual or organization—whose network of processes must be specified through the construction of a meaningful Causal Loop Diagram. The symptom is not "the" problem, because "the" problem lies in the structure of the system and its dynamics (Fig. 5.19).

FOURTH PRINCIPLE:

1. In finding solutions for systemic problems, do not be content with symptomatic solutions; look for systemic-structural levers that can produce the more incisive effect.

2. If there are several systemic levers, choose the most efficient and effective one, which produces the maximum effects with the minimum effort.

3. The choice of structural and decisional levers, as well as the intensity of the actions to modify their values, must follow from a careful construction, interpretation, and assessment of the system's causal map.

Peter Senge defines a definitive solution that exploits the potential of the system's structure and its loops—not limiting itself to symptomatic interventions on individual variables—as the *leverage effect*.

The advantage of systems thinking derives from the leverage effect – seeing in what way the actions and changes in the structures can lead to long-lasting, meaningful improvements (Senge 2006, p. 131).

We can derive the:

FIFTH PRINCIPLE: If you want to solve a problem by modifying the system structure or by introducing new programs that readjust the processes, then always remember the LAW OF DYNAMIC INSTABILITY (Sect. 1.7.4) and pay attention to:

1. The changes that can trigger dynamics of expansion or reduction. No expansion or reduction will be maintained for long; each reinforcing loop always produces a balancing loop that, in the long run, can also reverse the direction of the dynamics.

2. The changes that balance dynamics of expansion or reduction; every balancing loop generates contrasting processes that can prevail and restore the original dynamics, often in the reverse direction.

This last principle represents the logical conclusion to the application of the first principle.

SIXTH PRINCIPLE: Try not to commit the same mistake two (or more) times; if today's problems come from yesterday's "solutions" (Senge 2006, p.57), you must ensure that today's solutions, no matter how capable they are of solving the problem, do not create new problems tomorrow.

Although these rules are clear and acceptable, the problem solving process is not easy; the large number of variables, the different levels of influence, and the time dynamics represent problems in applying Systems Thinking to Problem Solving. A careful inspection of the structural map of the system that generates problems should normally allow us to identify *definitive solutions*. Often, however, the errors are so subtle that they are not perceived, and thus they become almost inevitable. In fact, the causal loop diagrams do not automatically *highlight* any solution; instead, they must be carefully studied in order to identify one or more linked *loops*—which are called *system*, or *structural levers*—which, by being acted upon, can cause the *leverage effect*. This represents the most difficult moment; if we cannot recognize the *system lever* that can bring about the *leverage effect*, then no decision can be said to be consciously taken and no problem definitively resolved.

5.12 Complementary Material

5.12.1 Multicriteria Decision-Making

In order to offer an example of the decision-making process in the simple case of a multicriteria decision under conditions of certainty (see: Fishburn 1967a, b; Miller and Starr, 1969; Bouyssou et al. 2006; Saaty 1990; Ho et al., 2010; Kabir et al. 2014), let us consider choosing a car by trying to satisfy the following five decision-making criteria, which we can also view as the "objectives" to achieve.

- o_1—minimum purchase cost,
- o_2—minimum operating cost,
- o_3—maximum speed,
- o_4—maximum comfort,
- o_5—maximum representativeness as status symbol.

Taking account of these decision-making criteria, we shall limit the choice to five models, which represent the courses of action.

- c_1—two-seat sports car,
- c_2—four-seat sports car,
- c_3—big-engined station wagon,
- c_4—diesel station wagon,
- c_5—minivan.

We must now specify the degree of preference for each alternative for each decision-making criterion (or objective).

- For the first three objectives we use the car manufacturers' sticker prices.
- For the objectives of comfort and representativeness we devise a score on a scale of 0–10, taking into account as well the "opinions" of the family members.

The *decision-making problem* can thus be represented by the *preference matrix* in Table 5.1. Written in the matrix squares for each line is the following:

- At the top center: the result for each alternative for each decision-making criterion
- In italics at the bottom left: the order of preference for the alternatives, on a simple scale of 1–5.

Table 5.1 Preference matrix

	o_1 = purchase cost €/000	o_2 = operating cost €/km	o_3 = speed	o_4 = comfort	o_5 = representative ness
c_1 – two-seat sports car	50	0.7	220	4	10
Preference	*2*	*3*	*1*	*5*	*1*
c_2 – four-seat sports car	60	0.8	210	6	9
Preference	*4*	*4*	*2*	*5*	*2*
c_3 – big-engined s-w	65	0.9	200	8	8
Preference	*5*	*5*	*3*	*2*	*3*
c_4 – diesel station wagon	55	0.6	170	7	6
Preference	*3*	*1*	*5*	*3*	*5*
c_5 – minivan	45	0.65	180	10	8
Preference	*1*	*2*	*4*	*1*	*4*

Observe that none of the alternatives (cars) is immediately preferable to the others; that is, there is no "dominant" alternative. The decision can be made using only a single decision-making criterion (objective) or all the criteria simultaneously. In the first case, the governance must specify the decision-making criterion adopted; in the second case, the *degree of relative importance* of the criteria.

This example immediately shows that every *multicriteria* decision has no immediately determinable solution, depending instead on the *decision-maker's preference regarding the decisional criteria*. Every decision is made up of other decisions.

If we assume that the decision maker is "young" and not particularly sportive and assign to the decision-making criteria the weight indicated in the upper row of the preference matrix in Table 5.2, we can calculate a weighted preference index to determine, for each alternative, the *weighted average of the preference indicators*. These *preference indices* are shown in the last column on the right of the preference matrix in Table 5.2.

Since the scores from 1 to 5 are inversely related to the preference (index 1, bottom, preferred to index 5, top), in order to choose it is necessary to identify the course of action corresponding to the lowest average preference index (right-hand column); in our case, the choice would be to purchase the c_1 = "two-seat sports car," which has an average preference index = 1.9. If we insatead assume that the choice must be made by the "head of family," who assigns to the preference criteria the weights indicated in the matrix in Table 5.3, the choice would be to buy the minivan.

Table 5.2 Preference matrix for a "young" decision maker

	o_1 = purchase cost €/000	o_2 = operating cost €/km	o_3 = speed	o_4 = comfort	o_5 = representativeness	Index values
Weight of objectives	0.3	0.1	0.2	0.1	0.3	1.0
c_1 – Two-seat sports car						
Preference	2	3	1	5	1	1.9
c_2 – Four-seat sports car						
Preference	4	4	2	5	2	**3.1**
c_3 – Big-engined s-w						
Preference	5	5	3	2	3	3.7
c_4 – Diesel station wagon						
Preference	3	1	5	3	5	3.8
c_5 – minivan						
Preference	1	2	4	1	4	**2.6**

Table 5.3 Preference matrix for a "head of family"

	o_1 = purchase cost €/000	o_2 = operating cost €/km	o_3 = speed	o_4 = comfort	o_5 = representativeness	Index values
Weight of objectives	0.3	0.1	0.2	0.1	0.3	1.0
c_1 – Two-seat sports car						
Preference	2	3	1	5	1	1.9
c_2 – Four-seat sports car						
Preference	4	4	2	5	2	**3.1**
c_3 – Big-engined s-w						
Preference	5	5	3	2	3	3.7
c_4 – Diesel station wagon						
Preference	3	1	5	3	5	3.8
c_5 - minivan						
Preference	1	2	4	1	4	**2.6**

5.13 Summary

We have learned the following:

1. How to recognize or design the logical structure of a control system by recognizing the variables to control [*Y*], the objectives [*Y**], and the control levers [*X*] (Sect. 5.1).
2. There exist two control variants: structural and symptomatic controls, which intervene in many circumstances when we cannot identify the lever that acts directly on *Y* (Sect. 5.2).
3. How to evaluate the effectiveness and efficiency of a control system (Sect. 5.3) and how to strengthen *undersized* control systems (Sect. 5.4).
4. What the risks are of failure of the control carried out by a control system (Sect. 5.5).
5. The existence of Risks of Failure of the Control Process Due to same Characteristics of the Variables to be Controlled (Sect. 5.6).
6. What the risks are of failure due to the erroneous choice of levers. In this regard, two important archetypes were presented (Sect. 5.7): the FIXES THAT FAIL (Fig. 5.3) and the SHIFTING THE BURDEN (Fig. 5.4); a variant is also presented (Fig. 5.8).
7. We have analyzed four control pathologies: discouragement, insatiability, persistence, and underestimation (Sect. 5.8), and the four archetypes that represent them (Figs. 5.9, 5.10, 5.11, 5.12, 5.13, 5.14, 5.15, 5.16, and 5.17).
8. Lastly, we have examined the *problem-solving* and *decision-making* processes and their relation to the control system (Sects. 5.9 and 5.10).

Part II
The Magic of the Ring

Chapter 6
The Magic Ring in Action: Individuals

Abstract Part 1 presented the *theory* of control systems. A control system can be thought of as a *Ring* that rotates several times to guide the variables [*Y*], toward the assigned objectives [*Y**], through the control levers [*X*]. The first five chapters produced, so to speak, the anatomy of *Rings*, presenting their logical structure, their *modus operandi*, the main types, and the risks of failure in the control. In Part 2 we now move on to the *discipline* of Control, that is, training ourselves to observe, recognize, model, and simulate the action of the *Rings* in every micro and macro "environment" where they carry out their regulating function. In this chapter, we shall try to recognize control systems in the domestic and civic environment.

Keywords Overhead control systems · Public lighting · Urban waste · Control of fine particles · "Gaia" and Daisyworld · Global warming · Gulf Stream · Earthquakes and tsunamis · Postural control systems · Control of needs and aspirations · Biological clocks

> In the arithmetic of life, one is always many. Many often make one, and one, when looked at more closely, can be seen to be composed of many (Lynn Margulis, cited by Habib 2011).

> Let me briefly explain. Homeostasis is the tendency of a complex system to run towards an equilibrial state. This happens because the many parts of the complex system absorb each other's capacity to disrupt the whole. Now the ultimately stable state to which a viable system may run (that state where its entropy is unity) is finally rigid—and we call that death. If the system is to remain viable, if it is not to die, then we need the extra concept of an equilibrium that is not fixed, but on the move. What causes the incipiently stable point to move is the total system's response to environmental change; and this kind of adjustment we call adaptation (Stafford Beer 1975, p. 426).

Part 1 presented the *theory* of control systems. A control system can be thought of as a *Ring* that rotates several times to guide the variables [*Y*], toward the assigned objectives [*Y**], through the control levers [*X*]. The first five chapters produced, so to speak, the anatomy of *Rings*, presenting their logical structure, their *modus operandi*, the main types, and the risks of failure in the control. In Part 2 we now move

P. Mella, *The Magic Ring*, Contemporary Systems Thinking,
https://doi.org/10.1007/978-3-030-64194-8_6

on to the *discipline* of Control, that is, training ourselves to observe, recognize, model, and simulate the action of the *Rings* in every micro and macro "environment" where they carry out their regulating function. It is not a question of overestimating the action of the *Rings* in our world but of realizing that the world itself is made up of *Rings* and that only their joint action can permit the birth and existence of living organisms, maintain the dynamics of the environmental variables, and produce evolution in nature as well as in society, collectivities, and organizations. It is no exaggeration to say that the *Rings* will not cease to amaze us and that we must recognize there is something *magical* in their existence. The *Ring* thus reveals its magic and becomes a *Magic Ring*. To accustom the reader to observing Rings, this chapter will take him on a mental journey through various "environments" characterized by the action of *Rings*. I do not claim that this treatment will be exhaustive; my objective is to indicate to the reader a path for recognizing *Rings* and their action which, through their *ubiquitous presence*, make all aspects of our world possible, in all its facets. By applying the first rule of systems thinking, which obliges us to *see the trees and the forests*, by zooming out and in, we shall try to recognize control systems in ever wider contexts, beginning with the domestic and civic environment and then moving on to the physical and social environment in our world. This journey, though presented in summary fashion, reveals, in whatever environment we explore, an impressive quantity of *Rings* of various types and sizes. For this reason I shall put off examining the delicate control systems regarding societies and organizations until Chaps. 7 and 9.

6.1 Magic *Rings* Operating on a Wonderful Day

The simplest way to begin our journey into the "magic" world of control systems is to rethink our typical day. Obviously I will present some moments from "my typical day," but I invite the reader to reflect on his own experience, which could even be quite different. In any event, in both cases we will find the same *Rings*, though perhaps at different moments.

I am sleeping. My brain, in order to allow me to fall asleep and remain sleeping, has for hours raised the threshold values of perception, Y^*, so that all the modest sensory stimuli do not keep me awake. While sleeping, my *attention* is "turned off," so to speak. Suddenly the ringing of the alarm clock, Y, exceeds the threshold values, producing a variance between the nocturnal quiet and the noise perceived by my sense of hearing, which I perceive as annoying and which moves me to "reactivate" my attentive capacity (signs of awakening = $Y^* - Y$) in order to activate some control lever to make the annoyance cease. I move my arm, X, to reduce the distance, Y, between my hand and the source of the sound, Y^*, and I try to reach the alarm clock (distance from the alarm clock = $Y^* - Y$). Guided by the *routines* developed over many years, the tactile sensors of my fingers, X, reach the button, Y^*, after first having touched several objects on the night stand; once I recognize it (difference between button and the other perceived objects = $Y^* - Y = 0$), I press it (alarm position = off) so that the ringing stops (alarm clock signal = 0). Thanks to these three *Rings* (reaching the source of the sound, recognizing the button, pressing the button) quiet is restored and drowsiness ceases. I am aware that describing the action of the various control systems by also indicating parenthetically the elements Y, Y^*, X, and E, which characterize these systems, weighs down considerably the text without adding further clarity; henceforth, when necessary I shall only indicate the type of control system (CS) that is at work.

Like a dove looking down from on high, as soon as I am awake I recognize the position of my slippers, and I quickly move my feet toward them (recognizing CS and attainment CS with a fixed objective). With my slippers on I assume a vertical position (dynamic objective), which is maintained by the control system for equilibrium (vestibular apparatus and leg and back muscles), and maneuver among the furniture in the room to reach the bathroom (tracking CS), where I begin a series of ritual activities that invariably are repeated each morning. After having completed the cycle of the Continuous Input-Point Output CSs (I shall leave this to the reader's imagination) I start the shower-shaving process. I shall not spend time describing the control system that allows me to shower but only mention that this is a multiobjective *Ring* (control of temperature, water flow, amount of shampoo and soap). The point is to emphasize how complex the control process for shaving is and the fundamental role played in this by the CSs for the positioning of the blade on the face and the control of the razor's movement (attainment CSs). The fingertips, running over every millimeter of my facial skin, act as sensors of the control system to recognize the imperfect shave and the presence of residual whiskers to be eliminated. The action of these *Rings* can be more easily perceived by considering how many times the hand is taken away from the face and then once again reapplied, with extreme precision, and how many times the hand runs over the face again and again during shaving.

At this point the clothing and dressing ritual begins. The recognizing CSs enable me to choose the clothes to wear and to properly match my shirt, tie, socks, and shoes to my suit. The postural control systems allow me to quickly dress and to control, almost without reflecting on it (thanks to the muscle routines), the buttoning of my shirt, and the size and position of my tie knot (qualitative CSs). Perhaps we do not reflect enough on how many *Rings* are activated to tie our shoes; we have to use both hands and a number of fingers as levers to move the shoelaces so as to form a knot with the proper tension.

There is no need for further detail about the problem of dressing; let us move now to the breakfast process, which is regulated by a number of repetitive control systems. There are many attainment CSs that allow us to grab our mug and spoon, pour the coffee, reach the sugar bowl, and pour the sugar, thereby eliminating the distance between the sugar bowl and mug while at the same time maintaining the spoon horizontal. The control system for the coffee's temperature leads us to activate the levers needed to bring the temperature to a level where we can drink the coffee without burning our lips. Several qualitative CSs are set off that enable us to regulate the quantity of sugar and cool down the coffee, without making the coffee too sweet or too cold.

Once breakfast is over the complex activity of making my way to work begins. After activating the recognizing CS to verify the elevator has arrived (in the contrary case, pushing the button to call it) I go down to the garage. In order to get in my car and leave the garage I activate a number of *Rings*. An attainment CS quickly operates several times to allow me to leave the garage by detecting the distances between the car and the right and left walls of the garage door (alternating objectives CSs). Then, after activating the recognizing systems for the presence of obstacles and pedestrians, I drive to my office. As we know, driving requires the activation of multiobjective and multilever CSs that entail the contemporary activation of a number of elementary *Rings*. While controlling the speed and distance of the cars ahead (variable objectives CSs), I recognize the signals from the traffic light and the cars ahead, accelerate, brake, turn the steering wheel, observe the lights, accelerate, brake, turn the steering wheel, brake suddenly, turn the wheel, stop, start moving again, turn left, gain speed, etc. This, in short, is how all the *Rings* needed to drive to my final destination act, dozens of times per minute. And let us not forget that usually the control systems of stress and anger are also activated; these are also needed to arrive at the office and start work under the proper psychological conditions. There are dozens of people to recognize, greet, accompany, and dozens of documents to recognize, read, file away, or respond to that require the activation of a multitude of recognizing CSs as well as the necessary attainment CSs. Likewise, it is easy to see how many attainment and recognizing CSs are needed to write at the computer.

All the activities so far described obviously require the operation of systems of biological control that make up our body, in particular all the CSs of vision, hearing, tactile capacity, as well as postural systems that modify the position of our limbs to allow us to remain balanced, walk, drive, etc. I shall let the reader imagine, by referring to his personal experience, the number and type of *Rings*, of the most varied nature, that are activated at other times during the day, when we speak with our spouse, children, friends, read the paper or relax in front of the TV.

Now it is nighttime. In bed, automatic mental control systems slowly deactivate our attention while also raising the threshold levels to inhibit the perception of sensory stimuli. Our eyes close and we start to fall asleep (non-REM followed by the REM phase), in anticipation of starting a new day, one of many thousands of future days in which millions of control systems will act, second after second, billions of times, to make our existence possible.

Now that we are aware that our lives as individuals would not be possible without the action of these many *Rings*, we can certainly realize, perhaps with some amazement, why we must speak of "magic" *Rings* as the basic instruments of our existence.

6.2 *Rings* Operating in the Domestic Environment

The previous *section* took the simplest route to begin a journey in the universe of the most evident and verifiable control systems which, day after day, allow us to continue down our path in life. However, the *Rings* are everywhere and make possible our behavior in life. The *Rings* do not only allow us to exist as individuals but operate in the most unthinkable places, situations, and circumstances; the fact we are not used to noticing them takes nothing away from their ubiquity. To accustom ourselves to the Discipline of Control it is useful to proceed by means of progressively zooming in or out to observe our environment: start with the simplest and most evident control systems—which occur daily in our home, work or town environments—and then extend our observation and reflection to those that operate at a wider or a narrower level.

The domestic environment is ideal for our first zooming activities regarding this discipline. Our household appliances are by now all built using automatic control systems. The fridge maintains the temperature objective when there are changes in the internal conditions (introduction of hot food) and external conditions (the room temperature increases). The microwave shuts off when the programmed time objective is reached, or when we open the door.

Along with automatic systems, where our intervention is limited to setting the objective—as in the case of the elevator in Fig. 2.19, where we simply push the button to indicate the objective that the "machine" then achieves without any further intervention on our part, or the refrigerator or air conditioner, where the objective is set through the appropriate analog or digital instruments—many others exist where we ourselves are components of the control process, either because we monitor and observe the objectives (temperature, levels, etc.) or because we regulate them (rotation of the levers, pressure on buttons, opening of doors and windows, etc.).

We become part of the control process at sundown, for example, when, no longer able to read our newspaper—despite the fact the control system for the eye has dilated the pupils as much as possible—we regulate the light in our room to maintain a visibility limit, turning on artificial light or opening wider the shutters. However, today on the market we can buy sensor devices to automatically regulate the light throughout the day, after setting the objective for the desired amount of light. Every time we pour wine into a glass (halt control system) or control the stock of food in the fridge (PI–CO control system), we are at the same time monitors (level of wine and stocks) and regulators (stop pouring and visit the hypermarket).

Our computer is a galaxy of control systems in terms of both the hardware and the various software we use; however, it could not function without our intervention; the same occurs with our gas stove, electric oven, air conditioner, television, washing machine, dishwasher, iron, radio alarm clock, and wrist watch. And what about the flow controls we daily carry out using the regulators for our various taps to control water levels, flows, and temperature? How can we cook a cake or make risotto if we do not regulate the level of the gas and how long it is on? Today ever more versatile kitchen robots are becoming common and android robots are already on the market, which represent true ambulatory control systems. We can also consider the control of our pantries and how we change our way of dressing with the changing seasons, the places we frequent, and the circumstances in which we present ourselves. In short, our entire domestic life is only possible thanks to the many *Rings* that permit us to achieve objectives of well-being or to respect the myriad of work and safety constraints.

In all domestic control systems, we play a principal role as users and managers–governors, and only our experience enables us to successfully achieve our objectives. Grandma's cake never burns because she checks how it is cooking minute by minute, while it is likely her granddaughter burns her cake from being distracted by the endless number of telephone calls she receives.

However, I can safely state—referring to Fig. 6.1, which derives from Fig. 3.1—that our domestic life is characterized by continual control processes in which, in varying ways, we play a role, either in a minimum way [① if we only set the objectives], an intermediate manner [① + ②) + ③ if we set the objectives, note down the error and decide the action to take on the control levers], or in global terms ① + ②) + ③ + ④) [if we form the entire chain of control; we are the control systems]. If you are not convinced of this, "look around."

If we even consider our car as part of our environment, we can see that while driving it we form together a formidable multiobjective and multilever control system—of which we are the governors–users (we take a trip and establish the policy indicating the priority of the objectives), the manager (we decide the strategy and regulate the control levers), even the apparatus in the chain of control (we record errors, turn the steering wheel, and activate all the technical apparatuses)—that permits us to reach our desired destination by controlling direction, speed, and travelling time, while taking into account speed constraints and external disturbances (e.g., pedestrians crossing the street).

If you are amazed by the actions of a "Formula 1" race-car driver, who controls his car (accelerator, brakes, and gears) to gain the maximum speed, without swerv-

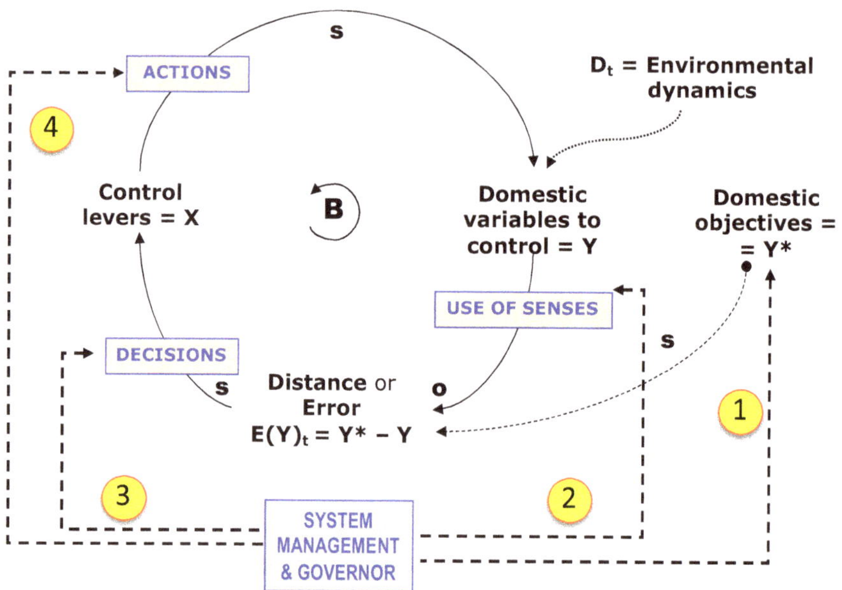

Fig. 6.1 General model of our intervention in domestic control systems

ing, while paying attention to the other drivers, do not underestimate your own abilities when, each morning, you go to work by car in the crowded city streets. The control systems you manage are more complicated than those of the professional driver, having to take into account traffic lights, pedestrians, bicycles, cars braking suddenly, etc., which force you to regulate your speed, lateral and frontal distances, curves, lights, look out for the traffic police, and so on, through continuous actions on the available levers. The car has hundreds of automatic control systems, most of which we ignore (me included), that allow us to drive without worrying about minute objectives and constraints that must be met by the various processors controlled by the automatic electronic devices in the propulsion mechanism, the mechanical structure, and the electrical system. In many car models today, control is becoming almost total: the lights turn on automatically when luminosity is reduced; the wipers start automatically when it rains and automatically go faster as the rain intensifies; all kinds of sensors tell us when the car is in gear, the doors are open, the tire pressure is low, what the petrol level is or the maintenance requirements are, and so on. Some models are equipped with systems to control (by preventing) nodding off by the driver, by detecting the time the eyelids are lowered and producing vibration in the steering wheel or some other algedonic signal. Recent models reduce speed—without the driver's intervention—when there is an obstacle in the road and even park automatically.

For the most part of domestic control systems are multilever and multilayer. Even the air conditioner achieves its temperature objective using many levers. The model shown in Fig. 6.2, which completes Fig. 2.6, involves a three-layer system.

In the first layer the temperature is controlled by two operational levers: the amount of time the air conditioner is on and the fan speed (if available). If these two levers are not enough, there is a structural lever available that modifies the action rate to increase the temperature of the air that is expelled. And if this too is insufficient, recourse can be made to a third layer, which increases the number of splits per room.

What is more interesting, however, is observing how many control systems, even functioning as autonomous systems, are interconnected—or interconnectable—so as to form control systems that are much more powerful and flexible in achieving their objectives. The control of the hot (radiators) and cold air (air conditioners) interacts to maintain a comfortable temperature in our homes year round. Yet even the control of the heat emissions from our appliances or our lighting system, or the control represented by the opening of our windows in summer and winter contributes to maintaining a comfortable temperature.

We have thus arrived at an initial conclusion: though not always possible, when we set an objective in our domestic life we can almost always "put in play" a multitude of *Rings* that form an integrated holonic super system of control, whose various component *Rings* represent the operating levers (Sect. 3.8).

It is not farfetched to state that the first forms of human progress at the dawn of civilization were precisely the result of the continual search for control processes and systems that make life possible; avoiding burns, producing and preserving food, finding protection from dangers, and so on, provided the stimulus for controlling

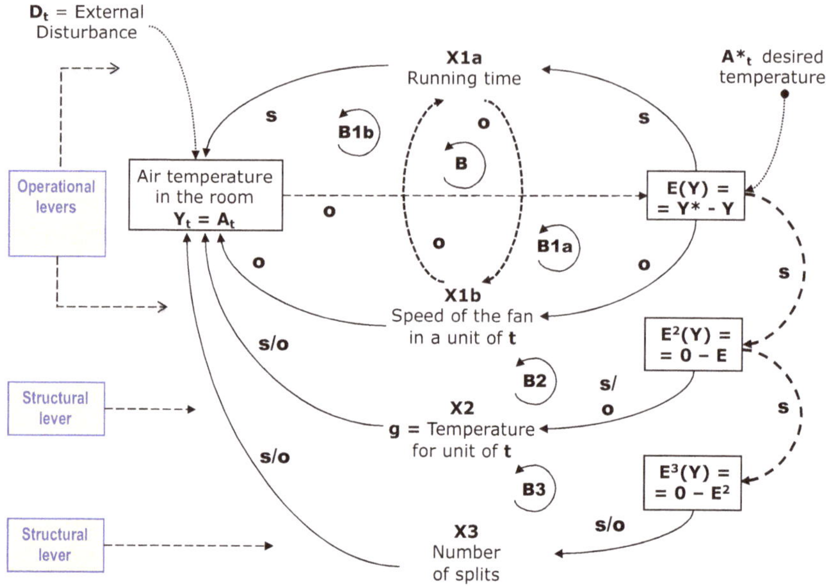

Fig. 6.2 Air conditioner as a three-layer control system

fire, raising animals, growing crops, and building homes. Since then man has come a long way, but even today the search for a comfortable life continues with many projects to aid man in his domestic chores. We can foresee that the continued progress in *domotics* will before long allow for the *remote control* of all our household electrical appliances, which will be equipped with automatic control systems to carry out their processes and limit human intervention, reaching the point where objectives are also set in the form of *standard procedures*.

Recognizing CSs is also part of electrical appliances. For example, vacuum cleaners become automatic robots that explore every angle in the house; washing machines recognize the weight and type of load and choose and measure out the detergent to use as well as the washing time. And recently a brand of refrigerator has been marketed whose models are equipped with ten sensors for the automatic and dynamic control of the temperature and humidity levels to permit the best preservation of our food by signaling the food that will expire next and that already past the expiry date. The "virtual chef" is a multiobjective recognizing, attainment, and halt control system that simultaneously regulates the many cooking variables for a variety of food. There are even *safety doors* available, equipped with sensors that display the temperature, which are automatically blocked when the error indicates a dangerous temperature.

Progress in computers has led to the creation of control systems to aid people with physical problems. "Emc" and "Paravan" now sell a "four-way" joystick that encapsulates the three basic control levers needed to drive a car. Through a single lever the joystick manages the steering wheel, gas pedal, and brakes, and it can be used by disabled people for driving. Soon it will probably also be equipped for all other drivers, since we all would like to control all the levers with the minimum fatigue and maximum safety when behind the wheel. Already commercially available are apparatuses that use voice commands to regulate lights, directional indicators, windshield wipers, horns, door locking, and convertible tops. From what we read in many newspapers and on web sites, progress in control systems for car transport does not stop there; there are now even completely automatically driven cars that arrive at the programmed destination without the need for a driver.

> The *Google driverless car* is a project by Google that involves developing technology for driverless cars. …The project team has equipped a test fleet of at least ten vehicles, consisting of six Toyota Priuses, an Audi TT, and three Lexus RX450h, each accompanied in the driver's seat by one of a dozen drivers with unblemished driving records and in the passenger seat by one of Google's engineers… The system drives at the speed limit it has stored on its maps and maintains its distance from other vehicles using its system of sensors. The system provides an override that allows a human driver to take control of the car by stepping on the brake or turning the wheel, similar to cruise control systems already found in many cars today. …In August 2012, the team announced that they have completed over 300,000 autonomous-driving miles (500,000 km) accident-free, typically have about a dozen cars on the road at any given time, and are starting to test them with single drivers instead of in pairs. Three U.S. states have passed laws permitting driverless cars as of September 2012: Nevada, Florida, and California (Google/Driverless, car project 2012, Online).

> The engineers from Google explain that the Prius utilizes a series of cameras and a roof-mounted, spinning laser to see what is going on around it. The result is a vehicle which might just be safer than one with a human behind the wheel. However, according to the report, the goal of the system is not to completely remove the driver from the equation; the system is pitched as more of a "super cruise-control" than a full auto-drive system. The theory is that it would be useful for traffic-filled commutes to and from work, and it might be a nice solution to eliminate or reduce distracted-driving. Get a phone call? Hit the Google button and let the car have the wheel while you take your call (Google/Prius 2010, Online).

Remaining on the topic of control systems for transport, for some years now a very sophisticated instrument known as "Segway" has been sold for individual, low-speed locomotion. "Segway" is a bicycle with two parallel wheels equipped with multiple MEMS, that is, solid gyroscopic sensors that can emulate human equilibrium and which permit both control of vertical stability for the passenger, even when the vehicle is stationary on two wheels, as well as total maneuverability, with forward and backward direction, since the "Segway" can easily navigate curves and attain speeds of around 20 km per hour. "Segway" can, in all regards, be considered a multilever and multiobjective control system, since stability is maintained through electric apparatuses that eliminate even the minimum error in equilibrium detected by the multiple sensors it is equipped with. Change in direction comes from shifting a vertical, inclinable steering column right or left, as if this were a man-sized joystick.

As a final example of automatic systems in the domestic environment, in Italy the Computer Vision Laboratory, Visilab, at the University of Messina, has created software for a blind assistant.

Blind Assistant is an attempt to realize a software that can help men and children that are blind or visual impaired. This software doesn't require a complex and dedicated hardware: in fact, it is designed to run on simple and cheap devices such as Sony Playstation Portable (PSP). Blind Assistant uses Nanodesktop technology for windows environment, it is able to speak with the user using the speech synthesis engine ndFLite, and it is able to recognize the commands spoken by the blind through the speech recognition engine ndPocketSphinx. In this way, the blind can control the program using the integrated voice interface. At the moment, Blind Assistant provides 7 functionalities: Face recognizer, Room recognizer, Optical chair recognizer, Color scanner, Mail reader, DMTX scanner, Clock/Calendar. The face recognition technology is provided using ndOpenCV, a porting of the original Intel OpenCV libraries that has been developed by the same author, and which is part of the Nanodesktop SDK. The system is able to recognize the name and the position of the people that are present in that moment in your room, using complex algorithms (such as Intel EigenFaces/PentLand PCA) […] (Visilab (2013, Online).

6.3 Overhead *Rings* in the External Microenvironment

We need only leave our houses and look around to understand that we could not survive long in our outside microenvironment—represented by our city and region—without a very large number and variety of *Rings* that show us how to move and adapt ourselves in this environment. I would also observe that we ourselves represent direct control systems that carry out a large number of control processes: for example, whenever we avoid running into other passersby; judge the best moment for crossing the street, despite the traffic; check if we need to stop for the traffic light; avoid puddles; reach our destination; and select the bell to ring to gain entrance. Moving around the city or countryside as control systems, we can achieve our quantitative and qualitative objectives and abide by the large number of constraints that limit our discretionary power.

However, the outside environment is rich in various kinds of control systems which we are not even able to perceive. Many of these are automatic, while others require the intervention of man (teams of operators), which are carried out above our heads, as silent as they are useful: these are *overhead* control systems.

Many essential *overhead* control systems are managed by *policy makers* and *public regulators* to provide public services or protect public health, in order to satisfy a multitude of stakeholders that make up the collectivity in a given territory. Figure 6.3 shows the general model of how overhead control systems operate in the external microenvironment to satisfy the needs of stakeholders through the provision of services meeting given quality standards. Among the many overhead control systems that exist, I shall only briefly mention the most useful ones.

We do not realize how many controls are needed to supply that extraordinary good, that is, the energy to illuminate streets and houses and to operate all the mechanisms that could not work without it. Whether or not it is produced through hydroelectric processes, some kind of combustion process, wind or solar power, or even nuclear energy, the production of electricity must be minutely controlled in all its phases in order to maintain constant the tension and ensure there are adequate flows

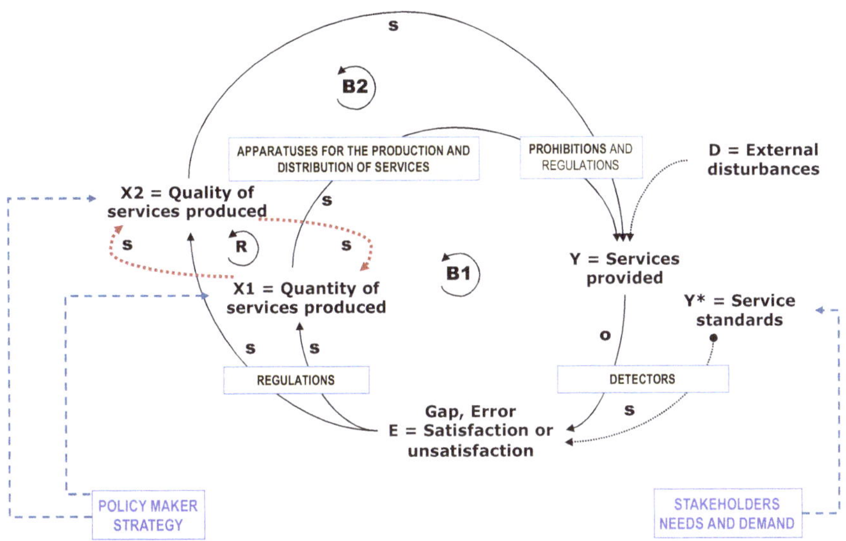

Fig. 6.3 Overhead control systems operating in the external microenvironment

by automatically importing energy from surplus producing countries to avoid black-outs during periods of intense use. Generally this involves automatic engineering controls on the electrical distribution network that, in any event, refer to well-known systems.

A multilever and multilayer control system we are not aware of is the one that guarantees regularity in the level of public lighting, with the automatic switching on of street lights when the outside light falls below a specific threshold and the turning off of the lights in the opposite case, as shown in the simplified model in Fig. 6.4.

Parallel to the control system for the quality of public lighting is the control system for *energy savings*, which determines the optimal period for turning on the lighting facilities in streets, squares, public buildings, stadiums, etc. Lastly, let us not forget the control system that allows us to pay for our energy use through meters and, in the event the meters do not permit an automatic reading (detection devices), personnel trained to read them (detection). The bills are then sent out and the payment verified.

What we have just observed about the production and distribution of electricity is equally true for *natural gas* for domestic or industrial use. We do not realize how many controls are there, not only in the initial phase when the gas is injected into the distribution network but also, and above all, in keeping the network working perfectly efficiently. A qualitative control system placed in the switchboard system in the main junctions of the distribution network allows maintenance teams to promptly intervene (control implementation) as soon as a malfunctioning is detected (detection devices) with respect to the set standards; our noses can even detect malfunctioning with respect to set standards when there is a gas leak in our town.

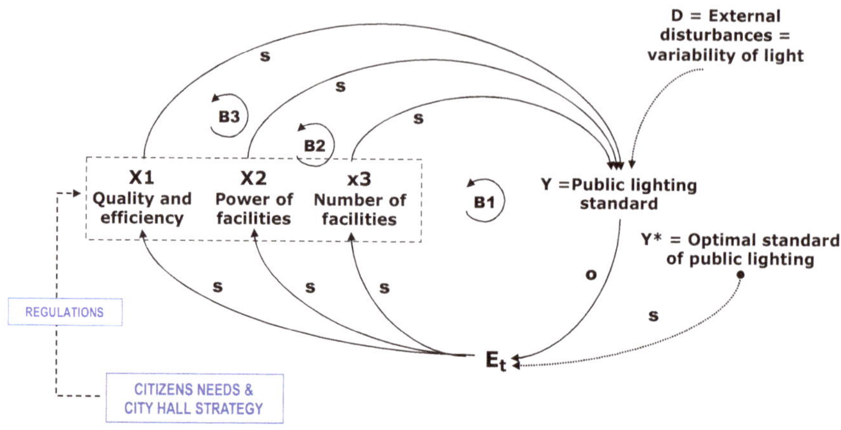

Fig. 6.4 Control of the level and quality of public lighting

Residential centers could not exist without the highly important control systems for liquid and solid wastes, which even the Romans were aware of: before any urban expansion they enlarged the sewer system. Today the production and distribution system for the necessity goods in life is also a powerful producer of solid wastes of all kinds that must be eliminated every day.

Sewers and systems for collecting and disposing of waste are vital for avoiding health catastrophes, but these systems must be guided by very sophisticated control systems.

We all know what happens when there is a malfunctioning in the control system that regulates the collection of solid waste: suffocating garbage piles disrupt normal activity in a city and lead to intervention by our man-run and civic hygiene control system that forces us, despite our protests, to become active participants in the authorities' efforts to activate or strengthen the control levers for the volume of waste.

The model in Fig. 6.5 indicates the levers that can be used to control solid urban waste and for street cleaning.

Figure 6.6 presents a model that illustrates how a control system for urban hygiene can operate through the sewerage system. In recent decades our cities have experienced another important problem: along with the elimination of liquid and solid waste, there is the problem of the elimination of *gaseous waste* from all types of harmful emissions. Particularly felt is the need to control *particulate matter*, or PM_{10}, which is so dangerous to our health because it is composed of a harmful *mix* of particulates, that is, solid and liquid particles suspended in the air (whose diameter is equal to or less than 10 μm) that contain dust, smoke, soot, and aerosol (a mix of micro drops of various liquid substances).

Particulate matter comes from various sources, among which hydrocarbon combustion by cars and heating plants, fires, volcanic emissions, tire wear, asphalt particles, micro-pollens, etc. It is not yet possible to control pollution levels with

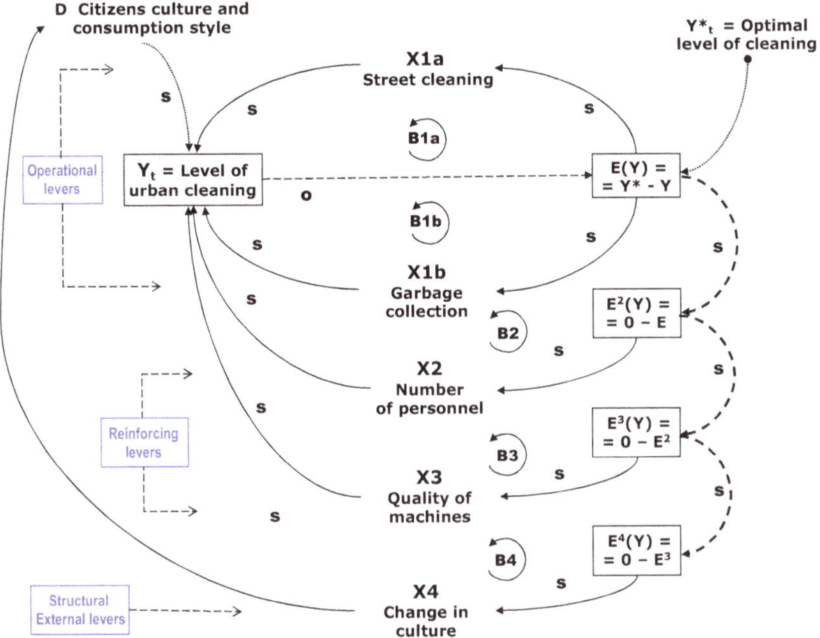

Fig. 6.5 Control system for solid urban waste

automatic control systems. However, some well-known levers can be adopted for the control of particulate matter such as traffic restrictions, time and temperature constraints on home furnaces, incentives to adopt solar energy for heating, and so on.

To carry out the necessary controls with concrete measures large cities have installed a network of detectors for pollution levels that allow authorities to detect deviations from the tolerance limits (constraints), which can lead to corrective measures, as long as the control bodies, which are governed by bureaucratic and political structures, are willing to intervene. It is no surprise that orders from the regional authorities to limit traffic in a certain region are never accepted by all the cities in the region and that, for certain reasons, some local authorities permit a continual increase in the pollution thresholds.

Figure 6.7 illustrates a simple model of a control system for *particulate matter*. Note the multiple levers that are used and their classification into operational, auxiliary, and structural levers based on the environment in question and the regulatory organization.

I should mention two other types of pollution: pollution from the intense *magnetic fields* produced by electric energy distribution systems and by systems that produce communications signals, and *acoustic pollution*, caused by noisy instruments or traffic. We must unfortunately admit that, unlike other types of pollution, the control systems for these two forms are quite tenuous. Although there are monitoring instruments to detect levels above the danger thresholds, the control system

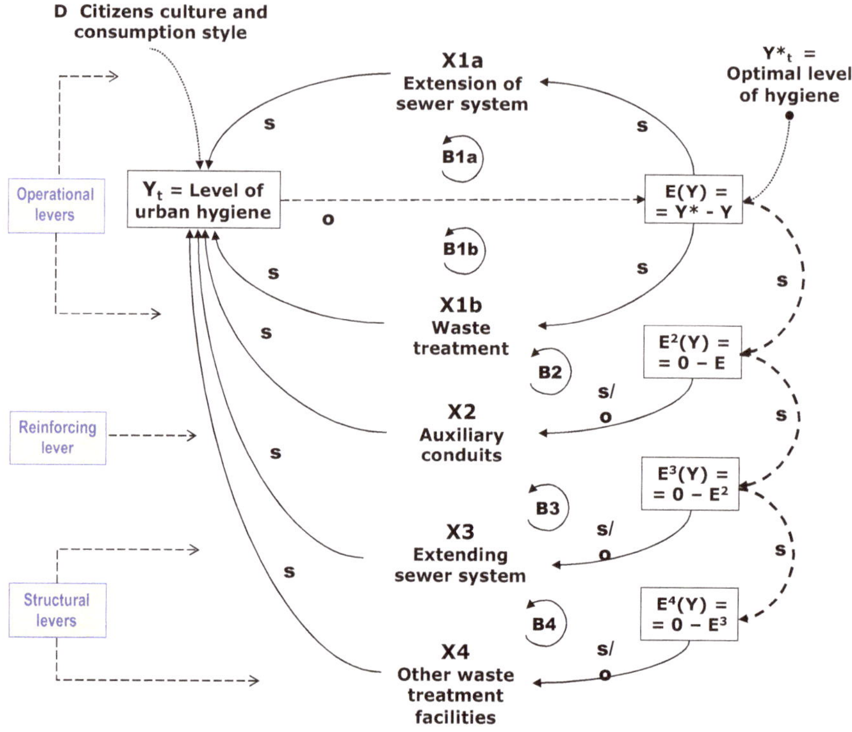

Fig. 6.6 System of control of liquid waste treatment

has two weak links: the scarcity of control levers to act on the sources of pollution—structures solidly "in place" in the residential areas in cities—and the lack of willingness on the part of public authorities to implement the few existing control levers, such as a ban on new plants, moving existing ones to other sites, or installing protective barriers.

A brief mention of a sore point: why can't we control the punctuality of *public* buses in large, crowded cities with choking traffic? The answer is twofold: on the one hand, there is a lack of effective control levers, and on the other the external disturbances to the scheduled route times are so great that control is quite problematic. What levers could the public regulators implement? More efficient route planning? The creation of preferential lanes? Making it easier for users of public transport to get on and off the vehicles? Further development of the public transport network? All of the above can be effective in theory, but in practice traffic disturbances thwart the efficacy of any lever.

The range of overhead control systems is not of course limited to the few examples above, but these nonetheless are sufficient to make the reader reflect on the urban environment and recognize there are many other *Rings* that make possible the world around and above us. In a few years' time we will have intelligent cities that integrate all the control systems we have so far examined—waste disposal, hygiene,

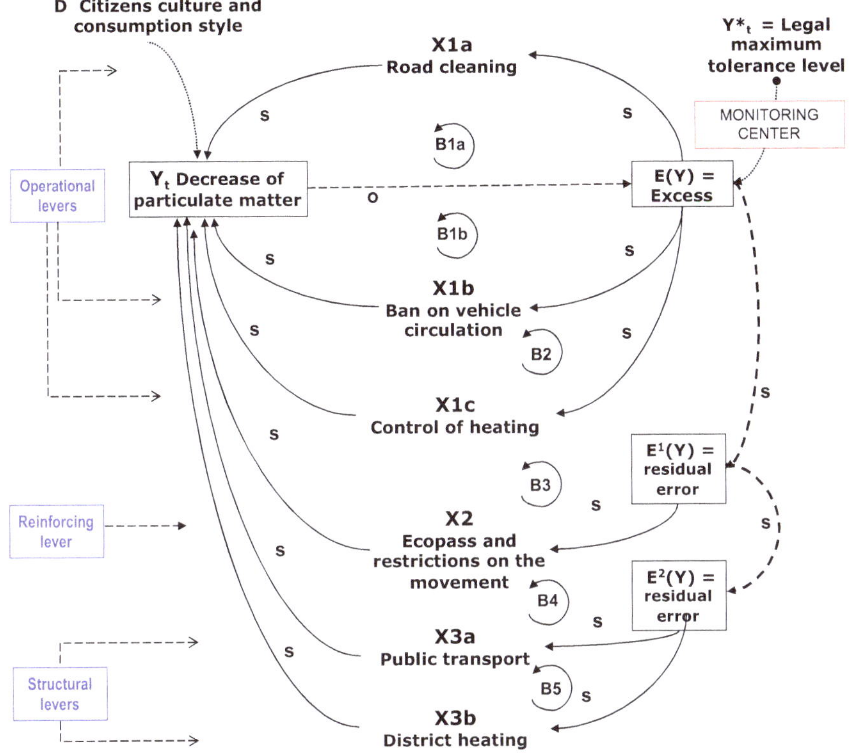

Fig. 6.7 Control system for fine particles

transport, pollution, etc.—to form a global control system that operates thanks to the regulation of a powerful central information processing system. Here is an example of a successful experiment.

> PlanIT Valley will combine intelligent buildings with connected vehicles, while providing its citizens with a higher level of information about their built environment than has been possible previously. Its efficiency will extend into the optimum control of peak electricity demand, adapted traffic management for enhanced mobility, assisted parking and providing emergency services with the capacity to have priority when needed in the flow of traffic.
>
> PlanIT Valley will enable the enhanced monitoring of the vital signs of urban life, the condition and performance of vehicles and infrastructure. As a result, managers will be able to optimize normal daily operations of the city and provide greater certainty in reacting to extraordinary events through real-time modeling and simulation. With a view to incorporating new developments, urban management control systems will be updated with the latest information and technology as these emerge.
>
> Urban mobility will be based on lightweight and fuel efficient electric or hybrid vehicles built with the use of new composite materials. When connected to the city and services, this will make better use of mobile computing capacities in vehicles, benefitting from rich wireless capabilities optimizing the use of bandwidth. These will enable citizens to fully exploit energy storage capabilities while driving, use electricity more efficiently in homes and sell excess capacity to the electric power grid (PlanIT Valley Project 2013, Online).

6.4 *Rings* Acting in the External Macro Environment

If we zoom out and broaden our observational horizon to the macro environment, we observe a large number of control systems operating in various contexts. However, it is not easy to identify these since we are not accustomed to thinking of macro phenomena as elements in some control process. With a bit of practice the discipline of control will enable us to better understand the ubiquitous presence of *Rings* in the macro environment as well.

Let us above all consider the significance of many of the structures we see around us: dams, canals, boundary walls, and observation towers, all testify to man's desire to control his hostile surroundings. These are levers to dike a river or sea, or to protect ourselves from the enemy hordes. The more frequent the floods were, the more embankments that were built; the larger the invasions, the more widespread the ramparts and walls for protection, as shown by the Great Wall of China.

Beyond these evident levers for man's control over nature, we must recognize that on a macro level "nature" itself is regulated by the action of a large number of "natural" *Rings*. These systems are not always easy to identify and represent, but their action is constant; the discipline of control must accustom us to recognizing and constructing models of them. The difficulty in recognizing such *Rings* comes from the fact that environmental dynamics—unlike the case with man-made control systems—do not achieve explicit *objectives* but rather avoid exceeding certain natural "*constraint values*," or even achieve "*limit values*" believed optimal in relation to certain environmental conditions. Natural control systems always behave in the same way, but, independently of the peculiar nature of "natural" limits or constraints, they can be described using the usual control system model, as shown in Fig. 6.8.

When a system is in an equilibrium state, $Y_t = Y^*$, with respect to the value of a "natural" variable considered optimal, or binding, in those conditions, and this equilibrium is changed by the action of environmental disturbances—often considered random because of our ignorance of the causal factors that give rise to them—there then emerge variances, $E(Y)$, which, though not immediately perceivable, trigger changes in some other "natural" variable, X_t, which brings the system back to equilibrium, thereby eliminating $E(Y)$.

There are three main difficulties in perceiving a similar control action. The *first* is simply the fact we do not always know the optimal values, Y^*, that the "natural" *Ring* tries to achieve and maintain. The *second*, which is more "fleeting," is that the optimal values occur when particular environmental conditions exist in the form of specific values of other accompanying variables that are not always known or easily knowable; these are not part of the system but belong to the external environment. The *third* difficulty is connected to the fact that, because of the dynamic nature of the accompanying variables, the optimal values change over time, even rapidly, with the result that Y^* is constant only for a brief period and can vary, often unpredictably, in subsequent periods.

To demonstrate this let us take the example of a very simple "natural" *Ring* we all can observe: the control system leading to the formation of "fair weather small cumulus clouds," those isolated small clouds in the shape of a small spherical pile

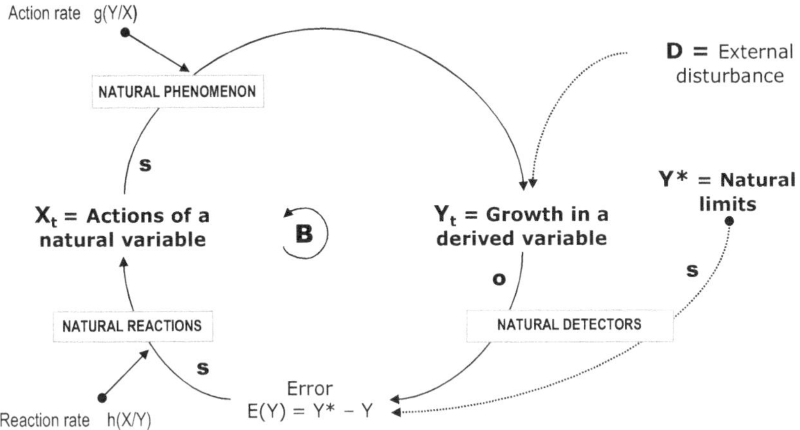

Fig. 6.8 General model of how "natural" control systems operate in the external macro environment

that, during the warm days of spring or summer, suddenly appear in the sunny sky above the plains, above the seas, or hovering over the mountain tops.

These small cumulus clouds can be interpreted as a *symptom*—that is, the error $E(Y)$—that occurs during a control process under way (invisible to us) that leads to a re-equilibrium in the density of small masses of air (commonly called "bubbles") through water vapor condensation, when the latter reaches a certain density at a given atmospheric pressure with respect to the surrounding air.

Referring to Fig. 6.9, we can briefly define the density, $D_t = H_t/T_t$, of a mass of air as a ratio between the humidity (H_t) and temperature (T_t) at a certain pressure. The higher the humidity and the lower the temperature, the more the increases in density.

We can imagine there exists a limit value, D^*, at which the mass of air is saturated and gives off the humidity it contains, at the same time lowering the temperature. D^* can be interpreted as a limit (or a constraint/objective) of the control system that produces small fair-weather cumulus clouds.

With the aid of Fig. 6.9—which is flipped over with respect to the general model in Fig. 2.2, in order to visualize the small cumulus cloud phenomenon—we can easily describe the process.

The irradiance increases the temperature (T) of an air mass that persists over an area with a particular shape, such as an area of heat-retaining *dark land* which is surrounded by light-colored land, resulting in the air mass heating up more than the surrounding air. At the same time, the irradiance favors evaporation that increases the humidity (H) of that same mass on the dark area. In most cases the increase in temperature is greater than that in humidity, so that the density $D_t = H_t/T_t$ decreases and the air mass becomes lighter, following the direction "**o1**."

Nothing happens until a chance event (external disturbance)—for example, a gust of wind, the movement of air from a passing freight train, or some other event—detaches the air bubble, which, being light due to the high temperature, due to Archimedes' buoyancy force, tends to rise and increase its distance from the ground

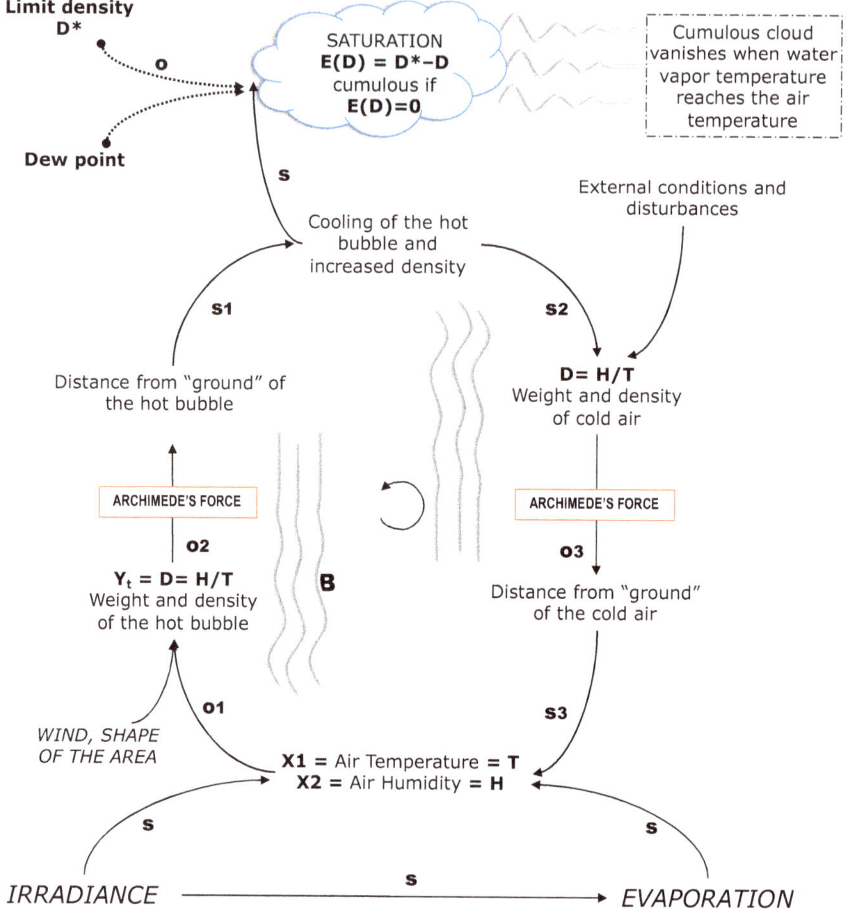

Fig. 6.9 The formation of fair-weather cumulus clouds and rising thermals

(direction "**o2**"). Thus we can think of *temperature* and *humidity* as two control levers (X1 and X2) in the process that involves the variation in air density ($Y_t = D_t$). Rising toward cooler regions, the bubble cools, thereby increasing (direction "**s1**") the density (D_t). This is the first phase in the control process.

We assume that at a given temperature, called the *dew point*—which varies according to a number of conditions—the density, (D_t), reaches the limit density, (D^*); the air bubble then becomes saturated and releases its humidity, which, as it condenses, becomes visible as a "fair-weather cumulus cloud" that releases its "latent heat from evaporation," probably further reducing its temperature. However, the process is not over yet. Once the rising air cools and releases its humidity it regains weight (direction "**s2**") and sinks toward the ground (**o3**), where, if the conditions are those of the previous cycle, it heats up again and gains humidity (**s3**). If the control system repeats loop [**B**] for several cycles (there are 3 "**os**"), various air bubbles are produced that rapidly follow on one another to create a true column of rising hot air, a rising thermal, that comes up against a flow of descending cold air. It is these opposing air flows that are a source of enjoyment for lovers of the air sports of gliding, hang gliding, and paragliding. Practitioners literally jump from one cumulus cloud to another in the search for updrafts (avoiding downdrafts), which take them higher and allow them to remain flying for hours. I do not know whether Fig. 6.9 provides an effective explanation, but one thing is certain: cumulus clouds form and represent a visible reality.

This raises an *initial question*: what meaning do we attribute to cumulus clouds? As we can see from Fig. 6.9, they are not the output of some process but the visible *symptom* of the action of natural invisible *Rings* when the rising air bubble reaches the limit value, D^*. They are not the effect of the control levers but of the trends in $E(D)$. It is enough to rotate Fig. 6.9 180° to clearly recognize the structure of the general model of control systems we are all familiar with, in which the small cumulus clouds become the revelatory sign of the elimination of the error regarding the saturation of the air.

A *second question*: why don't the fair-weather cumulus brighten up all our spring and summer days, but instead are only rarely visible, more of an exception than the rule? The answer lies in the peculiarities of the control process, which depends on very particular conditions regarding irradiance, evaporation, wind, air conditions above the ground and, above all, on the dew point and the limit density value, (D^*), which vary according to external circumstances and can be considered as disturbance factors.

Moreover, the model in Fig. 6.9 can be considered a general one for interpreting the formation of clouds as a consequence of a convective heat transfer resulting from shifting air flows that are heated at ground level and pushed toward high-altitude atmospheric strata; the temperatures of these air flows falls by around 6.5 °C for every 1000 m of altitude, as is well known to pilots flying at 10,000 m, where there is a risk of ice forming on the wings. All air masses heat up and grow moist during summer days, but if the hot air slowly and continually rises, if there is not much humidity, if the above-ground temperature is constant for hundreds of meters, and if D^* is very high, then the conditions do not exist for the air to release its humidity, and thus no cumulus will form. For this reason the "fair-weather cumulus" is a rare rather than a normal occurrence.

The explanation for the formation of cumulus clouds leads us to the common and general phenomenon of *cloud formation*. As the preceding process clearly indicates, clouds form in the same way: they are the result of a "natural" invisible *Ring* that causes the water vapor of vast air masses to condense. When these air masses slowly heat up and gain moisture over vast areas they produce clouds as symptoms that the dew point has been reached. When the densities of the water vapor and the air temperature are high, the cloud formation process can lead to the formation of large cumulus clouds, repeating on a large scale, with more marked dynamics, the control process that produces "small cumulus clouds." The air below the large cumulus clouds, pushed by the cold air, rises with impressive speed and force, dragging the lovers of gliding, caught unawares, to such high altitudes that the unfortunate victims suffer frostbite. While fair-weather cumulus form at relatively modest altitudes, where the air is only a few degrees colder, large cumulus, several kilometers high, encounter very low temperatures (the adiabatic vertical gradient corresponds to a fall of around 6.5 °C/km).

Finally, note that the formation of large cumulus clouds can arise from the coming together of masses of air with different densities and temperatures as a result of differences in atmospheric pressure. Since it is heavier, the shifting mass of cold air (or the cold front) normally wedges in below the lighter mass of hot air, thus producing a sudden cooling of the latter, which releases a large amount of vapor into the atmosphere. It is this vapor which, in fact, gives rise to large cumulus clouds that can even lead to thunderstorms.

The formation of fog at ground level derives from the same natural *Ring* that forms clouds in the atmosphere. When evaporation is uniform over a large area and there are no air bubbles with different densities, thus leading to a fall in air temperature toward the dew point, then humidity condenses at the ground level. Fog is nothing other than the symptom of the elimination of the difference between the effective temperature and the dew point; in other words, the symptom that the density has reached its saturation point, thus eliminating the density error and giving off water vapor. On cold winter days the same invisible *Ring* acts to form frost. Instead of falling to the dew point, the temperature must fall to the *frost point*, where the water vapor is transformed into ice. An entirely analogous phenomenon during cool summer evenings is the formation of dew.

6.5 Planetary *Rings*: "Gaia" and Daisyworld

If we broaden our observational horizon we can note that a large number of natural and invisible control systems are operating on our planet in a manner similar to that of the *Rings* we have examined in Sect. 6.4. These systems give rise to vast recurring phenomena at a planetary level; for example, the control system that regulates the water, or hydrologic cycle, the formation of permanent winds, trade winds, and monsoons, and the formation of ocean currents.

I shall examine the water cycle model in Sect. 6.11.1. Here I wish to look at the highly important system that controls the dynamics of *ocean currents*, which are in fact produced by the action of an *invisible Ring* that operates in a completely analogous manner as that which produces the fair-weather cumulus clouds in Fig. 6.9, though on a larger scale and in a different environment.

The oceans are crucial for the accumulation of heat and its distribution by means of the ocean currents:

> [The Oceans] possess a large heat capacity compared with the atmosphere, in other words a large quantity of heat is needed to raise the temperature of the oceans only slightly. In comparison, the entire heat capacity of the atmosphere is equivalent to less than three *metres* depth of water. That means that in a world that is warming, the oceans warm much more slowly than the atmosphere [...]. The oceans therefore exert a dominant control on the rate at which atmospheric changes occur (Houghton 1994, p. 94).

Here I shall only consider how the *Ring* operates that produces and maintains over time the *Gulf Stream* in the northern hemisphere. The currents produced in the southern hemisphere follow the same logic, though with dynamics that acts in a different direction and with variations that derive from this different orientation.

It is well known that the Gulf Stream originates in the Gulf of Mexico, follows the North American coast, crosses the Atlantic, and reaches Northern Europe, brushing the coasts of Portugal, Spain, France, and England, reaching as far as Scandinavia. From a physical point of view, the Gulf Stream is a giant flow of water kept distinct from the surrounding bodies of water due to its different temperature and saline density; it transports water at a rate of 74 million cubic meters per second at a speed of 1–2 m/s; in the middle Atlantic it is 400 km wide with an average temperature of 14 °C. From a climatic point of view it is a powerful means of transporting heat toward Europe, where it mitigates the climate by impeding the arrival of freezing arctic air.

The *Ring* in Fig. 6.10 illustrates how the Gulf Stream originates and is maintained. It is easy to see the similarity with the model in Fig. 6.9, with the obvious exceptions due to the fact it is a water current and not an air current, and that its direction of movement is determined not only by the heat–cold differential but also from the Coriolis effect that is a result of the Earth's rotation (see: http://science.yourdictionary.com/coriolis-effect).

The current flows continuously following a circular *logical* model—a cycle that repeats itself year after year—even if, in fact, its actual route is more irregular. Considering only the *logical system*, Fig. 6.10 clearly shows that the Gulf Stream is kept active by two sources: a heat source (solar irradiation) located in the Gulf of Mexico, and a cold source (arctic air) from the North Pole. The heat source heats the water, which, becoming saltier and lighter due to Archimede's buoyancy force, remains at the surface, where the winds can begin to move it north.

The Coriolis effect from the Earth's rotation directs the current toward the northeast, where the heated and salt-rich surface water begins to move north in the hemisphere, from the Central American coast toward the North European one. Proceeding northeast, the water encounters an area of colder air and starts to lose heat to the air (heating the air), thereby maintaining temperate the regions of Northern Europe. Once it has given off its heat, the increasingly colder water (due to the arctic climates of the northernmost regions) becomes even denser and heavier, so that the current must descend in depth due to Archimedes' force, subsequently moving southwest across the ocean as a result of the Coriolis effect. Moving down toward the equator in deep waters, the Gulf Stream runs into warmer water and regains sufficient heat to bring it to the surface in the Gulf of Mexico, where once again it is significantly heated before it again flows north, tracing a gigantic three-dimensional "8," after which the cycle starts again.

Now let us turn to other relevant "natural," invisible *Rings* that lead to the maintenance of acceptable temperatures on a planet populated by biological species that can alter temperatures simply through variables related to their numbers. Let us take as an example Daisyworld, an abstract world created by James Lovelock in his fundamental work, "Gaia" (written with the contribution of the biologist, Lynn

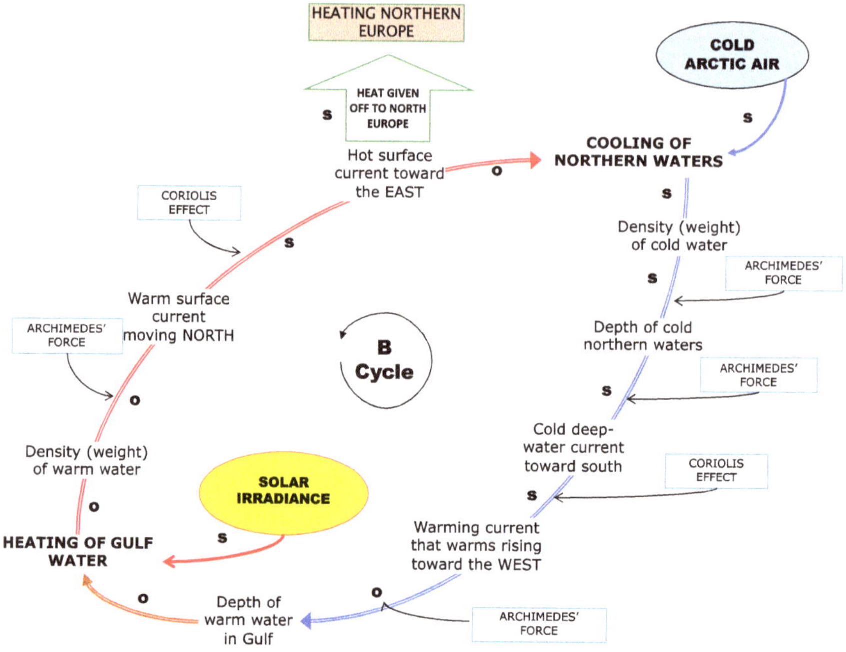

Fig. 6.10 The Gulf Stream *Ring*

Margulis). Lovelock sought to demonstrate how the Earth, as a unitary system made up of a multitude of inseparably connected physical and biological variables, can in principle be considered an evolving "living being" that maintains itself in equilibrium over time through a holarchy of reinforcing and balancing loops that, despite the fact they are invisible, operate on a planetary scale.

> We now see that the air, the ocean and the soil are much more than a mere environment for life; they are a part of life itself. Thus the air is to life just as is the fur to a cat or the nest for a bird [...] There is nothing unusual in the idea of life on Earth interacting with the air, sea and rocks, but it took a view from outside to glimpse the possibility that this combination might consist of a single giant living system and one with the capacity to keep the Earth always at a state most favorable for the life upon it. An entity comprising a whole planet and with a powerful capacity to regulate the climate needs a name to match. It was the novelist William Golding who proposed the name Gaia. Gladly we accepted his suggestion and Gaia is also the name of the hypothesis of science which postulates that the climate and the composition of the Earth always are close to an optimum for whatever life inhabits it (Lovelock 2011, Online).

Since one of the Earth's problems is the maintenance of the thermal equilibrium that allows various species to remain alive over time, Lovelock and Watson create Daisyworld, an imaginary ecosystem, to show that this equilibrium can be maintained—despite variations in insulation—thanks to the *Ring* based on the unconscious variation of biological masses.

> This book is the story of Gaia, about getting to know her without understanding what she is. Now twenty-six years on, I know her better and see that in this first book I made mistakes. Some were serious, such as the idea that the Earth was kept comfortable by and for its inhabitants, the living organisms. I failed to make clear that it was not the biosphere alone that did the regulating but the whole thing, life, the air, the oceans, and the rocks. The entire surface of the Earth including life, is a self-regulating entity and this is what I mean by Gaia (Lovelock 1979, Preface).

> Lynn Margulis, the coauthor of Gaia hypotheses, is more careful to avoid controversial figures of speech than is Lovelock. In 1979 she wrote, in particular, that only homeorhetic and not homeostatic balances are involved: that is, the composition of Earth's atmosphere, hydrosphere, and lithosphere are regulated around "set points" as in homeostasis, but those set points change with time (Environment 2013, Online).

Daisyworld is viewed as a simplified hypothetical world on a planet that orbits around a star that is similar to the sun. Daisyworld is covered in daisies (imaginative component), which belong to two specimen populations of opposite color: the *white* Daisies, which flourish in hot climates, since the white color allows them to reflect most of the sun's light and thus not die from scorching, and the *black* Daisies, which instead have adapted to living in cold environments, since their dark color absorbs a good part of the solar light, thereby supplying them with the energy to live.

Watson and Lovelock (1983) show how this simplified ecosystem can self-regulate, regarding both the thermal equilibrium and the biological masses it is composed of, and indefinitely resist random limited variations in the heat emitted by the sun-like star.

> When I first tried the Daisyworld model I was surprised and delighted by the strong regulation of planetary temperature that came from the simple competitive growth of plants with dark and light shades. I did not invent these models because I thought that daisies, or any other dark- and light-colored plants, regulate the Earth's temperature by changing the balance between the heat received from the Sun and that lost to space (Lovelock 1988, p. 37).

I wish to demonstrate that the control system by which Daisyworld maintains itself in thermal equilibrium is composed of the joint action of two *Rings*, as shown in the Causal Loop Diagram in Fig. 6.11.

We must assume above all that there are vital temperature intervals for the two daisy populations. For the white daisies (WD), which live in the heat, there must be a minimum temperature, $C_{\text{MIN-WHITE}}$, below which there is no guarantee of the vital conditions for life, and above which (until a maximum temperature) they reproduce/die out at a *vital action rate*, "g(WD/C)," for each degree of temperature (the rates of reproduction and extinction do not necessarily have to coincide and be constant). For the black daisies (BD), on the other hand, we assume there is a maximum temperature, $C_{\text{MAX-BLACK}}$, above which they cannot survive, but below which they reproduce/die out at a given *vital rate*, "g(BD/C)."

The mass of white and black daisies determines the reflecting surface according to the coefficients "m/r," which, if necessary, are separated into white and black daisies. Variations in the reflecting surface determine the variation in the temperature according to the coefficient "$h(C)$."

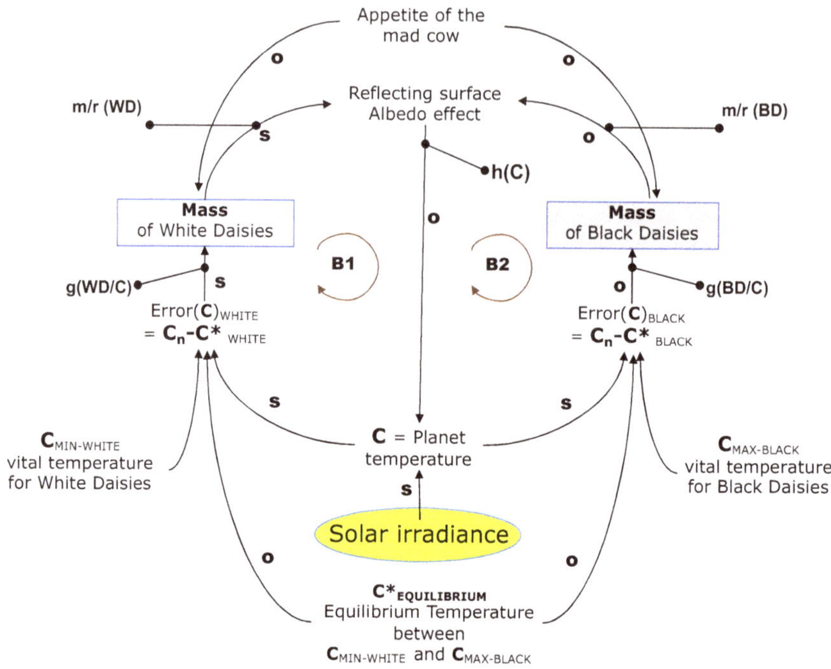

Fig. 6.11 Daisyworld and self-regulation (the model will be simulated in Sect. 6.11.2)

Let us agree that the system is temporarily in equilibrium at a temperature of $C^* = 40$ °C, and let us assume that, due to some accidental phenomenon, there is more sunlight, or that the sun, if possible, is hotter (a convenient hypothesis), for example $C_3 = 60$ °C at period $n = 3$ (see the simulation in Fig. 6.25).

What happens? Reading the model in Fig. 6.11 from left to right, we deduce that when the greater irradiance raises the planet's temperature, according to the link in direction "**s**," the white daisies have an error$(C)_{\mathrm{WHITE}} = C_n - C^*_{\mathrm{WHITE}} = 20$ °C with respect to the equilibrium temperature (note that, in order to make the description of the model more understandable, error(C) is calculated as the difference between $C_n - C^*_{\mathrm{WHITE}}$). This positive error produces an increase in the mass of white daisies at the rate "g(WD/C)," since they are used to living in the heat. However, this causes an increase in the reflecting surface based on the coefficient "m/r(WD)" (direction "**s**"), which reduces the planet's temperature according to the *action rate* "$h(C) < 0$" (direction "**o**"); this reduction lowers error$(C)_{\mathrm{WHITE}}$ and causes a reduction in their mass, thereby reducing the reflecting power and initiating a new balancing cycle. This represents the first balancing loop, [**B1**] (there is only a single "**o**").

Operating simultaneously is the loop [**B2**], shown on the right side, which refers to the black daisies (BD). The increase in the planet's temperature increases error$(C)_{\mathrm{BLACK}}$, which, because of the link in direction "**o**," reduces the mass of black daisies at the rate "g(BD/C)," since these are adapted to cold climates; the reduced mass increases the reflecting surface (direction "**o**"), and this causes a reduction in the planet's temperature according to "$h(C)$." Thus loop [**B2**] is formed (there are two "**os**") which, together with [**B1**], rebalances the temperature and the size of the biological masses (see the simulation in Sect. 6.11.2).

Obviously, if the temperature of the planet should rise above the maximum allowed for black daisies, $T_{MAX\text{-}BLACK}$, or fall below the minimum for white ones, $T_{MIN\text{-}WHITE}$, there would be a mass extinction of one of the populations; however, equilibrium could still be reached if the action and reaction rates of the surviving population permitted this by restoring the climate conditions that could favor the spontaneous regeneration of the extinct population.

Lovelock himself recognized the importance of identifying the survival limits of daisies.

> Very few assumptions are made in this model. It is not necessary to invoke foresight or planning by the daisies. It is merely assumed that the growth of daisies can affect planetary temperature, and vice versa. Note that the mechanism works equally well whatever the direction of the effects. Black daisies would have done as well. All that is required is that the albedo when the daisies are present be different from that of the bare ground. The assumption that the growth of daisies is restricted to a narrow range of temperatures is crucial to the working of the mechanism, but all mainstream life is observed to be limited within this same narrow range (Lovelock 1988, p. 57).

The automatic natural *Ring* that regulates the temperature on Daisyworld strikingly resembles the two-lever temperature control system of Fig. 4.3, which controlled the water temperature of a shower with two separate taps. The mass of white daisies corresponds to the flow of cold water; the mass of black daisies to the flow of hot water.

However, there is an important difference: in Fig. 4.3 the temperature objective is set by the governor–manager and the control system regulates the flows from the hot and cold water taps so as to ensure the objective is reached. The Daisyworld model, on the other hand, is basically a self-regulating system for the two daisy populations rather than a control system for the planet's temperature, which is maintained through self-regulation.

We can make Daisyworld exactly like the shower system with two taps by choosing as the controlled variable the planet's temperature and as the objective the temperature that leads to equilibrium between the daisy masses. In reality, systems thinking teaches us that, since the three variables (white and black daisies, and temperature) are linked in a single dynamic system, none of the three rules is regulated by the others; instead all three are interconnected variables in a single system.

Another analogy is also possible: the Daisyworld model acts similarly to the control system that leads to the stabilization of prices through variations in demand and supply, shown in simplified form in Fig. 4.26. Though Daisyworld operates in a similar way, it differs from demand–supply equilibrium in its effects, since the objective of the latter control system is to maintain demand and supply equal through price variations. The Daisyworld model, on the other hand, has the objective of restoring the equilibrium temperature, which can also be achieved through different quantities of white and black daisies.

6.6 Control System for Global Warming: The Myopia Archetypes in Action

Once we have understood the operational mechanism of the Daisyworld model, a question naturally arises: does the model for temperature regulation in Daisyworld also apply to our planet? The answer appears to be no. In fact, the Earth's temperature has continued to increase for decades, and it is difficult to imagine some form of automatic *Ring* that can return it to stable levels. Heating instead produces the opposite effect to that foreseen for Daisyworld, and the reason for this is easy to comprehend: the *albedo effect* which, in Daisyworld, was regulated by variations in the amounts of white daisies, is mainly produced on our planet by the extension of the polar ice cap. Moreover, on Earth the role of the black daisies is attributable to the amount of forests, woods and dark algae in the oceans. On Daisyworld an increase in temperature was automatically "balanced" by an increase in the amount of white daisies and a reduction in the amount of black daisies, leading to the forming of the two *balancing* loops shown in Fig. 6.11 and summarized in the left-hand side of the model in Fig. 6.12.

On our planet, on the other hand, an increase in temperature produces two *reinforcing* loops. As we can see on the right-hand side of Fig. 6.12, loop [**R1**] originates because the increase in temperature causes the surface area of the polar ice caps to shrink. Loop [**R2**] occurs because the increase in temperature makes the forests, woods and algae increasingly lush and extensive, thereby reducing the albedo effect rather than increasing it, as would be necessary to return temperature increases to an equilibrium state. The joint effect of the two reinforcing loops is to gradually increase global warming. Another difference between Daisyworld and Earth is that the latter is characterized by the *greenhouse effect*, to which human activity contributes. By amplifying the effects of solar radiation, the *global warming effect* is intensified.

The global warming phenomenon is produced by numerous causes, many of which are due to man's activities which, following the archetype of "short-term,

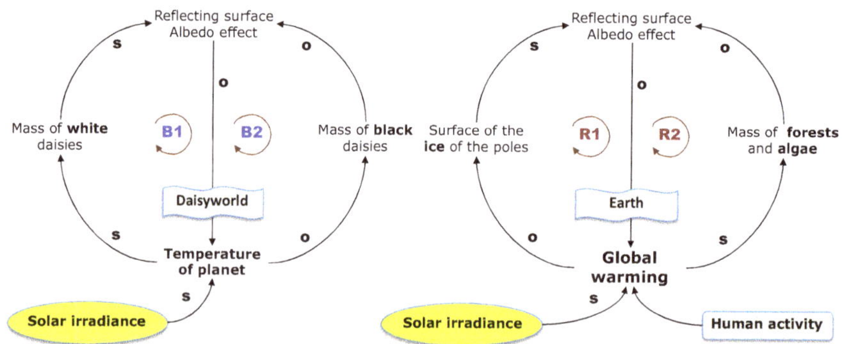

Fig. 6.12 Daisyworld and the Earth in comparison

local and individual preference" (Sect. 1.6 and Fig. 1.12), have increased *greenhouse gases* in the atmosphere (carbon dioxide, methane, nitrous oxide, hydrofluorocarbons, perfluorocarbons, and sulfur hexafluoride).

> Three major pollution issues are often put together in people's minds: global warming, ozone depletion (the ozone hole) and acid rain. Although there are links between the science of these three issues (the chemicals which deplete ozone and the particles which arc involved in the formation of acid rain also contribute to global warming), they are essentially three distinct problems. Their most important common feature is their large scale (Houghton 1994, Preface).

Among the many models available for understanding how global warming is produced, especially useful is the one by Peter Senge (1990, p. 345), which is presented in Fig. 6.13.

Senge's model is so clear there is no need for comment. The growth in global warming caused by loop [**R3**]—as a consequence of economic development as well as of the other sources of CO_2 emissions—seems unstoppable, and this produces and will continue to produce more and more damage due to unimaginable increases in sea levels, weather disturbances, the breakup of glaciers and drought, etc., in less time than predicted. For some (Maslin 2004) this damage—if we leave out population effects—could speed up economic development: houses to rebuild, entire cities to be moved, new products, etc., thereby increasing growth even more. Whatever its causes, this phenomenon can be regulated by a multilever control system (involving both artificial and natural levers), which, without claiming to be complete, is indicated in Fig. 6.14, which expands on Fig. 6.13.

The control system that regulates global warming uses three orders of levers, which are indicated in the lower lines in the figure. The operational and extraordinary levers are *artificial*; their action is specified in the dark boxes and their effects

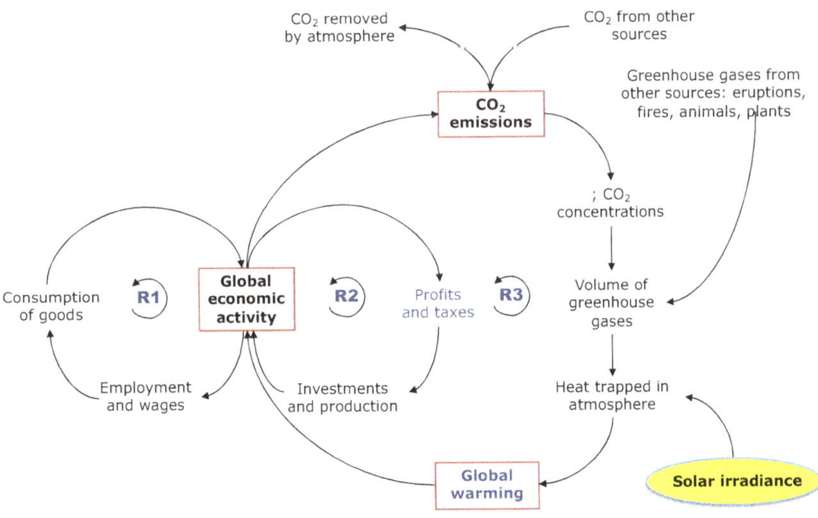

Fig. 6.13 A model for global warming

represented by the arrows with the broken lines. The *natural* lever is structural in nature; it is shown in the lower right, and its action is represented by the double-lined arrows.

To orient ourselves in the challenging model in Fig. 6.14, I suggest the following reading sequence. When *global warming* is perceived and its effect begins to become worrisome, a *social alarm* arises (shaded box), which becomes ever more intense ("**s**"), producing a political movement that leads supranational regulators to set tolerance *limits to greenhouse gases* ("**o**") through a series of international agreements, among which the Kyoto protocol—signed in Kyoto on December 11, 1997, by over 160 countries at the COP3 Conference during the United Nations Framework Convention on Climate Change (UNFCCC)—along with other subsequent international agreements. These treaties oblige industrialized countries to reduce emissions of polluting substances (by at least 5.2% from 2008 to 2012, with respect to emissions in 1990, which is taken as the base year). Recently the Bali agreement (2007) has established a drastic reduction (25–40% by 2020) in greenhouse gases.

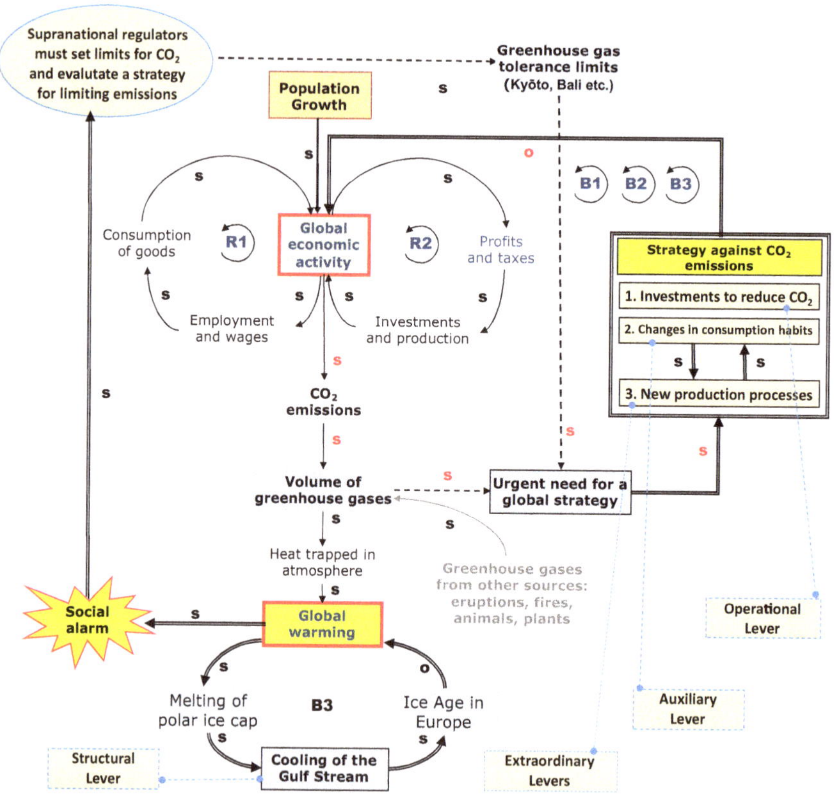

Fig. 6.14 Model of a control system for global warming (derived from Fig. 6.13)

In order to reduce the gap between actual levels of CO_2 and those set by the regulators, the nations that sign the treaties must come up with a strategy composed of various levers (shown in the square on the right-hand side). These strategies activate above all the operational levers (box 1), which consist of *investments* to finance measures to reduce as much as possible CO_2 emissions, in order to bring them below the tolerance limits, thereby producing loop [**B1**] (these measures are too great in number to be detailed in the model). However, the social alarm also activates two other extraordinary levers: spreading a culture to *change consumption habits* (box 2), trying to limit the consumption of goods requiring a greater production of greenhouse gases, and the consequent change in *production processes* (box 3), again with the aim of reducing emissions of CO_2 and other greenhouse gases, thus producing loop [**B2**] in Fig. 6.14. Levers 2 and 3 are independent but mutually reinforcing.

The *structural* lever complements the operation of the artificial levers and acts to modify the structure of the geographic environment in which global warming produces its negative effects. The overall action of this lever can be viewed as an autonomous *natural* and automatic control system integrated into the system that produces global warming. Such a natural automatic *Ring*—which produces loop [**B3**] in Fig. 6.14—is identified by some (Lemley 2004, Online; NASA 2004, Online; Melchizedek 2004, Online) as the cooling of the Gulf Stream due to global warming, which, by melting the polar glaciers, dumps a mass of fresh cold water, which is also poor in salt, into the Atlantic Ocean, thereby cooling the ocean currents.

Figure 6.15 adds elements to the Gulf Stream model in Fig. 6.10 to demonstrate in more detail how the natural *Ring* [**B3**] in Fig. 6.14 operates to control global warming. The mass of cold, fresh water that spills into the Atlantic Ocean from massive ice cap melting ("**s1**") prematurely cools the flow of the Gulf Stream ("**s2**"), which, once cooled, precipitately sinks toward the ocean floor ("**s3**" and "**s4**") and turns toward the equator in a southwesterly direction.

The gradual melting of the polar ice caps, caused by the global warming, produces masses of sweet, cold water that pour into the Atlantic Ocean, diminishing its salinity. This phenomenon interrupts the current's "motor," pushing the Gulf Stream to the bottom and slowing its velocity. This precipitant sinking and slowing down of the Gulf Stream causes a smaller mass of the current's warm water to reach the North Atlantic ("**o1**"), and thus a smaller transfer of heat ("**s5**"), thereby turning the water colder again, with the risk of new glacier formation ("**o2**"). This hypothesis is supported by a large amount of data, and even the study of past weather phenomena appears to show that the last ice age, some 8500 years ago, was caused in fact by the collapse of a glacier barrier in the Atlantic.

Glaciation in Northern Europe increases the *albedo effect* ("**s6**") because of two joint effects, similar to what we have witnessed in Daisyworld: on the one hand, the extension of the ice mass in Northern Europe (which corresponds to the increase in the mass of white daisies) and, on the other hand, the reduction of the green mass of northern forests (corresponding to the reduction in the mass of black daisies). These two effects reduce global warming ("**o3**"), at least temporarily, forming loop [**B1**].

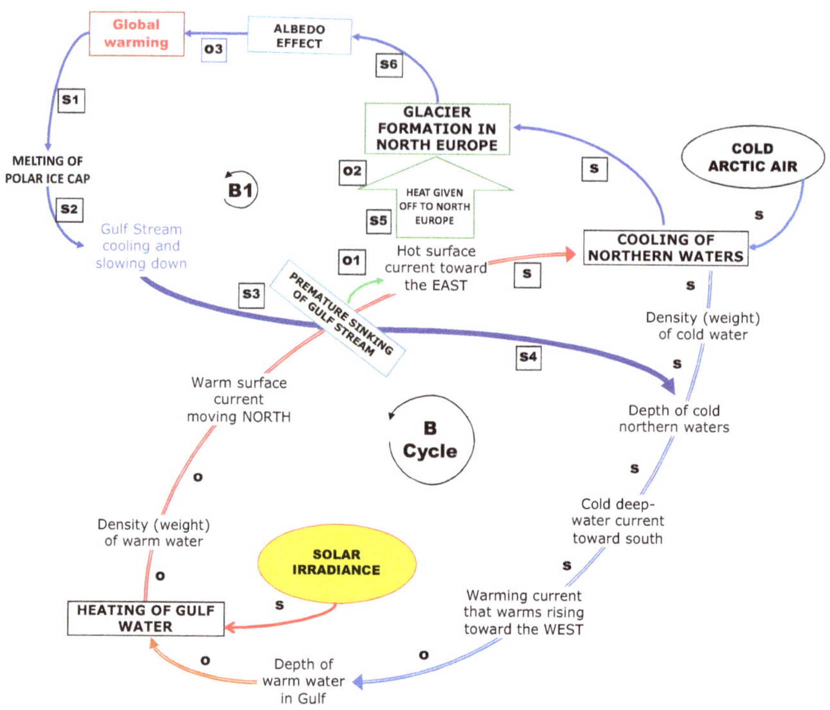

Fig. 6.15 The natural *Ring* to control global warming through the cooling of the Gulf Stream

These effects are summarized in Fig. 6.16. Loop [**B**] controls global warming by using the lever of the temperature in the northern regions, which depends on the Gulf Stream; this lever functions as an effector to regulate this temperature, as illustrated in the model in Fig. 6.15. The sinking of the Gulf Stream reduces the temperature in the north, which activates loop [**R**], which in turn accentuates the albedo effect by regulating the two levers that extend the ice masses and cover the forest/green zone with ice. The natural *Ring* just described is, in fact, a *magic Ring*, which, if it ever spontaneously self-activates, is capable of dealing with the global warming problem.

Attempts have also been made to control global warming using other *artificial auxiliary levers*, but not all of those conceived of, even though apparently rational, are effective.

An artificial lever, which is simple to produce with the appropriate means, is that which produces the albedo effect on the oceans by spraying a wide area of clouds hovering above the oceans with seawater spray; the clouds then turn white, since they contain microscopic salt particles. In this way, the clouds will reflect a greater amount of solar radiation to produce the desired albedo effect, and as a result the Earth will be heated up to a lesser extent.

The idea behind the marine cloud-brightening (MCB) geoengineering technique is that seeding marine stratocumulus clouds with copious quantities of roughly monodisperse sub-

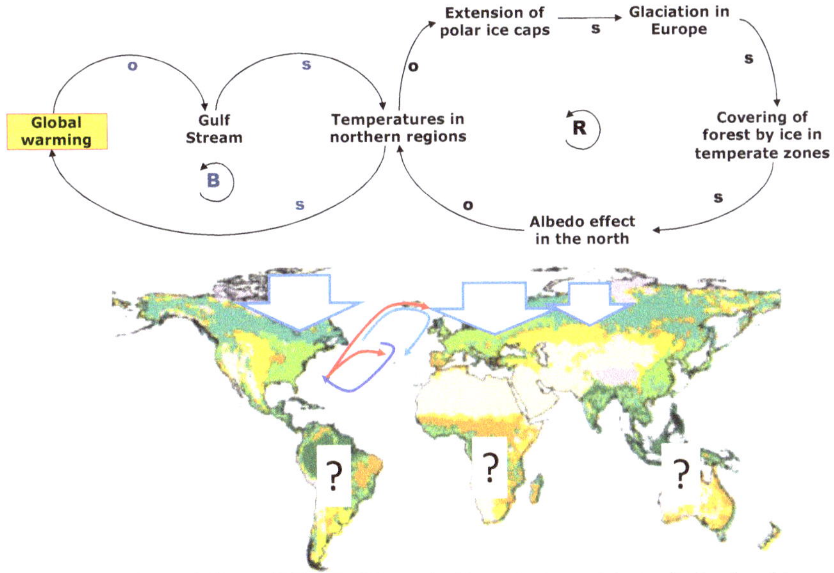

Fig. 6.16 The Daisyworld effect in the control of global warming

> micrometre sea water particles might significantly enhance the cloud droplet number concentration, and thereby the cloud albedo and possibly longevity. This would produce a cooling, which general circulation model (GCM) computations suggest could – subject to satisfactory resolution of technical and scientific problems identified herein – have the capacity to balance global warming up to the carbon dioxide-doubling point (Latham et al. 2012, p. 4217).

> But at this point an unwanted surprise appears. Researchers have realized that the final effect was an undesired increase of rain on land, even to the point of causing terrible monsoons. Such violent phenomena occur, in fact, when the air over land is warmer than that over the oceans; the latter air then cools and produces clouds which are very white (Caprara 2010, Online).

Figure 6.16 indicates how global warming can, at least potentially, be countered by the natural Gulf Stream cooling phenomenon caused by global warming itself, as part of a balancing loop that controls the growth of global warming. However, I must also mention four *reinforcement loops* caused by global warming that lead to its increase. These reinforcement loops, represented in the model in Fig. 6.17, can be summarized as follows:

(1) *Reduction of reflective surfaces*. The first harmful effect of global warming is the melting of the glacier ice on the snow-capped mountains, especially in the polar regions; the disastrous consequence is that the reduction of icy surfaces reduces the "albedo effect" resulting in a further increase in global warming.

> Unexpectedly rapid warming attributed to man-made climate change is stripping the Arctic Ocean of its mantle of ice. Predictions that the North Pole could

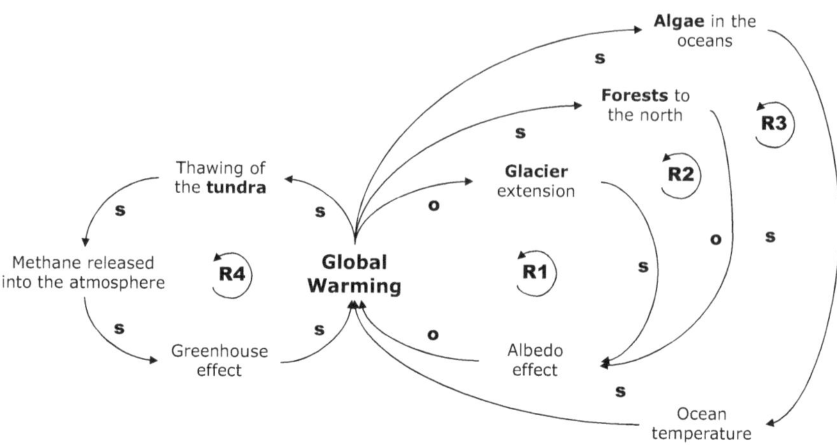

Fig. 6.17 Four harmful loops that reinforce global warming

be open water by the end of the century are looking conservative. Since 1979, a third of the pole's summer ice has disappeared. This year's melt reduced coverage to a record low of 1.7 million square miles, the U.S. National Snow and Ice Data Center at Colorado University reports. Rather than a specter far beyond our lifetimes, Arctic experts say, the accelerated melting could produce an ice-free summer Arctic by 2030. The environmental repercussions from such an extensive defrosting of the region are enormous. Open, dark water absorbs heat from sunlight, while an icy skin reflects much of it back into space. Hence, the more open ocean, the faster the melting of the remaining ice, creating a feedback loop that could eat away at the adjacent Greenland Ice Sheet (Chron 2018, online).

(2) *Increase in forest-covered areas*. In parallel with the melting of ice, global warming, despite man-made deforestation, favors the expansion of forests into increasingly larger areas previously covered in ice or formed by clear ground. The dark leaves of forest trees can not only trap heat, producing aerosols that interact with methane (Marshall, 2020; Unger, 2014), but in the long run also reduce the "albedo effect," thereby favoring the action of solar radiation and, consequently, increasing global warming.

Many scientists applaud the push for expanding forests, but some urge caution. They argue that forests have many more-complex and uncertain climate impacts than policymakers, environmentalists and even some scientists acknowledge. Although trees cool the globe by taking up carbon through photosynthesis, they also emit a complex potpourri of chemicals, some of which warm the planet. The dark leaves of trees can also raise temperatures by absorbing sunlight. Several analyses in the past few years suggest that these warming effects from forests could partially or fully offset their cooling ability. … At the same time, some researchers worry about publishing results challenging the idea that forests cool the planet. One scientist even received death threats after writing a commentary that argued against planting trees to prevent climate change (Popkin 2019).

Harmful algae usually bloom during the warm summer season or when water temperatures are warmer than usual. Warmer water due to climate change might favor harmful algae in a number of ways: ... Algal blooms absorb sunlight, making water even warmer and promoting more blooms (EPA 2019, online).

(3) *Expansion of dark algae*. The phenomenon in (2), which affects the land areas that are uncovered, has its counterpart in the spreading of dark algae in the Antarctic and in the oceans, which can increase global warming with similar results.

Warming in the Antarctic Peninsula has already exceeded 1.5 °C over pre-industrial temperatures, and current Intergovernmental Panel on Climate Change (IPCC) projections indicate further global increases. Set against a background of natural decadal temperature variability, climatic changes on the Peninsula are already influencing its vegetation. With the available area for plant colonisation on the Peninsula likely to increase by up to threefold due to this warming, understanding how snow algae fit into Antarctica's biosphere and their probable response to warming is critical to understanding the overall impact of climate change on Antarctica's vegetation. … Several studies have used satellite observations to investigate snow and ice algae on larger scales, implicating algal blooms as significant drivers for darkening and enhancing melt of the Greenland ice sheet (Gray et al. 2020).

(4) *Thawing of the tundra*. In parallel with the phenomenon in (1), global warming produces the thawing of the immense tundra, whose frozen soils unfortunately trap methane bubbles; melting releases large amounts of these methane bubbles, which increases the greenhouse effect with an intensity about 30 times greater than that produced by CO_2 (Macdonald 2016; Tollefson 2018; Mooney 2018). The prospects for humanity are not encouraging, but it is to be hoped these reinforcement loops will accelerate the intervention of the balancing loop provided by the Gulf Stream.

This extraordinary sight - in a video filmed of the tundra on remote Belyy Island in the Kara Sea off the Yamal Peninsula coastline - was witnessed by a scientific research expedition. Researchers Alexander Sokolov and Dorothee Ehrich spotted 15 patches of trembling or bubbling grass-covered ground. When punctured they emitted methane and carbon dioxide, according to measurements, although so far no details have been given. The reason is as yet unclear, but one possible explanation of the phenomenon is abnormal heat that caused permafrost to thaw, releasing gases. Alexander Sokolov said that this summer is unusually hot on the Arctic island, a sign of which is polar bears moving from the frozen sea to the island. Scientists have warned at the potential catastrophic impact of global warming leading to the release into the atmosphere of harmful gases hitherto frozen in the ground or under the sea. A possibility is that the trembling tundra on Bely Island is this process in action (Siberian Times reporter 2016, online).

6.7 *Rings* Acting on Earthquakes and Tsunamis

After this brief overview on Global Warming above, it is also important to reference a study of control systems for catastrophic events. Here I will just mention the studies to control great storms and tornadoes by bombarding them with appropriate chemical substances and those to control the route of asteroids and meteors if they should be on a collision course with Earth. The need for a control system to counter the collision of asteroid “2004 MN4,” forecast for April 13, 2029, is quite urgent (Britt 2004; Hecht 2007). Science fiction? Not any longer. When projects for control systems become reality, science fiction gives way to engineering (Fleming 2007). Among the harmful phenomena capable of producing catastrophic events, two particularly dangerous ones are deserving of mention: earthquakes and tsunamis.

The first question is: how do earthquakes originate? Referring to the technical literature (Stein and Wysession 2009; Adagunodo and Sunmonu 2015; Shearer 2019), the lithosphere is divided into tectonic plates, each of which is characterized by its own independent motion (uncontrollable by man) with respect to the adjacent lithosphere, probably produced by the convective motion of the magma in the mantle. The main plates are North America, South America, Europe, Africa, the Arabian Peninsula, India, Australia, Antarctica, the Pacific, Nazca, plus smaller ones such as Cocos, Juan de Fuca, and the Philippines. Near the points of contact between the tectonic plates, faults can originate that represent fractures in the Earth’s crust that places two plates in contact with each other due to the forces inside the Earth.

> A fault is a fracture or zone of fractures between two blocks of rock. Faults allow the blocks to move relative to each other. This movement may occur rapidly, in the form of an earthquake - or may occur slowly, in the form of creep (USGS, online).

The points of contact between the plates, called *margins*, can be of three main types (USGS online: Adagunodo and Sunmonu 2015).

(a) *Converging*, when the plates collide and overlap, causing the Earth’s crust to buckle and deform to create mountain ranges.
(b) *Diverging*, when the plates move away and separate, leaving an empty space that is filled by the magma of the mantle; if this happens in the oceans, the magma rises from the depths of the Earth’s mantle to the surface, forming volcanoes and mountains along the suture zones and producing a change in the ocean floor.
(c) *Transforming*, when two plates rub against each other (the San Andreas fault line in California is an example of this); the rubbing motion often triggers powerful earthquakes, like the one that devastated San Francisco in 1906.

The diverging and converging movement of the plates due to the convective motion in the mantle produces fractures in the Earth’s crust from which the magma, a mass of molten and glowing material below the Earth’s crust, escapes. This is one of the main causes of *volcano creation*. The movement of the plates, however, also produces earthquakes, which can be much more serious. An earthquake, or seism, is a rapid vibration of the Earth’s crust caused by contact between moving plates when *margins* are formed. Simply put, along the faults (margins) the rocks remain in tension, thus accumulating increasing energy, which increases the tension; when the rupture load is reached, that is, the resistance of the rocks is exceeded whether by compression, displacement, or friction, in a matter of seconds the rocks slide against each other and *release the accumulated energy*, causing an earthquake of greater or lesser intense vibrations that is transmitted to people and structures of all kinds (Kanamori and Brodsky 2004, p. 1438). The intensity of the quake will depend on the *type of margins*, the amount *of accumulated energy*, the *type of rocks involved*, and their *rupture load.*

The point on Earth where the sudden release of energy manifests itself is *the hypocenter*; the place on the Earth's surface placed on the "vertical axis" of the hypocenter is called the *epicenter*, which is generally the area most subject to earthquakes. The most common earthquakes are those associated with the movement of the ground along the faults, but there are also those associated with volcanoes and their activity, with human activities, explosions, and collapsing buildings or parts of mountains, or even nonterrestrial events, such as the impact of celestial bodies. Figure 6.18 presents the Causal Loop Diagram which, in the language of Systems Thinking (Chap. 1), describes the mechanism behind the formation of earthquakes and volcanic eruptions.

Note that the "engine" that provides energy for the accumulation of tension in the plates is represented, for both phenomena, by the *movement of the plates*, indicated as an exogenous variable in the system. Also note that both earthquakes and volcanic eruptions release energy, which reduces the energy accumulated in the plates. This is perfectly in line with our experience, which tells us that the two phenomena always last for a limited time only to cease and resume often centuries down the road. For this reason, those who study earthquakes and eruptions constantly monitor

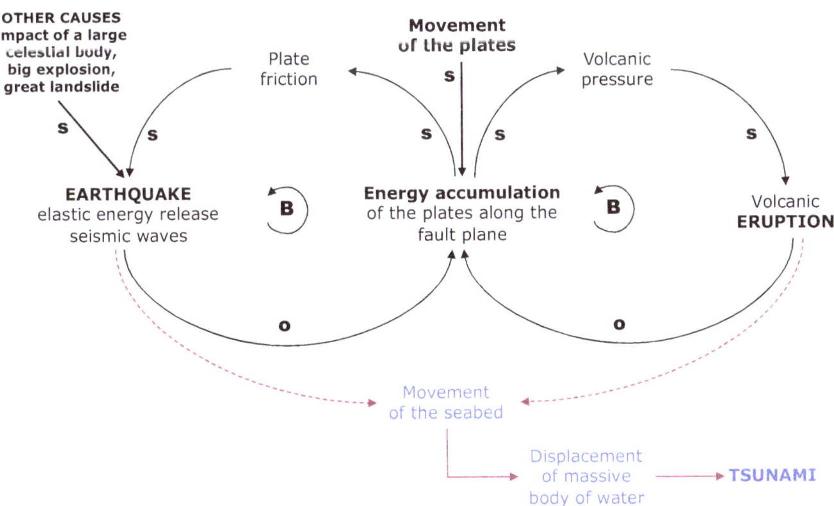

Fig. 6.18 The formation of earthquakes from volcanic eruptions and tsunamis

the situation of faults and "calderas," the magma chambers that formed as a result of a previous eruption where new eruptions are likely to arise. The bottom of Fig. 6.18 also highlights the mechanism behind the formation of tsunamis, which originate from movements of the seabed due to earthquakes and/or volcanic eruptions that, by shifting significant masses of water, generate the typical mega waves.

I would now like to make some observations on the "control" of earthquakes and tsunamis, meaning, of course, not the control of the phenomena itself but of their "disastrous effects." In fact, not only is the control of an earthquake impossible, but a scientific protocol has not yet been established for its prediction. The most effective damage *control levers* for earthquakes are:

Earthquake Early Warning (EEW): these exploit the differences in the propagation speed of two types of seismic waves: the "primary" waves and the "secondary" waves that make the ground vibrate horizontally and vertically, respectively, regarding their propagation direction. "Primary" waves propagate at a rate of between 1.5 and 8.0 km/second, while the "secondary" waves, the most harmful ones, are about 1.7 times slower than the others. EEWs take advantage of the different propagation speeds of "early" seismic waves, which represent a very short-term warning signal that arrives about 9 seconds before the devastating seismic waves. This very short amount of time is never enough to warn the population, but it allows for *automatic prevention systems* to be activated, including: the automatic shutdown of trains and subways and the activation of a mechanical blocking system for the rails to avoid overturning; automated systems for closing gas and electrical networks and the activation of auxiliary electric generators so as not to disrupt essential activities, such as surgeries, obstetrics, and the delivery of essential services; and the launching of an immediate general alarm through television outlets, audible alarms, and short message service (sms). Of course, the EEW system requires that a seismic monitoring network be created to detect waves and predict the scale of the event (Guerrero Gámez et al. 2017). Individual trains are also equipped with earthquake sensors to ensure the safety of all passengers through an immediate shutdown.

> Millions of people, however, could be travelling on Tokyo's railway and subway network when the quake hits. The Tokyo Metro says its infrastructure has been seismically reinforced, and that trains will make an immediate emergency stop in strong shaking; it advises passengers to hold tightly to handrails and straps (Hurst 2020).

The *Earthquake Early Warning* obviously does not prevent earthquakes; it works precisely because the earthquake has already begun.

Active protection, that is, the use of effective seismic building techniques pertaining to seismic engineering, such as seismic insulation; at present, these techniques can minimize the damage even of extremely powerful earthquakes and are widely used in some of the most seismic areas in the world. Skyscrapers are built from steel modules that withstand vibrations and often rest on elastic bearings capable of dampening the resonance produced by seismic waves that cause buildings to vibrate (Sun et al. 2018). Even "low-lying" dwellings must be built according to specific seismic standards to prevent the possibility of collapse and, above all, of fire and gas explosions in the event of a rupture of the pipes. Water and gas distribution networks must also be designed to withstand rupturing due to earthquakes of a certain magnitude (Heginbotham 2018).

Personal prevention of earthquakes, including all the measures that must be put in place daily by families and individuals to withstand an earthquake.

> For those trapped, all offices and many private houses in Japan have an earthquake emergency kits, including dry rations, drinking water, basic medical supplies. Offices and school also keep hard-hats and gloves for use in the event of a quake (Foster 2011, online).

In the case of entrapment in the home or evacuation, it is particularly useful to keep stocks of food and basic necessities on hand, in addition to first aid equipment, bottled water, food rations, gloves, face masks, insulation sheets, survival tools like torches, and even radios that broadcast regular updates.

> The Tokyo Metropolitan Government has compiled a manual called "Disaster Preparedness Tokyo" (Tokyo Bousai*) to help households get fully prepared for an earthquake directly hitting Tokyo and other various disasters.
>
> [. . .] If a major disaster strikes and disrupts infrastructure, it would even be difficult for the authorities to respond quickly. You should thus keep a stockpile of items that can allow you to live without relying on anyone for at least a week until relief arrives. This is the idea of emergency stockpiles… What you need when you leave home is an emergency bag containing the minimum requirements, which is compact enough to be carried to the evacuation center (Metro Tokyo, online).

Social prevention of earthquakes, including all forms of *Education About Prevention* through systematic teaching so that all citizens, especially students, can acquire a culture of *conscious and controlled reaction to earthquakes* resulting in rational behavior learned through periodic exercises: protect yourself, do not run away, wait for relief, help the injured, gather in designated safe areas, etc.

> Every schoolchild in Japan will be familiar with earthquake drills in which alarms sound and children retreat under their desks to shelter from falling debris. The drills take place every month, with the children being taught to go head-first under the desk and cling to table legs until the quake is over. If the children are out in the playground the rush to the centre of any open space to avoid being hit by falling debris. The local fire department also takes groups of children into earthquake simulation machines to familiarize them with the sensation of being in an earthquake. Schools with two storeys or more have evacuation chutes which children can slide down to safety. After a quake hits, children are required to stay in school until an adult comes to collect them, in case their homes are damaged or their family members aren't available to look after them (Foster 2011).

The action of the different earthquake control levers is represented in the Control System in Fig. 6.19.

Most of the above considerations also remain valid for the damage caused by tsunamis. The only effective way known at present to detect the actual generation of a tsunami from an underwater earthquake or a volcanic eruption is to directly measure sea level variation immediately after the event is detected. Under no circumstances, however, no matter how timely and refined, can theoretical forecasting models and systems measuring sea level variation through the generation of abnormal waves limit the damage and casualties produced by a tidal wave if this were triggered by a seismic phenomenon very close to the coast line, since it would not be possible to alert the population in time. In tsunami-prone areas, the only effective active prevention measure would be to stop the construction of settlements along the

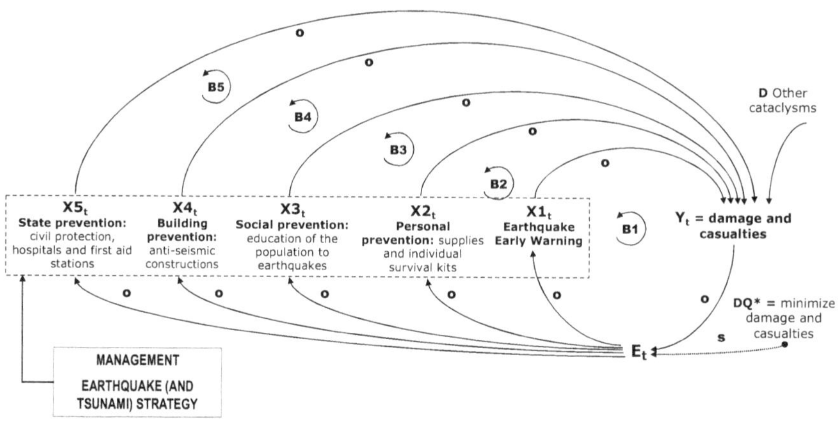

Fig. 6.19 Control Systems against damage from earthquakes and tsunamis

coasts and for a few kilometers inland. This solution could, if anything, apply to new settlements but could not be applied to cities built in past centuries.

6.8 *Rings* Acting on the Human Body

Now that we have observed the external world, it is appropriate to apply the discipline of control to our bodies, which, together with our minds, can be considered a psychophysical system that is kept alive by a network of natural and acquired control systems.

In fact, the same definition of life proposed by Herberto Maturana and Francisco Varela in their work, *Autopoiesis and Cognition* (1980) gains its operational validity from the existence of control systems as constituent structural elements of every living being. Living beings, man in particular, must be observed as closed structures due to their self-organization and conceived of as autopoietic and living machines or systems, that is, *homeostatic* systems which, thanks to a complex network of holonic multilever, multilayer control systems, manage to maintain at the intercellular, intracellular, and organic levels a delicate equilibrium over time that allows them to continually reproduce the organization that defines them. These ideas are expressed most clearly by a direct quote from the work of Maturana and Varela:

> Autopoietic machines are homeostatic machines. Their peculiarity, however, does not lie in this but in the fundamental variable which they maintain constant. *An autopoietic machine is a machine organized (defined as a unity) as a network of processes of production (transformation and destruction) of components which: (i) through their interactions and transformations continuously regenerate and realize the network of processes (relations) that produced them; and (ii) constitute it (the machine) as a concrete unity in space, in which they (the components) exist by specifying the topological domain of its realization as such a network.* It follows that an autopoietic machine continuously generates and specifies its own organization through its operation as a system of production of its own components,

> and does this in the endless turnover of components under conditions of continuous perturbations and compensation of perturbations. Therefore, an autopoietic machine is homeostatic (or rather a relations-static) system which has its own organization (defining network of relations) as the fundamental variable which it maintains constant (Maturana and Varela 1980, pp. 78–79).

This quote clearly shows that every *autopoietic machine* must regulate, or maintain constant, the *network of relations* that defines its own organization; in other words, without control systems that regulate the organizational variables, no system can be and remain autopoietic. Every autopoietic system thus has, in its *Rings* that regulate homeostasis, the vital condition for its existence.

On the basis of this definition the two authors also specify the concept of a living system:

> If living systems are machines, [then the fact] they are physical autopoietic machines is trivially obvious: they transform matter into themselves in a manner such that the product of their operation is their own organization. However, we deem the converse is also true: a physical system, if autopoietic, is living. In other words, we claim that the notion of autopoiesis is necessary and sufficient to characterize the organization of living systems (Maturana and Varela 1980, p. 82).

As a result, the theory of autopoiesis—that is, the theory of living systems—clearly shows that our biological structure cannot maintain itself and develop over time without a network of *natural automatic Rings* that continually regenerates the network of vital processes by maintaining homeostasis and permitting autopoiesis to occur. Such *homeostatic natural automatic* control systems, together with those of *postural control*, which can also be nonautomatic, are fundamental for our existence. In the words of Norbert Wiener:

> Our homeostatic feedbacks have one general difference from our voluntary and our postural feedbacks: they tend to be slower. There are very few changes in physiological homeostasis – not even cerebral anemia – that produce serious or permanent damage in a small fraction of a second (Wiener 1961, p. 115).

Not all relevant *Rings* that maintain homeostasis, even though in a disturbed environment, are known today, even if it is now clear that the levers used to regulate the various vital and cerebral parameters can also be chemical ones; biology and medicine are taking giant steps in uncovering the complex structure of microscopic and macroscopic *Rings*. The so-called *postural* control systems mentioned by Wiener are equally relevant in our daily lives, for us as well as for any animal characterized by articulation.

> In the human body, the motion of a hand or a finger involves a system with a large number of joints. The output is an additive vectorial combination of the outputs of all these joints. We have seen that, in general, a complex additive system like this cannot be stabilized by a single feedback. Correspondingly, the voluntary feedback by which we regulate the performance of a task through the observation of the amount by which it is not yet accomplished needs the backing up of other feedbacks. These we call postural feedbacks, and they are associated with the general maintenance of tone of the muscular system (Wiener 1961, p. 107).

Postural control is usually achieved through a *multilever* control system—one which is entirely similar to that illustrated in Fig. 4.10 for focusing on an object on a horizontal plane—since it requires a number of levers to eliminate the distance, as Wiener recognizes in the above quote: "a complex additive system like this cannot be stabilized by a single feedback." However, I would also note that, in order to achieve the *postural control* of our joints, for example, those that allow us to grasp an object, the control system not only is *multilever* but also has features of a *holonic* control system just like the one presented in Sect. 3.8 for the control of the shifting of a point, P_t, in a multidimensional space in order to reach a point, P^*, which is set as the objective.

In fact, if we think about it, the action of moving our hand to grasp a small object (a pencil, pipe or cigarette, using Wiener's examples) in a fixed position is, from the control point of view, nothing other than the shifting of the present position of the hand, P_t, toward the position of the object, P^*. This is a movement in three-dimensional space, along with other, more precise spatial movements of the fingers, which must contract at the same time and at the "right point" in order to grasp the object. However, the postural control needed to grasp an object is complicated by the fact that the movement of the hand (and, on a smaller scale, the fingers) requires the joint action of various articulations—among which, the entire body, the bust, shoulder, arm, forearm, wrist, and, lastly, the fingers—each of which must, in turn, move in different spatial directions by means of the muscles that serve as translators, as shown in Fig. 6.20.

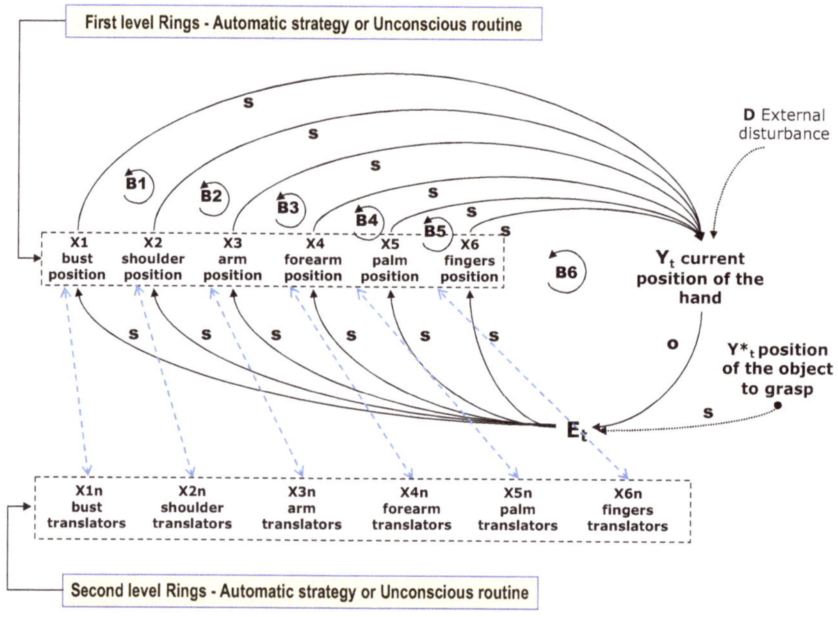

Fig. 6.20 Model of a holonic control system for grabbing a small object

In fact, the control systems for each articulation must, though autonomous, be coordinated to permit the overall control of the movement of the hand toward the object–objective.

Operating similarly is the control of the movement of the two fingers that must "grab" the small object. The control system is further complicated if we consider the fact that, in order to move the articulations, it is necessary to control a number of muscles, which must be appropriately taught to guarantee "adequate" strength with respect to the size and weight of the object to grasp. Thus, a second level of translator control systems is formed, which is necessary for the functioning of the first-level control systems. The model in Fig. 6.20 becomes in all respects a *holonic* control system as soon as we realize that the control of the translators requires at least three third-level control systems, each of which is needed to modify the position of the articulation on a spatial coordinate, that is, to modify the tension of the muscles that change the direction of the movement of each articulation. All control systems, at the various levels, are coordinated by a precise automatic strategy, that is, by *unconscious routines* which in part are innate and in part derive from our experiences beginning in our childhood.

We could zoom in even more, but even without considering in detail the control of the fascia and individual muscular fibers, which make up the effectors, a conclusion becomes immediately clear: grasping a small object, no matter how simple it may seem to us, requires a large number of *Rings* which are coordinated and cooperate with one another. The operation of the "*macro*" global *Ring* that eliminates the gap between the position of the hand and object requires, in turn, the action of a number of "*meso*" *Rings*, which control the variation in the position of the individual articulations and, in an even more detailed manner, the action of the "*micro*" *Rings* that control the effector muscles of the other higher-level *Rings*.

Generalizing the above discussion, the control of the movement from a given point that requires a *series of articulations* can be carried out by a holonic, multilever, control system (which is probably also multiobjective) made up of control systems of various levels. The further down we move to the sublevels of such systems, the more automatic and involuntary their operation becomes. As a result, such a holonic system must operate based on a strategy (and, when necessary, a policy as well) that is translated into *motor planes of action*, or in *physiological routines*—which we are thus not aware of—that regulate the levers in an *unconscious manner* to allow us to focus our *conscious* attention on the higher-level "macro" *Ring* needed to grasp the object. In this sense, we can clearly understand Wiener's words when he describes postural control systems with articulations:

> We have thus examples of negative feedbacks to stabilize temperature and negative feedbacks to stabilize velocity. There are also negative feedbacks to stabilize position, as in the case of the steering engines of a ship, which are actuated by the angular difference between the position of the wheel and the position of the rudder, and always act so as to bring the position of the rudder into accord with that of the wheel. The feedback of voluntary activity is of this nature. We do not will the motions of certain muscles, and indeed we generally do not know which muscles are to be moved to accomplish a given task; we will, say, to pick up a cigarette. Our motion is regulated by some measure of the amount by which it has not yet been accomplished (Wiener 1961, p. 97).

No comment could be clearer, but one thing is certain: without the action of the many *Rings* that operate "magically" in a coordinated and cooperative manner unbeknownst to us, postural control would not be possible. To better understand these conclusions, let us consider several of the many *Rings* that involve our motor and perceptual apparatuses that are operating at every moment in our daily lives. These apparatuses continually become honed starting from the first days of life, and we come to take no notice of them because their regulation becomes so automatic, thanks to the *motor plans of action* and *physiological routines* that control our limbs and muscles (at various levels), allowing us to concentrate on the voluntary controls. However, we can become aware of them through several simple experiments.

Try this very simple one. Place several objects in front of you (e.g., a sheet of paper, a book, pen, glass, or other small object), then close your eyes and try to grab the objects, one after the other, with your right hand. I am sure you will all succeed in grabbing the objects, even with your eyes closed, from remembering their position. Now repeat the experiment, but try to note the involuntary postural control and understand how your hand adapts the position of the fingers in a different and specific way depending on the object you are about to grab. Without our being aware of it, every day the magic *Rings* that control our fingers act automatically to allow us to do all the movements for grabbing, squeezing, releasing, etc. Once again with our eyes closed, if we try to turn on the light with the wall switch we realize how the magic *Rings* that control the position of our arm, hand and fingers direct our index finger to the switch we have to turn on. This is the power of the *instrumental* automatic postural control for the control system for achieving the objectives we consciously set for ourselves.

A second, equally simple and meaningful, experiment could also be useful. Sit at a desk, or a level surface, in front of a mirror with a piece of paper and pencil. With your right hand (if you are not left-handed) print your name in block letters on the paper, with letters 8–10 cm high, or draw a figure of your choice of similar size. Now try to trace the letters, but using your left hand. By observing the imprecision in your tracing, which covers the original writing or drawing with clear-cut errors, you will understand how the postural control system for articulation acts to allow us to control the position of the left hand in tracing over the original drawing. Proceeding with the experiment (using first the right hand and then the left one), trace again the letters (perhaps turning over the paper), but looking at the sheet of paper in the mirror and not directly. It is easy to see how difficult controlling the direction of the pencil becomes and how many errors occur in tracing the new lines over the original ones. The explanation for this is clear: an action so unusual as tracing lines on a page observed from a mirror impedes the action of the normal *physiological routines*, which are replaced by conscious regulations that are slowly modified. With patience, and trying the experiment several times, you will see that the errors decrease the more you repeat the movement; the hand will begin to move the pencil with greater security the more the *physiological routines* are readjusted.

This experiment teaches us many things; in particular, it confirms that under normal conditions the activities that require the use of our limbs are regulated by automatic control systems of which we are not aware. The second lesson is that such control systems have been routinized by "daily use"; we all remember how much effort it took in kindergarten or elementary school to correctly trace letters and form words. The third lesson is that the control systems for our movements and actions increase our precision through repetition, training and experience.

These evident facts are rich in consequences since, without entering into the complex "world" of learning, it lets us understand that this is in large part linked to the improvement in the efficiency of control systems through greater experience of the manager, or even from the consolidation of experience in *routines* and *standardized strategies* that make the regulation of the levers automatic and precise. To convince ourselves of this conclusion, we need only observe how children are taught to speak and move their limbs through the continual repetition of controls and the correction of the imprecise use of control levers; and to observe how much effort is required to teach them to remain erect and walk. Similarly, consider how activating the complex control system to remain vertical during a bus or subway trip, without resorting to hand straps or poles, has become for us an almost unconscious action. Also consider the effort a driving student must make to learn to guide a car or to park, control processes which, for experienced drivers, have become *routine*.

Finally, I must mention the very powerful system of multiple *Rings* that operate in a coordinated and cooperative manner to allow us to speak. This represents a form of control which differs from the postural one but which is no less important. The control systems for words must at the same time control the breathing in and out of air as well as its expulsion through the larynx and vocal chords, which must in turn be modulated to produce the desired sounds, phonemes, words, and sentences with the correct pronunciation (independently of meaning). Moreover, it is necessary to control the movements of the jaw and, above all, the tongue and lips.

These involuntary *Rings*, of which we have no awareness, are magic because they operate with very high precision and incredible speed. Learning to speak begins when we are infants, and it continues through our entire life. It is easy to observe, on the one hand, the effort and patience of teachers in teaching young children the right position of their mouths and tongues in order to transform their stuttering into phonemes and words; and, on the other hand, the many attempts by young children to imitate the sounds pronounced by their teachers. Moreover, all these facets of speaking are well known to those who, having to learn another language, must apply the control systems for words to regulate their pronunciation.

6.9 Control Systems for Survival as Psychophysical Entities

Man, as a living being, not only is an autopoietic system but can also be considered (to an outside observer) as a conscious cognitive system, since, with internal organs for memory, computation, and evaluation (preferences), he is able to compare objects, calculate information, and construct representations in order to couple himself successfully to the environment and survive, even by modifying his own structure in line with the variations permitted by his genetic and operative program (Walsh 1995; von Krogh and Roos 1996).

> (1) A cognitive system is a system whose organization defines a domain of interactions in which it can act with relevance to the maintenance of itself; the process of cognition is the actual (inductive) acting or behaving in this domain. Living systems are cognitive systems, and living as a process is a process of cognition. This statement is valid for all organisms, with and without a nervous system.
>
> (2) If a living system enters into a cognitive interaction, its internal state is changed in a manner relevant to its maintenance, and it enters into a new interaction without a loss of identity (Maturana and Varela 1980, p. 13).
>
> Observers know and create their environment through interactions with it. This interaction involves an explicit or implicit prediction about the environment (Uribe 1981, p. 51).

The preceding *sections* sought to demonstrate how the interactions of man (and of a wide variety of animal species) with his environment is made possible by the ubiquitous presence and efficient action of innumerable *Rings*, voluntary or automatic, which are necessary to recognize and obtain all the objects needed to survive, or to flee from harmful phenomena, based on models of acquired knowledge. I do not wish to return to this discussion, but rather to consider in this *section* several fundamental control systems that operate every day at the *macro level* to allow us to survive, such as psychophysical systems, in particular the *Rings* that enable us to search for energetic and metabolic inputs in the environment held to be useful to maintain the network of vital processes.

We are all aware that under normal conditions we do not eat to the point of indigestion or drink until we are ready to explode. We do not run until we are ready to drop, stay awake for days on end, look directly at the sun, or stay where the temperature is too low or too high. Control systems watch over us with appropriate levers so that satiety, fatigue, blinding light, sleep, heat, and cold reveal the state of error with respect to the normal physiological values of the macro variables of our existence, and activate the corresponding levers to restore the conditions for our existence.

Let us begin by recognizing that, as systems possessing physical and mental activities, we are equipped with control systems that produce equilibrium in our psychophysical states. The general model of such systems is illustrated in Fig. 6.21. Such control systems, through the internal perceptive capacities of the governor–manager, detect at every instant the distance, or error, $E(Y)$, of the organism's state, Y_t, with respect to the "normal" states Y^*; that is, the breakdown of the psychophysical equilibrium. The error is perceived as negative, unpleasant, in need of being eliminated, minimized, or avoided. These unpleasant errors are called *needs* in the language of economics.

To eliminate needs we require *goods* of a certain *quality* and *quantity*. The quantity and quality of goods represents the two types of basic levers; the consumption of goods, quantitatively and qualitatively defined, modifies the state of our organism. When a need is perceived, man makes decisions to procure the necessary quantity and quality of the good to satisfy it. The *Ring* that allows us to deal with and manage our needs, shown in the general model in Fig. 6.21 must be supplemented

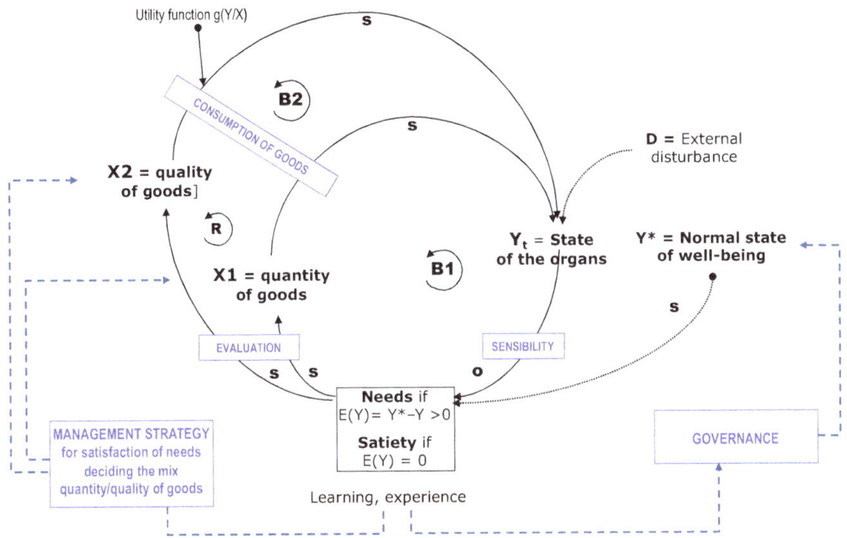

Fig. 6.21 General model of the control systems of needs

by specific *Rings* for the various types of needs; these *Rings* are normally multilever and multiobjective control systems. We are cold (deviation) and the need arises to find clothes (first lever) to protect us and a house (second lever) for the night where we can light a fire (third lever); we are hot and thirsty (deviations), and so there is need for an ice-cold drink (first lever), a refreshing shower (second lever), and a comfortable armchair in an air-conditioned room (third lever). We need to visit a friend who lives far away (objective), and in that moment we feel the need for transportation and choose to take our car (lever); we regularly feel the stimulus of hunger (deviation), and every time this occurs, the need for food arises (lever).

Needs are present in all aspects of our lives. It is useful to mention Maslow's (1943, 1970a) attempt to classify needs by proposing a scale, or pyramid, of needs (see also Katz 1964; Organ 1990):

1. *Biological and physiological needs*: in advanced industrial societies physiological needs (air, food, drink, shelter, warmth, sex, sleep, etc.) are not very motivating, since they are reasonably well satisfied.
2. *Safety needs*: even safety needs (security, order, law, stability, freedom from fear) are relatively well satisfied, thanks to the continuity in the level of resources and the insurance and social welfare systems.
3. *Belongingness and love needs* (friendship, intimacy, affection and love—from work group, family, friends, romantic relationships): these cannot be renounced and operate at the social behavioral level.
4. *Esteem needs:* these correspond to the aspirations for self-esteem and social status (achievement, mastery, independence, status, dominance, prestige, self-respect, respect from others); they join together to express the individual's need

for his own identity, to stand out in his environment; a favorable sense of self-identity is strengthened by the recognition and approval of others, by the prestige and status one enjoys.

5. *Need for self-actualization:* at the top of the hierarchy of human needs is that for self-actualization (realizing personal potential, self-fulfillment, seeking personal growth, and peak experiences). This is the need to realize one's own potential capacities, to develop and grow in an autonomous and constant manner, to be creative and full of ingenuity.

> It is quite true that man lives by bread alone — when there is no bread. But what happens to man's desires when there is plenty of bread and when his belly is chronically filled?
>
> At once other (and "higher") needs emerge and these, rather than physiological hungers, dominate the organism. And when these in turn are satisfied, again new (and still "higher") needs emerge and so on. This is what we mean by saying that the basic human needs are organized into a hierarchy of relative prepotency' (Maslow 1943, p. 375).

> The late Abraham H. Maslow, the father of humanist psychology, showed that human wants form a hierarchy. As a want of a lower order is being satisfied, it becomes less and less important, with a want of the next-highest order becoming more and more important. ... As a want approaches satiety, its capacity to reward and with it its power as an incentive diminishes fast. But its capacity to deter, to create dissatisfaction, and to act as a disincentive rapidly increases (Drucker 1977, p. 195).

During the decades 1960s and 1970s, Maslow (1970a, b) enlarged the original five-level Hierarchy proposing an eight-stage model (McLeod 2007):

1. *Biological and physiological needs*
2. *Safety needs*
3. *Love and belongingness needs*
4. *Esteem needs*
5. *Cognitive needs* (knowledge, meaning, culture, etc.)
6. *Aesthetic needs* (search for beauty, balance, form, etc.)
7. *Self-actualization needs*
8. *Transcendence needs* (helping others to achieve self-actualization)

> Less studied but knowable through common observation are the *cognitive needs* for sheer knowledge (curiosity) and for understanding (the philosophical, theological, value-system-building explanation need). Finally, least well known are the impulses to beauty, symmetry, and possibly to simplicity, completion, and order, which we may call *aesthetic needs*, and the needs to express, to act out, and to motor completion that may be related to these aesthetic needs (Maslow 1981, p. 2).

> In the history of theorizing about love relations as well as about altruism, patriotism, etc., much has been said about the *transcendence of the ego*. [. . .] Furthermore, it seems quite clear that this need to go out beyond the limits of the ego may be a need in the same sense that we have needs for vitamins and minerals, i.e., that if the need is not satisfied, the person becomes sick in one way or another. I should say that one of the most satisfying and most complete examples of ego transcendence is a healthy love relationship (Maslow 1981, p. 194).

We are easily aware of the fact that the quality of our lives depends on the number and types of needs and aspirations we can satisfy, and above all on the way in which we achieve this satisfaction. Until a few decades ago man could satisfy only his primary needs for food, clothing, and protection from disease, and only a few people managed to satisfy their aspirations for culture, art, and entertainment, which today are available to everyone. This is a general phenomenon: in all ages, all civilizations, and in every area of the world we observe a twofold transition: from low-level needs to high-level ones, and from needs to aspirations. And this leads to an increase in welfare and to the search for wealth as an opportunity and means of welfare.

Fortunately man not only does aim to eliminate the *unpleasant states* that the many *Rings* psychologically and mentally reveal to him but also perceives *pleasant states* he wishes to acquire, maintain, or increase through appropriate control systems of pursuit, whose general model is shown in Fig. 6.22.

I shall call these states *aspirations*, which are produced by the gap, or error, between a desired state of well-being—which increases with the consumption of

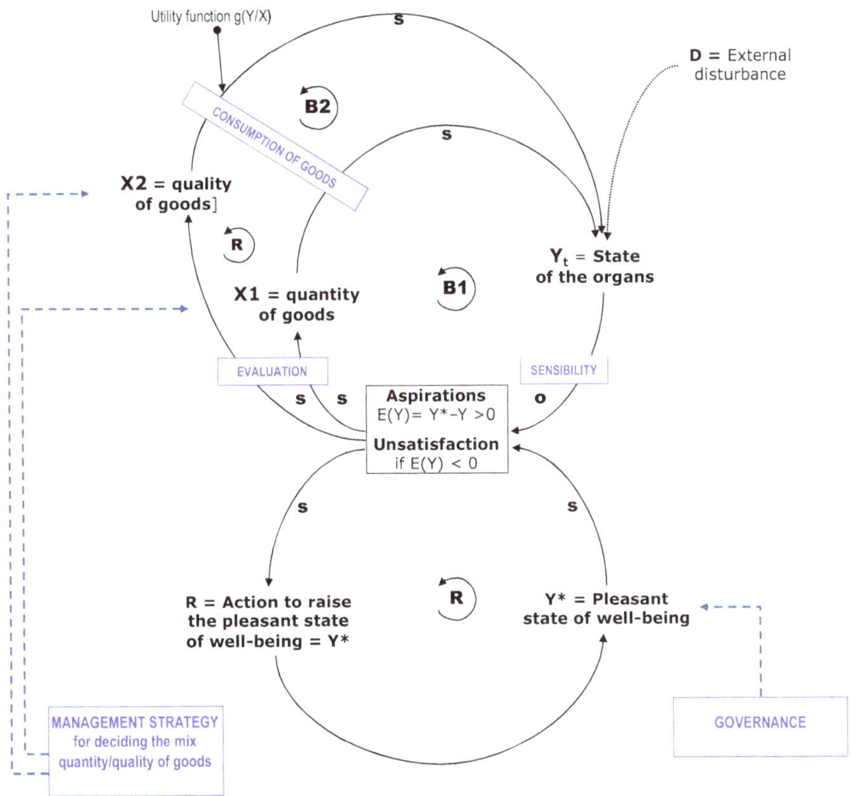

Fig. 6.22 General model of control systems for aspirations

goods—and the state deriving from the consumption of available goods. The desire for a fashionable article of clothing does not derive so much from the *need* to protect ourselves from the cold as from the *aspiration* to be admired (objective); we feel the *need* for a car (more precisely, we feel the need for transportation) but we *aspire* to have a Ferrari or Lotus (objective), even if many other cars are able to satisfy the same need. But each time an *aspiration* is satisfied its level increases, so that it becomes more intense, thereby triggering a control system of pursuit toward an objective that increases the more we try to achieve it, based on the typical archetype of "Insatiability" (Fig. 5.12).

The action of control systems for *needs* and *aspirations* is so evident that their operation can be considered to be ingrained in our biological and social programming. With reference to Jacques Monod in his famous work, *Le hazard et la nécessité* (1970), we are conscious of the fact that, thanks to the existence of such control systems, man can be defined as a *teleonomic system* that tends, like all other animals, to ensure the survival of the species, of the group it identifies with based on the family's and individual's territory (tribe, nation, etc.).

> All artifacts are the product of the activity of a living being that expresses in this way, and in a particularly evident manner, one of the fundamental features that characterizes all living beings, without exception: that of being an object endowed with a project which is represented within their structures and is carried out by means of their performance (for example, the creation of artifacts).
>
> Rather than refusing to accept this idea (as some biologists have tried to do), it is instead indispensable to recognize it as being essential for the definition of living beings themselves. We can say that the latter are distinguished from all other structures of systems in the universe by this property that we will denote as teleonomy.
>
> We arbitrarily choose to define the essential teleonomic project as consisting in the transmission, from one generation to the next, of the unvarying contents characteristic of the species. All the structures, all the performances, all the activities which contribute to the success of the essential project will thus be called "teleonomic" (Monod 1970, pp. 22–25).[1]

Teleonomy is achieved on the condition that the *Rings* that reveal the continual creation of *needs* and *aspirations* are active and that these lead to the search for levers to achieve the maximum satisfaction to ensure the maximum quality of life for the individual's survival, which is necessary to perpetuate the species to which it belongs. For this reason, in my opinion there is a close relationship between *tele-*

[1] Tout artefact est un produit de l'activité d'un être vivant qui exprime ainsi, et de façon particulièrement évidente, l'une des propriétés fondamentales qui caractérisent tous les êtres vivants sans exception: celle d'être des objets doués d'un projet qu'à la fois ils représentent dans leurs structures et accomplissent par leurs performances (telles que, par exemple, la création d'artefacts). Plutôt que de refuser cette notion (ainsi que certains biologistes ont tenté de le faire), il est au contraire indispensable de le riconnaître comme essentielle à la définition même des tre vivants. Nuos dirons que ceux-ci se distinguent de tiutes les autres structures de tous systèmes présents dans l'univers, par cette propriété que nous appellorons la téléonomie) (Monod 1970, p. 22) [...] Nous choisirons arbitrairement de définir le projet tèléonomique essentiel comme consistant dans la trasmission, d'une génération à l'autre, du contenu d'invariance caractéristique de l'espèce. Toutes les structure, toutes les performances, toutes les activitès qui contribuent au seccès du projet essentiel seront donc dites "téléonomiques" (Monod, ibidem).

onomy and *autopoiesis*, since teleonomy—understood as the attitude of a species, as a collectivity, to perpetuate itself—can be considered the phenomenology that corresponds to autopoiesis—understood as the self-production of species with respect to the individuals.

> In effect teleonomy is teleology made respectable by Darwin, but generations of biologists have been schooled to avoid 'teleology' as if it were an incorrect construction in Latin grammar, and many feel more comfortable with a euphemism (Dawkins 1982, cited by Barrows 2001, p. 705).

In fact, our way of life depends on the number and types of *needs* and *aspirations* we can satisfy, and above all on the way we achieve this satisfaction. These general comments make it easier to investigate and recognize the various *Rings* that maintain teleonomic conditions for the individual. Among the most evident are the control systems for food, water, fuel, electricity, and heating needs, those to regulate our health and hygienic conditions, to control the social relations of education, and to form thoughts about the self and others. The building of hospitals, hotels, vacation resorts, the continual increase in the size and efficiency of new firms, and the increase in highly gratifying jobs are all examples of action levers used in these control systems for needs and aspirations. However, we can easily see—as we have observed on several occasions above—that even our daily behavior is the result of the action of control systems that allow us to achieve our objectives in life. The action of *Rings* on our behavior is pervasive, and our behavioral life depends on these systems.

This should be clear, but it is useful nevertheless to reinforce this by presenting the interpretation of behavioral control provided by Perceptual Control Theory (PCT), a psychological-cognitive approach that bases the study of our behavior on the control systems that determine it. PCT focus on man—and more generally on cognitive systems (entities or agents)—trying to explain how such entities control what happens to them by investigating the relation between actions and objectives, perceptions and actions, and perceptions and reality based on the typical logic of control systems (Carver and Scheier 1981).

> Perceptual Control Theory (PCT) is a theory of how 'control' works to the level of detail that control can be modelled with precision. It regards life as a process of control and proposes that all living things are purposeful in doing what they do.
>
> To put it simply, people need a goal, a means, a resource and they need to pay attention to the results – which are the effects on our own experiences. These elements weave together in a closed circle that builds control. The goal is what the person can hold in mind and strive towards – to be a fireman, to be loyal, to be a good son or daughter. In PCT this is known as a reference value. The means are the behaviours, skills, work, emotional attunement to others and ways of thinking (outputs) that a person exercises to achieve this goal. The resource is the feature of the environment, including people around us, who can help us achieve this goal. Paying attention to our own experiences closes the loop because we perceive our results and compare them to our goal. If we have not reached our goal yet, we continue to develop the means to achieve it. When this cycle is allowed to prosper at a young age, a sense of purpose is built up that brings in the local environment and one's own community as part of one's own sense of purpose rather than as a force to be challenged and attacked (Mansell 2011, Online).

The guiding principle of PCT is that human behavior is not the output of a system of transformation of *stimuli* into *actions*—as in traditional behavioral theory—but the input (*X*) of a control system to achieve (*Y*) what man wants: his objectives, desires, and aspirations (*Y**). In other words, the stimulus to action is considered the perception of an error, *E*(*Y*), between the present position (*Y*) of the agent and the objective (*Y**) he wishes to achieve. In the web site of the Control Systems Group, this logic is summarized as follows:

> PCT is a theory about 'control' in living things, like people, cats and fish. It's about the control of what they perceive – what they notice, look at, feel, hear, taste and smell.
>
> The living thing is in the driver's seat. PCT is a theory from the person's own point of view – or the animal's own point of view…
>
> When things are working normally, the person gets to experience what they want to experience. It is 'just right' – like the perfect cup of coffee or tea…
>
> But how do they know this is happening?
>
> According to PCT, the person compares a 'standard' – what they want – with what they are experiencing right now – their perception.
>
> The difference between the two – the discrepancy or error is being measured. The bigger the error the more the effort the person makes to reduce it, until the error is zero – this means they get what they want (http://www.pctweb.org/whatis/whatispct.html).

I wanted to include this long citation because it clearly confirms that our behavior is nothing other than a continuous chain of actions to activate the control levers necessary to achieve an objective. Thus, according to PCT, our actions are the physical activities of the control system levers (*X*) that are activated based on the mental processes for detecting error (attention) and for regulating behavior aimed at the objective. If we add that the objective (*Y**) is the output (*Y*) to achieve (grab a pen, take a trip to a certain place, take the bus, etc.) and is generated by the same mental process that produces the actions to undertake (*X*) to activate the levers (output to carry out), then the following conclusion is valid: every human actión must be interpreted in the context of the control systems of which it is a part. This produces the following *corollary*: man is a *teleological system*. He does not act based on "cause → effects," "actions → reactions," "stimuli → responses," "disturbances → adaptations," but to achieve *objectives* of various kinds through *actions* and *behaviors* that represent the levers (*X*)—or inputs—of the control system that aims at eliminating the error perceived between the present position of the individual (*Y*) and the objective, (*Y**) in accordance with the general model of a control system.

William Powers, in his book *Behavior: The Control of Perception* (1973), proposed a more complex formulation of PCT. He starts from the general thesis that human behavior is activated by a control system to eliminate the effects of the external disturbances [*Dm*] on the *M* vital variables [*Ym*] (where *m* varies from 1 to *M*); these disturbances produce the errors *E*(*Ym*), which we perceive as the stimuli that activate our behavior (*Xn*). However, Powers observes that these perceptions can be divided into a hierarchical order covering nine levels: (1) Intensity, (2) Sensation or Vector, (3) Configuration, (4) Transitions, (5) Sequence, (6) Relationship, (7) Program, (8) Principles, and (9) System concepts. This order is not obligatory but rather only represents a proposal for further refinement; in fact, Powers subsequently introduced two new levels: 10 events and 11 categories, thereby also modifying the hierarchical order. What is important to note is that the perceptions at each level activate not only the behavior needed to control those perceptions but also the behavior to control the perceptions at the higher lever. Powers therefore recognizes that human behavior must be viewed as a typical holarchy of perceptions and behavior at various levels. To emphasize the idea of a hierarchy of perceptions and hierarchical controls by means of coordinated behavior, Powers defined his new formulation as the Hierarchical Perceptual Control Theory, or HPCT, precisely because the theory includes a hierarchy of perceptions and a parallel hierarchy of control. I shall not go beyond this brief description of Powers' work; however, I would underscore that he illustrated the role of memory and imagination as means for reorganizing perceptions. He introduced a second-level control defined as the *reorganizing system*, which operates to change the organization of the perceptual control hierarchy (Y) so as to remove potentially harmful conflicts, $E(Y)$, and improve (X) the control of perceptual signals at all levels.

If we consider all the observations from this *section*, we see that the Discipline of Control has clearly demonstrated that man, as a psychophysical system, can exist and continue living only to the extent he is equipped with a holonic network of *Rings* that maintains autopoiesis and permits him to behave in a teleonomic and teleological manner. Even Ludwig von Bertalanffy, considered the founder of the general theory of systems, clearly recognized that:

> So a great variety of systems in technology and in living nature follow the feedback scheme, and it is well-known that a new discipline, called Cybernetics, was introduced by Norbert Wiener to deal with these phenomena. The theory tries to show that mechanisms of a feedback nature are the base of teleological or purposeful behavior in man-made machines as well as in living organisms, and in social systems (von Bertalanffy 1968, p. 44).

Nevertheless, teleological behavior cannot be viewed as limited to activities needed for survival alone but must be extended to all those creative and artistic forms of behavior that denote man's innate tendency to achieve objectives involving his self-fulfillment and the improvement of his conditions in life. In order to fully understand how fundamental the above-mentioned action of the *Rings* are in guiding man's actions toward objectives with a strong *creative* intellectual content, we need only reflect on how such systems act to guide artists, painters, sculptors, musicians, architects, poets, etc., in producing their works.

Consider the number and type of magic *Rings* that have influenced the sensibility to sound, color, proportion, and shape, and that have guided the movement of the eyes and limbs to allow Leonardo to paint the Mona Lisa, Michelangelo to sculpt the Pietá, Bernini to build his cathedrals, Beethoven to compose his symphonies, and Dante to write the Divine Comedy. If the control systems that have guided these artists seem powerful, the cathedrals in Siena and Orvieto, with the splendor of their frescos, statues, paintings, inlaid marble and wood, stain-glassed windows and mosaics represent what is perhaps an unequalled sample of the joint action of billions of *Rings* that magically have guided the work of thousands of artists (and workers) toward the objectives conceived by their imagination. Every museum, architectural work, opera house, or library testifies to the incredible action of the magic *Rings* in leading man toward artistic creations of all kinds for his self-fulfillment.

6.10 *Rings* That Regulate Biological Clocks

To conclude my observations I would note that, like all living beings, our body has a number of "time" control systems that regulate most of our physiological functions at all levels. Our behavior, like that of the organisms of all living species, is conditioned by the variations in the rhythms of the sun and moon—during the year, the months, and the diurnal and nocturnal paths of the sun—as if such variations were controlled, with respect to given periods or instants memorized as objectives, by innate control systems, which are commonly called *biological clocks* and studied in chronobiology (Dunlap et al. 2003).

How do organisms "know" when it is time for a particular activity, say plants to detach the fruit from the branches; the chrysalis to emerge from its cocoon; penguins, other birds, and fish to migrate to lay their eggs? Why does a woman's pregnancy last 40 weeks while for other species the time period is different? Why do we feel the urge to eat, sleep, and wake up after an average number of hours?

The most intuitive answer is that all these cyclical events are regulated by biological clocks (Koukkari and Sothern 2006). In fact, we can assume that a *biological clock* is an *impulse* control system (Sect. 4.3) whose objective, Y^*, is the value of some type of *accumulation* variable Y, positive or negative, or the progressive elimination of some type of *dissipative* variable, as shown in the model in Fig. 6.23.

The gradual accumulation of values for Y continues until Y^* is reached. The gradual elimination of the dissipative variable continues until Y attains the value

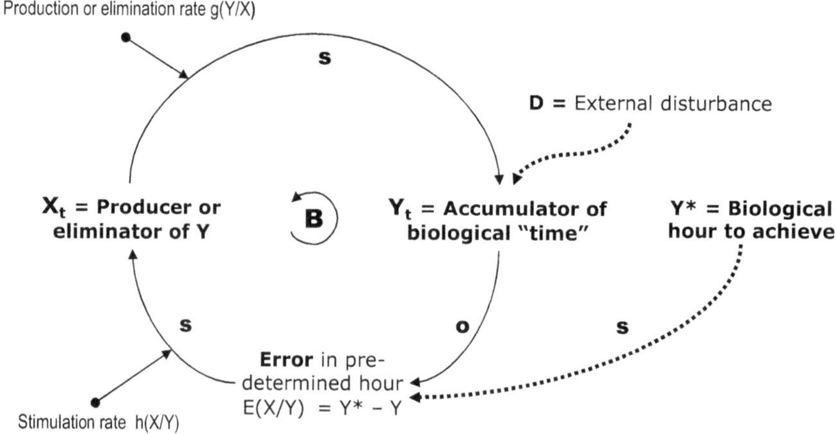

Fig. 6.23 The biological clock as an impulse control system

$Y^* = 0$. The $E(Y) = 0$ represents the signal that the physiological processes, until that moment inhibited, have been activated. The nature of the variable Y, whose values are accumulated or eliminated, depends on the type of biological clock and on the specific organism in which it operates. Chemical substances as well as frequency signals can be accumulated produced by the action variable X, for example, melatonin from the pineal glands, cortisol from the suprarenal glands, the circadian rhythm of intracellular potassium, and so on. In any event, it is clear that the term "biological clock" is used for convenience sake; however, there is no independent clock that "detects" the passing of time. Instead, there is a control system that "detects" the accumulation or the elimination of some flow variables, usually chemical, biochemical or physical, until they reach levels (upper or lower) contained in the genetic programming of the biological individual.

6.11 Complementary Material

6.11.1 The Water Cycle

Very briefly, the *water cycle* describes when water at ground level partially evaporates and is transferred to the atmosphere to form cloud masses which subsequently descend back to the ground in the form of rain, hail, or snow. The water cycle can be illustrated by the model in Fig. 6.24 (which does not claim to be complete) which, representing a continual, repetitive (though not closed) process, can in all respects be likened to an invisible planetary *Ring*.

It is well known that evaporation is a continuous phenomenon when the temperature is above zero, and that it is favored by air pressure and wind. Evaporation becomes quite intense as a result of solar irradiation over large areas, lands, forests and oceans. Unlike what occurs in small cumulus clouds that form due to convective movement, water vapor rises slowly, invisibly, increasing the density of large masses of air that can be moved by the wind. Clouds form due to water vapor condensation when the air masses encounter colder masses with different densities. If the temperature falls below the saturation temperature, the water vapor of the clouds generally *precipitates* in the form of rain; if the temperature falls further, this *precipitation* is in the form of snow.

6.11.2 Daisyworld Dynamics

Watson and Lovelock (1983) produced a model simulating Daisyworld, assigning appropriate values for the growth of daisies based on temperature variations, translating the variation in the masses of white and black daisies into variations in the albedo effect, as indicated in the model in Fig. 6.11. Usually the decision to use only

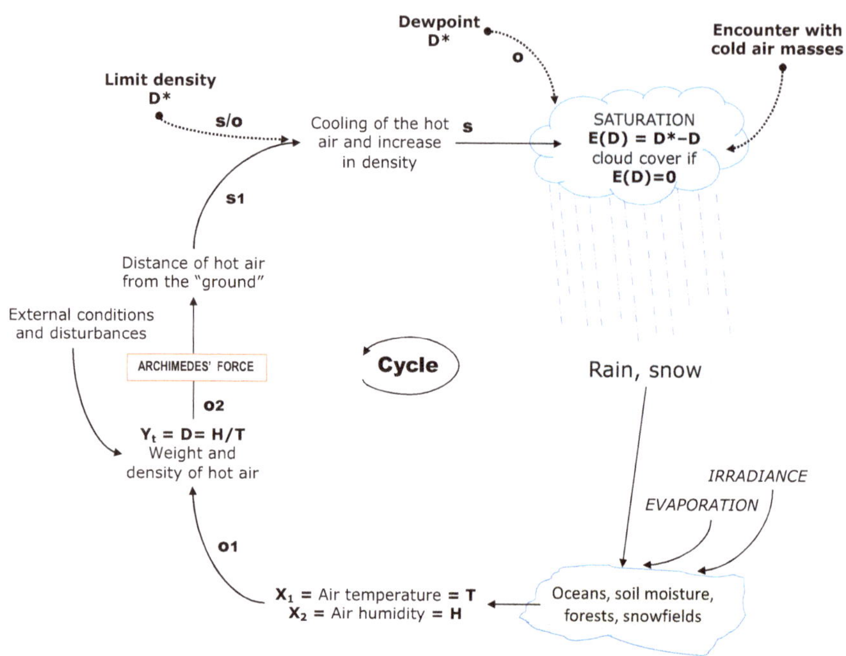

Fig. 6.24 The water cycle model

white and black daisies is based on the simplifying assumptions that the other daisy color gives no direct reproductive advantage and that the two extreme colors, black and white, make the feedbacks that regulate the entire system more effective. However, Lovelock also designed a model which includes grey daisies in order to make variations in planetary temperatures less drastic.

> With rare exceptions, biologists either ignored the models or remained as skeptical as ever. A persistent criticism from biologists was that, in a real world, daisies would have to use some of their energy to make pigment and therefore would be at a disadvantage compared with unpigmented gray daisies. In such a world no temperature regulation would take place. As they put it, "the gray daisies would cheat".
>
> Stimulated by their criticism I made a model with three species of daisies. All that the new model required was another set of equations to describe the temperature and growth of the gray daisy species. It was a matter of introducing sober-suited middle-management daisies to a world of colorful eccentrics (Lovelock 1988, p. 45).

In order to take into account the three daisy populations, each sensitive in a different way to temperature variations, the model in Fig. 6.11 would have had to have been extended by a third loop and have behaved similarly to The Law of Dynamic Instability (Sect. 1.7.4).

Even if there are a large number of simulation applets for Daisyworld, which the reader can easily find on the Internet, I feel it useful to undertake an Excel simulation, which in my view provides the immediate advantage of allowing us to grasp the dynamics of the daisy populations and of the temperature under differing

W_0	1000	B_0	1000	h(C)	-0.1
g(WD/C)	1.0%	g(BD/C)	1.0%	C_0 =	40
m/r WD	1	m/r BD	1	C* =	40
				tolerance	1

n	WHITE Daisies mass	BLACK Daisies mass	DELTA albedo WL	DELTA albedo BL	Total albedo	C_n = Daisyworld temperature	C* OBJECTIVE	D_n DISTURBANCES	ERROR C_n - C*	mad cow WL	mad cow BL
0	1000	1000	0.0	0.0	0.0	40.0	40.0		0.0		
1	1000	1000	0.0	0.0	0.0	40.0	40.0		0.0		
2	1000	1000	0.0	0.0	0.0	40.0	40.0		0.0		
3	1000	1000	0.0	0.0	0.0	60.0	40.0	20.0	20.0		
4	1010	990	10.0	10.0	20.0	58.0	40.0		18.0		
5	1020	980	10.1	9.9	20.0	56.0	40.0		16.0		
6	1030	970	10.2	9.8	20.0	54.0	40.0		14.0		
7	1041	961	10.3	9.7	20.0	52.0	40.0		12.0		
8	1051	951	10.4	9.6	20.0	50.0	40.0		10.0		
9	1062	941	10.5	9.5	20.0	48.0	40.0		8.0		
10	1072	932	10.6	9.4	20.0	46.0	40.0		6.0		
11	1083	923	10.7	9.3	20.0	44.0	40.0		4.0		
12	1094	914	10.8	9.2	20.1	42.0	40.0		2.0		
13	1105	904	10.9	9.1	20.1	40.0	40.0		0.0		
14	1105	904	0.0	0.0	0.0	40.0	40.0		0.0		
15	1105	904	0.0	0.0	0.0	20.0	40.0	-20.0	-20.0		
16	1094	913	-11.0	-9.0	-20.1	22.0	40.0		-18.0		
17	1083	923	-10.9	-9.1	-20.1	24.0	40.0		-16.0		
18	1072	932	-10.8	-9.2	-20.1	26.0	40.0		-14.0		
19	1061	941	-10.7	-9.3	-20.0	28.0	40.0		-12.0		
20	1050	951	-10.6	-9.4	-20.0	30.0	40.0		-10.0		
21	1040	960	-10.5	-9.5	-20.0	32.0	40.0		-8.0		
22	1030	970	-10.4	-9.6	-20.0	34.0	40.0		-6.0		
23	1019	979	-10.3	-9.7	-20.0	36.0	40.0		-4.0		
24	1009	989	-10.2	-9.8	-20.0	38.0	40.0		-2.0		
25	999	999	-10.1	-9.9	-20.0	40.0	40.0		0.0		
26	999	999	0.0	0.0	0.0	40.0	40.0		0.0		
27	999	999	0.0	0.0	0.0	40.0	40.0		0.0		
28	999	999	0.0	0.0	0.0	40.0	40.0		0.0		
29	999	999	0.0	0.0	0.0	40.0	40.0		0.0		
30	999	999	0.0	0.0	0.0	40.0	40.0		0.0		

Temperature Dynamics

Populations Dynamics

White daisies mass

Black daisies mass

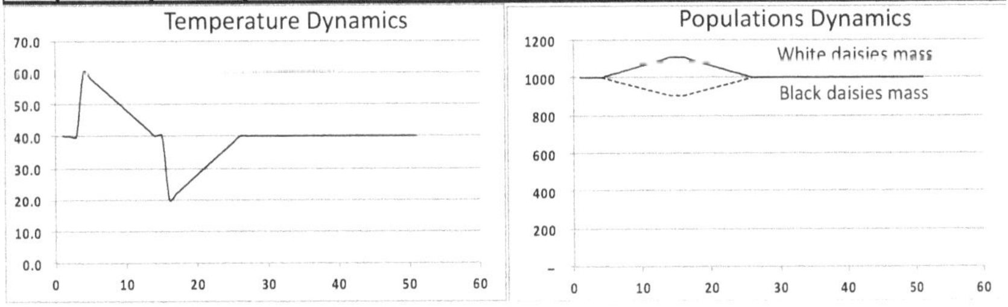

Fig. 6.25 First Daisyworld simulation

hypotheses. It is sufficient to construct the model based only on two populations of white and black daisies. Figure 6.25 shows the dynamics of the white and black daisy populations and of the temperatures in Daisyworld.

The action rates are indicated in the initial control panel, which assumes an equilibrium between the daisy masses at a temperature of $C^* = 40$ °C. At instant $n = 3$ there is a rise in the temperature of the C_3 disturbance to 20 °C. The second column clearly shows that starting from instant $n = 4$ the number of white daisies begins to increase at the rate of $g(\mathrm{WD}/C) = 1\%$ (as if a flow of cold water were triggered), while at the same time the number of black daisies begins to decline at the rate $g(\mathrm{BD}/C) = 1\%$ (as if a flow of hot water were shut off), which, for simplicity's sake, coincides with that of the white daisies. The joint dynamics of the two daisy populations increases the albedo effect with a consequent gradual decline in the planet's temperature. Beginning at instant $n = 13$ the dynamics stabilize around the value WDmass = 1105 and BDmass = 904, and the temperature in Daisyworld returns to $C_{13} = 40$ °C. A new equilibrium is achieved with different masses from the initial ones.

Let us further assume that at $n = 15$ a cold front occurs that lowers the temperature of the C_{15} disturbance to −20 °C. This lowering of the temperature in Daisyworld reduces the mass of white daisies, while that of the black daisies begins to increase, thereby producing a negative albedo effect that leads to an increase in the planet's temperature, which at $n = 25$ returns to $C_{25} = 40$ °C. Note that there is a change in the masses of white and black daisies with respect to the previous equilibrium attained at $n = 13$: the masses return to the initial level of WDmass = 1000 and BDmass = 1000, which includes a rounding off by one unit owing to the tolerance of 1 (control panel). The dynamics of the daisy masses and of the temperatures are shown in the graph in the lower part of Fig. 6.25.

Figure 6.26 presents a second simulation that more realistically provides for gradual cold impulses lasting 4 years, from $n = 4$ to $n = 7$, followed, after some time, by gradual cold impulses, from $n = 18$ to $n = 21$. From year $n = 10$ until $n = 13$ a random disturbance is introduced that strikes the white daisies, which are reduced in number due to the so-called "mad cow" effect, that is, a random reduction in the mass of daisies, as if they were eaten by a cow that randomly eats white daisies and black daisies.

As the model predicts, both the effects of the warm and cold impulses and the disturbances caused by the "mad cow" effect on the temperatures are always eliminated by the dynamics of the masses of the two daisy populations, as can clearly be seen in the graph at the bottom of Fig. 6.26.

6.12 Summary

We have learned that systems are everywhere, inside and outside us, in the biological, social, and physical world; we started an ideal journey among the systems that characterize the following:

(a) A typical day in our lives (Sect. 6.1)
(b) Our domestic environment (Sect. 6.2), discovering that various forms of direct intervention can be brought to bear on such systems (Fig. 6.1)
(c) The local external *micro* environment (Sect. 6.3), discovering the *overhead control systems* that pervade every moment of our town life (Fig. 6.3)

W_0	1000	B_0	1000	h(C)	-0.1
g(WD/C)	1.0%	g(BD/C)	1.0%	C_0 =	40
m/r WD	1	m/r BD	1	C* =	40
				tolerance	1

n	WHITE Daisies mass	BLACK Daisies mass	DELTA albedo WL	DELTA albedo BL	Total albedo	C_n = Daisyworld temperature	C* OBJECTIVE	D_n DISTURBANCES	ERROR C_n - C*	mad cow WL	mad cow BL
0	1000	1000	0.0	0.0	0.0	40.0	40.0		0.0		
1	1000	1000	0.0	0.0	0.0	40.0	40.0		0.0		
2	1000	1000	0.0	0.0	0.0	40.0	40.0		0.0		
3	1000	1000	0.0	0.0	0.0	40.0	40.0		0.0		
4	1000	1000	0.0	0.0	0.0	42.0	40.0	2.0	2.0		
5	1010	990	10.0	10.0	20.0	44.0	40.0	4.0	4.0		
6	1020	980	10.1	9.9	20.0	48.0	40.0	6.0	8.0		
7	1030	970	10.2	9.8	20.0	54.0	40.0	8.0	14.0		
8	1041	961	10.3	9.7	20.0	52.0	40.0		12.0		
9	1051	951	10.4	9.6	20.0	50.0	40.0		10.0		
10	1052	941	0.5	9.5	10.0	49.0	40.0		9.0	-10.0	
11	1052	932	0.5	9.4	9.9	48.0	40.0		8.0	-10.0	
12	1053	923	0.5	9.3	9.8	47.0	40.0		7.0	-10.0	
13	1053	914	0.5	9.2	9.8	46.0	40.0		6.0	-10.0	
14	1064	904	10.5	9.1	19.7	44.1	40.0		4.1		
15	1074	895	10.6	9.0	19.7	42.1	40.0		2.1		
16	1085	886	10.7	9.0	19.7	40.1	40.0		0.1		
17	1085	886	0.0	0.0	0.0	40.1	40.0		0.1		
18	1085	886	0.0	0.0	0.0	38.1	40.0	-2.0	-1.9		
19	1074	895	-10.8	-8.9	-19.7	36.1	40.0	-4.0	-3.9		
20	1063	904	-10.7	-9.0	-19.7	32.1	40.0	-6.0	-7.9		
21	1053	913	-10.6	-9.0	-19.7	26.0	40.0	-8.0	-14.0		
22	1042	922	-10.5	-9.1	-19.7	28.0	40.0		-12.0		
23	1032	932	-10.4	-9.2	-19.6	30.0	40.0		-10.0		
24	1021	941	-10.3	-9.3	-19.6	31.9	40.0		-8.1		
25	1011	950	-10.2	-9.4	-19.6	33.9	40.0		-6.1		
26	1001	960	-10.1	-9.5	-19.6	35.9	40.0		-4.1		
27	991	969	-10.0	-9.6	-19.6	37.8	40.0		-2.2		
28	981	979	-9.9	-9.7	-19.6	39.8	40.0		-0.2		
29	981	979	0.0	0.0	0.0	39.8	40.0		-0.2		
30	981	979	0.0	0.0	0.0	39.8	40.0		-0.2		

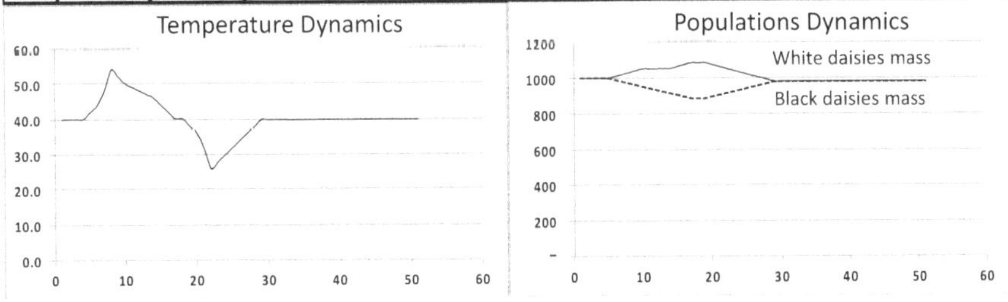

Fig. 6.26 Second Daisyworld simulation

(d) The external *macro* environment (Sect. 6.4), realizing that even many natural phenomena are governed by control systems in which only natural factors intervene (Fig. 6.8)

(e) The planetary environment (Sect. 6.5), referencing the model of planetary self-regulation in the Daisyworld model (Fig. 6.11); in particular the *Rings* that regulate the Gulf Stream (Fig. 6.10)

(f) The global warming problem (Sect. 6.6 and Figs. 6.13–6.17), thereby adapting the Daisyworld model to the reality of our planet (Fig. 6.12)
(g) Earthquakes and tsunamis (Sect. 6.7) (Figs. 6.18 and 6.19)
(h) Our body and articulations (Sect. 6.8) and the postural control systems (Fig. 6.20)
(i) Our psychophysical system (Sect. 6.9), in order to demonstrate that we exercise control over our lives in order to satisfy needs (Fig. 6.21) and aspirations (Fig. 6.22)
(j) Even the *biological clocks* that maintain the rhythm of our lives are, in fact, natural control systems (Sect. 6.10 and Fig. 6.23)

Chapter 7
The Magic Ring in Action: Social Environment and Sustainability

Abstract In Chap. 6, we headed off on an ideal journey to learn to recognize and model the action of multiple *Rings* operating in various environments. I hope the reader has become aware that without the action of the *Rings*, without their ubiquitous *magic* presence everywhere in a world made up of interconnected variables, life, order, and progress could not exist. The aim of this chapter is to guide the reader as he continues this journey through the world of human populations that interact to self-control their dynamics, and as he subsequently moves on to the action of the *Rings* in our ecosystems.

Keywords Social system · Political system · Control of coexistence · Social holarchy · Maintenance of autopoiesis · Tragedies of the commons · Complex adaptive systems · Non-renewable resource · Combinatory systems · Mother tongue · Change management · The PSC model · Stereotypes and gender discrimination

> The subject of this volume [The Social System] is the exposition and illustration of a conceptual scheme for the analysis of social systems in terms of the action frame of reference [...] The fundamental starting point is the concept of social systems of action. The interaction of individual actors, that is, takes place under such conditions that it is possible to treat such a process of interaction as a system in the scientific sense and subject it to the same order of theoretical analysis which has been successfully applied to other types of systems in other sciences (Talcott Parsons 1951, p. 3).
>
> 9.1 An organism or society is said to be in dynamic equilibrium if the Self-Assertive and Integrative tendencies of its holons counterbalance each other.
>
> 9.2 The term 'equilibrium' in a hierarchic system does not refer to relations between parts on the same level, but to the relation between part and whole (the whole being represented by the agency which controls the part from the next higher level) (Arthur Koestler 1967, p. 347).

In Chap. 6, we headed off on an ideal journey to learn to recognize and model the action of multiple *Rings* operating in various environments. I hope the reader has become aware that without the action of the *Rings*, without their ubiquitous *magic* presence everywhere in a world made up of interconnected variables, life, order, and

P. Mella, *The Magic Ring*, Contemporary Systems Thinking,
https://doi.org/10.1007/978-3-030-64194-8_7

progress could not exist. The aim of this chapter is to guide the reader as he continues this journey through the world of human populations that interact to self-control their dynamics, and as he subsequently moves on to the action of the *Rings* in our ecosystems. I shall then consider the action of control systems in social systems, understood as autopoietic units, which allow such systems to survive, develop their autopoiesis, and maintain in homeostatic equilibrium the relations among individuals and among the basic variables required for their survival. I shall examine in particular the logic of combinatory systems and that of complex adaptive systems (CAS) that support the dynamics within the social systems. The conclusion of this chapter is in line with that of the previous chapters: the biological and social environment cannot survive as a unit, nor can the individuals that compose this environment continue for long to survive, without the action of a countless quantity of interconnected "magic" *Rings*, nested over several levels and operating continuously. Chapter 8 will then examine organized systems in order to identify the *Rings* that permit the preservation of organizational relations and the interaction among organizations.

7.1 Social Environment

I define *society* (or *social system*) as a collectivity made up of one or more populations, even of different species, that live for a long time (even several generations) in the same territory according to *patterns* of stable *social relations* of coexistence imposed on them (systems of authority created by shared institutions) or which are self-generated (self-organization), thereby permitting coordinated and cooperative behavior for the survival of the society while also allowing its members to reproduce. In many societies, the members are *stratified* into homogeneous groups that do not always cooperate. In human societies, the members often also share their cultural systems and their expectations for behavior, which is useful and necessary for the general interest (the common good). The above definition is in line with the traditional one offered by sociologists and anthropologists. For example, Sandro Segre writes in his study on the sociologist Talcott Parsons (1951, 1971):

> A "system" is a stable set of interdependent phenomena, provided with analytically-established boundaries, which relates to an ever-changing external environment. A "social system" is a system of social interactions between reciprocally oriented actors. It consists of roles, collectivities, norms and values. The social system forms a system of societies, each characterized by relative autonomy, by its own territorial organization, and by its own sense of identity. In a social system, actors relate to one another by jointly orienting themselves to a situation through a language or other shared symbols. A social system comprises several subsystems or collectivities, all of which are functionally differentiated, interdependent and intertwined. Analysis of such a system focuses on the conditions in which interactions occur within particular collectivities (Segre 2012, pp. 2–3).

There are other forms of collective behavior that occur in combinatory systems and in CAS. I propose defining a *combinatory system*, or *simplex system*, as any collectivity formed by similar elements or agents (not organized by hierarchical relations or interconnected by a network or tree relationship) which displays an *analogous* micro behavior over time and produces similar micro effects. Together, the micro behaviors and micro effects give rise to a macro behavior (and/or macro effect) attributable to the collectivity as a whole. The collective behavior, though composed of the micro behaviors, conditions or directs the subsequent micro behaviors of the agents following a typical micro–macro feedback process.

Collectivities made up of heterogeneous elements that interact to form dynamic wholes, which are emerging and not predictable a priori, shall be defined as *complex systems*. The CAS approach, in particular (Kauffman 1993, 1996; Castelfranchi 1998; Gilbert 1995; Gilbert and Troitzsch 1999), studies how collectivities interact and exchange information with their environment to maintain their internal processes over time through adaptation, self-preservation, evolution, and cognition (in the sense of Maturana and Varela 1980, p. 13) and achieve collective decisions within a relational context of micro behaviors (Rao and Georgeff 1992; Wooldridge and Jennings 1994).

A collectivity of organized agents *specialized* according to function, functionality, functioning, and topology forms an *organization*, a social system (a *social machine* according to Maturana and Varela 1980), where the collective and individual behavior is determined by a network of stable relations, that is, the "organization" of the agents and processes for the purpose of creating a "stable structure".

> *Organization and Structure*. The relations between components that define a composite unity (system) as a composite unity of a particular kind constitute its organization. In this definition of organization the components are viewed only in relation to their participation in the constitution of the unity (whole) that they integrate. This is why nothing is said in it about the properties that the components of a particular unity may have other than those required by the realization of the organization of the unity (Maturana and Varela 1980, p. xix).

Chapter 9 will deal with the analysis of *organizations* and the *Rings* which allow the former to exist.

7.2 The *Rings* Regulating Social Systems: The Control of Coexistence

Even if the distinction between the *social*, *economic*, and *political* aspects of our behavior is relative, the *social environment* of which we all are a part is made up of a complex system of groups of individuals, of expanding size—individuals, couples, families, tribes, clubs, clans, villages, regions, populations, nations, associations, and organizations of various types and sizes—that mutually condition one another through a network of social and economic connections surrounding consumption and deaths (destruction), production and births (regeneration), and the exchange of persons, resources, labor, information, etc., which permits the replacement of elements (persons, resources, labor, information) that are no longer efficient with new elements, thereby indefinitely maintaining the network of connections.

Because there are no other factors that determine the maintenance of the social system apart from the network of connections which regenerates the elements forming those connections, this is a clear example of *autopoiesis* operating at a social level, similar to what Maturana and Varela indicated in their definition of a living system. In fact, Maturana and Varela stated that a social system can be defined as a structured system of autopoietic systems made up of individuals linked by social and economic relationships who seek to maintain their autopoiesis through social interaction.

> The question, 'What is a social system?' cannot be answered by simply describing a particular one because we do not know the significant relations that we must abstract when characterizing its organization. The question must be answered by proposing a system which, if allowed to operate, would generate a phenomenal domain indistinguishable from the phenomenal domain proper to a natural social system. Accordingly, I propose that a collection of autopoietic systems that, through the realization of their autopoiesis, interact with each other constituting and integrating a system that operates as the (or as a) medium in which they realize their autopoiesis, is indistinguishable from a natural social system [...]The structure of a society as a particular social system is determined both by the structure of its autopoietic components and by the actual relations that hold between them while they integrate it (Maturana and Varela 1980, pp. xxiv–xxv).

The authors also specify that if a social system can also include different autopoietic systems, each of these can be part of various social systems.

> An autopoietic system participates in the constitution of a social system only to the extent that it participates in it, that is, only as it realizes the relations proper to a component of the social system. Accordingly, in principle, an autopoietic system may enter or leave a social system at any moment by just satisfying or not satisfying the proper relations, and may participate simultaneously or in succession in many different ones (Maturana and Varela 1980, p. xxv).

When the *social system* is made up of multiple autopoietic systems, populations or social groups, these can *cooperate* in order to achieve the *common goal* of maintaining the collective autopoiesis, that is, the *common good* of the components of the *social system*. On the contrary, they can also be *in conflict* with other social systems perceived as a threat or obstacle to the achievement of individual autopoiesis. If the socioeconomic systems are as a whole *autopoietic* and made up of a multitude of component autopoietic systems that tend toward the *common good*, then it is clear the social systems are *homeostatic machines* "*...whose own organization (network of defining relations) is the fundamental variable that it maintains constant*." In other words, the systems must regulate or maintain constant the *network of relations* that defines its own organization.

> Thus small, closely knit communities have a very considerable measure of homeostasis; and this, whether they are highly literate communities in a civilized country or villages of primitive savages. Strange and even repugnant as the customs of many barbarians may seem to us, they generally have a very definite homeostatic value, which is part of the function of anthropologists to interpret (Wiener 1961, p. 160).

It is thus intuitive that, being *homeostatic* systems, *social systems* can produce their autopoiesis for a long time thanks to the action of the innumerable *Rings* which, at any moment, allow social interactions to develop, become coordinated, and constantly reproduce over time. It has long been recognized that control systems play an indispensable role in maintaining social systems. In his studies on social systems, the sociologist Kingsley Davis clearly recognizes that no social system can survive without a system of norms to which the individuals must conform:

> Always in human society there is what may be called a double reality – on the one hand a normative system embodying what ought to be, and on the other a factual order embodying what is... These two orders cannot be completely identical, nor can they be completely disparate (Davis 1949, p. 52).

The norms that lead to this factual order act as standards of behavior, or as true objectives of the social controls:

> [Norms] are controls; it is through them that human society regulates the behavior of its members in such ways that they perform activities fulfilling societal needs – even sometimes, at the expense of organic needs (Davis 1949, p. 52).

The logical consequence of the above aspects is that social systems must be characterized by *three structural elements*:

- A stable *system* of *social and economic rules of coexistence* (objectives, constraints, limits, even in the form of obligations and prohibitions) must circumscribe the behavior of the individuals and groups that make up the social system (in order to allow them to achieve their objectives without harming or destroying the homeostatic relationships that guarantee survival), so that the autopoietic relations are not altered, or, if altered, are restored, thereby allowing the groups and organizations to function as an autopoietic homeostatic machine. A social system made up of systems of individuals that share the same system of rules, including language, religion, traditions, and governance, takes the form of a *nation*.
- A stable *network* of *Rings* must make such a system of rules cogent, considering the rules as objectives, constraints, or limits which human behavior must achieve/respect and maintain. Control systems must identify the in behavior that goes beyond the rules and activate the most appropriate levers to bring behavior back in line with the rules.
- A stable *structure* of *social institutions* must guarantee the functioning of the stable network of *control systems*, acting as the governance that establishes the rules for coexistence and as the management that regulates the control levers. These institutions must guarantee the existence of efficient *chains of control*, that is, of effectors, detectors, and regulators that permit the concrete achievement of the *Rings* needed for the coordinated regulation of the rules which are "vital" for the long-lasting achievement of homeostasis. I propose defining these three elements—rules, control systems, and the institutions that produce them—as a *political system* which complements and completes the *social system*. The structure and the *modus operandi* of the *political system* depend on the cultural systems of the individuals or groups that make up the social system.

Taking account of the inevitable and fundamental economic interactions, I shall refer to the social system by the more complete term *socioeconomic political system*. When legalized in a stable manner, the *socioeconomic political system* referring to a given territory becomes a *state*, and the individuals living there, and who are legally subject to its norms, can be referred to succinctly as the *people* of the state. The logic of control systems which maintain individual behavior in line with the rules of coexistence can be outlined in the model in Fig. 7.1.

In this figure the *rules* ①, which discipline the individual behaviors of coexistence are usually translated by the *political system* into the "formal" laws of the legal system of the social system, or are represented by the "natural" laws codified as normal behavior or that derive from custom. The two types of rules interact in a network that is often difficult to identify; nevertheless, to make the action of control systems concretely possible, the institutions tend, where possible, to transform the "natural" rules into formal legal norms. It seems incredible when we consider how many laws, limits, bans, prescriptions, or obligations regulate our behavior at every moment of the day. For example, our home and the use of public services; traffic laws, regional and municipal regulations, as well as tax laws discipline our every movement in our towns, from the use of our car to that of public transport, from driving to parking regulations, automobile taxes to tolls. Norms, regulations, and customs of all types regulate how we enjoy our free time, imposing obligations and

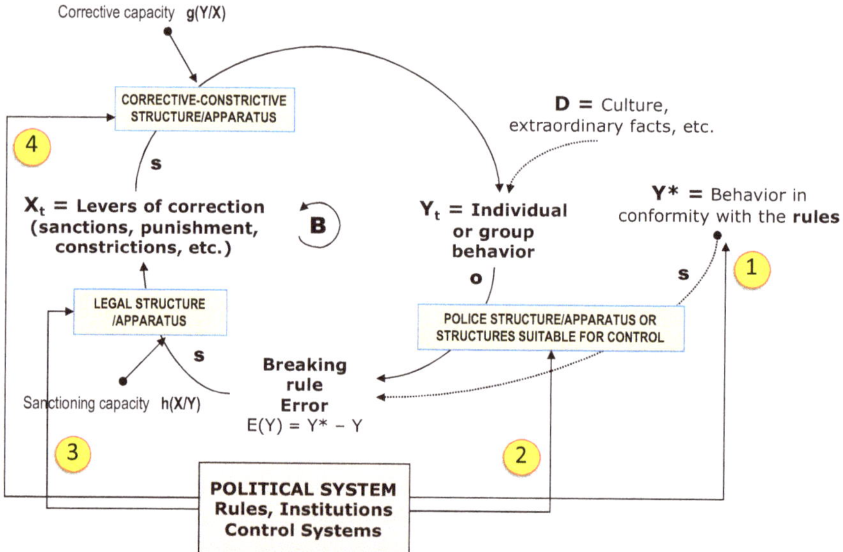

Fig. 7.1 The *Ring* for the control of the respect for the rules of coexistence in the social system

prohibitions in all areas: from the bar to the canteen, the church to the beach, and weddings to funerals. And norms regarding ceremonies and proper etiquette are no less stringent.

As shown in Fig. 7.1, in order to allow the social *Rings* to operate, the institutions must also create the following necessary apparatuses of the *chain of control*:

1. *Detectors*, ②, to identify and recognize behavior not in conformity with the rules: police apparatuses or other structures to check on the conformity of behavior in various areas: judges, arbitrators, head masters in schools, auditors, etc.
2. *Regulators*, ③, of the behavioral control levers (the legal apparatus and other apparatuses capable of deciding on types of sanctions)
3. *Effectors*, ④, which bring non-conforming behavior back in line or avoid future non-conforming behavior (corrective and, if necessary, constrictive apparatuses, even entailing limits on personal freedom)

The institutions that define the rules and produce the network of control systems are the *governance institutions*, which, within a *state*, represent the *government* structures. The *political institutions* that make up the *government* are normally holonic in nature, since they are arranged in various layers that form a more or less complex holarchy, in order to identify the *hierarchy of control systems* and the *managers* that will guarantee they function properly, as shown in the model in Fig. 7.2.

Such hierarchical forms of *government* arise from the need to control individuals or groups belonging to an increasingly larger territory. The holonic institutions can be either *ascending* or *descending*; if the base individuals (base holons) that form the "people" belonging to the social system produce first-level institutions which, in turn, form second-level ones, and so on, up to the maximum level (top holon), then the process is *ascending*. If the highest level (top holon) is formed, to which

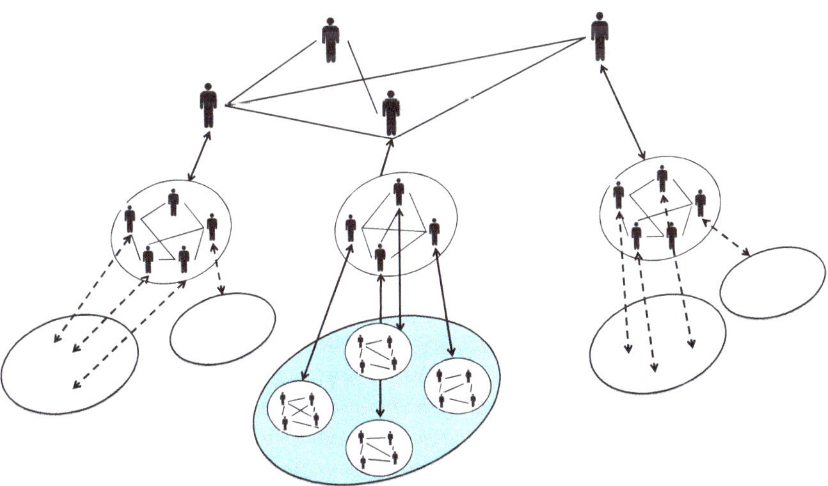

Fig. 7.2 Institutions over various layers, forming a holarchy

maximum authority is attributed, and this level then produces, through a division of power, the lower levels, the process is *descending*, as illustrated in the diagram in Fig. 7.3.

In general, we can state that the number of levels of institutions is higher the vaster is the social system (large territory) and the more the need is felt for a control of individuals' behavior (conflictuality); nevertheless, we can also observe a tendency for a reduction in the levels of the governance system and an increase in the number of individuals and in the size of the territory governed by the elements at each level. A social system comprised of a tribe has, as its highest institution, the tribal chief, who controls the behavior of the tribe's members in order to avoid homicides, thefts, infidelity, as well as to impose various types of tributes. The tribal head can be elected (ascending) or can dominate the "people" through physical or religious force (descending).

When the social system grows through the union of different tribes, a sovereign is needed who can control the tribal heads that, in some manner, must be maintained. When the system grows further, an emperor is required who can control the various sovereigns, however they are referred to. The formation of high-level governance structures (empires or the political union of states) is a result of the need for the control and governance of increasingly larger territories (the Roman Empire, the Chinese Empire, the Mogul Empire, the Ottoman Empire, the USA, USSR, EU, etc.). Many of these institutions are ascending in nature, elective (the tribes elect the tribal heads, who then elect the sovereign, etc.) but history teaches us that institutions

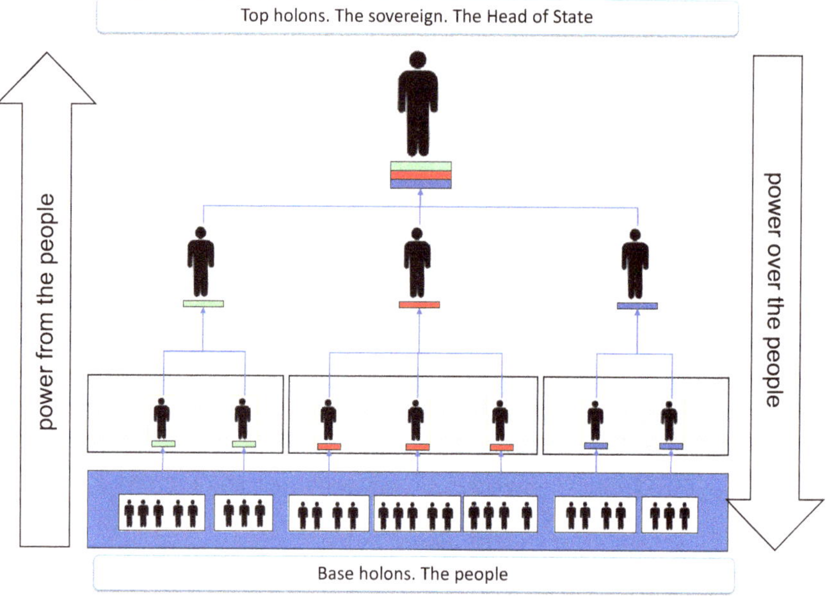

Fig. 7.3 Ascending and descending institutions. The different dimensions in the various levels indicate the concentration of power

that develop through an ascending process very often are transformed into descending institutions. The emperor is not elected but acquires authority and power from some "supreme authority," and in turn appoints the hierarchy of subordinate controllers, who have authority over individuals of increasingly smaller territories. Consider, for instance, the levels of the aristocratic hierarchy in Europe: emperor, king, prince, grand duke, archduke, landgrave, duke, marquis, count, viscount, knight, and noble.

Today the descending institutions of monarchies and dictatorial regimes are still with us; however, more numerous are the ascendant republics or parliamentary monarchies, which arise and are maintained in social systems where the "people" hold power, which they delegate in various forms and for more or less lengthy periods to their delegates. In many countries today, there is a hierarchy of delegated authority in cities, provinces, regions, nations, and supranational organizations. This delegated authority results in a *governance* that establishes the policies based on some rule of representation of the delegates; these rules can at times make the governance unstable and the setting of objectives and policies problematic. We can easily find numerous examples of this in the daily news. It is not beyond reason that, with progress in systems for the exchange of information through networks (the web), we may also someday witness a "flat" governance where the intermediate levels are eliminated, as some have already hypothesized.

> This allows us to conceptually divide the system into functional components responsible for the different stages of the processing of incoming matter and energy (metabolism), and for the processing of information needed to maintain cybernetic control over this mechanism (nervous system). As the system continues to evolve, on-going adaptation and division of labor lead to an increasingly diverse, complex, and efficient organization, consisting of ever more specialized components. This general model of complex, self-organizing systems can be directly applied to the present development of society. Since society is an organismic system consisting of organisms (individual people), it can be viewed as a "superorganism." Conspicuous trends such as globalization, automation, and the rise of computer networks can be understood as aspects of the general evolution towards increasing efficiency and interconnectedness which makes the superorganism ever more robust (Heylighen 2013, p. 32).

7.3 *Rings* That Maintain Autopoiesis in Social Systems

The control of the coexistence among individuals and groups—that is, the conformity of behavior to social rules—is necessary, though not sufficient, for the integrity of the "social fabric." Equally essential is the maintenance of *autopoiesis* in the social system as a whole. Thus, we must join other *macro* systems of *overall* control to the *micro Rings* that regulate individual relationships, guaranteeing in this way the regulation of the fundamental socioeconomic *macro variables* which characterize the *common good*. It is up to the *political system* to define the *objectives* of these autopoietic macro variables and determine the order in which the objectives must be pursued, thus performing the fundamental function of *governance* required to activate the macro *Rings* which, by controlling those variables, guarantee autopoiesis in the social system.

In order to control these socioeconomic *macro variables*, government institutions normally create appropriate *institutional macro structures* that are functionally specialized in managing *control systems* in order to achieve the *macro objectives* (the policy) defined by the governance. The totality of these structures represents the *operational apparatus* of the social system, on whose efficiency and size the operational efficiency of the control systems depends. The specialized organizations that permit the actual functioning of the *operational apparatus* make up the *bureaucratic-administrative apparatus*.

The *macro structures* are at the top of a holarchy because they are operationally divided into smaller structures with more circumscribed control powers, arranged in layers of *bureaucratic-administrative micro structures*, so that even the macro objectives can be holonically detailed as lesser objectives that form a holarchy of objectives, for whose achievement holarchies of control systems are formed.

In a modern social system, with ascending governance, the *bureaucratic-operational holarchies* are usually specialized for each of the social macro variables whose macro-objectives are set by the governance (policy), as shown in the model in Fig. 7.4.

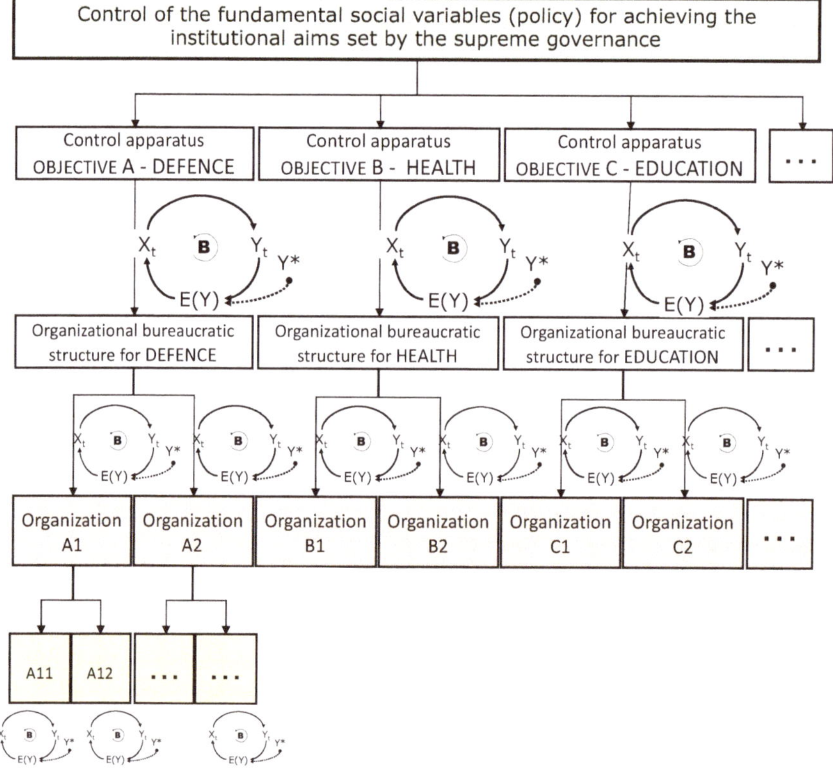

Fig. 7.4 Structure of the organizational bureaucratic apparatus that manages the *Rings* that control the autopoietic variables based on the governance policy

In every modern state, the *bureaucratic-administrative* structures at the highest level are the ministries (however denominated), each of which represents the management for the control of the social variable assigned to them. The objectives the ministries must achieve are divided into segments of objectives assigned, for example, to the ministry heads, who in turn control the increasingly smaller territorial units. In this way, offices, agencies, committees, etc., are created that provide the local governance and serve as elements in the chain of control of the various *Rings*.

In every society, we encounter the same higher-level "autopoietic variables" to which the governance *policies* assign objectives and priorities. In order to identify these variables, it is sufficient to make a list of the ministries that, though under different names, every state must set up. Limiting ourselves to only the most general and obvious ones, the fundamental variables that every state tends to control, and whose functions are assigned to the ministries, can be indicated in the Table 7.1.

Thus, for example, the macro variable "*public health*" is assigned general objectives by the general governance; these are usually qualitative objectives (health care for everyone, increase in the number of hospitals, a reduction in death rates by *y* %,

Table 7.1 Social variables necessary for the maintenance of autopoiesis

Social variables [sample list]	General objectives assigned to the social variables [sample description]
Defence	Maintain security in the face of external threats or aggression to the state
Public order and internal security	Guarantee adequate levels of internal security by controlling public order and dealing with emergencies and calamities of all kinds
Justice	Provide maximum levels of criminal, civil, and juvenile justice services along with adequate levels of prison facilities
Public health	Assure the highest levels of heath care and of treatment and prevention of illnesses
Education (schools, universities)	Assure the highest levels of educational services at all levels, from pre-school to doctoral
Economy	Provide incentives to, coordinate, and develop all forms of economic activity
Industry, commerce, agriculture, and tourism	Regulate and develop the main forms of production activity
Employment	Regulate the demand and supply of labor by increasing employment
Finance and treasury	Contain and maintain in equilibrium the flows and levels of tax revenue and expenditures (taxes, public debt, and public expenditures)
Families	Assure the creation and maintenance of families of all forms
Sport	Assure adequate access and support to all forms of sport
Protection of territories and seas	Further environmental protection, economic growth, and social cohesion in the territories
Infrastructure and transport	Provide adequate and modern systems of infrastructure and of public and private transport
Etc.	

etc.), which are then divided into territorial objectives assigned to the second-level governance, and so on for each successive territorial unit down to the cities and, even more specifically, each health center. For each of these objectives, in each territorial level, there are *Rings* operating to achieve them, whose management represents a segment of the *operational-bureaucratic apparatus*. For the macro variable "*instruction and education*," the ministry of education (which may be made up of specific ministries for schools, universities, and higher education institutes) sets up management specialized according to each type of sub-variable. Increasingly more detailed local bureaucratic apparatuses are then put in place, down to the individual educational institute (education managers, headmasters, deans, etc.).

This arrangement means that no social system can control the homeostasis of the social variables which are vital for the maintenance of its own autopoiesis without a multitude of *Ring* holarchies operating daily in every part of the territory. This control of the social *macro* variables operates parallel to the mixture of *micro* controls on the behavior of single individuals; these controls derive in large part from the control of the macro variables.

It is not easy to examine the autopoietic network of the *Rings* operating at the various levels of the social system, given their ubiquity and extreme variety. Two difficulties are immediately apparent. Above all, control systems that operate in the socioeconomic system are usually *multi-lever* and *multi-objective*; the difficulty in recognizing them is due to the fact that, on the one hand, many of the control levers—for example, those that act on social decay and on the violence in our stadiums—are not known or immediately perceivable, and on the other that there are interesting, varying objectives over time that present sudden shifts linked, for example, to unexpected political changes, population flows between regions, natural and social calamities, and so on. As a result it is difficult to detect the deviations from the objectives and to identify the control variables for their elimination. A general model of such control systems could be represented as that in Fig. 7.5.

A second difficulty comes from the fact the socioeconomic system is made up of a network of subsystems of varying types and sizes which are closely interconnected, so that any control of a system influences the other systems it is linked to. For example, the control of labor productivity and employment levels—which are variables from the economic environment—implies the control of consumer optimism for the future, which is a typical variable of the social environment. The control of the level of national income implies control of consumption and investment at a national level; but this imposes a control at the regional level, and thus at the municipal level, all the way down to the company level. The control of employment implies both a production and administrative control policy. Controlling the spread of drugs implies control of the effectiveness of the educational system and of the quality of the family's educational activities.

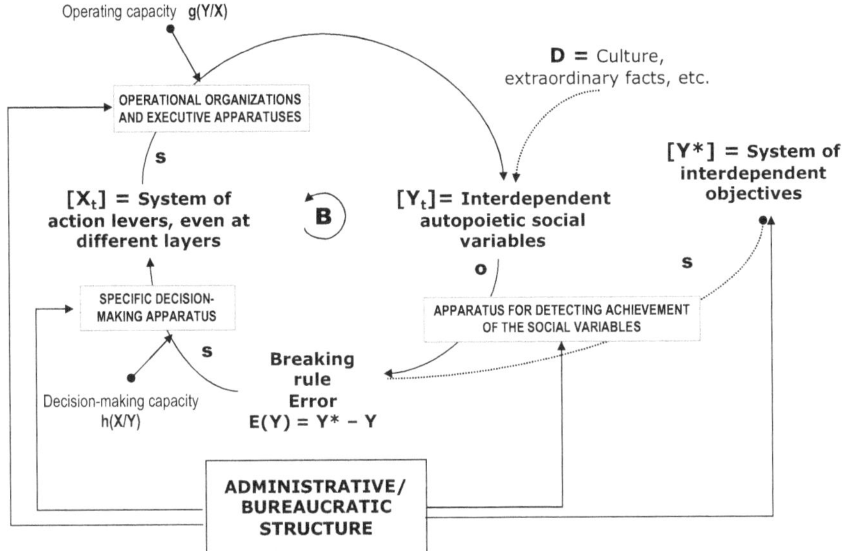

Fig. 7.5 General model of control systems for the autopoietic variables in social systems

7.4 Rings That Regulate Some Fundamental Variables in Social Systems

The preceding methodological observations, though brief, allow us to consider the typical control systems needed to regulate several macro variables fundamental for the preservation of social systems. I shall start with two particularly significant ones: the control of language and *traditions*, the control of *violence*, the control of *population size*, the control of *health*, *births*, *nutrition*, and many other equally relevant variables.

Language Surely no one will argue that the most deeply rooted tradition in any social system is the learning and maintenance of the *mother tongue* by the inhabitants of an area of varying size. This is a very powerful control system given that, as reported by SIL International the Ethnologue (Ethnologue 2013) site, there are at present 7105 languages, of which 2304 are in Asia and 2146 in Africa (Lewis et al. 2013, on line). The maintenance of spoken language—together with the written one, with its more narrowly codified syntax, which is less affected by practice—among the members of a community thus appears as one of the most powerful control system operating in human society. The linguistic control system teaches not only the rules of the new language but also the linguistic and dialectic inflections, setting off the levers to maintain them indefinitely by trying to use the instruction levers to eliminate speaker s (how often have scholars despaired over the many s in syntax and orthography!), according to the general model in Fig. 7.6.

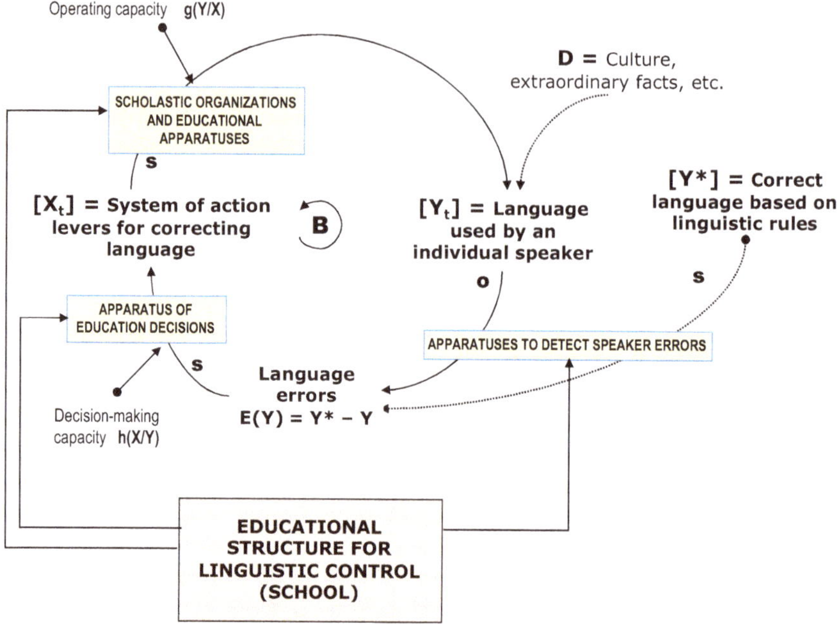

Fig. 7.6 *Ring* to control and preserve languages

At times, the family of the newborn knows only one language, which is different from those of the other members of the community (immigration, deportation, etc.); in this case, this language is passed on as the mother tongue of the newborn, but the linguistic control system acts to eliminate the language gap between the newborn and the language normally spoken by the other members of the society, as the newborn is forced to learn a second language that is dictated by the educational control system of the community. Usually this second language is held to be rewarding, and is thus passed on to the offspring (dialects are spoken today only by the elderly, for whom they represent the mother tongue, since the youth are taught the national language). At other times, in the same social environment several communities coexist, each characterized by its own language; in this case, there are as many control systems as there are different languages. If the languages are equally important with regard to interpersonal relations, and if marriages between people with different mother tongues are frequent, then the parents may pass on different mother tongues to their children at the same time (e.g., Switzerland). I will deal with this kind of control system in the Sect. 7.5, since the maintenance of the mother tongue is carried out by means of a combinatory system that operates automatically.

Violence and Wars Violence is deeply rooted in our competitive nature and often degenerates into highly destructive behavior. Precisely because violence represents behavior inherent to man and is manifested in different contexts (family, work, on the roads, in stadiums, etc.) and with differing intensities (verbal or physical violence, with or without weapons, robberies, thefts, etc.), the social system must keep the level of violence under the tolerance threshold—which is the control objective—

by using various multi-lever control systems acting at various levels through various types of levers: a system of prevention through education and job promotion, a system of punishment and repression by means of restrictive measures for violent individuals, the arrest of promoters of violence, the loss of civil rights, etc. Though apparently different in nature, organized crime, common delinquency, stadium violence, bullying at school, hazing in the barracks, mobbing at work, and so on, are all encompassed in such a control framework.

More complicated is the violence that takes the form of wars and ethnic conflicts of varying magnitude. Man is certainly not the only social being that engages in conflicts.

> One of the behaviors of ants most commonly observed in nature is overt aggression, often resulting in injury, death, and displacement of one colony by another. Auguste Forel was correct in observing that "the greatest enemies of ants are other ants, just as the greatest enemies of men are other men" […] Territorial fighting is very general in the ants, sometimes widespread over the foraging areas and sometimes confined to the immediate vicinity of the nest (Hölldobler and Wilson 1990, p. 398).

Today there are many ongoing conflicts in the world. Wikipedia (2020, Conflicts) lists more than 9 current wars and conflicts, as well as about 18 minor wars and conflicts.

Even if man, as a belligerent being, displays apparently unpredictable behavior, there is always some control system that stops wars from going on for too long, leading to the annihilation of one of the warring sides. This is testified to by the fact wars end sooner or later (even the Hundred Years' War only refers to a limited historical period). The main difficulty in identifying such control systems is identifying the constraints that oblige the belligerent parties to sit down at the peace table.

The so-called "religious" wars persist for a long time precisely because religions are deeply rooted in man's soul, especially in terms of prejudice and hate, rather than existing at an exclusively rational level. The outcomes of such wars are often horrific and their length unpredictable. Thus, the constraint that the war must end with the complete elimination of the "enemy," even if, rationally speaking, this appears too extreme, has occurred on many occasions. Other forms of wars of "conquest and defense" are motivated by the need for expansion (preservation) of the territory in which the social systems carry out their autopoiesis, due to the resources the territory offers as well as the need to subject the populations inhabiting it. In this form of conflict, while there are advantages that come with winning, there is still considerable damage even for the belligerent victor. A plausible limit to the continuation of war is linked to the trade-off between advantages and disadvantages. When the disadvantages exceed the advantages, it is highly likely, unless there are other non-economic motives pushing the sides to fight "to the last man," that the war will end due to discouragement on the part of one or both of the belligerents, similar to Richardson's model on the limits to arms growth (Sect. 1.7.1). More realistic constraints that tend to limit the damage from carrying on a war are the reduction in the enemy's offensive capabilities below a given threshold, thus minimizing the fear of the enemy, or a reduction in the defensive capabilities below limits perceived as dangerous for survival, so as to provide incentives to the enemy to take action to search for peace.

Population growth Two other fundamental control systems for autopoiesis in social systems are the control of *health* and *births*. The objective of the former is to increase the percentage of individuals who reach the limit of *life expectancy* at birth, that is, the average number of years an individual, from birth, can expect to live within a certain population. The system also aims to increase life expectancy itself. Since *life expectancy* depends on economic and cultural factors in the population, on the state of society, and on environmental conditions, the control system for this variable uses a certain number of control levers, among which fostering progress in medicine, improving physical hygiene, preventing disease through a system of early detection of infections and epidemics, diffusing dietary education, improving work conditions, and preventing and reducing the infant mortality rate due to work-related reasons or civil violence.

Births help maintain the population; however, too high or too low a birth rate with respect to the limit or objective necessary or useful for maintaining autopoiesis—also taking into account the death rate—necessitates a control system with multiple levers, which is not always easy to identify or hypothesize, since we do not know the optimal growth rate of the population. Births depend on psychological, economic, and social conditions whose effects are not always known; thus, the activation of appropriate control levers is often difficult. Such levers will impact lifestyles and work habits, family culture, the need for free-time, assistance for mothers and children, the condition of schools, and sports activities, requiring complex, long-term, and costly policies by the governance. Often favorable life conditions slow down births, while dangerous situations accelerate them. In some cases, the existence of appropriate facilities for the care of children (pediatric clinics, nursery schools, kinder-gardens, summer camps, sports centers, and so on) are useful, while in others they have no effect.

In cases of *declining* population, strategies were introduced to spur population growth by activating levers aimed at combatting infant mortality, improving living conditions and health, and providing various forms of economic incentives to produce large families; sometimes references to patriotic values can increase births, as occurred in Italy from the end of 1925 to the end of the Second World War and has even occurred in recent years. On the contrary, in some social systems during certain historical periods, where population has increased too much, strategies to reduce population growth were undertaken through restrictive measures sanctioned by law, such as in the People's Republic of China through the introduction of a one-child policy for the control of demographic growth (the laws introduced in 2001 that forbid families from having more than one child) and the tendency of Chinese families to favor male children, especially in rural areas (Von der Pütten 2008). This policy was revised in December 2013, and as of 2014 Chinese couples are now allowed to have two children. In short, every social environment must necessarily activate a hierarchy of coordinated control systems to control the birth and death rates, but the observation and planning of adequate control systems is anything but easy, as shown by recent history and daily news stories.

Nutrition and health I will only briefly mention another important intentional control system present in every socioeconomic environment. The autopoiesis in a social system depends on the fundamental control systems of *nutrition* or malnutrition in the population, as these influence the birth and death rates, work capacity, and the levels of defense against epidemic diseases. It is not always easy to design effective *Rings*, as testified to by the presence of endemic malnutrition, diseases, and infant death rates that are difficult to eliminate (sleeping sickness, leprosy, etc.) or by the recurring outbreak of flu epidemics. Normally the nutritional level of the population is a *symptom* whose elimination is almost always achieved through simple symptomatic or extraordinary levers (humanitarian aid, dietary help); it is more difficult to identify the *structural levers* that can reduce the variance between an acceptable level of dietary wellbeing and the effective level that falls below this standard.

Population longevity Ray Kurzweil (2005) states that man's ultimate aspiration is to achieve "immortality," or at least to increase individual and social longevity as much as possible. According to any dictionary, longevity is a term which, in biology, refers to the physiological ability of an organism to survive beyond the limit considered to represent the average life expectancy for the species to which it belongs. Longevity is not synonymous with aging but with healthy aging, in which individuals preserve their interest in the future. One indicator that expresses the dynamics of longevity is *life expectancy at birth*, a statistical measure that expresses the average number of years a person can expect to live at the time of his birth in a given country based on the mortality rates recorded on the date of measurement. This indicator is one of the most significant parameters of social, economic, and health conditions, thus setting itself up not only as a demographic indicator of expected longevity but also of the level of development of a country. Worldwide, women's life expectancy exceeds that of men due to both genetic factors and the different incidence of cardiovascular disease and other gender-type diseases. Therefore, to control longevity, life expectancy must be controlled/increased by activating a multi-lever control system that can be described according to the general levers indicated in Fig. 7.7.

Longevity control must, of course, be accompanied by that of other correlated variables whose dynamics create problems:

(a) *Social* problems that arise when, due to age limits, the inactive population will exceed the working population
(b) *Production* problems due to a lack of active work in many sectors of the economy; policies and strategies are needed to maintain the employment rate and increase productivity
(c) *Demographic problems* caused by the increasing gap between young and old
(d) *Health* problems, since the ageing of the population leads to an inevitable increase in the health care needs of citizens, which impacts health expenditures due to the increase in diseases typical of old age and the increase in health care treatments and consumption

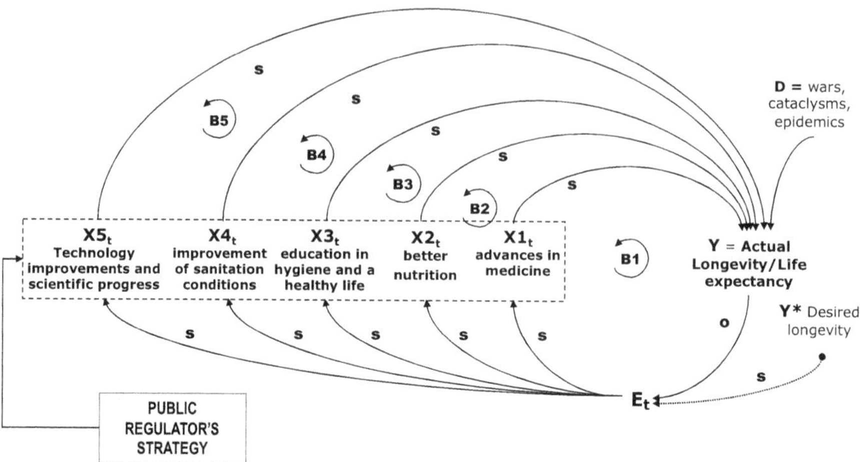

Fig. 7.7 Longevity control system

(e) Problems related to *health care assistance* and the need for its provision when families can no longer offer parental assistance; these problems are accompanied by a shift of employment both within health care facilities and in the elderly care facilities

(f) Problems concerning *the sustainability of the social security system;* if older workers are kept on at work, it could be more difficult for young people to enter the labor market

Economic variables. To conclude, I will mention several *Rings* that involve the economic variables, even when the economic system is so bound up in the social system that social variables are closely intertwined with economic ones. The trends in economic variables (consumption, production, savings, investment, loans, exports, imports, price and interest levels, the levels of transfers and taxes, etc.) representing the entire economy of the social system are determined by the individual behavior of productive and financial organizations and by consumers, which as a whole determine the level of the nation's wealth, which is related to the level of national income and the gross domestic product (GDP), which in turn depend on the level of production, consumption, and prices, and thus on investment, production costs, and capital costs, in a typical dynamic system. In order to control economic behavior, which at the aggregate level produces global economic flows, the governance of the socioeconomic system introduces a number of forms of *regulation*.

This term generally indicates government intervention in the market both through specific rules and by assuming the governance of the numerous economic *Rings* (activated by the economic control structures: ministries and other subdivisions of authority) identify the values to assign to the fundamental *control levers* in order to permit the *economic variables* to achieve the system's objectives or to bring these values back within the appropriate limits or respect the constraints of the economic policy the governance (government) feels appropriate for promoting the economic autopoiesis lying at the base of social autopoiesis.

> There are at least two large chunks of the economy that the competitive market model obviously does not describe or even purport to describe. These are the huge and growing public sector, the allocation of resources which is determined not by the autonomous market but by political decisions, and the public utilities [...].
>
> To be sure, the govemment influences the functioning of the private, competitive sectors of the economy as well in many ways—for example, by regulaling the supply and availability of money, enforcing contracts, protecting property, providing subsidies or tariff protection, prohibiting unfair competition, providing market information, imposing standards for packaging and product content, and insisting on the right of employees to join unions and bargain collectively. In principle, these influences, however pervasive, are intended to operate essentially at the periphery of the markets affected. ...In these sectors the govemment does not, or is not supposed to, decide what should be produced and how or by whom; it does not fix prices itself, nor does it control investment or entry on the basis of its own calculations of how much is economically desirable [...] In contrast, the government does do all these things with the public Utilities. Here the primary guarantor of acceptable performance is conceived to be (whatever it is in truth) not competition or self-restraint but direct governmental prescription of major aspects of their structure and economie performance...control of entry, price fixing, prescription of quality and conditions of service, and the imposition of an obligation to serve all applicants under reasonable conditions (Kahn 1988, p. 2/1).

I shall only briefly expand on the above discussion by observing that ensuring the growth rate objectives are achieved is the task of government economic policies; however, the widespread and continuing difficulties in producing the desired growth in national income and GDP testify to how difficult it is to identify and regulate the necessary control levers. Together with the growth in *national income* and GDP, *inflation* must also be kept under control. The fact there is still inflation in many countries and that governments are constantly concerned that it may rise again is the clearest evidence of the difficulty in planning an effective control system. It must be noted that the difficulties in controlling inflation also depend on the fact that, contrary to national income and GDP (which are variables that are to be controlled), *inflation is a symptom* of the lack of control of the socioeconomic system's most important variables—costs of particular factors (gold, oil, copper, iron scrap, etc.), production costs, productivity, etc.—whose trends depend on other variables which are not always economic in nature: the level of social and organizational conflict, confidence in the future, demand levels, the social security system, the propensity to save and to invest, and so on. In fact, in order to keep inflation under control, we must identify both the variables which are truly relevant in influencing inflation and the levers that can be adopted to produce values compatible with economic growth at low levels of price increase. I will just mention in passing the control systems for *unemployment*, *productivity*, and *tax evasion*, systems which are extremely important for every country's economic growth but which are difficult to design, since the dynamics of the variables to control depend on social and economic control levers which are not easy to identify.

The dynamics and control of other relevant variables observed in social systems will be interpreted using combinatory systems theory, which will be presented in the sections below.

7.5 *Rings* Within Collectivities as Combinatory Systems

A particular aspect that makes observing and designing control systems interesting but problematic is the fact the socioeconomic environment is made up not only of networks of systems and subsystems, which were discussed in the preceding *section*, but also of collectivities that I have defined as *simple social systems*—or simplex systems (Mella 2005a)—or, more generally, as combinatory systems.

In social environments, a combinatory system is a simple social system formed by a group of similar individuals, or agents (humans, animals, or other living units), forming the *base* of the system, not organized by hierarchical relations or interconnected by a network or tree relationship, that produces an analogous micro behavior (pedestrians walk along the sidewalk, elephants in the savanna) that can produce some observable micro effects (individual pedestrians leave garbage behind; individual elephants head toward other elephants). Combined together, the micro behaviors produce a macro behavior (general tendency to litter; a compact herd forms), which can produce an observable macro effect (a garbage pile forms; the herd opens a trail in the forest), which, in turn, conditions the subsequent micro behavior of the individuals (the garbage pile represents information about where one can leave garbage with impunity, the herd gives information that there is protection available).

Thus, on the one hand, the macro behavior of the system, as a collective unit, derives from the *combination* (appropriately defined) of the analogous micro behaviors of the individuals (from which the name combinatory systems derives); on the other hand, the macro behavior determines, conditions, or guides the subsequent micro behaviors. This reciprocal relation can be defined as micro–macro feedback. I assume that a *necessary* and *sufficient* condition for a collectivity (observable or assumed) to be considered a combinatory system is the existence of a feedback between the *micro* behavior of the individuals and the *macro* behavior of the collectivity constituting the system.

DEFINITION Combinatory System: A "combinatory system" is defined as any collectivity made up of a plurality of unorganized similar agents (or elements) producing analogous micro behaviors and showing, as a whole, a macro behavior and/or a macro effect, whose dynamics are created by a micro-macro feedback action. If, on the one hand, the macro behavior of the system, as a whole, derives from the combination, appropriately specified, of the analogous behavior (or effects) of its similar agents (hence the name combinatory system), on the other hand the macro behavior (or the macro effect) determines, conditions or directs the subsequent micro behavior, according to a feedback relation between the micro and macro behavior or effects (Mella 2017, p. 9).

The existence of micro–macro feedback thus leads to an essential consequence: the macro behavior (or effect) of the system cannot be considered a mere sum of the micro behavior (or effects) of its elements; the micro–macro feedback causes "emerging" types of macro behavior (or effects) which are not included in advance in the operating program of the agents' behavior but instead are attributable to the unit.

Combinatory systems are very common in the biological and social worlds, but we are not used to perceiving them (Mella 2005a). Examples of these are the systems that lead to the formation of groups of various objects, from cities to industrial districts (Mella 2006), or to the spread and maintenance of particular characteristics, from language to traditions, fashion to collective states of mind. If the perception of combinatory systems appears to be difficult, it becomes possible as soon as we apply the first rule of Systems Thinking, described as follows in its most complete form:

> To understand reality we must not limit ourselves to observing only individual objects, elements, or entities; it is necessary to "see" even the larger groupings that these compose, attributing to them an autonomous meaning. The converse process is also true: we cannot limit ourselves only to considering an object in its unity but must force ourselves "to see" its component parts. This rule, which is at the basis of Systems Thinking, can be translated as follows: if we want to broaden our intelligence we must develop the capacity to "zoom" between parts and wholes and between wholes and components (Mella 2012, pp. 9-10).

The action of combinatory systems occurs because the macro behavior—or the associated macro effect—translates into *global information* (of varying types) that represents a *global objective* individuals must or desire to achieve; this objective guides the choices of individuals who, noting a distance between their own micro behavior or micro state and that of the overall state of the system, try to eliminate the gap by adjusting their individual and collective behavior. The system's macro behavior, or macro effect, represents (or conditions) the *objectives* behind the individual choices.

It clearly follows from this that combinatory systems function due to the presence of *microRings* which, operating at the individual level, lead to uniform micro behavior by individuals in order to eliminate the (gap) with respect to the objective that is represented—or revealed—by the global information (macro behavior or effect). The above description can be better understood through the model in Fig.7.8.

For this reason, social combinatory systems can also be called *self-produced global information systems*, in order to distinguish them from *local information systems*, whose typical model is represented by complex systems simulated by cellular automata. When the actions of individuals are simultaneous, the agents of the system appear to *synchronize* their micro behavior. However, each micro behavior updates the global information, and this recreates a divergence that exerts even more influence on the individuals to conform to this information.

The mutual dependence between the *micro* and *macro* behavior (or their states or effects) represents the *micro–macro feedback* [**R**], which is maintained thanks to the individual control systems represented by the loops [**B**] in the lower portion of the model in Fig. 7.16, so that the combinatory system in principle repeats its cycles until the individual actions can update the self-produced global information which makes clear to each individual the distance, *E*(me), of the individual state (me) with respect to the state representing the objective (me*).

The feedback arises and is maintained by a set of *necessitating factors* which force the agents to adapt their micro behavior to the system's macro behavior or effect; the feedback is also maintained by the action of a set of *recombining factors*,

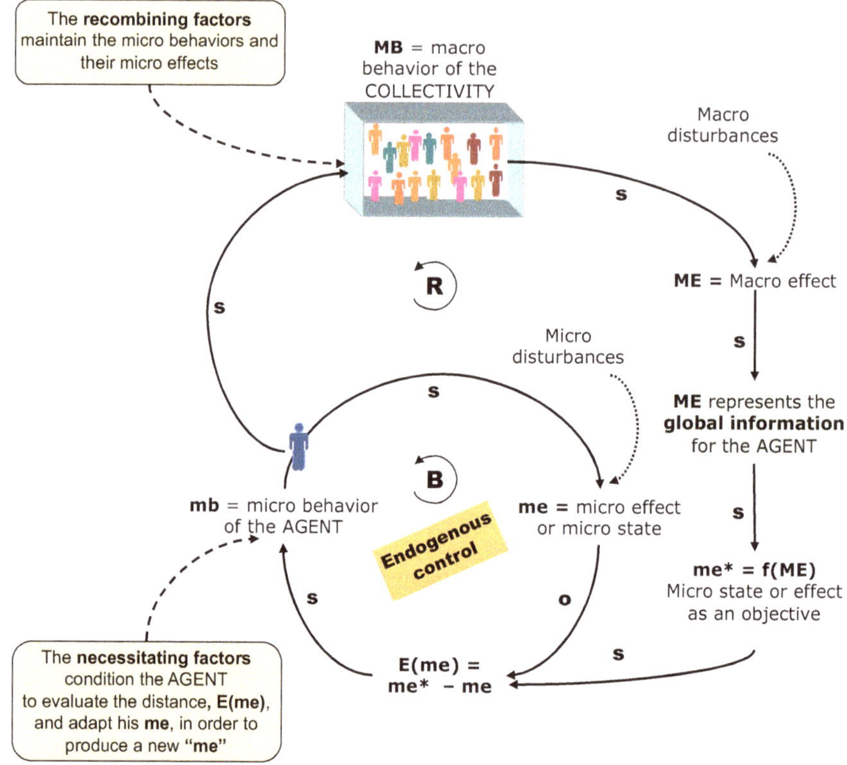

Fig. 7.8 Schematic general model of a combinatory system

which lead the collectivity as a whole to recombine the micro behaviors, or the micro effects, in order to produce and maintain the macro behavior, or the macro effect (see Sect. 7.9). Recognizing the existence of a *micro–macro feedback* and understanding the nature of both the *necessitating* factors and the *recombining* ones is indispensable for interpreting collective phenomena as deriving from a combinatory system.

The logic of combinatory systems allows us to identify a wide range of phenomena that depend on the individual control systems that lead to micro behavior compatible with the global information (macro behavior or macro effect) produced by collective action. To more easily understand the *modus operandi* of combinatory systems and the role of global information, let us consider the simple phenomenon, familiar to all of us, of a "*buzzing noise* arising in a crowded room."

What causes this buzz to form? It arises from the voice level (micro effect) of those present when they speak to each other (micro behavior), which is recombined by the shape of the room. But why do those present speak in a loud voice (macro behavior)? Because there is a buzzing noise (global information) which prevents them from being heard; this becomes the minimum constraint (objective) to exceed in order to be heard; if the voice level is not adequate (gap), it is not possible to communicate. Thus, if the buzz increases, those present, in order to be heard, must raise their voices. It seems they do this all together, as if the global information forces them to synchronize their micro behavior; but this causes the buzz to increase further, which obliges those present to raise their voices even more, which increases the buzz, which forces those present… etc., as part of a reinforcing loop that takes the buzz to the maximum bearable level (constraint); once this is reached, the individual control systems induce the speakers to be quiet, as we have all witnessed first-hand. The next *section* will provide a formal representation of the combinatory system I have described; for now

I will present the system that produces the process for "passing on a *language* within a population," which has briefly been dealt with in Sect. 7.10.2. Let us begin by stating that all parents pass on their mother tongue to their children (micro behavior), and that the children learn this language (micro effect). The population communicates (macro behavior) using the mother tongue (macro effect), which represents the global information that obliges families to teach that language to their children (constraint/objective) in order not to disadvantage them in their communication (s). The feedback is evident but—remembering the bad marks inflicted on us by our teachers because of syntax s—we can imagine there are also external control systems that both reinforce the main combinatory system, by detecting deviations between the language of the group members and the codified mother tongue (Eccles and Robinson 1984, p. 141), and try to eliminate these by using the traditional levers of academic teaching to correct the syntax s. In Italy, the Accademia della Crusca (National Academy for the Study and Preservation of the Italian Language) is the depository of "proper speech" in the Italian language. The existence of so many languages and equally numerous dialects show how powerful this combinatory system is and how efficient the individual and external control systems are. Thus, combinatory systems for the spoken language—to the same extent as those for the written language, with its more strictly codified syntax that is less affected by accepted practice (consider all the syntax and spelling s that have driven scholars crazy!)—are some of the most powerful systems operating in human society.

The process behind a "*fashion* spreading" appears different from that regarding the maintenance of a mother tongue; but the former is also produced by the action of a combinatory system and by the individual controls this entails. How do fashions arise, how do they cease, and why do some return while others die away forever? A fashion—whether clothing, watches, cars, toys, the use of a particular linguistic form, and so on—arises in a given environment through a novelty introduced (micro behavior) "by chance" by a given creator for the purpose of distancing himself from custom (objective to exceed). If the novelty is not pleasing, costs too much or has other disadvantages, the episode will remain an isolated one and no combinatory system will arise. If a sufficient number of individuals (minimum activation number) see the imitation of the innovation as an objective to achieve and follow the novelty in order to eliminate the deviation from the new desired state, then that innovation becomes "in fashion" and its spread represents the global information of the new "tendency." As a result a macro behavior is produced that refers to the collective system, which leads to an increase in the desire to imitate, thereby causing individuals, conditioned by the individual control systems, to undertake micro behaviors that imitate the novelty, with the objective of eliminating the between the desired and normal states. The resulting micro effects consist in the purchase of goods which are the object of imitation, a further increase in the desire to imitate, and the resulting macro effect of the further spread of the good that is "in fashion."

Similar to combinatory systems for the spread of a fashion are those that move individuals to "pursue *records* of whatever kind." Even if Pierre De Coubertin stated that "*it is important to participate, not to win,*" those who compete are not content merely to equal the record (objective to achieve), sparing no effort to eliminate the gap and exceed it, since the record, in any sporting event, is the global information that identifies the "absolute best." When the effort comes from a multitude of competitors, the record gradually improves, thus raising the level of the objective the individual competitors must achieve. It is easy to see in this mechanism the continuous action of individual control systems to eliminate the deviation between the record—which represents a dynamic objective—and the actual performance of the competitor, who represents the ARCHETYPE of "Strengthening Goals" (Fig. 5.12). It is normal today in national competitions for even the last-place runner in the 100-m dash to break the record from the 1896 Olympics, when the gold-medalist, Thomas Burke (USA), won with a time of 12 s; the current women's world record is held by Florence Griffith (USA): 10.49 s in 1988. The tributes for the record holder, the considerable prize money and social gratification are levers used by the collectivity to push competitors to improve their performance by activating control systems that, through the levers of training, improvement, the learning of techniques and tricks—and often the use of banned substances as well—seek to eliminate the gap between individual performance and the objective, despite accidents, victims and physical damage. Here, too, it is not difficult to see how this combinatory system is, in turn, reinforced by environmental control systems such as: the spread of sports education, government sports policies, and the building of more gyms, pools, and other sports facilities.

Let us move to another phenomenon, which is completely different but still produced by a combinatory system. Those who often travel on highways know how annoying and dangerous "*wheel* ruts" are; these are the parallel ruts that form on the pavement from the constant passage of heavy trailer trucks over the same trajectory. The trucks' wheels continually use the same section of pavement, especially on hot days when the pavement is softer, causing a typical subsidence that becomes a long wheel rut if observed over a long stretch on the pavement mantle. Wheel ruts occur especially on long straightaway sections of narrow lanes since, in order to stay within their lane and avoid invading the adjacent one, the drivers of large vehicles stick to the same strip of highway mantle. Ruts are less common along curves, since it is harder to keep the same trajectory in this case. Once formed, wheel ruts force trailer truck drivers to drive within the ruts on the highway so as to avoid dangerous swerving and route adjustments that could cause them to invade the adjacent lane. We can clearly see the action of the individual control system of the single drivers that try to eliminate s with respect to the trajectory imposed by the wheel ruts, which represent the global information that forces them to adopt a uniform micro behavior.

These few examples show that a large number of socioeconomic and sociopolitical phenomena can be convincingly explained using the model for combinatory systems. Without attempting a complete description, but merely to make order out of the many social phenomena produced by combinatory systems, I shall now present a typology composed of five classes, based on the macro effect produced.

1. *Systems of accumulation*, whose macro behavior leads to a macro effect which is perceived as the accumulation of objects, behaviors, or effects of some kind; this logic applies to quite a diverse range of social phenomena, among which the formation of urban or industrial settlements of the same kind, of industrial districts, the accumulation of garbage, graffiti, and writings on walls, and the clustering of stores of the same type in the same street (Mella 2006).
2. *Systems of diffusion*, whose macro effect is the diffusion of a trait or particularity, or of a "state," from a limited number to a higher number of agents of the system; systems of diffusion explain quite a diverse range of social phenomena: from the spread of a fashion to that of epidemics and drugs; from the appearance of monuments of the same type in the same place (the towers of Pavia, for example) to the spread and maintenance of a mother tongue or of customs.
3. *Systems of pursuit*, which produce a gradual shifting of the system toward a *dynamic objective*, as if the system, as a single entity, were pursuing a goal or trying to move toward increasingly more advanced states; this model can represent quite a different array of social behaviors: from the pursuit of records of all kinds to the formation of a buzzing in crowded locales; from the start of feuds and tribal wars in all ages to the overcoming of various types of limits.
4. *Systems of order*, which produce a macro behavior, or a macro effect, perceived as the attainment and maintenance of an ordered arrangement among the agents that form the system. Systems of order can be used to interpret a large number of social phenomena: from the spontaneous formation of ordered dynamics (to an observer) in crowded places (dance halls, pools, city streets, etc.) to that of groups that proceed in a united manner (herds in flight, flocks of birds, crowds, etc.); from the creation of paths in fields, wheel-ruts on paved roads and successions of holes in unpaved roads, to the ordered, and often artificial, arrangement of individuals (stadium wave, Can-Can dancers, Macedonian phalanx). Systems of order can also explain populations of insects, typically ants, which act by creating an "aromatic potential field" by spreading *pheromones* or other permanent messages; the increasing concentration of pheromone (global or macro information) increases the probability that each agent will move in the direction of that site. The micro–macro feedback is quite evident.
5. *Systems of improvement and progress*, whose effect is to produce progress, understood as an improvement in the overall state of a collectivity normally conceived of as a consequence of evolution.

Schematic and heuristic models of such systems are presented in Sect. 7.10.1. Examples of social variables that can be explained using combinatory systems are indicated in Sect. 7.10.2.

7.6 The Control of Combinatory Systems

As shown in Fig.7.8, combinatory systems can operate thanks to the micro control systems that allow agents to adapt their individual states to the macro state of the system as a whole. Nevertheless, combinatory systems can, in turn, also be controlled to favor or inhibit the attainment of certain individual and collective results considered to be harmful or useful.

Harmful behavior could be one that leads to the writing of "graffiti on a wall"; every instance of graffiti (micro effect) left on the wall (micro behavior) increases the mass of graffiti (macro effect), which in turn increases the probability that individual passersby will leave new graffiti, in a typical micro–macro feedback process. "Garbage piles" also influence passersby who want to get rid of their garbage; but each time garbage is left, the pile increases not only in size but also regarding its "attracting power" (global information), thereby producing an undesirable effect. Without an external control on the combinatory system, the effects of collective action based on self-produced global information could lead to harmful, even disastrous effects.

When, on the contrary, the combinatory system produces a useful macro effect, then the external control could accelerate the formation of the macro effect; we see this clearly, for example, in the formation of industrial districts (macro effect), which are formed when new firms are enticed to locate in a particular district (micro behavior), which makes the area even more advantageous by offering more employment and increasing the local wealth.

What happens if the combinatory system acts freely, without any external control? In some cases, a type of self-control is produced: no one can continue raising his voice to overcome the buzz in the crowded room; such a volume is reached that people spontaneously stop talking, and then the buzz diminishes (only to soon begin to rise again). In other cases, the combinatory system finishes operating when the maximum admissible density (or maximum level of the macro effect) is reached, making any further individual action impossible.

In medieval Pavia (northern Italy), a phenomenon occurred that was unique in the world. The "tower phenomenon" began around 1000, perhaps some decades before, and rapidly developed in 1100 and the following century, so that by around the year 1300 many had already been ruined. In 1570 the historian Breventano (1570) mentioned more than 170 towers, Spelta (1602) around 100, and Zuradelli (1888) counted 76, a very impressive number if we consider that a recent map of the town reveals 71 traceable towers within its walls, 7 of them almost intact. The towers had no useful function; they only reflected the desire of the noble families of Pavia to display their wealth. They were the visible icons attesting to the fact the family had attained wealth and could allow itself to make a show of this by using this symbol. I have carefully studied this by creating simulation software that provides results that conform to the historical data. The appearance of a swarm of towers is certainly the macro effect of a combinatory system that is easy to understand. The towers represent the global information that many families have achieved wealth. In order not to be any less appreciated, new families built (micro behavior) their own tower (micro effect), thereby making the global information more significant; the denser the "swarm of towers" became, the more those families without a tower felt inferior and were spurred to build their own tower.

Why did this phenomenon die out? The plausible explanation is the following: during that historical period the area inside the town's defensive walls extended over a surface area of about 900 × 900 m. If we consider that the towers were built in the corner of palaces belonging to noble and wealthy families, and if we assume that on average a palace was 30 × 30 m (which can be objectively verified), then there were no more than 300 palaces for rich and noble families in Pavia, if we take into account the surface area occupied by the less well-off townsfolk, as well as by the town's squares, roads, churches, and public buildings. We thus have a possible explanation: the towers had become so numerous (a density of one tower for every two palaces) that they ended up losing their power of attraction; the global information had changed from "if you don't have a tower you're not considered rich" to "if you build a new tower you're a parvenu." The phenomenon had come to an end.

Leaving aside the spontaneous cessation of combinatory systems, there is no doubt that very often *external controls* are carried out on combinatory systems using external control systems constructed expressly for that purpose. In order to understand how these external control systems for combinatory systems operate, we must recognize that the micro–macro feedback that produces the micro and macro behavior requires the contemporaneous presence of *necessitating* and *recombining* factors.

I define a *necessitating factor* as any element—a constraint, rule, condition, law, conviction, etc.—that "obliges" each individual in the collectivity to adapt his micro behavior to the macro behavior of the system by activating the individual control system that detects the gap between the individual's micro state and the macro state of the collectivity, leading to the activation of the control levers needed to eliminate this gap ($E(Y)$). The stronger the *necessitating* factor is, the more the micro behavior is activated. In combinatory systems made up of persons, the necessitating factors often originate from conscious motivations: necessity, convenience, opportunity, the desire not to be inferior, and so on. At other times, these factors can also be "natural" and act unconsciously, since they derive from the genetic or operational program of the individuals that form the *base* of the combinatory system.

To understand how *necessitating* factors operate, let us reflect on the highly powerful combinatory system involving the spread of drugs and ask ourselves what causes the addiction that moves the drug taker to buy and sell those substances (conscious need), thereby producing the collateral macro effects represented by the network of drug pushers and the cultivation and importation (macro effect) of such drugs. What moves the driver of a trailer truck to remain in his lane, if not the need to avoid dangerous swerving? What instincts incite individuals to perpetuate feuds and start wars?

The existence of one or more necessitating factors is indispensable though not sufficient; it is also necessary for the system—through some *recombining factor* (rule, convention, algorithm, nature of the environment the agents operate infect)—

to be able to recombine the micro behaviors (or micro effects) to produce the macro behavior (or macro effect) which, through the *micro–macro feedback*, is able to act on the necessitating factors. The recombining factors can also operate in various ways.

Consider how different the recombining effect is regarding the formation of the buzzing noise when the room is large as opposed to high-ceilinged or small; or flat as opposed to arched; or when there is a small or a large number of speakers. In the countryside, it is much simpler to scatter one's refuse haphazardly than to accumulate it. In cities, the systematic removal of garbage makes it almost impossible for the combinatory system that produces garbage piles to start up. In an area with a high population density, it is easier for a fashion to spread; even the action of advertising promotions, together with the abundance of sales outlets, makes the recombining of this system much stronger, faster, and more persistent.

From the above comments, it is easy to see that when the *macro* behavior or the *macro* effect of the combinatory system (ME_t) must be guided toward a desired objective or limit (ME*), set by some policy maker in the socioeconomic environment within which the combinatory system operates, specific external control systems can be activated whose control levers, [Xn_t], are represented by *reinforcing* or *weakening measures* (actions, provisions, constraints, limits, obligations, etc.) which, operating at a macro and/or micro level, modify the *recombining* and *necessitating* factors, influencing the *macro* and/or *micro* behaviors, thus directing the *macro* behavior of the combinatory system, as shown in Fig. 7.9.

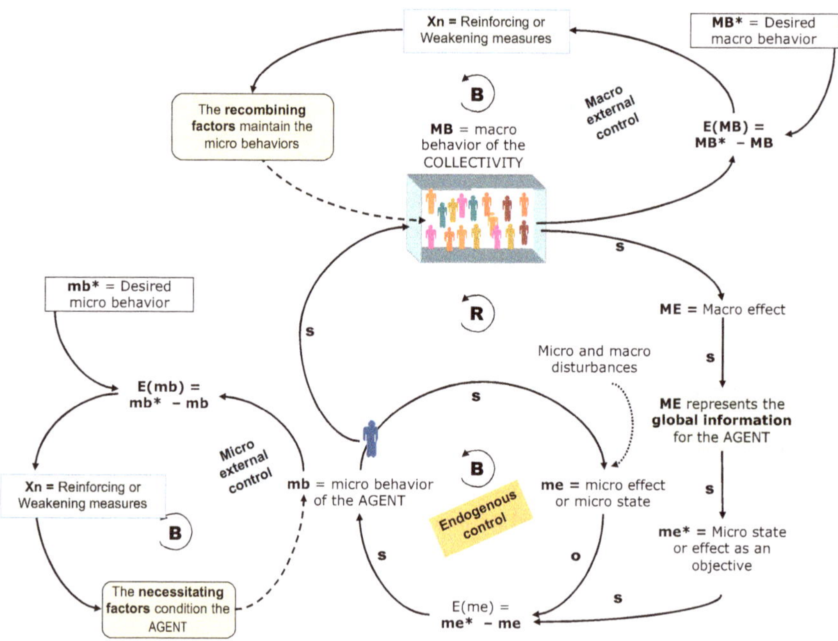

Fig. 7.9 External macro and micro controls in combinatory systems

As an example, let us apply all this to the combinatory system of *accumulation* that gives rise to industrial settlements or the formation of industrial districts (Mella 2006). When a certain area offers a positive differential in terms of economic efficiency compared to others (necessitating factor), then a certain number of entrepreneurs may decide to locate there (micro behavior) and set up production and commercial units (micro effect). The presence of a group of firms always produces economic advantages for future settlements (recombining factor), and this attracts new entrepreneurs that weigh these advantages against the disadvantages of locating in non-industrialized areas, in so doing activating a typical micro–macro feedback process. When the *minimum activation density* is reached (number of firms needed to produce a positive differential in economic efficiency), the combinatory system starts up and, in subsequent cycles, produces the collective inflow of other production units (macro behavior) and a rise and expansion in the typical industrial and commercial districts (macro effect). The system can also be controlled by both weakening actions (urban regulations, higher taxes and charges, etc.) and reinforcing actions (outright grants to locate in an area, the creation of infrastructures, etc.), which create new economic efficiency differentials. Reinforcements and weakenings are the levers [Xn_t] that control the combinatory system, and these can be applied both directly to the *macro* behavior—we shall define this as the *macro* control—or to the *micro* behavior, in which case the control will be defined as *micro* control.

The model of the combinatory system combined with the external control systems is shown in Fig. 7.9, in which we immediately see that the macro-level control is achieved through reinforcements and weakenings that act on the recombining factors, while the micro control requires reinforcements and weakenings to modify some necessitating factor. It should be clear that these are *external* control systems run by managers that are external to the combinatory system; these systems should not be confused with *internal* control systems, which are necessary to maintain the micro–macro feedback (Sect. 7.5). Also note that the micro control must not be confused with the *feedback* between the macro and micro behavior internal to the combinatory system.

I wish to proceed with the example of the productive settlements that form an industrial district or commercial zone, considering, however, how this phenomenon can be produced by a combinatory system of *diffusion* that operates in the socioeconomic environment to favor the *endogenous* spread of new firms.

Some successful firms that locate in a particular area and that are able to internally develop their employees, managers, and professional staff may decide, after acquiring the necessary competencies and pushed by the desire for personal profits (necessitating factor), to undertake an activity to exploit these competencies (micro behavior), thereby creating new businesses (micro effect). If those who undertake an autonomous business venture are successful in their attempt, then the combinatory system can start up and the area will see the growth of new businesses (macro effect) that *endogenously* develop over the territory. The process spreads and, after subsequent cycles, the group of workers is gradually transformed into a group of businessmen (macro behavior), thereby creating industrial settlements (macro effect), which probably will require bringing in new employees from other areas.

Since this spread of businesses is considered to be the source of well-being for the entire area, various forms of *macro* and *micro controls* can be carried out by activating several levers such as: creating professional schools that guarantee initial job entry (macro control); the availability of risk capital and loans (micro control); incentives to start new businesses through facilitating regulations to favor, for example, young entrepreneurs (macro control); and the creation of safety nets for failure (micro control). The creation of businesses exogenously, or through public authorities, which possess the appropriate business characteristics (even small-sized firms, the need for small-scale collateral production, the development of competencies) can artificially start up the system, on condition that the critical mass necessary to produce the necessitating and recombining factors that produce the micro–macro feedback is quickly achieved.

7.7 Sustainability of Social Behavior, Myopia Archetypes in Action, Population Growth, and Commons Depletion

The ARCHETYPE of the "three myopias," which reveals the incessant and widespread action of "short-term, local, and individual preference" (Sect. 1.6), produces worrisome appropriation phenomena which weaken the capacity of social systems and organizations to develop and even survive over time. This has given rise to and increased a "social alarm" that sees in the micro behavior of agents guided by "short-term, local, and individual preference" the main source of the problem concerning the *sustainability* of the macro behavior of social, economic, and environmental systems.

> Biologically, sustainability means avoiding extinction and living to survive and reproduce. Economically, it means avoiding major disruptions and collapses, hedging against instabilities and discontinuities. Sustainability, at its base, always concerns temporality, and in particular, longevity (Costanza and Patten 1995, p. 196).

The problem of sustainability is usually considered looking forward and interpreted as a problem of the *sustainable development* of the macro variables of collective systems. In social systems, sustainability is usually defined as:

> [...] a broad interpretation of ecological economics where environmental and ecological variables and issues are basic but part of a multidimensional perspective. Social, cultural, health-related and monetary/financial aspects have to be integrated into the analysis (Söderbaum 2008).

At the start of the 1970s, an important report entitled "The limits to growth" (Club of Rome 1972) initiated the international debate on how man should intervene to create a curve of logistic accommodation to resources in order to limit the exponential growth in population, food, industrial production, energy consumption, CO_2 emissions, etc., and thus avoid a catastrophe. The nature of the debate shifted from "limits to growth" to the concept of "sustainable development" with the publication of the report: "World Conservation Strategy" (1980) by the International Union for the Conservation of Nature and Natural Resources (IUCN). Referring to the definition by the "Brundtland Commission" (Brundtland 1987), it has been observed that:

> Over these decades, the definition of sustainable development evolved. ...This definition was vague, but it cleverly captured two fundamental issues, the problem of the environmental degradation that so commonly accompanies economic growth, and yet the need for such growth to alleviate poverty (Adams 2006, p. 31).

The most relevant and quoted definition of sustainability is the one from the report entitled, "Our Common Future," published by the World Commission on Environment and Development (WCED) (institutional version):

> Sustainable development is development that meets the needs of the present without compromising the ability of future generations to meet their own needs (Brundtland 1987, p. 43).

Nevertheless, in general, as Pearce has commented: "defining sustainable development is not a difficult issue. The difficult issue is in determining what has to be done to achieve sustainable development, assuming it is a desirable goal" (Pearce 1999, p. 69). More than a definition, this quote is a "plea" to undertake sustainable behavior to ensure that today's benefits do not become harmful for future generations. It is also possible to reformulate the hopes of the Brundtland Commission as follows: if, on the one hand, agents—men and organizations—have to continually develop, on the other they must avoid nonsustainable environmental, economic, and social behavior. If such behavior already exists, they should try to remedy the damage and above all avoid its repetition. In recent decades, the definition of sustainable development has evolved. This definition was vague but it cleverly captured two fundamental issues: the problem of the environmental degradation that so commonly accompanies economic growth and the need, nevertheless, for such growth to alleviate poverty (Adams 2006). Thus, today "there is wide consensus that the idea of sustainability figures as one of the leading models for societal development by indicating the direction in which societies ought to develop" (Christen and Schmidt 2012, p. 411).

The archetype of "short-term, local, and individual preference" (Fig. 1.19) reveals that the main problems of sustainability are caused by economic activity: individuals, as consumers, behave individualistically and with a short-term perspective due to their desire to consume goods in ever greater quantities and of ever higher quality; they do not know about, or are indifferent to, the long-run effects of their consumption. In fact, consumption requires heightened production activity, which, thanks to profit-motivated investment, which provides employment, earnings, and taxes to finance public services, makes immediately clear the short-term and almost always local individual benefits. This repeated and increasingly intense production activity, and economic activity in general, produces four evident problems regarding sustainability:

1. The increasing pollution, in particular global warming
2. The increase in the dynamics of the world population
3. The depletion of non-renewable resources
4. The depletion of disposal sites for waste and scraps of products that can no longer be used

Point (a) has been discussed in Sect. 6.6. This section will consider problems (b) and (c). Point (d) will be discussed in Sect. 7.10.4.

I would now like to examine point (b), which deals with the effects of the Myopia Archetypes on population dynamics, well aware of the highly sensitive nature of this topic, since it essentially involves the fate of all of humanity. I would reiterate that the Myopia Archetypes act at the collective and global levels to guide the behavior of individuals in a group or population viewed as a whole. In particular, the effects **of** the Individual Myopia archetype will be considered, which guides the action of the average person who, instinctively, feels the need to reproduce, thereby gaining the immediate *individual* advantages of joy and pride from the birth of a child, advantages that reduce the perception of the *global* disadvantages and harm to the environment, society and the economy from non-sustainable overpopulation. It is important to note that "population" dynamics is not only triggered by the innate instinct to reproduce but is also normally accentuated by the *culture of society*. Repetitive behavior in the short term triggers an increase in population; in cases of abundant resources, this increase is an advantage for society. Once the resource constraints have been exceeded, there is the risk of overpopulation and non-sustainability for the environment, society and the economy, as shown in Fig. 7.10.

The harmful consequences of overpopulation, which are heightened in the long run, are easy to imagine. However, it is useful to note the following unwanted harmful effects: the increase in the number of megacities, which today already number 40, the vulnerability to epidemics, the reduction in the availability of water resources, the non-sustainable exploitation of food sources, and the need for a continuous increase in global activity. The most serious problems of global non-sustainability are associated with the effects of the latter, since the continuous growth in global economic activity—which is needed for employment and the production of goods to satisfy the needs of a growing population—leads to:

1. The depletion of resources, especially non-renewable ones, above all those defined as "commons" due to the fact they are limited, of common use, and freely exploitable (Sect. 7.10.4)
2. The continuous increase in energy consumption
3. The increase in waste, especially solid and liquid wastes
4. The reduction or exhaustion of disposal sites for solid waste and the risk of uncontrolled waste disposal in the environment (plastic in the oceans and radioactive waste buried indiscriminately)
5. Increasing pollution and the dramatic effect of global warming and rising sea levels

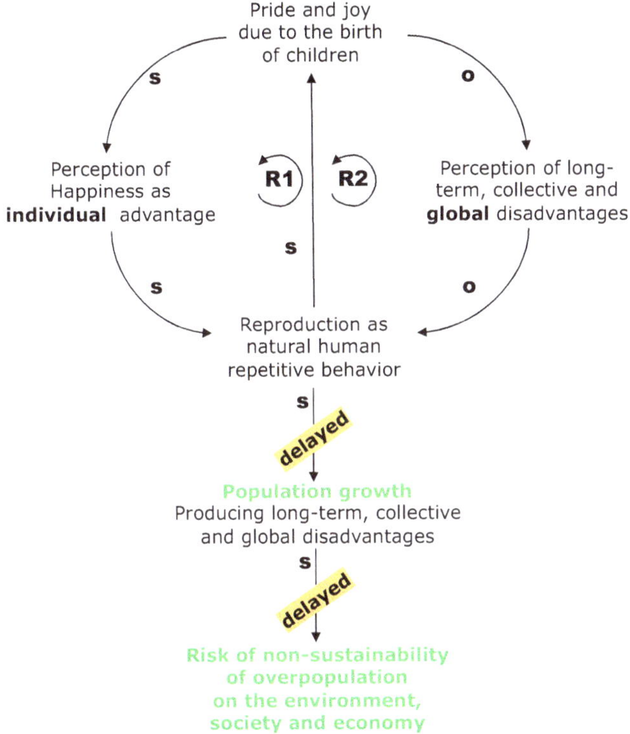

Fig. 7.10 How the archetypes of myopia produce the non-sustainability of "population" increases

Note that the archetype in Fig. 7.10 can also produce the opposite effect on the overall population. When an increasing number of individuals see a child as a burden, an obstacle to their career and individual freedom, the repeated behavior becomes less frequent and intense, then the global effect can be a reduction in the population, which can become non-sustainable when it does not allow the levels of economic growth, employment, and welfare to remain unchanged and alters the fiscal and pension system equilibriums, thereby making it more difficult for new generations to maintain their quality of life at the level attained by the previous generations. In some cases, the rationality of individual behavior is accompanied by a collective effect that leads to an improvement and to progress; in others to a negative, even catastrophic result.

A particular example of *negative collective effects* is the one that leads to the exhaustion of scarce, non-renewable resources that individuals can freely appropriate and use. This situation is normal whenever individuals, faced with a resource in danger of depletion, behave according to the Archetype of "short-term, local, and individual preference" (Sect. 1.6), which dominates the prevailing form of individual–agent, leading the latter to adopt individual, non-rational micro behaviors which result in disastrous collective macro effects (two other examples of which will be presented in Sect. 7.10.4). This relationship between individual, non-rational behavior, which is increasingly driven by the harmful collective effects, also makes the model of the combinatory systems of diffusion applicable (Sect. 7.10.1). This model is so common in social systems that Peter Senge considered it as a true systemic archetype, known as the Tragedy of the Common Resources, or the Tragedy of the Commons, knowledge of which is useful for understanding the causes of many types of extinction or exhaustion and for finding remedies (Sect. 7.10.4). This combinatory system can also be activated in non-human populations competing for a territory or a food source; if not stopped through exogenous controls, it can inevitably lead to the extinction of one of the populations concerned.

This combinatory system contains a certain number of individuals—the appropriators—who share a common resource available in limited quantities, called Common-Pool Resources, or simply "commons" (Hardin 1968, 1985; Ostrom 1990), which are not distributed based on some indicator of merit or need but freely exploited for individual activities, in order to achieve personal objectives whose outcome depends on the use of the resource. It is obvious that the more the resource is exploited individually the less is available collectively, so that even the activities that employ the resource are less efficient, and thus its use becomes increasingly "less economic." To achieve the objectives in a situation of lesser efficiency, the appropriators must intensify their activity, continuing to exploit the resource, no matter how scarce, until it is completely exhausted. The micro–macro feedback is evident: individual appropriation leads to the scarcity of the common good, but this macro effect leads individuals to intensify their individual appropriation. The result is a *tragedy*, since those who exploit the resource are forced to interrupt or reduce their activity because of the exhaustion. To better understand the action of this combinatory system and highlight the role of selfish and myopic individual decisions, it is useful to illustrate the simplified situation where there are only two agents, A and B, with external controls, as shown in Fig. 7.11, and where we assume the conclusions apply to any number of agents.

Technical Note There are different ways of representing this combinatory system. I prefer the model in Fig. 7.13 to that presented by Peter Senge (1990, p. 446), since the latter does not highlight the two individual control systems, that is, the fact that A and B have their own objectives even in the presence of diminishing returns from the use of the common resource. In the context of control systems, I feel that Senge's model might be completed as indicated in Fig. 7.13.

Referring to Fig. 7.11, the loops [**B1**] and [**B2**] are, in fact, individual control systems that show that the individual actions of A and B are undertaken to achieve a personal objective; if the common resource is abundant and the marginal performance is high, then these activities will produce a result in line with the objective. If there is no gap between the actual and desired result, A and B have no reason to intensify their resource-exploiting activities. Since we have assumed a limited availability of the common resource, the combined activities of A and B lead to a gradual reduction in the marginal performance of the resource, which is usually perceived with a (sometimes even considerable) delay.

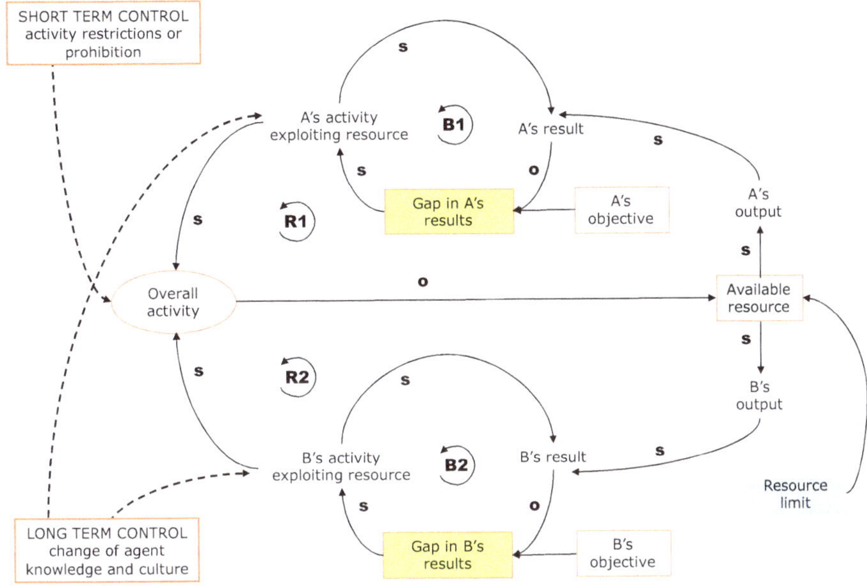

Fig. 7.11 Archetype of the Tragedy of the Common Resources. (*Source*: adapted from Mella 2012, p. 251)

Assuming they undertake the same activities, A and B see that their individual results clash and fail to achieve their objectives, which produces a distance, $E(Y)$, between expectations–objectives and results. This gap forces them to intensify their activities, which acts as a lever to achieve their expected outcomes. In any event, the efforts by A and B to pursue their objectives generate the reinforcing loops [**R1**] and [**R2**]—which correspond to the micro–macro feedback (divided into two loops to simplify the figure)—whose action inevitably leads to the exhaustion of the common resource, with negative effects for the individual and collective outcomes.

Figure 7.12 represents the whale hunting process (this example can be adapted to the hunting of elephants for ivory, seals for fur, etc.) under the assumption that the reproduction rate of the population of whales is lower than the hunting intensity, so that the simultaneous action of many whale hunters reduces the number of whales before they have had the chance to reproduce in adequate numbers, thereby diminishing the returns from hunting.

When the returns from hunting start to diminish, the whale hunters, in order to fill their oil and meat holds (the objective of every whale hunter), are pushed to intensify the hunt (remaining longer at sea), as shown in loops [**B1**] and [**B2**], until the whales reach the minimum admissible population, as shown in loops [**R1**] and [**R2**]. The two control systems, [**B1**] and [**B2**], can have different modes of action, all inevitably destined to result in the extinction of the scarce resource. The hunters could, for example, intensify their hunting activity not only by devoting more time to it but also by improving the quality of their equipment—radar, sonar, satellites,

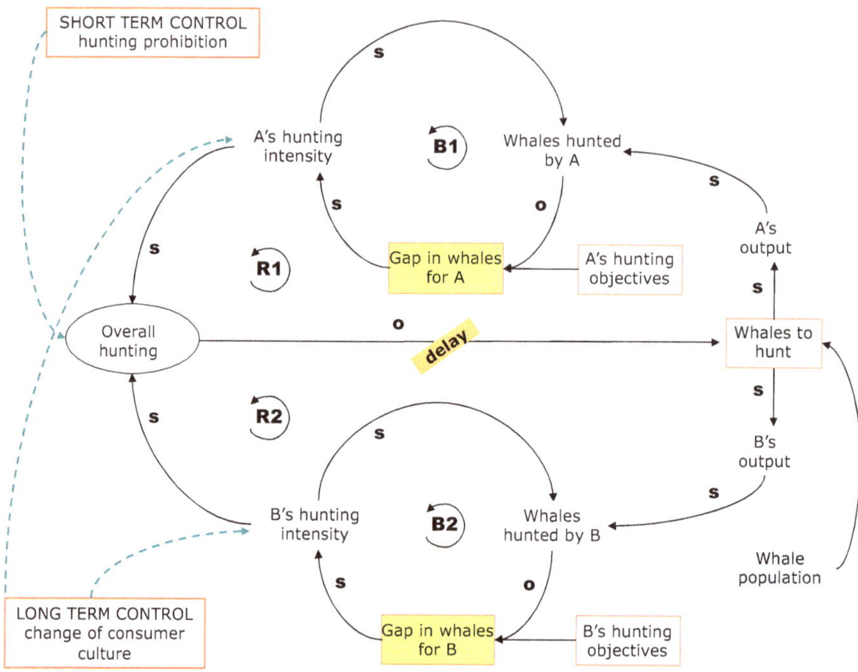

Fig. 7.12 Tragedy of the Common Resources in whale hunting. (*Source*: Mella 2012, p. 253)

harpoon cannons, or teleguided missiles—with the result that the whales, no matter how rare they become, are inexorably identified and hunted. The result? There is no way out for the whales, unless controls external to the combinatory system intervene.

The *external* macro control at the collective behavior level could operate by implementing several levers; for example, a ban on whale hunting by world authorities, or a ban on the consumption of whale meat and fat by individual countries. The external micro control could operate, for example, by granting subsidies to convert whaling ships to fishing vessels, or by initiating information campaigns to discourage consumers from eating whale meat, thereby making whale hunting less convenient from an economic point of view. In conclusion, in such combinatory systems individual interests guide the individual control systems that provide the impetus for the collective exploitation of scarce resources until they are depleted.

We can also more significantly represent the tragedy of the risk of extinction of whales by employing the model of the 3 myopias, illustrated in Fig. 7.13, which also highlights the opposition levers of the archetype (for a better understanding of the model, see Sect. 7.10.4).

Opposition to whale hunting The International Whaling Commission prohibits and/or places quotas on whale hunting and develops a common strategy of incentives and disincentives to control this hunting; consumption at high prices; the trade in meat as a food source; profits; demand by hunters; investments to exploit the hunting; incentives and information to change the production functions and the

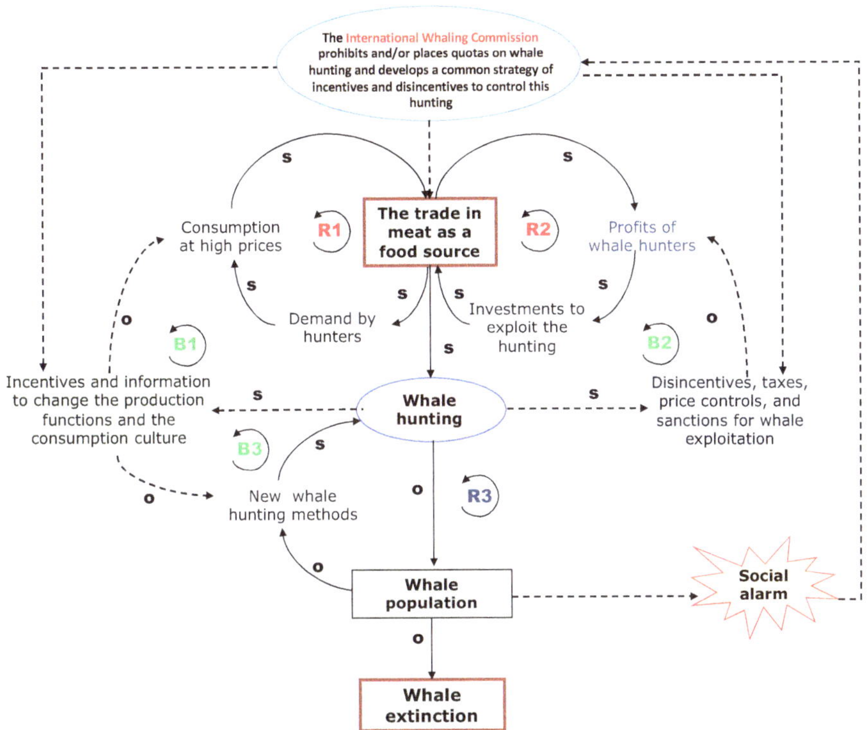

Fig. 7.13 The tragedy of the risk of extinction of whales

consumption culture; whale hunting; disincentives, taxes, price controls, and sanctions for whale exploitation; new whale hunting methods. In 1986, after years of struggle by Greenpeace and other associations, a worldwide ban on whale hunting was enacted. Japan ignored this ban under the pretext of "*scientific research,*" and in 2019 the Japanese government reinstated whale hunting for commercial purposes.

In The Tragedy of the Commons (1968), Garrett Hardin also emphasizes the fact that populations that share a common vital resource almost always head toward tragedy.

> Adding together the component partial Utilities, the rational herdsman concludes that the only sensible course for him to pursue is to add another animal to his herd. And another; and another... But this is the conclusion reached by each and every rational herdsman sharing a commons. Therein is the tragedy. Each man is locked into a system that compels him to increase his herd without limit—in a world that is limited. Ruin is the destination toward which all men rush, each pursuing his own best interest in a society that believes in the freedom of the commons. Freedom in a commons brings ruin to all (Hardin 1968, p. 1244).

If we observe the world around us, we see this archetype in operation in many situations: too many fishermen fishing in waters with limited populations of fish; the search for oil, minerals, and precious stones in the same area and, in general, the exploitation of limited natural resources by many users. We should not wonder if the tropical forests are reduced because their inhabitants try to transform them into arable land; if the aquifer levels in the desert fall because in many areas irrigation systems are built that take water from them; if rivers run dry because alongside them ever larger areas are transformed into rice fields; if the water-bearing strata in the deserts are being lowered because in many places irrigation plants are being built that take water from those strata; or if rivers are drying up because along their banks larger and larger areas are being turned into rice fields. It appears that even Easter Island was abandoned after its inhabitants destroyed all the island's wood resources in order to build their Moai. This archetype also operates in the business field, since its effects are completely analogous to those produced by "arms" growth, where firms go "hunting" for consumers in a market heading toward the saturation point.

The presence of limited common resources can cause conflicts that represent a *second tragedy* included in the tragedy of the commons. The clearest evidence of this are the numerous wars fought over water and oil, which are among the most limited and scarce vital goods on earth.

How can the action of the archetype be countered? There is only one answer: the managers/decision-makers must externally control the combinatory system that produces the harmful effects that must be avoided by specifying several useful levers to reduce conflicts and manage scarcity:

- Delimiting the boundaries of the common resource, restricting and possibly limiting the appropriators
- Specifying the technologies that can be used to exploit the resource, placing restrictions and quantitative and time limits on its allocation
- Providing incentives for resource saving and, when possible, for its regeneration
- Setting up an outside authority to accurately and continuously monitor individuals to ensure they are respecting the rules for resource allocation and sanctioning violations of the rules for resource exploitation
- Discouraging conflicts between resource users and seeking to resolve conflicts that have broken out in a peaceful manner
- Encouraging collective participation in the choices regarding the above points in order to favor agreements on collective exploitation, self-discipline, and shared social oversight

7.8 *Rings* Operating in Social Systems as Complex Adaptive Systems (CAS)

Social systems, as autopoietic systems, must not control the homeostasis of the fundamental variables for the survival of the individuals and groups that make up the system, nor limit themselves to creating the *Rings* for achieving the autopoietic macro objectives based on the *policy* decided by the governance (Sect. 7.3). Social systems can survive for a long time only if they react and adapt successfully to the disturbances (disasters, wars, radical technological innovations, etc.) and threats their structure can undergo, which the individuals in the system must face.

Every structural disturbance influences above all the autopoietic organization of the social system, the interpersonal relations and the governance; this obliges the individuals themselves, the groups, and the various coexisting autopoietic subsystems to react and adapt their mutual behavior and their relationships based on new forms of coexistence in order to survive, thereby also changing the political structure (rules and social control systems) and the economic structure (autopoietic objectives). The individual adaptive reactions lead to further changes in the social autopoietic structure which necessitates further individual adaptation, based on reciprocal control systems that also produce numerous cycles that end when the social system has absorbed the disturbances and found new homeostatic equilibriums between the individual behaviors and the basic variables for the system's preservation. These new equilibriums are maintained until new disturbances arise that require further adaptations.

Every social system, though composed of a number of agents with different individual features, social roles, economic motivations, different group memberships, and so on, represents a complex entity, of a *variable organizational nature*, capable of being "malleable" and thus adapting to disturbances (which do not overwhelm the system). In this regard, it can be considered a CAS with the capacity for *structural adaptation* and *organizational evolution*, whereby it produces new policies and strategies to control the cohesion and homeostasis of the autopoietic variables, as shown in this concise definition:

> Definition (1): A CAS consists of inhomogeneous, interacting adaptive agents. Adaptive means capable of learning.
>
> Definition (2): An emergent property of a CAS is a property of the system as a whole which does not exist at the individual elements (agents) level. Typical examples are the brain, the immune system, the economy, social systems, ecology, insects swarm, etc.
>
> Therefore to understand a complex system one has to study the system as a whole and not to decompose it into its constituents. This totalistic approach is against the standard reductionist one, which tries to decompose any system to its constituents and hopes that by understanding the elements one can understand the whole system (Ahmed et al. 2005, pp. 1–2).

According to Murray Gell-Mann (1992, 1994, 1995), the category of CAS should also include all the basic components of this system as well as individuals who are capable of surviving by adapting their behavior and producing new *schema of interaction* and coexistence that allow this behavior to be *predicted* and adapted to.

> Now how does a complex adaptive system operate? How does it engage in passive learning about its environment, in prediction of the future impacts of the environment, and in prediction of how the environment will react to its behavior?
>
> [...] The answer lies in the way the information about the environment is recorded. In complex adaptive systems, it is not merely listed in what computer scientists would call a look-up table. Instead, the regularities of the experience are encapsulated in highly compressed form as a *model* or *theory* or *schema*. Such a schema is usually approximate, sometimes wrong, but it may be adaptive if it can make useful predictions including interpolation

and extrapolation and sometimes generalization to situations very different from those previously encountered. In the presence of new information from the environment, the compressed schema *unfolds* to give prediction or behavior or both (Gell-Mann 1992, p. 10).

In addition to what Gell-Mann had to say, I would observe that CAS have been recently studied in the context of the theory of complex systems. Briefly, I would note that the CAS *approach* studies how complex systems interact and exchange information with their environment in order to maintain over time their internal structure and the network of vital processes and to develop a form of social cognition. The term CAS refers to a system (not exclusively of a social nature but of an organizational one as well) with the following properties:

- It is composed of a large number of primitive components, or "agents," different in nature (men, animals, plants, robots, scientific theories, neurons, etc.);
- Their number is not always fixed, so that the system can often be considered open; thus, it may be difficult or impossible to define system boundaries;
- It produces many types of different interactions among the agents and between the agents and their environment; these interactions are in the form of reinforcing and balancing loops, which make the interactions *non-linear* since, as we know, small actions by the agents can produce significant changes in the system;
- The agents are structurally coupled to other agents and to the environment, and they are subject to many environmental constraints;
- As a consequence of the interactions among the agents, the system's behavior evolves over time;
- Unanticipated *global properties or patterns emerge* as a result of often non-linear spatial–temporal interactions among a large number of component systems at different levels of organization.

There are three basic characteristics of a CAS which allow us to interpret and understand the capacities for self-control and lengthy survival in social systems:

- A CAS comprises *different types* of agents that try to adapt their behavior to that of the other agents they can observe adapting to one another "*…as a process of reciprocal selection of congruent paths of structural changes in the interacting systems which result in the continuous selection in them of congruent dynamics of state*" (Maturana and Guiloff 1980, p. 139);
- The agents' micro behavior is meant to adapt to disturbances only by following certain *local schema* (rules and strategies), not necessarily uniform across all agents, which can change over time when this favors adaptation; "*More generally, it is significant that any* CAS *is a pattern-recognition device that seeks to find regularities in experience and compress them into schemata*" (Gell-Mann 1994, p. 22);
- The processes of adaptation to environmental perturbations generates a succession of macro mutations in the system that enables it to *evolve* over time, thereby producing an evolutionary "history"; the dynamics of this evolution develop from the reciprocal local interactions according to local rules and strategies established over time or to variants produced by individual adaptation. Normally

the agents are not aware of the macro behavior of the entire system, adapting instead to that of their neighbors within a restricted range of knowledge.

> To break away from the paradigm of top-down modeling, we enter the science of complexity and its applications in complex adaptive systems theory. Instead of the top-down approach so entrenched in contemporary simulation and modeling, complex adaptive systems theory is based on the notion of model building from the bottom up. After three hundred years of dissecting everything into molecules and atoms and nuclei and quarks, they [scientists] finally seemed to be turning that process inside out. Instead of looking for the simplest pieces possible, they were starting to look at how those pieces go together into complex wholes (Waldrop 1992, p. 16).
>
> Complex systems are usually systems which have been created by evolution or an evolutionary process. Evolved systems which have a long historical background are nearly always complex. Complexity can be found everywhere where evolution is at work,

- in all living organisms which are subject to evolution
- in all evolving complex adaptive systems which have a long history
- in systems that have grown over a long period of time (CASG 2013, Online).

The term *complex evolving system* may be used to distinguish human from other CAS. In particular, *complex evolving systems* refer to those systems which are able to learn and which change their internal structure and organization over time, thus changing the behavior of the individual elements (Allen 1997). In social systems in particular, mutual adaptation among the individuals–agents not only is due to adaptation that is a reaction to the disturbances but derives from *forecasts*, *prospects*, and *expectations* regarding the behavior of the other individuals–agents, as is well stated by John Holland, the pioneer of CAS studies:

> Here we confront directly the issues, and the questions, that distinguish CAS from other kinds of systems. One of the most obvious of these distinctions is the diversity of the agents that form CAS. Is this diversity the product of similar mechanisms in different CAS? Another distinction is more subtle, though equally pervasive and important. The interactions of agents in CAS are governed by anticipations engendered by learning and long term adaptation (Holland 1995, p. 93).

It is easy to understand that the process of *individual adaptation* presupposes the action of innumerable *Rings* operating recursively, at different levels, among pairs of individuals and/or groups to allow the agents to recognize the dynamics and construct the *schema* (models which are stable but adaptable to the circumstances) to achieve a state in conformity with them, as shown in Fig. 7.14.

This model also enables us to see that the activity of each *complex adaptive agent* is conditioned by the agent's capacity to form and direct the evolution of the *schema* of adaptation on which the chosen coordination strategy depends, in order to adapt his behavior to the changing behavior of the other agents.

The repeated use of the innumerable *Rings* needed for adaptation and the continual modification of the *schema* of adaptation by all agents produces a form of *social learning*, which also makes increasingly more efficient and effective the action of control systems that operate recursively in order to allow the agents to achieve their individual objective through mutual adaptation in reaction to the observed behavior of the other individuals in the social system:

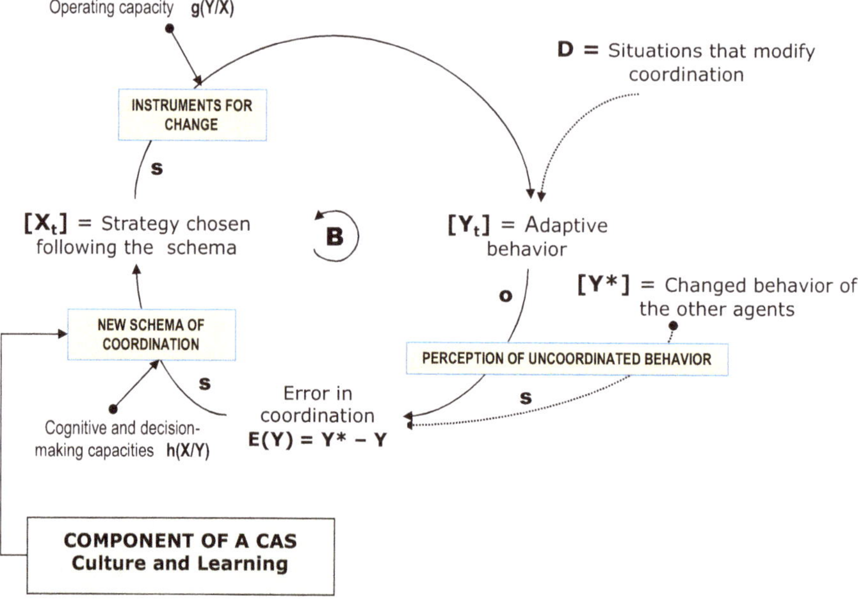

Fig. 7.14 The dynamic coordination of the agents in a CAS

> […] social learning is based upon either extrinsic or intrinsic factors or motivations. Intrinsic social learning occurs in social facilitation. It occurs when people are led by others to acquire new means (intrinsic) for achieving old goals. Extrinsic social learning occurs in imitation, when people learn through the observation of attractive and consistent social models. By observing their social models and recording when these apply reinforcing mechanisms (extrinsic), people learn to reinforce themselves (self-reinforcement) to do what others have reinforced, and abstain from doing what others have punished (Conte and Paolucci 2001, par. 2.2).

CAS operate in every context and provide interpretative models of the numerous varieties of phenomena in the world. It is useful to propose the excellent example of CAS presented by John Holland in the following extensive quote:

> On an ordinary day in New York City, Eleanor Petersson goes to her favorite specialty store to pick up a jar of pickled herring. She fully expects the herring to be there. Indeed, New Yorkers of all kinds consume vast stocks of food of all kinds, with hardly a worry about continued supply. This is not just some New Yorker persuasion; the inhabitants of Paris and Delhi and Shanghai and Tokyo expect the same. It's a sort of magic that everywhere is taken for granted. Yet these cities have no central planning commissions that solve the problems of purchasing and distributing supplies. Nor do they maintain large reserves to buffer fluctuations; their food would last less than a week or two if the daily arrivals were cut off. How do these cities avoid devastating swings between shortage and glut, year after year, decade after decade? The mystery deepens when we observe the kaleidoscopic nature of large cities. Buyers, sellers, administrations, streets, bridges, and buildings are always changing, so that a city's coherence is somehow imposed on a perpetual flux of people and structures. Like the standing wave in front of a rock in a fast-moving stream, a city is a pattern in time. No single constituent remains in place, but the city persists. To enlarge on the previous question: What enables cities to retain their coherence despite continual disrup-

> tions and a lack of central planning? There are some standard answers to this question, but they really do not resolve the mystery. It is suggestive to say that Adam Smith's "invisible hand" or commerce, or custom, maintains the city's coherence, but we still are left asking How? (Holland 1995, p. 1)

Holland's answer is that, naturally, the phenomenon described is an example of order, which can be interpreted as the action of a CAS. From these few observations we can easily see the basic differences between combinatory systems and CAS.

Firstly, combinatory systems do not necessarily present phenomena of adaptation but, generally, some form of self-organization due to the *micro–macro feedback*, which is the adaptation of agents to a synthetic variable produced by the macro behavior of the system. A *second* difference is observable also as regards the similarity of the agents. Combinatory systems are made up of similar agents, while, as Holland notes, CAS are composed of heterogeneous agents: "*Here we confront directly the issues, and the questions, that distinguish* CAS *from other kinds of systems. One of the most obvious of these distinctions is the diversity of the agents that form* CAS" (Holland 1995, p. 93).

The *third* main difference regards the absence of interactions among the agents; in combinatory systems, agents normally interact only with some macro variable and not each other. The *fourth* relevant difference is that the theory of CAS observes the macro effects of the system produced by the agents that follow a *schema* or change the schema previously followed. Any micro–macro feedback between the micro behaviors and the schema is considered as a relevant characteristic. *Finally*, ignoring the micro–macro feedback implies that CAS theory only focuses its attention on *necessitating* factors and ignores the *recombining* ones.

Taking account of these differences, we can conclude that combinatory systems can be viewed as a particular class of CAS. Nevertheless, combinatory systems can represent, precisely due to their simple logic, which can easily be observed in nature and easily described and simulated by using combinatory automata, a useful theoretical model to be applied in the study of populations of similar agents, though they are not applicable to complex social systems (Sect. 7.10.3).

7.9 Change Management in a Complex World: The PSC Model

Populations, societies, and collectivities are continually evolving complex systems made up of interconnected social, political, and economic entities (holons) that form a network (holarchy) of social, political, and economic processes that produce a continuous exchange of goods, knowledge, and value in order to regenerate the autopoietic social system (Sect. 7.3). In this interconnected world, the basic components of society are ceaselessly regenerated and the network continually redesigned; this continuous process of adaptation produces an inexorable *change* in the entire society.

> Change in the substance of any of the relationships affects the overall structure. Since a change in any relationship affects the position of those involved, the whole set of interrelated relationships is subject to change and that has consequences for the outcome of a relationship for those involved. A dyad, a relationship, is a source as well as a recipient of change in the network. […] (Håkansson and Snehota 1995, p. 41).

> We live in a changing world. People grow, business conditions fluctuate, personal relationships develop, major industries come and go, population increases, water shortages worsen, and scientific discoveries increase. Our principal concerns arise from change – growth, decay, and fluctuation (Forrester 1983, Foreword, p. IX).

Traditional processes are abandoned while other processes are added, spurred on by the *new technologies*, which change the way social units and organizations operate (Hayes and Jaikumar 1988) according to the model in Fig. 7.15. Not only do the cultural bases of individuals, societies, and organizations change, but the social units also evolve through modification of their structures, processes, and output, in the attempt to loosen the old restraints while setting new objectives and rewriting the programs for their achievement.

Change to the culture of a social system and of an organization assumes an individual and social *learning* process that not only depends on the motivations for change but also influenced by the "living" and the "personality," as Talcott Parson clearly states:

> The crucial point for the present is that the "learning" and the "living" of a system of cultural patterns by the actors in a social system, cannot be understood without the analysis of motivation in relation to concrete situations, not only on the level of personality theory, but on the level of the mechanisms of the social system.
>
> There is a certain element of logical symmetry in the relations of the social system to culture on the one hand and to personality on the other, but its implications must not be pressed too far. The deeper symmetry lies in the fact that both personalities and social sys-

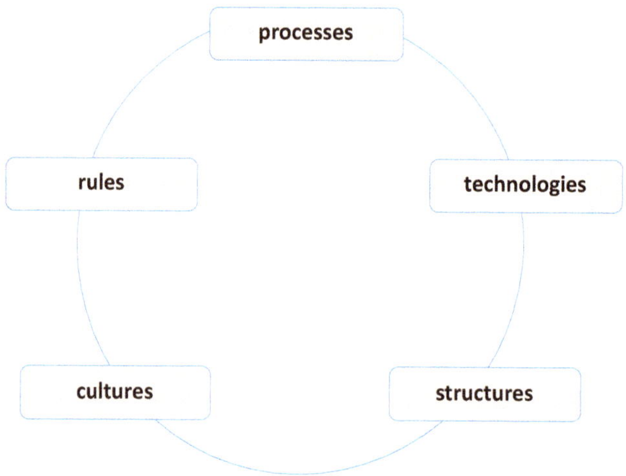

Fig. 7.15 The aspects of change in social environments (*Source*: Forrester 1983, Foreword, p. IX)

> tems are types of empirical action system in which both motivational and cultural elements or components are combined, and are thus in a sense parallel to each other (Parsons 1951, p. 17).

Without going too far back into economic history, we can observe that social units and structures, organizations in particular, must deal with change in their environment. Business organizations, in particular, must operate in a more competitive context of heightened change and growing globalization (Porter 2008).

Societies and organizations can, therefore, be conceived as systems in which the change process is incorporated (Liguori 2012). I suggest extending this concept to consider organizational change mechanisms alone as a control system, following the model shown in Chap. 2, Fig. 2.16, which allows researchers to understand and describe the qualities of change management (Robbins 1987). From this point of view, the change manager is seen as the one who, through a series of decisions, develops a strategy to identify the change drivers, "Xs," necessary to change the current state "Y_t." Thus, in order to survive in a dynamic and changing world, social units and organizations must develop their ability to quickly understand change and react to unfavorable triggers, developing an intelligent behavior; that is, they must transform themselves into learning organizations (Senge 2006) by adapting or innovating their own structures and, above all, their own cultures. The change management process is conceived of as:

> The systematic approach and application of knowledge, tools and resources to deal with change. Change management means defining and adopting corporate strategies, structures, procedures and technologies to deal with changes in external conditions and the business environment (SHRM 2020 Online).

This process becomes physiological, for individuals (Lewin 1947a, b), social groups, or organizations (Senge 2006), and it represents the natural approach for dealing with change both at the individual and organizational levels for all types of organization (Hiatt 2006).

For Kurt Lewin:

> A change towards a higher level of group performance is frequently short lived; after a 'shot in the arm,' group life soon returns to the previous level. This indicates that it does not suffice to define the objective of a planned change in group performance as the reaching of a different level. Permanency at the new level, or permanency for a desired period, should be included in the objective (Lewin 1947a, p. 228).

For this reason Lewin proposed a three-step model of change:

Step 1: *Unfreezing*. Unfreezing from a position of harmful equilibrium that impedes change in groups is necessary to create the motivation for change; however, this does not necessarily produce and control change. Unfreezing is the necessary condition for overcoming the barriers to organizational change and initiating change (Argyris 1993).

Step 2: *Moving*. Even when the situation to change is unfrozen, the directions of change are not always clear. The control process toward new desired states appears difficult, and for this reason Lewin recommends that the groups move forward gradually, recognizing all the forces at work and identifying and evaluating, on a trial and basis, all the available options. This is the perfect description of a control system for change, which produces a gradual shift to move a group from a situation that has been unfrozen to a new, evolving one. Change management is the management of this control system, which must move toward the objectives of change established by the governance.

Step 3: *Refreezing*. In this third step, the results of the changes judged to be satisfactory must be refrozen in order to stabilize the group at a new quasi-stationary equilibrium in order to maintain the new state relatively safe from regression.

The above considerations for group and organizations can refer, on a smaller scale, even to individuals. The survival of man, whether as an individual or as part of an organized group, depends on his ability to understand and dominate the changes in the variables that influence his choices and behavior, and he must develop the capability of being part of a network of knowledge and ideas. Therefore, such cognitive activities are very difficult today because change has become "complex"; individual, collective, and environmental dynamics interact, giving rise to reinforcing and balancing loops that form "complex" systems of variables that produce outcomes which are less and less understandable and predictable. In the face of environmental pressures, the capacity of the social systems to survive at length—that is, to maintain themselves as viable systems (Beer 1979) and develop teleonomy based on Jacques Monod's (1970) conception—depends on the ability of the managers of change to understand the change and regenerate the internal vital processes. According to Ross Ashby's well-known requirement of *necessary variety* (Sect. 5.3) in a world of increasing dynamics and complexity all self-directed social systems must be "reviewed" and strengthened through an effective action of *change management*.

Change management is not a simple process but involves several fundamental phases that can be listed as follows:

1. Ascertaining the level of dissatisfaction with the present situation and the need for change;
2. Recognizing the optimal situation to achieve (vision);
3. Identifying the paths of change;
4. Structuring a strategy of change;
5. Identifying the person in charge of the process and setting up the task force of change management;
6. Undertaking change and developing the actions of change;
7. Controlling the adequacy of the implemented changes in order to achieve the desired optimal situation;
8. Stabilizing the changes obtained if these are held to be satisfactory, or returning to phase 2.

Apart from the general method of the three steps proposed by Kurt Lewin, different strategies have been presented to direct the fulfillment of these steps—the ADKAR model (Hiatt 2006), the McKinsy "7S" model (Waterman et al. 1980), Kotter's model (Kotter 1996) as well as others—but all models show that the manager of change should have the ability to recognize and meet the demands of change, possess the will to make change happen, and gain the consent of the parties involved in the change. Interpreted from the control system view, these models lead to the recognition that change management is a general *control process* to *move the system to be changed* toward the objectives that allow them to survive.

The change management model proposed by John Kotter originates from his work *Leading Change: Why Transformation Efforts Fail* (1995), which describes the outcomes of numerous analyses he had undertaken over the years involving more than 100 different organizations during the implementation of change projects. The article describes *eight errors* the changing organization makes which can be fatal to the success of the process. The model is subsequently presented in the book titled *Leading Change* (1996, 2nd ed. 2012), in which Kotter modifies these eight errors, changing them into the *eight prescriptions* that underlie his model, each fundamental to moving on to the next one. The prescriptions can be summarized as follows.

1. *Establishing a Sense of Urgency*. In this early phase, it is very important to have a charismatic leader who knows how to exploit the dissatisfaction and discontent of the members of the organization to convince them of the need to support a change that improves the current condition. "*A paralyzed senior management often comes from having too many managers and not enough leaders. [...] Change, by definition, requires creating a new system, which in turn always demands leadership*"(Kotter, 1995, pg. 60). Projects often fail because managers who decide to support change fail to create a sense of urgency that is motivating enough for members of the organization who are inhibited by the fear of getting out of their comfort zone and changing their consolidated and routine habits into something new and unknown. Very often, however, it is the managers themselves who fear change because of what might happen to their position and privileges. "*When is the urgency rate high enough? From what I have seen, the answer is when about 75% of a company's management is honestly convinced that business-as-usual is totally unacceptable*" (Kotter, 1995).
2. *Creating the Guiding Coalition*. The change must be supported by a cohesive group of people who are sufficiently motivated to defend the project and carry it forward with conviction and perseverance. Initially few will join, but gradually, if a strong leader succeeds in taking advantage of the discontent properly, the number of participants will increase and the size of the group will grow considerably.
3. *Developing a Vision and Strategy*. "*In every succesful transformation effort that I have seen, the guiding coalition develops a picture of the future that is relatively easy to communicate and appeals to customers, stakeholders and employees,*" Kotter writes as he prepares to describe the third most common mistake: failure to come up with a vision. This step is very important to secure the support of the members of the organization, who are the ones the change is most directed at and who are most affected by it. The definition of a vision allows stakeholders to identify the objectives toward which their efforts should be directed, and for this reason it must be communicated clearly and accurately to make it understandable to all.

4. *Communicating the Change Vision*. All corporate communication systems must be used to communicate the vision, and managers should make the message more realistic to show that the objectives really exist and are not merely words.
5. *Empowering Employees for Broad-Based Action*. Managers must help individuals overcome the psychological blocks they are subject to, since the communication of the vision very often is not enough to eliminate the uncertainties. Most limitations arise from the fear among members of the organization that they are not up to the task or not prepared to get out of their routine.
6. *Generating Short-Term Wins*. In long-term projects, the lack of positive results in the short term can be a demotivating factor. "*Real transformation takes time, and a renewal effort risks losing momentum if there are no short term goals to meet and celebrate. Most people won't go on the long march unless they see compelling evidence within 12 to 24 months that the journey is producing expected results. Without short-term wins, too many people give up or actively join the ranks of those people who have been resisting change*" (Kotter, 1995, p. 65).
7. *Consolidating Gains and Producing More*. It is necessary to avoid the mistake of considering the process of change concluded as soon as the first "signs of victory" appear and the first positive results are observed, because this produces a decrease in engagement and the disappearance of the initial motivations.
8. *Anchoring New Approaches in the Culture*. In accordance with Kurt Lewin's concept of "refreezing," changes at the end of the change process must be incorporated into the organizational culture, which thus conforms to the positive results of change management. If the culture of the organization remains unchanged after the change management process, the old behavior and routines will return to guide the organization, thereby decreeing the failure of change in the long term. Therefore, the new, modified culture must ensure that new managers learn, share, and become the promoters of change themselves.

> I realize that in a short article everything is made to sound a bit too simplistic. […] But just as a relatively simple vision is needed to guide people through a major change, so a vision of the change process can reduce the error rate. And fewer errors can spell the difference between success and failure (Kotter, 1995, p. 67).

In his book "ADKAR: *a Model for Change in Business, Government, and Our Community*" (2006), Jeffrey Hiatt proposes a new model of change management called ADKAR, an acronym referring to the five objectives to achieve and complete to have success in a change project:

1. *Awareness*, that is, the awareness of those affected by the change regarding the reasons that made it necessary, the benefits it can bring, and the risks that would be faced if the process were not completed.
2. *Desire*, that is, the desire and will to participate in change.
3. *Knowledge*, which entails the knowledge, skills, techniques, or specific roles that every member of the organization must have to move change forward.

4. *Ability*, that is, the capacity to transform knowledge and skills into actions, and thus to implement change from a practical point of view.
5. *Reinforcement*, including the recognition in support of change provided by external factors (such as awards and celebrations) and internal ones (e.g., the personal satisfaction experienced by the proponents of change or the observation of the benefits deriving from change).

The sequencing of the various phases is not invertible since it represents the natural order in which an individual experiences and faces the change process.

Change is not always an accepted process within organizations. Rosabeth Moss Kanter identifies ten main reasons that make a subject hostile to change:

1. *Loss of control*. Change is perceived as something that could call into question independence in work, undermining that sense of self-determination needed in one's working and private life.
2. *Excessive uncertainty*. This is a particularly feared factor that pushes people to prefer a stable, even if non-optimal, situation to one that would bring only hoped-for benefits.
3. *Surprises*. Changes that occur too abruptly cause alarm among those concerned, who should instead have time to evaluate and accept change.
4. *Everything looks different*. Change undermines the routines that guide people's lives and make their job tasks familiar, and for this reason it must be limited to some necessary aspects, without changing the entire routine of the persons involved.
5. *Degradation*. Change could be perceived as the "correction" of a wrong situation and interpreted by those involved as the consequence of their own personal failure.
6. *Concerns about capabilities*. People often resist change because they feel inadequate and unable to support it, that they do not have the right skills to deal with it, or that it would take a great effort to acquire these skills.
7. *Increased workload*. Employees often fear that more effort may be required and that this will not be properly rewarded.
8. *Chain effect*. Any change in the current situation of the organization necessarily impacts other functions, some of which even distant from those directly affected; any possible resistance in this regard must be carefully evaluated.
9. *Old resentments*. There is often unresolved friction between colleagues in the organization that could resurface and create resistance when collaborative action is required.
10. *Sometimes the danger is real*. Change is not an advantage for everyone: for some, resistance is justified as change will cause redundancies or relocations.

Kanter points out that the main task of the leaders of the change process is not to make people feel totally comfortable with the change, which is practically impossible, but to minimize discomfort:

> Although leaders can't always make people feel comfortable with change, they can minimize discomfort. Diagnosing the sources of resistance is the first step toward good solutions. And feedback from resistors can even be helpful in improving the process of gaining acceptance for change (Kanter 2012, online).

The phases for carrying out a change management process are by now standardized and well-consolidated in doctrine, but I feel it is useful to reconsider the central phases in the process, which are indicated in points 3 and 4 (3. Identifying the paths of change; 4. Structuring a strategy of change), in order to place them in a dynamic pattern. With regard to those phases in the process, we must recognize that there are *three paths* for achieving change, each of which identifies different *control levers* that form the strategy of change and act at *different speeds*, as if they were interconnected *wheels* of an inexorable process of change which leads to change at different speeds as shown in the model in Fig. 7.16; the model takes into account the three fundamental "wheels of change," which act at different speeds but with a power of change inversely proportional to the speed itself.

1. The "wheel of change" in the internal operational programs and *processes*, understood also as norms, programs, regulations, instructions, etc.; this is the small wheel, easy to move but of limited effectiveness.
2. The "wheel of change" in the *structures* of systems; that is, in the organization of the organs and individuals; this is the medium-sized wheel, which is moved by the small one, though at a lower speed.
3. The "wheel of change" in the *culture*; that is, in the cognitive and behavioral models of the individuals or groups that participate in the social and organiza-

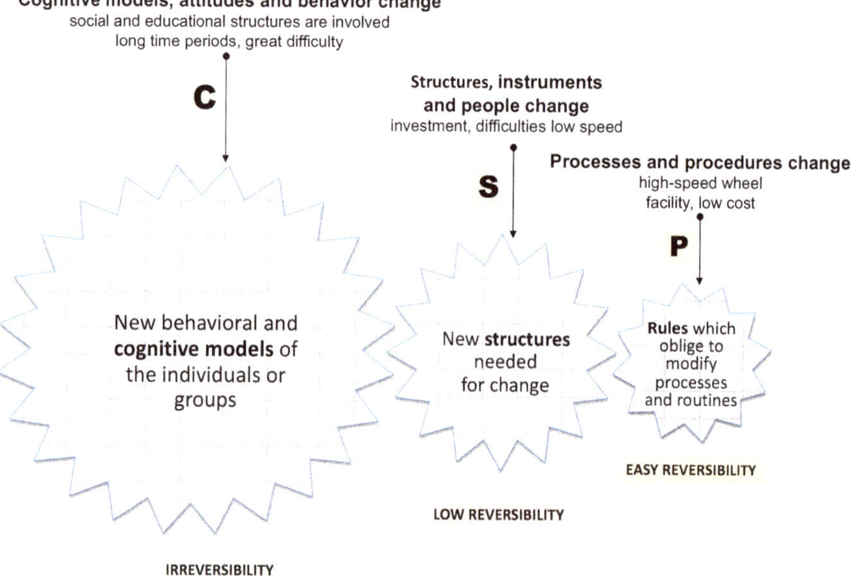

Fig. 7.16 The PSC model showing the wheels of change in social systems

tional structure; this is the largest and slowest wheel, with small movements that produce great changes.

I refer to model in Fig. 7.16 as the PSC Model (Processes, Structures, Culture). When a society or an organization succeeds in effectively influencing the culture of those belonging to it, then the new cognitive models (large and slow wheel) trigger a fast change in the structure (middle wheel), which, in turn, produces an ever faster adaptation in the programs and procedures (smallest wheel). Though slowly, the structure wheel of change turns the largest wheel, culture change, and this slow change produces the cultural innovation that modifies the foundation itself of society and organizations.

The concept of culture has been defined, in general terms, by many scholars, but it seems useful to touch on the precise contribution of Edgar Schein, whose thinking strongly focused on the concept of *organizational culture*.

> Culture can now be defined as (a) a pattern of basic assumptions, (b) invented, discovered, or developed by a given group, (c) as it learns to cope with its problems of external adaptation and internal integration, (d) that has worked well enough to be considered valid and, therefore (e) is to be taught to new members as the (f) correct way to perceive, think, and feel in relation to those problems (Schein 1990, p. 7).

He argued that the standards and values that characterize an organization are chosen from a range of options, each of which aim at the same objective and involve different levels of anxiety. Anxiety is defined as a symptomatic manifestation of stress, of which uncertainty is one of the main causes. In fact, it is precisely the uncertainty, and therefore the difficulty in predicting the outcome of a particular change, that triggers resistance to it.

> Every group and organization is an open system that exists in multiple environments. Changes in the environment will produce stresses and strains inside the group, forcing new learning and adaptation. [...] To some degree, then, there is constant pressure on any given culture to evolve and grow. But just as individuals do not easily give up the elements of their identity or their defense mechanisms, so groups do not easily give up some of their basic underlying assumptions merely because external events or new members disconfirm them (Schein 1990, p. 23).

On this line of thought, Peter Senge writes:

> Beliefs and assumptions, established practices, skills and capabilities, relationships, and awareness and sensibilities are five elements of culture that influence one another and impact the success of an organization … While these taken-for-granted ways of seeing the world (beliefs) are often invisible to those who hold them, they shape organizational practices, guide how people do things, and, in turn, determine what skills and capabilities people develop based on those organizational practices (Senge 2006, p. 285).

Organizational values, beliefs, and norms that shape people's behaviors—individually and in groups—which support the analysis, definition of acting mechanisms (levers of control) to deal with change for the overall benefit of the organization.

> Values are the social principals, goals, and standards that cultural members believe have intrinsic worth. They define what the members of a culture care about most and are revealed by their priorities (Hatch and Cunliffe 2013, p. 169).

> Values influence every aspect of our lives: our moral judgments, our responses to others, our commitments to personal and organization goals. Values set the parameters for the hundreds of decisions we make every day (Kouzes & Posner 2002, p. 48).

> Norm are expressions of values…While values specify what is important to the members of a culture, norms establish what sorts of behavior to expect from one another (Hatch and Cunliffe 2013, p. 169).

Corporate culture is conceivable as distributed on three levels: observable artifacts, values, and basic underlying assumptions (Schein 1992). Observable artifacts represent the visible part of the culture and include symbols, ceremonies, stories, slogans, physical layouts, dress codes, manners in which people address each other, as well as company records, products, and annual reports (Goffee and Jones 1998). They are the most visible, particularly to newcomers. In this level, and partly in the second one, we can find also organizational rules, procedures, and routines (ostensive parts) as external manifestations of a company's culture and knowledge (Pentland and Feldman 2008).

The second level, values, includes espoused and documented values, norms, beliefs, attitudes, feelings, ideologies, charters, and philosophies, which are at the first sight invisible, can be recognized through several survey instruments like interviews or questionnaires. Values determine how members interpret the environment and bond members to the organization. The most invisible level of culture, the third level, includes underlying and usually unconscious assumptions that determine employees' perceptions, thought processes, feelings, and behavior and are commonly taken for granted. They develop, when values become a part of members' mindset and affect their interpretation of a situation.

Hatch added a fourth element, *symbols*, to highlight the importance of *interpretation* and *meaning making* in culture; she also expanded Schein's model by adding the processes or interactions between the elements in Shein's model.

> In the domain of the top half of the model, assumptions and values shape activity such that artifacts of these influences are created and maintained within the culture. In the domain of the lower half of the model, organizational members choose some (but not all) of the artifacts available to them and use the selected artifacts to symbolize their meanings in communication with others (Poole and Van de Ven 2004, p. 206).

Referring to the PSC Model of Fig. 7.16, I observe that the rotation speed, the small wheel of *rules* and *programs* that regulate the processes is the fastest but also the least effective. It spins easily and quickly. Nothing is easier than changing a norm, a regulation, placing a ban, and limiting the alternatives. Any program change can be modified by a subsequent change. The speed of the small wheel's rotation assures a timely adjustment to environmental dynamics. However, the small wheel causes the medium wheel to spin, even though slowly. This form of speedy change, at relatively low cost, is typically symptomatic and lends itself to continual revisions, since it is easily reversible and more affected by the temporary equilibriums among the participants in the governance.

The change of an operational *program* imposes a change in the *processes*, which is often traumatic for the structure that is thus forced to adapt, slowly, in order to respect the modified programs; the structure learns the new changes and itself changes. Through a new regulation we speedily introduce new subjects in scholastic programs, modify the structure of university degree programs, place new bans or limits on the circulation of people and cars, limit the sale of certain goods, impose/ban on the use of additives, change the rules at work, forbid the turning on of boilers or air conditioners, forbid smoke in certain locales, etc. All these sudden changes produce a variation in the structures that must carry them out: schools must recruit the appropriate teachers, the universities must reorganize their faculties and departments, the traffic wardens must put up new signs, changes must be made in packaging and distribution processes, smoke ventilators must be installed and smoking corners set up, and so on. However, the changes in the structures that must adjust to the new processes proceed much more slowly than do the changes in the programs and processes themselves, since changes in the operational structure of the institutions and organizations are very costly; moreover, once carried out, structural changes are more difficult to change, becoming almost irreversible. The return to the *status quo* is nearly impossible, since the institutions and organizational system have been deeply modified (Watzlawick et al. 1974). Though slowly, the wheel of change in structures turns the largest wheel, one that of cultural change, and this slow change produces cultural innovation, which modifies the very foundations of society and of the organization. Once in place this process is nearly irreversible, since any change to it would require an equally long period for the change itself.

> Changing people's habits and way of thinking is like writing your instructions in the snow during a snowstorm. Every 20 minutes you must rewrite your instructions. Only with constant repetition will you create change (Donald Dewar, cited by Haines 2005, p. 53).

This third level of change also includes the change in social and organizational values, which Casey (2000) has defined as the third order of change. Cultural change can be quite inconvenient for social systems, since many individuals could find it difficult to accept, adapt to, and live the new cultural values. Thus, cultural change could produce a true evolution in the entire social system or in its large component parts, as well as a breaking up of the system into subsystems that adopt the old and new cultural and value structure. Cultural change represents the definitive solution that exploits the potential of informed and educated individuals. Without such change the normative efforts (small wheel) and the structural ones (medium-size wheel) run the risk of being merely symptomatic interventions on individual variables, and thus of not producing the *leverage effect* needed to deal with sustainability.

We arrive at these speculative conclusions by also considering the slowest wheel first. When a society or an organization succeeds in effectively influencing the culture of those belonging to it, then the new cognitive models (large and slow wheel) trigger a fast change in the structure (middle wheel), which, in turn, produces an ever faster adaptation in the programs and procedures (smallest wheel). In Fig. 7.17, it is easy to recognize the three wheels of change of Fig. 7.16, a typical system of multi-layer control.

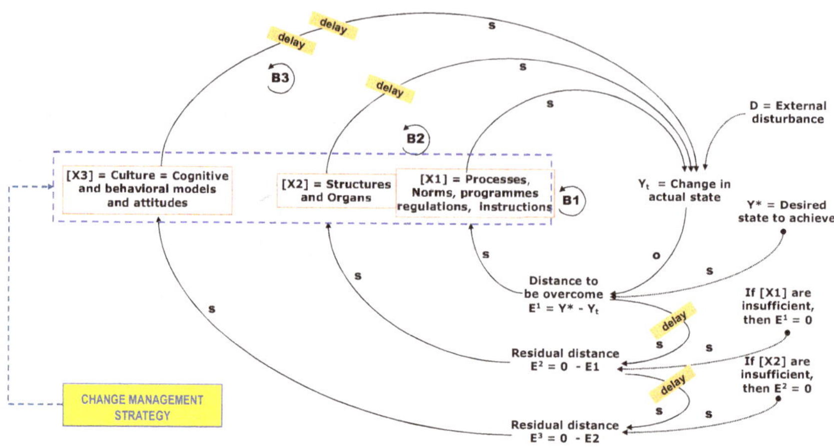

Fig. 7.17 The control of change through a multi-layer *Ring* representing the PSC model

The control process is realized through a *problem finding* and *problem solving* process (Nickols 2010, 2011). Considering change management as a particular form of creative problem solving, the two views, which we have called *top down* and *bottom up*, can, however, also be considered as true *strategies* of creative change (Osborn 1953):

- The *symptomatic strategy*, which proceeds top down, from fast to slow, contingent to permanent;
- The *permanent strategy*, which proceeds bottom up, from slow to fast, permanent to contingent.

Environmental changes, which are accelerated and interconnected, often affecting symbolic variables, produce unpleasant effects which the social systems interpret as *problems*, as *symptoms* to eliminate as quickly as possible. The symptomatic strategy (top down), which is reactive and quickly carried out, is considered an immediate, though contingent, remedy for dealing with unpleasant problems due to change. Change management must act rapidly, making the most urgent changes to the present situation, often without involvement of others in the organization, by acting on rapid levers such as programs, directives, and operational rules.

As the model in Fig. 7.18 shows, the symptomatic solutions, though fast, almost always lead to collateral effects that can worsen the problems arising from change. The *permanent* strategy (bottom up) is the most extensive, and for this reason it is slow and difficult to achieve, since it forces change on the cultural wheel by spinning the wheel of cultural change; but if it is successfully adopted, then the social system will have reacted to the change in a lasting way and have become able to deal with other rapid changes.

To quote Peter Senge:

> The advantage of systems thinking derives from the leverage effect – seeing in what way the actions and changes in the structures can lead to long-lasting, meaningful improvements. Often the leverage effect follows the principle of the economy of means, according

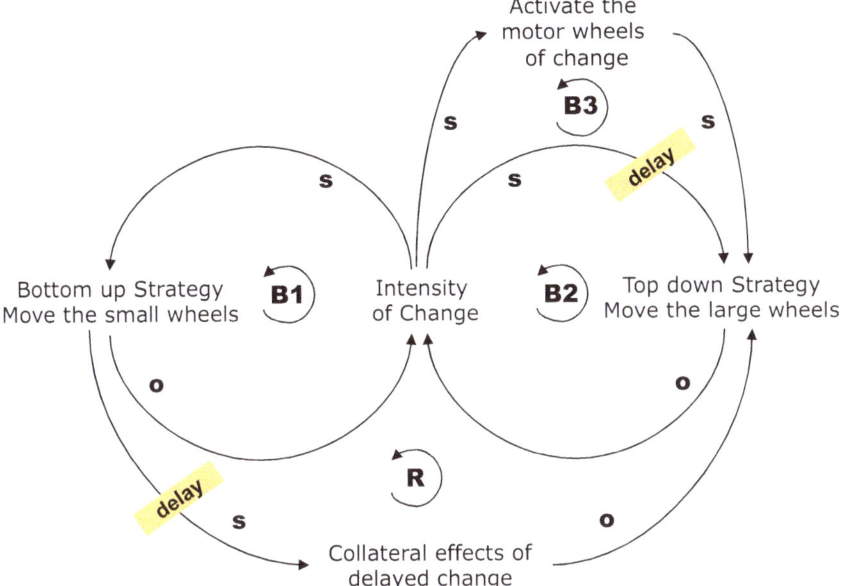

Fig. 7.18 The symptomatic and permanent solutions of change management

> to which the best results do not come from large-scale efforts but from well-concentrated small actions. Our non-system way of thinking causes significant specific damage because it continually leads us to concentrate on low leverage effect changes: we concentrate on symptoms of higher stress. We correct and improve the symptoms: but such efforts are limited, when things go well, to improving short-term factors, while worsening the situation in the long run (Senge 2006, p. 131).

In this way, the system not only can forbid alcohol and counter the spread of new drugs in order to reduce "Saturday night road deaths" but also can change the smoking and drug culture through a long-term strategy of education and the self-fulfillment of youth and not only limit, by law, toxic emissions or change the structure of productive systems but also introduce an appropriate cultural attitude to environmental problems.

> Change is a central concern in every human activity… But the processes of change have not been presented in an orderly way in our educational institutions. The dynamics of change have seldom been taught as a basic foundation that underlies all fields. The processes of change have not been organized so that they can be taught at all educational levels, even though a child, from his or her earliest awareness, begins to cope with change and to build an intuitive awareness of change (Forrester 1983, Foreword, p. IX).

Of the three paths of change in social systems, that which leads to a cultural adaptation appears, though slow to occur, the most necessary today, precisely due to its propulsive power regarding the other wheels of change in a world that is changing more and more rapidly and in a structurally complex manner.

Section 7.10.5 will consider the cultural change necessary to eliminate or mitigate a social phenomenon that is relevant and widespread in all social systems in the world: the influence of *stereotypes* in conditioning individual behavior. The action of stereotypes and their persistence over time can also be described using Combinatory Systems Theory.

7.10 Complementary Material

7.10.1 Models and Classes of Combinatory Systems

The typical logic of the combinatory systems outlined in Fig. 7.8 produces a wide class of collective phenomena; that logic can be generalized by introducing these simple control rules that direct the micro and macro behaviors:

1. Each agent is characterized by an *individual variable* of some kind (qualitative or quantitative) whose values—at any time t_h—represent the *individual states*, $x_i(t_h)$.
2. The collectivity is characterized by a *global variable* (qualitative or quantitative) whose values—at any time t_h—represent the *system state*, $Y(t_h)$.
3. Due to the presence of an opportune set of *recombining factors*, the system state—at any time t_h—derives from the *combination* (to be specified) of the individual states, following macro or *recombining rules*.
4. Each agent can perceive a gap (positive or negative) between his individual state and the state of the collectivity.
5. Due to the presence of an opportune set of *necessitating factors* each agent—at time t_{h+1}—decides, or is forced, to identify the most appropriate strategies to expand or reduce the perceived gap following the micro or *necessitating rules*.
6. As long as the necessitating and recombining factors are maintained, the *micro–macro feedback* can operate.
7. The agents are characterized by an initial state at time t_0, $x_i(t_0)$; in most cases, this initial state may be assumed to be due to *chance*.
8. The micro–macro feedback operates between the limits of the *minimum activation number* and the *maximum saturation number* of the agents who reveal the state maintaining the micro–macro feedback.
9. The sequences of state values over a period represent the macro and micro dynamics, or behaviors, of the system and the agents, as shown below in Fig. 7.19.

Combinatory systems can be represented by different models of increasing complexity. The simplest models are the *descriptive* ones that indicate in words—or by patterns analogous to Fig. 7.20 (energy inputs are not included)—the fundamental elements necessary for understanding the operative logic of systems that produce observable collective phenomena. The more powerful models are the *heuristic models* that try to simulate the system's dynamics by stating—or constructing ad hoc—a *set of rules* specifying: (a) the micro, or *necessitating* rules producing the micro

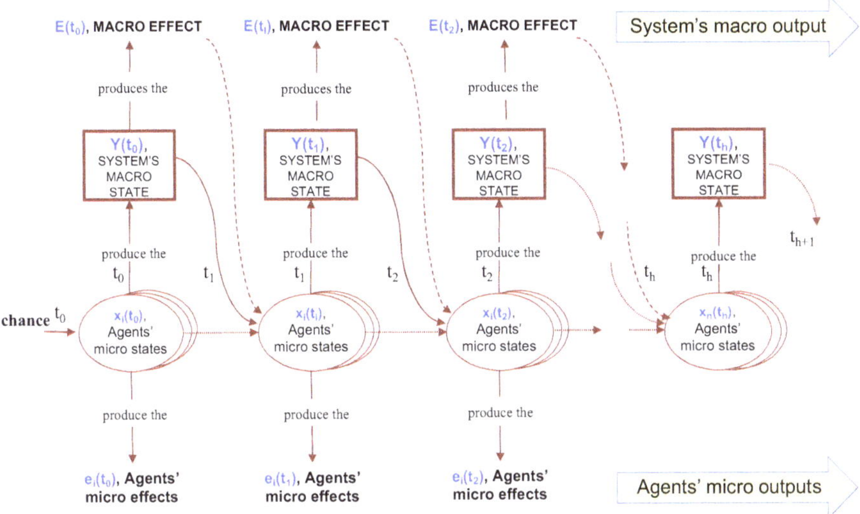

Fig. 7.19 The micro and macro dynamics and the micro–macro feedback

behaviors of agents as the consequence of the necessitating factors; (b) the macro, or *recombining* rules that produce the system's macro behavior, due to the presence of recombining factors; (c) the *micro–macro feedback* that allows the system to produce the observed phenomena; (d) the *strengthening*, *weakening* and *control levers*, when possible or admitted. Finally, we can build a combinatory automaton (Mella 2017) that specifies the mathematical and statistical simulation model that produces the micro and macro behaviors of the combinatory systems based on specific assumptions regarding the *recombining* and *necessitating* rules.

The *heuristic models* of the five important classes of combinatory systems (Sect. 7.5) can be briefly described as follows; a comparison between them clearly reveals the same *modus operandi* of the combinatory systems of the various heuristic models.

(a) Heuristic model for *systems of accumulation*.
  Necessitating rule: if you have to accumulate some object with others similar in nature (micro behavior), look for already-made accumulations, since this gives you an advantage or reduces some disadvantage (necessitating factor).
  Recombining rule: the environment preserves the accumulated objects, or is not able to eliminate them, thus maintaining the advantages of the accumulation; everyone accumulates (macro behavior), and an accumulation of some kind is created (macro effect).
  Micro–macro feedback: the accumulation is the macro effect of single micro behaviors; the larger the accumulation (macro effect), the more incentive (facility, probability) for accumulating (micro behaviors) objects (micro effects); the collective accumulation (macro behavior) leads to the maintenance or the increase of the accumulation.

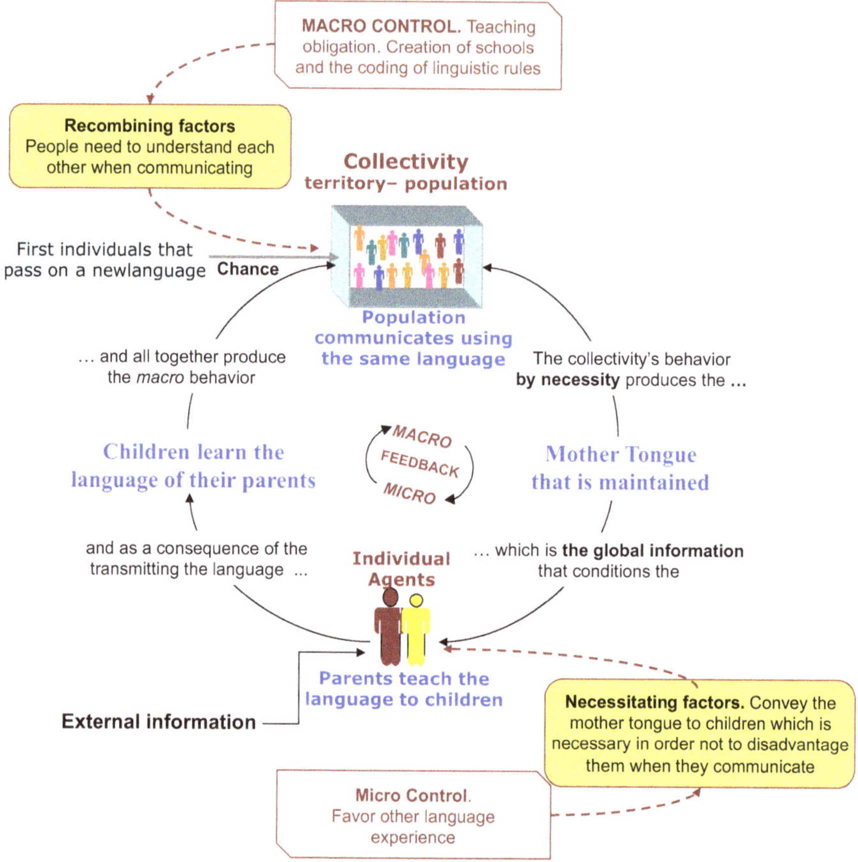

Fig. 7.20 Graphic model of the "maintaining-the-language" system

Strengthening, weakening, and control levers: a sign reading "accumulate here" or "accumulation prohibited" represents the best strengthening action. The prompt removal of the initial accumulation and a careful vigilance represent weakening factors. The external control acts on the accumulation and on the site where this forms; the internal control acts on the desire to accumulate *and* on the individuals' actions.

(b) Heuristic model for *systems of diffusion*.

Necessitating rule: if you see that an "object" is diffused, then it is "useful" for you to possess it or harmful not to possess it (necessitating factor), and you must try to produce or acquire it.

Recombining rule: the environment or the collectivity preserves the diffused objects and maintains the utility of possessing the object; the higher the individual's utility or need is to acquire the object, the more the object will spread throughout the collectivity.

Micro–macro feedback: a greater diffusion (macro effect) implies a greater desire to acquire the object (micro effect); the single acquisition (micro behavior) increases the collective diffusion (macro behavior).
Strengthening, weakening, and control levers: publicity and social gratification represent strengthening factors for the system; social disapproval and repression represent weakening factors; the macro control must act to limit the availability of purchasable goods; the micro control must act on individual preferences.

(c) Heuristic model for *systems of pursuit*.
Necessitating rule: if there is an objective, try to achieve it; if there is a limit, try to exceed it; if another individual overtakes you (negative gap), regain the lost ground; if you are even with someone, try to go ahead of him; if you are in the lead, try to maintain or increase your advantage (positive gap).
Recombining rule: the collectivity recognizes the validity of the objective and views limits in a negative way; the more individuals try to exceed the limit, the greater the chance of exceeding it, with a consequent advantage for those who succeed in doing so. This provides the incentive for the pursuit.
Micro–macro feedback: if everyone tries to go beyond the limit (macro behavior), then this limit is raised (macro effect), thereby eliminating the advantage for those who have already reached it (micro effect); this forces individuals to exceed the limit (micro behavior).
Strengthening, weakening, and control factors: if honor and prize money are awarded, then the system is strengthened due to the increase in gratification. If the achievement of the objective or the exceeding of the limit causes accidents, with victims, then there is a weakening. Another weakening factor involves the cost of competitions: the higher this is, the more difficult the search for the record becomes. Control factors for the system are represented, for example, by publicity concerning the record and the creation of venues where the sporting activity can be practiced. A typical micro control intervention is represented by a favorable culture for competitions.

(d) Heuristic model for *systems of order*.
Necessitating rule: there are advantages in maintaining a particular order and disadvantages in breaking it; if you want to gain advantages or avoid disadvantages, try to control your behavior so that you maintain or achieve the order that is indicated by the rules that establish it.
Recombining rule: the more the particular order is maintained, the greater the advantages from adjusting one's behavior to maintain it and the disadvantages from breaking it.
Micro–macro feedback: the order (macro effect) creates the convenience for individuals to maintain the arrangement and respect the rules (micro behaviors); everyone maintains a coordinated behavior (macro behavior).
Strengthenings, weakenings, and control: strengthenings to the system can be represented by awards for those who maintain the order and evident punishments for those breaking it. Introducing difficulties and constraints to the maintenance of order discourages individuals from respecting such maintenance.

Facilitating and removing constraints make it easier for agents to control their micro behavior. The macro control can be carried out through the supervision of a director who forces the dance pairs to dance in a rotating way. The prior instruction represents a form of micro control.

(e) Heuristic model for *systems of improvement and progress*.
Necessitating rule: if you perceive that the level of your improvement parameter is below the level of the system's progress parameter—that is, there is a negative gap between your state and that of the others—try to improve in order to reduce the gap and, if possible, to attain a positive gap; if you perceive there is a positive gap, do nothing or try to improve further in order to increase the favorable gap.
Recombining rule: the system must be able to notice the individual improvement and adjust the progress parameter to the average (or, more generally, to the combination) of the individual improvement measures.
Micro–macro feedback: individual improvement (micro effect) raises the parameter that measures collective progress (macro effect); this leads to the formation of positive and negative gaps that push the individuals to control their behavior in order to increase the gaps (if positive) or eliminate them (if negative).
Strengthenings, weakenings, and control: the macro control can be carried out through a supervisor or a manager who makes the agents aware of their inferior state and of the possibility of improving their position, or advises them on how to do so. The prior instruction of the dance pairs represents a form of micro control.

7.10.2 *The Heuristic Models Revealing the* Modus Operandi *of Some Relevant Social Phenomena Following the Combinatory Systems Model*

Maintaining the mother tongue Surely no one will argue that the most deeply-rooted tradition in any *social system* is the learning and maintenance of the mother tongue by the inhabitants of areas of varying size. Why does each of us—except for those lucky enough to be multilingual from birth—speak a certain language, which is considered the mother (natural) tongue, and learn other languages that are considered acquired? Why are there hundreds of different languages that are maintained even though a common language would obviously be useful to facilitate communication? Why do dialects exist and why are they maintained for the same language? Why in the case of two urban settlements, perhaps only a hundred meters apart, are there distinct dialects, so that we can unmistakably determine to which community someone belongs? The descriptive graphical model of this combinatory social phenomenon is shown in Fig. 7.20.

The rules to form the heuristic model that describes this combinatory system of order can be represented as follows:

Micro behavior and necessitating rule: while raising and educating your children, teach them the language of your forebears; consider as an error any syntactical or semantic deviation and try to eliminate these, since your children must be able to communicate within the collectivity.

Macro behavior and recombining rule: the educational apparatuses of the community preserve and diffuse the ancestral language (macro effect), which is spoken using the same structure, generation after generation; the language is codified and passed on without changes.

Micro–macro feedback. Chance and necessity: the language is preserved by being taught to newborns; it is thus the result of the combination of past *micro behaviors*, but it also conditions future ones. A new language, or a variant of the old one, can be introduced in a given territory "by chance"; if this new language, or new variation, comes out the loser in the competition with the pre-existing one, then no combinatory system is activated and the pre-existing language prevails. However, once the system is activated, "by necessity" it maintains the linguistic foundation through its transmission to newborns.

Strengthening, weakening, and control actions: the creation of schools and the cataloguing of linguistic rules in a written form (syntax) considerably strengthen the system; periodic waves of immigration can instead weaken it.

THE PRESERVATION OF RELIGIONS Among the many *social systems* with a functioning logic similar to the one we have just seen are those that lead to the preservation of the national religion and of local and territorial uses and customs.

THE MAINTENANCE OF STEREOTYPES Section 7.10.5 will consider the cultural change necessary to eliminate or mitigate a social phenomenon that is relevant and widespread in all social systems throughout the world: the influence of *stereotypes* in conditioning individual behavior. The action of stereotypes and their persistence over time can also be described using combinatory systems theory.

THE EXPANSION OF THE SIZE OF CITIES When one or more individuals need a house (necessitating factor) and feel they have found a place (environment) which represents a favorable place to live—for example, due to the presence of water (the great cities of antiquity grew up along the great rivers), or roads (many cities arose along important routes), or a hill that is useful for defense (how many fortified cities have been built on hills that overlook valleys or the sea?), or even simply a beautiful view—then the first dwellings (micro effects) are built (micro behavior). A city thus appears as the result of individual decisions to build a house in a favorable place. However, the presence of a city provides information that favorable living conditions have been found, which gives rise to powerful necessitating factors that influence the individual micro behavior; the city itself indicates the favorable conditions (recombining effect) that allow it to increase in size. In 1950, there were only two megacities: London and New York. According to Wikipedia (Megacities), today there are more than 40 megacities, and the largest cities are found on all continents. The heuristic model of this social behavior can be formulated as follows.

Micro behavior and necessitating rule: if you need to build a house, look for favorable conditions; if there is a city there already, it is assumed that favorable conditions exist. Leave your house to your descendants.

Macro behavior and recombining rule: the construction of new houses strengthens the urban settlement; the strengthening and growth of the city are signs that favorable conditions exist (opportunities, services, protection, etc.), which influences the micro behavior whereby new houses are added to the existing settlement. The older and larger the city is, the greater the incentive for new arrivals to the area to locate there.

Micro–macro feedback. Chance and necessity: a city is the result of individual decisions to build a house in a favorable place; but the presence of a city gives *information* that favorable living conditions have been found, and this influences the individual micro behavior (the city itself indicates the favorable conditions). A city arises "by chance" but, once begun, the phenomenon is "by necessity" maintained over time as long as the necessitating factor operates.

Strengthening, weakening, and control actions: overcrowding, an increase in the time needed to cross a town, the desire for solitude, and the need to preserve the surrounding areas: these all represent *weakening* factors, and where these prevail over the need for a community life, the "urban-settlement" combinatory system will not operate; instead a different system is activated that can be referred to as "maintain-the-territorial-division." Examples of *strengthening* factors are the danger of invasion, tourist attractions, tax incentives for construction, and the supply of attractive urban services. The macro control can act by means of urban planning; the micro control can influence the desire to live in a town, or the opposing desire to flee the crowds.

It is easy to see that the above model is also valid for interpreting the creation and expansion of Industrial Districts.

The spread of a fashion The spread of a fashion—whether this concerns clothing, clocks, cars, toys, monuments, a particular linguistic form, and so on—is one of the most easily recognizable social phenomena that can be explained by combinatory systems theory. The macro behavior referring to the fashion system leads to an increase in the desire to imitate (necessitating factor) and to the macro effect of the diffusion of the good that is the object of imitation and the advantage for those who possess it (recombining effect). The micro behavior is represented by the imitative actions regarding a novelty, conditioned by the individual need to set oneself apart from the norm: that is, to participate in the fashion. The micro effects consist in the acquisition of the good which is the object of imitation. The macro behavior (usually not directly observable) can be seen as an increase in the number of individuals imitating the innovation; the macro effect consists in the increase in the average density of the good in the environment and in the base. The heuristic model that describes this system can be represented as follows.

Micro behavior and necessitating rule: if you need to stand out, abandon the traditional way and follow an innovation; if you do not want to be considered as just a "somebody" but as someone different, then join in with a fashion and undertake an imitative micro behavior.

Macro behavior and recombining rule: if an individual distinguishes himself, he is appreciated and well-considered; the diffusion of goods in the environment, which represents the effect of the imitative micro behavior, increases the desire to adhere to a fashion. The greater the number of imitative behaviors, the more powerful is the fashion and the more pressing the desire to follow it, until the saturation point is reached.

Micro–macro feedback. Chance and necessity: the diffusion of an innovation is the result of the imitative micro behavior of the single individuals, but it also conditions future behavior. The innovative idea arises "by chance" and conditions the initial imitation. Once the imitative phenomenon has begun, it is maintained "by necessity" until the saturation density is reached.

Strengthening, weakening, and control actions: publicity and social gratification represent strengthening factors of the system; social disapproval (the fashion of driving at break-neck speed in cars, for example) and repression represent weakening factors. The macro control must act to limit the availability of purchasable goods, whereas the micro control must act on individual preferences.

The spread of epidemics The spread of an epidemic (Sect. 1.7.7) can impact all social systems on all continents. Combinatory systems theory explains how this phenomenon can develop. The epidemic begins when the carrier of a flu virus comes into contact with other individuals (limited information), thereby transmitting the virus (micro behavior). The higher the number of sick people (macro effect), the more likely it is that other people will become sick (macro behavior). In order to have an epidemic-like spread, the carrier must transmit the flu to other individuals, who must in turn pass it on to others, and so on. It is clear that if the first infected individual does not come into contact with anyone, then the combinatory system will not begin; but if by chance there is frequent contact with other individuals, then the system has a high probability of getting under way. When the number of infected persons exceeds a minimum activation number, the epidemic develops by necessity.

The spread of drugs The spread of drugs, in all its forms, represents one of the most devastating and difficult to control social phenomena. Combinatory systems theory can aid in interpreting the *modus operandi* of the system that generates this unstoppable phenomenon, which depends on the socio-cultural conditions of the collectivity and on its political and police organization. The diffusion of drugs develops where several individuals undertake micro behaviors that consist in individual drug-pushing actions that produce as micro effects the initiation of new addicts. When the number of addicts exceeds the minimum activation number, which usually is not very high, then an unstoppable and powerful combinatory system develops, producing the terrible effects we are all familiar with. The heuristic model that describes this system can be represented as follows.

Micro behavior and necessitating rule: if you need drugs, try to buy them; if you need money for drugs, obtain it by selling them by influencing others to use drugs.

Macro behavior and recombining rule: individual drug pushing spreads the use of drugs at the collective level, but the spread of the drug is the condition that favors drug pushing; the higher the number of drug addicts, the higher the number of pushers, which in turn leads to an increase in the number of addicts.

Micro–macro feedback. Chance and necessity: the spread of drugs (macro behavior) has as an effect the incentive to proselytize (micro behavior); but this in turn leads to the spread of drugs; the first addicts arrive "by chance," and if they are not rehabilitated then the macro behavior begins and is maintained "by necessity," with an increase in the number of new addicts.

Strengthening, weakening, and control actions: the strengthening of the police force, the destruction of distribution channels, the repression of drug selling organizations, the tendency to denounce drug pushers, and the creation of rehabilitation and detoxification centers are examples of *weakening factors* of the system. The strong presence of mafia-like organizations, the absence of social constraints, the proximity of import centers for drugs are examples of *strengthening factors*. The *macro control* must act upon the drug supply lines and the social conditions that favor drug addiction. The *micro control* must act by means of preventive campaigns that provide information about the dangers of drug use, along with rehabilitation facilities and, in general, social support and psychological measures that help drug addicts overcome their addiction.

7.10.3 The Combinatory Automaton to Simulate the Buzzing in an Indoor Locale

In order to simulate the behavior of the different classes of *combinatory systems* and their macro effects, it is useful to build a simple linear *combinatory automaton*, made up of a grid (or an array) of cells, each of which contains the value of the micro state of an agent of the combinatory system. At each instant, t_h, the automaton calculates the synthetic state of the entire system through an operation (addition, average, max value, min value, etc.) on the values of all the cells. At the subsequent instant, t_{h+1}, the individual cells are updated by calculating a new state that takes into account the overall system state, thus producing the dynamics shown in Fig. 7.21. In combinatory automata, the *recombining* and *necessitating* factors are translated into numerical probabilities that are assigned to the various cells and updated for each iteration by applying some rule.

The model in Fig. 7.21 shows a general schematic model of a probabilistic combinatory automaton (in simplified form), which formally represents the dynamic model of the combinatory system in Fig. 7.19. The most useful combinatory automata are the stochastic ones, in which a probability, p_i, is associated with the transition of state of each x_i; in the opposite case, they are *deterministic*. The *stochastic* automata form since, to carry out the simulations, it is necessary to translate the *necessitating factors*—which move each agent (each cell) to change its state—into a *probability of transition of state*.

The probabilities can also be different for each cell and change in subsequent moments. These form a *probability field*, $P(t_h) = [p_i]$, which is coupled to the state of the agents. The *social combinatory systems* that are most interesting and easiest to simulate are the *irreversible* ones (build a tower or not). In these systems, both the micro and macro behaviors produce permanent effects (towers built) that may be viewed as increasing or decreasing cumulative processes in which the probabilities are dependent on the synthetic state of the automaton at each transition. The *set of rules* specifying the initial state $X(t_0)$ and $P(t_0)$ and the functions F, f_i, G, and g_i represent the *operative program*, which produces the dynamics of the *combinatory automaton* and must be specified for each simulation.

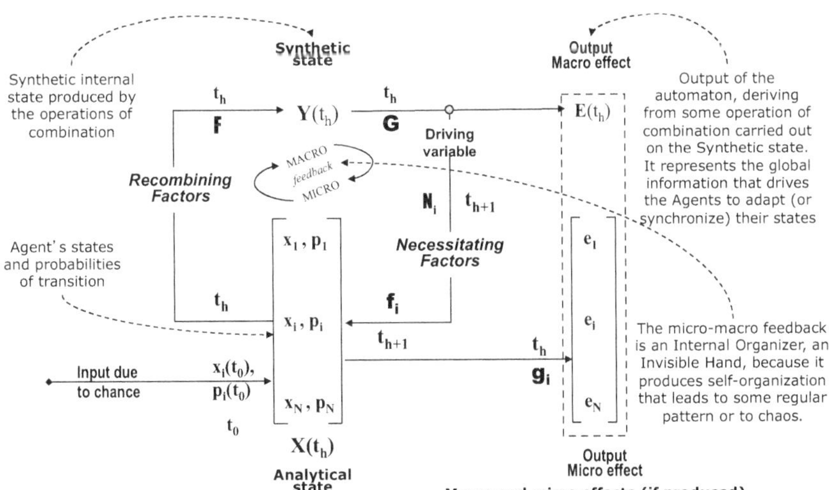

Fig. 7.21 The components of a stochastic combinatory automaton (for a more complete representation, see Mella 2017, p. 159)

Figure 7.21 allows us to easily produce the equations that characterize *combinatory automata.* In short, if we assume that the Social Combinatory System is formed by a base of N agents, "X_i," $1 \le i \le N$, and that at "time" "t_h" each agent has a state defined by "$x_i(t_h)$", then:

- The vector (or matrix, if the agents are arranged two-dimensionally):

$$X(t_h) = \left[x_i(t_h), 1 \le i \le N\right]$$

represents the analytical state of the automaton at "time" "t_h";

- The function "F", appropriately specified for each combinatory system, recombines the "micro" states to generate the "macro" state of the automaton:

$$Y(t_h) = F\left[X(t_h)\right]$$

- In the stochastic combinatory automata, the transition of state of the agents requires the specification of the *probability field,* $P(t_h) = [p_i]$, and of the transition of state functions "f_i," so that the state at "time" "t_{h+1}" can be determined based on the present state and the synthetic macro state $Y(t_h)$:

$$x_i(t_{h+1}) = f_i\left[x_i(t_h), p_i, Y(t_h)\right], 1 \le i \le N$$

- The G and g_i functions, appropriately specified for each phenomenon to be simulated, transform the *macro* and *micro* "states" into the *macro* and *micro* "effects";
- The entire combinatory automaton, with the micro–macro feedback, is described by the following system:

$$\begin{cases} Y(t_h) = F\left[X(t_h)\right] \\ x_i(t_{h+1}) = f_i\left[x_i(t_h), p_i, Y(t_h)\right] \end{cases}$$

A more analytical and in-depth description of a combinatory automaton is found in Mella, 2017, Ch. 3.

The combinatory automaton model of the combinatory system that produces buzzing noise in a crowded room is shown in Fig. 7.22, from which we can also derive the logic of the simulation model that describes the macro behavior of the system (production of noise) and the micro behaviors of the agents (speaking at a voice level that exceeds the noise).

Taking into account these assumptions, I can represent the combinatory system by the following simple probabilistic model:

Parameters of the model	symbols	A_1	A_2	A_3	...	A_{18}	A_{19}	A_{20}
Initial voice level of A_i	$v_i(t_0)$	1,00	1,00	1,00	...	0,00	0,00	0,00
Amplification coefficient	k =	0,8		a = 0	...			
External noise	Q =	10			...			
Position coefficient of A_i	w(n)	1,00	1,00	1,00	...	1,00	1,00	1,00
Decibels above R	$v_{i(min)}$	10	9	8	...	9	8	7
Decibels above R (random)	$v_{i(rnd)}$	2	3	4	...	3	4	1
Necessitating factor (probab.)	$s_{i\,[0,1]}$	60%	65%	70%	...	70%	80%	90%

Fig. 7.22 Control panel for the combinatory automaton that produces background noise

$$\text{Input of the initial state of the system}: v_i(t_0) \leftarrow \text{“chance”},\ 1 \le i \le N$$

$$R(t_h)] = \left\{ k\left[(1/N) \sum_{1 \le i \le N} v_i(t_h) \right] + Q\ r(t_h)_{[0,1]} \right\} (1-a), h = 0,1,2,\ldots$$

$$v_i(t_{h+1}) = \left\{ \left[w_i\ R(t_h) + v_{i(\min)} \right] + v_{i(rnd)}\ 1_i(t_h)_{[0,1]} \right\}\ s_{i[0,1]}\ b_{i[bol]}(t_h).$$

The descriptive model of the combinatory system producing noise in crowded locales is shown in Fig. 7.23. The basic variables and parameters that characterize this combinatory automaton can be summarized as follows:

- The group of speakers consists of N agents, A_i, with $1 < i < N$; their behavior is observed over discrete instants: t_h, $h = 0, 1, 2 \ldots$.
- $R(t_h) = k\,[(1/N) \sum_{1 \le i \le N} v_i(t_h)]$ represents the background noise, the Buzzing that arises from the combination of the speakers' voice levels; we assume this is determined by the arithmetic average of the speakers' voices, which represents the principal *recombining rule*; the average is rectified by the amplifying factor "*k*," whose value depends on the shape of the room.
- The background noise, $R(t_h)$, does not depend solely on the speakers' voice levels but on other factors as well, among which the outside noise, which represents a macro-level disturbance; this is indicated by $Qr(t_h)_{[0,1]}$, with Q being the disturbance and $r(t_h)_{[0,1]}$ the probability of this occurring at each instantly have also considered the possibility the background noise is externally controlled by sound absorbing panels by introducing a coefficient of sound-absorption, "a"; if $a = 1$ there is total sound absorption and $R(t_h) = 0$.
- $v_{i(\min)}$ indicates the number of decibels (which differs for each agent) each speaker must produce above $R(t_h)$ in order to be heard by the others.

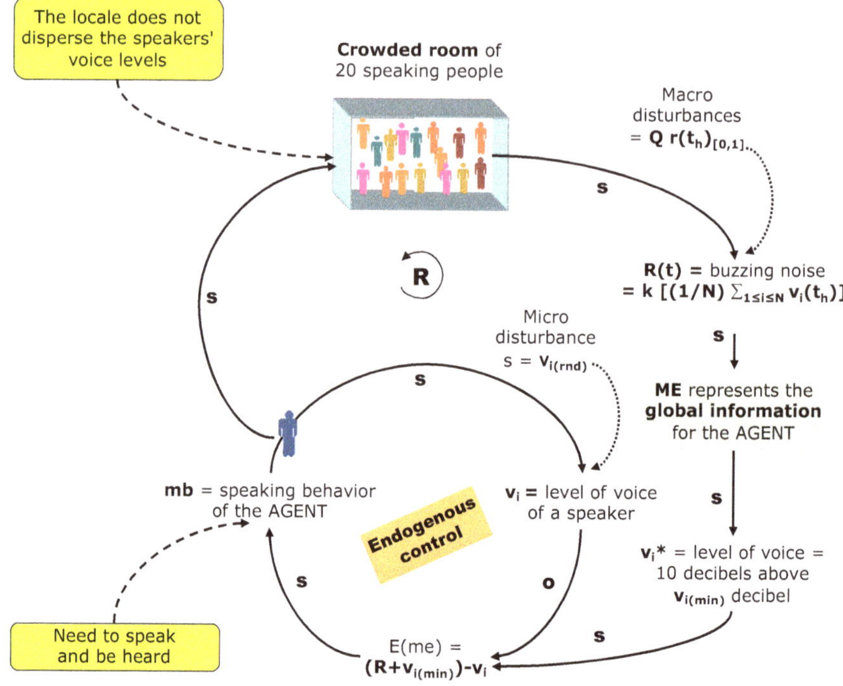

Fig. 7.23 Descriptive model for the combinatory system that produces background noise

- $v_i(t_{h+1}) = [w_i R(t_h) + v_{i(\text{min})}] + v_{i(\text{rnd})}$ represents the voice level of the agents above the buzzing level; the parameter w_i (usually set equal to 1) represents a synthetic measure of the position of each agent: if $w_i > 1$ the Buzzing is perceived as amplified by the speaker, who then further raises his voice.
- For each agent an anomalous variation of the voice level is added, indicated by $v_{i(\text{rnd})}$, to take into account random factors that oblige speakers to modify their way of speaking (emotional state, distance from other speakers, etc.).
- $s_{i[0,1]}$ indicates the probability of speaking; this is a very important variable in the model, since it synthetically expresses the complex of necessitating factors that push individuals to speak or remain silent; for simplicity's sake, this probability is assumed to be constant over time, even though it is different for each speaker.
- I have also introduced tolerance to noise, indicated by $b_{i(\text{bol})}(t_h)$; this is a Boolean variable that is equal to "0" when the speaker has reached the maximum level of toleration and stops talking.

The preceding rules have been translated into a combinatory automaton of $N = 20$ agents talking in a room, each of whom is characterized by individual parameters and probabilities of speaking specific to each agent, which are represented in the control panel in Fig. 7.22.

The simulation is presented in Fig. 7.24, which shows the dynamics in the voice levels (colored lines) and the buzzing noise (bold blue line) for a period of 30 cycles of the process (each cycle hypothetically lasts 2 s).

The *internal control* is provided by the individual speakers, who vary their voice levels when the buzzing varies; typical forms of external macro control are the placing of sound absorbing panels on the walls to reduce the action of the recombining factor that transforms the voices into a buzzing noise, thereby modifying the coefficient "k," or keeping macro-level external disturbances under control.

A form of *micro control* is the education and upbringing of the speakers, which affect the probability of speaking, which we have assumed includes all the necessitating factors. Another form is to assign a supervisor to the room (room director, usher) who, at regular intervals, invites those present to moderate their voice levels.

During a performance in a concert hall only rude people whisper, and they are immediately told to be quiet by those nearby or are even reproached by the usher. The dynamics produced by the combinatory automaton in Fig. 7.22, which simulates different situations, is shown in the upper model in Fig. 7.24. The noise (bold line) may be viewed as the output of the combinatory automaton constituting the collectivity considered as a whole.

7.10.4 Two Modern Tragedies of the Commons

I begin with the observation that economic activity, which in the short term and due to corporate, national, or regional (local) interests, has a growing need for non-renewable resources, becomes globally unsustainable in the long run and with a global dimension, because it depletes the non-renewable resources. The problem of the depletion of non-renewable resources arises when a large number of firms, referred to as "appropriators" (indicated concisely in the model in Fig. 7.25 by the term "Global Economic Activity") and located in different areas, use a Common-Pool Resource that is available in limited quantities (the "commons") which are not distributed among the appropriators based on some indicator of merit or need but freely exploited in competition for the individual activities of the firms, whose outcome (production of goods, profits, wages, etc.) depends on the use of the resource. The loops [**R1**] and [**R2**] in the model in Fig. 7.25 show how the economic activity of the appropriators produces advantages for both consumers and producers-investors; as long as these advantages persist, economic activity will continue to increase.

It is obvious that the more the resource is exploited individually, the less is available collectively, so that even the activities that employ the resource are less efficient, thus making its use increasingly "less economic." To achieve their final objectives in a situation of lesser efficiency the appropriators must intensify their activity, continuing to exploit the resource, no matter how scarce, until it is com-

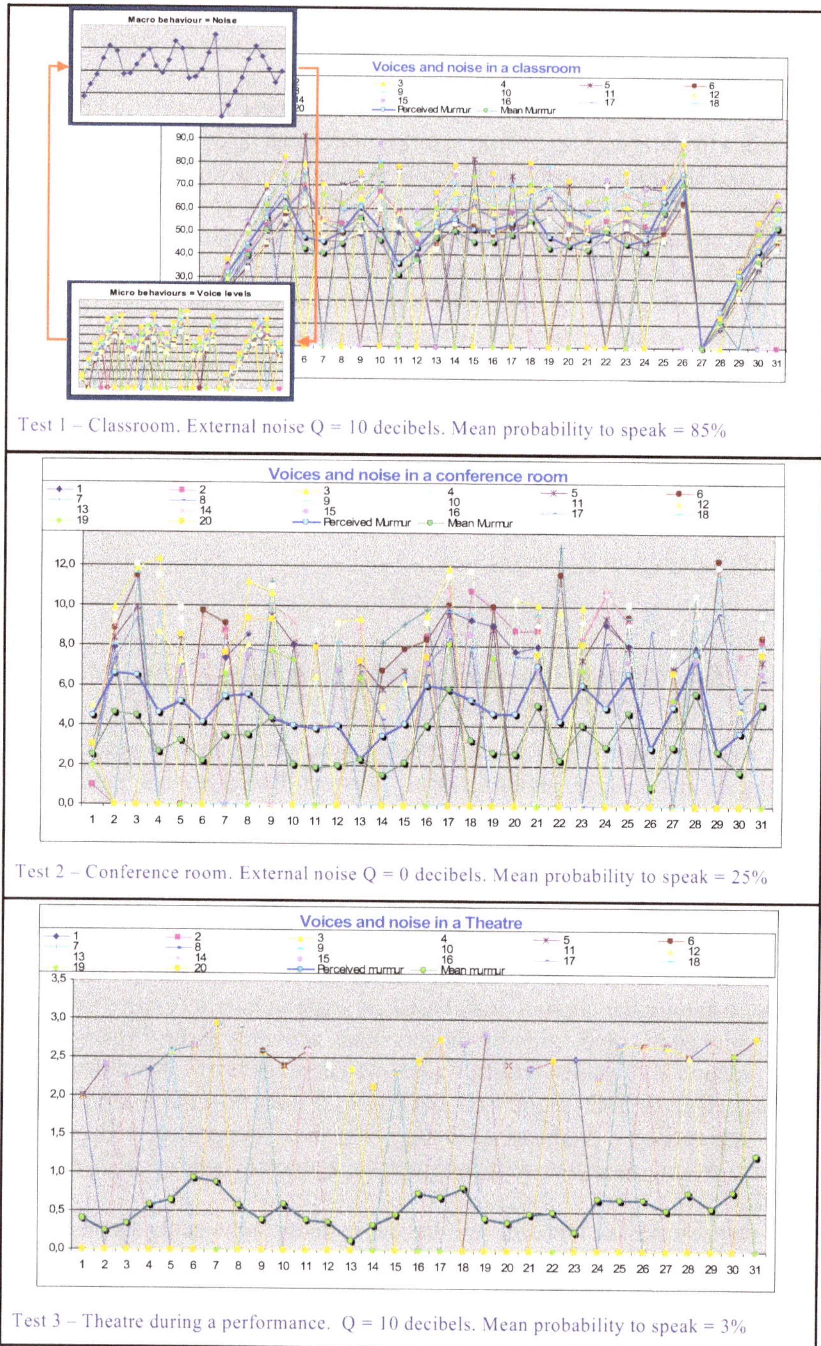

Fig. 7.24 Combinatory automaton simulating noise in different locales and situations (Powersim)

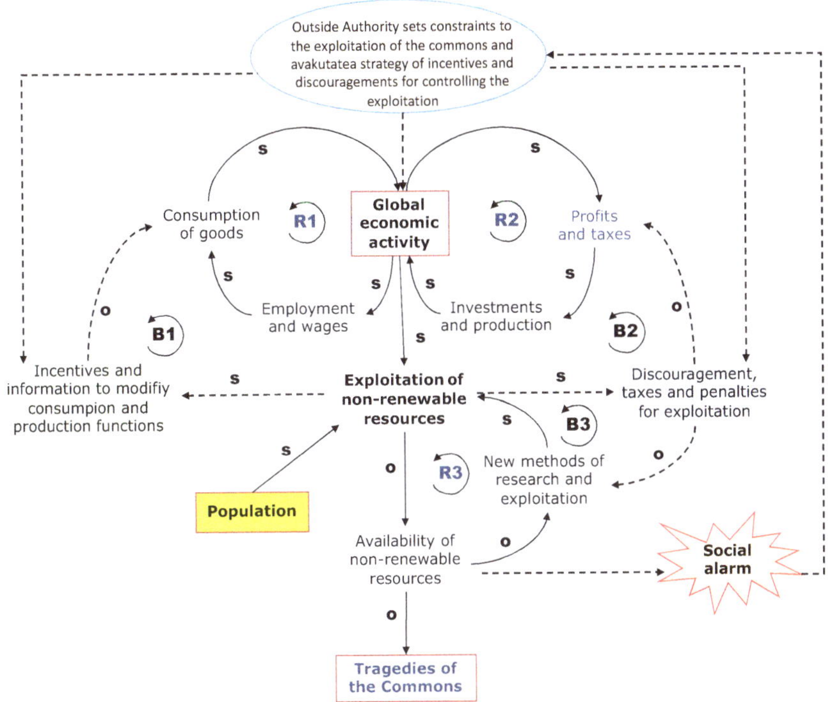

Fig. 7.25 Actions to counter the non-sustainability of the appropriation of the commons and to avoid the tragedies of the commons

pletely exhausted. The result is a "tragedy," since those who exploit the resource are forced to interrupt or reduce their activity because of its exhaustion. The efforts by the appropriators to pursue their objectives, which are individual and local and refer to a short-term time horizon, gives rise to more intense exploitation of the common-pool resources, as shown in loop [**R3**], whose action inevitably leads to the exhaustion of the common resource with negative effects for individual and collective outcomes.

If we observe the world around us we see this archetype in operation in many situations: too many fishermen fishing in waters with limited populations of fish; the search for oil, minerals, and precious stones in the same area; and, in general, the exploitation of limited natural resources by many users. In conclusion, when the individual agents—consumers and producers—that produce the global economic activity need to exploit non-renewable resources, without considering the sustainability of their activities, both with regard to consumers (eating whale meat may lead to the extinction of cetaceans) and producers (whaling vessels are now guided by satellite systems and can fire harpoons even from a distance of many kilometers in order to hunt down the few remaining whales), there is the inevitable risk such resources will be exhausted.

The action of the three archetypes summarized in Fig. 1.12 is inexorable; thus, educating producers and consumers regarding the question of long-term global sustainability has become increasingly more urgent, since the exploitation of limited common resources can cause conflicts that represent a "second tragedy." The clearest evidence of this is the numerous wars over water and oil, which are among the most limited and scarce vital goods on earth. How can the tragedies of the commons be countered? There is only one answer: when resources become very scarce a "social alarm" is sounded, and the public regulators, who become aware of the non-sustainability of the exploitation of limited resources, must externally control the behavior of consumers and entrepreneurs producing the harmful effects that must be avoided by specifying several useful actions to manage scarcity in order to reduce conflicts; some of these actions have been indicated in the model in Fig. 1.13:

(a) Weaken loop [**R1**] by activating loop [**B1**], thereby encouraging consumers to change their tastes, thus directing their preferences toward products that do not use up scarce resources. This action on consumers would also affect producers, who would find it even more necessary and urgent to restructure their production functions by substituting scarce resources with non-scarce ones;
(b) Weaken loop [**R2**] by activating loop [**B2**], thereby discouraging producers by creating disadvantages for those who take advantage of the exploitation of non-renewable resources. This action would force producers to pass on the greater costs to the prices of their products, which would discourage consumers from purchasing products using a large amount of scarce resources. The producers should modify their production function and substitute the scarce resources with non-scarce resources;
(c) Weaken loop [**R3**] by activating loop [**B3**], thereby introducing disincentives to discourage innovation and proposing a moratorium on further exploitation of non-renewable resources.

Other actions have been proposed alongside these three general ones, among which measures which directly impact the appropriators.

A second example that demonstrates the need to assess the sustainability of global economic growth concerns the problem of the disposal of waste, packaging material, and products that are no longer usable (cars, appliances, electronic goods, chemical materials, etc.). Residential centers could not exist without the control systems for liquid and solid wastes, of which even the Romans were aware: even before any urban expansion occurred, they began enlarging their sewer systems. Today the production and distribution system for the necessary goods in life is also a powerful producer of solid wastes of all kinds that must be eliminated daily.

The general problem can be represented in the model shown in Fig. 7.26. Economic activity inevitably produces waste and by-products that, together with human biological waste, pose a serious threat to sustainability for future generations, since waste cannot always be disposed of and, even when it can, this implies (especially for solid waste) the use of increasingly larger areas for landfills and waste treatment sites.

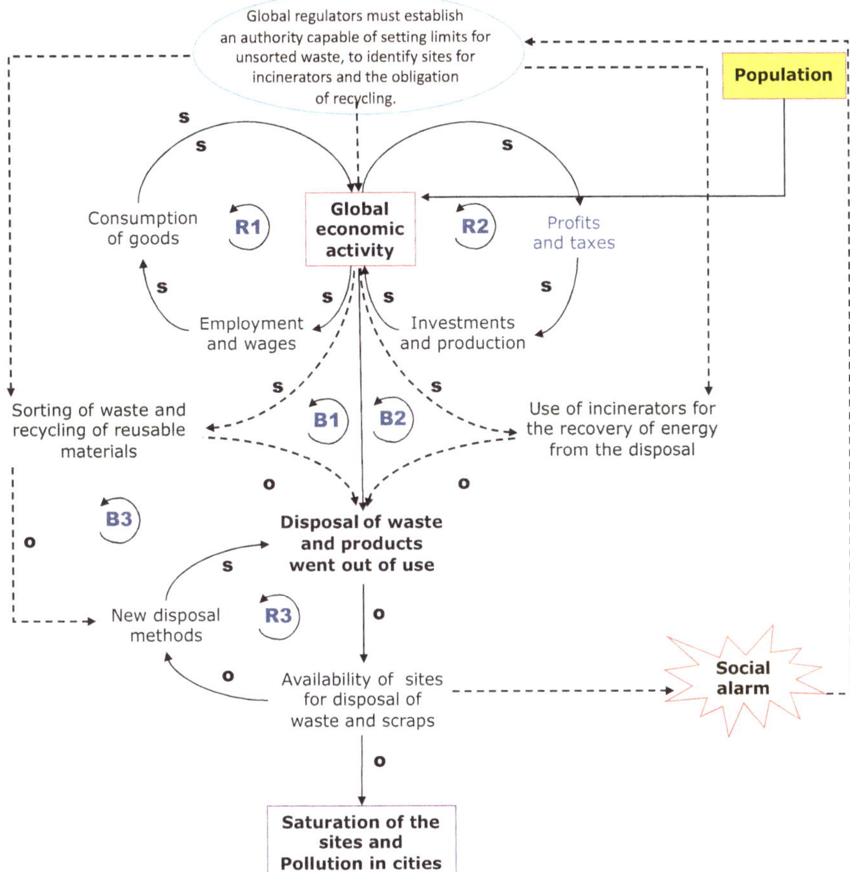

Fig. 7.26 The sustainability of pollution. Actions to counter the exhaustion of sites for the treatment of solid waste

The sustainability problem derives from the relative scarcity of such sites, which in turn results from the action of the archetype of local and individual preferences: communities with potential waste treatment sites refuse to accept the waste produced by inhabitants from other areas.

The areas housing landfills are in all respects Common-Pool Resources, and their gradual local exhaustion can give rise to the Tragedies of the Commons, exactly as described in the preceding section. Environmental disasters in the cities and geographical areas where the waste treatment sites are exhausted are ever more frequent, as well as the "wars" among inhabitants from differing areas who refuse to accept others' waste.

> Ruin is the destination toward which all men rush, each pursuing his own best interest in a society that believes in the freedom of the commons. Freedom in a commons brings ruin to

> all. [...] As a result of discussions carried out during the past decade I now suggest a better wording of the central idea: Under conditions of overpopulation, freedom in an "unmanaged commons" brings ruin to all (Hardin 1985, Online).

Global regulators consider the scarcity of waste treatment sites as a problem that gives rise to a social alarm. In order to avoid the sustainability problem arising from the waste produced by global economic activity, global regulators can carry out four types of intervention. They can:

(a) concentrate on the local problem of waste treatment and try to convince the inhabitants of other geographical areas to accept waste produced elsewhere;
(b) intervene in the waste production process through actions to discourage its production, taxes and penalties for polluting, and economic incentives for consumers and firms to induce them to modify consumption and production functions;
(c) provide incentives for the reutilization of waste or oblige individuals and firms to separate their waste collection and recycle their waste, thereby activating loops [**B1**] and [**B3**];
(d) modify the process of eliminating waste by providing incentives to, or obliging, individuals and companies to utilize waste in energy production and for heating purposes through incineration in increasingly more sophisticated, high-energy-efficiency incinerators; in this way, loop [**B2**] is activated along, once again, with [**B3**] (the arrow is not indicated in the figure).

Figure 7.26 illustrates the functioning of measures (c) and (d), since the others operate exactly the same as those in the example in the preceding section. Unfortunately, once again the archetype of local behavior can come into play, since the inhabitants of the areas where the sites would be located must be convinced to overcome their tendency toward localism and agree to separate waste collection and construct incinerators. Nevertheless, "*Not now and not in my backyard*" is still the most common response. Moreover, it is no surprise that orders from the regional authorities to limit traffic in a certain region are never accepted by all the cities in the region and that, for certain reasons, some local authorities permit a continual increase in the pollution thresholds.

7.10.5 The PSC Model Applied to Stereotypes and Gender Discrimination

Stereotypes can have a significant influence on people's behavior, and their spread and maintenance over time within social groups are a major phenomenon. The term "stereotype," already in use in typography, was introduced into social systems by journalist Walter Lippmann in his book *Public Opinion* (1922), with the meaning of rigid mental images, mental simplifications. The author used the concept of stereotype to provide a clear and effective analysis of the emergence of public opinion, how citizens' opinions become public opinion, and the role played in this process by the mass media.

> Real space, real time, real numbers, real connections, real weights are lost. The perspective and the background and the dimensions of action are clipped and frozen in the stereotype (Lippmann 1922, p. 100).

> What will be accepted as true, as realistic, as good, as evil, as desirable, is not eternally fixed. These are fixed by stereotypes, acquired from earlier experiences and carried over into judgment of later ones (ibidem, p. 107).

Stereotypes are usually defined as simplified and widely shared visions about a recognizable group of people who share certain characteristics, accompanied by the belief that an individual has certain characteristics only because he or she belongs to a group that generally possesses them (Valian 1999). A stereotype is formed when, within a social system, a characteristic of a not insignificant percentage of a category is extended to the totality of individuals. Stereotypes can, however, also refer to objects (if the black cat crosses in front of you, turn back), historical places, or eras, and they can be positive (e.g., Italians love the opera, the French are romantic), negative (e.g., Italians are all in the mafia, all rock musicians take drugs), or neutral (e.g., Christmas is not Christmas without snow and a crackling fire).

> From "girls suck at maths" and "men are so insensitive" to "he is getting a bit senile with age" or "black people struggle at university", there's no shortage of common cultural stereotypes about social groups (Kahn 2018, online).

Sometimes the stereotype is a caricature or reversal of some positive characteristics possessed by the members of a group, exaggerated to the point of becoming detestable or ridiculous, prompting people to *discriminate* and *reject* those who possess those characteristics. In any case, the stereotype gives one a coherent world view that makes individuals feel they are "*on the right side*." Stereotypes can help make sense of the world. They are a form of categorization that helps to simplify and systematize information (Wikipedia, Stereotypes), but at the same time they hinder knowledge of reality. "*Researchers often argue that stereotypes are tenacious; they tend to have a self-perpetuating quality that is sustained by cognitive distortion*" (Hentschel et al. 2019). According to Lippman:

> The systems of stereotypes may be the core of our personal tradition, the defenses of our position in society. They are an ordered, more or less consistent picture of the world, to which our habits, our tastes, our capacities, our comforts and our hopes have adjusted themselves (Lippmann 1922, p. 63).

Lippman acknowledges that stereotypes are pervasive throughout the social system, regardless of the age of the subjects, although they are particularly active in organizations, where judging and deciding on the basis of stereotypes often leads to unethical and usually even erroneous behavior. He thus identifies the cultural origin of stereotypes within organizations:

> The members of a hierarchy can have a corporate tradition. As apprentices they learn the trade from the masters, who in turn learned it when they were apprentices, and in any enduring society, the change of personnel within the governing hierarchies is slow enough to permit the transmission of certain great stereotypes and patterns of behavior. From father to son, from prelate to novice, from veteran to cadet, certain ways of seeing and doing are taught. These ways become familiar and are recognized as such by the mass of outsiders (Lippmann 1922, p.145).

There is a significant number of stereotypes, which vary according to the areas over which the social systems extend and in relation to the historically formed culture. Among the most common are the following (SPEEXX online):

- Gender stereotype
- Religious and racial stereotypes
- Stereotypes about parents
- Stereotypes about teachers and evaluators
- Stereotypes related to age and different physical abilities
- Stereotypes related to the professions
- Stereotypes related to substance addiction

Stereotypes are particularly harmful in organizations, where they are widespread because, whenever people need to be assessed, such as in recruitment processes, career advancement, task assignments, or the handing out of awards and punishments, they can lead to discriminatory behavior. Precisely because behavior is (often consciously) influenced by stereotypes, Peter Senge emphasized this danger in organizations and, while not explicitly mentioning stereotypes, pointed out the importance of expressing his Mental models (Sect. 1.7.9).

> 'Mental models' are deeply ingrained assumptions, generalizations, or even pictures or images that influence how we understand the world and how we take action (Senge 2006 p. 8).

Among the different types of stereotypes, I would like to briefly mention the gender stereotype, which defines how people "are," but also how they "should be," thereby creating expectations about the roles that men and women should take on as biological men and women.

> Gender stereotypes are generalizations about what men and women are like, and there typically is a great deal of consensus about them. According to social role theory, gender stereotypes derive from the discrepant distribution of men and women into social roles both in the home and at work (Hentschel et al. 2019, online).

> Gender stereotypes reflect normative notions of femininities and masculinities, women and men. Yet, like all aspects of gender, what constitutes stereotypical femininity or masculinity varies among cultures and over historical time. Gender stereotypes typically portray femininities and masculinities as binary opposites or dualisms, as, for example, between emotionality and rationality (Gendered Innovations 2020, online).

Stereotypical opinions on gender differences are not only present in the modern world but are rooted in history, and over the centuries they have produced a hierarchy between the status of man and that of woman (Schiebinger 2014). According to the gender stereotype, the characteristics attributed to the male are the ability to act, practicality, self-affirmation, competence, independence, strength, manhood, individuality, and a greater interest in social and public life, which is why only men were found in military careers until a few decades ago. The characteristics attributed to women, on the other hand, are emotional capacity, sociality, emotiveness, communicative skills, affectivity, altruism, interdependence, propensity for two-way relationships, sweetness, delicacy, softness, the tendency to manage one's own intimate world, as well as the tendency to favor motherhood, child care, and family care. For this reason, in management careers females have for the most part been discriminated against. According to the *Glass Ceiling Theory*, there is an invisible barrier that keeps women from rising in a hierarchy (Ferber and Nelson, 1993); therefore, *as the prestige of a job increases, the share of women in that position decreases*. According to *Tournament Theory*, hierarchical levels and salary differences should be based on productivity or merit, but when evaluators are influenced by stereotypes, especially of gender or race, such differences are hierarchical and economic and based on relative differences between individuals, so that the influence of stereotypes distorts the rationality of choices and leads to a waste of resources (Gilovich et al., 2002). These effects can be represented by the archetype of *Success to the Successful* archetype, as indicated in Fig. 7.27.

"Think Manager, Think Male" (for the role of secretary the opposite is true) is a typical gender stereotype that leads to discrimination in the evaluation and selection of managers (Schein et al., 1996) and reveals a negative feature of the cognitive process that distorts the use of information (Hudak, 1993), thereby producing systematic errors in the results obtained in both production organizations and at school (Biernat and Kobrynowicz, 1997).

A typical systematic error in the selection process is due to the *Pygmalion Effect:* stereotypes condition the evaluation and selection processes in such a way that the subject that most reflects the preconceived judgment (stereotype) of the evaluator is chosen (Rosenthal and Jacobson, 1968). In addition, if the evaluator is prejudiced regarding the abilities of the one who is being evaluated and expects the latter's performance to match these expectations, we have an instance of an interpersonal *self fulfilling prophecy*. The use of stereotypes by the evaluator results in an increase in discrimination, as he or she judges people by the group to which they belong. Discriminated employees can witness a fall in their self-esteem, and being less motivated they will be less committed to their work; therefore, their performance

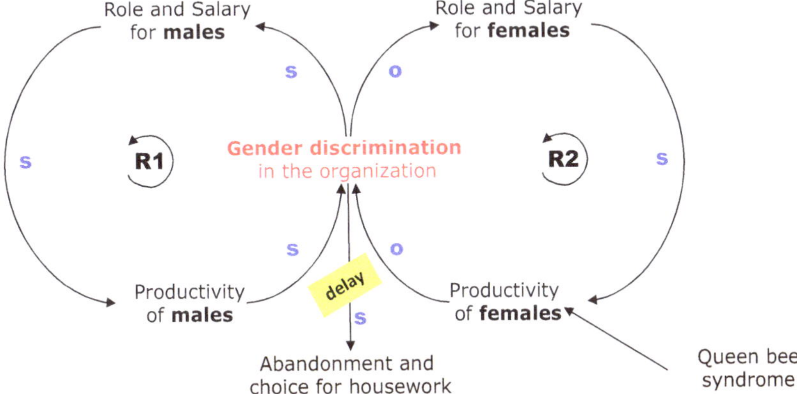

Fig. 7.27 Gender discrimination in the organization

will suffer, thus strengthening the core of truth contained in the stereotype, which will be considered even more of a valid yardstick for future evaluations. The *Pygmalion Effect* can be contrasted with the so-called *Galatea Effect,* which produces a *self-fulfilling prophecy* for the individual who evaluates himself. This effect occurs when an individual's high expectations of himself translate into high performance and the results achieved influence self-efficiency: the more such individuals are convinced of their ability to achieve a goal, the more likely it is to happen. Surprisingly, when a female excels at work she behaves according to the "Queen Bee Syndrome", in which a woman who has success in career advancement, and thus outshines other colleagues, does everything possible to hinder the potential professional growth of other women, favoring male employees to ensure that she is the only woman in power (Fig. 7.27).

Even in educational systems and regarding student performance the gender stereotype (as well as many other more dangerous ones) can lead to significant inequalities:

> Research shows that this threat dramatically depresses the standardized test performance of women and African Americans who are in the academic vanguard of their groups (offering a new interpretation of group differences in standardized test performance), that it causes disidentification with school, and that practices that reduce this threat can reduce these negative effects (Steele 1997, 613).

> Stereotypes are deeply embedded within social institutions and wider culture. They are often evident even during the early stages of childhood, influencing and shaping how people interact with each other. For example, video game designers designed a game platform for girls in pink because that is what the parents (who purchase the game) perceived their girls wanted. The girls themselves preferred darker metallic colors (Gendered Innovations (2020, online).

The stronger the stereotype, the more incisive its effects will be and the longer they are likely to last. The immediate recognition of certain stereotypes means they are widely used in the production of effective advertising and in sitcoms.

Since discrimination based on stereotypes of gender and, in particular, of religion and race are typically the result of widespread cultural distortion, the fight against stereotypes is above all a battle for change in cultures that judge, evaluate, and select on the basis of unfounded prejudices. To control the effect of discrimination due to the social spread of stereotyped judgments, it is necessary to rotate the "wheel of culture" (Fig. 7.16), establishing appropriate structures according to rules, regulations, and instructions (Sect. 7.9) that oppose the practice of discrimination in schools and organizations and in advertising messages of all kinds. If the evaluation distortions result from cultural factors, it is necessary to act against these stereotypes and produce a cultural change to identify and optimally allocate "talent." The identification of talent has social value because if it is not allocated properly, actual output will fall short of potential output. According to Tournament Theory (Lazear and Rosen 1981), social welfare is greater if the most gifted occupy the top positions. The model in Fig. 7.28 illustrates a control system for a Change Management strategy to fight stereotypes. It contains some abstract indications that need to be adapted to real-world situations.

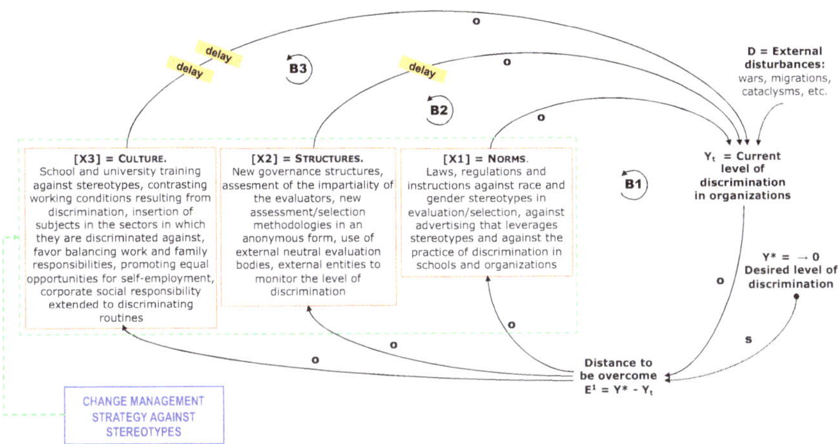

Fig. 7.28 Control system against the stereotype culture (reference to Fig. 7.17)

7.11 Summary

Continuing our ideal journey among the *Rings* that make our world possible, this chapter has examined the control systems that characterize the macro biological environment, in particular:

(a) The *Rings* that regulate the coexistence of individuals in social systems (Sect. 7.1 and Figs. 7.2, 7.3, and 7.4)
(b) The *Rings* for the control of autopoietic variables (Sects. 7.3 and 7.4) based on the governance policy (Figs. 7.1 and 7.5)
(c) The *Rings* that allow the combinatory systems which form in collectivities to be activated (Sect. 7.5) and controlled (Sect. 7.6) (Figs. 7.8 and 7.9)
(d) The *Rings* that regulate sustainability of social behavior, population growth and commons depletion (Sects. 7.7, 7.10.4 and Figs. 7.10, 7.11, 7.12, and 7.13)
(e) The *Rings* acting in social systems as CAS (Sect. 7.8)
(f) The *Rings* that produce change in social systems; the PSC model of change (three-wheels, three-speed) (Fig. 7.22) was presented (Sect. 7.9)
(g) The PSC model applied to stereotypes and gender discrimination (Sect. 7.10.5 and Figs. 7.27 and 7.28)

Chapter 8
The Magic Ring in Action: The Biological Environment

Abstract This chapter deals with the vast topic of population dynamics and control both from a quantitative view, the change in the number of individuals in a population, and a qualitative one, the variation or change in the phenotypes in evolutionary processes. Taking as a basis the Volterra–Lotka equations, Excel and Powersim (Powersim Software. 2018. http://www.powersim.com) simulations are presented for specific examples. The concept of "external" or "exogenous" control is also introduced, which is carried out by man through interventions aimed at increasing or reducing the size of one or more interconnected populations. Also considered is the evolution of populations of non-biological entities, particularly, those of robots and organizations that form the nodes of production networks.

Keywords Population quantitative dynamics · Population qualitative dynamics · Volterra · Lotka equations · Ecosystem · Population control · Theory of evolution · Evolution in non-biological populations · Evolution in production networks

This chapter deals with the vast topic of population dynamics both from a quantitative view, the change in the number of individuals in a population, and a qualitative one, the variation or change in the phenotypes in evolutionary processes. In a broader sense, what regards the populations living in a defined area or ecosystem is also applicable to species: that is, the set of individuals who show the same phenotypic characteristics. Quantitative dynamics are studied not only in an isolated population but also in the more interesting case of two or more populations interconnected in a simple *trophic* food chain in a prey–predator model, in the broadest sense of the term, thereby forming an *ecosystem*. This chapter introduces the concept of "population control." Two or more interconnected prey–predator populations regulate, in turn, their dynamics, even though unconsciously, exercising a control that has been defined as "natural" and "endogenous." The concept of "external" or "exogenous" control is also introduced, which is carried out by man through interventions aimed at increasing or reducing the size of one or more interconnected populations that are approaching dangerous minimum limits (populations of species in danger of

P. Mella, *The Magic Ring*, Contemporary Systems Thinking,
https://doi.org/10.1007/978-3-030-64194-8_8

extinction) or maximum ones (populations of invasive species) held to be advantageous to and congruous with the ecosystem. Also highlighted is the connection between *qualitative* and *quantitative* dynamics, in the sense that every mutation in the individuals of a population that increases the potency of the defense apparatus in prey, or the hunting apparatus in predators, necessarily modifies the rates of fecundity or extinction of that population, which necessarily affects the quantitative dynamics of all the interconnected populations. Taking as a basis the Volterra–Lotka equations, Excel and Powersim (2018) simulations are presented for specific examples in which population dynamics is treated under the assumption of unlimited resources, or resources that are limited but reproducible or limited until depletion. The case of environmental catastrophes that externally modify the dynamics of one or more populations is also touched upon. I will then adopt traditional Darwinian theory to consider a possible interpretation of evolution, understood as a natural process of *modification* and *enhancement* of the control systems of the individuals of a species, highlighting how evolution changes the efficiency of the effectors, detectors, and regulators that characterize the control systems of biological individuals. The chapter concludes with considerations regarding the dynamics and evolution of nonbiological species. In particular, production *networks* will be examined to highlight how their composite nodes (production and nonproduction organizations) condition, through their evolutionary dynamics, the evolution of the entire network toward increasingly efficient and effective structures.

8.1 Magic Rings in the Macro Biological Environment

Having already examined the *Rings* that permit individuals to exist, let us now examine control systems with regard to *biological environments*, considering above all how the dynamics of any type of biological collectivity can be controlled. As defined in Sect. 7.5, a *collectivity* is a plurality of *agents* that show an individual micro behavior over time—producing a micro effect—but, considered together, is capable of developing a macro behavior—and/or macro effect—which is attributed to the collectivity as a definable unit.

Collectivities can be composed of persons or animals, active biological organisms, or reactive ones (plants, microorganisms); in any case, if considered from a certain distance, from an endogenous perspective (von Bertalanffy 1968), collectivities can be observed as units (or macro agents) that appear distinct with respect to the individuals they are composed of and show an autonomous observable macro behavior (change in the state of a collectivity over time)—which can lead to observable macro effects—which derives from the interactions of the micro behaviors (change in the agents' states over time) or from the derived micro effects of the agents (Mella 2017). Collectivities can be observable (e.g., swarms, flocks, crowds, spectators at a stadium, students in a classroom, people talking in a crowded room, dancers doing the Can Can), or simply imaginable, for example, trailer trucks traveling a stretch of highway in a month, the noble families of Pavia who erected the 100 towers in the span of two centuries, a group of scientists who dedicate themselves to a branch of research, the consumers of a particular product during its entire life cycle, stockbrokers working on a certain day in world or European stock markets).

Collective phenomena are the essence of life at any level, both for human beings and animals, who operate in populations, societies (cities, nations, tribes, associations, teams, social units) and groupings of various kinds (crowds, hordes, flocks, flights, herds, schools, and so on).

I define a *population* as a collectivity of similar individuals (normally of the same species) that live in a given geographical area and can reproduce by crossbreeding with other individuals in the collectivity. A group of populations, even of different species, living contemporaneously in a given area and interacting for living space and food, form a biological community constituting an ecosystem. According to Capra, the languages of the early ecologists and of organismic biology were very similar; therefore, *biological communities* were compared to *organisms*. Subsequently, the plant ecologist A. G. Tansley introduced the term "ecosystem" to refer to animal and plant communities (Capra 1996b) The current concept of ecosystem refers to "a community of organisms and their physical environment interacting as an ecological unit."

> Since the language of the early ecologists was very close to that of organismic biology, it is not surprising that they compared biological communities to organisms. For example, Frederic Clements, an American plant ecologist and pioneer in the study of succession, viewed plant communities as "superorganisms". This concept sparked a lively debate, which went on for more than a decade until the British plant ecologist A. G. Tansley rejected the notion of superorganisms and coined the term "ecosystem" to characterize animal and plant communities. The ecosystem concept is defined today as "a community of organisms and their physical environment interacting as an ecological unit" (Capra 1996b, p. 33).

There are interesting forms of collective behavior that occur in combinatory systems (Sect. 7.5) and in CAS (Sect. 7.8). Following the definitions set out in Sect. 7.5, a *combinatory system*, or *simplex system*, must be understood as any collectivity formed by similar elements or agents (not organized by hierarchical relations or interconnected by a network or tree relationship) that displays an *analogous* micro behavior over time and produces similar micro effects. Together, the micro behaviors and micro effects give rise to a macro behavior (and/or macro effect) attributable to the collectivity as a whole. The collective behavior, though composed of the micro behaviors, conditions or directs the subsequent micro behaviors of the agents following a typical micro–macro feedback process (Mella 2017).

Collectivities made up of heterogeneous elements that interact to form dynamic wholes, which are emerging and not predictable a priori, shall be defined as *complex systems*. The CAS approach, in particular (Kauffman 1993, 1996; Castelfranchi 1998, Gilbert 1995; Gilbert and Troitzsch 1999) studies how collectivities interact and exchange information with their environment to maintain their internal processes over time through adaptation, self-preservation, evolution, and cognition (in the sense of Maturana and Varela 1980, p. 13) and achieve collective decisions within a relational context of micro behaviors (Rao and Georgeff 1992; Wooldridge and Jennings 1994).

A collectivity of organized agents *specialized* according to function, functionality, functioning, and topology forms an *organization*, a social system (a *social machine* according to Maturana and Varela 1980), where the collective and individual behavior is determined by a network of stable relations, that is, the organization of the agents and processes for the purpose of creating a stable structure:

> *Organization and Structure*. The relations between components that define a composite unity (system) as a composite unity of a particular kind constitute its organization. In this definition of organization the components are viewed only in relation to their participation in the constitution of the unity (whole) that they integrate. This is why nothing is said in it about the properties that the components of a particular unity may have other than those required by the realization of the organization of the unity (Maturana and Varela 1980, p. xix).

This *section* presents particular forms for the control of several fundamental variables regarding populations. I shall begin by observing that populations of individuals of a given species—or even an entire species, such as populations that existed in a given period—present at least two interesting trends: a *quantitative* dynamics (i.e., changes in the numbers of individuals over time) and a *qualitative* one (changes in the traits of the individuals over time).

The quantitative dynamics of population have always been an object of study in various disciplines, producing theories and systemic models, often mathematical, to describe and simulate the *quantitative* dynamics (that is, changes in the numbers of individuals over time) of a single population or of two or more interacting populations (Morris and Doak, 2002; Nowac, 2006). In particular, one of the major challenges of the twenty-first century is the understanding of population dynamics in complex ecosystems (Liow et al. 2011; Post and Palkovacs 2009; Schoener 2011).

However, population dynamics can also be studied from a *qualitative* point of view in terms of the evolution of the phenotypes (changes in the traits of the individuals over time), with mutations that are so relevant as to also lead to the transformation of an original population into one that is new and different from the preceding one. This is the typical area of study of evolution theories. In many cases, the two types of dynamics, *quantitative* and *evolutionary*, have a mutual relationship (Buchanan 2000): a population quantitatively in decline can change the direction of its dynamics when a relevant "mutation," which spreads rapidly among all the individuals, has the effect of strengthening its defense weapons, if the population is the *prey of another*, or its attack weapons, if the population *preys on others*. Since this mutated population interacts with others, its evolutionary changes will necessarily modify the dynamics of the interacting populations. The prey–predator system is a prime example of the importance of qualitative factors, in which any interaction with the qualitative traits of the populations is expressed by a quantitative oscillation (Kareiva et al. 1990).

The *first* objective is to demonstrate that in the world of living organisms—and also that of species, from a broader perspective—the populations of individuals that *interact* in the "struggle for survival" and "for reproduction," whether prey or predators, produce forms of "self-control" of their dynamics. This self-regulating behavior can be conceived of as the action of natural endogenous Control Systems, which allow populations to survive, develop, and maintain themselves in homeostatic equilibrium by regulating the relations among individuals and the environment and by activating the basic variables required for their survival. These natural Control Systems of the dynamics of biologically interconnected populations can, in principle, act on any number of populations that interact as prey and predator, thereby forming an *ecosystem*.

The *second* objective is to demonstrate that the "natural" self-produced *quantitative* dynamics—that is, those which are self-produced—that derive from the unconscious behavior of individuals of the same populations can be controlled by *external interventions* carried out to *support* or *inhibit* growth or to *interrupt* or *reverse* the decline in growth, as occurs in populations of invasive species (inhibiting growth) or populations of useful but endangered species (interrupting the decline in growth): cultivation and extirpation, of breeding and hunting that exist.

The *third* objective is to undertake an initial analysis of the *qualitative dynamics* of biological populations by proposing a model to represent their *evolutionary* dynamics and attempting to link these dynamics with the quantitative dynamics, thereby demonstrating that "variations" or "mutations" in a population (more generally, in a species) that alter its relationship with other populations produces an alteration, even one that is profound and fast, in its quantitative dynamics.

Chapter 7 examined the social systems and the forms of control characterizing them. Chapter 9 will deal with the analysis of *organizations* and the *Rings* that allow the former to exist.

8.2 Magic Rings That Regulate the Dynamics of Populations

I shall consider above all the *quantitative dynamics* of populations and try to reflect on the fact that the oceans are not only populated by sharks and the savannas by lions, even if these species are among the most efficient hunting machines in their environments; the seas abound in sardines and the savannas are populated by antelopes, even if both of these species are prey for large populations of predators. Above all, populations, and species in general, are normally self-balancing by means of some *natural* control system that stops them from increasing infinitely.

The model in Fig. 8.1 shows the main and general factors that determine the quantitative dynamics of any population.

Loop [**R**] in the model in Fig. 8.1 makes clear that every population increases naturally at a net *growth rate*, which is equal to the difference between the birth and death rates. When the birth rate falls below the death rate the population gradually decreases, running the risk of natural extinction. Loop [**B1**] causally links the dynamics of the population to that of the *prey* that provides its food. Under normal conditions, the more the population grows, the more the number of prey and quantity of food decreases, so that the population undergoes an initial balancing. If the prey belongs to another population, there is a joint dynamic between the two populations. If the reduction rate of the prey prevails over that of the natural rate of increase for this population, then the population may even risk extinction. Loop

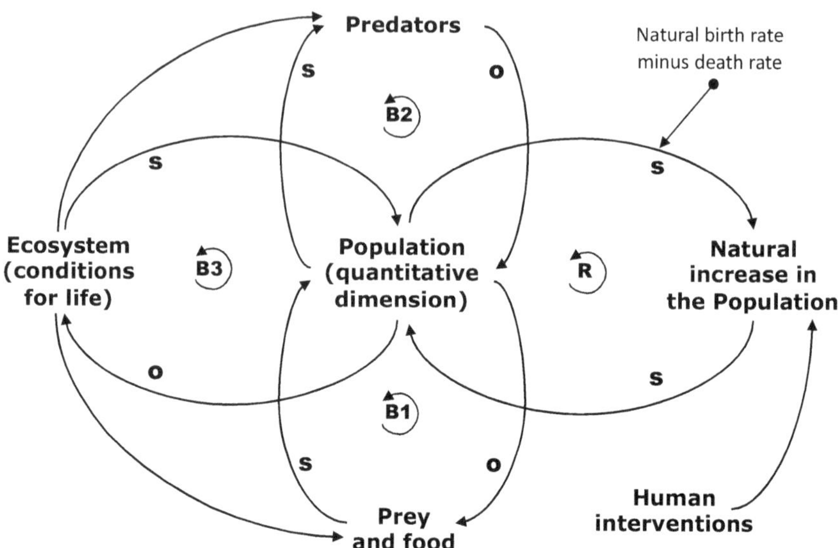

Fig. 8.1 Main factors in population dynamics

[**B2**] illustrates the same joint dynamics relationship; however, it considers the observed population as the prey of a population of predators. Taken together, loops [**B1**] and [**B2**] form an initial example of a *food chain*, in which the observed population represents the intermediate *Ring*. Finally, loop [**B3**] shows the relation between the dynamics of the population and the general conditions of life in the *ecosystem*. If the ecosystem offers excellent conditions of life, the population will increase, other circumstances held constant. Normally the opposite trend also holds: the more the population increases, the more the ecosystem deteriorates, as occurs in the ecosystem of the human population. A population that has suffered serious events (fires, floods, etc.) can have fewer possibilities for growth and may also represent easy prey for its predators, in turn encountering difficulties in finding prey or food.

The general model in Fig. 8.1, which can be adapted to every population, even the *vegetable kingdom*, does not illustrate any control system for population dynamics but rather the balancing factors that contrast the reinforcing loop that would lead to natural expansion. Describing and simulating these control systems are difficult tasks, since populations have dynamics that do not appear to approach some limit or "natural" value. Yet even the dynamics of biological collectivities seems to be determined by some control system, even though this is difficult to identify. For many populations man uses the levers available to him to set objectives and limits in order to control their dynamics in an evident manner. Proof of this lies in the hundreds, perhaps thousands, of forms of cultivation and extirpation, of breeding and hunting that exist. In other cases man radically influences the dynamics of populations through forms of protection and tolerance—which leads to overpopulation with respect to the environment (deer, rats)—as well as through actions that threaten their survival (sea turtles deprived of their eggs and whales hunted everywhere, as we shall see in Sects. 8.3 and 8.4). In Sect. 7.4, several social levers were considered—the battle against infant mortality, the improvement in living conditions and health, economic incentives to produce large families, laws and disincentives to discover new births, and so on—which can be used to control the population dynamics in some countries.

In other cases, populations are regulated by *Rings* that make use of "natural" levers. One such *Ring* includes a very potent lever: the natural instinct of adult males to eliminate a large number of newborns, especially if they are the offspring of other males, as occurs, for example, with hippopotamuses and lions. A similar mechanism can be observed in many other species, though with a variation: often it is the firstborn—for example, in a brood of birds—that stops the other eggs from hatching (Dawkins 1989, p.133). In both cases this control system, though seemingly cruel, ensures that population growth slows down when reaching an upper limit.

The topics below will be dealt with in the *sections* that follow:

1. Dynamics of a population without growth limits
2. Dynamics of a population with growth limits
3. Dynamics of two populations that interact as prey and predator
4. Exogenous control of the dynamics of two populations
5. Exogenous control of the dynamics of three or more populations comprising an ecosystem
6. Qualitative dynamics: the evolution or alteration of the quantitative dynamics

Excel will be used for simulations regarding a limited number of populations, as it is powerful as well as flexible tool, while Powersim will be used for three or more populations contemporaneously (Powersim Software).

8.3 Dynamics of a Population Without Growth Limits

When there are no limits to growth and no interaction with other populations, a population of size A_n, at "time *n*," with an *increasing* positive rate, "ri = rb – rd," representing the difference between the *birth* rate, "rb," and *death* rate, "rd," tends to grow infinitely, according to the simple *reinforcing loop* in Fig. 8.2, which assumes that the periodic increase in population is added to the pre-existing population based on the rate of *accumulation*, "racc," which expresses the percentage of newborns that survive to become adult individuals. For example, the survival rate of newborn "Caretta Caretta" turtles is very low (racc = 0.1%) because most of the small ones end up as victims of the large quantity of predators present in large numbers on beaches and in the coastal waters, while another percentage, disoriented by the lights of the houses, heads inland rather than into the sea (Sea Turtles, online). If there were no *limits to growth*, this would produce a population able to populate a very vast territory.

It is easy to derive the following formal model (Sect. 1.4):

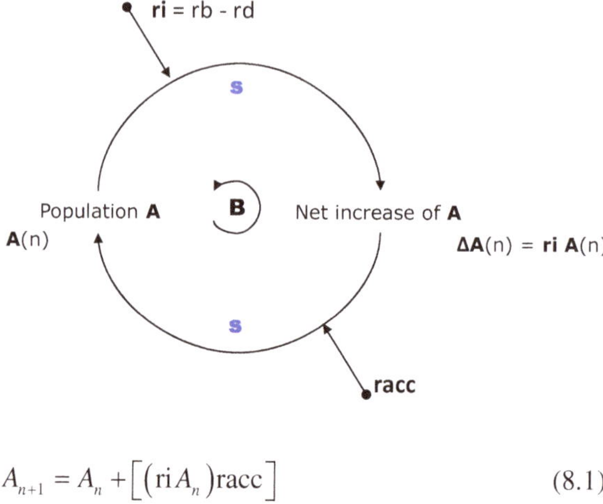

Fig. 8.2 Reinforcing loop producing an increase in a population without limits to growth

$$A_{n+1} = A_n + \left[\left(\mathrm{ri} A_n \right) \mathrm{racc} \right] \tag{8.1}$$

Setting racc = 1 and generalizing, after specifying the number of initial individuals, A_0, we obtain the well-known law of accumulation for each "*n*":

$$A_n = A_0 \left(1 + \mathrm{ri} \right)^n$$

Unlimited growth is rare since limiting factors normally intervene that are linked to the extension of territory, the availability of food, and the presence of competing populations. Only in very large areas with abundant food and a limited number of predators do we find examples that approximate unlimited, explosive growth. A striking case is the spread of rabbits in Australia, which has become a big ecological problem on that continent. Rabbits were imported to Australia at the end of the eighteenth century (probably in 1788) to provide a supply of meat to the inhabitants. Around the middle of the nineteenth century, 24 rabbits (along with other animals) were brought from England as prey for sport hunters. The imported rabbits, particularly strong and fast, crossbred with the rabbits that were already there, causing them to quickly proliferate, setting off a typical *population explosion* for rabbits. No attempt at control (hunting, the destruction of warrens, the infection of rabbits with the Myxoma virus, etc.) was successful (PestSmart Connect 2018, online; Queensland Country Life 1949).

This population explosion for rabbits follows the typical reinforcing *Loop* in Fig. 8.2, which produces an exponential increase clearly shown in the model in Fig. 8.3, for whose construction we have introduced the (didactic) assumptions a–f (note that the simulation model allows for these assumptions to be modified):

(a) For simplicity's sake, the population of rabbits is indicated by the variable X_n and its increase by Y_n; the growth rate of Y is indicated as a function of X by $g(Y/X)$.
(b) A period of 150 years; an initial population of $X_0 = 24$ pairs; each of the newly imported specimen pairs with one already present.

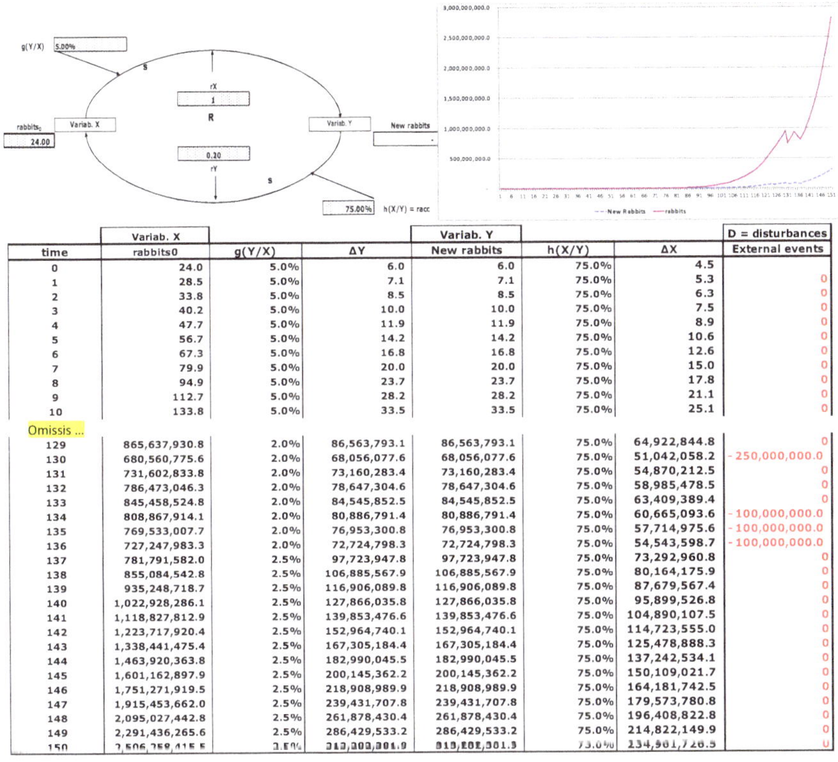

time	Variab. X rabbits0	g(Y/X)	ΔY	Variab. Y New rabbits	h(X/Y)	ΔX	D = disturbances External events
0	24.0	5.0%	6.0	6.0	75.0%	4.5	
1	28.5	5.0%	7.1	7.1	75.0%	5.3	0
2	33.8	5.0%	8.5	8.5	75.0%	6.3	0
3	40.2	5.0%	10.0	10.0	75.0%	7.5	0
4	47.7	5.0%	11.9	11.9	75.0%	8.9	0
5	56.7	5.0%	14.2	14.2	75.0%	10.6	0
6	67.3	5.0%	16.8	16.8	75.0%	12.6	0
7	79.9	5.0%	20.0	20.0	75.0%	15.0	0
8	94.9	5.0%	23.7	23.7	75.0%	17.8	0
9	112.7	5.0%	28.2	28.2	75.0%	21.1	0
10	133.8	5.0%	33.5	33.5	75.0%	25.1	0
Omissis ...							
129	865,637,930.8	2.0%	86,563,793.1	86,563,793.1	75.0%	64,922,844.8	0
130	680,560,775.6	2.0%	68,056,077.6	68,056,077.6	75.0%	51,042,058.2	-250,000,000.0
131	731,602,833.8	2.0%	73,160,283.4	73,160,283.4	75.0%	54,870,212.5	0
132	786,473,046.3	2.0%	78,647,304.6	78,647,304.6	75.0%	58,985,478.5	0
133	845,458,524.8	2.0%	84,545,852.5	84,545,852.5	75.0%	63,409,389.4	0
134	808,867,914.1	2.0%	80,886,791.4	80,886,791.4	75.0%	60,665,093.6	-100,000,000.0
135	769,533,007.7	2.0%	76,953,300.8	76,953,300.8	75.0%	57,714,975.6	-100,000,000.0
136	727,247,983.3	2.0%	72,724,798.3	72,724,798.3	75.0%	54,543,598.7	-100,000,000.0
137	781,791,582.0	2.5%	97,723,947.8	97,723,947.8	75.0%	73,292,960.8	0
138	855,084,542.8	2.5%	106,885,567.9	106,885,567.9	75.0%	80,164,175.9	0
139	935,248,718.7	2.5%	116,906,089.8	116,906,089.8	75.0%	87,679,567.4	0
140	1,022,928,286.1	2.5%	127,866,035.8	127,866,035.8	75.0%	95,899,526.8	0
141	1,118,827,812.9	2.5%	139,853,476.6	139,853,476.6	75.0%	104,890,107.5	0
142	1,223,717,920.4	2.5%	152,964,740.1	152,964,740.1	75.0%	114,723,555.0	0
143	1,338,441,475.4	2.5%	167,305,184.4	167,305,184.4	75.0%	125,478,888.3	0
144	1,463,920,363.8	2.5%	182,990,045.5	182,990,045.5	75.0%	137,242,534.1	0
145	1,601,162,897.9	2.5%	200,145,362.2	200,145,362.2	75.0%	150,109,021.7	0
146	1,751,271,919.5	2.5%	218,908,989.9	218,908,989.9	75.0%	164,181,742.5	0
147	1,915,453,662.0	2.5%	239,431,707.8	239,431,707.8	75.0%	179,573,780.8	0
148	2,095,027,442.8	2.5%	261,878,430.4	261,878,430.4	75.0%	196,408,822.8	0
149	2,291,436,265.6	2.5%	286,429,533.2	286,429,533.2	75.0%	214,822,149.9	0
150	2,506,[illegible]	[illegible]	[illegible]	[illegible]	75.0%	234,961,726.5	0

Fig. 8.3 Rabbit explosion (population without limits to growth)

(c) A rate of population increase (fecundity rate–mortality rate) at the start of the process of $g(Y/X) = 5\%$; for the following simulation, it is assumed that, partly as a result of the attempts by authorities to limit the phenomenon, the rate "g" diminishes with the increase in population: $g_{50} = 4\%$; $g_{74} = 3\%$; $g_{110} = 2{,}5\%$; $g_{124} = 2\%$; at $n = 137$, $g(Y/X)$ increases to $g_{137} = 2.5\%$.

(d) It is assumed that for each generation only a percentage racc $= h = 75\%$ of the new individuals contributes to increasing the population; the remaining ones are prey for the predators.

(e) We assume there are five pairings per year; this datum is inserted into the model by simply setting $r = 1/5 = 0.20$ to indicate that in 1 year the population undergoes a fivefold increase.

(f) Several catastrophic events have been hypothesized for the rabbit population; at $n = 130$, a cataclysm reduces the population by 250 million pairs in a single year. In the 3-year period from $n = 134$ to 136, the *Rabbit hemorrhagic disease* epidemic reduces the rabbit population by 100 million pairs.

The model shows that after 150 years—the equivalent of 750 generations—the rabbit population reaches about $X_{150} = 2.506$ billion pairs. Since Australia has a surface area of 7,617,930 km^2, after 150 iterations there will be an average population of 329 pairs of rabbits/km^2.

If we assume the authorities succeed in introducing new predators for rabbits, so that the number of newborns not yet at the reproductive age falls to $h = 50\%$ on average, even without the intervention of any cataclysms, then the population would reach a density of only one and a half pairs per square kilometer. Of course, the model is presented for the sole purpose of illustrating the action of the *reinforcing loop* in Fig. 8.2, representing a typical single population growth phenomenon; knowledge of the real birth and death rates of rabbits in Australia would allow us to refine the results of the model. Note that the population explosion model is valid for any type of "population" whose increase is exponential. The *reinforcing loop* in Fig. 8.2, which regulates this phenomenon, can also represent an annihilation process if the action rate "g" is negative and represents a rate of decrease that is self-reinforcing to the point that X is eliminated.

8.4 Dynamics of a Population with Limits to Its Available Resources: Malthusian Dynamics

The quantitative dynamics of a population may be driven by different factors (Steffan-Dewenter and Schiele 2008). In a bottom-up approach, resources act as density-independent factors that can influence the growth of the population (Eber 2004; Sibly et al. 2007). In a top-down approach, the natural enemies are the factors that can regulate dynamics (Berryman 2001). In particular, when coexisting species are competing for the same resources, food, and vital territory (Krizna et al. 2018), individuals more capable of obtaining resources have a greater chance of survival and reproduction (Malthus 1798). Therefore, the model in Fig. 8.2 is unrealistic and openly contradicts the "law of dynamic instability" (Sect. 1.5): no population grows infinitely, since growth is always balanced by at least *three* factors operating individually or jointly: (1) a maximum limit of available resources, (2) the presence of a prey–predator population, and (3) human external controls.

The first factor is particularly relevant and is always present in any population. When we assume that to survive the individuals consume resources that are limited, then population growth cannot be infinite; instead, it is constrained by the quantity of available resources, which represents a limit to growth, as shown in the model in Fig. 8.4. Going back to Malthus's work, *An essay of the principle of the population as it affects the future improvement of society* (Malthus 1798), the population is expected to grow only to the "stabilization point," which is limited by resources. The potential for population growth and quality of life is still represented by his fundamental idea, despite many modifications over time.

In Sect. 3.4 I have defined as "variable-objective systems" those control systems whose objective, Y_t*, changes over time, so that the system appears to follow a continually changing objective. Figure 8.5 presents a typical example of a *population dynamics* where there is a population—whose initial size (equal to 1000 units) is

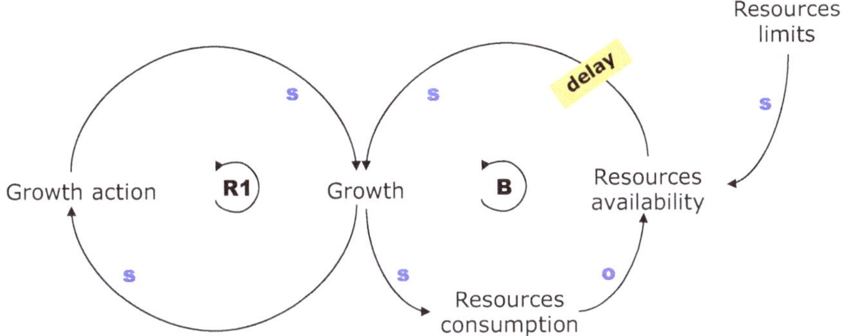

Fig. 8.4 Available resources limit growth

specified—which varies annually due to births and deaths. Since the birth rate is assumed to be greater than the death rate, year-to-year there is a gradual increase in population. However, there is also assumed to be a limited quantity of food (for convenience sake, food supplies that each year increase by the same amount; that is, food is a *renewable* resource, but it cannot increase beyond a maximum limit). Initially we assume that the amount of resources (7,500 units) is more than what is necessary for consumption, which is 5,000 units. Thus, the population can expand until the limit imposed by the availability of limited resources, which are completely used up at $t = 16$.

The simulation program also allows for the possibility that the population, having reached the resource saturation point, can initiate a *production process* to increase the resources and continue to expand; however, this growth will not continue for long since we assume there is a limit (10,000 units) to the expansion of the food due, for example, to the constraint from the need for other resources for its production (land, water, fertilizer, etc.). Therefore, the population expansion stops at $t = 16$. In short, with the data inserted in the control panel, the population will expand until $t = 9$ with the available food, at which point a production process for the resource will start, increasing the latter to 10,000 units. Subsequently, from $t = 19$ the population will stabilize at 2,062 units (column 2).

This simple simulation program allows for other hypotheses, in particular, that once the saturation point of the resource is reached there could be a change in the birth rate (downward) or the death rate (upward), in addition to the hypothesis of unit consumption (rationing). Figure 8.6 simulates the case where the process depends on the use of a *nonrenewable* productive resource available only in limited quantities (15,000 units).

If there were no limits to obtaining food, the population would continue to increase while producing more and more food, in an endless pursuit. By posting a limit to the reproducible resource needed for food production (20,000), the system tends gradually but inevitably to the above limit. Production starts at $t = 9$, when the food available in the system, which can be reproduced year after year (in the amount of 7,500 units), is completely utilized. Food is produced above 7,500 units gradually

		reaction time	1	
initial population	1,000	production	5.0	
birth rate	10%	if famine	9%	negative percentage
mortality rate	5%	if famine	6%	negative percentage
per capita food cons.	5.0	if famine	5.0	negative percentage
food as resource	7,500	maximum food	10,000	(about)

years	old populat. + births – deaths	birth rate	births	mortality rate	deaths	unit food consumption	food requirement	food available	humanitarian aid	productivity	production	food (surplus /deficit)	years
0	**1,000.0**	10.0%	100	5.0%	50	5.0	5,000	7,500		5.0	-	-2500	0
1	**1,050.0**	10.0%	105	5.0%	53	5.0	5,250	7,500		5.0	-	-2250	1
2	**1,102.5**	10.0%	110	5.0%	55	5.0	5,513	7,500		5.0	-	-1988	2
3	**1,157.6**	10.0%	116	5.0%	58	5.0	5,788	7,500		5.0	-	-1712	3
4	**1,215.5**	10.0%	122	5.0%	61	5.0	6,078	7,500		5.0	-	-1422	4
5	**1,276.3**	10.0%	128	5.0%	64	5.0	6,381	7,500		5.0	-	-1119	5
6	**1,340.1**	10.0%	134	5.0%	67	5.0	6,700	7,500		5.0	-	-800	6
7	**1,407.1**	10.0%	141	5.0%	70	5.0	7,036	7,500		5.0	-	-464	7
8	**1,477.5**	10.0%	148	5.0%	74	5.0	7,387	7,500		5.0	-	-113	8
9	**1,551.3**	10.0%	155	5.0%	78	5.0	7,757	7,757		5.0	256.6	0	9
10	**1,628.9**	9.0%	147	5.0%	81	5.0	8,144	8,144		5.0	644.5	0	10
11	**1,694.1**	9.0%	152	5.0%	85	5.0	8,470	8,470		5.0	970.3	0	11
12	**1,761.8**	9.0%	159	5.0%	88	5.0	8,809	8,809		5.0	1,309.1	0	12
13	**1,832.3**	9.0%	165	5.0%	92	5.0	9,161	9,161		5.0	1,661.4	0	13
14	**1,905.6**	9.0%	172	5.0%	95	5.0	9,528	9,528		5.0	2,027.9	0	14
15	**1,981.8**	9.0%	178	5.0%	99	5.0	9,909	9,909		5.0	2,409.0	0	15
16	**2,061.1**	9.0%	185	5.0%	164	5.0	10,305	10,000		5.0	2,500.0	305	16
17	**2,082.4**	9.0%	187	6.0%	207	5.0	10,412	10,000		5.0	2,500.0	412	17
18	**2,062.5**	9.0%	186	6.0%	186	5.0	10,312	10,000		5.0	2,500.0	312	18
19	**2,061.9**	9.0%	186	6.0%	186	5.0	10,309	10,000		5.0	2,500.0	309	19
20	**2,061.9**	9.0%	186	6.0%	186	5.0	10,309	10,000		5.0	2,500.0	309	20
21	**2,061.9**	9.0%	186	6.0%	186	5.0	10,309	10,000		5.0	2,500.0	309	21

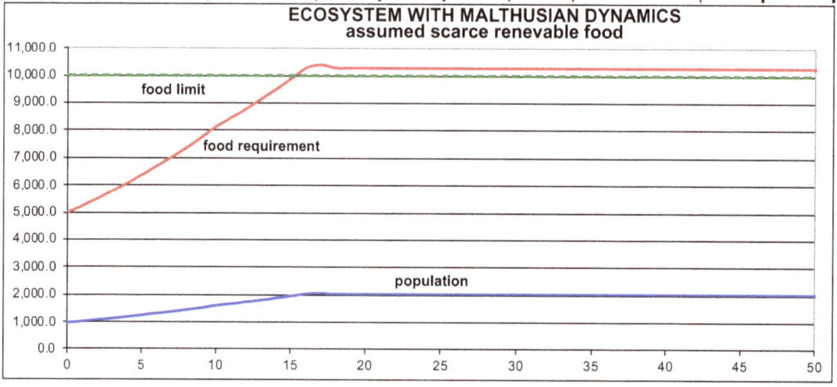

Fig. 8.5 Ecosystem with Malthusian dynamics and scarce renewable resources

until $t = 22$, when the scarce resource is completely utilized (last three columns of the table in Fig. 8.6). From then on, the production of food ceases and the population quickly falls to 1,579 units ($t = 26$).

Does not the possibility exist of humanitarian aid? Of course! However, the simulation shows that such aid—which would begin at $t = 35$ and cover a decade—would have a short-lived effect; when this aid ceases the *population*, at $t = 48$, returns to the *maximum allowed limit* of 1579 units. This second example highlights, though in a simplified manner, the fact that population growth, and the consequent growth in production, must consider the constraints imposed by the ecosystem's sustainability.

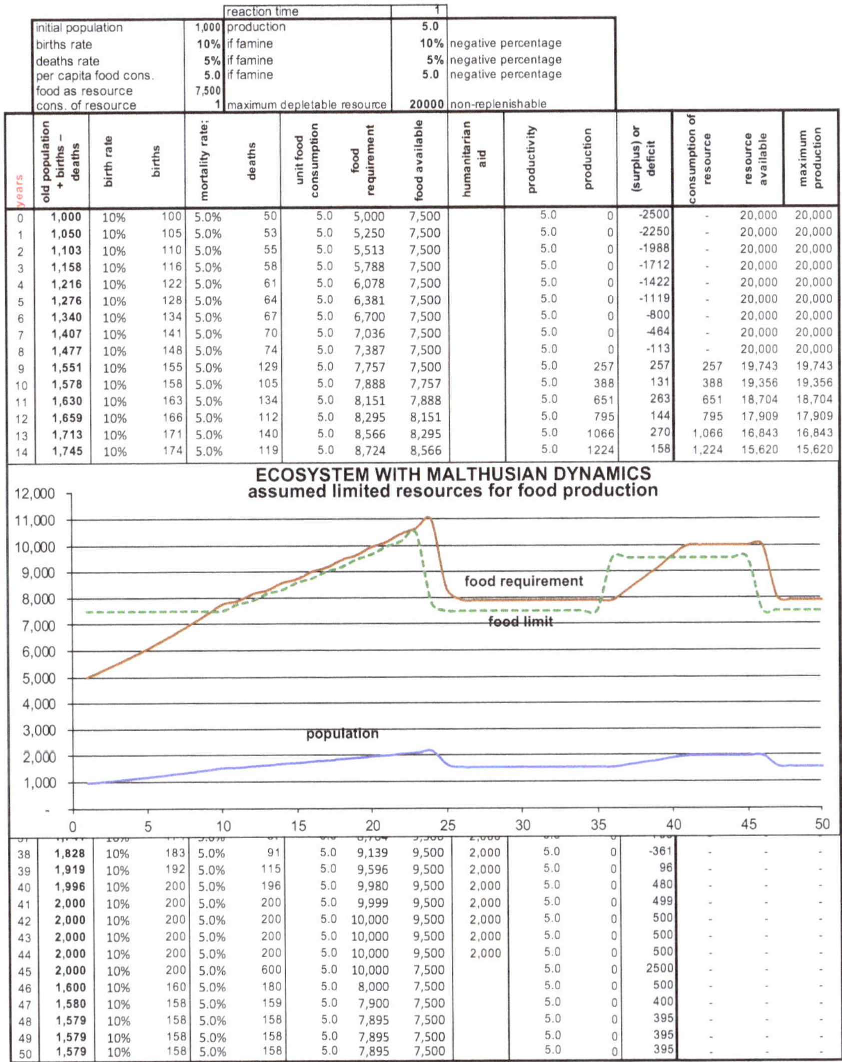

		reaction time	1	
initial population	1,000	production	5.0	
births rate	10%	if famine	10%	negative percentage
deaths rate	5%	if famine	5%	negative percentage
per capita food cons.	5.0	if famine	5.0	negative percentage
food as resource	7,500			
cons. of resource	1	maximum depletable resource	20000	non-replenishable

years	old population + births – deaths	birth rate	births	mortality rate;	deaths	unit food consumption	food requirement	food available	humanitarian aid	productivity	production	(surplus) or deficit	consumption of resource	resource available	maximum production
0	1,000	10%	100	5.0%	50	5.0	5,000	7,500		5.0	0	-2500	-	20,000	20,000
1	1,050	10%	105	5.0%	53	5.0	5,250	7,500		5.0	0	-2250	-	20,000	20,000
2	1,103	10%	110	5.0%	55	5.0	5,513	7,500		5.0	0	-1988	-	20,000	20,000
3	1,158	10%	116	5.0%	58	5.0	5,788	7,500		5.0	0	-1712	-	20,000	20,000
4	1,216	10%	122	5.0%	61	5.0	6,078	7,500		5.0	0	-1422	-	20,000	20,000
5	1,276	10%	128	5.0%	64	5.0	6,381	7,500		5.0	0	-1119	-	20,000	20,000
6	1,340	10%	134	5.0%	67	5.0	6,700	7,500		5.0	0	-800	-	20,000	20,000
7	1,407	10%	141	5.0%	70	5.0	7,036	7,500		5.0	0	-464	-	20,000	20,000
8	1,477	10%	148	5.0%	74	5.0	7,387	7,500		5.0	0	-113	-	20,000	20,000
9	1,551	10%	155	5.0%	129	5.0	7,757	7,500		5.0	257	257	257	19,743	19,743
10	1,578	10%	158	5.0%	105	5.0	7,888	7,757		5.0	388	131	388	19,356	19,356
11	1,630	10%	163	5.0%	134	5.0	8,151	7,888		5.0	651	263	651	18,704	18,704
12	1,659	10%	166	5.0%	112	5.0	8,295	8,151		5.0	795	144	795	17,909	17,909
13	1,713	10%	171	5.0%	140	5.0	8,566	8,295		5.0	1066	270	1,066	16,843	16,843
14	1,745	10%	174	5.0%	119	5.0	8,724	8,566		5.0	1224	158	1,224	15,620	15,620

years	old population + births – deaths	birth rate	births	mortality rate;	deaths	unit food consumption	food requirement	food available	humanitarian aid	productivity	production	(surplus) or deficit	consumption of resource	resource available	maximum production
38	1,828	10%	183	5.0%	91	5.0	9,139	9,500	2,000	5.0	0	-361	-	-	-
39	1,919	10%	192	5.0%	115	5.0	9,596	9,500	2,000	5.0	0	96	-	-	-
40	1,996	10%	200	5.0%	196	5.0	9,980	9,500	2,000	5.0	0	480	-	-	-
41	2,000	10%	200	5.0%	200	5.0	9,999	9,500	2,000	5.0	0	499	-	-	-
42	2,000	10%	200	5.0%	200	5.0	10,000	9,500	2,000	5.0	0	500	-	-	-
43	2,000	10%	200	5.0%	200	5.0	10,000	9,500	2,000	5.0	0	500	-	-	-
44	2,000	10%	200	5.0%	200	5.0	10,000	9,500	2,000	5.0	0	500	-	-	-
45	2,000	10%	200	5.0%	600	5.0	10,000	7,500		5.0	0	2500	-	-	-
46	1,600	10%	160	5.0%	180	5.0	8,000	7,500		5.0	0	500	-	-	-
47	1,580	10%	158	5.0%	159	5.0	7,900	7,500		5.0	0	400	-	-	-
48	1,579	10%	158	5.0%	158	5.0	7,895	7,500		5.0	0	395	-	-	-
49	1,579	10%	158	5.0%	158	5.0	7,895	7,500		5.0	0	395	-	-	-
50	1,579	10%	158	5.0%	158	5.0	7,895	7,500		5.0	0	395	-	-	-

Fig. 8.6 Ecosystem with Malthusian dynamics and scarce nonrenewable resources

8.5 Dynamics of Interacting Populations Forming a Trophic Food Chain: Volterra–Lotka Model

This *section* presents a particular form of control regarding two populations, prey and predators, carried out by the "natural" interaction of the two populations. Given the quantitative dynamics of populations, we must consider that the oceans are not only populated by sharks and the savannas by lions, even if these species are among the most efficient hunting machines in their environments; the seas abound in sardines and the savannas are populated by antelopes, even if both of these species are prey for large populations of predators. When two populations interact to form a trophic food chain (Post 2002; Post et al. 2000; Schoener 2011), in the sense they are linked by a prey–predator relationship, it is easy to intuit that their respective population dynamics cannot be independent of that of the other. Populations, and species in general, are normally self-balancing by means of some *natural* Control System that stops them from increasing infinitely, represented by conflicts with a second (or more than one) population connected to the first to form a food chain.

According to Nguyen and Yin (2016) interactions among species may occur by means of cooperation, predator–prey, or competition (Krizna et al. 2018). Consequently, due to interactions, the growth of one species is conditioned by the presence of the other. At this purpose, the coexistence of competitive species occur only when the *intraspecific* competition is stronger than the *interspecific* competition, and therefore when the competition for limited resources is stronger for members of the same species than for members of different species (Grover 1991; Macarthur and Levins 1967). When the individuals in a population are the prey of other predator populations, the "natural" Control System tends to carry out a reciprocal control of the number of individuals in both populations, often succeeding in producing an equilibrium state or an oscillating dynamics, which is, in fact, similar to that of the sardines and sharks in the Adriatic Sea observed by the zoologist Umberto D'Ancona. D'Ancona provided the occasion for Vito Volterra to formulate his well-known equations describing the coevolution between the two populations, showing how two prey and predator populations can have a dynamic equilibrium that takes various forms based on the variation rates for the populations, which are a function of the number of individuals in each population (Mella 2012). Subsequently, Vito Volterra (1931), and Alfred Lotka (1925), made Volterra's original model even more general, as we shall see shortly.

In order to make it easier to understand the reciprocal Control System between the number of individuals in populations (or species) connected in the food chain, we begin by examining the dynamics of only two species, which, for simplicity's sake, we indicate as A (*prey*) and B (*predators*), considering that, in effect, the evolution of the prey cannot occur without the simultaneous evolution of the predators, as shown in the simplified *causal loop diagram* in Fig. 8.7 describing the simplest system of equations proposed by Vito Volterra, which hypothesizes that prey and predator populations are linearly connected (Volterra, 1926).

Let us assume the observations originate in year $n = 0$, when the number of prey and predators is determined (we could also have indicated the pairs). Subsequently, the populations increase and decrease year after year at a fixed rate. The *prey* (**A**) increase by a certain yearly number according to the rate "a" (which takes account of the difference between the natural birth rate and the natural death rate), which is applied to the population of the previous year, $\mathbf{A}_n$, and decrease at a rate "b," which expresses the average voracity of the predators (**B**), taking into account the number of predators from the previous year, $\mathbf{B}_n$. The *predator* population **B** dies out from natural causes at a death rate "c" and increases by feeding on the prey at a reproduc-

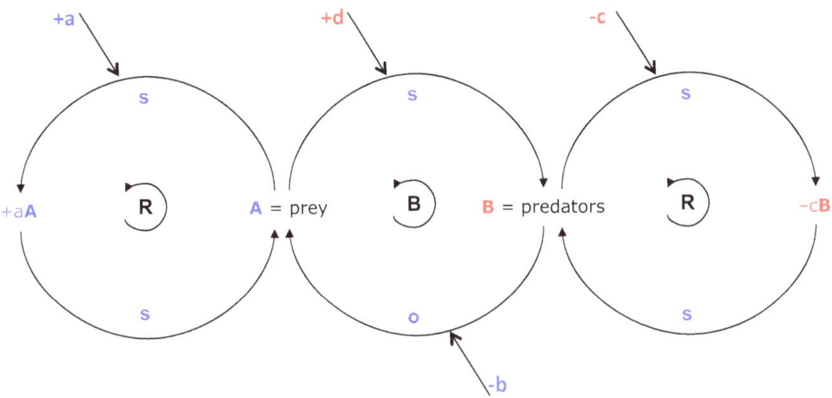

Fig. 8.7 Relationships between two interconnected populations

tion rate "*d*," which is tied to the food supply. Thus, the prey and predator populations at year "*n* + 1" are

$$\begin{cases} \mathbf{A}_{n+1} = \mathbf{A}_n + a\mathbf{A}_n - b\mathbf{B}_n \\ \mathbf{B}_{n+1} = \mathbf{B}_n - c\mathbf{B}_n + d\mathbf{A}_n \end{cases} \tag{8.2}$$

The equations in (8.2) form the natural system, producing an "internal control" of the two populations, each a function of the other. The respective dynamics over $N = 30$ periods are simulated in the graph in Fig. 8.8. When we insert the data into the control panel, in the form of a CLD, the model in Fig. 8.8 shows that in year $n = 15$, population "A = prey" reaches and surpasses population "B = predators," even though it began with fewer individuals ($\mathbf{A}_0 = 2{,}500$; $\mathbf{B}_0 = 4{,}000$). The reason is clear: there is a positive difference between the birth ($a = 5.5$) and death ($b = -1.5$) rates, and as a result, each year population **A** has a net increase of $(a + b) = 4\%$; **B**, on the other hand, has a lower rate of increase of $(d + c) = 1\%$. To make the model more realistic, Vito Volterra proposes a variant—introduced independently by Alfred Lotka as well in his research on the dynamics of the autocatalytic chemical reactions—where the rate of decrease of **A** and of increase of **B** are a function of the number of individuals in *both populations*. In fact, it is intuitively clear that the chance of being the victim of a predator depends not only on the number of predators but also on the number of prey, since a larger population means more likelihood of escaping from the hunt.

Moreover, for a predator the possibility of hunting a prey does not depend only on the number of prey but also on the size of the predator population, since the higher the number of predators, the more difficult it is to contend for the prey. In prey–predator interaction, the rate of prey capture and prey density are connected by a relationship that represents the functional response of a predator–prey system. In fact, as Staddon proposed in 2016, the risk to which the prey is exposed and the size of its population are strictly connected and depend on the predation rate (Staddon 2016).

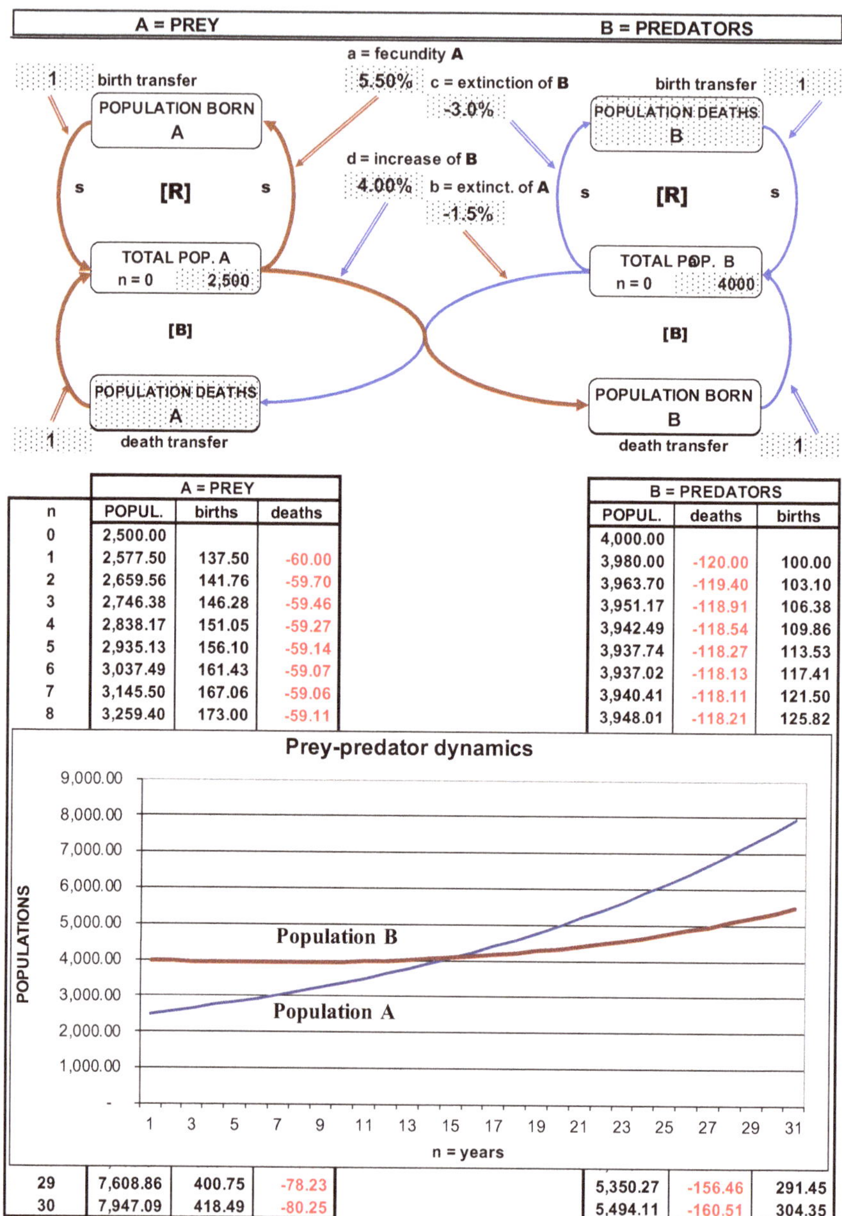

n	A = PREY POPUL.	births	deaths	B = PREDATORS POPUL.	deaths	births
0	2,500.00			4,000.00		
1	2,577.50	137.50	-60.00	3,980.00	-120.00	100.00
2	2,659.56	141.76	-59.70	3,963.70	-119.40	103.10
3	2,746.38	146.28	-59.46	3,951.17	-118.91	106.38
4	2,838.17	151.05	-59.27	3,942.49	-118.54	109.86
5	2,935.13	156.10	-59.14	3,937.74	-118.27	113.53
6	3,037.49	161.43	-59.07	3,937.02	-118.13	117.41
7	3,145.50	167.06	-59.06	3,940.41	-118.11	121.50
8	3,259.40	173.00	-59.11	3,948.01	-118.21	125.82
29	7,608.86	400.75	-78.23	5,350.27	-156.46	291.45
30	7,947.09	418.49	-80.25	5,494.11	-160.51	304.35

Fig. 8.8 Model of a control system of two populations based on Vito Volterra's equations

The variation in the model consists precisely in setting the number of deaths in population **A** and births in **B** at a "critical mass," which is given by the product of the number of prey and predator populations: "Critical Mass = $\mathbf{A}_n\ \mathbf{B}_n$." Thus, system (8.2) is modified as follows, producing the more realistic *endogenous natural*

Control System known as the Volterra–Lotka equations, which is simulated in the model in Fig. 8.9 for a period of 1000 iterations:

$$\begin{cases} \mathbf{A}_{n+1} = \mathbf{A}_n + a\mathbf{A}_n - b\mathbf{A}_n\mathbf{B}_n \\ \mathbf{B}_{n+1} = \mathbf{B}_n - c\mathbf{B}_n + d\mathbf{A}_n\mathbf{B}_n \end{cases} \quad (8.3)$$

In line with the model in Fig. 8.7, the more prey **A** increases in number, the greater the increase in predator **B** due to the abundance of food; this increase in **B** reduces the number of **A**, which causes a reduction in the number of **B**. If we exclude the case of initial values of zero, the populations have a mutual equilibrium when the initial values of each are proportionate to the rates of extinction and reproduction of the connected population. With the data from Fig. 8.9, the equilibrium values are determined by the following relations: $\mathbf{A} = c/d = 316.7$ and $\mathbf{B} = a/b = 112.5$ (Takeuchi 1995).

Note. In the simulations in Figs. 8.8 and 8.9, both the initial number of the two populations and the rates of increase "*a*" and "*d*" and decrease "*b*" and "*c*" were chosen arbitrarily. In constructing the simulation models referring to real populations, these rates must be quantified by studying the historical dynamics registered in the archives.

8.6 Natural Endogenous and Artificial External Controls of Two Interacting Populations

In Fig. 8.9, the "natural" endogenous control leads the two populations to maintain a reciprocal relationship that produces an oscillating increasing dynamics. It is possible to introduce other variations to "enrich" the system (for example, the competition for food among predators); since the variants do not modify in any way the basic logic of the system, those interested in further detail can consult specialist texts (Casti 1985; Flake 2001; Martelli 1999; Takeuchi 1995).

It is difficult to recognize a Control System in Fig. 8.9 (and in Fig. 8.8), since no "objective," "constraint," or "limit" is indicated. In fact, these are "natural" control systems—analogous to the one operating in Daisyworld (Sect. 6.5)—where the objective "reciprocal size of the two populations" depends on the type of population and on the environmental features. The populations "know" whether or not they can achieve equilibrium and what kind of equilibrium this would be:

> In spite of the extensive knowledge of self-reinforcing feedback in common folk wisdom, it played hardly any role during the first phase of cybernetics. The cyberneticists around Norbert Wiener acknowledged the existence of runaway feedback phenomena but did not study them any further. Instead they concentrated on the self-regulatory, homeostatic processes in living organisms. Indeed, purely self-reinforcing feedback phenomena are rare in nature, as they are usually balanced by negative feedback loops constraining their runaway tendencies. In an ecosystem, for example, every species has the potential of undergoing an exponential population growth, but these tendencies are kept in check by various balancing

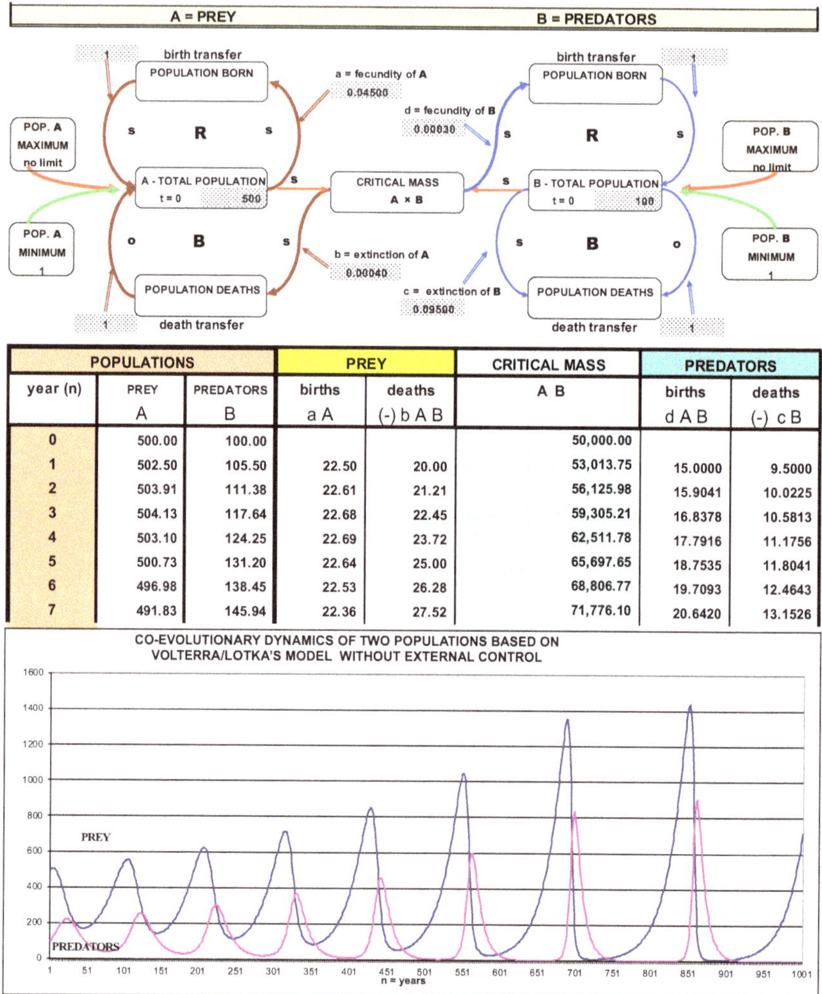

POPULATIONS			PREY		CRITICAL MASS	PREDATORS	
year (n)	PREY A	PREDATORS B	births aA	deaths (-) b A B	A B	births d A B	deaths (-) c B
0	500.00	100.00			50,000.00		
1	502.50	105.50	22.50	20.00	53,013.75	15.0000	9.5000
2	503.91	111.38	22.61	21.21	56,125.98	15.9041	10.0225
3	504.13	117.64	22.68	22.45	59,305.21	16.8378	10.5813
4	503.10	124.25	22.69	23.72	62,511.78	17.7916	11.1756
5	500.73	131.20	22.64	25.00	65,697.65	18.7535	11.8041
6	496.98	138.45	22.53	26.28	68,806.77	19.7093	12.4643
7	491.83	145.94	22.36	27.52	71,776.10	20.6420	13.1526

Fig. 8.9 Model of a natural endogenous control of two populations based on Volterra–Lotka equations without external control

> interactions within the system. Exponential runaways will appear only when the ecosystem is severely disturbed. Then some plants will turn into "weeds", some animals become "pests", and other species will be exterminated, and thus the balance of the whole system will be threatened (Capra 1996b, p. 63).

The law of dynamic instability (Sect. 1.5) is constantly operating. However, in the model in Fig. 8.9, it is possible to simulate the introduction of an *exogenous control*—for example, by assuming the possibility of varying the numbers in the population through outside intervention—when the population size exceeds certain maximum levels or is below given minimum levels. In the case of populations of fish or deer, for example, numbers can be controlled by fishing or hunting, if the number becomes excessive, or by repopulation, if the number becomes too low. Figure 8.10 starts from Fig. 8.9 and completes it by introducing *maximum* and *minimum limits* for the two populations.

In TEST 1, the *maximum* and *minimum* limits to the size of the two populations produce, after around 200 iterations, a regular dynamic for both populations in line with the limits. In TEST 2, *maximum* and *minimum limits* have been set for both populations; the maximum limit for the prey was increased to $\mathbf{A}_{max} = 800$ units, which impedes it from reaching the maximum, while the predator population is slightly higher than $\mathbf{B}_{max} = 500$ units.

The diagram in TEST 2 shows that the dynamics become irregular, as an environmental cataclysm reduces population **A** by 500 units over three periods: $n = 593$–595. Therefore, this demonstrates that from a theoretical point of view an *external control* can also be added to (8.3) by adding explicit control limits (objectives). Equation (8.3) will then turn into Eq. (8.4):

$$\begin{cases} \mathbf{A}_{n+1} = \mathbf{A}_n + a\mathbf{A}_n - b\mathbf{A}_n\mathbf{B}_n \\ \min\mathbf{A} \le \mathbf{A}_n \le \max\mathbf{A} \\ \mathbf{B}_{n+1} = \mathbf{B}_n - c\mathbf{B}_n + d\mathbf{A}_n\mathbf{B}_n \\ \min\mathbf{B} \le \mathbf{B}_n \le \max\mathbf{B} \end{cases} \tag{8.4}$$

To make the model even more realistic, we could also assume a modification in the birth and death rates of the populations when the maximum limits for population numbers are approached. It is also interesting to consider the natural limits to population expansion from limits to the available resources.

Note. It is necessary to give a meaning to the scales of the axes in the previous simulations. The scale adopted depends on the type of population in question and the research objectives. If, for example, one wanted to study the relations between sharks and sardines in the Mediterranean, a single iteration could represent a 3-month or 6-month period; however, this period would depend on the statistics in the archives of marine biologists.

Apparently, at least in theory the *external control* creates order in dynamics that were previously in disorder, thereby uncovering order in chaos. However, this introduces the important aspect of the effect of environmental disturbances, which not only create disorder in the dynamics of the affected populations but spread their effects through a chain of causes and effects that impact a much vaster ecosystem. For example, if a locale climate change or a disastrous explosion with intense smoke should reduce the number of bees in a territory, pollination would become difficult for many vegetal species; this would negatively impact other insects that are food for birds, who would then be forced to migrate, and so on. In an ecosystem comprising different populations, what affects the dynamics of even a single population produces increasingly disorderly dynamics in the interconnected species, thereby heightening the need for external controls.

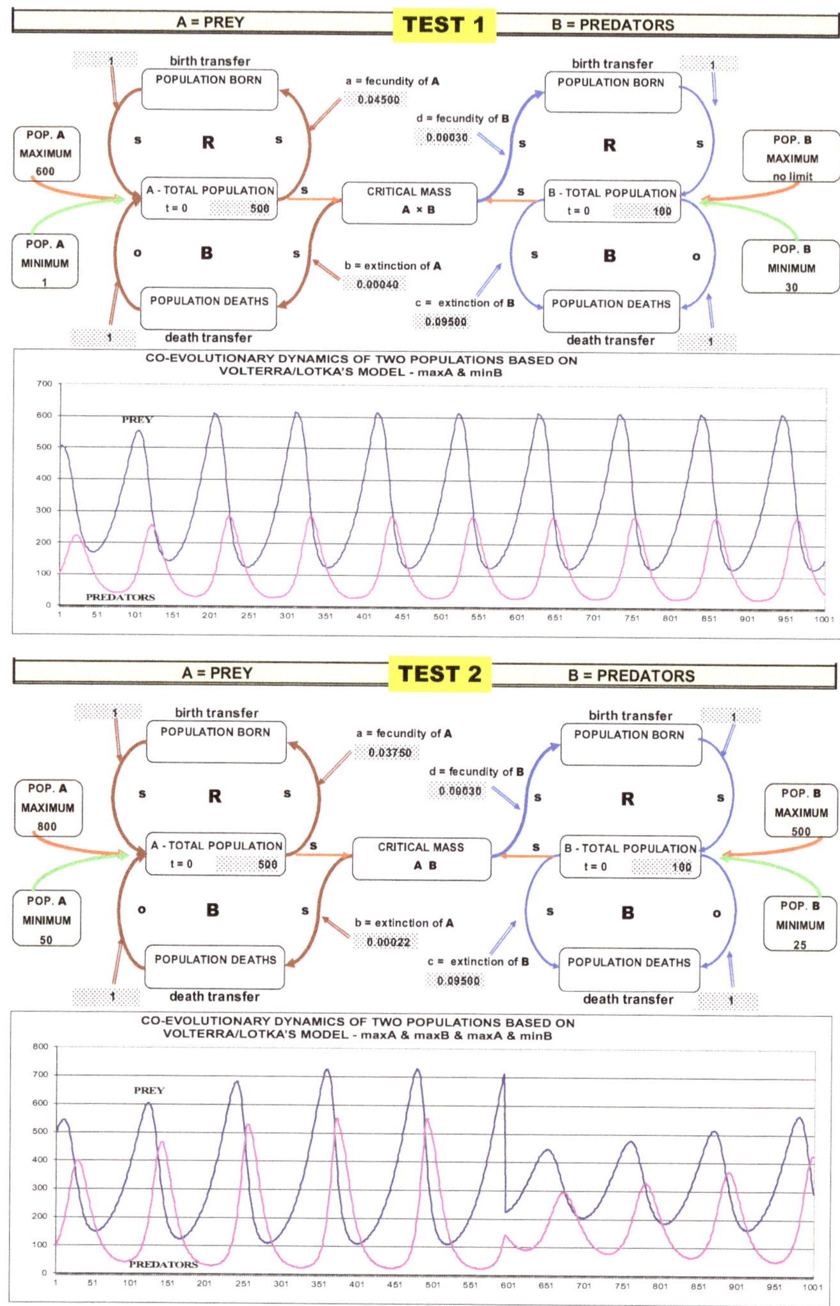

Fig. 8.10 Model of an external control of two populations based on Fig. 8.9

8.7 Magic Rings That Regulate the Dynamics of Three or More Populations Comprising an Ecosystem

The control system applied to the Volterra–Lotka model for two populations can be extended to any number of populations forming a food ecosystem. Some authors have identified limitations related to the case of only two populations (Paine 1966; Price et al. 1980), in particular, to the extent to which the populations are limited in achieving an equilibrium or limiting a cycle (Segel 1984). For this purpose, the same logic applied to the simulation of two populations can be useful for any number of populations forming a food ecosystem, after the rules of interaction between populations of prey and predators have been defined. Let us now consider the more general case of a simple ecosystem comprised of only three populations, which coexist in a prey–predator relationship (Freedman and Waltman 1984; Gakkhar and Gupta 2016; Hsu et al. 2015). In fact, while the classical ecological models have focused on two species (May 1973), the study of the dynamics between more than two populations may contribute to defining their interactions and to exploiting more phenomena that are exhibited in nature (Hastings and Powell 1991).

As an example, in Fig. 8.11 we consider three populations that form a food chain.

The first population of prey, **A**, serves as the first link in the food chain; the second population, **B**, feeds on the prey but represents, in turn, the prey fed on by the individuals in the third population, **C**, which is made up of pure predators that represent the final link in the food chain. Population **C** (pure predators) increases in relation to the available food (population **B**) and becomes extinct because of natural causes, since it is not a food source for other superordinate populations. This situation leads to an annual increase in population **A** at a rate determined by the difference between the birth and natural death rates; moreover, population **A** decreases in relation to the voracity of population **B**. The latter increases and becomes extinct as a function of populations **A** and **C** (prey and predators, respectively) at rates that are net of births and natural deaths.

Figure 8.11 also shows the complex oscillatory dynamics that result from the assigned parameters over 1,000 periods, after having also set *maximum* limits for the prey and *minimum* ones for the predators, limits that, representing merely an example, influence only the birth rates of the two populations. From the graph we can see that the dynamics of prey **A** conditions that of population **B**, which presents oscillations that tend toward stable values due to the control effect. We could easily set more elaborate limits that call for a maximum amount of resources available or territory size, referring to these to specify the birth and death rates and the outside interventions. The same logic applied to the simulation of three populations can be used for any number of populations. Figure 8.12 illustrates the Powersim model to simulate the dynamics of five populations for 5,000 periods that produces a more articulated trophic food chain.

Figure 8.13 presents the results of the simulation. The top graph shows the population dynamics determined only by the reciprocal control carried out by the populations forming the food chain. The lower graph presents the dynamics produced

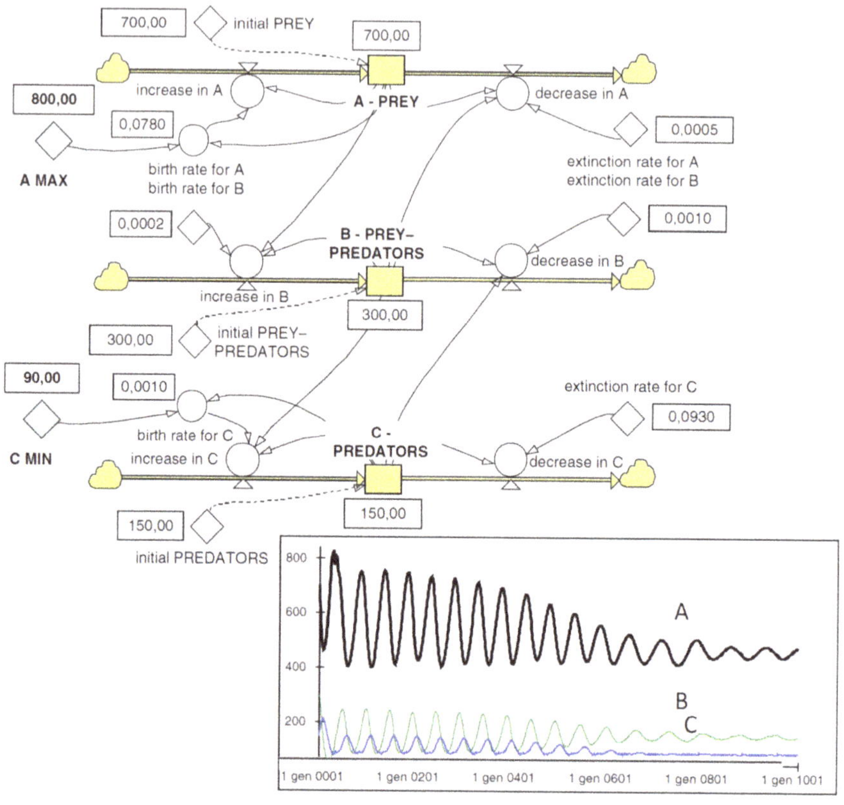

Fig. 8.11 Volterra–Lotka model for three populations, simulated using Powersim Software

when we impose (for clarity) maximum constraints on populations **A** and **B**, showing that the irregular, probably chaotic, dynamics that occur using only the reciprocal natural endogenous control (upper graph) becomes regular as soon as the simple external control hypothesized in the simulation is carried out. Only two limits have been introduced in the simulation for simplicity's sake: Max population **A** = 2000; Max population **B** = 800.

Despite the fact these models may capture many of the fluctuations in the ecosystems, we need to be aware that *ecosystems* are very difficult to simulate, since they can include thousands of populations of different species (for example, a rain forest or coral reef), whose reciprocal relations are so complex that it appears impossible to translate them quantitatively into birth and death rates and predation.

Ecologists have attempted to overcome the inadequacies of their simple models by including a structured hierarchy of species that are referred to as "food webs". In such a hierarchy there is a pyramid that is surmounted by the top predator, such as a lion, with the smallest numbers. The numbers increase as you go down each "trophic" level, until at the base of the pyramid are the most numerous primary producers, the plants, that provide food for the whole system. In spite of years of effort and computer time, the ecologists have made no real progress towards modeling a complex natural ecosystem such as a tropical rain forest or the three-dimensional ecosystem of the ocean. No models drawn from theoretical ecol-

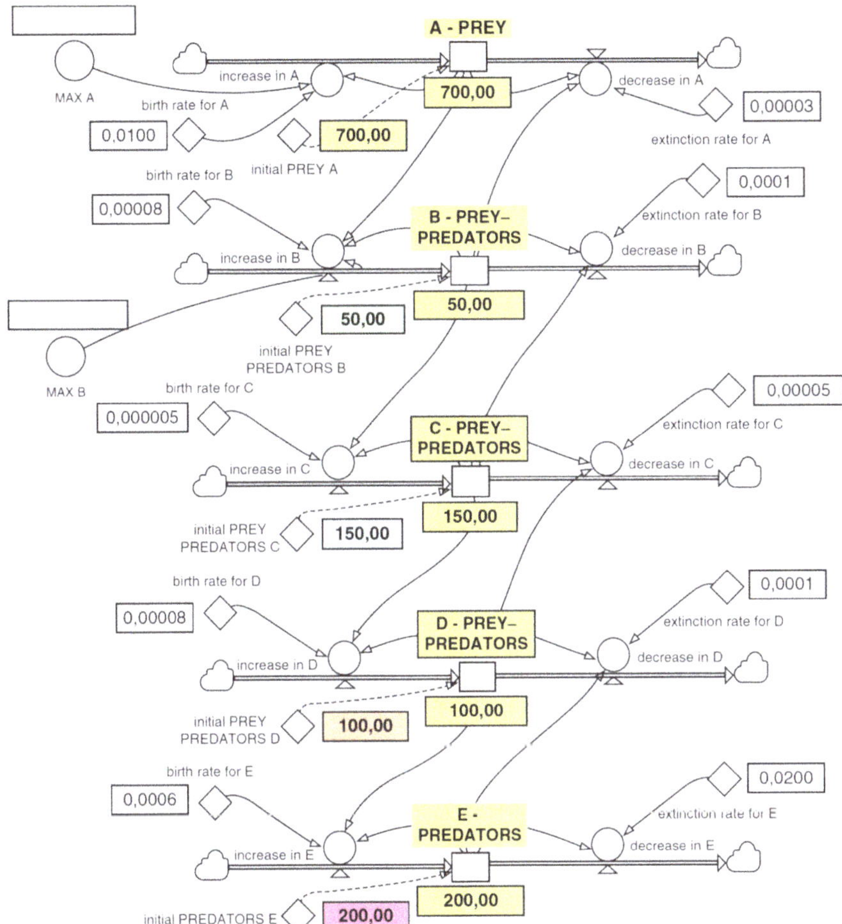

Fig. 8.12 Model of an ecosystem of five populations forming a food chain (simulated with Powersim)

> ogy can account in mathematical terms for the manifest stability of these vast natural systems (Lovelock 1988, p.49).

Four large classes of levers can be used to implement exogenous control:

- *Selective culling*: This lever is used to reduce the number of individuals in a population, animal or plant, to manage and coordinate the presence of specific species: for example, the culling of wild boars in cities, wolves in forests adjacent to the grazing land of farmed animals, etc.
- *Disinfestation*: This lever entails a general artificial reduction of a population in a given territory, for example, pest control against mosquitoes and snails.
- *Repopulation*: This is understood as activities to re-establish the presence of specific species, for example, chamois in the Alps or deer on the major Italian islands

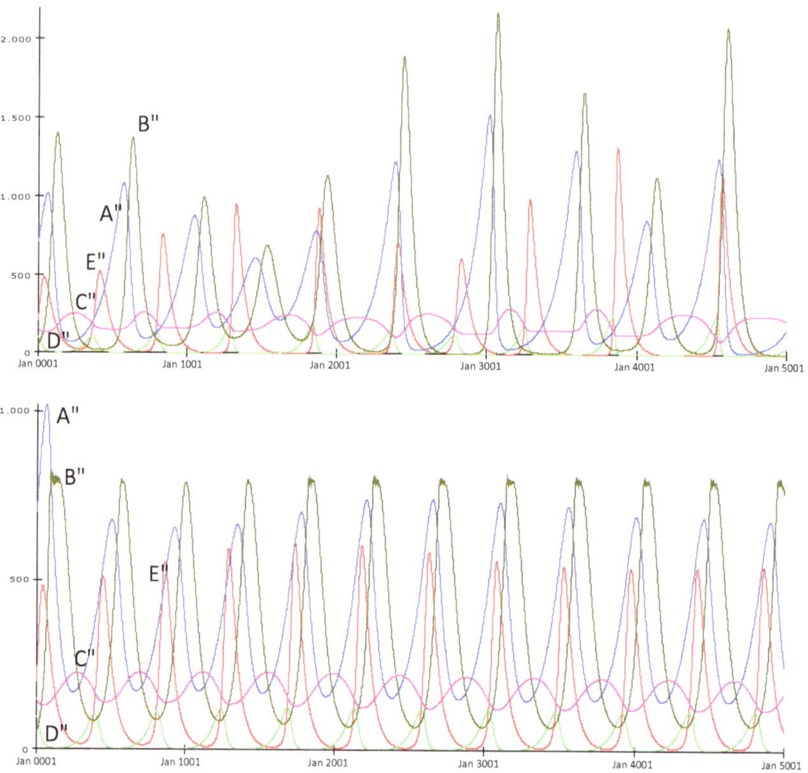

Fig. 8.13 Dynamics of five populations of an ecosystem, without (top graph) and with (lower graph) external control, simulated with Powersim (reference to Fig. 8.12)

of Sicily and Sardinia, through the "One Deer Two Islands" program financed by European funds from the LIFE+ program;

- The *introduction of new invasive species* to balance the uncontrollable expansion of other native species; for example, the introduction of the samurai wasp against the Asian bedbug.

It should be noted that many animal species (Asian bedbug, tiger mosquito, poppies in wheat fields, killer algae), whether they are introduced randomly or intentionally into a territory or are indigenous, seem to be uncontrollable regarding their reproduction, spread, and numbers. Examples of such species are the giant squid, jellyfish, and mosquitoes.

Here is a surprising example of an outside man-made *repopulation* control that has impacted the dynamics of many interconnected populations. In 1995, the introduction in Yellowstone National Park of 14 wolves, a species that had disappeared in 1926, had an unexpected and surprising effect on the entire ecosystem of the park. The new wolves, who were at the top of the food chain and undisturbed by predators, began to hunt the deer, thereby causing a rapid decline in the latter species, which even abandoned some of the park areas in which they were easy prey for the wolves. The reduction in the number of deer allowed several plant species that were food for the deer, above all poplars and weeping willows, to increase once again and flower, thereby leading to the spread of berries and numerous insect species, which attracted several species of birds. The wolves also reduced the coyote population, thus allowing red foxes, weasels, badgers, hawks and golden eagles to increase again (USP 2017). Deforestation, particularly of the tropical forests, instead represents an outstanding example of a larger alteration of a vast ecosystem caused by human intervention. The deforestation of tropical forests, normally through fires or the industrial cutting down of trees (more than 50,000 fires in 1991 alone) has reduced the forest cover around the world by over two million km^2. Since forests are a natural habitat for many thousands of life forms, their destruction has led to the extinction of numerous vegetal and animal species, with a resulting genetic impoverishment. For example, at risk today are the Sumatran orangutan, the Bornean orangutan (Voigt 2018), the green-eyed frog, the Angonoka tortoise, and the Manchurian leopard, to name but a few (Lindsey 2007).

Environmental disturbances to the quantitative dynamics of populations are not caused only by human interventions; natural disasters and extreme atmospheric events are an important source of the loss of biodiversity and increase the probability the ecosystem will collapse, to the point that several scientists maintain that the Holocene extinction, the sixth terrestrial phase of mass extinction, has already begun, perhaps leading to even worse consequences than the preceding ones.

> Therefore, although biologists cannot say precisely how many species there are, or exactly how many have gone extinct in any time interval, we can confidently conclude that modern extinction rates are exceptionally high, that they are increasing, and that they suggest a mass extinction under way—the sixth of its kind in Earth's 4.5 billion years of history (Ceballos et al. 2015, p. 3).

In any case, the models of quantitative dynamics show that no matter how large are the disturbances to populations caused by environmental disasters, there is always the chance for the interconnected populations to restore their dynamics by interconnecting to other populations to ensure their survival and that of the ecosystem (Sutter 2017).

Populations and species are *resilient* collective systems capable of adapting even to great shocks (Fabinyi et al. 2014) and returning to their original form even after an extremely dangerous environmental cataclysm (Laprie 2008; Lucini 2014).

> Resilience (from the Latin etymology resilire, to rebound) is literally the act or action of springing back. As a property, two strands can historically be identified: a) in social psychology [...], where it is about elasticity, spirit, resource and good mood, and b) and in material science, where it is about robustness and elasticity. The notion of resilience has then been elaborated [...] A common point to the above senses of the notion of resilience is the ability to successfully accommodate unforeseen environmental perturbations or disturbances (Laprie 2008).

Obviously, the deterioration in the global ecosystem also places the human species at risk; however, man, unlike the other animal and vegetal species, can defend himself through the use of elaborate instruments thanks to his intelligence.

8.8 Qualitative Dynamics of Populations—Evolution: The Framework

Let us now also consider the *qualitative dynamics* of populations—more generally, species—limiting ourselves to the control process carried out by *natural selection*, a powerful "natural" control system that leads to the gradual modification (in terms of improvement) of the *phenotypic traits* of the individuals in a population that give them the maximum chance of survival. That "evolution" is a natural form of a Control System was recognized by Alfred Wallace, who, along with Charles Darwin, was one of the founders of the theory of evolution, anticipating it 20 years before publishing in 1858, along with Darwin, the article titled *On the Tendency of Species to Form Varieties; and on the Perpetuation of Varieties and Species by Natural Means of Selection* (Smith 2005) In his paper *On the Tendency of Varieties to Depart Indefinitely from the Original Type* (1858) Wallace explicitly referred to the "centrifugal governor" in a typical system of elementary control for maintaining species stability (Sect. 3.9.1).

> The action of this principle [the feedback principle] is exactly like that of the centrifugal governor of the steam engine, which checks and corrects any irregularities almost before they become evident; and in like manner no unbalanced deficiency in the animal kingdom can ever reach any conspicuous magnitude, because it would make itself felt at the very first step, by rendering existence difficult and extinction almost sure soon to follow (Wallace 1858, online).

Not wishing to present too simplified a view of evolution, I shall limit myself here to referencing the concise thoughts of Darwin about the "natural selection process" based on the *struggle for survival* among individuals, which leads to the *qualitative evolution* of the species through the "*natural selection of the fittest*"—whatever interpretation we wish to give to this process, including Dawkins' version of a *selfish gene* (Dawkins 1989, 1st ed. 1976)). It is useful to note that there is not "only one" theory of evolution. I shall only mention the main theories of evolution that provide interpretations of the evolutionary phenomenon in the domain of life:

(i) Lamarckism, or the Theory of the Inheritance of Acquired Characters (Lamarck 1809), and Packard's Neo-Lamarckism (Packard 1901)
(ii) Darwinism, or the Theory of Natural Selection (Darwin and Wallace 1858; Darwin 1859)
(iii) De Vries' Mutation Theory (de Vries 1905)
(iv) Huxley's Neo-Darwinism (Huxley 1942), or the "modern theory" of evolution, supported by studies by Watson and Crick (1953) and by Monod (1970)
(v) Stephen Gould's Punctuated Equilibria Theory (Gould and Eldredge 1977)
(vi) The Selfish Gene Theory of Evolution, popularized by Richard Dawkins (Dawkins 1989)
(vii) Maturana and Varela's Autopoietic Theory of Evolutionary Processes (Maturana and Varela 1987)
(viii) Waddington's Epigenetics view of evolution (Waddington 1957)

(ix) Charles Thaxton's Intelligent Design as a challenge to Neo-Darwinism (Thaxton et al. 1984)

In a nutshell, the Darwinian theory of evolution (as expressed in the recent "modern theory") is based on two main concepts: the inheritance of mutated characteristics, or "descent with modification," and "natural selection," which are the main engines of the evolution of species. The *modus operandi* can be summarized by the following postulates:

- Individuals in a species show a wide range of variation and mutations caused by chance and environmental factors.
- The variations/mutations that affect the physical characteristics of an organism—the phenotype—can be passed on to the next generation.
- Individuals with characteristics "best suited" to their population and environment are more likely to survive and reproduce.
- Favorable mutations are maintained and transmitted to descendants in future generations.
- Individuals that are poorly adapted to their environment are less likely to survive and reproduce. This means that their mutations are less likely to be passed on to the next generation.

As a consequence, when a variation/mutation in the phenotype is more favorable for existence, it is transmitted to the offspring and, over a necessary period of time, this mechanism changes the entire population, sometimes producing a new species; otherwise the mutation tends to disappear. The core of Darwin's natural selection is the idea of "survival of the fittest" in the "struggle for existence," with a double meaning that a fitted phenotype living in a competitive environment has more chances to survive, as prey and predator:

- In the *struggle for life,* that is, in the search for food and in defense against predators (Darwin, 1859, p. 43)
- In the *struggle for reproduction*, in whatever form it takes place: asexual reproduction; blind mating, prevailing by force over sexual competitors; mating with sexual selection, attracting possible partners more easily

This process is explained in a masterful description by Darwin himself, which is worth reading in full:

> Owing to this struggle for life, any variation, however slight and from whatever cause proceeding, if it be in any degree profitable to an individual of any species, in its infinitely complex relations to other organic beings and to external nature, will tend to the preservation of that individual, and will generally be inherited by its offspring. The offspring, also, will thus have a better chance of surviving, for, of the many individuals of any species which are periodically born, but a small number can survive. I have called this principle, by which each slight variation, if useful, is preserved, by the term of Natural Selection, in order to mark its relation to man's power of selection (Darwin 1859, p. 61).

> Natural selection will modify the structure of the young in relation to the parent, and of the parent in relation to the young. In social animals it will adapt the structure of each individ-

ual for the benefit of the community; if each in consequence profits by the selected change. What natural selection cannot do, is to modify the structure of one species, without giving it any advantage, for the good of another species; and though statements to this effect may be found in works of natural history, I cannot find one case which will bear investigation (Darwin 1859, p. 87).

The mutation of certain phenotypic traits leads to greater attractiveness toward the partner. Therefore, individuals who have such mutations have an advantage in the *struggle for reproduction*. If the mutation is hereditary, it spreads to all individuals in the species; those who do not possess it are, in fact, *losers* who do not reproduce or reproduce with difficulty. Darwin named this form of competitive advantage Sexual Selection:

Sexual Selection. Inasmuch as peculiarities often appear under domestication in one sex and become hereditarily attached to that sex, the same fact probably occurs under nature, and if so, natural selection will be able to modify one sex in its functional relations to the other sex, or in relation to wholly different habits of life in the two sexes, as is sometimes the case with insects. And this leads me to say a few words on what I call Sexual Selection. This depends, not on a struggle for existence, but on a struggle between the males for possession of the females; the result is not death to the unsuccessful competitor, but few or no offspring. Sexual selection is, therefore, less rigorous than natural selection. Generally, the most vigorous males, those which are best fitted for their places in nature, will leave the most progeny [...] A hornless stag or spurless cock would have a poor chance of leaving offspring... (Darwin 1859, p. 86-87).

Recognizing the traits that increase the sexual attraction of some individuals as opposed to others is not simple since, as humans, we do not know the judgment criteria of animals and cannot evaluate which shapes, colors, scents, sizes, ceremonies, etc., are more attractive. We can only draw generalizations from observed behavior. Wallace himself observed that certain mutations changed certain phenotypic traits without, apparently, bringing advantages in the *struggle for life*. However, the logic of hereditary transmission of such seemingly useless mutations, which "*serve no material or physical purpose*," and their spread in a population or species must "obey" some principle of utility that, it is assumed, refers precisely to the struggle for reproduction:

Do you mean to assert, then, some of my readers will indignantly ask, that this animal, or any animal, is provided with organs which are of no use to it? Yes, we reply, we do mean to assert that many animals are provided with organs and appendages which serve no material or physical purpose. The extraordinary excrescences of many insects, the fantastic and many-coloured plumes which adorn certain birds, the excessively developed horns in some of the antelopes, the colours and infinitely modified forms of many flower–petals, are all cases, for an explanation of which we must look to some general principle far more recondite than a simple relation to the necessities of the individual (Wallace 1856, p. 30).

Take, for example, the mane of a lion. It has been shown, even through experimental observations, that the lioness mates more easily with lions with a dark mane because this characteristic is interpreted as an indicator of strength, aggression, and an abundance of testosterone:

Using life-sized dummies with detachable wigs, researchers in Africa have found that female lions favour mates with a dark head of hair (West and Packer 2002).

> Lions with the longest, darkest manes suffer from the heat more than their blonder or less shaggy peers but they do better in the hunt for mates. Scientists have found that the lion's mane works much like the peacock's tail; it tells females - and rivals - about a male's health and fitness (Highfield 2002, online).

> Not only does having a bigger, blacker mane mean more mates, but the lion is able to intimidate other adult males and better protect his cubs, says the study to be published Friday in the journal *Science*. "The results of these tests have been very exciting and suggest that the manes do play a role in sexual selection", said researcher Peyton West of the University of Minnesota, who is writing her thesis on the topic. The main reason lions with dark manes are attractive is that they have higher levels of testosterone. "It isn't surprising that females would prefer darker manes and males would be intimidating" (CBC News 2002, online).

A relevant question immediately arises: can natural selection justify the spontaneous genesis of such perfect organs, like the eye, so as to suggest a plan, a will, an intelligent design? Darwin offers this answer:

> To suppose that the eye, with all its inimitable contrivances for adjusting the focus to different distances, for admitting different amounts of light, and for the correction of spherical and chromatic aberration, could have been formed by natural selection, seems, I freely confess, absurd in the highest possible degree. Yet reason tells me, that if numerous gradations from a perfect and complex eye to one very imperfect and simple, each grade being useful to its possessor, can be shown to exist; if further, the eye does vary ever so slightly, and the variations be inherited, which is certainly the case; and if any variation or modification in the organ be ever useful to an animal under changing conditions of life, then the difficulty of believing that a perfect and complex eye could be formed by natural selection, though insuperable by our imagination, can hardly be considered real. ... (Darwin 1859, p. 167).

Obviously, Darwin could not have known that the source of the mutations in the phenotypes derived from mutations in genotypes. It was only in 1953 that James Watson and Francis Crick, discovering the structure of DNA, demonstrated the genetic mechanism of inheritance. Jacques Monod, in his famous *Chance and Necessity* (1970), clarified that evolution is due to "random" mutations produced in the genotype, which "by necessity" spread as a result of the invariant reproductive mechanism of cells, if such mutations bring advantages for existence; otherwise, they tend to disappear:

> A mutation represents in itself a microscopic, quantum event, to which, as a result, we apply the principle of indetermination: an event which thus is unpredictable by nature (Monod 1970, p. 97). … We call these events [mutations] accidental; we say they are random occurrences. And since they constitute the only possible source of modification in the genetic text, itself the sole repository of the organism's hereditary structures, it necessarily follows that chance alone is at the source of every innovation, of all creation in the biosphere. Pure chance, absolutely free but blind, at the very root of the stupendous edifice of evolution: this central concept of modern biology … is today the sole conceivable hypothesis … (Monod 1970, p. 112. The emphases are Monod's).

I would also suggest everyone meditate on the view of evolution proposed by Dawkins (1989, 2004), which places at the center of evolution the search for stability in the *genes* that transform the phenotypes of organisms (their *survival machines*) to ensure greater "*chances of survival*" of genes in their offspring in subsequent generations (Dawkins 1989, 2004).

> We are survival machines – robot vehicles blindly programmed to preserve the selfish molecules known as genes. This is a truth which still fills me with astonishment (Dawkins 1989, p. xxi).

Nevertheless, Darwin himself had recognized the role of "chance" in the evolutionary process:

> I have hitherto sometimes spoken as if the variations—so common and multiform in organic beings under domestication, and in a lesser degree in those in a state of nature—had been due to chance. This, of course, is a wholly incorrect expression, but it serves to acknowledge plainly our ignorance of the cause of each particular variation (Darwin 1859, p. 120).

Deserving of special consideration is Huxley's *Synthetic Theory* (1942), based on the ideas of Darwin, on Mendel's hybridization results, and on the central role of natural selection as the fundamental mechanism of evolution. In particular, Huxley recognized that the genes—located in a linear manner on the chromosomes—are units of both heredity and mutations. However, Huxley took another step forward, highlighting the importance of populations as units of the evolutionary process. For Huxley, natural selection does not operate through "survival of the fittest" but through the more comprehensive "process of adapting to the environment."

A brief mention of the *Intelligent Design Theory* may also be appropriate. Intelligent Design Theory begins with the observation that intelligent causes can do things that undirected natural causes cannot. Undirected natural causes can place scrabble pieces on a board, but they cannot arrange the pieces as meaningful words or sentences. To obtain a meaningful arrangement requires an intelligent cause. This intuition, that there is a fundamental distinction between undirected natural causes, on the one hand, and intelligent causes on the other, has underlaid the design arguments of past centuries. The theory of intelligent design holds that certain features of the universe and of living things are best explained by an intelligent cause, not by an undirected process such as natural selection. However, the dominant theory of evolution today is neo-Darwinism, which contends that evolution is driven by natural selection acting on random mutations, an unpredictable and purposeless process. It is this specific claim made by neo-Darwinism that intelligent design theory directly challenges.

8.9 Magic Rings That Regulate the Qualitative Dynamics of Populations Over Time

The model in Fig. 8.14 shows the joint evolutionary dynamics of a prey and predator population, clearly showing that while the life of a predator (attacker) depends on its offensive weapons for killing its prey, the life of the prey (defender) is linked to the predator's defensive capabilities. Predators with refined weapons (whatever the biological interpretation we choose to give to this term) have an advantage in their predation ($\mathbf{s}_1$), thereby reducing the chances for survival of the weaker prey ($\mathbf{o}_1$) (defenders); the lower number of weak prey increases the probability of procreation ($\mathbf{o}_2$) by the prey with strong defenses (hereditary transmission of traits), which leads to an increase ($\mathbf{s}_2$) in the effectiveness of the defenses for an increasingly higher number of prey. The descendants of the stronger prey have a high probability of inheriting the effective defense capabilities of the parents; thus, an ever-higher number of prey can escape from their predators ($\mathbf{s}_3$). The strengthened defense capacity of the prey reduces the probability of survival of the weaker predators ($\mathbf{o}_3$) (attackers); with more effective offensive weapons, the remaining predators can reproduce with greater frequency and with a good chance of passing on their weapons to their offspring ($\mathbf{o}_4$). On average, the predators have become stronger ($\mathbf{s}_4$) and can thus eliminate the weaker prey, thereby ensuring that those with more effective defense capabilities can procreate and transmit these capabilities to their descendants. The loop repeats itself, generation after generation (the system is repetitive), thereby producing a gradual improvement in the predators' weapons (a longer beak, more powerful jaws, greater speed of attack, etc.) and in the defense mechanisms of the prey (improved mimesis, thicker shells, greater speed of escape, etc.).

In the model in Fig. 8.14, two loops have been added to the main loop [**R**]: "$\mathbf{o}_5$" and "$\mathbf{o}_6$," and "$\mathbf{o}_7$" and "$\mathbf{o}_8$." These represent two forms of *external control* of the process which *weaken* the natural selection activity since, by "short-circuiting" the two relations that are delimited by these loops, they reverse the gradual expansion of the offensive and defensive weapons, thereby making both the prey and predators increasingly weaker and "undefended" in their struggle for survival. This weakening mechanism acts whenever human intervention tends to protect some species (first of all the human species) in the struggle for survival, thereby causing individu-

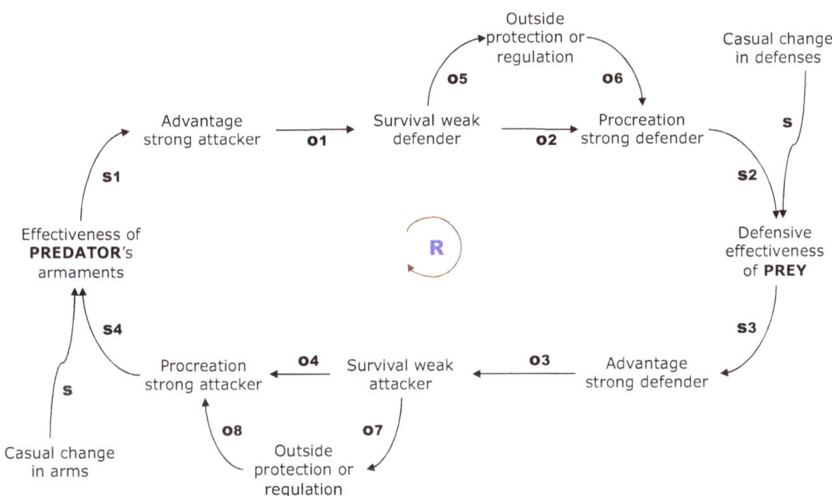

Fig. 8.14 Model of the evolution of phenotypes based on the process involving the struggle for life in populations of preys and predators

als to no longer be suited to competition; in fact, we can note the difficulties animals that have been held in captivity (protected) for a long time experience when they are reinserted into their natural habitat. Nevertheless, we must observe that the loops "$\mathbf{o}_5$" and "$\mathbf{o}_6$," as well as "$\mathbf{o}_7$" and "$\mathbf{o}_8$," can also act in the opposite way: when there is an absence of *external protection*, the winning traits are reinforced. This is the principle all breeders follow when they "control" and "guide" the reproduction of those individuals with the strongest traits, often leading to significant changes in animal populations and species.

Let us return to the main loop, [**R**], and ask whether, if left alone, evolution leads to a strengthening without limits in the attack and defense mechanisms—which also holds for the traits that provide an advantage in reproduction—or if, once again, some "*natural*" Control System intervenes to limit the excess numbers. The "law of dynamic instability" (Sect. 1.5) reminds us that nothing grows or remains stable endlessly; all growth—just as all declines—will be halted at a limit determined by the operation of some Control System. The "*natural*" Control System that sets a limit to the growth in attack and defense organs in the world of biology is represented—in simplified form—in Fig. 8.15.

When the weapons of the predator increase, there is soon a slowdown in the chances for further growth due to the presence of physiological limits (o_1), in accordance with loop [**B1**]; when the limit is reached, the individuals are incapable of surviving and therefore die, without offspring, thereby stopping any further transmission of their traits (s_4). The same process regulates the development of defenses in the prey, as shown in loop [**B2**].

One final observation: the models in Figs. 8.14 and 8.15 would invariably operate even if the conditions of the ecosystem in which the population is living should change.

> A major challenge for twenty-first century biology is to understand evolution in complex ecosystems [...]. How does a community of interacting species evolve in response to a new environment? How do evolutionary responses in turn affect ecosystem properties? These questions have numerous applications—ranging from managing gut bacterial communities for human health [...] to predicting species responses to climate change [...] —yet current evolutionary and ecological knowledge is insufficient to answer them (Barraclough 2015, p.26).

Let us assume that a sufficient number of predators and prey (for example, lions and gazelles) find themselves, due to a cataclysm, living in a different environment (for example, a mountain area without level ground). After a sufficient number of years, we would probably find gazelles and lions capable of climbing and jumping. Loop [**R**] would invariably operate to heighten these capacities in both populations; in the absence of mutual adaptation, one or both populations would become extinct (Brockhurst et al. 2003). It is intuitive that coevolution is faster for species with strong ecological interactions (Buckling and Rainey 2002) and that pairwise coevolution may change the direction of evolution with respect to the direction of the adaption in isolation (Schluter et al. 1985). The size of the environment is equally important. Neo-Darwinism's insight is relevant here; as mentioned above, Julian Huxley, after having highlighted the importance of populations as units in the evo-

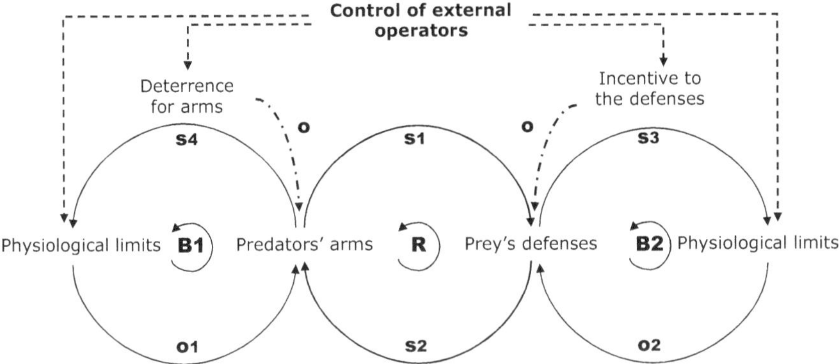

Fig. 8.15 Natural and "external" Rings regulating the evolution of phenotypes in populations of preys and predators

lutionary process, argued that natural selection operates through the process of "adapting to the environment" and that reproductive isolation favors "evolutionary divergence" due to "sexual communication," where there is a free flow of genes so that the genetic mutation, which by chance appears in certain individuals, gradually spreads throughout the entire population, and sometimes, over a long period, to species as well (Darwin 1859; Huxley 1942).

8.10 Combinatory Systems as a Tool for Simulating the Dynamics Regarding the Spread of a Favorable Mutation

It is possible to simulate the dynamics regarding the spread of a favorable mutation in a population by using a Combinatory Automaton, which is a tool for the quantitative modeling of phenomena that "spread" devised by Combinatory System Theory (Mella 2017, Chap. 3). In short, we should remember (Sect. 7.10.3) that a Combinatory Automaton is composed of a lattice, representing a population, each of whose cells contains the "state of an individual's phenotype" (or of one thousand, one million, one billion individuals, etc.). It is easy to understand that the value of each cell [1 = *mutated* or 0 = *not* mutated] at each moment depends on the total number of the mutated individuals who can spread their mutation, since a "micro–macro feedback" connects the "analytical (micro) values" of the cells and the "synthetic (macro) state" of the automaton (total number of mutated individuals). The most useful automata for simulating the Darwinian evolutionary mechanism are the *stochastic* and *irreversible* ones (Mella 2017, Sect. 3.2), since to simulate the spread of a mutation in a population it is necessary to express the *probability of transition* from a "normal" to a "mutated phenotype," referring to each individual (each cell).

Figure 8.16 shows the control panel and the results obtained from a Combinatory Automaton simulating the process of a "slow evolution" (gradualism) under the assumption that an advantageous mutation produces a superiority in "blind mating reproduction." For a didactic demonstration the following "probabilities of transition of state" of individuals have been hypothesized:

- *Fitness intensity of mating* = 1.5% (probability), which is a function of the neighboring individuals
- *Attractiveness for mating* = 0.05 (coefficient which is multiplied by the number of mutated individuals) under the assumption that the mutated individuals have a higher level of attractiveness to mating.
- *Minimum environmental probability of mating* = 0.20 (constant) under the assumption that the environment favors the reproduction of mutated individuals

The simulation presented in Fig. 8.16 shows a relatively slow but irreversible diffusion of the mutation.

The biologist Stephen Gould (Gould and Eldredge 1977), in systematically examining paleontological fossil records, argued that when relevant advantageous mutations occur, the evolution of a species may be relatively rapid, so that a new species appears after a relatively short period of time. He defined this model of evolution as a "Punctuated Equilibrium Pattern." Gould's hypothesis probably derives from de Vries' idea (1905) that evolution is a "discontinuous and jerky process" where new varieties and species are formed by discontinuous variations, and that mutations are the basis of evolution. Gould systematically examined paleontological fossil records, arguing that:

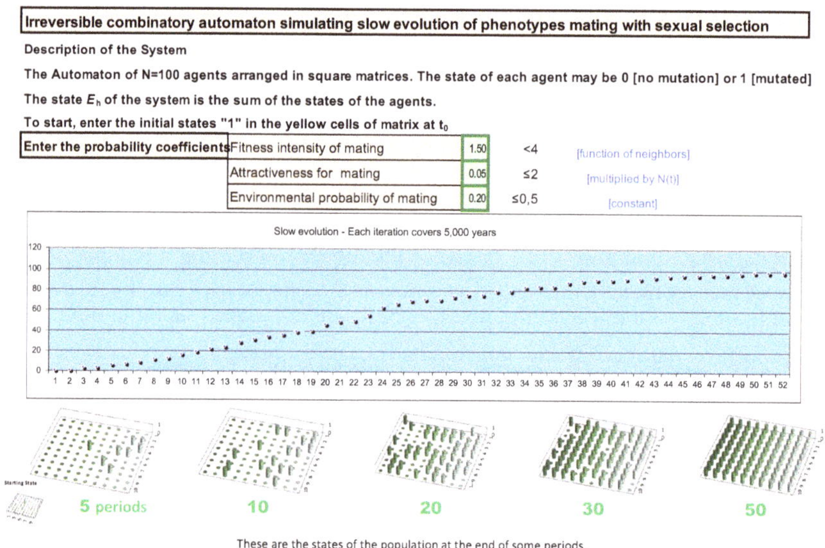

Fig. 8.16 Simulation of the spread of a mutation in a population of 100k individuals, with k equal to 1000 or a million, etc., of individuals

- Any species remains in a state of stability, called *stasis*, for a long period.
- When relevant evolutionary changes occur, the evolution of that species is relatively rapid, producing a division into two distinct species.
- The new species have a new *stasis* until other rapid evolutionary changes occur.
- In many cases, evolutionary dynamics reveal punctuated equilibria rather than generally smooth and continuous changes.

Figure 8.17 shows the graphical representation of a population dynamics showing a punctuated equilibrium. The graph gives the idea of an apparently instantaneous change—a revolution, requiring only 10,000 years—from which a new species is born.

Both "gradualism" and "punctuated equilibrium" describe the rates of speciation. For gradualism, changes in species are slow and gradual, occurring in small periodic changes in the gene pool, whereas for punctuated equilibrium, evolution occurs in spurts of relatively rapid change with long periods of nonchange. The simulation in Fig. 8.16 was based on the assumption of a relatively "slow" evolutionary process; Fig. 8.18 shows the control panel of the combinatory automaton that simulates an evolution with "discontinuity." This automaton differs from the previous one (Fig. 8.16) because I have introduced a "*Minimum Activation Number*," the number of mutated individuals that trigger quick evolution. Once the minimum has been reached, the probability of spread of the mutation becomes a function of the system state, that is, the number of mutated individuals, which makes evolution much faster, thus producing a discontinuity. In the model, the population dynamics are produced by the action of the following parameters:

- "*Fitness Intensity in Individuals*": this expresses the advantages the mutated individuals have regarding survival and reproduction.

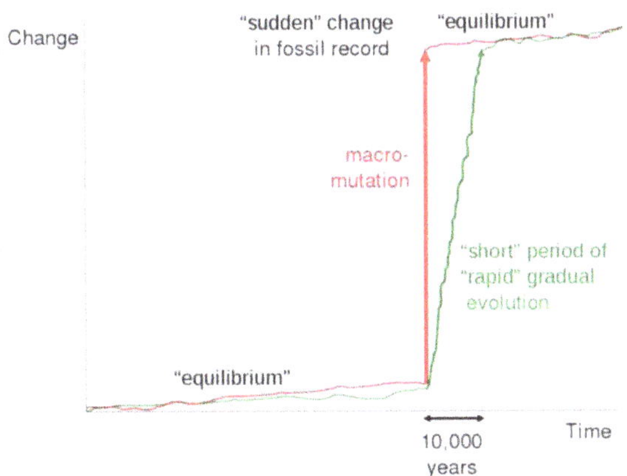

Fig. 8.17 Population dynamics showing a punctuated equilibrium (source: Wikipedia: Punctuated equilibrium)

Combinatory Automaton simulating the evolution with punctuated equilibrium in populations with sexual selection

Description of the System

The system is composed of 100 agents arranged in square matrices. Agent's state [0 or 1].

The state N(t) of the system is the sum of the states of the agents

minimum activation number	10	≤ 20
fitness premium for diffusion of mutation	1	≤ 5
fitness intensity in individuals	5	$1 \leq p_{min} \leq 10$
resistance to mutation	4	0 ≤ resist. ≤ 10

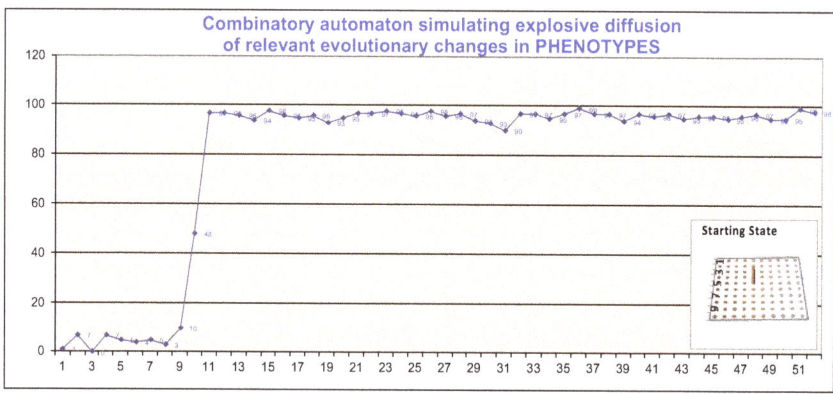

Fig. 8.18 Combinatory automaton that simulates an evolution with discontinuity

- "*Fitness Premium for Minimum Activation Number*": this determines the increase in the level of fitness for the mutated individuals when the minimum activation number has been reached.
- "*Resistance to Mutation*": this indicates the influence of environmental factors that may hinder the survival of mutated individuals.

This graph shows the dynamics of a population with a discontinuity due to an explosive evolution, under the assumption of "limited reversibility." We observe that the simulated population dynamics with the combinatory automaton perfectly represents the dynamics assumed by Gould (Fig. 8.17).

8.11 Qualitative Dynamics Interacting with Quantitative Dynamics

Whatever idea we have of the evolutionary process, the *qualitative* dynamics of populations and species also influences the *quantitative* dynamics; an alteration in some traits of a species can produce considerable changes in the number of individuals in that and other species connected to it in the ecosystem, at times with catastrophic effects (Buchanan, 2000).

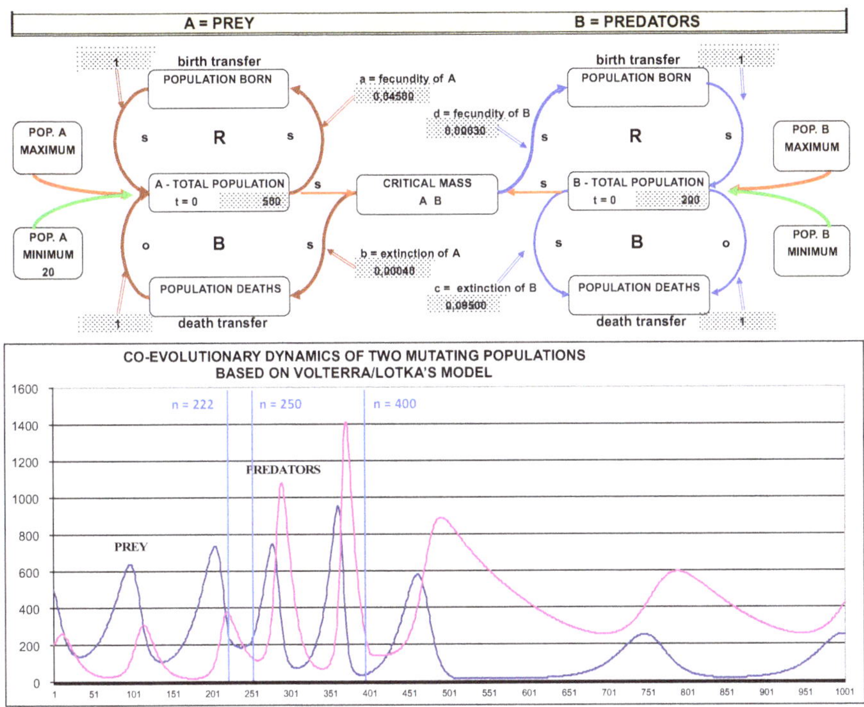

Fig. 8.19 Model of the natural endogenous control of two populations with mutations (based on Fig. 8.10)

Figure 8.19 uses the data from Fig. 8.10, with an initial minimum of $A = 20$ units. The somewhat "disorderly" dynamics in the diagram at the bottom derives from hypothesized mutations to the phenotypical features of the two populations, which impact the rates that determine the quantitative dynamics, according to the following hypotheses (didactic objectives):

- At time $n = 222$: a mutation occurs that strengthens the defense capabilities of population A, making predation by B more difficult; as a result, we assume that the extinction rate of A decreases from $b_{221} = 0.0004$ to $b_{222} = 0.0002$.
- At time $n = 250$: a new mutation in A modifies the rate of reproduction, which increases from $a_{249} = 0.045$ to $a_{250} = 0.080$.
- At time $n = 400$; population B undergoes a mutation, with its fecundity rate falling from $d_{339} = 0.0003$ to $d_{400} = 0.0001$.

8.12 Evolution of Non-biological "Species"

Some brief considerations are warranted here about the "future" evolution of a new, non-biological "species" that has recently appeared: the "robot species," in all its forms, whose *quantitative* dynamics and *qualitative* evolution are for the moment controlled externally by man. The entire world robot population reached 8.6 million at the end of 2010 (Guizzo 2010). However, in 2019 there were more than two million Industrial Robots and about 22.1 million units of Domestic Service Robots in use around the world, in factories, in our houses (IFR 2019, online), and in repetitive and dangerous work environments:

> The International Federation of Robotics (IFR) expects new installations of industrial robots to reach 200,000 units in 2014, and grow 12 percent annually between 2015 and 2017. That would boost the world's industrial robot population to more than 2 million in 2017 (ASME online).

Many industrial robots are evolving by connecting and collaborating to improve their performance. Robots that, while operating in different locations, perform the same task have become connected through the cloud and collaborative programs:

> Robots doing the same task connect across all global locations so performance can be compared and improved at the click of a button. Robots automatically download what they need to get started from a cloud library and then start to optimise through "self-learning" (IFR 2019, online)

It is not pure fantasy to believe the production of robots will be carried out by robots themselves as part of a close-looped technological chain that produces a positive feedback for continued technological improvement.

However, if Kurzweil's "Law of accelerating returns" (Kurzweil 2001) continues to hold, within a generation (about in the year 2.045/48) this species will become independent of man, with the risk it can self-determine its own quantitative and qualitative dynamics as well as those of all other forms of life. This is not science fiction, such as that found in "Terminator." A number of studies already exist on "evolutionary robotics" (Holland 1975; Koza 1992; Floreano and Nolfi 1997; Harvey et al. 2005; Floreano and Matiussi 2008; Floreano et al. 2008a, 2008b; Floreano and Keller 2010) that apply Darwin's principle of natural selection to the population of robots:

> The general idea of evolutionary robotics [...] is to create a population with different genomes, each defining parameters of the control system of a robot or of its morphology. The genome is a sequence of characters whose translation into a phenotype can assume various degrees of biological realism. For example, an artificial genome can describe the strength of synaptic connections of an artificial neural network that determines the behaviour of the robot. [...] At the beginning, robots have random values for their genes, leading to completely random behaviours. The process of Darwinian selection is then imitated by selectively choosing the genomes of robots with highest fitness to produce a new generation of robots. [...] This process of evolution can be repeated over many generations until a stable behavioural strategy is established. In some experiments this selective process has been performed with real robots whereas in other experiments physics-based simulations (Floreano and Keller 2010, online).

Aspects such as the alternation of Darwin's theory, the simulation of the quantitative dynamics of population mutation over time, the substantiation of the dynamics of continuing human population growth, and deeper consideration of the robot population can serve to indicate further hypotheses and insights in other fields: social behavior, languages, scientific and technical studies, evolutionary economics

(Boulding 1981), evolution in fashion, products, processes, Artificial Life (Levy 1992), and so on. For example, a *fashion*—whether this concerns clothing, clocks, cars, toys, a particular linguistic form, and so on—arises in a given environment from a novelty that is introduced (mutation) "by chance" by a given initiator (mutant) to get away from the usual routine. If the novelty (mutation) is not advantageous, not pleasing and costs too much, the mutation would remain isolated and no diffusion would develop. If the number of individuals that imitate the innovation (mutant) reaches the necessary activation density, then that mutation will spread and give rise to a "fashion."

It would be useful to explore the evolution of new scientific ideas when these compete among each other. For example, according to Thomas Kuhn, scientific progress is guided by paradigms—that is, by structures of conjectures, postulates and laws that, after a prescience phase, are accepted by a population of scientists, thereby giving rise to a period of normal science. A scientific revolution (mutation and diffusion of the mutated ideas) occurs when a paradigm is no longer able to solve scientific problems and is replaced by another, more suitable paradigm, with the formation of a new scientific community (Kuhn 1962):

> What are scientific revolutions, and what is their function in scientific development? Much of the answer to these questions has been anticipated in earlier sections. In particular, the preceding discussion has indicated that scientific revolutions are here taken to be those noncumulative developmental episodes in which an older paradigm is replaced in whole or in part by an incompatible new one (Kuhn 1962, p. 92).

The next *section* illustrates how Darwinian theory also acts in the field of production, leading to continuous improvements in production organizations and in the entire "productive machine," on which the fate of all humanity depends.

8.13 Evolution in Networks of Organizations: Production Networks

In this *section*, I will try to highlight how the production system that develops in an area—from the small municipal area to the national or even international one—is in fact a network of flows whose nodes are production units, companies, corporations, and public and nonprofit units. Each node receives input flows from upstream nodes and allocates the flows of its production to downstream nodes. Everything we consume is produced by a more or less vast network. To effectively observe production systems as production networks I have suggested a change of perspective (Mella 2019) by adopting the fundamental "dualistic variant" well described by Peter Senge:

> We all know the metaphor of being able to "pull back" from the details enough to "see the forest instead of individual trees" but, unfortunately, for most of us when we pull back we see only "a large number of trees" (Senge 1990, pp. 146-147).

I wish to investigate the "laws" that guide the evolution of the complex production systems that make available the goods and services from which—as if guided by a Smithian "miraculous" invisible hand—the continuity of the "subtle film of material called life" depends (Brown 1954 p. 3). In particular, I will try to demonstrate how production networks make their knots into units that evolve according to Darwinian logic.

Let me start by reminding you that Systems Thinking teaches us that if an observer wants to broaden his intelligence (Mella 2012) he must develop the capacity to "zoom out" from the parts to the whole and to "zoom in" from entities to components. This way of "connecting" the parts to the whole, and at the same time of identifying the relations between the "whole" and the "parts," is not an option but a necessary condition for knowing and thinking, as Edgard Morin tried to teach us:

> Since we have been domesticated by our education which taught us much more to separate than to connect, our aptitude for connecting is underdeveloped and our aptitude for separating is overdeveloped; I repeat that knowing, is at the same time separating and connecting, it is to make analysis and synthesis. We need to deeply reform all our way of knowing and thinking (Morin 2005, p. 21).

To understand the nature of production processes, we need to change our way of thinking about production, viewing it not as the result of the activity of individual production organizations but as a production flow obtained from a more or less widespread Production Network of interconnected production units (the nodes) all of which, consciously or not, are necessarily connected, interacting and cooperating in a coordinated way to combine and arrange, step by step, the factors—materials, components, manpower, machines, utensils, equipment, means of transport, infrastructure, and so on—to obtain flows of products—intermediate or final goods—and to sell these where there is a demand for them. To see how production activity develops through networks, let us consider a simple case, observing how our suit is produced.

Let us suppose that the suit we are wearing has only four components: the wool fabric, the interior cotton lining, the special thread for sewing and the buttons. Obviously, in order to obtain the wool fabric, we require sheep farmers who periodically, and with the appropriate instruments, sheer the flock. The raw wool must then be gathered, baled and transported in order to be washed, degreased, whitened, carded, and then spun. All these operations require machines of varying complexity produced by specialized companies, which, in turn, need electric motors, steel and plastic components, cables, monitors, safety systems, etc. The spun wool finally is sent to be dyed with colors that require producers of chemical components, metal or plastic packaging, thinning solutions, separators, and all other types of accessory. The colored wool is sent to the weaving mill with its modern robotic machinery produced by super-specialized companies. The pieces of fabric are finally packaged and purchased by the clothes manufacturer, who must cut and sew them using specialized machinery. The cuts of fabric are sewn together using a special thread, after which the interior lining is inserted. This final phase also requires another highly extensive production network. The reader will be left to combine these two networks with the one that intervenes to provide the thread and buttons. This is a short, abstract description of the production networks that allow thousands of people to wear a suit with jacket and pants like yours. What significance is there in producing and purchasing a suit? The production of the suit is normally, and simplistically, seen as the activity of a *terminal node* of several extensive *production networks* needed to obtain, transform, and combine the components of the flow of production. Purchasing the suit means benefitting from the output of the entire Production Network. And what about the completely different production networks for obtaining cars, shoes, wine, cheese, detergents, and hygienic goods? We only need enter a hypermarket to realize how many thousands of specific production networks operate to allow us to fill our shopping cart.

The production network perspective reveals a number of interesting characteristics, among which:

(a) The production of any good or service requires the existence of a production network. *No production is possible without a network.*
(b) The opposite relation is equally valid: *No production network can exist without production to be obtained* and purchased by a sufficient demand. Moreover, new production requires new networks or the adaptation of existing ones.
(c) The above characteristics clearly show that as consumers we need networks; however, the networks also need our consumption to repeat their transformation cycles over time.
(d) From the production networks perspective, it is relevant to note that when we purchase and consume a good we do not download the warehouse of a single production organization but download the whole production network; at the same time, we allow all the component modules to repeat their production cycles.
(e) The network functions only if *resources—labor* in particular—are available for production. If the network needs resources, the resources need the network to be included in the transformation cycles.
(f) In order to obtain the flow of its outputs, the network operates continuously over time for an unlimited period and does not have a precise reference space.
(g) Networks are a spatial and a temporal entities since they have no precise physical space of reference or a specific duration; the production units that compose them do not have the necessary spatial or temporal location but only the necessary interconnection of flows; wherever their nodes are located, the networks are units of connection that join the resource reservoirs, wherever these are located and from which they obtain their inputs, to the sources of demand, wherever located and to which their output flows.

Technical note. The term *network* is correctly preferred to the term *system*, or *structure*, since it brings out three aspects (Mella 1997a). *First*, that among the nodes—that is, the production units—there must always exist necessary and stable connections, which originate from the exchange and information processes. *Second*, that the network functions if all its component nodes act, according to their appropriate times and in an appropriate space, simultaneously and in a coordinated way, revealing a *selfish behavior* aimed at its survival within the network. *Third*, that the network operates to produce *flows* and not *individual products*; thus, the unitary activity of the network must be observed over a *meaningful time span*, in which we can identify the flows of interconnection among the nodes and among the external reservoirs.

> Despite differences in terminology, as well as in focus, between different researchers, there is a growing consensus around the idea that one of the most useful keys to understanding the complexity of the global economy—especially its geographical complexity—is the concept of the network (Coe et al. 2008: 2).

A *generalization* can be made here: "all" Production Networks represent an efficient system of "micro-local" transformation processes of resources to produce *flows* of goods or services to satisfy the demand for *final consumption goods*, which represent the *global* OUTPUT of these Networks.

To demonstrate how the nodes of production networks can be interpreted as vital units operating in a Darwinian environment, subject to the evolutionary logic of populations, it is necessary to identify six fundamental characteristics that distinguish production units when observed as nodes, or modules, that make up the networks. For a productive network to be able to operate and survive any disruptions, its modules, as holonic units, must have at least the following characteristics (Koestler 1967; Wilber 1995):

1. *Spontaneous genesis*: to an external observer, the nodes seem to arise spontaneously—based on a decision by some person or group—and can link up with upstream and downstream nodes; they require a *reservoir of demand*, defined as a number of units—individuals or organizations—that are often *territorially located* and overall represent a potential demand for goods.
2. *Intelligent behavior*: through their own management, the nodes behave rationally, seeking optimization and following the make-or-buy rule.
3. *Economic efficiency*: through the system of internal and social values, the nodes can evaluate their input and output flows and behave under the *survival condition* that the value of the output is not below that of the inputs in an appropriate period; and, if it is below this value, that the difference is not greater than the amount of the initial resource endowment (invested capital) (Chap. 9).
4. *Autonomy, self-affirmation*, and *durability:* the modules produce *autonomous cognitive processes* aimed at survival; once created, the production organizations tend to remain viable indefinitely by modifying their production processes and their connections; in this sense, they are *viable systems* according to Beer's definition (Beer 1979, 1981; Espejo and Harnden 1989), and autopoietic systems as defined by Maturana & Varela (1980).
5. *Specialization*: the network nodes tend to specialize their productive processes and products (Snow et al. 1992), limiting the range of possible processes and adopting only those required within the Production Network; the production units linked to the *consumption reservoirs* are *terminal production nodes*; the others, linked to these in an instrumental and specialized way, are *intermediate nodes*; as Coase (1937) and Williamson (1975, 1981) have demonstrated, production through modules specialized in the production of components, machinery, or services and assembled into finished products from different specialized modules is more efficient than that carried out by "monolithic organizations." For this reason, the network nodes mucst be equipped with expertise, research and development, and design (Heshmati 2003).

6. *Need to connect*: production *modules* cannot exist in isolation; their natural tendency is to *link up* with other production organizations because this is necessary for survival; in particular, they must be connected with a labor reservoir, which is not easily transferable. The fundamental characteristic of labor is that it is "idiosyncratic and place-bound"(Storper and Walker 1989, p. 155).
7. *Selfish Behavior:* the nodes, in order to maintain their autonomy and viability over time, must behave in a way consistent with their individual and local interests, while ignoring the long-run collective and global consequences their behavior may produce; *individualism, localism* and self-affirmation prevail in the node's behavior as a "natural" form of "selfishness."

> An entity… is said to be altruistic if it behaves in such a way as to increase another such entity's welfare at the expense of its own. Selfish behaviour has exactly the opposite effect. "Welfare" is defined as "chances of survival", even if the effect on actual life and death prospects is small […]. It is important to realize that the above definitions of altruism and selfishness are behavioural, not subjective (Dawkins 1989: pp. 4, 5).

This *seventh characteristic*, which recognizes the *selfish behavior* of the nodes of production networks, is very strong, and it must be mitigated when individual selfish behavior contrasts with the "common good." Production networks, even if they consist of "selfish nodes," must produce in a "sustainable world" to whose realization they must contribute. If we accept these seven *minimal* properties of network nodes, then we can recognize that production organizations take on both the meaning of *autonomous units*—if observed as individual production units—as well as the broader one of *interconnected units*—if considered as nodes of a network—since they are connected with other *upstream* organizations that provide them with factors of production and with other downstream organizations, to which their production flows. Therefore, we can immediately conceive of the knots of a network as *holons*, following Koestler and Wilber (Sect. 1.2) and the production network as a holonic network or reticular holarchy, understood as a network of networks (on the notions of holon, holarchy, holonic network, and reticular holarchy, see Mella 2009).

Above all, both Koestler and Wilber postulate that holons, like the nodes that form the Production Network, *arise spontaneously* and order themselves naturally in a holarchy or a holonic network (Mesarovic et al. 1970). A form of holonic network that is particularly interesting comes from Shimizu's idea (1987), which theorizes the Autonomic Cognitive Computer (ACC), a concept that interprets in holonic terms the processes of gradual informational synthesis through parallel processing by cognitive entities. In simplified terms, an ACC is made up of a parallel set of processors (machines, production departments, organizations and modules, decision points, and so on), arranged on various levels, which, at each level, carry out progressive synthesis of the inputs coming from the lower levels. Shimizu's construction presents us with two interesting productive applications in organizations that carry out complex processes: the *Holonic Manufacturing System* (HMS) and the *Bionic Manufacturing System* (BMS). An HMS (Adam et al. 2002; Kawamura 1997) is conceived of as a holarchy of modular production units—groups of similar machines (modules or cells) that carry out basic processes, together with groups of organizational units engaged in supply or selling activities and units of coordination—that constitute a complex process broken up into different levels through the successive syntheses of basic processes, in order to obtain a final product. A BMS considers a final product as a model to be achieved (Okino 1989; Tharumarajah et al. 1996), subdivided into autonomous segments to be obtained over various levels; it is not the processors that are considered as holons but rather the segments of the model to be achieved—called "modelons" (*model*s as ho*lons*)—which are produced from the gradual accumulation of previous segments in order to obtain the *final modelon*.

If we agree on the holonic vision of production networks, whose logic is the development of processes on several levels that combine to obtain finished products—then it immediately follows that we can interpret them as ACC and, in particular, as HMS or BMS. In general, the production network acts as Shimizu's ACC since, through the production units, it carries out the gradual combination of factors necessary to obtain the finished products. However, we can consider holons as production units, nodes of the network (that is, as processors) as well as their products, considered as models made by a certain set of those units. In the first case, the network is similar in its operational logic to an HMS whose operational modules develop the processes necessary to obtain the final consumer products, through the subsequent processing of materials, components, machinery, etc. In the second case, a holarchy of submodels is produced that has as a top modelon a final product and as a modelon at lower levels the components and other factors directly or indirectly included in the final modelon. The same logic is also valid if we consider the stratification of the values produced by each holon rather than the actual results (matter, materials, components, machinery, etc.). In both cases, the value of a final product is a top holon since it summarizes the value of all processing steps or sub models processed by subordinated holons.

8.14 The RULES of Selfish Behavior of Nodes (Modules) in Production Networks: Evolutionary Qualitative Dynamics of Nodes

By accepting the production network perspective, in which the nodes possess at least the aforementioned seven characteristics, we may also assume that each node, in following its tendency to strive for *self-affirmation and durability*, must follow these five RULES of "selfish behavior":

RULE (1)—*Reservoir of demand*: the *organization-node* identifies or creates a reservoir of demand compatible with its OUTPUT vector (volume, quality and price of the resulting production), under the condition of an *intelligent behavior*. Let us assume that the reservoir α for product A is characterized by the vector $\alpha A = [P_{max}, pP_{max}, pP_{min}, \mathrm{q}lP]_{A}$—in which the horizontal axis indicates the volume of potential demand (P_{max}) of the reservoir for the good produced by the

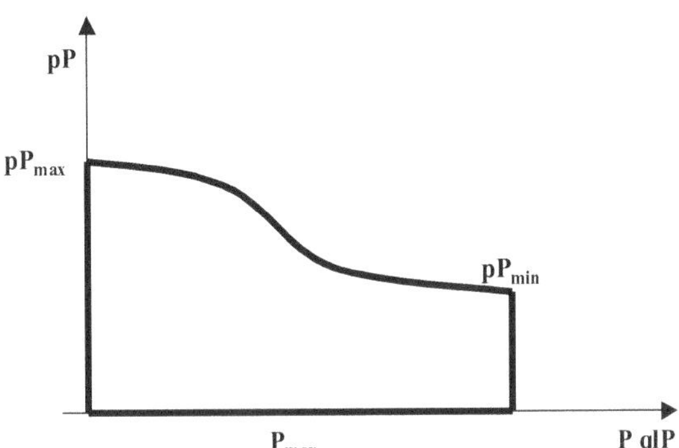

Fig. 8.20 Reservoir of demand for the good A

node at a value, *pP*, between a maximum and a minimum, with the quality level (ql*P*) assumed constant—and can be represented by a trapezoid like the one in Fig. 8.20.

RULE (2)—*Increase in size:* if it succeeds in connecting with a given reservoir of demand, the *organization-node* tries to attain the maximum size, in line with its INPUT vector (volume, quality and price of the utilized factors) and the available capital. This RULE has by now been accepted in the literature on the firm and managerial behavior, which sees the expansion in sales, and thus in production, as the main managerial objective. In his work, *On the Theory of Expansion of the Firm* (1962), William Baumol presents a model that shows the relation between the sales level and the profit level, compatible with the growth of the firm. In short, he introduces these hypotheses: first, the firm's objective, which derives from rational managerial behavior, is to achieve the maximum sales revenue, the maximum rate of growth in sales over time; second, profits derive from sales and represent the main form of self-financing.

Although the static theory of the firm is a helpful snapshot description of a system in motion, it is useful also to have an alternative construction of the kind which is described in this paper – another equilibrium analysis in which the rate of growth of output, rather than its level, is the variable whose value is determined by optimality considerations (Baumol 1962, p. 1078).

RULE (3)—*Readjustment of its* OUTPUT VECTOR *and its downstream connections:* if it cannot connect to a reservoir of demand—or if the reservoir to which it is connected is no longer compatible with the processes carried out—then the *organization-node* must try to modify its OUTPUT vector by adapting its internal processes, in accordance with the INPUT vector and the available capital.

RULE (4)—*Productive efficiency*: in any event, the *production module* must *always* try to *readjust* its INPUT vector to reduce the unit cost of production by increasing productive efficiency; that is, the continual search for higher technical returns from the factors and from labor productivity. This also implies the search for new *resource reservoirs* to reduce the unitary prices of factors and/or increase their quality.

RULE (5)—*Extinction*: if the management (cognition activity) cannot connect the *organization-node* in a convenient manner to a reservoir of demand or modify its internal processes to the extent necessary to maintain its autopoietic processes, the *organization* will be extinguished.

Regarding what is stated in RULES (1) and (2), rather than simply referring to the traditional notion of *demand* for a given good, the concept of *reservoir of demand* better represents the idea that the potential consumers or users can also have a geographic reference rather than a merely quantitative one, and that the *production nodes* may tend to connect with the reservoir rather than simply satisfy a certain stock of requests.

RULES (3) and (4) imply the concept of *resource reservoir* as being symmetric to that of a *demand reservoir*, with the difference that we can consider a resource reservoir both as a *site* where resources are present—a stretch of sea rich in tuna or seals, or an area rich in oil, water, gold-bearing metals, etc.—and as a *set of* nodes earlier in the process that can supply, competitively or as an alternative, materials, components and structural factors. A particular resource reservoir is represented by a *reservoir of labor*, which can be understood as an area with a certain quantity and quality of manpower availability at a given unit cost. Referring by analogy to Fig. 8.20, we can characterize resource and labor reservoirs by a vector that indicates the availability, unit value and quality of the available resources.

Rule (4) guarantees that even these elementary production organizations manifest the continual search for higher levels of *efficiency* through the progressive learning on the part of the organization. The spontaneous genesis makes it likely that, external to an organization, a new organization will be created capable of producing with greater efficiency, and thus at a lower cost, some components (materials or equipment) already internally produced by the original organization, which may then consider it convenient to connect serially to the newly generated organization to obtain the factors it needs at a lower cost or at a higher quality.

The improvement of production *modules* improves the performance of the entire *network*, which, in turn, provides the impetus for further improvements in the modules. It is precisely this characteristic of bidirectional influence, the interrelationship between *micro* and *macro*, that is between *whole* and *parts*, that gives rise to the fundamental *property of continuous improvement* that distinguishes every production network. In every production network, *reinforcement loops* are activated that enhance *self-organization* as well as produce a faster and more widespread *improvement in the performance* of modules at every level. Changes (mutations) that improve the production network are preserved and disseminated, and these can also generate new branches, while disadvantageous ones are eliminated or weakened; otherwise, the branch of the network in which the modifications occur is suppressed.

A new mobile chip that allows a new type of message to be sent will change all the production *downstream* of the new generation's cellphones; but that new chip will put pressure on component manufacturers *upstream* to improve the quality of their products, with new monitors, keyboards, microphones, etc. An improvement in the production of car engines will expand to all cars produced but will also require new research on the components of that engine. New crossbreeds in plants for cultivation or in farm animals will expand downstream to create new products or improve existing ones, but at the same time there will be upstream pressure for new feed, automatic stables, and other breeding tools. This evolutionary feature of nodes and networks is so obvious that it does not merit any further consideration; I would only point out that the technological, technical, and scientific progress of humanity is the consequence of the rapid spread of innovations throughout the production networks, in cases where an innovation improves the vector of INPUTS of *downstream modules* and the vector of the OUTPUTS of *the upstream modules*, expanding in the directions of the affected branches, often with *reinforcement loops*. Consider the case where an improvement of a component of a mix improves the quality of tires that improves the performance of racing cars. The increased performance in terms of road holding makes it possible to upgrade the engines; and more powerful engines require new braking systems and new steering control systems. This necessitates the study of new brake linings and more efficient gears, and in turn the development of new materials, etc., in an easily intuitive cycle of improvement (advantageous mutations). Now consider a new processor that allows for an improvement in the performance of an industrial robot that produces a qualitative improvement in the production of processors that improve the robots that will produce the processors of the next generation, in a very obvious reinforcement loop.

The cognitive ability of production modules, which must continuously develop creativity, research, and development, is therefore essential. Today, the great challenges concern the new materials, nanotechnologies, and the processes of alternative energy sources, as well as advances in robotics, biology and genetics. By focusing on the *micro level*, it is clear that these dynamics are produced by units that arise *spontaneously* for their own benefit and that, for their own survival, *connect* to each other at the local level on the basis of simple selfish rules. Being holonically connected *upstream* and *downstream* with other units, their existence and the meaning of their processes depend on the network that they themselves contribute to produce, developing forms of differentiation, innovation, organization, integration, and disintegration that represent the very essence of the *evolutionary dynamics* of the production networks. From a holonic point of view, *there is no network without nodes and no nodes without a network that connects them.*

The seven characteristics of the modules that form the nodes of the production networks, together with the five RULES just examined, lead to the conclusion that, at least from a logical–procedural point of view, within the networks there is a *competition* between nodes, which leads to a natural *selection process* that presents principles similar to those of Darwinian evolution.

> In social animals it will adapt the structure of each individual for the benefit of the community; if each in consequence profits by the selected change. What natural selection cannot do, is to modify the structure of one species, without giving it any advantage, for the good of another species; and though statements to this effect may be found in works of natural history, I cannot find one case which will bear investigation. [...] (Darwin 1859, pp. 86-87).

The network becomes an *environment* that produces *selective tendencies* that spontaneously arise in its nodes; requiring ever greater efficiency, the network favors *random creative mutations* in the production processes that raise the level of efficiency. Broadly speaking, the Production Network can be conceived of as the model that, from a theoretical point of view, justifies the idea that the economy must be interpreted as evolutionary economics, following the ideas of Joseph Schumpeter (1942), Armen Alchian (1950) and Kenneth Boulding (1981).

8.15 The Laws of Production Networks

The *cognitive* and *creative* processes that characterize *organizations* as *nodes* do not allow us to predict the actual evolution of production networks; nevertheless, if we assume that *organizations' nodes*—consciously or not—follow the selfish RULES (1)–(5) above (Sect. 8.14), then we can deduce five typical trends, or behavioral schema, referred to here as the laws *of Production Networks*, in order to highlight their apparent inevitability and cogency.

FIRST LAW: Production Networks tend to expand.

This law states that Production Networks tend to increase in depth (vertical expansion), in width (horizontal expansion), and in their ramification. The law is supported by the characteristics of the *organizations*—spontaneous genesis and tendency to connect—as much as by the basic RULES of behavior: if at any level of the network the *nodes* try to enlarge their demand reservoir, improve their own INPUT vector and increase in size, then it can be assumed there is always an increase in connections both along the boundaries of the network as well as internally, with the formation of increasingly more connected branches at ever greater levels of productive specialization. The expansion occurs in three ways:

(a) The network *expands its own boundaries* and new links are added to the borders to enable connection to new demand reservoirs for final products.
(b) The network *adds more levels*: productive specialization, creativity, and research lead to the spontaneous creation of *organizations* whose OUTPUTS are the specialized INPUTS of *successive* organizations; the production chain becomes longer as well as wider. Today, in modern economies, it is easy to recognize that even the smallest components of a product are acquired externally from newer and more specialized *organizations*.
(c) *Two or more networks merge*: many networks arise independently to produce distinct products aimed at differentiated demand reservoirs. When several final products are complementary and are obtained in an integrated manner from a single organization; when an autonomous product becomes a component of a final product; or when several intermediate processes are centered on antecedent hubs, the networks can also be considered merged.

SECOND LAW: Production Networks tend to increase the quality of their performance through a nonlinear cumulative process.

This law derives from the tendency of organizations to improve their input and output vectors and from the general property of Production Networks to spread their individual improvements. The network improves its performance even if a casual improvement occurs in the performance of only a single organization; but RULE (3) leads all organizations, in order to improve their input and output vectors, to produce innovations and to make discoveries and inventions that, if useful, spread simultaneously to all branches of the network, though with differing intensity, thereby involving the most distant and unforeseen ramifications (Schilling 2000). This law supports an important *corollary*: the improvement in the quality of the network's performance is permanent and cumulative, thus path-dependent and nonlinear, and in general exponential, producing an increasing return (in the sense of Arthur 1994) regarding the network's economic efficiency. In fact, as the improvements are transmitted to the network branches, they not only spread but, until substituted by other improvements, are preserved in time and space, producing a cumulative effect that leads to an acceleration in the progress of each sector. Each improvement derives from a creative or rational action based on a previous improvement. If this were not the case, then the monumental and accelerated progress in electronics, telecommunications, transport, war production and biology could not be explained.

THIRD LAW: Production Networks are resilient networks that tend to continue on as if they were living entities.

This law is explained by the natural tendency of *organizations*, as well as the branches that spread out from them in a *forward* direction, to survive through the adaptation of the INPUT and OUTPUT vectors, when this is necessary for their autonomy. Thus, when an *organization* is destroyed, the entire successive branch—which remains functional for some time—tries, in order to avoid extinction, to adapt by connecting to another *organization*; if this is not possible, the pressure to restore the antecedent branch is so intense as to make it likely that other *organizations* will be spontaneously created to substitute the destroyed one. If this spontaneous genesis does not occur, and if the *successive organizations* are also not able to internally produce the missing INPUTS, then it is likely that the INPUT vector will be modified so as to substitute various components for the latter. If this, too, is not possible, then the network will be broken. As a result, a Production Network is *resilient* since the ability of nodes to perceive, appreciate, transform, and utilize knowledge for adaptive purposes (Zahra and George 2002) enables the entire network to resist deformation or dynamic breakage and to return to its original form after deformation.

FOURTH LAW: Production Networks tend to combine with other networks in order to function.

Thus, production networks must avail themselves of other networks, with which they tend to combine. Here are the most relevant of such networks (which are touched on only):

(a) The *information network* (more generally, the network of Information and Communications Technology) that allows producers and consumers to exchange information about needs, products, and services. The Internet will soon be joined by Internet2, which will be faster, have more capacity, and will be more secure (D'Amours et al. 1999; Hesse and Coe 2006).
(b) The *logistics network*, which is needed so that the flow of resources can reach the producers and the machinery and facilities can be placed where they are most useful, and the products consumed where there is a demand for them (road and railway infrastructures, air and naval network, oil pipelines, oil tankers, warehouses, refrigerators, etc.) (Rodrigue and Hesse 2006).
(c) The *financial-insurance network*, which directs capital to those points in the network with the highest efficiency and quality (with the creation of electronic money, this network is arguably the first to reach a global level); despite the fact the Romans had constructed a respectable logistical network for those times, the production networks were quite circumscribed due to the lack of both technology as well as a financial network that would facilitate the transfer of funds.
(d) Last but not least, the very relevant *network of scientific and technological research*.

FIFTH LAW: Production Networks have their own inertia that always delays their adjustment to changing demand.

A long-lasting fall in consumption creates stress for networks. When an organization is forced to cease its activities, it is not only the different links in the network that are affected but all the links connected to it. Gaps and broken links form that cannot be repaired in a short time. Nevertheless, the network tends to survive, seeking to fix the gaps, substituting old links with new ones and expanding elsewhere (a reduction in the demand for flights—for example, after an airplane crash—can cause the collapse of many airline companies; the subsequent increase in demand for flights results in long waits before finding an available flight, leading to the birth of new companies that serve to restore the links in the network; in recent months overseas sellers have been battling over the lack of available transport ships, but the seagoing transportation network will adjust, though with a certain delay).

8.16 Complementary Material

8.16.1 The Nodes (Modules) Forming the Production Networks

Regardless of the specific characteristics of the *nodes* in each production network, we must assume that their INPUTS and OUTPUTS characterize their functionality in the network: that is, they are prepared to carry out production transformation processes (factor inputs transformed into production outputs), economic processes (input values transformed into output values and results), and financial processes (capital transformed into investments and profits), as detailed in Chap. 9. If we indicate by "node A," or "O_A," a generic "*module* A," then by taking into account these three typical transformations, we can represent a generic *autonomous and vital node* by the simple *standard module* in Fig. 8.21.

Sticking to the terminology of traditional business logic (Chap. 9), we observe that the INPUTS of "O_A" must at the very least be defined by a vector, CF_A, of the factor costs (material and components, energy and services, labor, and machinery and structure) that specifies the amounts (QF = amount of factors, according to the chosen production function and the appropriate time distribution), their unit value (pF_A = price, according to the supply markets), and the level of their quality (quality, according to the production specifications), with reference to a given time period (not indicated in the model). The production OUTPUTS is likewise defined by a vector, "RP_A" that indicates production volumes (QP, with the appropriate temporal distribution), their unit value (pP_A = price), and the level of their quality (quality). In current terminology, the sum of the costs of the factors used by "O_A" in production determines the cost of production (CP_A) which, in proportion to the amounts produced, quantifies the average unit cost of production "cP_A" over the observed period. It is also necessary to take into account the possible capital, "K" (a combination of Equity and Debt) needed for investments to start and maintain active production processes, both monetary, as in modern economies, and nonmonetary, represented by the supply of factors of various kinds.

Other, more elaborate models of *nodes* may be proposed, but the simple one presented here, in line with "business economics doctrine," is able, for our purposes, to highlight all the main variables that characterize the transformation processes of any production organization understood as a "module-holon."

In primitive economies, in which self production prevails, as well as in *nonbusiness organizations*—or utility companies—the value of production, "vP," is equal to its *usefulness* to the end consumer, or user; therefore, the cost of production is "charged" to the entire organization, since the products are sold without an explicit price.

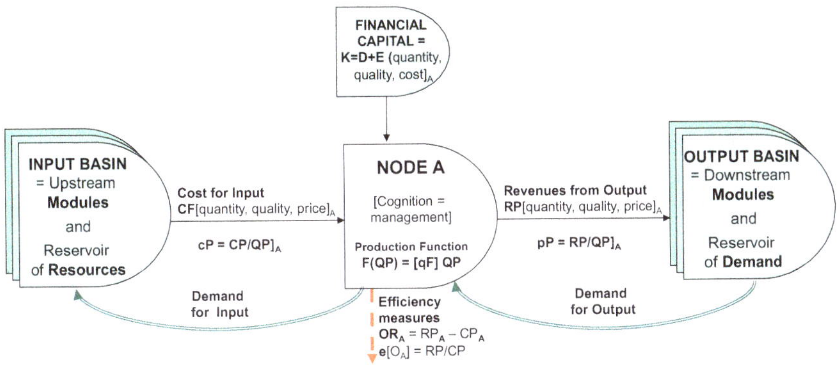

Fig. 8.21 The standard module of a productive network

In *business organizations*, "pP" represents a price. *For-profit organizations*, or companies, seek the maximum gap between "pP" and "cP" or, in equivalent terms, the maximum operating result. *Nonprofit organizations*, on the other hand, look for the minimum gap between "vP" and "cP," which is equivalent to producing with operating results tending toward zero.

Note. The *modules* show real, monetary, and financial inputs and outputs. Since this Chapter dealt with the production aspect, I felt it sufficient to characterize the standard *module* with economic flows and volumes defined in terms of values; only the financial input of capital was added to the real ones. The entire production network is therefore viewed as a network of economic flows. There are probably other forms of production organizations, but the model, in its essential aspects, can represent them.

8.16.2 The Genesis of Production Networks

A Production Network forms when several nodes, *production organizations* (Fig. 8.21) connect to one another through their INPUT and OUTPUT according to the RULES of *selfish survival*. In this sense the following considerations refer to *real* or *factual* Productive Networks and are not limited to *Rational Networks* that derive from voluntary agreements among firms to pool together resources and skills in order to form *holonic, virtual or extended enterprises* (Davidow and Malone 1992; Goldman et al. 1995; Huang et al. 2002).

Adopting a *functional perspective*, the nature of a Production Network implies that, for all production nodes that are not *primal* and final *nodes*, each output of a node is at the same time an input of some other nodes. Only the chain of connections, the network, takes on full significance as the system for the production of goods. The history and direct observation of primitive economies clearly shows that the first production nodes were final organizations, which we will call "O_A," spontaneously arising from self-production by seeking and transforming resources through the labor provided by the consumers themselves.

RULE (3) guarantees that even these elementary production organizations manifest the continual search for higher levels of efficiency through the progressive learning on the part of the organization (Sect. 8.14). The *spontaneous genesis* of nodes (Sect. 8.13) makes it likely that, external to "O_A," a new *organization*, "O_C," will be created, capable of producing with greater efficiency, and thus at a lower cost, some components (materials or equipment) already internally produced by "O_A." It can then be convenient for "O_A" to connect *serially* to "O_C" to obtain the factors it needs at a lower cost or at a higher quality [Model 1]. "O_A" is connected to the demand reservoir β, but at the same time it represents (is included in) the *demand reservoir* for "O_C." The OUTPUTS of "O_C" are INPUTS that "O_A" combines with other internal resources to obtain products.

[Model 1]

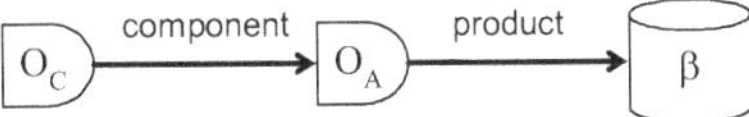

Ronald Coase's approach comes immediately to mind. It could be said that in transaction cost theory the management of the productive nodes is related to the control over the input costs that could, and should, be saved by the firm, compared to the market, in coordinating resources. Following this reasoning, Coase argued that:

> [the] entrepreneur has to carry out his function at less cost, taking into account the fact that he may get factors of production at a lower price than the market transactions which he supersedes, because it is always possible to revert to the open market if he fails to do this (Coase 1937, p. 392).

Referring to Coase's approach, Alchian and Demsetz (1972) argued that some issues related to *information* also play a fundamental role in defining the convenience for an organization in joining a Production Network.

The process shown in [Model 1] can repeat itself serially or in parallel. If a new *organization*, "O_H," should more efficiently produce several components needed by "O_C," then the branch may lengthen (serially) [Model 2].

[Model 2]

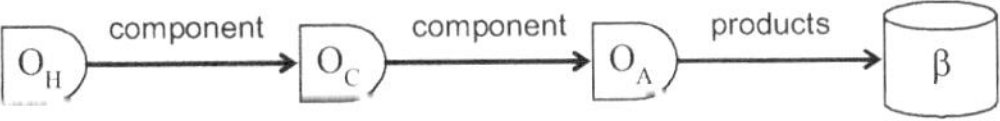

If an *organization*, "O_S," produces a high-quality utensil, tool or machine useful in the production process of "O_A," then we can have an enlargement in parallel of the *Productive Network*, with a ramification [Model 3].

[Model 3]

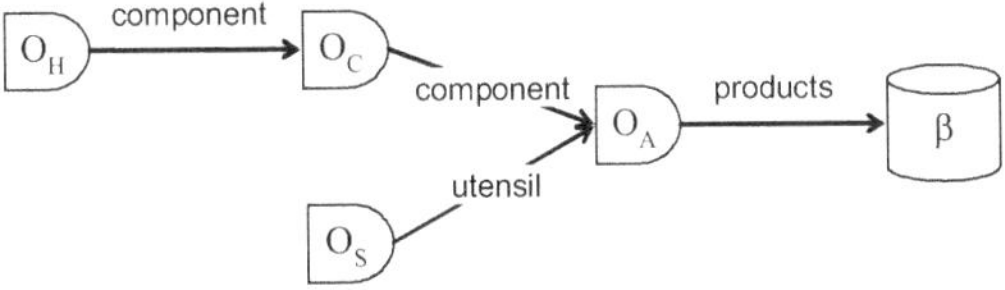

If "O_A" produces various products, P_1 and P_2 for example, it may be convenient to generate a specific autonomous *organization*, "O_B," parallel to "O_A," for production P_2, thus creating two independent branches of the overall network [Model 4].

[Model 4]

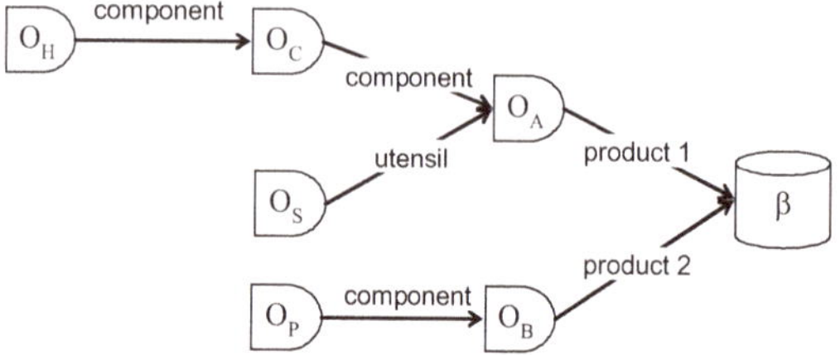

"O_A" and "O_B" connect to the same demand reservoir, creating *collaboration*, or *compete* with each other for connection to β. In fact, according to RULES (3) and (4) (Sect. 8.14), when it has to evaluate the adequacy of its own INPUT *vector*, each producer *organization* must always decide whether or not to *make* or *buy*. If the decision is to *buy*, then some specialized *organization* must exist *upstream* and connect with one *downstream* to supply the latter with the materials, components and machines it has given up producing internally.

There is also the reverse case: according to RULE (3), each *organization* must always evaluate the adequacy of its own OUTPUT vector, and it may be convenient for it to modify its own connection when this brings an improvement in the volume and/or in the unit value of the products. Let us suppose [Model 4] that "O_N," previously connected (downstream) to "O_A," disconnects from "O_A" to connect with "O_B." This decision can be considered as the shifting from "O_A" to "O_B" not only of "O_N" but also of the entire branch below it. As a result [Model 5], the branches of the network can vary their connections both *upstream* and *downstream*. Obviously the connection of "O_N" to "O_B" causes problems in the production process of "O_A"; if "O_A" is not able to replace "O_N," then it must modify its INPUT or OUTPUT vector; if this does not work, then "O_A" must disappear, with consequent difficulties for the entire antecedent branch formed by "O_H" and "O_M."

[Model 5]

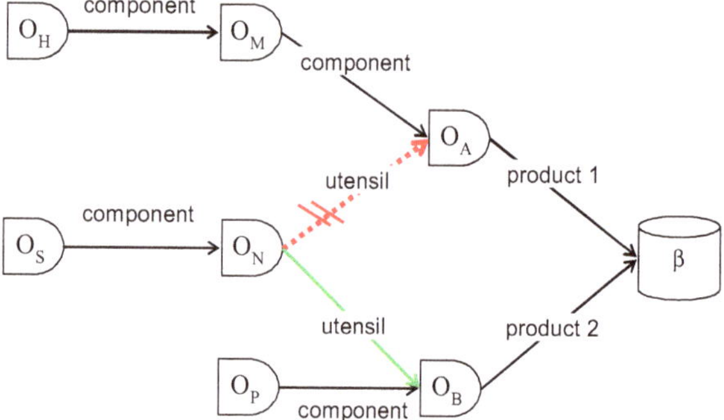

The genesis of new organizations capable of increasing the efficiency of the processes carried out by other already-connected organizations enables the Production Network to develop and extend itself in terms of size (parallel *organizations* that connect to the same demand reservoir) as well as depth (specialized *organizations* that connect serially). It is also possible for *organizations* to merge in order to create a larger production *node*.

Finally, let us consider the more complex case that occurs when, due to the specificity of its production, an *organization* employed in the production processes of several *organizations* or networks is connected at the same time to a number of other *organizations downstream*. Such an *organization* can be fully considered as a *hub*), in that it is the center of connection for many other *organizations* and branches that are variously situated in the network or that even make up different networks. The presence of *hubs* should not be considered a special case but rather the norm; according to RULE (2), each production *organization* must try to connect to the largest possible number of other organizations *downstream*, since these represent its demand reservoir. The more numerous are the *hubs* between different Production Networks, the more these networks become *integrated*, to the point of becoming *a single network*. Precisely due to the presence of *hubs*, the *Production Networks* can also become a complex behavioral network, as shown in [Model 6], since some *organizations* can have circular connections and generate dynamic or stable loops, even giving rise to evident paradoxes. This makes observation and modeling difficult; but this does not mean the *Production Network* loses its features as a modular system.

[Model 6]

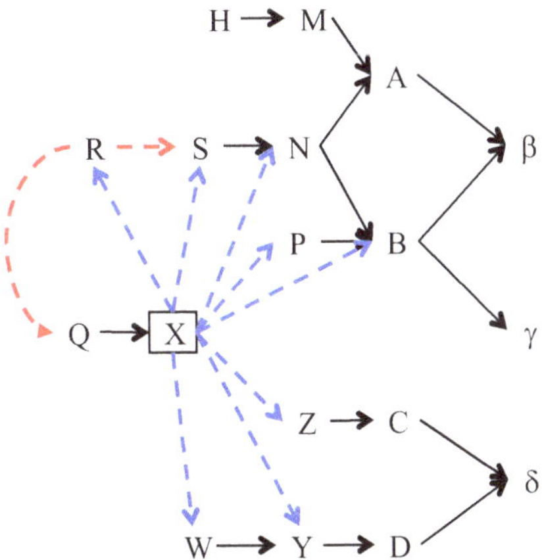

It is even more evident today that there is a tendency for modern economies to move toward production *specialization* and the expansion in network connections (Dyer 1997): *each* Production Network *that is not too elementary will be composed of organizations specialized in the production of materials, components, services, energy and machinery, all aiming at the production of final consumption goods*, as seen in the example in Sect. 8.14.

8.16.3 The Evolution of Production Networks: The Ghost in the Production Machine

Production networks are not a recent phenomenon; they have existed since the dawn of civilization, and from their earliest days began to expand, to improve their performance and to resist disturbances. We can outline the evolutionary trend toward ever vaster networks.

Local micro networks The first networks were territorial ones, which can be defined as *local micro networks*, which are characteristic of production in families, towns, and cities. In order to survive the town must produce food, shelter, furniture, cutlery, clothing, hunting and fishing tools, livestock, and crops. Taking on a specialized but interconnected form, such activities constitute an easily observable network. In such networks the *demand, resources* and *labor reservoirs* were found in a single circumscribed territory, so that the three elements of every production network—*resources, production* (the transformation of resources) and *consumption*—overlapped. Livestock were raised, land farmed, food and clothing produced, homes built with material from the territory, small exchanges carried out and learning acquired through experience and imitation.

Local macro networks The (inevitable) expansion of the *local micro networks* gave rise to the *local macro networks*, which extended across increasingly vaster areas: counties, provinces, regions; in all cases, though, the production of good and services for local consumption entailed local resources, with perhaps a slight separation between production and consumption areas, though these were always inside the territory of reference. Physical markets thus arose as places where the output of products for sale could be concentrated and supplied, with production knowledge accumulated and diffused orally and through apprenticeships. However, the problem of the transference of wealth was still very serious.

Bipolar micro networks The development of local networks soon led to the saturation limit of production potential due to a lack of resources. In order to survive networks had to expand beyond their original territories in order to find resources in other areas, thereby becoming bipolar networks: consumption and production referred now to a given demand reservoir confined to a particular territory, while the resources were "imported" from reservoirs located in other territories. This is how one should interpret Marco Polo and Christopher Columbus's journeys to discover new routes to locate far-off resources to import to the Continent. England's political expansion in the East or their colonial conquests in Africa and the Americas should be similarly viewed. Such a perspective makes clear Shell and Eni's reason (as well that of all oil companies) for extending their oil extraction networks to the remotest oil areas. The bipolar networks subsequently underwent a further change, which is still under way: according to the law of the improvement in economic efficiency (Second Law), production becomes increasingly detached from consumption and moves closer to the resources Production and resources are located in one area while consumption remains in another. It is not the resources that shift over space and time but the products. The great American livestock herds, the sugarcane cultivation, the transformation of rice in Lombardy and Northern Italy are all examples of this evolution in networks.

Multipolar networks We now come to the final evolution, that still under way: the networks become *multipolar*, where resources, production and consumption are found in geographically independent *reservoirs*, in areas that are separate but connected by a close network of exchanges carried out by other organizations serving as connectors, thereby forming the Production Networks. Not only are the *production* processes detached from *consumption* and *resources*, but *production* is segmented into thousands, millions of specialized production processes located in highly diverse areas: every product component is manufactured "elsewhere" (Bartlett and Ghoshal 1989), every machine to transform the resources is made "elsewhere." All the components of these machines are produced in thousands of different areas, and these networks inevitably become segmented due to strong production specialization. The multipolar networks become nonspatial and nontemporal.

International networks The multipolar networks that extend into different countries are international networks, which today dominate the global economy and expand thanks to the *connection* among the national networks. Today we can observe four large production networks, which have not yet merged: the North American network (U.S. and Canada); the Japanese network; the Chinese and Indian one; and the European network, with great potential but which unfortunately has been slow to take off, since EU national production networks have not yet completed the merger process. Other networks are in the process of being created, among which the South American and African ones.

Global network It won't be long that these networks will become connected, and this because the Laws that govern these networks theorize that this is necessary.

Global Big Machine When all networks are connected in a single large production network, when the world is a single Production Machine that provides consumer goods, then a true integrated "Big Production Machine" will have been achieved. If the RULES of selfish behavior of nodes (Sect. 8.14) and the laws of networks (Sect. 8.16.2) are valid, then the Big Production Machine will have autonomous existence as a superorganism, independent of the will of its modules and public regulators, and it will be resilient and increase in efficiency as we shall see in Chap. 10.

Paraphrasing Koestler (1967), it seems there truly is *a ghost in the production machine* whose *invisible hand*—acting on the individual nodes of the Productive

Network—determines increasing levels of productivity and quality; increases the quality and quantity of satisfied needs and aspirations; and reduces the burden of labor, thereby producing ever higher levels of progress in the entire Production Network. There is nothing metaphysical about this evolution: it is produced and governed by the laws of Production Networks.

8.17 Summary

This chapter has examined the *quantitative* (size) and *qualitative* (evolutionary processes) dynamics of populations of living beings and, from a broader perspective, of species.

Several models of quantitative population dynamics have been presented to simulate typical cases; above all, the dynamics of a population has been considered both in the case of an absence of constraints on growth and of growth limits due to the depletion of resources or a resource stock that is limited but can be increased through additional production. We then considered the dynamics of two or more populations connected by a prey–predator relationship or belonging to the same ecosystem. The idea here was to demonstrate that the systemic interconnection between two or more populations can produce ordered dynamics, as if an *endogenous*, automatic control regulated the size of these populations, as revealed in the Volterra–Lotka equations (applied to two populations). However, this model introduces the broader and more significant topic of *external*, artificial *control* of population dynamics in addition to the previously introduced *endogenous control*, which is achieved through various levers, ranging from repopulation to breeding, hunting or selective fishing, from targeted disinfestation to repopulating woodlands, etc. This chapter has also dealt with the topic of the *quantitative dynamics* of populations, which depend on *evolutionary phenomena* that, by modifying the phenotypes, favor adaptation to the new conditions of survival and modify the *birth* and *extinction* rates. Models have been presented to analyze the *qualitative* dynamics and to show how this can alter, even drastically, the quantitative dynamics. The chapter concluded by considering the dynamics and evolution of nonbiological species. In particular, the evolutionary phenomena that develop within *the production networks* were considered to highlight how the nodes that compose them (in production and nonproduction organizations) condition, through their evolutionary dynamics, the evolution of the entire network toward increasingly efficient and effective structures.

Chapter 9
The Magic Ring in Action: Organizations

Abstract Control systems also operate in production or consumption *organizations* which are unitary systems where the control is pervasive and essential. In particular, all production organizations are able to carry out their activities and survive over a long period of time precisely because they are preordained to undertake a constant control of the coordination of individuals and organs, the macro process they carry out as a single entity, and the network of micro processes of their organs. In this chapter, I shall examine organizations from five points of view: (1) as autopoietic and teleonomic systems, (2) as vital systems, (3) as systems for efficient transformation, (4) as systems subject to management control, both entrepreneurial strategic and managerial operational, and, finally, (5) as cognitive and explorative systems. Several instruments of control, such as budgeting and cost control, will be presented.

Keywords Organizations · Structural coupling · Viable system model · MOEST · Management control system · Strategic plans and programs · Short-term plans · Budget · Economic value of the firm · roe · roi · EBIT · Cost measurement · Cost control · Dashboards · Balanced scorecard · Position analysis · The Dupont Tree

> None of our institutions exists by itself and as an end in itself. Every one is an organ of society and exists for the sake of society. Business is no exception. 'Free enterprise' cannot be justified as being good for business. It can only be justified as being good for society (Peter Drucker 1973, p. 43).

> Although organizational learning occurs through individuals, it would be a mistake to conclude that organizational learning is nothing but the cumulative result of their members' learning. Organizations do not have brains, but they have cognitive systems and memories… Members come and go, and leadership changes, but organizations' memories preserve certain behaviors, mental maps, norms, and values over time (Bo Hedberg 1981, p. 6).

The discipline of control entails observing reality by filtering it through the models produced by the simple theory of control discussed in Part 1 of this book. Chapter 6 took the point of view of the individual observing the environment he or she lives

P. Mella, *The Magic Ring*, Contemporary Systems Thinking,
https://doi.org/10.1007/978-3-030-64194-8_9

in; Chap. 7 examined the *Rings* that operate in social systems, combinatory systems, and complex adaptive systems; Chap. 8 considered various forms of control operating within the Macro Biological Environment. Control systems also operate in production or consumption *organizations* which are unitary systems where the control is pervasive and essential. In particular, all production organizations are able to carry out their activities and survive over a long period of time precisely because they are preordained to undertake a constant control of the coordination of individuals and organs, the macro process they carry out as a single entity, and the network of micro processes of their organs. For this reason, organizations must necessarily create control systems at different levels (which are interconnected) in order to regulate all aspects of their performance, the internal organizational micro processes, and the relations with the market and external environment. As unitary wholes, organizations must themselves be considered control systems, whose lengthy existence depends, however, also on lower level control systems pervading every part of their structure, their processes, and, in the final analysis, their existence. In particular, we shall examine organizations from five points of view: (1) as autopoietic and teleonomic systems, (2) as vital systems, (3) as systems for efficient transformation, (4) as systems subject to management control, both entrepreneurial strategic and managerial operational, and, finally, (5) as cognitive and explorative systems. Once again we shall be able to see the action of the magic *Rings*, without which organizations could not exist but would instead become transformed into a group of agents operating in an individualistic and disorderly manner. Several instruments of control, such as budgeting and cost control, will be presented.

9.1 *Rings* That Allow Organizations to Exist as Autopoietic, Homeostatic, and Teleonomic Social Systems

It is now time to examine the action of the control systems we can observe, or design, in organizations, in particular *business* and *profit-oriented* production organizations.

Organizations are special *social systems* made up of individuals or *agents*, or groups of these, linked by a network of *stable organizational relations* or *structural constraints*—the "organization," in technical words—that specify, for each structural component, the following five elements: (1) a precise spatial and temporal *placement* in a hierarchy (topology), (2) a specialized *function* in relation to the macro function of the entire structure, (3) a specific *functionality* that delimits the admissible interactions of any agent with the others, (4) a set of standards of *functioning*, (5) a structure of control systems for the behavior and performance of agents and their processes.

The structural elements that conform to these five elements have no behavior or autonomous significance except in relation to the higher organizational level; they are *organs* of the system linked to the organization and to the functionality of all the organization's elements. Precisely by considering the organic structure of organizations as social systems, March and Simon proposed the analogy with biological systems, which, by definition, are organisms.

> Organizations are assemblages of interacting human beings and they are the largest assemblages in our society that have anything resembling a central coordinative system … the high specificity of structure and coordination within organizations—as contrasted with the diffuse and variable relations among organizations and among unorganized individuals—marks off the individual organization as a sociological unit comparable in significance to the individual organism in biology (March and Simon 1958, p. 4).

Through the network of stable relations among its organs, the organization is able to provide constraints to the behavior of its constituent agents in order to produce, at the various hierarchical levels, a network of coordinated and cooperative *processes* for the achievement of a common or an institutional goal; for this purpose the organization carries out the activities and operations needed to interact with its environment and achieve the objectives established by the governance in order to satisfy the interests of parties outside the organization: the *stakeholders*. In fact, each organization, as a unitary entity, exists within a social system and is structurally coupled with a more or less vast environment, thereby creating a network with other organizations. Many groups, other than the internal agents, can benefit or be harmed by the organization; these groups taken together make up the *stakeholders* that define the limits to the freedom of action of the organization, carrying out an external control of the dynamic processes leading to the achievement of the institutional goal.

An organization is created from the desire of a "founder" to achieve some goal; the "agents" willingly participate in the organization based on their *specific individual motivations*—usually a monetary compensation and a career path—accepting to limit their potential in order to become (or be a part of) *organs*. These agents thus *formally recognize* the constraints imposed by the *organizational relationships* and by the structural *ties* that oblige them to become "structurally coupled" with other "agents," giving rise to *specialized, coordinated*, and *cooperative* behavior. Thus, these agents agree to become part of an organizational structure and accept the objectives, plans, rules, and responsibilities as constraints on their behavior *inside the organization*, on the condition that the latter allows them to satisfy their *specific individual motivations* (Cyert and March 1963). Organizations differ from social systems in many ways, the most evident of which is the fact that the agents participate voluntarily in the organization because of the advantages it brings them and can also voluntarily cease to collaborate with it. In social systems the agents in most cases are obliged to participate and cannot decide not to do so, except with limited exceptions.

The *agents* and *organs* they constitute can be considered *processors* that allow the structure to produce a *network of recursive micro processes*—financing, investment, production of value, disinvestment and reinvestment, reimbursements and refinancing, etc.—which tend to maintain and perpetuate themselves over time, continually adapting to the environment. Such *micro* processes form larger processes, managed by higher level organs which tend to recursively regenerate themselves over time, to the point of forming a *macro* process (attributable to the organization as a whole) which transcends the *micro* processes produced by the organs, in that it represents the *emerging* result of the *network* of micro processes. The macro process produces the organization's outputs, the qualitative–quantitative results perceived by the environment, which taken together form the *outcomes* directed at the stakeholders. As a result of this, each organization, considered as an *organizationally closed system*, can be observed at three different levels, as shown in Fig. 9.1.

Starting from the bottom, Fig. 9.1 represents:

1. The *level of the organizational structure* of the agents, with the organizational relations and the behavioral constraints (lower level in Fig. 9.1).
2. The *level of the network of processes* regarding the activities and operations the agents must carry out (middle level in Fig. 9.1); the vertical upward arrows from level 1 indicate that the network of processes is activated by the behavior of the agents and the organs in level 1.
3. The *level of the macro processes* that produce the financial and economic flows which allow the organization to achieve the results, outputs, and outcomes necessary to attain its institutional goals and to exist for a long time (upper level in Fig. 9.1); the vertical upward arrows from level 2 indicate that the macro processes that produce the organization's outcomes derive from the micro processes on level 2 and thus the behavior of the agents on level 1.

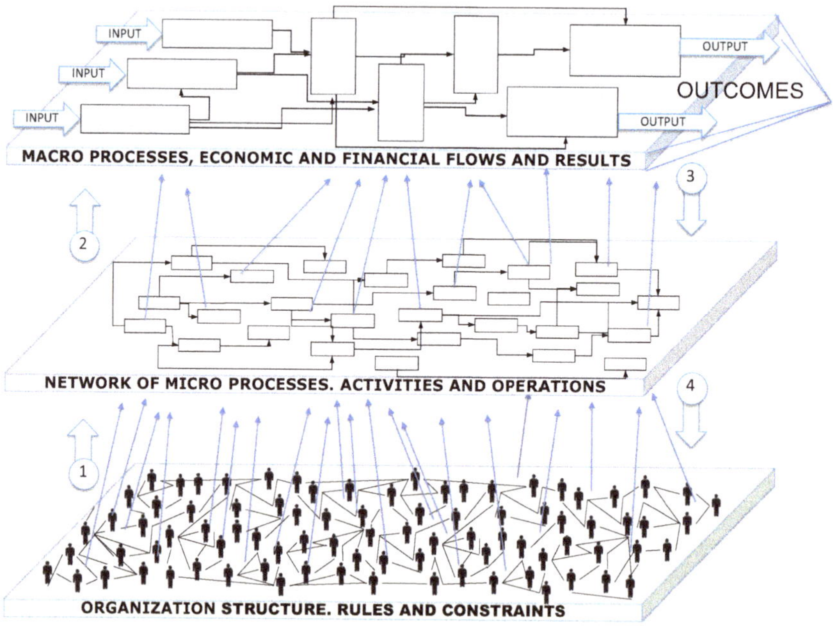

Fig. 9.1 The three interconnected levels of the organization

The three levels are thus interconnected by ascending and descending relations. The agents and organs in the organizational structure produce the activities and operations needed to carry out and maintain the *micro* processes (arrow ①); the *micro* processes are necessary to carry out the *macro* processes which, given the inputs, produce the outputs, results, and outcomes (arrow ②). The *macro* processes needed to achieve the results condition the composition of the network of micro processes (arrow ③), which in turn conditions the organizational structure (arrow ④).

Organizations survive as long as they can maintain their organizational constraints stable, replacing departing agents and inserting new ones that maintain the network of micro processes in order to activate the macro processes needed to achieve the goals. As long as they manage to repeat this cycle, organizations remain *vital*. In this sense organizations can, in all respects, be viewed as *autopoietic machines*, and thus *homeostatic* in line with Maturana and Varela's conception (Sect. 7.3), in which organizations tend to remain in existence for a long time by continually regenerating the *processors* (organs) and the *network of processes* which form the "organizational fabric." And as we know already: "[...] *an autopoietic machine continuously generates and specifies its own organization through its operation as a system of production of its own components, and does this in the endless turnover of components under conditions of continuous perturbations and compensation of perturbations*" (Maturana and Varela 1980, p. 79).

If we accept this view, then it is clear that the existence of every organization, as a single entity, is conditioned by *four* levels of control systems which, as shown in the model in Fig. 9.2, act on the elements on the three structural levels in Fig. 9.1 in order to maintain the autopoietic networks:

(a) The control systems of the structural coupling of the individuals in the organization based on the organizational constraints: I define these as the *Rings* needed for *autopoiesis*.
(b) The control systems for the efficiency, effectiveness, and equilibrium of the operations, activities, and micro processes that make up the autopoietic network the organization needs to produce the global macro process: I define these as the *Rings* necessary for *homeostasis*.
(c) The control systems to ensure the effectiveness of the macro processes that generate the inputs, outputs, results, and outcomes of the organization as a unit: I define these as the *Rings* needed for the *teleonomy* of the organization, since they function to guarantee that the outputs, results, and outcomes respond to the demands of the stakeholders, which represent the maximum *vital constraints* for the achievement of the institutional objectives.
(d) The control systems for the effectiveness of the *performance* of the agents and organs, which form the structure (lower level in Fig. 9.1) needed to activate and maintain the network of micro and macro processes: I define these as the *Rings* needed for the performance of the agents, organs, and entire organization.

These four types of *Rings* are in line with the traditional four forms of organizational control: *personnel* control, *action* control, *results* control, and *performance* control (Merchant 1985). The *Rings* for *autopoiesis* (type A) form a network of

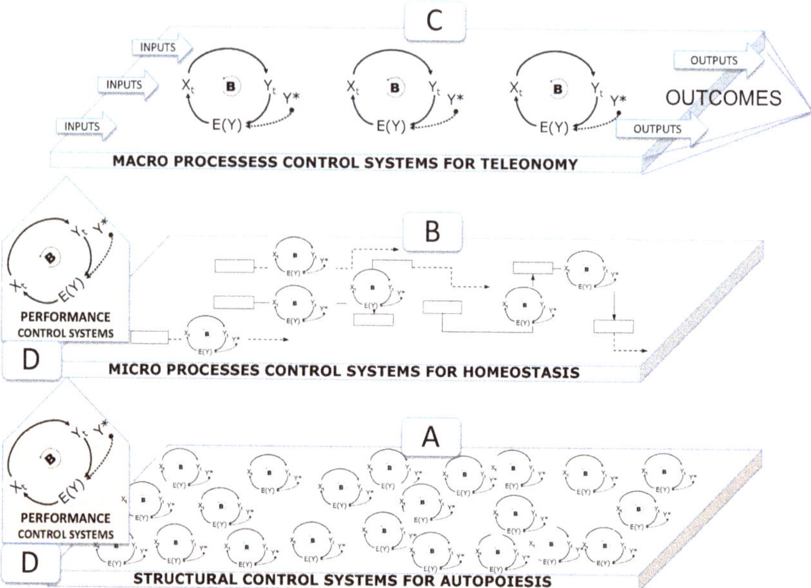

Fig. 9.2 The four levels of control systems for maintaining the organization as a unitary entity: [A] autopoiesis, [B] homeostasis, [C] teleonomy, [D] performance

organizational control systems aimed at producing an "alignment" among the agents in the organization and the various organs, so that their actions, functions, positions, tasks, and objectives will be *coordinated* and *cooperative over time*, which is a necessary condition for the long-lasting maintenance of the mutual *structural coupling* of agents and organs. I shall define this internal form of *control* of the structural coupling of agents and organs as a *structural self-control*, which is vital for any organization, and use the term *structural control systems* to refer to the *Rings* which achieve this; this form of control has also been defined as the control of *conformity* or bureaucratic control.

> Organizations require a certain amount of conformity as well as the integration of diverse activities. It is the function of control to bring about conformance to organizational requirements and achievement of the ultimate goals of the organization. The co-ordination and order created out of the diverse interests and potentially diffuse behaviors of members is largely a function of control [...] But regardless of how order is created, it requires the conformity of all or nearly all to organizational norms (Tannenbaum 1962, pp. 237–238).

The *structural self-control* also performs a *control* or a *surveillance of the position and behavior* of the various agents and organs, since each agent or organ in the structure has his or her/its own function to carry out within a specific field of action, according to shared standards, while respecting the external and internal norms based on a common ethical view.

> Cultural or normative controls which operate as alternatives to bureaucratic rules and direct supervision are new technologies of power developed within knowledge-intensive organi-

zations. The mobility and extensive exposure of … workers to a range of cross-cutting professional obligations coupled with the uncertainty and ambiguity of their tasks makes such control fragmented and contingent (McKinley and Starkey 1998, p. 10).

Structural self-control is made possible by the fact that the agents in the organization must accept the *responsibilities* that derive from their participation in the structure and respect the *authority* of the other agents and organs, which is established by the rules that discipline the structural links, as shown in Fig. 9.3.

The *structural coupling* control systems (lower level A in Fig. 9.2), as shown in Fig. 9.3, apply to every agent or organ (e.g., A) in order to maintain or restore the coordination and cooperation between the position of A, the assigned tasks, and the procedure for the functions and processes with respect to those organizationally assigned. These systems use a variety of control levers, [Xn], such as resources, information, orders, constraints, incentives, persuasion, awards, punishments, and career possibilities, to eliminate the gaps, $E = Y^* - Y$, which are "symptoms of non-coordination" and the "absence of cooperation," between the position/task/objective of A, Y(A), and the values required to maintain the *structural coupling* of A with B (Y^* = position/task/objective of A to allow coordinated and cooperative behavior with respect to B), taking into account the external disturbances (D), which alter behavior.

It is immediately clear that a large number of *Rings* are required to guarantee the structural coupling of all the agents or organs in the organization (lower level A in Fig. 9.2). *Structural self-control* must be carried out by *steering systems*, generally multi-lever, that continually corrects errors in the structural coupling, taking into account the fact that each change in the behavior of an agent A determines a conse-

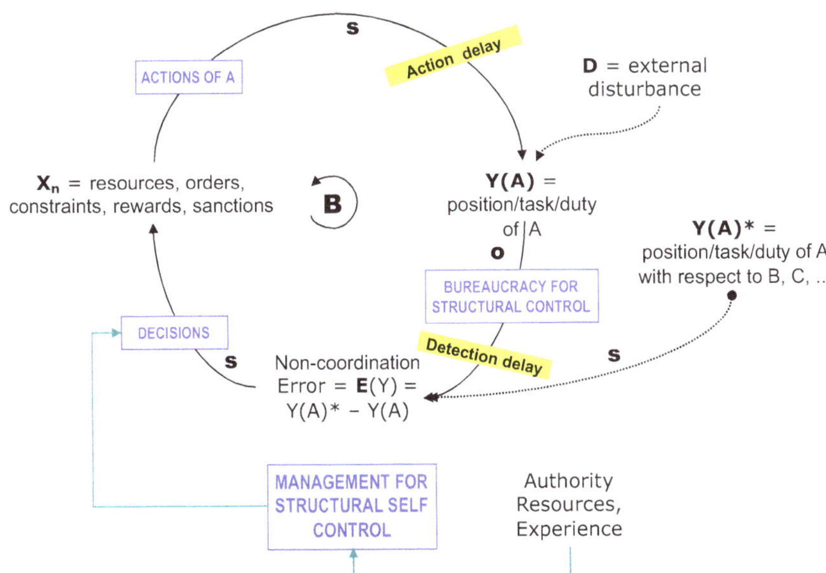

Fig. 9.3 Micro *structural control system* for the structural alignment of agents

quent change in the behavior of at least one agent B, thereby risking possible interference and oscillating behavior, according to the definition below of *structural coupling*:

> Autopoietic systems may interact with each other under conditions that result in behavioral coupling. In this coupling, the autopoietic conduct of an organism A [individual or organ] becomes a source of deformation for an organism B [individual or organ], and the compensatory behavior of organism B acts, in turn, as a source of deformation of organism A, whose compensatory behavior acts again as a source of deformation of B, and so on recursively until the coupling is interrupted. In this manner, a chain of interlocked interactions develops such that, although in each interaction the conduct of each organism [individual or organ] is constitutively independent in its generation of the conduct of the other, because it is internally determined by the structure of the behaving organism only, it is for the other organism, while the chain lasts, a source of compensable deformations which can be described as meaningful in the context of the coupled behavior (Maturana and Varela 1980, pp. 119–120).

Equally important is the function of the *Rings* that control the network of micro processes necessary to ensure *homeostasis* (type B in Fig. 9.2). These processes must guarantee that the *activities* and *operations* of the agents applied to the network of micro processes conform to the established standards of behavior and efficiency and produce the expected results in order for the higher level macro *processes* to be produced, as shown in the model in Fig. 9.4.

Clearly this form of control is not aimed at guaranteeing the structural coupling of agents (Fig. 9.3) but at ensuring that the behavior of the agents entails the operationality and the efficiency required to produce the *outputs* of the processes in which the agents are the processors. For this reason the control of the *network of processes*

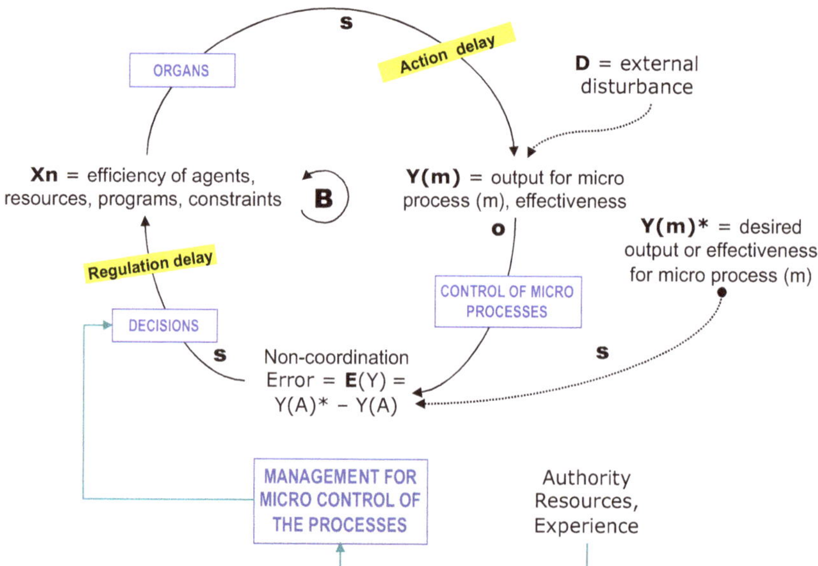

Fig. 9.4 Control systems for the control of *micro* processes

is, in fact, the control of the operationality and behavioral efficiency of the structure (agents and organs), and it seeks to align the *results* and *individual performance* (agents and organs) to the standards of quality, efficiency, and productivity set as constraints or objectives in order to ensure that the processes themselves operate according to the functional standards for the higher level network of macro processes.

> A distinction is drawn between two modes of organizational control, one based on personal surveillance, behavior control, and the other based on the measurement of outputs, output control. A study of employees over five levels of hierarchy shows that the two modes of control are not substitutes for each other, but are independent of each other. The evidence suggests that output control occurs in response to a manager's need to provide legitimate evidence of performance, while behavior control is exerted when means-end relations are known and thus appropriate instruction possible (Ouchi and Maguire 1975, p. 559).

Along with the *Rings* (types A and B above) operating *within* the organization at the *micro* levels of the individuals and organs, control systems must also operate at the *macro* level (type C in Fig. 9.2), *on* the organization as a *unitary system*, to guide its complex activities toward the institutional objectives. In order to accomplish this, such control systems regulate the macro processes to produce the *flows, inputs, outputs, results*, and *outcomes* necessary to satisfy the outside stakeholders' expectations, as shown in Fig. 9.5.

This incessant activity regarding the *production of outcomes* and the *search for inputs* to maintain the organization in existence shows that organizations, in addition to being autopoietic and homeostatic systems, also produce *teleonomic* behavior to achieve the "existential project" for which they were created, according to

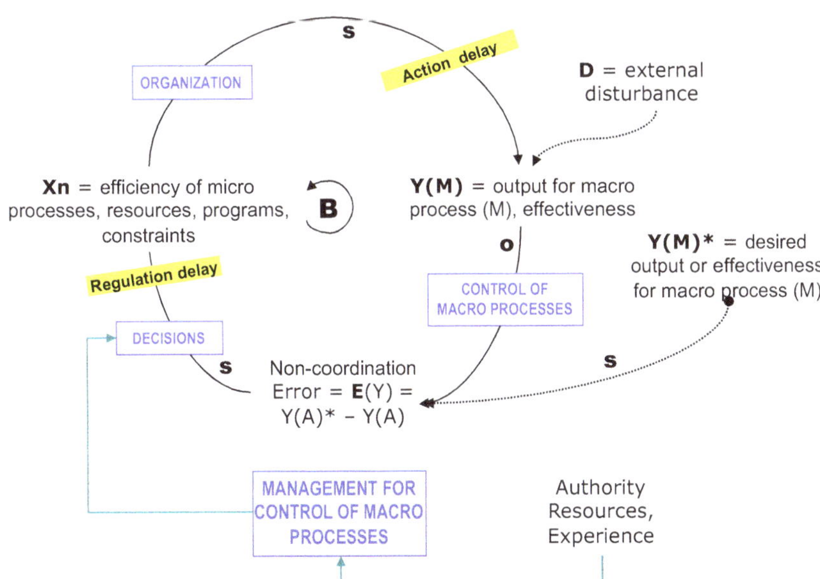

Fig. 9.5 Control systems for the control of *macro* processes

Jacques Monod's definition in Sect. 6.9. I shall refer to these *control systems* as the *Rings for teleonomy*, since they drive the macro processes toward the attainment of the outputs necessary to achieve the institutional goals (Fig. 9.5).

We immediately perceive the importance of control systems of performance (type D in Fig. 9.2). The structure of the agents and organs is instrumental for the workings of the network of operational *micro* processes necessary for the functioning of the *teleonomic macro* processes. Thus, the activity of the structure's components must conform to the *results* and *performance* standards required for the activities and processes to be carried out. The *Rings* for the *control of performance* (shown in Fig. 9.6) should activate the levers needed to bring the individual performance to the levels required by the tasks and functions, since the organization, as a unitary entity, must satisfy the needs of the *stakeholders*, thereby producing outputs, results, and outcomes which meet expectations.

If it succeeds in efficiently structuring the four networks of *Rings*—for *autopoiesis, homeostasis, teleonomy*, and *performance* of the agents and organs—the organization, as a unit, will manage to maintain a stable structural coupling *with the environment* not only by reacting to *environmental* disturbances through its own system of *Rings* but also by searching the environment for disturbances (variables, stimuli, inputs) which are useful for autopoiesis (Zeleny and Hufford 1992; Mingers 1995, 2002) while fleeing from those which are harmful.

In this sense the long-lasting organization displays a cognitive behavior aimed at survival, and it can be viewed as a living system that reproduces itself over time, along the lines of Maturana and Varela's analysis, a part of whose quote from Sect. 6.7 states: "*If living systems are machines, that they are physical autopoietic*

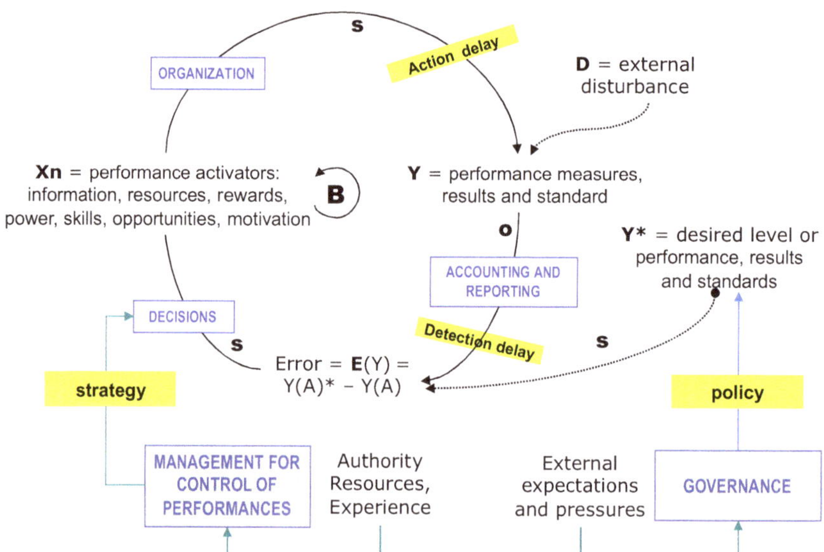

Fig. 9.6 Control systems for the control of *performance*

machines is trivially obvious [...] *However we deem the converse is also true: a physical system, if autopoietic, is living*" (Maturana and Varela 1980, p. 82).

The idea that the organization is a living system has been excellently described by Salvatore Vicari in a convincing book entitled *The Organization as a Living System* (Vicari 1991) and by Arie De Geus in his work *The Living Company: Habits for Survival in a Turbulent Business* (2002; see also 1988). De Geus clearly shows the importance of cognition and learning for an organization's teleonomy, especially large corporations, whose teleonomic activity can be interpreted only by assuming that the organization (company) is a living being and the decisions for organizational activities taken by this living being result from a learning process. It is not without significance that the *Forward* of this work was written by Peter Senge, who sums up the reasons organizations must be viewed as living beings and not as simple machines. Among these reasons, I find the following quite convincing:

> Seeing a company as a machine implies that its actions are actually reactions to goals and decisions made by management. Seeing a company as a living being means that it has its own goals and its own capacity of autonomous action.
>
> Seeing a company as a machine implies that it will run down, unless it is rebuilt by management. Seeing a company as a living being means that it is capable of regenerating itself, of continuity as an identifiable entity beyond its present members.
>
> Seeing a company as a machine implies that its members are employees or, worse, "human resources", humans standing in reserve, waiting to be used. Seeing a company as a living being leads to seeing its members as human work communities.
>
> Finally, seeing a company as a machine implies that it learns only as the sum of the learning of its individual employees. Seeing a company as a living being means that it can learn as an entity, just as the theater troop, jazz ensemble, or championship sport team can actually learn as an entity. In this book Arie argues that *only* living beings can learn (Senge 2002, pp. IX–X).

Francisco Varela holds that non-biological entities should not be defined as living in the traditional sense:

> Frankly, I do not see how the definition of autopoiesis can be directly transposed to a variety of other situations, social systems for example. It seems to me that the kinds of relations that define units like a firm (Beer) [concern] units [which] are autonomous, but with an organizational closure that is characterizable in terms of relations such as instructions and linguistic agreement (Varela 1981, p. 38).

In this sense, *if observed from the outside* organizations, in their relations with their environment, are dynamic systems capable of self-regulation which themselves can be considered to be macro control systems, as shown in the model in Fig. 9.7. Each time there is a gap between the *outcomes* (Y^*) the organization must provide its stakeholders to remain vital, and the *potential outcomes* (Y), *cognitive* and *operational levers*, are activated to search for the appropriate inputs and resources to produce the desired outputs and outcomes.

Whenever the organization is not able to produce a *macro* control to attain the desired *outcomes* for its stakeholders and to find the necessary *inputs*, it is incapable of reproducing its own organs (processors) and processes; the "organizational fabric" declines to a level such that the autopoietic network of vital processes is dismantled. The organization's *existence* comes to an end and the organization dies

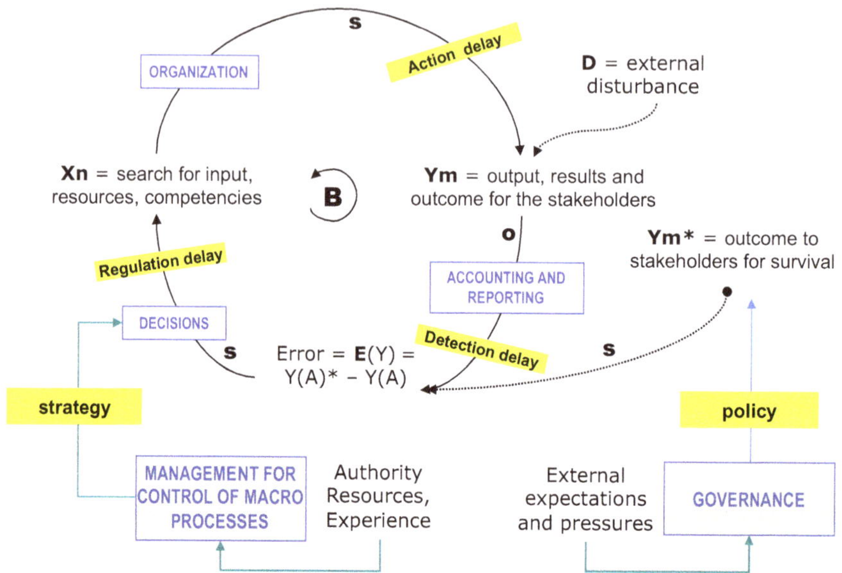

Fig. 9.7 The organization as a control system which regulates teleonomic behavior

out, or changes in an irreversible manner, losing its identity. In this sense *cognition* and *learning* are necessary for the existential success of the organization; as Peter Senge has masterfully shown, the organization must become a *learning organization* as soon as possible (Senge 2006).

There is thus an evident strengthening loop between the *macro control*, which permits the long-lasting structural coupling between organization and environment, and the *micro controls*, which guarantee coordination, cooperation, and efficiency of the performances of the agents and organs in the structure.

9.2 Viable Systems View

The above considerations lead us to a general conclusion: organizations of any kind, in particular *production, business-oriented* ones (firms), can, as autopoietic systems coupled to the environment, have a long-lasting existence if (1) at the *macro level* they operate as unitary control systems, equipped with an internal *chain of control* produced by the organizational structure, capable of achieving the organizational outcomes, and (2) at the *structural level* they are composed of micro *Rings* that guarantee *autopoiesis, homeostasis*, and *teleonomy*. There are specific theories and models which allow us to represent organizations–enterprises in general, firms in particular, as unitary control systems. In this *section* I will start from Stafford Beer's model (Beer 1979, 1981), universally known as the Viable System Model, or VSM, which is particularly significant (Fig. 9.8). The next *section* takes up a similar model, but one aimed at the objectives of control.

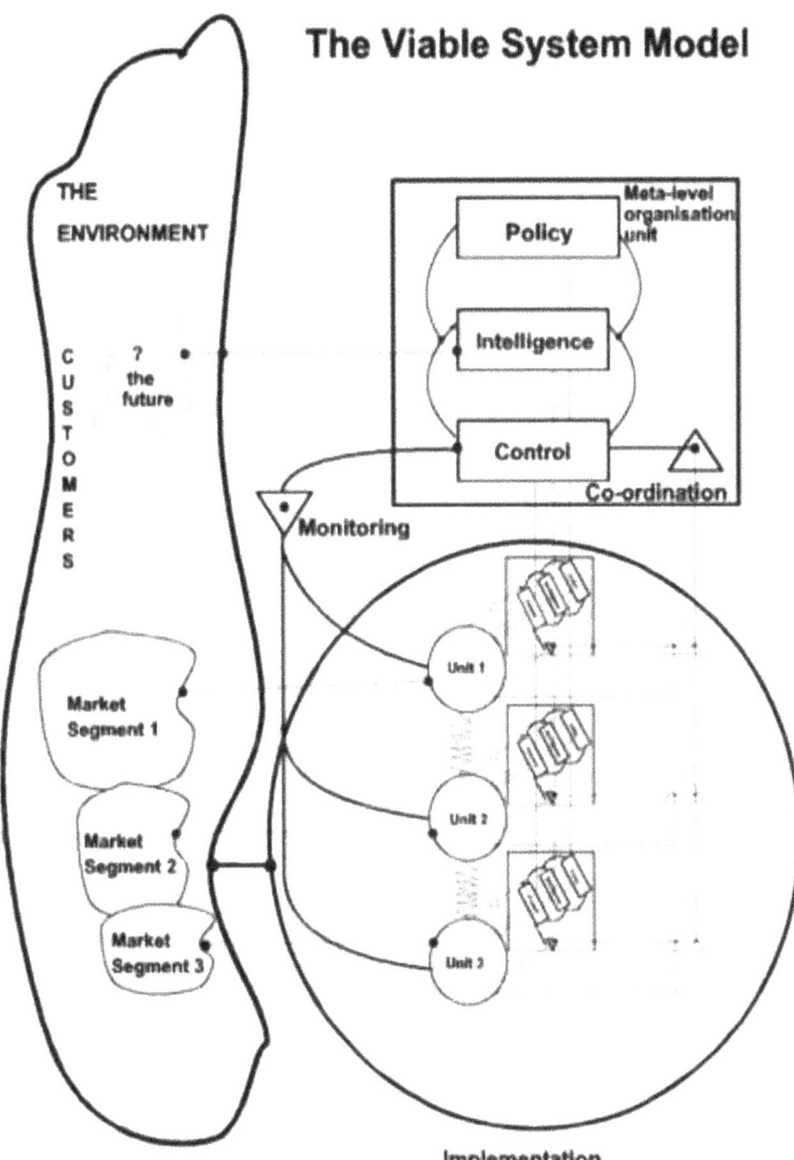

Fig. 9.8 The viable system model (*Source*: Beer 1981, p. 75)

Directing the reader to Beer's book for a detailed description, here it is enough to mention that Beer directly interprets organizations and firms, when observed from the outside, as *cybernetic systems*, and thus as unitary control systems that include management directly in the chain of control. In order for organizations to have a long-lasting existence, management must continually modify their behavior based on changes in the objectives and constraints set by the stakeholders and by outside events. Thus, Beer believes that organizations must be viewed as *viable systems* which, through their structure, which is capable of learning and cognition, can achieve an enduring structural coupling with the environment, continuing in this way to exist for a long time through continually adapting to the environment.

To maintain the conditions for *viability*, organizations, as control systems, internally determine the policies and activate the levers and strategies needed to eliminate the negative effects from environmental disturbances during the course of their existence, disturbances which cannot be foreseen at the moment the system is designed and created. In his book *Brain of the Firm*, Beer provides a definition of viability:

> This book has been wholly about the viable system. There must be criteria of 'independent' viability, even though any system turns out to be embedded in a larger system and is never completely isolated, completely autonomous or completely free (Beer 1981, p. 226).

> The object is to construct a model of the organization of any viable system. The firm is something organic, which intends to survive—and that is why I call it a viable system (*ibidem*, 75).

In this sense, even the VSM, in line with the *autopoietic* viewpoint, represents organizations as cognitive, adaptive, long-lasting (though the exact length of time is not predefined) systems coupled to the environment through systems of communication and exchange. However, the VSM takes a *macro* and *external* perspective, according to which organizations are entities that survive thanks to their cognitive and control structure and are able to communicate with the environment and acquire information needed to define and achieve, through the coordination of the operational units, their institutional objectives. From this point of view, and adopting Varela's analysis, the VSM represents the typical model of a *system open to the environment* through its own operations and communication activities, though *closed* regarding the processes of adaptation to environmental dynamics, as synthetically shown in my elaboration in Fig. 9.9.

However, conceiving of organizations as autopoietic systems (Sect. 9.1) adopts a mainly *micro* point of view, considering mainly the logic of the self-production and homeostasis of the "organizational fabric" and of the autopoietic *network* of processes considered to be *organizationally closed*, in the sense they form an unvarying structure that the system maintains stable and in homeostatic equilibrium over time (Bednarz 1988). For Varela (1979),

> […] a system that must carry out internal processes {P} as a condition for its existence, as typically occurs in human organizations (Sect. 9.1), has organizational closure if (1) the relationships among the processes {P} comprise a network {T} and not a simple hierarchy; (2) the processes in {T} are mutually interdependent for their own generation and realization; and (3) the set of processes {P} constitute the system as a unity "recognizable in the space (domain) in which the processes exist" (Varela 1979, p. 55).

Given these assumptions, the VSM appears to be a powerful instrument for understanding the *structural conditions* of vitality for the firm and at the same time can be interpreted using the same basic concepts adopted for autopoietic and cybernetic systems (Beer 1981).

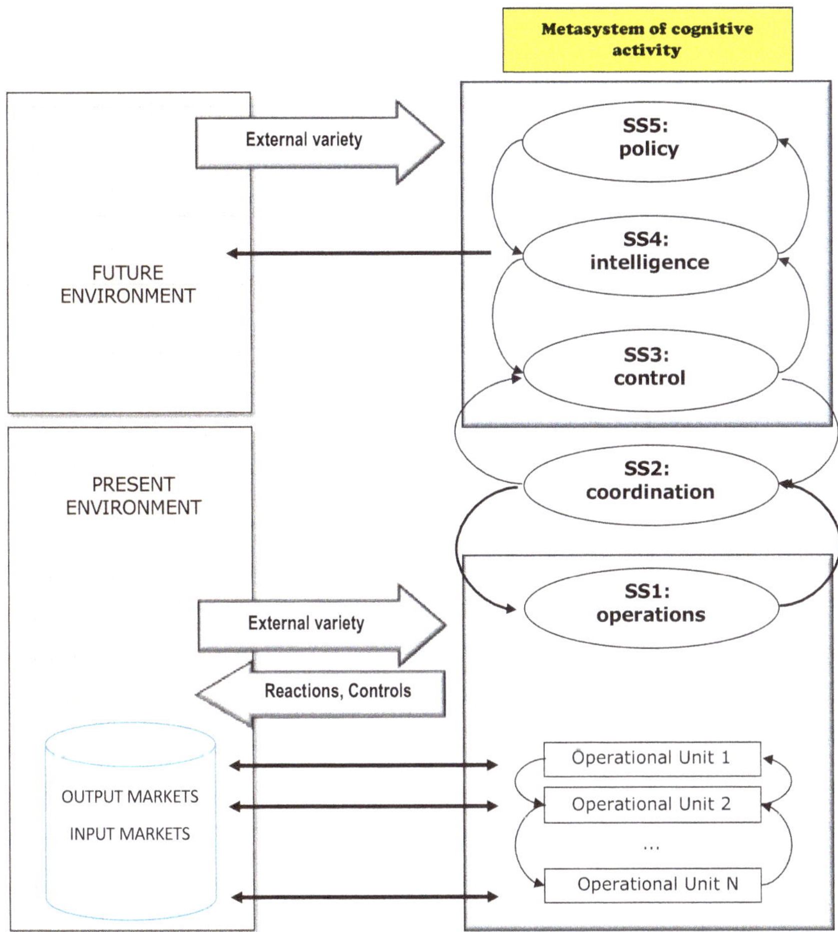

Fig. 9.9 A synthesis of the viable system model (my elaboration; reference: Fig. 9.8)

Beer's model is graphically very complex; however, in its essential schematic form it can be summarized as follows.

SS1: Operations This represents the *operational processes* of the organization, which are carried out by distinct *operational units* that can be observed in both their *vertical relations*—aimed at the achievement of the objectives of vitality and survival of the entire organization—and their *horizontal relations* involving interexchanges with the environment, to which, based on the autopoietic view, they are *structurally coupled* in order to achieve their own objectives to ensure survival. Thus, the operational units which make up SS1 are control systems directed at specific and particular objectives and constraints, internal as well as external.

SS2: Coordination Clearly, the operational units of SS1, since they employ common resources and can be in competition with the other units regarding objectives, are *interconnected* and normally *interfering*. As such they may be lacking in coordination and produce oscillatory dynamics in their local values which can cause inefficiency at the global system level. Precisely for this reason, SS2 is assigned the task of coordinating the interconnected operational units by undertaking control actions to avoid the harmful effects from interference and the lack of coordination. For this role SS2 tries to optimize the *structural coupling* between the operational units of SS1, thus operating as a control system in a way similar to what is depicted in Fig. 9.3, coordinating the operational units in order that these can achieve their individual objectives while respecting the common constraints.

SS3: Control The operational units of SS1, insofar as they are coordinated by SS2, must be directed as a whole toward the achievement of the higher order objectives, which refer to the overall organization. SS3 is entrusted with this task. By assigning specific programs of action to the various units of SS1, as part of a general *programing* accompanied by a constant *monitoring*, SS3 directs its processes to achieving the final objectives of the system. The term "*control*" itself used by Beer to refer to this subsystem clearly indicates that SS3 is a typical control system based on a particular programing. As it is able to activate a *range* of control levers, even at various levels, SS3 is assigned the task of formulating *strategies* regarding the use of the levers for the various objectives. Nevertheless, SS3 cannot separate itself from subsystems 4 and 5, which together form a higher level subsystem that produces *cognitive activity* of the organization.

SS4: Intelligence or research regarding information on the environment The survival capacity and the conditions of vitality do not depend solely on the present circumstances surrounding the environment, to which the organization is *structurally coupled*, but above all on future dynamics. It is thus vital for the organization to continually observe and forecast the "future environment" in order to enable SS3 to formulate programs of action and to adapt to these the units and activities of SS1. SS4 represents precisely the element of the *viable system* entrusted with proposing the vital objectives, based on predictable future scenarios, which are translated into programs of action, and with then identifying, gradually and in real time, the deviations from the programs, which are due to the actual environmental dynamics. Warning: "*The use of scenarios is also relevant to evaluation, particularly when dealing with negative evidence from the environment. [. . .]The use of scenarios might have identified the reactions to unfavorable forecasts prior to investing money on these forecasts*" (Armstrong 1983, p. 8).

SS5: Policy To complete the VSM Beer has perceptively noted that the objectives set by SS4 could be mutually incompatible or in competition with one another and thus not always suitable to maintaining future conditions of vitality. Thus, he holds that every organization must provide for a higher level activity—a true governance—to identify the *hierarchy* of objectives set by SS4, whose pursuit through the activities of SS1, coordinated by SS2 and controlled by SS3, represents the condition for the survival over time of the organization as a vital system. The introduction of SS5 is provided with a *unitary management*, an *entrepreneurial* as well as *managerial capacity* to define the *policies* for the achievement of the vital objectives, explicitly recognizing the fact that organizations, as unitary entities, are *multi-objective* control systems and that, as such, *strategies* for the control levers employed by lower level subsystems do not suffice; instead, a careful policy evaluation and rational ordering of objectives are essential.

In short, the VSM explicitly recognizes that every organization can be fully considered, in the joint action of its five subsystems, as a unitary, multi-lever, and multi-objective control system, as represented in Fig. 9.10. In this sense, the VSM (with the appropriate adjustments) can perfectly apply to organizations the heuristic schema used by the sociologist Talcott Parsons (1951, 1971) in order to interpret the general functioning of social systems, or society in general. This schema is called the "AGIL paradigm" ("scheme" or "pattern").

The AGIL paradigm is based, in fact, on the assumption that every social structure, as well as every organization considered as a social system with a stable closed structure (Sect. 7.3), seeks to remain vital and to guarantee its continuity with respect to its environment. To ensure its *vitality* every society must carry out four different functions or "functional imperatives" (the acronym AGIL derives from the initials of these functions).

Each of these imperatives corresponds to a particular subsystem of the social system.

A—*Adaptation* to the environment: in society the adaptation function involves the *economic subsystem*, that is, those structures that see to the production of the necessary resources (material goods, capital, knowledge, know-how, etc.) and the division of social labor, that is, the structuring and division of the tasks within the social structure to ensure adaptation.

G—*Goal attainment*, that is, the function needed for a society to attain its goals by seeking out the optimal use of resources to achieve society's main goals: In a society the goal attainment function pertains to the *political subsystem*, which must define and pursue those goals held to have priority.

I—*Integration*, that is, the function that integrates and makes coherent the choices and actions of the structures making up the social system (subsystems, groups, roles, etc.), thereby guaranteeing the integration of mutual expectations and the rational use of resources: Integration thus concerns the normative structures, that is, those which formulate and impose respect of the laws and sanction any violations in order to maintain the social fabric.

L—*Latency pattern maintenance*, which is a sort of cultural template that guarantees the passing on of values, cultural models, and roles that structure a society, thereby ensuring that these are interiorized by all the members of the social system: This function falls to the educational structures (school, training programs, family, rites, systems of meaning, etc.) which are predetermined to allow each member of a society to acquire the behavioral and value models suited to his role and function.

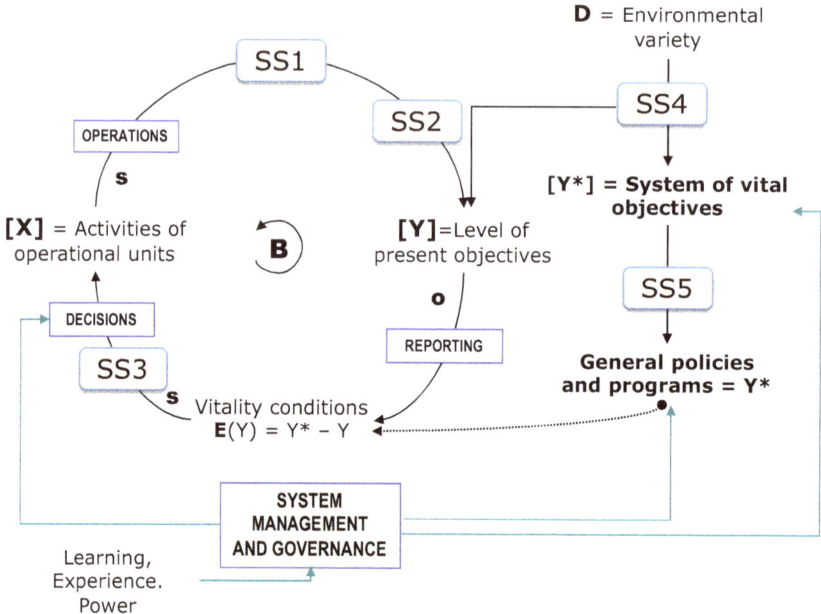

Fig. 9.10 The VSM as a multi-lever and multi-objective control system

Note. According to Beer, the viable system represents in all respects a holonic entity (Sect. 1.2), since every organization, while a complete unit, is in turn composed of smaller vital entities (operational units, departments, functions, divisions, etc.), with different levels of complexity, and at the same time part of a larger vital unit, as Beer clearly recognizes the holonic structure of his model, enunciating the following theorem, describing the concept of "cybernetic isomorphism":

> *Recursive System Theorem*. In a recursive organizational structure, any viable system contains, and is contained in, a viable system. There is an alternative version of the Theorem as stated in *Brain of the Firm*, which expressed the same point from the opposite angle: 'if a viable system contains a viable system, then the organizational structure must be recursive'. There is an alternative version of the Theorem as stated in Brain of the Firm, which expressed the same point from the opposite angle: 'if a viable system contains a viable system, then the organizational structure must be recursive' (Beer 1979, p. 118).

9.3 Organizations as Efficient Systems of Transformation

The preceding overview of Beer's viable system models refers to all organizations, independently of the nature of the processes they carry out. But what do production organizations and companies actually do to remain vital and effectively adapt to environmental changes?

To highlight the functions all organizations and firms must carry out to remain vital, I have proposed a specific model (in many respect parallel to the VSM) which considers organizations, and for-profit companies in particular, as *systems of transformation* which, in order to remain in existence over time, must carry out five interconnected vital *transformations*, each of which, operating with maximum efficiency, carries out a vital function similar to what is proposed in the VSM. Unlike the VSM, which represents organizations from the point of view of their *structural* synthesis, the Model of the Organization as an Efficient System of Transformation (MOEST) sees them from a *functional* viewpoint (Mella 2005b).

The MOEST can be applied both to *productive organizations* in general and to *enterprises* such as *business profit organizations* which, in *technically* producing goods and services, must also *economically* produce (and distribute) *value*. I shall apply the model to *capitalistic enterprises* which, in order to produce, require capital inputs in the form of *equity* and *debt*, with the constraint (obligation) of maintaining or increasing the value; producing remuneration, dividends, and interest (output); and paying back the invested capital.

The MOEST—represented in Fig. 9.11—shows, above all, how each firm must necessarily carry out three efficient "technical" transformations, so defined because they concern the productive, economic, and financial functions instrumental in allowing the organization to maintain its vitality in order to satisfy, in particular, the needs of its stakeholders and shareholders.

The complete model is very detailed; however, a synthesis would be useful to better understand the logic behind it. Hereafter referring to the MOEST in Fig. 9.11, I will begin with the last three transformations, which I have defined as "technical" to distinguish them from the first two, which I define as "cognitive."

TR1-P: Physical productive transformation, through which physical volumes of productive factors (F) are transformed into volumes of finished products (P): This is typically a transformation of utility: through the *operational units* that make up the *productive system*, the factors of production, having a given utility, are acquired (inputs) and transformed into products with a given level of quality capable of producing greater utility, which is sold to clients (outputs). The efficiency of the productive transformation is measured by two indicators: (1) *productivity* (π), understood as the capacity of the transformation to generate the maximum production output with the minimum inputs (consumption) of factors ($\pi = P/F$), and (2) *quality*, understood as the maximization of the *use function* (qlP) of the products.

TR2-E: Economic or market transformation, through which the organization, by activating a network of *economic* units and processes, businesses, and investment, tries to increase the value of the productive factors F (economic output), purchased at the input prices (pF), by using them to obtain production P that can be sold at remunerative prices (pP), thereby obtaining revenue (economic input): From the difference between revenues ($R = P \times pP$) and costs ($C = F \times pF$) the firm produces the *operating income* (operating profit) or EBIT: (economic inputs−economic outputs)=($R - C$) = EBIT, which is the synthetic value of the TR2-E. The function of this transformation is to negotiate as "advantageously" as possible the prices of the production and of the factors in order to produce the *maximum* revenues (input) with the minimum costs (output). The *economic efficiency* of this transformation, understood as the capacity to cover cost flows with revenue flows over a sufficiently long period of time, can be quantified by three main indicators: (1) the

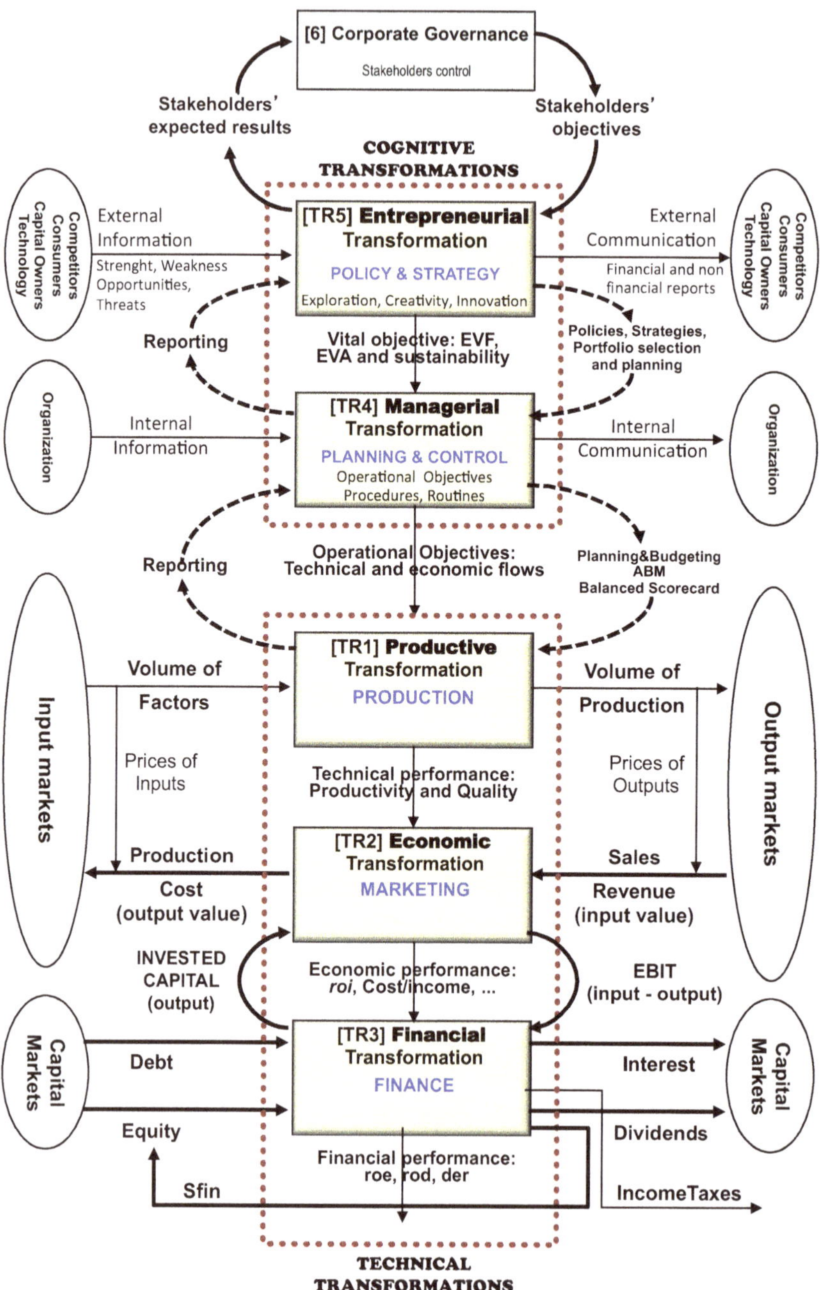

Fig. 9.11 The MOEST in brief

EBIT = $R-C$; (2) the ratio between EBIT and cost of production, that is, by the *return on cost* (roc); or (3) the ratio between revenues and costs (e = economic efficiency index) or its inverse ($1/e$ = cost/revenue ratio). We can write the economic efficiency index in the following form:

$$e = \frac{P \times pP}{F \times pF} = \pi \frac{pP}{pF} \tag{9.1}$$

We immediately observe that economic efficiency depends on productivity, which reflects productive efficiency, and on the ratio between the *average selling price* and the *average unit cost of production*, which represents *market efficiency*.

TR3-F: Financial transformation, through which the firm, in setting up its businesses, transforms the *acquired* capital (input)—in the form of equity (E) (underwritings, equity, stocks) or debt (D) (bonds, loans, and various forms of debt financing)—into *invested* capital ($I = E + D$), which is *used* (output) to finance TR2-E: By employing the invested capital (I) in TR2-E, the economic transformation is able to produce EBIT = $R - C$, which is then transformed into *flows* of interest payments (Int) for D and in dividends (Div) for E. Not all the EBT = (EBIT − I) becomes dividends, since a portion is assigned to taxes (Tax) and a portion is kept as self-financing (Sfin), which goes to increase equity for the growth of the firm. Only the residual amount is distributed as dividends. Thus, the financial transformation is typically a transformation of the risk of the capital investment in the hope of obtaining an adequate return. The *capitalistic firms* (business profit organizations) bear three types of correlated risks (Ruefli et al. 1999):

(a) *Technical*, or *production, risk* entails not being able to *attain* production goals
(b) *Economic*, or *market, risk* is the risk of not being able to *sell* the production obtained; there are two kinds of risk in this case:

1. *Demand risk* which derives from *consumer* freedom
2. *Competitive risk* which derives from the *freedom to take economic initiatives*

(c) *Financial risks*, connected to the impossibility of maintaining "*I*" and "*E*" *financially integral* (see the technical note below)

The efficiency of TR3-F is measured by the *yield on the capital*, which, as a first approximation, is quantified as the ratio between the *average return* on capital and the average capital amount, with reference to a given period. In detail, the main indicators of *financial performance* are (1) the return on equity (*roe* = EBT/E = R/E); (2) the return on debt (*rod* = I/D), taking into account the debt/equity ratio (*der* = D/E); and (3) the return on invested capital (*roi* = EBIT/I = OI/I). Note that *roi*, though it is an indicator of the overall financial efficiency of TR3-F, also expresses the economic efficiency of TR2-E. Thus, *roi* summarizes the efficiency of all three transformations.

Technical Note. A capital $K(t_0)$ that yields an income $R(T)$, with a $roi = R(T)/K(t_0)$, is kept *financially integral* at the end of period $T = [t_0, t_1]$ if $roi \geq roi^*$, where roi^* is the *opportunity cost* of $K(t_0)$ defined as the best roi^* of all the alternative available investments. *Proof*. The financial value K^F of a capital K that yields an income $R(T)$ can, for simplicity's sake, be set equal to the present value of $R(T)$ at rate "i"; that is: $K^F = R/i$. By definition, K is *financially integral* at the end of T if $K^F \geq K$. If $i = roi = R/K$, then $K^F = K$. If $i</>)roi$, then $K^F >/< K$. Setting $i = roi^*$ we obtain the following conclusion: in order to maintain a capital K financially integral the *roi* that is obtained must be greater than the opportunity cost. From the preceding definition we can assume that a necessary condition for a *capitalistic firm* to be created and continue to exist for period T is that $E^F \geq E(t_0)$, where $E^F = R°(T)/roe^*$ and roe^* is the minimum acceptable return for *equity holders* to maintain their capital invested in the firm. If $E^F \geq E(t_0)$, and if $R^*(T)$ is the net income that assures roe^*, then the difference $sfin = R°(T) - R^*(T)$ prepresent the *self-financing capital* that can be invested for the growth of the firm.

A necessary condition for the firm to activate the first three "technical" transformations is that two "cognitive" transformations also be carried out: the *entrepreneurial* (TR5-I) and *managerial* (TR4-M) transformations, whose function is to "control" the "technical" transformations. Though TR5-I is at the top of the model in Fig. 9.11, it is a preliminary requirement for TR4-M and for coherence sake will be examined first.

TR5-I: Entrepreneurial transformation, whose function is to monitor the *present* and *future* environments in order to (a) identify the survival conditions and define the *maximum objectives* that will guarantee an enduring vitality; (b) decide which *entrepreneurial policies* have priority in terms of the vital objectives; and (c) for each objective establish the *entrepreneurial strategies* to order the most effective control levers. To carry out this function, TR5-I manages the system mainly "transforming" external information and representations of the environment (sector, market, technology, etc.) into *policies, strategic objectives, and decisions* (Macintosh and Maclean 1999), preparing the long-term plans and programs and designing the *management control systems* that give rise to and regulate the three other transformations for the achievement of the quality, productivity, economic efficiency, and profitability objectives. I have referred to the transformation as "entrepreneurial," as it produces to the maximum extent possible the conceptual, creative, and innovative way of thinking (Christenson 1997; Deephouse 1999) by trying to change the *strategic position* of the firm in the environment (Nonaka and Takeuchi 1995; Mintzberg et al. 1998), in order to achieve the "max*roe*" necessary for maintaining the invested capital financially integral. Through TR5-I the firm, as a whole, can be viewed as a system in which present and future information can be transformed, by means of the economic calculation, into *political* and *strategic decisions* as well as into *regulations* for the purpose of achieving the vital objectives; the firm adapts itself for this purpose to the environment in order to carry out the three efficient "technical" transformations. The maximum vital objective in profit organizations is the production of value for shareholders, who normally represent the most relevant component of the stakeholders. The production of adequate levels of *economic value of the firm* (EVF), from which shareholder value derives, and the maintenance of the conditions of *sustainability* represent the maximum objectives imposed by TR5-I on TR4-M. However, the entrepreneurial transformation is, in turn, controlled by the stakeholders, who activate the corporate governance and set the maximum institutional objectives and the environmental restrictions for the survival of the organization as a vital system (Ashworth and James 2001).

TR4-M: MANAGERIAL TRANSFORMATION, which has a *control* function in order to allow the "technical" transformations to achieve, with the maximum efficiency, the *policies* and *strategies* decided by the entrepreneurial transformation: TR4-M undertakes five sub-functions: (1) it divides the vital objectives determined by TR5-I into operational objectives to be assigned to the organs (functions) and operational units; (2) it divides the *overall* entrepreneurial strategies drawn up by TR5-I into *functional* and *operational* strategies, which it assigns to the organs and operational units that carry out the "technical" transformations, and as a consequence (3) it draws up the operational programs and budgets that serve as the *operational objectives* for the control systems, which are required to achieve the maximum level of productive, economic, and financial efficiency; (4) it carries out the *managerial coordination* of the organs, operational units, and members of the organization that together represent the engines of the "technical" transformations; (5) it decides on the *operational regulations* which oblige the controlled units (organs, units, and individuals) to undertake the necessary actions to achieve the objectives (model in Fig. 9.2).

For convenience sake, I shall present in more concise form the system of the (macro) variables and of the basic values that regulate the MOEST to allow the reader to better understand the model in Fig. 9.11. The system of economic and financial values that represent the dynamics of the organization for any particular period can be summed up by the following balance sheet relations:

$$\begin{cases} L + I = D + E & [\text{financial position}] \\ C + \text{EBIT} = R & [\text{economic position}] \\ R = P \times pP & [\text{input of values}] \\ C = F \times pFP & [\text{output of value}] \\ \text{EBIT} = R - C & [\text{revenue production}] \\ \text{EBT} = \text{EBIT} - \text{Tax} & [\text{earnings before taxes}] \\ \text{EBIT} - Int - \text{Tax} = Div + \text{Sfin} & [\text{revenue distribution}] \end{cases} \tag{9.2}$$

In the above relations, L indicates liquidity (cash + receivables – trade payables), I is the invested capital, D and E represent debts and equity, Int = (D $i\%$) represents the interest paid on D, Tax = [(EBIT – Int) tax%] denotes taxes, and Div and Sfin indicate the dividends and the self-financing provisions.

The system of performance indicators can be summarized as follows:

$$\begin{cases} e = \frac{R}{C} = \pi \frac{pP}{pF} & \text{[economic efficiency]} \\ \pi = \frac{p}{F} & \text{[productivity = production efficiency]} \\ f = \frac{F}{P} = \frac{1}{\pi} & \text{[unit requirement of factors]} \\ cP = \frac{C}{F} & \text{[unit cost of production]} \\ roi = \frac{\text{EBIT}}{I = E + D} & \text{[economic and financial efficiency]} \\ roe = \frac{\text{EBT}}{E} & \text{[return of equity]} \\ rod = \frac{\text{Int}}{D} & \text{[return on debt]} \end{cases} \quad (9.3)$$

Figure 9.2, even if in extremely concise form, clearly shows that, even *functionally speaking*, every production organization must be viewed as a system of control systems that aims to achieve the objectives necessary for its survival. In fact, the MOEST includes three main forms of control: (1) *stakeholder control* (corporate governance in a broad sense) over the entrepreneurial transformation; (2) the *strategic control* of the entrepreneurial transformation, indicated in Fig. 9.12; and (3) the *operational control* of the managerial transformation, represented in the model in Fig. 9.13.

Figure 9.12 shows that the *entrepreneurial* transformation, TR5-I, identifies or receives from the governance the *vital objectives* for survival and determines the *policies* and *general programs* that become the *strategic objectives* the *managerial* transformation, TR4-M, must achieve through the control systems (normally defined as *strategic*), which act at the business and general function levels. Figure 9.13 indicates that the *managerial* transformation, TR4-M, translates the strategic objectives into *operational objectives*, to be achieved by means of a planning and budgeting program which is necessary for the *operational control* system to produce the necessary strategy to activate the available levers.

The *entrepreneurial* transformation, TR5-I, is, in turn, subject to an *institutional control* at an even higher level, carried out by the stakeholders, who represent the *Corporate Governance*; the amount of control TR5-I has in the organization depends, in fact, on the limits set by the governance. The models in Figs. 9.12 and 9.13 illustrate the role of the three technical transformations in implementing the control in order to achieve the vital objectives.

The operational units of SS1 as described by Beer (Fig. 9.9) correspond to the units that carry out the "technical" transformations of the MOEST. The "cognitive" transformations of the MOEST, both entrepreneurial and managerial, perfectly correspond to the activities assigned to the other four subsystems of the VSM.

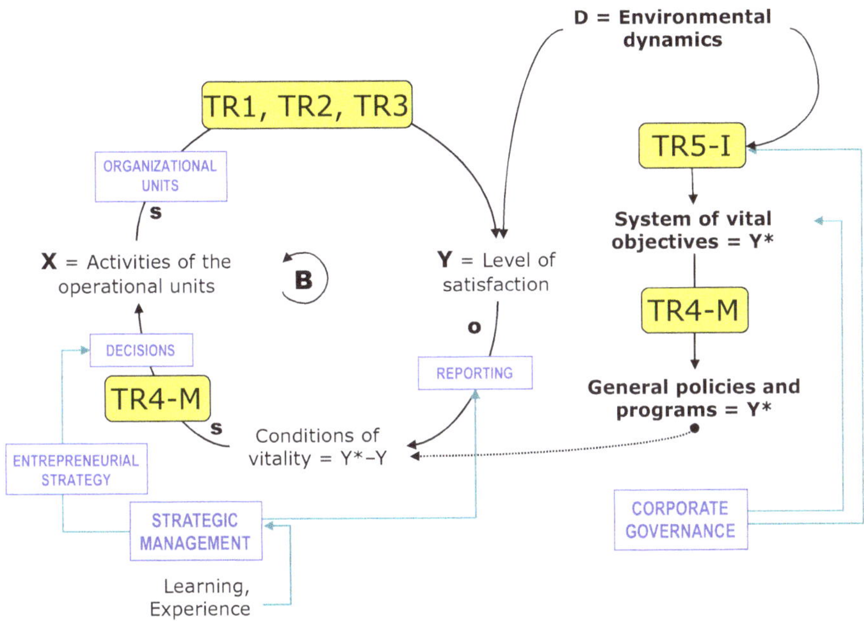

Fig. 9.12 The MOEST as a control system based on policies and strategies

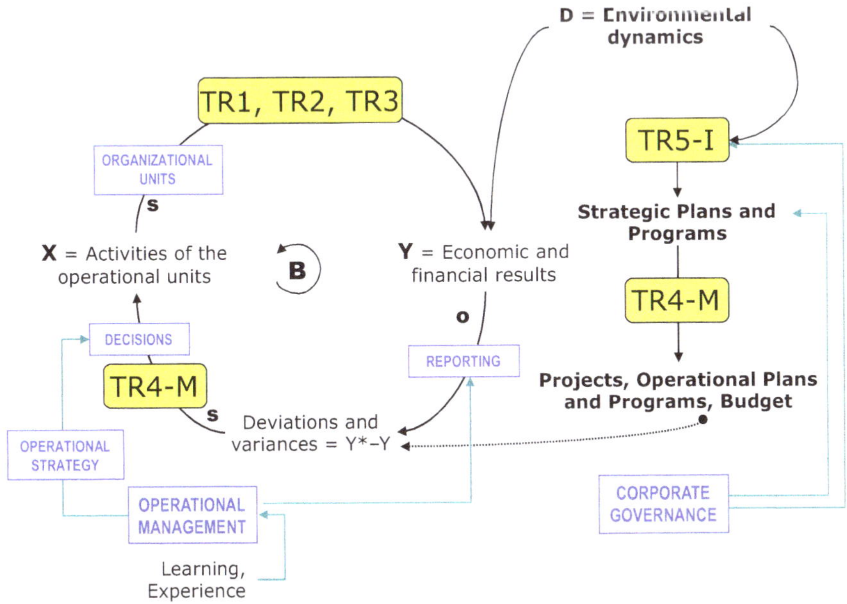

Fig. 9.13 The MOEST as a control system based on operational plans and budgets

In this sense the MOEST is in line with Talcott Parsons' view of the organization. Parsons distinguished between three organizational levels, each characterized by its own level of organizational control: the *technical*, the *managerial*, and the *institutional* levels. The *technical* level includes those activities necessary to achieve the organizational goals. At this level, organizational control involves the operational activities necessary to achieve the organizational objectives. The *managerial* level is entrusted with the administration of the organization; at this level the organizational control must see to the obtainment of the necessary resources for the technical level and for the satisfaction of the customers and users. The institutional level must align all the organizational goals with the aims of the stakeholders; at this level the organizational control must maintain the identification and the uniformity between the organizational and social objectives.

There are two differences between Beer's model and the MOEST. Above all, the latter explicitly presents the three different levels of control: *institutional, strategic*, and *operational*, each of which pursues objectives different in scope. Second, it emphasizes the possibility of creating a precise system of *performance indicators* represented by the efficiency indicators: productivity, economic efficiency, and profitability, indicated in (9.3) (in Fig. 9.11 these are only mentioned), all of which can be linked from the production of value point of view.

The MOEST shows us that the function of control systems, at the institutional, strategic, and operational levels, is to guarantee that the typical transformations achieve the performance objectives regarding value, economic results, and profitability throughout the organization.

Observing the MOEST in Fig. 9.11, we can deduce the fact that the *profit organization's autopoiesis* depends on its ability to:

1. Create a dynamic portfolio of businesses along a virtuous path from "question mark" to "cash cow" businesses (following the well-known Boston Consulting Group matrix 1970) by means of an effective *entrepreneurial function.*
2. Achieve the maximum exploitation of the present market and expand toward new markets in order to increase its production volume, P, and increase as much as possible the selling price, pP, through an efficient *marketing function.*
3. Contract the unit factor requirements while expanding the *quality* of products by means of an efficient production function, thereby increasing *productivity.*
4. Reduce the average factor prices through an efficient supply function which searches for supply markets where the factors have a higher quality and lower purchase prices.

The *autopoiesis of the capitalistic enterprise* is based on its capacity to regenerate its *financial* and *economic* circuits, which are made up of the value flows represented by the bold arrows in TR2-E and TR3-F in Fig. 9.11. The financial circuit is renewed if the capitalist firm succeeds in acquiring and preserving its invested capital (I)—which is necessary for structural investments—by means of an adequate *financial leverage* (debt/equity ratio, or *der*); but this requires that the suppliers of both debt and equity financial capital receive a fair remuneration, defined as one at least equal to their opportunity cost.

9.4 From MOEST to Management Control and Performance Management

In the preceding *sections* I have observed that, no matter how we choose to view them, production organizations, firms in particular, are control systems that, through their particular structure, succeed in remaining *living* and *vital* by adapting themselves to a changing environment, recreating the "organizational fabric" (*micro* level), identifying their vital objectives (*macro* level), and trying to achieve these through control systems of varying size operating at all organizational levels (Fig. 9.2) and regulated by the respective managers. For convenience sake, I shall define *organizational control* as the network of all control systems which, operating at different levels (individuals, organs, functions, departments, businesses, variously defined operational units, etc.), *guide* the organization to produce the maximum *efficiency* in its operations and processes and to achieve the maximum *performance* in terms of productivity, quality, economic efficiency, and profitability, in an ever-changing environment and respecting the limits set by the market and the stakeholders (Anthony 1998). The network of these *Rings*, which ensure efficiency and allow the operational objectives to be achieved, is normally defined as a *management control system*; this network, as a whole, results in the *management control* of the organizational performance of the firm as a unit.

This conclusion is, in fact, not new to Management Science, Organization Science, and Management Control Theory; in point of fact, these fields have a strong tradition of research in the disciplines and techniques of management con trol, which Robert Mocker defines as follows:

> Management control can be defined as a systematic effort by business management to compare performance to predetermined standards, plans, or objectives in order to determine whether performance is in line with these standards and presumably in order to take any remedial action required to see that human and other corporate resources are being used in the most effective and efficient way possible in achieving corporate objectives (Mockler 1970, p. 14).

Very concisely, keeping in mind the general model of a control system (Fig. 2.10), we can conceive of management control as the vast process that specifies and operationalizes the efficiency and the performance objectives of the organization and *activates* all the "apparatuses" of the control systems needed to achieve the objectives: the *effectors*, which correspond to the operational units; the *detectors*, which correspond to the monitoring, reporting, and performance-measuring centers; and the *regulators*, which correspond to the decision-making and responsibility units. Stafford Beer clearly illustrates this control system management role in organizations when he states that "*If cybernetics is the science of control, management is the profession of control*" (Beer 1966, p. 54). Even more clearly, in outlining the contribution of his book *Brain of the Firm* he states: "*This book is entirely concerned with the contribution which cybernetics, the science of control, can make to management, the profession of control*" (Beer 1979, p. 17).

It is not possible in this context to mention the thoughts of the many important management control scholars, not only due to the vastness of the bibliography but also because each author obviously follows his or her own conception of control and control system without, however, expressing this with an explicit definition. Nevertheless, it is useful to point out that, already in the 1960s, Robert Anthony presented a viewpoint of management control that reflected the different transformations indicated by the MOEST; in fact, he distinguished three types, or areas, of control systems: strategic planning, management control, and operational or task control.

> [Strategic planning is] the process of deciding upon objectives, on changes in these objectives, on the resources used to attain these objectives, and on the policies that are to govern the acquisition, use and disposition of these resources.
>
> [Management control is] the process by which managers assure that resources are obtained and used effectively and efficiently in the accomplishment of the organization's objectives.
>
> [Operational or task control is] the process of assuring that specific tasks are carried out effectively and efficiently (Anthony 1965, p. 16).

Referring to Fig. 9.11, it is clear that *strategic planning* is carried out by the *entrepreneurial* transformation in defining the time frame for the vital objectives indicated by the stakeholders and the overall strategy that indicates the strategic orientation for achieving these; that *management control* is carried out by the managerial transformation, which translates the strategic plans into specific control instruments for the operations of the "technical" transformations (Flamholtz 1996); and, finally, that operational control concerns the controls at the organizational structure level, which corresponds to the lower level in Fig. 9.2.

It is also useful to mention that the "management" concept itself has evolved to also include functions that differ from those related to the regulation of control systems. Already in 1916, Henri Fayol, in his book *Administration Industrielle et Generale* (translated in 1949 with the title *General and Industrial Management*), extended the concept of management to all activities of a firm, that is, to all the transformations indicated in the MOEST: "*technical, commercial, financial, security, accounting, managerial activities*." However, he also emphasized the control activities that characterize the organizational structure of the agents and organs; in fact, he stated that "*To manage is to forecast and plan, to organize, to command, to coordinate and to control,*" indicating the well-known 14 *principles of* management (Kramer 2010). In 1975, William Newman, in spreading the concept of control systems, though with his own system of symbols, clearly demonstrated the essential and pervasive nature of management control of processes, directly underscoring the relation between actions and objectives while stating that "*control is one of the basic phases of managing along with planning, organizing and leading*" (Zimmerman 1999). In his impressive book entitled *Management*, Peter Drucker emphasized above all the difficulties in appropriately using the term management while pointing out the empirical nature of management:

> The word "management" is a singularly difficult one. It is, in the first place, specifically American and can hardly be translated into any other language ... It denotes a function but

> also the people who discharge it. It denotes a social position and rank but also a discipline and field of study (Drucker 1973, p. 5).

> Management is a practice rather than a science. In this, it is comparable to medicine, law, and engineering. It is not knowledge but performance. Furthermore, it is not the application of common sense, or leadership, let alone financial manipulation. Its practice is based on knowledge and on responsibility (*ibidem*, p. 17).

He also recognizes that management operates at all levels, even that regarding entrepreneurial transformation:

> For three-quarters of a century management has meant primarily managing the established, going business. Entrepreneurship and innovation, while mentioned in many management books, were not seen as central from 1900 till today. From now on, management will have to concern itself more and more with creating the new in addition to optimizing the already existing. Managers will have to become entrepreneurs, will have to learn to build and manage innovative organizations (Drucker 1973, p. 31).

It should be clear that management control refers above all to the typical actions of the managers—the *performance makers* in the firm, according to the MOEST model—which involve setting objectives for the control of processes, defining the organizational actions, identifying gaps regarding their achievement, modifying the organizational activities themselves, implicitly assuming that the control is supported by the actions of the agents and organs in the organization (the effectors in the control systems); and actions, agents, and organs which themselves must be controlled.

The network of control systems activated by management control specifically to carry out the technical, economic, and financial transformations constitutes, as a whole, the *management control system*. This includes a system for the production of information and organizational rules and procedures needed to activate, maintain, and improve organizational behavior, which is achieved by acting on individual behavior, as set out in the following definition by Robert Simons and David Otley (Demartini 2014).

> Control in organizations is achieved in many ways, ranging from direct surveillance to feedback systems to social and cultural controls [...] Management control systems [are] the levers managers use to transmit and process information within organizations. For the discussion to follow, I adopt the following definition of management control systems: *management control systems are the formal, information-based routines and procedures managers use to maintain or alter patterns in organizational activities* [...] These information-based systems become control systems when they are used to maintain or alter patterns in organizational activities (Simons 1995, p. 5).

> Management control systems provide information that is intended to be used by managers in performing their jobs to assist organizations in developing and maintaining viable patterns of behavior (Otley 1999, p. 364).

This definition clearly emphasizes the role of the manager as described by Beer and as it appears in the control system model (Fig. 3.1): determine the performance objectives; obtain information from the performance measurement systems, which detect deviations from the established performance objectives; make decisions regarding the control in order to decide on a strategy; and send information on the levers activated by the effectors (agents or organs), which are the true passive subjects of the control. Management control systems are thus vital for maintaining the organizations as viable systems, since they continually adapt the organization to the environment to continually satisfy the objectives set by the stakeholders.

> In broad terms, a Management Control System is designed to help an organization adapt to the environment in which it is set and to deliver the key results desired by the stakeholder groups, most frequently concentrating upon shareholders in commercial enterprises. Managers implement controls, or sets of controls, to help attain these results and to protect against the threats to the achievement of good performance (Merchant & Otley 2007, p. 785).

Since for every activity the performance is related to the results, but the latter depend on behavior, it is necessary to undertake not only a control of the operations but also a control on the *motivations* behind the behavior which impacts the *performance* of the operational units as well as on the *emotions* of the personnel which influence decisions and the standards of behavior. This form of control represents the typical focus of *performance management* techniques, a term introduced by Aubrey Daniels in 1970 with this definition: "*Performance is the sum of behavior and results, and cannot be viewed as independent of either component. It is an outcome of effective management*" (Daniels and Daniels 2004). To measure and improve the performance of individual and group behavior it is necessary to link them to systems of rewards or incentives (control levers) (Fig. 9.6). The more instrumental operational concept of *performance management system* derives from that of *performance management*. The former represents the set of all control systems and their networks which allow the organization, its organs, and agents to achieve and improve their performance (for an ample bibliographical review see Chiara Demartini 2014).

> We view PMSs as the evolving formal and informal mechanisms, processes, systems, and networks used by organizations for conveying the key objectives and goals elicited by management, for assisting the strategic process and ongoing management through analysis, planning, measurement, control, rewarding, and broadly managing performance, and for supporting and facilitating organizational learning and change (Ferreira and Otley 2009, p. 264).

This concept is by now widespread in all types of organization. For example, from a link ("Performance Management") on its website, Berkeley University in California clearly indicates this concept with the following definition:

> The University of California, Berkeley is committed to a performance management system that fosters and rewards excellent performance.
>
> Effective performance management aligns the efforts of supervisors and employees with departmental and campus goals, promotes consistency in performance assessment, motivates all employees to perform at their best, and is conducted with fairness and transparency.
>
> The employee, the supervisor, and the University are critical members of a partnership that ensures performance planning, assessment, coaching, and development (Berkeley 2010, online).

Lastly, I would mention the concept of *administrative* control (or *work practice* controls), a general term indicating the *social aspect* of the control of agents as *individuals* who must operate in respect of the limits imposed by *responsibility* and *accountability*, which circumscribe the behavior of the agents and organs.

> Administrative controls develop control, since they direct employee behavior through the organizing of individuals and groups, the monitoring of behavior … and make employees accountable for their behavior … and the process of specifying how tasks or behaviors are to be performed or not performed (Green and Welsh 1988, p. 293).

> Procedures and policies, on the other hand, relate to the "bureaucratic approach to specifying the processes and behavior within an organization" (*ibidem*, p. 294).

The administrative control also takes on the significance of a control for the safety of the agents as workers in the organization, who must not suffer damage in carrying out their responsibilities. Finally, the term also has the bureaucratic meaning of *compliance control* regarding the operations and actions of the agents; this control is carried out by systems of recognition that verify that the acts are not contrary to the internal or the external rules of the organization.

Management control systems, however understood, are not always effective due to the fact that they control individuals in the organization who may not modify their behavior in the manner expected by the managers as a consequence of the action of the chosen control levers. Very often, no matter how well designed and structured a framework of management control systems in an organization may be, the results may not meet expectations, especially when some of the individuals subject to control band together (Hofstede 1978), in some cases bordering on a true rebellion, to sabotage the control actions. This situation has given rise to the expression "illusion of control":

> … the illusion fosters the belief among managers that conventional controls … accurately and validly measure, and thereby help determine, behavior. The illusion reflects a presumption that management can intervene when necessary and successfully effect change. Further, the illusion provides for the belief that, by changing a given mix of existing controls, managers make necessary and sufficient functional responses to internal or external change. To those managing with an illusion of control, negative consequences of managerial action often signify the necessity for more controls (Dermer and Lucas 1986, p. 471).

9.5 The *Rings* in *Macro* Management Control: The Objectives of the Strategic *Rings*: EVF and EVA

The MOEST model in Fig. 9.11 helps us understand that business profit organizations (firms) can be viewed not only as autopoietic and vital systems but also from two other points of view:

1. As *instrumental* systems through which people carry out the economic activities indispensable for production, consumption, savings, and investment of wealth

2. As *teleological* systems, since they are units with an autonomous existence with respect to the individuals operating within them; thanks to the managerial transformation they follow a trajectory in the production, economic, and financial space of which they are a part in order to achieve a system of vital objectives

The *ultimate aim* of organizations, which are viewed as *instrumental systems*, is the maximum achievement of the stakeholders' goals through a system of high-level objectives which I shall define as *institutional goals*. Satisfying the *institutional goals* represents, more than the *ultimate* objective of the organization, a true *condition of its existence*: the organization can continue to exist only if it achieves its *institutional goals*, thus leading to a long-lasting maintenance of the conditions of "unlimited life." In particular, the firm can continue to exist only if it produces "technical transformations" that can satisfy:

(a) The *clients*, through the production of prized goods or services (and thus possessing greater value with respect to the factors used to obtain them), of adequate quality, continually innovated, and at a price kept as low as possible with respect to competitors
(b) The *workers*, by offering them *monetary payments* that can lead to *cooperation* and *coordination* and a *work environment* that can best satisfy their professional aspirations and favor a *career* marked by commitment and loyalty toward the organization
(c) The *suppliers*, by guaranteeing stable supplies and punctuality in fulfilling orders
(d) The *suppliers of financial capital*, debt, by assuring a return in line with those offered by the market and, above all, security in the payment of interest and capital reimbursement
(e) The *suppliers of risk capital*, equity, by sufficiently increasing value through the production of high rates of return
(f) The entire *collectivity*, external stakeholders, by favoring employment, thereby guaranteeing a tax flow over time and safeguarding the environment by operating under conditions of sustainability

In order to achieve the *institutional goals* and respect the *constraints* set by the stakeholders, the TR4-M managerial transformation must provide the organization with specific control systems of various levels, which are *interconnected* to form a network and which often interfere with one another. The TR4-M must also define a system of coordinated *directional (managerial) objectives* that can be expressed quantitatively and represent a reference for control systems; the latter must ensure that the *directional objectives* are reached as well as maintain the desired *performance levels*, guaranteeing not only the *results* but also the operational *standards* of the processes and operational units. The *directional objectives* can be considered in all respects as *performance objectives* of the "technical" transformations. The system of control systems and directional objectives that must be achieved form the overall general management control system, which carries out two forms of control: *macro* and *micro* management control. These two aspects are distinguished not so much in a logical sense but in terms of the type of specific control system used, the objectives to achieve, and the time scale over which they act. *Macro* management control takes on the form of a *strategic control* as well as an *operational control*. We have already seen in Fig. 9.12 that the *strategic control*, after having received the *vital* objectives from the *entrepreneurial transformation*—which are conditioned by the policies and limits set by the *Corporate Governance* and the stakeholders, in particular the stockholders—formulates the *strategic plans* and *programs*, which specify the levers and strategies to pursue the vital objectives and the top priority policies.

Figure 9.14 specifies that these plans, which cover a multi-year period, represent for the strategic control system the maximum objectives management must achieve through the technical, economic, and financial *levers*.

Strategic reporting identifies mid-term variances between programs and implementations; deviations represent inputs from *regulatory* systems, which make strategic decisions regarding which levers to activate and at what reaction rates in order to take into account management urgency; on the basis of these decisions, the technical, economic, and financial levers are implemented, which the organization, through its management, will translate into production, economic, and financial outcomes. The cycle repeats itself until the organization has reached the optimal vitality levels.

Figure 9.15, which is derived from Fig. 9.14, provides a concise representation of an *operational control* operating through the budget. In the practical application of the *budgetary control*, the model in Fig. 9.15 is to be understood as referring to a multitude of analytical control systems, each of which must act to achieve *specific values for the objectives*, which are specified for each *segment* of the budget. Only through analytical control systems is it possible, for each specific objective value (e.g., Y^* = Sales of "alfa" product in "northern" area in the month of "April" for the sales division "Hamburg"), to identify the values attained, Y, and the deviations, $E(Y) = Y^* - Y$, and to decide on the actions, $[X]$, in order to eliminate $E(Y)$ or accept it. In this sense the control through the budget, even more so than the strategic control, requires the activation of a true *system* of coordinated parallel control systems to achieve the general objectives regarding planned profitability. The *continual reporting* detects the trajectory of the values produced by the organs, the results, and the performance of the operational management; subsequently, through a comparison with the budget, it determines the deviations, $E(Y)$, from which management gains information to make corrective decisions regarding the activities of the operational units. The budget structure in manufacturing firms and the role of reporting will be discussed in Sect. 9.9.1.

Though a detailed analysis of the system of strategic entrepreneurial *performance* objectives is not possible here, some further brief considerations are useful (for a more detailed discussion, see Mella 2005b, 2011). In small- and medium-sized firms, where *equity* is concentrated in the hands of a few "owners," the highest order directional objective can be considered to be the achievement of the *maximum flow of profits* from the most limited use of *equity* possible. The *roe* becomes the main indicator of *economic performance*, as long as it is in line with the institutional

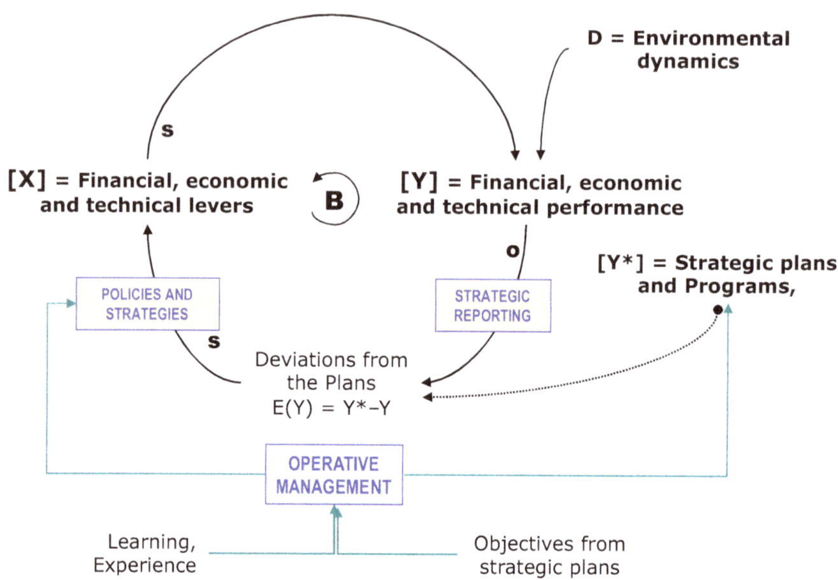

Fig. 9.14 General model of strategic entrepreneurial control of management in production organizations

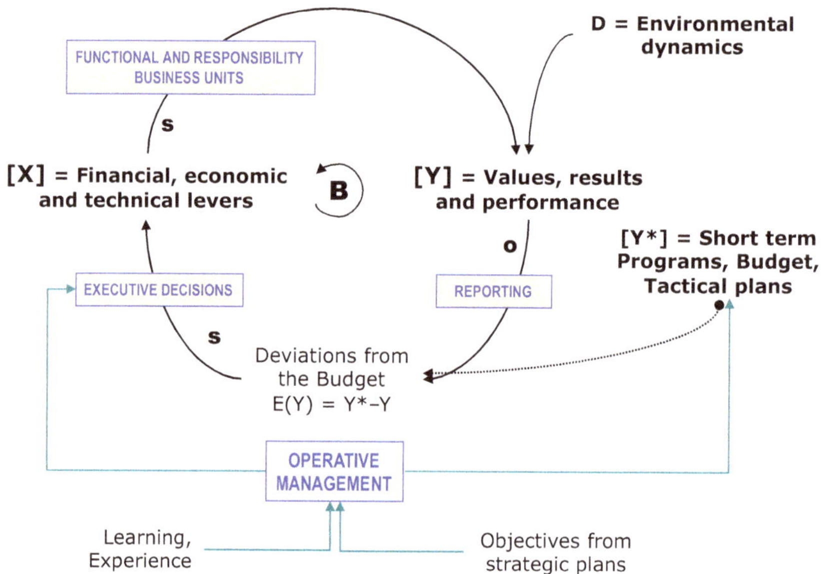

Fig. 9.15 General model of operational control of management in production organizations

goals and compatible with the *financial constraints* regarding the raising of the capital necessary for survival and organizational growth while also respecting the *monetary constraint* of a balanced cash management. For large corporations, where the capital is controlled by a large number of shareholders who can exchange their shares on the capital markets, the *entrepreneurial* transformation requires that the *managerial* transformation pursue *as* the *maximum* performance objective the *continual increase in shareholder value*; this objective translates into the managerial objective of obtaining *profitability* levels that maximize the EVF, since the EVF defines the theoretical value of share trading.

In short, recalling that the value of an amount of non-redeemable capital, K, corresponds to the present value of its future average income, R, discounted at the rate "i" based on the well-known formula for discounting a perpetual income stream ($K = R/i$), then, in a totally analogous manner, the EVF can be calculated as follows:

$$\text{EVF} = P^* / ce^* \tag{9.4}$$

where

- P^* indicates, for simplicity's sake, the expected future average profit of the firm before taxes (EBT), calculated with reliable and realistic multi-year plans.
- ce^* indicates the *cost of equity*, that is, the expected *financial interest* or *dividend yield* that would convince the shareholder to invest *equity* in the firm, a yield the company must guarantee in order to keep intact the shareholder's investment; thus, for the firm, ce^* theoretically represents the cost of attracting and maintaining equity.

Indicating by E the amount of the original *equity* investment which is assumed to produce P^*, then the ratio

$$roe^* = P^* / E \ (\text{return on equity}) \tag{9.5}$$

represents the highest order *managerial performance objective* of the firm, since it allows us to quantify EVF, as can be seen from the simple comparison of roe^* and ce^*. In fact, substituting P^* from (9.5) into (9.4) we get

$$\text{EVF} = E \left(roe^* / ce^* \right) \tag{9.6}$$

If $roe^* = ce^*$, the *entrepreneurial* transformation guarantees the monetary integrity of the equity invested, since EVF $= E$; only if $roe^* > ce^*$ does the firm produce value for the investor, since from (9.6) it follows that EVF $> E$. The difference $G = \text{EVF} - E$ is defined as the synthetic value of *goodwill*, representing the *capital gain* to the capitalist if the firm is sold.

From (9.6) we immediately observe that the highest order *managerial performance objective* is to obtain a roe^* greater than ce^*; this managerial objective is, in turn, more precisely defined in the additional *managerial* objective of producing revenues, R^*, and incurring costs, C^*, that enable the firm to produce P^*, which quantifies the roe^* necessary to increase EVF.

The calculation of *cost of equity* is not simple, but it should generally correspond, for example, to the average *roe* of the firms operating in the *same sector*, under conditions of equal risk, if we assume that the capitalist wants to invest *equity* in production firms of the same type, or correspond to the *risk-free* yield from *government securities*, if we assume the capitalist–entrepreneur is interested in the security of the investment more so than in the interest rate on the capital. However, the *risk-free* yields must be adjusted to take into account the different risks in investing in the firm (Mella 2005b).

Observing the MOEST (Fig. 9.11), we can consider the growth in EVF, or G, to be the maximum survival objective the *entrepreneurial* transformation sets for the *managerial* transformation in order to enable the latter to determine the *operational* objectives of R^*, C^*, and P^* for planning and budgeting. EVF (or G) can be viewed as the *global driver of the production of value*, since it serves as an objective to assess the overall performance levels of the four organizational transformations that succeed the entrepreneurial transformation.

Along with EVF, as a driver of the production of value, economic value added (EVA), or *residual income*, is also frequently considered. EVA was conceived of by Stern Stewart & Co., a global consulting firm, which launched it in 1989 (Stern et al. 1995). This measure shows in a simple manner the *residual economic result* obtained by subtracting the cost of debt and equity capital from NOPAT, that is, the *net operating profit after tax*:

$$\text{EVA}^{®} = \text{NOPAT} - \left(I + ce^{*}E\right) = \left(\text{EBIT} - \text{Tax}\right) - rod\ D - ce^{*}E \tag{9.7}$$

The innovation here is the fact that EVA is calculated by subtracting from NOPAT as the cost of capital not only the interest on debt but also the expected return to shareholders' equity (Ehrbar 1998). However, NOPAT does not correspond simply to after-tax EBIT before deducting the cost of capital but, except for particular cases, is determined by adjusting EBIT using a certain number of *key adjustments* to the book values in the income statement to take into account the costs that must be capitalized or the capitalized costs that should have been included in the calculation of the economic result.

> These adjustments aim to 1) produce an EVA figure that is closer to cash flows, and therefore less subject to the distortions of accrual accounting; 2) remove the arbitrary distinction between investments in tangible assets, which are capitalised, and intangible assets, which tend to be written off as incurred; 3) prevent the amortisation, or write-off, of goodwill; 4) eliminate the use of successful efforts accounting; 5) bring off-balance sheet debt into the balance sheets; and 6) correct biases caused by accounting depreciation (Young 1999, p. 8).

Taking into account (KA), NOPAT is calculated as follows (P indicates net profit):

$$\text{NOPAT} = \left(\text{EBIT} - \text{Tax}\right) + \text{KA} = \left(P + I\right) + \text{KA} \tag{9.8}$$

Ignoring KA, EVA then takes on the simpler form:

$$\text{EVA}^{®} = (roe\ E + rod\ D) - (rod\ D + ce\ E) = (roe - ce^*)\ E \tag{9.9}$$

From (9.9) we clearly see that EVA > 0 only if ($roe > ce^*$), that is, if actual *return on equity* is greater than the *expected return for shareholders*.

There are also other entirely equivalent ways to formulate EVA, but (9.9) suffices to show that, in fact, the maximization of EVF implies the maximization of EVA and vice versa (ignoring the *key adjustments*). In my opinion, therefore, though it is useful in assessing from year to year the actual balance sheet results, EVA does not represent an *explicit* managerial objective, as demonstrated by the following relation (Mella 2005b):

$$\text{EVF} = \frac{R^* + \text{EVA}}{roe^*} = E + \frac{\text{EVA}}{roe^*} = E + G \tag{9.10}$$

> EVA's top–down approach works well at higher levels of an organization but becomes more difficult to implement below the strategic business unit level. ABC was designed to overcome this deficiency. To implement EVA at the lowest level of *a* firm's operations, capital cost may be traced to activities and then to objects, using the principles of ABC (Kee 1999, p. 4).

9.6 The Objectives of the Operational *Rings*: *Roe** and *Roi**

Despite the fact *roe* is an autonomous indicator of financial performance, for the purposes of *management control* it represents a *performance objective* derived from other *managerial objectives*, in particular *return on investment* (*roi*), *return on debt* (*rod*), and *debt/equity ratio* (*der* = D/E), as clearly shown in Modigliani–Miller's (1958) well-known equation:

$$roe = \left[roi + (roi - rod)der\right](1 - \text{tax}) \tag{9.11}$$

Equation (9.11) is translated in the model in Fig. 9.16, which makes it clear that the control system for *roe** is typically a multi-lever system whose levers can act on:

(a) *roi*, that is, raising the return on invested capital above *rod*.
(b) *rod*, to produce a *spread* = (*roi* – *rod*) which, if positive, can raise *roe* above *roi*, taking into account the financial leverage obtained from *der*.
(c) *der*, increasing this ratio until the *roe* objective is achieved, with *roi* given and assuming *rod* is known; in this case it is necessary to act directly on the efficiency of the TR3-F, financial transformation, in determining the appropriate proportion between debt and equity.

(d) Self-financing (Sfin = EBT – taxes – dividends) to achieve an adequate *der* and limit the financial burdens.
(e) The *corporate tax rate* (tax) to reduce taxes.

The balancing loop (**B1**) in Fig. 9.16 clearly shows that the most important factor impacting *roe** is *roi*:

$$roi = \text{EBIT} / I = (R - C)/(D + E) \tag{9.12}$$

which represents a new *performance objective* even more significant than *roe**, since it represents at the same time the *maximum economic and financial performance of the firm*. As a result, "*roi**" becomes the most significant *managerial objective* in terms of planning and budgeting, since the other levers are subordinated to it. "*rod*" quantifies the cost of debt capital, which is assumed to depend on the credit market; "*tax*" indicates the average income tax rate as determined by each country's fiscal policy. The choice of levers to control *roi* represents the medium- to long-term strategy which *management control* must specify and the *performance management* must carry out in order to achieve the necessary "*roi**."

While *roe** is controlled by the multi-lever control system shown in Fig. 9.16, the control system for achieving the objective for *roi** requires, as (9.12) demonstrates, the activation of two main levers which allow for the control of EBIT (numerator) and invested capital, $I = D + E$ (denominator), whose values, in turn, become the *managerial* objectives of the economic transformation. Of the two levers, EBIT is the main one; the optimal value of EBIT* that achieves the objective for *roi** is, in the final analysis, the *maximum strategic performance objective* of the *economic* and *productive* transformation, which must be controlled for the survival of the firm. However, (9.12) clearly implies that the control of EBIT must be carried out by a control system that acts through two important levers: *sales revenues* (SR) and *production costs* (PC), as summarized in Fig. 9.17.

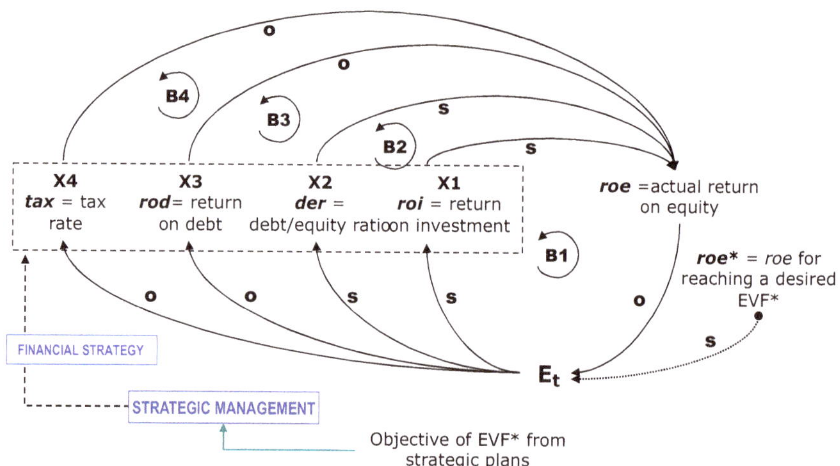

Fig. 9.16 General model of a control system for *roe*

The control system for EBIT carries out a *strategic management* control if it considers the EBIT flows for a multi-year future time period; it can be transformed into an *operational management* control if it covers only 1 year. In multi-business companies the control system for EBIT synthesizes a number of control systems which act in parallel fashion and all of whose objectives is the EBIT of each business.

A network of *lower level* control systems depends on these high-level control systems; the former operate in a coordinated and continual manner to control the *desired* levels of both *sales revenues*, SR*, and PC*, which ensure that the EBIT* objective will be met.

The control systems for achieving planned SR* and PC*, both globally and for each business, can activate various control levers in terms of product, market, price, technology, productivity, quality, distribution channel, etc. Therefore, the management for each control system must determine the appropriate activating *strategy*. Moreover, the various control systems for EBIT*, SR*, and PC* must achieve multiple, correlated (at times even often contrasting) objectives, so that it is necessary to establish not only a *strategy* but a control *policy* as well.

Figures 9.18 and 9.19 present in more detail the general models of the two control systems that operationally translate the revenue and cost levers indicated in Fig. 9.17. Such models can be specified for each business and each period of reference (year, quarter, etc.). In particular, Fig. 9.18 presents an example of a *multilayer* and *multi-level Ring* to control PC*.

The first level, operational control, is based on the two variables that can be impacted in the short run: the purchase price of factors of production and the unit factor requirements. Where there is a positive deviation $E(C)$ (greater-than-desired costs), management tries, on the one hand, to optimize supplies by reducing unit prices as much as possible and on the other to eliminate waste and redesign the

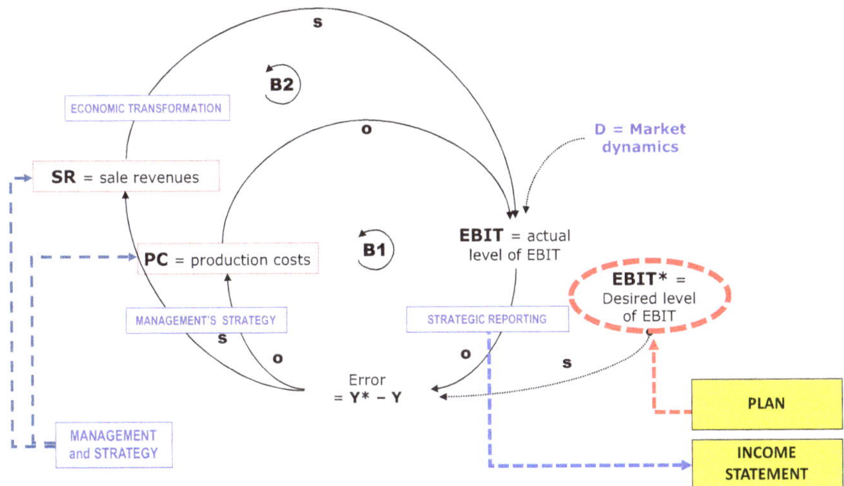

Fig. 9.17 General model of a control system for EBIT

product to reduce the unit requirements for material, labor, services, and other factors. The other measures (the order in which they appear in the model in Fig. 9.18 is merely indicative) are more of long term and structural. The investment in technology and process restructuring levers is needed to save on the use of factors of production; the lever for changing the pool of suppliers, placing them in competition with one another or seeking out more favorable contract terms, is needed to impact the level of supply prices. It is easy to imagine that these wide-ranging *structural* interventions most likely will not directly impact the *budget*; nevertheless, the model in Fig. 9.18 is so general as to be applicable also in cases of other forms of programing and control: *target costing* and *activity-based management*, and others still which I can only mention.

Figure 9.19 outlines the *multilayer* model of a *Ring* to achieve the desired level of revenues. This case also calls for various levels of intervention: the operational control entails the price and sales volume levers. The other action variables permit a medium- to long-term *structural* control. In particular (the order is merely indicative), the strengthening of the sales force and determination of monopolistic strategies (or, in any event, strategies capable of increasing the advantage from size) affect sales volumes; the marketing action, which entails the strengthening and

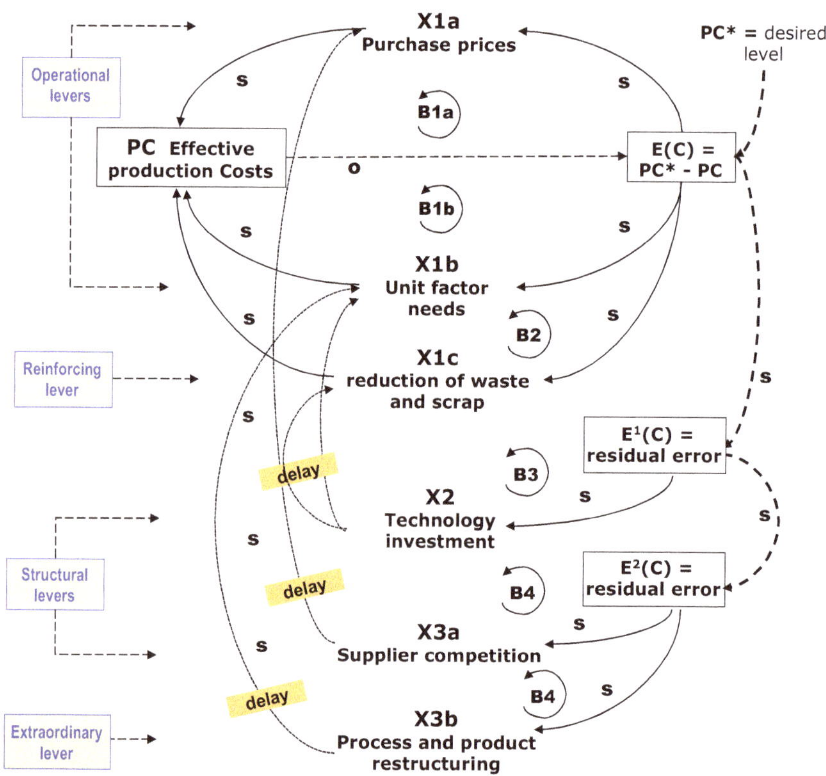

Fig. 9.18 General model of a multilayer system for controlling production cost

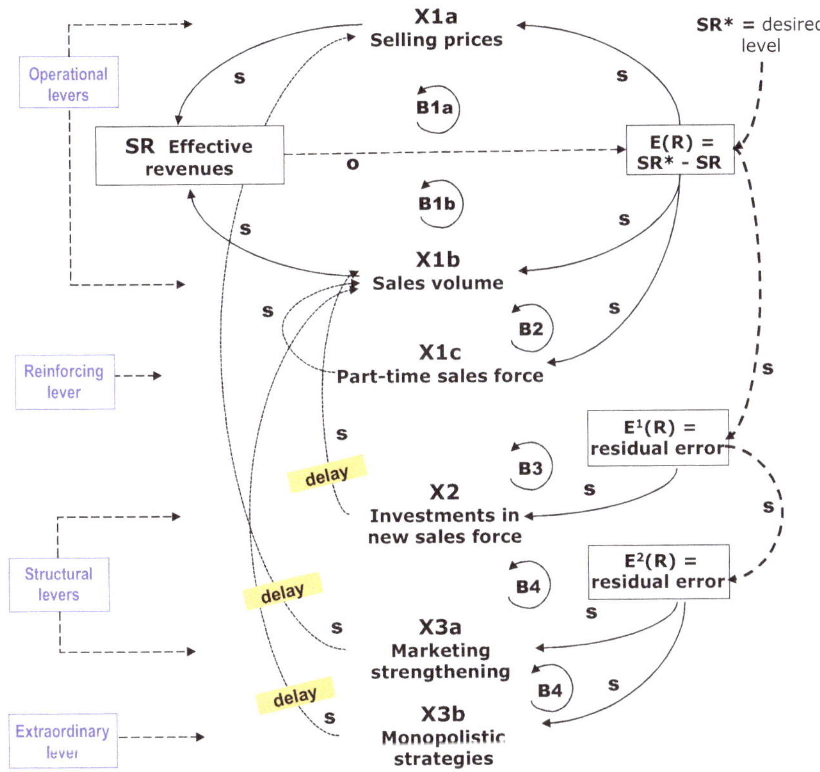

Fig. 9.19 General model of a multilayer system for controlling sales revenues

control of the respective variables (advertising, promotions, discounts, packaging, etc.), influences the selling prices.

Figure 9.20 combines the sales revenue and production cost *Rings*; it derives from Fig. 9.17, after substituting the synthetic PC and SR levers with the corresponding *Rings* in Figs. 9.18 and 9.19. Thus, the model in Fig. 9.20 represents the firm as a *multi-objective* control system in which management must try to simultaneously achieve the cost and revenue objectives, PC* and SR*, based on a well-delineated *policy* aimed at achieving the "goal" represented by the final *roe** and EVF* objectives. The *Ring* for the EBIT* in Fig. 9.20 is based on planning and programing, which is joined by the parallel operation of a process to *detect* performance and to evaluate it by identifying *variances* and deciding on the *strategic levers* for containing them.

We can clearly observe the similarity between the structures in the models in Figs. 9.18 and 9.19. This is no coincidence; in fact, it is easy to note that the two *Rings* are entirely similar and follow the pattern already outlined for the control system of a car's speed in Fig. 4.15. Moreover, the similarity between the model in Fig. 9.20 and the control system for the flight of an airplane in Fig. 4.16 is entirely evident. Even the symmetry between the models in Figs. 9.18 and 9.19 should not

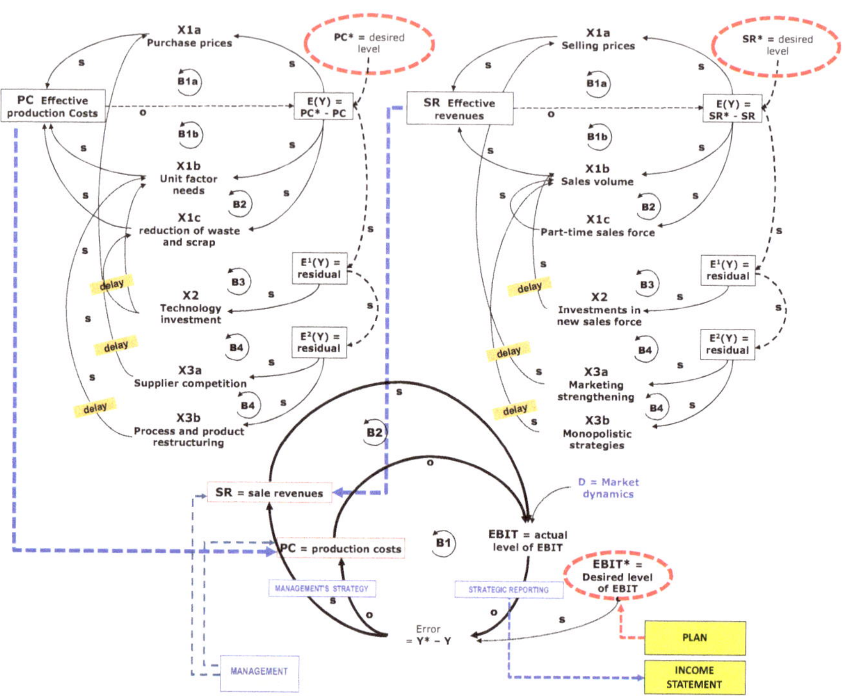

Fig. 9.20 General model of a multi-level, multi-objective control system for a firm's EBIT (see Figs. 9.17, 9.18 and 9.19)

surprise us; it is clear observing TR2-E in the MOEST (Fig. 9.11) that the control of revenue flows is (almost) symmetrical with respect to that of cost flows. We can intuitively deduce that the *operational control* of EBIT must achieve an even more analytical level by activating a network of increasingly more detailed *Rings* that operate in a coordinated and continual manner to control the level of the control levers in Figs. 9.18 and 9.19.

Figure 9.21 presents the control system for short-term *cash flow* (*K*), which originates with the cash inflow from *receipts* (KR variable) less the cash outflows from *payments* (KP variable).

The flows from revenues and payments represent the variables influenced by the current operational control of cash flow; if the restructuring of these flows were not sufficient to produce the desired volume of cash flow, then long-term and structural interventions would be called for that act on the other levers: investment, disinvestment, financing, etc.

In concluding this *section*, I would remind the reader that TR5-I, the *entrepreneurial* transformation (Fig. 9.11), sets objectives not only for profitability but also for the firm's *growth*, understood as a long-lasting growth in sales revenue and invested capital. The growth of the firm requires that:

(a) Sales can be increased with a consequent absorption by the markets.

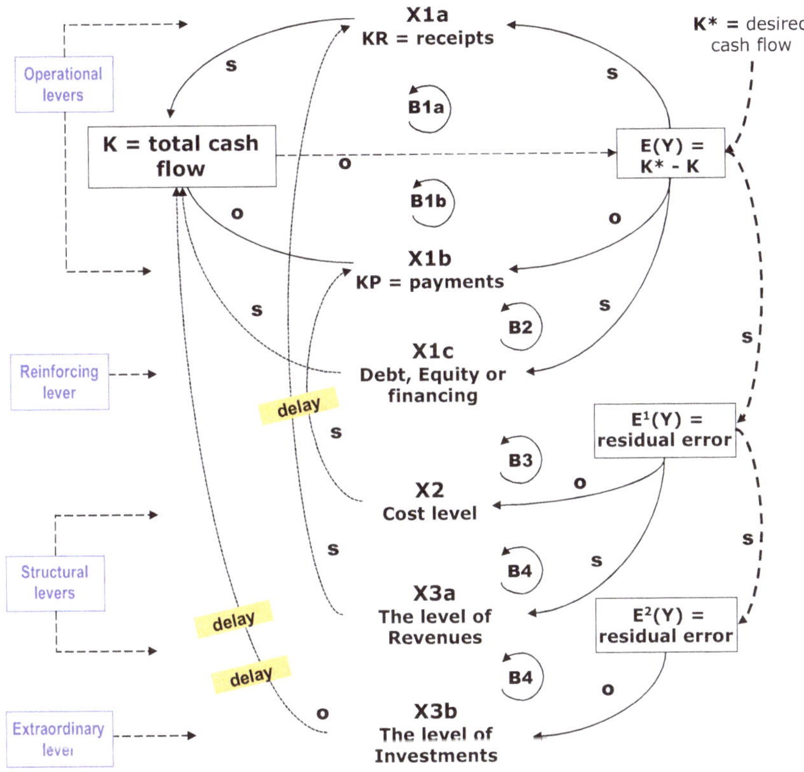

Fig. 9.21 General model of a control system for cash flow

(b) Along with the growth in sales there is an adequate growth in the production processes and thus in the capital invested in the firm.
(c) The increase in the size of the production processes is adequately financed.
(d) The growth in sales and invested capital leads to a proportionate growth in operating income as well.

Organizational growth requires the availability of financial resources, which can be obtained from *outside sources*—loans or increases in capital stock—or internal ones, typically *operating cash flow* (or gross self-financing and net self-financing). The generation of *net self-financing flows* over time is possible only if the firm's revenue is sufficiently high to allow for a fair amount of dividend distributions while at the same time allowing part of the profits to be kept as a reserve and thus as self-financing. Once again, the control of the performance of the *managerial* transformation can be further refined. It is not enough for $roe > roe^*$; it is also necessary for $(roe - roe^*)\ E \geq \text{sfin}^*$, where sfin* is the net self-financing needed to achieve the desired levels of growth.

For completeness sake, I would emphasize that the *managerial control* set up to *maximize shareholder value* and at the same time organizational growth is also referred to in modern management literature as value-based management, a new managerial approach aimed at the production of value for the shareholders (Mella and Pellicelli 2008).

Arnold's definition in this regard is significant:

> Value-based management is a managerial approach in which the primary purpose is long-term shareholder wealth maximization. The objective of a firm, its systems, strategy, processes, analytical techniques, performance measurements and culture have as their guiding objective shareholder wealth maximization (Arnold and Davies 2000, p. 9).

Copeland, Koller, and Murrin's definition is more specific:

> VBM is very different from 1960s-style planning systems. It is not a staff-driven exercise [...] Instead, it calls on managers to use value-based performance metrics for making better decisions. It entails managing the balance sheet as well as the income statement, and balancing long- and short-term perspectives (Copeland et al. 2000, p. 87).

Morin and Jarrel clearly refer to the double interpretation of VBM: a "mental attitude/selection of operational methods":

> [Value Based Management] is both a philosophy and a methodology for managing companies. As a philosophy, it focuses on the overriding objective of creating as much value as possible for the shareholders. ... As a methodology, VBM provides an integrated framework for making strategic and operating decisions (Morin and Jarrel 2001, p. 28).

> Once the company develops strategies, a number of operational drivers that are key to implementing the strategy have to be identified. By focusing on these operational drivers, the company's strategy is successfully implemented, which in turn improves the value drivers, creating aggregate value (*ibidem*, p. 343).

The control systems presented in this *section*, specifically MOEST, are appropriate for summarizing the logic of value-based management in capitalist enterprises. According to the MOEST, we can highlight the basic strategy of the control system that value-based management must activate to maximize EVF and shareholder value:

1. Select those investments having a $roi > min\ roi^*$—that is, the level of *roi* sufficient to achieve roe^*—for the entire firm; if there is more than one, it chooses the one having the max *roi* and the *minimum payback period*.
2. Choose the investments that, in any event, have $roi \geq rod$ and $roi \geq wacc$ (weighted average cost of capital) and which are sufficient to guarantee min roe^*.
3. Choose financing with min *wacc* and *min rod* (other conditions held constant).
4. If $rod < roi$, increase D and reduce E; turn to rule (1).
5. Substitute, when possible, investment I with J if $roi(J) > roi(I)$; in this way the average *roi* for the entire firm will increase.
6. Substitute, when possible, financing F with G if $rod(G) < rod(F)$ in order to reduce the average *rod* for the entire firm.

Trying to achieve the objective of increasing EVF means setting roe^* as the objective and using the levers shown in Fig. 9.16, which, in turn, represent lower level objectives whose attainment implies the activation of specific management *Rings*. In particular, roi^* becomes an objective for control systems for EBIT, which are presented in concise form in Fig. 9.20. Since net profit and EBIT derive from the "technical" transformations (productive, economic, and financial) indicated by the MOEST (Fig. 9.11), clearly it is necessary to consider a hierarchy of operational objectives for these three transformations, which will then be further detailed through operational goals to be achieved by the various business divisions in carrying out their operational processes. In particular:

(a) The *sales process* will be directed at increasing market share, price levels, percentage of market penetration, and so on.
(b) The *production* process will be aimed at objectives that specify the production and stock volumes, production cost levels, labor productivity, average production capacity employed, degree of flexibility of facilities and machines, degree of elasticity of machinery and equipment, replacement times, and so on.
(c) The *process for* purchases will be directed at specifying the volumes and rates of supplies, the minimum number of suppliers, the degree of independence toward certain markets, and so on.
(d) The *financing process* will aim at achieving objectives that determine the sources and cost of financing, the investment recovery period, and so on.

Thus, a "pyramid" of corporate objectives is formed, which is summarized in Fig. 9.22. In order to achieve these objectives, the entrepreneurial transformation must draw up a *strategy* which the managerial transformation translates into annual *budgets* to be met by the *micro management control.*

9.7 Operational Control of the Economic Transformation: From Cost Measurement to Cost Control

The *Economic Transformation* Control Systems in Fig. 9.11, referred to as TR2-E: Economic Transformation, are designed to control EBIT, or Operating Income, through the control of costs and revenues, according to models similar to those presented in Sect. 9.6, Figs. 9.17, 9.18, 9.19 and 9.20. This *section* considers the logic of cost determination and control. To *control costs,* rational *cost measurement* procedures must be adopted to obtain the "cost data" necessary for *cost control* (for more on this, see Mella 1997b, Chaps. 17, 18 and 19).

The *production cost*, C(P), of production "P" in the amount "QP" is defined as the sum of the costs of the "*N*" factors, $C(F_n, P)$, which are "consumed productively" to obtain QP:

$$C(P)=\sum_{n} C(F_n,P),\quad 1 \leq n \leq N \tag{9.13}$$

The production "QP" is called an *object of cost*; the costs of factors F_n, $1 \leq n \leq N$, used to obtain it—symbolized by $C(F_n, P)$—are called *costs elements* or *elementary costs.* If the costs of two (or more) productions, "QP" and "QP_2," are calculated at the same time, then the factors must be related to one or the other of the productions.

maximum EVF and maximum ROE

or, more realistically, a minimum level of profitability, ROE*, and a desired growth rate

objectives for ROE, ROI, ROD, DER, growth rate, afin

specified as

objectives for REVENUE (QP and Pp), VALUE ADDED, COST, MARKET SHARE, etc.

further broken down into objectives for

PURCHASES	PRODUCTION	FINANCE	SALES
optimize purchases and stock EOQ logistical optimization	increased return from reducing factors of production and production costs increased use of production capacity flexibility of machinery productivity and reduction of waste, etc. optimal production lot, etc.	reduction of financing requirements Debts Mix reduction of ROD increase of CIR optimization of DER	price increase/reduction increased market share and customer loyalty increased sales and customer satisfaction customer diversification, etc.

Fig. 9.22 The "pyramid" of corporate objectives

If there are "M" productions, then by extending the symbolism we obtain:

$$\begin{aligned} &C(P_1) = \sum_n C(F_n, P_1), \\ &C(P_2) = \sum_n C(F_n, P_2), \\ &\ldots \\ &C(P_m) = \sum_n C(F_n, P_m), \quad 1 \le m \le M. \end{aligned} \tag{9.14}$$

Cost calculations for the "M" productions are made easier by constructing a "*cost matrix*," or even an *F/P matrix*, to indicate that it is the matrix that correlates the F_n with the QP_m. The row indices in the matrix are the "N" factors and the column indices are the "M" productions. The cell corresponding to the intersection of each row with each column contains the factor costs, $C(F_n, P_m)$, as shown in Fig. 9.23.

An *important observation*: the production costs of the mth product must always be related to a *volume of production obtained* in a specified period "T," $QP_m[T]$. The average unit cost is divided by dividing the total cost by the production quantity in question:

$$c(P_m) = \frac{\sum_n C(F_n, P_m)}{QP_m[T]}, \quad 1 \le m \le M. \tag{9.15}$$

Expressions such as "The cost of an iMac is *X*," or "The cost of a ton of steel is *Y*" are meaningful only with reference to the quantity QP[*T*] of iMacs or steel to be produced (budgeted costs) or the quantity of products (final costs) in period *T*.

The construction of the *F/P matrix* requires that the following be specified precisely:

- The period "T" of reference for the calculation and the control
- The quantity of products relating to that period: $QP_m[T]$
- The value of the factors that must be entered in each cell of the matrix for each production: $C(F_n, P_m)$

These three elements define the *fundamental problem* of *cost measurement:* to "better correlate" the elementary costs of factors with the different productions included in the cost matrix in Fig. 9.23. To do so, the cost elements can be classified as follows:

A. *Special*, *direct*, or even (normally) *variable costs*: these relate to *operational factors*—such as materials, components, accessories, external production services, direct labor—which are purchased and used for individual units and are specifically attributable to a single cost object. The costs in this category can, therefore, be calculated and directly entered in each box corresponding to the production for which they were used; they are also referred to as *traceable costs* since they can be observed directly in the finished product. They are calculated by *multiplying three elements*:

   (i) The unitary factor requirement, qF_i
   (ii) The price, pF_i
   (iii) The quantity of production, QP_j

Products Factors	QP_1		QP_m		QP_M	totali
F_1						
F_n			$C(F_n, P_m)$			$CF_n = \Sigma_m C(F_n, P_m)$
F_N						
Totals			$C(P_m) = \Sigma_n (C(F_n, P_m)$			

Fig. 9.23 The cost matrix—F/P matrix

In other words:

$$C\left(F_i,P_j\right)=qF_i\times pF_i\times \text{QP}_j; \tag{9.16}$$

B. Costs that are common to different productions, indirect costs, or even (normally) fixed costs; these are costs incurred mainly due to structural factors, such as industrial buildings, machinery, plants, means of transport, furniture, patents, surveillance services, maintenance, etc., which by nature are usually used in common for multiple productions. For this reason, such costs must be "attributed" (divided, imputed) to the various QP_m; that is, divided into distribution shares, determined by referring to specific distribution drivers (or bases for imputation). If "Fk" is a common factor in space and/or time, used jointly for the productions "QP_m" and "$\text{QP}_{\text{m}+1}$" and with a common fixed cost equal to "$\text{C}(\text{F}_\text{k})$," it is necessary to divide this cost into two shares for separate entry in the corresponding cells of the two cost objects. The relation that connects the $\text{C}(\text{F}_\text{k})$ to the two cost objects has the following form:

$$\begin{aligned}
&\text{C}\left(\text{F}_\text{k}\right)=\text{C}\left(\text{F}_\text{k},\text{P}_\text{m}\right)+\text{C}\left(\text{F}_\text{k},\text{P}_{\text{m}+1}\right)\\
&\text{C}\left(\text{F}_\text{k},\text{P}_\text{m}\right)=\text{k}_\text{m}\text{C}\left(\text{F}_\text{k}\right)\\
&\text{C}\left(\text{F}_\text{k},\text{P}_{\text{m}+1}\right)=\text{k}_{\text{m}+1}\text{C}\left(\text{F}_\text{k}\right)\\
&\text{cP}_\text{m}=\text{C}\left(\text{F}_\text{k},\text{P}_\text{m}\right)/\text{C}\left(\text{F}_\text{k}\right)\\
&\text{cP}_{\text{m}+1}=\text{C}\left(\text{F}_\text{k},\text{P}_{\text{m}+1}\right)/\text{C}\left(\text{F}_\text{k}\right)
\end{aligned} \tag{9.17}$$

where $C(F_k, P_m)$ and $C(F_k, P_{m+1})$ are the shares of the common cost attributed to P_m and P_{m+1}, according to the *distribution coefficients* "k_m" and "k_{m+1}" applied to the common cost "$C(F_k)$" and determined using appropriate *cost drivers* (or *resource drivers*). If B indicates a chosen *cost driver* for the allocation of the common cost, and if we assume that B_m and B_{m+1} are the values of the *driver* for the two productions, then:

$$\begin{aligned}
k_m &= \frac{B_m}{B_m+B_{m+1}}\\
k_{m+1} &= \frac{B_{m+1}}{B_m+B_{m+1}}
\end{aligned} \tag{9.18}$$

There can be multiple *cost drivers*, whose choice can influence the distribution of cost. In general, they can be divided into two large categories:

(a) Physical or technical drivers: these are quantified as units of technical measures (hours of work, surfaces, number of pieces produced, km. traveled, weight, etc.).
(b) Value drivers or economic bases: these refer to monetary quantities (costs, revenues, added value, various economic margins).

Drivers can be used one at a time (*single-based* distribution) or jointly (*multiple-based* distribution). You can use the same *cost driver* for all indirect costs (*uniform-based* imputation), or different drivers for different indirect costs (*varied-based* imputation). In multi-product enterprises, characterized by complex production processes divided into well-distinguishable phases, *varied-based imputation* is normally applied, according to which the *drivers* are individually identified for each fixed cost to express the *causal relationship* between the drivers and the costs to be distributed (see example in Sect. 9.9.2).

Note: the *classification* between *direct* and *indirect* costs is never absolute; that is, there is no entirely *direct* or *indirect* cost due to the physical-technical nature of the factor; the classification depends on the *size of the cost object* being considered and on the *possibilities of correlating the costs with it.* Figure 9.33 demonstrates how an F/P *matrix* can be constructed with reference to a small leather business. The *cost measurement* system described so far is not the only one possible. In medium-large enterprises, two other methods, which will only be mentioned briefly, are normally employed (Mella 1997b; Goodwin et al. 2001):

1. The system that involves the localization of costs in operational or nominal cost centers, appropriately identified as, for example, departments, processing centers, service centers, which represent the cause of the consumption of factors, and therefore the elementary costs; at a subsequent stage, the costs of the centers are *transferred* to the different terminal productions according to the diagram in Fig. 9.24, which highlights that three categories of cost drivers are required for the *localization* procedure:
   (a) *Resource drivers*, which guide the allocation of factor costs to terminal productions, to products—if they are direct costs or *traceable costs*—as well as to the cost centers.
   (b) *Activity drivers*, which guide the allocation of costs among the centers; these generally refer to services provided by one center to another.
   (c) *Product drivers,* which guide the allocation to the finished products of costs located in the direct production centers when all the "cost transfers" have been completed.
2. The system known as *Activity Based Costing*: this costing system assumes that the costs of factors—excluding the variable direct costs relating to production—should be attributed to the *production activities*, direct and indirect, appropriately identified, which are the cause of the consumption of the factors, and therefore of the elementary costs (Cooper 1989; Cooper and Kaplan 1991). The costs of the activities are then attributed to the various terminal productions, using appropriate techniques (Moisello 2012).

Cost measurement is the premise for *cost management,* that is, the *cost control* process implemented to arrive at *rational decisions* regarding the following four fundamental problems, the solution to which will determine the efficiency of the economic transformation:

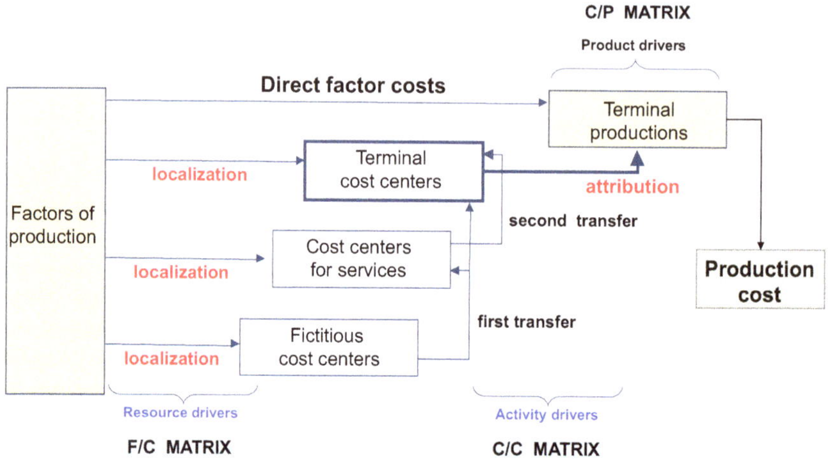

Fig. 9.24 General diagram of the localization procedure

1. *Whether to accept a sale price* proposed by a client; let us assume that a given client requests a batch of "QP = 2,000" units of a given product with specific features, for which he is willing to pay a maximum price of "$pP = 100$" per unit. Let us also assume that the economic transformation determines the average production cost of the "QP = 2,000" units, equal to $cP = 120$ per unit. In this case, since "$pP < cP$," the company will not begin production and will not satisfy the client's request, or it will make a counter-offer. Therefore, managing costs entails accurately analyzing the product and the production process to determine the maximum cost, which is set as the target (Sakurai 1989). This delicate control process comes under the broad concept of *target costing*, which is illustrated in the diagram in Fig. 9.25. An example is provided in Sect. 9.9.3.
2. *What price to set,* according to a process similar to the one in Fig. 9.26, which is applied, for example, when a company wants to produce bags of a given type and has to decide on the sales price, having set as a process goal a *return on cost* of roc = 30% (*cost-plus-pricing*). To set the sales price using this technique, it is necessary to quantify the *average unit cost* and then use this to derive the sales *price*, taking into account taxes. In reality, the *economic control management* sets the selling price, considering not only the *price-cost-quantity* relation but also the *price-demand* and *price-cost-quantity relations*, since a high price increases the risks of lower demand, thereby encouraging competition.
3. *Whether to produce internally or to purchase from an outside supplier*; a large car company buys the engines for its cars from an external supplier at price pM; since a large number of engines are purchased, the problem arises as to whether it is convenient to produce the engines internally rather than continuing to have them supplied from outside (in fact, the price paid includes the supplier's profit); the manager of the economic transformation must then calculate the (average) costs of production. If these are lower than the price paid to the supplier, it is

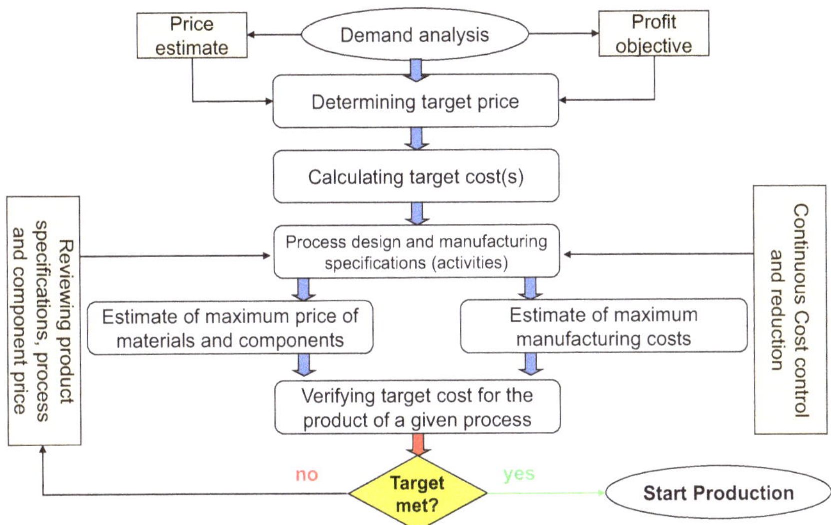

Fig. 9.25 Cost measurement and cost control in setting a target cost

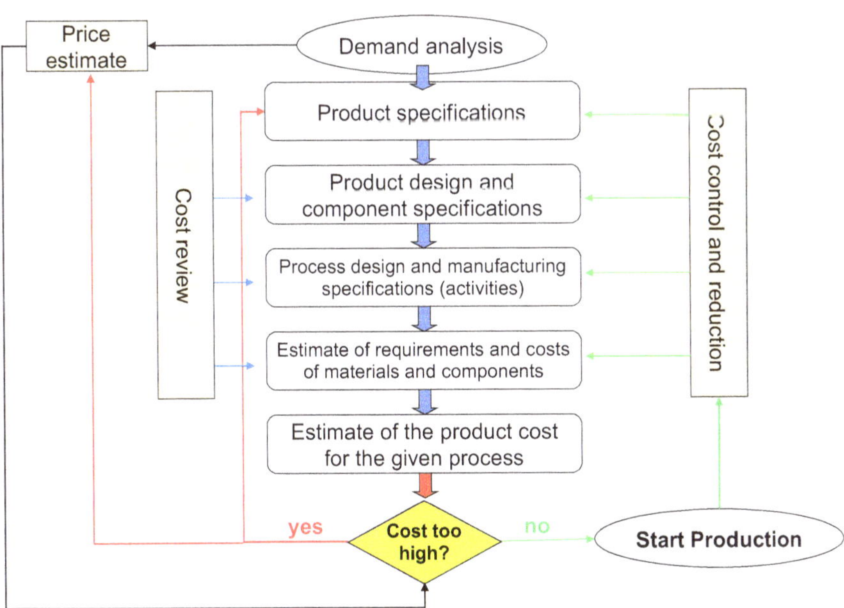

Fig. 9.26 Cost measurement and cost control to determine price

economically convenient for the company to produce its engines rather than to continue purchasing them from the supplier. Conversely, the same problem arises in deciding whether to internally produce a given component or to *outsource* its production (Pellicelli 2017).

4. How to *rationally correlate prices and costs* to choose the *production mix* that maximizes corporate RO. Specifically, how to check the cost-effectiveness of processes or products, choosing those processes to discontinue because they are cost-inefficient and which to reinforce (Moisello and Mella, 2020). An example is given in Sect. 11.20.3.
5. Other specific goals:

   (a) How to determine EOQ and EPQ for stock control (Sects. 11.6, 11.7, 11.8 and 11.9).
   (b) What *expected costs* are needed to prepare plans, programs, and budgets (Sect. 9.9.1). These costs are determined through an accurate forecast of supply price trends under certain assumptions regarding the desired structure of the product and the process phases.
   (c) How to *optimize the projects* in terms of time and cost of implementation (Sect. 11.18).
   (d) How to control the costs of the processes through a Continuous Control System whose aim is to keep *actual costs* below *standard* costs, that is, costs-objective determined ex ante, assuming specific levels of process efficiency. The latter are compared with the former, which are determined ex post, to detect any deviation, in which case corrective levers will be employed.

9.8 The Role of Planning in Transforming the Organization into a Cognitive, Intelligent, and Explorative System

The model in Fig. 9.27, which is derived from Fig. 9.13, highlights how *managerial control* translates multi-year strategic plans and programs into *operational programs*—and into *budgets*, in particular—that cover a one-year time horizon divided into smaller sub-periods. The *budget* thus acquires the significance of a system of short-term operational objectives, such as "path to follow" and "results to be achieved," or of a system of *constraints* assigned to the management of the operating units, which are variously organized and carry out the "technical" transformations. *Operational reporting* continuously tracks the trajectory, results, and performance of the operations management and, by comparing these with the budget, determines the deviations which provide management with information for operational decisions regarding the allocation of resources to the operating units and their processes.

In the practical application of *budget control*, the *budget summary*, although unitary, is broken down at a minimum into analytical *budgets* representing "segments" related to *functions*, *operating units*, and their *responsibilities*, *product lines*, and *operating areas*, defined, respectively, as functional *budgets*, *budget* of responsibilities, *budget* by *product line*, and budget by *geographic area*. Each analytical budget defines the objectives to be achieved in the different segments. Therefore, Fig. 9.27

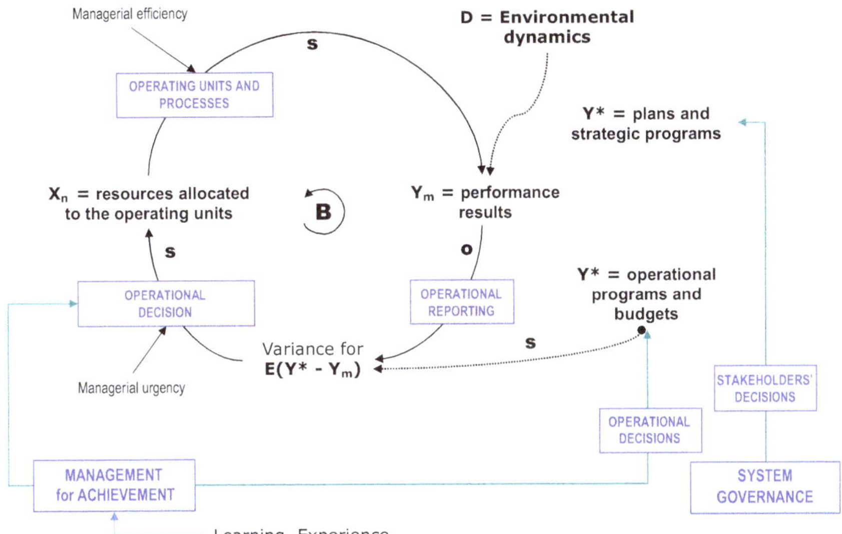

Fig. 9.27 General model of operational managerial control in production organizations

must be understood as the general model of a multiplicity of *analytical* control systems, each of which must aim to achieve *particular objective values* specified for each *segment* of the budget, separated into *function* (functional budgets for sales, production, procurement, etc.), *area* (geographic budget), *operating center* (center or department budget: printing, milling, painting, etc.), *organ* (responsibility budget assigned to the manager, official, employee, department head, etc.), *period* (period budget: week, month, quarter, etc.), and *product.*

In this sense, budget control requires activating a true *system* of Control Systems, each of which is coordinated with the others and has specific objectives that have been assigned to it; and, as a system, achieving the planned general objectives of economic, production, and financial efficiency (Anthony 1965).

From the above logic, it is clear that the purpose of *strategic business control* (multi-year plans and programs) and of *operational managerial control* (operational plans and budgets) is the same: to decide the specific *direction* the company will take in the *production, economic,* and *financial spheres* to achieve its objectives.

The plans, programs, and budget are, in effect, documents that, although they entail different degrees of analysis, detail the intra-period values (monthly, quarterly, and monthly in the budget) that must derive from the processes carried out by the functions and operating units (revenues, supply costs, labor costs, investment, cash flow, etc., although at different levels of analysis) necessary for the organization (company) to achieve the objectives and the performance of the various transformations indicated in the MOEST: EVF, monetary equilibrium, *roi, roe,* quality, productivity, etc.

The values indicated in the plans, programs, and budgets are therefore the formal representation of the *planned direction (trajectory)* in the production, economic, and financial spheres, defined in the different sub-periods of reference, required to achieve the objectives and the performance goals. Intra-period values represent *path* objectives, "Y_t*" (tracking objective), that the Control Systems must maintain over time by eliminating deviations, in order to achieve the desire outcomes and performance goals, Y*. If the environment were to be widely and suddenly disrupted, bringing considerable "disturbances" to the control systems, the objectives could become unrealistic; therefore, plans, programs, and budgets must be subject to continuous and timely *reviews* so that they represent plausible and motivating objectives for Management Control Systems. Just as the budget needs to be accurate and continuously reviewed, the *reporting* must be accurate and timely; it would be of no use to know the February sales data in June, or the production data after 3 months.

Strategic entrepreneurial control (multi-year plans and programs) and *operational managerial control* (operational programs and budgets) have the same function and aim: to determine a specific *direction* for the firm in the *productive, economic, and financial space.* In fact, since they specify in detailed manner the operations selected to be carried out to achieve the values of the final objectives (revenues, supply costs, investment, cash flow, etc.), the multi-year plans, programs, and budgets (though at different levels of analysis) are the formal representation of the *programmed direction (trajectory) of the firm system in productive, economic, and financial space*, defined for the various sub-periods of reference.

However, these documents represent not only the fundamental instrument for the system's *direction* but also the indispensable condition for the *management control*, which seeks to maintain the trajectory in the established *direction* through specific control systems. In particular, only a well-structured budget with continual and timely revision along with multi-year plans will allow a control system for management to exist. For an outside observer, such a system turns the firm into an *automatic control system* whose *management*, represented by the entrepreneurial and managerial transformation (Fig. 9.11), is an integral part that "learns from its own errors" in a continual and mutual interaction with the external environment. Reporting plays a fundamental role, as it allows the firm to detect the variances in the values produced with respect to the budgeted values and to accurately analyze these to determine the causes of such variance and decide on the control levers to activate. Just as the *budget* must be precise and continually revised, the *reporting* must be precise and timely; it would be of no use knowing February's sales data in June or the production data after 3 months.

Feedback control, which is typically *reactive* and based on eliminating "system errors" from the programmed path, cannot, and must not, be the only type of control. *Cognitive transformations* (entrepreneurial and managerial, (Fig. 9.11), through an accurate control to forecast the fundamental variables, must achieve a *feedforward control* by taking *proactive* actions to *prevent* variances, accurately deciding on the input levels that will produce the programmed output. In any event, *feedforward* control assumes a planning process which is continually updated and controlled through *feedback*. Planning and the budget are the *result* of the logic of *feedforward* control and represent the *instrument* for *feedback* control.

Planning in general, and the *budget* in particular, have other relevant functions, first and foremost as *instruments of cooperation and coordination* among the members, organs, functions, and processes of the organization, as foreseen by Beer's VSM (SS2) (Fig. 9.9) and by the MOEST (TR4-M) (Fig. 9.11). The budget is a tool for *coordination* since it is composed of functional budgets, which are divided into analytical budgets, all of which are coordinated by the economic calculation to produce a coherent set of values capable of verifying whether or not the objectives have been achieved. As a result, it is also an instrument of *cooperation*, since the achievement of the individual budgets by those in charge of them automatically entails the achievement of the final objectives of the firm.

The budget is therefore a basic *performance management instrument* since it allows the firm to control that the personnel and organs in the organization actually carry out their *responsibilities* and *tasks* in a coordinated and cooperative manner in order to achieve the various sections of the budget. However, for performance control the firm must be divided into *responsibility centers*, each of which with its own objectives to achieve, and include a system of incentives to ensure that personnel follow established levels of organizational efficiency. This form of control of responsibilities is also known as *task control*, which Robert Anthony defines as follows: "*Task control is the process of ensuring that specific tasks are carried out effectively and efficiently*" (Anthony 1988, p. 12).

To restore cooperation and coordination various control levels can be activated: from participating in decisions to a rewards/punishment system; from creating an efficient information system to improving working conditions; and from worker participation in the firm's capital to forms of identification of individuals in the mission of the organization they belong to. In order to carry out the management, coordination, cooperation, and organizational control functions the plans, programs, and budgets must not be the result of a simple bureaucratic process directed from the top (cognitive transformation) down (technical transformations) but rather of a democratic participatory process, from bottom up, so that decisions are accepted by the entire organization, since they are formalized through the consensual participation of the operational organs.

Faced with environmental change—whose visible features are the increasing speed of the production of "innovation"; the increasingly larger geographical areas over which the various organizations interact; and the more crowded network of interorganizational and intraorganizational relations that make it more difficult to interpret the overall behavior of each system—the capacity to achieve *teleonomic behavior* to ensure long-term survival depends on the ability of organizations and managers to understand change and to regenerate the cognitive processes, thereby producing a long-lasting autopoietic behavior.

If the firm, through the entrepreneurial and managerial control systems, learns from its errors, then the *plans, programs, and budgets become powerful tools of organizational learning*, since they will be based on:

- A careful analysis and evaluation of stakeholder expectations
- A process leading to a deep awareness of the organizational structure

- An accurate search for information, generally concerning future developments and forecasts regarding the environmental context, operating structure, and contingent situations
- A precise recognition of the external environmental constraints and the internal organizational ones
- The specification of the fundamental variables of interaction with the environment

As a result, the *planning* and *budgeting processes* move the firm to specify a system of dynamic objectives regarding productivity, economic efficiency, and profitability that guide management operations based on a typical *pull logic*; moreover, based on an equally typical *push logic* these processes help to gradually improve performance to permit the achievement of those objectives through performance levels continually adapted to a changing environment.

> The repertoire of tools that top managers use… is determined by their experiences. A dominant logic can be seen as resulting from the reinforcement that results from doing the right things … they are positively reinforced by economic success (Prahalad and Bettis 1986, pp. 490–491).

In this sense the activation of strategic and operational control systems represents the fundamental moment of reflection of the entire organizational activity: a continual critical rethinking of the mission, competitive position, and decisions to be made. Thanks to participation and sharing in discussing the underlying hypotheses of the plans and programs that set out the control objectives and create the conditions for coordination and cooperation, the entire organization becomes involved in a deep, collective learning process regarding the internal processes, constraints, limits, potentialities, and functional relations with the other processes and with the environment. If carried out with broad-based participation and motivation, this process will play a fundamental role regarding knowledge and result in the improvement at all levels of the organizational structure, leading the firm to behave according to the logic of *learning organizations*, which was briefly discussed in Chap. 1. From an instrument of *modeling to predict and control*, planning becomes a process of *modeling to learn* (Arie de Geus 1988, 1992):

> Institutional learning is much more difficult than individual learning. The high level of thinking among individual managers in most companies is admirable. And yet, the level of thinking that goes on in the management teams of most companies is considerably below the individual managers' capacities. […] Because high-level, effective, and continuous institutional learning and ensuing corporate change are the prerequisites for corporate success […] the critical question becomes, "Can we accelerate institutional learning?" (de Geus 1988, p. 70).

> The only relevant learning in a company is the learning done by those people who have the power to act (at Shell, the operating company management teams). So the real purpose of effective planning is not to make plans but to change the microcosm, the mental models that these decision makers carry in their heads. And this is what we at Shell and others elsewhere try to do (*ibidem*, p. 71).

> Our exploration into this area is not a luxury. We understand that the only competitive advantage the company of the future will have is its managers' ability to learn faster than their competitors. So the companies that succeed will be those that continually nudge their managers towards revising their views of the world. The challenges for the planner are considerable. So are the rewards (*ibidem*, p. 74).

As a system of efficient transformation (MOEST) the organization becomes in all respects a *cognitive*, *intelligent*, and *rational* economic *agent* with the capacity to *control* its own structure, processes, and dynamics in order to achieve ever-higher levels of efficiency. It is an *economic agent* because the organization–enterprise designs and follows a unique trajectory in the *productive*, *economic*, and *financial* space in which it operates, even though it is characterized by a multitude of organs, processes, and businesses, as VBM has shown.

The organization is an *intelligent cognitive agent* in that, as we have seen (SS4 and SS5 of the VSM, and TR5-I of the MOEST), the organization engages in a *cognitive* activity aimed at providing meaning to environmental stimuli, translating these into information and, through programing, structuring this into knowledge, adapting to the changing environment while maintaining its identity through a continual autopoietic process, as prescribed by the autopoietic approach. It is a *rational agent* in the sense that *cognitive* activity must lead to maximizing the *efficiency* of the vital transformations by searching for the best productive, economic, and financial performances (Fig. 9.11).

However, we can also consider that the organization–enterprise, being a *rational, cognitive agent*, is also an *explorative agent* (Kauffman and Levin 1987; March 1991; Lewin et al. 1999; Mella 2006) that, in continually searching to improve performance in any way possible (*rational* agent), *explores* its territorial environment which may be divided into *areas of interest* (continents, states, regions, provinces, etc.) and "moves" toward those *areas of greater attractiveness* (Scott and Bruce 1994; Gephart et al. 1996).

Let us assume that the organization–enterprise, based on its knowledge (*cognitive, intelligent* agent), can divide the "territory that can be reached through its processes" into significant areas and that for each of these it can determine the value of the performance indicators it wishes to maximize. In this way the "territory that can be reached" (for each area, and subarea, in which it is divided) is characterized by a *function of attractiveness* which indicates the average level of the performance indicators held to be significant, thereby forming an *attractiveness landscape* indicating which areas are most attractive and which less so in relation to the various performance indicators that were chosen. Based on the characteristics of each area, it is possible that the *attractiveness landscape* may reveal "valleys" of moderate attractiveness, "peaks" of high attractiveness, or undesirable "pits" (no attractiveness whatsoever) to be avoided.

Therefore, continuing to assume that we wish to choose *roe* and *roi* and their components (Sect. 9.6) as performance indicators, it is likely that an area rich in potential consumers and with few competitors will be highly attractive due to potentially high revenues, from the side of both quantity and price, and thus offer a high *roi*. On the other hand, an area with many competitors would not be very attractive since, because of competition in terms of price and quality, we can assume a lower *roi*. An area with a low tax inflow would, however, have a higher *roi* compared to the one with higher inflows, with all other factors held constant. An area with heavy pedestrian traffic could favor sales for a small retail firm, while the one with ample parking facilities could increase the economic and financial performance of a large retail trader. Note that, to an *explorative* agent, the areas of attractiveness should not be solely conceived of as areas in a geographical territory but also as areas in the economic-financial environment, which is made up of businesses, markets, distribution channels, technology, and legislation. The assumption that the organization–enterprise is a *rational* agent leads to the following *optimal strategy*:

1. Explore all accessible territories and areas that can be reached by its productive, economic, and financial processes.
2. Determine the most important factors *of attractiveness* or *undesirability*, and evaluate these in economic terms by forming the *fitness landscape*.
3. Translate the fitness landscape into the *attractiveness landscape*.
4. Shape, update, and continually explore, looking toward the future, fitness, and at*tractiveness landscapes*.
5. Choose the area(s) with the greatest *attractiveness*, that is, with favorable conditions for an increase in *roe* and *roi* (e.g., the ease with which new businesses can arise, greater sales volumes, expectations for better prices and supply costs, greater productivity and public subsidies, high levels of social protection, stimulating environment, abundance of infrastructures, lower tax burden).
6. Avoid the "pits," and attain the highest "peaks" of attractiveness, that is, those areas where the best performance can be achieved.

The formation process for the *fitness landscape* and the *attractiveness landscape* can be summarized in the table in Fig. 9.28.

Column [1] indicates the different *areas*, Ai, considered interesting by the firm or by one of its businesses. Column [2] includes the *factors of attractiveness* that characterize each area in relation to the type of firm or business. The vector [a_{11}, a_{12}, a_{13}, …] represents the fitness landscape of area Ai. Column [3] determines the value of the chosen *performance indicator*, calculated as the mix of values from column [2] carried out by the entrepreneurial transformation after careful evaluation. Column [4] specifies, if possible, a *value of risk* for each area (certain if $r = 1$; impossible if $r = 0$). The *index of attractiveness* for each area is determined taking into account *P* and *r*, contemporaneously; the simplest way is by multiplying the two indicators.

Areas of a chosen territory: (Ai), i = 1, 2...	Index of attractiveness (aij),	Performance indicators (Pi) evaluated for each area based on the index of attractiveness	Indicators of performance risk, (ri) for each area	Attractiveness landscape
[1]	[2]	[3]	[4]	[5]
A1	[f11, f12, f13, ...]	P1 = mix of [f11, f12, f13, ...]	r1	AL1 = P1 r1
A2	[f21, f22, f23, ...]	P2 = mix of [f21, f22, f23, ...]	r2	AL2 = P2 r2
A3	[f31, f32, f33, ...]	P3 = mix of [f31, f32, f33, ...]	r3	AL3 = P3 r3
A4	[f41, f42, f43, ...]	P4 = mix of [f41, f42, f43, ...]	r4	AL4 = P4 r4
...	...	...	...	...

Fig. 9.28 Process to determine the attractiveness landscape

The vector $[P_i \times r_i]$, $i = 1, 2,\ldots,$ indicated in column [5], represents the *attractiveness landscape*, Al_i, for the areas in the chosen territory.

Once again, even when viewed as an *explorative* agent, the organization–enterprise represents a control system, since the objective of achieving a given performance forces the organization–enterprise to continue exploring until it identifies the areas in which the objective is attained. The control system can act in two ways: the firm *explores* the environment to *achieve* a given performance level, or it undertakes exploratory behavior to *improve* the previous performance levels. In the first case (achieving a given performance level) the control system can be defined as one of *achievement* and in the second (improving previous performances) as one of *improvement*. Systems of *achievement* can be represented by the *Ring* in Fig. 9.29, which is adapted from Fig. 9.13 or Fig. 9.14.

Systems of *improvement* are typical *systems of pursuit* (Sect. 3.4) whose new objective for time "t + 1" is the best performance achieved at time "t," to which management adds a plausible increment, taking into account competitors and limits imposed by the stakeholders, as shown in Fig. 9.30.

9.9 Complementary Material

9.9.1 The Programing and Budgeting Process: The Role of Forecasting

The programing process as a moment of Control In Sect. 9.8 we saw that the whole *macro control process* that enables the organization, understood as a unitary system, to achieve the levels of performance that guarantee the greatest chances

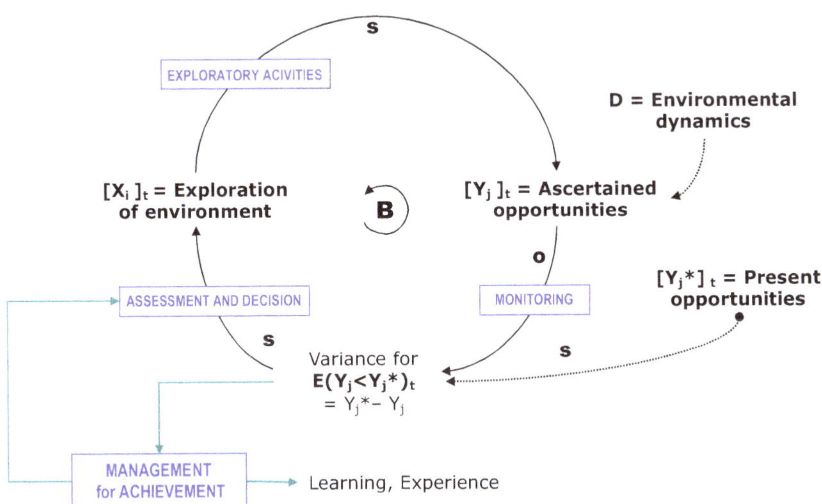

Fig. 9.29 Model of a firm as an exploratory system of achievement

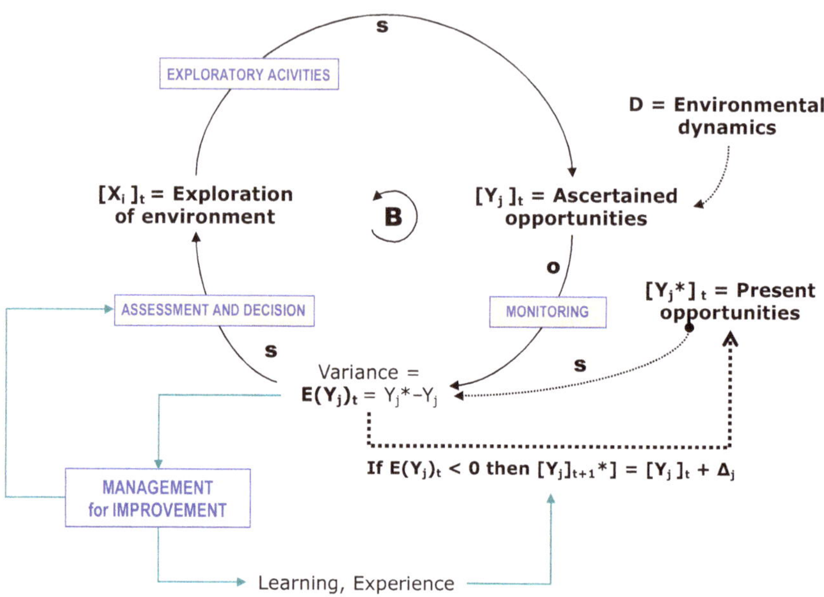

Fig. 9.30 Model of a firm as an exploratory system of improvement

of survival requires that long-term *plans,* short-term *programs,* and *budgets* be structured and their *values* become *objectives* that the control systems for the "technical-instrumental" transformation must achieve.

Multi-year (or long-term) *plans—strategic planning* in particular—typically cover a period of 3–5 years and even longer (depending on the reliability of long-term forecasts), and in a typical *feedforward control* these plans consider the effects the company's main strategic decisions and operations will have. They are designed to produce and maintain sufficient profitability to ensure continued growth in EVF; in fact, they consider the trends in the values of strategic, long-term variables, how to achieve a given market penetration, obtain a certain volume of turnover, reach the break-even point in a new production process within a defined number of years, modifying the range of products, activating a new distribution channel, changing the structure of the production processes, how to achieve an increase in productivity, an improvement in quality, a change in the organizational structure, and so on. Strategic plans must take into account the *conditions* of the external environment (markets, economic situation, etc.) and of the internal environment (available production capacities, input constraints, individual aspirations, organizational climate, etc.), as well as the strengths and weaknesses of competing companies, as part of a continuous effort to strengthen the company's position to meet the threats and opportunities of the external environment.

BUDGETING *Annual* (operational, financial statements) *programs* represent a segment of multi-year plans; they only cover *a single year* and provide detailed information for intra-annual periods (quarters and/or months). The annual program specifies the entire value system in terms of both the economic transformation (revenues, factor costs, depreciation, etc.) and the financial transformation (cash flow dynamics), and it sets short-term production and market performance targets. It determines sales for the subsequent 12 months, the quantity to produce, the material and labor requirements, the quantity to keep in stock, as well as the deferral periods to customers, the expected funding requirements to meet any cash deficits, and the forms of coverage. *Functional budgets* represent a detailed picture of the annual programs; taking into account the internal and external constraints, they specify, for each organizational function, operations manager, corporate "area" or even business, the objectives to be achieved in the different sub-periods throughout the year: sales revenues, the factor requirements and costs, and the investment and financing necessary to achieve the profitability and economic goals. The process of structuring the budget system involves the phases presented in Table 9.1.

Table 9.1 The planning and budgeting process

Phases of control	Activities in the various phases
1. Set the objectives	*roe*, *roi*, *roc*, P, pP, F, pF, value added, EBITDA, operating income, financial charges, etc., for each "subdivision" in the budget: intra-annual periods, functions, areas, production processes, products market segment, operating and responsibility centers, center operators, etc.
2. Search for information	This concerns all the internal organizational variables, quantitative and qualitative (economic, political, social, cultural, etc.) that constitute the state of the system.
3. Formulate a system of forecasts	This can concern the economic and operational context. Environmental scenarios are developed that predict how the system might evolve.
4. Identify constraints	Constraints can, for example, be technical, economic, financial, socio-political, managerial in nature.
5. Make the operating decisions	Rational decisions must be made regarding sales volumes and prices, factor requirements and purchase prices, unit costs, standard costs, financial transfers, levels of production efficiency (levels of productivity and quality), etc.
6. Formulate a hypothetical budget	An initial *hypothetical budget* is produced, which is checked for coherence and sustainability regarding the information at hand, the forecasts, and the constraints. Subsequent simulations lead to the choice of the *final budget*.
7. Assign responsibility for implementing the budget	The functional budget is divided into responsibility budgets for each center, organ, and individual in charge (effectors of the control system).
8. Implement the budget	The processes and operations are gradually implemented as indicated in the budget, in terms of quantity, quality, time, and values.
9. Prepare systematic reporting	*Reports* are prepared at the predetermined intervals to determine whether there is any variance or deviation between the planned and final situations.
10. Undertake analysis and control	The managers of the various Control Systems determine the *levers needed to eliminate any variance.*
11. Carry out continuous reviews	The budget is periodically *reformulated* at more frequent intervals to take into account current conditions.

The forecast One of the most important phases of the planning and budgeting process (phase 3) is the development of a consistent system to forecast the results of future operations. Most of the information needed to make rational decisions for defining plans, programs, and budgets is, in fact, information about the future dynamics of business events, obtained through *forecasting* processes.

> Forecasting methods, as defined here, are explicit procedures for translating information about the environment and the company's proposed strategy into statements about future results. What would be the results if the environment were favorable and we did A? What if it were unfavorable and we did A? What if it were unfavorable ap. nd we did B? (Armstrong 1983, p. 13).

Prediction is an intellectual activity through which we try to "make known beforehand" (from the Latin *praedicere*) a future event or the future unfolding of a phenomenon already in progress. Forecast does not mean *divination* or *clairvoyance.* The forecast is based on a knowledge of historical dynamics and scientific laws that, taking into account the current situation, allow us to infer the future dynamics of current phenomena. All other conditions equal, the best results will be achieved by the company that can formulate the most accurate and reliable forecasts, aided by all the statistical and simulation tools available (Chisholm et al. 1971; Hyndman and Athanasopoulos 2020). This ability depends on the theoretical and practical preparation of the people responsible for the forecasts and on the mass of useful information that can be collected and arranged. However, a forecast is not the same as planning; it is one aspect of planning, but between a *forecast* and a *plan* there must a *decision* to modify the forecasted future to obtain the desired values. Let us assume a firm predicts a 15% fall in global demand for its products. It may then set as an objective not losing more than 5% of demand, based on which it decides on and plans the appropriate levers. For example, it could develop an advertising campaign, review sales prices, change its packaging to make the product appear to have a higher functional quality (Sect. 10.3). To this end, it must decide on and plan the proper advertising means, the length of the campaign, the price reduction, the purchase of new machines for packaging, and so on.

The forecasts required to prepare the budget depend on the objects it refers to, which can be divided into three categories.

1. *Context forecasts:* these represent the fundamental forecasts about the environment the company will have to operate in; the forecasts refer to the dynamics and constraints regarding the fundamental variables (which often cannot be modified) presented by the internal and external environment.
2. *Operational forecasts*: these concern the trends in the variables that refer to the company, its processes, and its critical issues.
3. *Contingent forecasts*: these relate to the variables that define the particular operational "context" of the organization. This category includes, for example, forecasts about the average number of workers who will be active in the next month, the number of orders to be fulfilled and those that will arrive in the upcoming days, stock market listings, the price of raw materials, the inflation rate for the next month, and so on.

Long-term forecasts (the discovery of new energy sources, the international political situation, new inventions, the possible use of recent patents, etc.) are needed above all for long-term planning. *Short-term forecasts* are particularly important for planning and for the budget. It is intuitively clear that the reliability of forecasts is usually inversely proportional to the length of the forecast period, even if long-term forecasts, based as they are on the knowledge of stable evolutionary laws, can at times offer reliable outcomes. A forecasting system that offers an advance picture of a given environment represents an "environmental scenario." The professionals involved must hypothesize different environmental scenarios that reflect possible alternative forecast values. Planning and budgeting must choose the most likely scenario.

The structure of the budget in manufacturing firms As we know, a budget is a unitary document; however, structurally speaking, it is made up of a network of *functional* (or operational) budgets that relate to the individual "operational functions" of the "technical-instrumental" transformations of the organization–enterprise, which are linked together in the sequence indicated in Fig. 9.31. A brief description of the content of functional budgets is given in Table 9.2. As shown in the figure, in *market-oriented* companies, normally the *sales budget* serves as a guide for the other budgets. There are, however, particular situations in which the guiding budget for the formation of the general system of budgets is either the *production budget* or the *investment budget.* This normally occurs in *production-oriented* organizations—especially those that provide public services—which, due to constraints regarding production expansion or supplies, must subordinate sales volumes to production quantities, or to viable investments.

The budget time horizon The future period "covered" by the budget is called the budget *time horizon;* this is usually 12 months long, but the functional budgets mostly entail shorter time periods (monthly budgets, quarterly, etc.) representing the production, economic, and financial trajectories that control systems must ensure. Normally, the total sales budget comes from the summary of the analytical budgets that indicate the quantities to sell, prices, and revenues, which are specific for product, geographic region, customer type, form of distribution channel, etc. Sales figures normally refer to intra-annual periods: January, February sales, etc.; first quarter, second quarter sales, etc. The values of the individual functional budgets are summarized in a summary document called the *general operating budget*, which takes the form of a real balance sheet—balance sheet and income statement—prepared 12 months in advance with respect to the management activities in question, and corresponding to the annual program. The content of the different functional budgets is summarized in Table 9.2.

Reporting The control process entails continuously comparing the values programmed in the budget and those "achieved" by the management processes. The comparison assumes that reports are prepared by those who have been assigned a liability budget containing the final data, a comparison with the budget data, and the

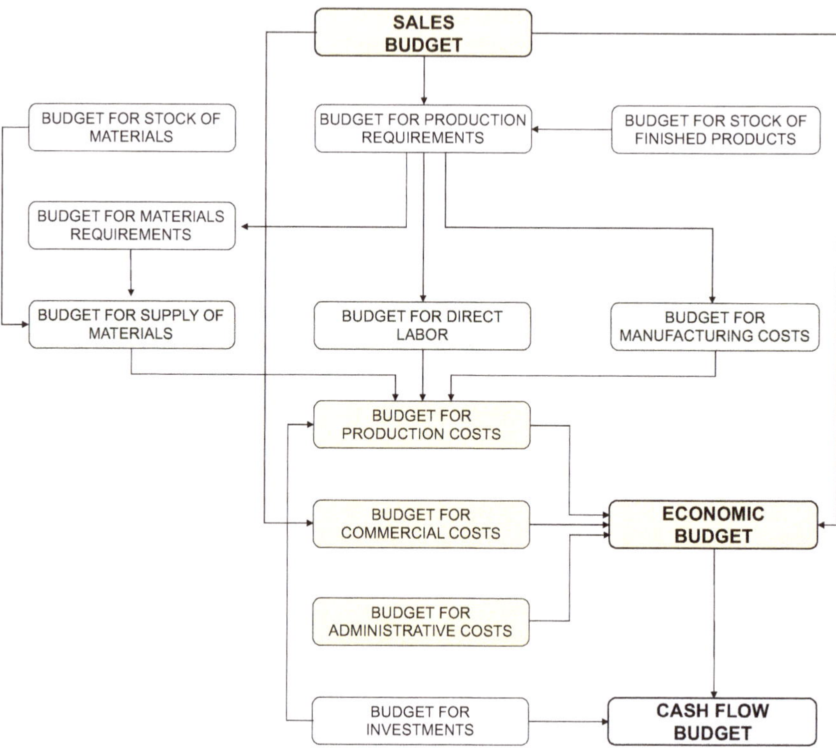

Fig. 9.31 The system of functional budgets

deviations (errors) uncovered, accompanied by an initial analysis to identify the causes. The *analytical reports* are grouped into one or more *summary reports* to obtain the synthetic data of the functional report, whose values are compared with those of the corresponding functional budget for that period (month, quarter, etc.). Therefore, assuming that a company has divided its production budget for two products into four factory budgets and that, for each factory, there are three division budgets and for each division six department budgets, the following number of reports could be drawn up:

- 72 department reports, to be drawn up by the department heads
- 12 division reports by the division managers
- 4 factory reports by the plant managers
- 2 product and factory reports by the production manager

The sliding report technique is often used. In addition to the data for the period in question—for example, February revenues—this report also indicates revenues for the preceding month, the last quarter, the last 12 months, February of a year ago, and any other summary data that allows the company to express a judgment on the extent to which the objectives have been achieved.

Table 9.2 The content of the functional budgets

Budget	Content
Sales	For each product, each geographic sales area, each type of client, this budget indicates the planned quantities month by month and the relative prices.
Stocks of materials, components, products	For each product, component, or material, this budget indicates the initial stocks and those planned for the end of the year, taking into account the pace of production.
Production	For each product, this budget calculates the quantity to produce month by month, taking into account the initial stocks and the final planned stocks.
Budget for materials requirements and direct services	For each component, material, and production service, this budget calculates the monthly requirements for each product, taking into account the production quantity.
Supplies	For each component, material, and production service, this budget determines the monthly supplies for each product, taking into account the market availability (seasonal) of the requirements and the stocks.
Direct labor	This budget determines the monthly direct labor (DL) requirements for each product, taking into account the production quantity.
Production cost (industrial costs)	This calculates the production cost as the sum of the costs of materials and services, the direct labor, and the depreciation of the multi-year costs of the structural factors.
Distribution costs (commercial costs)	This budget determines the commercial costs for supplies, royalties, the indirect costs of the seller, transport, and advertising.
Administrative costs (general costs)	This determines the general administrative costs: accounting, finance charges, rent, surveillance, and all other costs not pertaining to production or to product distribution.
Investment in structural factors	Taking into account the production volume, this budget determines the requirement regarding machine hours, plant services, and investment for new purchases.
Cash management and necessary financing	This budget identifies the cost-generated expenses and earnings generated by income; taking into account existing and reimbursed liquidity, debts, and credits, as well as the corresponding planned amounts; it also calculates incoming and outgoing cash flows for the next 12 months, highlighting any financial deficits and the financial sources to eliminate these.
Operating budget: economic and financial budgets	The various functional budgets represent items in the general operating budget, whose typical form is a budget drawn up 12 months in advance that calculates the efficiency indicators set as objectives: *roi, roe, roc*, etc.

9.9.2 Cost Measurement in a Small Leather Craft Firm

Figure 9.32 illustrates how production costs are estimated for two products manufactured by a small craft firm. The model can be generalized for any number of products and factors. As shown in the spreadsheet, taking the *total* of all the *elementary costs* —allocated directly or after their distribution —from the cells of *a single column* in the matrix (column 3 for Bags and column 4 for Shoes), we obtain the overall cost of production for each products, "QP_m"; in this example,

n [1]	F_n=Factors of Production [2]		P_1=BAGS [3]		P_2=SHOES [4]		Average and Total values [5]	
QP_m		units		1,000		2,000		
CM_1	Cost of Materials: **hides**			10,000		20,000		30,000
qM_1	Unit requirement	feet	4		4			
QM_1	Total requirement	feet	4,000		8,000		12,000	
pM_1	Purchase cost	euros	2.5		2.5		2.5	
cM_1	Unit cost	euros	10		10			
CM_2	Cost of Materials: **leather**			8,000		12,000		20,000
qM_2	Unit requirement	feet	2		1.5			
QM_2	Total requirement	feet	2,000		3,000		5,000	
pM_2	Purchase cost	euros	4		4		4	
cM_2	Unit cost	euros	8		6			
CL	Cost of **labor**			7,000		5,000		12,000
qL	Unit requirement	hours	3.5		1			
QL	Total requirement	hours	3,500		2,000		5,500	
pL	Purchase cost	euros	2		2.5		2.18	
cL	Unit cost	euros	7		2.5			
CS	Cost of **energy**			2,000		3,000		5,000
qS	Unit requirement	power	2		1.5			
QS	Total requirement	power	2,000		3,000			
pS	Purchase cost	euros	1		1			
cS	Unit cost	euros	2		1.5			
TOTAL DIRECT OPERATING COSTS				**27,000**		**40,000**		**67,000**
CI_1	Service cost: **transport means**	euros		10,000		10,000		20,000
d_1	Driver: items transported	units	500		500		1000	
$c_1$input	Allocation coefficient	euros	20		20			
$\%_1$input	Allocation percentage	%	50%		50%		100%	
CI_2	Machinery: **cutters**	euros		10,000		10,000		20,000
d_2	Driver: machine hours	hours	1,000		1,000		2000	
$c_2$input	Allocation coefficient	euros	10		10			
$\%_2$input	Allocation percentage	%	50%		50%		100%	
CI_3	Machinery: **sewing machines**	euros		6,000		4,000		10,000
d_3	Driver: machine hours	hours	1,200		800		2000	
$c_3$input	Allocation coefficient	euros	5		5			
$\%_3$input	Allocation percentage	%	60%		40%		100%	
TOTAL INDIRECT STRUCTURE COSTS				**26,000**		**24,000**		**50,000**
CP_m	**TOTAL PRODUCTION COST**	euros		**53,000**		**64,000**		**117,000**
cP_m	average unit cost	euros		53.0		32.0		
QP_m	Sales volume	units		1,000		2,000		
pP_m	Average prices	euros		55		40		
RP_m	**Total SALES REVENUE**	euros		**55,000**		**80,000**		**135,000**
OR_m	**Analytical results**	euros		**2,000**		**16,000**		**18,000**
ROC	Mark-up	%		3.8%		25.0%		15.4%
CPC	Percentage of cost	%		96.4%		80.0%		86.7%
E=1+ROC	Economic efficiency	Ratio		1.038		1.250		1.154

Fig. 9.32 Calculation of production cost using the F/P MATRIX

the production cost is a *total cost*, a *full cost*, and it refers to the quantities indicated in the second row, "QP_m":

$$CP_1(1000\,\text{Bags}) = 53{,}000$$
$$CP_2(2000\,\text{Shoes}) = 64{,}000.$$

Dividing the total production costs, "CP_m," by product quantity, "QP_m," the cost object, we obtain the *average unit costs*, indicated by cP_m:

$$cP_1(\text{Bags}) = 53.0$$
$$cP_2(\text{Shoes}) = 32.0$$

The "cP_m" represents fundamental information for the *business transformation* to control productions, decide on the product mix and the calculation of EPQ, and estimate the value of the products in stock. While for each of the operational factors: *hide, leather, labor,* and *energy* we indicate the values for pF_n and qF_n, which allow us to directly calculate the cost of each production process, for *cutters, sewing machines,* and *transport means*, which belong to the *structural factors* (which are fixed) category and provide their productive capacity to the production processes, the calculation is done by sharing the costs between the two productions. For these factors, we calculate the *cost of total capacity provided*, which is found in *column* [5], in correspondence to the green cells; the *cost drivers* the manager considers appropriate for expressing the *causal link* between the use of the factors and the production output are shown in lines d_1, d_2, and d_3.

To give a specific example, let us consider the factor *Sewing Machines*, whose capacity cost is CI_3 = 10,000 (column [5]), which supplies a total *functioning time* (driver) of d_3 = 2000 hours. We also assume that the total operating time of the machines is 1200 hours for Q_1 (the bags) = 1000 (many stitches) and only 800 hours for Q_2 (shoes) = 2000 pairs (few stitches). To distribute the cost CI_3 = 10,000, it is useful to calculate the *cost-sharing coefficients* (indicated in the figure as *%input*) as percentages of the hours of use for each product (1200 and 800, respectively), compared to the total hours supplied (2000):

$$\%\text{allocation}(\text{Bags}) = 60\% = (1200/2000)$$
$$\%\text{allocation}(\text{Shoes}) = 40\%$$

By applying these coefficients to the total cost of the sewing machines, CI_3 = 10,000, we get the allocation shares: CI_3(Bags) = 6,000; CI_3(Shoes) = 4,000.

The F/P *matrix* is also a useful instrument to calculate the analytical economic results by subtracting the total costs of each production from the sales revenue. This gives us the values in row OR_m.

9.9.3 Calculation and Control of Costs to Decide Whether or Not to Accept a Given Price

I now present a simple case to describe the cost *control* process to decide whether or not to accept a price proposed by the customer. The commercial manager of an Italian company specializing in the production of molded plastic products had been contacted by the procurement manager of a large supermarket chain regarding a supply of 200,000 units of a particular utensil. He was sent a sample and offered a price of $12 for *delivery ex warehouse buyer*. The sole administrator of the company summoned the *plant manager* and asked him to calculate the production costs (in euros) to verify whether the price was economically convenient. The plant manager examined the sample, which was a white, plastic tray with non-slip grooves for dishes that came in three different colors; nothing special, except for one detail: the tray had an upside-down V-shape and was completely perforated so that it could be inserted directly into most dishwashers on the market. It also had a hinge that allowed it to open horizontally, thereby doubling its load capacity in the event of normal manual use. Practically speaking, the product was decomposable into three parts:

A—lower rigid support
B—upper support
C—fastening hinge

From the technical point of view, the company's production structure would allow it to produce the item to specification; only the fastening hinge, made from anti-corrosive metal, would require some modification.

In his laboratory, the plant director produced a *concrete cast* of parts A and B, along with 50 injection tests, obtaining the following *average unit consumption* of raw materials:

	Part A	Part B	Tot. + 5%
Plastic	200 gr	100 gr	315
Thinner	10 gr	10 gr	21.0
Releasing liquid	20 gr	10 gr	31.5
Cleaning liquid	30 gr	20 gr	52.5
Dye	5 gr	5 gr	10.5

Since waste products averaged 4%, the above values were increased by 5% (last column). The manager had the hinge used in the sample tested in the laboratory, and the technicians said it could not be produced internally. He then instructed the procurements manager to find a supplier within 5 days. The plant manager personally went to a specialized company that produced plastic injection molds to request a quote for the two necessary *molds*. The mold for part A could be built at a cost of 12,000 euros and the one for part B at a cost of 10,000 euros. Production could take place with the injector already in operation, with an *average production capacity* of one piece every 15 seconds.

Next he calculated that, by working *two* 16-hour *shifts*, the order could be turned out in 52 days; instead, with *three shifts* it could take 35 days. He decided to calculate the costs based on the second hypothesis.

Meanwhile, the procurement manager chose the best offer for the *hinge*: € 0.20. The costs of the raw materials (in euros) were:

- plastic	3 per kg
- thinner	50 per kg
- releasing liquid	30 per kg
- cleaning liquid	10 per kg
- dye	60 per kg

Some elements were still needed to complete the calculation (all amounts are in euros):

(a) The *injection machinery* normally was depreciated in annual quotas of € 60,000 and was used on average for two shifts over 250 days a year.
(b) The machine *workforce* had an overall daily average cost as shown below (data in euros provided by the director of personnel):

- First shift	2000
- Second shift	1400
- Third shift	2200

(c) Packaging was in 50-unit plastic bags at a cost of € 0.20 per bag; the bags were then placed in a cardboard box that cost approximately € 1.
(d) The packaging machine was depreciated for 20,000 euros/year.
(e) Indirect labor cost per day was 1000 euros.
(f) Overall cost for the past year was 160,000 euros with an increase of 10% expected for the current year.
(g) Transport costs were 10 per package.
(h) Insurance costs were 1 per package.
(i) Design and production start-up costs were estimated at 5000.

Based on the above data, the production manager prepared the production cost prospectus shown in Fig. 9.33 (an F/P MATRIX with a single P); based on the results obtained after 5 days, he immediately informed the administrator that the average unit cost per tray was:

$$\text{Cost per unit} = 1{,}205{,}100 / 200{,}000 = €\,6.0255$$

which was rounded off to € 6.03. The managing director also met with the sales manager to find out if, in the meantime, he had had any other information about the company that requested the order. The sales manager reported that *banking information* revealed that the potential client was a sound business and punctual in its payments; he added, however, that the normal payment conditions were about 120 days.

The sole administrator carefully considered this additional information. The size of the order was such that Elastoplast would have had to resort to short-term bank lending to finance the purchase of the materials. It estimated that the cost of capital

	qF	pF	QP	C(F,P) = qF pF QP
Cost of materials and equipment				
PLASTIC	0.315	3	200,000	189,000
THINNER LIQUID	0.021	50	200,000	210,000
RELEASE AGENT	0.0315	30	200,000	189,000
CLEANING LIQUID	0.0525	10	200,000	105,000
DYE	0.0105	60	200,000	126,000
PACKAGING BOX	1	1.2	200,000/50	4,800
HINGE	1	0.2	200,000	40,000
Total cost				**863,800**
Cost of machinery and equipment				
MOULDS (1 × 12,000) + (1 × 10,000)				22,000
INJECTOR [60,000 / [(35 × 3) + (216 × 2)] × (35 × 3)				11,800
PACKAGING MACHINE				2,800
Total cost				**36,600**
Labor cost				
DIRECT LABOR	35	5,600		196,000
INDIRECT LABOR	35	1,000		35,000
Total cost				**231,000**
Commercial and administrative costs				
INSURANCE	200,000/50	1		4,000
TRANSPORT	200,000/50	10		40,000
GENERAL COSTS [(160,000 × 1.1) × (35/250)				24,700
PRODUCTION START-UP COSTS (not included in general costs);				5,000
Total cost				**73,700**
Total cost of the order				**1,205,100**

Fig. 9.33 F/P *matrix* for a single product

invested in production would be about 10%, which increased the average unit manufacturing cost (industrial cost) due to the financial charges.

As a result, the *total cost* became (in euros):

Average manufacturing cost = 6.030
Financial charges 0.1 × (120/360) × 6030 = 0.200 about
Total average unit cost = 6.230

The *exchange rate* was conservatively estimated at $/€ = 1.359 at the time of payment. Gross unit profit was calculated as follows:

Selling price (at the euro/dollar exchange rate) $12 /1.359 € = 8.830 €
Average total cost = (6.230) €
Gross average unit profit = 2.600 €

Taxes were assumed to be 40% of gross profits; therefore, net profit (in euros) was:

Gross average unit profit	= 2.600
Taxes on gross profit (t = 40%)	= (1.040)
Net unit profit	= 1.560.

The cost mark-up, a fundamental index of economic efficiency: approximately *roc* = 25.9%, was judged to be very satisfactory and compatible with the risks surrounding production, delivery, and receipt of payment. As a result, the managing director accepted the order for delivery in 35 business days, with a delivery time tolerance of 10%.

9.9.4 Standard Costing for Controlling Production Efficiency

As I observed in Sect. 9.7, in addition to its use for decision-making purposes, the calculation of production costs is also necessary to implement "cost control" through management microcontrol, which must ensure that costs are maintained within set "standards," that is, at precalculated levels. In this sense, cost control by means of standards is no longer part of the *cost measurement* process but the premise of *cost management* (Drury 1998, 2013). Since production costs are incurred to carry out production processes, keeping costs under control means, in fact, keeping the processes to which the costs are associated under control and improving their efficiency. As with any other form of control, *cost control* requires calculating "parameter costs," which are set as the objectives, and comparing these with the actual costs incurred. The "parameter costs" are determined by assuming *fixed levels of operational efficiency* in carrying out the production processes, assuming given *levels of performance of the factors*: materials, labor, and machinery. As a result, comparing an *actual cost*—calculated based on an *actual level* of operating efficiency—with a "parameter cost" highlights deviations, or errors, which indicate the need to control the operational efficiency of the processes. There are two types of parameter costs that are normally referred to for process control: *costs included in the budget* and *standard costs* (Hilton et al. 2003; Horngren et al. 2000; Eaton 2005).

> [standard cost] The planned unit cost of the product, component or service produced in a period. The standard cost may be determined on a number of bases. The main use of standard costs is in performance measurement, control, stock valuation and in the establishment of selling prices (CIMA 2005).

Budget costs are predetermined costs needed to draw up the production, commercial, and warehouse *budget*, as well as any other budgets that entail costs (Sect. 9.9.1), and to keep the production processes under control during the implementation of the budget. These costs are determined after accurate *forecasts* about the planned operational characteristics of the processes and, consequently, about the planned levels of operating efficiency. For example, to determine the budget for the labor force and production costs, it is necessary to forecast the average "absenteeism rate" for the whole year and for the individual quarters, even for the individual months to come (absenteeism may vary depending on the holidays and long weekends). This forecast must take into account planned interventions to control process efficiency.

Standard costs indicate the costs of a production process (cost center, or other object of reference) that are hypothetically determined, assuming that the object of reference is obtained at fixed levels of standard operating efficiency. *Standard costs* differ from *budget costs* because the latter refer to *actual expected and achievable efficiency levels* while *standard costs* refer to *assumed* efficiency levels. *Budget costs* can also be calculated only for productions that will actually be obtained, and which are thus planned; *standard costs* are calculated also for productions that are not obtained but whose achievement can only be simulated.

> Both Standard Costing and Budgetary Control are based on the principle that costs can be controlled along certain lines of supervision and responsibility, that focuses on controlling cost by comparing actual performance with the predefined parameter. However, the two systems are neither similar nor interdependent. Standard Costing delineates the variances between actual cost and the standard cost, along with the reasons. On the contrary, Budgetary Control, as the name suggest, refers to the creation of budgets, then comparing the actual output with the budgeted one and taking corrective action immediately (Key Differences 2015, online.

The calculation procedure for standard costs can be described as follows:

(a) Analyze the production process (or other object of reference) and set the desired levels of operating efficiency, or *standard objectives*; establishing set efficiency levels means assuming given levels of factor performance and, consequently, given utilization rates of both materials and labor.
(b) Translate the *objectives* into *standard operating practices,* that is, into a system of binding rules for implementing the production processes: that is, carry out the individual manufacturing operations in ways that allow the assumed levels of efficiency to be achieved.
(c) Given the assumed efficiency levels, determine the *standard quantities,* that is, the *standard consumption* of each factor (corresponding to qF in the formula in (9.16).
(d) Enhance *standard consumption* with a system of *standard unit values,* or *standard prices,* corresponding to pF in (9.16); the product of the standard amount and the standard prices is often referred to as the standard unit cost for the different factors.
(e) Assume a given *standard amount obtained* of the object of reference, corresponding to QP in (9.16).
(f) Calculate the *standard elementary cost* for each factor (direct and indirect) in relation to the previous *parameters* used to *determine* the standards: quantity, prices, and standard amount.
(g) Calculate the *standard cost of the object of reference* as the sum of the standard elementary costs, deriving from this, based on the standard amounts, the unit

standard cost (average cost). For a given product, the unit standard cost is calculated using the following function:

$$
\begin{aligned}
+\text{SCM} &= \text{Standard cost of materials for QP} \\
&= \text{Standard consumption} \times \text{Standard price} \\
+\text{SCL} &= \text{Standard cost of labor for QP} \\
&= \text{Standard hours} \times \text{Standard hourly cost} \\
+\text{SCI} &= \text{Standard share of industrial costs for QP} \\
&= \text{Standard industrial costs / Standard amount} \\
= \text{SC}(\text{P}) &= \text{Standard cost of production}\,(9.13) = \text{SCM} + \text{SCL} + \text{SCI} \\
&\text{Standard unit cost}\,(9.15) = \text{SC}(\text{P})/\text{QP};
\end{aligned}
$$

(h) Multiply the standard unit cost by a hypnotized or planned production, thereby obtaining the standard cost for that production volume.

The level of standard costs, as we have previously noted, depends on the level of *standard operational efficiency* (objective for the standard) assumed in the calculations; the higher the efficiency, the lower the costs. We can distinguish between the following:

(a) *Ideal standard costs*: these are determined using assumptions about absolute operational efficiency; they represent the lowest conceivable cost levels and are difficult to achieve. For example, the consumption of materials is calculated assuming the total absence of waste, the amounts of worker hours are quantified assuming a zero absenteeism rate.
(b) *Normal* or *realistic standard costs* (standard appraisal), which assume a higher than normal level of operational efficiency, though one which is not unattainable.
(c) *Current standard costs* (operating standard): these assume a level of efficiency similar to that found in the company.
(d) *Basic standard costs* (engineering standard), which express costs at a level of efficiency that should be considered normal, assuming current standard operating practices and a standard production volume; basic standards are kept fixed for a period of varying length, at the end of which they are subject to revision.

Current standards are used for the operating budget, while for cost control normal standards are also used, which provide an incentive for efficiency improvement. Variance analysis is the typical phase of "exception control" and applies to both budget costs (but also to revenues and other values) and *standard costs*. It is useful to briefly discuss the implementation for *standard costs*. After the standard cost of a given factor has been calculated, for example, the standard cost of materials (SCM), during the production process the error (total deviation or variance) is calculated with respect to the *actual cost* (realized or *final cost*) compared to the *standard cost* (cost objective), with a control system that operates continuously. The overall deviation from the standard cost comes from the influence of all the causes that may have affected actual production efficiency and made it diverge from the levels assumed in the calculation of standard costs. For control purposes, this figure is too concise and thus not yet useful; it is necessary to identify the "causes" that have led to changes in efficiency, and therefore to the deviation between costs. The search for these causes is the essence of "variance analysis" (Rushinek1983).

Let us assume that a comparison of the *standard costs* and the *final costs* of a given product in a given period reveals that there was an actual cost higher than the *standard* cost. Before identifying the *levers* use to eliminate the error, the manager will have to check whether the discrepancy is due to the use of defective materials or to reduced labor performance that ruined some components during the assembly phase. Let us also assume that for labor as well the *final cost* was considerably higher than the standard cost. The search for causes is crucial to structuring the control system. The error could mean that many of the hours paid by the company were "lost" due to increased absenteeism, illness, worker assemblies, etc., or that the workforce, due to insufficient preparation or reduced performance, took more hours than indicated in the standard practices (Sect. 10.14.7). On the other hand, let us assume that the costs of a given process—for example, maintenance and testing—are below the standard costs. The favorable error may indicate that, for the same number of units produced and examined, fewer defective units were found, resulting in less time required to search for and eliminate defects. However, it could also indicate a harmful reduction in the number of units that were to undergo maintenance and testing. These considerations also apply to *variance analysis* applied to budget data.

The use of standard values is not only intended to *control* the efficiency of processes but also to *improve* this efficiency by producing progressive learning on the part of workers. When the final values are aligned to the standards, this means, on the one hand, that the planned efficiency has been achieved, and on the other that it can be increased. The control system that is structured to achieve this dual purpose is, for all intents and purposes, a system that produces "insatiability," whose general structure is similar to that indicated in Fig. 9.34, which is derived from Fig. 5.12.

9.9.5 Dashboards in Performance Control Systems

In formulating the VSM (Sect. 9.2), Stafford Beer observed that at all levels of the organizational hierarchy managers serve as filters of the information from the lower levels, since they synthesize this information in order to transmit it to the higher hierarchical levels, which in turn filter it again by sending synthesized information to even higher hierarchical levels. In a complementary manner, the further one moves toward the operational base of the organization the more analytical the information becomes. For example, each head of sales for an area writes a weekly report summarizing the data from the salesmen in the area; these weekly reports are passed on to the head of area sales, who summarizes it and transmits the product sales

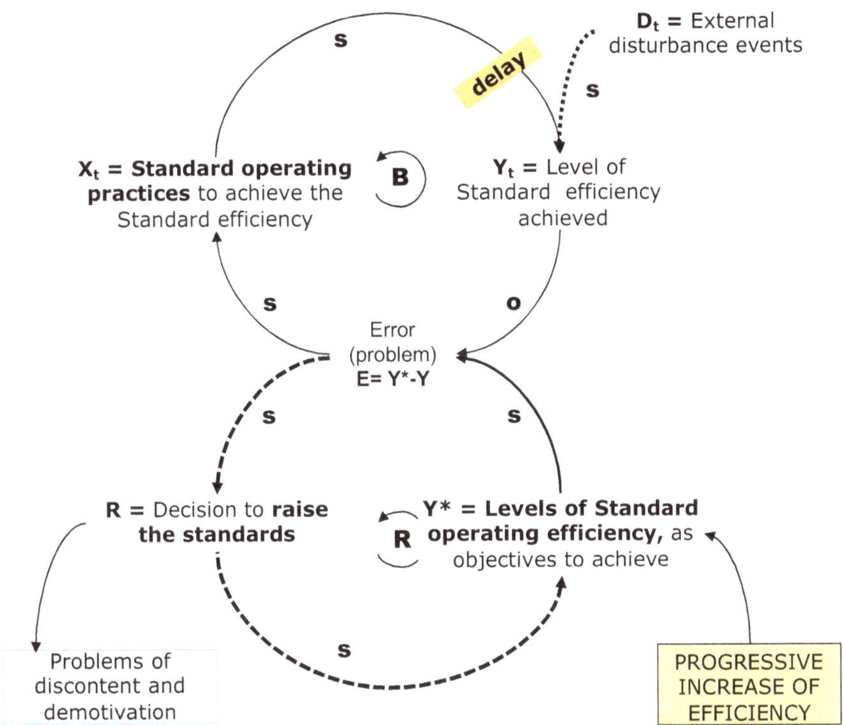

Fig. 9.34 The insatiability archetype acts to continuously improving efficiency by progressive learning on the part of workers

figures to the marketing head, who synthesizes the figures for sales for all products and transmits the results to the managing director, who, together with data for costs, drafts the reports of the gross results for the board of directors.

It is intuitively clear that in order to evaluate and control operational performance at the various organizational levels, the data from activities, operations, and processes that produce the economic and financial variables must be known in *real time*, so that the variances with respect to the objectives can be readily perceived and the necessary corrective measures can be taken. In order to carry out control in *real time,* the synthetic values for an entire period of observation (e.g.,, production, sales, monthly or quarterly stock) are not particularly significant, since they do not permit us to observe the dynamics regarding the underlying phenomena to be controlled (sales, receipts, and progressive losses), thus making it difficult to readily perceive the abovementioned variances.

For this reason the technique of constructing *dashboards* (or *management cockpits* or *tableau de bord*) to detect in real time the dynamics of the fundamental variables to control continues to spread (Mella 2012). For example, in hypermarkets dashboards are used to highlight the flows of the most relevant types of goods: stocks, resupplies, average supply times, average times goods are exposed on shelves, losses/damage, thefts, etc. In large hospitals dashboards are created for the daily control of the operational dynamics of the main clinics or divisions: numbers and types of patients admitted, length of waiting lists, average hospital stay, type of treatments, percentage of failures, multiple interventions, average amount of drugs administered, average consumption of medication materials, etc. In control theory, *corporate dashboards* must be considered in all respects as *continual reporting* instruments for monitoring a system of *performance objectives* and *standards* in order to allow for the control of operations and personnel at a specific operational level. The dashboard data can also be generally defined as *key performance indicators* (KPI), since they monitor the *performance* regarding the objective of the control process, as illustrated in the *Ring* in Fig. 9.35.

There is no set rule for constructing effective dashboards, even though several principles should be considered, such as:

- Creating above all an organizational culture of action aimed at achieving performance
- Integrating this performance culture with one characterized by an acceptance of performance control
- Creating dashboards to monitor the specific performance of certain personnel and defined functions (*key performance areas*)
- Including in the dashboard only those (few) KPIs that represent realistic performance objectives for the area in question
- Graphically representing the KPIs in an intelligible manner, using different characters, colors, and graphs to highlight them so that they can be immediately utilized
- Making clear, even graphically, the systematic links among the various KPIs, avoiding the use of colors and formats that can mislead perceptions of the important information for which the dashboard was designed in the first place

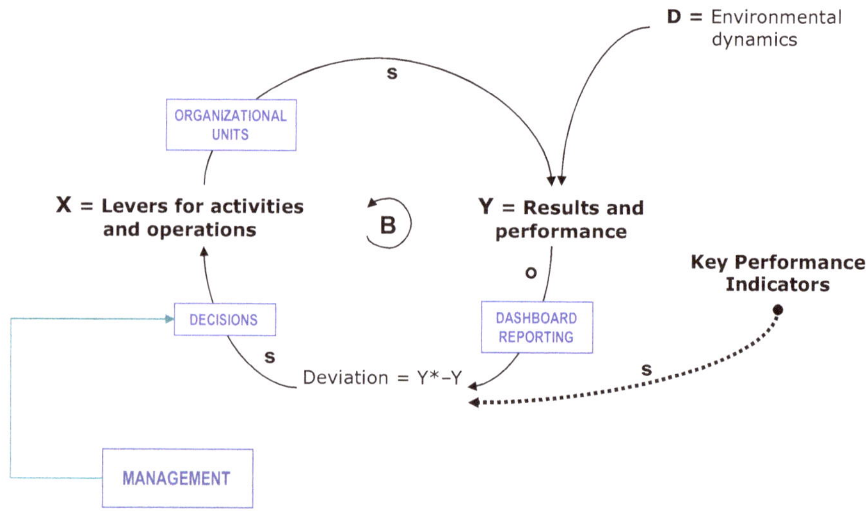

Fig. 9.35 The role of dashboards in performance control

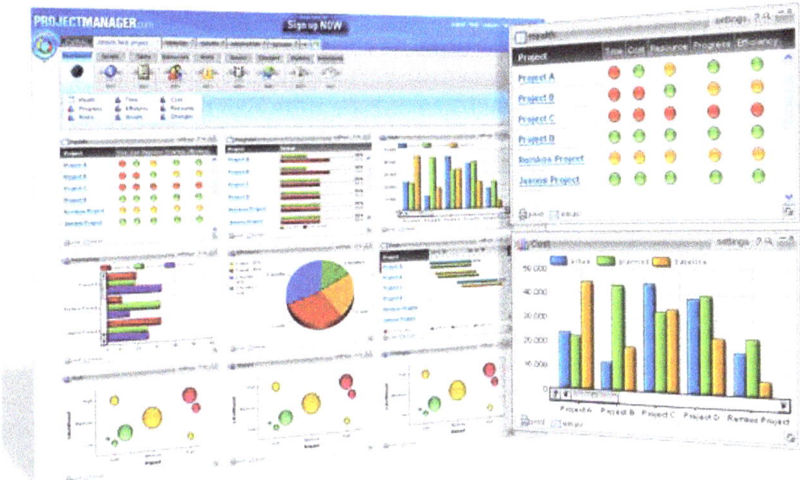

Fig. 9.36 An example of a dashboard (*Source*: http://www.projectmanager.com/project-dashboard.php

Figure 9.36 presents an example of an attractive graph of an organizational dashboard for top management.

9.9.6 The Balanced Scorecard in Performance Management Control Systems

Management performance control systems cannot be limited to merely controlling financial performance. A number of other interesting variables can serve as performance indicators. The *balanced scorecard* (BSC) is one type of *corporate dashboard*, created by Kaplan and Norton (1992, 1996), which provides top management with information for a continual evaluation of the performance of an entire firm (Thakkar et al. 2006). The BSC considers four *strategic variables* or *perspectives* (or focus) held to be fundamental for providing management with a "balanced" perspective of the *strategic performance* of the firm and an evaluation of the progress toward the objective of *shareholder value* creation imposed by *shareholders*.

1. *Financial perspective*: how the organization wishes to be viewed by its shareholders.
2. *Customer perspective*: how the organization wishes to be viewed by its customers.
3. *Internal business processes perspective*: through which processes must the organization develop its abilities in order to satisfy its shareholders and customers.

4. *Organizational learning and growth perspective*: which changes and improvements must the company make to implement its vision.

As a corporate dashboard designed to monitor corporate strategy, the BSC presents three unique features: it includes among the KPIs both financial measures and nonfinancial ones that can be adapted to the type of firm implementing them; it is directed mainly at top management and updated relatively slowly. The "balance" of the scorecard is reflected by the balance between outcome measures and performance driver indicators and that between financial and nonfinancial measures. In short, each perspective is represented in a *scorecard* and assigned a weight of relative importance; for each perspective a limited number of *performance measures* managers hold to be truly significant are included in the BSC, as shown in Fig. 9.37.

As an instrument of strategic control, the role of the BSC can be represented by a model entirely similar to the one in Fig. 9.38.

The following measures are particularly efficient in choosing each perspective:

(a) Measures for the financial perspective: value of the action, growth in profits, profit rate, ROI, EVA, ROE, operating costs, operating margin, corporate objectives, survival, profitability, growth, cost reduction, increase in ROI, cash flow, earnings, increase in earnings, profit rate of shares, and so on (Mella 2005b).
(b) Measures for the client perspective: service level, market share, new clients, new products, new markets, customer satisfaction, customer loyalty, product reliability, perceived quality of the product and/or collateral services, customer complaints, etc.
(c) Measures for the internal perspective: increase in efficiency, quality of processes, utilization rate of production capacity, stock storage period, waste, recycling rate of production waste, remanufacturing, lead time, average unit cost, employee morale, motivation, and so on.
(d) Measures for the learning and growth perspective: trend in value creation, product diversification, supplier diversification, increase in R&D, risk diversification, strengthening of internal control, development of new products, continual improvement, technological leadership, employee involvement, etc.

According to Kaplan and Norton, these perspectives and measures can, in particular contexts, be supplemented by others or enriched by new measures:

> The four perspectives could be viewed as a scheme of reference and not as a straightjacket. Many organizations use the BSC and establish relative weights for each of the scorecard measures. These relative weights are used to evaluate performance (Kaplan and Norton 1996, p. 34).

> If, as we have indicated, the scorecard could guide us in growing our business, then it is natural to believe it possible to change the number of perspectives, areas, or focusses (Olve et al. 1999, p. 120).

If also used as an instrument for determining strategy and not only for monitoring its implementation, the BSC takes on the dual role of an objective and a way of obtaining measurements, thus allowing the organization to determine the variances between the chosen strategy and the actual results.

Perspective	Number of Measures	Weight
Financial	5	22%
Customer	5	22%
Internal Business Processes	8 to 10	34%
Learning & Growth	5	22%

Fig. 9.37 Weights and measures of the BSC perspectives (*Source*: Kaplan and Norton 2001, p. 375)

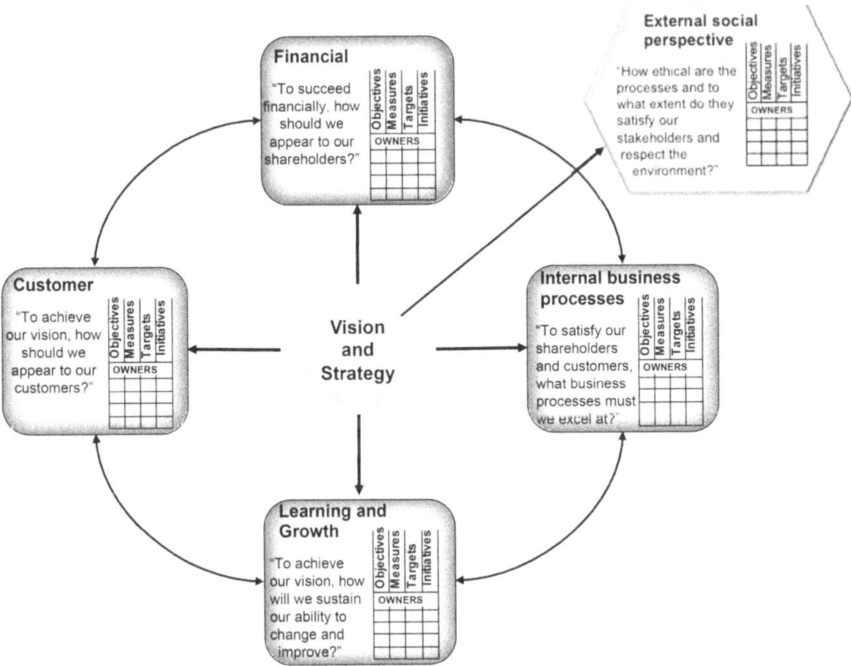

Fig. 9.38 The BSC as a strategy-forming instrument (*Source*: Mella 2012: p. 190)

9.9.7 External Control of the Company's Economic and Financial Efficiency: Position Analysis (Brief Outline)

The forms of management control presented in the previous *sections* are typically developed by internal corporate personnel who, with different characteristics in mind, mainly monitor the performance of the organization regarding the macro processes carried out at the third level (Fig. 9.1), where the costs, revenues, and economic results are determined. However, the control of the macro activity of production organizations can also be carried out by external parties based on data from the business transformation, [TR5], represented in Fig. 9.11, to express a judgment, an appreciation, of the system of values expressed in (9.2). *Appreciation* refers to a survey undertaken to express a judgment, favorable or unfavorable, an assessment, or a positive or negative acknowledgement regarding the company in operation, its management performance to date, and its fitness: that is, on the ability of the company to achieve future performance levels.

One of the established techniques widely used by individuals outside the organization to make their judgments is the Balance Sheet Analysis, which, based on the company's documents, as well as other information, uses descriptive and diagnostic tools to determine economic and financial indicators (Whittington 1980), whose values can be compared with similar values from *other companies* (spatial analysis), with previous *years* (time analysis), or with the aggregate *values* of all companies operating in the same sector (Mella and Navaroni 2012). Particularly useful is the third form of analysis, *position analysis,* since, by comparing specific indicators for a single company with the average results in the sector of reference, the company's "position" compared to other similar companies in the sector can be determined (Bernstein and Wild 1989). As a result, it is possible to express a positive or negative judgment about the company depending on whether its indicators are in a "better or worse position" compared to the sectoral average; the deviations that are calculated can therefore be used as a diagnostic tool for the appreciation of the company's performance.

For example, the company could have a higher return on investment (ROI) than the average company and thus be positioned more favorably; however, it may have a lower inventory turnover (Sect. 11.20.1) compared to the sectoral average, and this positioning may be an indication of stock mismanagement. With position analysis, the assessment of management is thus expressed by analyzing the deviations between each individual index of the company and the corresponding index of *competitors* or by examining the average sectoral index to determine an error that can uncover the causes of these differences and identify the strengths and weaknesses of the company vis-à-vis its *competitors*. Position analysis is therefore very effective and powerful in that it allows budgetary analyses to be used in a manner similar to the use of "medical analyses" in revealing anomalies, but only assuming the *correct calculation of the sectoral indices*, which must entail the same reclassification criteria used by the budget analyst (Salmi et al. 2005; Drake and Fabozzi 2008).

It should be noted that position analyses are very often implemented not only for individual indices, located above or below the average sectoral indices, but also for systems or clusters of significant indices (Mella 2005b). All indices can be appropriately linked, as revealed by the Dupont Tree, developed by Donaldson Brown in Dupont (Flesher and Previtis 2013) (Fig. 9.39).

> The basis of my report gauging the performance of the various operating departments was a simple mathematical formula: $R = T \times P$. The factor R represented the rate of return on capital invested, which is a final and fundamental measure of industrial efficiency in terms of management's primary responsibility. The T stood for the rate of turnover of invested capital, and P for the percentage of profit on sales. On the investment side T was broken down into components, embracing planL and other fixed investment items, as well as

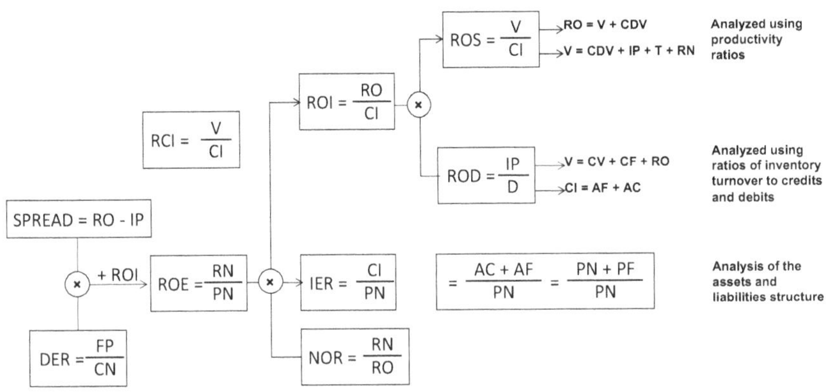

Fig. 9.39 The Dupont Tree represents the fundamental system of efficiency indicators with regard to profitability (my elaboration from Davis 1950, p. 7)

> amounts tied up in working capital in various categories such as raw materials, work in process, finished product, accounts receivable and required operating cash balances. ...
> A chart room was set up where these statistical data pertaining to each segment of the company's operations were displayed. Meetings were held regularly with department heads, and extended discussions were held regarding the possibility of improving specific cost and expense items, in relation to the end-result of return on invested capital …
> This approach resulted in a specific disclosure of causes and effects for the return on investment shown for each department. Effective control or the lack of it, for any item on either side of the equation could be identified, thus making possible efforts to improve conditions (Brown 1977, pp. 26–27).

In another form, the Dupont Tree analytically expresses the system of values [(9.2)]. Position analysis requires knowledge of the meaning of numerous budget indices, which it is not possible to further illustrate here.

9.10 Summary

After having identified the innumerable *Rings* that characterize individual behavior in our environment (Chap. 7), and *the* Rings that regulate the behavior of biological populations in ecosystems, collectivities, and societies (Chap. 8), I have tried to "see beyond" by examining control systems that influence *organizations*.

In particular, I have examined the *Rings* that allow organizations to exist:

(a) As autopoietic systems, along the lines envisioned by Maturana and Varela (Sect. 9.1, Fig. 9.2), and as teleonomic systems, following Monod's approach (Fig. 9.7)
(b) As vital systems, following Beer's VSM (Sect. 9.2, Fig. 9.8)
(c) As systems for efficient transformation, following the MOEST approach (Sect. 9.3, Fig. 9.10)

(d) As systems requiring management control (Sects. 9.4, and 9.5)—at both the strategic (macro control) and operational (micro control, Sects. 9.6 and 9.7) through the implementation of performance control systems and, in particular, of management cost control systems
(e) As cognitive, intelligent, and exploratory systems (Sect. 9.8)

We have examined several basic instruments for micro management control (Sect. 9.7):

1. The programming and budgeting process (Sect. 9.9.1)
2. Cost measurement. An example (Sect. 9.9.2)
3. Cost measurement to decide whether or not to accept a price (Sect. 9.9.3)
4. The use of Dashboards in performance control systems (Sect. 9.9.4, Fig. 9.34)
5. The use of the Balanced Scorecard in performance control systems (Sect. 9.9.5, Fig. 9.36)
6. The use of the position analyses based on The Dupont Tree model (Sect. 9.9.7 and Fig. 9.39).

Chapter 10
The Magic Ring in Action Explores Quality and Productivity

Abstract The discipline of control entails observing reality by filtering it through the models produced by the simple theory of control discussed in Part I of this book. Chapter 6 took the point of view of the individual observing the environment in which he or she lives; Chapter 7 examined the *Rings* that operate in social systems, combinatory systems, and complex adaptive systems; Chapter 8 considered various forms of control operating within the Macro Biological Environment; and Chap. 9 demonstrated how control systems operate in production or consumption *organizations* which are unitary systems where the control is pervasive and essential. This chapter presents the Rings which control the two fundamental variables—quality and productivity—that are vital for both organizations and the global economic system. I have also proposed the Law of Increasing Productivity and Quality: every individual, local, national, or global production system develops increasing productivity and higher levels of product quality over time.

Keywords Functional quality · Design-based quality · Environmental quality · Profit as a function of quality · Quality control · Quality assurance · Company-wide quality control · Quality drives loyalty, Confidence and reputation of the firm · Productivity control · Productivity levers · intrinsic drivers · Managerial drivers · Extrinsic drivers · Dynamics of productivity · The Law of Increasing Quality and Productivity · MAPI formula · total productive maintenance (TPM)

> Make your workplace into showcase that can be understood by everyone at a glance. In terms of quality, it means to make the defects immediately apparent. In terms of quantity, it means that progress or delay, measured against the plan, and is made immediately apparent. When this is done, problems can be discovered immediately, and everyone can initiate improvement plans (Taiichi Ohno, on line).

> MBWA (management by walking around, a term that I learned from Lloyd S. Nelson) is hardly ever effective. The reason is that someone in management, walking around, has little idea about what questions to ask, and usually does not pause long enough at any spot to get the right answer (Edwards Deming 1989, Chap. 2).

P. Mella, *The Magic Ring*, Contemporary Systems Thinking,
https://doi.org/10.1007/978-3-030-64194-8_10

As observed in Chap. 9, *micro management control* or *operational control* concerns the various functions (supply, production, sales, logistics, personnel, finance, etc.) and various operational units (plants, divisions, departments, individual facilities and/or workers, etc.) which, by activating various levels of operational levers, produce the values that must be achieved by those values chosen as objectives in short-term plans and in budgets. The micro control concerns the achievement of short-term objectives (producing at an average cost below a certain level; fulfilling an order at a given cost and within a given time period; not exceeding the maximum or the minimum warehouse stock; reducing the average cost of capital for the next quarter, etc.) and makes use of various and well-structured instruments to detect the *variances* and determine which control *levers* to adopt. This chapter examines the control processes that regulate the two most essential variables—quality and productivity—that drive both organization and the economic system, which is viewed as a network in which organizations represent the basic modules or nodes (Chap. 8).

The control of *product quality* and the quality of *production and sales processes* are fundamental to the *productive* transformation of the MOEST (Chap. 9), whether the firm produces goods or services or produces by job order or process. Quality must be ensured for the customer as a condition of survival for the organization-firm, since it influences considerably the efficiency levels of the TR1-P, the productive transformation, and the TR2-E, economic transformation (Fig. 9.11), both in terms of revenues, by influencing the selling price and demand level, and costs, since a variation in quality levels causes a variation in the cost of quality certification, protection, and restoration. There are numerous techniques for evaluating the *quality levels* and the *levers* to be used to control this variable, and these are complex and differ depending on the type of quality to be controlled. The control of the efficiency in the use of productive factors is equally fundamental to the efficiency of TR1-P and TR2-E (Fig. 9.11). Therefore, control of the productivity of work also becomes vital for the organization.

This chapter also tries to demonstrate the fundamental relevance of the control of *productivity*, the variable which is the basis of all productive systems, viewed as transformers of utility and value, since the search for maximum productive efficiency is necessary to reduce production costs and thus to produce value. What are the factors on which productivity depends and what levers can control this fundamental variable? After presenting a coherent frame of reference, I shall examine the drivers of productivity and then move on to discuss the consequences of the continual growth in productivity. The paper concludes with a simple but significant model that seeks to illustrate the relationship between productivity and employment. I have also proposed the Law of Increasing Productivity and Quality: every individual, local, national, or global production system develops increasing productivity and higher levels of product quality over time.

10.1 Quality Control: The Theoretical Framework—The Three Forms of Quality

The concept of *quality* is not easy to define (Plunkett and Dale 1987). It is "elusive," and understanding it is normally left to intuition. Several scholars have proposed giving up the attempt to precisely define the concept, viewing it as an intuitive, nondefinable, almost primitive, philosophical, even *metaphysical* notion derived from the concepts of differentiation, excellence, perfection, and consistency. For Barbara Tuchman (referring to publishable papers) quality is undeniably, though not necessarily, related to class, not in its nature but in its circumstances.

> … quality, as I understand it, means investment of the best skill and effort possible to produce the finest and most admirable result possible. Its presence or absence in some degree characterizes every man-made object, service, skilled, or unskilled labor—laying bricks, painting a picture, ironing shirts, practicing medicine, shoemaking, scholarship, writing a book. You do it well or you do it half-well. Materials are sound and durable or they are sleazy; method is painstaking or whatever is easiest. Quality is achieving or reaching for the highest standard as against being satisfied with the sloppy or fraudulent. It is honesty of purpose as against catering to cheap or sensational sentiment. It does not allow compromise with the second-rate … (Tuchman 1980, p. 38).

Robert Pirsig (1974), in his book *Zen and the Art of Motorcycle Maintenance: An Inquiry into Values*, introduces the idea of the Metaphysics of Quality, where quality is a direct experience independent of and prior to intellectual abstractions.

> Quality is the continuing stimulus which our environment puts upon us to create the world in which we live. All of it. Every last bit of it. Now, to take that which has caused us to create the world, and include it within the world we have created, is clearly impossible. That is why Quality cannot be defined. If we do define it, we are defining something less than Quality itself (Pirsig 1974, p. 114).

As stated in Sect. 9.3, quality must be ensured for the customer as a condition of survival for the firm, since it influences the efficiency of the production and market processes, both in terms of revenues, by influencing the selling price and demand level, and costs, since a variation in quality levels causes a variation in the cost of quality certification, protection, and restoration.

From a managerial perspective, the quality of goods or services produced by firms is defined in the UNI EN ISO 9000 as: "*Whatever the customer perceives good quality to be,*" referring not so much to the intrinsic features of the product as to consumer tastes and experiences. In 1983, the American Society for Quality Control (ASQC) proposed a more specific definition:

> A subjective term for which each person or sector has its own definition. In technical usage, quality can have two meanings: 1) the characteristics of a product or service that bear on its ability to satisfy stated or implied needs; 2) a product or service free of deficiencies. According to Joseph Juran, quality means "fitness for use"; according to Philip Crosby, it means "conformance to requirements" (ASQ 1991, online).

In addition to being general, these definitions are also "generic" (Russell and Miles 1998). Some definitions emphasize several characteristics of "quality" products or the extent to which these characteristics are perceived by the customers as being present in the products. Keith Leffler states:

> Quality is based on the presence or absence of a particular attribute. If an attribute is desirable, greater amounts of that attribute, under this definition, would label that product or service as one of a higher quality (Leffler 1982, p. 959).

For Armand Feigenbaum

> Quality is "best for certain customer conditions" (Feigenbaum 1951, p. 10). Quality is now and international business language and no longer travels under any single national passport or has any single geographic centerpoing or particular cultural characteristic (Feigenbaum 1997, p. 45).

In a similar vein, Robert Flood states that: "*Quality means meeting customer (agreed) requirements, formal and informal, at the lowest cost, first time every time*" (Flood 1993, p. 48). Sid Kemp views the customer as the main evaluator of quality, which represents: "*all elements of our product that add value for the customer or stakeholders, or are required for our product or service to meet relevant standards and regulations*" (Kemp 2006, p. 331). David Hoyle states that "*Quality is the extent to which a product or service successfully serves the purposes of the user during usage (not just at the point of sale)*" (Hoyle 2007, p. 10).

David Garvin's analysis (1984) goes deeper. In fact, Garvin highlights eight basic dimensions of quality: (1) Performance: a product's primary operating characteristics; (2) Features: the "bells and whistles" of the product; (3) Reliability: the probability a product will operate properly over a specified period under stated conditions of use; (4) Conformance: the degree to which physical and performance characteristics of a product match pre-established standards; (5) Durability: the amount of use one gets from a product before it physically deteriorates or until replacement is preferable; (6) Serviceability: the speed, courtesy, and competence of repair; (7) Aesthetic: how a product looks, feels, sounds, tastes, or smells; (8) Perceived quality: subjective assessment resulting from image, advertising, or brand name.

Taking into account these various meanings of "quality," in order to give coherent definitions, I propose identifying three distinct, though connected, notions that sum up the majority of definitions found in the literature and allow us to focus on quality from a threefold perspective: the customer, the product and the environment: *functional*, *design*, and *environmental* quality.

Functional (or market) quality defines the set of characteristics which, from the customer's point of view, make the *product* appropriate for *use*; that is, capable of satisfying a specific *use or utility function* of the good or service; in other words, the relationship between the user's needs and aspirations and the merchandising characteristics of the product—taking into account a desired standard of *reliability* (the product must provide use that is not interrupted due to imperfections) and *safety* (the product must be capable of use without damage or risks), as specified in ISO/IEC 9126:1991 (Band and Huot 1990; Band 1991; Oakland 1989). In this sense, for Ryall and Kruithof:

> Quality is consistently meeting the continuously negotiated needs and expectations of Customers, in the context of the needs and expectations of other interested parties, in ways that create value and satisfaction for all involved (2001, p. 20).

ISO 8402: 1994 (Quality Vocabulary) specifies the concept as "*the totality of characteristics of an entity that bear on its ability to satisfy stated and implied needs.*" In ISO 8402: 1986, version 3.1, the definition is: "*The totality of features and characteristics of a product or service that bear on its ability to satisfy stated or implied needs,*" very similar to that proposed by ASQC: "*The total features and characteristics of a product or a service made or performed according to specifications to satisfy customers at the time of purchase and during use.*" *Functional quality* is the basis for *customer satisfaction* (Wellemin 1990) as defined in ISO 9001:2000 (8.2.1), known as "Vision," which in fact emphasizes *functional quality*:

> … if before Quality meant satisfying all the Customer's expectations, now the concept of "expectations" is broadened and enriched through adjectives such as "explicit" and, above all, "implicit", with regard to both the external Customer – the recipient of the products – and the internal Customer – the department that receives the components and services of the other departments (ISO 9001 2000).

In this sense, quality refers to the concept of "fitness for use": that is, "*the extent to which a product successfully serves the purposes of the user*" (Juran and Godfrey 1999), and of "*attractive quality,*" which defines what the customer desires even though he is not yet aware of this (Kano et al. 1984).

DESIGN (OR INTRINSIC, PRODUCTIVE) QUALITY, or *conformance to specifications*, is the set of characteristics that, from an internal point of view (in terms of production processes) make all the product units *conform to a standard of reference* (prototype, sample, model, design), as stated by Philip Crosby (1979), who defined quality as "*conformance to requirements or specifications*" (Levitt 1972). To Crosby, any product that falls within its design requirements is a quality product, and for this reason the message to managers is "*do it right the first time (DIRFT). Requirements are the 'it'*" (Crosby 1979, p. 127). As a consequence, *the performance standard is zero defects*. "*What cost money are the unquality things – all the actions that involve not doing jobs right the first time*" (Crosby 1979, p. 2). The standard of reference can be represented by the unit of product introduced in the market in the past (temporal comparison).

Normally, the standard of reference is the initial project and the specifications provided therein for the production processes. Each company, whether large or small, must continually verify that the intrinsic quality of its products always conform to set qualitative standards or improves over time. For firms that produce on commission, these standards can even take the form of regulations inserted in the procurement contract that provide a detailed description of the work to be carried out accompanied by one or more technical designs or even a scale model. For mass production companies, the certification of quality over time is essential to avoid future complaints and returns from customers; and, in the final analysis, to avoid damaging the product image and losing market share (which is clearly linked to the control of the functional quality of the product). The need to at least distinguish between functional and design quality is explicitly recognized by Nelsen and Daniels, according to whom: "*quality has two meanings: 1. The characteristics of a product or service that bear on its ability to satisfy stated or implied needs; 2. A product or service free of deficiencies*" (2007, p. 54).

The *intrinsic quality* refers not so much to the product as to the *production flow*, that is to the flows of production units obtained during the production process. This is the notion that is considered in UNI EN ISO 9000:2000, which defines quality as the "*degree to which a set of inherent characteristics fulfils requirements.*"

ENVIRONMENTAL (OR CONTEXT) QUALITY is the set of characteristics which, from the point of view of external impact, make the product compatible with the environment, both in terms of pollution, waste disposal, environmental risks, or suitability for introduction into the context in question. The perception of environmental quality as defined in the ISO 14000 standards—which obliges firms to identify how the firm's activities, processes and products can have an "impact" on the environment—gives rise to the idea of the adequacy of the products relative to the environmental context in which they are employed during their life cycle. Today, in our highly interconnected society, no product can avoid a *judgment of adequacy*: some cars are too polluting; some homes scar the landscape; some packages are too cumbersome; some scooters too noisy; some motorcycles too dangerous; some railway lines too invasive, and so on (Claret 1981; Underwood 2003).

Moreover, these notions of quality consider the products as *fit for a purpose*: "*One of the possible criteria for establishing whether or not a unit meets quality, measured against what is seen to be the goal of the unit*" (Campbell and Rozsnyai 2002, 132). "*Fitness for purpose sees quality as fulfilling a customer's requirements, needs or desires*" (Juran and Godfrey1999; Crosby 1979).

Considering the product as "fit for a purpose," it seems clear that the three forms of quality outlined above apply to the product, including *packaging* and *distribution logistics* (Myers and Shocker 1981). A product with intrinsic quality may be judged negatively if its packaging makes its use less efficient and/or it is necessary to resort to inadequate logistics to acquire it. Donald Abbot (1989) points out that packaging has four fundamental roles: containment, protection, communication and utility, and thus plays a fundamental role in the value chain (Gofman et al. 2010; Young 2008). For example, an exquisite perfume in a bottle that is inconvenient or aesthetically unappealing, heavy, and difficult to dispose of can be viewed negatively on the whole from a functional and environmental point of view. In the same way, a point of sale considered to be of high design quality (attractive, providing a full range of products, can be functionally inadequate if the product must be purchased only at that place, which, however, is far from the town center and difficult to reach.

> Some of the services ultimately desired by consumers include bulk-breaking (as previously discussed), spatial convenience (being able to buy milk in the supermarket rather than having to drive out to a farmer to get it), timing of availability (having someone—the retailer and other channel members—plan to have toothpaste available in the store when the consumer needs it), and providing a breadth of assortment (the same store will carry different kinds of food and other merchandise from different suppliers (Perner 2018, online).

Similar considerations pertain to after-sales service (assistance, repairs, maintenance, etc.) when these are an integral part of the main "product."

> After sales service quality affects satisfaction, which in turn affects behavioural intentions. Hence, after sales services affect the overall offering, and thus the quality of the relationship with customers (Rigopoulou et al. 2008, p. 522).

10.2 The Relevance of Quality Control for the Economic Transformation

The control of the three forms of quality is essential for every organization because quality *represents* the set of product characteristics—function, design, and environmental impact—that *contributes in a relevant way to the purchase decisions, taking into account the selling price* and the general and specific economic conditions. Quality is fundamental because it represents the variable on which the *value perceived* by the customer depends, thereby assuring him the maximum *value for money function* (Wellemin 1990), understood as the maximum quality at the lowest price:

$$\text{Value perceived by the customer} = \mathit{Quality} \,/\, \mathit{product\ price} \qquad (10.1)$$

Quality (the numerator) conditions the *value*, with price being equal (denominator). However, a high level of quality allows production organizations to raise the price without reducing the value perceived by the customer. Therefore, since quality conditions the prices and the market, acting on economic efficiency mainly from the revenue side, it becomes a fundamental driver for value production (together with productivity, which conditions value from the cost side) since the level of quality perceived by the customer contributes in a relevant manner to purchasing decisions and the customer's propensity to pay the price (eq. 10.1).

For this reason, it is vital for any productive organization to set up a reliable system of *quality management* as a part of the general strategy for value production (Shetty 1987); this system should set appropriate "quality" objectives for the product and adopt a system to control the achievement and progressive surpassing of these objectives, as called for in the application of the Deming Cycle (Deming 1989, 2012).

> Practising quality control means developing, designing, producing, and supplying quality products and services which are the most economical and most useful for the consumer, providing him continual satisfaction (Ishikawa 1990).

As Fig. 10.1 shows, the perception of consumers regarding the *value* of the overall supply of goods and services (eq. 10.1) is directly dependent on the three forms of quality, which produce *loyalty*, increase *trust*, justify the *proposed selling price*, and sustain the level of demand by impacting *market share*.

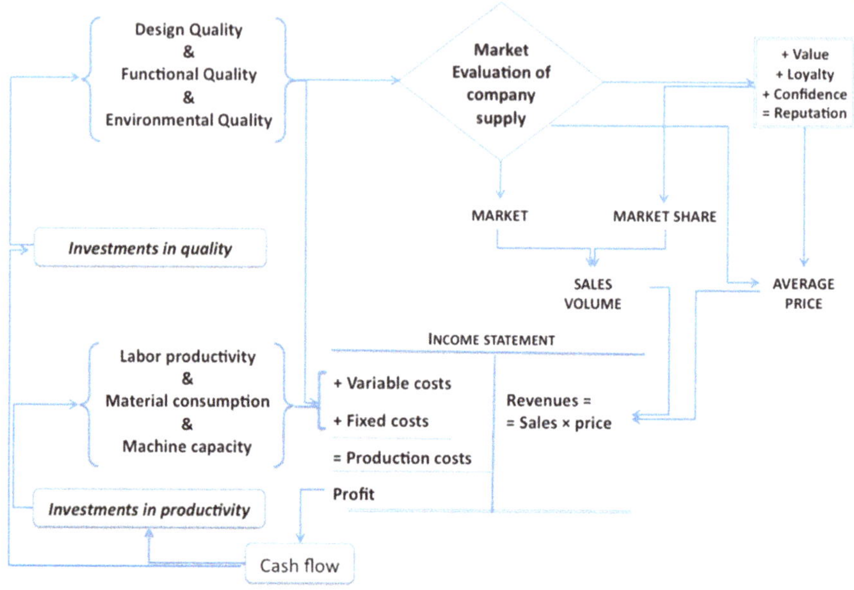

Fig. 10.1 Quality and productivity as a profit, and thus value, driver

Quality is also a fundamental driver of *marketing efficiency*, which is expressed by the following equation:

$$\text{Sales} = \text{Market} \times \text{Market share} \tag{10.2}$$

Quality influences the *market* itself by creating goods to satisfy new needs and aspirations or by varying the capacity of goods to satisfy high-level needs or aspirations in order to create, or increase the share in, an attainable market. However, quality also impacts *costs*, since the improvement in quality levels causes a reduction in *nonquality costs*, which are linked to certifying, preventing, or restoring quality (Sect. 10.5).

Profit creation generates *cash flow* that can favor new investments to improve quality and productivity. Investments for the improvement of quality should thus represent a relevant factor in the overall strategy through its effect on marketing and production. After these general considerations, let us examine the methods the production organization must follow to control the three forms of quality.

10.3 Control of "Functional" Quality

According to the concept of *functional quality*, every product has its own *use function*; that is, a *mix* of needs and aspirations—"expressed or implicit," agreeing with ISO 8402:1986—which a product can satisfy contemporaneously. The productive organization must manage quality by controlling that the product maintains, and if possible increases over time, its overall *use function*, which is the condition for controlling prices in some manner and producing economic value. Keeping functional quality under control is important since the *revenue curve* of the firm (Fig. 10.2c) depends on the form of activity (Ojasalo 2006). In fact, functional quality (ql) influences both the demand curve and prices (Fig. 10.2a). If we express demand as a function of both the price and functional quality, we observe how the curve varies with variations in this type of quality (Fig. 10.2b). Curve A in the graph on the left in Fig. 10.2b indicates the normal relation between demand (x-axis) and price (y-axis), assuming there is normal elasticity and the curve depends on the pricing maneuvers of the firm. The product of sales volume (indicated by S on the x-axis) and price level (indicated by p on the y-axis) expresses the revenue (R) attained for each (p, S) pair, which can be represented as the area of the rectangle whose vertex is at point (p, S) and whose opposite vertex is at the origin. The graph on the right in Fig. 10.2b shows the theoretical effect of increasing the level of functional quality, which causes the following:

(a) An upward shift of the demand curve, so that, at each price, the sales volumes are indicated on curve B, rising from S_A to S_B; total revenue (TR) thus becomes TR = R + R'.
(b) An increase in prices as well at each level of demand; this means that the slope of the demand curve is reduced and can be represented as the curve C, which shows that for each price level the quantity demanded is greater than what it would have been without the increase in quality. Total revenue is now TR = R + R' + R".

The trend in revenues as a function of the quality level can be shown by the curve in Fig. 10.2c, from which we can hypothesize that there is a minimum level of functional quality (ql min) below which revenues are not defined, as well as a maximum level (ql max) above which total revenues cannot increase as a result of further increases in quality (they can increase, of course, due to other causes that are not pertinent to a study on quality).

The successful firms today are those capable of anticipating variations in the use function of products, which thus allows them to achieve a timely *marketing strategy* that enables them to set higher prices due to the perception of a change in the products' use function. More specifically, the *levers* to control *functional quality* must certify that the product maintains (or increases) over time the overall *use function* required by the market. Firms carry out *functional quality* controls in the form of *Quality Function Deployment* (Bossert 1991) which proposes the development of new products or processes and the improvement of existing ones, as defined in the official site:

> Quality Function Deployment (QFD) was developed to bring this personal interface to modern manufacturing and business. In today's industrial society, where the growing distance between producers and users is a concern, QFD links the needs of the customer (target quality) with design, development, engineering, marketing, manufacturing, service and other organizational functions through its systematic deployment. 1. QFD seeks out spoken

Fig. 10.2 Functional quality, demand, and price. (**a**) Demand (D) and price (*p*) as functions of functional quality (qlP). (**b**) Variations in revenue (R) as a function of functional quality (plP). (**c**) Quality revenues as a function of functional quality (plP)

and unspoken customer needs from fuzzy Voice of the Customer verbatim; 2. QFD uncovers the "true" customer needs and "positive" quality that wows the customer; 3. QFD translates these into designs characteristics and deliverable actions; and 4.QFD builds and delivers a quality product or service by focusing the various business functions toward achieving a common goal: customer satisfaction (QFD 2019, online).

Quality Function Deployment

> … is a method for developing a design quality aimed at satisfying the consumer and then translating the consumer's demand into design targets and major quality assurance points to be used throughout the production phase. ... [QFD] is a way to assure the design quality while the product is still in the design stage (Akao 1990).

The phases of a QFD system can be summarized as follows (Woodhouse 1996; NPD solutions 2016):

1. Evaluating the strengths and weaknesses of the use functions of existing products and verifying consumer tastes in order to identify the consumer's real needs and aspirations, thereby creating, if possible, new forms of aspirations.
2. Searching in particular for the characteristics that can visibly distinguish the product from a quality perspective and identifying the priority decisions (deployment) for the product and the process that will maximize the customer's perception of value.
3. Designing a "customer focused" product manufacturing process while trying to match the attributes desired by the customers to the design specifications and production methods, even adopting *target costing*.
4. Coming up with new features that can visibly differentiate the product to create a product "image" that serves as an instrument for satisfying consumer aspirations (brand, trademark, imitation effect, etc.).

Among the instruments used to control functional quality, *Value Analysis* deserves mention. This instrument starts from the principle that the value of a product will be interpreted in different ways by different customers in terms of a high level of performance, capability, emotional appeal, style, etc., relative to its cost. This can also be expressed as maximizing the following function (10.1):

$$\begin{aligned} \text{Value} &= \left(\text{Performance} + \text{Capability}\right) / \text{Cost} = \text{Function} / \text{Cost} \\ &= \text{Functional quality} / \text{Cost} \end{aligned}$$

Value analysis consists of a series of techniques and studies aimed at finding the best product composition in terms of the form and structure of the components and the necessary manufacturing processes, with the aim of minimizing the drawbacks resulting from the product's use (Crow 2002).

> Value analysis is a problem-solving system implemented by the use of a specific set of techniques, a body of knowledge, and a group of learned skills. It is an organized creative approach that has for its purpose the efficient identification of unnecessary cost, i.e., cost that provides neither quality nor use nor life nor appearance nor customer features. When applied to products, this approach assists in the orderly utilization of better approaches, alternative materials, newer processes, and abilities of specialized suppliers. It focuses engineering, manufacturing, and purchasing attention on one objective-equivalent performance for lower cost (Miles 1989, p. 3).

This type of control, in particular *value analysis*, is carried out in the context of the marketing function, which must activate a multilever Control System to develop a strategy that includes the following *operational levers*, as shown in Fig. 10.3:

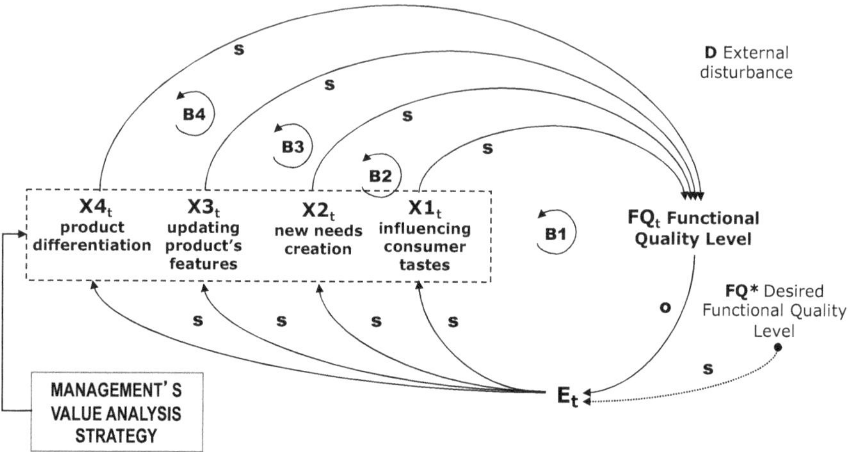

Fig. 10.3 Multilever control system for functional quality control

(a) Verifying and influencing consumer tastes (market research, consumer panels, pilot markets, etc.)
(b) Creating new needs and aspirations (advertising)
(c) Updating the product's technical features as the types of needs change (product design quality)
(d) Coming up with new features that can visibly differentiate the product so as to create a product "image" to serve as an instrument for satisfying consumer aspirations (brand, trademark, imitation effect, etc.).

While the analysis of value is a technique that encompasses the design and manufacturing phases, *quality circles* (Ishikawa 1980; Hutchins 1985; Akao 1990) are organizational instruments for the control and improvement of quality, especially the *design quality*, as part of the quality function deployment.

> [Quality Circle is] … a small group of people doing similar work who meet voluntarily and regularly, usually under the leadership of their supervisors; they identify and discuss their work problems (Gupta 2001, p. 771).

Who better than the workers involved in the production process to identify the problems in assembling, stocking, and distributing materials, component parts, subassemblies, and other items at any stage in the manufacturing process? Who better than the workers involved in the production process to identify the problems in assembling, stocking and distributing materials, component parts, subassemblies, and other items at any stage in the manufacturing process? Who better than these workers to suggest solutions to eliminate such snags? While the *quality circles* rely on the *opinions* of persons *inside* the firm, the *panel of consumers* technique directly seeks out the opinions of *customers*, forming a panel of consumers who are chosen based on statistical sampling techniques and then asked to try out the product—usually a new one—so that they can use it and express an opinion on its quality (and, in general, also on the price/quality relationship).

10.4 Control of "Design" Quality

Productive organizations must also verify that the *design quality* of the firm's production is at least equal to that of the competitors and remains constant or improves over time (Merrill 1997). *Design quality*, as far as it represents the product unit's conformity to a prototype or sample of reference, is mainly expressed as "*absence of defects*" or "*zero defects*" (Zemke 1990). Controlling design quality means:

(a) Maintaining a uniform technical standard in space and over time,
(b) Preventing or eliminating defects and maintaining or increasing reliability,
(c) Setting up a service to carry out: inspections, revisions, repairs, or substitutions.

The premise for effective design quality control is that the product be *properly designed* from the outset by setting rational standards that can be maintained and verified over time. For firms that produce job-orders, the standards are usually included in the contract in the form of detailed specification of the work to be carried out, accompanied by one or more technical designs or even a scaled model, in addition to Gantt and Pert Charts (Chap. 11). For firms with *mass production,* quality verification over time is essential to avoid future complaints and refunds, substitution or reparation costs, and, in the last analysis, to maintain the product's image and market share; this verification is combined with the control of functional quality (which typically concerns itself with effectiveness), though it remains a distinct activity.

The design and development process for a *new product* with high *design quality* standards can be summarized by the following features:

1. *Developing the idea and* assessing *the feasibility of the design.* This stage of product development is intended to develop the initial ideas about the new product. Is it technically achievable? Can it be achieved at a cost that is sustainable and convenient? This phase mainly requires the attention of the design and marketing areas and may take a long time.
2. *Detailed design.* If deemed feasible, the new product must undergo an analytical design to translate the original idea into the best form for efficient production.
3. *Prototype.* Usually, for products that are not too simple it is necessary to build a first version which is faithful to the detailed design in order to make any needed changes or even for product redesigning.
4. *Industrial prototype.* Taking into account the existing production equipment or that which has been built ad hoc, the product is evaluated for its concrete "manufacturability," introducing any changes in its design needed to simplify the manufacturing process (see Sect. 9.9.3). Raw materials can be replaced, adaptations made to equipment, verification undertaken to determine whether the new product can be inserted into some existing production line. At this stage, the suitability of existing production processes can be verified or new production processes tested in "pilot" plants. Production samples are obtained to simulate, as far as possible, all aspects of the manufacture of the product, so that all possible problems can be identified in advance, before the start of large-scale production.

5. *Large-scale production.* This phase coincides with giving full responsibility to the production for the manufacture of the product, which must be carried out using the existing plants, maintaining the characteristics of the initial design, and guaranteeing the quality standards.
6. *Variations in design after the marketing phase*. Many projects require modification, even when the product has already reached the production stage on a large scale. This may take into account suggestions or demands from clients or result from improvements to be made at the production level. These changes in the design may be minimal, but they nevertheless require the existence of a control system.

Design quality control is not limited to the design and production activities. In fact, most control processes concern the manufacturing and marketing processes and cover three areas, which are as follows:

1. *Control of supplies*, within the *Supplier Management Function* (Benoit et al. 2006; CIPS 2007): this aims at the *quality of materials and components*. It is necessary to undertake a process that, through defined control cycles, verifies that the input materials respect the product's required specifications; the outcome of the verifications determines the updating of the stock (unloading verification and warehouse loading) and the continual evaluation of the supplier's reliability, as stated in the *Quality Management System Section* of the ISO 13485 standard, which includes requirements for the process of supplier control (Gardner 2018).
2. *Control of the manufacturing process undertaken to produce and move the product*: this activity controls the manufacturing standards and decides whether to maintain the process or to modify the parameters which do not conform to the standards, and it involves the production function (Mrugalska and Tytyk 2015). We can further divide this control activity into:

   (a) *Production control*: this focuses on the *manufacturing processes* and, in turn, includes two phases:
      (i) *Monitoring* or *inspection*: gathers, files away, processes, and presents data on productivity, the utilization of the means of production, and the progress of production – number of pieces produced, excess production, production times, losses, etc.

   Being a key element of quality control, Product Inspections allow you to verify product quality on site at different stages of the production process and prior to its dispatch. Inspecting your product before it leaves the manufacturer's premises is an effective way of preventing quality problems and supply chain disruptions further down the line (QIMA, online).

      (ii) *Diagnostics*: to verify the state of the production process and the data collection, which serves to identify and signal functional anomalies in the plants and their causes; the control of the adequacy of the production process is fundamental, in the dual form of adequacy of the production system to the desired reliability standards and suitability of the individual workstations included in the system. The examination of the equipment should simulate the real production conditions, and the units produced should be thoroughly tested to ensure that the equipment or machinery is able to achieve and maintain the planned standards.

(b) *Logistics control*: this aims at verifying the punctuality and accuracy of the deliveries and whether or not the selling procedures have been respected; in short:

Logistics refers to the art of managing the flow of materials and products from the source to the user. The logistics system includes the total flow of materials, from the acquisition of raw materials, to the delivery of the finished product, to the ultimate users, and the related counter-flows of information that both control and record material movement (Virolainen 1991, online).

3. *Control and testing of the finished product*: this verifies that the products respect the standard specifications; for this control as well, the director of production or the product manager (if called for) is usually in charge, and the process involves two phases:

(a) *Testing*, a typical control of *efficiency*, when the functionality of the product is verified along with its suitability for providing the performances for which it has been designed; the outcome of the testing determines whether product quality is certified, whether the product is remanufactured, or whether it is discarded (Directive 75/318/EEC 1991, online);
(b) *Revision*, when consumer observations, complaints, and refunds are analyzed to identify the origin of the defects and prevent their future occurrence.

The Control of Design quality requires a multilever Control System like the one shown in Fig. 10.4.

At times the entire production must be controlled; at other times a sample control is carried out based on sophisticated statistical instruments (Smith 1998; Oakland 2014). This latter control is called the Statistical Quality Control, or Statistical Process Control (SQC or SPC) (Montgomery 2007; Raheem et al. 2016). A more detailed analysis of these controls can be found in Sect. 10.14.2 and in the ample bibliography in Industrial Statistics and Quality Control (Duncan 1986; Xie and Goh 1999; Ryan 2000).

The control systems for *design quality* verify, in particular, that the production processes maintain defects within the tolerance limit objectives (considered to be optimal) is shown in the simplified model in Fig. 10.5. The control of parameters normally uses *control charts* (Sect. 10.14.6), which are special charts used when it is necessary to control a continual process. The control charts indicate the checks carried out, the measures obtained, and the upper and lower limits of the maximum

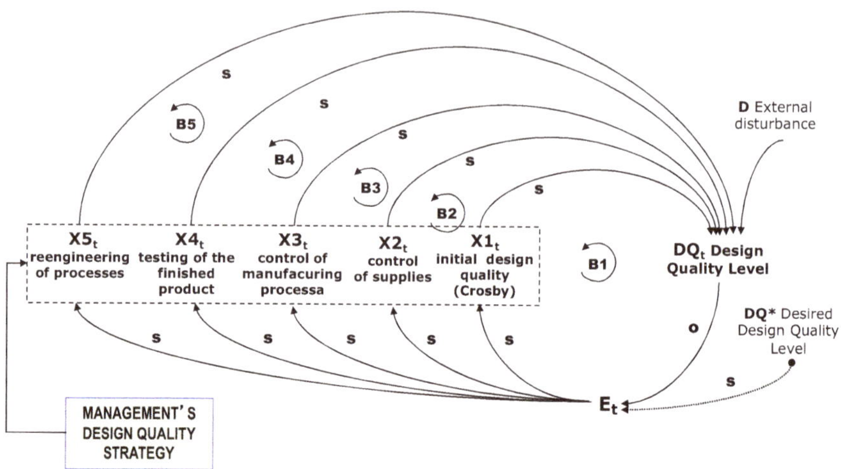

Fig. 10.4 Multilever control system for design quality control

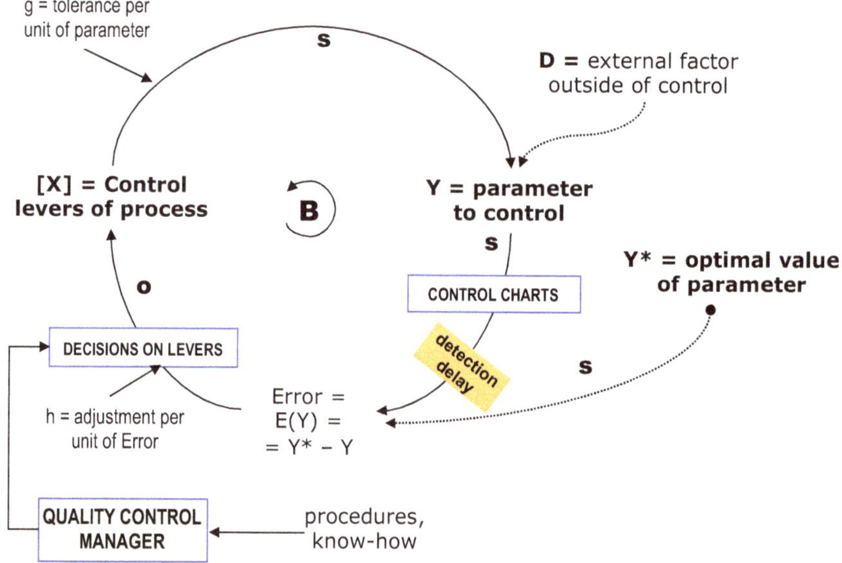

Fig. 10.5 General model of the control system of a parameter

tolerance measures for a given characteristic of the unit or for the production obtained from the process. The control charts allow for a simple and immediate diagnosis of the production process problems, on the condition that these problems are obtained *on line* (while the cycle is in progress) and in real time, that is, before the last moment has passed to intervene efficiently in the process.

In concluding this *section*, a brief mention is in order regarding the control systems that use structural levers through organizational restructuring and the *reengi-*

neering of processes, organs, and functions. These are long-term levers that lead to radical changes in the organization but which can increase the power of control systems for processes (Davenport and Stoddard 1994). Equally potent, but more gradual in moving toward the desired objectives, is the *continuous improvement* which leads to the gradual improvement in the performance of processes leading to the objectives (which normally involve quality) (Deming 1989).

10.5 Controlling the Costs of Quality

While *functional quality control* mainly impacts revenues, the control of design quality is of economic importance mainly from the *cost side* (Albright and Roth 1994), since there is a clear inverse relation between the control of design quality and defectiveness (Rosenfeld 2009). To control defectiveness, it is fundamental to structure an effective *defect prevention, appraisal and failure process* (Ferretti et al. 2013; Dhar and Nandi 2016) that will eventually result in the Six Sigma Quality Management control system (Bellows 2004) (Sect. 10.14.1).

If we assume we are controlling a given *parameter* that refers to finished products for all output units of the production process over a given interval, then we can become aware of the "design quality" of this same process in terms of *defectiveness*, by building a *control chart* like the one shown in Fig. 10.6; the left side reports the dynamics for the number of defective units; the right side shows the same process after the design quality control, which leads to an improvement that, on the one hand, reduces the standard average level of defectiveness, and on the other reduces the number and duration of "out of control points" (for details, see Sect. 10.14.6 and Fig. 10.15).

The *costs of design quality* concern both the products and the managerial system and can be divided into (Tang et al. 2004; Campanella 1999):

A. *Costs of internal and external nonconformities* (or full nonquality costs) linked to the *absence of control* (Schneiderman 1986);

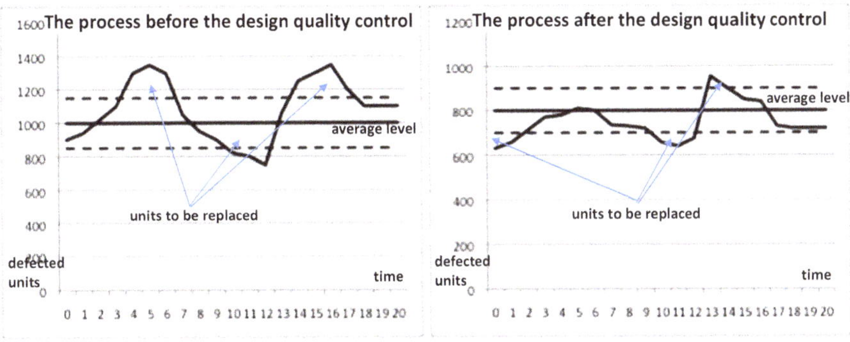

Fig. 10.6 Control charts showing the effects of design quality control

A.1 *Internal failure costs*: these derive from the need to eliminate anomalies and defects that occur in the production units *during* the manufacturing process;

A.2 *Lack-of-rationality costs*: these are connected to the lack of rationality in production, in the product, and in the marketing and post-sales organization due to the presence of defects;

A.3 *External failure costs*: these arise from the repair or substitution of the product *after* sale;

B. *Investment quality costs*, linked to the *procedures for implementing controls*;

B.1 *Prevention costs*: these arise from the attempt itself to *prevent* defects;

B.2 *Appraisal and assessment costs*: these arise from procedures to *verify* product defects;

C. *Opportunity quality costs* (*customer-incurred cost*, loss of productivity, *customer-dissatisfaction cost*, dissatisfaction shared by word of mouth, *loss-of-reputation cost*, customer perception of firm) (Harrington 1987), resulting in lower revenue due to the reduction in functional quality. In effect, design quality has a notable impact on the product's use function since, on the one hand, design quality damages and discourages the consumer (reduction in sales), and on the other forces the firm to lower prices; in general, the intangible costs are thus identified with the failed earnings caused by low levels of quality (Zemke 1990). *Indirect* costs and opportunity quality costs obviously derive from the lack of *verification* and *prevention* of defects. Taguchi (1986) considered quality in a "social loss approach" as "*the loss a product causes to society after being shipped, other than any losses caused by its intrinsic functions. This loss can be caused either by variability in the product's function or by adverse side effects*" (Taguchi 1986, p. 1). The two graphs in Fig. 10.7 clearly show the importance of *prevention* and *inspection* activities and their effect on defectiveness.

We can assume there is an optimal level of design quality that minimizes *total quality costs*, which is the sum of the direct and indirect costs, as shown in Fig. 10.8 (left). We can now modify the graph in Fig. 10.6 (left) by also adding "intangible costs" arising from the *lack of quality*. Such costs cause a shift to the right of the minimum point of the total cost curve, which indicates that the optimal level of quality is higher in the presence of intangible costs, as shown in Fig. 10.8 (right).

Figure 10.9 shows, in the same graph, the *total revenue curve* as a function of the level of *functional* quality and the *total quality cost* curve as a function of the level of *design* quality, where the quality levels are expressed on a uniform scale. This graph shows that the optimal level of overall quality—functional as well as design—can be determined as the level of overall quality that corresponds to the maximum profit, equal to the difference between only those revenues and costs connected to quality, which we assume can in principle be determined.

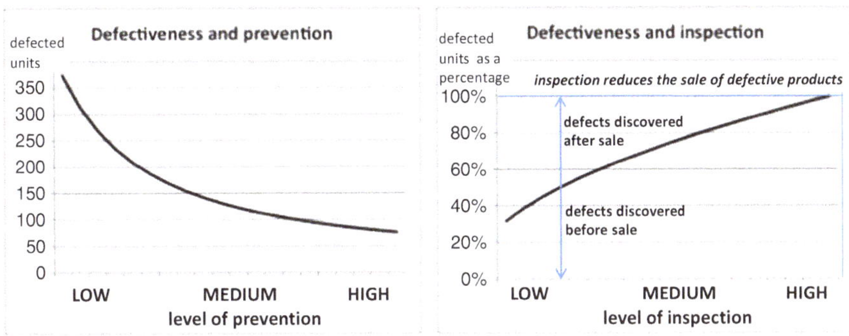

Fig. 10.7 Control of quality through prevention or inspection

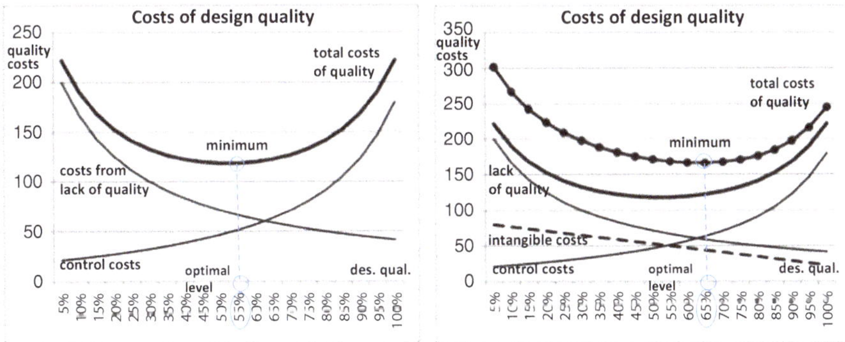

Fig. 10.8 Costs of design quality

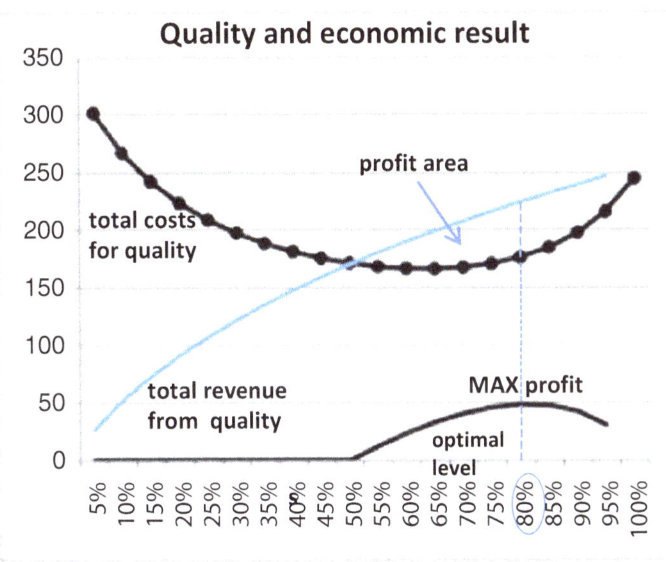

Fig. 10.9 Profit as a function of quality

10.6 Control of the "Environmental Quality"

The environmental compatibility of products must be continually monitored, taking account of the possibility that the search for functional and design quality is often in contrast with the objective of safeguarding the environment. The lack of environmental quality impacts revenues in the form of reduced demand and the costs needed to safeguard the environment. The *forms* of control of *environmental quality* are not simple to identify, since the standards of environmental compatibility are not absolute but depend on several variables:

(a) The state of *scientific and technical knowledge*. Up until the discovery of its terrible polluting power, asbestos was considered a key component of all fire-resistant material, providing such material with a high functional quality; and upon discovery of Mad-Cow Disease, even t-bone steaks, whose functional quality is not open to discussion, were temporarily banned;
(b) The *prevalent social culture*, since sensitivity toward environmental problems is not the same in all countries (the refusal to abandon the use of polluting products in several countries is indicative here). During the industrialization phase, the construction of residential "clusters" on the periphery of large cities was deemed essential for the internal migration of manpower; once this migration ended, each new "cluster" was, and is still, viewed negatively as a source of social degradation; and the energy crisis changed opinion toward nuclear power plants, which are widespread in industrialized countries; however, today this source of energy supply is increasingly rejected in favor of wind or solar energy;
(c) *Special economic needs*. Gas refrigerators worsen the hole in the ozone layer, and automobiles with polluting engines but low prices are viewed favorably in industrializing countries, even if their environmental quality is questionable.

Despite these difficulties, firms can control the environmental quality of their processes and products thanks to the use of increasingly sophisticated techniques in Environmental Impact Assessment and Sustainability Appraisal, whose application is obligatory for many production processes (Thérivel and Phillip Minas 2012).

10.7 From Quality Assurance to Company-Wide Quality Control

Firms cannot simply control quality by means of individual methodologies and instruments; they have to integrate methods and tools into a quality system to achieve the gradual involvement of all corporate functions in the control of quality, transforming the *Quality Control Function* into a *Quality Assurance Policy* with the aim of guaranteeing the total satisfaction of the final user (Vlăsceanu et al. 2004).

> Customer satisfaction determines the success of a new product and only products at high value meet needs of clients who expect them to perform correctly in their whole life cycle. In order to fulfill such requirements, the minimum of variation of parameters should be

> assured within the manufacturing processes and the product itself. From an elementary part to compound parts, they must be designed and manufactured on high quality level and be reliable and safe in use (Mrugalska and Tytyk 2015, p. 2730).

> Quality Assurance (QA) is a planned system of review procedures conducted by personnel not directly involved in the inventory compilation/development process. Reviews, preferably by independent third parties, are performed upon a completed inventory following the implementation of QC procedures. Reviews verify that measurable objectives were met, ensure that the inventory represents the best possible estimates of emissions and removals given the current state of scientific knowledge and data availability, and support the effectiveness of the QC programme (Winiwarter et al. 2006, p. 6.5).

This definition highlights the fact that firms must achieve the involvement of the entire organization in the Quality Assurance policy (Taguchi 1986). This necessitates:

> … a positive attempt by the organizations concerned to improve structural, infrastructural, attitudinal, behavioural and methodological ways of delivering to the end customer, with emphasis on: consistency, improvements in quality, competitive enhancements, all with the aim of satisfying or delighting the end customer (Zairi 2010, p. 74).

Quality Assurance cannot be only a production, marketing, or economic need but also must become the objective of a true *Total Quality Management* (TQM) strategy (Deming 1989) or one of *Company-Wide Quality Control* (CWQC) (Ishikawa 1985; Ishikawa and Lu 1989)—considered the Japanese version of TQM. The TQM strategy must involve the entire production, personnel and marketing structure (Oakland 2014) in the control and improvement of product quality and production, involving top management, directors, supervisors, and workers from all activity areas of the organizational group, such as market research, research and development, planning, purchases, sales management, production, inspections, selling, and personnel services (Sharma and Kodali 2008). Also of importance is the relationship with customers and suppliers, who are viewed as the engines for quality improvement; the customers make suggestions for the improvement of functional quality, and the suppliers must be the guarantors of design quality. The culture of quality spreads from the suppliers to their own suppliers, so as to permeate the entire production line and entire business sectors. Armand Feigenbaum (1991) has developed a ten-point framework to implement Total Quality Management, which is parallel to the 14 points proposed by Deming (1989) (see below).

A constant underlies the vision of these authors: the firm must create a system to manage quality which is aimed at customer service. It must reach every sector of the organization and be understood by all personnel, who then must truly believe in it and enthusiastically participate in its implementation by proposing the creation and spread of the *quality circles*. In Feigenbaum, Deming and Ishikawa's conceptual framework, quality is not achieved through a discrete process, through innovations, often temporary, in products and processes, but through a continual process of "small steps," the *kaizen* (Imai 1986; Tanaka 1994), which produces a considerable evolution even in the way of conceiving of the relationships among the various operational departments or centers within the firm. Each center must guarantee the maximum quality to the center downstream—its "customer"—and demand the maximum quality from the center upstream—its "supplier" (Chiarini et al. 2018).

In very general terms, it can be said that the Japanese vision of improving quality, which is now widespread around the world, stems from the peculiar conception of the factors of competitive success for companies (Sect. 10.14.1). Western companies, still tied to traditional patterns, interpret profit as a fundamental objective of business management (as is still taught today, even in university management courses in many Western universities), and quality, along with productivity, is the necessary condition for achieving profit. In Japanese companies, oriented toward functional and design quality as key factors of competitive success, the profit-quality nexus must be reversed: the real goal of the company is customer satisfaction, which is achieved through the quality of the products; profit is only a consequence of achieving customer satisfaction, if management is supported by adequate productivity values. This different view of the role of quality leads to different ways of *pursuing quality* (Deming 2012):

- The West holds the view that quality should be achieved through a *discrete* process, through (often occasional) innovations in products and processes;
- In the East, on the other hand, the pursuit of quality must be a *continuous* process, which takes place by taking small steps in a process called *kaizen*.

Table 10.1 compares certain elements of the quality improvement process that emerge from the discontinuous process in the West and the continuous *kaizen* process. In short, in the West, profit is the real objective of the company and quality is necessary to achieve it. In the East, on the other hand, quality is the real objective and profit is merely a consequence, a "premium" for achieving quality. This explains

Table 10.1 Two visions of quality

Western firm Quality must be controlled	Japanese firm Kaizen
The firm survives if it makes a profit (adequate profitability)	The firm survives only if the customer is satisfied (customer satisfaction)
Profit is the primary objective of the businessman	Quality is the primary objective; profits are the premium for competitive success
Quality affects revenues	Quality affects customer satisfaction
The search for quality entails costs	Quality has no cost if it "is arises" at the beginning
Quality must be controlled by a specific organ or center	Quality must involve the entire organization
Workers must be controlled; otherwise, there will be a decline in quality	Workers are the first controllers of quality
Suppliers must be controlled; otherwise, the supplies will be of low quality	Suppliers must be involved in the search for quality
The product must be controlled	The product must emerge already with the maximum quality and be improved through suggestions from the customers
Customers must be assisted after delivery	Customers must contribute to improving the product by pointing out defects or suggesting improvements

the importance of the relationship with customers and suppliers, which are seen as the engine of quality improvement. Customers give suggestions for improving functional quality, while suppliers are the guarantors of design quality.

William Edwards Deming, who is recognized as the "father" of the Japanese movement toward the "culture of quality," has tried to spread this culture even in the Western world by proposing the concept of Total Quality Management. To help management become aware of the need and the possibility to change traditional management culture, Deming has identified "14 *key principles for management,*" proposing a "system of Profound Knowledge" (Deming 2018) based on a "holistic" conception of the manager's function that involves four "pillars": appreciation of a system (i.e., Systems Thinking), knowledge of variation (to apply Control Theory), the theory of knowledge (i.e., thinking through models), and psychology.

> What we need is cooperation and transformation to a new style of management. The route to transformation is what I call Profound Knowledge (The Edwards Deming Institute 2020, online).

> The first step is transformation of the individual. This transformation is discontinuous. It comes from understanding of the system of profound knowledge. The individual, transformed, will perceive new meaning to his life, to events, to numbers, to interactions between people (ibidem).

Deming's "14 points" are quite complex and do not lend themselves here to an analysis, albeit a concise one. I shall only list them:

(1) Create constancy of purpose toward improvement; (2) Adopt the new philosophy; (3). Cease dependence on inspection to achieve quality; (4) End the practice of awarding business on the basis of price tag. Use a single supplier for any one item. (5) Improve constantly and forever the system of production and service; (6) Institute training on the job; (7) Institute leadership; (8) Eliminate fear. (9) Break down barriers between departments; (10) Eliminate slogans, exhortations and targets for the work force asking for zero defects and new levels of productivity; (11) Eliminate management by objectives; Substitute leadership; (11) Eliminate management by objective. Substitute workmanship; (12) Remove barriers to pride of workmanship; (13) Institute a vigorous program of education and self-improvement; (14) Put everyone in the company to work to accomplish the transformation. The transformation is everyone's work (Deming 1989, Chap. 2).

10.8 Controlling the Quality of the Organization to Generate Trust and Reputation

Firms must develop a policy to segment the three forms of quality within an organization and produce an improvement in the internal processes and products for the customer's benefit, thereby creating a holistic model of quality management (Srikanthan and Dalrymple 2005). Whatever the nature of the products offered, their *utility*, *value*, and *suitability* are the drivers of consumer *loyalty*, which, together with *trust*, build up the *reputation* of the firm, on which market share depends, which in turn impacts revenues, taking into account the role of design and functional quality on prices (Fig. 10.1). The first and most general action firms must undertake is to transfer the three notions of quality from the products to the organization—which, in turn, is the basis of market evaluation—by developing the following:

A. The *functional quality of the firm*, by adjusting the characteristics of the firm so that it is *judged* by the market as suitable for producing value for all the external stakeholders and assigned a value as a single entity; that is, the firm is characterized by the same "fit for purpose" concept that we have examined for products: *"fitness for purpose" is a definition of quality that allows institutions to define their purpose in their mission and objectives, so "quality" is demonstrated by achieving these"* (Woodhouse 1999, 29–30).
B. The *design quality of the firm*, by adjusting the characteristics of the organization to carry out production processes which conform to standards considered suitable (with respect to sanitary and safety norms at the workplace; secure and useful products); a high quality in carrying out the company processes (seriousness, consistency in the procedures, manufacturing uniformity, etc.) generates *reliability* for consumers (Robertson 1971; Krishnaiah and Rao 1988).
C. The *environmental quality of the firm*, by conforming the qualitative features of products to create the vision of a firm capable of playing a positive role in the environment, both from the social point of view (employment, housing improvement, consumer utility, etc.) and the environmental one (nonpolluting, respect of safety regulations, etc.). The perception of environmental quality leads to *appreciation* of the firm by both economic stakeholders and social ones (the political community, unions, territorial agencies, etc.) (Sewell and Brown 2009).

Figure 10.10 shows the joint effect of quality control product and company quality:

1. The three forms of *product* quality, expressed in terms of *utility, value,* and *suitability,* are the drivers of customer loyalty (Dick and Basu 1994) and brand store loyalty (Story and Hess 2006).
2. The three forms of *company* quality, in terms of *evaluation, reliability,* and *appreciation,* produce *consumer confidence* (Acemoglu and Scott 1994) and positive *customer sentiment* toward the firm as an operating entity in the environment (Bennett and Harrell 1975; Rust et al. 1999).

TYPE OF QUALITY	PRODUCT	FIRM	PRODUCT & FIRM
Functional	Value, Satisfaction	Evaluation	
Design	Utility, Image	Reliability	
Environmental	Suitability	Appreciation	
Effects	LOYALTY	CONFIDENCE	REPUTATION & EXCELLENCE

Fig. 10.10 The effects of product and corporate quality

3. *Loyalty*, together with *confidence*, enhance the firm's *reputation*, which represents the fundamental value driver on which market share, price level and, as a consequence, revenues and added value depend (Sweeney and Soutar 2001).

   According to this logic:

   > The leverage of a strong brand name can substantially reduce the risk of introducing a product in a new market by providing consumers the familiarity of and knowledge about an established brand. Moreover, brand extensions can decrease the costs of gaining distribution and/or increase the efficiency of promotional expenditures (Aaker and Keller 1990, p. 27).

Customer satisfaction has always been seen as a key indicator of quality policies. However, it must be accompanied, when possible, by so-called *customer delight*, which indicates *not only* the extent to which a customer's expectations are met, but also the extent to which the customer judges that the quality of the asset/service purchased significantly exceeds his initial expectations, thus creating a positive emotional reaction generating a positive "word of mouth" that affects market demand and extinguishing the company and its products/services from those of the competition. *Customer delight* is not synonymous with *customer satisfaction*: the latter occurs when a client purchases a product to satisfy his needs, uses it, and through this use receives the expected benefits.

> The customer can be delighted only if organisations know what the customer expectations are. "Delight is nothing other than exceeding customer expectations" (Popli 2005).

> Customer delight happens when you surprise a customer (or client) by exceeding expectations. When expectations are met, you have customer satisfaction. When expectations are exceeded, you achieve customer delight (Willis 2015 online).

With *customer delight*, the customer's benefit from the product is not limited to the satisfaction of his need but exceeds the buyer's expectations, thereby "making him happy."

There are three objectives for the company in pursuing customer delight:

1. Make customers loyal to its brand (Sewell and Brown 2009).
2. Increase the "value" perceived by the customer; when the product is considered to be of higher than expected quality, customers place less importance on the price.
3. Push customers, through "word of mouth," to become "testimonials" by speaking positively about the product, so that the brand or the company store acquires a proven reputation, supported by other experiences, since the reputation of the brand depends heavily on the corporate identity of the company.

10.9 The Control of Productivity: The Efficiency of Productive Transformation

As shown in Fig. 9.11, the *economic efficiency* of the TR2-E: ECONOMIC OR MARKET TRANSFORMATION depends on the production efficiency of the TR1-P: PHYSICAL PRODUCTIVE TRANSFORMATION. It is useful here to take up and expand on the considerations made in Sect. 9.3 to clarify the role of productivity in controlling EBIT or Operating Return, as summarized in Fig. 10.1. First of all, we must recall that production systems interface with input (supply) markets, where they purchase the factors of production from other organizations, called suppliers, as well as from workers, in short, and with output (or end) markets, where they sell their production to other organizations, called customers. The purchase prices (pL, pM, pP&M) allow *costs* to be quantified as factor values [C] (inputs); the selling prices (pP) quantify the revenues [R] as values of the production [P] (output). Through the valorization of the inputs and outputs by means of prices, the *productive* transformation becomes a *value transformation*, as shown in Fig. 10.11.

The EBIT can be quantified by the following equation:

$$\text{EBIT} = \left[\text{RP-}\left(\text{CL} + \text{CM} + \text{CR\&P}\right)\right] = \text{RP} - \text{CP} \quad (10.3)$$

where RP denotes the sales revenues and [CP = (CL + CM + CR&P)] denotes the overall production cost. The ratio between [CP] and [QP] is defined as the *average unit cost* of production [cP = CP/QP], which may be rewritten as follows:

$$\text{cP} = \left(\text{ql} \times \text{pL}\right) + \left(\text{qM} \times \text{pM}\right) + \left(\text{qP \& M} \times \text{pP \& M}\right) \quad (10.4)$$

We can transform [10.3] into a more significant expression for EBIT by writing the revenues and costs directly as a function of the selling prices [pP] and unit costs of production [cP] as calculated by equation [10.4]. We thus obtain:

$$\text{EBIT} = \text{RP} - \text{CP} = \left(\text{pP} \times \text{QP}\right) - \left(\text{cP} \times \text{QP}\right) = \left(\text{pP} - \text{cP}\right) \times \text{QP} \quad (10.5)$$

The previous expressions clearly show that the production organization—assuming it makes decisions rationally—can try to control EBIT by using four decision-making levers to:

(a) Expand the *volumes* of production [QP] through the *sale function.*
(b) Control the *selling prices* [pP] through the *marketing function* in order to expand QP and reduce the risks of competition.

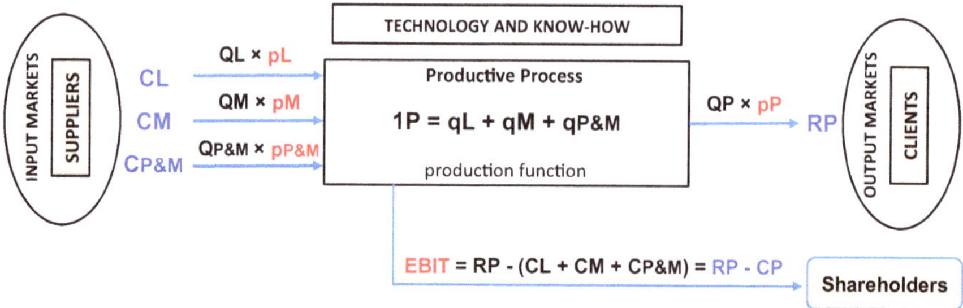

Fig. 10.11 The economic transformation (author's elaboration)

(c) Reduce the unit factor requirements [qL, qM, qP&M] which influence unit costs, through the *production function*, as shown in (10.4) (Syverson 2002), with an intensive use of new technologies (Stiroh 2005).
(d) Negotiate lower *factor supply prices* (for pL, pM, etc.) through the *supply function.*

The European Commission communication to the European Council and Parliament of 21 May 2002, whose object was: Productivity: the Key to Competitiveness of European Economies and Enterprises, provides this definition:

> In formal terms, labor productivity is the quantity of labor required to produce a unit of a specific product. In the macroeconomic context, labor productivity is measured as a country's gross domestic product (GDP) per capita of employed population. Productivity growth depends on the quality of physical capital, improvements in the skills of the labor force, technological advances and new ways of organising (Arnold and Dennis 1999). Productivity growth is the principal source of economic growth (COM 2002a).

If we let "πL" stand for the average productivity of labor (L), the COM definition indicates the *efficiency in the use of the labor factor*, which is directly quantified by the ratio:

$$\pi L = \frac{QP}{QL} = \frac{QP}{qL \times QP} = \frac{1}{qL} \tag{10.6}$$

where QP is the volume of production and QL is the quantity of labor.

More generally, I define *productive efficiency* as the capacity of the *productive system* to reduce the unit factor requirements and *productivity* as the capacity of a productive system to maximize the quantity produced with the minimum use of labor.

If we let "πF" stand for the average productivity of a generic factor "F" (used in a productive system for a defined period and under unchanging conditions) and generalize the above definition, then productivity is measured through the following relationship:

$$\pi F = \frac{QP}{QF} = \frac{QP}{qF \times QP} = \frac{1}{qF} \tag{10.7}$$

which immediately shows that *productivity* is entirely equivalent to *productive efficiency* (Nordhaus 1996). The search for increasingly *productive efficiency* levels (and thus *labor* productivity) is undertaken by all *production organizations* or *firms*, since:

(a) As (10.4) clearly shows, the more the *productive efficiency* increases, the more the *production costs* fall (the unit requirements, qF, fall), with supply prices held equal (David and Wright 1999a).
(b) Alternatively, as follows from (10.5), we can neutralize the increases in the *factor prices* in order not to reduce EBIT.

(c) The lower production costs allow the firms to obtain higher levels of EBIT even without increasing *selling prices*.
(d) Firms can maintain or improve the levels of EBIT also by reducing the *selling prices*, thereby undertaking a strategy for increasing demand and reducing potential competition (Sen and Farzin 2000).

For this reason, all *production organizations* pay particular attention to the trends in *labor productivity*.

> The ability to obtain more output from given inputs of labor and capital corresponds to growth in productivity. Productivity growth depends on the quality of physical capital, improvements in the skills of the labor force, technological advances and new ways of organising these inputs. Historically, productivity growth has been the principal source of economic growth. It has made possible an expansion of output, not just without concomitant increases in inputs, but with important reductions in hours worked over the medium term. In doing so, it has made a sustained rise in real incomes possible (COM 2002b: par. 2).

10.10 Explaining Productivity. The "Intrinsic" Drivers of Labor Productivity

There are two questions which arise when analyzing the increasing productivity of labor:

1. What does the increase in productivity in individual productive systems depend on? This leads us to search for the *factors of productivity*.
2. What do the trends in the productivity of the entire economic system depend on?

In order to answer the *first question*—on what does the increase in productivity in individual productive systems depend—I shall use the term *drivers of productivity* to refer to those elements, phenomena, causes or variables that, in the equation (10.6), can produce an increase in QP and/or a reduction in QL. Though varied, the *drivers of productivity* can be grouped into few categories which, though they interact, must be kept separate in order to facilitate our observation. Let us first consider the "*intrinsic*" *drivers* of productivity, which derive from Adam Smith's famous example of the manufacture of pins to illustrate the advantages of the division of labor.

> This great increase in the quantity of work, which, in consequence of the division of labor, the same number of people are capable of performing, is owing to three different circumstances; first, to the increase of dexterity in every particular workman; secondly, to the saving of the time which is commonly lost in passing from one species of work to another; and lastly, to the invention of a great number of machines which facilitate and abridge labor, and enable one man to do the work of many (Smith 1776, p. 5).

Generalizing Adam Smith, I propose this typology, and referring to (10.6):

A. *Passive drivers*: These increase QP with QL held constant; there is only one passive driver of productivity: *fertility*, in all its forms: the fertility of land, water, subsoil; *natural* (the banks of the Nile) or *artificial* fertility (irrigated and fertilized land).

B. *Active drivers*: These reduce the QL needed to produce with fertility held constant; there are three types of active drivers of productivity, which are as follows:

   1. *Skill*: It is easy to imagine that a skilled fisherman or hunter can obtain the same amount of goods in less time than other, less-skilled producers; obviously, applying the same amount of labor as the others, they can obtain a larger quantity of products.
   2. *Equipment:* From the first chipped rock to modern machines and digitally-controlled plants, equipment extends the capacities of «human hardware» (represented by our body, its limbs and by our brain), reducing the effort and danger associated with production and rewarding skill. Holding the amount of labor employed constant, we can greatly increase the amount of clothes produced by using a loom rather than a crochet needle; with labor hours held constant, we can increase beyond measure the amount of arable land by using a modern-day tractor, which uses a six-bladed plough, compared to what we could obtain with a single-bladed plough pulled by a pair of oxen. It is easy to imagine how few cars would be produced without the modern robotized assembly line; the construction of three pyramids with the limited equipment at the time required more human labor that the construction of all of Manhattan's skyscrapers put together. We can easily recognize the progress computers and robots have produced until now and will produce looking forward.
   3. *Specialization*: A characteristic of the organization is the functional division of labor; this means that each worker in the organization provides his skilled- and equipment-aided labor to undertake a specific activity; this specialization, together with the equipment available, further increases productivity.

C. *Endogenous or psychological drivers*: These are the psychological conditions that lead man to supply his labor to a given organization; these drivers can be divided into the following:

   1. *Motivation*: Man is willing to supply his labor only if adequately motivated and if there are expectations that this will satisfy his needs or motivations. Once the main motivation was paid in the form of wages, salary, or profit; today other motivations, intellectual in nature, have been added to those regarding monetary payments: a job that gives satisfaction is often preferable to a better paid but boring one. Motivation drives man not only to be more efficient but also to continually learn to improve his preparation. Motivation not only is linked to a worker's personal and family situation but also depends on individual "vision" (improving one's status) and the collective culture (advance the group, the homeland, etc.); this driver also includes the worker's ethical attitude toward his role in the production and toward time-wasting.

2. *Satisfaction*: Motivation provides work incentives to man; the initial *motivations must be followed by satisfaction, that is, the satisfactory achievement of* motivations.

Intrinsic drivers directly improve the worker's performance. Maximum productivity occurs when skilled, equipped, and organized work, properly motivated and satisfied, is supplied in a fertile environment, that is, when production is organized into productive systems within production firms and enterprises. The primary challenge to developed economies will be to create the endogenous drivers of productivity, that is, to motivate and satisfy workers (Baumol et al. 1989). Certainly, the problem for the future (which already today has become a crucial point) will be to improve productivity through the quality of both products and working conditions.

10.11 The "Managerial" Drivers of Labor Productivity

To check the labor productivity in an organization, the productivity factors must be applied by management through effective organizational tools, the most relevant of which include:

(a) The *quality* of labor in terms of skill (linked to learning) and the makeup of the workforce (gender, civil status, age, educational background, etc.); a continual updating of the skill levels.
(b) The degree of *rationalization* in labor employment, which depends on the following:

   1. A more rational analysis of working methods and procedures (this factor will be taken up in the next *section*)
   2. A more careful analysis of working times
   3. A better planning of work activities regarding the various production processes, taking into account the needs of the various operating centers in the company
   4. A variation in labor mobility, that is, in the company's ability to shift the labor force among various divisions to reduce the rate of excess labor and ensure a more rational distribution of resources

(c) The organized control of *labor performance* linked to the following:

   1. The introduction or improvement of incentives (motivation and satisfaction), both monetary (bonuses, excess incentives, etc.) and nonmonetary (job security, protection against risks and occupational diseases, indirect benefits, such as child care facilities, summer camps, organized vacations, subsidized housing, transport, etc.).
   2. Improvements in working conditions (lighting, noise, cafeteria service, hygiene, etc.). Thus, for example, with working time held constant, an expert

worker will produce more than an inexperienced one. A tired worker, or one forced to work under unfavorable physical and psychological conditions will, with all other conditions equal, produce a lower quantity than another who can work under better conditions.

(d) A more rational *organization of tasks*, a more prudent allocation of times and more balanced distribution of work among the production centers, so that any shortages in certain centers can be covered through the reassignment of times and workplaces of workers from other centers, and so on.
(e) An improvement in the quality of employed labor by influencing the productivity of the other factors, that is, materials (e.g., improvement in the quality of materials or the time needed for these to enter into the production processes) and equipment (e.g., an increase in mechanization, an improvement in layout, etc.). It is especially important to coordinate labor with the mechanical factors of production by making better use of machines and equipment.

10.12 The “Extrinsic” Drivers of Labor Productivity

Along with these traditional *intrinsic drivers* of productivity, we can identify other, *extrinsic* drivers, so called because they involve the organization of productive systems, the environment within which the work is carried out, and, in the final analysis, the firm’s policies regarding increased productivity. These differ depending on the type of productive sector and the type of production; among the most important are the following (for a more in-depth economic and historical treatment, see the vast literature on this subject-matter) (Schmitz Jr. 2001):

(a) *Continual mechanization*: Mechanization began with the origins of man himself, with the natural creative capacity of the worker to produce equipment, a capacity subsequently passed on to the organization as a unitary system. In recent years, this process has accelerated due to electronic and computer technologies.
(b) *Online automatic control systems for processes*: One of the factors that slows down work is the control function. Mechanization and electronic technologies have automatized the control of processes, with a noticeable increase in labor efficiency in the productive system.
(c) *Work environment—ergonomics:* A well-equipped and ergonomic work environment speeds up the execution of tasks to the advantage of productive efficiency.
(d) *Advanced information systems* and *dematerialization*: Making information gathering and access to data bases immediate, thereby increasing their quantity and variety, improves the decision-making process, facilitates the execution of tasks, and avoids errors that require control, with a consequent savings in labor time in every phase of the production process.

(e) *Standardization* and use of *new materials*: Designing products and processes based on unified criteria, utilizing materials and components with specific predefined (and thus interchangeable) standards conceived and designed to make the processes they are used in more efficient, speeds up assembly work and makes it more precise, and reduces quality restoration measures, thereby increasing the productivity of the entire production system.
(f) Increase in the *speed of processes* through streamlined production and the continual improvement and rationalization of the flow of component assembly through the search for *total quality* and the application of *just in time* (Demartini and Mella 2011; Brynjolfsson 1994; Brynjolfsson and Hitt 2000).
(g) Rationalization of *logistics* and of materials handling (Bowersox et al. 2005).
(h) Outsourcing and *decentralizing* the search for productivity to more efficient units (Mella and Pellicelli 2012).
(i) Progress in *scientific and technological research*, in particular the development of systems of energy supply and of the productive use of energy, as masterfully demonstrated by Carlo Maria Cipolla in *The Economic History of World Population* (1962).
(j) Improving the *organizational learning*. The faster the organizations learn how to develop motivation for individual and collective learning the more the entire organization is capable to increase production efficiency (Senge 2006).

10.13 The "Law" of Increasing Productivity and Quality

We can argue (as observation will prove) that in an *economic system* with a developed *network* of production firms (Sects. 8.13, 8.14, and 8.15), each productive organization has a higher probability of long-term survival if it produces at levels of productivity at least equal to those of other firms; in fact, to survive, each production firm must try to improve its efficiency, and thus its productivity levels. This general *behavior* of every productive organization, every node of a production network, results in the continual increase of overall productivity in the system, thereby triggering a positive *feedback* that leads to an increasing trend both in the firm's *micro* behavior as well as in the system's *macro behavior*. What I have just said about "productivity dynamics" also applies to "quality dynamics". The SECOND LAW OF PRODUCTION NETWORKS clearly states that Production Networks tend to increase the quality of their performance—mainly due to the increase *in* the productivity and quality of production—through a nonlinear cumulative process. This law applies to both network performance improvements and the individual production units that make up the nodes, since it is characteristic of nodes to improve their efficiency, along with that of the entire network, as indicated by their third feature (Sect. 8.14). Thanks to the action of production networks, the increase in *productivity* and quality levels becomes the *dominant phenomenon* in the entire economic scenario and ends up becoming *institutionalized*; so much so that I would not hesitate to describe it as a true *hypothesis*, or "Law", concerning how economic man behaves.

Law of increasing productivity and quality to successfully survive as nodes in production networks, firms tend to achieve increasing productivity and develop higher levels of product quality; this individual response leads to a general increase in the productivity of the system, which, in turn, conditions the successful behavior of the firm. Therefore, the network of production enterprises tends to achieve increasingly higher levels of productivity and quality, but it is governed in turn by this continual increase in productivity.

This "law" presents two "corollaries" that, in fact, represent the very foundations on which the Law of increasing productivity and quality is based.

First Corollary If a "logical" or "technological" innovation in a process can save time and capital, obtaining the same quality and quantity, then that innovation will be adopted and will spread to all similar processes.

Second Corollary If a "logical" or "technological" innovation in a process can improve the quality of the result achieved in a process, with the same time and capital requirements, then that innovation will be adopted and ill spread to all similar processes.

There is no need to study the history of technology to verify the action of these two corollaries; we need only reflect on our daily processes. If it took us 10 minutes to make mayonnaise by hand, with the discovery of the minipimer we reduce the time to a fraction of what was previously required and are slowly incentivized to use the minipimer for many other cooking processes. Imagine how much the productivity of potters and the quality of vases have increased as a result of the invention of the lathe? Are there still ships for transporting people and goods? The introduction of the engine (coal, oil, or nuclear) has reduced transport times, increased the size of ships, and made service comfortable and safe. These are only a few of the examples that could be cited.

The improvement in productivity leads to a reduction in the amount of work required for production not only at the level of the individual enterprise, but also at that of the entire economic system; this dynamic can easily be seen in the following phenomena:

(a) Increase in the age when people begin their working activity (raising of the age for compulsory schooling, increase in schooling, etc.) (Iacovou and Berthoud 2001)
(b) Reduction in the age when work activity ends (retirement)
(c) Reduction in the average annual working days (longer holidays)
(d) Reduction in the average daily work hours (shortened work week)
(e) Increase in nonemployment (unemployment, delayed employment, and layoff due to redundancy)

If the first few factors are viewed positively, the increase in nonemployment or in unemployment is worrisome today, even if the transition toward the *jobless economy* is inevitable (David and Wright 1999b; ASQC 1961) (Sect. 10.14.8).

I have tried to provide operating justifications for this hypothesis demonstrating how, from a typical holonic viewpoint (Mella 2009), this increase in productivity derives from the activities of production firms forming the global production network, the Big Machine (Sect. 8.16.3).

10.14 Complementary Material

10.14.1 Six Sigma Quality Management System

"Six Sigma" is a term coined by Bill Smith, an engineer at Motorola, who was the first to study the method ("Six Sigma®" is, in fact, a registered trademark of Motorola), which is part of the methods for the *quality assurance* of *design quality*, which aims at the greatest reduction possible in *average defectiveness* (Wheat et al. 2003; Taghizadegan 2006; Pyzdek 2003).

Until a few decades ago, when manufacturing processes had a large manual component, it was considered normal for the units of product, no matter how accurate the production process was, to have defects, or deviations from the standard size. For example, if a size 44 suit must have 52 cm long sleeves, because this is the measure established in the model (or prototype) developed by the designer (design quality), it is probable that the actual length of the sleeves are not likely to always be 520 mm for all units; in some cases it will be 521, in others 519, in still others 522, 518, 523, or 517, and so on.

It is intuitive to assume that if you perform, for example, 1000 measurements, most of the sleeves will have a length that differs only slightly from that of the prototype, while the number of sleeves with significant differences decreases the more the size of the defect increases. We may then conclude that the number of cases with defects is inversely proportional to the extent of the defect. For example, with a "population" of $N = 1000$ suits, we could discover defects with the frequencies shown in Table 10.2 (tolerance expressed in millimeters, with discrete measurements of 1 mm).

When compared to 1000 measurements, the data in the left-hand column expresses the frequencies of the sleeves that have an increasing defect (error) compared to the prototype. We can present the data in a chart as shown in Fig. 10.12. Despite the limited data, the curve here recalls (very roughly) the Bell curve typical of the Normal

Table 10.2 Results from a sample analysis of defectiveness

Number of cases	Measurement	Size of defect
13	490 mm or less	–30 mm
45	500 mm	–20 mm
100	510 mm	–10 mm
700	520 mm	0
85	530 mm	+10 mm
40	540 mm	+20 mm
17	550 mm or more	+30 mm

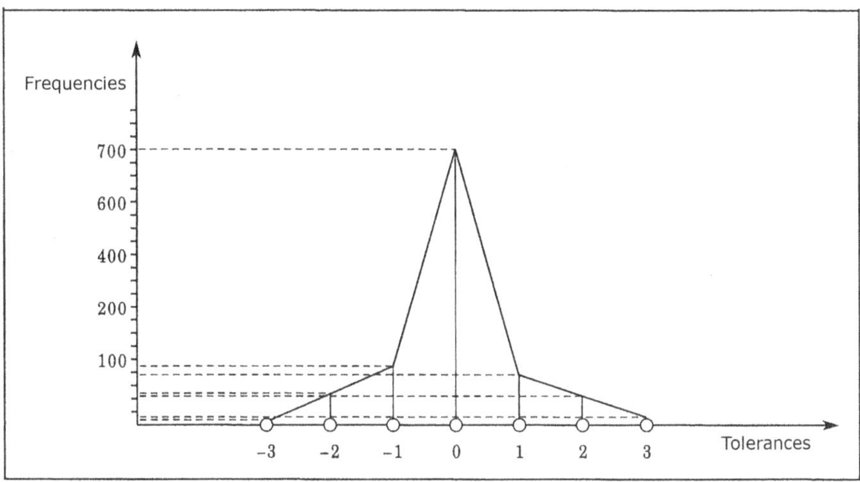

Fig. 10.12 Normal curve of the errors caused by defects found in a small amount of production

Distribution, thereby highlighting a reduction in the frequency of defects as the size increases which is (approximately) symmetrical to the maximum frequency.

If the designer who designed those clothes (producing the prototype) had established a maximum tolerance of more or less 20 mm, then all the clothes with sleeves from 500 to 540 mm in length would be judged "acceptable," since the length of the sleeves fell within the established tolerance; clothes with sleeves exceeding the range of acceptability would be considered "defective" since they do not conform to the model, and thus must be remade. This simple example shows us the innovative logic introduced by Six Sigma, which is based on a sample control of defects found in a "population" of manufactured units. The objectives of the control of defectiveness in flow production are defined by Motorola as follows:

> A defect rate of no more than 3.4 per million; statistically, allowing for some variation in mean, this approaches zero defects ... At Motorola, we actually have a measure for quality which we call "Six Sigma," and this literally affects everybody and everything we do, every minute, of everyday. Six Sigma is basically a target based on zero defects per million manufactured parts. At present, we are hitting 99.9996%, which is so close to perfection that we are now using a parts-per-billion measure for defects (Bellows 2004, p. 3).

Therefore, Six Sigma introduces as a general measure of defectiveness the number of defects per "million" units. A nearly perfect process, or at least one of excellent quality, is one that produces no more than 3.4 defects per million units; in other words, the production flows must have an admissible tolerance between the normal average value $\pm 6\sigma$, where "σ" is the standard deviation of the *normal distribution* (Fig. 10.13).

A defect rate of 3.4 units *per million* appears to be quite low, but we must judge it based on the number of elements that can present a defect. For example, according to recent data from Apple, around 1000 million iOS devices have been sold. Assuming we are observing a single component that can present "fatal" defects, according to the Six Sigma method, the production processes should be so accurate

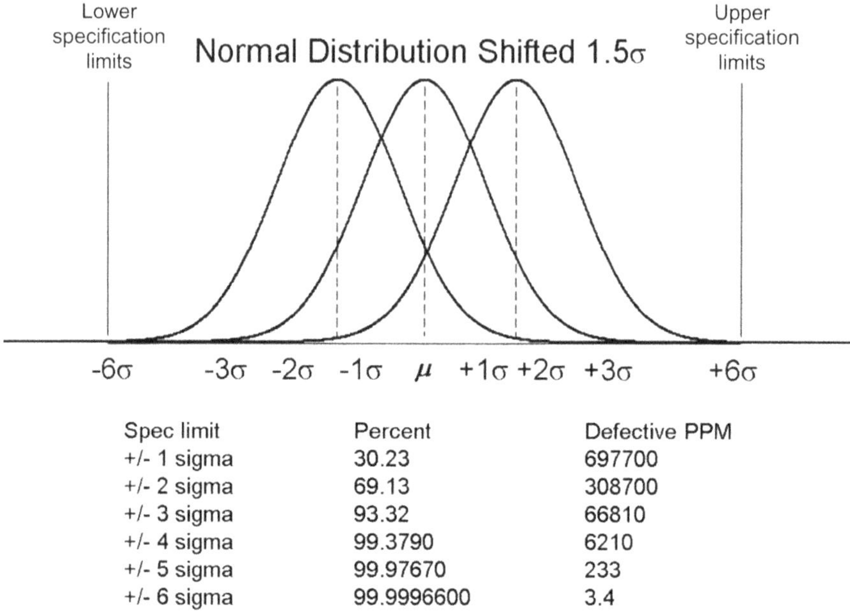

Spec limit	Percent	Defective PPM
+/- 1 sigma	30.23	697700
+/- 2 sigma	69.13	308700
+/- 3 sigma	93.32	66810
+/- 4 sigma	99.3790	6210
+/- 5 sigma	99.97670	233
+/- 6 sigma	99.9996600	3.4

Fig. 10.13 The statistical rules of Six Sigma (source: my elaboration from Montgomery and Woodall 2008, p. 332)

as not to leave more than 3400 defective units to be replaced. If we assume each device has 100 potentially defective components (keys, scroll bars, batteries, led lights, etc.), then according to Six Sigma the potential number of devices to replace would still be 340,000.

Still unsatisfactory is the data for mishandled baggage at airports around the world. According to data published in the 2018 ATI-Baggage Report, which refers to data published by SITA in 2017, the number of mishandled baggage is 5.57 bags per thousand passengers, hence 5557 per million.

> Mishandled bags represent a minority of the estimated 4.6 billion or so bags handled each year. Nevertheless, they cost the industry an estimated US$2.3 billion in 2017, which gives an indication of the business benefits that airlines can reap from their investment in-end-to-end bag tracking. Looking at the longer term trend, passenger numbers have soared by 64% since 2007. In this period, the mishandling rate per thousand passengers has reduced by 70.5% and there has been a 46.2% cut in the annual cost of baggage mishandling to the industry (ATI 2018, p. 15).

This figure reveals how difficult it is to control the airport baggage delivery, the quality of which, according to the data in Fig. 10.13, is just above 4 σ.

Six Sigma must not be considered as only a process of statistical measurement but above all as a well-tested set of *instruments* and sophisticated *techniques* aimed at reducing the variability (tolerance) and defectiveness of a product-process, that is, a powerful managerial tool to improve *overall quality* (Basu and Wright 2003). These instruments are referred to as *Lean Six Sigma* (George 2003; Näslund 2008;

Basu 2009; Voehl et al. 2016), and it entails five phases in the DMAIC form (Henshall 2017; Beemaraj and Prasad 2018), as described in Fig. 10.14, and in the DMADV form.

> The fundamental objective of the Six Sigma methodology is the implementation of a measurement-based strategy that focuses on process improvement and variation reduction through the application of Six Sigma improvement projects. This is accomplished through the use of two Six Sigma sub-methodologies: DMAIC and DMADV. The Six Sigma DMAIC process (define, measure, analyze, improve, control) is an improvement system for existing processes falling below specification and looking for incremental improvement. The Six Sigma DMADV process (define, measure, analyze, design, verify) is an improvement system used to develop new processes or products at Six Sigma quality levels. It can also be employed if a current process requires more than just incremental improvement (ISIXSIGMA 2019, online).

Fig. 10.14 DMAIC process and Lean Six Sigma. (Source: Ahmed 2019)

Although Six Sigma is an autonomous technique, its principles are embedded in TQM (Sheehy et al. 2002), and the two methods are compatible in many ways. Schroeder et al. (2008) has attempted to show that the introduction of Six Sigma to organizations that already have TQM (Sect. 10.7) leads to further improvements in the quality control process (Anderson et al. 2006), providing a strong structure for achieving greater process improvements. Salah et al. (2009) have demonstrated that Six Sigma and TQM (Sect. 10.7) have a common basis in terms of aims, principles, theory, philosophical approach, focus on process, and dependence on management support. We can generalize by stating that Six Sigma transforms TQM into a "holistic" TQM. The two philosophies are also integrated on the applied level: TQM can help Six Sigma and Six Sigma extends TQM; according to Yang (2004), the relation between Six Sigma and TQM can be expressed as follows:

$$\textit{Current business system} + \textit{Six Sigma} = \text{TQM}$$

I end this *section* by mentioning the Control Systems that rely on long-term structural levers in the form of organizational restructuring and the re-engineering of processes, organs and functions that can increase the potency of process Control Systems (Davenport and Stoddard 1994). Equally as powerful, though more gradual in its interventions to achieve the desired objectives, is the "kaizen," a term introduced by Masaki Imai (1986), in *Kaizen, The Key to Japan's Competitive Success,* indicating the concept of continuous improvement.

> KAIZEN means improvement. Moreover, it means continuing improvement in personal life, home life, social life, and working life. When applied to the workplace KAIZEN means continuing improvement involving everyone – managers and workers alike (Kaizen Institute 2020, online. https://www.kaizen.com/what-is-kaizen.html).

Kaizen is the gradual, ongoing improvement in the performance of processes through the involvement of workers at all the organizational levels to achieve the quality and productivity objectives (Deming 1989; Imai 1986; Tanaka 1994). In his 1997 book, *Gemba Kaizen: A Commonsense, Low-Cost Approach to Management,* Masaki Imai (2007) stated that Kaizen must be applied where "the real work" takes place to make real improvements to the organization of the company. With the term Gemba Kaizen, Imai identified the countless daily "little improvements" that are made to the place where "the real work" is done, for example, the assembly line in the production sector or the office where employees interact with customers in the service sector.

10.14.2 Time and Motion Study and Work Sampling

[1] Time and motion study In controlling work productivity and quality control, statistical techniques can be used which are as simple as they are effective, whose aim is to improve business *performance* by keeping track of the activity times of men and machines (Karger and Bayha 1987). A particularly effective method and technique for controlling and improving work performance is *work sampling* (WS), also referred to as activity analysis, or method for the *statistical sampling of work*, which aims at measuring the time of activity or inactivity of men and machines by operating with multiple instant surveys, at random intervals, on a sample of a population consisting of activities/inactivity (or similar events), thereby obtaining statistical measures that refer to the population. The WS was, in fact, designed to control labor performance, the factor that most explains the dynamics of the cost of production, within the context of the *performance management* paradigm (Otley 1999; Demartini 2014) and the *performance measurement* paradigm (Johnson and Kaplan 1987; Kaplan and Norton 1996, 2001). The definition of work sampling can be usefully taken from the DCAA. U.S. Department of Defense Audit Agency manual (2000), which states:

> Work sampling is broadly defined as the application of statistical sampling techniques to the study of work activities. […] work sampling is typically used to estimate the proportion of workers' time that is devoted to different elements of work activity. […] observed activities are grouped into either of two main classifications: working or nonworking (2000: I1).

Measuring and controlling the technical efficiency of labor have always been a sensitive issue of productivity control in performance management, and different methodologies have been developed for their implementation, among which it is worth remembering *Gross Time Study*—which aims at maximizing the ratio of gross production volumes, appropriately defined, and the overall time taken to obtain these—and the *Continuous Time Studies,* to optimize working time and minimize the duration of each phase of activity carried out by the worker (Gilbreth and Gilbreth 1919; Allderige 1954). More generally, we can also mention *Time and Motion Study,* a technique developed by Frederick Taylor, the Father of Scientific Management (Taylor 1919) and (Garvin 1984), the formalizers of the *Methods Engineering Approach*, also known as *Business Process Re-engineering*, part of the *Continuous Improvements* approach, which sought not only to optimize work time, but also to reduce the number of operations required to perform a task (Gilbreth 1914). Time and Motion Study calls for calculating the time it takes for a number of workers to do the same task, usually in order to assign standard execution times. With regard to these techniques, the following excerpt containing the considerations of Frederick Taylor is instructive:

> The CHAIRMAN. When you make a time study of a man at physical labor do you not always eliminate in that time study the pauses at his labor, so as to get at the accurate and actual time in which he performs the labor? …
>
> Mr. TAYLOR. To answer your question, we do both things. We take the gross time, the whole time which the man takes in doing the job, and then we make at the same time another study which includes the productive time alone, the time he is engaged in actual work. On this printed page there is a study of the gross time, and a study of the productive time as well.
>
> The CHAIRMAN. When you make a study of the productive time you eliminate in that study, and are able to do so by virtue of your stop watch, the periods in which whe workman is not engaged in productive work?

> Mr. Taylor. Yes, sir.
> The Chairman. How can you take a mental study of the productive time, the mental time that it takes to work out a problem? Would you be able in your time study to take the time of the mental pauses that occur during the time when the problem was being worked out?
> Mr. Taylor. The time during which the man stops to think is part of the time that is not productive (Wilson et al. 1912).

[2] The work sampling function These techniques are complemented by Work Sampling, a methodology for detecting the times—frequencies, or probabilities—of activity or inactivity of men and machines that was first applied in the 1930s in the U.S., based on the L.H.C. studies (Tippett 1935, 1947), albeit with a different terminology: the snap-reading method (Tippett 1947, p. 109). This work was taken up by R. L. Morrow (1957) to determine the ratio-delay, that is, the ratio of inactivity or rate of delays with respect to the total working hours assigned to a task, which led to an investigation of the causes of such delays. The initial analyses by Tippet and Morrow gave rise to a vast reference literature, testifying to the relevance of the application of the statistical sampling technique both in the manufacturing sector (Correll and Barnes 1950; Gustat 1955) and in different contexts (Stevenson et al. 1999).

Work sampling allows two types of investigation to be undertaken:

1. Measure the percentage of activity/inactivity of a given task carried out by a man or machine, investigating the causes of inactivity;
2. Establish working time standards to be assigned to the performance of certain types of production with a view toward optimizing the use of resources.

Precisely because it allows dichotomous characteristics present in a large number of elements of a population to be identified, this technique is proving to be applicable (and is increasingly being applied) even in quite different areas; in particular, in the production, commercial, marketing, quality control, health care, banking, transport, entertainment, and auditing sectors, among others.

[3] Principles of work sampling Statistical sampling based on work sampling follows two *basic principles*:

1. Simplify elementary surveys to the maximum, resulting, where possible, in the assessment of only two states, or parameters, of opposing value: active/inactive, yes/no, on/off, stationary/moving, time greater than 12 min./time less than 12 min., etc. Information about the activity of machines or men is, therefore, simplified as it is easy to describe two opposing *"states"* of activity simply by counting the number of times; the subject of the investigation was "observed" in that given situation with respect to the total number of observations.
2. The survey is carried out on a *sample* of the entire *observation universe* and the results obtained refer to the *entire observation universe,* that is, the set of all machines, workers, or activities of the company surveyed. "Sample" means a part, more or less large and of appropriate size, of that universe, chosen with random criteria according to statistical rules.

It is intuitive to realize that:

(a) The greater the sample size, the higher the *level of confidence* in the *accuracy* of the results; if we look at 500 workers out of 1000 and find that 15% of them are absent every Monday of each week, then we can reasonably assume that *about* 15% of the entire universe of workers are absent on Mondays; or that the probability of finding absent workers is 15%. With a sample of half the population, there is considerable confidence that the observation of 500 workers can also be extended to the other 500. If we look at only 50 workers out of 1000 and find that absences are 18% ($p = 18\%$), then the result (frequency of absenteeism) observed in 50 workers can only be extended to the other 950 workers with a lower degree of confidence;

(b) The smaller the sample size, the lower the *cost* of data collection.
For this reason, the fundamental problem to be solved in work sampling is the most *convenient sizing* of the sample to be observed in order to make the results as significant as possible at the lowest possible cost; sampling must be made as efficient as possible.

[4] Statistical rules for work sampling The WS is based on formulas for the specific statistical distribution of the binomial type (YES–NO) that perfectly fits the method of recording activities that can be analyzed through work sampling. In work sampling, choosing the number of elements to be included in the sample does not depend on the size of the population to be examined, but on three fundamental parameters that must be established in the research setup phase, taking into account the phenomenon being observed and the objectives to be achieved:

1. The desired *confidence interval* (F), which expresses the *level of confidence in the* results. *Interval* (F) indicates in percentage terms the probabilistic confidence in the acceptance of a result; it is measured in the *standard deviation*, (σ_P), of the distribution. It seems reasonable to expect that the value actually observed in the statistical survey belongs to the range consisting of the *average value* of the distribution, plus or minus the error, expressed by the *standard* deviation (σ_P); the standard deviation will be higher the higher the desired confidence level in the results. It is possible to set $F = 1, 2, 3$, or even several times the standard deviation. Specifically:

   $\pm 1\sigma_P$, equal to a level of confidence in the results of 68.27%
   $\pm 2\sigma_P$, equal to a level of confidence in the results of 95.45%
   $\pm 3\sigma_P$, equal to a level of confidence in the results of 99.72%

   For common use analyses, a level of 95.45% corresponding to $\pm 2\sigma_P$ – with $F = 2$ – is sufficient, but in the *six sigma methodology*, for example, it is set at six times the standard deviation, with a confidence level approaching 100%.

2. The *precision,* or *accuracy, of the measurement* (S), which indicates the percentage of error deemed permissible between the observed value and the estimate, that is, what percentage of error can be tolerated with respect to the exact measure. For common analysis, a precision of $2\% < S < 5\%$ is adopted.

3. The *probability*, or *frequency* of an event (P), whose existence must be ascertained. This represents the statistical probability (or frequency) of the data to be recorded in the population, that is, the probability the event surveyed will or will not occur. This probability can be estimated by using two methodologies:

   1. *Inductive method*: this consists in examining probability studies of similar populations for similar companies and sectors, referring to surveys in similar fields found in the academic literature, or to surveys of statistical bodies. For example, it is plausible to assume that the probability inactivity occurs during the operation of a press in a metallurgical company is almost analogous to that of a press of the same size in a metalworking company;
   2. *Deductive method*: this is based on ad hoc "pilot studies" to detect the probability of the event happening or not happening in a small-sized sample. This first approximation obtained from the pilot study is used to determine the sample size that can measure the same probability of first approximation even in the population.

[5] THE SAMPLE SIZE FOR WORK SAMPLING The *optimal sample size*, "n," is the number of observations to be made to achieve a result that has the established level of confidence and accuracy. Referring the reader to statistical textbooks for a complete demonstration, it is sufficient here to point out that to determine "n" we must first of all consider the following equation:

$$S \times P = F \times \sigma_P \tag{10.8}$$

where:

- S indicates the desired degree of final precision; this is expressed as $\pm S$
- P indicates the percentage of the activity to record, and it is expressed as P = (Number of positive observations/Number of observations carried out)
- F indicates the degree of confidence as the number of units of standard deviation σ_P ($F = 1$ if we desire a confidence level of 68%; $F = 2$ for a confidence level of 95%; $F = 3$ for a confidence level of 99%)
- σ_P indicates the standard deviation of the binomial distribution:

$$\sigma_P = \sqrt{\frac{(1-P) \times P}{n}} \tag{10.9}$$

- $F \times \sigma_P$ thus represents the average that contains the desired percentage of precision.

The number "n" of sample elements to observe for a confidence level of 95% is calculated using the following general equation, obtained by substituting (10.9) into (10.8) and setting $F = 2$:

$$S \times P = F \times \sqrt{\frac{(1-P) \times P}{n}} \tag{10.10}$$

From the preceding equations, it immediately follows that:

$$n = \frac{F^2 \times (1-P) \times P}{S^2 \times P^2} \tag{10.11}$$

$$S = \frac{F}{P} \times \sqrt{\frac{(1-P) \times P}{n}} \tag{10.12}$$

$$P = \frac{F^2}{(S^2 \times n) + F^2} \tag{10.13}$$

In the previous equation (10.10), there are two unknowns:

P is the percentage of activity (probability of occurrence of the data to be recorded) to be determined
n is the sample size.

In order to proceed in calculating (10.11), (10.12), and (10.13)), it is necessary to quantify an estimated value for P by performing a preliminary *pilot study* to determine a value of first approximation for P, and subsequently a more accurate estimate.

The previous formulas assumed that *work sampling* was applied to continuous processes, in progress, and that therefore the source population *was not known* by definition. We can remove this limitation and apply work sampling to *populations of known size*. In this case, the sample size is quantified by the following expression, instead of by (10.11):

$$n = \frac{F^2 \times N \times [P \times (1-P)]}{[S^2 \times (N-1)] + [F^2 \times [P \times (1-P)]} \tag{10.14}$$

In practice, without directly using the previous formulas, the size of the samples can be determined according to how the work sampling is to be used by establishing the number of observations to be made according to precalculated graphs or tables in which, for a given level of confidence, "n" can be determined based on a first approximation for S and P. If this approximate sizing is not used, it is necessary to apply (10.10) to quantify the sample size, assuming a value for P that is estimated or determined from the pilot survey. Once "n" is determined, the data is collected and processed. *During the survey*, at the end of each day (or subsequent stage of the survey) summaries are made of the operations performed. The systematic recording of this data is carried out on control cards whose limits are calculated on the basis of previously established elements. If the process is kept within the control limits, it continues on as normal; otherwise, the data that is beyond acceptable levels is examined. At *the end of the survey,* the summary calculations are carried out to obtain the final percentage of activity compared to the total number of observations. To check whether the analysis has produced results that fall within the desired level of precision, equation (10.5) above is used. If the accuracy *actually* obtained is greater than that established when setting up the process, the analysis may be considered completed; otherwise, additional analysis is required, since the number of observations is insufficient.

[6] HOW TO EXTRACT A RANDOM SAMPLE Equations (10.11) and (10.14) allow us to determine the sample size. However, in order to carry out the surveys in practice, it is necessary to select from the N elements of the population the "n" elements of the sample to be surveyed. The easiest way to obtain a random sample is to resort to random numbers, that is, numbers whose values have the same probability of occurring in the sequence. The following procedure can be used to do this manually:

(a) Progressively number every element of the population, from 1 to N.
(b) Generate *random numbers* containing a number of digits equal to the number of digits contained in the universe components, for example, four digits for a universe of 8679 units; if $N = 125{,}000$, then we require random numbers of six digits; and so on.
(c) Choose the random number from which to start the generation of the sequence.
(d) Progressively read off "n" random numbers that follow from the first number chosen.
(e) Select the items in the population that correspond to these numbers.

All statistical textbooks, even elementary ones, contain *tables of random numbers*: sequences of numbers whose values have the same probability of occurring in the sequence. You can generate a table of random numbers with different software, in particular with Excel, by selecting the appropriate function and specifying the range the random numbers should fall into. As many cells are generated as needed to cover the population numbering. In continuous processes, this procedure can be automated already during the production phase by indicating, for each unit produced, whether or not it should be included in the sample.

Once the elements to control have been randomly determined, it is necessary to determine the period of time in which to make the observations. If the observations refer to a sample of events that have already occurred, then the time frame of the observations is not relevant, and the operator can proceed in the time period and order the deems most appropriate. If the observations are related to processes that develop over a period of several days, a table of random numbers can again be used to randomly establish the days on which the observations are to be carried out. If these are to be done at different moments on the same day, the same technique can determine the time or minute at which to make the observations.

10.14.3 *Application of Work Sampling in the Control of Industrial Processes*

A useful application of work sampling is the detection and control of plant activity/inactivity. Suppose you want to detect the degree of operation of a homogeneous *fleet of machines* of a given company or department and that it is considered sufficient in this case to consider whether the machines are active or inactive during the day. Since there are many machines and it is not possible to directly observe the entire *machine fleet,* work sampling can be implemented after specifying the following parameters:

- A degree of precision of $S = 2\%$
- Percentage of machine activity which, from experience, is expected to be observed, estimated at $P = 80\%$
- A confidence interval equal to 95%, with $F = 2$

By entering this data in the formula (10.11), we get $n = 2500$, which corresponds to the number of observations needed to obtain an estimate of the operating time of the machinery. However, the observations must be made on the basis of some *random criteria of choice* regarding the machinery to be monitored at different points during a chosen time frame for the analysis. The value for the number of active machinery that is expected to be observed has been estimated at $P = 80\%$. This value must then be refined after an initial number of observations that act as a *pilot survey*. It is then decided to observe 20% of the machines, that is, 500 out of the 2500 total, and to *recalculate* P based on the results of those 500 observations, which are indicated in Table 10.3.

At the end of the pilot survey (500 out of 2500 observations), the activity status of the machines turns out to be $P = 85\%$, compared to 80% forecast before the start of the statistical sampling. With this information from the pilot study, it is now possible to proceed according to two alternatives:

1. Continue the surveys for all $n = 2500$; since the new value of P is greater than initially assumed, the significance objective is certain to be achieved.
2. Recalculate the number of observations to be made taking into account the new value of $P = 85\%$.

We proceed with the second alternative, since increasing P reduces the size of the survey sample, thus optimizing the time and cost of sampling. Replacing $P = 80\%$ with $P = 85\%$ in equation (10.11) we obtain $n = 1765$; with the new value of

Table 10.3 Results of the 500 pilot observations of the activity state of the machinery

Observed state	Number of observations	P = percentage in operation
Machines that are operating	425	85%
Machines that are inactive	75	15%
Total observations	*500*	*100%*

Table 10.4 Results of the 1800 observations of the activity state of the machinery

Observed state	Number of observations	P = percentage in operation
Machines that are operating	1359	85.5%
Machines that are inactive	261	14.5%
Total observations	*1800*	*100%*

$P = 85\%$, we no longer need to make 2500 observations, but only 1765. It is decided in the end to perform 1800 observations, the results of which are shown in Table 10.4.

To conclude the analysis of work sampling, it is necessary to check whether the results obtained satisfy the desired accuracy. Setting the final figure of $P = 85.5\%$ in equation (10.12) above, we obtain: $S = \pm 1.99\%$. Therefore, the accuracy of the sampling is fully within the desired limits, having set the target at $S = 2\%$.

10.14.4 Example of Work Sampling in the Quality Control of a Continuous Production Process

WS is also a useful technique for continuous production control, as the following case illustrates. BETA's production manager was concerned about the dangerous reduction in the weight of the honey dripping mechanically into the glass jars. Weight loss could be noticed and lead to complaints from wholesalers and retailers, as well as consumers. It was necessary to monitor the weight of each jar to recalibrate the electronic scale when the measured weight fell below 5% of the guaranteed weight on the package. He thus decided to implement a *work sampling plan* to statistically control the weight of the jars in such a way as to ensure that the *average weight* detected in the *sample* referred to the packaged units of the entire *production*.

Since the control had to be implemented *during* the manufacturing stage, it was not possible to know the size of the statistical universe. The sample size was determined by equation (10.11) based on the following objectives:

- Desired degree of accuracy: $S = 4\%$.
- Error percentage on average weight of 5%; therefore, the probability of finding full jars was $P = 95\%$.
- Confidence interval for the results (confidence level) equal to 95%, and thus $F = 2$.

The following value was obtained: $n = 132$ jars to check. After determining the number of observations, it was necessary to determine how to extract the random sample. Since 3600 jars of honey are packed in an hour, the manager, instead of making a random extraction with the help of random number tables (or similar procedures), decided on a constant-cycle check, controlling 1 out of every 28 jars ($28 = 3600/132$).

10.14.5 Application of Work Sampling in Marketing Control

Marketing can also usefully apply work sampling to carry out process and quality tests, as shown in the following case. The marketing director of the company "Chocolate S.A." had noticed, with regret, that the email inbox of the company's website contained an increasing number of emails with complaints from customers who had not found the "gift voucher" in their boxes of chocolates, despite the widespread publicity at the time the product was launched. The general manager was immediately informed and summoned the packaging manager along with the production manager, asking them to submit, within 24 hours, a plan to control the packaging process with an estimate of the relative costs. He then ordered the commercial director to estimate the economic damage resulting from the mishap.

The next day, the production manager submitted Table 10.5 with an estimate of the costs of the *control,* assuming a daily production of 20,000 boxes, taking into account both the cost of the workers involved and that of the destroyed material (boxes), net of the recoveries made.

Based on the data above, a sampling plan is drawn up as shown in Table 10.6 that determines the size of the sample using (10.14), under two hypotheses: high and average precision.

Table 10.5 Average cost of control for boxes of chocolate (daily production of $N = 20{,}000$ boxes)

Daily production	20,000	Boxes
Control cost: labor	20.50	€/box
Control cost: processed material	10.50	€/box
Total gross average cost	31.00	€/box
Material recovered	−2.00	€/box
Total net average cost	29.00	€/box

Table 10.6 Sampling plan (daily production of $N = 20{,}000$ boxes; average cost of control equal to € 29/package)

FIRST HYPOTHESIS: High precision	
Daily production	$N = 20{,}000$
Rate of acceptable error: 2%	$P = 0.02$
Desired precision: (±)1%	$S = 0.01$
Confidence interval: 95.5%	$F = 2$
Size of the work sampling	$n = 754$ boxes
Cost of survey under the high precision hypothesis	€ 29 × 754 = € 21,866
SECOND HYPOTHESIS: Medium precision	
Daily production	$N = 20{,}000$
Rate of acceptable error: 5%	$P = 0.05$
Desired precision: (±)2%	$S = 0.02$
Confidence interval: 95,5%	$F = 2$
Size of work sampling	$n = 464$ boxes
Cost of sampling under an average precision hypothesis	€ 29 × 464 = € 13,456

The commercial director estimated a 20% drop in demand if the defects continued. Since each package has a sale price of 19 euros and an average unit cost of production of 14 euros, the loss in margin is, therefore, equal to 5 euros for each unsold package, entailing a total commercial loss of 4000 units and 20,000 euros. The General Manager decided to go with the *medium-precision* control project and wait some time before assessing the quality improvements.

10.14.6 Control Charts

The *control cards,* or Shewhart cards (named after Walter Shewhart proposed them in the 1930s), or process-behavior charts, represent another important statistical tool for the *quality control of processes* by analyzing the dynamics they produce (Nelson 1984). Special perspectives are adopted in *quality control* when you want to monitor the dynamics of an ongoing process to identify the causes of the variation in results that can be due to *natural causes*, which give rise to random or natural variations in the results, and to *special causes*, which produce abnormal and specific variations. To identify the values that indicate the process is out of control, the control charts indicate the upper control limits, UCL, and the lower control limits, LCL, and at times even the *warning limits*, which represent the maximum measures of tolerance, or acceptability, of the characteristic to be controlled in the units or of the manufacturing processes, for example, a range, mean, or proportion.

Control charts can be used to control several characteristics. Normally, we distinguish between parameters that are expressed by continuous data (length, weight, temperature, thickness, etc.) and attributes expressed by the discrete data (number of defective units, defects per unit, etc.).

- R-Charts that control the ranges of the elements
- $\overline{X}$ -Charts that control the means
- P-Charts that control the proportions
- c-Charts that control the number of occurrences

In constructing a control chart, the time at which the inspections were carried out or the size of the samples analyzed is indicated on the *x-axis*, while the *y-axis* charts the measurements obtained. The graph thus specifies, *for each assessment,* the results obtained, so that even visually it is possible to identify the moment, or moments, when the *limits of tolerance* or *acceptability* are exceeded. At such moments, the process appears out of control and must be stopped to ascertain and remove the cause of malfunction of the equipment or of the workers. The *center line* (or target line) is usually determined by the average value of the data detected; the width of the interval between the UCL and LCL with respect to the center line is determined after deciding the range of the tolerated results: if the survey values are assumed to follow a normal distribution and, graphically, the data appears as a normal curve for the error, then the size of the control limits, taking into account the variability of the detected values, is determined using the *standard deviation*, "σ," within the limits indicated in Fig. 10.13. If "x_n" represents the measured values for the elements "x," then for the times/samples $n = 0\ 1, 2, \ldots N$ the standard deviation is determined as follows:

$$\sigma = \sqrt{\frac{\sum_n \left(x_n - \bar{X}\right)^2}{N-1}}$$

where $\bar{X}$ is the average value. If $F = 1, 2, 3, \ldots$ indicates the number of times chosen to calculate the *tolerance limits* given "σ," then "UCL = Fσ" and "LCL = −Fσ". It is immediately clear that as "F" increases so does the tolerance interval. With $F = 1$, only those results that fall within ±68.27% of the average values are deemed to be under control; when $F = 2$, the tolerance rises to ±95.45%; with $F = 3$, it becomes ±99.72%. Shewhart had determined that it was always convenient to choose $F = 3$, a rule still followed today (Shewhart 1931; Kazmier 2004). An examination of the Control Chart allows us to observe the dynamics over time of the measurements taken. If all the values examined are within the control limits, then the process is "under control," that is, "stable." Values that lie outside the tolerance limits indicate that when they were measured the process was out of control. When these values occur, the search for the *out of control causes* and their elimination or mitigation should begin. Control cards also allow an evaluation to be made regarding the dynamics of the values even for processes that are under control.

The Control Charts allow for a simple and immediate diagnosis of the drawbacks of the production process. However, the control system adopted must enable the "online" Control Charts to be obtained while the process is *under way*, that is, with detections carried out on the same process while the cycle is taking place, and in "*real time*," before the deadline has passed to effectively intervene in the process, in order to return it to the predetermined limits of efficiency before too many defects have occurred. When the Control Charts signal an anomaly in the production process, it is necessary to stop the process, investigate the causes of the malfunction, and intervene to eliminate the factor that has produced the variation. For this reason, it is recommended that job output controls be done over relatively short intervals. Lloyd Nelson, in a paper entitled *The Shewhart Control Chart—Tests for Special Causes* (1984), proposed a list of eight types of observable dynamics in the Control Charts. Nelson's work is widely cited, but others have also proposed specific models of dynamics (Noskievičová 2012; Smajdorova and Noskievicova 2017; Wikipedia 2020, Nelson Rules.) Nelson's models can serve as a useful guide.

A simple example will clarify the structure of the Control Charts and the operational information needed to control the monitored process. A chemical company creates a product using an ingredient of a certain density. To verify that the machinery maintains the mixture at the constant density of the ingredient, an online control process is carried out by systematically testing selected samples at constant intervals. Before the control of the process is undertaken, the company does a pilot study on 20 samples to determine the average density and acceptability limits, UCL and LCL. By keeping constant the calculated values for the first $N = 20$ surveys or samples, the process is checked. The data and control cards for 30 observations (20 from the pilot study and 10 from the actual control) are shown in Fig. 10.15. The data "x_n" in black was that used in the pilot study; the data in red was measured after the start of the online control. The Control Chart immediately shows that the process was out of control starting from the sample period n = 23, and that therefore, the mixing machinery must be recalibrated.

10.14.7 *Gross and Net Labor Productivity*

When calculating labor productivity, we must specify whether we are referring to the labor *assigned* to a given process—that is, labor theoretically supplied by all workers assigned to that process—or to the labor effectively *employed* only by those

CONTROL CHART FOR DENSITY OF AN ACTIVE INGREDIENT

"n"	x_n
1	0.15
2	0.153
3	0.144
4	0.142
5	0.163
6	0.175
7	0.166
8	0.139
9	0.155
10	0.138
11	0.165
12	0.139
13	0.144
14	0.16
15	0.152
16	0.168
17	0.141
18	0.22
19	0.148
20	0.146
21	0.158
22	0.189
23	0.2
24	0.225
25	0.228
26	0.3
27	0.31
28	0.31
29	0.32
30	0.33

Mean value = 0.1554
σ = 0.01875
UCL = 0.21
LCL = 0.10

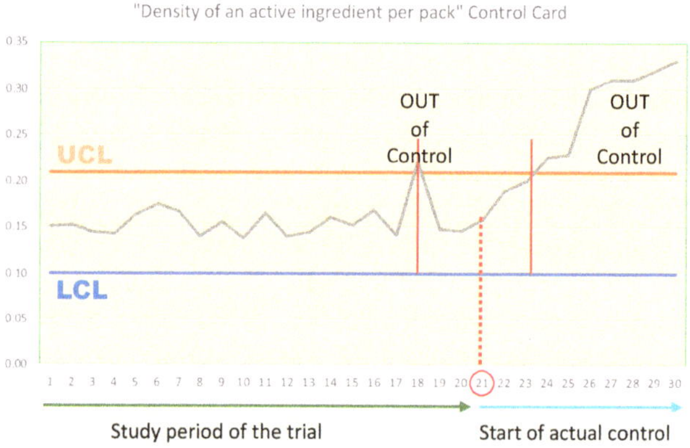

Fig. 10.15 Control chart for a mixing process

workers actually "operating" during the production cycle. The following expression links "gross" productivity of *assigned* labor to "net" productivity of *employed* labor:

$$\pi L_{assigned} = \frac{QP}{QL_{employed}} \times \frac{QL_{employed}}{QL_{assigned}} \quad (10.15)$$

From (10.15), we see that, since $QL_{employed} \leq QL_{assigned}$, "gross" productivity of *assigned* labor is usually less than that of *employed* labor, since it also depends on the *intensity of employed labor ratio* (second factor); that is, on the relationship between the *employed* labor (numerator) and the *assigned* labor (denominator). This ratio is less than and tends toward "one" when all the assigned labor is utilized. There are three main factors that influence the *intensity of employed labor ratio*:

1. The *absentee rate* (ta): this expresses the percentage of hours not worked due to worker "absences" of various kinds due to maternity leave, justified absences, unjustified absences, other types of authorized leave, etc.
2. The *rate of idleness of the production process* (ti): this expresses the hours of idle functioning of the production system due to programmed maintenance, breakdowns, etc.
3. The *rate of excess labor* (ts): this expresses the quantity of excess labor assigned to a given production center with respect to actual needs. This rate is an indicator of the improper allocation of labor.

If we express the quantity of *employed* labor, these three factors should be filtered out from *assigned* labor as follows:

$$QL_{employed} = QL_{assigned}(1 - ta)(1 - ti)(1 - ts) \quad (10.16)$$

Then, productivity becomes:

$$\pi L_{assigned} = \frac{QP}{QL_{employed}}(1 - ta)(1 - ti)(1 - ts) \quad (10.17)$$

Equation (10.17) allows us to express more clearly the productivity of *assigned* labor (the basis for calculating labor cost) in terms of the effective output of labor and of those factors—absenteeism, idleness and excess labor—that reduce this output with respect to the more significant output of *employed* labor.

Let us assume, a given production center is allocated 30 workers for 8 hours a day, and that only 25 workers are "needed"; that the absentee rate is 18%; and that in a month of 20 working days, 4 on average are lost due to various causes. The *production* from the center will then be equal to QP = 50,000 units.

The *labor* hours assigned to the center will be:

$$QL_{assigned} = 30 \times 8 \times 20 = 4{,}800 \text{ hours.}$$

The "gross" productivity of the *assigned* labor will be:

$$\pi L_{\text{assigned}} = 50{,}000 / 4{,}800 = 10.417$$

Since ta = 18%; ti = 20% = (20–16)/20; and ts = 16.6% = (30–25)/30, the quantity of *employed* labor will be:

$$\text{QL}_{\text{employed}} = 4{,}800 \times (82\%) \times (80\%) \times (83.33\%) = 2{,}624 \text{hours}$$

The net productivity of *employed* labor will increase to:

$$\pi L_{\text{employed}} = 50{,}000 / 2{,}624 = 19.055.$$

We can observe many examples of production organizations with excess labor and idle production facilities where an improvement in acceptable productivity levels requires eliminating the "surplus" and shutting down under-utilized operating units. Also clear are the effects from a reduction in absenteeism and the under-utilization of working times as important drivers for productivity increases.

10.14.8 Productivity, Consumption, and Employment: Toward the Jobless Economy?

Let us go back to (10.6) and specify the period of observation, T, by writing $\pi L(T)$ to indicate labor productivity for a period $T = 1$, equivalent to 1 year:

$$\pi \text{L}(T) = \frac{\text{QP}(T)}{\text{QL}(T)} \tag{10.18}$$

We can express QL as a variable that depends on QP(T) and πL(T):

$$\text{QL}(T) = \frac{\text{QP}(T)}{\pi \text{L}(T)} \tag{10.19}$$

Equation (10.19) makes it clear that, in order for QL(T) not to diminish over time, we can choose one of two alternatives:

- Attempt to increase QP(T)
- Slow down the growth in πL(T)

The second hypothesis, a slowdown in the growth of πL(T), is unrealistic regarding the network of enterprises which constitute the productive structure of the economic system; πL(T) can in all regards be considered a noncontrollable variable, subject to the Law of increasing productivity.

In order to analyze the other alternative, increasing QP(T), it is appropriate to recall that *final consumption products* can be divided into two large categories, in relation to the speed of consumption:

1. *Immediate consumption products*: food, general apparel, services, etc.; these will be indicated by IcP.
2. *Durable consumption products*: housing, furniture, cars, special apparel, etc.; these will be indicated by DcP.

Let's introduce the *simplifying hypothesis* that in period $T = 1$ year, only these two categories of goods, IcP and DcP, are produced and calculate how much of each category of good we need to produce in period T for a consumer.

1. Goods IcP: if we assume that the *average* per capita *unit consumption* in each instant "t" is equal $\kappa(IcP)$; then in period $T = 1$ year, the quantity to produce of *IcP is:*

$$\mathrm{Q}IcP(T) = \int_{\mathrm{T}} \kappa(IcP)\,\mathrm{dt} = \kappa(IcP)\int_{\mathrm{T}} \mathrm{dt} = \kappa(IcP)T \tag{10.20}$$

Example if everyone ate one sandwich a day, then in period $T = 1$ year Qsand (1 year) = 365 sandwiches.

2. Goods DcP: if we assume that in period $T = 1$ year, the *average unit* consumption is equal to $\theta(DcP)$ and that a good has an average life of d/T, then $r = T/d$ indicates the average *turnover* of each unit of good, so that the quantity to produce for a consumer is:

$$\mathrm{Q}DcP(T) = \theta(DcP)(T/d) = \theta(DcP)r \tag{10.21}$$

Example if everyone consumes one pair of shoes each month, then $d = 1/12$ and $r = 12$, so that in 1 year Qshoes (1 year) = 12 pairs. If we buy a new car once every 5 years, then $d = 5$ and $r = 1/5$, and Qauto (1 year) = 1/5.

If the economic system has M consumers and unit consumption is uniform (average consumption), then, considering (10.20) and (10.21) the overall quantity of production in T is:

$$\mathrm{QP}(T) = M[\mathrm{Q}IcP(T) + \mathrm{Q}DcP(T) = M\left[\kappa(IcP)T + \theta(DcP)r\right] \tag{10.22}$$

If πL(T) in the system is unique (an unnecessary but useful hypothesis), then by substituting (10.22) into (10.19) we get:

$$\mathrm{QL}(T) = \frac{M\left[\kappa(IcP)T + \theta(DcP)r\right]}{\pi\mathrm{L}(T)} \tag{10.23}$$

We can immediately deduce from this that, in order to contrast the increase in πL(T), the following alternatives must be evaluated:

(a) Increase the average consumption of immediate consumption goods, $\kappa(IcP)$. Example: more oranges in orange juice, more sugar in sweets, more meat in hamburgers, etc..
(b) Increase the average consumption of durable consumer goods, $\theta(DcP)$. Example: two cell phones per person, two cars per family, three pairs of shoes per month, ten shirts per year.
(c) Increase the turnover, $r = T/d$, of durable consumer goods (reduce their life or increase the speed of consumption). Example: new personal computers every 2 years; throw away the shoes after using them; change furniture every 3 years, etc.
(d) Increase the number of consumers, M. Example: a computer for everyone; a cell phone for children; a car at 14 years of age, etc.
(e) Invent new goods to satisfy aspirations or reinvent old products.

In "wealthy" economic systems, these alternatives are all practiced today, in various forms and to various degrees. Must the economy become *destructive*? Even aspirations remain the same only when goods are scarce or low in quality. If there is an abundance of goods, and all of high quality, then motivation is lost. Example: a better restaurant is justified if restaurants normally are of average quality. Furs, Rolex watches, Ferraris have a unique attraction only if possessed by a few people. But we cannot escape from the Law of the increase in productivity and quality.

If in equation (10.23), πL(T) in the system increases and *it is not possible to increase* QP(T), then QL(T) must *necessarily* decline. An alarming question immediately arises: *in a world of highly automated production systems and in a world populated by robots, will there be enough work for everyone?* If we assume there are N workers in the system, then:

$$\mathrm{QL}(T) = N\,\mathrm{hl} \tag{10.24}$$

where "hl" are the per capita labor hours necessary to achieve QL(T); substituting (10.24) on the left side of (10.23), we get:

$$N\,\mathrm{hl} = \frac{M\left[\kappa(IcP)T + \theta(DcP)r\right]}{\pi\mathrm{L}(T)} \tag{10.25}$$

From (10.25) clearly follows that if πL(T) increases and the numerator is unchanged:

(a) With "hl" constant, labor would be freed up, since N must fall; unemployment would become a structural factor.
(b) With N constant, the unit labor requirement, "hl," would fall; this decline in labor hours would be gradual.

In fact, this is what has always occurred! The increase in productivity makes *wealth more abundant*, and on the other hand, this has translated into a *gradual reduction in the labor* needed to produce (Blank and Shapiro 2001). If the increase in productivity over time represents a trend (Law of increasing productivity) and if the increase in wealth is viewed as progress, how then can we counter the fall in the quantity of labor needed for production and ward off the fear of the *jobless economy*?

1. It is estimated that between 35% and 50% of jobs that exist today are at risk of being lost to automation. Repetitive, blue- collar type jobs might be first, but even professionals—including paralegals, diagnosticians, and customer service representatives—will be at risk. This is n't not just science fiction, it is happening now (Marr 2017, online)
2. Cutting-edge robots with artificial intelligence will have almost entirely replaced human workers within just 20 years, a world-leading expert on the subject has claimed. To prepare—he says he need to rethink our jobs, and our society. [...] Within the next two decades the human race will have to "*rethink our society itself, rethink capitalism, we need to rethink the way we are as human beings*," thanks to the impact of artificial intelligence and robotics (Brown 2016, online)

> Every job which takes less than 5 seconds to think will be done by robots (Rouhiainen 2019, p. 103)
> Martin Ford, author of *Rise of the Robots*, also highlights the fact that the kinds of tasks that are routine and repetitive will be the first to be assigned to robots, saying: *I personally believe that, in the future, we could well get into a situation where jobs simply disappear. And it will be especially any kind of job that is routine or repetitive on some level. A lot of those jobs are going to disappear* (Rouhiainen 2019, p. 103).

3. The imagined future where humans may not have to work, as machines will be taking care of an ever-widening range of our needs and wants, is not assured, but is highly probable. We can debate the timeline and keep stuffing this difficult conversation into a can to kick it down the road, but what would be more constructive would be delving into this debate headfirst, trying out new policies, learning from one another and shaping our jobless future to minimize its discontents. Our kids, the Gen Zs, will thank us for it (Chakhoyan 2017, online)

There is even clearer evidence today, in modern economies dominated by production networks, of the tendency toward specialized production and the broadening of network links. The spread of production networks will result in the only firm in existence being the entire network, which, despite everything, will also be governed by the Law of Increasing Productivity and Quality, this fulfilling the prophecy of Ray Kurzweil.

> An analysis of the history of technology shows that technological change is exponential, contrary to the common-sense "intuitive linear" view. So we won't experience 100 years of progress in the 21st century—it will be more like 20,000 years of progress (at today's rate). The "returns," such as chip speed and cost-effectiveness, also increase exponentially. There's even exponential growth in the rate of exponential growth. Within a few decades, machine intelligence will surpass human intelligence, leading to The Singularity—techno-

> logical change so rapid and profound it represents a rupture in the fabric of human history. The implications include the merger of biological and nonbiological intelligence, immortal software-based humans, and ultra-high levels of intelligence that expand outward in the universe at the speed of light (Kurzweil 2001, online).

10.14.9 *Control of Plant Efficiency for Quality and Productivity. The Total Productive Maintenance (TPM)*

Among the quality costs, maintenance and plant efficiency costs play an important role. Poorly maintained plants become inefficient and reduce design quality, thereby increasing product noncompliance costs (Ebeling 1997). They also increase equipment costs and reduce the productivity of the functional labor force at the plants themselves; increase the costs of preventing and inspecting defects; and, above all, increase the risk that defective products will be manufactured which will need to be repaired or replaced and that nonreprocessable production waste will be produced. Machinery with poor maintenance can result in more or less prolonged downtimes that force other production units, both upstream and downstream, to shut down, resulting in significant costs for unproductive manpower, a failure to meet the terms of delivery, penalty payments, a loss of image, and even a loss of customers (Fig. 10.16).

> Equipment maintenance represents a significant component of the operating cost in transportation, utilities, mining, and manufacturing industries. The potential impact of maintenance on manufacturing performance is substantial. Maintenance is responsible for controlling the cost of manpower, material, tools, and overhead … (Ahuja and Khamba 2008, p. 710).

To avoid all these inconveniences, repairs ex-post to restore the functionality of inefficient plants is not enough. The company instead must implement a *policy of*

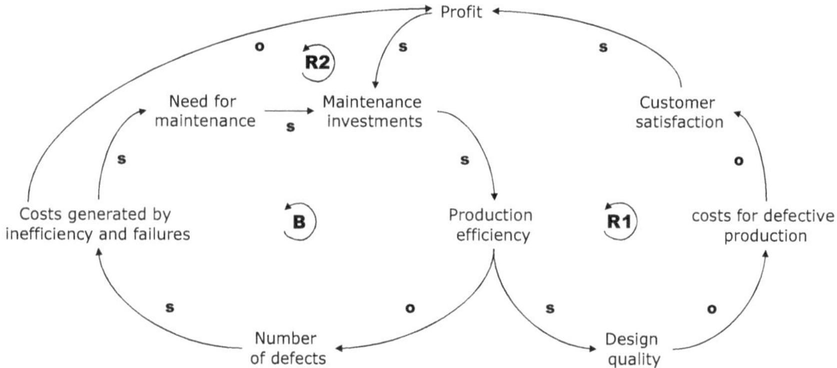

Fig. 10.16 Need for and utility of maintenance

prevention that, as much as possible, avoids the emergence of defects or abnormal functioning of the machines. This can be achieved by introducing a specific *Systematic Maintenance* business function to which *maintenance planning* is entrusted. According to UNI regulations, maintenance is defined, in general, as "*a combination of all technical, administrative and management actions, during the life cycle of an entity, intended to maintain or return it to a state in which it can perform the required function* (UNI 13306 2018, p. 2.1)." Normally, there is a distinction between *ordinary* and *extraordinary maintenance.* The main difference between routine maintenance and extraordinary maintenance lies precisely in their respective functions.

(a) *Ordinary maintenance:* this includes mainly restoration and corrective maintenance and routine preventive maintenance during the life cycle of the plant in order to (UNI 11063 2003, 4.1):

- Maintain the original integrity of the product
- Maintain or restore the efficiency of the good
- Contain normal wear and tear
- Ensure the useful life of the good
- Cope with accidents

(b) *Extraordinary maintenance*: this includes maintenance to upgrade equipment and significant preventive maintenance, which involves plant overhauls to improve its efficiency and life, since this type of maintenance:

- Can prolong the useful life of the equipment and/or improve its efficiency, reliability, productivity, maintainability, and inspectability;
- Can retain the original characteristics (plate data, sizing, building values, etc.) and the essential structure
- Does not involve changes in how the asset is used

We can distinguish four types of maintenance.

(a) *Corrective, ordinary, breakdown,* or *reaction maintenance*: this is carried out after a failure has occurred in an attempt to return a plant or equipment to the state that allows it to perform the requested function.

Run to failure is a maintenance strategy where maintenance is only performed when equipment has failed. Unlike unplanned & reactive maintenance, proper run-to-failure maintenance is a deliberate and considered strategy that is designed to minimize total maintenance costs (Dutschke 2014, online).

(b) *Preventive maintenance*: this is performed before a failure has occurred; it can be programmed at regular intervals, regardless of the state of the plant, or at intervals chosen based on predetermined parameters. It uses statistical and operational research models (McCall 1965; Kyriakidis and Dimitrakos 2006) and also relies on the opinion of experts (Wang and Zhang 2008).

[Preventive maintenance is] An equipment maintenance strategy based on replacing, or restoring, an asset at a fixed interval regardless of its condition.

Scheduled restoration tasks and replacement tasks are examples of preventive maintenance tasks (Gulati et al. 2010).

(c) *Condition-based maintenance* and *reliability-centered maintenance*: these fall under the category of preventive maintenance and include prevention measures subject to the existence of a certain condition of an asset, which can vary for metrics such as mileage, molding machine beats, operating time, thereby avoiding the disadvantages from excessive prevention (e.g., replacing a plant component when conditions of wear do not justify the intervention).

(d) *Predictive maintenance*: this reveals the presence of a progressive irregularity through the discovery and interpretation of weak warning signals regarding the final failure; this form of maintenance has spread thanks to increased monitoring capacity and the collection and analysis of data "supplied" by the machine, along with the ability to develop technological tools capable of learning and providing new information by processing the data obtained (Yam et al. 2001).

Figure 10.17 highlights a possible system to control the state of efficiency of a machine through various forms of maintenance.

In today's world of highly computerized and automated production, modern maintenance is no longer an episodic reaction to failures and malfunctions, but an activity that must be conceived of as a new business service whose features increasingly extend throughout the organization and whose characteristics can be outlined as follows:

- Maintenance work is transformed into programmable work.
- The maintenance function is transformed into an accountable unit.
- Maintenance is carried out to schedule, coordinate, and control.
- Staff training plays a key role.

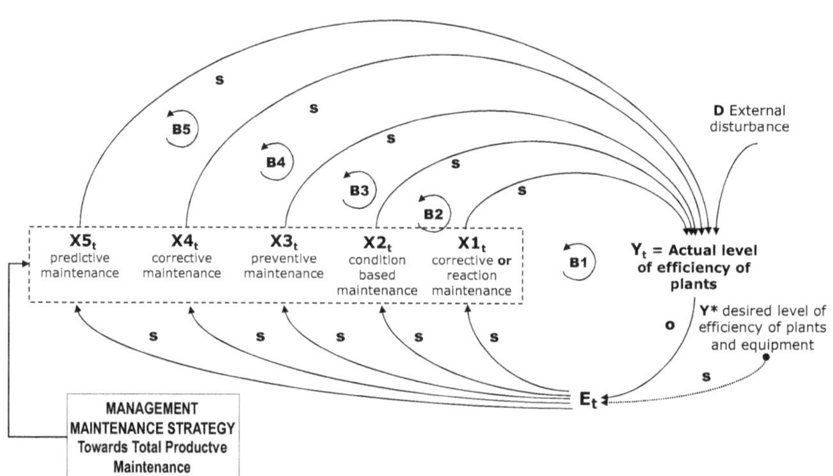

Fig. 10.17 Maintenance strategy toward total productive maintenance

- Modern maintenance is also asked to improve and maintain the systems for the safety and health of the worker. Accountability is needed.

This new maintenance role is called Total Productive Maintenance (TPM).

> TPM is a unique Japanese philosophy, which has been developed based on the Productive Maintenance concepts and methodologies. This concept was first introduced by M/s Nippon Denso Co. Ltd. of Japan, a supplier of M/s Toyota Motor Company, Japan in the year 1971. Total Productive Maintenance is an innovative approach to maintenance that optimizes equipment effectiveness, eliminates breakdowns, and promotes autonomous maintenance by operators through day-to-day activities involving total workforce … (Ahuja and Khamba 2008, p. 715).

> It can be considered as the medical science of machines. Total Productive Maintenance (TPM) is a maintenance program which involves a newly defined concept for maintaining plants and equipment. The goal of the TPM program is to markedly increase production while, at the same time, increasing employee morale and job satisfaction (Venkatesh 2007, p. 3).

> TPM was introduced to achieve the following objectives. The important ones are listed below.

- Avoid wastage in a quickly changing economic environment.
- Producing goods without reducing product quality.
- Reduce cost.
- Produce a low batch quantity at the earliest possible time.
- Goods send to the customers must be nondefective (ibidem).

TPM combines the "Western method" of preventive maintenance with the "Japanese" method of Total Quality Management (Sect. 10.7), which involves personnel at every business level (Miyake and Enkawa 1999). In fact, the adjective "total" takes on three meanings:

(a) Total efficiency: which indicates the pursuit of economic and financial efficiency;
(b) Total maintenance system: this includes reactive, corrective, preventive, and proactive maintenance;
(c) The participation of all employees in the autonomous maintenance activities carried out by workers in small groups.

According to the Japan Institute of Plant Maintenance, TPM has five fundamental objectives (Nakajima 1989):

1. Maximize the overall effectiveness of the plants;
2. Establish an accurate system of preventive maintenance during the life of the plant;
3. Hold all the factory functions accountable;
4. Involve all staff;
5. Promote production maintenance through "motivating management," with the work carried out by small autonomous groups.

TPM also considers *predictive maintenance* as essential; therefore, it emphasizes the fundamental *monitoring* function, which is indispensable for the realization of an effective and efficient *predictive* and *condition-based* maintenance. Monitoring is made possible thanks to a sophisticated "sensor" that detects at any time, and in real time the condition of the plant, thereby allowing maintenance intervention to be activated only in the presence of a potential failure, regardless of the scheduled maintenance plans. The application of TPM, which involves the whole organization, makes it necessary to spread *the maintenance culture* to all levels: from the machine operator to the process manager, who also becomes the head of maintenance. The workers become the executors of Autonomous Maintenance who transfer the activities of preventive maintenance at the first, or routine level (inspections, cleaning, controls, replacements, disassembly, small repairs, etc.) to the production workers. The status of Autonomous Maintenance is achieved through check-lists for verification and comparison specifically prepared for the TPM context.

In conclusion, maintenance is not synonymous with "repair" but with a process aimed primarily at understanding the technical and human causes that can prevent failures and maximize the efficiency and effectiveness of the service to ensure the overall economic optimization of the production system. Therefore, considerable importance is being attributed not only to the preparation of staff, but also to their team spirit. In fact, in the maintenance area it is necessary to combat individualism, which allows neither for collective growth nor development. This problem can be solved through greater communication and greater involvement in the processes that govern the company, which is precisely what lies behind the TPM approach.

10.14.10 Control of Plant Efficiency, Replacement, the MAPI Formula

The maintenance processes presented in the previous *section* keep the machinery running efficiently but cannot extend its useful life beyond a certain time limit because, even if technical performance is not reduced, the machinery undergoes an inevitable process of obsolescence. New plants that perform the same functions but with quality and cost advantages make it convenient sooner or later to replace those in operation. Therefore, it is necessary to check when the optimal "time" has arrived to replace an old operating system with a new one. This form of control is known as the replacement problem, which specifically refers to calculating the optimal date for replacing the industrial plants and equipment (Connor and Evans 1972; Fabrycky et al. 1972). Management must thus continually evaluate with even more care the convenience of their plant investments; once the investments are undertaken, they must monitor the prospects for future use in order to determine the economically useful life cycle, determining the optimal period in which to discontinue operating the plant. Such calculations estimate the period of time beyond which the continued utilization of a plant, already installed and operational, is no longer economically viable and should thus be replaced or shut down.

The calculation of the *optimal operating period* for a single plant, not included in a chain of renewals, is a typical economic–financial problem (Massé 1962): management must to determine how much time is needed to cover the purchase and operating costs for the plant with the sales revenue of the production obtained from its use, with the desired operating result. The problem could be solved calculating year-to-year the financial present value of the future costs and revenues. The replacement year is determined when the present value of costs exceeds that of revenues. It is also necessary to consider that the length of the operating period of the plant itself conditions the costs and revenues that derive from it, in that:

- The purchase cost of a plant equivalent to the one to be replaced presumably varies from year–to-year since, due to technological progress, the retired plant is replaced with one technically more advanced and with greater productivity.
- The cost and revenue flows are sensitive to the period of reference, because of both the volume of demand and the prices, if the more modern plant allows the same products to be obtained at lower costs, or products of higher quality and with more attractiveness for the market.
- The disposal value of the retired plant varies according to the year it is shut down.

These considerations make it clear that the choice of the replacement year of the machinery in use affects both the calculations of economic convenience for the new machinery and, consequently, similar calculations for the replacement of the latter in future. When the renewal is delayed the firm will most likely see its production costs increase at an increasing rate and will not be able to remain price competitive with respect to its competitors, the more so if the latter use more modern and efficient plants. If the delay of the renewal is too long, the business connected to the plant will enter its declining phase, becoming a "dog" (in the product portfolio matrix of The BCG, Henderson Inst. 1970), so that it will have increasing difficulty raising the necessary capital for renewal, which is more urgently required due to economic necessity. Similar problems would be encountered by anticipating the replacement since, due to the contraction of the pay-back period, the cycle of economic utilization of the previous investment would not be completed, and the capital that is invested and not yet written off through the depreciation process cannot easily be fully recovered through the sale of the retired plant. From the above considerations, it follows that more than referring to an individual operation, renewal must be viewed more broadly to include the concept of "chain of renewals" in which the replacement of each link in the chain, due to technical progress, inevitably ends up conditioning all the subsequent links and, as a result, the entire chain of subsequent substitutions.

In 1949, the Machinery Allied Production Institute in Washington set up a study committee chaired by George Terborgh to come up with a new evaluation system for investments aimed at the equipment-replacement problem, which led to a general rule, usually referred to as the MAPI formula (Terborgh 1949, 1956, 1958), that allows for the determination at a given moment of the opportuneness of replacing an existing plant, *the defender plant*, with a *new plant* that has come on the market, called the potential *challenger plant*.

> The present work ... is an attempt to rethink fundamentally the underlying theory of equipment policy. . . . it is an attempt to integrate into this theory a recognition of the phenomenon of obsolescence, strangely ignored i^as to future obsolescence at least) by existing theory. It is thus a piece of intellectual pioneering (Terborgh 1949, p. vii)

Even if the MAPI formula has been the focus of criticisms (Dean and Smith 1955; Leung and Tanchoco 1983; Abbott and Ring 1983), it offers a general solution to the equipment-replacement problem that covers a wide range of cases.

> Essentially, it provides a quick test of current replaceability on the basis of a prefabricated structure of assumptions and projections into which the analyst can insert the necessary stipulations and estimates for the case in hand (the amount of the investment, the income tax rate, the interest rate, the service life, the terminal salvage ratio, etc.). It is an elaborate gadget to simplify an otherwise almost insoluble problem (Terborgh 1956, p. 138).

Terborgh assumes continual technological evolution, which posits that progress is continuous and capable of making available to the firm a plant perfected at every future instant (in an abstract sense). There is thus the assumption of a continued operational superiority of future plants with regard to those the market offers and will offer; or, conversely, a continual operational inferiority of the plant to be replaced with regard to the new ones. Plant E_t in use at instant "t" is called the *defender*. The new plant, E_{t*}, available at "t*", is called the *current challenger*. Compared to the challenger, the defender has an *operating inferiority* that manifests itself in the following values:

- Gross revenues, E_t, less than those obtainable from E_{t*}
- Annual costs, E_t, greater than those required by E_{t*}
- Consequently, an annual margin, E_t, less than that obtainable from E_{t*}; this margin has a monotonic nondecreasing trend for each future $t > t^*$
- If the defender's margins have a *monotonic nondecreasing* trend year-to-year, then its *operating inferiority* will also have a monotonic nondecreasing trend

From the theoretical point of view, if the cost of capital was zero then the optimal renewal policy would obviously consist in replacing the plant every year (even at every instant, given the hypothesis of the continuity of technical progress) in order to have the best plant at every moment (Terborgh 1958). If we abandon this unrealistic hypothesis and assume that every investment entails a cost for the invested capital, then it is realistic to assume that the cost of the capital invested in the defender plant decreases as its period of utilization increases. Regarding the challenger, as the utilization period increases there are two opposing trends at work: on the one hand, the cost of the invested capital declines, but on the other the *operating inferiority* increases. The *optimal solution* is to find the point where the sum of the values of these trends is at a minimum—Terborgh defines this as the "adverse minimum"—that is, the minimum of all the alternative magnitudes, each of which is adverse.

Two fundamental hypotheses are posited to formulate the evaluation criterion:

1. All the future plants (the succession of challenger plants) have an adverse minimum equal to that of the best plant at the time of the survey. This is equivalent to hypothesize not only the continuity of technical progress but also its constancy

and uniformity over time. This serves to narrow the analysis to the data regarding the defender plant and to the best plant available for substitution at that moment, without worrying about what would instead happen if, due precisely to technical progress, the successive challenger should have very different economic-technical improvements.
2. The operating inferiority of the defender increases linearly over time, that is, both the costs regarding the physical deterioration of the plant (costs for maintenance and lower productivity) and those linked to the plant's obsolescence growing linearly.

These hypotheses reduce the analysis of the opportuneness of an investment in plant substitution to a comparison of the adverse minimum of the most technically advanced plant (the challenger) and the plant in operation at that moment (the defender).

Technical note To calculate the *adverse minimum*, we can indicate by:

- E_0 the defender plant.
- E_1, E_2, E_3, etc., the *challengers* available for substitution in the subsequent years.
- m_0, m_1, m_2, m_3, etc., the annual operating margins (revenues less operating costs) that would occur under the assumption that after t_0 the best new plant is always utilized. For convenience sake, we assume that the margins are discrete and achievable at the end of each year.

If we set "t" as the instant in which the plant E_0 begins operating, and setting $h = 1, 2, 3, \ldots$, and then the initial annual margins for each *new* and improved plant will be:

- $m_1(t + 1)$, $m_2(t + 2)$, $m_3(t + 3)$, … $m_h(t + h)$, and, based on our hypotheses:
- $m_1(t + 1) < m_2(t + 2) < m_3(t + 3) < \ldots < m_h(t + h)$.

If the firm keeps plant E_1 in operation in the year subsequent to year one, when technically better plants are instead available which, through technical progress, can offer increasingly greater initial annual margins, it will achieve lower annual margins; in other words, the firm would sustain costs due to the obsolescence of the plant with respect to new ones. If the firm should thus maintain E_1 in operation past the first year, it would achieve lower annual margins, even while plants with higher annual margins would be available. If we indicate by:

- $m_1(t + 1)$ the annual margin achievable with plant E_1 1 year after it has entered into service at instant t
- $m_1(t + 2)$ the margin achievable with E_1 2 years after it has become operational ...
- $m_1(t + h)$ the margin attainable with E_1 h years after its adoption by the firm,

then we obtain the following relation:

$$m_1(t+1) > m_1(t+2) > m_1(t+3) > \ldots m_1(t+h-1) > m_1(t+h)\ldots$$

The above relation indicates that if the firm should continue to utilize E_1 for periods subsequent to the first year, it would suffer disadvantages from lower margins owing to the gradual deterioration of the plant, together with the following:

- An increase in the operating costs due to a reduction in the technical productivity of the plant
- A reduction in sales revenue from products which, due to the wear and tear on the plant, may have lower quality
- An increase in maintenance costs

We thus define the *operating inferiority* of E_1, *for each subsequent year of utilization* $(t + h)$, as the difference between the net margin of a new plant and that of the plant in consideration is:

$$m_{\mathrm{h}}(t+h)-m_1(t+h)=g_1(t+h)$$

The *net present value* of the operating inferiority for a duration of "n" years of plant utilization after adding the purchase cost C_1 and subtracting the revenue from scrap after "n" years,$S_1(n)$ at a chosen given discount rate, "i," takes on the meaning of the *total opportunity cost* of the plant for "n" years:

$$B_1(n)=C_1-S_1(n)v^n+\sum_{h=1}^{n}g_1(t+h)v^h \tag{10.26}$$

where $v^h = (1 + i)^{-h}$ is the present value coefficient at discount rate "i."

Using (10.26), it is more significant to determine the average annual burden of service of the plant in use for "n" years, in the form:

$$b_1(n)=B_1(n)\frac{i}{1-v^n} \tag{10.27}$$

If we take "n" as a variable, we can calculate (10.27) for periods of increasing length, $n = 1, 2, 3, ..,$ and determine the period "n^*" that corresponds to the *minimum average annual burden of service* (10.27). This value, $b_1(n*) = min$, represents the *adverse minimum* of $E_1(n^*)$. However, since we are looking for the desirability of substitution of the *defender* plant E_0, we must compare the adverse minimum of the challenger E_1—that is, $b_1(n*) = min$ – with the *annual replacement cost* of the defender E_0, called the *adverse cost*, which is calculated as follows:

$$\Delta_0(t)=\left[S_0(t)-S_0(t+1)\right]+S_0(t)i+g_0(t+1) \tag{10.28}$$

where:

- $[S_0(t) - S_0(t + 1)]$ represents the loss in value of the *scrap* due to an additional year's utilization of E_0
- $S_0(t)$ i is a year's lost interest on the *terminal value*

- $g_0(t+1)$ is the operating inferiority of E_0 following its prolonged use for another year

The defender E_0 should be replaced with the challenger E_1 when:

$$\Delta_0(t) > b_1(n*) \tag{10.29}$$

If this optimality condition is not satisfied then it is economically more useful to utilize E_0 for another year, at the end of which a new comparison will be needed, but this time between E_0 and E_2 for the following year. If the inequality $\Delta_0(t+1) > b_2(n*)$ does not still exist, then it will be necessary to utilize E_0 for another year and then compare E_0 and E_3. The process should be repeated every year until the optimality conditions exist. The procedure presented to this point is difficult to apply in practice due to the difficulties in estimating the *adverse minimum* of the challenger plant relative to the length of utilization of the defender. In fact, a too forward-looking knowledge would be required, which is almost impossible to satisfy.

Rather than calculate, for each year of the probable life of the challenger plant, the annuality equivalent to *total opportunity cost*, as is the case in (10.27). Terborgh, in order to calculate the *adverse minimum*, provides the following formula, which is easier and more practical to apply:

$$b_1(n*) = \frac{1}{2}\left[(C_1 i) - g_1\right] + \sqrt{2C_1 \ g_1} \tag{10.30}$$

- Assuming "zero" as the elimination value and "i" the interest rate
- C_1 is the purchasing cost of the challenger plant
- g_1 is the average, constant over time (in case of the linearity of technical progress), of the *operating-cost inferiority gradient*; this is assumed to be equal to that of the plant in current operation

The adverse minimum calculated with (10.30) is compared with the replacement cost of the defender plant, as shown in (10.28). We thus return to the optimality condition (10.29), which determines whether or not renewal is appropriate.

After this long technical note, a numerical example will clarify the application of the MAPI Formula given in (10.30).

Let us suppose the following data:

- The defender is operating for 12 years and has a residual value of 20,000,000 USD.
- After another year of operation, the residual value is estimated to be 12,000,000 USD.
- The challenger, possessing greater technical efficiency, has an initial cost of 200,000,000 USD.
- The interest rate chosen is 10%.
- The challenger plant allows the firm to achieve, for the following year, the economic advantages indicated in Table 10.7, which compares the alternative minimums of the *defender* and *challenger* plants.

The value of the sum of the alternative minimums of the challenger plant is calculated by applying the MAPI Formula (10.30):

$$b_1(\text{n}*) = \frac{1}{2}\left[(C_1 i) - g_1\right] + \sqrt{2C_1 g_1}$$

Assuming that the residual value of the scrap is zero, a lower operating efficiency rate that is constant and equal to that of the defender plant, and inserting the following values:

- $C_1 = 200{,}000{,}000$
- $g_1 = 60{,}000{,}000/12 = 5{,}000{,}000$
- $i = 0.10$

we obtain:

$$b_1(n*) = \frac{1}{2}\left[(200{,}000{,}000 \times 0.10) - 5{,}000{,}000\right] + \sqrt{((2200{,}000{,}0005{,}000{,}000)}$$
$$= (7{,}500{,}000 + 44{,}721{,}400 +) = 52{,}221{,}400.$$

Since 52,221,400 < 70,000,000, and thus the optimality condition (10.29) exists, it is absolutely appropriate to replace the plant in operation with the new one.

For further development of the MAPI formula, see Mella et al. (2010), which also presents a new numerical example.

Table 10.7 Application of the MAPI formula: an example

	Advantages for following year if the challenger plant is installed (USD)	Advantages for following year if the defender plant is maintained (USD)
Lower maintenance costs	3,500,000	
Lower labor costs	34,500,000	
Lower general production costs	25,000,000	
Lower insurance costs		3,000,000
Total	**63,000,000**	**3,000,000**
Inferior operating efficiency of the *defender* plant compared to the *challenger* plant for the following year		60,000,000
Loss in value of old plant		8,000,000
Passive interest on the residual value of defender plant		2,000,000
Sum of alternative minimums of defender plant		**70,000,000**

10.15 Summary

There are two fundamental variables—*quality* and *productivity*—which drive both the organization and the economic system observed as a network in which organizations represent the basic modules or nodes (Chap. 8). Controlling these variables is of the utmost importance because both the efficiency of the entire organization and improvements in the well-being of the entire society depend on them. At the organizational level, *quality* control is essential both to improve customer and market relations (Sect. 10.2) and to avoid or reduce nonquality costs (Sect. 10.5).

With regard to quality control, the chapter began by presenting three meanings of the term "quality": functional quality, design quality, and environmental quality. There are numerous and complex techniques for evaluating the *quality levels* and the *levers* to be used to control this variable, which differ depending on the type of quality to be controlled. *Section* 10.3 illustrated the levers for the control of functional quality; Section 10.4 indicated how to control design quality; and the levers for environmental quality control were presented in Sect. 10.6.

After considering *quality* control, *productivity* control was examined. The latter is the variable which is the basis for all productive systems, since the search for maximum productive efficiency is necessary to reduce production costs and thus to produce value, following the directions outlined in Chap. 9. After defining the concept of productivity as a variable that indicates a reduction in the unit quantities of factors utilized in the production processes and discussing the factors on which productivity depends, I then examined the drivers of productivity followed by a discussion of the consequences of the continuous growth in productivity. Since quality and productivity are variables whose levels are continuously increasing, the chapter concluded by presenting the Law of Increasing Productivity and Quality (Sect. 10.13), followed by interesting Complementary Materials that include the following:

- Six Sigma Quality Management (Sect. 10.14.1)
- Time and Motion Study and Work Sampling (Sect. 10.14.2)
- Control Charts (Sect. 10.14.6)
- Productivity, Consumption, and Employment (Sect. 10.14.8)
- Equipment maintenance and Total Productive Maintenance (TPM) (Sect. 10.14.9)
- Optimal operating period for a single plant. The MAPI Formula (Sect. 10.14.10)

Chapter 11
The Magic Ring in Action: Control of Production and Stocks

Abstract The control of the warehouse and production processes, typical forms of management control aimed at the processes of production and economic transformation (TR1-P and TR2–3) described in Sect. 9.3, is also indispensable for the formation of the production and purchasing budget since the quantities to be produced and purchased in a given period depend on the planned stock of materials and products at the end of the budget period. A number of calculation models have been developed for warehouse and production control. Some simple models assume knowledge of the amount of demand for stocked goods; others, more sophisticated, assume that demand is known only in terms of the distribution of probabilities. The logic of warehouse control will be presented, including the calculation of the optimal purchasing and production batches even under the assumption of production constraints. An examination of the logic of the "global" control methods will follow: MRP, JIT, OPT, concluding with FMS and HMS. The final part of the chapter addresses the problem of *project control* by examining PERT/CPM and ALTAI not only as *network programming techniques* but also as optimization and control tools.

Keywords Coupling and uncoupling function of stocks · ABC method · Cost of preservation, or storage costs (sc) · Acquisition/ordering costs, or setup costs (ac) · Economic batch of supply · Wilson's hypotheses · Optimal production batch · Linear programming · Optimal production batches under the hypothesis of production constraints · Production control: "pull" and "push" method · MRP · JIT · OPT · FMS · HMS · CIM

> I say an hour lost at a bottleneck is an hour out of the entire system. I say an hour saved at a non-bottleneck is worthless. Bottlenecks govern both throughput and inventory (Eliyahu Goldratt, in Goldratt and Cox 1984)

> The repertoire of tools that top managers use… is determined by their experiences. A dominant logic can be seen as resulting from the reinforcement that results from doing the right things … they are positively reinforced by economic success (Prahalad and Bettis 1986, pp. 490–491).

P. Mella, *The Magic Ring*, Contemporary Systems Thinking,
https://doi.org/10.1007/978-3-030-64194-8_11

> Why not make the work easier and more interesting so that people do not have to sweat? The Toyota style is not to create results by working hard. It is a system that says there is no limit to people's creativity. People don't go to Toyota to 'work' they go there to 'think' (Taiichi Ohno, quoted by Ballé 2015, online).

Management control, or *operational control*, concerns the various functions (supply, production, sales, logistics, personnel, finance, etc.) and various operational units (plants, divisions, departments, individual facilities, and/or workers, etc.) which, by activating various levels of operational levers, produce the values that must be achieved by those values chosen in the budget objectives. The *operational control* is oriented toward the achievement of short-term objectives (producing at an average cost below a certain level; fulfilling an order at a given cost and within a given time period; not exceeding the maximum or the minimum warehouse stock; optimization of the production mix for a given period), making use of various and well-structured instruments to detect the *variances* and determine which control *levers* to adopt.

Although it is not possible to examine here the many *Rings* that make operational control possible, it is nevertheless useful to take a brief look at some of these control systems which are particularly significant and carry out a *micro-management control* of the *productive transformation*: *stocks and production*, which are "technical" variables—together with quality and productivity, as discussed in Chap. 10—that require detection instruments and control levers of a "technical" nature. I will start by presenting some classic inventory control models, in particular, quantity control by calculating the Economic Batch of Supply of the Optimal Production Batch, based on Wilson's hypotheses. I will then address the programming and control of *production with constraints*, presenting the linear programming method. Today, companies have a large number of control tools that rely heavily on IT; such tools provide complete control of production and storage, thereby greatly increasing process efficiency. Given the breadth of the issues dealt with, I will not be able to provide an in-depth treatment of the various topics. However, even a concise presentation of these topics is useful. After the presentation of the technical aspects of batch size and programming, I will consider material control and production systems, making a summary presentation of the global methods of control: MRP, JIT, OPT, and concluding with FMS and HMS. The last part of the chapter will address the problem of project control by examining the *network techniques* of PERT/CPM programming and the problem of assigning workloads in the event of the simultaneous execution of different projects.

11.1 Politics of Stocks and Production: The Coupling and Uncoupling Function of Stocks

From a *technical* point of view, in manufacturing companies stocks represent *volumes* of materials (primary and ancillary), semi-processed goods and components, or finished products—or goods and packaging in commercial companies—called the average stable stock level. These goods are stored in the warehouse for a given period. From an *economic* point of view, they represent monetary investments with a more or less high turnover, depending on the type of goods in stock. *Functional stocks* are not mere instant stocks or purchases for the purpose of price speculation (speculative buying stocks) but the result of a well-controlled storage process, based on accurate economic calculations that attempt to reduce the disadvantages due to the asynchronies of the business rhythms (*use* and *production*) compared to those of the markets (*purchases* and *sales*) and to the risks from the shutting down of processes (*minimum security stocks*).

Referring to the *functional stocks*, in industrial companies these simultaneously fulfill a *coupling* and *decoupling* function to ensure the elasticity of the processes of supply, production, and sales. They fulfill a coupling or anticipatory function in bridging the time gaps between the procurement process, the production process, and the sales process. As a result, production and sales requirements can be met regardless of the pace of production and of the acquisition of raw materials and the pace of the sales market. At the same time, stocks fulfill a *decoupling function*; raw material stocks make the production process partly independent of supplies; the stocks of products decouple the production process from that of sales (Fornaciari and Galassi 2012). The two functions cannot be considered separately: the inventory fulfills a decoupling function precisely because it serves a coupling function, allowing for the elimination of *asynchronies* between purchasing processes, production, and sales (Fornaciari and Galassi 2012), thereby making the pace of production processes stable; this stability promotes both the achievement and maintenance of higher levels of production quality (Eroglu and Ozdemir 2007) and an increase in productivity by facilitating the application of the control techniques examined in Chap. 10. The objective of independence between the rhythms of production and sales is important since a relatively uniform pace of production is reflected in the company's procurement policy, which is able to schedule its purchases on a regular basis without necessarily being subjected to supplier prices. The stability of the supply of a *downstream* company has a positive impact on the stability of production of *upstream* suppliers; this gradual rationalization extends to all companies that are part of the supply chain (Mentzer et al. 2001; Gardner 2018).

Finally, it is important to note that the warehouse optimization policy assumes that *functional stocks* are considered to be *physiological*; they do not represent a *redundancy* but an operational necessity, precisely due to their decoupling and coupling function. Therefore, *it is necessary to optimize their flow and level over time*. This concept is typical in Western companies created during the first industrial revolution, which were rigidly structured to achieve uniform production rhythms in an economy that was not *saturated* or sufficiently computerized, and with a non-advanced logistics system. Alongside (and perhaps opposed to) this is the approach of Japanese companies (now widespread, however, throughout the productive world)—that have had to face the industrial problems of the *saturated* economy (Chiarini 2010), according to which stocks are the consequence of the *rigidity* of production and can (indeed, must) be avoided. It is necessary to eliminate them through a policy of *flexible manufacturing*, which tends to produce with *zero stocks* (Browne et al. 1984; Raouf and Ben-Daya 1995).

When stocks are *functional,* and therefore play a coupling and decoupling role, it is clear that the control of physiological stocks is still interrelated to the size and rhythm of use of the plants; the dynamics of stocks depend, therefore, on the policy of *the plants*.

These stock/plant relationships should be examined by distinguishing between two typical business situations:

- The company produces goods whose trend in demand can, to a certain extent, be predicted constantly over time.
- The demand for the goods produced is subject to strong fluctuation, which are more or less recurrent over time.

In the first situation (constant demand) the company can schedule the size of the plants, and therefore the available production capacity, on the basis of the constant volume of demand. In this case, *functional stocks* of products may be very limited and able to cope with only small fluctuations in short-term demand or with temporary production interruptions. Note, however, that, even in the presence of constant demand, it may be economically convenient for companies to buy plants with more productive capacity than is sufficient to meet demand, in anticipation of future sales expansion. The most effective size of the plants is determined by an appropriate calculation of the economic benefits of investment, which will take into account not only the present situation but also, and above all, the expected future one, both in terms of demand and the period of use and costs of management and renewal. In the second situation, that is, where demand *fluctuates periodically* (cyclically or *seasonally),* management must make a choice between different plant sizes and stocks, quantities that must be decided within the framework of a *common production and storage policy* (Lorenc and Lerher 2019). The most cost-effective solution must be searched for between two extreme solutions (Fig. 11.1):

(a) The plants are sized to the *average demand* to obtain a constant production over time to produce stocks of products during periods of low sales (Fig. 11.1a) for use in coping with demand at the high end of the cycle. The productive capacity of the plants is maintained at the highest level and on a regular basis; on the other hand, at alternate periods the company finds it has stocks with consequent storage costs (Song and Zipkin 1993).
(b) Plants are sized to *maximum demand,* with variable production and the absence of stocks (Fig. 11.1b); the installed plants must have a production capacity equal to the maximum levels of demand, and therefore higher purchase and maintenance costs; on the other hand, they can be used at varying rates with the advantage of no inventory and storage costs.

The two alternatives for the size of the plants and stocks are not always discretionary; there are particular situations in which plant policy must necessarily be subject to the size of demand, as occurs for example:

1. In the case of companies making products with time-limited shelf life.

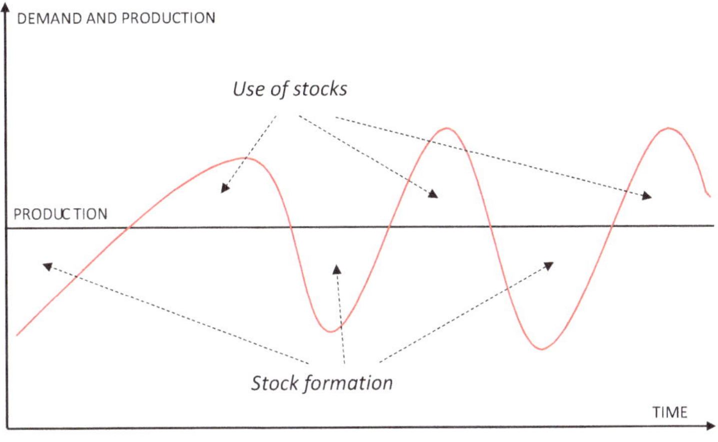

(a) **Production capacity sized to the average level of demand**

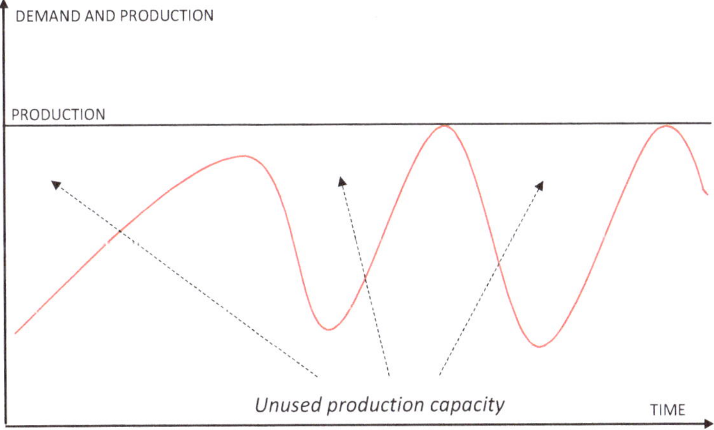

(b) **Production capacity sized to the maximum level of demand**

Fig. 11.1 Production and inventory policy as a function of demand

2. When products, even though physically preservable, are not economically preservable, as their demand is greatly influenced by changes in *fashion* and *consumer preferences*, resulting in a high *risk of depreciation* of stocks.
3. In the case of companies whose products cannot be warehoused since they provide *services*.

Similar is the case of companies that, since they transform raw materials only available seasonally (agricultural enterprises in general, sugar mills, fruit canning companies, etc.), have to size their productive capacity and rates not so much to the demand for the finished product but to the availability of materials.

11.2 The *Rings* in Management Control of Stocks—The Concentration of the Warehouse: The ABC Method

The operational control of the warehouse is the result of a particular economic-technical calculation whose main objective is to determine:

(a) The optimal quantities of stock
(b) The replenishment quantities
(c) The ordering times

to achieve the higher objective, determined by the economic transformation, of minimizing the cost of storage and/or the amount of capital invested in stocks (Tersine 1988). The three previous controls are defined together as the "technical control of the warehouse," which represents one of the most useful forms *of operational control,* both in commercial and industrial companies, precisely because, allowing for a reduction in the capital invested in stocks, it raises ROI, assuming the same operating result, and reduces the cost of capital, assuming, as is normally the case, that the warehouse is financed with loan capital. Keeping stocks under control means minimizing the cost of inventories while maintaining a balance between the investments required and the risks associated with the lack of inventory. Inventory management is the most useful means of controlling and optimizing inventory.

The warehouse control should take in all the items in stock: finished and semi-finished products, materials, components, tools, goods, and packaging. In companies that have many items in stock (hypermarkets, component sellers, etc.), a control that extends to all items could become problematic since the cost to implement it would be excessive compared to the cost savings benefit; therefore, the control is usually aimed only at *items/categories of goods* in stock whose value represents the largest part of the entire value of the stock. To identify the *items* on which to carry out the check, the simple "ABC" technique, or 80/20 method, can be used. This refers to the Lorenz-Pareto principle that highlights the degree of *concentration* of the warehouse (Ackerman 2015; Juran 2019). It is conceptually simple in that it requires that the annual consumption of each item in the categories of goods be valued and that all the categories be listed in order of decreasing annual value. Analyzing the *average composition* of the goods moved during the year generally reveals that a very small number of items, about 10–30%, determines a high percentage of the total annual value, about 75–80%; the following 10–15% of items represent about 10–15% of the value of the annual inventory movements, while the remaining items correspond to a value of 5–10% of the total value of moved items.

Implementing the ABC method is very simple (Mella 1997a, b, Sect. 19.7). We indicate by *N* the number of items in the warehouse on which to carry out the analysis and by "C_i" (*i* 1, 2, 3,..., *N*) the value of the "consumption"—appropriately defined—of the i-*th* article (items of the same kind can be accumulated). The first step in the procedure is to sort the *N* items in descending order based on the value of the "C_i." As an easy to understand example, let us assume that there are 20 categories of goods in the warehouse, for simplicity's sake labeled from "*a*" to "*t*." The accounting office calculates the value moved during the year for each category. Using an Excel file, a *distribution* can easily be calculated, as in Table 11.1, where the total value of the articles moved and the percentages per item and the cumulative amounts are calculated, which are equal to $N = 976$ units of value. Subsequently, in column 4 of the table the percentage ratio between the value of each individual item and the total value is determined. Column 5 calculates the cumulative sums of the percentage ratios, and column 6 shows the categories that are considered significant along with the number of items included in each class.

By placing the values from columns 1 and 4 on the x- and y-axes, respectively, we obtain the Pareto curve in Fig. 11.2, which will be further away from the bisectors of the axes the more concentrated the value is. The division of the stored items

Table 11.1 Division into categories of the items moved during the year and their division into categories using the "ABC" technique

Items	SALES per YEAR (thousand Eur)	Nr of codes per item	% on total sales	% cumulative	CLASSES	Nr. of items	Nr. of codes
[1]	[2]	[3]	[4]	[5]	[6]	[7]	[8]
a	200	10	20.49%	20.49%	*A*	5	90
b	180	20	18.44%	38.93%			
c	160	30	16.39%	55.33%			
d	150	10	15.37%	70.70%			
e	100	20	10.25%	80.94%			
f	35	40	3.59%	84.53%	*B*	4	170
g	30	40	3.07%	87.60%			
h	25	30	2.56%	90.16%			
i	20	60	2.05%	92.21%			
j	12	80	1.23%	93.44%	*C*	11	740
k	11	70	1.13%	94.57%			
l	10	50	1.02%	95.59%			
m	9	70	0.92%	96.52%			
n	8	60	0.82%	97.34%			
o	7	80	0.72%	98.05%			
p	6	50	0.61%	98.67%			
q	5	70	0.51%	99.18%			
r	4	80	0.41%	99.59%			
s	3	70	0.31%	99.90%			
t	1	60	0.10%	100.00%			
TOTAL	976	1000	100.00%	100%		20	1000

CLASSES	SALES per YEAR	Nr of codes per CLASS	% on total sales	% cumulative	% Nr. of items
A	790	90	80.94%	80.94%	5
B	110	170	11.27%	92.21%	4
C	76	740	7.79%	100.00%	11
Total	976	1000	100%		20

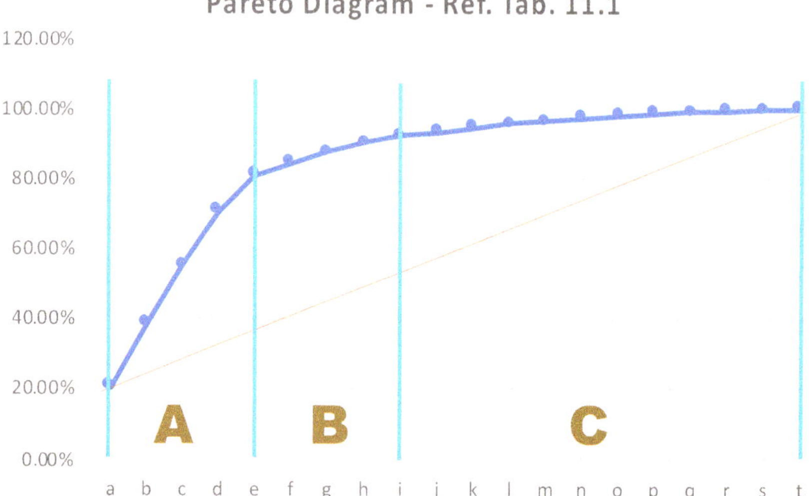

Fig. 11.2 "ABC method." Pareto diagram for the example in Table 11.1

in the ABC categories can be used to prepare appropriate alpha-numeric categories based on which it is possible to optimize the most relevant items using specifically designed calculation programs (Farné 2010). It will also be feasible to establish a procurement strategy, taking into account the control objectives and the size of the warehouse, based on which it is possible to determine which categories should be subject to individual control and which, instead, to a simpler control. We can consider not only the overall value of the various items but, when they indicate a type of good, also the number of categories within each item. In the example in Table 11.1, the item "a" is given a total valuation of Ca = 200; however, "a" could consist of a single product or na = 10 products, 20 products, etc., each of which could be subjected to a specific individual check.

Let us now assume that the items in Table 11.1 include a total of 1000 articles whose annual consumption is valued at € 976,000. If we assume that the restocking policy is to place 3 orders per year for each type, equal to 4 months of consumption, the company must then place 3000 orders per year, 1000 per quarter, and the value of each order represents the requirement for one quarter, that is, about € 325,000. If we assume that consumption is evenly distributed over time, then the average value of the stock is equal to half of the ordered quantity, in our case 2 months of consumption, which, expressed in monetary terms, is about € 163,000. We note from the bottom of Table 11.1 that the ABC method has revealed a warehouse inventory divided into the following categories:

- Category A = € 790,000, with an average inventory of 790/6 = (approximately) 132
- Category B = € 110,000, with an average inventory of 110/6 = (approximately) 18

- Category C = € 76,000, with an average inventory of 76/6 = (approximately) 13
- For a total of € 796,000, with an average inventory of 790/6, or about 163 per year

Let us now assume that management has decided to apply a different ordering policy for the three categories A, B, C, namely:

- Category A items: one order per month (12 orders per year)
- Category B items: one order every 4 months (3 orders per year)
- Category C items: one order per year

Recalculating the new average stock value, we see that this value has been significantly reduced, almost halved:

- Category A = € 790,000, with an average stock of about 790/24 = (approximately) 33, with the number of orders quadrupling (from 3 to 12)
- Category B = € 110,000, with an average inventory of about 110/6 = (approximately) 18, remaining at 3 orders per year
- Category C = € 76,000, an average inventory of 76/2 = (approximately) 38, or 1 order per year
- For a total of € 976,000, an average inventory of about 89 over the entire year
- These conclusions, of course, need to be combined with an analysis of the correct calculation *of the optimal purchase batch*, according to the techniques that will be presented in the *sections* below. In any event, the ABC method allows us to highlight the articles on which to focus control.

11.3 Cost of Preservation or Storage Costs (SC)

The example at the end of the previous *section* shows that the inventory value of a given *item* depends on the number of replenishing orders needed to cover its total annual requirement. Stocks would be drastically reduced if reordering was frequent and for low volumes. A decrease in the number of reorders would, on the contrary, increase the average value of stocks. However, each ordering process has fixed, unavoidable costs. The greater the increase in the number of reorders, even if this leads to lower investment in stocks and a reduction in the average investment costs, the greater the increase in fixed reordering costs.

(a) *The fundamental problem of inventory control* is immediately clear: to determine *the optimal economic level, physiologically necessary*, of the stocks of each significant item (according to the chosen ABC method) that minimizes the costs of the storage process while maintaining the level of storage considered compatible with the coupling and decoupling function (Bartmann and Beckmann 1992; Fornaciari and Galassi 2012).
(b) The stocks of *raw materials* and *consumables* must be sufficient to feed the production process: the objective *must be to ensure that stock shortages do not*

result in delays in the planned production rates and in machinery and worker downtime.

(c) Stocks of *finished products* must be sufficient to feed the sales process and prevent *the lack of inventory from causing delays in ordering and possibly a loss of customers,* to the benefit of competing companies.

With the above objectives in mind, the optimal stocking level depends on three factors:

1. The *need* for materials or products during the supply period or, which is the same thing, the *curve representing the outflow* of raw materials or finished products from the warehouse.
2. The *supply time,* or *lead time*: that is, the interval between the issuance of a purchase order (raw materials) or of a production order (finished products) and when what is ordered becomes available.
3. The *cost* of *supplying* and preserving inventory, which affects the *optimal quantity of supply.*

If we assume we know the factors in points (a) and (b), then the level of inventory depends essentially on the economic convenience of storage and of the procurement process, and thus on the balance between the advantages and disadvantages, in terms of costs and risks, of stocking. The advantages of stock formation have been sufficiently considered in the previous *sections*; we now turn our attention to the disadvantages, above all, in terms of costs and risks, which can be grouped into three basic categories (Swamidass 2000; Durlinger and Paul 2012; Azzi et al. 2014):

A. *Storage costs* or *holding* or *conservation costs* (SC): that is, the variable costs of conserving or maintaining stock; these are *variable costs* relative to the quantity of the stock
B. *Acquisition/ordering costs,* or *supply batch* or *setup costs* (AC):that is, the costs of moving or procuring batches of the various items; these are *batch costs* that are *fixed* relative to the quantity of the stock but vary with changes in the number of batches
C. *Storage risk costs* (RC): that is, costs connected to storage risks.

The *storage costs* (SC, category A) that inevitably arise are those related to *the preparation of warehouses for the storage of inventory,* namely:

(a) The costs of preparing the space physically occupied by the stored goods
(b) The costs of purchasing and renting the facilities necessary for the storage of the stocked goods
(c) The costs of the plants and equipment to permit the most efficient arrangement of the goods regarding their movement and maintenance (shelving units and shelves where goods can be easily placed, and their quantity and status of preservation checked)
(d) Labor costs for maintaining the warehouse
(e) The accounting costs for the warehouse

(f) Financial charges. As we know, storage entails a capital investment that results in financial charges linked not to the physical size of the stocks transferred over time but to the monetary resources invested in the inventory. The cost of the capital invested in inventories depends on three factors: (i) the value of the capital invested in a unit of stock, (ii) the time the unit is kept in the warehouse, and (iii) the calculated or paid interest rate for each unit of capital invested in stock.

The amount of these costs depends on the physical characteristics of the stocked goods; it should be noted that many costs are common to the different stocks and that, as a result, the problem of how to distribute these costs among the different stocks must be solved. It is therefore necessary to find a criterion for measuring the degree of use of the plants, and often it is assumed, sometimes incorrectly, that the incidence of the cost is proportional to the space occupied. Generally, a careful study of the warehouse leads to determining an *average daily* (or weekly or monthly) *cost per stocked* unit, which in all respects represents an *average unit full cost of storage* for each unit based on a verified, assumed, or expected normal volume of stock.

We indicate by "sc" the *average unit full cost, which represents the precise determination of the storage costs* per unit of time; if, in a given year, it is necessary to utilize or produce a total quantity of Q^*, then the quantity "$Q = Q^*/\mathrm{r}$" must be kept in stock, which depends on the *inventory turnover* "r": that is, on the moving average stock level for the year. We can then calculate the *average storage costs* for the average stock very simply as:

$$\mathrm{SC}\left(Q^*, r, T\right) = \mathrm{sc}\frac{\mathrm{Q}^*}{2\mathrm{r}}T \tag{11.1}$$

(11.1) highlights how the storage costs for the *average stock*, "Q/2r," are inversely proportional to the turnover "r" and directly proportional to the quantities Q^* and the retention periods T.

It is clear that, if we assume that during a period of $T = 12$ months there is an estimated requirement of $Q = 1200$ units and, for simplicity's sake, we assume a turnover of "$r = 1$," with a single restocking of $Q^* = 1200$ units at the beginning of T, or the entire quantity Q^*, which will be used gradually over period T and reach zero at the end of the 12 months (point input-continuous output), then the *average stock* will be "$Q^*/2r = 600$" and the *average storage costs* for 12 months will be (Fig. 11.3a):

$$\mathrm{SC}\left(Q^* = 1{,}200, r = 1, T = 1\right) = \mathrm{sc} \times 600 \tag{11.2}$$

If we instead assume that "$r = 2$" and there are two restockings of $Q = 600$ units each at the start of two 6-month periods with $T = 12/2$, then the average stock for each period will be $Q^*/2r = 300$ units. The increase in the turnover from "$r = 1$" to "$r = 2$" will halve the average stock from 600 to 300 units for all of T, and thus also halve the yearly inventory costs (Fig. 11.3b):

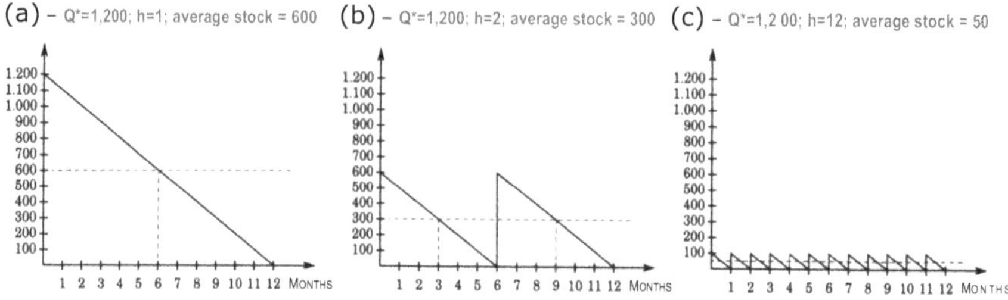

Fig. 11.3 Average storage as a function of the inventory turnover

$$\mathrm{SC}\left(Q* = 1,200, r = 2, T = 1/2\right) = \mathrm{sc} \times 300 \tag{11.3}$$

If, in order to satisfy demand, there are "$r = 12$" replenishments of 100 units each at the start of twelve equal periods of $T = 1/12 = 1$ month, then the *average stock* for each subperiod will be 1200/24 = 50 units, which will also be the value of the average storage for the entire period $T = 12$ months. Therefore, the increase in the turnover from "1" to "12" leads to a reduction in the average storage from $Q = 600$ to $Q = 50$ *for the entire period* T, and as a result a reduction in the storage costs (Fig. 11.3c):

$$\mathrm{SC}\left(Q* = 1,200, r = 12, T = 1/12\right) = \mathrm{sc} \times \frac{1200}{12 \times 2} = \mathrm{sc} \times 50 \tag{11.4}$$

11.4 Supply Batch Costs or Setup Costs (AC) and the Costs of Storage Risk

The *supply batch costs* (AC), that is, the fixed costs of moving or procuring batches of the various items, represent the second category of costs related to the formation of stocks: that is, the costs related to the order and entry of the goods in the storage facilities. They are mainly a function of the number, location, structure, and organization of the company's warehouses, and they typically include:

(a) The cost of periodic examination of existing warehouse availability
(b) The cost of the procedure for issuing procurement requests
(c) The cost of issuing the actual order to the supplier or the production order
(d) The cost of receiving and testing the goods

Specific mathematical models allow for the *optimal location* of warehouses to be estimated in a company with several supply and storage centers. In any event, wherever the warehouse is located, collecting the goods and bringing them to the warehouse involves the use of plants and equipment, material, and labor, whose costs *are not proportional* to the quantity of the goods stored (Q) or the duration of storage (T) but *vary only* in relation to the *number of replenishments effected.*

Procurement costs are, therefore, *fixed costs* for each restocking; if the rate of turnover "*r*" also indicates the number of restockings, then the higher "*r*" is, with reference to period *T*, the higher the overall costs will be for the collection and placement of the goods.

Finally, let us consider the *storage risk costs* (RC), which depend on:

(a) The possibility of stocked goods suffering physical deterioration and decline
(b) The possibility of economic aging, that is, obsolescence

The qualitative and quantitative changes in the goods in stock are hardly subject to strict and defined laws, although physical deterioration can be considered as an expression of the nature of stocks in relation to a given preservation criterion. The damage that the company faces may be a lower selling price of non-performing or obsolete products (for finished products) or deteriorating conditions of use (for raw materials), assuming it is still possible to use the damaged stocks. In relation to these harmful events, the company can follow two policies:

1. Adopt processes to prevent the occurrence of the event or to mitigate its harmful effects
2. Enter into contracts with insurance companies and pay a premium to ensure compensation for any damages caused by the events stipulated in the contract.

The adoption of the first criterion involves costs not strictly dependent on the stock quantity, whereas in the second case the insurance premiums are generally proportional to the value of the stock and the duration of the transfer. Therefore, while the costs for storage risks are a distinct category compared to the other costs, for the purposes of the overall cost calculations they can be included among either *storage costs*, if they vary with *Q* and/or *T*, or ordering costs, if related to "*r*."

Hereafter, they will not be considered as a distinct category but always as part of one of the latter two categories of cost.

11.5 Determination of the Economic Supply Batch: Wilson's Hypotheses

Increasing "r" does not always bring economic benefits. While this entails a reduction in the cost of capital and the risks of technical and economic obsolescence, it also implies an increase in the frequency of supply orders, which inevitably leads to an increase in *fixed costs* deriving from the additional procurement requests and an increased risk of undersupply. It is therefore necessary to seek forms of rational calculation to directly optimize the amount of the *batches*.

We define:

- "Supply batch" or "order quantity," "*Q*," simply as the quantity of a given good subject to a supply operation
- "Economic Supply Batch" or "Economic Order Quantity" (EOQ) as the *purchase order quantity* "q" for replenishment that minimizes total inventory costs

There is a vast literature on calculating EOQ (Whitin 1954; Veinott 1966; Clark 1972; Nahmias 1982; Silver 1981; Silver et al. 1998; Wagner 1980). A number of models have been presented for determining EOQ (Chu et al. 1998; De Toni et al. 1988; Goswami and Chaudhuri 1992; Cheng 1991); some assume the parameters contained in the calculation are fully known, while others that only their probabilities are known. Let us consider the simplest of all models: Wilson's model (or formula), also known as the *Harris-Wilson model*—since it was proposed by Ford Whitman Harris in 1913, and later developed by R. H. Wilson (1934)—and apply it to calculating the economic supply batch of a raw material.

In its *deterministic* form, the model is based on the following assumptions:

1. The *need for raw materials* is constant over time and supposedly known a priori with certainty, as it derives, for example, from the budget for the materials requirement
2. The *optimum quantity* to order is the one that allows for a cost-effective balance between the fixed *acquisition costs* and the *holding costs,* which are typically variable in proportion to the quantity of the batch purchased
3. The *storage costs* increase as the size of the lot increases; their dynamics are assumed to be directly proportional to the quantity purchased
4. The *acquisition costs*, being *lot* costs and not quantity costs, are a decreasing proportion of total costs as the quantity of the batch purchased increases
5. The company makes its purchases exclusively from outside suppliers; the *unit purchase price* is known in advance and does not vary depending on the quantity supplied; that is, it remains constant whatever the quantity ordered for each batch.

On the basis of these simple assumptions, we can produce Fig. 11.4, including in it the trends of SC and AC; the sum of the increasing and decreasing segments of these curves represents the trend of the total unit *cost* as the size of the economic supply lot Q changes.

From an examination of Fig. 11.4, we note that the *minimum total cost* for the quantity "Q = EOQ" is where the incidence of *acquisition costs* equals the unit *cost of storage*. Therefore, EOQ represents the quantity to be ordered that minimizes the overall cost of storage. It can be observed that the total cost in proximity ("around") EOQ decreases and increases slightly, so that small errors in the determination of EOQ result in irrelevant variations in the total cost of the batch.

Figure 11.5 illustrates the general model of the control system for *stocks* of a component or a material purchased externally; this is typically a system operating through impulses (Fig. 4.7). An example of a calculation of EOQ is undertaken in Sect. 11.20.2.

Impulse control systems, which we examined in Sect. 4.3, can be used to control the supply of goods, materials, and components in production organizations based on the general model of control systems for warehouses illustrated in Fig. 11.5. This is not a technical control of stocking and un-stocking, which depends solely on the size of the warehouse, since it is necessary to also introduce the *costs* relative to the storage system:

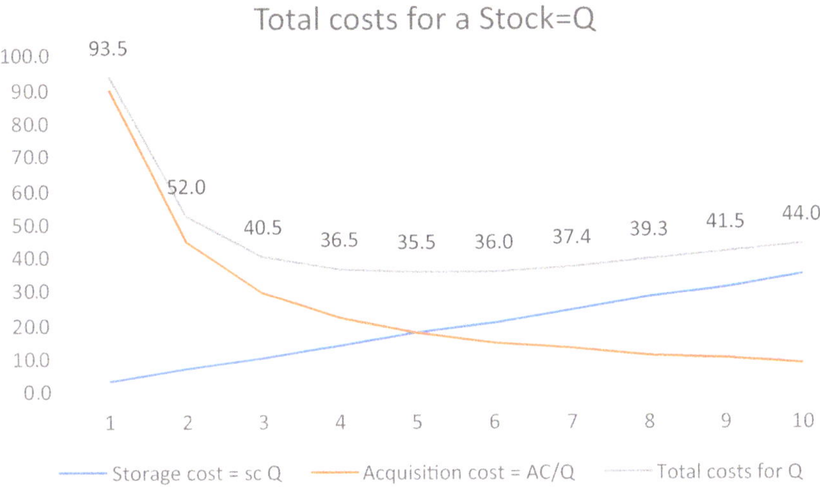

Fig. 11.4 Total storage costs

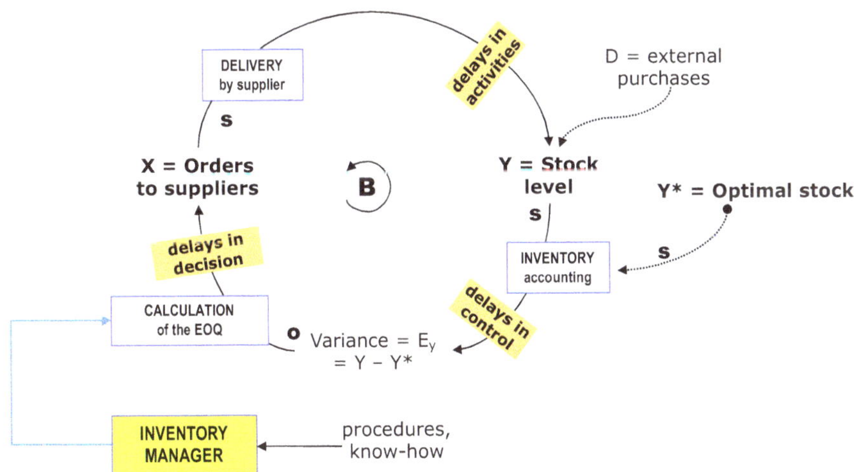

Fig. 11.5 Simplified model of a control system for the stock of a component

A. The average *storage cost* (sc) per unit of stock (e.g., \$2 per day per piece); the overall storage costs will thus be proportional to the total value of the average stock, *Q*/2, which arises from the shipment purchase.

B. The *supply cost* (cs) for a single shipment, which is assumed fixed, whatever the size (e.g., \$100 for each order).

11.6 Model of a Warehouse Control System Based on Wilson's Formula

There are several methods to calculate EOQ, but the easiest is to use Wilson's formula in its simplest version (Wilson 1934), where it is assumed that the value of stored goods has no influence on EOQ (Ballou 1967; Neal 1962). Under this simple assumption, EOQ is determined based only on an analysis of the storage costs (Wilson 1977); in other words, the purchase price is not included in the calculation; for example, Waters (1992); Arsham (2006); Woolsey and Maurer (2005); and Axsäter (2015) present a vast collection of calculation models based on assumptions that are alternatives to Wilson's formula.

- "Q^*," the total annual requirement (or that related to period T)
- "T," the reference period of the calculations; $T = 365$ days (solar year), or another convenient period (e.g., a semester, or 250 work days)
- "Q = EOQ," the *unknown* size of the supply batch
- "$Q/2$," the average stock (zero initial and final quantities)
- "sc," the *unit storage cost* per unit of stock in terms of value and time (e.g., € 50 per piece per day, or 10% of the weekly purchase cost); the total storage costs will then be proportional to the value of the average stock, $Q/2$, which is determined by the size of the purchased batch
- "SC," storage *cost* for the whole year
- "ac," the unit *supply batch cost* for each *individual batch;* "ac" is assumed fixed, whatever the batch size
- "AC," the total *acquisition cost* for all batches in a year
- "t," the interval during which the quantity supplied "Q" runs out; at the end of t, a new order must be made because "q" has been used up
- "$n = Q^*/Q = T/t$" is the number of batches required during the year; it corresponds to the rate of inventory turnover under the assumption of zero initial and final stocks;
- "TC = SC + AC" is the total cost of stock in period T.

To analytically determine EOQ, it is sufficient to calculate TC as the sum of the SC and AC cost functions and determine the size of Q that minimizes this cost.

The *storage cost*, "SC," depends on both the average stock, $Q/2$, and the overall period T (see Fig. 11.4). Recalling that $Q = Q^*/\mathrm{n}$, we can write:

$$\mathrm{SC} = \frac{Q}{2}\,\mathrm{sc}\,T = \frac{Q^*}{2\mathrm{n}}\,\mathrm{sc}\,T \tag{11.5}$$

The cost "AC" of the "n" batches purchased in T is obtained by multiplying the *supply cost* of a batch, "ac," by the number of batches ordered, n = Q*/Q:

$$\mathrm{AC} = \frac{Q^*}{Q}\mathrm{ac} \tag{11.6}$$

As shown in Fig. 11.4, we immediately see from (11.5) and (11.6) that as "Q" increases, SC increases and AC decreases. The optimal supply batch is the quantity "Q" that gives the minimum storage cost "TC":

$$\mathrm{TC} = \mathrm{SC} + \mathrm{AC} = \frac{Q}{2}\mathrm{sc}\,T + \frac{Q*}{Q}\mathrm{ac} \tag{11.7}$$

EOQ can be obtained in two ways: analytically and geometrically. *Analytically*, by calculating the first derivative of TC with respect to Q and verifying that the second derivate is always positive, and by following some simple steps (for positive values of Q), we obtain the desired value of the optimal batch (with $T = 1$):

$$\mathrm{EOQ} = Q = \sqrt{\frac{2\,\mathrm{ac}\,Q*}{\mathrm{sc}\,T}} = \sqrt{\frac{2\,\mathrm{ac}}{\mathrm{sc}}}\sqrt{Q*} \tag{11.8}$$

Setting the second root equal to a constant K, (11.8) immediately shows that the amount of EOQ increases less than proportionally to the increase in the annual need for $Q*$.

$$\mathrm{EOQ} = K\sqrt{Q*} \tag{11.9}$$

EOQ could also have been obtained *geometrically* by directly calculating the minimum point on the TC curve in Fig. 11.4 corresponding to the point Q that satisfies the equality SC = AC:

$$\frac{Q}{2}\mathrm{sc}\,T = \frac{Q*}{Q}\mathrm{ac} \tag{11.10}$$

Setting $T = 1$, with a few simple steps we once again obtain the value for EOQ in (11.8).

Once EOQ has been calculated, we can then determine:

(a) The number of shipments to order:

$$n = \frac{Q*}{\mathrm{EOQ}} \tag{11.11}$$

(b) The average duration of a shipment, that is, the number of days "t" between the replenishment and its depletion, during which we get the daily output, "d," and at the end of which we require a new input:

$$t = \frac{T}{n} = \frac{365}{n} \tag{11.12}$$

d = daily need	40	ca = supply c.	100.00	d = daily need	40.00
T = days in the year	250	cs = storage c.	2.00	riordino Y_0 =	1000.00
Q = total need for period	10000	r = delay	5.00	safety stock Y* =	200.00
				t = length of stock	25.00

$$EOQ = \sqrt{\frac{2\ ca\ Q}{cs}} = \sqrt{\frac{2\ ca}{cs}} * \sqrt{Q}$$

t	Δ X1	Δ X2	Δ Y	Y* = objective	Δ objective	E = disturbances	E(Y) = Y* - Y
0	0.0	0.0	1000.0	200.0	0.0	0.0	-800.0
1	-40.0	0.0	960.0	200.0	0.0	0.0	-760.0
2	-40.0	0.0	920.0	200.0	0.0	0.0	-720.0
3	-40.0	0.0	880.0	200.0	0.0	0.0	-680.0
4	-40.0	0.0	840.0	200.0	0.0	0.0	-640.0
5	-40.0	0.0	800.0	200.0	0.0	0.0	-600.0
6	-40.0	0.0	760.0	200.0	0.0	0.0	-560.0
7	-40.0	0.0	720.0	200.0	0.0	0.0	-520.0
8	-40.0	0.0	680.0	200.0	0.0	0.0	-480.0
9	-40.0	0.0	640.0	200.0	0.0	0.0	-440.0
10	-40.0	0.0	600.0	200.0	0.0	0.0	-400.0
11	-40.0	0.0	560.0	200.0	0.0	0.0	-360.0
12	-40.0	0.0	520.0	200.0	0.0	0.0	-320.0
13	-40.0	0.0	480.0	200.0	0.0	0.0	-280.0
14	-40.0	0.0	440.0	200.0	0.0	0.0	-240.0
15	-40.0	0.0	400.0	200.0	0.0	0.0	-200.0
16	-40.0	0.0	360.0	200.0	0.0	0.0	-160.0
17	-40.0	0.0	320.0	200.0	0.0	0.0	-120.0
18	-40.0	0.0	280.0	200.0	0.0	0.0	-80.0
19	-40.0	0.0	240.0	200.0	0.0	0.0	-40.0
20	-40.0	0.0	200.0	200.0	0.0	0.0	0.0
20	0.0	1000.0	1200.0	200.0	0.0	0.0	-1000.0
21	-40.0	0.0	1160.0	200.0	0.0	0.0	-960.0
22	-40.0	0.0	1120.0	200.0	0.0	0.0	-920.0
23	-40.0	0.0	1080.0	200.0	0.0	0.0	-880.0
24	-40.0	0.0	1040.0	200.0	0.0	0.0	-840.0
25	-40.0	0.0	1000.0	200.0	0.0	0.0	-800.0

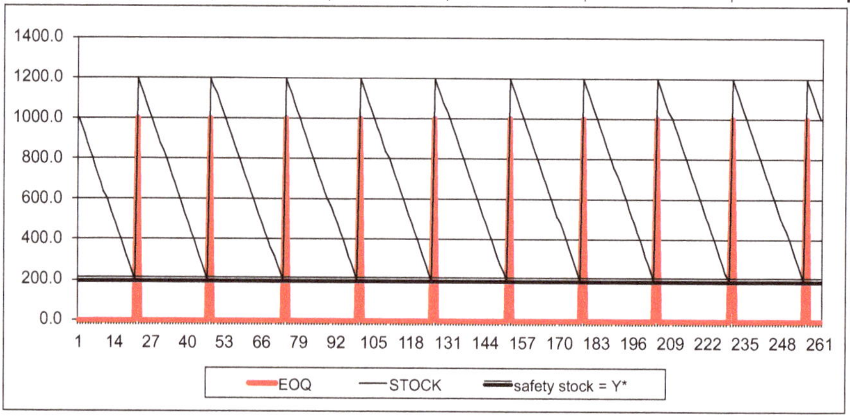

Fig. 11.6 Impulse control system of a warehouse using Wilson's formula for calculating the optimal reordering shipment. The red lines indicate the dynamics for each stock without delivery delays ($t° = 0$); the black lines, the dynamics assuming $t° = 5$ days

Figure 11.6 provides a numerical demonstration of the application of the warehouse control based on the data inserted in the initial control panel. Note in particular that we have assumed an annual period of $T = 250$ days under the realistic assumption that the average work period is 5 days over 50 weeks in the year. Applying Wilson's formula, the optimal shipment is EOQ = 1000 units. Since the daily consumption is $d = 40$ units, the warehouse must be resupplied every $t = 25$ days.

We can also introduce the assumption that the orders made on a given day will arrive with a constant delay, "$t°$," and that, in order to avoid delay risks, the warehouse manager desires a safety stock, "ss = $t°\ d$." If, for example, the delivery delay was on average "$t° = 5$" days and "$d = 40$" pieces per day, then a safety stock of "$S^* = 200$" pieces would be needed. For the warehouse manager, this simply means moving up the initial supply by 5 days with respect to the shipment duration period, "$t = 25$" days.

Since we have assumed a safety level of "$S^* = 200$" units, the initial supply takes place on day 20 instead of day 25. Figure 11.6 also presents the stock dynamics (continuous consumption and discontinuous supply) for the entire duration of the control period, which is assumed to be "$T = 250$" work days.

11.7 The Restocking Point and the Safety Stock Under Probability Assumptions

Taking up the considerations at the end of the previous *section*, note that the calculation of the EOQ for warehouses of materials or goods allows us not only to calculate the *optimal quantity to order* but also to know on what *date to place the orders*, assuming a constant withdrawal of the goods in stock and that, once exhausted, deliveries take place at the same time the order is sent to the supplier. The date of the EOQ supply order would be the same date the stock from the previous batch is depleted. Therefore, once we have calculated the number of units of EOQ, "n," to be purchased to meet the annual requirement "Q^*," and thus determined the duration of a batch, "$t = 360/n$," starting from the date the zero stock situation is noted for the first time, an order once every "t" days must be made. Under our assumptions, it is certain that when the stock of the previous batch falls to zero, the next batch will be delivered. This situation is, however, *unrealistic*, and so two different hypotheses need to be considered:

- There is a *lead time* of constant duration between the date the order goes out and the date on which the goods arrive and are verified to comply with the order, or are tested.
- Withdrawals from the warehouse are never perfectly regular.

In terms of the first complication, note that if the batches are equal—for example, "$Q = 300$" units and "$t = 15$" days—then under the assumption of a *constant withdrawal*, we can assume a daily requirement of "$d = Q/t = 20$" units per day. If the *average replenishing time* is "$t° = 6$," it is necessary to send out an order 6 days before the end of "t"; this can also be expressed in terms of the *minimum stock level* at which the order must be sent, which is defined as the *reorder point.* Under our assumptions, the *reorder point* would be "$Q° = 6$ days × 20 units = 120 unit," corresponding to a level of stock that would be withdrawn during the 6 days of lead time.

For the purposes of *warehouse control,* this means there is no need to control the *flow of time* (order 6 days before depletion) *but the flow of stocked goods*, as the reordering must take place when the goods in stock fall to "$Q° = 120$" units. If the new batch arrives on time after 6 days, then the stock is replenished at the very moment it reaches zero, as can easily be verified in Table 11.2.

Let us now turn to the second complicating factor: *non-constant demand* while the batch lasts. If the withdrawal of goods from the warehouse is not constant, two cases can exist:

- If "$Q° > 120$ units," there is a *stock breaking,* with clear-cut *damage* to the *sub-stock.*
- If the restocking requirement is instead "$Q° < 120$ units," then the replenishing may be delayed to the *benefit* of the storage cost.

Therefore, to correctly determine the *order point* it is necessary to know the average expected materials requirement during the lead time period. This requirement can be assumed to be a random variable that, with a given probability, takes on values below, equal, or greater than the theoretical value, assuming a constant withdrawal. Based on experience (repeated observations), we might, for example, find the values presented in Table 11.3.

From this random variable $S(n)$, which has the typical Gaussian form, we can calculate the *expected average value*, which represents the quantity to be normally kept in stock during the period in question:

$$S = \sum_{n=1}^{N} S(n)\,p(n) \tag{11.13}$$

Table 11.2 Level of reorder assuming a constant withdrawal of 20 units/day

Day in period "*t*"		Stock level	
1		300	
2		280	
…			
ß			
8		160	
9		140	
10 ←	Order date (time control)	**120** ←	Order point (quantity control)
11		100	
12		80	
13		60	
14		40	
15		20	
16	Arrival of order	**300** ←	New Stock

as well as the probability variance, which indicates the degree of uncertainty about the average value:

$$\sigma(S)^2 = \sum_{n=1}^{N} \left[S(n) - S\right]^2 p(n) \tag{11.14}$$

The average squared amount of stock:

$$\sigma(S) = \sqrt{\sum_{n=1}^{N} \left[S(n) - S\right]^2 p(n)} \tag{11.15}$$

indicates the number of units that on average exceed the average requirement (in this case the stock is broken and sub-stock must be used) or are lower than the average requirement (in this case there is an abundance of stock in that period).

We indicate with "$F°$" the *Safety Stock,* that is, a quantity of stock with respect to the average levels that, as a precautionary measure, should be kept in stock to avoid the sub-stock risk, which would cause damage (production shutdown or loss of customers). From Sect. 10.14.1 we know that if the safety stock is "$F^* = \sigma(S)$," there is an even higher probability of a stock break (32%); if the safety stock is "$F^* = 2\ \sigma(S)$," then the probability is drastically reduced (5%); with a safety stock of "$F^* = 3\ \sigma(S)$," it is almost certain there will be no stock break, as the probability of this occurring is less than 1%. In other words, stating that for a given raw material there is the probability (e.g., 95%) of *no* stock break means that if you receive 100 requests for materials from production during the supply period, 95 of them will be in smaller quantities than that at which a stock break would occur. Conversely, 5 withdrawal orders may be large enough to produce a situation where the company is out of stock.

Therefore, the safety stock (SS) is equal to:

$$\mathrm{SS} = k\sigma(S)$$

Table 11.3 The withdrawal requirement as a random variable

Requirement for 6 days	Probability—$p(n)$
$Q° = S(n)$	
100	2%
105	3%
110	8%
115	15%
120	42%
125	18%
130	8%
135	3%
140	1%
	100%

As a result, the *reorder point* is:

$$Q^{\circ} = S + k\sigma(S) \tag{11.16}$$

The factor "*k*" is referred to as *the safety factor*, and it is decided on by warehouse policy. The higher "*k*" is, the higher the storage costs of the safety stock and the lower the costs associated with the stock break. In theoretical terms, therefore, "*k*" should be set at a value that makes the storage cost (disadvantage of too high a safety stock) equal to the cost of breaking the stock (disadvantage for a safety stock that is too low). In practice, to calculate the probability of $S(n)$ it is necessary to do an accurate survey of consumption during the reordering period over a sufficient number of intervals and to calculate the frequencies of the various withdrawal amounts. The frequencies of the withdrawals can be taken as approximations *of probabilities.* Safety stock calculations can be carried out much more easily by replacing the average squared stock by the *mean absolute deviation* (MAD), which avoids having to calculate squares and square roots.

$$\text{MAD} = \frac{1}{N}\sum_{n=1}^{N}\left|S(n) - S\right| \tag{11.17}$$

Statistical reliability is not compromised by adopting MAD; it is always possible to shift from one procedure to another according to the relationship that binds the two quantities, in the belief that, with good approximation and with errors having a normal distribution, the average squared deviation will equal "1.25 MAD". Holding a safety stock of "SS = 1 MAD" means the probability of not having a stock break is 78.81%; with a safety stock of "SS = 2 MAD," the probability rises to 94.52%, and so on. An example is shown in Sect. 11.20.2.

11.8 Calculating the Replenishing Batch Assuming Proportionally Varying Prices

Until now, it has been assumed the unit price is constant at "*p*," irrespective of the ordered quantity "*Q*," and therefore it has not been considered in the calculations. As is well known, very often suppliers charge different prices depending on the purchase quantity. There are two possible forms of discount:

(a) Discounts granted according to the total quantity purchased by the company over a certain period of time, regardless of how total quantity is divided into individual orders
(b) Discounts applied on the basis of the quantity ordered from time to time.

The first type of discount does not directly affect the purchaser's stock as the price remains fixed until the minimum quantity is reached, after which the discount is applied. I will therefore consider the *discount applied to each batch purchased* using the simplified assumption that the unit price *continuously* decreases in proportion to the order size. To calculate EOQ we must minimize a function that includes, in addition to the ordering and storage costs considered in the previous models, the purchase costs as well, by *explicitly introducing prices* in the model.

In addition to the symbols introduced in Sect. 11.6, let be:

- "*p*—aQ" is the variable purchase price, which is assumed to be decreasing; it consists of two elements: "*p*" is the list price and "a" the discount factor per quantity ordered; "–aQ" is the reduction in price, which is proportional to the batch amount.

The total cost function (11.7) to minimize must then also include the *purchase cost* of the batch: "PC = (*p*-aq) *Q*":

$$\mathrm{TC} = \mathrm{AC} + \mathrm{SC} + \mathrm{PC} = \mathrm{ac}\frac{Q*}{Q} + \mathrm{sc}\frac{Q}{2}T + (p - \mathrm{aq})Q* \tag{11.18}$$

As usual, we calculate the first derivative of "TC" and set it to zero; then we calculate the value of "*Q*" from the resulting equation, choosing the values "$Q > 0$"; assuming a positive denominator, we find the solution formula, which provides real solutions if "sc *T*—2a *Q**)>0":

$$\mathrm{EOQ} = Q = \sqrt{\frac{2\,\mathrm{ac}\,Q*}{\mathrm{sc}\,T - 2\,\mathrm{a}\,Q*}} \tag{11.19}$$

As an example, let us assume that the annual requirement of some material is "*Q** = 10,000 units," that unit *supply batches* are estimated at "ac = $500" per batch, that the *storage cost* is "sc = $3" *per unit*, and that the purchase price is "*p* = 250-0.05 *Q**" per unit purchased.

Using this data, from (11.19) we obtain:

$$\mathrm{EOQ} = Q = \sqrt{\frac{2 \times 500 \times 10{,}000*}{(3 \times 365) - (2 \times 0.05 \times 10{,}000)}} \approx 324$$

The *number of orders* is *n* = (10.000/324) = (approximately) 31.

The *average duration* of a batch is *t* = (365/31) = (approximately) 11.5.

11.9 The Calculation of the Optimal Production Batch

Wilson's formula can also be used to calculate the *optimal production batch,* also called the *economic production quantity*, EPQ (Arsham 2006). In this case, for the units of finished products to be kept in storage, it is necessary to know both the *storage costs*, "sc" (considered fixed for each product batch), and the *average unit production cost*, "pc," calculated after choosing a costing system (Sect. 9.9.2). We indicate with:

- –"Q*" the total annual production requirement as presented, for example, in the production budget;
- "pc" the *average unit* industrial cost of production—net of supply costs—which is assumed fixed for each quantity produced (it can be both a *direct cost* and a *full cost*, but will naturally vary the meaning of the results achieved);
- "PC" the cost of producing the batches in period T
- "Q = EPQ" the *unknown* quantity of the production batch
- "ac" the unitary *supply batch cost*, assumed fixed for any volume "Q"; "ac" could include, for example, the costs of equipping the machinery or of handling the batch in the warehouse
- "AC" the yearly *supply batch cost* for all batches
- "sc" the *unit storage cost*, calculated based on the *average batch value* (e.g., 3% of the average production cost)
- "SC" the yearly *storage cost*
- "$Q/2$" the *average stock* (zero initial and final quantities)
- "t" the interval over which "Q" is used, at the end of which a new order must be made because "Q" is depleted
- Q_n = Q*/Q the yearly number of batches required
- PCQ the total cost of production, which includes both the industrial production costs and the procurement and storage costs

We can then immediately calculate:

(a) The *production cost of a batch*: "pc Q"
(b) The production cost of "n" batches: "PC = pc Q n"
(c) The *supply batch cost* for the "n" batches: "AC = ac n"
(d) The *storage cost of a batch*, which is proportional to its value; the *storage cost* depends on the unit production cost, "pc" (to which "ac" can eventually be added), and on the *unit storage cost*, "sc," multiplied by the unit value, "pc," the *average stock* and the duration of a batch "t":

$$\text{Storage Cost of a Batch} = \text{sc pc} \frac{Q}{2} t$$

(e) The *total storage cost* of the "n" batches is:

$$\text{SC} = \left(\text{sc pc} \frac{Q}{2} t \right) n$$

(f) The cost for production and storage of the n batches is:

$$\mathrm{CPQ} = \mathrm{PC} + \mathrm{AC} + \mathrm{SC} = \left(\mathrm{pc}\,Q\,n\right) + \left(\mathrm{ac}\,n\right) + \left(\mathrm{sc}\,\mathrm{pc}\frac{Q}{2}t\right)n \qquad (11.20)$$

Recalling that "$(Q\ n) = Q*$," "$(t\ n) = T$," and "$n = Q*/Q$," the previous equation becomes:

$$\mathrm{CPQ} = \left(\mathrm{pc}\,Q*\right) + \left(\mathrm{ac}\frac{Q*}{Q}\right) + \left(\mathrm{sc}\,\mathrm{pc}\frac{Q}{2}T\right) \qquad (11.21)$$

As usual, here we have to calculate the quantity "Q" that minimizes the previous total cost function CPQ. With a few simple steps, we obtain the value of EOQ by deriving "CT" from "Q" and calculating the value of "Q" that makes the first derivative zero; since "(pc sc) > 0," and considering as admissible only values of "$Q > 0$," and noting that the second derivative of "CQ" with respect to "Q" is always positive, EOQ represents a minimum value:

$$\mathrm{EOQ} = \sqrt{\frac{2\,\mathrm{ac}\,Q*}{\mathrm{pc}\,\mathrm{sc}}} \qquad (11.22)$$

Note the close similarity between the results here and those from calculating the optimal supply batch in (11.8); however, to calculate the *optimal production batch*, we must also take into account the production cost; we can see that the batch size *decreases* the more the average unit production cost, "pc", *increases*.

Important observation: (11.22) above can also be used to calculate the economic *supply batch* from outside suppliers when it is necessary to take into account the purchase price as well. In this case, it is enough to repeat the same logic, replacing the unit cost, "cp," by the unit price, "p," to obtain the following formula that expands on [13]:

$$\mathrm{EOQ} = \sqrt{\frac{2\,\mathrm{ac}\,Q*}{p\,\mathrm{sc}}} \qquad (11.23)$$

Technical note: calculating the optimal supply batch with Wilson's formula was based on the assumption of a *constant* requirement, "$Q*$," of materials or products and their regular use over time; under that hypothesis, the requirement of each subperiod represented a proportional fraction of the total requirements. There are, however, cases where purchase or sales processes entail requirements that are not evenly distributed, so that, for example, the January requirement may be quite different from that of February, which in turn might differ from that of March, etc. In this case, Wilson's model becomes difficult to apply, so that other procedures must be sought to calculate the quantities to be stored in the different subperiods "T" is divided into. Even in the case of requirements for materials or products that are not evenly distributed over the different subperiods, the optimal EOQ quantity should be equal to the sum of a given number of periodic requirements. The solution for EOQ can be found using *dynamic programming techniques*, one of the simplest of which is the so-called *Part-Period Balancing method*, whose logic is to "balance," step by step, *the ordering and maintenance costs*. Without going into detail, this method calculates the *storage cost* for ever larger batches that include the cumulative requirements of a number of subperiods. The overall cost of storage obviously increases as the number of requirements included in the calculation increases; the rule is: *it is economically convenient* to make an order when *the storage cost exceeds the supply cost*.

11.10 The Control of Production with Constraints. Direct Costing Rules and the Linear Programming Method

To control stocks, we have seen that it is essential to know the overall volume of goods to be kept in stock for the year. This implies that the manager must know the production levels on which both the consumption of the materials and the availability of the stored products depend. *Production control* is mainly a technical problem that requires the use of specialized software to manage processes, schedule operations, and deal with the logistics of the materials and products obtained. In multi-product companies, controlling production becomes an economic problem in all cases where production takes place at the same time, in period "T," but there are constraints on production capacity. When the production of the different products occurs under "conditions of limited resources," and the production of one product inevitably absorbs resources at the expense of another, it is necessary to determine the optimal quantities to be produced while also taking into account the need to maximize profits, or margins, compatible with the limitations of the availability of scarce factors of production—the production capacity of machinery and limited availability of materials and labor—that need to be shared among the different production processes. Only after determining the overall quantity to produce of the various products will it be possible to determine the most economically feasible production batch using the techniques presented in the previous Chapter.

When production is subjected to a single capacity constraint, the optimal production mix can be determined using the *direct costing method* decision-making rules, which assume that each unit produced and sold must necessarily *cover* the variable costs "$VC_n = vc_n\, qP_n$" for the $n = 1, 2, \ldots N$ production processes and offer an overall *contribution margin* $CM = \sum CM_n$ that can absorb the *fixed costs* needed for production; or, in any event, as much of these as possible.

With *direct costing* there is no need to quantify earnings or losses per unit of product; *for those goods produced jointly by the company with the involvement of all the fixed costs,* only the total profits or margins must be calculated. The explanatory capacity of *direct costing* and its rationale derive from the fact it *focuses* on the unit contribution margin, "$cm_n = p_n - cv_n$," and the overall margin, "$CM = \sum cm_n\, qP_n$."

This method leads to the following *operational rule* for the economic calculation of the unit price/cost comparison (Moisello and Mella 2020):

RULE 1): A product with price p that is greater than the direct cost cv can be produced, since its contribution margin covers *at least* a share of the fixed costs of production; as a result, the following relationship must always hold:

$$p \geq \mathrm{vc}$$

RULE 2): If the company *has no production capacity constraints*, where two products have different unit margins the company must favor the production of the one with the larger unit contribution margin

RULE 3): If the company *has a capacity constraint*, in that it has a limited quantity of a factor needed for the different products, it must favor the product that has the *highest ratio between the unit contribution margin and the quantity of the limited factor*; more specifically, if the company produces ALFA and BETA with a constraint on the quantity of the materials available to both, and if qM(ALFA) and qM(BETA) indicate the unit quantities of raw materials used by the two products, then the company must favor the product with the highest value for the following ratios:

$$\frac{\mathrm{cm}(\mathrm{ALFA})}{\mathrm{qM}(\mathrm{ALFA})} \text{ and } \frac{\mathrm{cm}(\mathrm{BETA})}{\mathrm{qM}(\mathrm{BETA})}$$

RULE 4): *When there are two or more production capacity constraints*, the optimal *mix* is obtained by setting up an *optimal constraint problem* through *linear programming*, setting as the objective function of the maximization of the overall operating result, calculated as the difference between the total contribution margins of all the products and the total fixed costs.

When Rule 4 applies, a technique that has been adopted for decades and is simple to apply in general is the *Linear Programming Method* (LPM), proposed by George Bernard Dantzig (1963), which can be usefully applied where there are problems of multiple continuous-flow production, or similar problems. I cannot go into the mathematical details here but will instead limit myself to presenting the fundamental characteristics of the method. An example of an application is presented in Sect. 11.20.3 together with the method-solving guidelines (a detailed treatment of the mathematical aspects can be found in Dantzig 1963; Pierre 1987; Brickman 2012; Gass 2010, and in Operations Research texts). It is worth noting that for the concrete solution of linear programming problems effective software applications are also available.

Very briefly, since the availability constraints of one or more factors of production means it is impossible to determine the optimal quantity of a product without at the same time taking into account the effects on the production of other products, LPM calculates the *optimal mix* of products to maximize an *economic function* that depends on the quantity of the different products obtained; this function could be a function that maximizes the overall revenues, gross sales margins, or the company's net profits. However, it could also be the production costs function that has to be minimized. The *economic function* is called *the objective function.* The quantities of production on which the objective function depends are subjected to a *linear equality* and *linear inequality constraints system.* The quantities produced by the different production processes cannot be negative: the *condition of non-negativity of production* is also necessary. Therefore, *linear programming* is a method that allows a limited set of resources to be optimally distributed among a number of competing productions in order to optimize an *objective function.* In what follows we will assume that the production policy has already been defined and that the quantities to be produced are independent of the stocks of finished products at the beginning and at the end of the period T on which the programming extends. We define an "acceptable solution" a set of "N" values that can be assumed by the N variables (the "qP_n," for the $n = 1, 2, \ldots$ N, production processes) and meet the constraints and the condition of non-negativity.

George Dantzig has proven the following theorems:

1. The admissible set of acceptable solutions is convex.
2. The *optimal solution*, that is the set of "n" optimal values, belongs to the frontier of admissible sets.
3. If the problem admits more than one solution, it has endless solutions given by the convex linear combinations of the different optimal solutions.

The *simplex method* represents an iterative method that, by subsequently analyzing the "n" solutions that belong to the frontier of the admissible set, identifies the "n" values that *optimize the objective* function. The simplex method could be implemented with "manual" calculations, but there are now numerous tools that allow for the automatic calculation of the optimal solution (Wolfram-Alpha's Mathematica Linear Programming Solver 2020, online; Sierksma and Zwols' Linear Optimization Solver 2015a, online). I will present an example of production programming whose solution is determined by the very simple model developed by Sierksma and Zwols (2015b), which will demonstrate how easy it is to find the set of "*n*" optimal solutions.

A company normally manufactures products "A," "B," and "C" (for simplicity I will also use these names to indicate the quantities of the variables). While A and B are completely separate products, C is economically complementary to the others: two units of C are sold for each unit of A and one unit of C for each unit of B. From the sale of the products in "week h," the following unit contribution margins are expected: cm(A) = 10,000; cm(B) = 6000; cm(C) = 5000. The production services of three machines must be used to manufacture the products: M1, which can be active for 40 hours per week, M2, available for 20 hours per week, and M3, which can be used for 40 hours per week. For "week h," "3" M1 machines, "4" M2 machines, and "9" M3 machines are available. To produce A, 60 minutes of M2 are required and 480 minutes of M3; product B requires 252 minutes of M1, 30 minutes of M2, and 50 minutes of M3; C requires 36 minutes of M1 and 160 minutes of M3. These data are arranged in Table 11.4.

Table 11.4 Availability and capacity constraints

Machine type	Number of machines	Minutes to produce A	Minutes to produce B	Minutes to produce C	Available machine minutes
M1	*3*	0	252	36	$3 \times 40 \times 60 = 7200$
M2	*4*	60	30	50	$4 \times 20 \times 60 = 4800$
M3	9	480		160	$9 \times 40 \times 60 = 21{,}600$
Unit Margins		10,000	6000	5000	

From the data in Table 11.4 we can set out this linear programming problem:
Objective function

$$Z\max = 10{,}000\,A + 6{,}000\,B + 5{,}000\,C$$

System of constraints

$252\ B + 36\ C < 7{,}200$	constraint for M1
$60\ A + 30\ B + 50\ C < 4{,}800$	constraint for M2
$480\ A + 160\ C < 21{,}600$	constraint for M3
$C = 2\ A + B$	complementarity constraint
$A, B, C > 0$	non-negativity condition

The first three constraints represent the production capacity limitations of the three machines, which are available for the number of hours specified above. The fourth, the complementarity constraint, expresses the economic links between the three production processes. Finally, the last constraint represents the non-negativity conditions for the production amounts.

Sierksma's and Zwols's model (2015b) allows us to set out the problem using the following symbols, where x1, x2, and x3 represent the production of A, B, and C, respectively.

PROGRAM START

```
var x1 >= 0;
var x2 >= 0;
var x3 >= 0;
maximize z:   10,000*x1 + 6,000*x2 + 5,000*x3;
subject to c11:            252*x2 + 36*x3 <= 7,200;
subject to c12:   60*x1 +   30*x2 + 50*x3 <= 4,800;
subject to c13:  480*x1 +           160*x3 <= 21,600;
subject to c14:    2*x1 +       x2        = x3;
end;
```

END OF PROGRAM

By clicking "Run-Solve" the model produces, almost instantaneously, the following optimal values, as can be verified by inserting them in the program:

```
[A = 20, B = 20, C = 60]
[Optimal objective value = 620,000]
[hours of machine use: M1 = 7,200; M2 = 4,800; M3= 19,200]
```

Section 11.20.3 will present another example of production under linear constraints, solved using Wolfram Mathematica.

11.11 From Programming to the Control of Production, The "Push" Method, The Materials Requirements Planning System

We shall now consider modern forms of production transformation control, which, due to their power, pervasiveness, and effectiveness, simultaneously enhance the control systems for warehouses, productivity, and quality, while also allowing an increase in the efficiency of the economic transformation. I will briefly present the principles of the techniques known as:

- MRP (or MRPI), Material Requirements Planning,
- CRP, capacity requirements planning,
- DRP, distribution requirements planning,
- MRPII, manufacturing resources planning,
- JIT, Just in time and Toyotism,
- OPT, Optimized Production Technology,
- FMS, Flexible Manufacturing System,
- HMS, Holonic Manufacturing Systems,
- CIM, Integrated Manufacturing Systems.

The principles that inspire these techniques emphasize "time" as the main variable to be controlled, which implies the regulation of the *sequence* and *timing* of production and storage activity (Ptak 1991; Schmenner 1993; Neely 2007). Controlling "time" at all stages of the production transformation process means having as objectives *time compression* (Tuttle 1994; Towill 1996; Mason-Jones and Towill 1999) and the *shortening of time-to-market*, in all its forms, including controlling production lead times, facilitating the *synchronization* of production lines, ensuring the *flexibility* (production mix) and *elasticity* (quantities) of production processes, and making efficient the processing of material and component requirements (Lei & Goldhar 1991; Plenert 1999). These controls are implemented using levers that can also be traced back to *material handling* activities. Careful management of *supplies* and of production *stocks* is perhaps the simplest time compression measure, considering that storage activities are parallel to those of production orders; therefore, procurement lead times can be controlled so as to make available only the necessary materials.

When determining production schedules by calculating the production quantities, the *production control* phase must be followed; this refers to the set of decisions and controls necessary to determine *when* (start and end dates of production), *where* (manufacturing centers), *how* (processing cycles), *with what resources* (materials or production capacity), and *to which destination* (immediate delivery or storage) production plans should be implemented, thus coordinating the planning and implementation processes. This means controlling the request for and progress of orders to processing centers, specifying how much each plant must produce, and sorting the outputs.

Production control is based on the control of the *progress of orders,* that is, assigning processing priorities to those orders related to parts of the finished product that must be manufactured in the same work center. These priorities are subject to constant change due to the fact customers can change the specifications or quantities of orders, anticipate or delay the requested delivery date, or even cancel the order. In addition, anomalous situations can occur within the factory, such as machine failures, an unexpected increase in defective parts, and quality problems. The problem, however, is mainly informational since, without accurate and timely information on the current state of production, it is impossible to obtain adequate activity levels consistent with the programs to be carried out, which inevitably leads to the accumulation of unfilled orders, a climate of disorder, and a constant state of emergency. These problems highlight the distinction between *push* and *pull.*

In a *traditional production control system*, the aim is to optimize the performance of each *operations center* so that it minimizes downtime and produces what is needed at the appropriate time for downstream centers. It is therefore necessary to coordinate the production "from downstream to upstream," starting from sales forecasts for a specific time period, from which a main production schedule is derived indicating the ordered sequence of the individual products or models to be produced in the assembly phase. The upstream manufacturing phases are informed of the pace and quality of parts to be manufactured through the issuance of several specific production programs enough in advance of the expected completion date so that the departments have time to prepare manpower and machinery, manufacture a given component, and send it to the next work center. This method is known as *Materials Requirements Planning*, MRP, a computerized system of planning and control for the planning of "materials" through the calculation of *requirements*, for determining the quantities and types of materials required and the *timing* of flows, and for synchronizing supplies with the *rhythms* of the production processes (the correct acronym for this technique is MRPI, to distinguish it from Manufacturing Resource Planning, MRPII and MRPIII, which will be briefly examined in Sect. 11.14).

Calculating the requirements for each assembly, sub-assembly, part, and component of the different products is carried out starting from the *main production program,* which must be developed in advance to define the planned product *quantities* and *delivery times* within a given time frame; the unit requirement data of each component, or "item/category," are applied to the quantities. In order to determine production schedules, it is necessary above all to estimate both customer demand and the time the production departments will take to complete the necessary components and assemble them. In an unplanned market characterized by elements of uncertainty, forecasts are not always accurate, and thus the simultaneous modification of all the production programs is required.

Given the *production schedule*, the fundamental document becomes the *bill of materials*—that is, the structured list of all materials or parts needed to produce a specific finished product—as a tool to divide the main production schedule into raw material *purchase orders* and *production orders* to be sent to the various departments, obviously taking into account the quantity in stock of the different components. From this information, the MRP system refers to the necessary quantities of each "category," or *gross requirement,* to meet the requirements contained in the production schedule. From the *gross requirements,* possibly supplemented by the size of the safety stock, the net *requirement* is determined, taking into account the warehouse quantities and the quantities for which a purchase or production order has already gone out. From the *net requirement*, the supply plan (order plan) is derived, which indicates exactly the date on which the production or purchase order of the materials should be issued; in this way the "category" will be purchased or produced only when they are actually requested. The availability of the materials is planned with precise timing, so that the supplies ensure the components arrive at the warehouse just in advance of when the withdrawal takes place, thus reducing the average stock, with significant savings to storage costs.

If the request phase has been done correctly and in accordance with the production capacity of the production structure (company, plant, department, etc.), and if no external events occur to disturb the scheduled sequence, all components of a given product will arrive at the assembly line at the same time. Upon completion, the products (which are not ready to be delivered) will first be sent to a warehouse and then delivered according to the actual orders of the customers. It is important that changes in schedules be communicated instantly to the processing centers to avoid generating stocks of components, semi-finished or finished products that are difficult to re-use. As it is difficult to manage frequent changes, companies often decide to maintain a high inventory between the various phases to ensure greater overall system responsiveness.

Considering the frequency and methods of *reprogramming*, MRP techniques can operate in two different ways (Schmitt 1984):

(a) *Regenerative systems* that periodically recalculate to fully explode the requirements based on the most recent production schedule. This procedure is suitable in the case of infrequent rescheduling and involves the cancellation of the previous schedule, which is replaced in full by a new program. These systems have the advantage of updating the schedule to any changes in the database (technical changes, change of parameters and production units, etc.), but they are particularly costly because of the large amount of information that has to be dealt with.
(b) *Incremental systems or net change systems* that reprogram based on *partial explosions* of the requirements as a result of partial changes in the production program. In this case, only the requirements for the categories affected by the change are recalculated, while the remainder of the schedule and the time horizon remain unchanged from the last schedule. Incremental systems are effective in the case of recalculations that are close in time, since they reduce waiting times.

The application of an MRP system involves a high commitment of organizational resources; the complexity of the method and the need for IT systems to make the programming more efficient are sometimes costly, but the result is significant benefits in terms of the accuracy and timeliness of the entire production information system.

11.12 The "Pull" Method

To avoid the inconveniences of the *push* method, an alternative system can be adopted which is based on the *pull* method. In general, in both push and pull systems the production schedule extends over a time horizon equal to the production time; however, the magnitude of this horizon changes, becoming reduced in the pull case. A push system occurs when the production and procurement lead time, P, is longer than the period in which the products are to be delivered, D, making it necessary to provide and plan in advance for the arrival of materials in the factory as well as for the processing orders so as to reduce the production lead time to a minimum. The push system is thus based on accurate predictions. A pull system, on the other hand, is based on those orders in the portfolio whose delivery times have been determined to be greater than the production and procurement lead times; therefore, orders "pull" production and supplies, and no forecasts are needed, since the production and delivery times are known. Of course, even the pull system always requires ensuring the necessary production capacity: that is, the availability of the plants and the workforce that must be acquired sufficiently in advance to make them available at the time of use.

Briefly, in this case the materials and components are "recalled" by the user centers and, consequently, made available only when necessary; there is no longer the need to "decompose" the main production schedule into so many processing orders to be sent to individual departments; it is sufficient that the final assembly line receive the sequence of the components of the models to be produced, in the quantity updated at the last moment using the actual customer orders, so that the necessary parts can be taken in the right quantity and at the right time. This operational method works its way "backwards" throughout the production system, since each manufacturing stage takes only what it needs from the immediately preceding one, which, in turn, merely produces what has just been requisitioned. This production control system is also an inventory control system since, based as it is on replacing what has been consumed, it eliminates the need to keep buffer stocks, which in a push system could be induced not only by demand forecasting errors, but also by asynchronies between two stages that are linked.

The *pull* method requires that materials and components, once processed, be held in small outbound storage points rather than *pushed* on to the next department. This is an excellent *visual error signaling device* to control the production of each department since, if no one comes to pick up the parts in question, the workers in the department can immediately see that their production must stop, therefore avoiding stock accumulation. Such a system can be thought of as a flow channel that combines the producer department with the user department, within which the stock cannot exceed a limited quantity (limited size of the storage point). If you take a piece from one end, the rest of the material advances into the flow channel and leaves an empty space at the extreme feeding point. The rate at which stocks are withdrawn for use "downstream" adjusts the rate of feeding by the "upstream" department.

Since the system flows backwards, that is, from the end point to the initial point of production, only the former must be made aware of any changes to the production plan. Demand fluctuations are rapidly reflected backwards throughout the production system, without the need for a central planning office to process and transmit modified production schedules to processing centers, as is the case in a *push* system. The difference with respect to a traditional *push method* is that with a pull system there is no advanced production; the system, responding to precise traction signals, provides only those products that, based on orders updated in real time, are likely to be sold.

In order for a *pull system* to work, certain characteristics are required of the production system:

1. First of all, it must be possible to produce in *continuous flow*. It is necessary to revise the layout of the machinery so that each manufacturing center has an inbound and outbound storage point to ensure a minimum accumulation of *functional buffer stock*; assembly lines will have more inbound storage points, as many as there are work stations that bring their components together.
2. Each central production unit (plant, department) must have only one supply point per component or material, even if that component or material is then used in different operations centers; it is also necessary to precisely plan the path the material must take through the production system.
3. The *leveling of production* should be implemented as much as possible, with reduced re-tooling time. Leveling production means making production in minimal batches, leaning more toward the unit, thereby making production flexible (Liker and Morgan 2006). This helps to balance all operations that feed into the final assembly, ensuring maximum fluidity without interruptions or stoppages.
4. The production process must be characterized by repetitiveness of operations even if this condition is less in evidence in the case of production-to-order. The *pull* system is very effective in the manufacture of pre-coded products: that is, those with a *bill of materials* and an easily definable product structure.

11.13 The Importance of Scheduling, Leveling, and Balance in Production

Whatever the method adopted for the control of production, the *scheduling* of the production control is important in order to optimize the production sequences to maximize the performance of the production factors and obtain a regular and timely flow of products throughout the various manufacturing stages manufacture. Correct scheduling is necessary for the configuration of the *manufacturing cycle*, that is, the ordered sequence of operations involving the handling and processing of materials and components that allow the necessary products for delivery and sale to be delivered to the warehouses for finished products within the established time period and in the correct amounts. The cycle begins by removing materials from the raw materials warehouse and ends with the final assembly of the finished product and its storage in the warehouse. This type of activity is essential in the case of discontinuous production, which consists of products that involve the same machines or operations centers. Scheduling implies that production volumes, times, and processes have been defined, and that the orders to fulfill and the resources available are already known. The decisions to be made concern the *production sequence* of the different orders or the operations related to all the orders for the purpose of:

(a) Reducing as much as possible the asynchronies and time lags between the work of the different production centers
(b) Avoiding wasted time associated with unnecessary work stoppages or excessive stocks
(c) Providing regular work to the workforce
(d) Meeting the pre-arranged delivery deadlines
(e) Reducing costs: since time is cost-related, the best sequence is the one that requires the least machine re-tooling and production time. In scheduling, therefore, the time period between the start of production and the delivery of the finished product is set so that the agreed time with the customer is met and the production costs minimized.

The *scheduling techniques* that can be used depend on the type of organization that characterizes the production process. It is possible, in fact, to obtain particular benefits from one or another system of production organization through appropriate standard scheduling procedures without requiring a general reorganization of the plants. There are two extreme types of production organization: *functional* and *product line*. In the first, the production centers are organized according to the different types of machinery or operations. Materials pass through the departments in separate batches corresponding to individual manufacturing orders. Although this type of organization operates with a lower plant production capacity, it entails significant costs related to the continuous need to re-tool the machines for continuous changes of production and to the need for more versatile, and therefore probably higher-paid, personnel. Therefore, each unit to be produced should be directed through the system and assigned to the various work stations.

In the organization of production by *product line*, on the other hand, the plants are specialized to manufacture a set of similar products for which all operations are combined, with the risk of having difficulty distinguishing between the individual orders. Therefore, organization by product line is characteristic of large-scale industries where demand is stable and production limited to a few basic products. While this type of organization has the advantage of a faster flow of materials through the various manufacturing phases, lower re-tooling costs, and lower stock during manufacture, it also presents high rigidity and necessitates extensive investment in plants. This activity requires scheduling limited to the need to determine the speed of production. In most manufacturing companies, elements of both types of system can be found.

In *batch production*, where production lines are re-tooled from time to time to process, in successive batches, different products in large numbers, *scheduling* determines the *priorities* regarding the manufacture of the different batches, considering both the delivery time constraints and the manufacturing time; difficulties can occur when a change in the priority of some orders occurs, resulting in the need to change the sequence of operations previously established. One form of scheduling used in these cases is the ALTAI method, which requires preliminary network programming (PERT), which will be examined in Sec. 11.18.

The *leveling of production* is a fundamental aspect of *flow production*; instead of obtaining a product in a single annual batch, it is preferable to produce it with a constant, monthly, or even weekly output, since this prevents the company from storing an entire batch equal to the annual or monthly requirements and allows it to launch production on the basis of actual needs and not on the annual forecasts, producing in the short term the *production mix* necessary for deliveries. For example, leveling on a weekly basis means producing all categories of products in a week with weekly production batches. *Leveling* has significant advantages in the case of production characterized by "cascading" departments, since the quantities of products planned for monthly (weekly, or daily) leveling must be leveled in all departments in the cascading hierarchy so that the outgoing production is actually what has been requested. Therefore, the control of the leveling of the production requires setting up a *levelling program* consisting of a more rational distribution of the requirements for materials and manpower, so as to obtain a harmonic distribution of each manufactured product while production is under way.

Controlling the *balance*, or *synchronization of production,* is also critical for the production flow organization. In order to always have the entire production activity under control so it can adapt to sudden changes in demand, the entire manufacturing process must maintain a regular pace and act in a repetitive and consistent manner at different stages of the process. If synchronization is lacking, different parts of the production line could behave autonomously, thus causing delays and also increasing manufacturing costs. It is not useful for a department, in order to optimize the use of its production capacity, to produce large quantities if this involves accumulations of work-in-process materials due to a lower operating speed in some downstream department. For synchronization, however, it may be appropriate to slow down the whole line of production, thereby lowering the efficiency of some machines but guaranteeing greater fluidity and control of the production process. For this reason, synchronization must ensure that each flow, each machine, or workplace proceeds at the same speed, determined by the slowest department, which represents the *bottleneck* in the process (Tuttle 1994).

11.14 The Evolution of MRP, MRPII, and MRPIII: Toward Just-In-Time (JIT)

For production control managers, the MRP system favors planning guidance and, thanks to the natural tendency of managers to improve control systems, also the search for more complex control systems aimed at planning the entire set of production resources and factor of production requirements. MRP systems have therefore gradually evolved, and the first form of evolution can be traced back to the emergence of so-called *closed-loop* MRP in which the system is able to verify the existence of available capacity in each department and to keep it under control by making a connection between MRP, which deals with the materials, and CRP, the *capacity requirements planning* that deals with production capabilities; the result is that, in addition to the confirmed orders, those planned by the MRP can also be translated by the CRP into capacity requirements and into relative departmental loads, which are appropriately leveled and compatible with existing capacity limits. The *closed-loop* MRP is thus able to plan the availability of three key resources: materials, production capacity, and time.

Just as the *closed-loop* MRP permits downstream control of the entire production program, there is an attempt upstream to adapt the planning of the distribution system requirements with production planning, leading to the configuration of the DRP system, *distribution requirements planning*, which coordinates the flow of information coming from the market. DRP emulates the operating mechanism of the MRP, considering the data from the peripheral warehouses or retail outlets in the market, planning the flow of demand from the periphery to the center and taking into account the respective availability, company policies, and the reassortment times of each distribution channel segment.

The informational integration of the areas involved in the planning and management of physical flows has recently led to the success of *manufacturing resources planning,* known as MRPII, which, while maintaining the underlying logic of MRPI systems, expands the planning process to all production resources: production capacity, investment, and personnel. Thanks to the progress of IT, hardware, and software tools, including the new bands and the expansion of broadband, which facilitates the speed of transmission of a large amount of necessary data, MRPII has also been expanded to include maintenance management, quality control, and cost accounting control in programming (Wight 1995; Jiang and Han 2009); MRPIII has become an advanced production management system, covering almost all management aspects of production

The term Just-In-Time, JIT, became popular in Japan in the early 1960s in shipyards, where, as a result of the high productivity of the steel mills, steel stocks were reduced, thereby achieving much reduced order handling times thanks to the possibility of obtaining batches equal to a few days of demand; in effect, therefore, "just-in-time" compared to usage. The practice has spread rapidly to other industrial sectors and to the internal production phases, and Toyota has applied it in all its factories, conceptualizing its various aspects and spreading them to suppliers. The JIT that has developed in the West is precisely the one developed by Toyota.

Limits of space do not allow for a discussion of the historical and economic reasons that led to this form of production control. It is appropriate, nevertheless, to recall that in the West, just-in-time was initially received with an opposing perspective since, more than an organizational-management model, it was seen as the return of an industrial policy made possible by the different cultural and social contexts in Japan, where labor management did not present the conflicts and problems found in Europe and the United States. A second aspect that initially helped to curb the introduction of this production model in Western enterprises was the belief that the MRP and JIT approaches were incompatible. However, there are no irreconcilable differences between the two systems; the fundamental difference between the two systems is the time horizon they adopt. MRP starts from a production program that covers a time interval greater than the lead-time of the upstream phases that are to be managed and planned; as a result, all the internal tasks thus entail backward scheduling. The JIT ties the single phase to the one downstream, not in terms of planned needs but of immediate needs. This way of proceeding eliminates many of the drawbacks that characterize control using MRP: in particular, the need for *functional* and *safety stocks*, which increases as the time horizon considered widens. While MRP is a typical *push* system, JIT is a *pull* system in which materials do not advance in the production process but are requisitioned by the downstream department when they are required. From a logical point of view, the *optimum* situation for a *pull* system is achieved when the logistical flow consists of transformation and handling operations performed on the *individual item* at the time it is required downstream. For JIT, this represents the ideal goal.

The essence of just-in-time, however, is much more complex, as it aims to *continuously improve* the use of available resources and *eliminate all waste*. According to Taiichi Ohno, the father of Toyota's production system (Ohno 1988), Western companies cannot adequately deal with the elimination of waste precisely because they follow the logic of large-scale production, with labor specialization (the Taylor model) and mass production assembly chains (Fordism) (Blackburn 1991; Wakamatsu 2009). He pointed to "Seven Wastes" which he considered the *mortal sins* in the production process (Smith and Reinartson 1991):

1. Delay, waiting, or time spent in a queue with no value being added
2. Producing more than you need
3. Over-processing or undertaking nonvalue-added activities
4. Transportation
5. Unnecessary movement or motion
6. Inventory
7. Product defects

According to JIT, all waste must be eliminated; however, most of the effort must be concentrated on *stocks* because they stem from the asynchrony between upstream operations and those downstream. When there is a significant amount of *asynchrony* the waste is obvious, and the situation is made worse by the risk of obsolescence. Products that remain as stock do not increase in value; when the stock is not strictly essential for the smooth functioning of the production operations, which add value, the stock represents a waste, as does also the space it occupies. This gives rise to the fundamental ideas behind the just-in-time system:

1. Companies should produce only *the amount* required by commercial needs, thus eliminating any waste from over-production
2. Companies should produce at the moment demand arises, so that the stocks of components and materials are eliminated along with any wasted space
3. There should be immediate delivery to eliminate the stocks of finished products.

Hence the custom of referring to the JIT system as *stockless production policy*. JIT, however, does not just emphasize *reducing inventories*, which are seen as *indicators of inefficiency* in a production system, but also *reducing the production lead time*, defined as the average time it takes for a production batch to complete all processing steps and be available for delivery. Monitoring and controlling the *lead time* is crucial because only a limited part of this is employed in activities that add value to the product, while a more or less lengthy part of the lead time is absorbed by operations that do not add value to the product but can generate unnecessary costs: handling and transport, queues and waiting times for manufacturing processes, material delivery times, machine downtime, etc., all operations that are unproductive and as a result must be minimized or eliminated. In short, the Japanese system overturned traditional theories because it required a shift in approach from maximizing the use of *installed* resources (traditional approach) to minimizing *utilized* resources to respond to changes in demand and, above all, minimizing stocks and lead time. The *flexibility* and *elasticity of production*, which ensure the direct link between market trends and production flow, are *another priority objective of just-in-time*.

As mentioned above (Sect. 10.14.1), the JIT was developed in Toyota, by Taiichi Ohno, giving its name to a new production policy called "Toyotism"—synonymous with flexibility and stock reduction—to distinguish it from "Fordism," synonymous with rigidity and constant production volumes. In extreme synthesis, the basic principles of Toyotism are as follows (Imai 1986; Benton and Shin 1998; Liker and Meier 2005; Liker and Morgan 2006):

1. *Jidoka*, which means "automation of machinery with a human mind": This requires maximum *automation* of the plants accompanied by *autonomous* operation and, above all, by *human intervention* to test the quality of operations online. This means automatically controlling the defects throughout the production process, since limited downtime to re-tool a machine is preferable to a waste of materials and the production of defective parts. In fact, *jidoka* requires that production line be stopped when anomalies occur, and to do this human discernment is indispensable, as well as the possibility to rely on the intervention of automatic devices.
2. *Shojinka*, that is, flexibility in the employment of labor.
   (a) The worker must be able to do many simple tasks and a number of complex ones; in this way he will be *employed in a flexible manner on several machines*
   (b) The amount of work should vary as production levels change; *workloads must be leveled and idleness eliminated*
3. *Soikufu*, that is, inventiveness and creativity: The experience and creativity of each production component must be used to involve each worker in the goal of improving *productivity* and *quality*
4. *Just-in-time*, the basic principle that gives the name to the entire philosophy: Try to reduce to zero every *lead time* in the production process, upstream, during, and downstream. This means maximizing production efficiency by minimizing

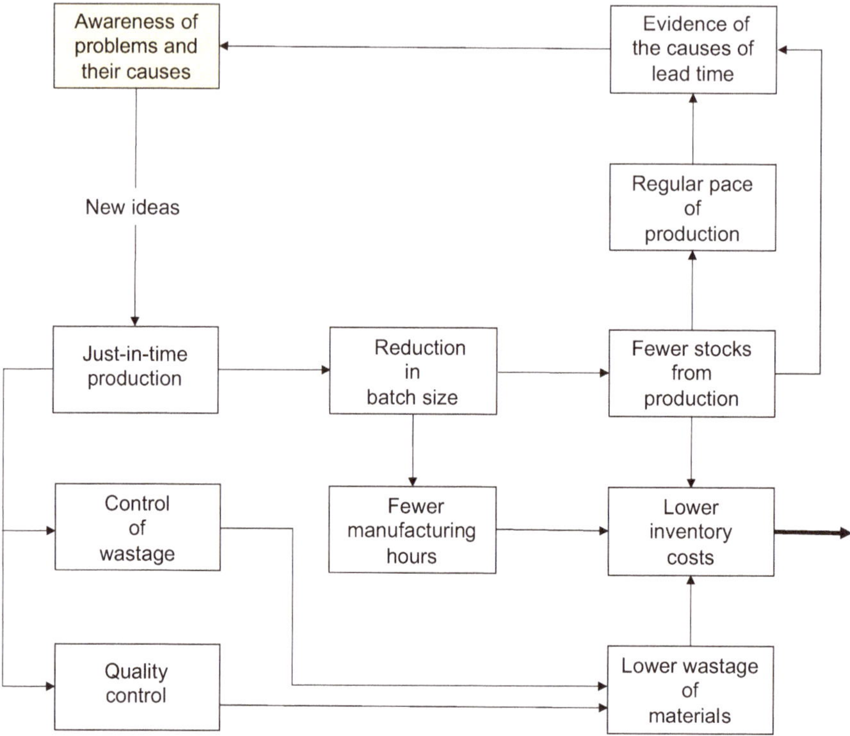

Fig. 11.7 JIT to control stocks, lead times, and processes

the ratio of "dead time" to "active time"; the presence of stocks can make control of the production process less urgent and lead to increased production inefficiency. In fact, JIT is not only a technique for controlling stocks and production but a philosophy/technique whose broad objective is the search for total quality, and as such it is part of the Total Quality Management approach (Sect. 10.7), as summarized in Fig. 11.7.

11.15 Just-In-Time and Stockless Production

Just-in-Time is not only a principle but a consolidated structure of engineering techniques designed to continually improve the use of the available resources by seeking flexibility and elasticity in production in order to guarantee timeliness in every phase of organizational processes. It is useful to mention some measures to implement a program of stockless production:

1. Reduce the setup time
2. Develop good relations with suppliers
3. Redesign the layout

4. Employ multipurpose workers
5. Eliminate defects: excellent quality
6. Produce modular products
7. Level production
8. Focus on customer satisfaction: The quality of production is central to customer satisfaction and should be improved by adding to the traditional product a series of complementary services (Hall and Jackson 1992; Hum and Sim 1996). For more on this point, see Sect. 10.7.
9. Associate with these measures the *Kanban* method.

Reduce set-up time In the just-in-time production system, setup time is particularly important because all steps taken to produce without stock have as a priority reducing setup time. Setup time is the time between the production of the last unit of an item and the production of the first unit of another item. The setup activities include both direct interventions on the means of production, such as the preparation of a plant or assembly line, as well as operations relating to the organizational aspects, and therefore the preparation of raw materials, semi-finished products, and all the documentation relating to the activities themselves (machine time record, the design of the detail to be started up in production, the control order of the finished detail). Reducing setup time allows for a faster turnover of the product categories over a given period of time for a plant, line of production, or an entire department. This makes it convenient to schedule shorter manufacturing cycles, and therefore lead to a reduction in the work in process stocks, producing a similar effect on stock levels in warehouses. It is important to note that the reduction in the setup time of machinery, in addition to measures regarding the product and therefore the project, permits the so-called leveled production; that is, a production program that, in a certain period, is constant in quantity and variable in its mix, as will be discussed below. It was, in fact, Taiichi Ohno who believed that Toyota needed a drastic reduction in re-tooling time, and at the beginning of the 1970s the company achieved considerable reductions in the setup times; for the presses for molding car sheet parts, the re-tooling was reduced over a 5-year period by about an hour and 12 minutes. This allowed the company to work toward an even more ambitious goal, the single setup: that is, a re-tooling time in single digits, and therefore less than ten minutes. In some cases, Toyota achieved times below one minute, also referred to as one touch setup, or instant re-tooling (Smith and Reinartson 1991).

Developing supplier relationships Apart from changes to the production process and the management of the production process, there can be no stockless production unless the purchasing policies are properly acted on, because the supply conditions often limit in a crucial way the level of raw material inventories. JIT has, of necessity, led to a review of the relationships with suppliers, involving them in the short- and medium-term satisfaction of the client company, with the stated aim of developing a climate of mutual trust and collaboration in the management of materials. The main objective in a just-in-time supply relationship is the reliability of deliveries and their synchronization with the production program of the client company. Suppliers are regarded in the same way as external departments and therefore as production units, to improve production and reduce stock levels as well as reduce the overall cost of supply. Inventory control is no longer entrusted only to one company but to all the companies that form the value chain (Rich and Hines 1997). The traditional customer–supplier relationship therefore tends to be modified to achieve the so-called *partnership in profit*, a relationship according to which customer and supplier then become "business partners," leading both to assess the appropriate cost reductions along the entire logistical-production chain. Another development that directly engages suppliers in the production process of the customer enterprise is the outsourcing of the quality control of materials and components from the purchasing company to the supplier itself, even in the face of an increase in price, since the detection of defects in the inputs is much less expensive the earlier they are discovered.

Layout redesign The overall performance of an operating system is the result of interactions between its components: labor, facilities, and materials. The most common organizational solution adopted in the design and implementation of production structures is the *functional* or *process layout*, which is based on the specialization of functions and the maximum productivity specific to the individual operations. The general orientation of just-in-time in the design of the layout is the so-called *organization by product*, which is based on the identification of production units, departments, and plant lines with the product they produce. The work required to redesign the layout of a company varies in amount and covers the technical area regarding component manufacturing, the assembly process, and the entire plant. The technique used by just-in-time in the design of the factory layout is *group technology*, which entails the transition from a functional job shop organization to a factory-cell organization. Group technology identifies the groups of plants or machines that will then be dedicated to the production of a specific family of products. The most obvious advantage of a manufacturing cell organization is the significant simplification of the flow of materials, which results in a significant reduction in the production lead-time due to the elimination of waiting times and the decrease in handling time (Smith and Reinartson 1991).

Multipurpose workers A JIT-type production system aims at reducing labor costs and relying on the versatility of workers. The first objective is pursued by automating operations and through small group work. The aim of *multipurpose workers* is to improve the flexibility of the production system by making it easier to shift workers to different plants and employ them for different tasks, which necessarily implies the periodic turnover of work positions to make each worker learn how all the installed plants work and to form a better understanding of the stages of production and a thorough knowledge of the process and the product. From an organizational point of view, this makes it possible to assign the same worker to manage multiple plants at the same time, resulting in a reduction in waiting times while the machines are engaged in the production process. An additional advantage associated with flexibility in assigning workers to different tasks is faster adaptation of production capacity to changes in demand, especially when such changes relate to the product mix rather than the quantities.

Elimination of defects: excellent quality Since, as we know (Chap. 10), quality control is fundamental to the economic efficiency of the company, JIT requires that it should not be limited to the control of finished products but focused on controlling the whole process, which is capable of associating a specific cause to any type of defect at the time it occurs, so that it can be quickly eliminated during (and not at the end of) the processing cycle.

Modular products In order to reduce production costs, it is necessary to act at a technical level, especially during the planning phase, through standardization and modularization. *Standardization* is an organizational process for achieving a high degree of uniformity in products or components. It can be divided into: simplification, whose goal is to reduce the excessive variety of products, and unification, searching for the most rational ordering of components through the regulation of common product features and compliance requirements. *Modularization* is a technique that seeks to structure the product as a set of modules, each of which is assigned a certain function. Each module can, in turn, be thought of as a product to be decomposed into submodules. The advantage of modularization is twofold: on the one hand, a simpler production flow can be designed, and on the other it is easier to continuously update the product when technological improvements occur without having to modify the whole project, but simply by replacing the obsolete module with an updated one with the same requirements.

Production leveling In the sales field, just-in-time is applied to make the product available for sale only in the quantity required by the market. This can be achieved by a production process that adapts promptly to market developments. The instrument for adjusting production to changes in demand is the *leveling* of production. If production is leveled, a production line will no longer have to produce only one type of product in large batches; on the contrary, it will have to manufacture many varieties of products every day depending on the variability of demand. In this way, production is always up-to-date and stocks are reduced.

The Kanban method The just-in-time production system is typically based on the "pull" method, as production is driven by final demand. JIT for production control uses the Kanban system, a Japanese term that means "tag or card," which is affixed on a sliding container of pieces of materials, components, or products.

> Kanban is a Japanese term: 'kan' means 'visual' and 'ban' means card, so roughly translated, it means 'card you can see. [. . .]... Kanban helps create responsive work sites in factories and other organizations, facilitate quality control, and design work sites that support human dignity and allows workers to realize their full potential. The 'Toyota Way' or Kanban has been adopted and is still being evolved by companies in the car business, and in business and services including healthcare, governments, and Agile software development - in fact, any organization that wants to continuously improve (Smartsheet, online).

1. The *Kanban withdrawal* (movement card) authorizes the withdrawal of a certain type and quantity of material by downstream manufacturing; it is always associated with a standard container and circulates between the external storage point and the internal storage point of the user production center.

2. The *production order Kanban* authorizes production for a quantity of units, equal to a standard container, to replace a withdrawn container; it is used only within the production center. As soon as the pieces to be manufactured are removed from the container at the user center, they are detached and put in a collection container. When a fixed number of Kanbans have been accumulated, or at predetermined time intervals, a worker moves them to the upstream processing phase, which authorizes the removal of other pieces. JIT entails an information system which aims at regulating the flow of materials within the production process, limiting the level of expenditure and at the same time meeting market demand by adapting the level of production to the fluctuations that occur in the product mix and quantity. Two observations are necessary: first, while JIT production can be developed even without a Kanban system, the latter is useful as it can be applied only under conditions of low re-tooling times and very small batches; second, today, where the Kanban system still applies, transport trolleys have evolved into automated guided vehicles, AGVs, which are semi-autonomous moving robots, and Kanbans, functioning as information cards, have become displays written by the control computer.

11.16 Optimized Production Technology: OPT

Optimized Production Technology (Jacobs 1983; OPT 2000) is a form of production improvement that aims at reducing production time and costs. The technique was introduced by Eliyahu Goldratt as the *Theory of Constraints* (TOC) in his book *The Goal* (Goldratt and Cox 1984), which quickly gained popularity (Balderstone and Mabin 1998), followed by the book titled *What Is This Thing Called Theory of Constraint and How It Should Be Implemented* (Goldratt 1990).

> The TOC analyses the constraints according to an organizational perspective, focusing on the consequences that removing the constraint would have on company results. The TOC encourages managers to channel their efforts into the reduction of bottlenecks, and constraints in general, by investing resources into their elimination, starting from the biggest constraints. The continuous process of removing constraints tends to affect ever larger company sub-systems, thus stimulating cross-functional coordination and cooperation (Moisello 2012).

The OPT has the dual goal of:

(a) Planning the optimal sequence of operations for the different departments and operations centers, to balancing flows and minimizing or eliminating bottlenecks, taking into account both the existing production capacity, the capacity constraints, and the priorities of the different operations
(b) Identifying the optimal use of critical resources and facilities, as well as how to minimize the stock of work in process (Neely 2007)

Contrary to MRP and JIT, OPT considers it fundamental to plan production by prioritizing both existing production capacity and the priorities of operations to be implemented, allowing for the optimization of the use of *critical resources* and workloads, the reduction of human and plant work time, and the minimization of the work in process warehouses and their related costs. OPT presents itself as a program implemented on a mainframe that addresses the problem of overall production optimization through a series of interconnected partial programs, each responsible for finding the optimal performance in a specific production area by analyzing the critical resources or the potential bottlenecks represented by personnel or machines.

OPT programming develops a master program that attempts to achieve a global optimization of production through local optimizations: in a few minutes, the local daily schedule for many operations centers can be obtained. The procedure considers, sequentially and at fixed intervals of time, how available factor inputs should be used to comply with requests, after priorities have been set for each operation using weighted functions of a number of variables, such as:

- The assortment of desired products
- The deadlines to meet
- The safety stock to maintain
- The use of resources that present bottleneck risks

Using the prioritized list of products to be obtained, the program allocates resources, starting or continuing with the production of products with higher priority and continuously considering the availability of production inputs in the time interval considered. Optimizing workloads, reducing inventory, shortening work time, and reducing relative costs thus become an effect of this Control System, not a goal, as with MRP and JIT. This overall optimization through local optimizations requires only local information, with the advantage of reducing the amount of information to be processed, thereby increasing the timeliness of the program's responses. In addition, analytical programming is not required for all departments but only for *critical* departments where there are "bottlenecks." The simulation of the system, however, can be linked with MRP, which allows for coordination of the plans generated by the OPT, with the programming of non-problematic resources making optimization flexible as conditions change, with significant benefits to productivity and overall cost-effectiveness. Another advantage of OPT is that it does not require total involvement of the organization as other methods do, although the effect is to increase the efficiency of workers.

The setting of the time intervals the analysis is segmented into are based on the minimum amount of work management believes should be scheduled (e.g., a minimum of 4 hours of work for each machine or worker); each minimum interval is converted into a minimum amount of output that must be obtained or transferred for each operation. Using the list of products to be obtained, arranged according to their priority, OPT allocates resources with the aim of continuing or starting up production of the products with higher priority, considering the availability of production inputs in the interval in question. Therefore, the OPT program tests the existing workload and highlights the bottlenecks in plant capacity or the inadequate use of resources due to operations that generate bottlenecks.

Overall, OPT allows programs to be developed in a short time: in a few minutes you can get the local daily schedule for many operations centers. The programming speed is higher than that achieved with MRP, but normally the OPT system is more efficient in the case of a few core products with large batches, each requiring, among other things, a limited number of operations. Moreover, since the program does not consider management costs, it should not be used to make long-term strategic decisions such as the purchase of new equipment or the dismissal/recruitment of workers; instead, such decisions should be based on analyses and calculations not included in the program (OPT 2000).

11.17 The Flexible Manufacturing System (FMS) and the Holonic Manufacturing Systems (HMS)

The Flexible Manufacturing System, FMS, which makes wide use of the possibilities offered by robots and systems equipped with *artificial intelligence* in industrial processes, represents a further development of production control techniques (Lee 1993; Boer and Krabbendam 1992; Kusiak 1986; Genus 1995; Christopher 2000). Thanks to high automation and previously unattainable computing power, FMS integrates and rationalizes many of the most important production functions permitting the simultaneous programming of:

- The loading, unloading, and storage of components and subassemblies
- The automatic substitution of equipment and tools
- The sequencing of machine manufacturing
- Data processing for manufacturing processes
- Personnel coordination

The FMS also produces fully automated storage and production lines by integrating the programming and control of operating machines into a computerized system of integrated control that leads to optimal programming of manufacturing and batches and the minimization of material and product lead time within the system. The integration of control produces considerable flexibility in the production capabilities of the automated assembly line, which can allow a wide variety of different parts or components to be obtained that fulfill the production needs in the optimal quantity and time with a rationalization of time, processing, costs, and an increase in productivity and quality. To overcome the production bottlenecks, a production plan is developed that takes full advantage of computerized techniques. The extreme evolution of FMS is the automated factory, in which all physical flows and all manufacturing are programmed in an integrated form that makes use of some typical tools (Buzacott and Yao 1986).

Numerically controlled tool machines The numerical control allows a machine to perform a predetermined sequence of features and also controls for the proper execution of the operations performed at each workstation. A numerically controlled machine –through sensors, vision systems, code readers, or magnetic platelets – is generally able to identify incoming parts from a given center in preparation for performing the correct processing program. With such devices it is also able to transmit to the central computer information about the number of parts produced, the state of the tools, and the processing errors. Keeping the wear and tear of the tools under constant control, the machine also automatically determines their placement for the correct execution of the tasks.

Industrial robots These are machines of automatic manipulation (robots) that are increasingly improved for both precision and program repeatability; they can perform a series of operations with elevated flexibility and governability and represent a special class of technologically advanced machine that can perform scheduled operations in a fully automated production system. Robots allow for high levels of flexibility since they can interact with each other and with other synchronized peripheral equipment by relying on a "library" of programs shared with other machines, thereby integrating the operation of the production process.

Automated handling systems An automated handling system of parts and tools allows for greater flexibility and reduced crossing times, resulting in a reduction in the stock of products being manufactured. FMS makes extensive use of AGV (automated guided vehicles) that are guided by infrared or induction systems to perform autonomously and at controllable speeds all loading and unloading operations for which routes from all the computers in the factory can be programmed. The AGV, therefore, confer flexibility to transport, being able to establish and rearrange complex routes with great ease, in a short time and at a low cost.

Automated warehouses In flexible production, a variety of parts can be stored and tools with different degrees of wear can be used, but the system must know at all times the location and status of the individual parts stored. The automated storage system provides appreciable benefits in terms of reducing human labor, the amount of available space occupied, and the size of stocks. An automated warehouse is typically managed by a computer connected with a database, which provides real-time information about the status of each stored unit available. This information may relate to the storage cells or to individual units; in the first case, it is possible to know the number of pieces in stock, the saturation level, the position of the individual items, and the blank cells; in the second, the information relates to the nature and status of each component occupying the cell, which allows data to be aggregated and turnover indices for each item to be obtained and the risk of depletion assessed.

Holonic Manufacturing Systems (HMS) (Adam et al. 2002; Kawamura 1997) represent an evolution of FMS. These are the prototypes of *agile manufacturing systems* (Christopher 2000; Leitão and Restivo 2006); that is, automized, highly flexible production systems—making wide use of machines, robots, work-cells, and labor units—that can deal with the rapid changes faced by all the mechanized-production manufacturing enterprises, flow or batch order: variety and uncertainty of demand, changes in tastes, reductions in the life cycle, and the need to reduce time to market (Mason-Jones and Towill 1999).

> Agile Manufacturing is primarily a business concept. Its aim is quite simple – to put our enterprises way out in front of our primary competitors. In Agile Manufacturing, our aim is to combine our organisation, from this integrated and coordinated whole for competitive advantage, by being able to rapidly respond to changes occurring in the market environment and through our ability to use and exploit a fundamental resource – knowledge (Kidd 2000; see also: Sanchez and Nagi 2001) people and technology – into an integrated and coordinated whole. We will then use the agility that arises from this integrated and coordinated whole for competitive advantage, by being able to rapidly respond to changes occurring in the market environment and through our ability to use and exploit a fundamental resource – knowledge (Kidd 2000; see also: Sanchez and Nagi 2001).

Holonic Manufacturing Systems (HMS) consist of relatively autonomous operations centers, called *holons*, that can interact with each other and the external environment to launch and manage orders to cover production requirements and to interact with the customers targeted by production.

> [...] the three fundamental functions of a Holon. These are Planning, Execution and Monitoring. They have to be performed by all Holons in order to be able to act autonomously. Thus they have to be able to compile their own plans and execute them while monitoring their progress. [...] Furthermore, the functionality is attributed to the Holons, which also consist of the physical shop floor equipment and materials. Thus Order holons also have this functionality, which allows the planning of the production flow to be performed in co-ordination by both material and resources in contrast to traditional systems... (Stylios et al. 2000).

In order to study the HMS, an association has been created which has defined the technical, informational, and operational specifications necessary for a network of machines to be considered an HMS (Mařík et al. 2011, p. 206). The "specific techniques" of the HMS *Consortium* provide the following definitions (Adam et al. 2002):

> *Holon*: An autonomous and cooperative building block of a manufacturing system for transforming, transporting, storing and/or validating information and physical objects. The holon consists of an information processing part and often a physical processing part. A holon can be part of another holon.
>
> *Autonomy*: The capability of an entity to create and control the execution of its own plans and/or strategies.
>
> *Cooperation*: A process whereby a set of entities develops mutually acceptable plans and executes these plans.

A set of *blocks* that in parallel produces materials or similar services forms a module; several modules can constitute a higher ordered holon which, in turn, can be included in other blocks of even higher levels. Holons are characterized by technical and informational attributes that allow for the planning and execution of their functions as well as for coordination with other holons. Through its component blocks, the Holonic Manufacturing System integrates the entire range of manufacturing activities, from planning to supply, from production to marketing to logistics. In its simplest configuration (Kanchanasevee et al. 1997), an HMS for a manufacturing company that is market-oriented includes three types of holons: the product

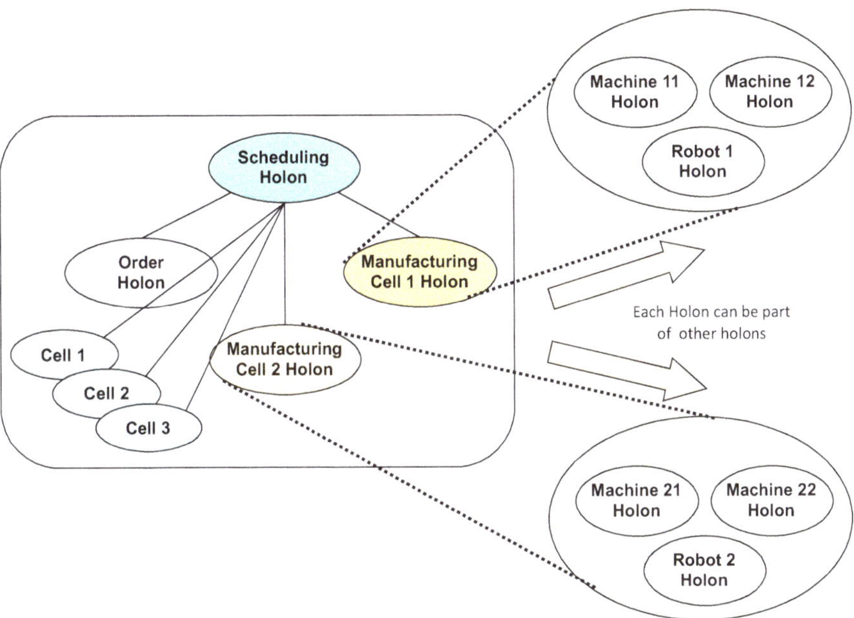

Fig. 11.8 Holonic modules of a Holonic manufacturing system

holons, which indicate the products in the catalogue and their components (sub-holons); the resource holons, which specify the resources available for production; the order holons, which identify the market demand. The holons representing the production processes of an HMS are shown in Fig. 11.8.

The FMSs and the HMSs can be considered real Computer-Integrated Manufacturing (CIM) Systems; that is, a fully computer-controlled production system, whose programs can regulate every moment in the flow of production, from its design to the supply phases, the handling of materials during manufacturing, storage, cost control, and industrial accounting, even in cases of multiple simultaneous processing: for example, in the automotive, aviation, space, and navigation companies (Lei and Goldhar 1991).

It is appropriate to mention the more pervasive and common information system for increasing the efficiency of business control systems: Enterprise Resource Planning (ERP), a computer system capable of collecting, processing, and improving all the information necessary for business management and control. The ERP system is "multipurpose" in that it contains modules that support as many business processes as possible: from processes to manage materials and goods – procurement, storage, and usage scheduling – to sales and delivery processes; from production control – production times, flows, assembly, and costs of each processing phase – to marketing control; from the management of product and materials data to documentation and structured reporting; from production, purchasing, and sales planning to the control of the corresponding processes and flows; from stock control to logistics; from accounts receivable, payables, banks, receipts, and payments to financial and industrial accounting.

An important function of ERP systems is to carry out *simulations* of the above processes even before accepting orders and starting production, thereby allowing the company to quickly identify and *eliminate* any *constraints* on the use of the machines, any bottlenecks, and any errors at all stages of production and logistics before they occur. The ERP system also allows for the scheduling of operations, logistics times, and the efficient use of the means of transport; it is therefore essential for the efficiency of production, the use of human resources (payroll, shifts, and measurable performances), and the supply chain, allowing the company to develop "dashboards" (Sect. 9.9.5) that provide managers with an overview of the financial data useful in highlighting the areas of success and the problem areas.

Finally, it is also worth noting that ERP systems are the most efficient tool for implementing "Business Process Reengineering" (BPR) and for supporting "Change Management" processes, which lead to greater production and organizational efficiency. The term Business Process Reengineering (BPR) was introduced by Hammer & Champy (1990, 1993) and Davenport & Short (1990) as a methodology for managing business transformations in order to analyze processes and discover new ways to increase performance (Davenport, 1993). For Hammer and Champy (1993) *"Reengineering, properly, involves the fundamental rethinking and radical redesign of business processes to achieve dramatic improvements in critical, contemporary measures of performance, such as cost, quality, service, and speed"* (p. 32). Davenport and Short (1990) also posit that BPR can be interpreted not only as the *"analysis and design of processes in order to maximize customer value while minimizing the consumption of resources required for delivering their product or service"* (p. 5), but also as a method that *"encompasses the envisioning of new work strategies, the actual process design activity, and the implementation of the change in all its complex technological, human, and organizational dimensions"* (p. 7).

For this reason, BPR could be thought of as part of a Change Management process that, in times of rapid change, becomes physiological for individuals (Lewin, 1947a, b), social groups, and organizations (Senge, 2006), representing a natural approach to addressing change both individually and organizationally for all types of organization (Hiatt, 2006).

> A change towards a higher level of group performance is frequently short lived; after a 'shot in the arm,' group life soon returns to the previous level. This indicates that it does not suffice to define the objective of a planned change in group performance as the reaching of a different level. Permanency at the new level, or permanency for a desired period, should be included in the objective. (Lewin 1947a, p. 228).

11.18 Control of Projects Through Multiobjective Control Systems: PERT/CPM

All organizations have projects, more or less complex, but some are specialized in contracting out production, including:

- The construction of specific and unique products: for example, publishing a book, making a film, constructing a machine, and setting up an assembly line

- The production of large works (construction) such as skyscrapers, stadiums, bridges, ships, and missiles
- The study of a prototype and the launch of its production
- Undertaking an advertising or a marketing campaign

Every project requires large investments which are affected by the length of the project and the quality of the work required. Therefore, a project must be carefully controlled from three points of view (Archibald 2003):

1. *Quality efficiency* to guarantee that it is appropriate for use by the client (functional objective)
2. *Time efficiency* to *verify* that the project is carried out in the minimum time allowed, given the technical length of the various activities, the user's needs, and the constraints from the machine capacity of the producer
3. *Economic efficiency* to *verify* that the project is achieved with the minimum cost, given the efficiency objectives and the coordination of needs regarding other projects under way at the same time.

When the organization must carry out complex projects, with a precise starting and ending date, it turns to special control instruments, among which *scheduling and time programming*. In order to control and optimize projects, various techniques have been developed to represent the projects, the most widespread of which are the Gantt diagrams, developed by Henry Gantt around 1910 to build sequential models, or *chrono programs*, of the project; the CPM, developed in 1957 by Du Pont to solve scheduling problems (Kelley and Walker 1959); and the PERT, developed in 1958 by the U.S. Navy for the Polaris missile program (Pinedo 2009; Malcolm et al. 1959; Roman 1962; Levin and Kirkpatrick 1966).

The latter two techniques build *network models* of the project, or *activity maps*, which provide a view of the links between the activities rather than their chronological order. Today these three techniques are joined in a single *graph technique* called PERT/CPM, which is integrated by the Gantt diagrams; this grid is produced by Microsoft Project™ and can be found in other software readily available on the Internet (Critical Tools 2020).

The control of projects through CPM and PERT is based on the idea that the duration of an activity, which requires the use of resources over time, depends on the quantity of resources needed to carry out the activity, and thus on the cost, according to an inverse relationship. To reduce the duration, it is necessary to increase the quantity of resources per unit of time; vice versa, a reduction in resources leads to an increase in the duration of any activity. It is thus possible to control the *duration* by modifying the *cost* or to control the *cost* by increasing the activity's *duration*. The basic difference between these two techniques lies in the method for estimating the duration of the activities. CPM produces a deterministic estimate, while PERT a probabilistic one that identifies three durations, pessimistic, normal, and optimistic, taking their weighted average, with the normal estimate having a weight four times that of the pessimistic and optimistic ones (Conde 2009).

Very briefly, *temporal economic control* entails the following steps:

1. Analyzing the project and scheduling of the component activities, specifying for each of these the links with activities upstream and downstream and their duration (in days, weeks or months, in relation to the project's overall scheduling), taking into account available resources and the cost for the assigned duration
2. Building the project grid, that is the network of activities for the project, which is a *model* that presents in a reciprocal relationship the various activities that constitute the project, precisely specifying the *order relationships* among the activities themselves and determining the *initial* and *final* event of the entire project
3. Estimating the length of time and cost of the various phases
4. Calculating the project's duration, or *critical path*, since if one of its activities starts late and/or its duration is prolonged there is a delay in the completion of the entire project
5. Estimating the cost of the entire project
6. Calculating the *float* (*slack*) of the activities of the other non-critical chains, which for each activity specifies how long the start can be delayed or how much longer its execution can be without compromising the project's completion date
7. Optimizing the activities to reduce the length of time and costs
8. Scheduling and releasing of the Gantt diagrams, which places all the activities on a precise temporal scale, ordering them in a logical sequence and specifying for each the effective start date, duration, and completion date.
9. Optimizing the workloads of the personnel and the machines.

Figure 11.9, which represents the *map* of the phases (in summary form) of a project to launch a new product in the form AOA (Activity on Arrow), allows us to observe the basic elements of grid planning (this example is in Mella 2012):

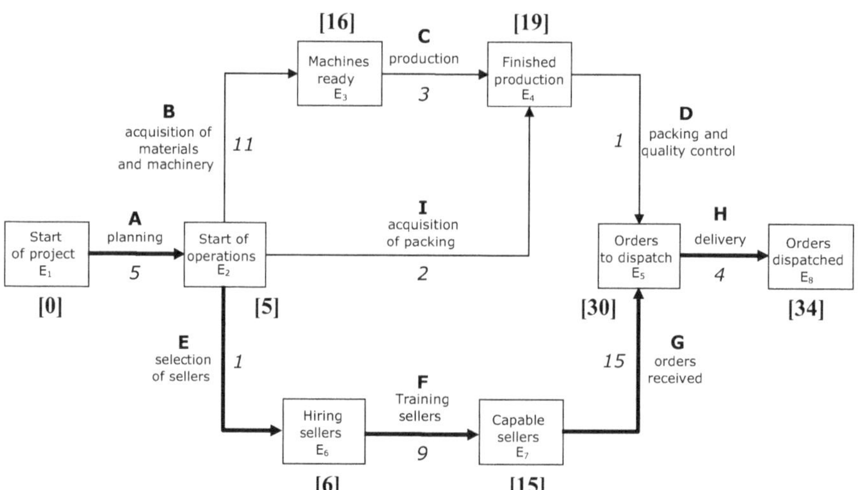

Fig. 11.9 The network for a product-launching project

(a) The eight blocks – indicated by E_1, E_2, … E_8 – show the *events* that represent the states of progress of the project.
(b) Each event (state of progress) is marked by a *date* that places it sequentially with respect to the events that precede it; by convention, the *initial event* is assigned the date [0].
(c) The nine arrows – indicated by A, B, … I – represent the (macro) activities that produce progress in the implementation (state of progress) of the project.
(d) Each arrow (activity) is characterized by a *length of implementation* – in specific time units (days, weeks, months, etc.) – written in italics next to the name of the activity.
(e) The date of each event (state of progress) of the project (block) is determined by the sum of the durations of the activities, the succession of which forms a *chain* that, starting from the initial event, leads to that event.
(f) When the same event is produced by several *chains* of activities, then the date of the event is determined by the duration of the longest chain that reaches it; for example, E_4 is dated [19] since the date is determined by the longer of the two chains that reach it: [A → B → C], with duration 19, and [A → I], with duration 7.
(g) Through this procedure we determine the date of the *final event*, which represents the date for the conclusion of the *entire project*, that is, its duration.
(h) The chain of activities (shown by the thicker arrows) that determines the final date – in this case, [A → E → F → G → H], with duration [34] – is called the *critical path*, since lengthening the execution by only a single time unit of one of its activities would lengthen the overall length of the project; vice versa, to reduce the duration of the project, the length of execution of some critical activities is necessary.
(i) The execution of the *critical activities*, which form the critical path, must thus be specially controlled.
It is clear that the activities that lie on the non-critical chains are subject to possible *shifting* as regards the initial and ending dates and, above all, the *length of* execution; PERT defines as *slack* or *float time* the amount of time that a *non-critical activity* can be delayed without causing a delay in the entire project; the *non-critical activities* can be carried out more slowly, thereby reducing their float time and providing them with fewer resources per unit of time (Lamberson and Hocking 1970).
(j) If we do not want to extend the duration of a non-critical task, we can postpone its execution; for example, the activity "I" has a duration of "*2*" time units, but it can also start after [5], but not after [17], because otherwise it would postpone the date of event E_4. The last date when a non-critical task can start is defined as a Late Event Time (LET), and it is calculated by subtracting the length from the date of the final event: for example, for activity "I":
LET(I) = date of E4 - duration of I = [19] - *2* = 17;
this means that "I" cannot start before the dates of E2 = [5] or after LET(I) = 17. Similarly, LET(D) = [30] - 1 = 29. When a non-critical activity starts at its LET, it becomes critical.

(k) For such simple projects, the calculations can be carried out in an equally simple manner. However, PERT/CPM provides concrete algorithms to automatically quantify length, costs, and slack times, even for projects with hundreds or thousands of activities.

With knowledge of the resources used for the various activities, the consequent length of the activities, and the qualitative standards of the work undertaken, it is possible to control the project through a *system* of control systems that reflect the project model itself. For each activity, a *micro* control system is developed to monitor the *length, cost*, and *quality* of execution; a broader system concerns the various *chains* in the project; with even greater synthesis, we can construct the *macro* control system for the entire project.

This *macro* control system is not only multilevered (Sect. 4.5) but also multiobjective in a true sense (Sect. 4.6), since the contracting firms must verify *contemporaneously* the achievement of the objectives regarding *length, cost*, and *quality* in agreement with the commissioning firm; usually, the control of one of these objectives interferes with that of the others. As can be seen in the project in Fig. 11.10, a shortening of the overall duration can lead to greater cost and often to a decline in the level of quality.

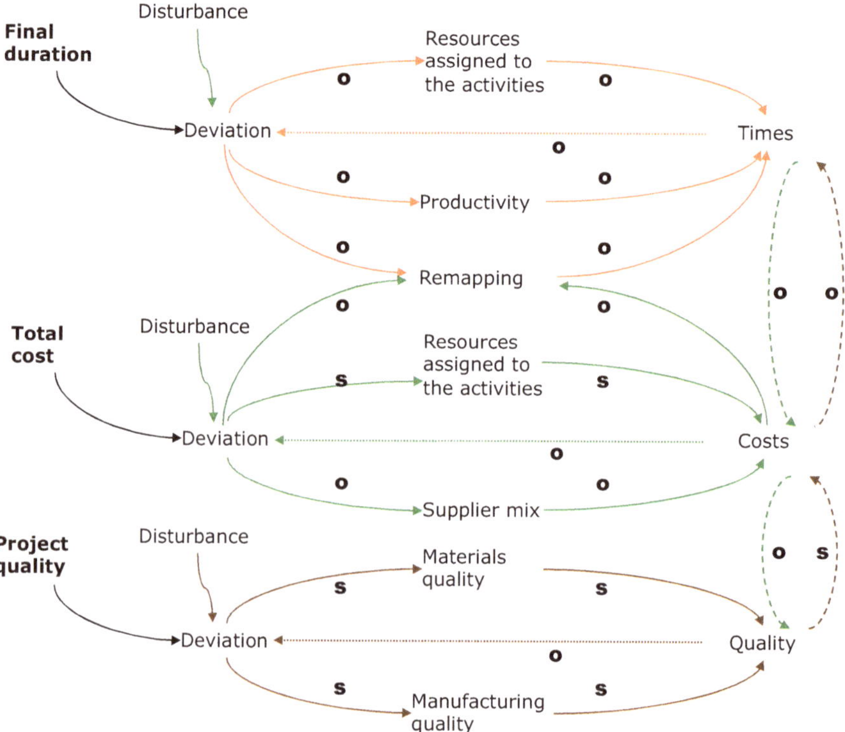

Fig. 11.10 Multiobjective system of control for projects

Thus, it is fundamental to prioritize the objectives. In typical business projects all three objectives are relevant, and a *policy* establishing the priorities must be set at the beginning. To carry out the control, several classes of control *levers* can be used, which involve:

1. Varying the quantity/quality of the resources assigned to each activity to control the duration and quality of the critical and non-critical activities (operational lever)
2. Varying the starting date of the activities, especially the *non-critical* ones, by optimizing the *machine load* and *capacity* needed for manufacturing, with appropriate *algorithm assignments* (operational lever)
3. Acting on the *float times* and on the *non-critical activities* to free up resources for the *critical* activities (operational lever)
4. Selecting appropriate suppliers and productivity increases to control the costs and quality of execution (extraordinary lever)
5. Redesigning the structure of the project (remapping) to modify the *critical* and *non-critical* chains, thereby eliminating or merging certain activities (structural lever)

If the project extends over a vast spatial horizon (activities carried out in various places) and a lengthy one (long duration), there is greater probability that external disturbance events can alter the time, cost, and quality of the activities. When such events cause errors that cannot be eliminated through the control levers, it is necessary to recognize the impossibility of achieving the original objectives and to prepare a *revision* of the project for those parts that must still be achieved.

11.19 Multiprojects: The Control of the Priorities and Workload

It may happen that multiple projects can be implemented and advanced at the same time, the activities of which can compete for execution times and common resources. Since the activities of the projects can be, and normally are, carried out by common operational centers, it is necessary to solve the problem of scheduling all the activities of every simultaneous program to optimize the time, cost, and quality of each. An algorithm is therefore needed to assign the various tasks to the work centers that are supporting them, so as not to cause (or minimize) underloads or overloads at each center and, at the same time, not to cause (or minimize) delays in the maximum planned duration of the projects.

Even in this complex case, it might be advisable to use PERT to form a single large network comprising all the various activities related to the individual projects. This technique, however, would be inefficient and burdensome to implement, since there would be a need to continuously review the grid whenever the start date of an activity had to be changed to conform it to the production capacity of the work centers. Moreover, even theoretically speaking such a solution would be difficult to implement because the duration of one activity would depend on that of other simultaneous activities competing with it to be served by the work center; it would not be possible to determine the specific duration if the grid of other activities was not completed, but this would require that the start and end times of the tasks to be allocated were established in advance.

To overcome these drawbacks of applying PERT to multiple contemporary projects, the ALTAI method (Analysis, Leveling, Integrated Automatic Timing) was developed. This is a technique for assigning tasks to operations centers to time the actual start dates of all the activities of all projects. ALTAI is used when, through PERT, all simultaneous programs have been examined, analyzed, and timed individually and separately, resulting in the preparation of the grids and the calculation of all the relevant parameters. The *optimal allocation criterion* provided by ALTAI is to minimize the sum of the costs and charges of overutilization and underutilization of the operating centers and the costs and charges associated with the delay in the completion of the projects.

Let us assume that "*N*" projects are scheduled at the same time, for convenience numbered progressively. We indicate with "Tf_n" the final date calculated with PERT for the nth project and with "ΔTf_n" the delay from the end date; let us also assume that "Cd_n" represents the costs and unit charges per delay on the final date of the nth project (for example, in USD per day of delay in terms of a penalty, lost interest on late receipts of fees, etc.). The total costs and charges, "TCd," for the delay of the "*N*" projects—if a delay were to occur on the completion date—would then be:

$$TCd = \sum_{n}^{N} \Delta Tf_n \, Cd_n$$

If the company can use "*M*" operating centers (jobs, machinery, service stations, operating stations, etc.), indicating with "P_m" the mth center, with "Cs" the costs and charges of overloads or underloads (supposedly equal, although nothing prevents their being considered separately), and with "S_m" the overall areas of overloading and underloading at the mth center, then the *overall costs and charges*, the "TCs," resulting from the *non-normal* use of the operating centers would be:

$$\text{TCs} = \sum_{m}^{M} \text{S}_m \text{Cs}$$

The optimal distribution of the project activities among the various centers is:

$$\min \text{TC} = \text{TCd} + \text{TCs}.$$

In order to obtain the previous minimum cost condition, however, it would be necessary to try all the different possible and compatible arrangements of all the different activities at the different operations centers. However, as can be easily imagined, this would take a long time and a high computing capacity, normally incompatible with project times. ALTAI proposes an approximate solution which, although not optimal, is in practice satisfactory, consisting of applying the following rules:

(a) All activities are placed on the same critical level and are assigned to the various centers according to their Late Event Time (LET).
(b) The activity that has the lowest LET should be prioritized, a criterion that applies to all the others, until all the tasks have been assigned. By using LET to evaluate whether or not to assign tasks, all tasks become critical, thereby respecting the fundamental condition.

Among the methods that lead to approximate results for the allocation problem, the ALTAI solution (Ricciardi 1973) appears efficient, as it allows for the exploitation of all possible slippages related to the different tasks to be assigned. However, this criterion, while satisfactory, is not optimal. If one operations center must be assigned two tasks of considerably different duration, and the one with the longest duration must be served first, then the servicing of the other will be postponed. Since the latter activity is critical, because it was ordered based on the final date, there will be a delay in the completion of its program. In such a case, therefore, it may be more convenient to service the activity of shorter duration rather than the other one, even if this was pre-ordered with respect to the final date, for the logical reason that *a small delay in a lengthy activity is preferable to a lengthy delay in a short activity*.

Let us assume that two activities are sent to the same operations center:

- Activity A with LET(A) = 10 and length of execution of 100 time units
- Activity B with LET/B) = 15 and length of execution of 5 time units

Based on the rules of ALTAI, activity A should be carried out first, since LET(A) < LET(B); however, the execution of A would result in a delay of 95 time units; since both are critical, the entire project B is included in would be delayed by 95 units, in which case it would be preferable to begin B and delay the project containing A by only 10 units.

The use of ALTAI criterion to schedule activities according to the final date can be fairly easily applied:

1. Set up N projects
2. Review the activities of all the projects and arrange them based on LET
3. Calculate the production capacity of the M operations centers
4. Assign the activities to the centers based on LET, trying, however, to eliminate inconsistencies and incompatibilities

If the capacity of a center is saturated but there are other activities to be served, one of the following alternatives could be adopted:

(a) Postpone the start of other activities until the center has freed up production capacity; this would result in a shifting of the final date of the projects which include the activities; this could also cause knock-on effects for other centers that could have operated at full speed if such activities had been carried out rather than delayed

(b) Expand the performance capacity of the work center to which the activities are assigned by transferring staff and resources from other unsaturated centers, thereby increasing the production efficiency of the center by lengthening the working time; all these solutions would, of course, increase the unit servicing cost of the center
(c) Re-examine the projects and try to change the final dates of the activities and/or their duration, compatible with the cost and scheduling objectives of the firm.

When none of these possibilities can be implemented, all that remains is to postpone when the first of the activities will be undertaken for the saturated center; the date it actually gets under way represents the "actual start date."

If there is a different *degree of urgency* regarding when the different projects are to be carried out—which must necessarily be respected—priority could be given to the implementation of the activities of the most urgent projects by establishing appropriate *priority coefficients*, "prc," expressed in time units, which are higher the greater the damage caused by the delay. Each task is assigned a *priority last date* (PLD) calculated by subtracting "prc" from the LET:

$$\mathrm{PLD} = \mathrm{LET} - \mathrm{prc}$$

In this way, the final dates are moved up depending on the urgency of the activity and the risk from any delay. Projects that cannot be delayed can be marked by a code that assigns "absolute priority" to the execution of their activity.

Note. In *Computer-Integrated Manufacturing Systems* the optimal allocation of the activities at the various operations centers can be automatically implemented using optimization programs.

11.20 Complementary Material

11.20.1 Methods for Calculating the Inventory Turnover and the Average Storage Period

As we saw in Sect. 11.3, when the inventory must be controlled it is useful to know the inventory turnover, a number that indicates the ideal number of times the inventory should be replenished and depleted during a period T, usually 1 year, relative to a given article or category of articles (materials, components, products, etc.). The inventory turnover, "r," can result from company policy or simply from a calculation; in the latter case, it can be determined by *physical quantity* or by *value*. Calculating "r" by *physical quantity* is very useful since it provides timely information on the turnover of each item in stock, or of a given category of items, expressed in the same unit of measure. However, the calculation necessarily implies keeping a *warehousing accounting system* that can record the physical movements of each item. The *determination by value*, which can be done even for the entire inventory considered as a unitary amount, implies a less sophisticated *warehouse accounting* system but provides less useful information, since the results are affected by the trends in prices or supply costs, in addition to the trends in the physical movements.

The turnover index of stock *by physical quantity* is determined by the following ratio:

$$r = \frac{\text{total output of period } T}{\text{average annual stock in } T} \tag{11.24}$$

The calculation of the output, the numerator above, can be carried out:

- *directly*, using inventory withdrawal data supplied by *warehouse accounting*;
- *indirectly*, using the following model:

$$\text{output in } T = \text{initial stock} + \text{input in } T - \text{final stock} \tag{11.25}$$

The *indirect* determination is simpler since the data to include in (11.25) can be easily obtained by the *physical inventory* of the initial and final stocks, while the input in T must be obtained by the purchase invoices, in the case of materials or goods, or a bill of lading in the case of finished products. A more significant problem is determining the *average annual stock* to include in the denominator to calculate "*r*." Here, too, there are two solutions:

1. If other forms of supply control are implemented which lead to the calculation of the *economic supply batch*, then the average stock can be set equal to half the economic lot plus the safety stock (Sect. 11.7).
2. If stock inventory data, $G_1, G_2, \ldots G_K$, was gathered for "K" subperiods of equal length—for example, at the end of each month, every 2 months, every quarter—so that their sum is equal to T, then the *average annual stock* can be set equal to the arithmetic mean of the existing quantity at the end of each of these periods, adding to this the initial stock, G_0, and dividing by K + 1:

$$\text{average annual stock} = \frac{G_0 + G_1 + G_2 + \ldots G_K}{K + 1} \tag{11.26}$$

Another useful datum for inventory control is the *average storage period*, whose significance is complementary to that of "*r*"; this measure indicates the length of the period (months, weeks, days) during which the company is ensured the availability of stock in the amount of the *average annual stock*. The *average storage period* can be easily calculated by dividing the entire period T (in the time units of measurement it is expressed in) by "*r*." If the period of observation is "T = 1 year = 365 days," then the *average storage period* in days is given by the following ratio:

$$\text{average storage period} = \text{inventory turnover} = \frac{365}{r} \tag{11.27}$$

Table 11.5 Monthly warehouse movements

January 1		Input	Output	Stocks
Opening stock – January, 1	30			
January		15	20	25
February		20	25	20
March		30	25	25
April		20	30	15
May		35	25	25
June		30	20	35
July		30	25	40
August		5	15	30
September		20	25	25
October		25	30	20
November		20	20	20
December		25	30	15
Total	30	275	290	295
Average value				25

In addition to using the preceding equation to quantify the *average storage period*, the following entirely equivalent ratio can be used:

$$\text{average storage period} = \frac{\text{average annual stock}}{\text{total output of period } T / 365} \tag{11.28}$$

An example will clarify the above calculation. Let us assume the inputs (loading) and outputs (unloading) of a given warehouse item, recorded on a monthly basis, are as indicated in Table 11.5.

Using (11.25) to calculate the *average annual stock*, we get

$$\text{average annual stock} = \frac{30 + 295}{13} = 25\,\text{units}$$

Applying (11.24), we can calculate the inventory turnover:

$$r = \frac{290}{25} = 11.6\,\text{times}$$

(11.27) allows us to calculate the *average storage period*:

$$\text{average storage period} = \frac{365}{11.6} = 31.5\,\text{days}\,(\text{about})$$

This result tells us that the annual inventory turnover will ensure a replenishment for around 31.5 days. From (11.28) we obtain the same result:

$$\text{average storage period} = \frac{25}{290/365} = 31.5\,\text{days}\left(\text{approximately}\right)$$

The inventory turnover, "r," can be calculated by considering the values rather than the physical quantities, which is especially useful when the calculation concerns the entire inventory or a heterogeneous range of goods or products. In this case, we can use the following simple equation:

$$\text{inventory turnover} = r = \frac{\text{Value of output in period}\,T}{\text{Value of average annual stock}} \tag{11.29}$$

For finished products or goods, the output data correspond to the sales values, which can be obtained by the general accounting process or directly from invoices. In its simplicity, (11.29) offers an approximate and scarcely significant measure of "r" since the numerator and denominator have no homogeneous expression: the first represents revenues and the second is quantified on the basis of production costs. In order to make the two terms homogeneous, it would be necessary to include in the numerator not sales but the *cost of goods sold*; the simplest procedure is to remove from sales the *average mark-up* percentage, which is equivalent to the return on sales (ROS). Knowing the ROS, we can calculate the *cost of the sale* simply by reducing sales by the ROS that is determined by the sales.

The inventory turnover, "r," calculated using data expressed in values, refers to "how many times the capital invested in stock is renewed during the year." An increase in "r," ceteris paribus, reduces the incidence of the cost of invested capital, in addition to that of inventory costs and the risks from obsolescence. It is immediately clear that if the inventory turnover equals "1" during the year, then \$100 invested in that inventory must be financed for a year. If the cost of the invested capital were 5% per year, the company would have a cost of \$5 to cover the investment in inventory; if, however, "r" were 2, then the cost of the invested capital would decrease to \$2.5; if "r" increased to 3, 4, or 5, then the cost of capital would decrease to 2, 1.5, and 1, respectively.

11.20.2 Control of the Stock of Material: Calculating EOQ

A company estimates its need for a given material for an entire year at "$Q^* = 3{,}820$" units. Table 11.6 shows the needs "$S(n)$" per week (n) for that material, in addition to the absolute difference with respect to the average. Analytical accounting provides the following unit *supply costs* (ac):

1. Cost of personnel: € 6
2. Handling and transport: € 4
3. Telephone, printing, and postal costs: € 3
4. Administration and computer costs: € 3

Total (ac): € 16

The purchase cost of each unit of stock is "price = 10 €/unit" and the *storage cost* per unit of stock is 30% of the price per day, which amounts to "sc = 3 €/day." To calculate the optimal purchase batch, we can apply (11.8) using the data in Table 11.6; setting "$T = 1$" we obtain:

$$\mathrm{EOQ} = \sqrt{\frac{2\,\mathrm{ac}\,\mathrm{Q}*}{\mathrm{sc}}} = \sqrt{\frac{2 \times 16 \times 3{,}820}{3}} = 201.86 \approx 202\,\mathrm{units}$$

The number of reorders (n) and the average storage period (t) is determined as follows:

$$n = \frac{\mathrm{Q}*}{\mathrm{EOQ}}\frac{3{,}820}{202} \approx 19\,\mathrm{orders}$$

$$t = \frac{T = 365}{n} = \frac{365}{19} \approx 19\,\mathrm{days.}$$

The cost of each supply batch is:

$$\mathrm{TC} = \left[(10 + 3)202\right] + 16 = €\,2{,}642.$$

We then calculate the *safety stock*, whose probability of being understocked is less than 5%, assuming a *lead time* of 1 week; this value is subject to change once the reorder period is known. To have a safety stock aligned with the desired probability, the factor "k" *applied to the* MAD is "$k = 2$." It is then necessary to calculate the average weekly withdrawal "$S(n)$" and the specification of the MAD. Taking the weekly consumption data in Table 11.6, we calculate the arithmetic mean:

$$F = \frac{3{,}820}{52} \approx 73$$

Therefore:

$$\mathrm{MAD} = \frac{600}{52} \approx 12\,\mathrm{units}$$

We now calculate the *order point*, after careful data gathering has determined that the reorder period is 2 weeks. The average demand is twice that of the average weekly demand: "SS = 2 MAD = 24," and thus represents the *reorder point*. "$Q°$" is equal to:

$$Q^{\circ} = 2\,\mathrm{F} + \mathrm{SS} = 2\,\mathrm{F} + 2\,\mathrm{MAD} = 146 + 24 = 170.$$

Table 11.6 Weekly requirements and deviations with respect to the average

weeks	S(n)	deviation	*weeks*	S(n)	deviation	*weeks*	S(n)	deviation
1	70	3	*19*	71	2	*37*	63	10
2	65	8	*20*	74	1	*38*	100	27
3	60	13	*21*	67	6	*39*	97	24
4	64	9	*22*	70	3	*40*	93	20
5	53	20	*23*	62	11	*41*	63	10
6	70	3	*24*	79	6	*42*	101	28
7	58	15	*25*	76	3	*43*	72	1
8	59	14	*26*	64	9	*44*	81	8
9	65	8	*27*	82	9	*45*	96	23
10	68	5	*28*	79	6	*46*	56	17
11	70	3	*29*	61	12	*47*	91	18
12	80	7	*30*	55	18	*48*	88	15
13	74	1	*31*	84	11	*49*	63	10
14	63	10	*32*	92	19	*50*	90	17
15	62	11	*33*	96	23	*51*	70	3
16	60	13	*34*	85	12	*52*	99	26
17	81	8	*35*	61	12	**Total**	**3820**	**600**
18	65	8	*36*	52	21	**average**	**73**	**12**

This figure signifies that the warehouse manager will make a new order of "EOQ = 202" units when the stock falls to "Q° = 170 units." This will ensure a service level close to 95% based on the inventory requirements. The maximum inventory will be:

$$\text{EOQ} + \text{SS} = 202 + 24 = 226\,\text{units}.$$

11.20.3 *Linear Programming for Multiproduction with Several Capacity Constraints: Example*

Let us assume that in a given month a company can produce and sell two units of products A and B given the prices and costs in Table 11.7 in order to maximize the operating results, "OR," as the sum of the total contribution margins, "CM," from which the fixed costs specific to each product, "FC," and those common to the two production processes, "FCT," are deducted. This produces the following objective function using the *direct costing* method:

$$\max \text{OR} = \left\{\left[\text{CM}(\text{A}) - \text{FC}(\text{A})\right] + \left[\text{CM}(\text{B}) - \text{FC}(\text{B})\right]\right\} - \text{FC}_{\text{T}}$$

$$\max \text{OR} = \{[5{,}000\text{A} - 2{,}000{,}000] + [4{,}000\text{B} - 4{,}000{,}000]\} - 10{,}000{,}000$$

$$\max [\text{OR} + 16{,}000{,}000] = 5{,}000\text{A} + 4{,}000\text{B}.$$

Simplifying the above:

$$Z = \frac{[\text{OR} + 16{,}000{,}000]}{1{,}000}$$

The objective function is thus simply:

$$\max Z = 5\text{A} + 4\text{B}:$$

There are no capacity constraints regarding the machinery, but the company must take into account the available raw materials. As shown in Table 11.7, "2" units of material G and "6" of H are needed to produce two units of A, while for B, "4" units of both G and H are needed. There are only 16,000 units of G and 18,000 of H available. In constructing the linear programming *model*, given the simplicity of the problem, it is convenient to indicate by A and B not only the names of the products but also their variable quantities.

Table 13 easily provides us with the *system of constraints*:

$$\begin{cases} 2\text{A} + 4\text{B} \leq 16{,}000 \text{ constraint of material G} \\ 6\text{A} + 4\text{B} \leq 18{,}000 \text{ constraint of material H} \\ \text{A}, \text{B} \geq 0 \text{ condition of non} - \text{negativity.} \end{cases}$$

Using *Wolfram-Alpha's Mathematica* (2020) we obtain Fig. 11.11, which shows the area of the admissible solutions (in gray) and the straight line (from the series of straight lines) of the objective function, which is tangent to the area. The theorems mentioned in Sect. 11.10 indicate, again using Mathematica, the following optimal solution, which corresponds to the coordinates of the point of tangency:

$$[\text{product A} = 500 \text{ units;,product B} = 3\text{;,}750 \text{ units}],$$

Table 11.7 Prices and costs of two products under capacity constraints

Values	Symbol	Product A	Product B	Total
Unit prices	p	16,000	18,000	
Unit variable cost	vc	11,000	14,000	
Unit contribution margin	cm	5000	4000	
Specific fixed costs	FC	2,000,000	4000,000	6000,000
Common fixed costs	FC_T			10,000,000
Constraints				
Material G	qG	2	4	≤ 16,0000
Material H	qH	6	4	≤ 14,000

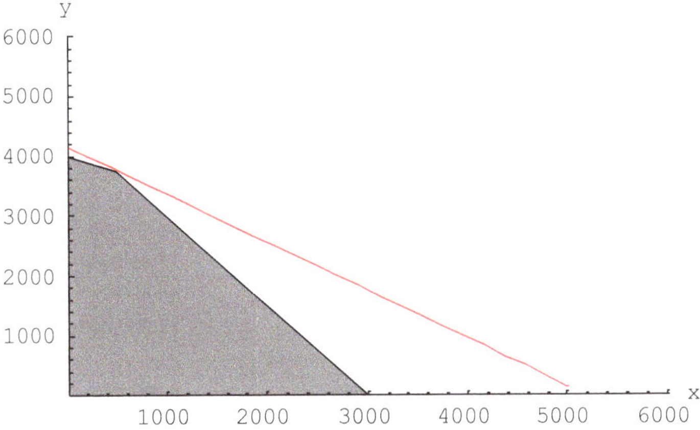

Fig. 11.11 The optimal solution using the linear programming method

which represent the *optimal production program* that maximizes the company's expected operating results. Corresponding to this *mix*, the objective function, in its simplified form Z, has the following maximum value:

$$Z = (5 \times 500) + (4 \times 3{,}750) = 17{,}500$$

By inverting the above simplification, we can calculate the following OR:

$$\text{OR} = 17{,}500(1{,}000) - 16{,}000{,}000 = 1{,}500{,}000.$$

After calculating the optimal solution, we can construct the *analytical accounting matrix* in Table 11.8 to demonstrate that the results are indeed optimal.

We immediately observe the erroneous conclusions that could have been reached if we had based the production program on the full-costing method rules and decided to eliminate production of A; in this case, analytical accounting reveals a negative operating result with fc(A) > p(a). Under the assumption that production of A is suspended, the following materials would be freed up:

material G available for B = 500 (2) = 1000
material H available for B = 500 (6) = 3000
Since 4 units of both G and H are needed for each unit of B, the new availability after production of A is suspended would at most lead to 250 additional units of B; in fact:

material G available for B/qG(B) = 1000/4 = 250 = ΔQB
material H available for B/qG(B) = 3000/4 = 750 = ΔQB (quantity not available)

By increasing the production of B by 250 units, the overall total rises to 4000, giving us the *analytical accounting matrix* in Table 11.9. The operating result in

Table 11.8 Calculation of analytical results with the optimal *mix*, under conditions of capacity constraints

Values		Product A		Product B	Total
Quantity	500		3750		
Prices	16,000		18,000		
Earnings		***8,000,000***		***67,500,000***	***75,500,000***
Unit variable costs	11,000		14,000		
Variable costs		*5,500,000*		*52,500,000*	*58,000,000*
Contribution margin		**2,500,000**		**15,000,000**	**17,500,000**
Specific fixed costs		2,000,000		4000,000	6000,000
Net operating margin		**500,000**		**11,000,000**	**11,500,000**
Imputation percentage	11		89		100
Imputed fixed costs		***1,100,000***		***8,900,000***	***10,000,000***
Total cost		**8,600,000**		**65,400,000**	**74,000,000**
Full cost	17,200		17,440		
Operating result		**- 600,000**		**2,100,000**	**1,500,000**

Table 11.9 Calculation of the analytical results with changes to the optimal mix

Values	Product A	Product B	Total
Quantity	0	4000	
Prices	16,000	18,000	
Earnings	**0**	**72,000,000**	**72,000,000**
Unit variable costs	11,000	14,000	
Variable costs	0	56,000,000	56,000,000
Contribution margin	**0**	**16,000,000**	**16,000,000**
Specific fixed costs	2,000,000	4000,000	6000,000
Net operating margin	0	**12,000,000**	**10,000,000**
Imputation percentage	0	100	100
Imputed fixed costs	**0**	**10,000,000**	**10,000,000**
Total cost	0	**70,000,000**	**72,000,000**
Full cost	0	17,500	
Operating result	**- 2,000,000**	**2,000,000**	**0**

Table 11.9 is lower than that in the previous table, and the solution is sub-optimal. In reading Table 11.9, we must keep in mind that the *specific fixed costs* for A are understood to apply even if A is not produced. The objective function would not take into consideration the situation in which the specific fixed costs could be saved through a halt in production.

11.21 Summary

This chapter has examined the logic of warehouse control and production, which is essential to ensuring the efficiency of the production and economic transformation (Sect. 9.3). The main focus was on the connection between the inventory and production strategies, highlighting the coupling and decoupling function of stocks (Sect. 11.1). It also emphasized the opportunity to apply the ABC method to identify the most important materials and products on which to focus control (Sects. 11.2 and 11.20.1).

After specifying the two main types of costs related to stock formation, *storage costs* and *supply batch costs* (Sects. 11.3–11.4), the logic behind calculating stocks based on Wilson's hypothesis (Sect. 11.5) was discussed, with a presentation of the analytical models for the calculation of the Economic Batch of Supply, the Restocking Point, the Safety Stock, and the Optimal Production Batch (Sects. 11.5, 11.6, 11.7, 11.8, and 11.9 and 11.20.2), even in the presence of constraints on production resources (Sects. 11.10 and 11.20.3).

Sections 11.11, 11.12, 11.13, 11.4, 11.5, 11.6, and 11.17 examined the modern techniques regarding the integration of inventory and production controls, known by the acronyms: MRPI, MRPII, MRPIII, JIT, OPT, FMS, and HMS. The chapter concluded by discussing the reticular programming techniques, PERT/CPM, to control projects, with a look at the problem linked to assigning workloads when several projects are being implemented simultaneously (Sects. 11.18 and 11.19).

Chapter 12
The Magic Ring Explores Cognition and Learning

Abstract This chapter concludes the journey we have undertaken by examining Control Systems that make learning and knowledge possible. I have limited the examination to those *Rings* needed for the conscious "mind" to function, deliberately choosing as a guide Gregory Bateson's model of the "mind" as a calculator of various levels of differences, presenting a model of how knowledge is formed through the perception and systematization of differences. I have therefore examined the problem of the formation of signs and the use of language to reveal and communicate knowledge. The reader is invited to read the initial sentences in Sect. 12.1, in which I point out the difficulties in dealing with the topic of knowledge by using mainly the sense of sight and employing a written language.

Keywords Bateson's conjecture · Cognitive system · Technical description of objects · Comparing objects · Composite objects · Technical definition of concepts · Meaningful technical definitions · Languages and signs · Models · The explanation process

> I have said that what gets from territory to map is transforms of differences and that these (somehow selected) differences are elementary ideas. But there are differences between differences. Every effective difference denotes a demarcation, a line of classification, and all classifications are hierarchic. In other words, differences are themselves to be differentiated and classified. In this context I will only touch lightly on the matter of classes of difference, because to carry the matter further would land us in the problems of Principia Mathematica (Gregory Bateson 1972, pp. 463–464).

> On the other hand, this reluctance to adopt a conceptual framework in which apparently separable higher mental faculties as, for example, "to learn," "to remember," "to perceive," "to recall," "to predict," etc., are seen as various manifestations of a single, more inclusive phenomenon, namely, "cognition," is quite understandable. It would mean abandoning the comfortable position in which these faculties can be treated in isolation and thus can be reduced to rather trivial mechanisms. Memory, for instance, contemplated in isolation is reduced to "recording," learning to "change," perception to "input," etc. In other words, by separating these functions from the totality of cognitive processes one has abandoned the original problem and now searches for mechanisms that implement entirely different func-

P. Mella, *The Magic Ring*, Contemporary Systems Thinking,
https://doi.org/10.1007/978-3-030-64194-8_12

> tions that may or may not have any semblance with some processes that are, as Maturana pointed out, subservient to the maintenance of the integrity of the organism as a functioning unit (von Foerster 2003, p. 172).

> The belief that one's own view of reality is the only reality is the most dangerous of all delusions (Watzlawick 1976, p. XIII).

This chapter concludes the journey we have made until now by examining Control Systems that make learning and knowledge possible. Since the topic of knowledge is incredibly vast and cannot be scaled down to fit in with the content and size of this book, I have limited the examination to those *Rings* needed for the conscious "mind" to function. I have deliberately chosen to use as a guide Gregory Bateson's model of the "mind" as a calculator of various levels of differences, undertaking to formally operationalize this model. I hope to succeed in presenting the reader with a convincing model of how knowledge is formed through the perception and systematization of differences, following a typical constructivist view (Mella 2020), which is reinterpreted starting from Bateson's model, which tries to demonstrate that the mental construction of individual "reality" is made possible by the action of numerous Rings. Knowledge that is structured "in" the mind can only be revealed through behavior produced by the cognitive system. As Jean Piaget observed (1937), knowledge should be considered as the highest form in the "world" of adaptation of a complex organism.

I have therefore examined the problem of the formation of signs and the use of language to reveal and communicate knowledge. I have only touched the problem of the relation between knowledge and the "brain" and the problem of conscious knowledge. The topic of learning is only examined with regard to several aspects that involve the construction of models and the explanation process. Once again, the action of *Rings* is fundamental to the development and transmission of knowledge. I invite the reader to read the initial sentences in Sect. 12.1, in which I point out the difficulties in dealing with the topic of knowledge by using mainly the sense of sight and employing a written language.

12.1 Dimensions of Knowledge

This chapter will demonstrate that Control Systems are indispensable for the "knowledge process" and for the formation of the "content of knowledge" through which "cognitive individuals" are structurally coupled to the external environment, trying to survive by reacting to disturbances.

It is not simple to analyze in a book, "as humans," the knowledge phenomena in living beings equipped with a nervous system. The *first difficulty* is entirely clear: dealing with knowledge through a book means speaking with a *written language* (which belongs to a specific linguistic code (English)) which represents the result of a highly complex and evolved mental activity in human beings. Using this language may make it difficult to describe the elementary mental processes that precede the production of the language; for this reason I shall try to integrate the lexical language (written English) with more abstract graphic models. The *second difficulty concerns* the almost obligatory use of examples that make use of the *sense of sight*, thus making it difficult to speak and write about stimuli and sensations which derive from other senses without being able to represent them nonlinguistically. While I can say: "Observe these figures…," it is difficult for me to say: "Try these flavors … or smell these scents…." Flavors and scents must be indicated through words; this assumes the reader already has knowledge of their meaning.

The *third most important difficulty* is *methodological in nature*, since, in my view, there are at least four dimensions for observing mental phenomena, all of which are interconnected, even though they can be examined separately:

The "brain" dimension (in general, the central nervous system) as a "physical system" that produces basic neuronal processes from which knowledge is built as the computation of neuronal stimuli. The connections among the neurons and the various types of schema that occur through the ordering of combinations of perceptions produce the mental sensations corresponding to colors, flavors, smells, etc. Such stimuli are not in the environment and the brain does not perceive them as the result of specific stimuli; they derive instead from the interconnection of stimuli from multiple sensors operating in parallel, which are combined to produce the variety we perceive (Fodor and Pylyshyn 1988). Simulation through cellular automata seeks to demonstrate the hypothesis of the *connectionist* functioning of the brain:

> The network contains "word nodes" each of which standardly becomes activated as a result of the presentation of a particular word, and each of which represents the presence of that word. These word nodes are activated by "letter nodes" each of which represents and standardly responds to the presence of a particular letter at a particular position in the word. Finally, each letter node is activated by "feature nodes" each of which represents and standardly responds to the presence of a certain feature of the shape presented at a particular position: a horizontal bar, a curved top, etc. (Lormand 1991, online).

1. The "mind" dimension, as an *emerging* system and feature of the brain's neuronal activity, which produces knowledge as a process necessary for the survival of living beings; this dimension cannot be directly observed but it can be deduced from observing behavior. Therefore, without behavior we cannot infer the existence of knowledge. As humans we can all directly verify this dimension through our awareness; we know nothing about how our brain functions, with its 100 billion neurons, but we feel heat and cold, we see color, we distinguish signs and try to understand their meaning.
2. The "awareness" dimension, as *conscious* knowledge (thought-containing content, including every form of language), which is a capacity presumably limited to man only.
3. The "intelligence and learning" dimension, as processes in the formation and modification of the content of knowledge.

In a chapter dedicated to Control Systems, it is not possible to treat these four aspects of knowledge, which have vast implications in many disciplines. Even a summary examination would be beyond the scope of this chapter. I will thus only present and attempt to operationalize Gregory Bateson's theory on the functioning of the mind (point 2 above), which produces knowledge through the elaboration of differences at multiple levels. My objective is to demonstrate that knowledge, even at the level of the "mind," assumes an incredible number of Control Systems.

The "brain dimension" (point 1) and its relation to the cognitive and behavioral process has been studied by neuroscience:

> Scientists still have not uncovered the full extent of what the brain can do. This single organ controls every aspect of the body, ranging from heart rate and appetite to emotion and memory. [...] Neuroscientists specialize in the study of the brain and the nervous system. They are inspired to try to decipher the brain's command of all its diverse functions. Over the years, the neuroscience field has made enormous progress. Scientists continue to strive for a deeper understanding of how the brain's 100 billion nerve cells are born, grow, and connect. They study how these cells organize themselves into effective, functional circuits that usually remain in working order for life (Society for neuroscience, web page).

We do not "observe" knowledge in the brain (more generally speaking, in the central nervous system), but only neuronal phenomena as the condition for knowledge to emerge, in the sense that according to the rules of implication: "knowledge emerges from a functioning brain" since "if there is knowledge then there is a functioning brain,," and "if the brain does not function, then knowledge will not emerge." The brain is conceived of as a "computational machine" that functions as a closed system whose inputs are its own states:

> Some of us consider the Central Nervous System to be a fundamentally closed system with its basic organization geared toward the generation of intrinsic images (thoughts or predictions), where inputs *specify* internal states rather than inform "homuncular vernacles" (Llinás and Paré 1996, p. 4).

This perspective is typical of *functionalism* and *connectionism*, according to which the mind produces processes of knowledge that depend on (are produced by or correlated with) neuronal physical processes. For the neurobiologist Gerald Edelman, who has developed a theory of neuronal group selection, or Neural Darwinism (see the quote below), as a model for brain functioning, ... [The] *mind is a special kind of process depending on special forms of matter* (Edelman 1992, p. 6). But then, Edelman goes on, we can say that mind and matter are both processes, i.e., interactive and not static. Are these processes the same? Or, are they similar to one another; and if they are similar, in what ways are they similar? Or, are they different from one another, so different that conflation would be in grave error? An examination of the vast and well-developed field of neuroscience, which is, moreover, highly technical and specialized, goes beyond the limits of this book; however, I would note that Thomas Henry Huxley had already, in the second half of the nineteenth century, and even without any sophisticated neuroscientific knowledge, observed the relations between the nervous system and behavior with his famous "example of the frog,," which led him to conclude (erroneously from a contemporary perspective) that animals are, or behave as if they were, automata:

> As I have endeavoured to show, we are justified in supposing that something analogous to what happens in ourselves takes place in the brutes, and that the affections of their sensory nerves give rise to molecular changes in the brain, which again give rise to, or evolve, the

> corresponding states of consciousness. Nor can there be any reasonable doubt that the emotions of brutes, and such ideas as they possess, are similarly dependent upon molecular brain changes. Each sensory impression leaves behind a record in the structure of the brain—an "ideagenous" molecule, so to speak, which is competent, under certain conditions, to reproduce, in a fainter condition, the state of consciousness which corresponds with that sensory impression; and it is these "ideagenous molecules" which are the physical basis of memory. It may be assumed, then, that molecular changes in the brain are the causes of all the states of consciousness of brutes (Huxley 1874, pp. 239–240).

If we assume the *connectionist hypothesis* is valid, then the development of knowledge by the functioning brain through the network of its connections is simulated by computational techniques that rely on multiple models of neural networks:

> The computational approach to these levels, which we call functional reconstruction of neuronal systems, strives to perform realistic simulations while giving adequate representation to the particularities of neuronal morphology, the various voltage-dependent and chemically mediated ion currents in membrane-bound channels, neurosecretory activity [...] Flexibility and versatility are the keynotes to this approach which consequently uses a compartmental model of the neuron (Perkel 1993, p. 40).

> The large gaps in our understanding of biological memory schemes and mechanisms, the limited applicability of cellular mechanisms and of connection-modification rules in hardware devices, together with the emergence of feature-detection properties in parallel distributed networks constitute the greatest challenge for computational neuroscience (ibidem, p. 44).

The "awareness" dimension (point 3 above) is encompassed in the field of study of various disciplines (Neuman 2003), first and foremost psychology, which investigates conscious and unconscious mental processes, the latter, in my view, deriving from nonconscious, innate, or acquired knowledge. According to Gerald Edelman:

> Primary consciousness may thus be briefly described as the result of the ongoing discrimination of present perceptual categorizations by a value-dominated self-nonself memory (Edelman 1989, p. 102).

> My main focus was on perceptual categorization as it related to memory and learning. I proposed that these functions could be understood in terms of "neural Darwinism"—the idea that higher brain functions are mediated by developmental and somatic selection upon anatomical and functional variance occurring in each individual animal (ibidem, p. xvii).

> I proposed that this ability depended critically on two of the most striking features of the brain, its variability and its re-entrant connectivity (ibidem, p. xviii).

The "intelligence and learning" dimension (point 4 above) is vast and includes the formation of experience, voluntary or guided, even in living beings without a nervous system but still coupled to the environment, and the formation of thought, linguistic and nonlinguistic content, which is needed to improve the existence of cognitive individuals. Various disciplines operate in this field of research, among which cognitivism, pedagogy, psychology, and artificial intelligence. "Artificial Intelligence is one of the newest fields in science and engineering. Work started in earnest soon after World War II, and the name itself was coined in 1956" (Russell and Norvig 2009, p. 4):

> Artificial intelligence is a rapidly evolving field of engineering with an ultimate objective to build machines capable of acting and thinking like human beings. The early phase of AI was concerned with developing programs for theorem proving and game playing. Modern AI encompasses various tools and techniques for humanlike reasoning, learning, planning, language and pattern recognition. Artificial intelligence is probably one of the most successful branches of a broad area of computing (Kumar 2008, p. 1).

The next few *sections* will thus be limited to a treatment of the "mind" dimension and the "intelligence" dimension, not so much to describe the content of mental and cognitive processes but to demonstrate how basic mental processes are based on a multitude of *Rings* acting so quickly as to remain entirely hidden from our awareness.

12.2 Bateson's Model: The "Mind" as a Calculator and Transformer of Differences

Following the *constructivist* (*epistemological*) *view* (e.g., Kelly 1955; Piaget 1937; Watzlawick 1976; Maturana and Varela 1980, 1992; Lewin 1935; von Foerster 1984, 1990, 2003; Luhmann 1990, 1997; Lee 2000; Bateson 2000, 2002: Ceccato 1969, 1974; Wittgenstein 1922; and many others), "reality" is seen as a mental construction composed of the perception and processing of a variety of stimuli that arrive in the form of atomic differences through *sense* organs (sight, hearing, touch, etc.) and general physical sensibility (position of the limbs, sensations in the internal organs, etc.). Such stimuli are continually selected, ordered, and classified by our mind in a continuous process of "cognition," which constitutes our "knowledge."

In his excellent book, *Mind and nature: A necessary unity* (2002, 1st ed. 1979) Gregory Bateson proposes an epistemological theory of knowledge based on the simple yet convincing model of "mind" as the capacity of a cognitive system, or individual, to form a representation (map) of the world (territory) through the perception and ordering, even at successive levels, of differences:

> 1. Mind is an aggregate of interacting parts or components. 2. The interaction between parts of mind is triggered by difference and difference is a non-substantial phenomenon not located in space or time. 3. Mental process requires collateral energy. 4. Mental process requires circular (or more complex) chains of determination. 5. In mental process the effects of difference are to be regarded as transforms (that is, coded versions) of the difference which preceded them. 6. The description and classification of these processes of transformation discloses a hierarchy of logical types immanent in the phenomena (Bateson 2002, p. 92).

Bateson does not take into consideration the brain, but views the *mind* as the "processor" of knowledge, an open system that acts as a "machine"—with multiple inputs—capable of calculating differences, memorizing and comparing these, and finding analogies. Moreover, Bateson—by adopting a simple metaphor—distinguishes between knowledge and what is known, comparing knowledge to a map, what is known to a territory: "The map is not the territory, and the name is not the thing named" (Bateson 1979, p. 30). The map—that is, knowledge—is formed by taking account of the differences the observer perceives in the territory represented; these differences and their transforms are "elementary ideas … and these differences are themselves to be differentiated" (Bateson 2000, p. 463). Our "mind," as a "processor" of knowledge, carries out a continual function of discovering relationships in the patterns of differences, and this process leads to the emergence of a hierarchy of differences based on which all knowledge is constructed (Bateson 2000, pp. 454–471; Bateson 2002, p. 106).

Note: Some of the contents of this chapter have been taken from and expanded on in Mella, P. (2020), *Constructing Reality. The "Operationalization" of Bateson's Conjecture on Cognition*, Springer Nature. The present work includes a broad bibliography on the constructivist context in which Bateson theorized that cognition is based on the abstract notion of "difference" as an element through which a cognitive system is able to achieve the "mental construction of the world."

Even if Bateson does not explicitly state this, I suggest we recognize that the "mind" is not a logical-abstract entity but the principal source of knowledge for *cognitive systems*, in particular "*conscious cognitive systems*." I define the latter as autopoietic systems which, through the mind, succeed in giving *significance* to *environmental stimuli*, transforming these into reactions (which an observer describes as macro behavior or processes interacting with the environment) in the *search* for positive stimuli for survival, or, alternatively, the *flight* from negative stimuli. In order to produce Bateson's model we must assume that cognitive systems must possess:

(a) *Sensor organs* that can perceive differences (in the form of stimuli, impulses, disturbances, etc.).
(b) *Computational* capacities, through which differences are ordered and these differences themselves are differentiated.
(c) *Memory capacity*, which reconstructs or preserves the computed differences through changes in state (of greater or lesser duration) of the internal processes of memory.
(d) The *evaluation capacity*, needed to give "weight" to the dynamics of the internal states which have favored or hindered the search for positive stimuli or the flight from negative ones.
(e) *Effector organs*, which give rise to environmental behavior involving the search for positive stimuli or the flight from negative ones; more specifically, the effector organs (mimicry, sound emission, signs, etc.) must be able to express thought content.

The mind functions through the action of all the components of the cognitive system, even if Bateson believes *differences* are not substantial phenomena since they do not result from a comparison involving the values of variables the cognitive subject is capable of (must be capable of) determining and comparing in order to produce new differences for subsequent comparison. Rules 5 and 6 above, which define the logic of the mental process, clearly show that Bateson views the "mind" as operating in a holonic fashion by forming a holarchy, that is, "a hierarchy of logical types immanent in the phenomena" (Bateson 1979, p. 122).

To apply Bateson's model we must postulate that, in order to undertake a recursive process to determine successive levels of differences, the "mind-processor of knowledge" must, as a minimum, be "structured" to carry out four basic operations, all of which are necessary:

1. *Distinction*, or perception; that is, identifying a *primary difference* with reference to a "null" internal state.
2. *Comparison*, which leads to the identification of *differences of differences* by forming *differences* of various levels (through some sense organ), and their inclusion in various types of *dimensions*.
3. *Perception of analogies* (through some form of memory), which leads to the identification of ranges of variation between *differences* held not to be "too different" from each other.
4. *Association*, which leads to considering a certain number of *differences* as a "different unit," as if they constituted a single *object* of observation.
   In fact, wherever information—or comparison—is of the essence of our explanation, there, for me, is mental process. Information can be defined as a difference that makes a difference (Bateson 2002, p. 91).

As *rule* 5 of "Bateson's conception of mind" clearly indicates, the differences are produced at various levels. Figure 12.1 shows the *Ring* needed to produce the *primary differences* in a cognitive system with a *sensor* capable of distinguishing between the outside disturbances. Referring to Fig. 12.1, I shall define as *basic*, or *primary differences* those differences formed by the cognitive system when the internal mental state is "null" (this is indicated by s_n = "Ø") and the environment produces a disruption or disturbance (e_n = "y") to some sensor organ.

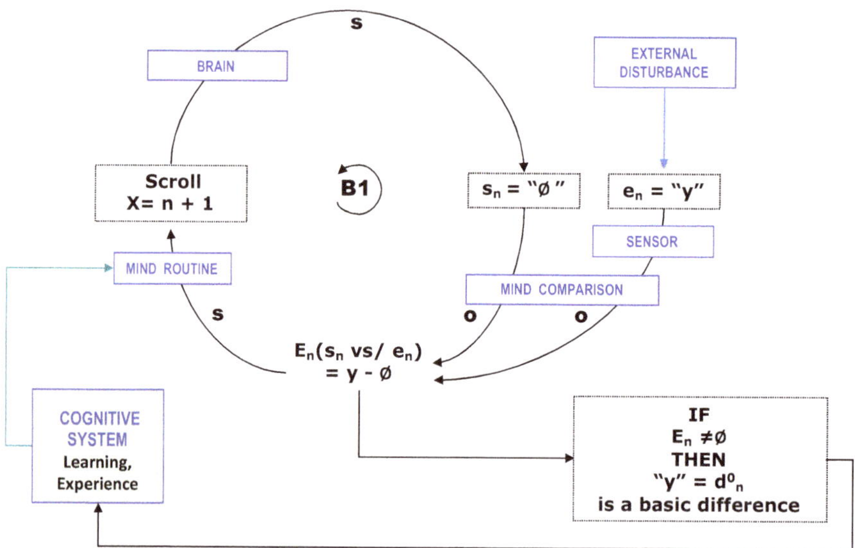

Fig. 12.1 The formation of basic differences through a sensor

The "mind" carries out a basic comparison to determine the primary difference, which I shall indicate as d_n^0 ="y"; the subscript "*n*" indicates that the *Ring* operates during a general phase "*n*" of the process, or at the general instant "*n*" when it perceives the difference d_n^0 ="y." In other words, the mind makes a comparison with the "null state"; a basic difference is perceived as the error produced by this comparison. The mind should be able to memorize all the differences, even those which are repeated, and distinguish these from the null state; for example, d_1^0 ="j," d_2^0 ="k," d_3^0 ="w," d_4^0 ="k," d_5^0 ="j," d_6^0 ="w," etc.

Obviously, if the cognitive system had numerous *sensors* of the same type, each capable of distinguishing similar differences supposed to exist at the same instant (hundreds of contemporaneous perceptions by means of hundreds of spinal hairs on an insect), or had different types of sensors that could distinguish specific differences (visual, auditory, in terms of flavor, etc.), we would have to assume an equal number of *Rings*, similar to those in Fig. 12.1, acting contemporaneously. Adopting "human" terminology, we could say that the cognitive system "sees" light, "tastes" a flavor, "hears" a sound, etc., when it detects a difference with respect to a "null" state.

Figure 12.2 illustrates the *first-level differences* from a comparison of *basic differences* held to be comparable by the cognitive system; for example, various intensities of light, differences in flavor, color, sound, etc. For Bateson, these first-level differences at instant "*n*," which I indicate as d_n^1, can be thought of as a transformation

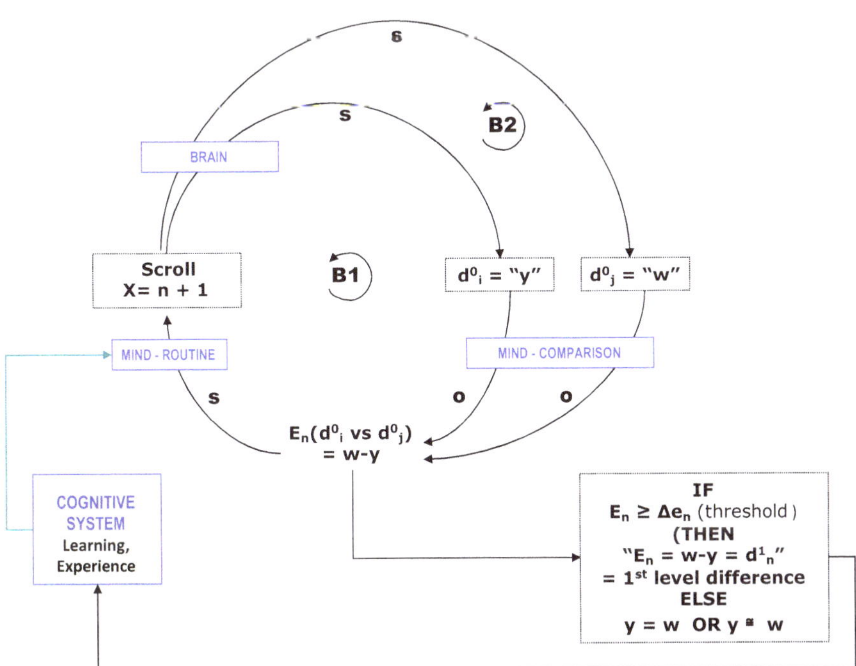

Fig. 12.2 The formation of first-level differences with a sensor

of "zero-level" differences, when the error arising from the comparison exceeds a tolerance threshold, ΔE_n, as shown in Fig. 12.2.

I hold that it is not possible to compare basic differences that derive from different types of sensors: sounds with colors, temperatures with flavors, etc. Colors, flavors, smells, tactile sensations, and so on (expressed in our human language) are examples of basic differences; the color stimuli cannot be compared to flavor stimuli, even if the mind recognizes two stimuli as two colors and two stimuli as two flavors. If the cognitive system has different sensors and *memory*, it can activate a large number of *Rings* to provide the "mind" with a "catalogue" of differences of the same level and an "experience" of perceiving differences. By increasing the level of comparison regarding comparable differences, the mind of the cognitive system can create a "vast catalogue" of differences of various levels which, according to Bateson, can be considered as the result of the transformations of differences of previous levels through the detection of error by the *Rings* undertaking the comparisons.

Figure 12.2 also illustrates the notion of "equality," which cannot be directly perceived but derives instead from the absence of differences ($E_n = 0$) between differences of the same level, so that there is no significance in identifying differences of a subsequent level. We can all verify this by comparing, for example, two flavors, colors, or sounds; if we do not find any differences we conclude that those flavors, colors, and sound are "equal." If there are minimal differences (below the tolerance threshold), those flavors and sounds are judged to be "entirely similar."

In order to understand how the mind can detect differences at higher levels, it is useful to turn to the idea of *dimensional distance*, which for the mind represents a "*step*" between similar perceptions. If several differences of order zero have been perceived, then a comparison of these can allow different "steps" to be identified which form a dimensional scale of distance. In this way various dimensional categories are formed. The identification of distances can be repeated by in turn considering the distances as perceptions. Thus, higher-order differences are formed, as illustrated, for example, in the following Table 12.1.

12.3 Bateson's Model Expanded: The "Mind" as a Calculator of Similarity and Analogy: The Construction of Dimensions

Even if Bateson does not touch on this, I observe that thanks to the *mind's* capacity to compare differences the *cognitive system* can, in addition to detecting higher-level differences or ascertain equalities, carry out two further operations: detecting similarity and analogy.

Two primary differences, "*h*" and "*k*" (etc.), though different, can be considered *similar* if they derive from the same sensor organ at a different moment and under different conditions and cannot be compared through primary differences, "*v*" and

Table 12.1 Higher-order differences

Perceptions	Dimensions	Distance	Variation	ETC
Differences of order zero: d_n^0	Differences of order one: Δ_1	Differences of order two: Δ_2	Differences of order three: Δ_3	Δ_k
Yellow	Color	Chromatic scale	Variations in tone	
Red				
Green, etc.				
Sour	Flavor	Scale of flavors	Intensity	
Sweet				
Bitter, etc.				
High	Height	Scale of heights	Variations in height	
Higher				
Higher still,				
etc.				

"*w*" (etc.), derived from other sensor organs. For example, [d_n = "green" ≠ d_m = "yellow"] and [d_i = "sweet" ≠ d_j = "savory"] but [d_n and d_m] cannot be compared to [d_i and d_j], since they derive from the action of different sensors; nevertheless, d_n and d_m are differences which are similar to each other, in the same way that d_i and d_j are similar. I shall define as basic dimensions: D_A, D_B, … D_M …, etc., the set of all primary differences, $\{d_{A\cdot n}\}$, $\{d_{B\cdot n}\}$, … $\{d_{M\cdot n}\}$, …, etc., the "mind considers *similar* because they can be compared.

By generalizing, we can assume that after having perceived the primary differences the mind manages, thanks to its *memory*, to order these as distinct dimensional elements. If we assume (again using our human minds) that the mind perceives the primary differences d_r = "blue" and d_s = "humid," it can conclude that:

$$d_r = \text{"blue"} \in D_{\text{COLOR}}$$
$$d_s = \text{"humid"} \in D_{\text{TACTILE}},$$

and that d_r and d_s cannot be compared. Secondly, we can assume that when it perceives a difference, for example, d_n = "dark blue," the mind can immediately determine that "dark blue" belongs to the dimension D_{COLOR}. The mind forms the dimensions as *scales of differences* held to be *similar*. A third hypothesis is to assume that once it has identified the difference $d_{\text{COLOR } q}$ = "dark blue," the mind is also able to compare this difference with other differences contained in D_{COLOR} by identifying second-level differences (light blue, navy blue, grayish, etc.) included in second-level dimensions, such as $D_{\text{SCALES OF BLUE}}$, $D_{\text{SCALES OF RED}}$, etc. (Table 12.1). Only through *analogies* can the mind create concepts, and thus develop knowledge. We can recognize the capacity to identify *differences* and construct *analogies* in a mental experiment proposed by Bateson

> Let me invite you to a psychological experience, if only to demonstrate the frailty of the human computer. First note that differences in texture are different (a) from differences in colour. Now note that differences in size are different (b) from differences in shape.

> Similarly ratios are different (c) from subtractive differences. Now let me invite you… to define the differences between "different (a)", "different (b)", and "different (c)" in the above paragraph (Bateson 2000, pp. 463–464).

The experiment proposed by Bateson refers to differences that can be perceived by different sensors, but it is impossible not to note how many times we apply the magic *Rings* in Figs. 12.1 and 12.2. The capacity of the cognitive system to calculate and compare differences obviously depends on the sensors which are available. Consider how reality has been expanded due to the invention of the microscope or telescope, instruments that, as we know, have increased the capacity of the eye to perceive differences. In my view, the two Rings shown in those figures can also be valid for much more basic cognitive systems.

Clearly these basic operations are possible only if we assume that the *structure of the cognitive system* can set off a large number of superfast *Rings* (magic) operating, for example, as indicated in the model in Fig. 12.1 or 12.2.

Once Bateson had affirmed his epistemological principle—the "mind" constructs knowledge through differences—he did not produce an operational-logical process to derive concepts, ideas, and meanings from differences (of differences).

Strictly following the direction outlined by Bateson, I propose three steps:

1. Construct *descriptions* of objects and *definitions* of concepts only by making use of "primitive" operations involving *comparison*, the *identification* of differences, and *analogy*;
2. Define the process of *denomination* through which the "mind" manages to represent objects (descriptions) and concepts (definitions) through *signs* (descriptions of signs) and *signifiers* (definitions of signs), thereby forming *semes*.
3. To the concepts of *description, definition* and *denomination* to operationally deal with the problem of *truth* as correspondence, making use of a reliable *process of determination*.

12.4 The First Step Toward Knowledge: The Process of the *Description* of Objects

Let me start by saying that in Bateson's model, the mind does not *observe* objects but compose them through operations involving the perception of differences and the search for similarities. A certain number of *differences* at various *levels*, even recognized as belonging to different dimensions—which the mind, through *memory*, recognizes as being associated in some way or deriving from the same source—represent the *description* of an "object." As intelligent human cognitive systems, when we observe "the pen on the desk," "the neighbor's house," "the just-purchased candy," "our aunt," we have no doubt we are "observing" an "object"—possessing color, weight, hardness, flavor—even if we in fact perceive "sensorial stimuli" (visual, tactile, auditory, etc.) which our mind "associates" in a unitary system: "the pen," "the house," "the candy," and "the person." I have written "the pen on the desk" in quotes to indicate reference to a particular observed item; that is, a particular pen. The need to use language to describe the observed operations entails the need as well to distinguish between "words" to indicate the observed objects and the concepts (Sect. 12.6). Nevertheless, we are accustomed to also considering as "objects" a window, a hand, a road, an intersection, even if it is difficult to perceive the weight of a window, the height of a hand, the flavor of a road, or the color of an intersection.

The mind carries out the same basic activities for "objects" that it does with perceptions: it perceives the objects, already present at an unconscious level, and compares them to uncover differences or equalities. This comparison among objects and the search for differences or equalities among objects represents the *base mental operations* of conscious cognitive activity. The mind forms at an unconscious level the observed dimensions (with any eventual dimensional distances) as well as the objects of observation. The dimensions and the objects represent the fundamental elements of conscious observation for the mind.

Using a simple formal notation, in order to operationally translate Bateson's difference-based epistemology, "describing" an object "A" means:

(a) Identifying or choosing a convenient number N of *dimensions*—or *scales of differences*—D_1, D_2, ..., D_n, ..., D_N, that denominate *observable dimensions*, which depend on the structure of the "mind-processor." The ordered set of observable dimensions forms the *observed universe* (UN) adopted by the mind for its descriptions:

$$\mathrm{UN} = \left[D_1, D_2, \ldots, D_n, \ldots, D_N\right]. \tag{12.1}$$

(b) Identifying the differences observed in (associated with) object "A"—and which constitute "A"—by specifying the state, $d_n(\mathrm{A})$, that each D_n assumes in object "A," obtained through a precise process of *qualitative* or *quantitative determination* (according to the limits of the "mind-processor"). The vector of the determinations (differences) $d_n(\mathrm{A})$ for each $D_n \subset \mathrm{UN}$ forms the *technical description* of "A," which I shall represent with the following notation:

$$\left[\text{des A}\right] = \left[d_1(\mathrm{A}), d_2(\mathrm{A}), \ldots, d_n(\mathrm{A}), \ldots, d_N(\mathrm{A})\right] \tag{12.2}$$

We must assume that each object "A"—for the "mind-processor" of a given cognitive system—corresponds to a *description* assuming the form in (12.2) achieved through the differentiation process indicated in (12.1). We can apply (12.2) for the technical descriptions of some of the following objects:

boy **b o y** BoY BObO BoA BOa B **O** **a** boA A **a** B *B* a

Let us assume the sensors allow us to construct the following UN:

UN = [D_1 = "letter," D_2 = "up/low case," D_3 = "color," D_4 = "font," D_5 = "size," D_6 = "distance from next object," D_7 = "is part of a word"].

The descriptions of some of the objects are represented by the following vectors:

[des **b**] = [b, lowercase, black, times, 10 pt., none, "is" part of a word],
[des **b**] = [b, lowercase, black, Arial black, 10 pt., 3 pt., "may be part" of a word],
[des B] = [b, uppercase, red, times, 10 pt., none, "may be" part of a word],

[des **a**] = [a, lowercase, red, Arial, 20 pt., 3 pt., is not part of a word], etc.

The *definition* process is always *relative*, since it depends on the *number* of dimensions considered in UN and the *precision* of the specification of their state, which the "mind" can produce through some process of qualitative or quantitative *determination*.

The *descriptions* can be the following:

- *Punctual*, if they concern a single object, or *general*, if they concern the dimensions common to the objects of a given set.
- *Static*, if the variables are observed at an instant in time, or independent of the time.
- Variable, or *dynamic*, if they also concern the variation in the state of the variables with respect to another variable that the "mind" conceives as "time"; the *dynamic descriptions* assume the search for differences among the differences observed at different moments, which in turn assumes the use of memory.

The *dynamic punctual technical description*, with reference to period *T*, can assume the following structure:

$$\left[\text{des A}(t)\right] = \left[d_1(\text{A},t), d_2(\text{A},t), \ldots, d_n(\text{A},t), \ldots, d_N(\text{A},t)\right], t \text{ in} T. \quad (12.3)$$

I have defined (12.2) and (12.3) as *technical descriptions* since they are independent of any linguistic representation; they are the result of a "mental process." However, they can be translated into *linguistic* descriptions using a chosen language (Sect. 12.10). Without entering into overly complex cognitive problems, but only in order to justify Bateson's idea of mind, I would note that for humans [des A] seems to derive directly from object "A." However, this is true only for objects we can perceive by sight. Try to identify an object in the dark using the many differences perceived through touch; or to identify food by its taste with your eyes covered, or aroma and the many differences perceived with the sensors on the tongue. You will realize the object is not the source of your perceptions but instead derives from these, since the mind "associates" them by forming a perceptual unit:

> We create the world that we perceive, not because there is no reality outside our heads, but because we select and edit the reality we see to conform to our beliefs about what sort of world we live in (Bateson 2000, p. vii).

Some further considerations will clarify the *human* "potency" of the mental operation of constructing *technical descriptions*, even if what we have learned during our lifetimes and through the use of language oblige us to make an effort not to think directly of the objects but to concentrate on their dimensions, D_n, and their states, d_n.

Above all, I would observe that we are used to defining *dimension* (connotation, characteristic, parameter, or way an object manifests itself) as any *association of differences* judged to be similar, which the mind can *perceive* in defined and given spatial–temporal contexts and which, as humans, we associate with "objects." Operationally speaking, the *dimensions* can thus be viewed as *generalized modes of perception* (weight, color, flavor, etc.) or of *association* (name, relative position, etc.) regarding certain *differences* (perceivable, defined, or imaginary) that a given observer judges to belong to the same species even though he places them in different objects. These dimensions are not "observable" but are "constructed" by the mind in order to compare various objects in relation to the states taken on by the dimensions created for purposes of comparison. Any further calculation of differences regarding the objects absolutely requires that the mind be able to determine in some way the state taken on by the dimensions "in" the various objects.

I shall define as *determination* the mental operation by which the mind, in making comparisons or employing first-level dimensional states which have already been *memorized*, specifies the state of each dimension of the technical description of an object in a given observable universe. To make this aspect even clearer, let us look at the example in Fig. 12.3, which, unfortunately, must necessarily refer to the human mind.

Through the operation of sensorial (sight, touch, etc.) and mental (memory, concentration) "organs," based on given modalities which we shall not examine here, the mind is able to recognize Fig. 12.3 in its "entirety" as a *unitary object* through the perception and arrangement of the differences found (traces of ink of various colors) in comparison to the uniformity present (blank white sheet of paper). The mind can therefore consider the *unitary object* (i.e., the overall figure) as a composite object (Sect. 12.5) and *isolate* in it a variety of more *circumscribed* objects; it can focus attention on the hatched box at the bottom, which, observing its relative (dif-

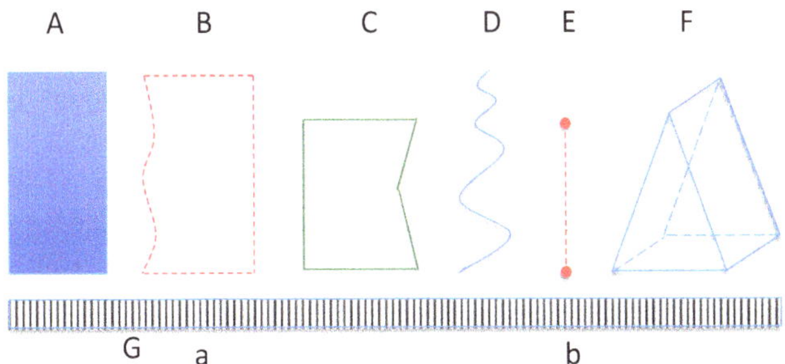

Fig. 12.3 Differences and dimensions

ference) position can be interpreted as a "plane of reference." The other figures "rising up" from this plane of reference can then be observed, in addition to the symbols we recognize as letters of the alphabet.

Concentrating our observation on the figures "rising up" from the base, the mind can perceive differences in each with respect to the others; for example, all are linked to other figures (which we recognize as capital letters) placed above them, and each figure is "erected" differently with respect to the plane of reference and "rests" on it in a different manner. We can thus clearly grasp the process involving the "construction" of the *dimensions*, understood as generalizations of the perception of differences which are distinct yet held to be of the same type. Through extrapolation the mind generalizes that the "name" represents a characteristic common to all the figures, and that for each of these this dimension presents a particular "state"; in the first figure on the left (according to our human perspective), the dimension takes on "state" A, in the first on the right, "state" F, and so on.

Perceiving in all the figures a different "*upward*" movement with respect to the designation G (plane of reference) leads us to generalize the "*height dimension*" of those figures; perceiving a difference in the area on which the figures rest on the plane of reference leads to the generalization "*width dimension*"; through association with other types of differences the mind perceives in all the specific figures, which it considers to be homogeneous, the mind makes other generalizations regarding the dimensions, for example, "color," "shape," "perimeter," "surface" (total, lateral, base), "position," and so on, according to the mind's ability to find similarities among the dimensional states perceived in the particular figures. To conclude this example, I would maintain that in Bateson's model the *dimensions* are created through the mental process involving the *generalization of* specific *differences* perceived in the observed objects, which are considered similar with respect to a common and unvarying *point of reference*. Thus, this example allows us to conclude that the relation [states → dimensions → objects] can be inverted: [objects → states → dimensions], without getting away from Bateson's logic. I do not think the idea would have "entered into the mind" of any human observer (with vast experience) of also associating with the specific objects in Fig. 12.3 the dimensions of *weight, specific weight*, or *brand of computer with which the figures were traced*. In fact, in Fig. 12.3 none of these dimensions manifest particular "perceivable states" that, through generalization, allow one to create those dimensions. Perhaps only the figure "designated" as F, which is observed as being three-dimensional, could have a "state" denominated as "weight" and one as "volume." The figures "denominated" D and E manifest the dimension "length," but not that of "surface," which instead characterizes the figures denominated A, B, and C. On the other hand, a mind could attribute to the figure "denominated" G the dimension "length," as proof that the dimensions are constructed by the mind through the generalization of quantifiable specific differences in the objects. Until now I have ignored the two signs "a" and "b" corresponding to the figures denominated "B" and "E," below the figure denominated "G." In which dimensions can they be compared? Perhaps the mind, with the obvious dimension *capital letter/lowercase letter* excluded, could consider "a" and "b" as "denominations" for only the two objects whose state is "dotted" in the *type of line* dimension. However, I do not exclude the possibility some minds might compare these using other dimensions.

The above example shows, even if the distinction does not possess absolute validity, the possibility of considering separately the *observed* dimensions with respect to those *associated* with a particular object of observation. The dimensions *observed* represent differences the human observer feels can be detected in the object: color, weight, volume, function, shape, flavor, material of construction, composition, structure, observation date, etc. The dimensions *associated* with objects, on the other hand, are determined by the observer himself, if necessary by developing dimensions from another category. The relative position of the object with respect to the observer, the name, specific weight, average weight, prevalent color, code number, name of builder, date of construction, weight per square decimeter of the base, and so on, represent examples of associable dimensions. The distinction is relative and it is not always easy to allocate the dimensions to one or the other category. If we consider that all observations are extremely subjective, then all the dimensions are relative to the mind of the observer, and for this reason can be considered *associated* with the object. If, on the other hand, we consider the object as a unit that is directly describable, then not only would the dimensions held to "naturally" emerge from the object's existence be *detectable* but so too would extension, mass, electrical charge, as well as any difference inherent in the object, without which the object "would not be" observable. An object "has no" weight until a mind produces it by calculating a difference within a given observational universe (12.1).

The above considerations imply that, at least *in man's case*: (1) the technical descriptions of the objects of observation depend on the number, type, and value (state) of the dimensions characterizing them; (2) there is no limit to the number of dimensions by which objects can be described; this number does not depend on the nature of the object but on the number of sensors and the experience (objectives, education, limits) of the observer; (3) to make the technical descriptions operationally efficient, we must *limit* the number of dimensions taken into account for the actual [des A] in (12.2).

12.5 Comparing Objects

When the "mind" applies the process for recognizing the *differences of differences* to the technical descriptions, it is then able to distinguish between objects that are *equal* or *different.*

I define the *equality* of "objects" by writing:

$$\begin{aligned}&\mathrm{A}=\mathrm{B}\ \text{in}\ [\mathrm{UN}]\ \text{if, for each}\ D_n\subset[\mathrm{UN}],\ \text{we have}: d_n(\mathrm{A})=d_n(\mathrm{B});\\&\text{that is, if}\ [\text{des A}]=[\text{des B}].\end{aligned}\tag{12.4}$$

I define the *difference* of objects by writing:

$$\begin{aligned} &\text{A} \neq \text{B in } \left[\text{UN}\right] \text{ if, } \textit{for at least one } D_n \text{, we have } d_n\left(\text{A}\right) \neq d_n\left(\text{B}\right); \\ &\text{that is, if } \left[\text{des A}\right] \neq \left[\text{des B}\right] \end{aligned} \tag{12.5}$$

Comparing A and B through their technical descriptions, [des A] and [des B], assumes the use of *Rings* operating over many cycles and at high frequency, similar to those shown in Fig. 12.4.

It is important to observe that *equality* or *difference* in objects can be affirmed only after we have defined the *observable universe*, [UN]. Many errors in observation occur because a [UN] of reference is not specified.

Comparing the *technical descriptions* of objects also leads to distinguishing between *simple and composite objects.* An object is *simple* if, for the "mind" and/or for a specific observation, it is held to by *unitary*; that is, it cannot be broken up into other elementary objects, B, C, etc., so that there is no sense in holding that [des A] can be broken up into autonomous [des B], [des C], etc. An object A is *composite* if it can be considered to be composed of (or broken up into) autonomous components (parts), B, C, etc., which can be independently described; in other words, if it makes sense to obtain [des A] = [des B] + [des C].

The mind also distinguishes between *separate* and *united* objects. The objects A and B are *separate* if, *for each* $D_n \subset$ [UN], [des A] can be formed through determinations which are independent of those necessary to form [des B]. If there is a D_n such that we can observe only d_n(A&B), and it is not possible to observe d_n(A)

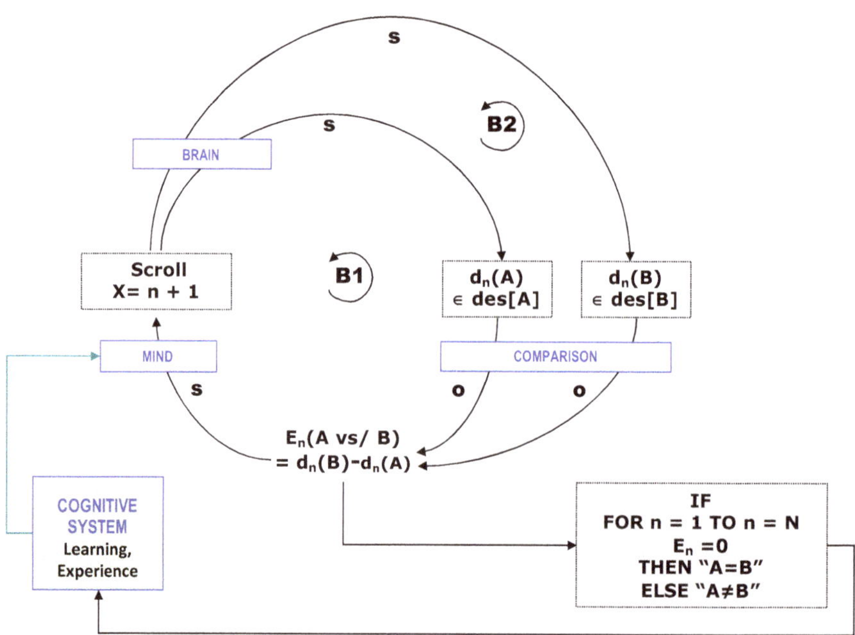

Fig. 12.4 A comparison of objects through their technical description

separately from d_n(B), then the objects A and B are *united* in (for) that dimension. Arm, hand, and fingers are *united* objects but hand, wedding ring, and watch are *separate* objects.

Two objects, A&B, though *united* in terms of the values of the dimension D_n, can be *separable* if, through some *conventional procedure*, we can determine the values d_n(A) and d_n(B) and include these separately in [des A] and [des B]. If there is at least one D_n which does not admit any conventional procedure to independently determine the states d_n(A) and d_n(B), then the A&B are *not separable* in the absolute sense. If a pair of oxen are pulling a cart it is not possible to observe as two objects the individual oxen each pulling half a cart. Two *separable* objects, A and B, can be *united* to form the object A&B if it is possible to form [des A&B] so that for each "*n*": $d_n(\text{A} \,\&\, \text{B}) = d_n(\text{A}) + d_n(\text{B})$. Two *united* objects, A and B, are *autonomous* if they can be described by different dimensions, leaving out only those dimensions for which they prove to be united. A finger is an object we all recognize as *autonomous* with respect to the palm of the hand, even if it is *united* to it, because we believe the *finger* and *palm* to be different in structure and function and consider it irrelevant to also include in the two *technical descriptions* the color and weight of both the finger and palm. A tree trunk can be independently described with respect to its foliage and roots, even if it is united to these.

The objects which can only be *described* through *differences* detected with the sensors (detected states of dimensions) can be defined as *real objects* for the observer. Referring to the "human mind," the use of our *associative capacities* (Sect. 12.2) often leads us to include in [des O] dimensional states which are not observable or not pertinent, thereby creating imaginary objects deriving from the mental activity that, using new terms or based on unusual forms, "associates" values (differences) of dimensions *not observable* in those objects, as occurs, for example, when we associate a color to an electron, describe a three-headed dog, a unicorn, a winged-steed, or a flying carpet. As they cannot be observed, *imaginary objects*, though they can be *enunciated*, can also be called *nominal*. Some of the *imaginary objects* which observation has not yet shown to exist, but neither demonstrated do not exist, can be defined as *hypothetical* (e.g., antimatter, mathematical "strings," soul, etc.).

12.6 The Second Step Toward Knowledge: The Process of *Definition*: From Objects to Concepts

Technical descriptions represent the first step in acquiring knowledge. The second step is that by which, through *analogy*, the "mind" transforms the "technical descriptions" of objects into "technical definitions" from which we form *concepts* (ideas) of objects. A "technical definition" is a *"general" description* that specifies the *limits* of variability in the "technical descriptions" of *objects* in order that these can be considered *analogous;* that is, all belonging to the same *open set* represented by the definition (genus, species, type, class, etc.).

The *process of defining concepts* thus implies an *analogical generalization* through which the differences found in various "technical descriptions" of objects are evaluated by the "mind" (calculating differences of differences) in order to "conserve" the *analogies* and to form a *general class of objects* that we can then associate with a *generic object.*

In formal terms, if, for each $D_n \subset$ UN, the symbol $\Delta d_n(\mathrm{O}^*)$ indicates *the range of admissible variation* of the dimensional states in order for an *object* "O" to be part of the *concept* "O*" (idea, intuition, general significance, abstraction, generalization, image, etc.), then the "technical definition" of "O*" is represented by the vector:

$$\left[\text{def O}^*\right]=\left[\Delta d_1\left(\mathrm{O}^*\right),\Delta d_2\left(\mathrm{O}^*\right),\ldots,\Delta d_n\left(\mathrm{O}^*\right),\ldots,\Delta d_N\left(\mathrm{O}^*\right)\right]\subset \mathrm{U}\left(\mathrm{N}\right). \quad (12.6)$$

I can state that for the knowing "mind" the *object* "O" is part of the *concept* "O*" if [des O] ⊆ [def O*]. More simply put, each specific object can be considered an "example" of the "idea" of a "generic object." I describe "*that*" particular chair but I define "chair"; "*that*" chair is an example of "chair."

The *definition* can be formulated using two processes:

1. *Denotative* or *extensive*, if the process identifies the set (denotation or extension) of the descriptions [des O_1], [des O_2], [des O_3], etc., that can be included in [def O*].
2. *Connotative* or *intensive*, if it identifies the *dimension* and *range of variation* within which the *states* d_n(O) identified in the objects (in their descriptions) must be included in order for them to be indicated by the definition (connotative or intensive). The connotative definition is provided in (12.6).

Thus, for example, all the objects denominated A to F in Fig. 12.5 are undoubtedly different from one another, given the technical descriptions the reader can make using (12.2) and considering the following simplified UN:

UN = [D_1="number of sides," D_2="shape of sides," D_3="length of sides," D_4="vastness of area," D_5 = "surface color," D_6="direction of vertex (point)," D_7 = "name," D_7 = "position," etc.]

The generalizing activity could hold, for example, that the figures A, C, and D are *analogous*, since they have the same "number of sides," independent of their

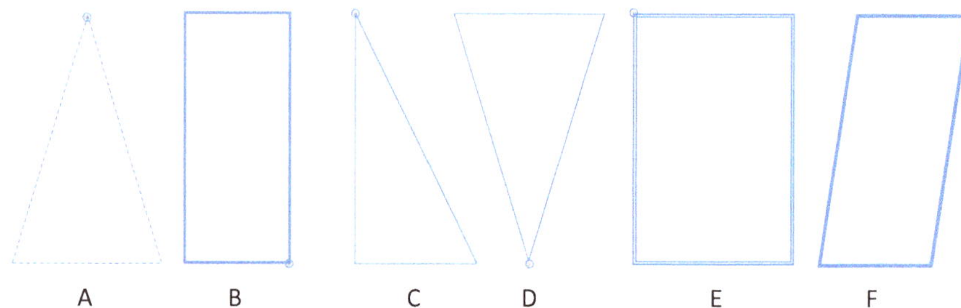

Fig. 12.5 Producing a technical definition

"shape," "height," "breadth," "width," and "surface color." Figures B, E, and F could also be considered *analogous* in terms of "number of sides." However, another mind might have considered as *analogous* those figures whose sides are a certain thickness, independently of the number of sides, while a third mind could have viewed as *analogous* those figures whose vertices (indicated by a solid circle) are pointing north. Analogical activity, being mental, is always subjective.

We can generalize by going back to the example in Fig. 12.5 and keeping Eq. (12.6) in mind: two objects, A^I and A^J, are considered *analogous* if, though their technical descriptions differ, the individual differences in quality and quantity are held to be included within the *range of admissible variation* characterizing "the" object A* in general": that is, the *concept* A* (for the moment we have to resist the urge to use language to denominate A*). The mind can now *generalize* as follows: the objects A, C, and D from the previous example, whose technical descriptions differ, are considered *analogous* because the states (differences) of the variables (scales of differences) from their technical descriptions (association of differences) revealed by the sensors are not "too different" from one another; using *analogical generalization*, the three objects can be thought of as special cases of a general object: "triangle," having the following technical definition (referring always to the shapes in Fig. 12.5):

[def TRIANGLE*] =
=["number of sides" = three and only three (no range of variation)
"shape of sides" = any whatsoever (infinite range of variation)
"length of sides" = any whatsoever, as long as it is positive (unless we are considering figures with negative measurements for the sides with respect to the origin) (infinite range of variation)
"width of area": correlated with the length of the sides
"surface color": any whatsoever
"direction of highlighted vertex": any whatsoever].

We can reverse the process: given the *technical definition* of the general object A*, how can we verify if the specific objects, A^I and A^J, for which the mind has calculated [des A^I] and [des A^J], are similar? The answer is simple: we need to verify that each *value* for the technical descriptions is included in the *admissible range of variation for the existence of the analogy*. Only if the values for all the dimensions, $d_n(A^I)$ and $d_n(A^J)$, of the two objects are included in $\Delta d_n(A^*)$ are the objects *analogous* and can be considered as "examples" ("→" means "example of") of the concept [def A*]. Therefore, indicating analogy with the symbol "≈," we can write the result of the comparison of the similar objects in one of the following ways (Fig. 12.6):

$A^I \approx A^J$, that is $[A^I, A^J] \rightarrow A^*$, that is
[des A^I] ≈ [des A^J], that is
{[des A^I], [des A^J]} ⊂ [def A*], that is
$d_n(A^I) \in \Delta d_n(A^*)$ and $d_n(A^J) \in \Delta d_n(A^*)$, for each "n."

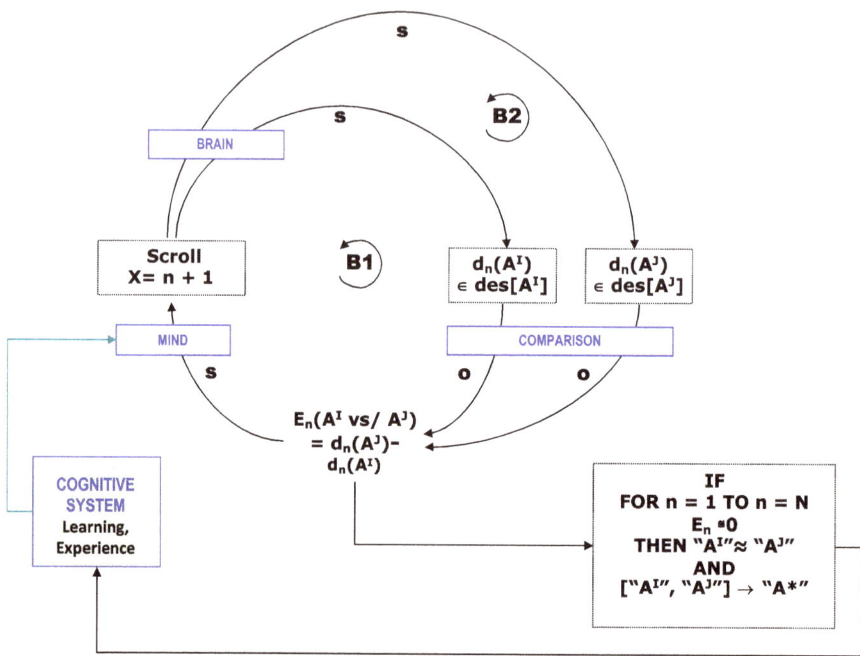

Fig. 12.6 Verifying the similarity of the objects and their belonging to a "concept"

The *general object* "the triangle" is a *conceptual abstraction* connected to the *analogical* and *generalizing* activities of the mind. In fact, the *general object* represents the set (defined intensively) of all specific objects (the individual triangles that can be observed or imagined) for which an analogy exists in the mind of a given cognitive system; that is, the set of specific objects whose technical *description* is included in the technical *definition* of the general object.

Objects are described; *concepts* are defined.

I shall define "a thing" as that which the mind does not define; this is a concept that includes every describable object. The technical definition of "a thing" is so general as to include as particular cases any specific object of observation. Formally, we can write:

$$\left[\text{def THING}^*\right] = \left[\Delta d_i\left(\text{THING}^*\right) = \text{" whatever," for each "}i\text{"}\right]..$$

This concept is in line with the use we make of the concept of "thing" when (through language) we want to designate unrecognized or unrecognizable objects of observation or to indicate any object or concept that is not described or defined, for example: "Some*thing* strange was seen in the sky"; "There are many *things* on the table; bring me all of them"; "The specific weight can be determined for all *things* that have a volume and a weight"; "All the visible and invisible *things* have been created by God," etc.

12.7 The Three Paths of Knowledge in a Two-Dimensional Cognitive Universe

Following Bateson's assumptions, we can now verify how the mind-processor—by means of the four basic mental operations *distinction*, *comparison*, *perception of analogies* and association (Sect. 12.2)—can gain *knowledge* by developing *three minimal cognitive processes* at both the conscious and unconscious levels (Fig. 12.7):

1. *Cognition, or perception, of objects*: adopting the mental processes of *perception* and *comparison*, and employing constitutive memory, the mind forms *descriptions* of objects—[des A], [des B], [des C], etc.—distinguishing these from their *background*.
2. *Understanding*: after having described several objects, the mind uses *analogical abstraction* to develop [def A*], [def B*], [def C*], etc.; that is, it finds in the *observational universe* a *zone* that corresponds to [def A*], a second zone corresponding to [def B*], a third zone corresponding to [def C*], and so on.
3. *Recognition*: in the presence of a specific object "O," the mind carries out the [des O] and determines in which [def O*] [des O] is included. The object "O" then becomes a *point* in [def O*]. If the *recognition* is successful, the mind *recognizes* (*knows*) "object O" as belonging to "concept O*." In the opposite case, it determines a new [def O*].

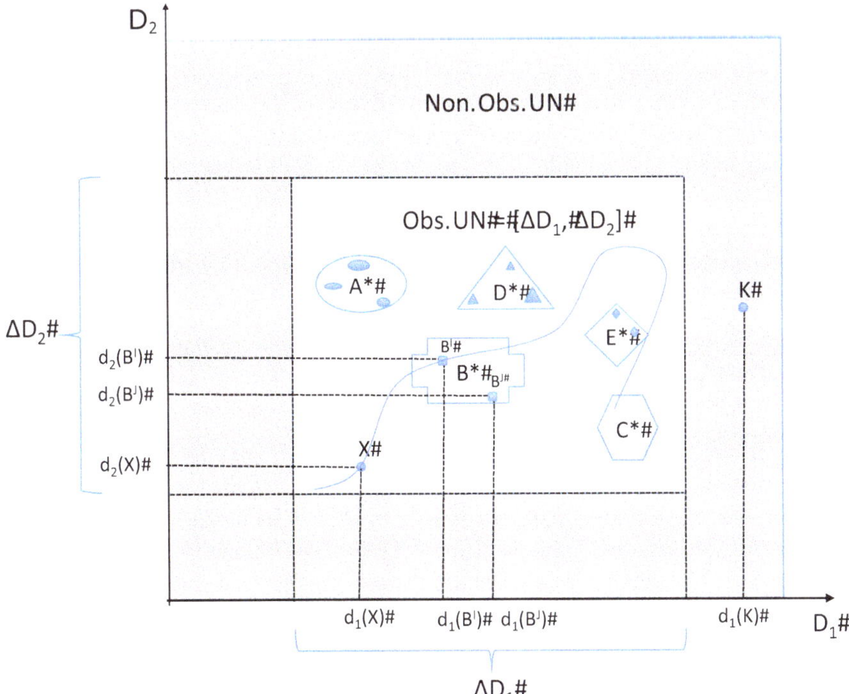

Fig. 12.7 Observable universe, descriptions and definitions

"Knowledge" is acquired through a continuous cycle of "*cognition*," "*understanding*" and "*recognition*," which together form our *minimal thoughts*.

To make the continuous action of the three moments of knowledge clearer, let us consider a "minimum mind" which receives the differences from only sensor organs and thus is only able to distinguish differences belonging to only two dimensions, D_1, D_2, representing a *minimal universe*: UN = [D_1, D_2], shown as the axes in Fig. 12.7. Let us further define ΔD_1 and ΔD_2 as the segments of the two dimensions that form UN, whose values can be distinguished by the sense organs that send the differences to the mind. The area determined by the segments of such dimensions which are knowable to the observer through the available instruments represents the *observable universe*: Obs-UN = [ΔD_1, ΔD_2], which, in Fig. 12.7, corresponds to the white rectangle. The grey area represents the universe which cannot be observed by the mind.

Referring to Fig. 12.7, each point of the Obs-UN represents the technical description of an *observable object*. The areas A*, B*, etc., represent *technical definitions*, that is [def A*], [def B*], etc. An object "X" is the point described by the coordinates d_1(X) and d_2(X), which the mind uses to construct [def X] = [d_1(X), d_2(X)]. Since the object X does not belong to any of the areas indicated in Obs-UN, the mind can describe X but not understand it, since the concept (the technical definition) to represent it is missing. Completely different is the situation regarding objects B^I and B^J, which can be described by the mind but also understood as objects representing examples of the concept [def B*]. This is represented graphically in Fig. 12.7 by points B^I and B^J included in the area-concept B*.

In general, the mind can *know* any object O^H if it describes it in Obs-UN (cognition); it can *understand* it if it detects a similarity with other objects which are "not too different" from it, thus forming the *concept* [def O*] (understanding). When the mind distinguishes another object, O^K, as being similar to O^H, it can *recognize* O^K as belonging to the concept [def O*] (recognition). More generally speaking, any object X *known* (distinguished) by the mind through [des X] is *recognized* if the mind succeeds in including the [des X] in some memorized *technical definition*. It can then recognize that X is A* if [des X] $\subset$ [def A*]; X is B* if [des X] $\subset$ [def B*], etc. This recognition can fail in two cases:

1. When the mind is not able to (adequately) determine the *technical description* of the object, perceiving "things," not objects.
2. When it is not able to determine the *technical definition* that encompasses that description.

The surface of Obs-UN occupied by the *technical definitions* forms the *known universe*; the set of points that corresponds to objects included in the known universe forms the *observed universe*. If [des X] does not belong to any of the definitions in Obs-UN, then a *new definition* arises represented solely by the point indicating the object: [def X*] = [des X]. Knowledge is increased because the known universe is enriched by the new definition. The curved line in Fig. 12.7 represents an ideal *path of knowledge* of a mind which, through its sensors, explores Obs-UN and, after having perceived X, continues its exploration, perceiving B^I and recognizing this as a point in B*, subsequently moving on to point C*, which it can recognize. I would note, even though it may seem obvious, that the "path" shown in Fig. 12.7 is not in a *spatial* territory but in a *mental* one, which is made up of concepts "mapped" in an *observable universe* (Fig. 12.8).

I would propose a mental experiment which is very difficult for adult humans, since our mind has already constructed all its technical definitions, that is, its concepts, and has associated a specific name to each concept. We must make an effort to simulate a mind that explores a given Obs-UN without already possessing any concepts or a language to name them, but which instead creates the concepts using the technical definitions.

Imagine a baby in a carriage led around by his mother for the first time in the fruit and vegetable section of the supermarket. While the mother pushes the carriage around the various food stands the baby's mind observes, as we can imagine, many differences in color and size that he arranges into "objects." There are many round objects (the baby is still unfamiliar with the concept of "sphere") and different colored ones: yellow, red, green, green and red, red and yellow, etc. (adopting the terms we already know). His mind can recognize similarities and he forms the concept (I must express it in words) of *large round fruit* (which to us is an apple). He then observes many other rounded objects, small but only red, with some variants and with stems: considering them to be similar, he includes them in the concept of *small round fruit with loops* (cherries). Other objects are red and small as well, but without stems and heart-shaped (I use this dimension only for purposes of this example), which he joins together under the concept *red heart-shaped fruit* (strawberry).

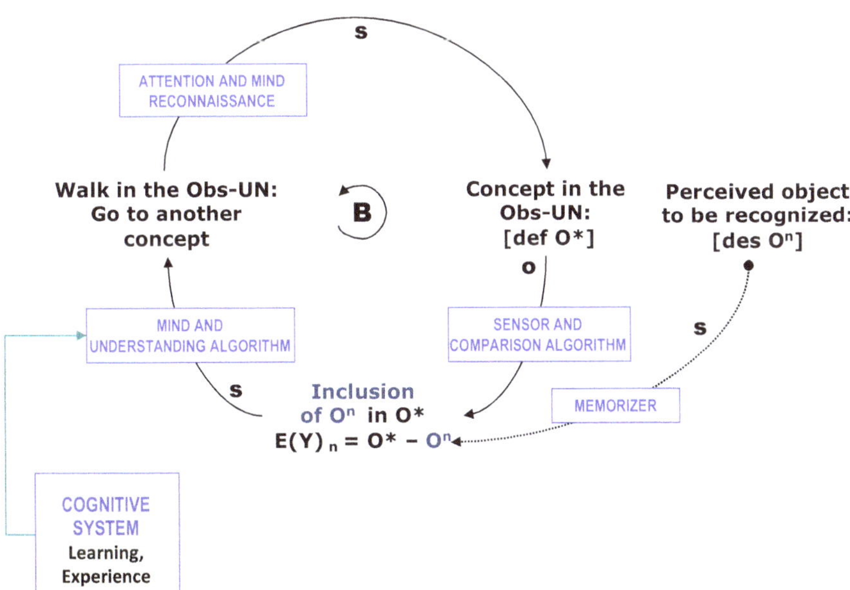

Fig. 12.8 The *Ring* for recognizing and identifying objects

Taking the example further, the long yellow objects are placed under the concept *long yellow fruit* (banana); the pear-shaped ones are *pot-bellied pointed fruit*; the dark colored, hairy cylindrical objects are *hairy fruit* (kiwis), etc. The baby is not aware that the objects he observes can be grouped under the concept *fruit*, but I have used this term to make the example easier to understand. The baby returns home and sees on the table a *long yellow fruit*, several *small round fruit* and some *large round fruit*. Soon after the grandmother (concept) arrives with a bag containing some *small round fruit* and some *pot-bellied pointed fruit*.

The above process is repeated week after week, until the baby's mind is able to "denominate" the objects and the "concepts"; that is, he begins to perceive signs–words, form analogies, for example, between the word "pear" pronounced by his mother, father, grandmother, and sister and various *pot-bellied pointed fruit*. He hears "vocal signals" that are different because pronounced by different people that he, nevertheless, judges similar in designating those objects. Finally, 1 day, again at the supermarket, he points at a *pot-bellied pointed fruit* and his mother says "pear" (pardon my conciseness here). Then he indicates a *large round yellow fruit* and his mother says "apple"; a *large round red fruit*, which the mother again refers to as "apple; a *small round fruit*, which is referred to by the sign "cherry"; the *round heart-shaped* fruit is referred to as "strawberry, while a pointed pot-bellied brown one elicits the sign "pear" from the mother. After a certain number of iterations of the process of perception the baby—thanks to the powerful memory all babies possess—forms his *understanding*; that is, he manages to give a defining content to the technical descriptions of many objects, thereby forming an *observed universe* of concepts to help him recognize various objects of that type.

Once again the grandmother arrives and takes out a *long yellow fruit* from the bag, which the baby recognizes as "banana," saying "*babanana*" as a sign to indicate the concept. His grandmother corrects him: "Not '*babanana*,' "*banana*." The baby recognizes a *small round fruit without a loop* and calls out "cherry." However, the grandmother makes him understand in some way that he has made a mistake. The technical description of that *small round fruit* does not belong to the concept of *cherry* but to that of "cherry tomato." The baby introduces the new concept, thus widening his *observed universe*. The grandmother then takes out a *large round bluish-violet fruit*, causing the baby to exclaim: "apple." Once again the grandmother patiently disrupts the baby's observed universe by saying: "eggplant," thus producing a new concept.

The importance of the *observed* universe is clear. Changing the UN means that the knowledge processes must also be restructured. To show this, let us continue with our example and assume that 1 day the baby *tastes* a slice of apple. The grandmother asks him: "What fruit is that?" The baby cannot avoid becoming aware of the new perception in the "flavor" dimension, but obviously he cannot recognize the fruit since its flavor was not a dimension he used to describe different fruits and to recognize them by means of their respective concepts. Once he has tried several times a sample of all the known fruits (whose definition he possesses) and associated the new states of the "flavor" dimension to the perceived objects, his knowledge is broadened and he is able to recognize a particular fruit either by looking at it or by tasting it. In fact, before tasting them the baby observed the fruits in the three-dimensional universe:

$$\text{UN} = \left[D_1 = \text{" shape," } D_2 = \text{" dimension," } D_3 = \text{" color"} \right].$$

After having distinguished the flavors and with the "flavor" dimension thus added, the universe of observation becomes four-dimensional:

$$\text{UN} = \left[D_1 = \text{" shape," } D_2 = \text{" dimension," } D_3 = \text{" color," } D_4 = \text{" flavor"} \right].$$

This broadening of the UN noticeably increases the capacity to form concepts and "ideas" and, above all, to carry out recognitions. The "idea" of the *general object* (also referred to as *concept, notion, generalization, abstraction, intension*, etc.) corresponds to the object O* of the *analogical generalization* specified by [des O*].The translation of the symbolic structure proposed for the terms *image, configuration, idea*, and *understanding* is summarized in Fig. 12.9.

The broad context in which the object of observation is identified can be referred to as "landscape." A *landscape* is not a composite object since its components are distinctly observed even though connected in terms of belonging to that scenario. Nevertheless, a *landscape* (or scenario) can be considered a composite object whenever the observation involves the entire *landscape* as a unit.

If M objects: O^1, O^2, …, O^M, constitute a *landscape* "L," then we can write:

$$[\text{L}] = \left\{ \left[\text{des O}^1 \right] \cup \left[\text{des O}^2 \right] \cup \ldots \left[\text{des O}^M \right] \right\}. \tag{12.7}$$

If, for all the objects composing a scenario [L], we emphasize the time dimension "duration" (as perceived by humans), defined in meaningful terms, then the entire scenario can also be considered to be observed in the time dimension. This scenario *evolves* through changes in the number and type of objects it is made up of at any moment. I shall refer to "environment," [E], as a scenario in which the duration of observation, *T*, is also defined. Therefore, to an observer every object can be part of a landscape and be observed in an historical, present, or future environment.

MENTAL OPERATION	FORMALIZATION	SYMBOLISM
Image of O	Distinction in the Obs-UN	O
Configuration of O	Technical description	[des O]
Idea, concept, abstraction of O*	Technical definition	[def O*]
Understanding	Meaningful technical definition	[min def O*]

Fig. 12.9 Mental representations of objects of observation

12.8 The Human Mind Constructs Meaningful Technical Definitions

The mind would not be efficient in operating as a calculator of differences if, in addition to the activity of *perception* and *analogy*, it did not also carry out *simplifying* activities. The continual search for "simplicity" (even though this term is subjective) pushes man toward "simplification," without which any generalization would be become extremely challenging in terms of efficient observation. Analogy is efficient only through simplified perceptions, descriptions, and definitions, since the mind tends to ignore anything appearing "superfluous" so as to concentrate on the perception of dimensions thought to be "relevant" for knowledge. Thus, above all in formulating descriptions or definitions, the minds of human beings tend to ignore many dimensions of the observable universe, since their inclusion would make the descriptive and definition process so complex and redundant as to render it inefficient. For this reason, there is an attempt to reduce to a minimum the number of dimensions considered in the descriptive and defining activities in the knowledge process in order to form *minimal* descriptions and definitions, which we shall call "*meaningful descriptions and definitions*" (or *basic* concepts): [min des O] and [min def O*].

Many *meaningful descriptions* and *definitions* even include a single dimension held to be particularly representative; these are called *elementary*. Below are some of these types of definitions, expressed, for simplicity's sake, in a linguistic form (the typology is also valid for descriptions):

(a) *Ostensive* definition: this lets the observer form the analogy by merely indicating several examples of the object of the analogical abstraction ("what is an ant?"; "any insect similar to those you see on the tree!").
(b) *Extensive* definition: this lists "all" the objects (extension) that must be contained in the definition ("what is a sisar?"; "any stellar object whose definition is listed on p. 22 of the Atlantis of the Sky").
(c) *Genetic* definition: this highlights the origins of the objects to be included in the definition ("an American is any individual born in the USA.").
(d) *Historical–geographic definition*: this considers the place and time the objects can be observed.
(e) *Structural* definition: this brings out the structure of the defined objects ("a hand is a limb composed of the following elements …").
(f) *Functional* definition: this brings out the function of the objects in the definition.
(g) *Modal* definition: this indicates the composition of the objects contained in the definition (materials, color, shape, etc.).
(h) *Instrumental* definition: this brings to light the possibilities for using the defined objects.
(i) *Teleological* definition: this considers the objectives of the objects of observation.

(j) *Operational* definition: this specifies the operations needed to identify or recognize the objects of observation in the definition. Operational definitions are particularly effective in defining abstract or composite objects.
(k) *Stipulative* definition: a definition created ad hoc for a specific purpose; for example, a specific communication or argumentation.

The meaningful definitions commonly used in human knowledge derived from combinations of a limited number of the preceding types. Some further considerations will clarify the different types of meaningful technical definitions.

Ostensive definitions are not true definitions, since they do not, in fact, highlight a specific dimension. Through ostension we can merely indicate one or a few objects which exemplify the definition we wish to form, leaving it to the cognitive subject to apply the analogy to the technical descriptions of the objects by identifying the meaningful dimensions. Just as in the example with the baby, ostensive definitions are usually formed as an answer to a question. To the question: "What is a pineapple?" the mother indicates several pineapples on the supermarket stand, letting the baby make the generalization "large rounded fruit with protruding needles," or something similar. As can be deduced, *ostensive definitions* are generally very meaningful, but only for the mind that has created them; they are not easily generalizable since the dimensions chosen are completely subjective.

Genetic definitions do not directly consider the dimensions of shape, weight, color, etc., but instead focus on the moments, causes, factors, and procedures that have given rise to the objects observed. Thus, for example, we could define "bread" as the product obtained from a baker; "pizza" as the product eaten in pizzerias, etc. The genetic definition of a prime number could refer to the procedure for verifying this: "a prime number is any whole number divisible only by 1 or itself." These definitions can be called genetic since they illustrate the way the objects that can be included in the definitions are produced.

Historical–geographic definitions are those that point out the historical or geographical location of the objects that must be included in them. For example, "Eskimos" are "individuals from the extreme north of Alaska, Canada and Greenland." "Crocodiles" are "reptiles that live in rivers, lakes and swamps of Central and South America, mainly in the tropical band." Obviously such definitions are not always meaningful, and for this reason they are associated with other types of definitions.

Much more significant are the *structural* definitions that present the dimensions consisting of the "structural conformation" of the objects of observation; that is, their system of components, parts, organs, essential elements, etc., together with their "organization." The "arm" is "each of the upper limbs of a human, composed of two parts joined by the elbow; one part begins at the shoulder and the other ends at the wrist." This structural definition is somewhat approximate though very significant, since it allows the observer to immediately identify the examples after having made the technical description.

If we wish to extend the definition of "arm" to that of "robot" as well, we can construct the *functional definitions*, which indicate several possible and evident ways the objects of observation function. "Arm" is "a structure, biological or mechanical, with a long and relatively thin shape that extends from a trunk and that, through two articulations joined by an elbow, permits a prehensile mechanism (a hand, a claw, an artificial limb) to move." This definition is very meaningful and general, since it also includes the biological structures of animals or of other mechanisms.

I shall term *modal definitions* as those that include the visual or tactile dimension of the objects or elements in the definition or those characteristics that highlight the differing fundamental modalities. "Lemon" is any yellow fruit with a rind and a sour taste." "Muscat grape wine" is "any sweet, fizzy dessert wine."

The *instrumental definition* is also very significant. This definition is applied to objects of observation instrumental to a particular objective. In fact, this definition brings out the different admissible alternatives, procedures, modalities, etc., regarding the use of the objects. A "chair" is "an instrument used to sit down." This definition does not refer to the place where the objects can be observed (homes, cafes, schools, etc.), their structure (with or without wheels, with an inclinable or rigid back, etc.), or the material they are made of (wood, steel, Plexiglas, etc.), only to their use. In many cases, these definitions are sufficiently significant, in others they must be aided by recourse to other dimensions, for example, structure (the seat must not have arms, otherwise it is an armchair; it must have a backrest, otherwise it is a stool) and height (a chair that is too low is a footstool). It is not always easy to distinguish between *functional definitions* and *instrumental definitions*, since the function of both can be linked to what the object of observation can be used for.

Objects of observation can be systems of various kinds which the mind considers characterized by independent objectives that are achieved by means of teleological behavior. Often objectives are attributed to the objects. *Teleological* definitions are those that aim at emphasizing the objectives that characterize or are attributed to the objects of observation. For example, we can define "enterprises" as organizations existing in a market economy and whose ultimate objective is to create profit and stockholder value through the production or exchange of goods or services.

Meaningful definitions, though they allow us to identify the objects of observation, often do not place the observer in a position to actually search for these objects. *Operational definitions* are those that include elements—procedures, heurisms, algorithms, etc.—that permit the observer to search for and identify the objects of observation. When is a number a "perfect square?"; when this number is obtained by multiplying a whole number by itself." What is a "physical field?"; roughly speaking, it is "the set of values for each point in space of a given physical, scalar or vectoral quantity which is determined through an appropriate procedure and using appropriate instruments."

12.9 The Signification Process: Signs, Denomination, Signifier, Seme, and Sememe

In order to analyze the process by which the *mind* forms knowledge—that is, the *thought process*—it was necessary to express myself through a *language*. However, it should be clear that until now, though having to express myself linguistically, I have described *basic mental operations* deriving from the mind's ability to calculate, compare, and associate differences, assuming the mind has not yet developed a *language* to share its knowledge with other minds; that is, to *communicate* its *thought content* with other cognitive subjects. The *mental knowledge* developed by a cognitive subject is, so to speak, a "private" phenomenon which can be perceived from "outside the mind" only if that subject is able to produce some form of *behavior*, in particular a *meaningful behavior*, to communicate the knowledge it has acquired.

The perception of danger and fear in a dog can be deduced by observing its "tail between its legs"; the pleasure felt by a male peacock upon seeing a female one can be deduced by observing how it displays its tail; the lion's desire to attack a prey can be deduced by its attack position, while the horns pointed toward the lion reveal the mental activity of the buffalo, which is turned to its defense on perceiving the predator. All "higher-order" animals, in particular quadrupeds, cetaceans, fish, birds, etc., even crustaceans and mollusks, have developed some form of signaling behavior to express to other individuals some thought content. I am certain the reader will agree that the "whale's song," that is, the catalog of sounds produced by whales, dolphins, and porpoises, is nothing other than behavior from the emission of sound waves through appropriate effectors in order to communicate to other members of the species. Humans have developed a very powerful communicative form of behavior through the formation and use of numerous *languages* which makes it possible to efficiently communicate the thought content the mind has developed (Sect. 12.10).

The analysis of the structure and function of languages goes beyond the scope of this book. In order to complete this brief "operational translation" of Bateson's model, it is nevertheless useful to at least examine the *denomination* process through which the "mind" associates to its *mental content* (i.e., the *objects of observation, descriptions* and *definitions*) other *conventional objects of communication*: the intentional "signs" produced through its own behavior (speaking, gesturing, writing, pressing keys, etc.), using these "signs" to then communicate to other minds its own "private" *mental objects* by transmitting the signs in the form of *signals* emitted by some specialized effector.

In short, the *process of communicating* thought content, T_A, from a cognitive subject A (transmitting subject) to subject B (recipient) assumes that:

1. A associates to T_A the sign ST_A it produces; that is, it *denominates* T_A by the sign ST_A.
2. A transforms ST_A into a signal, $\sigma T_{A,}$ obtained from one of its effectors (indicating with the hand, transmitting sound, etc.).

3. The signal σT_A reaches B.
4. B, through σT_A, recognizes the sign ST_A that A intended to transmit.
5. B, through ST_A, forms in its own mind the thought content *denominated* by ST_B.
6. $ST_B = ST_A$.

Condition (6) states that the communication is successful when B "recreates in its mind" the same thought content that A intended to transmit to it. In the simplest cases the success of the communication can be perceived both by A (transmitter), evaluating the behavior of B (appropriate response by B to A's question; appropriate execution by B of A's order, etc.), and by B (recipient), evaluating the correct effects from its own behavior, which is carried out by interpreting ST_A.

This interpretation of the communication process, as an act involving the transmission of thought content by the transmitter (in particular *technical descriptions* and definitions), through "signs" that allow the mind of the receiver to reconstruct the former's thought content, is perfectly in line with the autopoietic perspective of cognitive systems that produce linguistic interactions:

> Due to the nature of the cognitive process and the function of the linguistic interactions, we cannot say anything about that which is independent of us and with which we cannot interact; to do that would imply a description and a description as a mode of conduct represents only relations given in interactions. [...] it follows that reality as a universe of independent entities about which we can talk is, necessarily, a fiction of the purely descriptive domain, and that we should in fact apply the notion of reality to this very domain of descriptions in which we, the describing system, interact with our descriptions as if with independent entities. This change in the notion of reality must be properly understood. We are used to talking about reality orienting each other through linguistic interactions to what we deem are sensory experiences of concrete entities, but which have turned out to be, as are thoughts and descriptions, states of relative activity between neurons that generate new descriptions [...] we recognize that we, as thinking systems, live in a domain of descriptions, as has already been indicated by Berkeley, and that through descriptions we can indefinitely increase the complexity of our cognitive domain (Maturana and Varela 1980, pp. 52–53).

It is evident that the *denomination* process is fundamental for all communication. Recalling Bateson's motto: "The map is not the territory, and the name is not the thing named," and also Saussure, for whom "The linguistic sign unites, not a thing and a name, but a concept and a sound-image" (Saussure 1916, p. 66), I shall define *denomination* as the process by which a *sign* (a conventionally accepted name, in particular) is assigned to a concept:

> Some people regard language, when reduced to its elements, as a naming-process only—a list of words, each corresponding to the thing that it names. ... This conception is open to criticism at several points. It assumes that ready-made ideas exist before words; it does not tell us whether a name is vocal or psychological in nature ...; finally, it lets us assume that the linking of a name and a thing is a very simple operation—an assumption that is anything but true. But this is a rather naive approach [...] The linguistic sign unites, not a thing and a name, but a concept and a sound-image. The latter is not the material sound, a purely physical thing, but the psychological imprint of the sound (Saussure 1916, pp. 65, 66).

> The bond between the signifier and the signified is arbitrary. Since I mean by sign the whole that results from the associating of the signifier with the signified, I can simply say: *the linguistic sign is arbitrary* (ibidem, p. 67).

Only when the baby's grandmother in the example in Sect. 12.7 pronounces the vocal sign "pear" by pointing at several *pears* does she (ostensibly) *denominate* using this vocal sign "pear," a fruit the baby has mentally observed; that is, which he has described several times, thereby forming a concept. When the baby wants a pear (*definition*) as fruit he will use the sign "pear" he has heard, *trying to use the correct pronunciation*. Observing his mother's smile, he will note her "sign" of joy in hearing her son denominate that type of fruit, and the child will thus be convinced that using the vocal sign "pear" is very effective for transmitting to others his thought content (*definition*) regarding the nature of the desired fruit.

Signs can be used in three types of *denomination*:

(a) *Proper denomination* matches a *sign* to a *technical description* of a single object "O," and that sign becomes the *proper name* of the *described* object, the only one which can be denoted by that description and which represents the *signified* of S:

$$\textit{Proper denomination} \text{ of } \left[\text{S denoting"O"}\right] = \left[\text{des O}\right]. \tag{12.8}$$

(b) *Intensive common denomination* matches a sign to a *technical definition*. That sign becomes the *common name* for all those objects, denotable by that definition, which constitute the *signified* of S:

$$\textit{Intensive common denomination} \text{ of } \left[\text{S denoting } \text{O}^*\right] = \left[\text{def } \text{O}^*\right]. \tag{12.9}$$

(c) *Extensive common denomination*, which explicitly indicates all objects that can be denominated by S:

$$\textit{Extensive common denomination of} \left[\text{S denoting } \text{O}^*\right] = \left[\text{des } \text{O}_1, \ldots \text{des } \text{O}_N\right]. \tag{12.10}$$

The objects which a sign S can denominate and indicate—individual objects or concepts—form the "meaning of S":

$$\textit{Meaning of} \left[\text{S denominating O,or } \text{O}^*\right] = \left[\text{des O}\right] \text{ or } \left[\text{def } \text{O}^*\right]. \tag{12.11}$$

The *meaning* is conventional for a group (or social context) and refers not to the *objects indicated* but to the *indicator signs*. Imagine the baby's surprise when, during a trip to Turkey, he asks for a pear using the vocal sign "pear," but without being understood.

When I pronounce the vocal sign "university" it is totally different from the one pronounced by my daughter, my grandchild, my neighbor, or any of the readers. We all use different inflections, tones, and we all have different pitches, length of pronunciation, expressions, being able to pronounce the vocal sign either by smiling or crying, with clenched teeth or an open mouth, with surprise or anger, etc. And yet our mind, when considering the different signs for "university," as *objects* that have been perceived several times, compares them and quickly finds the similarity by creating the *technical definition* of the objects–signs–vowels: "university." The same process occurs reading the graphical signs (the underlined words are originally colored, while they appear to the reader as black and white):

University, UNIVERSITY, University, **university**, *University*, university, **University**, UNIVERSITY etc.

These are all different *signs*, but our mind (due to our basic school instruction) has soon become able to form a *generalizing analogical abstraction* in which, and despite the evident differences in color and font, all the graphical signs contain the same letters and are associated with the same content we are all familiar with.

The set of all "signs" having the same *signified* mental object (description or definition) is the *signifier of the sign* and corresponds to the *technical definition* of the sign:

$$\textit{Signifier of the sign}\left[S^*\right] = \left[\text{def } S^*\right]. \tag{12.12}$$

As a result, every sign which is part of a *signifier* is appropriate for indicating any of the components of the corresponding *signified*; likewise, any element of a *signified* can be indicated by any sign of a corresponding *signifier.* The correspondence (denoted by ↔) of the *signifier* (12.12) to what is *signified*, (12.8), (12.9), or (12.10), represents a *seme*, in Prieto's sense of the term (1966), and the convention adopted by a collectivity to achieve this correspondence is the *semic code*:

$$\textit{Seme} = \left[\text{def } S^*\right] \leftrightarrow \left[\text{def } O^*\right], \text{by means of a } \textit{semic code}. \tag{12.13}$$

The *sememe* of a sign is defined as the set of all *interpretants* of the *meaning* of the sign in a specified language (Prieto 1966).

$$\textit{Sememe of}\left[\text{def } S^*\right] = \text{Interpretants of } \textit{Meaning of}\left[\text{S denominating O,or } O^*\right]. \tag{12.14}$$

In order to better specify the three concepts, signifier, seme, and sememe, note that the graphical sign "dog" denotes a *concept* (technical definition) which can indicate different specific animals that the mind recognizes as dogs:

["Fido," "Buck," "Smurf," "Ball," "Black," "Samson," etc.] = Extensive common denomination = *extensive meaning* of the sign "dog."

The same meaning (the set of the observed dogs) can also be evoked by the signs:

[dog, *dog*, *dog*, DOG, *dog*, DOG, **dog**, DOG, dOg, DoG, d o g , D O G, **D O G**, etc.] = *Signifier* of the sign "dog".

The *sememe* of "dog" may be the set.

["our most faithful friend," "the faithful guardian of our home," "the most intelligent of pets," in addition to other expressions that designate "dog"] = *sememe* of [def S*].

It is clear that the introduction of *signs* and *semi* require the use of multiple Recognizing and Identifying Control Systems (Sect. 3.2) that operate at very high speed and over a great many cycles. Figure 12.10 presents the *Ring* for "constructing" the correct graphical form of a *written sign*, S, or the correct pronunciation, if S is a vocal sign, assuming that the mind has already formed the *Signifier of* [S*], as defined in (12.12).

When the cognitive system that arises from applying the Ring in Fig. 12.10 succeeds in forming and communicating the *sign*, learning to do so more and more quickly, the receiver must be able to *identify* and *interpret* it. The identification of the sign represents the mental operation that identifies the *Signifier of the sign* = [def S*], to whom the sign belongs, as illustrated in (12.12). The Identifying Control Systems are thus activated, which are similar to those in the model in Fig. 12.11.

Once the sign is identified, the recipient must *interpret* it; that is, determine its meaning. From (12.8) we know that, in order to determine the significance of the sign S representing a proper name, each recipient must recognize or identify the object determined by the [des O] or [min des O]; if the sign S indicates a concept, such as in (12.9), then every recipient must identify or determine the correct [def O*] or [min def O*], as defined in Sect. 12.8.

Having learned from earliest infancy to manage the Recognizing and Identifying *Rings*, it is not easy for us to realize how intensively and quickly the *Rings* operate

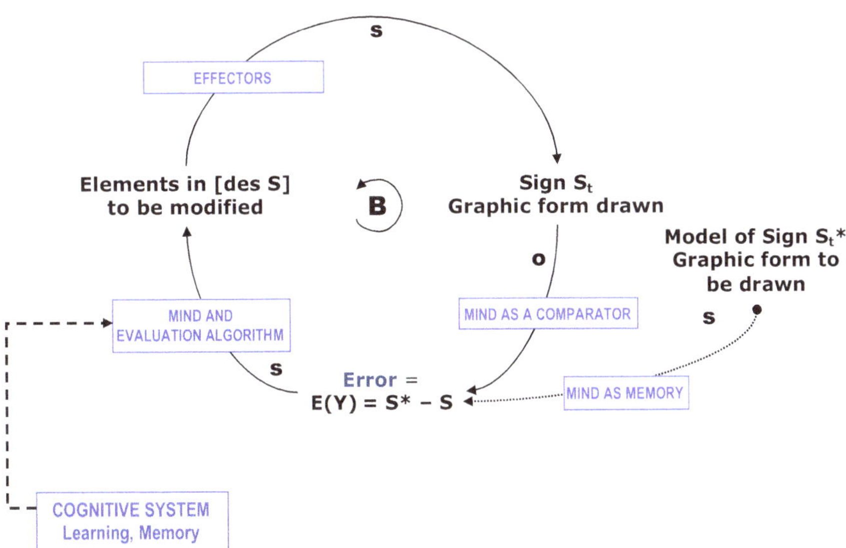

Fig. 12.10 *Ring* for controlling the correct form of a sign

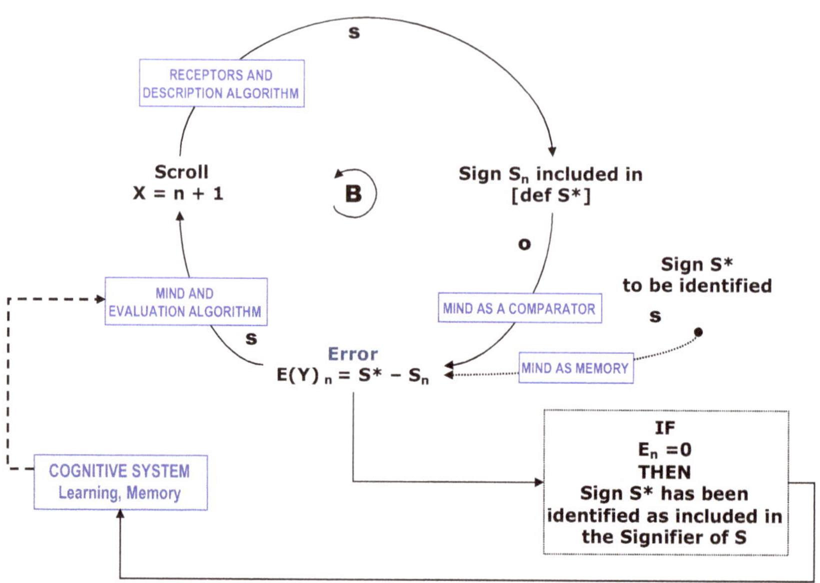

Fig. 12.11 The *Ring* to identify a sign belonging to a signifier

that choose the correct meaning from our Obs-UN. One such *Ring* is illustrated in the model in Fig. 12.12.

The above considerations can be easily understood if we turn again to the example of the baby who knows the fruit and vocal signs that designate the concepts (technical definitions) of the various fruit, recognizing their similarity even though they are described differently. Thus, consider the case regarding the recognition of the signs of the following propositions (strings of signs):

これらは兆候です , 이러한 징후입니다

I am not able to identify the signs, and thus I cannot interpret them, even if I am sure the mind of a Japanese or Korean person would immediately recognize the *signifier* of the signs and interpret them by forming the sentence: "these are signs."

12.10 Languages and Signs

Following Prieto, *semiology* is the science that studies human behavior regarding the attribution of *semes*, which represents a broader meaning than the original one attributed by Saussure. The field of *semiology* intersects with both the *linguistics*

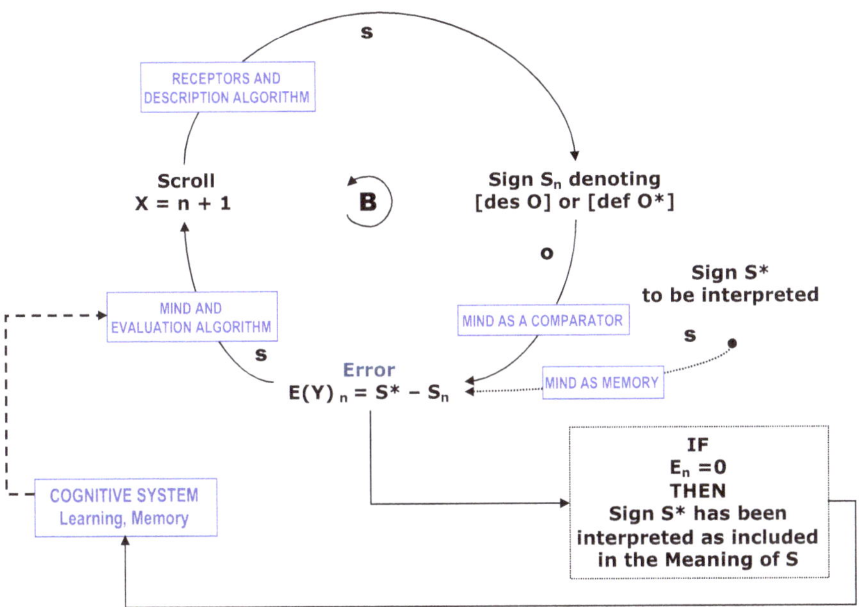

Fig. 12.12 *Ring* for interpreting the meaning of an identified sign

field, which studies linguistic signs and languages, and *semiotics*, the discipline that studies linguistic and nonlinguistic "signs" (Nöth 1995):

> [...] without doubt verbal language is the most powerful semiotic device that man has invented; [...] nevertheless other devices exist, covering portions of a general semantic space that verbal language does not. So that even though this latter is the more powerful, it does not totally satisfy the effability requirement; in order to be so powerful it must often be helped along by other semiotic systems which add to its power. One can hardly conceive of a world in which certain beings communicate without verbal language, restricting themselves to gestures, objects, unshaped sounds, tunes, or tap dancing; but it is equally hard to conceive of a world in which certain beings use only words; when considering … the labor of mentioning states of the world, i.e. of referring signs to things (in which words are so intertwined with gestural pointers and objects taken as ostensive signs), one quickly realizes that in a world ruled only by words it would be impossible to mention things (Eco 1976, p. 174).

I propose to define *language* as a system of *intentional semes* and *sememes*, codified by a collectivity, through which men can attempt *linguistic communication* involving any thought content whatsoever, by producing a system of *semic actions*. There are various types of languages, which differ according to the signs they use and the "codes" through which the *denomination* process unfolds. Some are specialized languages, employing, for example, logical, mathematical, graphical, or iconic signs. The most extensive language is the *natural verbal* one, through which the human mind can communicate any type of thought to another mind:

> [This can be expressed] … by saying that every theory of signification and communication has only one primary object, i.e. verbal language, all other languages being imperfect

> approximations to its capacities and therefore constituting peripheral and impure instances of semiotic devices. Thus verbal language could be defined as the *primary modeling system*, the others being only "secondary," derivative (and partial) translations of some of its devices .... (Eco 1976, p. 172).

Every language can be viewed in instrumental terms. A fundamental function of languages, no matter what their form (written, spoken, gestural, mimicking, ritual, iconic, and so on) is to enable communication to occur between individuals in order for them to share thought content.

In his view of social systems as autopoietic systems, Niklas Luhmann assigns language a fundamental role, since it permits communication among individuals and the generation of orders and progress (Luhmann 1995). In his book, Theory of Society (1997), he pays particular attention to language, claiming that the words (of the spoken language) have no precise significance if considered alone, but rather serve the function of producing sentences with meaning and, above all, increasingly newer sentences:

> Speech possesses its very own *form*. As a form with two sides it emerges from the difference between *sound* and *meaning* ... Spoken communication is the processing of meaning within the medium of sound. ... In order to construct the difference between medium and form within speech itself, the medial substrate of speech, the difference between sound and meaning, must be *underspecified*. Without under specification there would be nothing left to say because everything would already have been said. This problem is solved with the difference between words and sentences. Words are also constellations of sounds with meaning; but they do not determine the sentences that they can produce when combined.... With the help of speech, one may say *something that has never been said before* (italics in original) (Luhmann 1997, pp. 213–215; translation by Daniel Lee 2000, p. 326).

Thought contains basic content—[des O] and [def O*]—and composite content, which derive from a union of basic content in time and space. Thought is continuous over time and discontinuous regarding its content. It is continuous in that during wakefulness, it flows uninterruptedly; it is discontinuous in that thought content can suddenly change, shifting from one mix of objects and concepts to a new one when called for by communication and the survival of the cognitive subject. Language not only tries to express thought content by using appropriate *semes* that adapt to the flow and necessities of the cognitive subject, but it also tries to express such content more precisely. Therefore, language is revealed as a *flow of signs* denoting successions of *thought content*, even in a memorized form. Despite this variety, the thought content language can transmit can be reduced to the following categories:

1. *Expressions of judgments* and *emotions*, understood also as expressions of opinions and evaluations, impressions, indications of kindness, beauty, amazement, annoyance, happiness or unhappiness, and so on ("What a beautiful rose!").
2. *Declarations of intentions* to undertake certain actions or types of behavior, positive and negative ("I desire to pick the rose before it withers").
3. *Orders*, understood as manifesting a desire, or a necessity, to a subject that he behaves in a certain way ("You go pick the rose, but be careful not to leave the stem too short").
4. *Questioning*, expressed as requests for thought content and answers ("Should I also pick the tulips?").

5. *Answers to questions* and *orders*; that is, the revelation of thought content that has been requested ("Ok. I'll go then and pick the rose"; "No, leave the tulips where they are").
6. *Descriptions* and *procedures*; that is, the results from observing objects and portions of reality and the procedures for obtaining certain objects as the result of the application of the rules established by the procedure.
7. *Information*, or *specific data* useful in carrying out operations or activities.
8. *Argumentation*, through which the cognitive unit tries to judge the truth of certain statements (answers, descriptions or information), and seeks the explanation of the phenomena observed (Sect. 12.14.2).

Functions 1–5 mainly communicate the "internal states" of the cognitive subject; functions 6 and 7 communicate the "states of the world," that is, the states of the variables that describe or define objects assumed observable by both parties involved in the communication. Function 8 tries to arrive at the *conviction* that there exists a correspondence between the significance inferable from a statement and the "states of the world" the statement indicates; in particular, it tries to derive *explanations* of the "states of the world," deriving these from other "states of the world" which are observed or assumed or from a theory on the states of the world. Example: information: "Our friends expect us for dinner in 20 minutes"; expression: "Aren't you ready yet?!?"; information: "It takes 20 minutes to get there by car"; judgment: "Come on! You're always late because you're too slow"; question: "You want to get me in trouble?"; order: "Hurry up then!"; argumentation: "Otherwise we'll arrive too late"; answer: "I'm not the one whose slow; it's you who are stressing me out!"; argumentation: "And it's not true it will take us 20 minutes, because we'll take the ring road!."

Having distinguished among these basic types of thought content leads us to conclude that the fundamental function of languages, that of communicating thought content, can be divided into *specific functions*: the *expressive, declarative, imperative, inquisitive, replicative, descriptive, informative, and argumentative* functions.

The classification proposed conforms to the language functions analyzed in terms of transmissible thought content, and can be traced to the famous linguistics classification proposed by the scholar in *phonology*, Roman Jakobson (1970). Jakobson analyzes the functional aspects of linguistic acts in the *context of the communication process*, distinguishing among:

1. The "expressive" or "emotive" function: "This tends to give the impression of a certain emotion, true or simulated."
2. The "conative" function: "The conative function is expressed in its purest grammatical form in the vocative and imperative case which, from a syntactical and morphological, and even often phonological point of view, diverges from the nominal and verbal categories."
3. The "phatic" function: "There are messages that serve essentially to establish, extend, prolong, or interrupt communication, to check if the circuits are working, attract the attention of the interlocutor or to ensure it does not fade."

4. The "metalinguistic" function: "Whenever the sender and/or recipient deem it necessary to check whether or not they are properly using the same code, the discourse centers on the code: it fulfills a metalinguistic function."
5. The "poetic" function: "The poetic function is not the sole function in the art of language, merely the dominant, decisive one, despite the fact that in other verbal activities it only plays a subsidiary, accessory role."
6. The "referential" function: informative propositions that communicate contextual facts, or "states of the world."

It is not difficult to relate these six functions to the eight types of thought content that can be communicated.

Language can also be used *pragmatically* (phatic function) even simply for *ceremonial aims* ("Good morning," "Welcome," "Pleasure to see you," etc.), and can carry out a *self-referencing* function by indicating to the receiving subject that the sender wants to start, continue, or stop communicating with him ("Hey!," "What rainy weather," "Over and out"). Usually communication serves to transmit relatively complex thought content and to allow interactions among the various subjects or groups of subjects. Apart from simple cases ("Stop," "Be careful," "Enough!," etc.) several *semic* actions are needed to transmit systems of signs variously arranged, even when they are chosen by universes of different signs (e.g., the oral word can be accompanied by signs in the miming-gestural language).

The semiological and linguistic study of the way systems of signs are arranged and form in languages can be undertaken following three approaches:

1. The *semantic* study, which specifically studies the relation between signs and meanings for both the senders and recipients.
2. The *syntactic* study, which analyzes the relations between signs and the possibility of coherently combining them.
3. The *pragmatic* study, which considers the ways in which the linguistic signs are used in practice within a given "community" of individuals characterized by a given culture in order to "do something" to achieve results or objectives. For example, the sign "stop" written in words or on a sign at the end of the street, the sign "Come at 9 a.m.," or the sign "I understand. I'm coming" clearly are sent to produce or declare the effects of behavior and they produce or declare the effects of behavior if the meaning of the sign is recognized by those receiving it. This ability of signs (and thus of languages) is known as *performativity*, a term introduced by John Langshaw Austin, a language philosopher who has studied the pragmatic effects of signs and languages:

> It is worthy of note that, as I am told, in the American law of evidence a report of what someone else said is admitted as evidence if what he said is an utterance of our performative kind: because this is regarded as a report, not so much of something he said, as which it would he hear-say and not admissible as evidence, but rather as something he did, an action of his. This coincides very well with our initial feelings about performatives (Austin 1962, in Bial 2004, p. 145)

It is evident that signs are evaluated in the pragmatic sphere in a community that shares the *semic code*. In particular, *performativity* characterizes juridical language, which by definition is normative.

I would like to conclude this brief synthesis by considering the types of signs that must characterize all languages in order for these to carry out their communicative function, thus allowing the thought content to be *enunciated* through a succession of signs forming a *proposition*. A *proposition* is a sequence of elementary signs with a complete meaning, in that the signs serve to enunciate, express, and transmit the content of a defined thought (expressions, orders, questions, information, argumentation, etc.). Let us assume that those signs are determined by a unique *semic code* that is conventionally accepted by both the speaker and listener. In order to express complex thought content we use a *statement* made up of a sequence of propositions that express a system of meanings representing the overall meaning. The meaning of a proposition or a statement represents the *message* the sender wishes to transmit to the recipient, which falls under one of the eight functions examined above.

Without entering into a typological distinction, which can be found in any dictionary and grammar book in any language, the signs (in particular, the vocables, words, literary terms) that make up the propositions and statements can be divided into two *main* classes and two *accessory* ones, according to their main function.

Above all, there are the signs of a given *signifier*, whose main function is to denote objects (nouns) or actions (verbs) and their characteristics (attributes) according to how the observer forms these in his mind. For simplicity's sake, considering only the signs–words in the written language, note that the sign "Piero" indicates the *idea* of a specific individual with this name that is the[des PIERO]. For example, "Mella," to indicate the state of the "surname" dimension; "male" to indicate the state of the dimension "sex"; "born in Italy," to indicate the state of the dimension "place of birth"; "tall …," "hair color …," etc. For a *meaningful* technical description it could be enough to produce the phrase "Piero Mella, professor at Pavia University, in Italy." If my son were to tell a friend that he was referring to [des PIERO] he would simply say: "my father."

Though we seldom realize it, most of the words we use indicate *dimensions* (height, speed, weight, color, license plate, etc.) or attributes and dimensional states (180 cm, 60 km/h, 80 k, green, JK 335 OP, etc.). Even *actions* expressed by signs–verbs are part of the descriptions or definitions: "Piero took a taxi," "Mary runs 2 miles every day," "I ate a steak with fries," etc., which describe observed objects. We can also include in this category the signs that indicate an intention or judgment: "I'm going home," "I don't think I'll take a taxi," "It's been a terrible day and I'm tired." Even questions and orders must be expressed with signs that indicate objects of thought referring to descriptions or definitions: "Go get the pen on my table," I can't find it," "Are you blind? It's right in front of your eyes!."

In order to transmit thought content through propositions and statements, composed of the signs from the preceding category, a *second* category of signs is necessary which functions in order to *give orders to the mind*, in particular to its attention capacity, so as to evoke the "mental operations" needed to arrange the *objects of thought* in an appropriate "order" in the context of the linguistic statements. I am referring to the *ordering* or *correlating* signs for the meanings of other signs, or, in general, to the signs that evoke the *mental operations* that give "orders" to the meanings of other signs from the same class.

In written English, for example, the sign "...on..." does not *denominate* any object of observation but merely *evokes* the *mental operation* "overlaying" needed to order and correlate the *objects of thought* indicated by "...." For example, the proposition "the bottle is *on* the floor" has a complete meaning, since it allows us to state an acceptable mental object; "the floor is *on* the bottle," on the other hand, forces us to evoke a concept whose meaning is unacceptable. There are many and varied linguistic signs that designate *mental orders* in order to allow the mind to form the objects of though indicated by the language. I shall leave the reader to recall the numerous classes of signs of this type; here I shall mention only the most obvious ones:

1. The numerous types of signs of *conjunction*: "...and...," "...or...," "...either...," "...neither...," "and...and...," "either...or...," "neither...nor...," "both... and...," "...also...," etc.
2. The *articles*, "the..." and "a..."; in many languages these are specified by grammatical category: male or female, singular or plural, relative to the words that express mental objects.
3. The *prepositions*, "of...," "to...," "from...," "in...," "with...," "on ...," "between...," etc.
4. All the various types of *adverbs*, including *adverbs of time* ("still," "later," "yesterday," etc.); *adverbs of quantity* ("many," "little," "too much," etc.); *adverbs of quality* ("slowly," "carefully," "suddenly," etc.); *adverbs of place* ("above," "below," "in front of," "behind," "right," "left," etc.).
5. The signs to indicate singular or plural, male or female; for example, in English "words" order the mind to identify a group of words; the statement "my cars is red" and "my child are at school" stop someone receiving them (outside of any particular context) from identifying any meaning, since the plural and singular expressions contrast with the meaning of the verb.

In order to see how the signs of the preceding categories send *orders* to our mind, consider what is evoked by the following statements: "Carla grabbed the pencil and placed it above...," "Carla grabbed the pencil and placed it above the ...," or "Carla grabbed the pencil and placed it above our ...." I could continue on with this example, but a moment of thought will surely enable the reader to see that every sentence spurs the mind to complete the image the speaker is creating. The role of the signs sending orders to the mind and "creating" the meaning of a statement was pointed out by Silvio Ceccato, an Italian linguist and cybernetician who, in his book entitled *Course in Operational Linguistics* (1969) demonstrated that the *ordering* signs are finite in number and, through appropriate *sequences of attentional acts*, their operational meaning in the construction and interpretation of linguistic expressions can be defined. Ceccato proposes a simple experiment: place the tips of the fingers of one hand on the edge of a desk. Reflect on what you "feel" if I tell you: "you are touching *the front part* of the desk," or alternatively: "you are touching *the back part* of the desk." Your attention is diversely affected by each sentence and, in fact, you think you are touching what is "ordered" by the signs "front part" and "back part" with a shift in attention.

Languages usually include a third class of signs: the so-called signs of *punctuation* (".", ",", ":", "/", etc.), whose function is to separate or order the linguistic signs arranged in sequence in order to allow the mind receiving the message to correctly identify the succession of meanings of the words that make up the statement or proposition. A fourth class includes the *functional signs* (e.g., "?," "!," "!?," etc.), whose function is to evoke the linguistic subfunction chosen by the sender in transmitting a statement. The signs of the last two classes are considered *accessory* since their use can be avoided by replacing them with the signs from the main classes; for example, "Come here!" can be interpreted in the written language as "I order you to come to me." A proposition (and thus a statement) will be composed of signs from the four classes indicated above, arranged according to the conventional order set by the code that attributes meaning to the terms used:

> I shall reconsider human knowledge by starting from the fact that we can know more than we can tell. This fact seems obvious enough; but it is not easy to say exactly what it means. Take an example. We know a person's face, and can recognize it among a thousand, indeed among a million. Yet we usually cannot tell how we recognize a face we know. So most of this knowledge cannot be put into words (Michael Polanyi 1967, p. 4).

12.11 Factual Truth and Falseness

Among the eight functions of languages, of particular importance is the *argumentative* one, since it carries out linguistic transmissions on the "state of the world" a mind observes and desires to transmit to another mind through language. Following Copi and Cohen (2008), we shall define a *declarative proposition*, or *statement*, in any language as a sequence of basic signs capable of expressing a thought content that can be ascertained to be "true" or "false" using some *conventional procedure*:

> We know that every statement is either true or false. Therefore we say that every statement has a truth value, where the *truth value* of a true statement is *true*, and the truth value of a false statement is *false*. Using this concept, we can divide compound statements into two distinct categories, according to whether the truth value of the compound statement is determined wholly by the truth values of its components, or is determined by anything other than the truth values of its components (Copi and Cohen 2008, p. 290).

Verification and *falsification* are the cognitive operations that *verify* whether or not a *declarative statement* regarding differences, descriptions, definitions, or cognitive procedure (Sect. 12.10) has an observable or derivable meaning which is in accordance with the observations on the "states of the world." Thus, the truth does not refer to the "states of the world" but to the *statements* on the states of the world. As *truth* derives from a *mental process* of verification or falsification, it is always relative, at least as regards the instruments for observing the state of the world and the language used to construct the statement.

I wish to hypothesize about how the "Batesonian mind" can ascertain the truth. Let us suppose that a declarative proposition, *E*, transmitted from Alfa to Beta in a given language, asserts that the difference: $d_n(\text{A}^*) \in [\text{def A}^*]$ is *true;* for example,

"Snow is white" states that in [def A*] = [def "snow*"], d_n(A*)="color of snow=white." The proposition "Snow is white," composed and transmitted by Alpha, is *true* for Beta if the latter can, using his own procedure (which is not necessarily similar to that used by Alfa), construct [def A*], thereby determining d_n(A*) and verifying it belongs to the definition; otherwise the proposition is *false*:

> But one might object that, ultimately, truth is a matter of using and accepting a sentence as an adequate description of a state of affairs. Thus, "Snow is white" is true if and only if we are prepared to use and accept that sentence to describe a property that snow in fact has. "Snow is black" is a misassignment; it is false, because we are not prepared to use and accept that sentence as a description of snow [*more correctly: a description of a dimensional state of any object we denote as "snow"* (author's note)]. But with "Snow is marble" we may begin to hesitate; perhaps in certain circumstances, it *is* a true metaphorical description (Ankersmit and Mooij 1993, p. 78).

Adopting Bateson's concept of mind, let us consider Alfred Tarski's rule (1944) for truth: "'Snow is white', is true if and only if snow is white" (Tarski 1944, p. 342). The formal statement "Snow is white" is *true* for Beta if Beta *observes* that, in objects where [des snow] ⊆ [def snow*], the dimension D_{COLOR} always assumes the state [d_{COLOR}(snow)= "white"] ∈ [def white*]. According to the procedure examined in Sect. 12.7, *truth* thus depends on the processes of *distinction* (white?) *understanding* (snow? color?) and *recognition* ("this" is snow and its color is white?) and assumes a reliable *process of determination* (What does "snow" mean? What does "white" mean in a chromatic scale?). These observations are perfectly in line with Tarski's general definition of truth: "A sentence is true if it is satisfied by all objects and false otherwise" (Tarski 1944, p. 353).

Alfa's statement to Beta that [Δd_n(O*)] ⊂ [def O*] (e.g., "men are mortal") or that [des O_m] ⊂ [def O*] (e.g., "Piero is a man") is *true* if Beta is able to undertake a cognitive procedure that can construct [def O*], that is, "man," and can recognize that [Δd_n(O*)], that is, "mortal," belongs to it and identify [des O_m] = Piero as an element of [def O*]. In conclusion, the statement "Piero is a man" is *true* if [des PIERO] ⊂ [def MAN].

The statement "The fourth root of 100,000 is 17,7827941" is true if the [def SQUARE ROOT] includes a calculation procedure that, when applied, provides the declared value.

The concept of truth and falsehood, made operational by using Bateson's model, is *factual* in nature because the truth is conceived of as a *correspondence* between the meaning of the proposition and what the mind can observe. However, the *argumentative function* of language also tries to determine the truth or falsehood of *composite statements* through the relations of truth and falsehood in the specific, or "atomic" statements, taking account of the meaning of the connective signs that link them together. I cannot deal here with the topic of the search for formal truth and falsehood, since this topic is typical of the logic, in particular the formal logic, which adopts specialist formal language to demonstrate the logical truth of composite statements, determining their truth by the component atomic statements.

I shall conclude by reminding the reader that the truth or falseness of a statement (concerning what we have perceived and translated into linguistic signs) depends on the sensory organs of a person "A" who must interpret that proposition and compare it to the states of the world or with other propositions on these states. This implies that a proposition by this person "A," characterized by a given set of sensory organs, affirming the truth of a des(o) or a def(o*) may not be verifiable by another individual, "B," with a different set of sensory organs, unless the latter can "translate" the statements by "A" into a language that constructs statements that can be verified by "B." An example of this is the quote below which, though referring to humans, can be extended to any communication context involving subjects equipped with different sensors:

> The linguist manipulates the syntactic, phonological, and semantic forms and judges and/or asks native speakers to judge whether the consequences are a well-formed sentence in the language, an ambiguous string or any one of an array of numerous other possibilities. The relevant reference point by the very nature of the research is internal to the bearer of the internal grammar—the native speaker himself. To put the matter in a somewhat different form, suppose that we succeeded in constructing an instrument that purportedly arrived at the same judgments for visual inputs as those possessed by normally sighted people.
>
> How would we know whether the instrument worked? The answer clearly is that we would accept the instrument as accurate if and only if the responses of the instrument matched those of normally sighted people. In other words, we would calibrate the instrument by using precisely the same set of judgments (intuitions) reported by the people involved that we presently use in the absence of such an instrument. Thus in fields where the patterning under scrutiny is patterning of the behavior of human beings, the reference point and the source of the judgments will necessarily be the human being (Bostic St. Clair and Grinder 2001, p. 76).

12.12 Scientific Laws and Theories as Definitions and Conjectural Models of the World

Bateson's model can be extended and operationalized to also take in the important topic of *scientific explanations*. In fact, human thought is not limited to the identification and description of objects and the positioning of technical definitions. Instead such operations represent the premise for more advanced mental activities through which the conscious (and intelligent) cognitive system succeeds in giving explanations in order to understand certain phenomena in the world and to forecast these, in order, where possible, to control events.

"Advanced" mental activities are those directed contemporaneously at a group of objects, examples of which are:

A. SET FORMATION: mental activity whereby different objects are considered as belonging to a set, with regard to specific D_n (books with a cover), d_n (books with a red cover), or Δd_n (books by Italian or English authors). "Sets" are formed as objects resulting from set formation, which itself appears to be an "essential" mental activity: I do not describe the object "set" but I form it as an independent object of observation composed of *simple, separate objects*: that is, the elements of the set. Sets are mental objects resulting from the application of a technical definition to distinct objects:

$$[\mathrm{SET}] = \left\{ \left[\mathrm{des}\ \mathrm{O}^1 \right] \cup \left[\mathrm{des}\ \mathrm{O}^2 \right] \cup, \ldots, \cup \left[\mathrm{des}\ \mathrm{O}^M \right] \right\}.$$

B. CLASSIFICATION: this is the operation of forming sets (or that of technical definition) applied several times to form subsets, or classes, of objects of observation. Each class is characterized by one or more dimensions and/or one or more ranges of admissible variation. More precisely, each class is characterized by a specific "technical definition for the class." The operation of "identification" (recognition) is associated with classification (understanding).

C. SYSTEMATIZATION: the activity that takes place, as we know, when we observe objects linked by certain relations, with the result that these objects are interrelated and lose their observational individuality, becoming elements in a new entity: the "system"; a system is a composite object formed by *simple, unified objects* including a network of relations:

D. LEGALIZATION: the activity which searches for and identifies, in the observed universe, various types of regularity among the objects of defined sets. Various types of observation repeated with *regularity*, such as *technical descriptions* of events, phenomena, and objects, including their dimensional states, even under different observational conditions, lead the "mind-processor" to *legalization*—that is, to *generalizing analogical abstraction*—which results in those regularities; these regularities are *defined* as *empirical laws* and constitute valid models of knowledge for all the observed objects. Every "law" derives from an *inductive* process; that is, from an *abstraction* whereby the mind *generalizes* for "all the objects x" (in formal logic we write (x), or $\forall x$):

- *Specific dimensions* (all flying animals must have wind sensors)
- *Specific dimensional states* (all arachnids have eight legs)
- *Specific relations* (all blue-winged butterflies have antennae as long as their forewings)

The mind observes "a finite number, N, of objects x"; the inductive generalization expressing a relation R between two classes of objects, x_n and y_n, can be expressed, for example, as follows:

$$\begin{array}{ll} \text{IF} & [x_n \mathrm{R} y_n] = \text{"TRUE" for } n = 1, 2, \ldots, N \ (\text{observation}) \\ \text{THEN} & (x)\ (y)\ [x\ \mathrm{R}\ y] = \text{"TRUE"} \ \ (\text{inductive generalization}) \end{array}. \tag{12.15}$$

(12.14) leads the mind to make the following deduction:

$$\begin{array}{ll} \text{IF} & (x)\ (y)\ [x\ \mathrm{R}\ y] = \text{"TRUE"} \\ \text{THEN} & x_{N+1} \mathrm{R}\ y_{N+1} = \text{"TRUE"} \end{array}. \tag{12.16}$$

The validity of this inference process is based on a number N of specific observations held to be sufficiently adequate to make the generalization plausible. Each new observation which confirms the relation (12.15) observed until that moment "validates" the generalization. Only a single contrary observation is needed to reject the generalization, unless it is transformed into a probabilistic generalization based on frequencies. In its most elementary form:

$$\begin{aligned}&\text{IF } \left[x_n \mathrm{R}\ y_n\right] = \text{"TRUE" for } n = 1,2,\ldots,M < N,\\&\text{THEN } (x)\,(y)\,\left[x\ \mathrm{R}\ y\right] = \text{"TRUE" with probability } (M/N)\end{aligned} \qquad (12.17)$$

When a scientist says that the evidence is sufficiently good to establish a scientific hypothesis, he means that the evidence is such that the hypothesis could be inferred from it according to one of the recognized principles of scientific inference. His interest is in the adequacy of the evidence for this purpose, for this is what would be in dispute if a fellow scientist questioned the 'well-groundedness' of the hypothesis. [...].

Thinking only of the considerations which are relevant to a scientist, a philosopher will be quite right in taking the question as to whether or not an inductive inference is valid and the belief in a scientific hypothesis yielded by it is a reasonable belief as being a question as to whether or not the empirical evidence is such that the hypothesis can be obtained from it by following recognized 'rules of procedure' (Braithwaite 1953, p. 262).

For an *empirical law* (*norm* or *generalization*) to be defined as *scientific* it must:

1. Be presented as the following equivalent type of statement (12.15): "if [des A] then [des B], always"; (x) (y) ([des x] → [des y]); ∀[des A], then we observe [des B], always; [des B] does not exist without [des A], etc.
2. Have empirical content; without empirical content it can at most be a formal law.
3. Present relations between objects belonging to *open sets* that are connotatively defined and whose extension is not finite or entirely known; this is thus valid not only for observed objects but also for all those objects that have the connotations that define a set, even if they have not yet been observed.
4. Not present a relation that derives from conventions or the application of a procedure.
5. Be *verifiable* or *falsifiable*; that is, confirmed by favorable cases, or positive examples, or confuted by unfavorable cases, or contrary evidence.
6. Be *coherent* with other accepted scientific laws and permit *deductions* when included in deductive argumentation.

E. THEORIZATION: the activity of an evolved mind whose aim is to *interpret* and *explain* the perceived reality, even going beyond known scientific laws. Theorization is needed to give "closure" to scientific explanation (Sect. 12.14.2). Through this activity scientific theories are formed; that is, formalized systems of coherent hypotheses or conjectures, linked together, which aim to justify, and possibly explain, the existence of "laws" and their connections, thus forming a unitary descriptive and predictive corpus. Having identified regularities and *laws* for a given observed universe, the observer/mind then tries to understand the *reason* for their existence. He thus presents *theories*—that is, hypotheses or conjectures (systems of hypotheses)—that could justify the affirmed regularities. The theories (as well as the single laws) can be interpreted as *formal hypothetical definitions* of the *observed universe* presented by man to completely *describe* that universe. Theories, as *explanatory hypotheses*, must not only contribute to *explaining observed* facts but also permit *forecasts* about *observable* facts. In their so-called realistic conception of knowledge, Stephen Hawking and Leonard Mlodinow, recognize that all we can know about "reality" consists of networks of world pictures, or general models, expressed even through mathematical language; only the systematic connection of a set of observations accompanied by a conceptual model and by rules connecting the concepts of the model to the observations enables the mind to explain those observations by connecting them by rules to concepts defined in models:

In the history of science we have discovered a sequence of better and better theories or models, from Plato to the classical theory of Newton to modern quantum theories. It is natural to ask: Will this sequence eventually reach an end point, an ultimate theory of the universe, that will include all forces and predict every observation we can make, or will we continue forever finding better theories, but never one that cannot be improved upon? We do not yet have a definitive answer to this question … (Hawking and Mlodinow 2010, p. 8).

Multiple, equally valid, world pictures exist; therefore, science requires multiple models to encompass existing observations:

Like the overlapping maps in a Mercator projection, where the ranges of different versions overlap, they predict the same phenomena. But just as there is no flat map that is a good representation of the earth's entire surface, there is no single theory that is a good representation of observations in all situations (Hawking and Mlodinow 2010, p. 9).

12.13 Intelligent Cognitive Systems: The Power of Models

Having demonstrated Bateson's hypotheses that the mind is nothing other than a perceiver, calculator, and analyzer of various levels of differences, we can now define a *conscious* cognitive system as an *autopoietic system* that through the "mind" is able to distinguish differences as external disturbances and develop knowledge by adapting to these disturbances, thus forming concepts that allow the mind to link its own mental representations to form an "external world" to which it has no direct access.

Following Bateson, the "mind" thus represents the computational system capable of constructing knowledge and transforming the autopoietic system into an *observer system* that can *describe* objects and produce *definitions* of concepts from which a *knowledge process* can be derived:

What we know is generally considered to be the result of our exploration and understanding of the real world, of the way things really are. After all, common sense suggests that this objective reality can be discovered. …How we know is a far more vexing problem. To solve it, the mind needs to step outside itself, so to speak, and observe itself at work; for at this point we are no longer faced with facts that apparently exist independently of us in the outside world, but with mental processes whose nature is not at all self-evident (Watzlawick 1984, editor's Foreword).

The world, as we receive it, is our own invention (von Foerster 1984, online p. 45).

I shall define a *cognitive system* as *intelligent* when, through the "mind," it not only "constructs" the world but also produces a codification process of the difference information, creating a system of *semes*, a *language* through which the *cognitive system* links up with other *cognitive systems* in *formal communication processes* that allows it to form *scientific laws* and develop *argumentations*. It thus follows that a necessary and sufficient condition for a *cognitive system* to also be *intelligent* is that at the same time it is *autopoietic* and capable of developing formal *communications behavior* with other systems it is linked to. In the end, this is the ultimate meaning of Turing's Test (1950), which we are all familiar with. This is the basic hypothesis of *functionalism*, one of the most important research fields in cognitive science (Clark 1991). Extrapolating from Andy Clark's view of the theory of symbols (signs possessing conventional meaning):

The functionalist is in many ways the natural bedfellow of the proponent of the physical-symbol-system hypothesis. For the physical-symbol-system hypothesis claims that what is essential to intelligence and thought is a certain capacity to manipulate symbols (Clark 1989, p. 19)

A physical symbol system, according to Newell and Simon (1976, pp. 40–42) is any member of a general class of physically realizable systems meeting the following conditions:

1. It contains a set of symbols, which are physical patterns that can be strung together to yield a structure (or expression).
2. It contains a multitude of such symbol structures and a set of processes that operate on them (creating, modifying, reproducing and destroying them according to instructions, themselves coded as symbol structures).
3. It is located in a wider world of real objects and may be related to that world by designation (in which the behavior of the system affects or is otherwise consistently related to the behavior or state of the object) or interpretation (in which expressions in the system designate a process, and when the expression occurs, the system is able to carry out the process).

In effect, a physical symbol system is any system in which suitably manipulable tokens can be assigned arbitrary meanings and, by means of careful programming, can be relied on to behave in ways consistent (to some specified degree) with this projected semantic content (Clark 1989, pp. 4–5).

With our powerful minds and our capacity to construct instruments to expand the limits of our sensors and effectors, we humans are able to expand our intelligence by constructing *descriptions* and *definitions* relative to ever wider "portions of the world," by observing and understanding the "increasingly smaller," the "increasingly larger" and the "increasingly more distant." The phenomena observed are many, interconnected, variable and multidimensional, giving rise to all types of differences. Bateson's mind finds it difficult, if not impossible, to mentally describe such a vast "world," and as a result it is more and more difficult to communicate to other persons the results of our observations; ordinary language, spoken or written, is not always adequate to form and efficiently transmit complex representations, which is why man has always sought to define specific signs to construct efficient forms of representation, even in alternative languages. The signs and languages of geometry, logic, and mathematics greatly expand the mind's capacity for synthesis and analysis. Moreover, thanks to the continual advancements in these languages man has succeeded in understanding (defining) the unobservable (nondescribable) objects: from atoms to gravity, genes to evolution, complex dynamics to chaos, etc.

In short, the human mind can efficiently form descriptions and concepts by using appropriate *signs* and *operators*, defined in a specific manner, in order to construct increasingly more efficient models of the observable universe. For example, all the models in this book allow us to define the control process using the graphical tools I have adopted; however, the use of differential equations and integral calculus, or the use of Powersim, MatLab, and so on, would add other descriptions and definitions. And then there is the written language of music which, with its unique notation, allows us to codify the sounds–notes (primary differences), their pitch, intensity, timbre (higher-order dimensions) in order to form musical phrases which together form well-structured models (sonatas and symphonies, for example) and to transmit these in space and time so that they can be newly translated into sounds by reconstructing in the minds of future musicians and listeners the same model as that of the composer. If today we can listen a thousand times to Beethoven's 9th symphony, this is thanks to the system of musical signs that allowed the composer to leave behind his musical score, which a thousand interpreters can now newly translate into sounds.

In fact, the universes of observation, being composed of a group of objects, all of which multidimensional and often closely interrelated, are almost always too complex to lend themselves to observation and comprehension. Therefore, observational activities could not effectively and efficiently be undertaken if the observed did not formalize the results of such observations using appropriate *meaningful semic representations* that constitute *models*. As such, models are "depictions" of the results of observational activity and are instrumental in efficiently communicating the observational results for reasoning, argumentation, and explanation (Sect. 12.14.2).

We can distinguish between *mental models* and *formal models*. The former are the models the *intelligent* cognitive system creates in its "private" mental sphere to represent the content of the meaning it wishes to transmit (if it is the sender) or that it must interpret (if it is the recipient), which it uses to evaluate "the world" and make instinctive decisions. *Formal models* are made up of systems of concrete signs, linguistic or symbolic, which formalize and transfer a meaning and are used to communicate and enable weighty decisions to be made (Sect. 12.14.1).

Though approximate, *mental models* indispensable for *intelligent* thought, as they are the basis for an understanding of reality. The psychologist Philip Johnson-Laird analyzed their cognitive function in his book *Mental Models: Towards a Cognitive Science of Language, Inference and Consciousness* (1983), an analysis that was taken up again in many of his subsequent works:

The psychological core of understanding, I shall assume, consists of having a "working model" of the phenomenon in your mind. If you understand inflation, a mathematical proof, the way a computer works, DNA, divorce, then you have a mental representation that serves as a model of an entity in much the same way as, say, a clock functions as a model of the earth's rotation. .. Many of the models in people's minds are little more than high-grade simulations, but they are none the less useful provided that the picture is accurate (Johnson-Laird 1983, pp. 2, 4).

Peter Senge, in The Fifth Discipline, deals with mental models on more than one occasion, offering these definitions:

> "Mental models" are deeply ingrained assumptions, generalizations, or even pictures or images that influence how we understand the world and how we take action. Very often, we are not consciously aware of our mental models or the effects they have on our behavior. For example, we may notice that a co-worker dresses elegantly, and say to ourselves, "She's a country club person." About someone who dresses shabbily, we may feel, "He doesn't care about what others think" (Senge 1990, p. 8).

> Our 'mental models' determine not only how we make sense of the world, but how we take action. Mental models can be simple generalisations such as 'people are untrustworthy' or they can be complex theories, such as my assumptions about why members of my family interact as they do. But what is important to grasp is that mental models are active—they shape how we act. If we believe people are untrustworthy, we act differently than if we believed they were trustworthy" [...] Why are mental models so powerful in affecting what we do? In part, because they affect what we see. Two people with different mental models can observe the same event and describe it differently, because they've looked at different details (Senge 1990, p. 160).

Jay Forrester pointed out this "mental model of a mental model":

> The mental image of the world around us that we carry in our heads is a model. One does not have a city or a government, or a country in his head. He has only selected concepts and relationships, which he uses to represent the real system (Forrester 1971, p. 213).

Mental models are the result of, and at the same time the basis for, the cognitive activity we are continuously producing and represent the primary form of representation of the world and the relations we observe in it. Nevertheless, mental models, precisely because they are produced "in the private sphere," have a clear limit: not only can they be imprecise and vague but also they are often *erroneous* and, if not corrected through subsequent verifications or educational processes, can produce seriously harmful results. There are an endless number of examples of this, some of which it is useful to present here.

Before Isaac Newton the prevailing mental model for gravity was, in simple words, that: "The apple breaking off from the branch falls toward the ground, attracted by the earth, following a trajectory directed toward the center of the globe." Newton changed the mental model of gravity as follows: "Apples are attracted to the earth as much as the earth is attracted to apples, following a trajectory determined by their mutual centers of gravity." Another prevailing mental model was: "All bodies in motion are destined to stop, unless a force intervenes to keep them in motion." Newton's First Law of Motion changed this mental model by introducing the principle of inertia: All bodies will continue moving in a uniform straight line (or remain still) unless some other force causes it to change this state." Another obvious erroneous mental model, which, unfortunately, is widespread even today, even among educated persons, is the following: "When you have a cold your nose runs; I see that your nose is running, so I must stay away from you because I don't want to catch your cold." The logical mistake is clear as soon as we consider that my nose could be running because I have just finished crying my eyes out or because I suffer from hay fever.

Precisely to overcome these limits and risks due to reasoning and decisions based solely on *mental models*, man has always sought to accompany these, even at times replace them, with the more efficient *formal models* that allow him to represent and share, through specific language and using various types of signs and symbols, all thought content (Sect. 12.14.1), thereby allowing him to create science and improve the *explanatory* processes (Sect. 12.14.2).

Mental and *formal* models can represent objects of observation through their *descriptions* and technical *definitions*, as well as sets, systems, laws and theories, and entire universes of observation. The approximation to the cognitive "reality" that is to be represented is a necessary feature of every model, even operationally speaking, since it may be the only means of investigating a reality which is too complex, or of studying certain aspects of a single reality that cannot be manipulated and directly investigated. However, this approximation must not be overly vague but instead achieve maximum utility for an efficient and effective representation, communication, and understanding. If the approximation has such utility, then the model can be a substitute for the particular reality and the universe of observation in further observational research. The differences and dimensions which the mind identifies or constructs in the model become direct objects of the observational activity; the signs of the model replace the mental objects in the processes regarding knowledge and recognition, the formation of laws, and the construction of hypotheses.

I shall conclude by going back to Sect. 1.2. I am convinced that intelligence depends on the ability to construct coherent and sensible models in order to acknowledge and understand the "world." I believe that:

> If the capacity to *see* and not simply look at [the world] depends on the ability to construct models to understand, explain and simulate the world, then the most useful and effective models to strengthen our intelligence are the systems ones based on the logic of Systems Thinking (Mella 2012, p. 6), since these models allow us to represent complexity, dynamics, and change (Sect. 1.5).

12.14 Complementary Material

12.14.1 Models

It is useful to mention several important types of models, emphasizing that it is possible to combine these types to obtain highly structured classes of models, such as:

(a) Models in lexical or specialist languages
(b) Iconic and formal models
(c) Models of objects and relations
(d) Concrete and abstract models
(e) Descriptive and operational models
(f) Scale and analogic models
(g) Static and dynamic models
(h) Qualitative and quantitative models
(i) Deterministic and stochastic models

Models in lexical or specialist language Lexical language (English, for example) represents the basic instrument for translating thought content into transmissible models. If I have to describe my sister's wedding to a friend I have to use lexical language, transforming the mental *descriptions* of the ceremony into a flow of words following an informational path of my choosing (I first describe the setting, the spouses or the guests, etc.). On the other hand, if I must represent string theory, then lexical language is not suitable and I will have to use the same mathematical language theoretical physicists use to construct their models on the theory of the universe.

Models of objects and models of relations Here the distinction concerns the thought content to which we give priority in representing the models. Models of objects depict the objects of observation or individual *mental models*, simple or composite (a firm, landscape, etc.); models of relations instead represent the relations that bind together several objects, or the components of a structured object. Models of sets and systems belong to this class of models.

Concrete and abstract models This dichotomy refers to "material" the model is composed of. Concrete models are made up of materials whose properties concretely represent the dimensions of the objects they depict (statue, scale model of a machine, plastic in a building structure, etc.); abstract models instead represent the objects through signs (words in a language, signs and symbols, designs, graphs, etc.).

Iconic and formal models. This classification regards the "form" of representation of the original. Iconic models are constructed so as to "resemble" the original (photograph, films, a square, a design, a wax statue of Napoleon, a reduced-scale airplane model with the same colors as the original, and so on). Formal models instead represent the original in "schematic" terms and do not allow us to recognize the original thing being represented, though we can perceive the dimensions intended to be represented.

Within *formal* models we can distinguish among: *verbal, symbolic, logical-mathematical*, and *schematic* models. *Verbal* models are expressed using an everyday language, such as in the reasoning process: "If all men are mortal and Greeks are men, then Greeks are mortal." *Symbolic* models instead employ a symbolism based on an agreed upon and recognizable code. *Logical-mathematical models* rely on the symbolism typical of the language of logic and mathematics. For example, let us assume we agree on the following use of symbols: if H stands for Men, G for a Greek man, m for a mortal, x for a variable object, then: "any x which belongs to H also belongs to m; any x that belongs to G also belongs to H; any x that belongs to G must be an 'm'." Obviously to express models of dynamic and Control Systems, *schematic* models may be more appropriate, since these immediately depict the object to represent through graphical signs and not linguistic ones in the strict sense of the term. For dynamic systems the language of differential equations or difference equations could be more appropriate, accompanied by graphical representations of dynamic systems.

The advantage of *iconic* models is that they are able to immediately perceive the dimensions of shape, color, weight, etc., which often cannot be easily expressed in formal terms: "What color was the bride's bouquet?" To answer this question a photograph would be a better iconic model. Abstract models, though possessing the disadvantage of not being able to perceive similarities, have the advantage of being able to perceive analogies and commonly accepted conventions, and they are more easily constructed and transmitted. More specifically, linguistic models, literary, logical or mathematical, allow for a nearly unlimited representation of the depicted objects. Schematic models, though having a more limited range of possible representations, offer the advantage of immediacy and the possibility of highlighting relations in structural terms. Let us suppose we wish to represent the model of a system expressing the spatial positioning of 5 objects: A, B, C, D, and E. A verbal language would allow us to describe the system, for example, using the following annotation: "A is to the left of B; A and B are above C and D; C is to the left of D; E is below C and D; A is linked to B and C; C is also linked to D and E; E is also interrelated with D." Thanks to the use of the signs "left," "above," "below," etc., which order our mind to undertake the mental operations for perceiving the arrangement, our mind can form a mental model, perhaps with some difficulty, given the number of objects and the interrelationships. The schematic model in Fig. 12.13, where the arrows indicate relations and the positions are visually apparent, is much simpler and more significant. Note that almost all the models of systems presented in this book are schematic.

Descriptive and operational models These classifications are determined by the operational objectives the subject wishes to achieve through the model. Descriptive models serve to describe or explain the object they represent by presenting the rel-

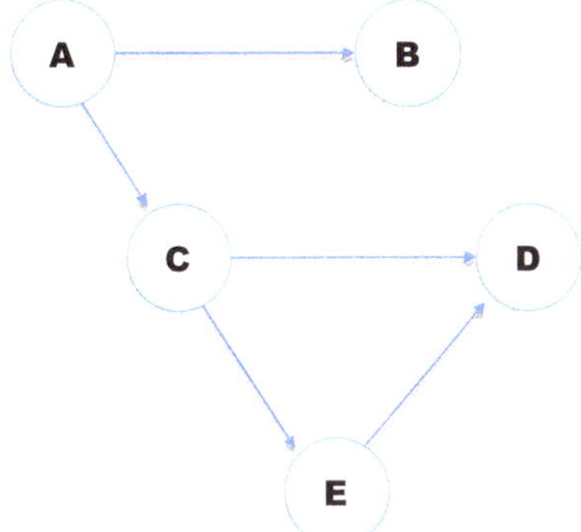

Fig. 12.13 Schematic model of interrelated objects

evant dimensions through some language or schema; they can represent a technical description (a red-sided triangular figure; a green surface; a 5 cm base; 4 cm in height; 42° vertex angle, etc.) or a technical definition (if AB, BC, and CD are linked segments forming a figure lying in a plane, then ABC is a triangle). *Operational* models are structured to simulate the "behavior" or dynamics of the phenomenon being represented (a mechanism to simulate earthquakes; systems of differential equations to simulate growth in an economic system; Excel programs to simulate the dynamics of Control Systems, etc.).

Scale and analogic models These classifications are based on the type of approximation the model introduces in the represented phenomena. In *scale* models all the dimensions of the model are maintained, but some dimensional measurements are presented with a different scale through enlargement or reduction. Analogic models, on the other hand, modify the observed dimensions and represent the thought object with dimensions having similar features. In other words, while scale models maintain the initial quantitative dimensions and the ratios between them (changing only the distance), analogic ones depict the original dimensions through different dimensions while respecting the ratios (and/or relations and/or distances) already observed in the represented object. Since the construction of scale models implies the conversion of the measurements based on scale, which changes the units of measurement with respect to those used in quantifying the original objects, it is necessary to avoid the mistake of considering the original dimensions as all being equally and arbitrarily modifiable. In fact, some dimensions can be reduced in scale only up to certain levels; for example, the surface tension of water molecules can have no influence on a model of a ship with a scale of 1:100, but it would undoubtedly be relevant in cases of greater reductions in scale: a needle will float on water, but a needle whose weight is increased tenfold will inevitably sink. Even with the construction of *analogic* models it is necessary to proceed carefully. For example, if it is not possible to simulate the flight of an airplane by constructing a model that can actually fly, we can build a similar and operational model by simulating the flight in a wind tunnel.

Static and dynamic models The former depict structural relations among the components of a system that do not evolve over time (a company's balance sheet is a typical static model of values). Dynamic models, on the other hand, consider time (or space) as a fundamental variable; every dynamic dimension of the original must be appropriately depicted in the model (e.g., models that represent self-controlled systems are dynamic).

Qualitative and quantitative models The distinction here refers to the type of dimensions held to be fundamental more so than to the results achieved. Models that, irrespective of their conformation, represent quantitative dimensions from the original are *quantitative*. Qualitative models instead highlight the qualitative dimensions. The schematic model in Fig. 12.13 is typically qualitative, since it emphasizes the *name* of the elements and the order relations among them.

Deterministic and stochastic models In this case the distinction regards the way the relations represented in the model are determined, typically when the model concerns dynamic systems. Probabilistic models hypothesize that the inputs and/or outputs are random variables whose probability distribution is assumed known.

12.14.2 The Explanation Process and Its Operational Closure

When the *mind* of man is faced with phenomena or events he cannot *understand*, a situation of *ignorance* exists that gives rise to *curiosity*, which leads to a *question* that requires an *explanation*; that is, guidance toward the construction of a model from which those phenomena or events could derive. Technically speaking, *explaining* a phenomenon F (*explanandum*) means *constructing a model for it based on some other knowledge*, thereby identifying a *succession* of statements; that is, constructing a deductive *argumentation* (*explanans*) from which it is possible to derive F as a *valid* conclusion (Hempel 1965). Every *explanandum* requires an appropriate *explanans*. Knowledge progresses due not only to new observations and mental models but also to subsequent explanations beginning from models that are well known or assumed to be so. Nevertheless, explanation is a *backward-recursive process*: every *explanans* of an explanation becomes, in turn, an *explanandum* that can require a subsequent explanation, and so on up the line, just like the "game of *why*" that all children love to play. Any ignorance is explained with some piece of knowledge; but this, in turn, is based on other knowledge (Keil 2005).

I indicate with $E(n)$ the *explanans* and with F the *explanandum*; the recursive explanation then has the following structure:

- F because $E(n)$,
- $E(n)$ because $E(n-1)$,
- $E(n-1)$ because $E(n-2)$,
- …
- $E(0)$ End.

The *explanation process* is *closed* when we reach $E(0)$; that is, an explanans with no further explanandum. I define $E(0)$ as the *operational closure* of the explanation, which represents the *point of ignorance*, the limit to knowledge. $E(0)$ can denote two possible forms of ignorance:

- $E(0)$ = "That which is not yet known." This represents *reversible* ignorance, which can take on the form of: (1) *temporary ignorance* ("research in progress …," "we are checking," etc.): (2) *disinterest*, that is, satisfaction with $E(1)$ ("I'm not interested," "Enough already!"); (3) *secretiveness*, that is, the desire to hide knowledge ("I can't explain it to you").
- $E(0)$ = "That which we can never know." In this situation, *operational closure* represents *irreversible ignorance*, which can take various forms: (1) *permanent ignorance*, that is, that which *reveals itself* incapable of *ever* being known (Heisenberg's indetermination principle, Gödel and Turing's theorems, and others that are similar); (2) the *mysterious*, that is, the *unknowable* (ineluctable principles, myths, God, metaphysics or agnosticism, acts of faith, etc.); (3) the *postulate*, that is, a nondemonstrable, only assumed, origin from which to derive the explanation ("We hypothesize that …," "Given that …"; "We assume that …," etc.); (4) *chance*, in all its manifestations (quantum randomness, genetic mutations, initial impulses of combinatory systems, etc.); (5) *necessity*, that is, the self-sufficiency of $E(0)$ ("It must be like this," "It can't be otherwise," "The immobile engine," etc.). In these cases operational closure is an *explanatory dogma*.

In very concise terms, an *explanation* is given below:

(a) *Complete*, if it also includes its own operational closure; otherwise it is *partial*.
(b) *Global*, if its operational closure is a *dogma* and does not allow for further explanations; otherwise it is *intermediate*.
(c) *Coherent*, if the explanans does not include elements that contradict other explanations.
(d) *Valid*, if it derives from a correct inference.
(e) *Truthful*, if the $E(t)$ do not contain false elements that have nevertheless been declared to be true.
(f) *Logical*, if it allows us to derive the explanandum through a satisfactory operational closure on the basis of a certain stock of knowledge and in a given social context.
(g) *Efficient*, if the explanans is composed of a minimum number of premises.
(h) *Effective*, if it is complete, logical and allows for verification or forecasts.

The coherence, validity, and truthfulness of an explanation are, according to the laws of science, always verifiable; logic, efficiency, and effectiveness are instead subjective. Every attempt at a global explanation is based on irreversible ignorance. After having outlined the main features of the explanation process, we will examine five types of explanation processes that I consider the most relevant and which rely on different premises.

Common Sense, Descriptive, or Contingent Explanations The so-called common (simplistic, unsophisticated, or descriptive) explanation is the one we daily apply to justify some phenomenon or to answer some question. In the common explanation the phenomenon or question to explain is derived from models that usually describe what we must explain without looking for the underlying scientific causes; "the light has come on because Aldo has flipped the switch"; "he got a fever because he was bit by an insect," etc. The common explanation at times relies on intentions ("the light has come on because Aldo had to go into the garage"), chance ("I met Aldo because he happened to be passing by"), experience ("it's raining because I went out without my umbrella"), regularity or common sense laws ("this is a rainy spring because the winter was dry"). The same phenomenon can thus give rise to a number of common explanations that depend on the most convincing factors and that usually allow for one or at most two levels ("you made a mistake because you were distracted. What are you thinking of?"; "I'm thinking of what I'm going to do tonight"). This type of explanation is usually expressed descriptively and does not make use of sophisticated qualitative models (Hempel 1965). Included in this typology are historical "explanations" as well as "attribution theory," in order to take into account one's own and others' behavior, attributing this to the most likely "causes," such as justification (Heider 1958). Another theory included in this typology is the 5-Ws—Who, What, When, Where, and Why—which accounts for most of the explanatory needs in our daily lives. The general structure of the common explanation is as follows:

Explanans	P = descriptive, contingent or simplistic, unsophisticated Premises: – functions, – aims, – intentions, – causes, – chance, etc.
Explanandum	F = phenomenon to explain
Conclusion	F occurs because the premises P have occurred
Operational closure	This is the last of the premises that recursively explain F

The contingent explanations can easily become tautological—and thus nonexplanatory—when one tries to explain a phenomenon simply through its definition: "Inflation exists because all prices are rising"; "internationalization is an inevitable process because socio-political barriers are overcome," etc.

Classical Scientific Explanations The classical explanations are those that are used in any scientific context where the explanans is a model which, in addition to the *initial conditions* of the phenomenon to explain, C, also includes the *causal laws* or functional relations, L, as well as the *scientific theories* or assumptions that can take into account the explanandum, according to the following schema:

Explanans	T = scientific theories and postulates L = scientific laws or statistical generalizations C = initial conditions
Explanandum	F = phenomenon to explain
Conclusion	F is observed because, given the initial conditions C, it follows from the laws L, if we accept theory T
Operational closure	The fundamental theories and postulates represent the operational closure of the explanation

By varying the theoretical context and/or the laws adopted and/or the initial conditions, we also vary the explanation (Cupples 1977). This form of explanation has become recognized as the scientific explanation *par excellence*:

> Forty years ago a remarkable event occurred. Carl G. Hempel and Paul Oppenheim published an essay, "Studies in the Logic of Explanation," which was truly epoch-making. It set out, with unprecedented precision and clarity, a characterization of one kind of deductive argument that, according to their account, does constitute a legitimate type of scientific explanation. It came later to be known as the deductive-nomological model. This 1948 article provided the foundation for the old consensus on the nature of scientific explanation that reached its height in the 1960s (Salmon 1990, 3).

> Hempel claimed that "there are two types of explanation, what he called 'deductive-nomological' (DN) and 'inductive-statistical' (IS) respectively." Both IS and DN arguments have the same structure. Their premises each contain statements of two types: (1) initial conditions C, and (2) law-like generalizations L. In each, the conclusion is the event E to be explained [...] The only difference between the two is that the laws in a DN explanation are universal generalizations, whereas the laws in IS explanations have the form of statistical generalizations (IEP online, pp. 2-3).

Procedural Scientific Explanations The procedural is a very common type of explanation, even though it has not received proper attention in the literature. It is used whenever a phenomenon does not derive from a particular model but rather appears to be the result of some elaboration or calculus, or the application of some algorithm, procedure, or program (Gibbon 1998). When we ask ourselves why the solution to extracting the square root of an expression does not correspond to the answer in our textbook, why our ticket was not drawn in the lottery, or why we ran into the bumper of the car in front of us, we must look for the answer in the procedure followed for the calculation or lottery drawing, or we must examine our parking attempt. The procedural explanation can take on the following structure.

Explanans	P = procedure whose application produces F C = conditions for applying the procedure
Explanandum	F = phenomenon to explain
Conclusion	F derives from P applied under the conditions C
Operational closure	The procedure represents the operational closure of the explanation

The procedural explanation comes under the generalized heading of systemic explanation.

Systemic Scientific Explanations The systemic explanation must be used when the explanandum cannot be reduced to a model that includes laws and theories (classical explanation) or results from the application of a procedure; instead it must be considered a phenomenon connected to the dynamics of some system process whose model we are trying to uncover. The systemic explanation must highlight the processes P that have generated F; the systemic structure S that supports those processes; the programs Π that guide the latter; and the environment E that conditions them. The model has the following structure:

Explanans	S = systemic structure that generates the processes that give rise to explanandum F Π = programs that make the structure generate the processes P = processes generated by the structure S through the operations of the programs Π E = environment to which the system is coupled
Explanandum	F = phenomenon to explain
Conclusion	F derives from P produced by S in E by means of Π
Operational closure	The system (environment–structure–programs–processes) represents the operational closure of the explanation

The systemic explanation is thus more powerful, as it can take into account and justify any phenomenon whatsoever, from global warming to population dynamics, the deviation of the route of a space probe to the spread of epidemics. Most of the explanations presented in this book by using the Control Systems model belong to the class of systems explanations. This class is not an alternative to other explanations but rather encompasses them, becoming a type of even more general explanation.

Teleological Explanation Systems explanations can be kept separate from *teleological explanations*, which try to take into account the behavior of a system (usually a biological one) by using the notion of goal (Lennox 1992). Such explanations regard systems with objective, or "goal-directed" systems to which are applied the Control Systems presented throughout the book:

> The necessity of teleological explanation for biology has been questioned by the formalist tradition. We have seen that the ideal of explanation for the formalists is the deductive model. Many attempts have been made to incorporate teleological explanation into the deductive model. Indeed such an incorporation is necessary if there is to be a formal reduction of biology to physics and chemistry. The basic line of argumentation is to subsume biological "goal-directed" systems under the wider category of "directively organized" systems which can apply to both living and inorganic systems (Plamondon 1979, p. 153).

Teleological explanations are evolutionary explanations that justify certain biological phenomena linked to individuals and populations and which derive from some necessary aim: a type of *backward causation*. As an example of a teleological explanation, let us assume we wish to explain "why the winter coat of the hare becomes white." The evolutionary teleological explanation could be presented as follows:

1. Differences in color exist in all natural hare populations.
2. If in winter a hare has a white coat, it is better able to camouflage itself and improve its chances of survival.
3. The greater its chances of survival are, the greater are those that it will successfully procreate.
4. Thus, on average hares with a lighter-colored winter coat will have more offspring compared to those with darker coats.
5. The color of the winter coat is a hereditary feature.
6. Thus, the following generation will see a greater number of hares with a white winter coat.
7. After many generations, only hares with white winter coats will exist.

The model of the teleological explanation has the following structure:

Explanans	S = systemic structure that generates the processes that give rise to explanandum F G = Goal that S must achieve C = Control process carried out by S E = environment to which the system is coupled
Explanandum	F = dynamics of S to explain
Conclusion	F derives from S in E because C directs S toward G
Operational closure	The Control System directing S to G represents the operational closure of the explanation

The Best Explanation While the classical explanation is particularly useful in the context of the experimental sciences, systemic explanation can also be used to explain individual, nonrepeatable events involving unique facts deriving from a system's behavior. The former bases its validity on the initial conditions of the phenomenon in question and on a system of laws and theories that represent the final closure of the explanation process. The systemic explanation is based on the search for the most appropriate system model for deriving the explanandum F and has as its operational closure the boundaries of the system and the extension and organization of the network of processes that constitute its functioning.

The best explanation is the one considered most likely; that is, the one that best satisfies the need for the individual or scientific community to take account of the explanandum. I do not believe anyone would turn to physiological or psychological laws and theories or to logistical processes to explain "why there is no more mayonnaise."

A good explanation must permit accurate forecasts. *Forecasting* phenomenon F means developing an argumentation which, based on laws and specific conditions, allows us to deduce F before we observe it (independently of that observation). It is immediately clear that the structure of the argumentation which produces the *explanations* or *forecasts* is entirely analogous; the difference is only in the time dimension. If the explanandum is "dated" (t_1) and the explanans (t_2), we have an explanation; if, on the other hand, the premises refer to the instant (t_1) and the conclusion to instant (t_2), the argumentation is in the nature of a forecast. While the excellence of the explanation depends on the extent to which the premises justify the conclusion, that of forecasts depend on the extent to which the conclusions are justified by the premises. Thus, scientific and systemic explanations are preferred since these argumentative structures are considered to be the best even in providing accurate forecasts of the occurrence of F.

In many circumstances a *realistic approach* must be followed; as Hawking and Mlodinow have stated (Sect. 12.12), it is necessary to constantly search for new forms of explanation that can explain the observations of facts, thus enriching in an increasingly coherent manner the networks of world pictures that can provide explanations and understanding. The explanatory process can thus be linked to the theory of Control Systems, since it develops dynamically, as shown in the model in Fig. 12.14.

When there are two equally likely explanations, the one whose explanans includes the theories, procedures, and systems the scientific community holds to be the simplest and most accredited is generally considered the better one. If a given scientific community is divided into groups that accept different operational closures (laws, theories, structures, procedures, programs, etc.), then alternative explanations for a given phenomenon are possible, all of which are equally accredited. The explanation becomes unequivocal only when progress in that scientific field leads to one operational closure prevailing over the others. Often those explanations

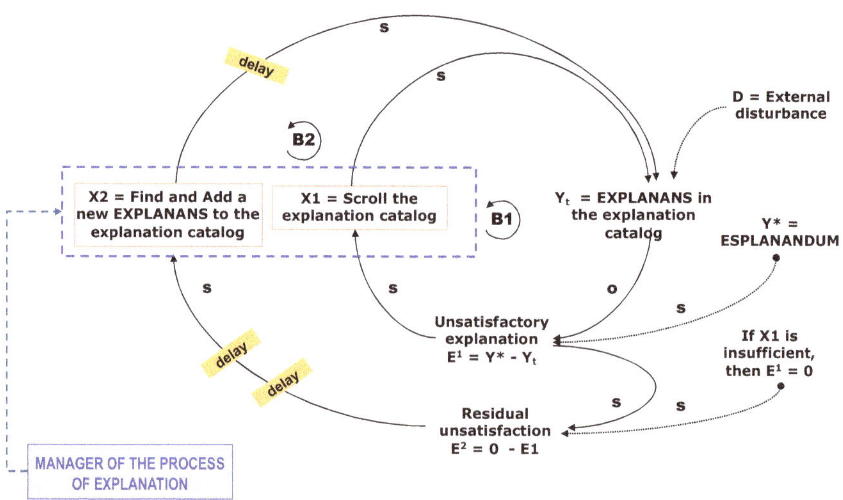

Fig. 12.14 The *Ring* representing the explanation process

are considered best that include among their premises only (or only a few) causal laws. However, I believe I have demonstrated that the inclusion of causal laws is often impossible when the phenomenon to explain derives from a system; in that case the systemic explanation is more convincing.

Finally, I would note that in order to understand and explain reality, Systems Thinking and System Dynamic approaches, jointly applied, allow us to identify the best systemic explanations by constructing models of dynamic, repetitive, and interconnected systems and simulating their operation. Objects observed from a static vision, nonrepetitive systems, individual phenomenon, simple causes, simple effects, a lack of memory: these are the errors Systems Thinking tries to eliminate.

12.15 Summary

To conclude this brief journey in the world of *Magic Rings*, this chapter has dealt with, even though with reference to a limited set of arguments, the important topic of knowledge in order to demonstrate that even thought processes, understood as emerging activities of the "mind" made possible by the processes of the brain, can take place only thanks to a large number of *Rings* operating at very high speed and in such a routine manner as to escape our attention. This chapter has focused on the following:

(a) Presenting Bateson's model of the "mind" as a calculator and transformer of differences (Sect. 12.2 and Figs. 12.1 and 12.2).
(b) Hypothesizing how the "mind" calculates similarities and analogies by constructing observational dimensions (Sect. 12.3).
(c) Indicating how the "mind" can describe objects (Sect. 12.4).
(d) Hypothesizing the process through which the mind compares objects through their technical descriptions (Sect. 12.5 and Fig. 12.4)).
(e) Presenting the *definition* process through which the mind generalizes and makes abstractions by moving from objects to *concepts* (Sect. 12.6 and Fig. 12.6).
(f) Constructing a model of a two-dimensional cognitive universe (Fig. 12.7) that represents the three paths of knowledge: *cognition of objects*, *understanding* concepts and *recognition* of objects (Fig. 12.8) as belonging to a particular concept (Sect. 12.7).
(g) Indicating the ways in which the human mind constructs meaningful technical definitions (Sect. 12.8).
(h) Presenting the signification process, through which the mind attributes meaning to the signs in order to represent and transmit its thought content to other minds (Sect. 12.9); to do so, the mind must avail itself of a number of *Rings*:

    - That control the correct form of a sign (Fig. 12.10)
    - That identify a sign belonging to a signifier (Fig. 12.10)
    - That interpret the meaning of an identified sign (Fig. 12.12)

(i) Undertaking a brief discussion of languages and the signs they are composed of, along with the eight fundamental functions of languages (Sect. 12.10).
(j) Translating Bateson's model into the concepts of factual truth and falseness (Sect. 12.11).
(k) Trying to demonstrate how Bateson's model can be extended and operationalized to even take in the important topic of scientific explanations; scientific laws and theories as definitions and conjectural models of the world are produced by advanced mental processes (Sect. 12.12).
(l) Arriving at a definition of an intelligent cognitive system as one equipped with a mind that operates according to Bateson's hypotheses (Sect. 12.13).

The chapter ends with a brief summary of the explanation process and its operational closure (Sect. 12.14.2).

Chapter 13
Concluding Remarks: Toward a General Discipline of Control

Abstract Understanding the theory of control systems is necessary to attain a true "control discipline". To introduce this "discipline", Part II guides the reader on an ideal journey in learning to recognize and model the action of multiple Rings which are observed, or even only imagined, to be operating in various environments. This chapter concludes the journey and presents some final considerations to stimulate further reflection. In the first *section* below, I shall attempt to outline several fundamental general *hypotheses* in order to propose a "control discipline." In the second *section*, I touch on the *human aspects* of control. A reflection on the content and limits of this study is presented as FAQs in the final pages of the chapter.

Keywords Control Systems discipline · The human aspects of control · Self-control · Control of chaotic dynamic · The world viewed as a single control system · Control system in religions · Variant structure control systems · Transformation systems · Production systems · Neural networks

> An analysis of the history of technology shows that technological change is exponential, contrary to the common-sense "intuitive linear" view. So we won't experience 100 years of progress in the twenty-first century—it will be more like 20,000 years of progress (at today's rate). The "returns", such as chip speed and cost-effectiveness, also increase exponentially. There's even exponential growth in the rate of exponential growth. Within a few decades, machine intelligence will surpass human intelligence, leading to The Singularity—technological change so rapid and profound it represents a rupture in the fabric of human history. The implications include the merger of biological and nonbiological intelligence, immortal software-based humans, and ultra-high levels of intelligence that expand outward in the universe at the speed of light (Kurzweil 2001, online).

In Part I of this book I have tried, using the language of systems thinking to present the structure and typology of the control systems that allow man to create a livable world and to aspire to progress. Understanding the theory of control systems is necessary to attain a true control discipline. To introduce this "discipline", Part II guides the reader on an ideal journey in learning to recognize and model the action of multiple Rings which are observed, or even only imagined, to be operating in

P. Mella, *The Magic Ring*, Contemporary Systems Thinking,
https://doi.org/10.1007/978-3-030-64194-8_13

various environments. This chapter concludes the journey into the logical world of control systems. At this point, it is useful to present some final considerations to stimulate further reflection. In the first *section* below, I shall attempt to outline several fundamental general *hypotheses* in order to propose a control discipline. These *hypotheses* lead to the conclusion that "*The world is conceivable as a Control System*" since its existence, as our own, is subordinated to a network of control systems, based on a typical multilevel holonic arrangement, which can be *viewed* as, and in fact *acts* as, a global control system. In concluding this work, I also feel that it is necessary to touch on the *human aspects* of control. Control systems are useful instruments for man, who in most cases is the manager-user of such systems. We cannot ignore the human attitude to control; thus, I have left these considerations until the concluding *section*. A reflection on the content and limits of this study is presented as FAQs in the final pages of the chapter.

13.1 A Possible Control Systems Discipline

The exploratory path into the world of *Rings* undertaken in Part II (following the *systems thinking* point of view) allows me to suggest several generalizations, or conjectures, which, somewhat emphatically, serve to outline the structure of a hypothetical control systems discipline. In order to understand and justify these conjectures, we must always keep in mind that the aims of a control systems discipline should be logical constructions—not physical apparatuses—which are not always immediately visible or understandable. A constant intellectual application is needed to master control systems, especially when these concern micro and macro environments too distant from human experience.

Let me start with the idea that if the world (as observed by systems thinking) is made up of repetitive and recursive systems of variables (Chap. 1) and actually exists and evolves in an orderly manner, except for pockets of disorder that man tends to eliminate, then it should be clear, even to a nonspecialist observer, that all the variables must be subject to some form of control. The variables of which the world is composed have limits or some objective value—determined by man or by the natural characteristics of the world—toward which they tend and to which they return when some disturbance moves them away from these values. A world without objectives, limits, or constraints that eliminate the disturbances to the component variables, both at the micro and macro levels, would be one dominated by chance, without any equilibrium or permanence. When disturbances move the dynamics of the variables away from the objectives, limits, or constraints, an adequate number of efficient *Rings* must exist to restore equilibrium, order, and stability and allow the world to exist in the form familiar to us all. Disorder and chaos are innate in many meteorological and geological phenomena, as well as in complex systems, but where possible man must control these dynamics and create order or at least protect himself from the consequences of disorder.

When this is not possible, it is not a question of the lack of *logical* control systems but of the difficulty in their practical realization through apparatuses sufficiently powerful and complex to deal with extreme dynamics. We need to only reflect on climate variables (Chap. 6) and on the disturbances to these from human activity in order to understand which wide-ranging *Rings* can regulate these equilibriums and how man tries to employ other control systems to eliminate or reduce these disturbances or seek refuge from the harm produced by the disturbed dynamics. If he is unsuccessful in these control actions and the climate variables attain other equilibriums, the world as we know it today could alter in an extremely damaging direction for mankind. New variables—heat, cold, sea levels, ozone and ultraviolet ray levels, etc.—must be controlled through systems that are increasingly vaster and interconnected. And what about the internal physiological controls and the external behavioral ones that account for our daily behavior?

From these premises I suggest the following conjectures, which go back in a systematic way to the concepts developed in the previous chapters:

FIRST GENERAL CONJECTURE: *The world is composed of control systems.*

According to the typical systems thinking perspective, the world is made up of variables, [Y], *from whose dynamics we can always identify or fix objectives or constraints,* [Y*], *whose achievement is made possible by the action of control variables,* [X]*—connected or unconnected—in balancing* loops. Ergo: *The world is made up of control systems interconnected in various ways.*

If we accept this first supposition and focus our attention on any type of control system, we realize that it is structured in a given way, since it belongs to a definite observational and operational context. Since the variables that make up the world are interconnected, by varying this context—zooming in or out—every control system is interfered with by other systems and the control must also extend to these interconnected systems. The control then becomes *multiobjective*; as a rule all multiobjective control systems must be strengthened by adding new levers. Thus, if we wish to increase crop yields and the birth rate in stock farming it is necessary to act on the availability of water by exploiting the naturally available quantities, thereby altering the flow of watercourses to create artificial reservoirs or drilling in underground strata. But this control system must take into account the drying up of surrounding areas. Thus the original control system acquires a new objective—not to dry up the surrounding areas—and must make use of other levers to provide the amount of water required by the arable land, for example, through aqueducts and desalinization plants.

Every *multiobjective* system is usually a *multilever* one as well. Each new lever that is added can, in turn, represent a variable to control for other connected systems; thus, the primitive control system is enlarged to take into account the necessary control on the levers that have been gradually added.

The *first conjecture*, together with these observations, allows me to derive a new conjecture:

SECOND CONJECTURE: *Control systems tend to become multilever and multiobjective.*

The more we extend the boundaries of the reality to control, the more numerous and interdependent are the variables [Y]—*taking into account the system of objectives* [Y*], *the control levers* [X], *and the disturbance variables* [E]. Ergo: *The more we zoom in, the more the control systems become connected and interactive; normally, it is possible to set new objectives, add other levers, and identify new disturbance variables.*

At a given level of observation control systems are usually viewed as autonomous entities; however, we must accustom ourselves to thinking of them as interacting with other systems at the same level or at a superordinate or a subordinate level. The easiest means of interaction is the single-directional one, which occurs when the dynamics of Y in system A become a disturbance for system B or the bidirectional one when, simultaneously, A influences B and B influences A (interacting systems as discussed in Chaps. 2 and 4), with the emergence of *causal loops* of varying complexity.

Apart from the specific interactions, we can view control systems as classes of observed entities placed at different hierarchical levels. This can easily be seen when we consider the fact that many control systems, though independent, are necessary for the functioning of others that, though independent as well, permit others at a higher level to function. In other words, there appears to be a holonic hierarchy (holarchy) among control systems (Sect. 3.8).

We can realize the different levels of control systems that exist as soon as we consider the fact that we can control our body's actions (movement, eating, etc.) because our organs function properly thanks to their control systems; and our organs, in turn, function correctly thanks to the cellular control systems that make up our tissue. It is equally clear that I can control my fingers for typing this sentence because I control my arm, shoulder, and entire skeletal muscle system, together with my sight organs; these words can flow from the keyboard because the neural system of the brain is under control; but this implies that the areas of the brain entrusted with the various cerebral functions are under control along with the individual neurons that have to control the impulses that arrive from "upstream" in order to pass on other impulses "downstream."

It is also clear that the control of the direction and speed of our car is possible because the various car systems and components, such as the gas level in the tank, turbo action, electrical system, and tires, are connected to specific control systems. Even the control system that tries to keep pollution emissions below certain thresholds implies, at the global level, the control of polluting companies, which limit individual emissions by controlling the energy consumption of their facilities and the maintenance state of their factories. How can infant deaths be controlled if we do not activate controls on health structures (obstetric and neonatal wards) and families (hygiene and diet)? How can we control health structures or families without controlling the level of preparation and education of the personnel? One of the problems in rich countries today is the control of Saturday night road accidents involving youth. It is clear that controlling the trend in such accidents implies controlling driving speed and the consumption of drinks and drugs. But to control speed and the alcohol level we need appropriate control systems, etc., which are part of a clear-cut hierarchy of controls.

In short, without taking anything away from the circular interconnections among control systems, it is clear that every control system at a certain level usually implies the existence of a lower level of control. These considerations lead me to the following conjecture:

THIRD CONJECTURE: *Control systems form a holarchy.*

The world can be viewed as a holarchy of control systems at various levels. The systems at each level have their own properties, new or emerging; however, they are also influenced by those at a lower level while in turn influencing those at a higher level. Thus, *the world can exist because it is made up of a holarchy of control systems.*

As mentioned in Sect. 3.8, the concept of *holarchy* comes from that of a *holon*, which was coined by Arthur Koestler (1967) to indicate any object or concept that can be observed from three points of view: (1) as an *autonomous* and *independent* entity that acts according to its own "canon" of behavior; (2) as a *superordinate* entity with respect to a certain number of component parts that it transcends and which displays emerging properties; and (3) as a *subordinate* entity that is part of a vaster whole that conditions it (for a more in-depth treatment, see Mella 2009). Koestler considered in particular the class of holons formed by *vital systems* that display the dual tendency of self-survival and integration:

> Every holon has the dual tendency to preserve and assert its individuality as a quasi-autonomous whole; and to function as an integrated part of a larger whole. This polarity between the Self-Assertive and Integrative tendencies is inherent in the concept of hierarchic order; a universal characteristic of life (Koestler 1967, p. 343).

Holons can be arranged in a "natural way" according to interconnected *levels* that make up a hierarchical ordering, which can be either vertical (holons arranged *above* and *below*) or horizontal (holons arranged *before* or *after*). The vertical arrangement is called a holarchy and the horizontal one a holonic network. Koestler (1967, p. 344) defines *output hierarchies* as those that operate according to the *trigger-release principle*. In this case, the *top holon* produces important processes that send signals to the base holons in order to condition their behavior. He defines *input hierarchies* as holarchies that operate based on the logic of successive filters; these produce progressive syntheses from the subordinate levels up to the superordinate ones, as if at each level the holons filtered or synthesized the inputs from the subordinate holons. It is clear that control system holarchies are simultaneously *input holarchies* and *output holarchies*, as shown in Sect. 3.8.

From an *ascending* observational approach—gradually zooming out—we can see that a miniscule pixel functions because a micro control system provides it with a certain state, since a superordinate control system makes that state necessary in order to adjust the state of the monitor. But the state of the monitor derives, in turn, from the control objectives of the software that is operating at that moment, which in turn responds to the control needs of the operator who is using it, who in turn is influenced by the control needs of the group he or she operates in, which in turn is affected by the control system represented by the operational center where the group carries out its activities; this is part of a larger control system made up of the department, division, and entire organization, which in turn is controlled by other superordinate systems. The same applies for all organizations that can be considered control systems with regard to the achievement of the institutional goals for which they were created; the macrolevel control directs (triggers) and depends on (filters) the control carried out by the organizational organs themselves, which direct (trigger) and depend on (filter) the control of the individual members, the processes, the operations, the movements, etc. This process involves a *descending* observational approach that zooms in from a broader to a narrower perspective.

We must now ask if observing control systems from a holonic perspective enables us to derive some generalizations. In my opinion, the following supposition immediately arises.

FOURTH CONJECTURE: *Control strategies and policies are necessary.*

The higher the level we reach in the holarchy of control systems that make up the world, the more we encounter multilever, multiobjective, and interdependent control systems. The lower down we go in the holarchy, the more we find single-lever, single-objective, and independent control systems. Thus, *every control system that is not single-levered and single-objective requires the definition and implementation of appropriate control strategies and policies.*

In fact, the examples we have proposed show that the more control systems operate at a minute level—the base level (base holons)—the more their functioning is regulated by a single lever or by a limited number of levers. The further we move away from such base systems, the more they are structured with various control levers.

1. Holons, on successively higher levels of the hierarchy, show increasingly complex, more flexible, and less predictable patterns of activity, while on successive lower levels we find increasingly mechanized, stereotyped, and predictable patterns.
2. All skills, whether innate or acquired, tend with increasing practice to become automatized routines. This process can be described as the continual transformation of "mental" into "mechanical" activities (Koestler 1967, p. 346).

While a pixel, a radar gun, or a boiler in a condominium has only one objective and one control lever, the work of a computer operator, transport minister, or project manager involves controls that make use of multiple levers to achieve a number of objectives. We cannot control the monitor if we do not take into account the program and work environment. To reduce the number of highway accidents, speed controls are not enough; we must take into account the road system, the right to enjoy oneself, the needs of night clubs and of families, etc. We cannot control emissions only through controls on home boilers in a city; we must also consider other possible levers (traffic restrictions, replacement of obsolete facilities, congestion charges for cars, and so on). We also need to consider the cost and length of operation of the various levers, in addition to other objectives, such as minimum heating levels, commuter needs, and, lastly, social discontent.

The conclusion is clear: the higher up we go in the holarchy of control systems, the more necessary it is to specify a control *policy* as well as a coherent *strategy*. This conclusion, together with the *second conjecture of control*, implies that as the number of objectives to achieve increases so, too, does the number of variables to control; therefore, the levers that are employed must increase to allow for the control, and as a result the structure of the control system expands and becomes increasingly complex. What can we infer about the relation between the efficacy and structure of these control systems?

The fundamental principle of cybernetics gives us the following supposition:

FIFTH CONJECTURE: *It is always possible to strengthen the power of a control system by adding new structural control levers. Structural levers are preferable to symptomatic ones.*

Given a realistic objective, the control levers must be quantitatively matched *to the controlled variables and to the disturbances that condition the latter's behavior.* Thus, *when the levers are not able to achieve the objective, given the disturbances, the control system can be improved either by changing the* capacity *of the existing levers (section and/or velocity) or by adding other control levers.*

The inadequacy of control systems in achieving their objectives does not always represent a structural defect in planning the number and capacity of the control levers; in many cases, the system only apparently fails when unachievable objectives are set. In this case, rather than an overhaul of the system it is necessary to set more realistically attainable values for the objectives. The problem of the adequacy of the system with regard to the objectives clearly emerges when the objective, in addition to its quantitative value, also has a precise time reference that may not be achievable by means of the normal logical and technological knowledge available at that moment. It is useless to try to sink a ship by hitting it with a slingshot; the control lever (kinetic energy of the rock) is too small with respect to the variable to control (ship's mass); it is equally useless to hit a fly (mass of the target) with a cannon (energy of the projectile).

However, we must also consider that the choice of control lever depends on both the type of objective and the magnitude of the error that arises when the variable to control is far from the objective and the significance the manager attributes to this error. We must consider that not all possible control levers are also useful ones. Many systems become multilever when we add increasingly more efficient levers to substitute the *symptomatic* ones.

In conclusion, if the control system cannot take on the states of reality to be controlled through an adequate set of *structural* levers, then the control is destined to fail as soon as the real states can no longer be represented by the system. This leads to a further supposition.

SIXTH CONJECTURE: *Control systems tend to improve their efficiency.*

The holarchy of control systems is itself a control system that continuously produces new control systems that are more powerful and precise in their logical *and* technical *structure, thereby improving the quantitative and qualitative performance.* Ergo: *Inefficient control systems inevitably tend to be replaced by more efficient ones through which new, more ambitious* objectives *can be achieved and stronger* limits *reached.*

We know that progress in control systems derives in large part from improvements in the effector, detector, and regulator apparatuses that make up the physical structure of the systems. However, equally important is the contribution to the improvement of control systems from the continuous search for new logical structures of control that stem from the gradual improvement in our capacity to observe the world.

One obvious example of this is the strengthening and improvement in the control levers regarding infectious diseases from the discoveries of Edward Jenner, the English physician and scientist who, in 1796, was the first to understand the possibility of immunization against smallpox by means of the inoculation of a milder form of the smallpox vaccine (Riedel 2005). I do not know if Jenner was aware of the existence and understood the functioning of the powerful control systems of recognition-identification represented by the immune system (Sect. 3.2); in any case, control by means of the vaccine represented not only a strengthening of the pre-existing immune system but, with respect to the knowledge available at that time, a new control system, more logical than biological, that was championed and popularized by Louis Pasteur and systematically studied by Robert Koch, the pioneer of modern bacteriology, who, through his causal laws, enabled this important class of control system to be formalized. These laws are as follows:

1. The pathogenic agent must be found in all cases of disease.
2. It must be possible to isolate the pathogenic agent in the host and grow it in a culture, that is, in the laboratory.
3. When we transfer the pathogenic agent to a noninfected host, we reproduce the same disease.
4. The same pathogenic agent must be isolatable in the new infected host.

And what about the new range of control systems that have been conceived of and developed after the discovery of Alexander Fleming, the biologist and pharmacologist who, in 1928, thought to attribute antibacterial capabilities to the *Penicillium notatum* mold, thereby deriving the famous penicillin that is one of the most powerful levers for controlling infectious diseases. Today the control of diseases in individuals has become common; nevertheless, man tends to set even loftier objectives, aspiring to a global control of infectious diseases while moving toward the objective of "zero disease" over the entire continents. In fact, cholera, plague, leprosy, and malaria have almost disappeared from our world, being confined to areas not yet reached by any control system.

Let us now consider another important chapter in the strengthening of control systems: inventions and progress in mechanics and computer science and their incorporation into robotic control systems. We have come very far indeed from Leonardo da Vinci's "lion robot," of which there is no physical trace, though there are references in his manuscripts together with suggestions for realizing his project for a robot soldier! According to historians (Rosheim 2006), this lion robot was commissioned from Leonardo to celebrate the entry into Lyon of the French king, Francis I. Perhaps in homage to this city, Leonardo designed, and according to historians constructed, a semimovable lion programmed to walk toward the king, stop in front of him, and rise up on its hind legs while opening its chest with its front paws to display a lily (symbol of France) in place of its heart (Atlantic Codex, f. 812; Madrid Codex I f. 90 v).

And who can ignore the qualitative leap in mechanical control systems due to the invention of the chip that permits digital control. Who could have imagined the progress that has taken man from the initial woodturners to modern robotic turners with laser control? Finally, who could have imagined the power and improvement in control systems that have permitted space probes to leave the solar system or the Hubble telescope (which will be replaced by the James Webb Space Telescope) to identify the most distant galaxies?

These considerations lead us to a new *conjecture.*

SEVENTH CONJECTURE: *The holarchy of control systems produces a continuous improvement in controlling our world.*

The holarchy of control systems that make up our world produces a continuous improvement, allowing us to increase the variety and strength of the possible controls and the scope of the attainable objectives. Ergo: *From a systems thinking perspective, the perpetual improvement in control systems permits the control of every micro and macro variable that makes up our world.*

We can view the holarchy of control systems as a control system that produces progress. There is no global desire to produce global progress; continuous improvement is not the result of decisions by some supreme authority but derives from the invisible hand of the unknown action of control systems that improve other control systems, which in turn improve other control systems as part of a circular causal chain that in the end produces the global improvement we are experiencing.

Echoing Koestler, we can thus consider the holarchy of control systems as an *open hierarchic system*, a *machine* made up of a holarchy of holons interacting at different levels that can survive, thanks to the action of control systems.

> I have tried to explain in it the general principles of a theory of Open Hierarchic Systems (O. H. S.), as an alternative to current orthodox theories. It is essentially an attempt to bring together and shape into a unified framework three existing schools of thought—none of them new. They can be represented by three symbols: the tree, the candle and the helmsman. The tree symbolizes hierarchic order. The flame of a candle, which constantly exchanges its material, and yet preserves its stable pattern, is the simplest example of an 'open' system. The helmsman represents cybernetic control. Add to these the two faces of Janus, representing the dichotomy of partness and wholeness, and the mathematical sign of the infinite … and you have a picture-strip version of O. H. S. theory (Koestler 1967, pp. 220–221).

This system of control systems produces general progress in life through the 2D improvement—upward and downward—in the holons-control systems, as if there were a *Ghost in the Machine*, thereby producing an inevitable evolutionary process of improvement (Banathy 2000).

This leads to the final supposition:

EIGHTH CONJECTURE: *The world is conceivable as a control system.*

The holarchy of control systems inevitably tends to transform itself into a network of multidimensional, multilevel control systems that act as a resistance to disorder and disturbances. Ergo: *The world in its unity and totality can be viewed as a single control system.*

Control systems tend to be holonically interconnected in an holonic network of relations that, by favoring local control, produce global resilience of the network as a whole. The network functions if all its Rings (in the nodes) act, according to their appropriate times and in an appropriate space, simultaneously and in a coordinated way, revealing a selfish behavior aimed at their survival within the network. The network can withstand damaging events and a lack of resources, replace local control systems with other ones, and repair the damages caused by external disturbances in the form of natural calamities. The network tries to repair the damage or restore the functionality of the destroyed control system by replacing those parts that do not improve efficiency. In other words, the network tries to survive (Arthur et al. 1997; Barabási 2002).

More inefficient control systems cease their activities, and others are created produce a higher level of control, extending control to multiple objectives. Political and legislative superstructures favor the creation of new and efficient systems that improve and expand the holonic network. In other words, the network tries to survive by eliminating every form of disturbance that can endanger the homeostasis of our autopoietic variables.

> We define survivability as the capability of a system to fulfill its mission, in a timely manner, in the presence of attacks, failures, or accidents. We use the term system in the broadest possible sense, including networks and large-scale systems of systems (Ellison et al. 1997, 97).

The holonic network of control systems is, from a theoretical point of view, capable of achieving higher objectives (improvement), as part of an unending interaction among individual lower level control systems, which are connected hierarchically and in a reticular manner to systems at vaster levels. It is precisely due to this global structure of the network of control systems that the local interactions among systems are increasingly intense and robust and produce global homeostasis acting as an "emergent algorithm."

> An emergent algorithm is any computation that achieves formally or stochastically predicable global effects, by communicating directly with only a bounded number of immediate neighbors and without the use of central control or global visibility (Fisher and Lipson 1999, p. 6).

> An emergent algorithm may produce results that are distributed globally but do not exist locally, that are entirely local, or that are represented both locally and globally (*ibidem*, p. 4)

Increasingly concrete and evident is the risk that the local control systems that, at various levels, constitute this network, will be managed in a conflictual manner, thus leading to the achievement of inopportune objectives. There is also the risk of a dearth of managers capable of dealing with the challenges posed by the complexity of structural interactions. I shall say no more, since any further comment would no longer be technical in nature but political.

13.2 The Human Aspects of Control

As we know from Sect. 3.1, in many circumstances man not only represents an element of the control systems (artificial or natural) that surround him but also normally interacts with the *control processes* those systems carry out. In artificial automatic systems, the interaction is minimal: it is manifested by a simple positioning of the objectives, with man playing the role of *governor* of the system. The maximum interaction occurs when man is an integral part of the entire *chain of control* and, using his limbs, senses, and decision-making capacity, replaces the effector, detector, controller, and communication systems.

Moreover, as a *detector*, man recognizes the *errors* in the vital variables with respect to the internal homeostatic constraints, attributing to these *errors* the meaning of unpleasant *sensations*: pain, thirst, hunger, cold, etc., which lead him to make urgent control decisions, often aimed at eliminating the symptom (unpleasant sensation), rather than to intervene on the appropriate variable to control (Sect. 5.6). Even more direct is human intervention in many social systems and in organizations, as will be illustrated in Chaps. 7 and 8.

There are many aspects that characterize the *interaction* between man, as manager–governor, and control systems:

A. Fallibility
B. Learning
C. Inventive and innovative capabilities
D. The search for active control
E. Intolerance toward passive control

A. *Fallibility* is the first and most evident characteristic of man's interaction with control systems. Understood as the human trait of making mistakes and committing errors, thereby compromising the control system's operation and the success of control processes, fallibility takes on different expressions and nuances. It can derive from an *inability* to understand how to regulate the control levers, recognize the correct variables to control, or measure the error, as occurs with children when they do not know how to move their effectors in order to use a pencil to draw a shape or with people who are driving a car for the first time and do not know how to regulate the steering wheel to guide the car or park it. Fallibility due to "ignorance" is overcome with practice and thus with *learning* and the gaining of experience. Fallibility can also be the result of various *impediments*, such as disease, disabilities, and distraction, which limit man's possibility of action regarding the apparatuses that form the system's chain of control. "I could not hear the alarm because I was listening to the stereo"; "My eyes were tearing up, so I couldn't see I was about to hit the wall"; "I had drunk too much and my reflexes were too slow"; "I was on the telephone and couldn't turn off the tap in the bathroom, so the house was flooded"; "Too tired, too distracted, too fast, too little light, too tiring, too changeable, etc." are all examples of factors that hinder efficient control.

I will mention other examples. The Greek general (manager) orders the standard bearer, in the heat of battle, to display the flags (which are codified) that give the cavalry the order to attack. However, because of the noise the standard bearer, who is in charge of communications, mistakenly orders the infantry to retreat from battle, thereby causing the army's defeat. Spotting an enemy submarine approaching, the commander orders the helmsman to "steer to starboard," but the latter has been injured and carries out the maneuver too slowly, thus causing the ship to be hit by an enemy torpedo. The head oarsman of a trireme (regulator), who calls out the rhythm to the rowers, receives the order from the commandant (manager) to increase the pace of the strokes in order to avoid a collision (detection of distance and speed) with the enemy vessel; however, the rowers (effectors) are exhausted and cannot keep up with the pace, and thus the ship is rammed. The motorcyclist (manager) encounters a very tight curve but, due to his or her speed, is not able to regulate the motorcycle's inclination (control lever); thus, he does not succeed in (effector) negotiating the curve, ending up off the road.

These few simple examples clearly show that human fallibility can intervene in an evident or subtle manner in all phases of the control system to produce an inefficient functioning of all the apparatuses in the chain of control, including the system for the transmission of information. There are three basic lines of action to mitigate the fallibility of man's intervention in control systems.

The *first* is obviously the proper *instruction* of the manager–governor regarding the features of the various elements in the control system. The preceding *section* of this chapter indicated the control system features that can represent a source of failure; we must know what these are and avoid them. The *second* line of action is to provide the manager–governor with the necessary support to facilitate his or her interventions: typically, the *error signals* (e.g., acoustic alarms for distance, rear video cameras for cars, speed limit controls, and detectors of moving obstacles), *calculation systems* for calculating the optimal regulation, automatism in the use of the effector, etc. and the other *algedonic signals* that warn about the impossibility of achieving the objectives. The *third* and most drastic form of reduction in human fallibility is the complete substitution of man in control systems by transforming the latter into automatic systems (errors by surgeons are avoided by robot surgeons; driving errors are avoided by automatic systems; braking errors are avoided by automatic braking systems; pilot errors are avoided by an automatic pilot; etc.).

B. To compensate for his *fallibility*, man possesses a *learning capacity*; he is able to *learn*, that is, to *control* himself by transforming himself into a true control system of his own actions in order to eliminate errors committed as the *agent* of the control. *Learning* by the manager is manifested, in successive cycles of a control system, as the *ability to become more and more effective in eliminating the error* through more precise and timely regulation decisions. Learning increases the manager's *experience* and improves his or her capacity to detect, regulate, and implement the operations needed to control the dynamics of the passive variables, as shown in the model in Fig. 13.1, which refers to the general model in Fig. 3.1, to which has been added the control systems of the managers themselves.

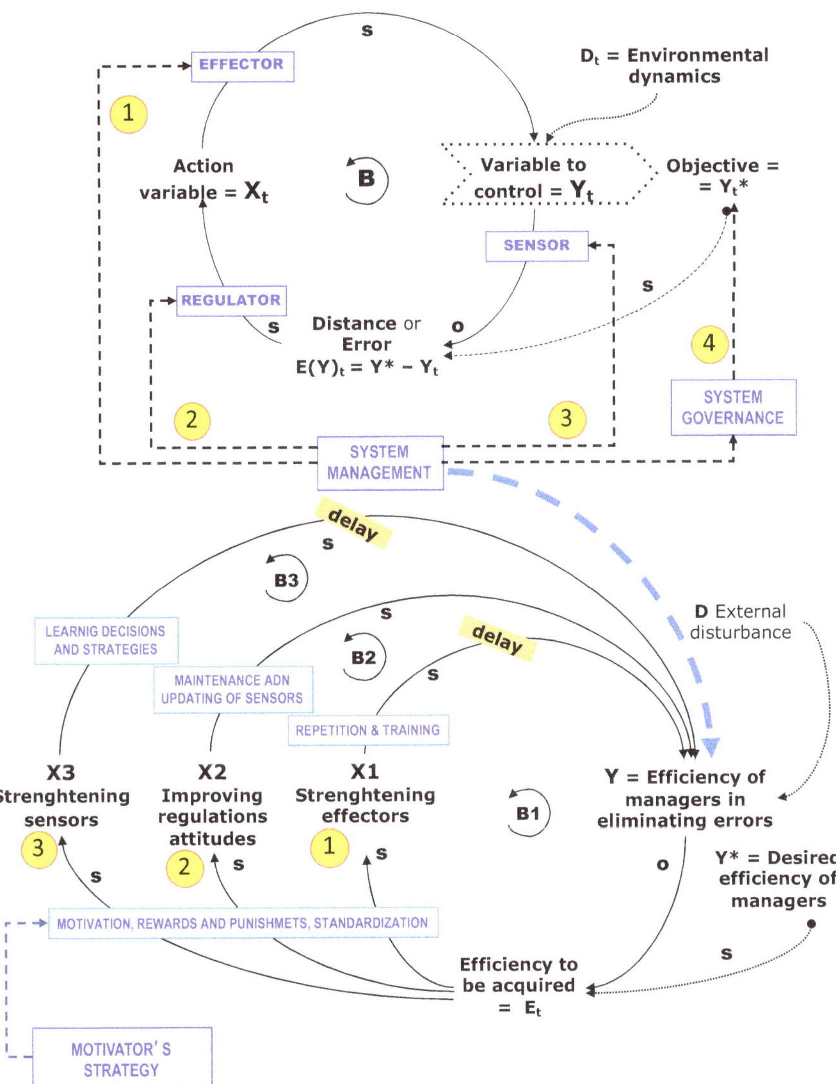

Fig. 13.1 Managerial learning for the efficiency of control (starting from Fig. 3.1)

Whenever control is carried out by man, the ability to correctly maneuver the control levers is essential for an increasingly more precise and effective control. When man is part of the *effector* apparatus, learning is in the form of *training*, mental and muscular improvement, to acquire *automatism* in actions. When man is part of the *detection* apparatus, he must prevent or mitigate the decay in the error perception organs by correcting focusing problems through proper lenses, maintaining the hearing and olfactory apparatuses, and so on. When man intervenes as a *regulator*, he must avoid alterations in judgment, a decline in the capacity to calculate, by resisting the taking of drugs that can inhibit these capacities or by using drugs that can block the decline in the cognitive functions needed for regulation. There is no need to provide abstract examples; everyday life testifies to the importance of learning and improvement in controlling our driving, handwriting, gesturing, postures, and relationships with our neighbors.

To facilitate learning and consolidate the influence of experience in human control, the most effective line of action is the standardization of the means of man's intervention through the formalization of procedures, strategies, and, more generally, *routines*. Section 4.5 introduced the concept of *routine* (or plan of action) to indicate a consolidated strategy the manager feels is efficient to apply to the manual control of a control system that involves numerous cycles or to regulate similar control systems in all similar control contexts. The consolidated and formalized *routines* facilitate learning on the part of various individuals involved in managing similar control systems with regard to the variables to control [*Y*] and the levers to use [*X*]. In learning and applying the *routines*, even the manual control tends to become similar to an automatic control for the manager–governor, since the routine indicates to the manager how to regulate the lever without having to calculate and decide anew, on each occasion, the variations in [*X*] and the maneuvers of the apparatuses that form the chain of control.

I can still remember how difficult it was, during my first driving lessons, to guide that formidable system of multiobjective and multilever controls in the car. Regulating the wheel to turn right, but not before regulating the directional indicator, and at the same time regulating the speed in adequate proportion to the driving maneuver, regulating braking by judging how much room there is to stop. How many uncertainties and repetitive actions there were! Gradually, by following the *routines* recommended by the driving instructor and learning how to apply them, and at times improve them, I learned to drive a car, and now, after years of driving, it does not seem possible that I had to learn to apply so many *routines*. Subsequently, I decided to get a pilot's license for a lightweight airplane. The control of even a small plane requires the application of a myriad of routines. There is a routine for every regulation, every lever, every objective, and every external disturbance. Check the runway and wind, listen to the instructions from the control tower, take off at the right speed, check to see the sky is clear, go at full throttle, gain speed, pull back on the control stick, rise to 450 feet, fold in the landing gear, give feedback to the control tower, rise to the proper altitude, avoid the off-limit flying zones, regulate the engine revs, check the direction of flight with the compass and correct your route, correct the lateral movement if there is lateral wind, check the temperature of the cylinders, check the sky up ahead, gain altitude, descend if you are too high up, etc. For each control lever there are smaller routines that indicate how many levers to use at the same time, which maneuvers have priority, etc.; all these controls represent the normal flight routine, which has to be repeated for each flight, and most of which must be repeated at regular intervals, many times, for each flight.

Which routines do we follow each morning to get ready to go to work? Which routines do chefs follow to prepare the various dishes ordered by customers at the same table? There is probably no need for other examples to understand that routines are the result of the *learning* needed to reduce the *fallibility* that characterizes man's interaction with control systems.

Learning is typical in most animals; many of their controls are guided by innate routines; for example, all spiders know which levers to activate and in which order to individually weave a spiderweb; all squirrels know the routine to collectively follow in building a small dam. Many routines in the animal kingdom produce unstoppable control automatisms. The lowly dung beetle pushes balls of excrement following an innate routine that obliges it to use its back legs by walking backward; it continually repeats this innate routine by always moving its legs in the same way, simulating the pushing movement even when the excrement ball is taken away from it (Russell and Norvig 2009, p. 39).

Normally, there is no learning in mechanical or physiological automatic control systems. On the contrary, in such systems there is usually a decline in performance over time. To combat this, there is need for *maintenance* of the efficiency of all the apparatuses that make up the chain of control or at least for *prevention* to slow down the gradual decline in the system's operation. Nevertheless, learning can also be found in artificial automatic systems, especially in recognition control systems (Sect. 3.2), where the regulation is carried out through neural networks that "learn" how to maneuver the control levers based on the frequency and type of error, reacting increasingly more efficiently. A typical example is the modern video game, where a human player competes against the routines carried out by a computer which, with its "artificial intelligence," learns to react with increasing precision, thereby forcing the player to also learn by devising game strategies, that is, his or her own personal routines.

C. A third characteristic distinguishes human intervention in control: *conceptual* and *innovative* behavior, in other words, man's capacity to identify and conceive of new control levers to strengthen systems, thereby further increasing his experience and laying the groundwork for new forms of learning, as part of a reinforcing *loop* that is continually acting to produce the *progress* we can all observe in the world.

There is no comparison between the rudimentary control levers of primitive sailboats with rudder bars and the modern servomechanisms of the giant cruise ships that carry 5000 passengers and punctually control room service, meals, entertainment, etc. Think about the leap in quality in the control of picture taking. The first cameras had completely manual controls, with imprecise, difficult-to-regulate levers. Today's modern reflex cameras have made man the *governor* who decides when and where to take photos. The *manager* has become automatic and regulates focusing, shutter speed, shutter openings, and other levers with the utmost precision, thereby always guaranteeing quality photos. Consider also the vastly improved quality in modern plasma and LCD television technology compared to the first sets with manual controls; and how far we have come from wood fireplaces and stoves to the modern thermoconvectors that regulate the desired temperature for each room in our house separately. It is easy to recognize the improvements in the levers for the technological, social, biological, and medical controls that make our lives easier in an increasingly crowded and complex world. All these improvements are the result of man's capacity to continually invent more sophisticated control systems and of the continual innovation in the apparatuses that allow these systems to operate.

D. After this rapid analysis of the interactions between man and control systems, several considerations about the emotional aspects of human intervention in *control processes* are in order. Many control systems in which man is the manager–governor are designed to control the behavior of other men, in the broadest sense of the term. We can thus distinguish between *controller* (or active) *subjects* and *controlled* (or passive) ones. *Controller subjects* are individuals, groups, and organizations that represent the *active elements* in the control processes; as governors they set the objectives of the control and design and produce the control systems in order to control the behavior of the other *controlled subjects*, according to a predetermined policy. *Passive subjects* of control are those that must be controlled in, among other things, their actions, behavior, and interactions while respecting the constraints in the achievement of their objectives.

Applying the first rule of systems thinking and zooming in or out, it becomes clear that man, when observed as a *controller subject*, wants to extend his control (at times, his *dominion*) to cover every controllable variable regarding the behavior of other men but also, and more generally, regarding his own body, environment, the animal, and vegetal world as well as the physical environment in the broadest sense. Through his conceptual and learning capacities man can continually identify new control systems and increasingly more powerful control levers in order to travel to deep space and control the weather, desertification, global warming, disease, social groups, minutest parts of his cells, and even biological clocks that determine his longevity.

I propose the following *first supposition*: Thanks to his intelligence, there appears to be a "natural" (even innate) tendency in man to control every variable that impacts his existence as an individual, a member of a group, and as a species. We can paraphrase John von Neumann's quote at the beginning of Chap. 2 and state that if "*All unstable processes we shall control,*" it is also true that "Man *can*, and *must* control all the unstable processes in his world," particularly those that can benefit his existence and well-being.

Through science and technology man can conceive of, design, and produce evermore powerful and precise chains of control of all kinds. Aided by mathematics, quantum physics, space, nuclear, genetic and computer technology, nanotechnologies, materials science, and all other forms of invention and discovery, man can design multilever and multiobjective systems that are increasingly vast, complex, and flexible in order to control not only himself and his own environment but also the infinitely small and infinitely large elements in his world.

E. If we consider man as a *passive subject* of control, the postulate in the first supposition appears to become inverted: man displays a natural aversion to being controlled. Rules are followed only if forced on him, and, in any event, they are not accepted willingly. Man is clearly "rebellious" toward all forms of coercion, and it is also evident that he wants to be the *controller* of the behavioral variables, not *controlled* by them. This attitude comes out strongly in revolutions or in wars against an enemy that wants to subjugate a population; but it can also be affirmed by all those that unwillingly accept undergoing periodic medical checkups, submit tax returns, or obey the rules set by organizations and social groups. And it is the attitude even of the most observant religious individuals who must follow the precepts imposed by their church.

In light of these observations, we can posit the following *second supposition*: Man displays a "natural" *resistance* to being a passive subject of control processes. Thus, the aforementioned quote by John von Neumann could be completed as follows: "Man *wants* to control all the unstable processes in his world, but he possesses a natural *resistance* to being the passive subject of the control processes that concern his own behavioral sphere."

A final observation of the human aspects of control regards *self-control*. I have presented numerous examples of *self-regulation* regarding variables related to man as a psychophysical entity: the control of equilibrium, temperature, needs, aspirations, handwriting, etc. However, now I wish to deal with *self-control*, a term I use to indicate the control exerted by man over his behavior, lifestyle, moral and ethical choices, etc. as well as the capacity to achieve given standards of behavior or directly desired objectives chosen from other possible objectives. Self-control is not an innate attitude; man experiences instincts and reacts to these often in an unconscious manner. Self-control modifies human behavior to produce desired actions and reactions, which are often imposed on man. The defensive instinct leads most of us to turn away from the sight of blood and to avoid touching snakes, mice, and even insects. And yet millions of persons have activated forms of self-control to modify their conscious behavior: doctors, nurses, and first-aid personnel treat wounds every day; biologists, researchers, and even children will themselves do not fear contact with certain animals.

Many of us instinctively react physically and verbally to violence and oppression. A good deal of self-control is needed to forego such instinctive reactions and rely instead on legal remedies. Many people are content with their present state, while others try to improve their situations even if this entails great sacrifice. Not to mention the self-control of practitioners of Yoga and meditation, who regulate the rhythms of their breathing, move their limbs into incredible positions, and even control their states of consciousness. In conclusion, self-control enables man to consciously modify his state and the quality of his behavior for a host of reasons, as the following quote clearly illustrates:

> Self-control refers to the capacity for altering one's own responses, especially to bring them into line with standards such as ideals, values, morals, and social expectations, and to support the pursuit of long-term goals. Many writers use the terms self-control and self-regulation interchangeably, but those who make a distinction typically consider self-control to be the deliberate, conscious, effortful subset of self-regulation. [...] Self-control enables a person to restrain or override one response, thereby making a different response possible (Baumeister et al. 2007, p. 351).

Clearly, self-control follows the general model of every control process. It is necessary to define the behavioral variables to control, Y, and the objectives, Y^*, as

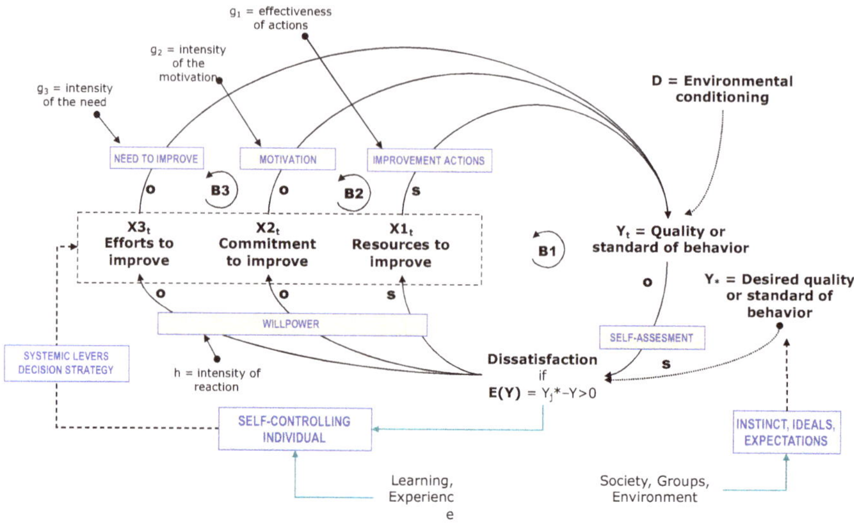

Fig. 13.2 The general model of self-control

well as to specify the apparatuses that form the chain of control. Figure 13.2 proposes a simple model of a system that carries out the self-control on a single behavioral variable. There can be multiple and varied control levers. The model in Fig. 13.2 indicates three such levers: the resources, commitment, and effort acceptable as levers through which the individual can implement the self-control. Each lever could be individually specified and have different effectors that provide different actions for improvement with different action functions, g. When the self-assessment detects a gap between the desired quality of behavior and the actual behavior, thus producing *insatisfaction*, the individual must then vary the state of the levers through a reaction function h, which represents the level of *motivation* and the *cultural state* of the individual; this reaction function can modify his or her *willingness* to provide greater resources, commitment, and effort to improving his or her state.

It is not only athletes, scientists, researchers, religious personalities, or ascetics that produce continuous forms of self-control; anyone possessing self-esteem and the esteem of others will wind up reflecting on the quality of his or her behavior, attitude, and relations and undertake self-control, even though other alternatives are possible that often are more convenient and easily realizable, though less stringent and imperative.

As a conclusion to the above observations, I would suggest completing the first two hypotheses on man's attitude toward control with the following auspicious ending: "if man *wants* to control all the unstable processes in his world but possesses a natural *resistance* to being the passive subject of the control processes that concern his own behavioral sphere, he should at least be capable of controlling himself."

13.3 Conclusion: Frequently Asked Questions (FAQ)

After this long journey, I have recommended the reader undertake to help him or her explore the magic world of control systems, it is useful, as an aid to further reflection, to go back and present several basic themes as recurring questions that anyone might ask himself or herself or me.

Are the control systems we have studied in this book different from those presented in books on the theory of regulation and control in engineering?

Absolutely not, even if they seem different, since I have employed a different language and different diagrams to represent them. As shown in Sect. 2.16.10, engineering books generally use rectangular boxes that contain the processes, or functions, that transform the inputs, *X*, into the outputs, *Y*, which are indicated on the arrows. Moreover, in engineering books on control systems, the objective, *Y**, is usually considered as the fundamental informational input of some "technical machine" that must achieve the objective, to which the *deviation* is added via the feedback line. Our model, on the other hand, emphasizes the relation between the control levers, *X*, and the controlled variables, *Y*; the objective, *Y**, enters into the calculation of the deviation and is represented on the right, usually next to *Y*. Finally, engineering books deal with real systems, mechanical, electrical, hydraulic, etc., that is, they deal with *technical* control systems made up of actual apparatuses. This book instead presents control systems as *logical models* which must be constructed in the real world by a "control chain" of technical apparatuses. I have considered some concrete control systems the radio, shower, bathtub, car, submarine, etc. in order to indicate models that are easy to understand and remember.

Are they then different from those presented in mathematics books on systems theory?

Here, too, the answer is no. In this book I have presented the *logic* of control without recourse to formal mathematics, preferring simple linear systems (when possible) with constant coefficients or systems similar to these. System theory texts—or those on system science—deal with abstract systems that are usually based on the theory of differential equations that connect *X* and *Y*. Math books that examine systems also consider a third variable, the *internal state of the system*—whose variations are intermediate to those of the inputs and outputs—by studying the relations [$X \rightarrow$ internal state] $\rightarrow$ [internal state $\rightarrow Y$]. The internal states of systems are not explicitly considered in this book since the relations between *X* and *Y* have been directly described by means of the simplest *transfer functions*.

The text presents models of very simple real systems. Are these typical models, that is, prototypes?

The answer is yes. The aim of this book is not to explain control systems that are designed or constructed in concrete situations occurring in the physical, biological, and social worlds, but to propose several simple models to aid the reader in observing and understanding the magical functioning, the wonderful complexity, and the incredible variety of control systems that enable our world to exist and progress.

How do control systems function, and what is their significance? These are the questions I hope the material in this book will provide answers to. Precisely for this reason I have chosen to utilize quite elementary, at times trivial, models of control systems, but ones which everyone knows and has experienced. Each of these elementary systems has a unique characteristic that influences the control process. Thus, they can be considered prototypes of elementary though general systems. The *control of the level of liquids*—in the example I used, a cistern—is the prototype of control systems with dependent and constrained levers. The *control of velocity*—in our example, referring to a car—represents the prototype of multilever and multilevel control systems that require a control strategy. Finally, the control of *3D movements*—in our case, a traditional airplane and a Cartesian robot—can be the prototype of multiobjective systems that require a control policy. The numerous variants, together with the other examples (including those that concern the interconnections among systems), can always be linked to the preceding prototypes.

Why is the book divided into two parts?

Part I, entitled "Discovering the Ring", presents the logic of control systems and proposes a minimal typology. Part II, "The Magic of the Ring", tries to illustrate the significance of the ubiquitous presence of control systems in all observational contexts and at every possible level we might zoom in on. Borrowing Charles Morris' (1938) three dimensions to describe languages (syntactic, semantic, and pragmatic), I suggest interpreting Part I as the *syntax* of control systems and Part II as the *semantics* of these systems. The *pragmatics* of control systems, understood as the answer to the question "How do we do things with control systems?" can be perceived by intuition throughout the book or can be recognized by observing the real world. In fact, the *Rings* are instruments for achieving an objective, respecting a constraint or staying within a limit. They exist and are constructed to allow for the ordinary operations of any kind of process.

Can the control systems presented explain everything?

The answer is yes, but under the condition that the reader succeeds in applying the models to both the macro and micro environments presented in Part II, following the general rule of systems thinking that recognizes the importance of zooming out and in to observe and understand the world. Though elementary in nature, the prototypes I have proposed are useful models for understanding the logic of control as well as for identifying the means by which this logic can be translated into concrete observations, more so than into concrete realizations. They are not operational models and cannot explain everything, but they do offer a general and broad vision of the problem of control from a systems thinking point of view.

The text also presents simulation programs. Are these in any way operational?

The simulation programs are above all illustrative and not immediately operational. The models are elementary just like the control systems they simulate; they do not take into account the numerous complications and disturbances that can present themselves in the real world. Their aim is simply to numerically verify the coherence of the models presented. For "small" systems I have preferred to build the models using Excel—which I think everyone is familiar with, at least its elementary commands, which were the only ones adopted—and Powersim, one of the many available simulators for "large" systems. I admit that the reader might prefer setting up my simple programs in a different manner to make them more elegant and complete. Nevertheless, the more expert reader could easily simulate the control systems by building ad hoc programs using any personal program language available today (from the immortal C to C++ to Java). Even *Mathematica*, "*The world's ultimate application for computations,*" as written in the Wolfram Research Web site (http://www.wolfram.com/mathematica/), allows us to simulate highly elaborate control systems by using differential equations and differences at any level of complexity.

What is innovative about this book?

The book presents several innovative features. Formally, the most evident is that it sets the control process within the framework of the general theory of systems, which is presented following the systems thinking approach. In fact, the book borrows its terminology and models from systems thinking. In terms of substance, the book seeks to launch a message more than to present a technique.

What is the message of the book?

Even though we are not accustomed to seeing them, control systems dominate our world, our life, and our very existence. "Seeing" control systems is not difficult, but we are not trained to think in terms of control. The control process takes shape in one or more balancing *loops* that are produced by control systems of every type, in every context, and for every purpose. To observe them we must accustom ourselves to systems thinking; but, as many have pointed out, teaching people to think in terms of *loops* is not widespread in our educational system, which is instead oriented—in terms of both science and techniques—toward teaching how to identify chains of open causal relations, along the lines of Kantian thought. The *message is therefore* if you truly want to understand the world, use systems thinking and, above all, think in terms of control systems. Search for the *loops* as well as the objectives, constraints, or limits toward which the variables tend. Finally, try to understand which variables are the control levers and how they operate to reduce the distance from the objectives.

Is control theory a science or a discipline?

Picking up from the previous message, here is another piece of advice: Do not be content only to observe control systems; you must get used to *thinking* in terms of *control processes*. To do this, you must train yourself, apply the concepts you have learned on a daily basis, and broaden your observational experience. In short, you must consider the observation of the world—in terms of control systems—as a *discipline of control* that will help you improve your intelligence and expand and improve your capacity to create models to understand and to explain and forecast phenomena. The simpler the basic instruments are, the more we will understand the control mechanisms; the more we understand the control systems, the more we will apply ourselves to the control discipline; the more advanced we are in the discipline, the more desire we feel to learn sophisticated instruments of understanding and simulation from books on engineering, math, physics, biology, economics, sociology, etc.

Is this why the text favors simpler control systems?

Yes. You cannot undertake the discipline of gliding by trying to pilot a glider without instructions and proper training. Similarly, unless you are an engineer or a mathematician, it is not possible—or, if possible, certainly not profitable—to attempt to understand the control methodology in terms of differential equations and complex feedback. I have preferred an elementary, intuitive, visual, gradual, nonmathematical, and logical approach to control, using the language of systems thinking, The reader who studies the control discipline "will know" when it is the right moment to move on to more sophisticated techniques.

Are there other topics the book does not deal with that should be studied?

The answer is certainly "yes." In Part II of the book, I have tried to present a large variety of control systems that surround us and appear in the most varied environments: within ourselves as individuals, in our homes, in our towns, in animal populations, in society, and in organizations. I have even quickly delved into the world of weather phenomena. My aim was to explore control systems that allow man as a psychophysical being to exist, briefly examining the control systems that could serve to operationalize Bateson's model of the mind. Though I have tried to examine various topics, many have had to be omitted due to the difficulty in entering into technical details as well as the vastness of the control problems that I would have had to deal with. For example, I have not considered the physiological control systems of the human and animal body in general and its organs in particular. Any physiology book on humans and animals is so voluminous as to make it difficult to choose the topics to treat. Neither have I taken up, even in part, those control systems activated by the endocrine system; apart from some brief comments in the preface, I have also avoided examining control systems that regulate the functioning of our nervous system and brain to produce thought and behavior. The knowledge regarding chemistry, electrochemistry, and neuronal and endocrinal physiology is so specialized that translating this into control system models would not have been very effective. The control systems chosen for presentation in this book therefore do not exhaust the types of systems or the possible observational contexts. There is much more also to be written on social and organizational control systems as well as on physical and biological ones. Finally, having chosen a didactic, nonmathematical approach, several important aspects have been neglected, three in particular: the problem of the stability of several multilever control systems that are interconnected (interfering showers, for example); the behavior of nonlinear control systems; and the relation between control theory and problem solving. Though I have simply noted this last point, the topic deserves more detailed treatment.

The text does not mention the problem of controlling the future, that is, the process through which the present produces the future but the future conditions the present. Can we control the future?

This is a perplexing question. Perhaps I have not explicitly written that there is a mutual relationship between the present and the future, but was there any need to?

The entire control discipline assumes this relationship; in fact, it is based on this. Let us reflect a moment: The idea itself of *objective* indicates our capacity and will to orient our actions toward future states we desire to achieve; the idea itself of a control lever assumes the possibility of acting "today" to produce the dynamics that help achieve "tomorrow's" objectives. Asking if we can control the future makes no sense in the context of control theory; by definition, control always involves the future (objectives), which conditions the present (control levers), which produces the future (movement of Y toward Y^*). In any event, in many parts of the book I have explicitly considered the processes of *change* and *decision-making* (and thus of the desirable or the necessary future) as control systems. Therefore, the control concept is intrinsically linked to the concept of future.

With extreme synthesis, what is the essence of the control discipline?

As I have observed several times in the book, control systems are at the basis of the order, and thus the existence and progress, of the world, which is understood here as the system of interconnected and interacting variables that must be subjected to some control system. Very succinctly, the discipline of control proposes three main teachings.

First: If we truly want to understand the world it is necessary but not sufficient to observe it in terms of systems of interconnected and interacting variables (systems thinking); we are also obliged to recognize and understand the operational logic of the control systems that condition the movements of those variables toward the achievement of some objective, constraint, or limit.

Second: Control systems occur at multiple levels and in ever broader contexts; the higher up we go in the hierarchy of control systems, the more powerful these become. Starting from single-lever and single-objective systems—typical of the more minute biological automatic control systems—we see that these progress, at higher levels, to artificial multilever and multiobjective systems that are increasingly more powerful and flexible. For this reason, we can think of control systems as elements of a typical dual-directional holarchic arrangement.

Third: There is a close relation between problem solving—which concerns the malfunctioning of the system—and control systems, since, in general, problems derive from the lack of control systems, the inadequacy of the existing ones, or their malfunctioning. When we do not correctly apply control systems to the variables in our life, then we are the cause of our problems. If we encounter problems we must always ask the following: Which control system did not function and why? Which control systems can we apply? How can we improve the control process?

For these reasons, the discipline of control must accompany us all through life in order to accustom us to identifying the objectives (constraints, limits), Y^*, and the dynamics in the Y variables that make up our life, and to help us understand which levers, X—guided by the amount of error $E = Y^* - Y$—can determine the dynamics of any system, taking into account the external disturbances. Only in this way can we control our future.

13.4 Complementary Material

13.4.1 Control System in Religions: Approaches in Catholicism and Buddhism

It is not easy to find a shared definition of "religion." However, taking a cue from ancient Western tradition, Cicero's definition appears acceptable to me: "Religion is all that concerns the care and veneration directed at a superior being whose nature we define as divine" ("Religio est, quae superioris naturae, quam divinam vocant, curam caerimoniamque effert" Cicerone. De inventione. II,161).

In general, all religions can be distinguished on the basis of three variables (the three "C"): the "creed," that is, the salvific end religion is thought to allow us to achieve; the "cult," that is, the practices adopted to carry out the salvific behavior that guarantees we will achieve the goal of salvation; and the "Church," that is, the set of structures (places of worship) and individuals (the clergy: priests, gurus, imams, etc.) that in various ways control the diffusion of the "cult," which allows the faithful to achieve their "creed" of salvation. I hold that every religion is structured in such a way that the "believers" must and are able to control their behavior so that it is in keeping with the optimal salvific behavior specified by the "creed" and controlled through the "cult" supplied by the "Church." This logic also applies to all kinds of "salvific practices," even those of a nonreligious nature, as, for example, in Buddhist practice.

Every religion and salvific practice thus includes (or in itself can be considered as) a multilever control system whose general model includes a *structural* control and also at times a *symptomatic* one. In such a control system, the objective is a "minimal (sufficient) state of perfection" and the variable to control is the "actual state of existence"; the gap between these two states represents the "imperfection" to be eliminated (defined qualitatively and often also quantitatively), and this can be interpreted and denominated in several ways in the differing religions: sin, sorrow, disobedience, blindness, offense to God, etc.

When the individual is aware of being in a "present state of imperfection," he or she must try to eliminate the "imperfection" (error) by means of control levers established (or suggested) by the "cult" proposed by his "Church." These control levers are normally modifications in the quality (and often in the quantity) of the present behavior in order to bring the "present state of existence" closer to the "minimum state of perfection." We can distinguish between levers that act in direction "**o**," or *anti-evil levers*, in that these try to eliminate the negative characteristics of behavior that inhibit the "present state of existence" from reaching the "state of perfection" set as the objective; and levers that act in sense "**s**," or *levers of good* (broadly speaking and to be specified on each occasion), which improve the "present state of existence" not only by eliminating the imperfection but also by bringing the present state to levels that are superior to the objective, thereby producing a "state of holiness" (grace, illumination, inner peace, etc.).

Figure 13.3 describes the general model I have outlined. The system is normally multilever, since it relies on a multiplicity of *structural* levers of the X1 type (in favor of good) and the X2 type (against evil). It is likely that once the error (imperfection) is eliminated the manager–believer will continue to activate the "levers of good" by giving $E(Y)$ a negative value. In this case the error takes on the meaning of "perfection," and the "state of perfection" produces some form of "holiness," whose meaning is specified differently in each religion.

Many religions have "rules of the Church" that permit recourse to *symptomatic levers* in order to *quickly* eliminate the error without activating the structural levers "for good" and "against evil" (which are difficult to activate, as their action is very *slow*). For example, the Catholic religion has the sacrament of Penitence (remorse, confession, and good intentions), which leads the believer to temporarily eliminate his or her state of sin (imperfection). In Hinduism, bathing in the sacred rivers in the sacred cities is permitted (on special occasions), the Ganges being the most famous, as a way of obtaining forgiveness for one's sins and as an aid in achieving salvation. Similarly, going on a pilgrimage to Ayodhya, an ancient city on the right bank of the Ghaghra River, and bathing in the river can annul even the worst acts imaginable (e.g., killing a Brahmin).

Without pretending to be complete, but only to suggest a guide for the reader, I will try to apply the model in Fig. 13.3 to outline the control system inherent in the Catholic religion and in Buddhist practice.

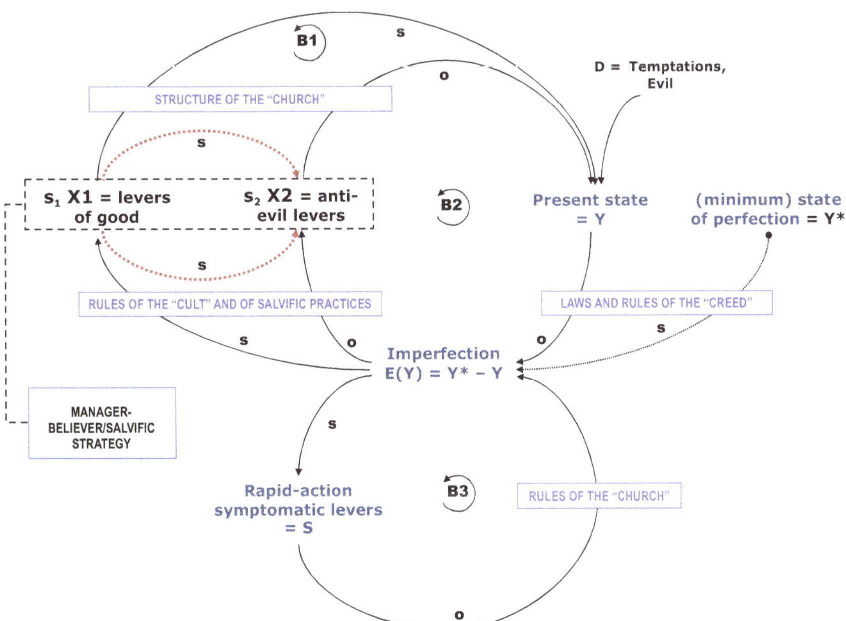

Fig. 13.3 General model of a control system of a religion

In my opinion, the Catholic religion is based on four principles well known to the believer: (1) sin exists (as does evil); (2) one must be "healed" from sin and maintain oneself in a state of grace in order to merit eternal life in God's presence; (3) there are ways to redeem oneself from sin and acquire and maintain a state of grace; (4) the Catholic church represents the "vehicle" (teachings, practices, and structures) for eliminating sin from the life of the faithful and for allowing him or her to gain eternal life. A few brief considerations are in order here (I apologize for the summary and partial nature of what follows).

The first principle should be entirely evident: Sin is an offense to God. We are born with original sin, and Jesus has taken on Himself the sins of all mankind, redeeming the world through his death on the cross, thereby reconciling man to His Father: "*God the Father, 'sending his own Son in the likeness of sinful flesh and for sin, he condemned sin in the flesh'*" (Encyclical Letter "Dominum et Vivificantem" of the Supreme Pontiff JEAN-PAUL II 1986, Catechism, online). Sin appears to be unavoidable:

> Man cannot avoid having at least venial sins, as long as he is alive. Nevertheless, he must not give too much weight to these sins, which are defined as venial. They seem insignificant when you commit them, but how frightening when we add them up! Many little things put together form weighty things: many drops fill up a river, just as many grains of sand a heap. What hope remains then? First and foremost, Confession should be made … ["*Non potest homo, quamdiu carnem portat, nisi habere vel levia peccata. Sed ista levia quae dicimus, noli contemnere. Si contemnis quando appendis; expavesce, quando numeras. Levia multa faciunt unum grande: multae guttae implent flumen; multa grana faciunt massam. Et quae spes est? Ante omnia, Confessio* …."] (Sanctus Augustinus, In epistulam Iohannis ad Parthos tractatus, 1, 6: PL 35, 198).

The second principle recognizes that the ultimate end of the believer in the Catholic religion is to attain and maintain a state of grace in order to merit eternal life in the presence of God, who sent Jesus Christ, his only son, to take upon Himself the sins of the world for the salvation of mankind and to die upon the cross: "*He will come again to judge the living and the dead*" (The Apostles Creed); "*He will come again in glory to judge the living and the dead, and his kingdom will have no end*" (The Nicene Creed): "*Immortality" in a Christian sense means the survival of man after his terrestrial death, for the purpose of eternal reward or punishment*" (Encyclical Letter "Mit Brennender Sorge" of the Supreme Pontiff Pius XI 1937, Catechism, online).

> But to live always, without end—this, all things considered, can only be monotonous and ultimately unbearable. This is precisely the point made, for example, by Saint Ambrose, one of the Church Fathers, in the funeral discourse for his deceased brother Satyrus: "Death was not part of nature; it became part of nature. God did not decree death from the beginning; he prescribed it as a remedy. Human life, because of sin … began to experience the burden of wretchedness in unremitting labor and unbearable sorrow. There had to be a limit to its evils; death had to restore what life had forfeited. Without the assistance of grace, immortality is more of a burden than a blessing". A little earlier, Ambrose had said: "Death is, then, no cause for mourning, for it is the cause of mankind's salvation" (Encyclical Letter "Spe Salvi" of the Supreme Pontiff Benedict XVI 2007, Catechism, online).

The third principle is that there is a remedy for sin; there are types of behavior to *avoid* because they lead to sin and behavior to *pursue* because it leads to the annulment of sin and to the state of grace. The fourth principle states that Jesus, through his Church, teaches people the "paths" to take to avoid sin or to be "cured" from sin in order to acquire and maintain a state of grace.

It is not easy to list these "paths," but generally speaking they concern behavior to be avoided or improved in order to avoid sin, along with behavior to undertake and maintain to achieve a state of grace (Catechism, online). Above all, it is necessary not to disobey the Ten Commandments (loop [**B1**] in Fig. 13.4) and to avoid the six sins against the Holy Ghost and the four sins that call out for revenge in the presence of God (FrancoBampi, online). Moreover, it is necessary to avoid the seven deadly sins: (1) pride, (2) greed, (3) lust, (4) wrath, (5) gluttony, (6) envy, and (7) sloth (loop [**B2**]). There is a multitude of types of virtuous behavior (state of grace); an initial list is indicated in the formation of loops [**B3**] and [**B4**]. Loops [**B5**] and [**B6**] elevate the state of the soul through the typical practices of the Church: the sacraments, precepts, and liturgy.

Figure 13.4 only shows the main loops, since many other ways of gaining salvation can be found in the teachings of the Church and in the behavior of the saints. It may happen that a believer not only has no sins and avoids behavior that could lead to sin but also behaves in such a virtuous manner as to lead his or her soul to a state above and beyond what is required for eternal salvation. In such cases, the error is in fact negative and denotes a state of holiness.

The model of the control system in Fig. 13.3 can be applied not only to religions but also to the "salvific practices" that did not originally represent a religion in the

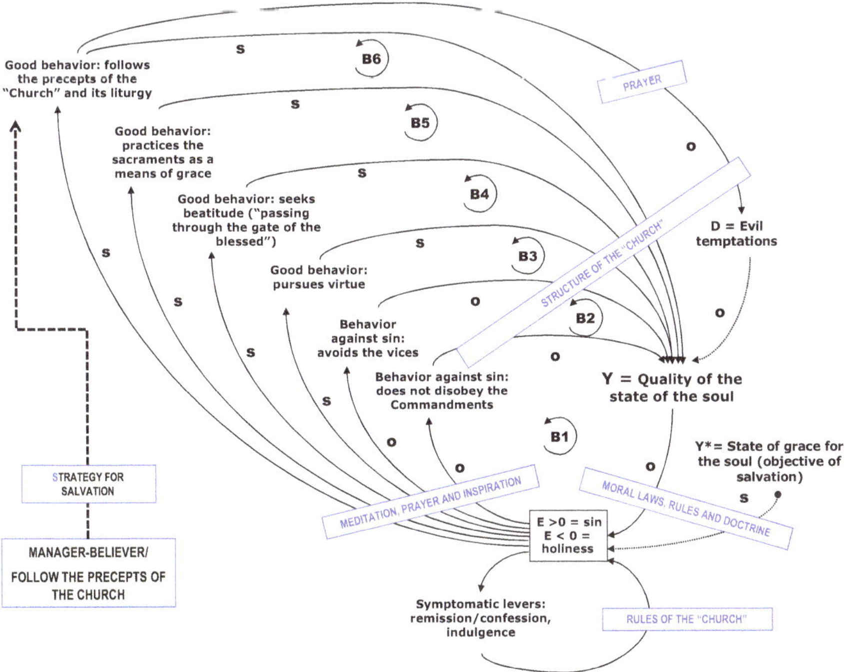

Fig. 13.4 The pluri-lever control system of (in) the Catholic religion (a simplified model)

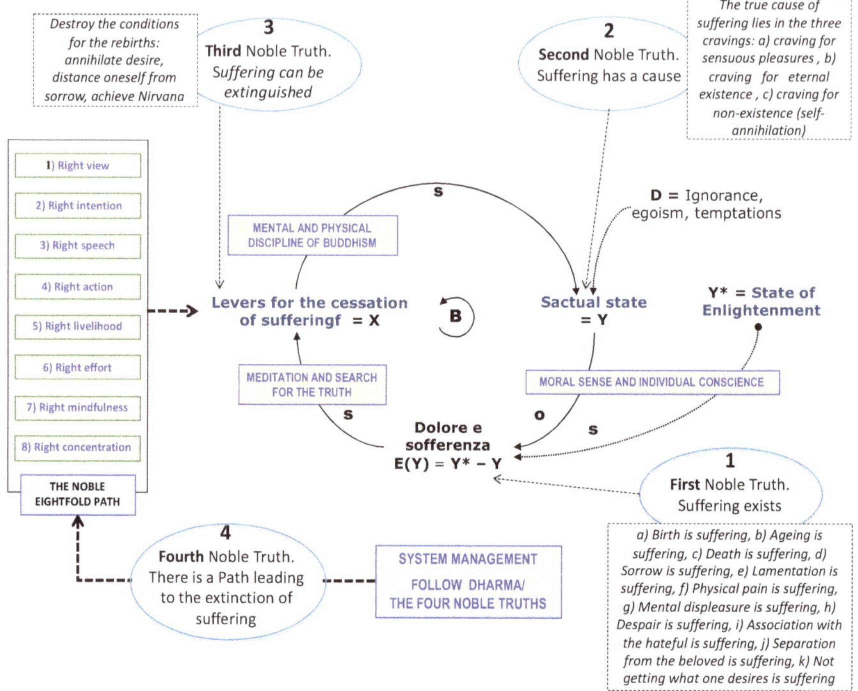

Fig. 13.5 The pluri-lever control system (of) in Buddhism (a simplified model)

strictest sense. I shall try to apply it to Buddhism in Fig. 13.5, asking indulgence of the reader for the summary nature of the treatment. The very heart and core of Buddhism represent the logical schema of the "Four Noble Truths," put forth by Buddha in 588 B.C., which represent the essence of the teaching (Dhamma or Dharma). They are considered as "truths" because they are *real* and form an *incontrovertible* fact of life. We may be aware or unaware of them, but they really exist in the world; Buddha simply revealed them to mankind (see The Dalai Lama 1998 online).

In order to better understand the significance of the "Four Noble Truths" we must keep in mind that Buddhism did not arise as a religion but as a life practice within Hinduism. The "Four Noble Truths" are (His Holiness The Dalai Lama 1998) as follows:

1. *Suffering exists*; this is a universal truth: (a) birth is suffering; (b) aging is suffering; (c) death is suffering; (d) sorrow is suffering; (e) lamentation is suffering; (f) physical pain is suffering; (g) mental displeasure is suffering; (h) despair is suffering; (i) association with the hateful is suffering; (j) separation from the beloved is suffering; (k) not getting what one desires is suffering.
2. *Suffering has a cause;* the origin of suffering is craving, which passes on from birth to birth; accompanied by pleasure and greed, it finds increasingly renewed

delight in the here and now. Craving is of three kinds, namely, craving for sensual pleasure, craving for eternal existence, and craving for nonexistence.

3. *Suffering can be extinguished* by cutting off its causes through the complete fading away and extinction of its origin, craving. The extinction of suffering is obtained through the complete elimination of desire, the separation from longings. The state in which the complete cessation, or extinction, of craving, ignorance, suffering, and the cycle of involuntary rebirths (saṃsāra) is achieved is Nirvana (or Nibbana). Once Nirvana is attained the chain is broken: life and death no longer have any meaning once the state of enlightenment is reached. The Buddhist conceives of Nirvana not only as the liberation of the soul from rebirth after death but also as the state of perfect peace and happiness and of imperturbable serenity and the absence of sorrow. For this reason, achieving Nirvana is for the believer the ultimate aim in life, the supreme state of enlightenment in which liberation from sorrow is achieved and the cycle of reincarnations is interrupted;
4. *The path leading to the extinction of suffering* is the "Noble Eightfold Path" composed of eight steps or factors: (1) right view, (2) right intention, (3) right speech, (4) right action, (5) right livelihood, (6) right effort, (7) right mindfulness, and (8) right concentration (Bhikkhu 1984).

> Considered from the standpoint of practical training, the eight path factors divide into three groups: (1) the moral discipline group, made up of right speech, right action, and right livelihood; (2) the concentration group, made up of right effort, right mindfulness, and right concentration; and (3) the wisdom group, made up of right view and right intention. These three groups represent three stages of training: the training in the higher moral discipline, the training in the higher consciousness, and the training in the higher wisdom (Bodhi 2011, online, Chap. 2).

13.4.2 Control of Chaos and of Chaotic Dynamics

Figure 8.12 (Chap. 8) presented five populations forming an ecosystem whose chaotic dynamics became regular and periodical as a result of an artificial control that set *maximum* or *minimum* constraints for the number of individuals. These constraints can be respected through selective hunting to reduce the number of individuals in excess of the maximum or through planned repopulation to increase the number of individuals when they fall below the minimum. Figure 8.10 allows us to reflect on the possibility of extending control systems to chaotic dynamics in general. I would suggest that chaos is a widespread phenomenon, at least in theory; however, we usually are not aware of this since we tend to eliminate or reduce its effects with appropriate controls. I shall first clarify what notion of chaos I am referring to.

If we set aside the mythological or the cosmogonical notion of chaos—as the disordered primordial universe from which our known ordered universe was formed—then the term chaos was first formally used by the American mathematicians, James Yorke and Tien Lien Li, in a 1975 work entitled Period 3 Implies Chaos, which dealt with differential equations, in particular the logistics function (Kellert 1993). We can thus limit ourselves to the notion commonly used in the mathematical theory of systems, according to which deterministic chaos is the characteristic of a system's dynamics that, although regulated by relatively simple recursive functions, such as a quadratic equation of the type

$$x_{n+1} = cx_n\left(1 - x_n\right), \tag{13.1}$$

has a temporal dynamics apparently devoid of cyclicity, or regular occurrence, which is sensitive to the initial conditions so that modest variations in these conditions can produce a temporal dynamics that is totally different and unpredictable.

This definition of chaos as a characteristic of the dynamics of a recursive system describes the so-called deterministic chaos (Flake 2001; Gleick 2008). A similar definition can be used for multiagent systems where individual behavior is dependent on a change of state individual probability function. The dynamics of this system—derived from the synthesis of individual behaviors—is unpredictable, without particular cyclicity or recurrences. In this system the chaos is defined as probabilistic, and it is revealed because the individual behaviors are regulated by specific probability functions, one of which is the so-called tent map (Mella 2007; Mella 2017, Sect. 3.4.3). Figure 8.13 shows a chaotic system in population dynamics, since the simulation model immediately demonstrates that modest variations in the initial population conditions (initial numbers, rates of increase and decrease, even for a single population) produce entirely different dynamics.

If we refer to Fig. 13.6, which indicates the chaotic dynamics of the function (13.1), and assign certain initial values to "x_0" and "c," we can clearly see the dependence of the dynamics on the initial conditions. When the conditions $c = 3.920$ and $x_0 = 0.889$ change slightly, becoming $c = 3.925$ and $x_0 = 0.891$, the dynamics are completely different. Therefore, for certain initial values the quadratic function is chaotic, since it is "sensitive" to these conditions and does not display any particular cycles (at least for the period considered in the simulation).

The mathematician and meteorologist, Edward Lorenz, became aware of this feature of chaotic dynamics almost by chance when, simulating the dynamics of a quadratic function using various pairs of initial values, he observed that for some pairs the dynamics were in fact chaotic, thereby concluding that meteorological processes, being sensitive to initial conditions (pressure, humidity, temperature, etc.), can have chaotic dynamics for particular values of the initial state parameters. Lorenz introduced the term *butterfly effect*, in 1972, at a meeting of the *American Association for the Advancement of Science*, in his speech entitled "*Does the flap of a butterfly's wings in Brazil set off a tornado in Texas?*" Lorenz provocatively stated that if the theories of complex systems and chaos were correct, then the fluttering wings of a butterfly in Brazil would be enough to alter climate patterns, even permanently, and generate typhoons in the Caribbean, or suffice to alter climate patterns, even permanently. However, Lorenz also stated that: "*If the flap of a butterfly's wings can be instrumental in generating a tornado, it can equally well be*

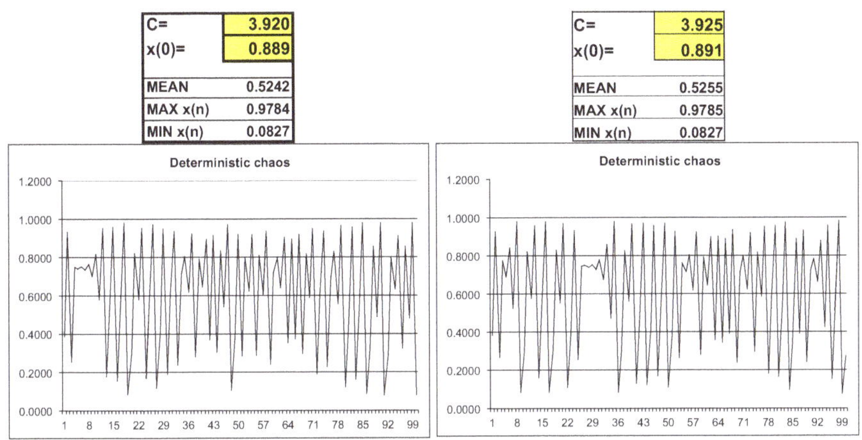

Fig. 13.6 The chaotic dynamics of a quadratic function with two different initial conditions

instrumental in preventing a tornado" (Lorenz 1972, p. 1). Alan Turing expressed a similar idea in saying that shifting a single electron at a given time by a billionth of a centimeter could, thanks to a very long cause and effect chain, give rise to very diverse events, such as the death of a man a year later due to an avalanche or his being saved (Turing 1950). For this reason, the butterfly effect is also known as the Turing Effect (Sect. 5.6). Dynamical systems that show extreme sensibility to initial conditions are chaotic systems. This means that infinitesimal variation in the initial situation leads to large scale variation in the systems dynamics. The evolution of a chaotic system can be predicted with a high degree of probability in the short term, but beyond that, its behavior becomes extremely difficult to predict, and had already been pointed out by Jules Henri Poincaré:

> A very small cause which escapes our notice determines a considerable effect that we cannot fail to see, and then we say the effect is due to chance [...] But it is not always so; it may happen that small differences in the initial conditions produce very great ones in the final phenomena. A small error in the former will produce an enormous error in the latter. Prediction becomes impossible, and we have the fortuitous phenomenon (Poincaré 1908, p. 3).

Figure 13.6 shows the maximum and minimum values of the chaotic function for the period of reference. It is natural to then test how the dynamics of the function would vary if we introduced a simple control: for example, setting the condition that the function cannot exceed the *maximum* value $x_{max} = 0.950$ or, alternatively, the *minimum* value $x_{min} = 0.088$. With these very limited maximum and minimum controls (for simplicity's sake applied one at a time), the dynamics soon become cyclical and the chaos vanishes, as Fig. 13.7 clearly shows.

It is precisely the ease with which we can keep chaos under control, even with loose constraints, which allows us to understand why chaos has almost disappeared from our daily lives, remaining only where the controls are not applied (for more, see Flake, 2001: Chap. 13). Thus, it would be easy to eliminate chaos in the whirlpool of a stream by simply building a bank along the sharpest curves. The chaos in the behavior of pedestrians that run into each other on a crowded sidewalk (for a simulation, see Legion, online) could be kept under control by appropriately indicating separate paths for both directions using signs. The chaos of a crowded bus stop, the entrance to crowded locales, or the departure gates at airports could be eliminated by a ticket system indicating the order for those who want to gain access.

13.4.3 Variant Structure Control Systems

The control systems presented in the previous chapters were always of the *invariant structure* or *invariant action* kind: the operating logic behind both the number of levers as well as the technical apparatuses (Fig. 2.10) remained unchanged during the control process. In other words, the *structural characteristics* of the "technical system" (Sect. 2.9) were kept unchanged; these characteristics are summarized in the *action* function, $g(Y/X)$, and the *reaction* function, $h(X/Y)$. These functions were assumed to be unchanged during the control cycle. To make the hypothesis of invariance more evident, these functions were in fact given the most elementary form, that of the "action rates" and "reaction rates," "g" and "h," respectively. In other words, they were constants independent of X, Y, and t. Nevertheless, in Sect. 5.4 we saw that the structure of a control system can be modified through the internal or the external strengthening of the system (Sect. 5.4).

These structural modifications make control possible even when the manager cannot directly act on the variable X to modify the value of Y and eliminate the error, $E(Y) = Y^* - Y$, but must necessarily modify the action and reaction rates. Thus, we

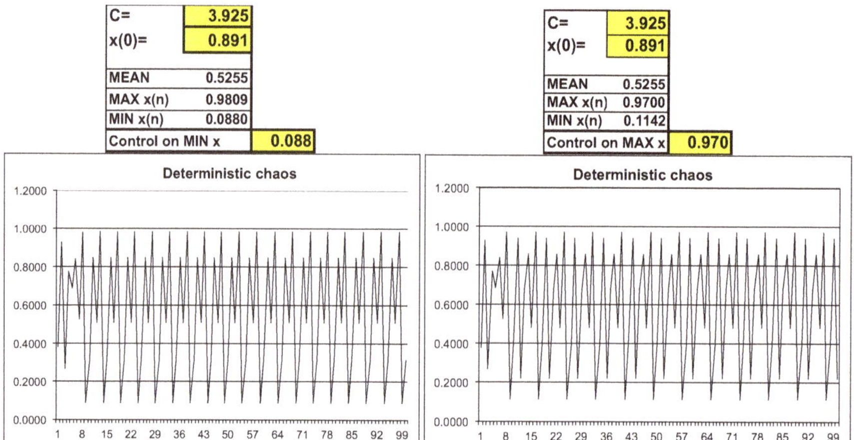

Fig. 13.7 The nonchaotic dynamics of the quadratic function with a maximum or a minimum control

have a different form of control in which, since X cannot be manipulated, the manager must try to eliminate $E(Y) = Y^* - Y_t$ by acting only on the rates "g" and "h" (more generally, on the structure of the action and reaction functions), so that, even with the values of X unvaried, the desired values of Y_t are consistent with Y^*. We can refer to such systems as *variant action* (or *structure*) control systems, due precisely to the fact that the values of Y depend not on variations of X but on values of the rates "g" and "h," which are varied to allow the varying processes (black boxes) to carry out the control in the presence of a flow of X that cannot be modified. The general model of a single-lever invariant action control system, presented in Chap. 2, Fig. 2.2, can be reformulated as shown in Fig. 13.8, where the action rate "g" itself becomes the *control lever* to be modified (assuming that the rate "h" has been readjusted); X is no longer the control lever but instead represents an *external variable* such as D.

Figure 13.8 can be reformulated in the more significant and understandable manner shown in Fig. 13.9, which, while appearing quite different than Fig. 13.8, is in fact entirely equivalent. The advantage of transforming Fig. 13.8 into Fig. 13.9, by isolating the variable X, allows us to see more clearly that X is a variable considered as an independent "input" of the system and no longer as a control lever. In order to ensure that the input X will produce a Y that tends toward Y^*, the control system must act to suitably vary the action rate $g(Y/X)$. The advantage of Fig. 13.9 is even clearer when we have to construct dual-lever control systems whenever Y depends on two or more external variables, $X1$ and $X2$, acting in parallel based on their respective action rates, g_1 and g_2. The model in Fig. 13.9 would then expand to become the model in Fig. 13.10.

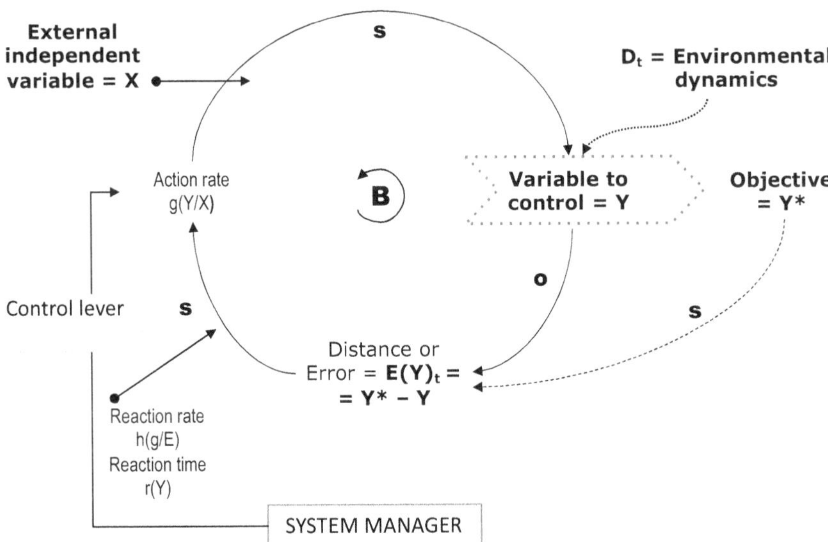

Fig. 13.8 An elementary variant action control system (reformulation of Fig. 2.2)

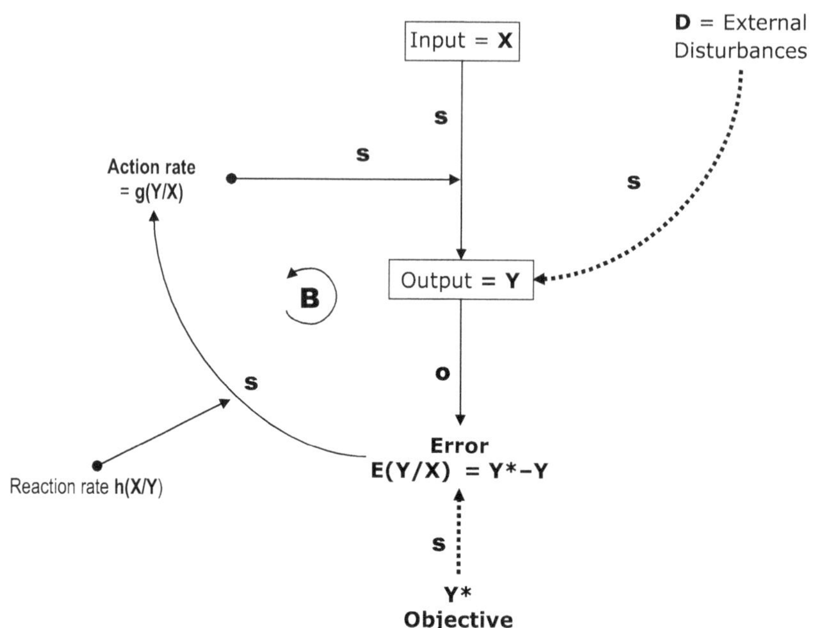

Fig. 13.9 General model of a single-lever variant action control system (reformulation of Fig. 13.8)

The model in Fig. 13.10 represents the variant structure of a typical dual-lever control system whose levers can act jointly (Sect. 4.1) or separately (Sect. 4.2). Thus, in systems where the two levers act "in parallel" there are three possibilities for controlling Y (i.e., for eliminating the error): acting on g_1, on g_2, or on both. The choice obviously depends on the type of situation the model represents; in any event, a precise control *strategy* must be defined.

We can formulate a third standard model that assumes that the systems operate "in series," or "sequentially," as indicated in Fig. 13.11, which clearly shows that we are again presenting a dual-lever control system since, during the initial phase, the input X is transformed into the output Y1 according to a variant action rate, g_1. Subsequently, Y1 becomes the input that the system transforms into the final output Y2, according to the specific variant action rate $g_2(Y_2/Y_1)$ for that phase. The error is still calculated with reference to Y2, but its elimination entails the threefold possibility of acting on g_1, g_2, or both; as always, the order of the reaction defines a control *strategy*. As I have already observed, in the models in the figures above, the reaction rate "*h*" could actually be modified to allow the control.

By observing the models in the figures above, we clearly see that, beyond the exchange of the variables that represent the control levers (X in the models in the previous chapters, the rate "*g*" in the models in this *section*), the control logic is the same: acting on the *control levers* to move Y toward the objective, thereby eliminating the error. The fundamental problem which the system "designer" must face is

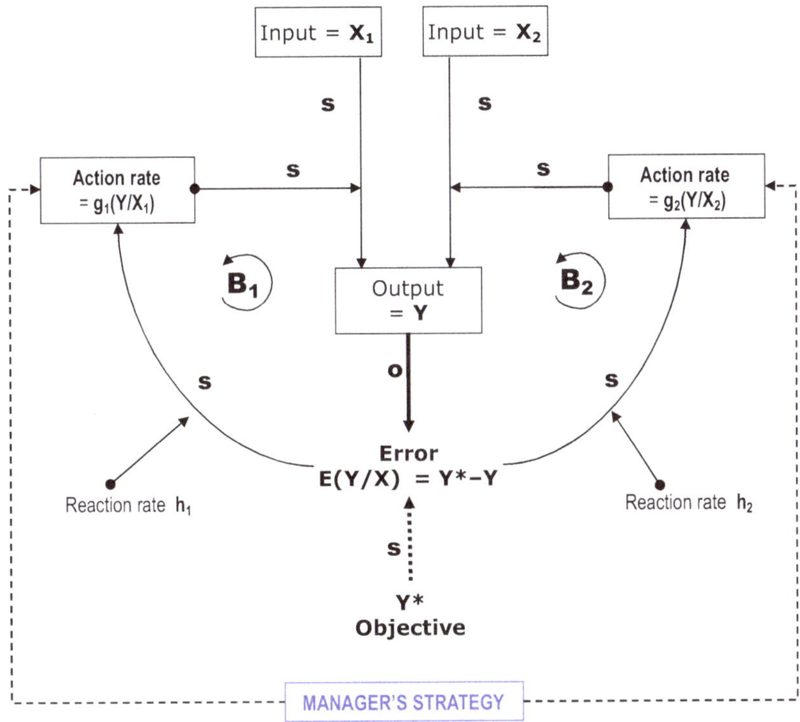

Fig. 13.10 General model of a "parallel" dual-lever variant action control system

determining the action rate "*g*" needed to eliminate the error without being able to intervene on the quantity of flow of the *X* variables. The solution to this problem can be found relatively easily for the basic systems in the models in the above figures; however, for more complicated structures, such as systems of *composite* transformation with more than two inputs and numerous links, "in series" and "parallel" at the same time and over various layers, the regulation of the values of "*g*" can turn out to be quite difficult.

I hold that variant action control systems constitute perhaps the simplest models for representing a wide class of physical or logical systems, among which:

A. Transformation systems
B. Production systems
C. Neural networks

A. *Transformation systems.* Though we cannot give due consideration here to this topic, some brief observations are nevertheless in order. I define *transformation systems* as all those apparatuses that transform an input variable, *X*, into an output variable, *Y*, with the "convention" that *Y* represents the result of some *qualitative* transformation (pieces of meat versus minced meat; dirty dishes versus clean dishes, etc.) or *quantitative* transformation (a lamina with thickness "*a*" versus a lamina

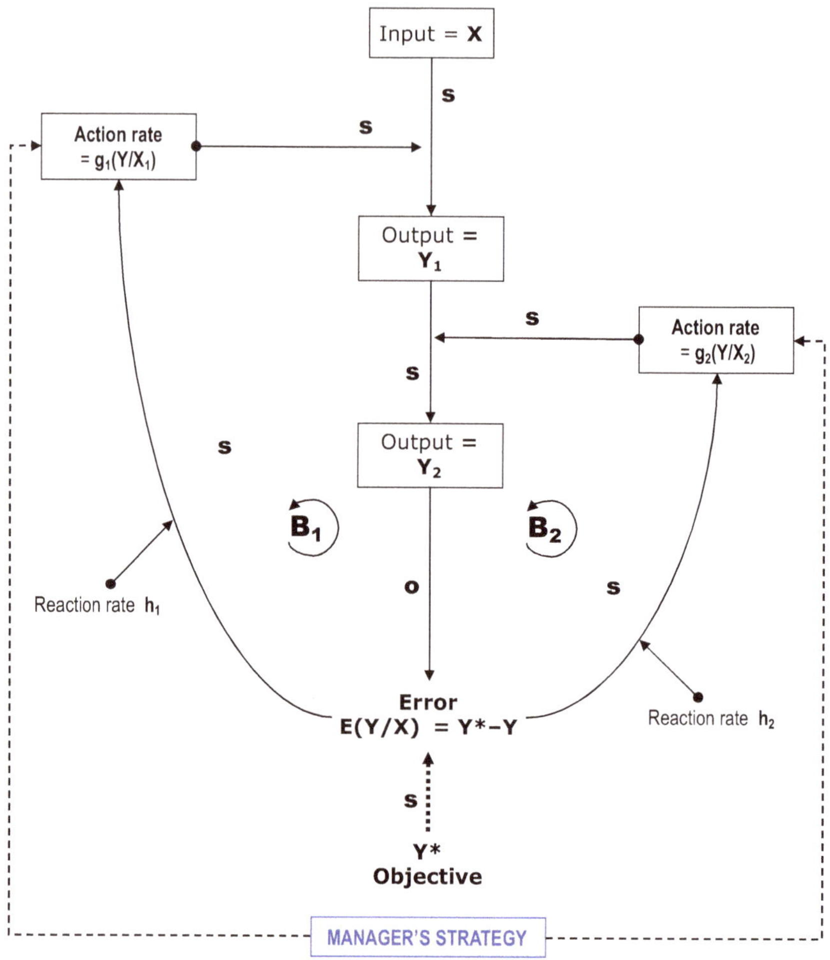

Fig. 13.11 General model of an "in series" dual-lever variant action control system

with thickness "*b*"; investment of an amount of capital "A" and disinvestment of capital in the amount of "*A* + *I*," etc.) of X itself, based on action rate "*g*." In many cases the control system cannot act on *X*, the value (or variable) that must be transformed, which represents an external input, but instead needs to modify the action rate "*g*" so that the transformation produces the desired *Y*. The rate "*g*" takes on the significance of "transformation coefficients."

B. *Production systems* can be viewed as a particular class of transformation system. Through appropriate production apparatuses they produce the various combinations of factors of production flows, *X*, in order to obtain flows of finished products, *Y*, which can be considered in all respects as the result of the transformation of factors according to the productivity rate "*g*." When the control system cannot achieve the production objectives by maintaining constant the productivity rates and modifying the factor amounts (control levers) X, which arrive at irregular intervals (e.g., consider sick people in a hospital), the productivity rates must be modified by *varying* the productive structure.

C. *Neural networks* can also be viewed as particular transformation systems—which act through modules that are linked and structured over various layers—which, through successive processes ("in series" and "parallel") modify the values of the independent inputs, X, into values of output, Y, by carrying out operations on X or on the sublevels of Y through appropriate coefficients of transformation known as "weights" (which normally are between 0 and 1). These weights can be fully equated with the action rate "g." If the neural network has an objective Y^* and, through the inputs X and the "propagation" process, the network produces a value Y which differs from Y^*, then the transformation coefficients (weights) must be modified, since it is not possible to act on the inputs to the input modules in order to gradually move Y toward Y^*. In this sense the neural network learns to "recognize" Y^* given the entry values X, which cannot be directly controlled. Therefore, even the neural networks, no matter how they are designed, can in principle be conceived of as variant structure control systems; in this case they can learn through the feedback processes, which modify the weights of the links among the upstream and downstream nodes. Today the neural networks normally do not have a control system for the weights, which instead are determined randomly based on reiterated "backpropagation" processes.

Transformation systems are usually *composite* since their structure has more than two inputs arranged over various layers. In order to implement the structural control for *composite* transformation systems it is necessary, as always, to define a strategy, among which we can mention the following:

(a) *Simple strategies*, which act on a single lever represented by a chosen transformation coefficient; the aim is to try and identify a single "g" such that its modification immediately eliminates $E(Y)$. This strategy is not always possible since there may not exist a g_i which, by itself, and taking into account the range of admissible variations, succeeds in bringing Y toward Y^*. The easiest strategy is to act on the final layer—let us say, M—considering as inputs of layer M the outputs of layer M-1. Alternatively, we can maintain the transformation coefficients of layer M and act on one of the layers of the preceding levels. It is fundamental that the manager act on only a single g_i, choosing the one which eliminates the error, no matter the layer it pertains to.

(b) *Composite, or multilever, strategies*. The manager can act on several levers at the same time, trying to simultaneously modify several transformation coefficients (action rates). A composite strategy can be identified fairly easily by operating on a chain or a path, that is, on an uninterrupted sequence of arrows which, starting from any input X_h, reaches Y. Setting the final output as $Y = Y^*$ and proceeding backwards, the manager tries to identify for modification one or more Y_h from the preceding layer in order for Y to assume the value Y^*. Values for the "g_h" coefficients (if admissible) are sought which allow Y to be attained at the end of the causal chain.

(c) *Global strategies*, which seek to modify the values of all the transformation rates, g_i, for all the variables that precede Y. It is difficult for the manager to make perfect calculations outside of extremely simply situations.

A simple global strategy (usually used in neural networks without direct feedback) is to effect a random change of the g_i in the following manner:

1. Assign rectifiers, h_i, to all the g_i coefficients whose values are randomly (or intuitively) set by the control system manager.
2. Calculate the new value for Y1; if the error increases, assign new correctors, again randomly (or intuitively) chosen.
3. Repeat these operations until $E(Y)$ is eliminated or the error cannot be further reduced.

A general strategy that can always be applied, which is a simplification of the above strategy, is shown in the general model in Fig. 13.12. This model recalculates, for each layer, the coefficients needed to achieve the objective, $Y^* = 150$.

Assigning to the Xi the input values shown in the cells in the first row ($X_1 = 20$, $X_2 = 10$, and $X_3 = 5$) we follow the arrows that indicate the causal links that produce the transformations. Each cell next to the arrows indicates the theoretical values of the "old g_i" coefficients, which allow the values of the "old Y" in the various layers to be calculated through the values of X. Using the "old g_i," the "old Y" value output of 270 is obtained. Since the objective has been set at $Y^* = 150$, the error is equal to the "old $E(Y)$" or -120. Therefore, it is necessary to again determine the "old $_g$" values for all the transformations in order to find a correct "new Y" value equal to 150, which produces a "new $E(Y)$" = 0. The *automatic controller* in the lower part of Fig. 13.12 enables the "new g_i" to be calculated (following the strategy which is apparent by examining the figure)—which produces the "new Y" in the various layers—and the "old $E(Y)$" to be eliminated, thus achieving the objective and maintaining stable the initial input values of X.

The model in Fig. 13.12 is flexible since it even allows the system manager to adapt various coefficients, thereby facilitating his or her task. In fact, when an "original" coefficient in the model is replaced by a "new" one, all the old "g_i" are automatically recalculated so that the manager can then proceed to the recalculations based on any hypothesis of change whatsoever, choosing the alternative he or she holds to be most effective and to represent the preferred strategy. It is possible to modify a single transformation coefficient or several coefficients. For each choice, the model recalculates all the coefficients. Obviously the *automatic calculator* is programmed specifically for the transformation system in Fig. 13.12, but the principle can also be applied to more complex structures.

In order to choose the best strategy, a cost/benefit analysis is required, as described in Sect. 4.7. In fact, rectifying the "old g_i" coefficients likely will entail a cost for the *technical* intervention on some apparatus, thus necessitating rectifications that are more efficient in terms of both implementation time and cost.

In order to clarify the above concepts, let us assume that the model in Fig. 13.12 represents a hydraulic network which, receiving liquid at various amounts of

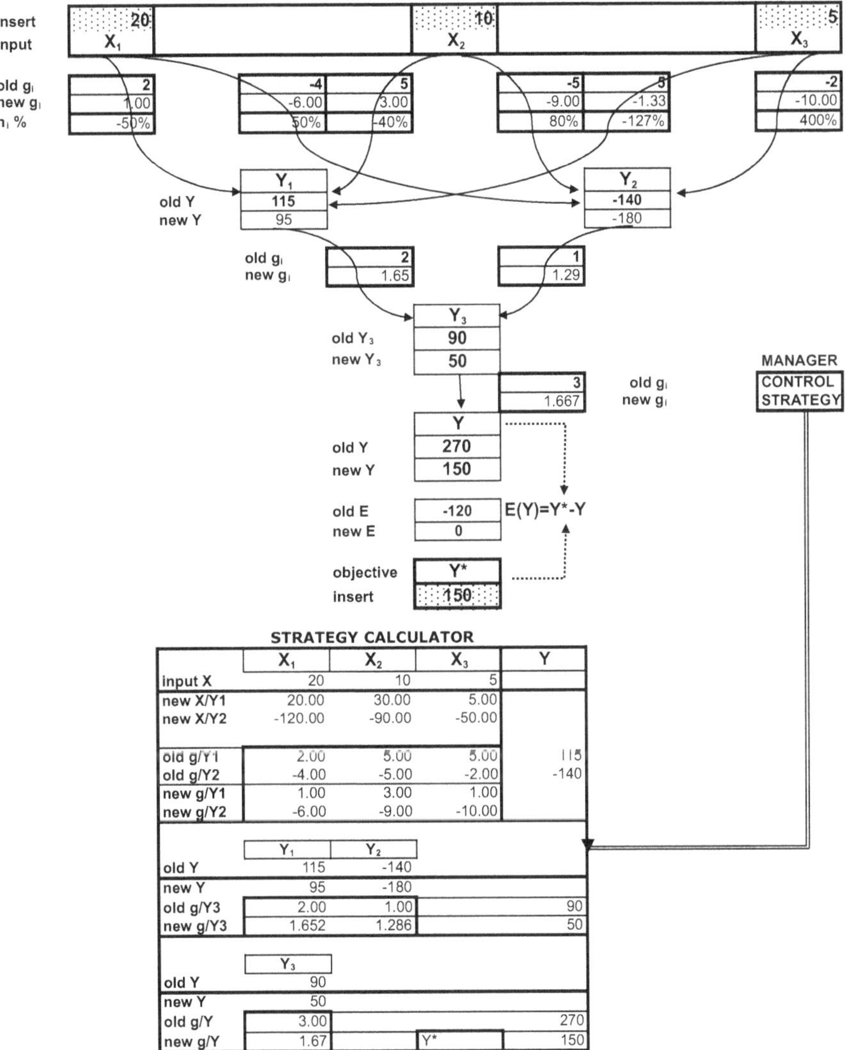

Fig. 13.12 Control of a multilayer transformation system

pressure from three sources (X_1 = 20, X_2 = 10, and X_3 = 5), varies, at successive stages, the overall pressure based on the transformation coefficients. The value Y indicates the pressure obtained, while Y^* indicates the desired pressure. If the incoming pressure is not varied, which is shown alongside the X_i, it becomes necessary to modify the outgoing pressure by means of appropriate pumps and valves, shown alongside each Y_i preceding Y (taking into account the various layers). We can intuitively understand from this example that if the manager were "near" the outflowing liquid, Y, then the control of the system, achieved by changing the first-level flow coefficients (from X to Y_1 and Y_2), would entail a high cost, since the control values are numerous and far from the operator. In this case, it would be simpler to adjust only those valves that regulate the coefficients on the curved line that connects Y_3 to Y. Nevertheless, if this control should require a nonadmissible variation of the installed regulator and it were necessary to substitute this with a more powerful regulator, with high energy consumption, then the steep cost of this strategy would make it more efficient to regulate the mechanisms located "near" the inputs.

Thus, we can say in general that the most effective strategy is the one that minimizes the cost of the control in terms of both time and money, assuming equal advantages in reducing $E(Y)$.

13.5 Summary

In this final chapter, I have presented some concluding considerations to stimulate further reflection:

1. In Sect. 13.1, I have outlined several fundamental general *hypotheses* in order to propose a *Control Systems Discipline*.
2. In Sect. 13.2, I have discussed the *human aspects* of control: fallibility, learning attitude, inventive and innovative capabilities, the search for active control, intolerance toward passive control.
3. In Sect. 13.3, I have used FAQs to investigate the content and limits of this book.
4. In Sect. 13.4, I have suggested that the Magic Rings also act in contexts we might never imagine:

(a) In the control carried out by (in) religions and regarding life practices (Sect. 13.4.1).
(b) In the control of the chaotic dynamics produced in our environment, as shown by the control of deterministic chaos (Sect. 13.4.2).
(c) In the control of *variant structure* systems (Sect. 13.4.3), where the X variables cannot be modified and the control is achieved through appropriate values of the "g" and "h" rates. Three fundamental classes of variant action systems (which were mentioned only in passing) are (1) transformation systems, (2) production systems, and (3) neural networks.

I have only briefly discussed these types of control systems, which nevertheless deserve a broader analysis which goes beyond the scope of this book. Thus, I have only presented transformation systems and a general algorithm to illustrate the control strategies.

Symbols (minimal)

X_t → values of the control lever

Y_t → values of the variable to be controlled

Y^* → fixed objective

Y_t^* → variable objective

D_t → external disturbance values

$\Delta(Y)_t = Y^* - Y_t$ (or $E(Y)_t$) → deviation, gap, error for systems [**s** → **o** → **s**].

$\Delta(Y^*)_t = Y_t \quad Y^*$ (or $E(Y^*)_t$) → deviation, gap, error for systems [**s** → **s** → **o**]

$g(Y/X)$ → action rate

$h(X/Y)$ → reaction rate

$r(X, Y)$ → reaction time

$\underset{+/-}{X} \xrightarrow{\mathbf{s}} \underset{+/-}{Y} \quad \underset{+/-}{X} \xrightarrow{\mathbf{o}} \underset{-/+}{Y}$ → direction of the variations

$X \xrightarrow{\mathbf{s/o}} Y \xrightarrow{\mathbf{s/o}} Z$ → open causal chain

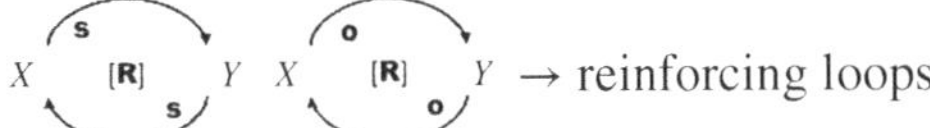

→ reinforcing loops

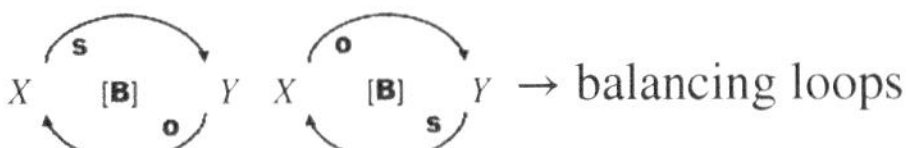

→ balancing loops

P. Mella, *The Magic Ring*, Contemporary Systems Thinking,
https://doi.org/10.1007/978-3-030-64194-8

Glossary

Action rate "$g(Y/X)$" Indicates the variation of Y for each unit of variation of X.

Active variable, Xt In the general model of the control system, any variable X capable of modifying Y. See **Levers**.

Apparatuses See **Chain of control**.

Archetypes Models or patterns that occur continuously, also known as **generic** structures. "*One of the most important, and potentially most empowering, insights to come from the young field of systems thinking is that certain patterns of structure recur again and again. These "systems archetypes" or "generic structures" embody the key to learning to see structures in our personal and organizational lives*" (Senge 2006:93).

Causal loop diagram (CLD) Model of a system built following systems thinking using arrows to indicate the direction of the causal relationships among the variables. A CLD consists of a system of loops in which all variables are linked by arrows, without there being an initial and final variable. All the variables are connected. By connecting a number of variables and determining the direction of variation, we can build models of every dynamic system, keeping in mind that we must zoom in order to analyze the processes in more detail, in order to identify and connect other important variables.

Chain of control The fundamental "machines" or "apparatuses" that represent *physical control systems* and whose functioning produces the dynamics in the active and passive variables and determines the variance. Effector: produces a variation in X_t into the corresponding variation in Y_t Detector (sensor, or comparator): measures the value of Y, compares this with the objective Y^* (or the constraint $Y°$), and determines the deviation $E(Y)$ Regulator (or compensator): "activates" the lever X taking account of $E(Y)$ Information transmission: represents the "real" *chain of control* that produces the "formal" control system

Control apparatuses See **Chain of control.**

P. Mella, *The Magic Ring*, Contemporary Systems Thinking,
https://doi.org/10.1007/978-3-030-64194-8

Control discipline This is based on the hypothesis that control systems occupy a preeminent place among all types of systems. Even though we are not accustomed to "seeing them," they are all around us, and only their presence makes our world, life, society, and our very same existence possible, producing an ordered and livable world, erecting barriers against disorder, and guiding the dynamics of the system toward equilibrium states. The control discipline teaches us how to recognize control systems, build models, simulate their behavior, and use these to improve our knowledge and our behavior. I would propose naming the control discipline as the Sixth Discipline, or the *control discipline of the individual, the collectivity, and the organizations in the ecosystem*; this is the discipline of the present and future of our world.

Control objective Each specific value, Y^*, or each trajectory of values, Y_t^*, however formed, which are set as values that the **objective variable**, Y_t, must achieve or maintain. The control objective can also mean the *constraint* which is not to be exceeded or the *limit* to be respected. If Y^* is a vector of values, $[Y^*]$, then the control system is multi-objective and the system manager must establish a **control policy**. The objectives can be both quantitative and qualitative.

Control process See **Control system (concept)**.

Control system (concept) Any set of apparatuses, *logical* or *technical* (algorithm or machine, rule or structure, etc.), which, for a set of instants, perceives $E(Y)_t$, calculates and assigns the values X_t, and produces the appropriate Y_t to gradually eliminate, when possible, the error $E(Y)_t = Y^* - Y_{t^*}$ at instant t^*. The control system is repetitive and functions by means of action (X acts on Y) and reaction ($E(Y)$ acts on X through Y), activating a closed-loop or feedback control. With a certain number of iterations on the control lever, it tries to achieve the objective (goal-seeking systems) or to respect the constraints or limits (constraint-keeping systems) by gradually eliminating the deviation $E(Y)$. Control systems have a *logical structure* and a *technical structure*. The former consists of the logical relationships between variables, always arranged in one or more balancing loops in order to develop feedbacks between $[X]$ and $[Y]$; the latter is formed by the **chain of control.**

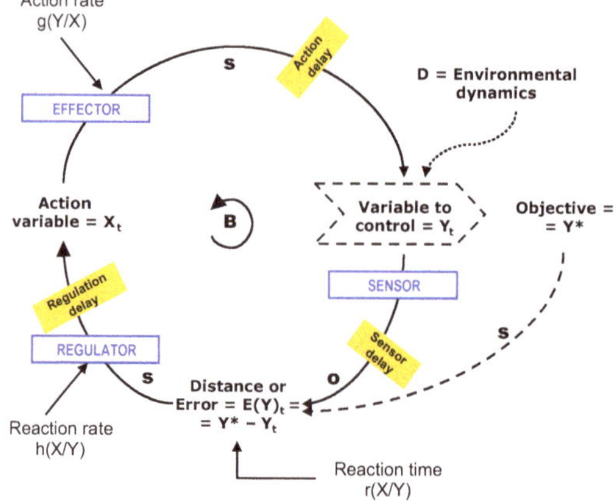

Control system (logical structure) The general model of a control system is a balancing loop, which consists of the elements shown in the model.

Control system (technical structure) The set of apparatuses that constitute the *physical control systems*. See **Chain of control**.

Control variable Synonym for **lever**.

Cybernetic control system An automatic control system guided toward a fixed (or even variable) objective determined by an outside governor, but guided by a manager within the chain of control.

Delay Abnormal or unexpected length of action of *X* on *Y*. There are three types of delay: *Action delay* (or response delay), which slows the response of *Y* to an impulse from *X*; this depends on the effector *Detection delay* (or informational delay), which acts on our perception and on the measurement of the error *Regulation* (decisional) *delay*, which occurs when the regulator does not respond promptly to the error

Deviation see Error, ***E***.

Direction of the causal link The arrow that indicates which variable is the *cause* and which the *effect* in a causal relation typical of systems thinking. The *cause* variable (*X* = input) is written at the tail of the arrow; the *effect* variable (*Y* = output) at the head (first arrow). If the causal link shows the opposite direction (second arrow), then *X* is the effect and *Y* the cause.

$$X \longrightarrow Y$$

$$X \longleftarrow Y$$

Direction of the variations This indicates the concordance in the signs of the variations between two variables, *X* and *Y*, which are linked in the same direction. Variations of *X* are linked to those of *Y* in the same direction (**s**) if *Y* increases/diminishes when *X* increases/diminishes (first arrow). Variations of *X* are linked to *Y's* variations in the opposite direction (**o**) if *Y* diminishes/increases when A increases/diminishes, thus presenting the opposite sign of variation (second arrow).

$$\underset{+/-}{X} \xrightarrow{\ s\ } \underset{+/-}{Y} \quad \underset{+/-}{X} \xrightarrow{\ o\ } \underset{-/+}{Y}$$

Discipline Following Peter Senge: "A discipline is a developmental path for acquiring certain skills or competencies. **[…]** To practice a discipline is to be a lifelong learner. You "never arrive"; you spend your life mastering disciplines" (Senge 2006:10).

Distance See Error, ***E***(***Y***).

Disturbances, *D* In the general model of the control system, each variable, D_t, external to the system, that alters the values of *Y*, regardless of the values of *X*. Disturbances can affect all the **apparatuses** that make up the **control chain**.

Error, ***E*(*Y*)**, (distance, deviation, gap, variance) In the general model of the control system, the variable "$E(Y)_t = \Delta(Y)_t = Y^* - Y_t$," which represents the *distance* or *deviation* between the values of the objective Y^* and those of "Y_t."

Feedback control systems Control system in which the control of Y_t is achieved through decisions to vary X_t over a succession of repetitions of the control cycle.

Feedforward control systems, also called *decision* or *one-shot control systems* Control system which try to reach the objective through a single cycle, and thus through a single *initial decision* made based on precise calculations; however, when an error is detected in the attempt to achieve Y^*, the manager cannot correct this through other regulating decisions to eliminate the error. They are typically systems that "fire a single shot" to achieve the objective.

Forrester, Jay Founder of **system dynamics**, in his fundamental work *Industrial Dynamics* (1961).

Gap See **Error, *E***.

Goals In the general model of the control system, any specific value, Y^*, or any trajectory of values, Y_t^*, however defined, that must be reached and possibly maintained by the control variables. If Y is a vector $[Y^*]$, then the control system is a multi-objective system and the manager must establish a **policy** of control. Goals can be quantitative or qualitative.

Governance of the control system The process through which a subject, the governor, determines the objective Y^*, or the vector $[Y^*]$ that management must achieve.

Interfering control systems Two or more control systems interfere with each other if they are interrelated, so that the values of Y_A of a control system A determine and are determined by the values of Y_B of other interfering control systems. The interference can also occur in both directions.

iThink, Stella See Simulations **softwares**.

I/O systems See Type **of control systems**.

Levers In a control system, each control variable, X, which – through some process – acts on the *variable to control*, Y, in order to eliminate $E(Y)$. If X is a vector of variables $[X]$, then the system is multi-lever and the manager of the system should establish a **strategy** for the control.

Loop A closed causal chain formed by a circular link between two variables, X and Y, which can be linked by two opposite directions with respect to the causal link. Loops can be basic, when there are only two variables, or compound, when more than two variables are joined in a circular link. There are only two basic types of loop: *reinforcing loops* [**R**], which produce a reciprocal increase or reduction – in successive repetitions of the system's cycle – in the values of the two variables having reciprocal causal links; and *balancing loops* [**B**], which maintain relatively stable the values of the connected variables.

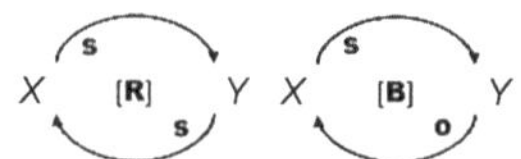

Manager of the control system   The subject that maneuvers the control system in order to achieve the goals established by the governance, by adjusting the control levers, and deciding on the **strategy** and **policy** to be adopted. The subject can be external to the control system or internal to the system itself, replacing some apparatus.

Multi-lever control systems   See Type **of control systems**.

Multi-objective control systems   See Type **of control systems**.

Objective variable, Y Synonym for **Variable to control**.

ON/OFF systems   See Type **of control systems**.

Open causal chain   Connection among a succession of several variables entailing some causal link.

$$X \xrightarrow{\textbf{s/o}} Y \xrightarrow{\textbf{s/o}} Y$$

Passive variable, Y_t Synonym for **Variable to control**.

Period of control   The number of intervals, over the typical time frame of the system, which are necessary to stably achieve the objective, in the absence of outside **disturbances**.

Pluri-lever   Synonym for **multi-lever**.

Pluri-objective   Synonym for **multi-objective**.

Policy (in control)   In multi-objective systems, policy represents the order of priority in achieving different objectives.

Powersim   See Simulation **softwares**.

Problem solving (PS)   The process through which we seek the solution of any problem. The PS can be interpreted as a control system, because a problem can be interpreted as an error or deviation, $E(Y)$, between two states – current (Y) and desired (Y^*) – of a variable, Y (which we assess as useless or harmful), and that we want to eliminate or reduce as far as possible by acting on the variables X that can change Y. In PS, solving the problem means designing the control system in order to identify the lever X_t, which, by producing the values Y_t – taking into account the states of nature, D_t – enables you to eliminate $E(Y)$. Notice that the deviation is not the problem but the symptom that is revealed. The problem lies in the malfunctioning of the system that generates the symptom.

Reaction rate "$h(X/Y)$"   Indicates the variation in X for each unit of variation in Y; usually "$h = 1/g$" if the system is perfectly symmetrical to the control.

Reaction time "$r(X/Y)$" (time to eliminate the error)   Indicates the speed with which the control system moves toward the objective; a reaction time of $r = 1$ indicates an immediate but sudden control; a reaction time $r > 1$ makes the control slower but smoother.

Regulation   In a broad sense, this is synonymous with control. In a strict sense, it is a process that assigns values to the **control levers**, based on the **deviation** and the **reaction rate**.

Senge, Peter   He is considered to be the popularizer of systems thinking, with his fundamental text: *Fifth Discipline: The Art and Practice of the Learning Organization* (Senge 1990).

Simulation softwares Software programs (simulation tools) created to simulate the behavior of systems with a high number of variables. I would mention: Powersim (www.powersim.com) MyStrategy and SYSDEA (www.strategydynamics.com/mystrategy/) ithink e stella (www.iseesystems.com/index.aspx) Vensim (www.vensim.com) Excel Software (www.excelsoftware.com/)

Sixth discipline See **Control discipline**.

Stock and flow diagram (SFD) or level and flow structure (LFS) System dynamics theory assumes that a dynamic system can normally be viewed as composed of *stock* or *level* variables, and of *flow* or *rate* variables that change the amount of stock. The flow variables that increase the stock (in various ways to be defined) can be considered as inputs; those that decrease it, as OUTPUTS. Several **Simulation softwares** are based on this logic.

Strategy (in control) Control with two independent control variables *always* implies a *strategy* that defines an *order of priorities* regarding action on the control "levers."

Symptom In **problem solving**, understood as a process of control, this indicates the meaning that the manager gives to the deviation $E(Y)$.

System (from a systems thinking perspective) Systems thinking defines a "system" as a unitary set of interconnected variables possessing its own autonomy, which is capable of producing emerging macro-dynamics that do not coincide with any of the micro-dynamics of the individual variables or their partial subsystems.

System dynamics The technique for translating the qualitative models (causal loop diagrams) into quantitative (stock and flow diagrams) ones that, by quantifying the initial values of the temporal variables and specifying the variation parameters, are able to generate the dynamics of those variables. This technique was developed as industrial dynamics by **Jay Forrester** in the 1960s.

Systems archetypes See **Archetypes**.

Systems thinking Discipline that recommends that we observe the world as a system of dynamic systems, recursive and repetitive – often with memory – each composed of interacting and interconnected variables. Systems thinking was presented by Peter Senge in *The Fifth Discipline: The Art and Practice of the Learning Organization* (Senge 1990). Systems thinking does not only represent a specific technique for constructing models but also a mental attitude, an approach, a logic, and a language. The systems are represented through qualitative models called **causal loop diagrams**.

Systems thinking laws There are two fundamental laws that derive from systems theory: 1. *The Law of Structure and Component Interaction*: In order to understand and control the dynamics in the world, it is necessary to identify the systemic structures that make up this world. Corollary: In observing a dynamic world, the ceteris paribus assumption is never valid. 2. *Law of Dynamic Instability*: Expansion and equilibrium are processes that do not last forever; they are not propagated ad infinitum. Sooner or later stability is disturbed. Sooner or later the dynamics are stabilized.

Type of control systems (CSs) minimal typology: ***Artificial*** and ***natural*** CSs: control systems designed and constructed by man are artificial or manmade control systems. ***Manually controlled*** systems, ***automatic*** CSs, or *cybernetic* systems: Systems which involve continuous manual control by a human operator are manually controlled control systems; control systems which incorporate the manager in their chain of control can be defined as automatic control systems. ***Quantitative*** CSs: The control is *quantitative* if the variable Y_t represents a *quantitative measure* and the objective is a value that Y_t must reach (*set reference*) or a set of values to be achieved (*track reference*); ***qualitative*** CSs: Systems in which both the variable Y_t and the objective Y^* are qualitative in the broadest sense of the term (objects, colors, flavors, forms, etc.). ***Attainment*** CSs: I shall define *attainment control systems* as those control systems that act to "attain" a *quantitative* objective, independently of the control context. ***Recognition*** and ***identification*** CSs: A general class of qualitative control system where the qualitative variable to control, *Yn*, represents an "object" that the system *scans* in order to recognize or identify a "model-objective," Y^*. ***Steering*** CSs (also called "two-way" systems): Systems acting to achieve the objective through positive or negative adjustments of *Yt* that converge toward Y^*, independently of the initial value of *Y*. ***Halt*** CSs (also called "one-way" systems): systems in which Y^* is by nature a non-exceedable limit or constraint. ***Collision***, *anti-collision,* and *alignment* CSs: Two interfering systems whose objectives are dynamic and tend to coincide are defined as "collision systems." If the objectives tend *not* to coincide, they are defined as "anti-collision systems." If the objectives maintain a constant distance, they are defined as "alignment systems." ***Goal-seeking*** CSs: Those in which Y^* represents a quantitative objective – to achieve a result or to attain or maintain operating standards. ***Constraint-keeping*** CSs, in which Y^* represents a constraint to respect or a limit not to be exceeded; these are usually halt systems. ***Regulator*** CSs, where Y^* (objective or constraint) is a specific value (goal, constraint, or limit) that Y_t must reach and maintain over time. ***Tracking*** CSs, or *path systems*, where Y^* (goal, constraint or limit) is a *trajectory*, that is, a sequence of values, Y^*_t, however formed, which Y_t must follow. **Multi-lever** or pluri-lever CSs, if the control is via a vector of levers $[X]$; if $[X] = X$, the system is single lever. ***Independent levers*** CSs: Multi-lever systems whose levers can be adjusted independently; dependent lever systems: if the levers allow only variations in the opposite direction. ***Multi-objective***, or pluri-objective CSs: If the control is a vector of objectives $[Y^*]$; if $[Y]^* = Y^*$, the system is single objective, in which case $[Y^*]$ must be equivalent to $[Y]$. ***Independent objectives*** CSs: Multi-objective systems whose objectives can be achieved independently; bound objective systems if a goal impedes the achievement of other goals. CSs with or without **delays**. CSs of ***direct*** or ***indirect*** *control*: A control system is direct if *X* and *Y* vary in the same direction "**s**"; otherwise, it is indirect. ***Autonomous***, or ***interfering*** CSs: Interfering systems are those whose values of *Y* are influencing each other. CSs with ***fixed*** or ***variable objectives*** (or systems of pursuit"): In the former, the variable must achieve a passive constant, Y^*; in the latter, the target Y_t^* represents a variable that depends on "*t*," *Y*, and/or *X*.

On/off CSs: Halt systems that achieve Y^* by activating [on] the X_t lever for a set time, T^*, until the system stops [off], only to start up again when the disturbances, D, once again produce an error of a set amount, ΔE^*. **I/O** CSs: Particular *steering* systems that try to achieve the objective Y^* by "turning on" [I] the X lever once or several times and for a given length of time, which is decided on each occasion by the manager, in order to obtain a fixed value ΔY, after which the lever is "turned off" [O], thereby eliminating $E(Y)$. The lever is once again turned on when D produces another error. ***Tendential*** CSs. In these systems, Y_t represents a time series (for example, the scores over time in an archery competition by participants trying to improve their performance) which, through some statistical procedure (trend, moving average, etc.), aims to ensure that Y_t tends toward a value-objective, Y^*, through a statistical measure. ***Combinatory*** CSs: Systems in which each Y_t represents a synthetic value, $[Y_{tN}]$, derived from the combination (to be specified in some manner) of the values of a vector of N variables, each of which is controlled by its own control variable.

Variable to control Any variable Y controlled by X in order to achieve a value Y^* set as an objective.

References[1]

Aaker, D. A., & Keller, K. L. (1990). Consumer evaluations of brand extensions. *The Journal of Marketing, 54*(1), 27–41.

Abbott, D. A. (1989). *Packaging perspectives*. Dubuque, Kendall: Hunt Publishing.

Abbott, R. A., & Ring, E. A. (1983). The MAPI method—Its effects on productivity: An alternative is needed. *Journal of Manufacturing Systems, 2*(1), 15–30.

Abdul-Rahman, H. (1993). Capturing the cost of quality failures in civil engineering. *International Journal of Quality & Reliability Management, 10*(3), 20–32.

Acemoglu, D., & Scott, A. (1994). Consumer confidence and rational expectations: Are agents' beliefs consistent with the theory? *The Economic Journal, 104*(422), 1–19.

Ackerman, K. (2015 online). *Practicing Pareto practically in the warehouse*. https://exclusive.multibriefs.com/content/practicing-pareto-practically-in-the-warehouse/distribution-warehousing

Ackoff, R. L., & Sasieni, M. V. (1968). *Fundamentals of operations research*. London: Wiley.

Adagunodo, T. A., & Sunmonu, L. A. (2015). Earthquake: A terrifying of all natural phenomena. *Journal of Advances in Biological and Basic Research, 1*(1), 4–11.

Adam, E. E., & Ebert, R. J. (1992). *Production and operations management*. Englewood Cliffs: Prentice Hall.

Adam, E., Mandiau, R., & Kolski, C. (2002). *Une Methode de modelisation et de conception d'organizations Multi-Agents holoniques*. Paris: Hermes.

Adams, W. M. (2006). *The future of sustainability: Re-thinking environment and development in the twenty-first century*. http://iucn.org

Ahmed, S. (2019). Integrating DMAIC approach of lean six sigma and theory of constraints toward quality improvement in healthcare. *Reviews on Environmental Health, 34*(4), 427–434.

Ahmed, E., Elgazzar, A. S., & Hegazi, A. S. (2005). *An overview of complex adaptive systems*. arXiv:nlin/0506059 [nlin.AO]. http://arxiv.org/abs/nlin/0506059v1

[1]We know nothing at all. All our knowledge is but the knowledge of schoolchildren. The real nature of things we shall never know (Albert Einstein).

As this book deals with Control Systems considered from various points of view, a complete bibliography may appear to be boundless. Here I have only listed the works cited in the book. All the sites mentioned have been visited in April 2014. In order to make it easier for the reader to directly examine

P. Mella, *The Magic Ring*, Contemporary Systems Thinking,
https://doi.org/10.1007/978-3-030-64194-8

Ahuja, I. P. S., & Khamba, J. S. (2008). Total productive maintenance: Literature review and directions. *International Journal of Quality & Reliability Management, 25*(7), 709–756.

Akao, Y. (1990). *Quality function deployment: Integrating customer requirements into product design*. Cambridge, MA: Productivity Press.

Albright, T. L., & Roth, H. P. (1994). The measurement of quality costs. *Accounting Horizons, 6*(2), 15–27.

Alchian, A. A. (1950). Uncertainty, evolution, and economic theory. *Journal of Political Economy, 58*(3), 211–221. http://institutoamagi.org/download/Alchain-Armen-Uncertainty-Evolution-and-Economic-Theory.pdf

Alchian, A. A., & Demsetz, H. (1972). Production, information costs, and economic organization. *The American Economic Review American Economic Association, 62*(5), 777–795.

Allderige, J. M. (1954). Work sampling without formulas. *Factor Management and Maintenance, 112*(3), 136–138.

Allen, P. M. (1997). *Cities & regions as self-organizing systems: Model of complexity*. Amsterdam: Gordon and Breach.

Alleyne, R. (2010). Titanic sunk by steering blunder, new book claims. *The Telegraph*. www.telegraph.co.uk/culture/books/booknews/8016752/Titanic-sunk-by-steering-blunder-new-book-claims.html

Anderson, V., & Johnson, L. (1997). *Systems thinking basics. From concepts to causal loops*. Waltham: Pegasus Communications.

Anderson, R., Eriksson, H., & Torstensson, H. (2006). Similarities and differences between TQM, Six Sigma and Lean. *The TQM Magazine, 18*(3), 282–296.

Andrews, A. (2011). *Butterfly effect: How your life matters*. Nashville: Thomas Nelson.

Ankersmit, F. R., & Mooij, J. J. A. (1993). *Knowledge and language* (Metaphor and knowledge) (Vol. III). Dordrecht: Kluwer.

Anthony, R. N. (1965). *Planning and control systems: Framework for analysis*. Boston: Harvard University.

Anthony, R. N. (1998). *The management control functions*. Boston: HBS Press.

Arbib, M. A. (1987). *Feedback and feedforward*. http://www.answers.com/topic/feedback-and-feedforward

Archibald, R. D. (2003). *Managing high-technology programs and projects*. Hoboken: Wiley.

Argyris, C. (1993). *Knowledge for action. A guide to overcoming barriers to organizational change*. San Francisco: Jossey Bass.

Argyris, C., & Schön, D. (1974). *Theory in practice. Increasing professional effectiveness*. San Francisco: Jossey Bass.

Arnold, G., & Davies, M. (2000). *Value-based management. Context and application*. London: Wiley.

Armstrong, J. S. (1983). Strategic planning and forecasting fundamentals. Marketing Papers, 123. http://repository.upenn.edu/marketing_papers.

Arnold, R., & Dennis, R. (1999, April, 7–12). Perspectives on productivity growth. *Business Economics*.

Arsham, H. (2006). *Economic order quantity and economic production quantity*. London: Wiley.

Arthur, W. B. (1973). *Optimal control theory with time delay*. Ph.D. Thesis, University of California, Berkeley, Operations Research Center, Report No. ORC 73–27.

Arthur, W. B. (1994). *Increasing returns and path dependence in the economy*. Ann Arbor: University of Michigan Press.

Arthur, W. B., Durlauf, S. N., & Lane, D. A. (1997). *The economy as an evolving complex system II*. Reading: Addison-Wesley.

Ashby, R. W. (1957). *An introduction to cybernetics* (2nd ed.). London: Chapman & Hall.

Ashworth, G., & James, P. J. (2001). *Value-based management: Delivering superior shareholder value*. Englewood Cliffs: Financial Times, Prentice Hall.

ASQ. American Society for Quality. (1991). https://asq.org/quality-resources/quality-glossary/q

ASQC. Quality Costs Technical Committee. (1961). *Quality costs – what and how?* (1st ed.). Milwaukee: American Society for Quality.

ATI. (2018). *Baggage Report 2018: published by SITA*. https://www.sita.aero/resources/type/surveys-reports/baggage-report-2018
Sanctus Augustinus. In *Iohannis epistulam ad Parthos tractatus* PL 35 et CC SL, 37, 198 (W. J. Mountain, textus adhuc ineditus).
Austin, J. L. (1962). *How to do things with words*. Cambridge, MA: Harvard University Press. In Bial, H., & Schechner, R. (2004). *The performance studies reader*.
Axsäter, S. (2015). *Inventory control*. Cham: Springer International Publishing Switzerland.
Azzi, A., Battini, D., Faccio, M., Persona, A., & Sgarbossa, F. (2014). Inventory holding costs measurement: A multi-case study. *The International Journal of Logistics Management, 25*(1), 109–132.
Bakshi, U. A., & Bakshi, V. U. (2009). *Automatic control systems*. Pune: Technical Publications.
Balderstone, S. J., & Mabin, V. J. (1998). A review of Goldratt's theory of constraints (TOC)–lessons from the international literature. In *Operations Research Society of New Zealand 33rd Annual Conference*.
Ballé, M. (2015, online). *Organizing for learning is critical to sustaining your kaizen efforts and improving your company.* https://planet-lean.com/learning-organization-kaizen-stick/
Ballou, R. H. (1967). Improving the physical layout of merchandise in warehouses. *Journal of Marketing, 31*(3), 60–64.
Banathy, B. (2000). *Guided evolution of society. A systems view*. New York: Springer.
Band, W. A. (1991). *Creating value for customers: Designing and implementing a total corporate strategy*. London and New York: Wiley.
Band, W. A., & Huot, R. (1990). Quality and functionality equal satisfaction. *Sales and Marketing Management in Canada, 31*(3), 4–5.
Barabási, A. (2002). *Linked: The new science of networks*. Cambridge, MA: Perseus.
Barraclough, T. G. (2015). How do species interactions affect evolutionary dynamics across whole communities? *Annual Review of Ecology, Evolution, and Systematics, 46*(1), 25–48.
Barrows, E. M. (2001). *Animal behavior desk reference: A dictionary of animal behavior, ecology and evolution*. Boca Raton: CRC Press.
Bartlett, C. A., & Ghoshal, S. (1989). *Managing across Borders: The transnational solution*. Boston: Harvard Business School Press.
Bartmann, D., & Beckmann, M. J. (1992). *Inventory control. Models and methods* (Lecture notes in economics and mathematical systems). Berlin/Heidelberg: Springer.
Basu, R. (2009). *Implementing six sigma and lean – A practical guide to Tools and techniques*. Oxford: Butterworth Heinemann.
Basu, R., & Wright, J. N. (2003). *Quality beyond six sigma*. Oxford: Butterworth Heinemann.
Bateson, G. (1972). See Bateson, G. (2000).
Bateson, G. (1979). See Bateson, G. (2002).
Bateson, G. (2000). *Steps to an ecology of mind*. Chicago: University of Chicago Press. Excerpts, Google Books.
Bateson, G. (2002, 1st ed. 1979). *Mind and nature: A necessary unity*. Cresskill: Hampton Press. (Originally published by Bantam, 1979). Excerpts, http://www.oikos.org/mind&nature.htm
Baumeister, R. F., Vohs, K. D., & Tice, D. M. (2007). The strength model of self-control. *Current Directions in Psychological Science, 16*(5), 351–355.
Baumol, W. J. (1962). On the theory of expansion of the firm. *The American Economic Review, 52*(5), 1078–1087.
Baumol, W. J., Blackman, S. A., & Wolff, E. N. (1989, 1st ed. 1972). *Productivity and American Leadership. The long view*. Cambridge, MA, MIT Press. Wiley.
BCG, The Boston Consulting Group. (1970). Corporate portfolio management: Theory and practice. *Applied Corporate Finance, 23*(1), 63–76.
Bednarz, J. (1988). Autopoiesis: The organizational closure of social systems. *Systems Research, 5*(1), 57–64.
Beemaraj, R. K., & Prasad, K. A. (2018). Six sigma concept and DMAIC implementation. *IJBMR, 3*(2), 111–114.

Beer, S. (1966). *Decision and control*. London: Wiley.
Beer, S. (1975). *Platform for change*. Chichester: Wiley.
Beer, S. (1979). *The heart of enterprise*. London: Wiley.
Beer, S. (1981, 1st ed. 1972). Brain of the firm. Wiley, London and New York.
Beer, S. (1992). Preface. In H. R. Maturana, & F. J. Varela (1992).
Beer, S. (1995). *Diagnosing the system for organizations*. London: Wiley.
Bellinger, G. (2004). *Archetypes, Musings*. http://www.systems-thinking.org/arch/arch.htm
Bellows, W. J. (2004). Conformance to specifications, zero defects and six sigma quality–a closer look. *International Journal of Internet and Enterprise Management, 2*(1), 82–95.
Bennett, P. D., & Harrell, G. D. (1975). The role of confidence in understanding and predicting buyers' attitudes and purchase intentions. *Journal of Consumer Research, 1*(2), 110–117.
Benoit, S., Fassnacht, M., & Klose, S. (2006). A framework for supplier relationship management (SRM). *Journal of Business-to-Business Marketing, 13*(4), 69–94.
Benton, W. C., & Shin, H. (1998). Manufacturing planning and control: The evolution of MRP and JIT integration. *European Journal of Operational Research, 110*(3), 411–440.
Berkeley University of California. (2010). *Guide to managing human resources*. https://www.csusm.edu/hr/performance_management/index.html
Bernstein, L. A., & Wild, J. J. (1989). *Financial statement analysis: Theory, application, and interpretation*. Homewood: Irwin.
Berryman, A. A. (2001). Functional web analysis: Detecting the structure of population dynamics from multi-species time series. *Basic and Applied Ecology, 2*(4), 311–321.
Bhikkhu, B. (1984). *The noble eightfold path. The way to the end of sufffering*. http://www.thesecularbuddhist.com/nep_02.php
Bial, H., & Schechner, R. (2004). *The performance studies reader*. New York: Routledge.
Bianchi, M. (2019). Beyond the structural modelling for the analysis of organizational performances in the resilience management. *Economia Aziendale Online, Special Issue, 10*(1), 63–74.
Biernat, M., & Kobrynowicz, D. (1997). Gender-and race-based standards of competence: Lower minimum standards but higher ability standards for devalued groups. *Journal of Personality and Social Psychology, 72*(3), 544.
Blackburn, J. D. (1991). Time-based competition: The next battleground. In *American manufacturing*. Homewood: Irwin.
Blackwell, D., & Girshich, M. A. (1954). *Theory of games and statistical decisions*. London: Wiley.
Blank, R. M., & Shapiro, M. D. (2001). Labor and the sustainability of output and productivity growth. In A. B. Krueger & R. M. Solow (Eds.), *Prepared for sustainable employment*. New York: Russell-Sage Foundation/Century Foundation, in process.
Bodhi, B. (2011). Noble eightfold path: Way to the end of suffering. Pariyatti Publishing. https://www.accesstoinsight.org/lib/authors/bodhi/waytoend.html.
Boer, H., & Krabbendam, K. (1992). Organizing for manufacturing innovation. The case of Flexible Manufacturing Systems. *International Journal of Operations & Production Management, 12*(7/8), 41–56.
Bossert, J. L. (1991). *Quality function deployment: a practitioner's approach, 21*. Milwaukee: ASQC Quality Press.
Bostic-St. Clair, C., & Grinder, J. (2001). *Whispering in the wind*. Scotts Valley: J & C Enterprises.
Boulding, K. E. (1956). General system theory, management science. The skeleton of science. *Management Science, 2*(3), 197–208.
Boulding, K. E. (1981). *Evolutionary economics*. Beverly Hill: SAGE Publications.
Bouyssou, D., Marchant, T., Pirlot, M., Tsoukias, A., & Vincke, P. (2006). *Evaluation and decision models with multiple criteria*. New York: Springer Sciences + Business Media.
Bowersox, D., Closs, D., & Bixby, M. (2005). *Supply chain logistics, management*. Boston: McGraw-Hill Irwin.
Braithwaite, R. B. (1953). Scientific explanation. A study of the function of theory, probability and law in science. Tarner Lectures. CUP Archive, Cambridge University Press.

Braun, W. (2002). *The system archetypes*. http://www.albany.edu/faculty/gpr/PAD724/724WebArticles/sys_archetypes.pdf

Breventano, S. (1570). *A history of antiquity, nobility, and of notable things of the city of Pavia* [original: Istoria della antichità, nobiltà, et delle cose notabili della città di Pavia]. Pavia: Hieronimo Bartholi, MDLXX. http://books.google.it/books?id=_CQ8AAAAcAAJ&printsec=frontcover&hl=it&source=gbs_ge_summary_r&cad=0#v=onepage&q=torri&f=false

Brickman, L. (2012). *Mathematical introduction to linear programming and game theory*. Heidelberg: Springer Science & Business Media.

Britt, R. R. (2004). *Asteroid with chance of hitting earth in 2029 now being watched 'very carefully'*. www.space.com/scienceastronomy/asteroid_risk_041224.html

Brockhurst, M. A., Morgan, A. D., Rainey, P. B., & Buckling, A. (2003). Population mixing accelerates coevolution. *Ecology Letters, 6*(11), 975–979.

Brown, D. (1977). Reminiscences of an Industrialist. Easton, PA: Hive Publishing Company.

Brown, H. (1954). *The challenge of man's future*. New York: Viking Press, (quoted by, Cipolla, 1962).

Brown, A. (2016). *Your job won't exist in 20 years: Robots and AI to 'eliminate' ALL human workers by 2036*. https://www.express.co.uk/life-style/science-technology/640744/Jobless-Future-Robots-Artificial-Intelligence-Vivek-Wadhwa

Browne, J., Dubois, D., Rathmill, K., Sethi, S. P., & Stecke, K. E. (1984). Classification of flexible manufacturing systems. *The FMS magazine, 2*(2), 114–117.

Brun, H. (1970). *Technology and the composer*. http://www.herbertbrun.org/techcomp.html

Brundtland, G. H. (1987). *Our common future*. United Nations General Assembly. World Commission on Environment and Development (WCED)—The "Brundtland Commission", Future. Oxford, UK: Oxford University Press, United Nations.

Brynjolfsson, E. (1994). *The productivity paradox of information technology: Review and assessment*. Center for Coordination Science, MIT Sloan School of Management, Cambridge, Massachusetts at: http://ccs.mit.edu/papers/CCSWP130/ccswp130.html

Brynjolfsson, E., & Hitt, L. (2000). Beyond computation: Information technology, organizational transformation and business performance. *Journal of Economic Perspectives, 14*(4), 23–48.

Buchanan, K. L. (2000). Stress and the evolution of condition-dependent signals. *Trends in Ecology and Evolution, 15*(4), 156–160.

Buckling, A., & Rainey, P. B. (2002). The role of parasites in sympatric and allopatric host diversification. *Nature, 420*(6915), 496–499.

Burgiel, S. W. (2020). The incident command system: A framework for rapid response to biological invasion. *Biological Invasions, 22*(1), 155–165.

Buzacott, J. A., & Yao, D. D. (1986). Flexible manufacturing systems: A review of analytical models. *Management Science, 32*(7), 890–905.

Campanella, J. (1999). Principles of quality costs: Principles, implementation, and use. In *ASQ (American Society for Quality) World Conference on Quality and Improvement Proceedings*, 507.

Campbell, C., & Rozsnyai, C. (2002). *Quality Assurance and the Development of Course Programmes*. Papers on Higher Education Regional University Network on Governance and Management of Higher Education in South East Europe. Bucharest: UNESCO.

Capra, F. (1996a). The rise of systems thinking. In F. Capra (Ed.), *The web of life*. http://www.maaber.org/sixth_issue/deep_ecology_1e.htm#_ftnref35

Capra, F. (1996b). *The web of life: A new scientific understanding of living systems*. New York: Anchor Books.

Caprara, G. (2010, July 1). White cloud brings rain [original: Nuvola bianca, porta pioggia]. *Corriere della Sera*. http://www.corriere.it/scienze/10_luglio_01/nuvole-bianche_93fb9560-8537-11df-a3c2-00144f02aabe.shtml

Cartwright, D. (Ed.). (1947). *Field theory in social science: Selected theoretical papers (188–237)*. London: Tavistock.

Carver, C. S., & Scheier, M. F. (1981). *Attention and self-regulation: A control-theory approach to human behavior*. New York: Springer.

Casey, W. (2000). *Project tools for leading organizational change*. Wheat Ridge: Executive Leadership Group.

CASG, Complex Adaptive Systems Group. (2013). http://www.cas-group.net

Castelfranchi, C. (1998). Through the minds of the agents. *Journal of Artificial Societies and Social Simulation, 1*(1), 1–5. http://jasss.soc.surrey.ac.uk/1/1/5.html

Castera, L. (2016). *The law of accelerating returns in the shipping industry*. https://blog.octopi.co/2016/11/16/the-law-of-accelerating-returns-in-the-shipping-industry/

Casti, J. L. (1985). *Nonlinear system theory*. Orlando: Academic Press.

CBC News. (2002). *The mane attraction: prefer dark-haired lions*. https://www.cbc.ca/news/technology/the-mane-attraction-lionesses-prefer-dark-haired-lions-1.305212

Ceballos, G., Erlich, P. R., Barnosky, A. D., Garcia, A., Pringle, R. M., & Palmer, T. M. (2015). Accelerated modern human-induced species losses: Entering the sixth mass extinction. *Science Advances, 1*(5), e1400253–e1400253.

Ceccato, S. (1969). *Course in operational linguistic [italian: Corso di linguistica operativa]*. Milano: Longanesi.

Ceccato, S. (1974). The third cybernetics. For a creative and responsible mind [Original: La terza cibernetica. Per una mente creativa e responsabile]. Milan, Italy: Feltrinelli.

Chakhoyan, A. (2017). The jobless world and its discontents. *World Economic Forum*. https://www.weforum.org/agenda/2017/01/jobless-world-and-its-discontents/

Champernowne, D. G. (1969). *Uncertainty and estimations in economics*. Edinburg: Oliver e Boyd.

Chandler, A. D., Jr. (1962). *Strategy and structure: Chapters in the history of the American Industrial Enterprise*. Cambridge, MA: MIT Press.

Checkland, P. (1999). *Systems thinking, systems practice*. London: Wiley.

Cheng, T. C. E. (1991). An economic order quantity model with demand-dependent unit production cost and imperfect production processes. *IIE Transactions, 23*(1), 23–28.

Chiarini, A. (2010). *Lean organization for excellence*. Milan: Franco Angeli.

Chiarini, A., Baccarani, C., & Mascherpa, V. (2018). Lean production, Toyota production system and kaizen philosophy. *The TQM Journal, 30*(3). https://doi.org/10.1108/TQM-12-2017-0178.

Christen, M., & Schmidt, S. (2012). A formal framework for conceptions of sustainability—A theoretical contribution to the discourse in sustainable development. *Sustainable Development, 20*(6), 400–410.

Christenson, C. (1997). *The innovator's dilemma*. Cambridge, MA: Harvard Business School Press.

Christopher, M. (2000). The agile supply chain. Competing in volatile markets. *Industrial Marketing Management, 29*(1), 37–44.

Chron. (2018). Arctic defrosting: Melting of polar ice pits environmental alarm against desire to exploit newly accessible energy sources. *Hearst Newspapers, LLC*. https://www.chron.com/opinion/outlook/article/Arctic-defrosting-Melting-of-polar-ice-pits-1551037.php

Chu, P., Chung, K. J., & Lan, S. P. (1998). Economic order quantity of deteriorating items under permissible delay in payments. *Computers & Operations Research, 25*(10), 817–824.

Churchman, C. W., & Ackoff, R. L. (1954). An approximate measure of value. *Operations Research, 2*(2), 172–187.

Churchman, C. W., Ackoff, R. L., & Arnoff, E. L. (1957). *Introduction to operations research*. London: Wiley.

CIMA – Chartered Institute of Management Accountants. (2005). *Official terminology*. Oxford: CIMA.

Cipolla, C. (1962). *The economic history of world population*. Harmondsworth: Penguin Books.

CIPS, Chartered Institute of Purchasing and Supply. (2007). *How to appraise suppliers*. https://www.cips.org/Documents/Resources/Knowledge%20How%20To/How%20to%20Appraise%20Suppliers.pdf

Claret, J. (1981, May 24–26). *Never mind the quality*. Management Accounting.

Clark, A. J. (1972). An informal survey of multi-echelon inventory theory. *Naval Research Logistics (NRL), 19*(4), 621–650.

Clark, A. (1989). Microfunctionalism: Connectionism and the scientific explanation of mental states. In A.

Clark, A. (1991). *Microcognition: Philosophy, cognitive science, and parallel distributed processing*. Cambridge: MIT Press.

Club of Rome. (1972). *The limits to growth*. A Potomae Associates Book.

Coase, R. H. (1937). The nature of the firm. *Economica, 4*(16), 386–405.

Coe, N. M., Dicken, P., & Hess, M. (2008). Global production networks: Realizing the potential. *Journal of Economic Geography, 8*(3), 271–295.

COM, European Commission communication to the European Council and Parliament. (2002a online). http://europa.eu/legislation_summaries/enterprise/industry/n26027_en.htm

COM, European Commission communication to the European Council and Parliament. (2002b online). http://eur-lex.europa.eu/LexUriServ/LexUriServ.do?uri=CELEX:52002DC0262:EN:HTML

Conde, E. (2009). A Minmax regret approach to the critical path method with task interval times. *European Journal of Operational Research, 197*(1), 235–242.

Connor, J., & Evans, J. B. (1972). *Replacement investment: A guide to management decisions on plant renewal*. Epping: Gower Publishing Company, Limited.

Conte, R., & Paolucci, M. (2001). Intelligent social learning. *JASSS Journal of Artificial Societies and Social Simulation*. http://jasss.soc.surrey.ac.uk/4/1/3.html

Continuous Improvement Associates. (2003). *Systems thinking archetypes*. http://www.exponentialimprovement.com/cms/uploads/ArchetypesGeneric02.pdf

Cooper, R. (1989). The rise of activity based costing—Part three: How many cost drivers do you need and how do you select them? *Journal of Cost Management, 2*(4), 34–46.

Cooper, R., & Kaplan, S. (1991, May–June). Profit priorities from activity-based costing. *Harvard Business Review, 69*(3), 130–135.

Copeland, T., Koller, T., & Murrin, J. (2000). *Valuation: Measuring and managing the value of companies* (3rd ed.). London: Wiley. (2nd ed., 1996).

Copi, I. M., & Cohen, C. (2008, last ed.). *Introduction to logic* (14th ed.). Edinburgh: Pearson.

Copi, I. M., & Cohen, C. (2011). Introduction to logic (14th ed.). Edinburgh, TX: Pearson.

Correll, D. S., & Barnes, R. M. (1950). Industrial application of the ratio-delay method. *Operational Research Quarterly*, 69–70.

Costanza, R., & Patten, B. C. (1995). Defining and predicting sustainability. *Ecological Economics, 15*(3), 193–196.

Coyle, R. G. (1977). *Management system dynamics*. London: Wiley.

Critical Tools. Project Planning Software. (2020). http://www.criticaltools.com/pertchartexpert-software.htm

Crosby, P. B. (1979). *Quality is free*. Boston, New York: McGraw-Hill.

Crow, K. (2002). *Value analysis and function analysis system technique*. http://www.npd-solutions.com/va.html

CSS, Complex Systems Society. (2013). *About complex systems*. http://www.complexssociety.eu/aboutComplexSystems.html

Cupples, B. (1977). Three types of explanation. *Philosophy of Science, 44*(3), 387–408.

Cyert, R. M., & March, J. G. (1963). *A behavioral theory of the firm* (2nd ed.). Englewood Cliffs: Prentice Hall.

D'Amours, S., Montreuil, B., Lefrançois, P., & Soumis, F. (1999). Networked manufacturing: The impact of information sharing. *International Journal of Production Economics, 58*(1), 63–79.

Daniels, A. C., & Daniels, J. E. (2004, 1st ed. 1992). *Performance management: Changing behavior that drives organizational effectiveness* (4th ed.). Atlanta: Performance Management Publications.

Dantzig, G. B. (1963). *Linear programming and extensions*. Princeton: Princeton university Press. https://doi.org/10.7249/R366.

Darwin, C. (1859). *On the origins of species by means of natural selection*. London: Murray. http://darwin-online.org.uk/content/frameset?itemID=F373&viewtype=text&pageseq=1

Darwin, C., & Wallace, A. R. (1858). On the tendency of species to form varieties; and on the perpetuation of varieties and species by natural means of selection. *Zoological Journal of the Linnean Society, 3*(9), 46–62.
Davenport, T. H., & Short, J. E. (1990). The new industrial engineering: information technology and business process re-design. Sloan Management Review, 31, 411–427.
Davenport, T. H., & Stoddard, D. B. (1994). Reengineering: Business change of mythic proportions? *MIS Quarterly, 18*(2), 121–127.
David, P. A., & Wright, G. (1999a). *Early twentieth century growth dynamics: An inquiry into the economic history of our ignorance*, SIEPR discussion paper no. 98-3. Stanford.
David, P. A., & Wright G. (1999b). *General purpose technologies and surges in productivity: Historical reflections on the future of the ICT revolution.* http://economics.ouls.ox.ac.uk/12488/1/a4.pdf
Davidow, W. H., & Malone, M. (1992). *The virtual corporation.* New York: Harper Business.
Davis, K. (1949). *Human society*. New York: Macmillan.
Davis, T. C. (1950). How the DuPont Organization Appraises Its Performance. Financial Management Series, 94. New York: American Management Association.
Dawkins, R. (1989 1st ed. 1976). The selfish gene. Oxford: Oxford University Press.
Dawkins, R. (2004). *The ancestor's tale: A pilgrimage to the dawn of evolution.* Boston: Houghton Mifflin.
DCAA. U.S. Department of Defense. (2000). *Work sampling*. Audit Agency manual (Appendix 1).
de Carvalho, J. L. (1993). *Dynamical systems and automatic control*. Englewood Cliffs: Prentice Hall.
de Geus, A. (1988). Planning as learning. *Harvard Business Review, 66*(2), 70–74.
de Geus, A. (1992). Modelling to predict or to learn? *European Journal of Operational Research, 59*(1), 1–4.
de Geus, A. (2002). *The living company: Habits for survival in a turbulent business*. Boston: Harvard Business Review Press.
de Saussure, F. (1916). *Course in general linguistics* [original: Cours de linguistique Générale]. Charles Bally and Albert Sechehaye. New York: Philosophical Library. Excerpt, http://faculty.smu.edu/dfoster/cf3324/saussure.htm
De Toni, A., Caputo, M., & Vinelli, A. (1988). Production management techniques: Push-pull classification and application conditions. *International Journal of Operations & Production Management, 8*(2), 35–51.
de Vries, H. D. (1905). *Species and varieties: Their origin by mutation: Lectures delivered at the University of California by Hugo de Vries*. Chicago: Open Court Publishing.
Dean, J., & Smith, W. (1955). Has MAPI a place in a comprehensive system of capital controls? *The Journal of Business, 28*(4), 261–274.
Deephouse, D. L. (1999). To be different, or to be the same? It's a question (and theory) of strategic balance. *Strategic Management Journal, 20*(2), 147–166.
Demartini, C. (2014). *Performance management systems. Design, diagnosis and use*. New York: Springer.
Demartini, C., & Mella, P. (2011). Time competition. The new strategic frontier. *iBusiness, 3*(2), 136–145.
Deming, W. E. (1989, 1st ed. 1982, 2nd ed. 1986). *Out of the crisis. Quality, productivity and competitive position.* Cambridge, MA: MIT Center for Advanced Engineering Study: MIT Press.
Deming, W. E. (2012). *The essential Deming: Leadership principles from the father of quality*. New York: McGraw Hill Professional.
Deming, W. E. (2018). *The new economics for industry, government, education*. MIT press.
Deming Institute. (2020). The Edwards Deming Institute: *The Deming System of Profound Knowledge*. https://deming.org/explore/sopk/
Dempsey. L. (2015). *System archetypes.* http://www.fashionforafiniteplanet.com/system-archetypes/
Dermer, J. D., & Lucas, R. G. (1986). The illusion of managerial control. *Accounting, Organizations and Society, 11*(6), 471–482.

Dewan, E. M. (1969). *Cybernetics and the management of large systems*. New York: Spartan Books.
Dhar, S., & Nandi, S. (2016). A quantitative approach to measure effectiveness of defect prevention process. *IJIRCCE, 4*(8), 15301–15310.
Dick, A. S., & Basu, K. (1994). Customer loyalty: Toward an integrated conceptual framework. *Journal of the Academy of Marketing Science, 22*(2), 99–113.
Directive 75/318/EEC. (1991). https://www.ema.europa.eu/documents/scientific-guideline/specifications-control-tests-finished-product_en.pdf
Dowling, A. M., MacDonald, R. H., & Richardson, G.P. (1995). Simulation of systems archetypes. In *Proceedings of the 1995 international system dynamics conference*, 2, pp. 454–463.
Drake, P. P., & Fabozzi, F. J. (2008). *Financial ratio analysis. Handbook of finance*. Wiley.
Drucker, P. F. (1973, last ed. 1986). Management. Tasks, responsibilities, practices. New York: Harper & Row.
Drucker, P. (1977). *People and performance*. New York: Harper & Row Pub (last ed. Routledge, 2013).
Drury, C. M. (1998). *Standard costing*. Cengage Learning Publisher.
Drury, C. M. (2013, 3rd ed.). *Management and cost accounting*. Springer (1st ed. 2007).
Du Pont De Nemours, E. I., & Company. (1961). Executive committee control charts: A description of the Du Pont Chart system for appraising operating performance. *American Management Association Bulletin, 6*.
Duncan, A. J. (1986, 1st ed. 1974). Quality control and industrial statistics. Homewood: Irwin.
Dunlap, J. C., Loros, J., & De Coursey, P. J. (2003). *Chronobiology: Biological timekeeping*. Sunderland: Sinauer.
Durlinger, P. P. J., & Paul, I. (2012). *Inventory and holding costs*. Durlinger Consultant. p. 1.
Dutschke, J. (2014). *Run to failure: make it part of your maintenance planning*. https://www.fiixsoftware.com/blog/run-failure-make-part-maintenance-planning/#:~:text=Run%20to%20failure%20is%20a,to%20minimize%20total%20maintenance%20costs
Dyer, J. H. (1997). Effective Interfirm collaboration: How firms minimize transaction costs and maximise transaction value. *Strategic Management Journal, 18*(7), 535–556.
Dyson, F. (1988). *Infinite in all directions*. New York: Harper and Row.
Eaton, G. (2005). *Management accounting official terminology*. Burlington: Elsevier.
Ebeling, C. E. (1997). *An introduction to reliability and maintainability engineering*. Boston, New York: McGraw-Hill.
Eber, S. (2004). Bottom-up density regulation in the Holly leaf-miner Phytomyza Ilicis. *Journal of Animal Ecology, 73*(5), 948–958.
Eccles, J. C., & Robinson, D. N. (1984). *The wonder of being human. Our brain & our mind*. New York: Free Press.
Eco, U. (1976). *A theory of semiotics*. Bloomington: Indiana University Press.
Edelman, G. (1989.) *The remembered present by a biological theory of consciousness* (review by Bobby Matherne 1997). New York: Basic Books.
Edelman, G. (1992). *Bright air, brilliant fire: On the matter of the mind*. New York: Harper Collins, Basic Books.
Edwards, M. (2003). *A brief history of Holons*. http://www.integralworld.net/edwards13.html
Ehrbar, A. (1998). *EVA: The real key to creating wealth*. New York: Wiley.
Ellison, B., Fisher, D. A., Linger, R. C., Lipson, H. F., Longstaff, T., & Mead, N. R. (1997). *Survivable network systems: An emerging discipline*. Survivable Network Technology Team CERT® (revised May 1999).
Encyclical Letter. (1937). *"Mit Brennender Sorge" of the Supreme Pontiff Pius XI*. http://www.vatican.va/content/pius-xi/en/encyclicals/documents/hf_p-xi_enc_14031937_mit-brennender-sorge.html
Encyclical Letter. (1986). *"Dominum et Vivificantem. On the Holy Spirit in the Life of the Church and the World" of the Supreme Pontiff JEAN-PAUL II*. http://www.vatican.va/content/john-paul-ii/en/encyclicals/documents/hf_jp-ii_enc_18051986_dominum-et-vivificantem.html

Encyclical Letter. (2007). *"Spe Salvi" of the supreme Pontiff Benedict XVI*. http://www.vatican.va/content/benedict-xvi/en/encyclicals/documents/hf_ben-xvi_enc_20071130_spe-salvi.html
Environment. (2013). http://www.environment.gen.tr/gaia/70-gaia-hypothesis.html
EPA – United States Environmental Protection Agency. (2019). *Climate change and harmful algal blooms*. https://www.epa.gov/nutrientpollution/climate-change-and-harmful-algal-blooms
Eroglu, A., & Ozdemir, G. (2007). An economic order quantity model with defective items and shortages. *International Journal of Production Economics, 106*(2), 544–549.
Espejo, R., & Harnden, R. (1989). *The viable system model*. New York: Wiley.
Ethnologue, Languages of the World. (2013). http://www.ethnologue.com/ethno_docs/distribution.asp?by=area
Evans, J. S. (1991). Making fast strategic decision in high velocity environments. *Academy of Management Journal, 32*, 542–576.
Ezekiel, M. (1938). The cobweb theorem. Quarterly Journal of Economics, 52, 255–280.
Fabinyi, M., Evans, L., & Foale, S. J. (2014). Social-ecological systems, social diversity, and power: Insights from anthropology and political ecology. *Ecology and Society, 19*(4). https://doi.org/10.5751/ES-07029-190428.
Fabrycky, W. J., Ghare, P. M., & Torgersen, P. E. (1972). *Industrial operations research*. Englewood Cliffs: Prentice-Hall.
Fayol, H. (1949, 1st ed. 1916). *General and industrial management* [original: Administration Industrielle et Générale]. New York: Pitman Publishing Corporation.
Feigenbaum, A. V. (1951). *Quality control: Principles, practice, and administration: An industrial management tool for improving product quality and design and for reducing operating costs and losses*. Boston, New York: McGraw-Hill.
Feigenbaum, A. V. (1991, 1st ed. 1951). Total quality control. Boston, New York: McGraw-Hill.
Feigenbaum, A. V. (1997). Changing concepts and management of quality worldwide. *Quality Progress, 30*(12), 45–48.
Ferber, M., & Nelson, J. (1993, last ed. 2020). Feminist economics today: Beyond economic man. Chicago, University of Chicago Press.
Ferreira, A., & Otley, D. (2009). The design and use of performance management systems: An extended framework for analysis. *Management Accounting Research, 20*(4), 263–282.
Ferretti, S., Caputo, D., Penza, M., & D'Addona, D. M. (2013). Monitoring systems for zero defect manufacturing. *Procedia CIRP, 12*, 258–263.
Fishburn, P. C. (1964). *Decision and value theory*. London: Wiley.
Fishburn, P. C. (1967a). Additive utilities with incomplete product sets: Application to priorities and assignments. *Operations Research, 15*(3), 537–542.
Fishburn, P. C. (1967b). Methods of estimating additive utilities. *Management Science, 13*(7), 435–453.
Fisher, D. A., & Lipson, H. F. (1999). Emergent algorithms—A new method for enhancing survivability in unbounded systems. *The 32nd Hawaii International Conference on System Sciences*. http://citeseerx.ist.psu.edu/viewdoc/download?doi=10.1.1.105.8956&rep=rep1&type=pdf
Flake, G. W. (2001). *The computational beauty of nature*. Cambridge, MA: The MIT Press.
Flamholtz, E. G. (1996). *Effective management control: Theory and practice*. New York: Springer.
Fleming, J. R. (2007). The climate engineers. *Wilson Quarterly, 31*(2), 46–60. http://www.colby.edu/sts/climateengineers.pdf
Flesher, D. L., & Previtis, G. J. (2013). Donaldson Brown (1885–1965). The power of an individual and his ideas over time. *Accounting Historians Journal, 40*(1), 79–102.
Flood, R. L. (1993). *Beyond TQM*. London and New York: Wiley.
Flood, R. L. (1999). *Rethinking the fifth discipline*. New York: Routledge.
Floreano, D., & Mattiussi, C. (2008). Bio-Inspired Artificial Intelligence. Theories, Methods, and Technologies. Cambridge: The MIT Press.
Floreano, D., & Keller, L. (2010). Evolution of adaptive behaviour in robots by means of Darwinian selection. *PLoS Biology, 8*(1), e1000292. https://doi.org/10.1371/journal.pbio.1000292.

Floreano, D., & Nolfi, S. (1997). God save the red queen! Competition in co-evolutionary robotics. In J. R. Koza, K. Deb, M. Dorigo, D. B. Fogel, M. Garzon, et al. (Eds.), *Genetic programming: Proceedings of the second annual conference* (pp. 398–406). San Francisco: Morgan Kaufmann.

Floreano, D., Dürr, P., & Mattiussi, C. (2008a). Neuroevolution: From architectures to learning. *Evolutionary Intelligence, 1*(1), 47–62.

Floreano, D., Husbands, P., & Nolfi, S. (2008b). Evolutionary robotics. In B. Siciliano & O. Khatib (Eds.), *Springer handbook of robotics* (pp. 1423–1451). Berlin: Springer.

Fodor, J. A., & Pylyshyn, Z. W. (1988). Connectionism and cognitive architecture: A critical analysis. *Cognition, 28*(1–2), 3–71.

Fornaciari, L., & Galassi, D. (2012). *Strategies and economic-financial control for the point of sale* [original: Strategie e control-lo economico finanziario per il punto vendita]. Milan: IPSOA.

Forrester, J. W. (1961). *Industrial dynamics*. Waltham: Pegasus Communications.

Forrester, J. W. (1970). Urban dynamics, IMR. *Industrial Management Review, 11*(3), 67.

Forrester, J. W. (1971). *Counterintuitive behavior of social systems*. Collected papers of J. W. Forrester, (pp. 211–244). Cambridge, MA: Wright-Allen Press.

Forrester, J. W. (1983). Foreward. In N. Roberts, D. Andersen, M. S. Garet, W. A. Shaffer, & R. M. Deal.

Forrester, J. W. (1991). *System dynamics and the lessons of 35 years*. Sloan School of Management, M.I.T. ftp://nyesgreenvalleyfarm.com/documents/sdintro/D-4224-4.pdf

Forrester, J. W. (1999). *System dynamics: The foundation under systems thinking*. Cambridge, MA: Sloan School of Management, MIT 02139. http://clexchange.org/ftp/documents/system-dynamics/SD2011-01SDFoundationunderST.pdf

Foster, P. (2011). Japan earthquake: Country better prepared than anyone for quakes and tsunamis. *The Telegraph*. https://www.telegraph.co.uk/news/0/japan-earthquake-country-better-prepared-anyone-quakes-tsunamis/

Freedman, H. I., & Waltman, P. (1984). Persistence in models of three interacting predator-prey populations. *Mathematical Biosciences, 68*(2), 213–231.

Gakkhar, S., & Gupta, K. (2016). A three species dynamical system involving prey-predation, competition and commensalism. *Applied Mathematics and Computation, 273*, 54–67.

Gardner, M. (2018). *Step-by-step guide for managing suppliers*. https://quality.eqms.co.uk/blog/iso-90012015-managing-suppliers-step-by-step-guide

Garvin, D. (1984). What does "product quality" really mean? *Sloan Management Review, 26*(1), 25–43.

Gass, I. S. (2010). *Linear programming: Methods and applications* (5th ed.). Dover: Dover Publications.

Gell-Mann, M. (1992). Complexity and complex adaptive systems. In *The evolution of human languages. Santa Fe Institute Studies in the Sciences of Complexity, 11* (pp. 3–18). Redwood City: Addison-Wesley.

Gell-Mann, M. (1994). *The quark and the jaguar*. New York: W. H. Freeman.

Gell-Mann, M. (1995). What is complexity? *Complexity, 1*(1), 16–19.

Gendered Innovations. (2020, online). *Stereotypes*. Stanford: Stanford University. https://genderedinnovations.stanford.edu/terms/stereotypes.html

Genus, A. (1995). *Flexible strategic management*. London: Chapman & Hall.

George, M. (2003). *Lean six sigma*. Boston, New York: McGraw-Hill Education.

Gephart, M. A., Marsick, V. J., Van Buren, M. E., & Spiro, M. S. (1996). Learning organizations come alive. *Training & Development, 50*(12), 35–45.

Gibbon, D. (1998). *Procedural explanations*. http://coral.lili.uni-bielefeld.de/Classes/Summer98/PragEngDialogue/pragengdialogue/node18.html

Gilbert, N. (1995). Simulation: An emergent perspective. In *Conference on new technologies in the social sciences* (pp. 27–29). Bournemouth.

Gilbert, N., & Troitzsch, K. G. (1999). *Simulation for the social scientist*. Buckingham: Open University Press.

Gilbreth, F. B. (1914). *The primer of scientific management*. New York: D. Van Nostrand.
Gilbreth, F. B., & Gilbreth, L. M. (1919, 1st ed. 1917). Applied motion study: A collection of papers on the efficient method to industrial preparedness. New York: Macmillan.
Gilovich, T., Griffin, D., & Kahneman, D. (Eds.). (2002). *Heuristics and biases: The psychology of intuitive judgment*. Cambridge University Press.
Gleick, J. (2008). *Chaos: Making a new science (Revised edition)*. London: Penguin.
Goffee, R., & Jones, G. (1998). *The character of a corporation: How your company's culture can make or break your business*. London: Harper Business.
Gofman, A., Moskowitz, H. R., & Mets, T. (2010). Accelerating structured consumer-driven package design. *Journal of Consumer Marketing, 27*(2), 157–168.
Goldman, S. L., Nagel, R. N., & Preiss, K. (1995). *Agile competitors and virtual organizations*. New York: Van Nostrand Reinhold.
Goldratt, E. M. (1990). *What is this thing called theory of constraint and how it should be implemented*. Croton-on-Hudson: North River Press.
Goldratt, E. M., & Cox, J. (2016, 1st ed. 1984). *The goal: A process of ongoing improvement*. London: Routledge.
Goodwin, G. C., Graebe, S. F., & Salgado, M. E. (2001). *Control system design*. Englewood Cliffs: Prentice Hall.
Google, Driverless Car Project. (2012). http://en.wikipedia.org/wiki/Google_driverless_car
Google, Prius. (2010). http://www.autoblog.com/2010/10/14/video-abc-news-gets-taken-for-a-spin-in-googles-self-driving-t/
Gopal, M. (2002). *Control systems: Principles and design*. New Delhi: Tata McGraw-Hill Education.
Goswami, A., & Chaudhuri, K. S. (1992). An economic order quantity model for items with two levels of storage for a linear trend in demand. *Journal of the Operational Research Society, 43*(2), 157–167.
Gould, S. J., & Eldredge, N. (1977). Punctuated Equilibria: The tempo and mode of evolution reconsidered. *Paleobiology, 3*(2), 115–151.
Gray, A., Krolikowski, M., Fretwell, P., Convey, P., Peck, L. S., Mendelova, M., & et al. (2020). Remote sensing reveals Antarctic green snow algae as important terrestrial carbon sink. Nature Communications, 11(1), 1–9. https://doi.org/10.1038/s41467-020-16018-w.
Green, S. G., & Welsh, M. A. (1988). Cybernetics and dependence: Reframing the control concept. *Academy of Management Review, 13*, 287–301.
Grover, J. P. (1991). Resource competition in a variable environment: Phytoplankton growing according to the variable-internal-stores model. *The American Naturalist, 138*(4), 811–835.
Guerrero Gámez, S., Darido, G., Sánchez d'Ocon, I., & Shibuya, N. (2017). *How to protect metro systems against natural hazards. Countries look to Japan for answers*. Transport for Development. https://blogs.worldbank.org/transport/how-protect-metro-systems-against-natural-hazards-countries-look-japan-answers
Guizzo, E. (2010). *World robot population*. https://spectrum.ieee.org/automaton/robotics/industrial-robots/041410-world-robot-population
Gulati, R., Kahn, J., & Baldwin, R. (2010). *The professional's guide to maintenance and reliability terminology*. Reliabilityweb.com Incorporated.
Gupta, R. C. (2001). *Statistical quality control* (6th ed.). Delhi: Romesh Chader Khanna.
Gustat, G. H. (1955). Applications of work sampling analysis. In *Proceedings of the tenth time study and methods conference*. New York: SAM-ASME.
Habib, C. (2011). *As above, so below: The worldview of Lynn Margulis*. http://realitysandwich.com/128651/above_so_below_worldview_lynn_margulis/
Haines, S. (2005). *Enterprise-wide change: Superior results through systems thinking*. San Francisco: Pfeiffer-John Wiley.
Håkansson, H., & Snehota, I. (1995). *Developing relationships in business networks*. London: Routledge.
Hall, D., & Jackson, J. (1992). Speeding up. New product development. JIT can help put products in customer hand faster. *Management Accounting, 74*(4), 32–36.

Hardin, G. (1968). The tragedy of the commons. *Science, 162*, 1243–1248.
Hardin, G. (1985). *An ecolate view of the human predicament.* www.garretthardinsociety.org/articles/art_ecolate_view_human_predicament.html
Harrington, H. J. (1987). *Poor-quality cost.* Milwaukee: ASQC Quality Press.
Harris, R. A. (1998). *Introduction to decision making.* www.virtualsalt.com/crebook5.htm
Harris, R. A. (2002). *Creative problem solving: A step-by-step approach.* Los Angeles: Pyrczak Publishing.
Harvey, I., Di Paolo, E., Wood, R., Quinn, M., & Tuci, E. (2005). Evolutionary robotics: A new scientific tool for studying cognition. *Artificial Life, 11*(1–2), 79–98.
Hastings, A., & Powell, T. (1991). Chaos in a three-species food chain. *Ecology, 72*(3), 896–903.
Hatch, M. J., & Cunliffe, A. L. (2013). *Organization theory: Modern, symbolic and postmodern perspectives.* Oxford: Oxford University Press.
Hawking, S. W., & Mlodinow, L. (2010). *The grand design.* Bantam. Excerpt, http://bookre.org/reader?file=65691
Hayes, R. B., & Jaikumar, R. (1988). Manufacturing's crisis: New technologies, obsolete organizations. *Harvard Business Review, 66*(5), 77–85.
Heaven & Earth Incorporated. (2000). What are system archetypes? Cambridge, MA. http://www.heartlandcenters.slu.edu/kmoli/assignments/Archetypes.pdf
Hecht, J. (2007). *NASA analysis of asteroid risk deeply flawed, critics say, NewScientist.com news service.* http://space.newscientist.com/article/dn11901-nasa-analysis-of-asteroid-risk-deeply-flawed-critics-say.html
Hedberg, B. (1981). How organizations learn and unlearn. In P. Nystrom & W. Starbuck (Eds.), *Handbook of organizational design, 1* (pp. 3–27). New York: Oxford University Press.
Heginbotham, C. (2018). *How Japan deals with earthquakes (and why It's the safest country in the ring of fire).* Medium, https://medium.com/@claireheginbotham/how-japan-deals-with-earthquakes-and-why-its-the-safest-country-in-the-ring-of-fire-42925f22f1ff
Heider, F. (1958). *The psychology of interpersonal relations.* New York: Wiley.
Hempel, C. G. (1965). *Aspects of scientific explanation and other essays.* New York: The Free Press.
Henderson Institute. (1970). *The Product Portfolio.* https://www.bcg.com/publications/1970/strategy-the-product-portfolio.aspx
Henshall, A. (2017). *DMAIC: The complete guide to lean six sigma in 5 key steps.* https://www.process.st/dmaic/
Hentschel, T., Heilman, M. E., & Peus, C. V. (2019). The multiple dimensions of gender stereotypes: A current look at men's and women's characterizations of others and themselves. *Frontiers in Psychology, 10*, 11. https://www.frontiersin.org/articles/10.3389/fpsyg.2019.00011/full
Heshmati, A. (2003). Productivity growth, efficiency and outsourcing in manufacturing and service industries. *Journal of Economic Surveys, 17*(1), 79–112.
Hesse, M., & Coe, N. M. (2006). Making connections: Global production networks, standards and embeddedness in the mobile telecommunications industry. *Environment and Planning, 38*(7), 1205–1227.
Heylighen, F. (2013). *The global superorganism: An evolutionary-cybernetic model of the emerging network society.* www.txtshr.com/computers/11212/index.html
Heylighen, F., & Joslyn, C. (2001). The law of requisite variety. *Principia Cybernetica.* http://pespmc1.vub.ac.be/REQVAR.html
Hiatt, J. M. (2006). *ADKAR: A model for change in business, government and our community.* Loveland: Prosci Research.
Highfield, R. (2002). *Lionesses don't prefer blonds - in the mane.* https://www.telegraph.co.uk/news/science/science-news/4768860/Lionesses-dont-prefer-blonds-in-the-mane.html
Hilton, R. W., Maher, M. W., & Selto, F. H. (2003). *Cost management: Strategies for business decisions.* New York: McGraw-Hill/Irwin.
Ho, W., Xu, X., & Dey, P. K. (2010). Multi-criteria decision making approaches for supplier evaluation and selection: A literature review. *European Journal of Operational Research, 202*(1), 16–24.

Holland, J. H. (1975). *Adaptation in natural and artificial systems: An introductory analysis with applications to biology, control, and artificial intelligence*. Ann Arbor: University of Michigan Press.
Holland, J. H. (1995). *Hidden order*. Cambridge, MA: Perseus Books.
Hölldobler, B., & Wilson, E. O. (1990). *The ants*. New York: Springer.
Horngren, C. T., Foster, G., & Datar, S. M. (2000). *Cost accounting: A managerial enphasis*. Upper Saddle River: Prentice-Hall.
Houghton, J. T. (1994). *Global warming: The complete briefing*. Cambridge, UK: Cambridge University Press.
Hoyle, D. (2007). *Quality management essentials*. Amsterdam: Elsevier.
Hsu, S. B., Ruan, S., & Yang, T. H. (2015). Analysis of three species Lotka-Volterra food web models with Omnivory. *Journal of Mathematical Analysis and Applications, 426*(2), 659–687.
Huang, B., Gou, H., Liu, W., Li, Y., & Xie, M. (2002). A framework for virtual enterprise control with the Holonic manufacturing paradigm. *Computers in Industry, 49*(3), 299–310.
Hudak, M. A. (1993). Gender schema theory revisited: Men's stereotypes of American women. *Sex Roles, 28*(5–6), 279–293.
Hum, S. H., & Sim, H. H. (1996). Time-based competition: Literature review and implications for modelling. *International Journal of Operations & Production Management, 16*(1), 75–90.
Hurst, D. (2020). *Guardian Tokyo week*. https://www.theguardian.com/cities/2019/jun/12/this-is-not-a-what-if-story-tokyo-braces-for-the-earthquake-of-a-century
Hutchins, D. C. (1985). *The quality circles. handbook*. New York: Pitman Press.
Huxley, T. H. (1874). On the hypothesis that animals are automata, and its history. *Fortnightly Review, 16*, 555–580.
Huxley, J. (1942). *Evolution: The modern synthesis*. London: George Allen and Unwin.
Iacovou, M., & Berthoud, R. (2001). *Young peoples' lives: A map of Europe*. Colchester: University of Essex, Institute for Social and Economic Research. https://www.iser.essex.ac.uk/files/iser_reps/pdf/002.pdf.
IEP, Internet Encyclopedia of Philosophy. (2014). *Theories of explanation*. http://www.iep.utm.edu/explanat/
IFR. International Federation of Robotics. (2019 online). *Industrial and service Robots*. IFR Press Conference. https://ifr.org/downloads/press2018/IFR%20World%20Robotics%20Presentation%20-%2018%20Sept%202019.pdf
Imai, M. (1986). *Kaizen, the key to Japan's competitive success*. New York: McGraw-Hill.
Imai, M. (2007). *Gemba Kaizen. A commonsense, low-cost approach to management*. New York: Mcgraw-Hill.
Imboden, D. M., & Pfenninger, S. (2013). *Introduction to systems analysis*. New York: Springer.
Insight-Maker. (2017). *Systems archetype*. https://insightmaker.com/tag/Systems-Archetype
Ishikawa, K. (1976). *Guide to quality control*. Tokyo: Asian Productivity Organization.
Ishikawa, K. (1980, 1st ed. 1970). *QC circle KORYO: General principles of the QC circle*. Tokyo: Japan: Union of Japanese Scientists and Engineers.
Ishikawa, K. (1985). *What is total quality control? The Japanese way*. Prentice Hall: Englewood Cliffs.
Ishikawa, K. (1990). Japan and the challenge of Europe 1992. *Futures, 22*, 664–665.
Ishikawa, K., & Kondo, Y. (1969). Education and training of quality control in Japanese industry. *Reports of Statistical Application and Research, 16*, 85–104.
Ishikawa, K., & Lu, D. J. (1989). How to apply companywide quality control in foreign countries. *Quality Progress, 22*(9), 70–74.
ISIXSIGMA. (2019). *What is six sigma*? https://www.isixsigma.com/new-to-six-sigma/what-six-sigma/
Jacobs, F. R. (1983). The OPT scheduling system: A review of a new production scheduling system. *Production and Inventory Management, 24*(3), 47–51.
Jakobson, R. (1970, 1st ed. 1963). Course in general linguistics [original: *Essais de linguistique générale*]. Les Editions de Minuit. http://nedelcu.ca/documents/Jakobson-Theo-de-la-communication.pdf

Jiang, W., & Han, J. (2009). The methods of improving the manufacturing resource planning (MRP II) in ERP. *2009 International Conference on Computer Engineering and Technology, 1*, 383–389.

Johnson, H., & Kaplan, R. (1987). *Relevance lost – The rise and fall of management accounting*. Boston: Harvard Business School Press.

Johnson, G., & Scholes, K. (2002). *Exploring corporate strategy* (6th ed.). London: Prentice Hall.

Johnson, R. A., Kast, F. E., & Rosenweig, J. E. (1963). *The theory and management of systems*. New York: McGraw-Hill.

Johnson-Laird, P. N. (1983). *Mental models: Towards a cognitive science of language, inference, and consciousness*. Cambridge, MA: Harvard University Press.

Juran, J. M., & Godfrey A. B. (1999, 1st ed. 1951). Quality control handbook. Boston: McGraw-Hill.

Juran Knowledge Base. (2019). *Pareto principle (80/20 rule) & Pareto analysis guide*. https://www.juran.com/blog/a-guide-to-the-pareto-principle-80-20-rule-pareto-analysis/

Kabir, G., Sadiq, R., & Tesfamariam, S. (2014). A review of multi-criteria decision-making methods for infrastructure management. *Structure and Infrastructure Engineering, 10*(9), 1176–1210.

Kahn, A. E. (1988). *The economics of regulation: Principles and institutions*. Cambridge, MA: The MIT Press.

Kahn, S. (2018). The terrifying power of stereotypes – And how to deal with them. *The conversation*. https://theconversation.com/the-terrifying-power-of-stereotypes-and-how-to-deal-with-them-101904

Kanamori, H., & Brodsky, E. E. (2004). The physics of earthquakes. *Reports on Progress in Physics, 67*(8), 1429.

Kanchanasevee, P., Biswas, G., Kawamura, K., & Tamura, S. (1997). Contract-net-based scheduling for holonic manufacturing systems. *Architectures, Networks, and Intelligent Systems for Manufacturing Integration, 3203*, 108–115.

Kano, N., Seraku, N., Takahashi, F., & Tsuji, S. (1984). Attractive quality and must-be quality. *Journal of Japanese Society of Quality Control, 14*(2), 39–48.

Kanter, R. M. (2012). Ten reasons people resist change. *Harvard Business Review, 74*. https://hbr.org/2012/09/ten-reasons-people-resist-chang

Kaplan, R. S., & Norton, D. P. (1992). The balanced scorecard. Measures that drive performance. *Harvard Business Review, 70*, 71–79.

Kaplan, R. S., & Norton, D. P. (1996). *The balanced scorecard. Translating strategy into action*. Boston: Harvard Business School Press.

Kaplan, R. S., & Norton, D. P. (2001). Transforming the balanced scorecard from performance measurement to strategic management. *Accounting Horizons, 15*(2), 147–160.

Kareiva, P., Mullen, A., & Southwood, R. (1990). Population dynamics in spatially complex environments: Theory and data. *Philosophical Transactions of the Royal Society B: Biological Science, 330*(1257), 175–190.

Karger, D. W., & Bayha, F. H. (1987). *Engineered work measurement: The principles, techniques, and data of methods-time measurement background and foundations of work measurement and methods-time measurement, plus other related material*. New York: Industrial Press.

Katz, D. (1964). Motivational basis of organizational behavior. *Behavioral Science, 9*(2), 131–146.

Kauffman, S. A. (1993). *The origins of order: Self organization and selection in evolution*. New York: Oxford University Press.

Kauffman, S. A. (1996). *At home in the universe: The search for laws of complexity*. London: Penguin.

Kauffman, S. A., & Levin, S. (1987). Towards a general theory of adaptive walks on rugged landscapes. *Journal of Theoretical Biology, 128*, 11–45.

Kawamura, K. (1997). Holonic manufacturing systems: an overview and key technical issues. *4th IFAC workshop on intelligent manufacturing systems: IMS'97, Seoul, Korea*, 33–36.

Kazmier, L. J. (2004, 1st ed. 1976). *Schaum's outline of theory and problems of business statistics*. New York: McGraw-Hill.

Kee, R. C. (1999). Using economic value added with ABC to enhance production-related decision making. *Journal of Cost Management, 13*(7), 3–15.

Keil, F. C. (2005). Explanation and understanding. *Annual Review of Psychology, 57*, 227–254. http://www.jsu.edu/dept/psychology/sebac/fac-sch/rm/Ch1-2.html

Kellert, S. H. (1993). *In the wake of chaos: Unpredictable order in dynamical systems*. Chicago: University of Chicago Press.

Kelley, J. E. Jr., & Walker, M. R. (1959). Critical-path planning and scheduling. *Proceedings of the eastern joint computer conference, Boston*, 160–173.

Kelly, G. (1955). *The psychology of personal constructs*. New York: Norton.

Kemp, S. (2006). *Quality management demystified*. Boston, New York: McGraw-Hill.

Key Differences. (2015). *Difference between standard costing and budgetary control*. https://key-differences.com/difference-between-standard-costing-and-budgetary-control.html

Kidd, P. T. (2000). *Agile manufacturing: Forging new frontiers*, Cheshire Henbury. http://www.cheshirehenbury.com/agility/agilitypapers/paper1095.html

Kim, D. (1999). *Shifting the burden: Moving beyond a reactive orientation. The systems thinker*. Waltham: Pegasus Communication.

Kim, D. H. (2000). *Systems archetypes I: Diagnosing systemic issues and designing high-leverage interventions*. Waltham: Pegasus Communications.

Kim, D. H., & Anderson, V. (2011). *Systems archetype basics*. Waltham: Pegasus Communications. https://thesystemsthinker.com/wp-content/uploads/2016/03/Systems-Archetypes-Basics-WB002E.pdf

Klir, G. J. (1969). *An approach to general systems theory*. New York: Van Nostrand Reinhold.

Klir, G. J. (1991). *Facets of systems science*. New York: Plenum Press.

Koestler, A. (1967). *The ghost in the machine*. London: Arkana.

Koestler, A. (1972). *The roots of coincidence* (2nd ed.). London: Hutchinson.

Koestler, A. (1978). *Janus; a summing up*. New York: Random House.

Kotter, J. P. (1995). Leading change: Why transformation efforts fail. *Harvard Business Review, 86*, 97–103. https://hbr.org/1995/05/leading-change-why-transformation-efforts-fail-2

Kotter, J. P. (1996, 2nd ed. 2012). *Leading change*. Boston: Harvard Business School Press.

Koukkari, W. L., & Sothern, R. B. (2006). *Introducing biological rhythms*. New York: Springer.

Kouzes, J. M., & Posner, B. Z. (2002). *The leadership challenge*. San Francisco: Jossey-Bass Wiley.

Koza, J. R. (1992). *Genetic programming: On the programming of computers by means of natural selection*. Cambridge, MA: The MIT Press.

Kramer, D. (2010). *Management theoretician. Henri Fayol's 14 principles discussed*. http://david-kramer.wordpress.com/2010/01/21/management-thinker-henri-fayol-14-principles-discussed/

Krippendorff, K. (1986). *A dictionary of cybernetics*. Unpublished report. http://pespmc1.vub.ac.be/ASC/ALGEDO_REGUL.html

Krishnaiah, P. R., & Rao, C. R. (1988). *V.7: Quality control and reliability*. Amsterdam: North-Holland Ed.

Krizna, M., Tahara, T., Tainaka, K., Ito, H., Morita, S., Ichinose, G., Togashi, T., Nagatani, T., & Yoshimura, J. (2018). Multi-species coexistence in Lotka-Volterra competitive systems with crowding effects. *Scientific Reports, 8*(1), 1–8.

Kuhn, T. S. (1962). *The structure of scientific revolutions* (2nd ed., 1970). Chicago: University of Chicago.

Kumar, E. (2008). *Artificial intelligence*. New Delhi: I. K. International Publishing House.

Kurzweil, R. (2001). *The law of accelerating returns*. www.kurzweilai.net/the-law-of-accelerating-returns

Kurzweil, R. (2005). *The singularity is near: When humans transcend biology*. Penguin Books.

Kusiak, A. (1986). *Flexible manufacturing systems. Methods and studies*. North-Holland, Amsterdam: Elsevier.

Kyriakidis, E. G., & Dimitrakos, T. D. (2006). Optimal preventive maintenance of a production system with an intermediate buffer. *European Journal of Operational Research, 168*(1), 86–99.

Lama, H. D. (1998). *The four noble truths: Fundamentals of Buddhist teachings*. Kinddle Edition.

Lamarck, J. B. (1809). *Philosophie zoologique, ou, Exposition des considérations relative à l'histoire naturelle des animaux*. Hachette Livre BNF (reprint, 2012).

Lamberson, L. R., & Hocking, R. R. (1970). Optimum time compression in project scheduling. *Management Science, 6*(10), B597–B606.

Laprie, J. C. (2008). From dependability to resilience. In *38th IEEE/IFIP international conference on dependable systems and networks* (pp. G8–G9).

László, E. (1983). *Systems science and world order. Selected studies*. Oxford, UK: Pergamon Press.

Latham, J., et al. (2012). Marine cloud brightening. *Philosophical Transactions. Series A, Mathematical, Physical, and Engineering Sciences, 370*(1974): 4217–4262. https://www.ncbi.nlm.nih.gov/pmc/articles/PMC3405666/

Lazear, E. P., & Rosen, S. (1981). Rank-order tournaments as optimum labor contracts. *Journal of Political Economy, 89*(5), 841–864.

Lee, C. Y. (1993). A recent development of the integrated manufacturing system: A hybrid of MRP and JIT. *International Journal of Operations & Production Management, 13*(4), 3–17.

Lee, D. (2000). The Society of Society: The grand finale of Niklas Luhmann. *Sociological Theory, 18*(2), 320–330.

Leffler, K. B. (1982). Ambiguous changes in product quality. *The American Economic Review, 72*(5), 956–967.

Lei, D., & Goldhar, J. D. (1991). Computer-integrated manufacturing (CIM): Redefining the manufacturing firm into a global service business. *International Journal of Operations & Production Management, 11*(10), 5–18.

Leigh, J. R. (2004). *Control theory*. London: The Institution of Electrical Engineers.

Leitão, P., & Restivo, F. (2006). ADACOR: A holonic architecture for agile and adaptive manufacturing control. *Computers in Industry, 57*(2), 121–130.

Lemley, B. (2004). A new ice age. Could global warming trigger a big freeze? *Discover magazine*. https://www.cbsd.org/cms/lib/PA01916442/Centricity/Domain/1622/Article%20E%20-%20A%20New%20Ice%20Age.pdf

Lennox, J. G. (1992). Teleology. In E. Fox Keller & E. A. Lloyd (Eds.), *Keywords in evolutionary biology* (pp. 324–333). Cambridge: Harvard University Press.

Leung, L. C., & Tanchoco, J. M. (1983). Replacement decision based on productivity analysis—An alternative to the MAPI method. *Journal of Manufacturing Systems, 2*(2), 175–187.

Levin, R. I., & Kirkpatrick, C. A. (1966). *Planning and control with PERT*. New York: McGraw-Hill.

Levitt, T. (1972). Production-line approach to service. *Harvard Business Review, 50*(5), 41–52.

Levy, S. (1992). *Artificial life: The quest for a new creation*. London: Penguin Books.

Lewin, K. (1935). On the structure of the mind. In *A dynamic theory of personality – Selected papers*. In Read Books Ltd, (2013). http://gestalttheory.net/archive/lewinp.pdf

Lewin, K. (1947a). Frontiers in group dynamics. In D. Cartwright (Ed.), *Field theory in social science: Selected theoretical papers* (pp. 188–237). London: Tavistock.

Lewin, K. (1947b). Group decisions and social change. In T. M. Newcomb & E. L. Hartley (Eds.), *Readings in social psychology* (pp. 197–211). New York: Holt.

Lewin, A. Y., Long, C. P., & Carroll, T. N. (1999). The coevolution of new organizational forms. *Organization Science, 10*(5), 535–550.

Lewis, M. P., Simons, G. F., & Fennig, C. D. (Eds.). (2013). *Ethnologue: Languages of the world*. Dallas: SIL International. http://www.ethnologue.com/ethno_docs/distribution.asp?by=area

Lienhard, J. H. (1994). I sell here, sir, what all the world desires to have. POWER. *Energy Laboratory Newsletter, 31*, 3–9. http://www.uh.edu/engines/powersir.htm.

Liguori, M. (2012). The supremacy of the sequence: Key elements and dimensions in the process of change. *Organization Studies, 33*(4), 507–539.

Liker, J. K., & Meier, D. L. (2005). *The Toyota way fieldbook (A practical guide for implementing Toyota's 4Ps)*. New York: McGraw-Hill Education.

Liker, J. K., & Morgan, J. M. (2006). The Toyota way in services: The case of lean product development. *Academy of Management Perspectives, 20*(2), 5–20.

Lindsey, R. (2007). Tropical deforestation. *NASA's Earth Observatory*, 508. https://earthobservatory.nasa.gov/Features/Deforestation [15 January 2018].

Liow, L. H., Valen, L. V., & Stenseth, N. C. (2011). Red queen: From populations to taxa and communities. *Trends in Ecology and Evolution, 26*(7), 349–358.

Lippmann, W. (1922, online). *Public opinion*. https://books.google.it/books?id=fqpEWJHEzIkC&printsec=frontcover&dq=Lippmann+public+opinion&hl=en&sa=X&ved=2ahUKEwj13YHA9KHqAhVHmYsKHVlzD8oQ6AEwAHoECAAQAg#v=onepage&q=Lippmann%20public%20opinion&f=false

Llina's, R., & Paré, D. (1996). The brain as a closed system modulated by senses. In R. Llinas & P. S. Churchland (Eds.), *The mind-brain continuum*. Cambridge, MA: MIT Press.

Lorenc, A., & Lerher, T. (2019). Effectiveness of product storage policy according to classification criteria and warehouse size. *FME Transactions, 47*(1), 142–150.

Lorenz, E. (1972). *Predictability: Does the flap of a Butterfly's wings in Brazil set off a tornado in Texas?*. Paper delivered at the American Association for the Advancement of Science, Washington. http://static.gymportalen.dk/sites/lru.dk/files/lru/132_kap6_lorenz_artikel_the_butterfly_effect.pdf

Lormand, E. (1991). *Connectionist languages of thought*. http://www-personal.umich.edu/~lormand/phil/cogsci/clot.htm

Lotka, A. J. (1925). *Elements of physical biology*. Baltimore: Williams & Wilkins.

Lovelock, J. E. (1979). *GAIA, a new look at life on earth*. Oxford, UK: Oxford University Press.

Lovelock, J. E. (1988). *The ages of Gaia: A biography of our living earth*. Oxford, UK: Oxford University Press.

Lovelock, J. E. (2011). *What is Gaia*? http://www.ecolo.org/lovelock/what_is_Gaia.html

Luce, R. D., & Raiffa, H. (1967). *Games and decision*. London, UK: Wiley.

Lucini, B. (2014). *Disaster resilience from a sociological perspective: Exploring three Italian earthquakes as models for disaster resilience planning*. Springer Science & Business.

Luhmann, N. (1990). The cognitive program of constructivism and a reality that remains unknown. In W. Krohn et al. (Eds.), Selforganizalion. Portrail of a scientific revolution (pp. 64–85). Dordrecht: Kluwer Academic Publishers.

Luhman, N. (1995). *Social systems*. Stanford: Stanford University Press (translation of the original book written in German language).

Luhmann, N. (1997). *Theory of society*. Stanford: Stanford University Press (translation of the.

Macarthur, R., & Levins, R. (1967). The limiting similarity, convergence, and divergence of coexisting species. *The American Naturalist, 101*(921), 377–385.

Macdonald, F. (2016). Something's making Siberia's tundra literally bubble under People's feet. Science alert. https://www.sciencealert.com/something-s-making-siberia-s-tundra-literally-bubble-under-people-s-feet

Macintosh, R., & Maclean, D. (1999). Conditioned emergence: A dissipative structures approach to transformation. *Strategic Management Journal, 20*(4), 297–316.

Malcolm, D. G., Roseboom, J. H., & Clark, C. E. (1959). Application of a technique for Research and Development program evaluation. *Operations Research, 7*(5), 646–669.

Malthus, T. R. (1798). *Essay on population as it affects the future improvement of society*. London: J. Johnson.

Mansell, W. (2011). *Control and purpose. A scientific and practical statement in response to the UK Riots 2011*. http://www.pctweb.org/LondonRiots.pdf

March, J. G. (1991). Exploration and exploitation in organizational learning. *Organization Science, 2*(1), 71–87.

March, J. G., & Simon, H. A. (1958). *Organizations*. London: Wiley.

Mařík, V., Vrba, P., & Leitão, P. (Eds.). (2011). *Holonic and multi-agent systems for manufacturing. HoloMAS 2011, Toulouse, France, proceedings*. Springer.

Marr, B. (2017). *The 4th industrial revolution and a jobless future - A good thing?* https://www.forbes.com/sites/bernardmarr/2017/03/03/the-4th-industrial-revolution-and-a-jobless-future-a-good-thing/#2b0d9cee44a5

Marshall, M. (2020). Planting trees doesn't always help with climate change. *BBC Future*. https://www.bbc.com/future/article/20200521-planting-trees-doesnt-always-help-with-climate-change

Martelli, M. (1999). *Introduction to discrete dynamical systems and Chaos*. London: Wiley.
Maslin, M. (2004). *Global warming*. New York: Oxford University Press.
Maslow, A. H. (1943). A theory of human motivation. *Psychological Review, 50*, 370–396.
Maslow, A. H. (1970a). *Motivation and personality*. New York: Harper & Row.
Maslow, A. H. (1970b). *Religions, values, and peak experiences*. New York: Penguin. (Original work published 1966).
Maslow, A. H. (1981). *Motivation and personality*. Prabhat Prakashan.
Mason-Jones, R., & Towill, D. R. (1999). Total cycle time compression and the agile supply chain. *International Journal of Production Economics, 62*(1), 61–73.
Massé, P. (1962). *Optimal investment decisions*. Englewood Cliffs: Prentice-Hall.
Maturana, H. R., & Guiloff, G. (1980). The quest for the intelligence of intelligence. *Journal of Social and Biological Structures, 3*(2), 135–148.
Maturana, H. R., & Varela, F. J. (1980, 1st ed, 1972). Autopoiesis and cognition. The realization of living. Boston: Reidel Publishing.
Maturana, H. R., & Varela, F. J. (1987). *The tree of knowledge: The biological roots of human understanding*. New Science Library/Shambhala Publications.
May, R. M. (1973). Stability and complexity in model ecosystems. *Monographs in Population Biology, 6*, 1–235.
Mayer, R. E. (1992). *Thinking, problem solving, cognition* (2nd ed.). New York: W. H. Freeman and Company.
McCall, J. J. (1965). Maintenance policies for stochastically failing equipment: A survey. *Management Science, 11*(5), 493–524.
McCurri, J. (2018). Japan shrinking as birthrate falls to lowest level in history. *The guardian*. https://www.theguardian.com/world/2018/dec/27/japan-shrinking-as-birthrate-falls-to-lowest-level-in-history
McKinley, A., & Starkey, K. (Eds.). (1998). *Foucault, management and organization theory*. London: Sage.
McLeod S (2007). *Maslow's hierarchy of needs*. http://www.simplypsychology.org/maslow.html
Meadows, D. H. (1982). Whole earth models and systems. *Co-Evolution Quarterly, 34*, 98–108.
Melchizedek, D. (2004). *Dry/ice: Global warming revealed*. http://www.spiritofmaat.com/announce/ann_dryice.htm
Mella, P. (1997a). *From systems to systems thinking: to understand the systems and systems thinking* [original: Dai sistemi al pensiero sistemico: per capire i sistemi e pensare con i sistemi]. Milano: Franco Angeli.
Mella, P. (1997b). *Management control* [original: Controllo di gestione]. Turin: Utet.
Mella, P. (2005a). Observing collectivities as simplex systems: The combinatory systems approach. *Nonlinear Dynamics, Psychology, and Life Sciences, 9*(2), 121–153.
Mella, P. (2005b). Performance indicators in business value-creating organizations. *Economia Aziendale Online, 2*, 25–52.
Mella, P. (2006). Spatial co-localisation of firms and entrepreneurial dynamics: The combinatory system view. *The International Entrepreneurship and Management Journal, 2*(3), 391–412.
Mella, P. (2007). Combinatory systems and automata: Simulating self-organization and chaos in collective phenomena. *The International Journal of Knowledge, Culture, and Change Management, 7*(2), 17–28.
Mella, P. (2009). *The holonic revolution. Holons, holarchies and holonic networks. The ghost in the production machine*. Pavia: University Press.
Mella, P. (2011). Managing business value-creating organizations. *International Journal of Knowledge, Culture, and Change management, 10*(9), 1–19.
Mella, P. (2012). *Systems thinking: Intelligence in action*. New York: Springer.
Mella, P. (2017). *The Combinatory Systems theory*. New York: Springer.
Mella, P. (2019). The ghost in the production machine: The laws of production networks. *Kybernetes, 48*(6), 1301–1329.
Mella, P. (2020). *Constructing reality: The "operationalization" of Bateson's conjecture on cognition*. Cham: Springer Nature.

Mella, P., & Navaroni, M. (2012). *Financial statements analysis* [original: Analisi di bilancio]. Santarcangelo di Romagna: Maggioli Editore.

Mella, P., & Pellicelli, M. (2008). The origin of value-based management: Five interpretative models of an unavoidable evolution. *International Journal of Knowledge, Culture, and Change management, 8*, 23–32.

Mella, P., & Pellicelli, M. (2012). The strategies of outsourcing and offshoring. *American International Journal of Contemporary Research, 2*(9), 116–127.

Mella, P., Pellicelli, M., & Meo, C. C. (2010). The replacement-renewal of industrial equipment. The MAPI formula. *Analele Universitatii din Oradea Stiinte Economice, 1*, 667–680.

Mentzer, J. T., et al. (2001). Defining supply chain management. *Journal of Business Logistics, 22*(2), 1–25.

Merchant, K. A. (1985). *Control in business organizations*. Boston: Pitman Publishing.

Merchant, K. A., & Otley, D. T. (2007). A review of the literature on control and accountability. In C. S. Chapman, A. G. Hopwood, & M. D. Shields (Eds.), *Handbook of management accounting research* (pp. 785–804). Amsterdam: Elsevier Press.

Merrill, P. (1997). *Do it right the second time: Benchmarking best practices in the quality change process*. Milwaukee: ASQC Quality Press.

Mesarovic, M., Macko, D., & Takahara, Y. (1970). *Theory of hierarchical, multi-level systems*. New York: Academic Press.

Metro Tokyo. (online). https://www.metro.tokyo.lg.jp/english/guide/bosai/index.html

Meyerhoff, J. (2005). The 20 tenets. In *Bald ambition: A critique of Ken Wilber's theory of everything, Ch. 1(B)*. Frank Visser Pub. http://www.integralworld.net/meyerhoff-ba-1a.html

Miles, L. D. (1989). *Techniques of value analysis and engineering* (3rd ed.). http://wendt.library.wisc.edu/miles/book/preface.pdf

Miller, D. W., & Starr, M. K. (1969). *Executive decisions and operations research*. Englewood Cliffs: Prentice-Hall.

Mingers, J. (1995). *Self-producing systems: Implications and applications of autopoiesis*. New York: Springer.

Mingers, J. (2002). Can social systems be autopoietic? Assessing Luhmann's social theory. *The Sociological Review, 50*(2), 278–299.

Mintzberg, H., Ahlstrand, B., & Lampel, J. (1998). *Strategy safari. A guided tour through the wilds of strategic management*. New York: Prentice Hall.

Miyake, D. I., & Enkawa, T. (1999). Matching the promotion of total quality control and total productive maintenance: An emerging pattern for nurturing of well-balanced manufactures. *Total Quality Management & Business Excellence, 10*(2), 243–269.

Mockler, R. J. (1970). *Readings in management control*. New York: Appleton-Century-Crofts.

Modigliani, F., & Miller, M. (1958). The cost of capital, corporation finance, and the theory of investment. *American Economic Review, XLVIII*(3), 261–297.

Moisello, A. M. (2012). Costing for decision making: Activity-based costing vs. theory of constraints. *International Journal of Knowledge, Culture & Change in Organizations: Annual Review, 12*, 1–13.

Moisello, A. M., & Mella, P. (2020). Matching revenues and costs: The counter-intuitive rationality of direct costing. *International Journal of Business and Management, 15*(1), 202–222.

Monod, J. (1970). *Chance and necessity: Essay on the natural philosophy of modern biology* [original: Le hazard et la nécéssité]. New York: Vintage Books (1972). (original: Paris: Seuil).

Montgomery, D. C. (2007). *Introduction to statistical quality control*. London and New York: Wiley.

Montgomery, D. C., & Woodall, W. H. (2008). An overview of six sigma. International Statistical Review/Revue Internationale de Statistique, 329–346.

Mooney, C. (2018). Arctic Cauldron. Arctic Lakes that don't Freeze. *The Washington Post*. https://www.washingtonpost.com/graphics/2018/national/arctic-lakes-are-bubbling-and-hissing-with-dangerous-greenhouse-gases/

Morgan, G. (1988). *Riding the waves of change: Developing managerial competencies for a turbulent world*. San Francisco: Jossey-Bass.

Morin, E. (2005). *Restricted complexity, general complexity*. Proceedings (2008), Intelligence de la complexité: epistemologie et pragmatique, Cerisy-La-Salle, France, June 26th, 2005. https://arxiv.org/pdf/cs/0610049.pdf
Morin, R., & Jarrel, S. (2001). *Driving shareholder value*. New York: McGraw-Hill.
Morris, C. W. (1938). *Foundations of a theory of signs*. http://www.scribd.com/doc/51866596/Morris-1938-Foundations-of-Theory-of-Signs
Morris, W. F., & Doak, D. F. (2002). *Quantitative conservation biology: Theory and practice of population viability analysis* (pp. 325–372).
Morrow, R. L. (1957). *Motion economy and work measurement*. New York: The Ronald Press Company.
Mrugalska, B., & Tytyk, E. (2015). Quality control methods for product reliability and safety. *Procedia Manufacturing, 3*, 2730–2737.
Myers, J. H., & Shocker, A. D. (1981). The nature of product-related attributes. *Research in Marketing, 5*(5), 211–236.
Nagrath, I. J., & Gopal, M. (2008). *Control systems engineering*. Kent: Anshan.
Nahmias, S. (1982). Perishable inventory theory: A review. *Operations Research, 30*(4), 680–708.
Nakajima, S. (1989). *TPM development program: Implementing Total productive maintenance*. Portland: Productivity Press.
NASA. (2004). A chilling possibility. *Science*. http://science1.nasa.gov/science-news/science-at-nasa/2004/05mar_arctic/
Näslund, D. (2008). Lean, six sigma and lean sigma: Fads or real process improvement methods? *Business Process Management Journal, 14*(3), 269–287.
Neal, F. L. (1962). Controlling warehouse handling costs by means of stock location audits. *Transportation and Distribution Management, 2*, 31–33.
Neely, A. (2007). Measuring performance: The operations management perspectives. In A. Nelley (Ed.), *Business performance measurement. Unifying theory and integrating practice* (2nd ed., pp. 64–81). Cambridge: Cambridge University Press.
Nelsen, D., & Daniels, S. E. (2007). Quality glossary. *Quality Progress, 40*(6), 39.
Nelson, L. S. (1984). The Shewhart control chart—Tests for special causes. *Journal of Quality Technology, 16*(4), 237–239.
Neuman, Y. (2003). *Processes and boundaries of the mind*. New York and Berlin: Springer.
Newell, A., & Simon, H. A. (1976). Computer science as empirical enquiry: Symbols and search. *Communications of the Association for Computing Machinery, 19*(3), 113–126.
Nguyen, D. H., & Yin, G. (2016). Coexistence and exclusion of stochastic competitive Lotka-Volterra models. *Journal of Differential Equations, 262*(3), 1192–1225.
Nicholson, M. (1999). Lewis Fry Richardson and the study of the causes of war. *British Journal of Political Science, 29*(3), 541–563.
Nickols, F. (2010). Change management 101. *A primer*. http://www.nickols.us/change.pdf
Nickols, F. (2011). *Change management as problem finding and problem solving*. www.odnetwork.org/publications/seasonings/samples/article_nickols.php
Nonaka, I. (1994). A dynamic theory of organizational knowledge creation. *Organization Science, 5*(1), 14–37.
Nonaka, I., & Takeuchi, H. (1995). *The knowledge-creating company: How Japanese companies create the dynamics of innovation*. Oxford: Oxford University Press.
Nordhaus, W. D. (1996). Do real-output and real-wage measures capture reality? The history of lighting suggests not. In *The economics of new goods* (pp. 27–70). Chicago: University of Chicago Press.
Noskievičová, D. (2012), Analysis of nonrandom patterns in control chart and its software support. *Metal, 2012*, 23. http://metal2013.tanger.cz/files/proceedings/02/reports/402.pdf
Nöth, W. (1995). *Handbook of semiotics*. Bloomington: Indiana University Press.
Nowac, M. A. (2006). *Evolutionary dynamics*. Harvard University Press.
NPD Solutions. (2016). *Customer-focused development with QFD*. http://www.npd-solutions.com/qfd.html

Oakland, J. S. (1989). *Total quality management*. London: Heinemann Professional Publishing.
Oakland, J. S. (2014). *Total quality management and operational excellence: Text with cases*. London: Routledge.
Ogata, K. (2002). *Modern control engineering*. Englewood Cliffs: Prentice Hall.
Ohno, T. (1988). *The Toyota production system; beyond large-scale production*. Portland: Productivity Press.
Ohno, T. (2020). *AZQuotes.com, Wind and Fly LTD, 2020*. https://www.azquotes.com/author/44597-Taiichi_Ohno
Ojasalo, J. (2006). Quality for the individual and for the company in the business-to-business market: Concepts and empirical findings on trade-offs. *International Journal of Quality & Reliability Management, 23*(2), 162–178.
Okino, N. (1989). Bionical manufacturing systems. In T. Sata (Ed.), *Organization of engineering knowledge for product modelling in computer integrated manufacture*. Netherlands: Elsevier.
Olve, N. G., Roy, J., & Wetter, M. (1999). *Performance drivers: A practical guide to using the balanced scorecard*. London: Wiley.
OPT – Optimized Production Technology. (2000). In P. M. Swamidass (Ed.), *Encyclopedia of production and manufacturing management*. Boston: Springer.
Organ, D. W. (1990). The motivational basis of, organizational citizenship behavior. *Organizational Behavior, 12*(1), 43–72.
Osborn, A. F. A. (1953). *Applied imagination: The principles and procedures of creative thinking*. New York: Scribner's Sons.
Ostrom, E. (1990). *Governing the commons. The evolution of institutions for collective action*. Cambridge, UK: Cambridge University Press.
Otley, D. (1999). Performance management: A framework for management control systems research. *Management Accounting Research, 10*(4), 363–382.
Ouchi, W. G., & Maguire, M. A. (1975). Organizational control. Two functions. *Administrative Science Quarterly, 20*(4), 559–569.
Packard, A. S. (1901). *Lamarck, the founder of evolution. His life and work with translations*. Longmans, Green and Co: New York/London.
Paine, R. T. (1966). Food web complexity and species diversity. *The American Naturalist, 100*(910), 65–75.
Parsons, T. (1951). *The social system*. New York: Free Press.
Parsons, T. (1971). *The system of modern societies*. Englewood Cliffs: Prentice Hall.
Pearce, D. (1999). *Economics and environment: Essays on ecological economics and sustainable development*. Cheltenham: Edward Elgar Publishing Ltd.
Pentland, B. T., & Feldman, M. S. (2008). Designing routines: On the folly of designing artifacts, while hoping for patterns of action. *Information and Organization, 18*(4), 235–250.
Perkel, D. H. (1993). Computational neuroscience. Scope and structure. In E. L. Schwartz (Ed.), *Computational neuroscience* (pp. 38–45). Cambridge, MA: MIT Press.
Perner, L. (2018). *Channels of distribution*. University of Southern California. http://consumerpsychologist.com/distribution.html
PestSmart Connect. Centre for Invasive Species Solutions. (2018). https://www.pestsmart.org.au
Piaget, J. (1937). *The construction of the real in the child* [original: La construction du réel chez l'enfant]. Geneva: Delachaux et Niestlé,
Pichler, F. (2000, 1st ed. 1980). On the construction of A. Koestler's holarchical networks. In R. Trappl (Ed.), *Cybernetics and systems*.
Pierre, D. A. (1987). *Optimization theory with applications*. New York: Dover Publications.
Pinedo, M. L. (2009). *Planning and scheduling in manufacturing and services*. New York: Springer.
Pirsig, R. M. (1974). *Zen and the art of motorcycle maintenance: An inquiry into values*. New York: Bantam Books.
Plamondon, A. L. (1979). *Whitehead's organic philosophy of science*. Albany: State University of New York Press.
PlanIT Valley Project. (2013). http://www.living-planit.com/design_wins.htm

Plenert, G. (1999). Focusing material requirements planning (MRP) towards performance. *European Journal of Operational Research, 19*(1), 91–99.

Plunkett, J. J., & Dale, B. G. (1987). A review of the literature on quality-related costs. *International Journal of Quality and Reliability Management, 4*(1), 40–52.

Poincaré, J. H. (1908). *Chance, science and method, part 1, Ch 4*. https://www.stat.cmu.edu/~cshalizi/462/readings/Poincare.pdf

Polanyi, M. (1967). *The tacit dimension*. Chicago: University of Chicago Press.

Poole, M. S., & Van de Ven, A. H. (2004). Theories of organizational change and innovation processes. In *Handbook of organizational change and innovation* (pp. 374–397).

Popkin, G. (2019). How much can forests fight climate change?. *Nature, 565*(7739), 280–282. https://www.nature.com/articles/d41586-019-00122-z

Popli, S. (2005). Ensuring customer delight: A quality approach to excellence in management education. *Quality in Higher Education, 11*(1), 17–24.

Porter, M. E. (2008). *On competition* (Updated and expanded edition). Boston: Harvard Business School Press.

Post, D. M. (2002). Using stable isotopes to estimate trophic position: Models, methods, and assumptions. *Ecology, 83*(3), 703–718.

Post, D. M., & Palkovacs, E. P. (2009). Eco-evolutionary feedbacks in community and ecosystem ecology: Interactions between the ecological theatre and the evolutionary play. *Philosophical Transactions of the Royal Society B: Biological Sciences, 364*(1523), 1629–1640.

Post, D. M., Pace, M. L., & Halrston, N. G. (2000). Ecosystem size determines food-chain length in lakes. *Nature, 405*(6790), 1047–1049.

Powers, W. T. (1973). *Behavior: The control of perception*. Chicago: Aldine.

Powersim Software. Powersim. (2018). http://www.powersim.com

Prahalad, C. K., & Bettis, R. A. (1986). The dominant logic: The new linkage between firm diversity and performance. *Strategic Management Journal, 7*(6), 485–501.

Price, P. W., Bouton, C. E., Gross, P., McPheron, B. A., Thompson, J. N., & Weis, A. E. (1980). Interactions among three trophic levels: Influence of plants on interactions between insect herbivores and natural enemies. *Annual Review of Ecology and Systematics, 11*(1), 41–65.

Prieto, L. J. (1966). Messages et signaux. Paris: Presses Universitaires de France.

Proctor, T. (1999). *Creative problem solving for managers*. New York: Taylor & Francis.

Ptak, C. A. (1991). MRP, MRPII, OPT, JIT, and CIM. Succession, evolution, or necessary combination. *Production and Inventory Management Journal, 32*(2), 7–11.

Pyzdek, T. (2003). *The six sigma handbook*. Boston, New York: McGraw-Hill.

QFD Institue. (2019). *What is QFD*? http://www.qfdi.org/what_is_qfd/what_is_qfd.html

QIMA. Product Inspection Services. (2019). *What is a product inspection*? https://www.qima.com/quality-control-services/product-and-manufacturing-inspections

Queensland Country Life. (1949). *Politics saved the rabbit*!. https://trove.nla.gov.au [15 January 2018].

Raheem, M. A., Gbolahan, A. T., & Udoada, I. E. (2016). Application of statistical process control in a production process. *Science Journal of Applied Mathematics and Statistics, 4*(1), 1–11.

Rangone, A., & Mella, P. (2019). Obstacles to managing dynamic systems. The systems thinking approach. *International Journal of Business and Social Science, 10*(8), 24–41.

Rao, A., & Georgeff, M. (1992). Social plans: Preliminary report. In E. Werner & C. Castelfranchi (Eds.), *Decentralized AI 3, proceedings of MAAMAW'91*. Amsterdam: Elsevier North Holland.

Raouf, A., & Ben-Daya, M. (1995). *Flexible manufacturing systems: Recent developments*. Amsterdam: Elsevier.

Reynolds, K. (2013). Strategic thinking for today's project managers. *PMI® global congress 2013—North America*. New Orleans: Project Management Institute. https://www.pmi.org/learning/library/strategic-thinking-today-project-managers-5910

Ricciardi, M. (1973). *Pert, Altai and other networking programming techniques* [original: Il Pert l'Altai e le Altre Tecniche Reticolari di Programmazione]. Milan: Etas Kompass.

Rich, N., & Hines, P. (1997). Supply-chain management and time-based competition: The role of the supplier as- sociation. *International Journal of Physical Distribution & Logistics Management, 27*(3/4), 210–225.

Richardson, L. F. (1949). *Arms and insecurity*. London: Stevens & Sons (published posthumously, 1960).
Richardson, G. P. (1991). *Feedback thought in social science and system theory*. Philadelphia: University of Pennsylvania Press.
Richmond, B. (1991). *Systems thinking. Four key questions*. Watkinsville: High Performance Systems.
Richmond, B. (1993). Systems thinking: Critical thinking skills for the 1990s and beyond. *System Dynamics Review, 9*(2), 113–133.
Richmond, B. (1994). *System dynamics/systems thinking: Let's just get on with it*. Sterling: International Systems Dynamics Conference. Retrieved from http://webspace.webring.com/people/ah/himadri_banerji/pdf/systhnk.pdf
Riedel, S. (2005). Edward Jenner and the history of smallpox and vaccination. *Proceedings of Baylor University Medical Center, 18*(1), 21–25. http://www.ncbi.nlm.nih.gov/pmc/articles/PMC1200696/
Rigopoulou, I. D., Chaniotakis, I. E., Lymperopoulos, C., & Siomkos, G. I. (2008). After-sales service quality as an antecedent of customer satisfaction: The case of electronic appliances. *Managing Service Quality: An International Journal, 18*(5), 512–527.
Robbins, S. P. (1987). *Organization theory: Structure, design, and applications*. Englewood Cliffs: Prentice-Hall.
Roberts, N. (1978). Teaching dynamic feedback systems thinking: An elementary view. *Management Science, 24*(8), 836–843.
Roberts, N., Andersen, D., Garet, M. S., Shaffer, W. A., & Deal, R. M. (1983). *Introduction to computer simulation: A system dynamics modeling approach*. Reading: Addison Wesley.
Robertson, A. G. (1971). *Quality control and reliability*. London: Pitman.
Robertson, S. I. (2001). *Problem solving*. Hove: Psychology Press.
Rodrigue, J. P., & Hesse, M. (2006). Global production networks and the role of logistics and transportation. *Growth and Change, 37*(4), 499–509.
Roman, D. D. (1962). The PERT system: An appraisal of program evaluation review technique. *The Journal of the Academy of Management, 5*(1), 57–65.
Rosenfeld, Y. (2009). Cost of quality versus cost of non-quality in construction: The crucial balance. *Construction Management and Economics, 27*(1), 107–117.
Rosenthal, R., & Jacobson, L. (1968, expanded ed. 2003). *Pygmalion in the classroom*. Crown House Pub.
Rosheim, M. (2006). *Leonardo's lost robots*. New York: Springer.
Rouhiainen, L. (2019). Artificial Intelligence: 101 Things You Must Know Today About Our Future. Cindy Estra.
Ruefli, T. W., Collins, J. M., & Lacugna, J. R. (1999). Risk measures in strategic management research: Auld lang syne? *Strategic Management Journal, 20*(2), 167–194.
Rushinek, A. (1983). Control aspects of standard costing: Variance analysis, inflation adjustment, the learning curve and their computer applications. *Managerial Finance, 9*(1), 14–18.
Russell, G. R., & Miles, M. P. (1998). The definition and perception of quality in ISO-9000 firms. *Review of Business, 19*(3), 13.
Russell, S., & Norvig, P. (2009). *Artificial intelligence, a modern approach* (3th ed.). Englewood Cliffs: Prentice-Hall.
Rust, R. T., Inman, J. J., Jia, J., & Zahorik, A. (1999). What you Don't know about customer-perceived quality: The role of customer expectation distributions. *Marketing Science, 18*(1), 77–92.
Ryall, J., & Kruithof, J. (2001). *The quality systems handbook*. Australia: Consensus Books.
Ryan, T. P. (2000). *Statistical methods for quality improvement*. London and New York: Wiley.
Saaty, T. L. (1990). How to make a decision: The analytic hierarchy process. *European Journal of Operational Research, 48*(1), 9–26.
Sakurai, M. (1989). Target costing and how to use it. *Journal of cost management, 3*(2), 39–50.
Salah, S., Carretero, J. A., & Rahim, A. (2009). Six sigma and total quality management (TQM): Similarities, differences and relationship. *International Journal of Six Sigma and Competitive Advantage, 1*(3), 237–250.

Salmi T, Nikkinen J, & Sahlström P. (2005). *The review of the theoretical and empirical, basis of financial ratio analysis revisited.* http://lipas.uwasa.fi/~ts/ejre/ejre.html
Salmon, W. C. (1990). *Four decades of scientific explanation.* Minneapolis: University of Minnesota Press.
Sanchez, L. S., & Nagi, R. (2001). A review of agile manufacturing systems. *International Journal of Production Research, 39*(16), 3561–3600.
Sandquist, G. M. (1985). *Introduction to system science.* Englewood Cliffs: Prentice Hall.
Schein, E. H. (1990). Organizational culture. *American Psychological Association, 45*(2), 109.
Schein, E. H. (1992, last ed. 2010). Organizational culture and leadership. San Francisco: Jossey-Bass Wiley.
Schein, V. E., Mueller, R., Lituchy, T., & Liu, J. (1996). Think manager—Think male: A global phenomenon? *Journal of Organizational Behavior, 17*(1), 33–41.
Schiebinger, L. (2014). Gendered innovations: Harnessing the creative power of sex and gender analysis to discover new ideas and develop new technologies. *Triple Helix, 1*(1), 1–17.
Schilling, M. A. (2000). Toward a general modular systems theory and its application to interfirm product modularity. *Academy of Management Review, 25*, 312–334.
Schluter, D., Price, T. D., & Grant, P. R. (1985). Ecological character displacement in Darwin's finches. *Science, 227*(4690), 1056–1059.
Schmenner, R. W. (1993). *Production/operations management: From the inside out.* Englewood Cliffs: Prentice-Hall.
Schmitt, T. G. (1984). Resolving uncertainty in manufacturing systems. *Journal of Operations Management, 4*(4), 331–345.
Schmitz, J. A. Jr. (2001). *What determines labor productivity? Lessons from the Dramatic Recovery of the U. S. and Canadian Iron-ore Industries.* Federal Reserve Bank of Minneapolis Research Department Staff Report 286/2001/03.
Schneiderman, A. M. (1986). Optimum quality costs and zero defects: Are they contradictory concepts? *Quality Progress, 19*(11), 28–31.
Schoener, T. W. (2011). The newest synthesis: Understanding the interplay of evolutionary and ecological dynamics. *Science, 331*(6016), 426–429.
Schroeder, R. G. (1993). *Operations management. Decision making in the operations function.* New York: McGraw-Hill.
Schroeder, R. G., Linderman, K., Liedtke, C., & Choo, A. S. (2008). Six sigma: Definition and underlying theory. *Journal of Operations Management, 26*, 536–554.
Schumpeter, J. (1942). *Capitalism, socialism and democracy.* New York: Harper (1975).
Scott, S. G., & Bruce, R. A. (1994). Determinants of innovative behavior: A path model of individual innovation in the workplace. *The Academy of Management Journal, 37*(3), 580–607.
Sea Turtles. (2020, online). https://www.seeturtles.org/baby-turtles
Segel, L. A. (1984). *Modeling dynamic phenomena in molecular and cellular biology.* Cambridge: Cambridge University Press.
Segre, S. (2012). *Talcott parsons: An introduction.* Lanham: University Press of America.
Sen, S., & Farzin, R. (2000). Downsizing, capital intensity, and labor productivity. *Journal of Financial and Strategic Decisions, 13*(2), 73–81.
Senge, P. (1990). *The fifth discipline: The art and practice of the learning organization.* New York: Doubleday/Currency.
Senge, P. (2002). *Foreword to A. De Geus.*
Senge, P. (2006, 1st ed. 1990). *The fifth discipline: The Art and practice of the learning organization.* New York: Doubleday/Currency (last edition, revised and enlarged).
Senge, P., & Lannon-Kim, C. (1991). The systems thinking approach. *The Systems Thinker Newsletter, 2*(5). Cambridge, MA: Kendall Square.
Senge, P., Kleiner, A., Roberts, C., Ross, R., & Smith, B. (1994). *The fifth discipline fieldbook.* New York: Doubleday.
Senge, P., Smith, B., Kruschwitz, N., Laur, J., & Schley, S. (2008). *The necessary revolution. Working together to create a sustainable world.* New York: Broadway Books.
Sewell, C., & Brown, P. B. (2009). Customers for life: How to turn that one-time buyer into a lifetime customer. Currency.

Shaikh, S., & Leonard-Amodeo, J. (2005). The deviating eyes of Michelangelo's David. *Journal of the Royal Society of Medicine, 98*(2), 75–76. http://www.ncbi.nlm.nih.gov/pmc/articles/PMC1079389/

Shapiro, S. (2011). *Website*. www.steveshapiro.com/innovation/page/7/

Sharma, M., & Kodali, R. (2008). TQM implementation elements for manufacturing excellence. *The TQM Journal, 20*(6), 599–621.

Shaw, M. C. (2001). *Engineering problem solving: A classical perspective*. Norwich, New York: Noyes Publications.

Shearer, P. M. (2019). *Introduction to seismology*. Cambridge: Cambridge University Press.

Sheehy, P., Navarro, D., Silvers, R., Keyes, V., & Dixon, D. (2002). *The black belt memory jogger*. Salem: Goal/QPC and Six Sigma Academy.

Shetty, Y. K. (1987). Product quality and competitive strategy. *Business Horizons, 30*(3), 46–52.

Shewhart, W. A. (1931). *Economic control of quality of manufactured product*. New York: Van Nostrand.

Shibata, H. (1998). *Problem solving: Definition, terminology, and patterns*. www.mediafrontier.com/Article/PS/PS.htm

Shimizu, H. (1987). A general approach to complex systems. In R. Graham & A. Wunderlin (Eds.), *Bioholonics' in lasers and Synergetics*. Berlin: Springer.

SHRM. (2020). *Managing organizational change*. https://www.shrm.org/resourcesandtools/tools-and-samples/toolkits/pages/managingorganizationalchange.aspx

Siberian Times Reporter. (2016). *Trembling tundra – the latest weird phenomenon in Siberia's land of craters*. http://siberiantimes.com/ecology/others/news/n0679-trembling-tundra-the-latest-weird-phenomenon-in-siberias-land-of-craters/

Sibly, R. M., Barker, D., Hone, J., & Pagel, M. (2007). On the stability of populations of mammals, birds, fish and insects. *Ecology Letters, 10*(10), 970–976.

Sierksma, G., & Zwols, Y. (2015a). *Linear and integer optimization: Theory and practice*. Taylor and Francis: CRC Press.

Sierksma, G., & Zwols, Y. (2015b). *Linear optimization solver*. https://online-optimizer.appspot.com/?model=builtin:default.mod

Silver, E. A. (1981). Operations research in inventory management: A review and critique. *Operations Research, 29*(4), 628–645.

Silver, E. A., Pyke, D. F., & Peterson, R. (1998). *Inventory management and production planning and scheduling*. New York: Wiley.

Simon, H. A. (1962). The architecture of complexity. *Proceedings of the American Philosophical Society, 106*(6), 467–482.

Simon, H. A., & Associates. (1986). *Decision making and problem solving*. Washington, DC: National Academy of Sciences, National Academy Press. http://dieoff.org/page163.htm

Simons, R. (1995). *Levers of control: How managers use innovative control systems to drive strategic renewal*. Boston: Harvard Business School Press.

Skyttner, L. (2005). *General systems theory: Problems, perspectives, practice*. Singapore: World Scientific Pub. Online Google Books.

Smajdorova, T., & Noskievicova, D. (2017). Parametric versus nonparametric statistical process control. In B. Katalinic (Ed.), *Proceedings of the 28th DAAAM international symposium* (pp. 0944–0949). Vienna: Published by DAAAM International. https://www.researchgate.net/publication/321663539_Parametric_Versus_Nonparametric_Statistical_Process_Control. Accessed 09 May 2020.

Smartsheet. (2020, online). *Everything you need to know about Kanban Cards*. https://www.smartsheet.com/everything-you-need-know-about-kanban-cards?iOS=

Smith, A. (1776). *An inquiry into the nature and causes of the wealth of nations*. London: Methuen. http://www.gutenberg.org/catalog/world/readfile?fk_files=3275283

Smith, G. (1998). *Statistical process control and quality improvement*. Columbus: Prentice Hall.

Smith, C. H. (2005). Alfred Russel Wallace, past and future. *Journal of Biogeography, 32*(9), 1509–1515.

Smith, B. S. (2013). Zeno's paradox of the tortoise and achilles. *PRIME*. http://platonicrealms.com/encyclopedia/Zenos-Paradox-of-the-Tortoise-and-Achilles/
Smith, P. G., & Reinartson, D. G. (1991). *Making products in half the time*. New York: Van Nostrand Reinhold.
Snow, C. C., Miles, R. E., & Coleman, H. J., Jr. (1992), Managing 21st century network organizations, Organizational Dynamics, 20(3), 5–20.
Society for Neuroscience. (2019). *Web Page*. https://www.brainfacts.org/sitecore/content/Home/SfN/About/About-Neuroscience
Söderbaum, P. (2008). *Understanding sustainability economics: Towards pluralism in economics*. London: Earthscan.
Song, J., & Zipkin, P. (1993). Inventory control in a fluctuating demand environment. *Operations Research, 41*(2), 351–370.
Sontag, E. D. (1998). *Mathematical control theory: Deterministic finite dimensional systems* (2nd ed.). New York: Springer.
SPEEXX. on line. https://www.speexx.com/it/speexx-blog/stereotipi-italiani/
Spelta, A. M. (1602). *History written by Antonio Maria Spelta, citizen of Pavia* [original: Historia d'Antonio Maria Spelta, Cittadino Pavese]. Pavia: Pietro Bartoli, MDCII (Library of the University of Pavia).
Srikanthan, G., & Dalrymple, J. (2005). Implementation of a holistic model for quality in higher education. *Quality in Higher Education, 11*(1), 69–81.
Staddon, J. E. (2016, 1st ed. 1983). Adaptive behavior and learning. Cambridge, MA, Cambridge University Press.
Steele, C. (1997). A threat in the air: How stereotypes shape intellectual identity and performance. *American Psychologist, 52*(6), 613–629.
Steffan-Dewenter, I., & Schiele, S. (2008). Do resources or natural enemies drive bee population dynamics in fragmented habitats? *Ecology, 89*(5), 1375–1387.
Stein, S., & Wysession, M. (2009). *An introduction to seismology, earthquakes, and earth structure*. New York: Wiley.
Sterman, J. D. (2000). *Business dynamics: Systems thinking and modeling for a complex world*. New York: McGraw-Hill.
Sterman, J. D. (2001). System dynamics modeling: Tools for learning in a complex world. *California Management Review, 43*(4), 8–25.
Stern, J. M., Bennett Stewart, G., III, & Chew, D., Jr. (1995). The EVA financial system. *Journal of Applied Corporate Finance, 3*(2), 38–55.
Stevenson, J., Caverly, S., Srebnik, D., & Hendryx, M. (1999). Using work sampling to investigate staff time allocation in community mental health centers. *Administration and Policy in Mental Health, 26*(4), 291–295.
Stiroh, K. J. (2005). Information technology and the world economy. *The Scandinavian Journal of Economics, 107*(4), 631–650.
Storper, M., & Walker, R. (1989). *The capitalist imperative: Territory, technology and industrial growth*. Oxford: Blackwell.
Story, J., & Hess, J. (2006). Segmenting customer-brand relations: Beyond the personal relationship metaphor. *Journal of Consumer Marketing, 23*(7), 406–413.
Stylios, C., Langer, G., Iung, B., Hyun, Y. T., Sorensen, C., Weck, M., & Groumpos, P. (2000). Discipline research contributions to the modelling and design of intelligent manufacturing systems. *Studies in Informatics and Control, 9*(2), 111–132.
Sun, X., Zhang, H., Meng, W., Zhang, R., Li, K., & Peng, T. (2018). Primary resonance analysis and vibration suppression for the harmonically excited nonlinear suspension system using a pair of symmetric viscoelastic buffers. *Nonlinear Dynamics, 94*(2), 1243–1265.
Sutter, J. D. (2017). Sixth mass extinction: The era of biological annihilation. *CNN*. https://edition.cnn.com/2017/07/11/world/sutter-mass-extinction-ceballos-study/index.html [15 January 2018].

Swamidass, P. M. (Ed.). (2000). *Inventory holding costs. Encyclopedia of production and manufacturing management*. Springer Science & Business Media.
Sweeney, J. C., & Soutar, G. (2001). Consumer perceived value: The development of a multiple item scale. *Journal of Retailing, 77*(2), 203–220.
Syverson, C. (2002). *Output market segmentation and productivity*. Washington, DC: WP, Center for Economic Studies, U.S. Census Bureau.
Taghizadegan, S. (2006). *Essentials of lean six sigma*. Amsterdam: Elsevier.
Taguchi, G. (1986). *Introduction to quality engineering: Designing quality into products and processes*. The Organization: Asian Productivity Organization. Un. of Michigan.
Taguchi, G. (1988). The development of quality engineering. *The ASI Journal, 11*(1), 5–29.
Takeuchi, Y. (1995). *Global dynamical properties of Lotka-Volterra systems*. Singapore: World Scientific.
Tanaka, T. (1994). Kaizen budgeting: Toyota's cost-control system under TQC. *Journal of Cost Management for the Manufacturing Industry*, 56–62.
Tang, S. L., Aoieong, R. T., & Ahmed, S. M. (2004). The use of Process Cost Model (PCM) for measuring quality costs of construction projects: Model testing. *Construction Management and Economics, 22*(3), 263–275.
Tannenbaum, A. S. (1962). Control in organizations: Individual adjustment and organizational performance. *Administrative Science Quarterly, 7*(2), 236–257.
Tarski, A. (1944). The semantic conception of truth: And the foundations of semantics. *Philosophy and Phenomenological Research, 4*(3), 341–376.
Taylor, F. (1919). *The principles of scientific management*. New York: Harper & Brothers.
Teal & Co. (2016). *The frog in boiling water: The myth and the metaphor*. http://tealandco.com/the-frog-in-boiling-water-the-myth-and-the-metaphor/
Terborgh, G. W. (1949). *Dynamic equipment policy*. New York: McGraw Hill.
Terborgh, G. W. (1956). Some comments on the Dean-Smith article on the Mapi formula. *The Journal of Business, 29*(2), 138–140.
Terborgh, G. W. (1958). *Business investment policy, A MAPI study and manual*. Washington: Machinery and Allied Products Institute and Council.
Tersine, R. J. (1988). *Principles of inventory and materials management*. New York: North Holland.
Thakkar, J., Deshmukh, S. G., Gupta, A. D., & Shankar, R. (2006). Development of a balanced scorecard: An integrated approach of Interpretive Structural Modelling (ISM) and Analytic Network Process (ANP). *International Journal of Productivity and Performance Management, 56*(1), 25–59.
Tharumarajah, A., Wells, A. J., & Nemes, L. (1996). Comparison of the bionic, fractal and holonic manufacturing system concept. *International Journal of Computer Integrated Manufacturing, 9*(3), 217–236.
Thaxton, C., Bradley, W., & Olson, R. (1984). *The mystery of life's origin*. Philosophical Library.
Thérivel, R., & Phillip Minas, P. (2012). Ensuring effective sustainability appraisal. *Impact Assessment and Project Appraisal, 20*(2), 81–91.
Tippett, L. H. C. (1935). A snap-reading method of making time studies of machines and operatives in factory surveys. *Journal of Textile Institute, 26*(75), T13–T51.
Tippett, L. H. C. (1947). The study of industrial efficiency, with special reference to the cotton industry. *Journal of the Royal Statistical Society, 110*(2), 108–122.
Tollefson, J. (2018). Satellite spies methane bubbling up from Arctic Permafrost. *Scientific American*. https://www.scientificamerican.com/article/satellite-spies-methane-bubbling-up-from-arctic-permafrost/
Towill, R. (1996). Time compression and supply chain management. A guided tour. *Logistics Information Management, 9*(6), 41–53.
Tuchman, B. W. (1980, November 2). The decline of quality. *New York: Times Magazine*, 38–47.
Turing, A. (1950). Computing machinery and intelligence. *Mind, 59*(236), 433–460. http://www.abelard.org/turpap/turpap.htm

Tuttle, G. (1994). High-speed management and new product development. In S. S. King & D. P. Cushman (Eds.), *High-speed management and organizational communication in the 1990s* (pp. 195–218). Albany: State University of New York Press.
Underwood, R. L. (2003). The communicative power of product packaging: Creating brand identity via lived and mediated experience. *Journal of Marketing Theory and Practice, 11*(1), 62–76.
Unger, N. (2014). To save the planet, don't plant trees. *NY Times*. https://www.nytimes.com/2014/09/20/opinion/to-save-the-planet-dont-plant-trees.html

UNI EN ISO 9000:2000 (2000). Quality management systems - Requirements. https://www.iso.org/standard/21823.html
UNI 11063 2003 (2003) Maintenance - Definitions of ordinary and extraordinary maintenance. http://store.uni.com/catalogo/uni-11063-2003
UNI 13306 2018 (2018). Maintenance - Maintenance terminology. http://store.uni.com/catalogo/uni-en-13306-2018
Uribe, R. (1981). Modeling autopoiesis. In M. Zeleny (Ed.), *Autopoiesis: A theory of living organization*. New York: Elsevier-North Holland.
USGS, US Geological Survey. (online). https://www.usgs.gov/faqs/what-a-fault-and-what-are-different-types?qt-news_science_products=0#qt-news_science_products
USP – The Untamed Science Portals. (2017). *Did yellowstone wolves really save the Park's ecosystem?* http://www.untamedscience.com/biology/ecology/ecology-articles/wolf-reintroduction-yellowstone/ [15 January 2018].
Valian, V. (1999). *Why so slow?: The advancement of women*. Cambridge, MA: MIT Press.
Vanecek, A., & Celikovsky, S. (1996). *Control systems: From linear analysis to synthesis of chaos*. Englewood Cliffs: Prentice Hall.
Varela, F. J. (1979). *Principles of biological autonomy*. Boston: Kluwer Academic.
Varela, F. J. (1981). Describing the logic of living. In M. Zeleny (Ed.), *Autopoiesis: A theory of living organization*. New York: Elsevier.
Varmah, K. R. (2010). *Control systems*. New Delhi: Tata McGraw-Hill Education.
Veinott, A. F., Jr. (1966). The status of mathematical inventory theory. *Management Science., 12*(11), 745–777.
Velentzas, J., & Broni, G. (2011). Cybernetics and autopoiesis theory as a study of complex organizations. A systemic approach. In *ICOAE, international conference on applied economics*, pp. 737–747.
Venkatesh, J. (2007). *An introduction to total productive maintenance (TPM)*. The Plant Maintenance Resource Center, 3–20. http://www.plant-maintenance.com/articles/tpm_intro.shtml
Vicari, S. (1991). *The firm as a living system* [original: L'impresa vivente]. Milan: Etas Ed.
Virolainen, V. M. (1991). *Control systems for logistics performance*. Laxenburg: International Institute for Applied Systems Analysys. http://pure.iiasa.ac.at/id/eprint/3577/1/CP-91-003.pdf
Visilab. (2013). *Blind assistant project*. http://visilab.unime.it/~filippo/BlindAssistant/BlindAssistant.htm
Vlăsceanu, L., Grünberg, L., & Pârlea, D. (2004). *Quality assurance and accreditation: A glossary of basic terms and definitions*. Bucharest: UNESCO-CEPES, Papers on Higher Education.
Voehl, F., Mignosa, C., Harrington, H. J., & Charron, R. (2016). *The lean six sigma black belt handbook: Tools and methods for process acceleration*. Cambridge, MA: Productivity press.
Voigt, M. (2018). *Dramatic decline of Bornean Orangutans*. Max Planck Institute for Evolutionary Anthropology. https://www.mpg.de [15 January 2018].
Volterra, V. (1926). Variations and fluctuations of the number of individuals in animal species living together [original: Variazioni E Fluttuazioni Del Numero D'individui in Specie Animali Conviventi]. *Memorie Della R. Accademia Dei Lincei, 6*(2), 31–113.
Volterra, V. (1931). *Lessons on the mathematical theory of struggle for life* [original: Leçons esur la théorie mathématique de la Lutte pour la vie]. Paris: Gauthier-Villars.

von Bertalanffy, L. (1968). *General system theory: Foundations, development, applications.* New York: Braziller.
von der Pütten, J. C. (2008). *Moral issues and concerns about China's one-child policy: A cosmopolitan perspective.* Norderstedt: GRIN.
von Foerster, H. (1984). Principles of self-organization in a socio-managerial context. In H. Ulrich & G. J. B. Probst (Eds.), *Self-organization and Management of Social Systems* (pp. 2–24). Berlin: Springer.
von Foerster, H. (1990). Ethics and second-order cybernetics. *Cybernetics & Human Knowing.* http://www.imprint.co.uk/C&HK/vol1/v1-1hvf.htm
von Foerster, H. (2003). *Understanding understanding: Essays on cybernetics and cognition.* New York, Berlin: Springer. https://www.pangaro.com/Heinz-von-Foerster/Heinz_Von_Foerster-Understanding_Understanding.pdf
von Foerster Page. (online). Maintained and copyright © by Alexander Riegler 1996–2020. http://www.univie.ac.at/constructivism/HvF.htm
von Krogh, G., & Roos, J. (1996). The epistemological challenge: Managing knowledge and intellectual capital. *European Management Journal, 14*(4), 333–337.
von Neuman, J., & Morgenstern, O. (1953, 1st ed. 1944). *Theory of games and economic behavior* (5th ed.). Princeton: Princeton University Press.
Waddington, C. H. (1957). *The strategy of the genes.* London: George Allen & Unwin.
Wagner, H. M. (1980). Research portfolio for inventory management and production planning systems. *Operations Research, 28*(3-part-i), 445–475.
Wakamatsu, Y. (2009). *The Toyota mindset: The ten commandments of Taiichi Ohno.* Enna Products Corporation.
Waldrop, M. M. (1992). *Complexity. The emerging science at the edge of order and chaos.* New York: Simon & Shuster.
Wallace, A. R. (1856). On the habits of the orang-Utan of Borneo. *Annals and Magazine of Natural History, 18*(103), 26–32.
Wallace, A. R. (1858). *On the tendency of varieties to depart indefinitely from the original type.* The Alfred Russel Wallace Page hosted by Western Kentucky University (Archived from the original on 29 April 2007). http://rpdata.caltech.edu/courses/Evolution_GIST_2013/files_2013/articles/Ternate_1858_Wallace.pdf
Walras, L. (1874). *Elements of pure economics: Or the theory of social wealth* (translated in French in 1926 and then published in English). Homewood: Irwin, 1954.
Walsh, J. P. (1995). Managerial and organizational cognition: Notes from a trip down memory lane. *Organization Science, 6*(3), 280–321.
Wang, W., & Zhang, W. (2008). An asset residual life prediction model based on expert judgments. *European Journal of Operational Research, 188*(2), 496–505.
Wasson, C. S. (2006). *System analysis, design, and development.* London: Wiley.
Waterman, R., Jr., Peters, T., & Phillips, J. R. (1980). Structure is not organization. *Business Horizons, 23*(3), 14–26.
Waters, C. D. (1992). *Inventory control and management.* Chichester: Wiley.
Watson, J. D., & Crick, F. H. (1953). Molecular structure of nucleic acids. *Nature, 171*(4356), 737–738.
Watson, A. J., & Lovelock, J. E. (1983). Biological homeostasis of the global environment: The parable of Daisyworld. *Tellus B, 35*(9), 286–289.
Watzlawick, P. (1976). *How real is real?: Confusion, disinformation.* Communication: New JorkVintage Books.
Watzlawick, P. (1984). *The invented reality: How do we know what we believe we know?* (Contributions to Constructivism). New York: Norton & Co. https://www.academia.edu/26596026/The_Invented_Reality_How_Do_We_Know_What_We_Believe_We_Know_Contributions_to_Constructivism_
Watzlawick, P., Weakland, J. H., & Fisch, R. (1974). *Change: Principles of problem formation and problem resolution.* New York: Norton.

Wellemin, J. H. (1990). *Customer satisfaction through quality*. Chartwell-Bratt (Publishing & Training) Ltd.
West, P. M., & Packer, C. (2002). Sexual selection, temperature and the lion's mane. *Science, 297*(5585), 1339–1343.
Wheat, B., Mills, C., & Carnell, M. (2003). *Leaning into six sigma: A parable of the journey to six sigma and a lean enterprise*. Boston, New York: McGraw-Hill.
Whitin, T. M. (1954). Inventory control research: A survey. *Management Science., 1*(1), 32–40.
Whittington, G. (1980). Some basic properties of accounting ratios. *Journal of Business Finance and Accounting, 7*(2), 219–223.
Widrow, B., Gupta, N. K., & Maitra, S. (1973). Punish/reward: Learning with a critic in adaptive threshold systems. *IEEE Transactions on Systems, Man, and Cybernetics, 5*, 455–465.
Wiener, N. (1954). *The human use of human beings: cybernetics and society*. Boston: Houghton Mifflin (Original work published in 1949).
Wiener, N. (1961, 1st ed. 1948). Cybernetics: Or control and communication in the animal and the machine. Cambridge, MA: MIT Press. Google Books.
Wight, O. (1995). *Manufacturing resource planning: MRP II: Unlocking America's productivity potential*. Wiley.
Wikipedia. (2020). *Nelson rules*. https://en.wikipedia.org/wiki/Nelson_rules
Wikipedia.com. (2020a). *List of ongoing armed conflicts*. https://en.wikipedia.org/wiki/List_of_ongoing_armed_conflicts
Wikipedia.com. (2020b). *Megacities*. https://en.wikipedia.org/wiki/Megacity
Wikipedia.com. (2020c). *Stereotype*. https://en.wikipedia.org/wiki/Stereotype
Wilber, K. (1995). Sex, *ecology, spirituality: The spirit of evolution* (2nd ed., 2000). Boston & London: Shambhala Publications.
Wilber, K. (2000, 1st ed. 1995). Sex, ecology, spirituality: The spirit of evolution. Boston: Shambhala.
Wilber, K. (2001, 1st ed. 1996). A brief history of everything. Boston: Shambhala.
Williamson, O. E. (1975). *Markets and hierarchies*. New York: Free Press.
Williamson, O. E. (1981). The economics of organization: The transaction cost approach. *The American Journal of Sociology, 87*(3), 548–577.
Willis, S. (2015). *What is customer delight? (and why should you care?)*. https://www.impactbnd.com/blog/what-is-customer-delight
Wilson, R. H. (1934). A scientific routine for cost control. Harvard Business Review, 13, pp. 116–128.
Wilson, H. G. (1977). Order quantity, product popularity, and the location of stock in a warehouse. *AIIE transactions, 9*(3), 230–237.
Wilson, W. B., Redfield, W. C., & Tilson, J. Q. (1912). *The Taylor and other systems of shop management: Hearings before to investigate the Taylor and other systems of shop management*, under authority of H. Res. 90, volume 3, U.S. Government Printing Office. https://babel.hathitrust.org/cgi/pt?id=nyp.33433004205641&view=1up&seq=7
Winiwarter, W., Ajavon, N. A., & Woodfield, M. (2006). *Quality assurance/quality control and verification*. IPCC Guidelines for National Greenhouse Gas Inventories. www.ipcc-nggip.iges.or.jp/public/2006gl/pdf/1_Volume1/V1_6_Ch6_QA_QC.pdf
Wittgenstein, L. (1922). *Tractatus Logico-Philosophicus* (Translated by C. K. Ogden). EBook #5740, Release Date: October 22, 2010. https://www.gutenberg.org/files/5740/5740-pdf.pdf
Wolfram|Alpha. (2020). *Mathematica linear programming solver*. https://www.wolframalpha.com/widgets/view.jsp?id=daa12bbf5e4daec7b363737d6d496120
Wolstenholme, E. F., & Corben, D. A. (1993). Towards a Core set of archetypal structures in system dynamics. In E. Zepeda & J. A. E. Macuca (Eds.), *The role of strategic modelling in international competitiveness*. System Dynamics Society.
Woodhouse, D. (1996). Quality assurance: International trends, preoccupations and features. *Assessment & Evaluation in Higher Education, 21*(4), 347–356.

Woodhouse, D. (1999). *Quality and quality assurance. Quality and internationalisation in higher education* (pp. 29–44). Paris: OECD.

Wooldridge, M., & Jennings, N. R. (1994, August). Toward a theory of cooperative problem solving. In *Proceedings of the 6th European workshop on modelling autonomous agents in a multi-agent world*. Odense.

Woolsey, R., & Maurer, R. (2005). *Inventory control*. Marietta: Lionheart Publishing.

Xie, M., & Goh, T. N. (1999). Statistical techniques for quality. *The TQM Magazine, 11*(4), 238–242.

Yam, R. C. M., Tse, P. W., Li, L., & Tu, P. (2001). Intelligent predictive decision support system for condition-based maintenance. *The International Journal of Advanced Manufacturing Technology, 17*(5), 383–391.

Yang, C. (2004). An integrated model of TQM and GE-six sigma. *International Journal of Six Sigma and Competitive Advantage, 1*(1), 97–111.

Young, S. D. (1999). Some reflections on accounting adjustments and economic value-added. *Journal of Financial Statement Analysis, 4*(2), 7–19.

Young, S. (2008). New and improved indeed: Documenting the business value of packaging innovation. *Quirk's Marketing Research Review, 22*(1), 46–50.

Zahra, S. A., & George, J. M. (2002). Absorptive capacity: A review, reconceptualization, and extension. *Academy of Management Review, 27*(2), 185–203.

Zairi, M. (2010). *Benchmarking for best practice*. London: Routledge.

Zeleny, M., & Hufford, K. D. (1992). The application of Autopoiesis in systems analysis: Are Autopoietic systems also social systems? *International Journal of General Systems, 21*(2), 145–160.

Zemke, R. (1990). Cost of quality: Yes, you can measure it. *Training, 27*(8), 62–63.

Zhao, W., & Chellappa, R. (2006). *Face processing. Advanced modelling and methods*. Boston: Academic Press.

Zhong, J. (2006). *PID controller tuning: A short tutorial*. http://saba.kntu.ac.ir/eecd/pcl/download/PIDtutorial.pdf

Zimmerman, J. L. (1999). *Accounting for decision making and control*. London: McGraw-Hill Higher Education.

Zuradelli, C. (1888). *The towers of Pavia* [original: Le torri di Pavia]. Pavia: Tipografia F.lli Fusi (Library of the University of Pavia).

Index

P. Mella, *The Magic Ring*, Contemporary Systems Thinking,
https://doi.org/10.1007/978-3-030-64194-8

H

I

J

K

L

N

S

www.ingramcontent.com/pod-product-compliance
Ingram Content Group UK Ltd.
Pitfield, Milton Keynes, MK11 3LW, UK
UKHW021832270726
14058UKWH00001B/100

* 9 7 8 3 0 3 0 6 4 1 9 6 2 *